Reinforced Concrete

Prentice Hall International Series
in Civil Engineering and Engineering Mechanics
William J. Hall, Editor

REINFORCED CONCRETE
Mechanics and Design

THIRD EDITION

JAMES G. MACGREGOR

University Professor Emeritus
Department of Civil Engineering
University of Alberta

PRENTICE HALL, Upper Saddle River, New Jersey 07458

293072

Library of Congress Cataloging-in-Publication Data

MacGregor, James G. (James Grierson)
    Reinforced concrete : mechanics and design / James G. MacGregor. –
–3rd ed.
      p. cm.
    Includes bibliographical references and index.
    ISBN 0–13–233974–9
    1. Reinforced concrete construction. 2. Reinforced concrete.
  I. Title.
TA683.2.M34   1997
624.1'8341—dc20   96–14925
                                    CIP

Editor-in-chief: *Marcia Horton*
Acquisitions editor: *Bill Stenquist*
Managing editor: *Bayani Mendoza DeLeon*
Project manager: *Jennifer Wenzel*
Cover director: *Amy Rosen*
Cover designer: *Joe Sengotta*
Interior: *Meryl Poweski*
Manufacturing buyer: *Julia Meehan*
Editorial assistant: *Meg Weist*
Cover photo: Provided by *James
          MacGregor* and is a photo of the Olympic
          Stadium in Montreal, Canada, under construction.

©1997, 1992, 1988 by Prentice-Hall, Inc.
Upper Saddle River, New Jersey 07458

The author and publisher of this book have used their best efforts in preparing this book. These efforts include the development, research, and testing of the theories and design procedures to determine their effectiveness. The author and publisher make no warranty of any kind, expressed or implied, with regard to the theories and design procedures contained in this book. The author and publisher shall not be liable in any event for incident or consequential damages in connection with, or arising out of, the furnishing, performance, or use of these theories and design procedures.

Printed in the United States of America

10

ISBN 0-13-233974-9

Prentice-Hall International (UK) Limited, *London*
Prentice-Hall of Australia Pty. Limited, *Sydney*
Prentice-Hall Canada Inc., *Toronto*
Prentice-Hall Hispanoamericana, S.A., *Mexico*
Prentice-Hall of India Private Limited, *New Delhi*
Prentice-Hall of Japan, Inc., *Tokyo*
Prentice-Hall Asia Pte. Ltd., *Singapore*
Editora Prentice-Hall do Brasil, Ltda., *Rio de Janerio*

Contents

Contents

Contents

CHAPTER 19   **DESIGN FOR EARTHQUAKE RESISTANCE**   **820**

Preface

Reinforced concrete design is both an art and a science. Since the 1960s, the emphasis on the science has increased as codes became more complex and as computers came to be used to design and detail concrete members. Today, contractors complain that current designs are difficult to build. Designers, faced with the myriad of rules, code clauses, and equations, feel that reinforced concrete design is a mystical science that few understand. This book presents the theory of reinforced concrete as a direct application of the laws of statics and the behavior of reinforced concrete. In addition, it emphasizes that a successful design not only satisfies the design equations, but also is capable of being built at a reasonable price.

The various topics—flexure, shear, columns, and so on—are presented at two levels in this book. Each subject starts with a basic presentation suitable for undergraduate university courses on reinforced concrete. It then moves to more advanced topics not normally found in American textbooks, including, for example, unsymmetrical beams and columns, strain compatibility solutions of beams, P–Δ analyses of frames, and the design of deep beams and column-beam joints. The latter concepts make this book a useful reference volume in design offices and a suitable text for graduate courses.

Particular emphasis has been placed on the logical order and completeness of the design examples. The examples are done in a step-by-step order and every step is worked out completely from first principles, at least once. Designers used to using design aids will recognize places where they can shorten the calculations. The examples have been chosen to illustrate the effects of unequal spans and other situations normally encountered in design but not in textbooks. Guidance is given in the text and in the examples to help students to make the many judgment decisions required in reinforced concrete design.

Chapter 1 sets the stage for the volume by providing definitions and giving illustrations of the various types of members built from reinforced concrete. A brief history of concrete, reinforced concrete, and codes for reinforced concrete is included.

Chapter 2 continues the introductory material with a discussion of the goals of structural design based, in part, on the limit states design concept. Limit states design is simply the traditional engineering approach of anticipating all of the ways that things can go wrong and taking steps to ensure that they don't. Considerable emphasis is placed on this throughout the book because, since the introduction of strength design in the 1963 ACI Code, concrete structures have become more and more slender and more apt to exhibit excessive cracking, deflections, or vibrations. Chapter 2 also contains a brief introduction to safety theory, a brief review of the loads considered in design, and a discussion of design for economy.

The significant properties of concrete and reinforcement are presented in Chapter 3 as a basis for developing the flexural theory, discussing time-dependent deflections, and so on. This chapter is also intended to serve as a ready reference source for information on the structural aspects of concrete technology.

Chapters 4 through 8 and 11 through 13 deal with the theory and design for various ultimate limit states, such as flexure, shear, anchorage, and so on. In each case, the discussion starts with a review of the behavior of concrete members and uses statics and mechanics to explain this behavior. Practical aspects of design and construction are introduced to explain code limitations and detailing rules. Appendix B of the 1995 ACI Code introduced a new method of setting an upper limit on the reinforcement in beams. This procedure has been used rather than the balanced reinforcement ratio because it is easier to apply.

Chapter 7 on torsion has been extensively revised to incorporate the 1995 ACI Code revisions to torsion design.

Chapter 8 on development has been revised extensively to incorporate the new development and detailing requirements of the 1995 ACI Code. A systematic method of carrying out the development calculations has been used in the examples.

The serviceability limit states, particularly deflection and crack control, are the subject of Chapter 9. The discussion includes the calculation procedures for checking deflections and crack widths, the limits that should be placed on these values, and why.

Chapters 10 and 13 and through 15 deal with the design of continuous slabs and beams, and two-way slabs. Chapter 13 starts with an overview of moments in slabs and the shear strength of two-way slabs and then goes on to present the direct design method with examples. Section 13–12 deals with the calculation of slab deflections. Chapter 14 covers the equivalent frame method. Elastic analysis of slabs, yield line theory, the strip method, and a simplified presentation of the advanced strip method are presented in Chapter 15 with examples.

Chapter 11 discusses the behavior and design of columns. The calculation of ϕ for tension failures is based on Appendix B of the 1995 ACI Code. Chapter 12 on slender columns has been revised in light of the extensive revisions of the slenderness provisions in the 1995 ACI Code.

Footings are discussed in Chapter 16. Chapter 17 presents shear friction, horizontal shear transfer and a new section on concrete composite beams.

The concept of D-regions, regions adjacent to discontinuities, is introduced in several places. Chapter 18 is devoted entirely to a detailed presentation of the theory and design of D-regions and contains several new examples.

Design for earthquake effects is presented in Chapter 19 which has been enlarged to include comprehensive examples of the design of a beam, a column, and a joint in a building in a high seismic region.

Appendix A presents 34 design tables and 14 design charts referred to throughout the book. These are gathered together for easy reference and make it possible to use the text in courses or in a design office without the need for a handbook. Ten tables in SI units have been added. The column interaction diagrams have been redrawn with ϕ based on ACI Code Appendix B.

A one-semester undergraduate course on reinforced concrete might cover Secs. 2-1 to 2-4 and 2-6 to 2-8 on the basis for design, safety factors, loads, and design for economy; Secs. 3-2, 3-3, and 3-9 on material properties; Chap. 4 and Secs. 5-1 through 5-3 on flexure; Secs. 6-1 to 6-3 and 6-5 on shear; Chap. 8 on anchorage; Secs. 9-1 to 9-5 on serviceability; Chap. 10 on continuous slabs and beams; Secs. 11-1 to 11-5 on columns and Secs. 16-1 to 16-5 on footings. A subsequent course might cover Chap. 7 on torsion, Secs. 12-1 to 12-4 on slender columns, Chap. 13 on two-way slabs, and Sec. 17-2 on shear friction. Chapters 18 and 19 would be optional. The prerequisites for these courses are a course in statics and a course in mechanics of materials.

A one-quarter undergraduate course on reinforced concrete might cover Secs. 2-1 to 2-4 and 2-6 to 2-8 on the bases for design; Secs. 3-2, 3-3, and 3-9 on material properties; Chap. 4 and Secs. 5-1 to 5-3 on flexure; Secs. 6-1 to 6-3 and 6-5 on shear; Secs. 8-1 to 8-7 on anchorage; Secs. 10-3 and 10-4 on one-way slabs, and Secs. 11-1 to 11-5 on columns. A subsequent one-quarter course might cover Secs. 9-1 to 9-5 on serviceability; the rest of Chap. 10; Secs. 16-1 to 16-5 on footings; Chap. 7 on torsion; Secs. 12-1 and 12-2 on slender columns and Secs. 13-3 to 13-10 on two-way slabs.

The text makes frequent reference to the 1995 ACI Code and assumes that the reader will have a copy of this code.

Although the foot-pound-second system of units is the main system of units throughout the book, eight examples in basic topics are repeated completely in SI (metric) units. Four of these are new in this edition. A number of metric design charts are given in Appendix A.

My sincere thanks to my friends, colleagues, and reviewers of the text for their suggestions for improvement, discussions of general approach, and the other assistance. In particular I wish to thank C. P. Siess; J. E. Breen, who initiated the idea of this book and had many helpful suggestions; R. Green, and J. K. Wight, whose comments and critiques of the first edition were invaluable; D. J. MacGregor, who drew many of the figures; and to my wife, Barb, whose continuing support, help, and encouragement have made all three editions possible.

I urge readers who have questions, suggestions for improvements, or clarifications, or who find errors, to write to me. I thank you in advance for taking the time and interest to do so.

I dedicate this book to the memory of my son, Robert J. G. MacGregor, PhD, P.E.

Reinforced Concrete

1

Introduction

1–1 REINFORCED CONCRETE STRUCTURES

Concrete and reinforced concrete are used as building materials in every country. In many, including the United States and Canada, reinforced concrete is a dominant structural material in engineered construction. The universal nature of reinforced concrete construction stems from the wide availability of reinforcing bars and the constituents of concrete, gravel, sand, and cement, the relatively simple skills required in concrete construction, and the economy of reinforced concrete compared to other forms of construction. Concrete and reinforced concrete are used in bridges, buildings of all sorts (Fig. 1–1), underground structures, water tanks, television towers, offshore oil exploration and production structures (Fig. 1–2), dams, and even in ships.

1–2 MECHANICS OF REINFORCED CONCRETE

Concrete is strong in compression but weak in tension. As a result, cracks develop whenever loads, or restrained shrinkage or temperature changes, give rise to tensile stresses in excess of the tensile strength of the concrete. In the plain concrete beam shown in Fig. 1–3b, the moments about O due to applied loads are resisted by an internal tension–compression couple involving tension in the concrete. Such a beam fails very suddenly and completely when the first crack forms. In a *reinforced concrete* beam (Fig. 1–3c), steel bars are embedded in the concrete in such a way that the tension forces needed for moment equilibrium after the concrete cracks can be developed in the bars.

Alternatively, the reinforcement could be placed in a longitudinal duct near the bottom of the beam, as shown in Fig. 1–4, and stretched or *prestressed*, reacting on the concrete in the beam. This would put the reinforcement into tension and the concrete in compression. This compression would delay cracking of the beam. Such a member is said

1

Fig. 1–1
City Hall, Toronto, Canada.

The Toronto City Hall consists of two towers, 20 and 27 stories in height, with a circular council chamber cradled between them. These structures and the surrounding terraces, pools, and plaza illustrate the degree to which architecture and structural engineering can combine to create a living sculpture. This complex has become the trademark and social hub of the city of Toronto. The council chamber consists of a reinforced concrete bowl containing seating for the council, press, and citizens. This is covered by a concrete dome. The wind resistance of the two towers results largely from the two vertical curved walls forming the backs of the towers. The architectural concept was by Viljo Revell, of Finland, winner of an international design competition. Mr. Revell entered into an association with John B. Parkin Associates, who developed the design and also carried out the structural design. The structural design is described in Ref. 1–1. (Photograph used with permission of Neish Owen Rowland & Roy, Architects Engineers, Toronto.)

Fig. 1–2
Glomar Beaufort Sea 1
(CIDS) being towed through
the Bering Straits to the
Beaufort Sea, Alaska.

This concrete island oil drilling structure consists of a steel mud base, a 230-ft-square cellular concrete segment, and a deck assembly with drilling rig, quarters for 80 workers, and supplies for 10 months. The structure is designed to operate in 35 to 60 ft of water in the Arctic Ocean. Forces from the polar sea ice are resisted by the thick walls of the concrete segment. Design was carried out by Global Marine Development Inc. Engineering and construction support to Global Marine was provided by A. A. Yee Inc. and ABAM Engineers Inc. (Photograph courtesy of Global Marine Development Inc.)

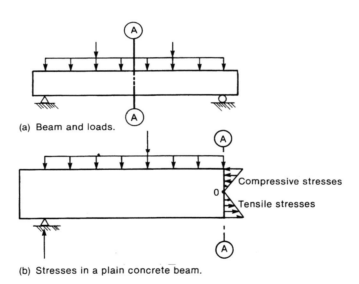

(a) Beam and loads.

(b) Stresses in a plain concrete beam.

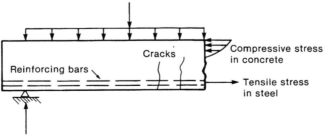

Fig. 1–3
Plain and reinforced concrete
beams

(c) Stresses in a reinforced concrete beam.

to be a *prestressed concrete* beam. The reinforcement in such a beam is referred to as *prestressing tendons* and must be high-strength steel.

The construction of a reinforced concrete member involves building a form or mold in the shape of the member being built. The form must be strong enough to support the weight and hydrostatic pressure of the wet concrete, and any forces applied to it by workers, concrete buggies, wind, and so on. The reinforcement is placed in this form and held in place during the concreting operation. After the concrete has hardened, the forms are removed.

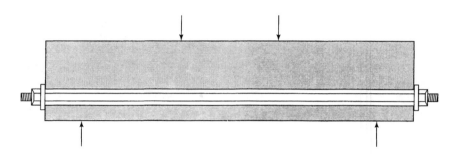

Fig. 1–4
Prestressed concrete beam.

1–3   REINFORCED CONCRETE MEMBERS

Reinforced concrete structures consist of a series of individual "members" that interact to support the loads placed on the structure. The second floor of the building in Fig. 1–5 is built of concrete joist–slab construction. Here a series of parallel ribs or *joists* support the load from the top slab. The reactions supporting the joists apply loads to the beams, which in turn are supported by columns. In such a floor, the top slab has two functions: (1) it transfers load laterally to the joists, and (2) it serves as the top flange of the joists, which act as T-shaped beams that transmit the load to the beams running at right angles to the joists. The first floor of the building in Figure 1–5 has a slab-and-beam design in which the slab spans between beams, which in turn apply loads to the columns. The column loads are applied to *spread footings*, which distribute the load over a sufficient area of soil to prevent overloading of the soil. Some soil conditions may require the use of pile foundations or other deep foundations. At the perimeter of the building, the floor loads are supported either directly on the walls as shown in Figure 1–5, or on exterior columns as shown in Fig. 1–6. The walls or columns, in turn, are supported by a basement wall and wall footings.

The slabs in Figure 1–5 are assumed to carry the loads in a north–south direction (see direction arrow) to the joists or beams, which carry the loads in an east–west direction to other beams, girders, columns, or walls. This is referred to as *one-way slab* action and is analogous to a wooden floor in a house, in which the floor decking transmits loads to perpendicular floor joists, which carry the loads to supporting beams, and so on.

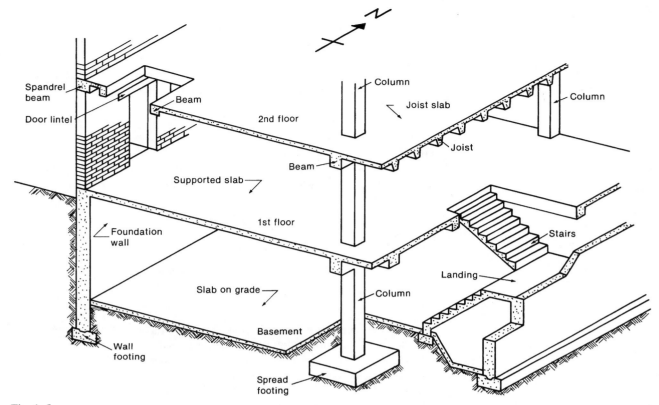

Fig. 1–5
Reinforced concrete building
elements. (Adapted from Ref. 1–2.)

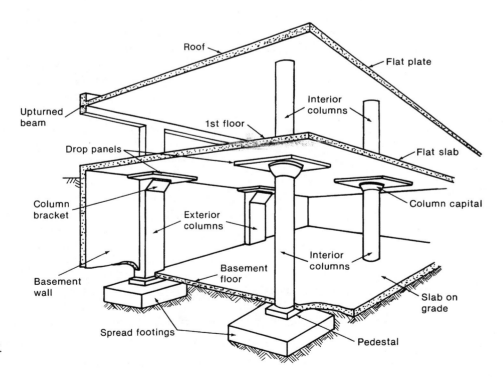

Fig. 1–6
Reinforced concrete building
elements. (Adapted from Ref.
1–2.)

1–3   Reinforced Concrete Members

5

The ability to form and construct concrete slabs makes possible the slab or plate type of structure shown in Fig. 1–6. Here the loads applied to the roof and the floor are transmitted in two directions to the columns by plate action. Such slabs are referred to as *two-way slabs*.

The first floor in Fig. 1–6 is a *flat slab* with thickened areas called *drop panels* at the columns. In addition, the tops of the columns are enlarged in the form of *capitals* or *brackets*. The thickening provides extra depth for moment and shear resistance adjacent to the columns. They also tend to reduce the slab deflections.

The roof of the building shown in Fig. 1–6 is of uniform thickness throughout without drop panels or columns capitals. Such a floor is a special type of *flat slab* referred to as a *flat plate*. Flat-plate floors are widely used in apartments because the underside of the slab is flat and hence can be used as the ceiling of the room below. Of equal importance, the forming for a flat plate is generally cheaper than that for flat slabs with drop panels or for one-way slab-and-beam floors.

1–4   FACTORS AFFECTING CHOICE OF CONCRETE FOR A STRUCTURE

The choice of whether a structure should be built of concrete, steel, masonry, or timber depends on the availability of materials and on a number of value decisions.

1.   Economy. Frequently, the foremost consideration is the overall cost of the structure. This is, of course, a function of the costs of the materials and the labor and the time necessary to erect them. Concrete floor systems tend to be thinner than structural steel systems because the girders and beams or joists all fit within the same depth, as shown in the second floor in Fig. 1–5, or the floors are flat plates, as shown in Fig. 1–6. This produces an overall reduction in the height of a building compared to a steel building, which leads to (a) lower wind loads because there is less area exposed to wind, and (b) savings in cladding and mechanical and electrical risers.

Frequently, however, the overall cost is affected as much or more by the overall construction time since the contractor and the owner must allocate money to carry out the construction and will not receive a return on their investment until the building is ready for occupancy. As a result, financial savings due to rapid construction may more than offset increased material and forming costs. The materials for reinforced concrete structures are widely available and can be produced as they are needed in the construction, as opposed to structural steel, which must be ordered and partially paid for in advance to schedule the job in a steel fabricating yard.

Any measures the designer can take to standardize the design and forming will generally pay off in reduced overall costs. For example, column sizes may be kept the same for several floors to save money in form costs, while changing the concrete strength or percentage of reinforcement to allow for changes in column loads.

2.   Suitability of material for architectural and structural function. A reinforced concrete system frequently allows the designer to combine the architectural and structural functions. Concrete has the advantage that it is placed in a plastic condition and is given the desired shape and texture by means of the forms and the finishing techniques. This allows such elements as flat plates or other types of slabs to serve as load-bearing elements while providing the finished floor and ceiling surfaces. Similarly, reinforced concrete walls can provide architecturally attractive surfaces in addition to having the ability to resist gravity, wind, or seismic loads. Finally, the choice of size or shape is governed by the designer and not by the availability of standard manufactured members.

3.   Fire resistance. The structure in a building must withstand the effects of a fire and remain standing while the building is evacuated and the fire is extinguished. A concrete building inherently has a 1- to 3-hour fire rating without special fireproofing or other details. Structural steel or timber buildings must be fireproofed to attain similar fire ratings.

4. Rigidity. The occupants of a building may be disturbed if their building oscillates in the wind or the floors vibrate as people walk by. Due to the greater stiffness and mass of a concrete structure, vibrations are seldom a problem.

5. Low maintenance. Concrete members inherently require less maintenance than do structural steel or timber members. This is particularly true if dense, air-entrained concrete has been used for surfaces exposed to the atmosphere, and if care has been taken in the design to provide adequate drainage off and away from the structure.

6. Availability of materials. Sand, gravel, cement, and concrete mixing facilities are very widely available, and reinforcing steel can be transported to most job sites more easily than can structural steel. As a result, reinforced concrete is frequently used in remote areas.

On the other hand, there are a number of factors that may cause one to select a material other than reinforced concrete. These include:

1. Low tensile strength. As stated earlier, the tensile strength of concrete is much lower than its compressive strength (about $\frac{1}{10}$), and hence concrete is subject to cracking. In structural uses this is overcome by using reinforcement, as shown in Fig. 1–3c, to carry tensile forces and limit crack widths to within acceptable values. Unless care is taken in design and construction, however, these cracks may be unsightly or may allow penetration of water.

2. Forms and shoring. The construction of a cast-in-place structure involves three steps not encountered in the construction of steel or timber structures. These are (a) the construction of the forms, (b) the removal of these forms, and (c) propping or shoring the new concrete to support its weight until its strength is adequate. Each of these steps involves labor and/or materials which are not necessary with other forms of construction.

3. Relatively low strength per unit of weight or volume. The compressive strength of concrete is roughly 5 to 10% that of steel, while its unit density is roughly 30% that of steel. As a result, a concrete structure requires a larger volume and a greater weight of material than does a comparable steel structure. As a result, long-span structures are often built from steel.

4. Time-dependent volume changes. Both concrete and steel undergo approximately the same amount of thermal expansion and contraction. Because there is less mass of steel to be heated or cooled, and because steel is a better conductor than concrete, a steel structure is generally affected by temperature changes to a greater extent than is a concrete structure. On the other hand, concrete undergoes drying shrinkage, which, if restrained, may cause deflections or cracking. Furthermore, deflections will tend to increase with time, possibly doubling, due to creep of the concrete under sustained loads.

1–5 HISTORICAL DEVELOPMENT OF CONCRETE AND REINFORCED CONCRETE AS STRUCTURAL MATERIALS

Cement and Concrete

Lime mortar was first used in structures in the Minoan civilization in Crete about 2000 B.C. and is still used in some areas. This type of mortar had the disadvantage of gradually dissolving when immersed in water and hence could not be used for exposed joints or underwater joints. About the third century B.C. the Romans discovered a fine sandy volcanic ash which, when mixed with lime mortar, gave a much stronger mortar that could be used under water.

The most remarkable concrete structure built by the Romans was the dome of the Pantheon in Rome, completed in A.D. 126. This dome has a span of 144 ft, a span not exceeded until the nineteenth century. The lowest part of the dome was concrete with aggregate consisting of broken bricks. As the builders approached the top of the dome they used lighter and lighter aggregates, using pumice at the top to reduce the dead-load moments. Although the outside of the dome was, and still is, covered with decorations, the marks of the forms are still visible on the inside.[1-3]

While designing the Eddystone Lighthouse off the south coast of England just before A.D. 1800, the English engineer John Smeaton discovered that a mixture of burned limestone and clay could be used to make a cement that would set under water and be water resistant. Owing to the exposed nature of this lighthouse, however, Smeaton reverted to the tried-and-true Roman cement and mortised stonework.

In the ensuing years a number of people used Smeaton's material, but the difficulty of finding limestone and clay in the same quarry greatly restricted its use. In 1824, Joseph Aspdin mixed ground limestone and clay from different quarries and heated them in a kiln to make cement. Aspdin named his product Portland cement because concrete made from it resembled Portland stone, a high-grade limestone from the Isle of Portland in the south of England. This cement was used by Brunel in 1828 for the mortar used in the masonry liner of a tunnel under the Thames River and in 1835 for mass concrete piers for a bridge. Occasionally in the production of cement the mixture would be overheated, forming a hard clinker which was considered to be spoiled and was discarded. In 1845, I. C. Johnson found that the best cement resulted from grinding this clinker. This is the material now known as portland cement. Portland cement was produced in Pennsylvania in 1871 by D. O. Saylor and about the same time in Indiana by T. Millen of South Bend, but it was not until the early 1880s that significant amounts were produced in the United States.

Reinforced Concrete

W. B. Wilkinson of Newcastle-upon-Tyne obtained a patent in 1854 for a reinforced concrete floor system that used hollow plaster domes as forms. The ribs between the forms were filled with concrete and were reinforced with discarded steel mine-hoist ropes in the center of the ribs. In France, Lambot built a rowboat of concrete reinforced with wire in 1848 and patented it in 1855. His patent included drawings of a reinforced concrete beam and a column reinforced with four round iron bars. In 1861, another Frenchman, Coignet, published a book illustrating uses of reinforced concrete.

The American lawyer and engineer Thaddeus Hyatt experimented with reinforced concrete beams in the 1850s. His beams had longitudinal bars in the tension zone and vertical stirrups for shear. Unfortunately, Hyatt's work was not known until he privately published a book describing his tests and building system in 1877.

Perhaps the greatest incentive to the early development of the scientific knowledge of reinforced concrete came from the work of Joseph Monier, owner of a French nursery garden. Monier began experimenting about 1850 with concrete tubs reinforced with iron for planting trees. He patented his idea in 1867. This patent was rapidly followed by patents for reinforced pipes and tanks (1868), flat plates (1869), bridges (1873), and stairs (1875). In 1880–1881, Monier received German patents for many of the same applications. These were licensed to the construction firm Wayss and Freitag, which commissioned Professors Mörsch and Bach of the University of Stuttgart to test the strength of reinforced concrete and commissioned Mr. Koenen, chief building inspector for Prussia, to develop a method of computing the strength of reinforced concrete. Koenen's book, published in 1886, presented an analysis which assumed that the neutral axis was at the midheight of the member.

The first reinforced concrete building in the United States was a house built on Long Island in 1875 by W. E. Ward, a mechanical engineer. E. L. Ransome of California experimented with reinforced concrete in the 1870s and patented a twisted steel reinforcing bar in 1884. In the same year, Ransome independently developed his own set of design procedures. In 1888 he constructed a building having cast-iron columns and a reinforced concrete floor system consisting of beams and a slab made from flat metal arches covered with concrete. In 1890, Ransome built the Leland Stanford, Jr. Museum in San Francisco. This two-story building used discarded cable car rope as beam reinforcement. In 1903 in Pennsylvania he built the first building in the United States completely framed with reinforced concrete.

In the period from 1875 to 1900, the science of reinforced concrete developed through a series of patents. An English textbook published in 1904 listed 43 patented systems, 15 in France, 14 in Germany or Austria–Hungary, 8 in the United States, 3 in the United Kingdom, and 3 elsewhere. Most of these differed in the shape of the bars and the manner in which the bars were bent.

From 1890 to 1920, practicing engineers gradually gained a knowledge of the mechanics of reinforced concrete, as books, technical articles, and codes presented the theories. In an 1894 paper to the French Society of Civil Engineers, Coignet (son of the earlier Coignet) and de Tedeskko extended Koenen's theories to develop the working stress design method for flexure, which was used universally from 1900 to 1950. During the past seven decades extensive research has been carried out on various aspects of reinforced concrete behavior, resulting in the current design procedures.

Prestressed concrete was pioneered by E. Freyssinet, who in 1928 concluded that it was necessary to use high-strength steel wire for prestressing because the creep of concrete dissipated most of the prestress force if normal reinforcing bars were used to develop the prestressing force. Freyssinet developed anchorages for the tendons and designed and built a number of pioneering bridges and structures.

Design Specifications for Reinforced Concrete

The first set of building regulations for reinforced concrete were drafted under the leadership of Professor Mörsch of the University of Stuttgart and were issued in Prussia in 1904. Design regulations were issued in Britain, France, Austria, and Switzerland between 1907 and 1909.

The American Railway Engineering Association appointed a Committee on Masonry in 1890. In 1903 this committee presented specifications for portland cement concrete. Between 1908 and 1910 a series of committee reports led to the *Standard Building Regulations for the Use of Reinforced Concrete* published in 1910[1-4] by the National Association of Cement Users which subsequently became the American Concrete Institute.

A Joint Committee on Concrete and Reinforced Concrete was established in 1904 by the American Society of Civil Engineers, American Society for Testing and Materials, the American Railway Engineering Association, and the Association of American Portland Cement Manufacturers. This group was later joined by the American Concrete Institute. Between 1904 and 1910 the Joint Committee carried out research. A preliminary report issued in 1913[1-5] lists the more important papers and books on reinforced concrete published between 1898 and 1911. The final report of this committee was published in 1916.[1-6] The history of reinforced concrete building codes in the United States was reviewed in 1954 by Kerekes and Reid.[1-7]

1–6 BUILDING CODES AND THE ACI CODE

The design and construction of buildings is regulated by municipal bylaws called *building codes*. These exist to protect the public's health and safety. Each city and town is free to write or adopt its own building code, and in that city or town, only that particular code has legal status. Because of the complexity of building code writing, cities in the United States generally base their building codes on one of three model codes: the *Uniform Building Code*,[1-8] the *Standard Building Code*,[1-9] or the *Basic Building Code*.[1-10] These codes cover such things as use and occupancy requirements, fire requirements, heating and ventilating requirements, and structural design.

The definitive design specification for reinforced concrete buildings in North America is the *Building Code Requirements for Reinforced Concrete (ACI–318–95)*,[1-11] which is explained in a *Commentary*.[1-11]

This code, generally referred to as the *ACI Code*, has been incorporated in most building codes in the United States and serves as the basis for comparable codes in Canada, New Zealand, Australia, and parts of Latin America. The ACI Code has legal status only if adopted in a local building code.

The ACI Code undergoes a major revision every six years. ACI–318–95 is the revision published in 1995. An interim revision or supplement is published halfway between the major revisions. This book refers extensively to the 1995 ACI Code. It is recommended that the reader have a copy available.

The term *structural concrete* is used to refer to the entire range of concrete structures from *plain concrete* without any reinforcement; through ordinary reinforced concrete, reinforced with normal reinforcing bars; through *partially prestressed concrete*, generally containing both reinforcing bars and prestressing tendons; to *fully prestressed concrete*, with enough prestress to prevent cracking in everyday service. In 1995 the title of the ACI Code was changed from *Building Code Requirements for Reinforced Concrete* to *Building Code Requirements for Structural Concrete* to emphasize that the code deals with the entire spectrum of structural concrete.

The rules for the design of concrete highway bridges are specified in the *Standard Specifications for Highway Bridges*, American Association of State Highway and Transportation Officials, Washington, D.C.[1-12]

Each nation or group of nations in Europe has its own building code for reinforced concrete. The *CEB-FIP Model Code for Concrete Structures*,[1-13] published in 1978 and revised in 1990 by the Comité Euro-International du Béton, Lausanne, is intended to serve as the basis for future attempts to unify European codes. This code and the ACI Code are similar in many ways.

2

The Design Process

2–1 OBJECTIVES OF DESIGN

The structural engineer is a member of a team whose members work together to design a building, bridge, or other structure. In the case of a building, an architect generally provides the overall layout, and mechanical, electrical, and structural engineers design individual systems within the building.

The structure should satisfy four major criteria:

1. Appropriateness. The arrangement of spaces, spans, ceiling heights, access, and traffic flow must complement the intended use. The structure should fit its environment and be aesthetically pleasing.

2. Economy. The overall cost of the structure should not exceed the client's budget. Frequently, teamwork in design will lead to overall economies.

3. Structural adequacy. Structural adequacy involves two major aspects.

(a) A structure must be strong enough to safely support all anticipated loadings.

(b) A structure must not deflect, tilt, vibrate, or crack in a manner that impairs its usefulness.

4. Maintainability. A structure should be designed to require a minimum of maintenance and to be able to be maintained in a simple fashion.

2–2 DESIGN PROCESS

The design process is a sequential and iterative decision making process. The three major phases are:

1. Definition of the client's needs and priorities. All buildings or other structures are built to fulfill a need. It is important that the owner or user be involved in determining

11

the attributes of the proposed building. These include functional requirements, aesthetic requirements, and budgetary requirements. The latter include first cost, rapid construction to allow early occupancy, minimum upkeep, and other factors.

2. Development of concept of project. Based on the client's needs and priorities, a number of possible layouts are developed. Preliminary cost estimates are made and the final choice of the system to be used is based on how well the overall design satisfies the prioritized needs within the budget available.

During this stage the overall structural concept is selected. Based on approximate analyses of the moments, shears, and axial forces, preliminary member sizes are selected for each potential scheme. Once this is done, it is possible to estimate costs and select the most desirable structural system.

The overall thrust in this stage of the structural design is to satisfy the design criteria dealing with appropriateness, economy, and to some extent, maintainability.

3. Design of individual systems. Once the overall layout and general structural concept have been selected, the structural system can be designed. Structural design involves three main steps. Based on the preliminary design selected in phase 2, a *structural analysis* is carried out to determine the moments, shears, and axial forces in the structure. The individual members are then *proportioned* to resist these forces. The proportioning, sometimes referred to as *member design*, must also consider overall aesthetics, the constructability of the design, and the maintainability of the final structure. The final stage in the design process is to prepare construction drawings and specifications.

2–3 LIMIT STATES AND THE DESIGN OF REINFORCED CONCRETE

Limit States

When a structure or structural element becomes unfit for its intended use, it is said to have reached a *limit state*. The limit states for reinforced concrete structures can be divided into three basic groups:

1. Ultimate limit states. These involve a structural collapse of part or all of the structure. Such a limit state should have a very low probability of occurrence since it may lead to loss of life and major financial losses. The major ultimate states are:

(a) Loss of equilibrium of a part or all of the structure as a rigid body. Such a failure would generally involve tipping or sliding of the entire structure and would occur if the reactions necessary for equilibrium could not be developed.

(b) Rupture of critical parts of the structure, leading to partial or complete collapse. The majority of this book deals with this limit state. Chapters 4 and 5 consider flexural failures; Chap. 6, shear failures; and so on.

(c) Progressive collapse. In some cases a minor localized failure may cause adjacent members to be overloaded and fail, until the entire structure has collapsed. Progressive collapse is prevented or slowed by correct structural detailing to tie the structure together and to provide alternative load paths in case of a localized failure.[2-1,2-2] Since such failures may occur during construction, the designer should be aware of construction loads and procedures. A structure is said to have *general structural integrity* if it is resistant to progressive collapse. Ways of providing structural integrity are discussed in Sec. 1.4 of the Commentary of Ref. 2–2.

(d) Formation of a plastic mechanism. A mechanism is formed when the reinforcement yields to form plastic hinges at enough sections to make the structure unstable.

(e) Instability due to deformations of the structure. This type of failure involves buckling and is discussed more fully in Chap. 12.

(f) Fatigue. Fracture of members due to repeated stress cycles of service loads may cause collapse.

2. Serviceability limit states. These involve disruption of the functional use of the structure but not collapse per se. Since there is less danger of loss of life, a higher probability of occurrence can generally be tolerated than in the case of an ultimate limit state. The major serviceability limit states include:

(a) Excessive deflections for normal service. Excessive deflections may cause machinery to malfunction, may be visually unacceptable, and may lead to damage to nonstructural elements or to changes in the distribution of forces. In the case of very flexible roofs, the deflections due to the weight of water on the roof may lead to increased deflections, increased depth of water, and so on, until the capacity of the roof is exceeded. This is a ponding failure and in essence is a collapse brought about by a lack of serviceability.

(b) Excessive crack width. Although reinforced concrete must crack before the reinforcement can act, it is possible to detail the reinforcement to minimize the crack widths. Excessive crack widths lead to leakage through the cracks, corrosion of the reinforcement, and gradual deterioration of the concrete.

(c) Undesirable vibrations. Vertical vibrations of floors or bridges and lateral and torsional vibrations of tall buildings may disturb the users. Vibration has rarely been a problem in reinforced concrete buildings.

Design for serviceability is discussed in Chap. 9.

3. Special limit states. This class of limit states involves damage or failure due to abnormal conditions or abnormal loadings and includes:

(a) Damage or collapse in extreme earthquakes

(b) Structural effects of fire, explosions, or vehicular collisions

(c) Structural effects of corrosion or deterioration

(d) Long-term physical or chemical instability (normally not a problem with concrete structures).

Limit States Design

Limit states design is a process that involves:

1. Identification of all potential modes of failure (i.e., identification of the significant limit states)

2. Determination of acceptable levels of safety against occurrence of each limit state

For normal structures this step is carried out by the building code authorities, who specify the load combinations and check factors to be used. For unusual structures the engineer may need to check whether the normal levels of safety are adequate.

3. Consideration by the designer of the significant limit states

Frequently, for buildings, a limit states design is carried out starting by proportioning for the ultimate limit states followed by a check of whether the structure will exceed any of the serviceability limit states. This sequence is followed since the major function of structural members in buildings is to resist loads without endangering the occupants. For a water tank, however, the limit state of excessive crack width is of equal importance to any of the ultimate limit states if the structure is to remain watertight. In such a structure the design might start with a consideration of the limit state of crack width, followed by a check of the ultimate limit states. In the design of support beams for an elevated monorail, the smoothness of the ride is extremely important, and the limit state of deflection may govern the design.

Basic Design Relationship

Figure 2–1a shows a beam that supports its own dead weight, w, plus some applied loads, P_1, P_2, and P_3. These cause bending moments, distributed as shown in Fig. 2–1b. The bending moments are obtained directly from the loads using the laws of statics, and for a given span and combination of loads w, P_1, P_2, and P_3, the moment diagram is independent of the composition or size of the beam. The bending moment is referred to as a *load effect*. Other load effects include shear force, axial force, torque, deflection, and vibration.

Figure 2–2a shows flexural stresses acting on a beam cross section. The compressive stresses and tensile stresses in Fig. 2–2a can be replaced by their resultants, C and T, as shown in Fig. 2–2b. The resulting couple is called an *internal resisting moment*. The internal resisting moment when the cross section fails is referred to as the *moment capacity* or *moment resistance*. The word "resistance" can also be used to describe shear resistance or axial load resistance.

The beam shown in Fig. 2–2 will safely support the loads if at every section the resistance of the member exceeds the effects of the loads:

$$\text{resistances} \geq \text{load effects} \qquad (2-1)$$

To allow for the possibility that the resistances may be less than computed, and the load effects may be larger than computed, *strength reduction factors*, ϕ, less than 1, and *load factors*, α, greater than 1, are introduced:

$$\phi R_n \geq \alpha_1 S_1 + \alpha_2 S_2 + \cdots \qquad (2-2)$$

where R_n stands for nominal resistance and S stands for load effects based on the specified loads. Written in terms of moments, Eq. 2–2 becomes

$$\phi_M M_n \geq \alpha_D M_D + \alpha_L M_L + \cdots \qquad (2-3a)$$

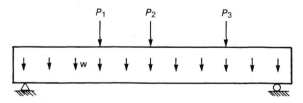

(a) Beam

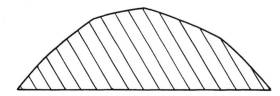

(b) Load effect — bending moment

Fig. 2–1
Loads and load effects.

(a) Stresses acting on a cross section.

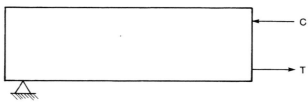

Fig. 2–2
Internal resisting moment.

(b) Internal couple.

where M_n is the *nominal moment resistance*. The word "*nominal*" implies that this resistance is a computed value based on the specified concrete and steel strengths and the dimensions shown on the drawings. M_D and M_L are the bending moments (load effects) due to the specified dead load and specified live load, respectively; ϕ_M is a strength reduction factor for moment; and α_D and α_L are load factors for dead and live load, respectively. The strength reduction factors are sometimes referred to as *resistance factors*.

Similar equations can be written for shear, V, or axial force, P:

$$\phi_V V_n \geq \alpha_D V_D + \alpha_L V_L + \cdots \qquad (2\text{–}3b)$$

$$\phi_P P_n \geq \alpha_D P_D + \alpha_L P_L + \cdots \qquad (2\text{–}3c)$$

Equation 2–1 is the basic limit states design equation. Equations 2–3 are special forms of this basic equation. Equation 11–1 of the ACI Code, for example, is the same as Eq. 2–3b except that in that equation, the group of terms $(\alpha_D V_D + \alpha_L V_L + \cdots)$ is expressed as V_u, which is defined as the *factored shear force*. Throughout the ACI Code, the symbol U is used to refer to the combination $(\alpha_D D + \alpha_L L + \cdots)$. This combination is referred to as the *required strength* or the *factored loads*. The symbols M_u, V_u, T_u, and so on, refer to *factored load effects* calculated from the factored loads, U, hence the subscript u.

2–4 STRUCTURAL SAFETY

There are three main reasons why some sort of safety factors, such as load and resistance factors, are necessary in structural design:

1. Variability in resistance. The actual strengths (resistances) of beams, columns, or other structural members will almost always differ from the values calculated by the designer. The main reasons for this are:[2–3]

(a) Variability of the strengths of concrete and reinforcement

(b) Differences between the as-built dimensions and those shown on the structural drawings

(c) Effects of simplifying assumptions made in deriving the equations for member resistance

A histogram of the ratio of beam moment capacities observed in tests, M_{test}, to the nominal strengths computed by the designer, M_n, is plotted in Fig. 2–3. Although the mean strength is roughly 1.05 times the nominal strength in this sample, there is a definite chance that some beam cross sections will have a lower capacity than computed. The variability shown here is due largely to the simplifying assumptions made in computing the resisting moment, M_n.

2. Variability in loadings. All loadings are variable, especially live loads and environmental loads due to snow, wind, or earthquakes. Figure 2–4a compares the sustained component of live loads measured in a family of 151-ft^2 areas in offices. Although the average sustained live load was 13 psf in this sample, 1% of the measured loads exceeded 44 psf. For this type of occupancy and area, building codes specify live loads of 50 psf. For larger areas the mean sustained live load remains close to 13 psf but the variability decreases, as shown in Fig. 2–4b. A transient live load representing unusual loadings due to parties, temporary storage, and so on, must be added to get the total live load. As a result, the maximum live load on a given office will generally exceed the 13 to 44 psf quoted above.

In addition to actual variations in the loads themselves, the assumptions and approximations made in carrying out structural analyses lead to differences between the actual forces and moments and those computed by the designer.[2-3] Due to the variabilities of resistances and load effects, there is a definite chance that a weaker-than-average structure may be subjected to a higher-than-average load. In extreme cases failure may occur. The

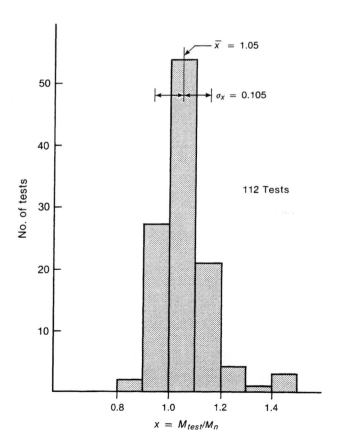

Fig. 2–3
Comparison of measured and computed failure moments based on all data for reinforced concrete beams with $f'_c > 2000$ psi, Ref. 2–4.

The Design Process

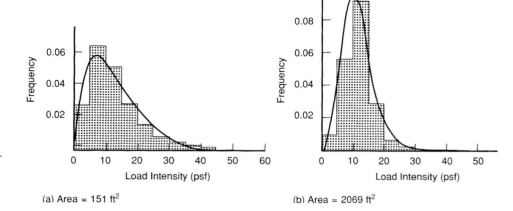

(a) Area = 151 ft²

(b) Area = 2069 ft²

load factors and resistance factors in Eqs. 2–2 and 2–3 are selected to reduce the probability of failure to a very small level.

A third factor that must be considered in establishing the level of safety required in a particular structure is:

3. Consequences of failure. A number of subjective factors must be considered in determining an acceptable level of safety for a particular class of structure. These include such things as:

(a) Cost of clearing the debris and replacing the structure and its contents.

(b) Potential loss of life. It may be desirable to have a higher factor of safety for an auditorium than for a storage building.

(c) Cost to society in lost time, lost revenue or indirect loss of life or property due to a failure. For example, the failure of a bridge may result in intangible costs due to traffic jams, and so on, which could approach the cost of the damage.

(d) Type of failure, warning of failure, existence of alternative load paths. If the failure of a member is preceded by excessive deflections, as in the case of a flexural failure of a reinforced concrete beam, the persons endangered by the impending collapse will be warned and will have a chance to leave the building prior to failure. This may not be possible if a member fails suddenly without warning, as may be the case with a tied column. Thus the required level of safety may not need to be as high for a beam as for a column. In some structures, the yielding or failure of one member causes a redistribution of load to adjacent members. In other structures, the failure of one member causes complete collapse. If no redistribution is possible, a higher level of safety is required.

2–5 PROBABILISTIC CALCULATION OF SAFETY FACTORS

The distribution of a population of resistances, R, of a group of similar structures is plotted on the horizontal axis in Fig. 2–5. This is compared to the distribution of the maximum load effects, S, expected to occur on those structures during their lifetimes, plotted on the vertical axis in the same figure. For consistency, both the resistances and the load effects are expressed in terms of a quantity such as bending moment. The 45° line in this figure corresponds to a load effect equal to the resistance. Combinations of S and R falling above this line correspond to $S > R$ and, hence, failure. Thus load effect S_1 acting on a structure having strength R_1 would cause failure, whereas load effect S_2 acting on a structure having resistance R_2 represents a safe combination.

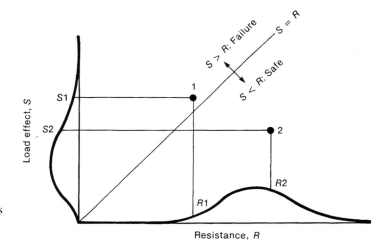

Fig. 2–5
Safe and unsafe combinations
of loads and resistances.
(From Ref. 2–6.)

For a given distribution of load effects, the probability of failure can be reduced by increasing the resistances. This would correspond to shifting the distribution of resistances to the right in Fig. 2–5. The probability of failure could also be reduced by reducing the dispersion of the resistances.

The term $Y = R - S$ is called the *safety margin*. By definition, failure will occur if Y is negative, shown shaded in Fig. 2–6. The *probability of failure*, P_f, is the chance that a particular combination of R and S will give a negative value of Y. This probability is equal to the ratio of the shaded area to the total area under the curve in Fig. 2–6. This can be expressed as

$$P_f = \text{probability that } [Y < O] \qquad (2\text{–}4)$$

The function Y has a mean value $\overline{Y}$ and a standard deviation σ_Y. From Fig. 2–6 it can be seen that $\overline{Y} = 0 + \beta\sigma_Y$, where $\beta = \overline{Y}/\sigma_Y$. If the distribution is shifted to the right by making $\overline{Y}$ larger, β will increase, and the shaded area, P_f, will decrease. Thus P_f is a function of β. The factor β is called the *safety index*.

If Y follows a standard statistical distribution, and if $\overline{Y}$ and σ_Y are known, the probability of failure can be calculated or obtained from statistical tables as a function of the type of distribution and the value of β. Thus if Y follows a normal distribution and β is 3.5, then $\overline{Y} = 3.5\sigma_Y$ and from tables of the normal distribution, P_f is $1/9091$ or 1.1×10^{-4}. This suggests that roughly 1 in every 10,000 structural members designed on the basis of $\beta = 3.5$ will fail due to excessive load or understrength sometime during its lifetime.

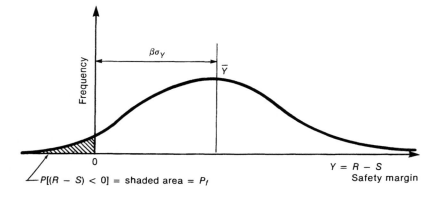

Fig. 2–6
Safety margin, probability of
failure, safety index. (From
Ref. 2–6.)

The appropriate values of P_f and hence β are chosen bearing in mind the consequences of failure. Based on current design practice, β is taken between 3 and 3.5 for ductile failures with average consequences of failure and between 3.5 and 4 for sudden failures or failures having serious consequences of failure.[2-6, 2-7]

Because the strengths and loads vary independently, it is desirable to have one factor or series of factors to account for the variability in resistances, and a second series of factors to account for the variability in load effects. These are referred to, respectively, as *resistance factors, ϕ*, and *load factors, α*. The resulting design equations are Eqs. 2–2 and 2–3.

The derivation of probabilistic equations for calculating values of ϕ and α is summarized and applied in Refs. 2–6, 2–7, and 2–8.

The resistance and load factors in the ACI Code were based on a statistical model which assumed that if there were a $1/1000$ chance of an "overload" and a $1/100$ chance of "understrength," the chance that an "overload" and an "understrength" would occur simultaneously is $1/1000 \times 1/100$ or 1×10^{-6}. Thus the ϕ factors were originally derived so that a strength of ϕR_n would be exceeded 99 out of 100 times. The ϕ factors for columns were then divided by 1.1 since the failure of a column has serious consequences. The ϕ factors for tied columns that fail in a brittle manner were divided by 1.1 a second time to reflect the consequences of the mode of failure. The original derivation is summarized in the appendix to Ref. 2–6.

2–6   DESIGN PROCEDURES SPECIFIED IN THE ACI BUILDING CODE

The 1995 ACI Building Code allows two alternative design procedures. The one most commonly used involves load and resistance factors and is referred to as *strength design*. This procedure is essentially limit states design except that primary attention is always placed on the ultimate limit states with the serviceability limit states being checked after the original design is completed.

ACI Secs. 9.1.1 and 9.1.2 present the basic limit states design philosophy of that code.

9.1.1—Structures and structural members shall be designed to have design strength at all sections at least equal to the required strengths calculated for the factored loads and forces in such combinations as are stipulated in this code.

The term *design strength* refers to ϕR_n, and the term *required strength* refers to the load effects calculated from factored loads, $\alpha_D D + \alpha_L L + \cdots$.

9.1.2—Members also shall meet all other requirements of this code to insure adequate performance at service load levels.

This clause refers primarily to control of deflections and excessive crack width.

Alternatively, *working stress design* can be used. Here design is based on *working loads*, also referred to as *service loads* or *unfactored loads*. In flexure, the maximum elastically computed stresses cannot exceed *allowable stresses* or *working stresses* of 0.4 to 0.5 times the concrete and steel strengths. The 1995 ACI Code refers to this procedure as the *alternate design procedure*. It is permitted in ACI Sec. 8.1.2. and details are given in ACI Appendix A.

The working stress design method assumes that the ultimate limit states will automatically be satisfied by the use of allowable stresses. Depending on the variability of the materials and loads, this is not necessarily so. ACI Sec. A.1.4 requires the designer to consider the deflection limit state and the crack-width limit state.

The drawbacks of working stress design are discussed in Refs. 2–6 and 2–7. The most serious drawbacks stem from its inability to account properly for the variability of the resistances and loads; lack of any knowledge of the level of safety; and its inability to deal with groups of loads where one load increases at a different rate than the others. This last criticism is especially serious when a relatively constant load such as dead load counteracts the effects of a highly variable load such as wind, as illustrated in Fig. 2–7. Here a 20% increase in the wind causes a 20% increase in the maximum flexural stresses (from 500 psi to 600 psi) as expected, but causes a 100% increase in the stresses at point A in Fig. 2–7d.

Plastic design, also referred to as *limit design* (not to be confused with limit states design) or *capacity design*, is a design process that considers the redistribution of moments as successive cross sections yield, forming *plastic hinges* which lead to a plastic mechanism. These concepts are of considerable importance in seismic design, where the ductility of the structure leads to a decrease in the forces that must be resisted by the structure.

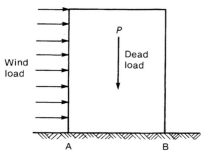

(a) Structure.

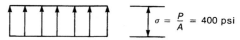

$$\sigma = \frac{P}{A} = 400 \text{ psi}$$

(b) Dead load stresses.

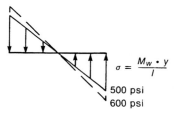

$$\sigma = \frac{M_w \cdot y}{I}$$

500 psi
600 psi

(c) Wind load stresses.

200 psi
100 psi

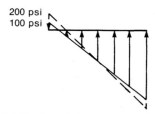

Fig. 2–7
Stresses due to counteracting loads. (From Ref. 2–7.)

(d) Combined load stresses.

Factored Loads, Required Strength

The ACI Code presents two different sets of load factors and load combinations. The load factors and combinations introduced in the 1963 ACI Code and amended in the 1971 ACI Code are presented in the body of the code in ACI Secs. 9.2.1 to 9.2.7. In 1982, the American National Standards Institute published *Minimum Design Loads for Buildings and Other Structures*, ANSI A58.1–1982 (recently issued as ASCE 7–95), which contained a unified set of load factors and load combinations, developed in Ref. 2–7, for use in the design of steel, timber, brick, and concrete structures. These load factors and load combinations differ from those in the ACI Code. Appendix C of ACI 318–95 presents the ASCE 7–95 load factors and load combinations in ACI Commentary Sec. RC.1 and presents compatible resistance factors in ACI Code Sec. C.1. The use of these is permitted when the structure has mixed construction, such as steel frames and concrete shear walls, for example. Design is to be based entirely on the load and resistance factors in ACI Chap. 9 or entirely on those in ACI Appendix C. When the load factors from ACI Appendix C are used, the resistance factors, ϕ, in ACI Appendix C must be used. In this book, the load factors in ACI Chap. 9 will be used throughout.

ACI Secs. 9.2.1 to 9.2.7 present a series of load factors and combinations of factored loads to be used in calculating the load effects. In the code the symbol U refers to a combination of factored loads; and the symbols M_u, V_u, T_u, and so on, refer to factored load effects (moments, shears, torques) calculated from U. The subscript u is reserved for load effects calculated from the factored loads, U. The ACI Code uses the term *required strength* to refer to factored load effects.

In the design of buildings that are not subjected to significant wind or earthquake forces, or for members unaffected by wind or earthquakes, the factored loads are computed from

$$U = 1.4D + 1.7L \qquad (2\text{--}5)$$
$$\text{(ACI Eq. 9--1)}$$

where D and L are the specified dead and live loads.

If wind loads do affect the design, ACI Sec. 9.2.2 requires that three combinations of loads be considered and the design based on the largest values of U of either sign at each critical section:

1. Where the load effects due to wind add to those due to dead or live loads:

$$U = 0.75(1.4D + 1.7L + 1.7W) \qquad (2\text{--}6)$$
$$\text{(ACI Eq. 9--2)}$$
$$U = 0.75(1.4D + 1.7W) \qquad (2\text{--}7)$$

2. Where the effects of dead loads stabilize the structure against wind loads, as in Fig. 2–7:

$$U = 0.9D + 1.3W \qquad (2\text{--}8)$$
$$\text{(ACI Eq. 9--3)}$$

But for any combination of D, L, and W, the required strength shall not be less than as given by Eq. 2–5 (ACI Eq. 9–1).

Similar load combinations are given in ACI Sec. 9.2.3 for earthquake loadings, Sec. 9.2.4 for lateral earth pressure, Sec. 9.2.5 for fluid pressures (in tanks, etc.), Sec. 9.2.6 for impact loads, and Sec. 9.2.7 for differential settlement, creep, shrinkage, and temperature change.

Equation 2–8 would be used to compute the stresses at point A in Fig. 2–7. At point B the most severe of Eqs. 2–5 to 2–7 would apply.

In the analysis of a building frame, it is frequently best to elastically analyze the structure three times, once each for 1.0D, 1.0L, and 1.0W and to combine the resulting moments, shears, and so on, for each member according to Eqs. 2–5 to 2–8. (Exceptions to this are analyses of cases in which linear superposition does not apply, such as second-order analyses of frames. These must be carried out at the factored load level.) The procedure used is illustrated in Example 2–1.

EXAMPLE 2–1 Computation of Factored Load Effects

Figure 2–8 shows a beam and column from a concrete building frame. The loads per foot on the beam are dead load, $D = 1.58$ kips/ft, and live load, $L = 0.75$ kip/ft. The moments and shears in a beam and the columns over and under the beam due to 1.0D, 1.0L, and 1.0W are shown in Fig. 2–8b to d.

Compute the required strengths using Eq. 2–5 through 2–8. For the moment at section A:

$$\text{(a)} \quad U = 1.4D + 1.7L$$
$$= 1.4(-39) + 1.7(-19) = -86.9 \text{ ft-kips}$$
$$\text{(b)} \quad U = 0.75(1.4D + 1.7L \pm 1.7W)$$
$$= 1.05(-39) + 1.275(-19) \pm 1.275(84)$$
$$= -172.3 \text{ ft-kips and } +41.9 \text{ ft-kips}$$

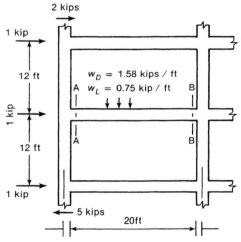

(a) Frame.

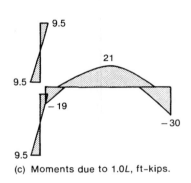

(c) Moments due to 1.0L, ft–kips.

Fig. 2–8
Moment diagrams—Example 2–1.

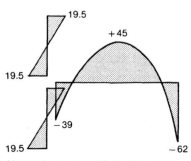

(b) Moments due to 1.0D, ft–kips.

(d) Moments due to 1.0W, ft–kips.

The Design Process

The positive and negative values of wind load moments are due to winds alternatively from the two sides of the building.

$$\begin{aligned}
\text{(c)} \quad U &= 0.75(1.4D \pm 1.7W) \\
&= 1.05 \times 39 \pm 1.275 \times 84 \\
&= -148.1 \text{ ft-kips and } +66.2 \text{ ft-kips}
\end{aligned}$$

It is not necessary to check the fourth load case because this problem does not involve uplift or overturning. Thus the required strengths, M_u, at section A–A are +66.2 ft-kips and −172.3 ft-kips. ∎

This computation is repeated for a sufficient number of sections to make it possible to draw shearing force and bending moment envelopes for the beam. (Bending moment envelopes are discussed in Sec. 10–3.) The solution of the four equations given above can easily be programmed for a programmable calculator with D, L, and W as input values and the seven values of U and/or the maximum positive and negative values of the factored load effect as output.

Factored Resistance, Design Strength

In the basic limit states design equations 2–2 and 2–3, the left-hand side (ϕR_n, ϕM_n, etc.) is referred to as the *factored resistance*. ACI Sec. 9.3 uses the term *design strength* to refer to *factored resistance*. The resistance factors, ϕ, are given in ACI Sec. 9.3.2, where they are called *strength reduction factors*. The following values are specified:

1. Flexure, with or without axial tension $\phi = 0.90$
2. Axial tension   0.90
3. Axial compression, with or without flexure:
    (a) Members with spiral reinforcement conforming to ACI Sec. 10.9.3   0.75
    (b) Other reinforced members   0.70
    Note that ϕ may be increased for very small axial forces as explained and illustrated in Sec. 11–4.
4. Shear and torsion   0.85
5. Bearing on concrete   0.70

Although not explained above, ACI Sec. 9.3.2.2 specifies a transition from $\phi = 0.90$ for flexure or axial tension to $\phi = 0.75$ or 0.70 for axial compression with or without flexure. Appendix B of ACI 318–95 presents a different definition of this transition than ACI Sec. 9.3.2.2. The method in Appendix B unifies the concepts used in making this transition for reinforced and prestressed concrete, and for beams and columns. For this reason ACI Appendix B will be used to define the ϕ factors in this book. The concepts will be discussed and illustrated in Secs. 4–3 and 11–4. It should be noted that if any part of ACI Appendix B is used, all of it must be used.

In regions of high seismic activity, lower strength reduction factors are used for shear in some cases; see ACI Sec. 9.3.4 and Sec. 19–5 of this book.

2–7   LOADINGS AND ACTIONS

Direct and Indirect Actions

An *action* is anything that gives rise to stresses in a structure. The term *load* or *direct action* refers to concentrated or distributed forces resulting from the weight of the structure and its contents, or pressures due to wind, water, or earth. An *indirect action* or *imposed*

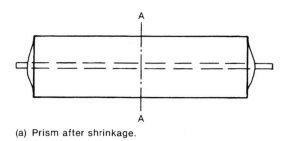

(a) Prism after shrinkage.

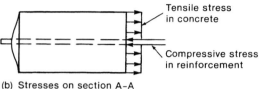

Tensile stress
in concrete

Compressive stress
in reinforcement

(b) Stresses on section A–A

Fig. 2–9
Self equilibrating stresses due
to shrinkage.

deformation is a movement or deformation which does not result from applied loads, but which causes stresses in a structure. Examples are uneven support settlements of continuous beams, and shrinkage of concrete if it is not free to shorten.

Because the stresses due to imposed deformations do not resist an applied load, they are generally *self-equilibrating*. Consider, for example, a prism of concrete with a reinforcing bar along its axis. As the concrete shrinks, its shortening is resisted by the reinforcement. As a result, a compressive force develops in the steel and an equal and opposite tensile force develops in the concrete, as shown in Fig. 2–9. If the concrete cracks due to this tension, the tensile force in the concrete at the crack is zero and for equilibrium, the steel force must also disappear at the cracked section.

Classifications of Loads

Loads may be described by their variability with respect to time and location. A *permanent* load remains roughly constant once the structure is completed. Examples are the self-weight of the structure, and soil pressure against foundations. *Variable* loads such as occupancy loads and wind loads change from time to time. Variable loads may be *sustained loads* of long duration, such as the weight of filing cabinets in an office, or loads of *short duration*, such as weight of people in the same office. Creep deformations of concrete structures result from the permanent loads and the sustained portion of the variable loads. A third category is *accidental loads*, which include vehicular collisions, and explosions.

Variable loads may be *fixed* or *free* in location. Thus the loading in an office building is free since it can occur at any point in the loaded area. A train load on a bridge is not fixed longitudinally but is fixed laterally by the rails.

Loads are frequently classed as *static loads* if they do not cause any appreciable acceleration or vibration of the structure or structural elements, and as *dynamic loads* if they do. Small accelerations are often taken into account by increasing the specified static loads to account for the increase in stresses due to such accelerations and vibrations. Larger accelerations such as those that might occur in highway bridges, crane rails, or elevator supports are accounted for by multiplying the live load by *impact factors,* or dynamic analyses may be used.

Three levels of live load or wind load may be of importance. The load used in calculations involving the ultimate limit states should represent the maximum load on the structure in its lifetime. Wherever possible, therefore, the specified live, snow, and wind loadings should represent the mean value of the maximum lifetime load. In checking the serviceability limit states, it may be desirable to use a *frequent* live load, which is some fraction of the mean maximum lifetime load (generally, 50 to 60%) and for estimating sustained load deflections it may be desirable to consider a *sustained* or *quasi-permanent* live load, which is generally between 20 and 30% of the specified live load. This differentiation is not made in the ACI Code, which assumes that the entire specified load will be the load present in service. As a result, service load deflections and creep deflections of slender columns tend to be overestimated.

Loading Specifications

Cities in the United States generally base their building code on one of three model codes: the *Uniform Building Code*,[2-9] the *Standard Building Code*,[2-10] or the *Basic Building Code*.[2-11] These three codes tend to be similar in many aspects of live loadings but differ considerably in the area of wind loadings. The loadings specified in the three model codes are based in large part on the loads recommended in *ASCE Minimum Design Loads for Buildings and Other Structures* (ASCE 7–95), formerly ANSI A58.1.

It should be emphasized that the basic structural design equation 2–2 implies that if the loads S_1, S_2, and so on, differ from code to code, then the load factors α_1, α_2, and so on, must also differ.

In the following sections, the types of loadings presented in ASCE 7–95 will be reviewed very briefly. This review is intended to describe the characteristics of the various loads. For specific values, the reader should consult the code in effect in his or her own locality.

Dead Loads

The *dead load* on a structural element is the weight of the member itself, plus the weights of all materials permanently incorporated into the structure and supported by the member in question. This includes the weights of permanent partitions or walls, the weights of plumbing stacks, electrical feeders, permanent mechanical equipment, and so on. Tables of dead loads are given in ASCE 7–95.[2-2]

In the design of a reinforced concrete member, it is necessary initially to estimate the weight of the member. Methods of making this estimate are given in Chaps. 4 and 10. Once the member size has been computed, its weight is calculated by multiplying the volume by the density of concrete, taken as 145 lb/ft^3 for plain concrete and 150 lb/ft^3 for reinforced concrete. (5 lb/ft^3 is added to account for reinforcement.) For lightweight concrete members, the density of the concrete must be determined from trial batches or as specified by the producer. In heavily reinforced members, the density of the reinforced concrete may exceed 150 lb/ft^3 when the weight of stirrups and longitudinal steel are included. In extreme cases design should be based on an estimate of the density for the members in question.

When working with SI units (metric units) the weight of a member is calculated by multiplying the volume by the mass density of concrete and the gravitational constant, 9.81 N/kg. In this calculation it is customary to take the mass density of normal density concrete containing an average amount of reinforcement (roughly, 2% by volume) as 2450 kg/m^3, made up of 2300 kg/m^3 for the concrete and 150 kg/m^3 for the reinforcement. The weight of a cubic meter of reinforced concrete is thus 1 m^3 × 2450 kg/m^3 × 9.81 N/kg/1000 = 24.0 kN and its weight density would be 24 kN/m^3.

The dead load referred to in Eqs. 2–5 to 2–8 is the load computed from the dimensions shown on drawings and the assumed densities. It is therefore close to the mean value of this load. Actual dead loads will vary from the calculated values because the actual dimensions and densities may differ from those used in the calculations. Sometimes the materials for the roof, partitions, or walls are chosen on the basis of a low bid, and their actual weights may be unknown at the time of the design. Tabulated densities of materials frequently tend to underestimate the actual dead loads of the material in place in a structure.

Some types of dead load tend to be highly uncertain. These include pavement on bridges, which may be paved several times over a period of time, or where a greater thickness of pavement may be applied to correct sag or alignment problems. Similarly, earth fill over an underground structure may be up to several feet thicker than assumed and may or may not be saturated with water. In the construction of thin curved shell roofs or other lightweight roofs, the concrete thickness may exceed the design values and the roofing may be heavier than assumed, leading to overloads.

If dead load moments, forces, or stresses tend to counteract those due to live loads or wind loads, the designer should carefully examine whether the counteracting dead load will always exist. Thus dead loads due to soil or machinery may not be applied evenly to all parts of the structure at the same time, leading to a critical set of moments, forces, or stresses under partial loads.

It is generally not necessary to checkerboard the self-weight of the structure by using dead-load factors of $\alpha_D = 0.9$ and 1.4 in successive spans because the dead loads in successive spans tend to be highly correlated. On the other hand, it may be necessary to checkerboard the superimposed dead load using load factors of $\alpha_D = 0$ or 1.4 in cases where counteracting dead load may be absent at some stages of construction or use.

Live Loads Due to Use and Occupancy

Most building codes contain a table of design or specified live loads. To simplify the calculations, these are expressed as uniform loads on the floor area. In general, a building live load consists of a sustained portion due to day-to-day use (see Fig. 2–4), and a variable portion generated by unusual events. The sustained portion changes a number of times during the life of the building when tenants change, the offices are rearranged, and so on. Occasionally, high concentrations of live loading may occur during periods when adjacent spaces are remodeled, office parties, temporary storage, and so on. The loading given in building codes is intended to represent the maximum sum of these loads that will occur on a small area during the life of the building. Typical specified live loads are given in Table 2–1.

In buildings where nonpermanent partitions might be erected or rearranged during the life of the building, allowance should be made for the weight of these partitions. ASCE 7–95 specifies that provision for partition weight should be made whether or not partitions are shown on the plans, unless the specified live load exceeds 80 psf. It is customary to represent the partition weight with a uniform load of 20 psf, or a uniform load determined from the actual or anticipated weights of the partitions placed in any probable position. ASCE 7–95 considers this as a live load because it may or may not be present in a given case.

As the loaded area increases, the average maximum lifetime load decreases because, although it is quite possible to have a heavy load on a small area, it is unlikely that this would occur in a large area (Fig. 2–4). This is taken into account by multiplying the specified live loads by a *live-load reduction factor*. In the 1988 edition of ASCE 7–95, this factor is based on the *influence area*, A_I, for the member being designed. To determine the influence area of a given member, one imagines that the member in question is raised by a unit amount, say 1 in. The portion of the loaded area that is raised when this is done is called the influence area, since loads acting anywhere in this area will have a significant effect on the load effects in the member in question. This concept is illustrated in Fig. 2–10 for an interior floor beam and an edge column. For the beam A_I is twice the tributary area

TABLE 2–1   Typical Live Loads Specified in ASCE 7–95

Apartment buildings	
Residential areas and corridors serving them	40 psf
Public rooms and corridors serving them	100 psf
Office buildings	
Lobbies and first-floor corridors	100 psf
Offices	50 psf
Corridors above first floor	80 psf
File and computer rooms shall be designed for heavier loads based on anticipated occupancy	
Schools	
Classrooms	40 psf
Corridors above first floor	80 psf
First-floor corridors	100 psf
Stairs and exitways	100 psf
Storage warehouses	
Light	125 psf
Heavy	250 psf
Stores	
Retail	
Ground floor	100 psf
Upper floors	75 psf
Wholesale, all floors	125 psf

Source: Based on *Minimum Design Loads for Buildings and Other Structures*, ASCE Standard ASCE 7–95, with the permission of the publisher, the American Society of Civil Engineers.

of the beam. For a column it is four times. Since two-way slab design is based on the total moments in one slab panel, the influence area for such a slab is defined by ASCE 7–95 as the panel area.

ASCE 7–95 allows reduced live loads, L, to be used in the design of members with an influence area, A_I of 400 ft^2 or more, given by

$$L = L_0\left(0.25 + \frac{15}{\sqrt{A_I}}\right)$$
(2–9)

where L_0 is the unreduced live load.

The live-load reduction applies only to live loads due to use and occupancy (not for snow, etc.). No reduction is made for areas used as places of public assembly, for garages, or for roofs. In ASCE 7–95, the reduced live load cannot be less than 50% of the unreduced live load for columns supporting one floor or for flexural members, and not less than 40% for other members.

For live loads exceeding 100 psf, no reduction is allowed by ASCE 7–95 except that the design live load on columns supporting more than one floor can be reduced by 20%.

The reduced uniform live loads are then applied to those spans or parts of spans that will give the maximum shears, moments, and so on, at each critical section. This is illustrated in Chap. 10.

The ASCE document requires that office and garage floors and sidewalks be designed to safely support either the reduced uniform design loads, or a concentrated load of 2000 to 8000 lb depending on occupancy, spread over an area of 30 in. by 30 in., whichever causes the worst effect. The concentrated loads are intended to represent heavy items such as office safes, pianos, car wheels, and so on.

The live loads are assumed to be large enough to account for the impact effects of normal use and traffic. Special impact factors are given in the loading specifications for supports of elevator machinery, large reciprocating or rotating machines, and cranes.

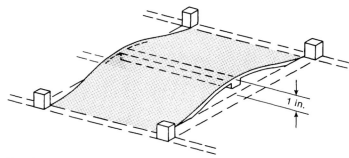

(a) Interior floor beam.

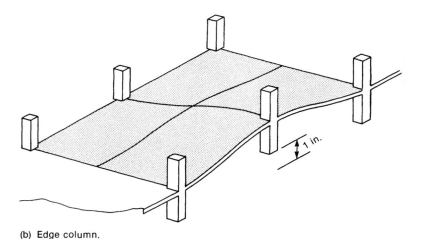

Fig. 2–10
Influence areas.

(b) Edge column.

Classification of Buildings for Wind, Snow, and Earthquake Loads

The ASCE 7–95 requirements for design for wind, snow and earthquake get progressively more restrictive as the level of risk to human life in the event of a collapse increases. These are referred to as *use categories*, and are:

I. Buildings and other structures that represent a low hazard to human life in the event of failure, such as agricultural facilities.

II. Buildings and other structures that do not fall into categories I, III, or IV.

III. Buildings or other structures that represent a substantial hazard to human life in the event of failure, such as assembly occupancies, schools, colleges, jails, and buildings containing significant quantities of toxic or explosive substances.

IV. Buildings and other structures designated as essential facilities, such as hospitals, fire and police stations, communication centers, and power-generating stations and facilities.

Snow Loads

Snow accumulation on roofs is influenced by climatic factors, roof geometry, and the exposure of the roof to the wind. Unbalanced snow loads are very common due to drifting or sliding of snow or due to uneven removal of snow by workers. Large accumulations of snow will often occur adjacent to parapets or points where roof heights change. ASCE 7–95 gives detailed rules for calculating snow loads to account for the effects of snow drifts. It is necessary to design for either a uniform or an unbalanced snow load, whichever gives the worst effect.

Snow load is considered to be a live load when applying the ACI load factors. A live-load reduction factor is not applied to snow loads.

Roof Loads

In addition to snow loads, roofs should be designed for certain minimum loads to account for workers or construction materials on the roof during erection or when repairs are made. Consideration must also be given to loads due to rainwater. Since roof drains are rarely inspected to remove leaves or other debris, ASCE 7–95 requires that roofs be able to support the load of all rainwater that could accumulate on a particular portion of a roof if the primary roof drains were blocked.

If the design snow load is small and the roof span is longer than about 25 ft, rainwater will tend to form ponds in the areas of maximum deflection. The weight of the water in these regions will cause an increase in the deflections, allowing more water to collect, and so on. If the roof is not sufficiently stiff, a *ponding failure* will occur when the weight of ponded water reaches the capacity of the roof members.[2-12]

Construction Loads

During the construction of concrete buildings, the weight of the fresh concrete is supported by formwork which frequently rests on floors lower down in the structure. In addition, construction materials are often piled on floors or roofs during construction. ACI Sec. 6.2.2 states that:

> No construction loads exceeding the combination of superimposed dead load plus specified live load shall be supported on any unshored portion of the structure under construction, unless analysis indicates adequate strength to support such additional loads.

Wind Loads

The pressure exerted by the wind is related to the square of its velocity. Due to the roughness of the earth's surface, the wind velocity at any particular instant consists of an average velocity plus superimposed turbulence, referred to as *gusts*. As a result, a structure subjected to wind loads assumes a basic deflected position due to the average velocity pressure and vibrates from this position due to the gust pressure. In addition, there will generally be deflections transverse to the wind due to vortex shedding as the wind passes the building. The vibrations due to the wind gusts are a function of (1) the relationship between the natural energy of the wind gusts and the energy necessary to displace the building, (2) the relationship between the gust frequencies and the natural frequency of the building, and (3) the damping of the building.[2-13]

Three procedures are specified in ASCE 7–95 for the calculation of wind pressures on buildings. These include the normal "analytical" calculation based on tabulated coefficients, a detailed calculation for tall slender buildings or flexible buildings based on the natural frequency and size of the building, and finally, a recommendation that in unusual cases, a more detailed analysis be carried out, possibly including a wind tunnel investigation.

In the analytical procedure the basic equation for computing the wind pressure on a building is

$$p = qGC_p \qquad (2\text{--}10)$$

where q is either the pressure q_z at height z above ground on the windward wall, or the pressure q_h at the mean roof height h on the roof, wide walls, and leeward wall. Sometimes it is necessary to allow for the effects of internal pressures. This is not generally the case when considering the main wind-resisting system in multistory buildings.

1. **Design pressure, p.** The design pressure is an equivalent static pressure or suction in psf assumed to act perpendicular to the surface in question. On some surfaces it varies over the height; on others it is assumed to be constant.

2. **Velocity pressure, q.** The wind velocity pressure, q psf, is the pressure exerted by the wind on a flat plate suspended in the wind stream. It is calculated as

$$q_z = 0.00256 K_z K_{zt} V^2 I \qquad (2\text{--}11)$$

where

V = basic 3-sec gust wind speed in miles per hour at a height of 33 ft (10 m) above the ground in open terrain

K_z = velocity pressure exposure coefficient, which increases with height above the surface and reflects the roughness on the surface terrain

K_{zt} = allows for wind speed up over hills

I = importance factor, which is a function of the building category

The constant 0.00256 reflects the mass density of the air and accounts for the mixture of units in Eq. 2–11.

Prior to 1995, V was based on the "fastest mile wind," which had a chance of 1 in 50 of being exceeded in any one year. This was the velocity corresponding to the time it took for a 1-mile-long piece of air to pass the wind gauge. In ASCE 7–95 the definition of V was changed to the velocity of a 3-sec gust, which has a 1 in 50 chance of being exceeded in any one year. The 1995 definition gives a much higher value of V than the earlier definition. However, since the gust effect has largely been accounted for by using the 3-sec gust speed, the gust factor, G, in Eq. 2–10 is close to 1.0. The overall result is relatively little change in the design pressure, p.

Maps and tables of V are given in the standard. Special attention must be given to mountainous terrain, gorges, and promontories subject to unusual wind conditions and regions subject to tornadoes. The importance factor, I, ranges from 0.87 for building use category I, 1.0 for normal buildings (building use category II), to 1.15 for building use categories III and IV. These values correspond to mean recurrence intervals of 1 in 25 years for use category I buildings, 1 in 50 for use II buildings, and 1 in 100 years for use category III and IV buildings.

At any location, the mean wind velocity is affected by the roughness of the terrain upwind from the structure in question. At a height of 700 to 1500 ft, the wind reaches a steady velocity as shown by the plots of K_z in Fig. 2–11. Below this height, the velocity decreases and the turbulence, or gustiness, increases as one approaches the surface. These effects are greater in urban areas than in rural areas, due to the greater surface roughness in built-up areas. The factor K_z in Eq. 2–11 relates the wind pressure at any elevation z feet to that at 33 ft (10 m) above the surface for open exposure. ASCE 7–95 gives tables and equations for K_z as a function of the type of exposure (urban, country, etc.) and the height above the surface.

3. **Gust response, factor, G.** The gust factor, G, in Eq. 2–10 relates the dynamic properties of the wind and the structure. For flexible buildings it is calculated. For most buildings, which tend to be stiff, it is equal to 0.8 for exposures A and B, and 0.85 for exposures C and D.

4. **External pressure coefficient, C_p.** When wind blows past a structure, it exerts a positive pressure on the windward wall and a negative pressure (suction) on the leeward wall, side walls and roof as shown in Fig. 2–12. The overall pressures to be used in the de-

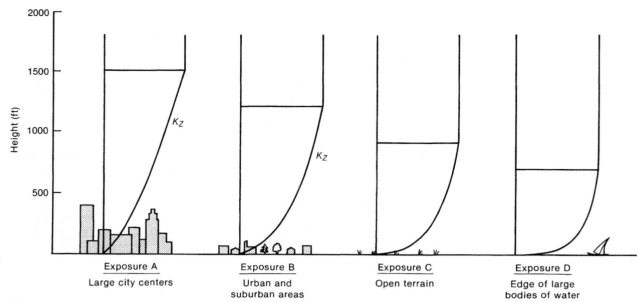

Fig. 2–11
Profiles of velocity pressure exposure coefficient, K_z, for differing terrain.

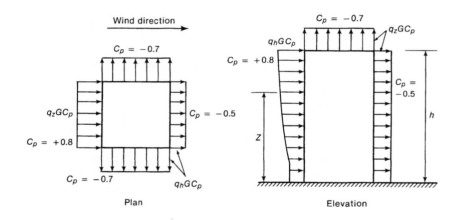

Fig. 2–12
Wind pressures and suctions
on a building.

sign of a structural frame are computed using Eq. 2–10, where C_p is the sum of the pressure coefficients for the windward and leeward walls. Values of the pressure coefficients are given in the loading standard. Typical values are shown in Fig. 2–12 for a building having the shape and proportions shown.

Earthquake Loads

Earthquake loads and design for earthquakes are discussed in Chap. 19.

Other Loads

ASCE 7–95 also gives soil loads on basement walls, loads due to floods, and loads due to ice accretion.

A major aim of structural design is economy. The overall cost of a building project is strongly affected by both the cost of the structure and the financing charges, which are a function of the rate of construction.

In a cast-in-place building, the costs of the floor and roof systems make up roughly 90% of the total structural costs. The cost of a floor system is divided between the costs of building and stripping the *forms*; providing, bending, and placing the *reinforcement*; and providing, placing, and finishing the *concrete*. Table 2–2 lists typical *relative* costs per square foot for several floor framing systems. The floor systems are listed in order of increasing complexity of form construction. Two things are noticeable: (1) the amount of materials goes up as the column spacing increases and as a result the cost increases, and (2) the cost of the forms is the biggest single item in the total costs, comprising 40 to 60% of the total. The major differences between the systems then come from increased amounts of materials as spans increase, and increased costs of forming as the complexity of the forms increases. In the case of the one-way joist floor, a portion of the form cost was for rental of prefabricated forms.

The cost data given in Table 2–2 suggest that floor forming costs should be a major consideration in the layout of the structural system. Formwork costs can be reduced by reusing the forms from area to area or floor to floor. Beam, slab, and column sizes should be chosen to allow the maximum reuse of the forms. It is generally uneconomical to try to save concrete and steel by meticulously calculating the size of every beam and column to fit the loads exactly, because although this may save cents in materials, it may cost dollars in forming costs.

Furthermore, changing section sizes often leads to increased design complexity, which in turn leads to a greater chance of design error and a greater chance of construction error. A simple design that achieves all the critical requirements saves design and construction time and generally gives an economical structure.

Wherever possible, haunched beams should be avoided. If practical, beams should be the same width as the columns into which they frame. Deep spandrel beams make it difficult to move forms from floor to floor and should be avoided if possible. In one-way joist floors it is advisable to use the same depth of joist throughout rather than switching from deep joists for long spans to shallow joists for short spans. The saving in concrete due to such a change is negligible and generally is more than offset by the extra labor of materials required, plus the need to rent and schedule two different sizes of joist forms. In joist floors, the beams should be the same depth as the joists.

If possible, a few standard column sizes should be chosen, with the same column size being used for three or four stories or the entire building. The amount of reinforcement and the concrete strength used can vary as the load varies. Columns should be aligned on a regular grid if possible and constant story heights should be maintained.

Economies are also possible in reinforcement placing. Complex or congested reinforcement will lead to higher per pound charges for placement of the bars. It is frequently best, therefore, to design columns for 1.5 to 2% reinforcement and beams for no more than one-half to two-thirds of the maximum allowable reinforcement ratios. Grade 60 reinforcement is almost universally used for column reinforcement and flexural reinforcement in beams. In slabs where reinforcement quantities are controlled by minimum reinforcement ratios, there may be a slight advantage in using grade 40 reinforcement. The same may be true for stirrups in beams if the stirrup spacings tend to be governed by the maximum spacings. However, before specifying grade 40 steel the designer should check whether it is available locally in the sizes needed.

Since the flexural strength of a floor is relatively insensitive to concrete strength, there is no major advantage in using high-strength concrete in floor systems. An exception to this would be a flat-plate system where the shear capacity may govern the thickness. On

TABLE 2–2 Breakdown of Floor Construction Costs[a]

	20 ft × 20 ft Panel		25 ft × 25 ft Panel	
	Relative Cost/ft^2	Percent	Relative Cost/ft^2	Percent
Flat plate				
Forms	0.43	43	0.44	38
Concrete	0.32	32	0.38	33
Reinforcement	0.25	25	0.33	29
	1.00		1.15	
One-way joists[b]				
Forms	0.62	58	0.61	54
Concrete	0.27	25	0.29	26
Reinforcement	0.19	17	0.23	20
	1.08		1.13	
One-way slab and beams				
Forms	0.58	48	0.62	45
Concrete	0.29	25	0.33	24
Reinforcement	0.32	17	0.42	31
	1.19		1.37	

[a]The costs of the various floor systems are expressed relative to the cost of a 20 ft × 20 ft panel flat plate. These cost ratios are based on a particular set of designs[2-14] for a common live load and a given set of costs.[2-15]

[b]The panel sizes considered are 20 ft × 20 ft and 20 ft × 30 ft.

the other hand, column strengths are directly related to concrete strength, and the most economical columns tend to result from the use of high-strength concrete.

2–9 HANDBOOKS AND DESIGN AIDS

Since a great many repetitive computations are necessary to proportion reinforced concrete members, handbooks containing tables or graphs of the more common quantities are available from several sources. The American Concrete Institute publishes its *Design Handbook* in several volumes, [2-16, 2-17, 2-18] and the Concrete Reinforcing Steel Institute publishes the *CRSI Handbook*.[2-19]

Once a design has been completed, it is necessary for the details to be communicated to the reinforcing bar suppliers and placers and the construction crew. The *ACI Detailing Manual*[2-20] presents detailing drafting standards and is an excellent guide to field practice. ACI Standard 301, *Specifications for Structural Concrete for Buildings*,[2-21] indicates the items to be included in construction specifications. Finally, the ACI publication *Formwork for Concrete*[2-22] gives guidance for form design.

The ACI *Manual of Concrete Practice*[2-23] collects together most of the ACI committee reports on concrete and structural concrete and is an invaluable reference on all aspects of concrete technology. It is published annually in hard copy and on a CD-ROM. The 1994 edition included 154 committee reports.

2–10 CUSTOMARY DIMENSIONS AND CONSTRUCTION TOLERANCES

The selection of dimensions for reinforced concrete members is based on the required size for strength, plus other aspects arising from construction considerations. Beam widths and depths and column sizes are generally varied in increments of 1, 2, 3, or 4 in. and slab thicknesses in $\frac{1}{2}$-in. increments.

The actual as-built dimensions will differ slightly from those shown on the drawings, due to construction inaccuracies. ACI Standard 347[2-22] on formwork gives the accepted tolerances on cross-sectional dimensions of concrete columns and beams as $\pm\frac{1}{2}$ in. and on the thickness of slabs and walls as in $\pm\frac{1}{4}$ in. For footings they recommend tolerances of $+2$ in. and $-\frac{1}{2}$ in. on plan dimensions and -5% of the specified thickness.

The lengths of reinforcing bars are generally given in 2-in. increments. The tolerances for reinforcement placing concern the variation in the depth, d, of beams, the minimum cover and the longitudinal location of bends and ends of bars. These are specified in ACI Secs. 7.5.2.1 and 7.5.2.2. ACI Committee 117 has published a comprehensive list of tolerances for concrete construction and materials.[2-24]

2-11 ACCURACY OF CALCULATIONS

In structural design, the loads, with the exception of dead load or fluid loads in a tank, are rarely known to more than two significant figures. Thus, although calculations should include three significant figures, it is seldom necessary to record more than this. Care should be taken in problems where load, forces, or stresses offset each other since the final value may be the difference of two large similar numbers.

Most mistakes in structural design arise from three sources: errors in looking up or writing down numbers, errors due to unit conversions, and failure to understand fully the statics or behavior of the structure being analyzed and designed. The latter type of mistake is especially serious since failure to consider a particular type of loading or the use of the wrong statical model may lead to serious maintenance problems or collapse. For this reason, designers are urged to use the limit states design process to consider all possible modes of failure and to use free-body diagrams to study the equilibrium of parts or all of the structure.

2-12 SHALL BE PERMITTED

Throughout the ACI Code, the word "may" has been replaced with the phrase "shall be permitted" or something equivalent, at the request of the three model codes. The phrase "shall be permitted" implies that the designer is permitted to use the alternative material or design method mentioned in the section in question.

3

Materials

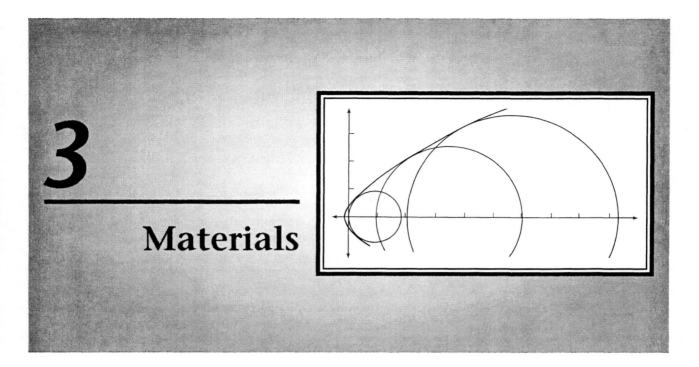

3-1 CONCRETE

Concrete is a composite material composed of aggregate, generally sand and gravel, chemically bound together by hydrated portland *cement*. The aggregate generally is graded in size from sand to gravel, with the maximum gravel size in structural concrete commonly being $\frac{3}{4}$ in., although $\frac{3}{8}$-in. or $1\frac{1}{2}$-in. aggregate may be used.

3-2 STRENGTH OF CONCRETE

Mechanism of Cracking and Failure in Concrete Loaded in Compression

Concrete is a mixture of cement paste and aggregate, each of which has an essentially linear and brittle stress–strain relationship in compression. Brittle materials tend to develop tensile fractures perpendicular to the direction of the largest tensile strain. Thus when concrete is subjected to uniaxial compressive loading, cracks tend to develop parallel to the maximum compressive stress. In a cylinder test the friction between the heads of the testing machine and the ends of the cylinder prevents lateral expansion of the ends of the cylinder and in doing so restrains the vertical cracking in those regions. This strengthens conical regions at each end of the cylinder. The vertical cracks that occur at midheight of the cylinder do not enter these conical regions and the failure surface appears to consist of two cones.

Although concrete is made up of essentially elastic, brittle materials, its stress–strain curve is nonlinear and appears to be somewhat ductile. This can be explained by the gradual development of *microcracking* within the concrete and the resulting redistribution of stress from element to element in the concrete.[3-1] Microcracks are internal cracks $\frac{1}{8}$ to $\frac{1}{2}$ in. in length. Microcracks that occur along the interface between paste and aggregate are called *bond cracks*; those that cross the mortar between pieces of aggregate are known as *mortar cracks*.

There are four major stages in the development of microcracking and failure in concrete subjected to uniaxial compressive loading:

1. Shrinkage of the paste during hydration and drying of the concrete is restrained by the aggregate. The resulting tensile stresses lead to *no-load bond cracks*, before the concrete is loaded. These cracks have little effect on the concrete at low loads and the stress–strain curve remains linear up to 30% of the compressive strength of the concrete, as shown by the solid line in Fig. 3–1.

2. When concrete is subjected to stresses greater than 30 to 40% of its compressive strength, the stresses on the inclined surfaces of the aggregate particles will exceed the tensile and shear strengths of the paste–aggregate interfaces and new cracks known as *bond cracks* will develop. These cracks are stable and propagate only if the load is increased. Once such a crack has formed, however, any additional load that would have been transferred across the cracked interface is redistributed to the remaining unbroken interfaces and to the mortar. This redistribution of load causes a gradual curving of the stress–strain curve for stresses above 40% of the short-time strength. The loss of bond leads to a wedging action, causing transverse tensions above and below the piece of aggregate.

3. As the load is increased beyond 50 or 60% of ultimate, localized *mortar cracks* develop between bond cracks. These cracks develop parallel to the compressive loading, due to the transverse tensile strains. During this stage, there is stable crack propagation; cracking increases with increasing load but does not increase under constant load. The onset of this stage of loading is called the *discontinuity limit*.[3-2]

4. At 75 to 80% of the ultimate load, the number of mortar cracks begins to increase and a continuous pattern of microcracks begins to form. As a result, there are fewer undamaged portions to carry the load and the stress–longitudinal strain curve becomes even more nonlinear. The onset of this stage of cracking is called the *critical stress*.[3-3]

If the lateral strains, ϵ_3, are plotted against the longitudinal compressive stress, the dashed curve in Fig. 3–1 results. The lateral strains are tensile and initially increase as expected from Poisson's ratio. As microcracking becomes more extensive, these cracks contribute to the apparent lateral strains. As the load exceeds 75 to 80% of the ultimate compressive strength, the cracks and lateral strains increase rapidly, and the volumetric strain (relative increase in volume), ϵ_v, begins to increase as shown by the broken line in Fig. 3–1.

The critical stress is significant for several reasons. The ensuing increase in volume causes an outward pressure on ties, spirals, or other confining reinforcement, and these in turn act to restrain the lateral expansion of the concrete, thus delaying its disintegration.

Equally important is the fact that the structure of the concrete tends to become unstable at loads greater than the critical load. Under stresses greater than about 75% of the short-time strength, the strains increase more and more rapidly until failure occurs. Figure 3–2a shows the stress–strain–time response of concrete loaded rapidly to various fractions of its short-time strength, with this load being sustained for a long period of time or until failure occurred. As shown in Fig. 3–2b, concrete subjected to a sustained axial load greater than the critical load will eventually fail under that load. The critical stress is between 0.75 and $0.80f'_c$.

Under cyclic compressive loads, axially loaded concrete has a *shake-down* limit approximately equal to the onset of significant mortar cracking at the critical stress. Cyclic axial stresses higher than the critical stress will eventually cause failure.

As mortar cracking extends through the concrete, less and less of the structure remains. Eventually, the load-carrying capacity of the uncracked portions of the concrete

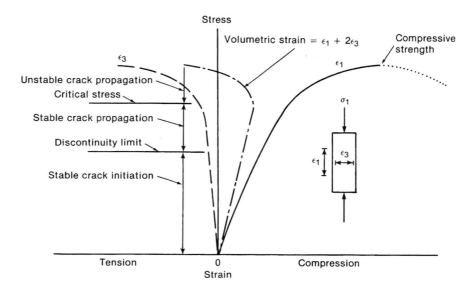

Fig. 3–1
Stress–strain curves for concrete loaded in uniaxial compression. (From Ref. 3–2.)

reaches a maximum value referred to as the *compressive strength* (Fig. 3–1). Further straining is accompanied by a drop in the stress that the concrete can resist, as shown by the dotted portion of the line for ϵ_1 in Fig. 3–1.

When concrete is subjected to compression with a strain gradient, as would occur in the compression zone of a beam, the effect of the unstable crack propagation stage shown in Fig. 3–1 is reduced because as mortar cracking softens the highly strained concrete, the load is transferred to the stiffer, more stable concrete at points of lower strain nearer the neutral axis. In addition, continued straining and the associated mortar cracking of the highly stressed regions is prevented by the stable state of strain in the concrete closer to the neutral axis. As a result, the stable crack propagation stage extends almost up to the ultimate strength of the concrete.

Tests[3–5] suggests that there is no significant difference between the stress–strain curves of concrete loaded with or without a strain gradient up to the point of maximum stress. The presence of a strain gradient does appear to increase the maximum strains that can be attained in the member, however.

The dashed line in Fig. 3–2c represents the gain in short-time compressive strength with time. The dipping solid lines are the failure limit line from Fig. 3–2b plotted against a log time scale. These lines indicate that there is a permanent reduction in strength due to sustained high loads. For concrete loaded at a young age the minimum strength is reached after a few hours. If the concrete does not fail at this time, it can sustain the load indefinitely. For concrete loaded at an advanced age the decrease in strength due to sustained high loads may not be recovered.

The *CEB-FIP Model Code 1990*[3–6] gives equations for both the dashed curve and the solid curves in Fig. 3–2c. The dashed curve (short-time compressive strength with time) can also be represented by Eq. 3–5, presented later in this chapter.

Under uniaxial tensile loadings, small localized cracks are initiated at tensile strain concentrations and relieve these strain concentrations. This initial stage of loading results in an essentially linear stress–strain curve during the stage of stable crack initiation. Following a very brief interval of stable crack propagation, unstable crack propagation and fracture occur. The direction of cracking is perpendicular to the principal tensile stress and strain.

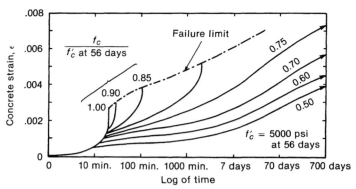

(a) Strain–time relationship.

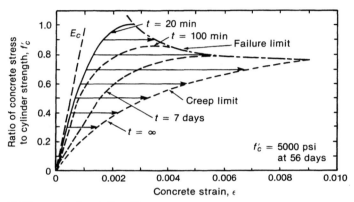

(b) Stress–strain relationship.

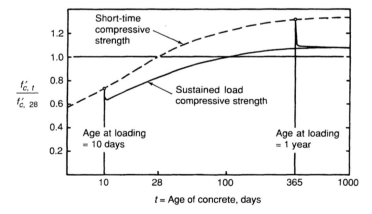

Fig. 3–2
Effect of sustained loads on behavior of concrete in uniaxial compression. (From Ref. 3–4.)

(c) Strength–time relationship.

Compressive Strength of Concrete

Generally, the term *concrete strength* is taken to refer to the uniaxial compressive strength as measured by a compression test of a standard test cylinder, because this test is used to monitor the concrete strength for quality control or acceptance purposes. For con-

Materials

venience, other strength parameters, such as tensile or bond strength, are related to the compressive strength.

Standard Compressive Strength Tests

The standard acceptance test for measuring the strength of concrete involves short-time compression tests on cylinders 6 in. in diameter by 12 in. high made, cured, and tested in accordance with ASTM Standards C31 and C39.

The test cylinders for an acceptance test must be allowed to harden in their molds for 24 hours at the job site at 60 to 80°F, protected from loss of moisture and excessive heat, and then must be cured at 73°F in a moist room or immersed in water saturated with lime. The standard acceptance test is carried out when the concrete is 28 days old.

Field-cured test cylinders are frequently used to determine when the forms may be removed or when the structure may be used. These should be stored as near the location of that concrete in the structure as practicable, and should be cured as closely as possible to the same manner as that of the concrete in the structure.

The standard strength "test" is the average of the strengths of two cylinders from the same sample tested at 28 days or an earlier age if specified. These are tested at a loading rate of about 35 psi per second, producing failure of the cylinder at $1\frac{1}{2}$ to 3 minutes. For high-strength concrete, acceptance tests are frequently carried out at 56 or 90 days because some high-strength concretes take longer than normal concretes to reach their design strength.

Statistical Variations in Concrete Strength

Concrete is a mixture of water, cement, aggregate, and air. Variations in the properties or proportions of these constituents, as well as variations in transporting, placing, and compaction of the concrete, lead to variations in the strength of the finished concrete. In addition, discrepancies in the tests will lead to apparent differences in strength. The shaded area in Fig. 3–3 shows the distribution of strengths of a sample of 176 concrete strength tests.

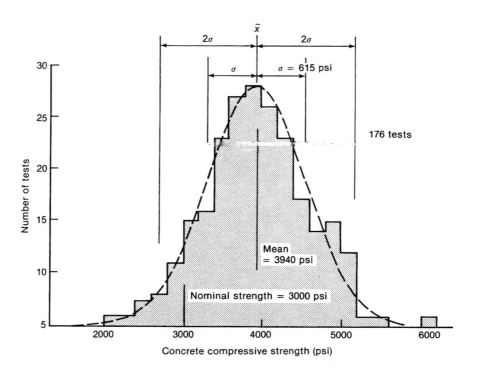

Fig. 3–3
Distribution of concrete strengths.

The mean or average strength is 3940 psi, but one test has a strength as low as 2020 psi and one is as high as 6090 psi.

If more than about 30 tests are available, the strengths will generally approximate a normal distribution. The normal distribution curve, shown by the curved line in Fig. 3–3, is symmetrical about the mean value, $\bar{x}$, of the data. The dispersion of the data can be measured by the *sample standard deviation*, s, which is the root-mean-square deviation of the strengths from their mean value:

$$s = \sqrt{\frac{(x_1 - \bar{x}) + (x_2 - \bar{x})^2 + (x_3 - \bar{x})^2 + \cdots + (x_n - \bar{x})^2}{n - 1}} \quad (3\text{–}1)$$

The standard deviation divided by the mean value is called the *coefficient of variation*, V:

$$V = \frac{s}{\bar{x}} \quad (3\text{–}2)$$

This makes it possible to express the degree of dispersion on a fractional or percentage basis rather than an absolute basis. The concrete test data in Fig. 3–3 have a standard deviation of 615 psi and a coefficient of variation of 615/3940 = 0.156 or 15.6%.

If the data correspond to a normal distribution, their distribution can be predicted from the properties of such a curve. Thus 68.3% of the data will lie within 1 standard deviation above or below the mean. Alternatively, 15.9% of the data will have values less than $(\bar{x} - s)$. Similarly, for a normal distribution, 10% of the data, or 1 test in 10, will have values less than $\bar{x}(1 - aV)$, where $a = 1.282$. Values of a corresponding to other probabilities can be found in statistics texts.

Figure 3–4 shows the mean concrete strength, f_{cr}, required for various values of the coefficient of variation if no more than 1 test in 10 is to have a strength less than 3000 psi. As shown in this figure, as the coefficient of variation is reduced, the value of the mean strength, f_{cr}, required to satisfy this requirement can also be reduced.

Based on the experience of the U.S. Bureau of Reclamation on large projects, ACI Committee 214 [3–7] has defined various standards of control for moderate-strength concretes. A coefficient of variation of 15% represents *average control* (see Fig. 3–4). About one-tenth of the projects studied had coefficients of variation less than 10%, which was termed *excellent control*, and another tenth had values greater than about 20%, which was termed *poor control*. For low-strength concrete, the coefficient of variation corresponding to average control has a value of $V = 0.15 f_c'$. Above a mean strength of about 4000 psi, the standard deviation tends to be independent of the mean strength, and for average control s is about 600 psi.[3–8]

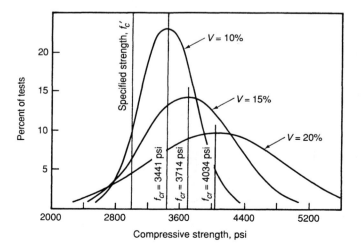

Fig. 3–4
Normal frequency curves for coefficients of variation of 10, 15, and 20 percent. (From Ref. 3–7.)

Materials

Building Code Definition of Compressive Strength

The *specified compressive strength*, f'_c, is measured by compression tests on 6 in. $\times$ 12 in. cylinders tested after 28 days of moist curing. This is the strength specified on the construction drawings and used in the calculations. As shown in Fig. 3–4, the specified strength is less than the average strength. The required mean strength of the concrete, f_{cr}, must be at least (ACI Sec. 5.3.2)

$$f_{cr} = f'_c + 1.34s \qquad (3\text{--}3)$$

or

$$f_{cr} = f'_c + 2.33s - 500psi \qquad (3\text{--}4)$$

where s is the standard deviation determined in accordance with ACI Sec. 5.3.1. Special rules are given if the standard deviation is not known.

Equation 3–3 gives the lowest average strength required to ensure a probability of not more than 1 in 100 that the average of any three consecutive strength tests will be below the specified strength. Alternatively, it ensures a probability of not more than 1 in 11 that any test will fall below f'_c. Equation 3–4 gives the lowest mean strength to ensure a probability of not more than 1 in 100 that any individual strength test will be more than 500 psi below the specified strength. Lines indicating the corresponding required average strengths, f_{cr}, are plotted in Fig. 3–4. In these definitions, a test is the average of two cylinder tests.

Factors Affecting Concrete Compressive Strength

Among the large number of factors affecting the compressive strength of concrete, the following are probably the most important for concretes used in structures.

1. Water/cement ratio. The strength of concrete is governed in large part by the ratio of the weight of the water to the weight of the cement for a given volume of concrete. A lower water/cement ratio reduces the porosity of the hardened concrete and thus increases the number of interlocking solids. Air voids introduced by air entrainment tend to reduce the strength. Voids due to improper compaction will tend to reduce the strength below that corresponding to the water/cement ratio. A water/cement ratio of 0.45 corresponds to 28 day strengths of 4000 to 5000 psi for air-entrained concrete and 5000 to 6500 psi for non-air-entrained concrete. For a water/cement ratio of 0.65, the corresponding ranges are 2500 to 3300 psi and 3300 to 4500 psi.

2. Type of cement. Five basic types of cements are produced:

Normal, Type I: used in ordinary construction where special properties are not required

Modified, Type II: lower heat of hydration than Type I; used where moderate exposure to sulfate attack exists or where moderate heat of hydration is desirable

High early strength, Type III: used when high early strength is desired; has considerably higher heat of hydration than Type I cement

Low heat, Type IV: used in mass concrete dams and other structures where heat of hydration is dissipated slowly

Sulfate resisting, Type V: used in footings, basement walls, sewers, and so on, exposed to soils containing sulfates

Figure 3–5 illustrates the rate of strength gain with different cements. Concrete made with Type III, high early strength cement gains strength more rapidly than does concrete made with Type I, normal cement, reaching about the same strength at 7 days as a corresponding mix containing Type I cement would reach at 28 days. All five types tend to approach the same strength after a long period of time, however.

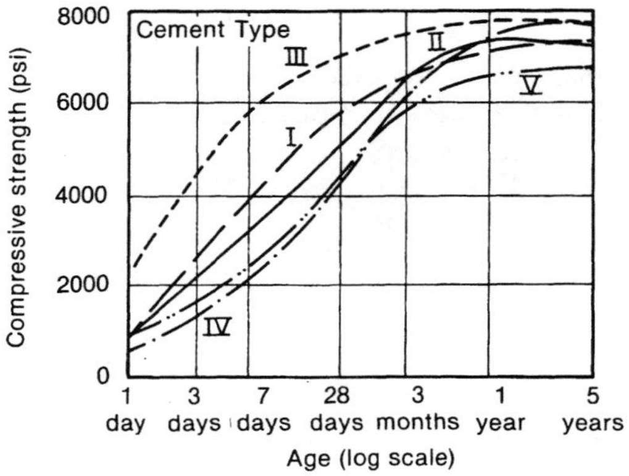

Fig. 3–5
Effect of type of cement on strength gain of concrete (moist cured, water–cement ratio = 0.49). (From Ref. 3–9 copyright ASTM; reprinted with permission.)

3. Supplementary cementitious materials. Sometimes a portion of the cement is replaced by materials such as fly ash, ground granulated blast furnace slag, or silica fume to achieve economy, reduction of heat of hydration, and depending on the materials, improved workability. Such materials are referred to as *pozzolans*, which are defined as siliceous, or siliceous and aluminous materials, which in themselves possess little or no cementitious properties, but which will, in the presence of moisture, react with calcium hydroxide to form compounds with such properties. When supplementary cementitious materials are used in mix design, the water/cement ratio, *w/c*, is restated in terms of the *water/cementitious materials ratio*, *w/cm* where *cm* represents the total weight of the cement and the supplementary cementitious materials, as defined in ACI Sec. 4.1.1. Upper limits on the amounts of fly ash, slag, or silica fume are given in ACI Sec. 4.2.3 for concretes exposed to freeze–thaw conditions. The design of concrete mixes containing supplementary cementitious materials is explained in Ref. 3–10.

Fly ash, precipitated from the chimney gases from coal-fired power plants, frequently leads to improved workability of the fresh concrete. It may slow the rate of strength gain of concrete but generally not the final strength, and, depending on composition of the fly ash, may reduce the durability of the hardened concrete.[3–11]

Ground granulated blast furnace slag tends to depress the early age strength and heat of hydration of concrete. Strengths at older ages will generally exceed those for normal concretes with similar *w/cm* ratios. Slag tends to reduce the permeability of concrete and its resistance to certain chemicals.[3–12]

Silica fume consists of very fine spherical particles of silica produced as a by-product in the manufacture of ferrosilicon alloys. The extreme fineness and high silica content of the silica fume make it a highly effective pozzolanic material. It is used to produce low-permeability concrete with enhanced durability and/or high-strength concrete.[3–10]

4. Aggregate. The strength of concrete is affected by the strength of the aggregate, its surface texture, its grading, and to a lesser extent by the maximum size of the aggregate. Strong aggregates such as felsite, traprock, or quartzite are needed to make very high strength concretes. Weak aggregates include sandstone, marble, and some metamorphic rocks, while limestone and granite aggregates have intermediate strength. Normal-strength concrete made with high-strength aggregates fails due to mortar cracking with very little aggregate failure. The stress–strain curves of such concretes tend to have an appreciable declining branch after reaching the maximum stress. On the other hand, if aggregate failure precedes mortar cracking, failure tends to occur abruptly with a very steep declining branch. This occurs in very high strength concretes (see Fig. 3–17) and in some lightweight concretes (see Fig. 3–25).

Concrete strength is affected by the bond between the aggregate and the cement paste. The bond tends to be better with crushed, angular pieces of aggregate.

A well-graded aggregate produces a concrete that is less porous. Such a concrete tends to be stronger. The strength of concrete tends to decrease as the maximum aggregate size increases. This appears to result from higher stresses at the paste–aggregate interface.

Some aggregates react with alkali in cement, causing a long-term expansion of the concrete that destroys the structure of the concrete. Unwashed marine aggregates also lead to a breakdown of the structure with time.

5. Moisture conditions during curing. The development of the compressive strength of concrete is strongly affected by the moisture conditions during curing. Prolonged moist curing leads to the highest concrete strength, as shown in Fig. 3–6.

6. Temperature conditions during curing. The effect of curing temperature on strength gain is shown in Fig. 3–7 for specimens placed and moist cured for 28 days under the constant temperatures shown in the figure, and then moist cured at 73°F. The 7- and 28-day strengths are reduced by cold curing temperatures, although the long-term strength tends to

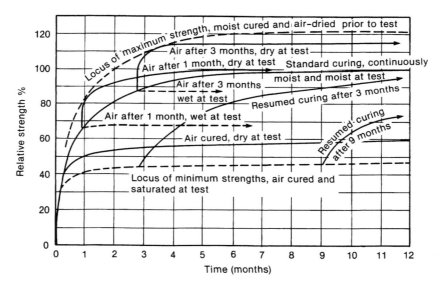

Fig. 3–6
Effect of moist–curing conditions at 70°F and moisture content of concrete at time of test on compressive strength of concrete. (From Ref. 3–13.)

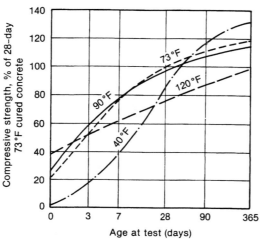

Fig. 3–7
Effect of temperature during the first 28 days on the strength of concrete (water/cement ratio = 0.41, air content = 4.5 percent, Type I cement, specimens cast and moist-cured at temperature indicated for first 28 days. All moist-cured at 73°F thereafter). (From Ref. 3–14.)

3–2 Strength of Concrete

43

be enhanced. On the other hand, high temperatures during the first month increase the 1- and 3-day strengths but tend to reduce the one-year strength. The temperature during the setting period is especially important. Concrete placed and allowed to set at temperatures greater than 80°F will never reach the 28-day strength of concrete placed at lower temperatures.

Occasionally, control cylinders are left in closed boxes at the job site for the first 24 hours. If the temperature is higher than ambient inside these boxes, the strength of the control cylinders may be affected.

7. Age of concrete. Concrete gains strength with age, as shown in Figs. 3–5 to 3–7. Prior to 1975, the 7-day strength of concrete made with Type I cement was generally 65 to 70% of the 28-day strength. Changes in cement production since then have resulted in a more rapid early strength gain and less long-term strength gain. ACI Committee 209 [3-15] has proposed the following equation to represent the rate of strength gain for concrete made from type I cement and moist cured at 70°F.

$$f'_{c(t)} = f'_{c(28)}\left(\frac{t}{4 + 0.85t}\right) \tag{3-5}$$

where $f'_{c(t)}$ is the compressive strength at age t. For Type III cement the coefficients 4 and 0.85 become 2.3 and 0.92.

Concrete cured under temperatures other than 70°F may set faster or slower than indicated by these equations, as shown in Fig. 3–7.

8. Maturity of concrete. The summation of the product of the curing temperature and the time the concrete has cured at that temperature is called the *maturity* [3-16] of the concrete as defined by

$$\text{maturity} = M = \sum_{i=1}^{n} (T_i - 10)(t_i) \tag{3-6}$$

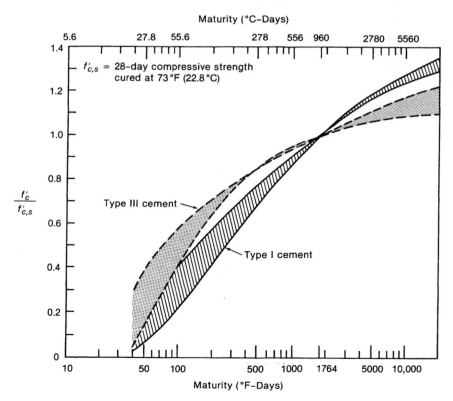

Fig. 3–8
Normalized compressive strength versus maturity.
(From Ref. 3–17.)

where T_i is the temperature in Fahrenheit during the ith interval and t_i is the number of days curing at that temperature. Figure 3–8 shows the form of the relationship between maturity and compressive strength of concrete. Although no unique relationship exists, Fig. 3–8 can be used for guidance in determining when forms can be removed.

9. Rate of loading. The standard cylinder test is carried out at a loading rate of roughly 35 psi per second and the maximum load is reached in $1\frac{1}{2}$ to 2 minutes, corresponding to a strain rate of about 10 microstrain/sec. Under very slow rates of loading, the axial compressive strength is reduced to about 75% of the standard test strength, as shown in Fig. 3–2. A portion of this reduction is offset by continued maturing of the concrete during the loading period.[3-4] At high rates of loading the strength increases, reaching 115% of the standard test strength when tested at a rate of 30,000 psi/sec (strain rate of 20,000 microstain/sec). This corresponds to loading a cylinder to failure in roughly 0.10 to 0.15 second and would approximate the rate of loading experienced in a severe earthquake.

Core Tests

The strength of concrete in a structure (*in-situ strength*) is frequently measured using cores drilled from the structure. These are capped and tested in the same manner as cylinders. ASTM C42–90 *Standard Method of Obtaining and Testing Drilled Cores and Sawed Beams of Concrete* specifies how such tests should be carried out. Core test strengths show a great amount of scatter because core strengths are affected by a wide range of variables.

Core tests have two main uses. The most frequent use of core tests is to assess whether concrete in a new structure is acceptable. ACI Sec. 5.6.4.2 permits the use of core tests in such cases and requires three cores for each strength test more than 500 psi below the specified value of f'_c. If the structure is dry in service, the cores are air dried for 7 days before testing and are tested dry. If the structure is wet in service, the cores are immersed in water for at least 40 hours before testing and are tested wet. ACI Sec. 5.6.4.4 states that concrete evaluated using cores has adequate strength if the average strength of the cores is at least 85% of the specified strength and no single core shows a strength of less than 75% of f'_c. This is just an acceptance rule. Because the 85% value tends to be less than the actual ratio of core strength to cylinder strength, taking the in-situ strength equal to (core strength)/0.85 overestimates the in-situ strength.

The second use of core test data is to determine the in-situ strength of concrete which is equivalent to the f'_c used in the design equations given in the code. This is referred to as the *equivalent specified strength* and is used when evaluating the strength of an existing member or structure. Bartlett and MacGregor[3-18] suggest the following procedure for estimating the equivalent specified strength of concrete in a structure using core tests.

1. Plan the scope of the investigation. The regions that are cored must be consistent with the information sought. That is, either the member in question should be cored, or, if this is impractical, the regions that are cored should contain the same type of concrete, of about the same age, and cured in the same way as the suspect region. The number of cores taken depends, on one hand, on the cost and the hazard from taking cores out of critical parts of the structure, and on the other hand, the desired accuracy of the strength estimate. If possible, at least six cores should be taken from a given grade of concrete in question. It is not possible to detect outliers (spurious values) in smaller samples and the penalty for small sample sizes (given by k_1 in Eq. 3–8) is significant. The diameter of the core should not be less than 3 times the nominal maximum size of the coarse aggregate and the length of the core should be between 1 and 2 times the diameter. If possible, the core diameter should not be less than 4 in. because the variability of the core strengths increases significantly for smaller diameters.

2. Obtain and test the cores. Use standard methods to obtain and test the cores as given in ASTM C 42–90. Carefully record the location in the structure of each core, the condition of the cores before testing, and the mode of failure. This information may be useful in explaining individual low core strengths. A load–stroke plot from the core test may be useful in this regard. It is particularly important that the moisture condition of the core correspond to one of the two standard conditions prescribed in ASTM C 42–90 and be recorded.

3. Convert the core strengths, f_{core}, **to equivalent in-situ strengths,** f_{cis}. This is done using

$$f_{cis} = f_{core}(F_{\ell/d} \times F_{dia} \times F_r)(F_{mc} \times F_d) \qquad (3\text{--}7)$$

where the factors in the first set of parentheses correct the core strength to that of a standard 4-in. diameter core, with length/diameter ratio equal to 2, not containing reinforcement.

$F_{\ell/d}$ = correction for length/diameter ratio as given in ASTM C 42–87

   = 0.87, 0.93, 0.96, 0.98, and 1.00 for ℓ/d = 1.0, 1.25, 1.50, 1.75, and 2.0, respectively

F_{dia} = correction for diameter of core

   = 1.06 for 2-in. cores, 1.00 for 4-in. cores, and 0.98 for 6-in. cores

F_r = correction for the presence of reinforcing bars

   = 1.00 for no bars, 1.08 for one bar, and 1.13 for two

The factors in the second set of parentheses account for differences between the condition of the core and that of the concrete in the structure.

F_{mc} = accounts for the effect of the moisture condition of the core at the time of the core test

   = 1.09 if the core was soaked before testing, and 0.96 if the core was air dried at the time of the test

F_d = accounts for damage to the surface of the core due to drilling

   = 1.06

4. Check for low outliers in the set of equivalent in-situ strengths. Reference 3–18 gives a technique for doing this. If an outlier is detected using a statistical test, one should try to determine a physical reason for the low strength.

5. Compute the equivalent specified strength from the in-situ strengths. The *equivalent specified strength,* f'_{ceq}, is the strength that should be used in design equations when checking the capacity of the member in question. To calculate it, one first computes the mean, $\bar{f}_{cis}$, and sample standard deviation, s_{cis}, of the set of equivalent in-situ strengths, f_{cis}, which remains after any outliers have been removed. Bartlett and MacGregor [3-18] present the following equation for f'_{ceq}. It uses the core test data to obtain a lower bound estimate of the 10% fractile of the in-situ strength.

$$f'_{ceq} = k_2\left[\bar{f}_{cis} - 1.282\sqrt{\frac{(k_1 s_{cis})^2}{n} + \bar{f}_{cis}^2(V_{\ell/d}^2 + V_{dia}^2 + V_r^2 + V_{mc}^2 + V_d^2)}\right] \qquad (3\text{--}8)$$

where

k_1 = a factor dependent on the number of core tests, equal to 2.40 for 2 tests, 1.47 for 3 tests, 1.20 for 5 tests, 1.10 for 8 tests, 1.05 for 16 tests, and 1.03 for 25 tests

k_2 = a factor dependent on the number of batches of concrete in the member or structure being evaluated, equal to 0.90 and 0.85, respectively, for a cast-in-place member or structure that contains one batch or many batches; and 0.90 for a precast member or structure

n = number of cores after removal of outliers

$V_{\ell/d}$ = coefficient of variation due to length/diameter correction, equal to 0.025 for $\ell/d = 1$, 0.006 for $\ell/d = 1.5$, and zero for $\ell/d = 2$

V_{dia} = coefficient of variation due to diameter correction, equal to 0.12 for 2-in-diameter cores, zero for 4-in. cores, and 0.02 for 6-in. cores

V_r = coefficient of variation due to presence of reinforcing bars in the core, equal to zero if none of the cores contained bars, and to 0.03 if more than a third of them did

V_{mc} = coefficient of variation due to correction for moisture condition of core at time of testing, equal to 0.025

V_d = coefficient of variation due to damage to core during drilling, equal to 0.025

The individual coefficients of variation in the second term of Eq. 3–8 are taken equal to zero if the corresponding correction factor, F, is taken equal to 1.0 in Eq. 3–8.

EXAMPLE 3–1 Computation of an Equivalent Specified Strength from Core Tests.

As a part of an evaluation of an existing structure, it is necessary to compute the strength of a 6-in.-thick slab. To do so it is necessary to have an equivalent specified compressive strength, f'_{ceq}, to use in place of f'_c in the design equations. Several batches of concrete were placed in the slab.

1. **Plan the scope of the investigation.** From a site visit it is determined that five cores can be taken. These are 4-in.-diameter cores drilled vertically through the slab, giving cores that are 6 in. long. They are taken from randomly selected locations around the entire floor in question.

2. **Obtain and test the cores.** The cores were tested in an air-dried condition. None of them contained reinforcing bars. The individual core strengths were 5950, 5850, 5740, 5420, and 4830 psi.

3. **Convert the core strengths to equivalent in-situ strengths.** From Eq. 3–7

$$f_{\text{cis}} = f_{\text{core}}(F_{\ell/d} \times F_{\text{dia}} \times F_r)(F_{mc} \times F_d)$$

The ℓ/d of the cores was 6 in./4 in. = 1.50. From ASTM C42–90, $F_{\ell/d} = 0.96$.

$$f_{\text{cis}} = f_{\text{core}}(0.96 \times 1.0 \times 1.0)(0.96 \times 1.06)$$

$$= f_{\text{core}} \times 0.977$$

The individual strengths, f_{cis}, are 5812, 5715, 5607, 5295, and 4720 psi.

4. **Check for low outliers.** Although there is quite a difference between the lowest and second-lowest values, we shall assume that all five tests are valid.

5. **Compute the equivalent specified strength.**

$$f'_{\text{ceq}} = k_2\left[\bar{f}_{\text{cis}} - 1.282\sqrt{\frac{(k_1 s_{\text{cis}})^2}{n} + \bar{f}_{\text{cis}}^2(V_{\ell/d}^2 + V_{\text{dia}}^2 + V_r^2 + V_{mc}^2 + V_d^2)}\right] \qquad (3\text{–}8)$$

The mean and sample standard deviation of the f_{cis} values are $\bar{f}_{\text{cis}} = 5430$ psi and $s_{\text{cis}} = 422$ psi, respectively. Other terms in Eq. 3–8 are $k_1 = 1.20$ for 5 tests, $k_2 = 0.85$ for several batches, $n = 5$ tests. Because no correction was made in step 3 for the effects of core diameter or reinforcement in the core (F_{dia} and $F_r = 1.0$), V_{dia} and V_r are equal to zero. The terms under the square-root sign in Eq. 3–8 are

$$\frac{(k_1 s_{\text{cis}})^2}{n} = \frac{(1.20 \times 442)^2}{5} = 56,265$$

$$\bar{f}_{\text{cis}}^2(V_{\ell/d}^2 + V_{\text{dia}}^2 + V_r^2 + V_{mc}^2 + V_d^2) = 5430^2(0.006^2 + 0.0^2 + 0.0^2 + 0.025^2 + 0.025^2)$$

$$= 37,918$$

$$f'_{\text{ceq}} = 0.85(5430 - 1.282\sqrt{56,265 + 37,918})$$

$$= 4281 \text{ psi}$$

The concrete strength in the slab should be taken as 4281 psi when calculating the capacity of the slab. ■

Strength of Concrete in a Structure

The strength of concrete in a structure tends to be somewhat lower than the strength of control cylinders made from the same concrete. This difference is due to the effects of different placing, compaction, and curing procedures; the effects of vertical migration of water during the placing of the concrete in deep members; the effects of difference in size and shape; and the effects of different stress regimes in the structure and the specimens.

The concrete near the top of deep members tends to be weaker than the concrete lower down, probably due to the increased water/cement ratio at the top due to upward water migration after the concrete is placed and by the greater compaction of the concrete near the bottom due to the weight of the concrete higher in the form.[3-19]

Tensile Strength of Concrete

The tensile strength of concrete varies between 8 and 15% of the compressive strength. The actual value is strongly affected by the type of test carried out to determine the tensile strength, the type of aggregate, the compressive strength of the concrete, and the presence of a compressive stress transverse to the tensile stress.[3-20]

Standard Tension Tests

Two types of tests are widely used. The first of these is the *modulus of rupture* or flexural test (ASTM C78 or C293), in which a plain concrete beam, generally 6 in. × 6 in. × 30 in. long, is loaded in flexure at the third points of a 24-in. span until it fails due to cracking on the tension face. The flexural tensile strength or modulus of rupture, f_r, from a modulus of rupture test is calculated using the following equation, assuming the concrete is linearly elastic:

$$f_r = \frac{6M}{bh^2} \tag{3-9}$$

where

> M = moment
> b = width of specimen
> h = overall depth of specimen

The second common tensile test is the *split cylinder* test (ASTM C496), in which a standard 6 in. × 12 in. compression test cylinder is placed on its side and loaded in compression along a diameter as shown in Fig. 3-9a.

In a split cylinder test, an element on the vertical diameter of the specimen is stressed in biaxial tension and compression, as shown in Fig. 3-9c. The stresses acting across the vertical diameter range from high transverse compressions at the top and bottom to a nearly uniform tension across the rest of the diameter, as shown in Fig. 3-9d. The splitting tensile strength, f_{ct}, from a split cylinder test is computed as

$$f_{ct} = \frac{2P}{\pi \ell d} \tag{3-10}$$

where

> P = maximum applied load in the test
> ℓ = length of specimen
> d = diameter of specimen

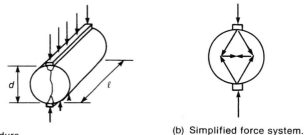

(a) Test procedure.

(b) Simplified force system.

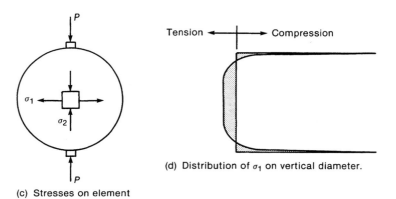

(c) Stresses on element

(d) Distribution of σ_1 on vertical diameter.

Fig. 3–9
Split cylinder test.

Various types of tension tests give different strengths. In general, the strength decreases as the volume of concrete that is highly stressed in tension is increased. A third-point loaded modulus of rupture test on a 6-in.-square beam gives a modulus of rupture strength f_r that averages $1.5 f_{ct}$, while a 6-in.-square prism tested in pure tension gives a direct tensile strength that averages about 86% of f_{ct}.[3–21]

Relationship between Compressive and Tensile Strengths of Concrete

Although the tensile strength of concrete increases with an increase in the compressive strength, the ratio of tensile strength to the compressive strength decreases as the compression strength increases. Thus the tensile strength is approximately proportional to the square root of the compressive strength. The mean split cylinder strength, $\bar{f}_{ct}$, from a large number of tests of concrete from various localities has been found to be[3–8]

$$\bar{f}_{ct} = 6.4\sqrt{f_c'} \tag{3–11}$$

where f_{ct}, f_c' and $\sqrt{f_c'}$ are all in psi. Equation 3–11 is compared to split cylinder test data in Fig. 3–10. It is important to note the wide scatter in the test data. The ratio of measured to computed splitting strength is essentially normally distributed.

Similarly, the mean modulus of rupture, $\bar{f}_r$, can be expressed as[3–8]

$$\bar{f}_r = 8.3\sqrt{f_c'} \tag{3–12a}$$

Again there is scatter in the modulus of rupture. Raphael[3–20] discusses the reasons for this. The distribution of the ratio of measured to computed modulus of rupture strength approaches a log-normal distribution.

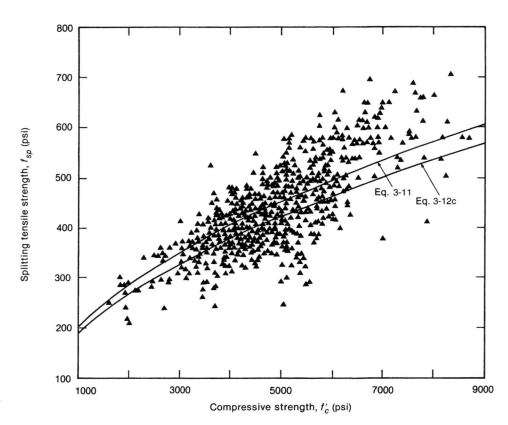

Fig. 3–10
Relationship between splitting tensile strengths and compression strengths. (From Ref. 3–8.)

ACI Sec. 9.5.2.3 defines the modulus of rupture for use in calculating deflections as

$$f_r = 7.5\sqrt{f_c'} \tag{3–12b}$$

A lower value is used in strength calculations (ACI Sec. 11.4.2.1):

$$f_r = 6\sqrt{f_c'} \tag{3–12c}$$

Factors Affecting the Tensile Strength of Concrete

The tensile strength of concrete is affected by the same factors as the compressive strength. In addition, the tensile strength of concrete made from crushed rock may be up to 20% greater than that from rounded gravels. The tensile strength of concrete made from lightweight aggregate tends to be less than for normal sand and gravel concrete, although this varies widely depending on the properties of the particular aggregate under consideration.

The tensile strength of concrete develops more quickly than the compressive strength. As a result, such things as shear strength and bond strength, which are strongly affected by the tensile strength of concrete, tend to develop more quickly than the compressive strength. At the same time, however, the tensile strength increases more slowly than would be suggested by the square root of the compressive strength at the age in question. Thus, concrete having a 28-day compressive strength of 3000 psi would have a splitting tensile strength of about $6.4\sqrt{f_c'} = 350$ psi. At 7 days this concrete would have compressive strength of about 2100 psi (0.70 times 3000 psi) and a tensile strength of about 260 psi (0.75 times 350 psi). This is less than the tensile strength of $6.4\sqrt{2100} = 293$ psi that one would compute from the 7-day compressive strength. This is of importance in choosing form removal times for flat slab floors, which tend to be governed by the shear strength of the column–slab connections.[3–22]

Strength under Biaxial and Triaxial Loadings

Biaxial Loadings

Concrete is said to be *loaded biaxially* when it is loaded in two mutually perpendicular directions with essentially no stress or restraint of deformation in the third direction as shown in Fig. 3–11a. A common example is shown in Fig. 3–11b.

The strength and mode of failure of concrete subjected to biaxial states of stress varies as a function of the combination of stresses as shown in Fig. 3–12. The pear-shaped line in Fig. 3–12a represents the combinations of the biaxial stresses, σ_1 and σ_2, which cause failure of the concrete. This line passes through the uniaxial compressive strength, f'_c, at A and A' and the uniaxial tensile strength, f'_t, at B and B'.

Under biaxial tension (σ_1 and σ_2 both tensile stresses) the strength is close to that in uniaxial tension, as shown by the region $B–D–B'$ (zone 1) in Fig. 3–12a. Here failure occurs by tensile fracture perpendicular to the maximum principal tensile stress as shown in Fig. 3–12b, which corresponds to point B in Fig. 3–12a.

When one principal stress in tensile and the other is compressive, as shown in Fig. 3–11a, the concrete fails at lower stresses than it would if stressed uniaxially in tension or compression.[3–23] This is shown by regions $A–B$ and $A'–B'$ in Fig. 3–12a. In this region, zone 2 in Fig. 3–12a, failure occurs due to tensile fractures on planes perpendicular to the principal tensile stresses. The lower strengths in this region suggest that failure may be governed by a limiting tensile strain rather than a limiting tensile stress.

Under uniaxial compression (points A and A' and zone 3 in Fig. 3–12a), failure is initiated by the formation of tensile cracks on planes parallel to the direction of the compressive stresses. These planes are planes of maximum principal tensile strain.

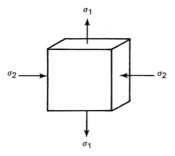

(a) Biaxial state of stress.

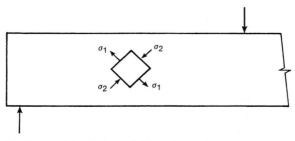

Fig. 3–11
Biaxial stresses.

(b) Biaxial state of stress in the web of a beam.

3–2 Strength of Concrete

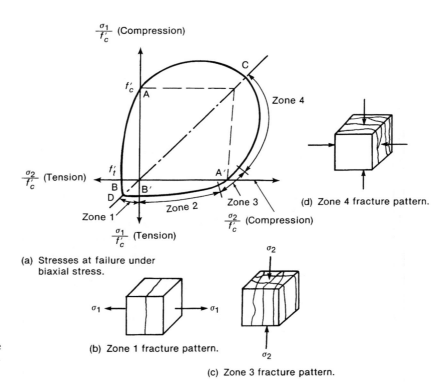

Fig. 3–12
Strength and modes of failure
of concrete subjected to biaxial stresses. (From Ref. 3–2.)

(a) Stresses at failure under biaxial stress.

(b) Zone 1 fracture pattern.

(c) Zone 3 fracture pattern.

(d) Zone 4 fracture pattern.

Under biaxial compression (region *A–C–A'* and zone 4 in Fig. 3–12a) the failure pattern changes to a series of parallel fracture surfaces on planes parallel to the unloaded sides of the member as shown. Such planes are acted on by the maximum tensile strains. Biaxial and triaxial compression loads delay the formation of bond cracks and mortar cracks. As a result, the period of stable crack propagation is longer and the concrete is more ductile. As shown in Fig. 3–12, the strength of concrete under biaxial compression is greater than the uniaxial compressive strength. Under equal biaxial compressive stresses, the strength is about 107% of f'_c, as shown by point *C*.

In the webs of beams, the principal tensile and principal compressive stresses lead to a biaxial tension–compression state of stress as shown in Fig. 3–11b. Under such a loading the tensile and compressive strengths are less than they would be under uniaxial stress, as shown by the quadrant *AB* or *A'B'* in Fig. 3–12a. A similar biaxial stress state exists in a split cylinder test as shown in Fig. 3–9c. This explains in part why the splitting tensile strength is less than the flexural tensile strength.

In zones 1 and 2 in Fig. 3–12, failure occurred when the concrete cracked, and in zones 3 and 4, failure occurred when the concrete crushed. In a reinforced concrete member with sufficient reinforcement parallel to the tensile stresses, cracking does not represent failure of the member because the reinforcement resists the tensile forces after cracking. The biaxial load strength of cracked reinforced concrete is discussed in the next subsection.

Compressive Strength of Cracked Reinforced Concrete

If cracking occurs in reinforced concrete under a biaxial tension–compression loading and there is reinforcement across the cracks, the strength and stiffness of the concrete under compression parallel to the cracks is reduced. Figure 3–13a shows a concrete element which has been cracked due to horizontal tensile stresses. The natural irregularity of the shape of the

Materials

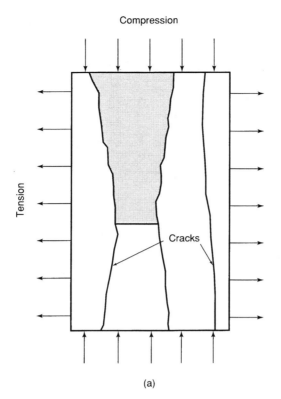

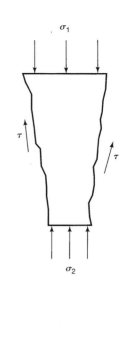

Fig. 3–13
Stresses in a cracked concrete element.

(a)   (b)

cracks leads to variations in the width of a piece between two cracks, as shown. The compressive stress acting on the top of the shaded portion is equilibrated by compressive stresses on the bottom and by shearing stresses along the edges, as shown in Fig. 3–13b. When the crack widths are small, the shearing stresses transfer sufficient load across the cracks that the compressive stress on the bottom of the shaded portion is not significantly larger than that on the top, and the strength is unaffected by the cracks. As the crack widths increase, the ability to transfer shear across them decreases. For equilibrium, the compressive stress on the bottom of the shaded portion must then increase. Failure occurs when the highest stress in the element reaches the uniaxial compressive strength of the concrete.

Tests of concrete panels loaded in-plane shear, carried out by Vecchio and Collins,[3–24] have shown a relationship between the transverse tensile strain, ϵ_1, and the compressive strength parallel to the cracks, f_{2max}:

$$\frac{f_{2max}}{f_c'} = \frac{1}{0.8 + 170\epsilon_1} \tag{3–13}$$

where ϵ_1 is the average transverse strain measured on a gauge length that includes one or more cracks. Equation 3–13 is plotted in Fig. 3–14. An increase in the strain ϵ_1 leads to a decrease in strength. The same authors[3–25] have suggested a stress–strain relationship, f_2–ϵ_2, for transversely cracked concrete:

$$f_2 = f_{2max}\left[2\left(\frac{\epsilon_2}{\epsilon_c'}\right) - \left(\frac{\epsilon_2}{\epsilon_c'}\right)^2\right] \tag{3–14}$$

where f_{2max} is given by Eq. 3–13, and ϵ_c' is the strain at the highest point in the compressive stress–strain curve, which the authors took as 0.002. The term in brackets describes a parabolic stress–strain curve with apex at ϵ_c' and a peak stress that decreases as ϵ_1 increases.

3–2 Strength of Concrete   **53**

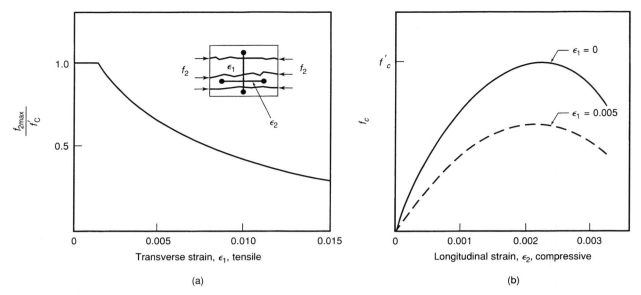

(a)

(b)

Fig. 3–14
Effect of transverse tensile
strains on the compressive
strength of cracked concrete.

Triaxial Loadings

Under triaxial compressive stresses, the mode of failure involves either tensile fracture parallel to the maximum compressive stress and thus orthogonal to the maximum tensile strain if such exists, or a shearing mode of failure. The strength and ductility of concrete under triaxial compression exceed those under uniaxial compression, as shown in Fig. 3–15. This figure presents the stress–longitudinal strain curves for cylinders subjected to a constant lateral fluid pressure $\sigma_2 = \sigma_3$, while the longitudinal stress, σ_1, was increased to failure. These tests suggested that the longitudinal stress at failure was

$$\sigma_1 = f_c' + 4.1\sigma_3 \tag{3–15}$$

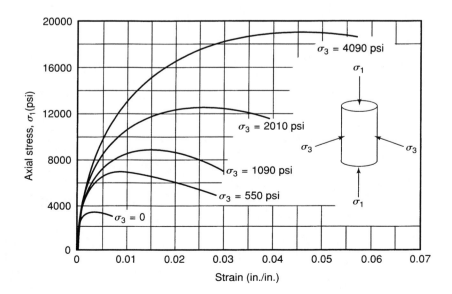

Fig. 3–15
Axial stress–strain curves
from triaxial compression
tests on concrete cylinders,
unconfined compressive
strength, $f_c' = 3600$ psi.
(From Ref. 3–3.)

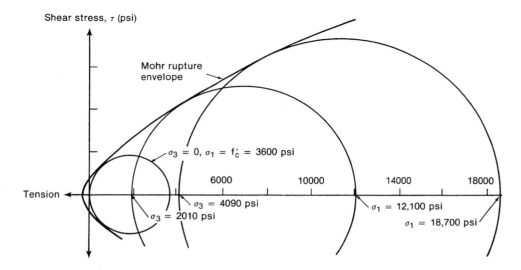

Fig. 3–16
Mohr rupture envelope for concrete tests from Fig. 3–15

Tests of lightweight and high-strength concretes,[3-26, 3-27] suggest that their compressive strengths are less influenced by the confining pressure, with the result that the coefficient 4.1 in Eq.3–15 drops to about 2.0.

The strength of concrete under combined stresses can also be expressed using a *Mohr rupture envelope*. The Mohr's circles plotted in Fig. 3–16 correspond to three of the cases plotted in Fig. 3–15. The Mohr's circles are tangent to the Mohr rupture envelope shown with the outer line.

In concrete columns or in beam–column joints, concrete in compression is sometimes enclosed by closely spaced hoops or spirals. When the width of the concrete element increases due to Poisson's ratio and microcracking, these hoops or spirals are stressed in tension, causing an offsetting compressive stress in the enclosed concrete. The resulting triaxial state of stress in the concrete enclosed or *confined* by the hoops or spirals increases the ductility and strength of the confined concrete. This is discussed in Chap. 11.

3–3 MECHANICAL PROPERTIES OF CONCRETE

The behavior and strength of reinforced concrete members is controlled by the size and shape of the members and the stress–strain properties of the concrete and the reinforcement. The stress–strain behavior discussed in this section will be used in subsequent chapters to develop relationships for the strength and behavior of reinforced concrete beams and columns.

Stress–Strain Curve for Normal-Weight Concrete in Compression

Typical stress–strain curves for concretes of various strengths are shown in Fig. 3–17. These curves were obtained in tests lasting about 15 minutes on specimens resembling the compression zone of a beam.

The stress–strain curves in Fig. 3–17 all rise to a maximum stress, reached at a strain of between 0.0015 and 0.003, followed by a descending branch. The shape of this curve results from the gradual formation of microcracks within the structure of the concrete, as discussed earlier in this chapter.

The length of the descending branch of the curve is strongly affected by the test conditions. Frequently, an axially loaded concrete test cylinder will fail explosively at the point of maximum stress. This will occur in axially flexible testing machines if the strain energy released by the testing machine as the load drops exceeds the energy that the specimen can absorb. If a member loaded in bending, or bending plus axial load, the descending branch will exist since, as the stress drops in the most highly strained fibers, other less highly strained fibers can resist the load, thus delaying the failure of the highly strained fibers.

The stress–strain curves in Fig. 3–17 show five properties used in establishing the mathematical models shown in Fig. 3–18 for stress–strain curve of concrete in compression:

1. The initial slope of the curves (initial tangent modulus of elasticity) increases with an increase in compressive strength.

The modulus of elasticity of the concrete, E_c, is affected by the modulus of elasticity of the cement paste and that of the aggregate. An increase in the water–cement ratio in-

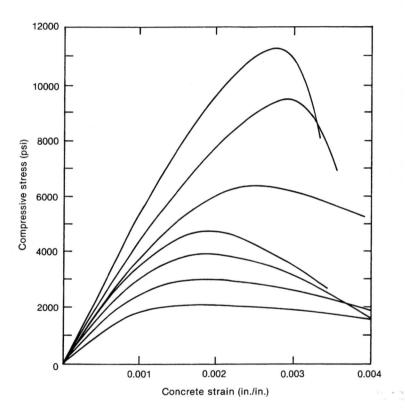

Fig. 3–17
Typical concrete stress–strain curves in compression. (From Refs. 3–28 and 3–29.)

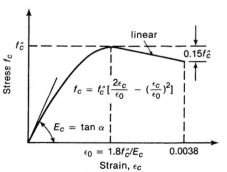

(a) Modified Hognestad. (From Ref. 3–31)

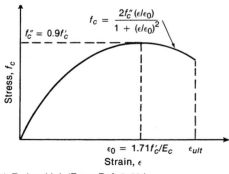

(b) Todeschini. (From Ref. 3–32.)

Fig. 3–18
Analytical approximations to the compressive stress–strain curve for concrete.

Materials

creases the porosity of the paste, reducing its modulus of elasticity and strength. This is accounted for in design by expressing E_c, as a function of f'_c.

Of equal importance is the modulus of elasticity of the aggregate. Normal-weight aggregates have modulus of elasticity values ranging from 1.5 to 5 times that of the cement paste. Because of this, the fraction of the total mix that is aggregate also affects E_c. Lightweight aggregates have modulus of elasticity values comparable to the paste, and hence the aggregate fraction has little effect on E_c for lightweight concrete.

The modulus of elasticity of concrete is frequently taken as given in ACI Sec. 8.5.1:

$$E_c = 33(w^{1.5})\sqrt{f'_c} \text{ psi} \tag{3-16}$$

where w is the weight of the concrete in lb/ft³. This equation was derived from short-time tests on concretes with densities ranging from 90 to 155 lb/ft³ and corresponds to the secant modulus of elasticity at approximately $0.50f'_c$.[3-30] The initial tangent modulus is about 10% greater. Because this equation ignores the type of aggregate, the scatter of data is very wide. Equation 3–16 systematically overestimates E_c in regions where low-modulus aggregates are prevalent. If deflections or vibration characteristics are critical in a design, E_c should be measured for the concrete to be used.

For normal-weight concrete with a density of 145 lb/ft³, ACI Sec. 8.5.1 gives the modulus of elasticity as

$$E_c = 57,000\sqrt{f'_c} \text{ psi} \tag{3-17}$$

2. The rising portion of the stress–strain curve resembles a parabola with its vertex at the maximum stress.

For computational purposes the rising portion of the curves is frequently approximated by a parabola.[3-31] This curve tends to become straighter as the concrete strength increases.[3-29]

3. The strain, ϵ_0, at maximum stress increases as the concrete strength increases.

4. The slope of the descending branch of the stress–strain curve tends to be less than that of the ascending branch for moderate strength concretes. This slope increases with an increase in compressive strength.

5. The maximum strain reached, ϵ_{cu}, decreases with an increase in concrete strength.

The descending portion of the stress–strain curve after the maximum stress has been reached is highly variable and is strongly dependent on the testing procedure. Similarly, the maximum or limiting strain, ϵ_{cu}, is very strongly dependent on the type of specimen, type of loading, and rate of testing. The limiting strain tends to be higher if there is a possibility of load redistribution at high loads. In flexural tests values from 0.0025 to 0.006 have been measured (see Sec. 4–1).

The two most common representations of the stress–strain curve consist of a parabola followed by the sloping line shown in Fig. 3–18a, terminating at a limiting strain of 0.0038, or a parabola followed by a horizontal line terminating at a limiting strain of 0.003 or 0.0035, which is widely used in Europe.[3-6] The stress–strain diagram in Fig. 3–18a is referred to as a modified Hognestad stress–strain curve.[3-31]

The stress–strain curve shown in Fig. 3–18b is convenient for use in analytical studies because it is a continuous function. The highest point in the curve, f''_c, is taken to equal $0.9f'_c$ to give stress block properties similar to the rectangular stress block of Sec. 4–2 when $\epsilon_{ult} = 0.003$ for f'_c up to 5000 psi. The strain ϵ_0, corresponding to maximum stress, is taken as $1.71f'_c/E_c$. For any given strain ϵ, $x = \epsilon/\epsilon_0$. The stress corresponding to that strain is

$$f_c = \frac{2f''_c x}{1 + x^2} \tag{3-18}$$

3–3 Mechanical Properties of Concrete

The average stress under the stress block from $\epsilon = 0$ to ϵ is $\beta_1 f''_c$, where

$$\beta_1 = \frac{\ln(1 + x^2)}{x} \tag{3-19}$$

The center of gravity of the area of the stress–strain curve between $\epsilon = 0$ and ϵ is at $k_2\epsilon$ from ϵ, where

$$k_2 = 1 - \frac{2(x - \tan^{-1}x)}{x^2\beta_1} \tag{3-20}$$

where x is in radians when computing $\tan^{-1}x$. The stress–strain curve is satisfactory for concretes with stress–strain curves that display a gradually descending stress–strain curve at strains greater than ϵ_0. Hence it is applicable to f'_c up to about 5000 psi for normal-weight concrete and about 4000 psi for lightweight concrete.

As shown in Fig. 3–15, a lateral confining pressure causes an increase in the compressive strength of concrete, and a large increase in the strains at failure. The additional strength and ductility of confined concrete are utilized in hinging regions in structures in seismic regions. Stress–strain curves for confined concrete are described in Ref. 3–33.

When a compression specimen is loaded, unloaded, and reloaded it has the stress–strain response shown in Fig. 3–19. The envelope to this curve is very close to the stress–strain curve for a monotonic test. This, and the large residual strains that remain after unloading, suggest that the inelastic response is due to damage to the internal structure of the concrete as suggested by the microcracking theory presented earlier.

Stress–Strain Curve for Normal-Weight Concrete in Tension

The stress–strain response of concrete loaded in axial tension can be divided into two phases. Prior to the maximum stress, the stress–strain relationship is slightly curved. The diagram is linear to roughly 50% of the tensile strength. The strain at peak stress is about 0.0001 in pure tension and 0.00014 to 0.0002 in flexure. The rising part of the stress–strain curve may be approximated either as a straight line with slope E_c and a maximum stress equal to the tensile strength, f'_t, or as a parabola with a maximum strain, $\epsilon'_t = 1.8f'_t/E_c$, and a maximum stress, f'_t. The latter curve is illustrated in Fig. 3–20a with f'_t and E_c based on Eqs. 3–11 and 3–17.

After the tensile strength is reached, microcracking occurs in a *fracture process zone* adjacent to the point of highest tensile stress, and the tensile capacity of this concrete drops very rapidly with increasing elongation. In this stage of behavior, elongations are concen-

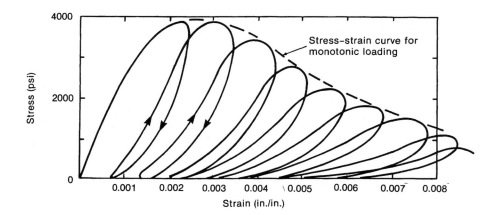

Fig. 3–19
Compressive stress–strain
curves for cyclic loads.
(From Ref. 3–34.)

Materials

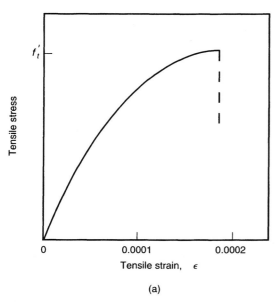
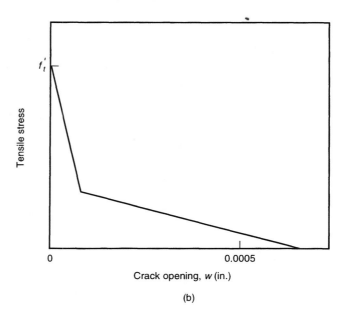

(a)

(b)

Fig. 3–20
Stress–strain curve and
stress–crack opening curves
for concrete loaded in tension.

trated in the fracture process zone while the rest of the concrete is unloading elastically. The unloading response is best described by a *stress versus crack opening diagram* as shown in Fig. 3–20b. The crack widths shown in this figure are of the right magnitude. The actual values depend on the situation. The tensile capacity drops to zero when the crack is completely formed. This occurs at a very small crack width. A more detailed discussion is given in Ref. 3–35.

Poisson's Ratio

At stresses below the critical stress (see Fig. 3–1) Poisson's ratio for concrete varies from about 0.11 to 0.21 and usually falls in the range 0.15 to 0.20. Based on tests of biaxially loaded concrete, Kupfer et al.[3-23] report values of Poisson's ratio for 0.20 for concrete loaded in compression in one or two directions, 0.18 for concrete loaded in tension in one or two directions, and 0.18 to 0.20 for concrete loaded in tension and compression. Poisson's ratio remains approximately constant under sustained loads.

3–4 TIME-DEPENDENT VOLUME CHANGES

Concrete undergoes three main types of volume change which may cause stresses, cracking, or deflections which affect the in-service behavior of reinforced concrete structures. These are shrinkage, creep, and thermal expansion.

Shrinkage

Shrinkage is the shortening of concrete during hardening and drying under constant temperature. The amount of shrinkage increases with time as shown in Fig. 3–21a.

The primary type of shrinkage is called *drying shrinkage* or simply *shrinkage*. It occurs due to the loss of a layer of adsorbed water from the surface of the gel particles. This

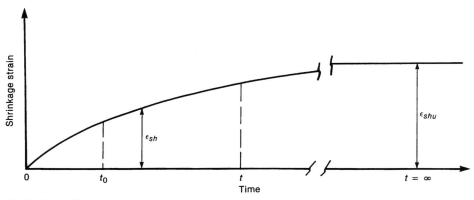

(a) Shrinkage of an unloaded specimen.

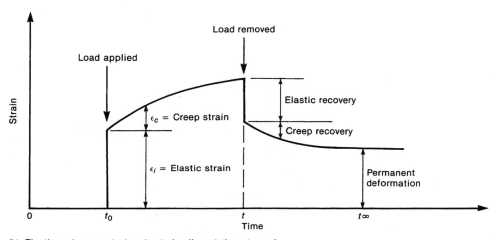

Fig. 3–21
Time-dependent strains.

(b) Elastic and creep strains due to loading at time, t_0, and unloading at time, t.

layer is roughly one water molecule thick or about 1% of the size of the gel particles. The loss of free unadsorbed water has little effect on the magnitude of the shrinkage.

Shrinkage strains are dependent on the relative humidity and are largest for relative humidities of 40% or less. They are partially recoverable on rewetting the concrete, and structures exposed to seasonal changes in humidity may expand and contract slightly due to changes in shrinkage strains.

The magnitude of shrinkage strains also depends on the composition of the concrete. The hardened cement paste shrinks, whereas the aggregate does not. Thus the larger the fraction of the total volume of the concrete that is made up of hydrated cement paste, the greater the shrinkage. The aggregates act to restrain the shrinkage. There is less shrinkage with quartz or granite aggregates than with sandstone aggregate, because the quartz has a higher modulus of elasticity. The water–cement ratio affects the amount of shrinkage because high water content reduces the volume of aggregate, thus reducing the restraint of the shrinkage by the aggregate. The more finely a cement is ground, the more surface area it has, and as a result, there is more adsorbed water to be lost during shrinkage and hence more shrinkage.

Drying shrinkage occurs as the moisture diffuses out of the concrete. As a result, the exterior shrinks more rapidly than the interior. This leads to tensile stresses in the outer skin of the concrete and compressive stresses in the interior. For large members, the ratio of volume to surface area increases, resulting in less shrinkage because there is more moist concrete to restrain the shrinkage. The shrinkage develops more slowly in large members.

A secondary form of shrinkage called *carbonation shrinkage* occurs in carbon dioxide–rich atmospheres such as those found in parking garages. At 50% relative humidity the amount of carbonation shrinkage can equal the drying shrinkage, effectively doubling the total amount of shrinkage. At higher and lower humidities the carbonation shrinkage decreases.

The ultimate drying shrinkage strain, ϵ_{shu}, for a 6 in. × 12 in. cylinder maintained for a very long time at a relative humidity of 40% ranges from 0.000400 to 0.001100 (400 to 1100 × 10^{-6} strain), with an average of about 0.000800. Thus, in a 25-ft bay in a building, the average shrinkage strain would cause a shortening of about $\frac{1}{4}$ in. in unreinforced concrete. In a structure, however, the shrinkage strains will tend to be less for the same concrete because:

1. The ratio of volume to surface area will generally be larger than for the cylinder, and as a result, drying takes place much more slowly.

2. A structure is built in stages and some of the shrinkage is dissipated before adjacent stages are completed.

3. The reinforcement restrains the development of the shrinkage.

The Euro-International Concrete Committee (CEB)[3-6] and the American Concrete Institute[3-15] have both published procedures for estimating shrinkage strains. The CEB method is more recent than the ACI procedure and accounts for member size in a better fashion. It will be presented here. The equations that follow apply only to the longitudinal shrinkage deformations of plain or lightly reinforced normal-weight concrete elements.

The axial shrinkage strains, ϵ_{cs}, occurring between times t_s at the start of shrinkage and t in plain concrete can be predicted using the formula

$$\epsilon_{cs}(t,t_s) = \epsilon_{cso}\beta_s(t,t_s) \tag{3-21}$$

where ϵ_{cso} is the basic shrinkage strain for a particular concrete and relative humidity, given by Eq. 3–22, and $\beta_s(t,t_s)$ is a coefficient given by Eq. 3–25 to describe the development of shrinkage between time t_s and t as a function of the effective thickness of the member.

$$\epsilon_{cso} = \epsilon_s(f_{cm})\beta_{RH} \tag{3-22}$$

where

$$\epsilon_s(f_{cm}) = \left[160 + \beta_{sc}(9 - f_{cm}/f_{cmo})\right] \times 10^{-6} \tag{3-23}$$

where f_{cm} is the mean compressive strength at 28 days, psi. This can be taken equal to f_{cr} as given by Eq. 3–3. For concrete with a standard deviation, s, equal to $0.15f'_c$, f_{cm} would be $1.20f'_c$. We shall use this value. Shrinkage is not a function of compressive strength per se. It decreases with decreasing water/cement ratio and decreasing cement content. The strength f_{cm} in Eq. 3–23 is used as an empirical measure of these quantities.

$f_{cmo} = 1450$ psi

$\beta_{sc} =$ coefficient that depends on the type of cement

$\quad = 50$ for Type I cement and 80 for Type III cement

$\beta_{RH} =$ coefficient that accounts for the effect of relative humidity on shrinkage

For RH between 40 and 99%;

$$\beta_{RH} = -1.55\left[1 - \left(\frac{RH}{RH_0}\right)^3\right] \tag{3-24}$$

For RH equal to or greater than 99%;

$$\beta_{RH} = + 0.25$$

RH relative humidity of the ambient atmosphere in percent

$RH_0 = 100\%$

The effect of relative humidity on the total shrinkage is illustrated in Fig. 3–22. More shrinkage occurs in dry ambient conditions.

The development of shrinkage with time is given by

$$\beta_s(t,t_s) = \left[\frac{(t - t_s)/t_1}{350(h_e/h_0)^2 + (t - t_s)/t_1} \right]^{0.5} \tag{3–25}$$

where h_e is the effective thickness in inches to account for the volume/surface ratio and is given by

$$h_e = 2A_c/u \tag{3–26}$$

and

A_c is the area of the cross section, in.^{2}

u is the perimeter of the cross section exposed to the atmosphere, in.

$h_0 = 4$ in.

t is the age of the concrete, days

t_s is the age of the concrete in days when shrinkage or swelling started, generally taken as the age at the end of moist curing

$t_1 = 1$ day

The development of shrinkage with time predicted using Eq. 3–25 is shorn in Fig. 3–23 for effective thicknesses of 4 in. and 24 in. Shrinkage develops much more rapidly in thin members because moisture diffuses out of the concrete more rapidly.

Because β_{RH} is negative, the computed shrinkage strain is also negative, implying that the concrete shortens due to shrinkage. In atmospheres with relative humidities greater than 99%, β_{RH} is positive, indicating the concrete swells in such environments.

The shrinkage predicted by Eq. 3–21 has a coefficient of variation of about 35%, which means that 10% of the time the shrinkage will be less than 0.55 times the predicted

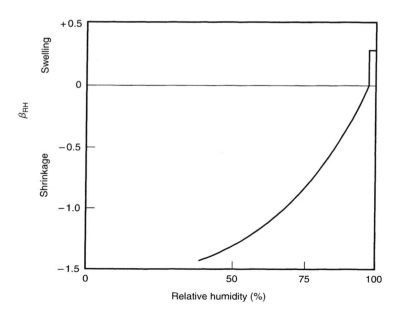

Fig. 3–22
Effect of relative humidity on shrinkage.

Materials

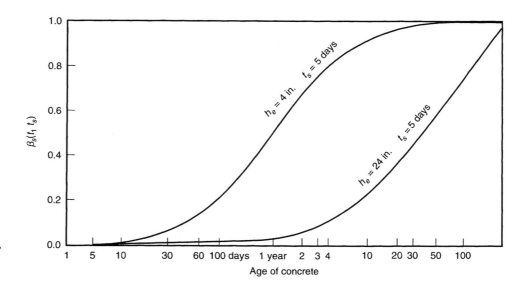

Fig. 3–23
Effect of effective thickness, h_e, on the rate of development of shrinkage.

shrinkage and 10% of the time it will exceed 1.45 times the predicted value. This is a very large spread. If shrinkage has a critical effect on a given structure, shrinkage tests should be carried out using the concrete in question.

If a lower level of accuracy is acceptable, the following values are representative of the shrinkage that would occur in 70 years in normal-weight structural concrete having strengths between 3000 and 7500 psi.

Dry atmospheric conditions or inside, RH = 50%:

Effective thickness, h_e = 6 in., $\epsilon_{cs}(70y)$ = −0.000560 strain

Effective thickness = 24 in., $\epsilon_{cs}(70y)$ = −0.000470 strain

Humid atmospheric conditions, RH = 80%:

Effective thickness = 6 in., $\epsilon_{cs}(70y)$ = −0.000310 strain

Effective thickness = 24 in., $\epsilon_{cs}(70y)$ = −0.000260 strain

EXAMPLE 3–2 Calculation of Shrinkage Strains

A lightly reinforced 6-in.-thick floor in an underground parking garage is supported around the outside edge by a 16-in.-thick basement wall. Cracks have developed in the slab perpendicular to the basement wall at roughly 6 ft on centers. The slab is 24 months old and the wall is 26 months old. The concrete is 3000 psi, made from Type I cement, and was moist cured for 5 days in each case. The relative humidity is 50%. Compute the width of these cracks assuming that they result from the restraint of the slab shrinkage parallel to the wall by the basement wall.

FLOOR SLAB

1. **Compute the basic shrinkage strain, ϵ_{cso}.**

$$\epsilon_{cso} = \epsilon_s(f_{cm})\beta_{RH} \tag{3–22}$$

$$\epsilon_s(f_{cm}) = \left[160 + \beta_{sc}(9 - f_{cm}/f_{cmo})\right] \times 10^{-6} \tag{3–23}$$

where

$$\beta_{sc} = 50 \text{ for Type I cement}$$
$$f_{cm} = \text{mean concrete strength} \approx 1.20f'_c$$
$$= 3600 \text{ psi}$$
$$f_{cmo} = 1450 \text{ psi}$$

$$\epsilon_s(f_{cm}) = \left[160 + 50(9 - 3600/1450)\right] \times 10^{-6}$$

$$= 486 \times 10^{-6} = 0.000486 \text{ strain}$$

$$\beta_{RH} = -1.55\left[1 - \left(\frac{RH}{RH_0}\right)^3\right] \tag{3-24}$$

where $RH = 50\%$, $RH_0 = 100\%$, and

$$\beta_{RH} = -1.55\left[1 - (50/100)^3\right]$$

$$= -1.356$$

Therefore, the basic shrinkage is:

$$\epsilon_{cso} = 0.000486 \times -1.356$$

$$= -0.000659 \text{ strain}$$

2. Compute the coefficient for the development of shrinkage with time.

$$\beta_s(t,t_s) = \left[\frac{(t - t_s)/t_1}{350(h_e/h_0)^2 + (t - t_s)/t_1}\right]^{0.5} \tag{3-25}$$

where h_e is the effective thickness $= 2A_c/u$. Consider a 1-ft-wide strip of slab exposed to the air on the top and bottom:

$$h_e = \frac{2 \times (6 \times 12)}{2 \times 12} = 6 \text{ in.} \qquad h_0 = 4\text{in.}$$

$$t = 730 \text{ days} \qquad t_s = 5\text{days} \qquad t_1 = 1\text{day}$$

$$\beta_s(t,t_s) = \left[\frac{(730 - 5)/1}{350 \times (6/4)^2 + (730 - 5)/1}\right]^{0.5}$$

$$= 0.692$$

This means that after two years 69% of the slab shrinkage will have occurred.

3. Compute the shrinkage strain, $\epsilon_{cs}(t,t_s)$.

$$\epsilon_{cs}(t,t_s) = \epsilon_{cso}\beta_s(t,t_s) \tag{3-21}$$

$$= -0.000659 \times 0.692$$

$$\epsilon_{cs}(t,t_s) = -0.000456 \text{ strain}$$

BASEMENT WALL

1. Compute the basic shrinkage strain, ϵ_{cso}. This will be the same as for the floor slab since f_{cm}, β_{sc}, and RH are the same.

2. Compute the coefficient for the development of shrinkage with time. Compute the effective thickness, $h_e = 2A_c/u$. Again considering a 1-ft-wide strip of wall. It is exposed to air only on the inside face.

$$h_e = \frac{2 \times (16 \times 12)}{12} = 32\text{in.} \qquad h_0 = 4\text{in.}$$

$$t = 791\text{days} \qquad t_s = 5\text{days} \qquad t_1 = 1\text{day}$$

$$\beta_s(t,t_s) = \left[\frac{(791 - 5)/1}{350 \times (32/4)^2 + (791 - 5)/1}\right]^{0.5}$$

$$= 0.184$$

3. **Compute the shrinkage strain,** $\epsilon_{cs}(t,t_s)$

$$e_{cs}(t,t_s) = -0.000659 \times 0.184$$
$$= -0.000121 \text{ strain}$$

RELATIVE SHRINKAGE AND CRACK WIDTH

Thus the relative shrinkage strain between the slab and the wall when the slab is one year old is $-0.000456 - (-0.000121) = -0.000335$. If the average crack spacing is 6 ft, the shortening between cracks will be about $6 \times 12 \times -0.000335 = -0.024$ in. Hence the cracks will be about 0.024 in. wide on average.

This calculation does not allow for the effect of the reinforcement in restraining the shrinkage strains. The actual shrinkage would be 75 to 100% of the calculated values. ∎

Creep

When concrete is loaded, an instantaneous elastic strain develops as shown in Fig. 3–21b. If this load remains on the member, creep strains develop with time. These occur because the adsorbed water layers tend to become thinner between gel particles which are transmitting compressive stress. This change in thickness occurs rapidly at first, slowing down with time. With time, bonds form between the gel particles in their new position. If the load is eventually removed, a portion of the strain is recovered elastically and another portion by creep, but a residual strain remains, as shown in Fig. 3–21b, due to the bonding of the gel particles in the deformed position.

The creep strains, ϵ_c, are on the order of one to three times the instantaneous elastic strains. Creep strains lead to an increase in deflections with time; may lead to a redistribution of stresses within cross sections; cause a decrease in prestressing forces; and so on.

The ratio of creep strain after a very long time to elastic strain, ϵ_c/ϵ_i, is called the *creep coefficient*, ϕ. The magnitude of the creep coefficient is affected by the ratio of the sustained stress to the strength of the concrete, the humidity of the environment, the dimensions of the element, and the composition of the concrete. Creep is greatest in concretes with a high cement paste content. Concretes containing a large aggregate fraction creep less because only the paste creeps, and that creep is restrained by the aggregate. The rate of development of the creep strains is also affected by the temperature, reaching a plateau about 160°F. At the high temperatures encountered in fires, very large creep strains occur. The type of cement (i.e., normal or high early strength cement) and the water–cement ratio are important only in that they affect the strength at the time when the concrete is loaded.

For creep, as for shrinkage, several calculation procedures exist.[3-6, 3-15] The method given here is from the *CEB-FIB Model Code 1990.*[3-6] It is applicable for concretes up to compressive strengths of about 10,000 psi subjected to a compressive loading up to about $0.40f'_c$ at an age t_0, exposed to relative humidities of 40% or higher and mean temperatures between 40° and 90°F. For stresses less than $0.40f'_c$, creep is assumed to be linearly related to stress. Beyond this stress, creep strains increase more rapidly and may lead to failure of the member at stresses greater than $0.75f'_c$, as shown in Fig. 3–2a. Similarly, creep increases significantly at mean temperatures in excess of 90°F.

The total strain, $\epsilon_c(t)$, at time t in a concrete member uniaxially loaded with a constant stress $\sigma_c(t_0)$ at time t_0 is

$$\epsilon_c(t) = \epsilon_{ci}(t_0) + \epsilon_{cc}(t) + \epsilon_{sc}(t) + \epsilon_{cT}(t) \tag{3–27}$$

where

$\epsilon_{ci}(t_0) =$ initial strain at loading $= \sigma_c(t_0)/E_c(t_0)$

$\epsilon_{cc}(t) =$ creep strain at time t where t is greater than t_0

$$\epsilon_{cs}(t) = \text{shrinkage strain at time } t$$

$$\epsilon_{cT}(t) = \text{thermal strain at time } t$$

$$E_c(t_0) = \text{modulus of elasticity at the age of loading}$$

The stress-dependent strain at time t is

$$\epsilon_{c\sigma}(t) = \epsilon_{ci}(t_0) + \epsilon_{cc}(t) \tag{3–28}$$

For a stress σ_c applied at time t_0 which remains constant until time t, the creep stain ϵ_{cc} between time t_0 and t is

$$\epsilon_{cc}(t,t_0) = \frac{\sigma_c(t_0)}{E_c(28)}\phi(t,t_0) \tag{3–29}$$

where $E_c(28)$ is the modulus of elasticity at the age of 28 days, given by Eq. 3–16 or 3–17, and $\phi(t,t_0)$ is the creep coefficient, given by

$$\phi(t,t_0) = \phi_0\beta_c(t,t_0) \tag{3–30}$$

where ϕ_0 is the basic creep given by Eq. 3–31, and $\beta_c(t,t_0)$ is a coefficient to account for the development of creep with time, given by Eq. 3–35.

$$\phi_0 = \phi_{RH}\beta(f_{cm})\beta(t_0) \tag{3–31}$$

where

$$\phi_{RH} = 1 + \frac{1 - RH/RH_0}{0.46(h_e/h_0)^{1/3}} \tag{3–32}$$

$$\beta(f_{cm}) = \frac{5.3}{(f_{cm}/f_{cmo})^{0.5}} \tag{3–33}$$

$$\beta(t_0) = \frac{1}{0.1 + (t_0/t_1)^{0.2}} \tag{3–34}$$

where h_e, h_0, RH, RH_0, f_{cm}, f_{cmo}, and t_1 are as defined in connection with Eqs. 3–21 to 3–26.

The basic creep coefficient, ϕ_0, is actually a function of the relative humidity, the composition of the concrete, and the degree of hydration at the start of loading. The last two of these are expressed empirically in Eqs. 3–33 and 3–34 as functions of the mean 28-day strength, f_{cm}, and the age at loading, t_0.

The development of creep with time is given by

$$\beta_c(t,t_0) = \left[\frac{(t - t_0)/t_1}{\beta_H + (t - t_0)/t_1}\right]^{0.3} \tag{3–35}$$

with

$$\beta_H = 150\left[1 + \left(1.2\frac{RH}{RH_0}\right)^{18}\right]\frac{h_e}{h_0} + 250 \leq 1500 \tag{3–36}$$

The effects of the effective thickness and age at the time of loading on the creep co-efficient $\phi(t,t_0)$ are illustrated in Fig. 3–24. The creep coefficient is about half as big for concrete loaded at one year as it is for concrete loaded at 7 days. The effective thickness has less effect than it did on shrinkage (Fig. 3–23), reducing the value of $\phi(t,t_0)$ by about 20% for the example shown.

When compared to creep test data, the creep coefficient $\phi(t,t_0)$ computed in this way has a coefficient of variation of about 20%.[3–6] Ten percent of the time the actual value of $\phi(t,t_0$ will be less than 75% of the computed value and 10% of the time it will exceed 125% of the computed value. If creep deflections are a serious problem for a particular structure, consideration should be given to carrying out creep tests on the concrete to be used.

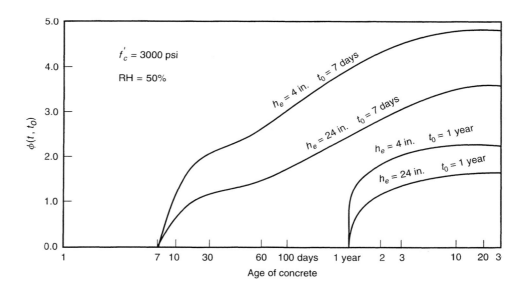

Fig. 3–24
Effect of effective thickness, h_e, and age at loading, t_0, on creep coefficient, $\phi(t,t_0)$

TABLE 3–1 Creep Coefficient, $\phi(70y,t_0)$ for Normal-Weight Concrete after 70 Years of Loading

Age at Loading, t_0 (days)	Dry Atmospheric Conditions (RH $= 50\%$)		Humid Atmospheric Conditions (RH $= 80\%$)	
	Effective Thickness, h_e (Eq. 3–26)			
	6 in.	24 in.	6 in.	24 in.
1	4.8	3.9	3.4	3.0
7	3.3	2.7	2.4	2.1
28	2.6	2.1	1.8	1.6
90	2.1	1.7	1.5	1.3
365	1.6	1.3	1.1	1.0

Source: Ref. 3–6.

In cases where a lower level of accuracy is acceptable, the creep coefficient at 70 years, $\phi(70y,t_0)$, can be taken from Table 3–1.

The total shortening of a plain concrete member at time t due to elastic and creep strains resulting from a constant stress σ_c applied at time t_0 can be computed using Eq. 3–28, which becomes

$$\epsilon_{co}(t) = \sigma_c(t_0)\left[\frac{1}{E_c(t_0)} + \frac{\phi(t,t_0)}{E_c(28)}\right] \qquad (3\text{–}37)$$

The term in brackets is the *creep compliance function*, $J(t,t_0)$, representing the total stress-dependent strain per unit stress.

EXAMPLE 3–3   Calculation of Unrestrained Creep Strains

A plain concrete pedestal 24 in. $\times$ 24 in. $\times$ 10 ft high is subjected to an average stress of 1000 psi. Compute the total shortening in 5 years if the load is applied 2 weeks after the concrete is cast. The properties of the concrete and the exposure are the same as in Example 3–2.

1. **Compute the basic creep coefficient, ϕ_0.**

$$\phi_0 = \phi_{RH}\beta(f_{cm})\beta(t_0) \tag{3-31}$$

$$\phi_{RH} = 1 + \frac{1 - RH/RH_0}{0.46(h_e/h_0)^{1/3}} \tag{3-32}$$

where

$$RH = 50\% \qquad RH_0 = 100\% \qquad h_0 = 4\text{in.}$$

$$h_e = 2A_c/u = 2 \times 24 \times 24/4 \times 24 = 12\text{in.}$$

$$\phi_{RH} = 1 + \frac{1 - 50/100}{0.46(12/4)^{1,3}} = 1.754$$

$$\beta(f_{cm}) = \frac{5.3}{(f_{cm}/f_{cmo})^{0.5}} \tag{3-33}$$

where

$$f_{cm} = 1.20f'_c = 3600\,\text{psi} \qquad f_{cmo} = 1450\text{psi}$$

$$\beta(f_{cm}) = \frac{5.3}{(3600/1450)^{0.5}} = 3.364$$

$$\beta(t_0) = \frac{1}{0.1 + (t_0/t_1)^{0.2}} \tag{3-34}$$

where

$$t_0 = 14\,\text{days} \qquad t_1 = 1\,\text{day}$$

$$\beta(t_0) = \frac{1}{0.1 + (14/1)^{0.2}} = 0.557$$

Thus

$$\phi_0 = 1.754 \times 3.364 \times 0.557 = 3.29$$

The basic creep coefficient is 3.29.

2. **Compute the development of creep with time, $\beta_c(t,t_0)$.**

$$\beta_c(t,t_0) = \left[\frac{(t - t_0)/t_1}{\beta_H + (t - t_0)/t_1}\right]^{0.3} \tag{3-35}$$

where:

$$\beta_H = 150\left[1 + \left(1.2\frac{RH}{RH_0}\right)^{18}\right]\frac{h_e}{h_0} + 250 \le 1500 \tag{3-36}$$

$$= 150\left[1 + \left(1.2\frac{50}{100}\right)^{18}\right]\frac{12}{4} + 250 = 700$$

$$t = 5 \times 365 = 1825\,\text{days} \qquad t_0 = 14\,\text{days} \qquad t_1 = 1\,\text{day}$$

$$\beta_c(t,t_0) = \left[\frac{(1825 - 14)/1}{700 + (1825 - 14)/1}\right]^{0.3} = 0.907$$

This indicates that 90.7% of the total creep has occurred at the end of 5 years.

3. **Compute the creep coefficient, $\phi(t,t_0)$.**

$$\phi(t,t_0) = \phi_0\beta_c(t,t_0) = 3.29 \times 0.907 \tag{3-30}$$

$$= 2.98$$

4. Compute the total stress-dependent strain, $\epsilon_{c\sigma}(t,t_0)$

$$\epsilon_{c\sigma}(t) = \sigma_c(t_0)\left[\frac{1}{E_c(t_0)} + \frac{\phi(t,t_0)}{E_{ci}}\right] \tag{3-37}$$

where

$$\sigma_c(t_0) = 1000\text{psi}$$

$$E_c = 57{,}000\sqrt{f'_c} \tag{3-17}$$

$E_c(t_0)$ is modulus of elasticity at 14 days, where the concrete strength at 14 days is given by Eq. 3–5 with $t = 14$ days:

$$f'_c(t) = f'_c(28)\left(\frac{t}{4 + 0.85t}\right)$$

$$= 3000\left(\frac{14}{4 + 0.85 \times 14}\right) = 2830\text{psi}$$

$$E_c(t_0) = 57{,}000 \times \sqrt{2830} = 3{,}032{,}000\text{psi}$$

E_{ci} is the modulus of elasticity at 28 days.

$$E_{ci} = 57{,}000\sqrt{3000} = 3{,}122{,}000\text{psi}$$

$$\epsilon_{c\sigma}(5y) = 1{,}000\left(\frac{1}{3{,}032{,}000} + \frac{2.98}{3{,}122{,}000}\right)$$

$$= 0.000330 + 0.000955 = 0.001285 \text{ strain}$$

The creep strain is almost three times the instantaneous strain.

5. Compute the total shortening.

$$\Delta\ell = \ell \times \epsilon_{c\sigma}(5y) = 120 \times 0.00128$$

$$= 0.154 \text{ in.}$$

The pedestal would shorten by 0.154 in. in 5 years. ∎

In an axially loaded reinforced concrete column, the creep shortening of the concrete causes compressive strains in the longitudinal reinforcement, increasing the load in the steel and reducing the load, and hence the stress, in the concrete. As a result, a portion of the elastic strain in the concrete is recovered and, in addition, the creep strains are smaller than they would be in a plain concrete column with the same initial concrete stress. A similar redistribution occurs in the compression zone of a beam with compression steel.

This effect can be modeled using an *age-adjusted effective modulus*, $E_{caa}(t,t_0)$, and an *age-adjusted transformed section* in the calculations[3-36 to 3-38] where

$$E_{caa}(t,t_0) = \frac{E_c(t_0)}{1 + \chi(t,t_0)[E_c(t_0)/E_c(28)]\phi(t,t_0)} \tag{3-38}$$

where $\chi(t,t_0)$ is an aging coefficient that can be approximated by[3-39]

$$\chi(t,t_0) = \frac{t_0^{0.5}}{1 + t_0^{0.5}} \tag{3-39}$$

The axial strain at time t in a column loaded at age t_0 with a constant load P is

$$\epsilon_c(t,t_0) = \frac{P}{A_{traa} \times E_{caa}(t,t_0)} \tag{3-40}$$

3–4 Time-Dependent Volume Changes

where A_{traa} is the age-adjusted transformed area of the column cross section. The concept of the transformed sections is presented in Sec. 9–2. For more information on the use of the age-adjusted effected modulus, see Ref. 3–37 and 3–38.

EXAMPLE 3–4 Computation of the Strains and Stresses in an Axially Loaded Reinforced Concrete Column

A concrete column 24 in. $\times$ 24 in. $\times$ 10 ft high has 8 No. 8 longitudinal bars and is loaded with a load of 630 kips at an age of 2 weeks. Compute the elastic stresses in the concrete and steel at the time of loading and the stresses and strains at an age of 5 years. The properties of the concrete and the exposure are the same as in Examples 3–2 and 3–3.

Steps 1, 2, and 3 are the same as in Example 3–3. The following quantities are computed:

$$f_c'(14) = 2830 \text{ psi} \qquad f_c'(28) = 3000 \text{ psi}$$

$$E_c(14) = 3{,}032{,}000 \text{ psi} \qquad E_c(28) = 3{,}122{,}000 \text{ psi}$$

$$\phi(t,t_0) = 2.98$$

4. Compute the transformed area at the instant of loading, A_{tr}. (Transformed sections are discussed in Sec. 9–2.)

$$\text{Elastic modular ratio} = n = \frac{E_s}{E_c(14)} = \frac{29{,}000{,}000}{3{,}032{,}000}$$

$$= 9.56$$

The steel will be "transformed" into concrete, giving the transformed area

$$A_{tr} = A_c + (n-1)A_s = 576 \text{ in.}^2 + (9.56 - 1) \times 6.32 \text{ in}^2$$

$$= 630 \text{ in.}^2$$

The stress in the concrete is 630,000 lb/630 in.^{2} = 1000 psi. The stress in the steel is n times the stress in the concrete = 9.56 $\times$ 1000 psi = 9560 psi.

5. Compute the age adjusted effective modulus, $E_{caa}(t,t_0)$, and the age-adjusted modular ratio, n_{aa}.

$$E_{caa}(t,t_0) = \frac{E_c(t_0)}{1 + \chi(t,t_0)[E_c(t_0)/E_c(28)]\phi(t,t_0)} \qquad (3\text{–}38)$$

where

$$\chi(t,t_0) = \frac{t_0^{0.5}}{1 + t_0^{0.5}} = \frac{14^{0.5}}{1 + 14^{0.5}} \qquad (3\text{–}39)$$

$$= 0.789$$

$$E_{caa}(t,t_0) = \frac{3{,}032{,}000}{1 + 0.789 \times \dfrac{3{,}032{,}000}{3{,}122{,}000} \times 2.98}$$

$$= 923{,}400 \text{ psi}$$

$$\text{Age-adjusted modular ratio, } n_{aa} = \frac{E_s}{E_{caa}(t,t_0)} = \frac{29{,}000{,}000}{923{,}400}$$

$$= 31.4$$

6. Compute the age-adjusted transformed area, A_{traa}, stresses in concrete and steel, and shortening. Again the steel will be transformed to concrete.

$$A_{traa} = A_c + (n_{aa} - 1)A_s = 576 \text{ in.}^2 + (31.4 - 1) \times 6.32 \text{ in.}^2$$

$$= 768 \text{ in.}^2$$

$$\text{Stress in concrete} = f_c = \frac{P}{A_{traa}} = \frac{630{,}000 \text{ lb}}{768 \text{ in.}^2}$$

$$= 820 \text{ psi}$$

$$\text{Stress in steel } n_{aa} \times f_c = 31.4 \times 820 \text{ psi}$$

$$= 25{,}750 \text{ psi}$$

$$\text{Strain} = \frac{f_c}{E_{caa}} = \frac{820}{923{,}400}$$

$$= 0.000888 \text{ strain}$$

$$\text{Shortening } \epsilon \times \ell = 0.000888 \times 120 \text{ in.}$$

$$= 0.107 \text{ in.}$$

The creep has reduced the stress in the concrete from 1000 psi at the time of loading to 820 psi at 5 years. During the same period, the steel stress has increased from 9560 psi to 25,750 psi. A column with less reinforcement would experience a larger increase in the reinforcement stress. To prevent yielding of the steel under sustained loads, ACI Sec. 10.9.1 sets a lower limit of 1% on the reinforcement ratio in columns.

The plain concrete column in Example 3–3, which had a constant concrete stress of 1000 psi throughout the 5-year period, shortened 0.154 in. The column in this example, which had an initial concrete stress of 1000 psi, shortened two-thirds as much. ∎

Thermal Expansion

The coefficient of thermal expansion or contraction, α, is affected by such factors as composition of the concrete, moisture content of the concrete, and age of the concrete. Ranges from normal-weight concretes are 5 to 7×10^{-6} *strain*/°F for those made with siliceous aggregates, and 3.5 to 5×10^{-6}/°F for concretes made from limestone or calcareous aggregates. Approximate values for lightweight concrete are 3.6 to 6.2×10^{-6}/°F. An all-around value of 5.5×10^{-6}/°F may be used. The coefficient of thermal expansion for reinforcing steel is 6×10^{-6}/°F. In calculations of thermal effects it is necessary to allow for the time lag between air temperatures and concrete temperatures.

As the temperature rises, so does the coefficient of expansion and at the temperatures experienced in building fires, it may be several times the value at normal operating temperatures.[3-40] The thermal expansion of a floor slab in a fire may be large enough to exert large shear forces on the supporting columns.

3–5 HIGH-STRENGTH CONCRETE

Concretes with strengths in excess of 6000 psi are referred to as *high-strength concretes*. Strengths of up to 18,000 psi have been used in buildings. Reference 3–27 presents the state of the art of the production and use of high-strength concrete.

Admixtures such as superplasticizers, silica fume, and supplementary cementing materials such as certain fly ashes improve the dispersion of cement in the mix and produce workable concretes with much lower water–cement ratios than possible previously. The resulting concrete has a lower void ratio and is stronger than normal concretes.

Coarse aggregates should consist of strong fine-grained gravel with a rough surface. Smooth river gravels give a lower paste–aggregate bond strength and a weaker concrete.

Mechanical Properties

As shown in Fig. 3–17, the stress–strain curves for higher-strength concretes tend to have a more linear loading branch and a steep descending branch. High-strength concrete exhibits less internal microcracking for a given strain than does normal concrete. In normal concrete, unstable microcracking starts to develop at a compressive stress of about $0.75f'_c$, referred to as the critical stress (see Sec. 3–2). In high-strength concrete the critical stress is about $0.90f'_c$. Failure occurs by fracture of the aggregate on relatively smooth planes parallel to the direction of the applied stress. The lateral strains tend to be considerably smaller than for lower-strength concrete. One implication of this is that spiral and confining reinforcement may be less effective in increasing the strength and ductility of high-strength concrete column cores.

Equations 3–16 and 3–17 overestimate the modulus of elasticity of concretes with strengths in excess of about 6000 psi. Reference 3–27 proposes that

$$E_c = 40,000\sqrt{f'_c} + 1.0 \times 10^6 \tag{3-41}$$

As noted ealier, E_c varies as a function of the modulus of the coarse aggregate.

The modulus of rupture of high-strength concretes ranges from 7.5 to $12\sqrt{f'_c}$. A lower bound to the split-cylinder tensile test data is given by $6\sqrt{f'_c}$.

Shrinkage and Creep

Shrinkage of concrete is approximately proportional to the percentage of water by volume in the concrete. High-strength concrete has a higher paste content, but the paste has a lower water/cement ratio. As a result, the shrinkage of high-strength concrete is about the same as that of normal concrete.

Test data suggest that the creep coefficient, ϕ, for high-strength concrete is considerably less than that for normal concrete.

3–6   LIGHTWEIGHT CONCRETE

Structural lightweight concrete is concrete having a density between 90 and 120 lb/ft^3 containing naturally occurring lightweight aggregates such as pumice; artificial aggregates made from shales, slates, or clays which have been expanded by heating; or of sintered blast furnace slag or cinders. Such concrete is used when a saving in dead load is important. Lightweight concrete costs about 20% more than normal concrete. The terms "all-lightweight concrete" and "sand-lightweight concrete" refer to mixes having lightweight sand or natural sand, respectively.

The modulus of elasticity of lightweight concrete is less than that of normal concrete and can be computed from Eq. 3–16.

The stress–strain curve of lightweight concrete is affected by the lower modulus of elasticity and relative strength of the aggregates and cement paste. If the aggregate is the weaker of the two, failure tends to occur suddenly in the aggregate and the descending branch of the stress–strain curve is very short or nonexistent, as shown by the upper solid line in Fig. 3–25. On the other hand, if the aggregate does not fail, the stress–strain curve will have a well-defined descending branch, as shown by the curved lower solid line in this figure. As a result of the lower modulus of elasticity of lightweight concrete, the strain at which the maximum compressive stress is reached is higher than for normal-weight concrete.

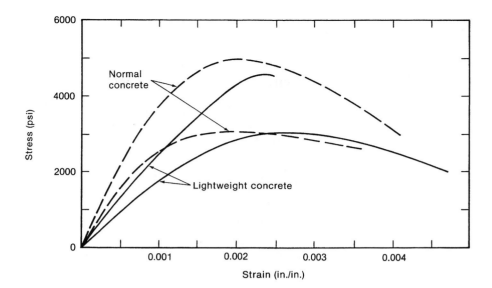

Fig. 3–25
Compressive stress–strain
curves for normal-weight and
lightweight concretes,
$f'_c = 3000$ and 5000 psi.
(From Ref. 3–41.)

The tensile strength of all-lightweight concrete is 70 to 100% of that of normal-weight concrete. Sand-lightweight concrete has tensile strengths in the range of 80 to 100% of those of normal-weight concrete. The fracture surface of lightweight concrete tends to be smoother than for normal concrete because the aggregate fractures.

The shrinkage and creep of lightweight concrete are similar or slightly greater than those for normal concrete. The creep coefficients computed using Eq. 3–30 can be used for lightweight concrete.

3–7 DURABILITY OF CONCRETE

The durability of concrete structures is discussed in Ref. 3–42. The three most common durability problems in concrete structures are:

1. Corrosion of steel in the concrete. Corrosion involves oxidation of the reinforcement. For corrosion to occur there must be a source of oxygen and moisture, both of which diffuse through the concrete. Typically, the pH value of new concrete is on the order of 13. The alkaline nature of concrete tends to prevent corrosion from occurring. If there is a source of chloride ions, these also diffuse through the concrete, decreasing the pH of the part the chloride ions have penetrated. When the pH of the concrete adjacent to the bars drops below about 10 or 11, corrosion can start. The thicker and less permeable the cover concrete is, the longer it takes for moisture, oxygen, and chloride ions to reach the bars. Shrinkage or flexural cracks penetrating the cover allow these agents to reach the bars more rapidly. The rust products that are formed when reinforcement corrodes are several times the volume of the metal that has corroded. This increase in volume causes cracking and spalling of the concrete adjacent to the bars. Factors affecting corrosion are discussed in Ref. 3–43.

ACI Sec. 4.4 attempts to control corrosion of steel in concrete by requiring a minimum strength and a maximum water to cementitious materials ratio to reduce the permeability of the concrete and by requiring at least a minimum cover to the reinforcing bars. The amount of chlorides in the mix is also limited. Epoxy-coated bars are sometimes used to delay or prevent corrosion.

Corrosion is most serious under conditions of intermittent wetting and drying. Adequate drainage should be provided to allow water to drain off structures. Corrosion is seldom a problem for permanently submerged portions of structures.

2. Breakdown of the structure of the concrete due to freezing and thawing.
When concrete freezes, pressures develop in the water in the pores, leading to a breakdown of the structure of the concrete. Entrained air provides closely spaced microscopic voids, which relieve these pressures. ACI Sec. 4.2.1 requires minimum air contents to reduce the effects of freezing and thawing exposures. The spacing of the air voids is also important and some specifications specify spacing factors. ACI Sec. 4.2.2 sets maximum water/cementitious materials ratios of 0.40 to 0.50 and minimum concrete strengths of 4000 to 5000 psi for concretes, depending on the severity of the exposure. These may give strengths higher than would otherwise be used in structural design. Thus a water/cement ratio of 0.40 will generally correspond to a strength of 4500 to 5000 psi for air-entrained concrete. This additional strength can be utilized in computing the strength of the structure.

Again, drainage should be provided so that water does not collect on the surface of the concrete. Concrete should not be allowed to freeze at a very young age and should be allowed to dry out before severe freezing.

3. Breakdown of the structure of the concrete due to chemical attack.
Sulfates cause disintegration of concrete unless special cements are used. ACI Sec. 4.3.1 specifies cement type, minimum water/cementitious materials ratios, and minimum compressive strengths for various sulfate exposures. Geotechnical reports will generally give sulfate levels.

Some aggregates containing silica react with the alkalies in the cement, causing a disruptive expansion of the concrete, leading to severe random cracking. This *alkali silica reaction* is counteracted by changing the source of the aggregate or by using low-alkali cements.[3-44] It is most serious if the concrete is warm in service and if there is a source of moisture.

Reference 3–45 lists a number of chemicals that attack concretes.

ACI Chap. 4 presents requirements for concrete that is exposed to freezing and thawing, deicing chemicals, sulfates, and chlorides. Examples are pavements, bridge decks, parking garages, water tanks, and foundations in sulfate-rich soils.

Sulfates cause disintegration of concrete unless special cements are used. ACI Table 4.3.1 gives special requirements for concrete in contact with sulfates in soils or in water. In many areas in the western U.S. soils contain sulfates.

3–8 BEHAVIOR OF CONCRETE EXPOSED TO HIGH AND LOW TEMPERATURES

When a concrete member is exposed to high temperatures such as occur in a building fire, for example, it will behave satisfactorily for a considerable period of time. During a fire, high thermal gradients are established, however, and as a result the surface layers expand and eventually crack or spall off the cooler, interior part of the concrete. The spalling is aggravated if water from fire hoses suddenly cools the surface.

The modulus of elasticity and the strength of concrete decrease at high temperatures, whereas the coefficient of thermal expansion increases. The type of aggregate affects the strength reduction, as shown in Fig. 3–26. Most structural concretes can be classified into one of three aggregate types: carbonate, siliceous, or lightweight. Concretes made with carbonate aggregates, such as limestone and dolomite, are relatively unaffected by temperature until they reach about 1200 to 1300°F, at which time they undergo a chemical change and rapidly lose strength. The quartz in siliceous aggregates, such as quartzite, granite, sandstones, and schists, undergoes a phase change at about 800 to 1000°F, which causes

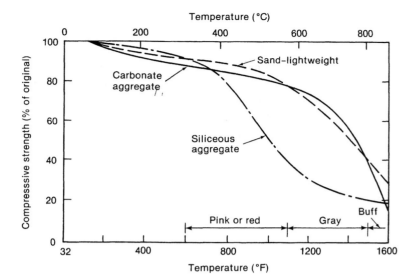

Fig. 3–26
Compressive strength of concretes at high temperatures. (From Ref. 3–40.)

an abrupt change in volume and spalling of the surface. Lightweight aggregates gradually lose their strength at temperatures above 1200°F.

The reduction in strength and the extent of spalling due to heat are most pronounced in wet concrete and, as a result, fire is most critical with young concrete. The tensile strength tends to be affected more by temperature than does the compressive strength.

Concretes made with limestone and siliceous aggregates tend to change color when heated, as indicated in Fig. 3–26, and the color of the concrete after a fire can be used as a rough guide to the temperature reached by the concrete. As a general rule, concrete whose color has changed beyond pink is suspect. Concrete that has passed the pink stage and gone into the gray stage is probably badly damaged. Such concrete should be chipped away and replaced with a layer of new concrete or shotcrete.

In cold temperatures, the strength of hardened concrete tends to increase, the increase being greatest for moist concrete.

3–9   REINFORCEMENT

Because concrete is weak in tension, it is used together with steel bars or wires that resist the tensile stresses. The most common types of reinforcement for nonprestressed members are hot-rolled deformed bars and wire fabric. In this book only the former will be used in examples, although the design principles apply with very few exceptions to members reinforced with welded wire mesh or cold-worked deformed bars.

The ACI code requires that reinforcement be steel bars or steel wires. Significant modifications to the design process are required if materials such as fiber-reinforced-plastic rods are used for reinforcement, because such materials are brittle and do not have the ductility assumed in the derivation of design procedures.

Hot-Rolled Deformed Bars

Steel reinforcing bars are basically round in cross section with lugs or deformations rolled into the surface to aid in anchoring the bars in the concrete (Fig. 3–27). They are produced

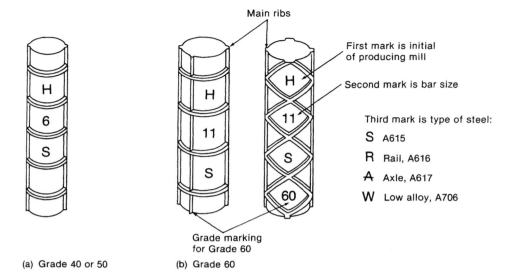

Fig. 3–27
Standard reinforcing bar markings. (Courtesy of Concrete Reinforcing Steel Institute.)

Main ribs

First mark is initial of producing mill

Second mark is bar size

Third mark is type of steel:

S   A615

R   Rail, A616

A   Axle, A617

W   Low alloy, A706

Grade marking for Grade 60

(a) Grade 40 or 50

(b) Grade 60

according to the following ASTM specifications, which specify certain dimensions and certain chemical and mechanical properties:

1. ASTM A 615: *Standard Specification for Deformed and Plain Billet-Steel Bars for Concrete Reinforcement.* This specification covers the most commonly used reinforcing bars. They are available in sizes 3 to 18 in Grade 60 (yield strength 60 ksi) plus sizes 3 to 6 in Grade 40 and sizes 6 to 18 in Grade 75. The specified mechanical properties are summarized in Table 3–2. The diameters, areas, and weights are listed in Table A–1 in Appendix A. The phosphorus content is limited to $\leq 0.06\%$.

2. ASTM A 616: *Standard Specification for Rail-Steel Deformed and Plain Bars for Concrete Reinforcement.* This specifies bars rolled from discarded railroad rails. This type of steel is less ductile and less bendable than A 615. A 616 steel is not widely available. It is rolled in sizes 3 to 11 in Grade 60.

3. ASTM A 617: *Standard Specification for Axle-Steel Deformed and Plain Bars for Concrete Reinforcement.* This covers bars rolled from discarded train car axles. It is produced in sizes 3 to 11 in Grades 40 and 60 and is less ductile than A 615 steel. It is not widely available.

4. ASTM A 706: *Standard Specification for Low-Alloy-Steel Deformed Bars for Concrete Reinforcement.* This specification covers bars intended for special applications where weldability, bendability, or ductility is important. As indicated in Table 3–2, A 706 requires a larger elongation at failure and a more stringent bend test than A 615. ACI Sec. 21.2.5.1 requires the use of A 706 bars or A 615 bars meeting special requirements in seismic applications. A 706 limits the amounts of carbon, manganese, phosphorus, sulfur, and silicon and limits the carbon equivalent to $\leq 0.55\%$. These bars are available in sizes 3 through 18 in Grade 60. There are both a lower and an upper limit on the yield strength.

Reinforcing bars are available in four grades, with yield strengths of 40, 50, 60, and 75 ksi, referred to as Grades 40, 50, 60, and 75, respectively. Grade 60 is the steel most commonly used in buildings and bridges. Other grades may not be available in some areas. Grade 75 is used in large columns. Grade 40 is the most ductile, followed by Grades 60, 75, and 50, in that order.

TABLE 3–2 Summary of Mechanical Properties of Reinforcing Bars from ASTM A 615 and ASTM A 706

	A 615			A 706
	Grade 40	Grade 60	Grade 75	Grade 60
Minimum tensile strength, psi	70,000	90,000	100,000	80,000
Minimum yield strength, psi	40,000	60,000	75,000	60,000
Maximum yield strength, psi	—	—	—	78,000
Minimum elongation in 8 in. gauge length, %				
No. 3	11	9	—	14
No. 4 and 5	12	9	—	14
No. 6	12	9	7	14
No. 7 and 8	—	8	7	12
No. 9, 10, and 11	—	7	6	10
No. 14 and 18	—	7	6	10
Pin diameter for bend test, where d = nominal diameter				
No. 3, 4, and 5	$3.5d$	$3.5d$	—	$3d$
No. 6	$5d$	$5d$	—	$4d$
No. 7 and 8	—	$5d$	$5d$	$4d$
No. 9, 10, and 11	—	$7d$	$7d$	$6d$
No. 14 and 18	—	$9d$	$9d$	$8d$

Grade 60 deformed reinforcing bars are available in the 11 sizes listed in Table A–1. The sizes are referred to by their nominal diameter expressed in eighths of an inch. Thus a No. 4 bar has a diameter of $\frac{4}{8}$ in. or $\frac{1}{2}$ in. The nominal cross-sectional area can be computed directly from the nominal diameter except for the No. 10 and larger bars, which have diameters slightly larger than $\frac{10}{8}$ in., $\frac{11}{8}$ in., and so on. Size and grade marks are rolled into the bars for identification purposes, as shown in Fig. 3–27. Grade 40 bars are available only in sizes No. 3 through No. 6. Grade 75 steel is only available in sizes 6 to 18.

ASTM A 615 and A 706 also specify metric (SI) bar sizes. These are sized so that the areas are even multiples of 100 mm^2 and the bars are referred to using nominal bar size numbers equal to their theoretical diameter rounded off to the nearest 5 mm. Thus a No. 10M bar has an area of 100 mm^2, corresponding to a theoretical diameter of 11.3 mm, while a No. 20M bar has an area of 300 mm^2, corresponding to a theoretical diameter of 19.5 mm. The diameters, areas, and weights of SI bar sizes are listed in Table A–1M in Appendix A.

ASTM A 615 defines three grades of metric reinforcing bars, Grades 300, 400, and 500, having specified yield strengths of 300, 400, and 500 MPa, respectively.

When this book was in press, the Concrete Reinforcing Steel Institute announced that it was reconsidering the metric bar sizes and may ask ASTM to revise its reinforcing bar standards. Two changes were contemplated: (1) the existing inch bar sizes would be retained, designated by a bar number equal to the diameter taken to the nearest whole millimeter, and (2) yield strengths of 300, 420, and 520 MPa are being considered. Because the change has still to be agreed to, the tables and examples in this book are all based on the reinforcing bars defined in the ASTM standards in existence at the end of 1995.

Mechanical Properties

Idealized stress–strain relationships are given in Fig. 3–28 for Grade 40 and Grade 60 reinforcing bars, a representative high-strength bar, and welded-wire fabric. The initial tangent modulus of elasticity, E_s, for all reinforcing bars can be taken as 29×10^6 psi. Grade 40 bars display a pronounced yield plateau, as shown in Fig. 3–28. Although this plateau is generally present for Grade 60 bars, it is typically much shorter. High-strength bars generally do not have a well-defined yield point.

Figure 3–29 is a histogram of mill test yield strengths of Grade 60 reinforcement with a nominal yield strength of 60 ksi. As shown in this figure, there is a considerable variation in yield strength, with about 10% of the tests having a yield strength equal to or

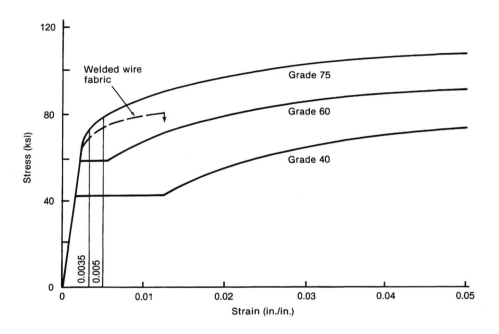

Fig. 3–28
Stress–strain curves for reinforcement.

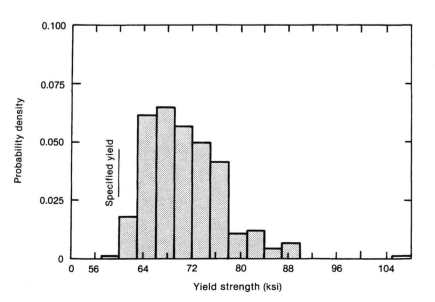

Fig. 3–29
Distribution of mill test yield strengths for Grade 60 steel. (From Ref. 3–46.)

greater than 80 ksi, 133% of the nominal yield strength. The coefficient of variation of the yield strengths plotted in Fig. 3–29 is 9.3%.

ASTM Specifications base the yield strength on *mill tests* which are carried out at a high rate of loading. For the slow loading rates associated with dead loads or many live loads, the *static yield strength* is applicable. This is roughly 4 ksi less than the mill test yield strength.[3-46]

Fatigue Strength

Some reinforced concrete elements, such as bridge decks, may be subjected to a large number of loading cycles. In such cases, the reinforcement may fail in fatigue. Fatigue failures of the reinforcement will occur only if one or both of the extreme stresses in the stress cycle is tensile. The relationship between the range of stress, S_r, and the number of cycles is shown in Fig. 3–30. For practical purposes there is a fatigue threshold or *endurance limit* below which fatigue failures will normally not occur. For straight ASTM A 615 bars this is about 24 ksi and is essentially the same for Grade 40 and Grade 60 bars.

The fatigue strength of deformed bars decreases as the stress range (the maximum stress in a cycle minus the minimum stress) increases, as the level of the lower (less tensile) stress in the cycle is reduced, and as the ratio of the radius of the fillet at the base of the deformation lugs to the height of the lugs is decreased. The fatigue strength is essentially independent of the yield strength.

The fatigue strength is strongly reduced by bends or tack welds in the region of maximum stress. These will reduce the fatigue strength by about 50%.

For design, the following rules can be applied. If the deformed reinforcement in a particular member is subjected to 1 million or more cycles involving tensile stresses, or a combination of tension and compression stresses, fatigue failures may occur if the difference between the maximum and minimum stresses under the repeated loading exceeds 20 ksi. In the vicinity of bends or in locations where auxiliary reinforcement has been tack welded to the main reinforcement, fatigue failures may occur if the stress range exceeds 10

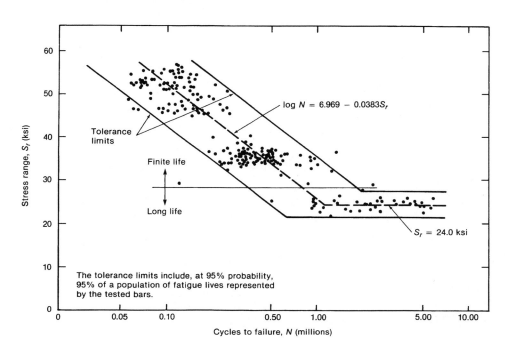

Fig. 3–30
Test data on fatigue of deformed bars from a single U.S. manufacturer. (From Ref. 3–47.)

ksi. Further guidance is given in Refs. 3–47 and 3–48. Design of reinforced concrete for fatigue is discussed in Sec. 9–8.

Strength at High Temperatures

Deformed reinforcement subjected to high temperatures in fires tends to lose some of its strength as shown in Fig. 3–31. When the temperature of the reinforcement exceeds about 850°F, the yield and ultimate strengths drop both significantly. One of the functions of concrete cover on reinforcement is to prevent the reinforcement from getting hot enough to lose strength.

Welded-Wire Fabric

Welded-wire fabric is a prefabricated reinforcement consisting of smooth or deformed wires welded together in square or rectangular grids. Sheets of wires are welded in electric-resistance welding machines, in a production line. This type of reinforcement is used in walls or slabs where relatively regular reinforcement patterns are possible. The ability to place a large amount of reinforcement with a minimum of work frequently makes welded-wire fabric economical.

The wire for welded-wire fabric is produced in accordance with the following specifications: ASTM A 82 *Standard Specification for Steel Wire, Plain, for Concrete Reinforcement*, and ASTM A 496 *Standard Specification for Steel Wire, Deformed, for Concrete Reinforcement*. The deformations are typically two or more lines of indentations of about 4 to 5% of the bar diameter rolled into the wire surface. Wire sizes range from about 0.125 in. diameter to 0.625 in. diameter and are referred to as W or D, for plain or deformed wires, respectively, followed by a number that corresponds to the cross-sectional area of the wire in 0.01-in.^{2} increments. Thus a W2 wire is a smooth wire with a cross-sectional area of 0.02 in.^{2} Diameters and areas of typical wire sizes are given in Table A–2a.

Welded-wire fabric corresponds to the following specifications: ASTM A 185 *Standard Specification for Steel Wire Fabric, Plain, for Concrete Reinforcement*, and ASTM A 497 *Standard Specification for Steel Wire Fabric, Deformed, for Concrete Reinforcement*.

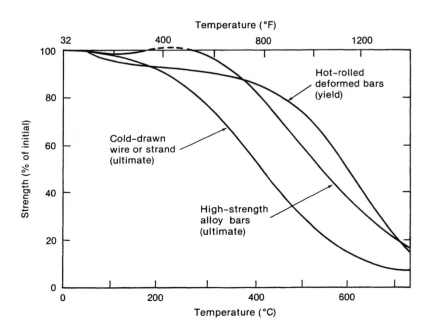

Fig. 3–31
Strength of reinforcing steels at high temperatures. (From Ref. 3–40.)

Deformed welded-wire fabric may contain some smooth wires in either direction. Welded-wire fabric is available in standard or custom patterns referred to using a style designation such as $6 \times 6 - W4 \times W4$. The numbers in the style designation refer to: (spacing of longitudinal wires $\times$ spacing of transverse wires–size of longitudinal wires $\times$ size of transverse wires). Thus a $6 \times 6 - W4 \times W4$ fabric has W4 wires at 6 in. on centers each way. Areas and weights of common welded-wire fabric patterns are given in Table A–2b.

Welded smooth wire fabric depends on the cross-wires to provide a mechanical anchorage with the concrete, while welded deformed wire fabric utilizes both the wire deformations and cross-wires for bond and anchorage. In smooth wires, two cross-wires are needed to mechanically anchor the bar for its yield strength.

Smooth and deformed fabric are made from wires ranging from about 0.125 in. to 0.625 in. in diameter. The minimum yield and tensile strength of smooth wire for wire fabric is 65 ksi and 75 ksi. For deformed wires, the minimum yield and tensile strengths are 70 ksi and 80 ksi. According to ASTM A 497, these yield strengths are measured at a strain of 0.5%. ACI Secs. 3.5.3.5 and 3.5.3.6 define the yield strength of both smooth and deformed wires as 60 ksi except that if the yield strength at a strain of 0.35% has been measured, that value can be used.

One problem with welded-wire fabric is the ductility of the wires. Frequently, such materials break at elongations of 1 to 3%, considerably less than required by the ASTM Specifications for reinforcing bars (see Fig. 3–28). The cold-working process used in drawing the small-diameter wires causes strain hardening of the steel. This eliminates the yield plateau of the stress–strain curve. The smaller the diameter of the wire, the more brittle the wire tends to be.

PROBLEMS

3–1   What is the significance of the "critical stress"

(a) with respect to the structure of the concrete?

(b) with respect to spiral reinforcement?

(c) with respect to strength under sustained loads?

3–2   A group of 43 tests on a given type of concrete had a mean strength of 3622 psi and a standard deviation of 421 psi. Does this concrete satisfy the requirements of ACI Sec. 5.3.2 for 3000-psi concrete?

3–3   The concrete containing Type I cement in a structure is cured for 3 days at 70°F followed by 6 days at 40°F. Use the maturity concept to estimate its strength as a fraction of the 28-day strength under standard curing.

3–4   Use Fig. 3–12a to estimate the compressive strength σ_2 for biaxially loaded concrete subjected to

(a) $\sigma_1 = 0$.

(b) $\sigma_1 = 0.75$ times the tensile strength, in tension.

(c) $\sigma_1 = 0.5$ times the compressive strength, in compression.

3–5   The concrete in the core of a spiral column is subjected to a uniform confining stress σ_3 of 800 psi. What will the compressive strength σ_1 be? The uniaxial compressive strength is 4000 psi.

3–6   What factors affect the shrinkage of concrete?

3–7   What factors affect the creep of concrete?

3–8   A structure is made from concrete containing Type I cement. The average ambient relative humidity is 70%. The concrete was moist cured for 4 days. $f'_c = 4000$ psi.

(a) Compute the unrestrained shrinkage strain of a rectangular beam with cross-sectional dimensions 8 in. $\times$ 20 in. at 3 years after the concrete was placed.

(b) Compute the stress dependent strain in the concrete in a 20 in. $\times$ 20 in. plain concrete column at an age of 2 years. A load of 400 kips was applied to the column at an age of 60 days.

4

Flexure: Basic Concepts, Rectangular Beams

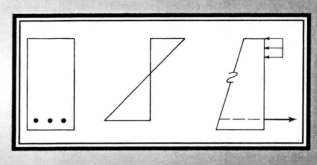

4–1 INTRODUCTION

In this chapter the stress–strain curves for concrete and reinforcement from Chap. 3 are used to develop a theory for flexure. This theory is applied to rectangular beam cross sections, and examples of the design of beam cross sections are given. In Chap. 5 this is extended to T beams and beams with compressive steel.

Because the complete design of a beam requires an understanding of shear, bond, and other aspects, a complete design example is deferred until Chap. 10, where all these aspects are considered. In Chap. 11 the flexure theory is extended to combined bending and axial load, to permit the design of columns.

Most reinforced concrete structures can be subdivided into beams and slabs subjected primarily to flexure (bending), and columns subjected to axial compression accompanied in most cases by flexure. Typical examples of flexural members are the slab and beams shown in Fig. 4–1. The load, P, applied at point A is carried by the strip of slab shown shaded. The end reactions from this slab strip load the beams at B and C. The beams, in turn, carry the slab reactions to the columns at D, E, F, and G. The beam reactions cause axial loads in the columns.

The slab in Fig. 4–1 is assumed to transfer loads in one direction and hence is called a *one-way slab*. If there were no beams, the slab would carry the load in two directions. Such a slab is referred to as a *two-way slab* (see Chaps. 13 to 15).

B-Regions and D-Regions

Through most of the length of a beam or column, a straight-line distribution of strains will exist and the normal flexure theory can be applied. Such regions are referred to as *B-regions*, where the B stands for beam or for Bernoulli, who first postulated the straight-line strain distribution. Adjacent to discontinuities, concentrated loads, holes, or changes in cross section, the strain distribution is not linear and different types of analyses must be used. Such regions

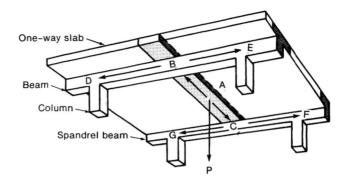

Fig. 4–1
One-way flexure.

are referred to as *D-regions*, where the D stands for discontinuity or for disturbed. Chapters 4 to 15 deal primarily with B-regions. D-regions are discussed in Chap. 18.

Analysis versus Design

Two different types of problems arise in the study of reinforced concrete:

1. *Analysis.* Given a cross section, concrete strength, reinforcement size, location, and yield strength, compute the resistance or capacity.

2. *Design.* Given a factored load effect such as M_u, select a suitable cross section, including dimensions, concrete strength, reinforcement, and so on.

Although both types of problems utilize the same principles, the procedure is different in each case. Analysis is easier since all the decisions concerning reinforcement, beam size, and so on, have been made and it is only necessary to apply the strength calculation principles to determine the capacity. Design, on the other hand, involves the choice of beam sizes, material strengths, and reinforcement to produce a cross section that can resist the loads and moments. Because the analysis problem is easier, most sections in this and other chapters start with the analysis to develop the fundamental concepts and then move to consider design.

Required Strength and Design Strength

The basic safety equation for flexure is:

$$\text{factored resistance} \geq \text{factored load effects} \tag{4–1a}$$

or

$$\phi M_n \geq M_u \tag{4–1b}$$

where M_u is the *moment due to the factored loads*, which the ACI Code refers to as the *required ultimate moment*. This is a load effect computed by structural analysis from the governing combination of factored loads given in ACI Sec. 9.2. The term M_n refers to the *nominal moment capacity* of a cross section computed using the nominal dimensions and specified material strengths. The factor ϕ is a *strength reduction factor* (ACI Sec. 9.3) to account for possible variations in dimensions and material strengths and possible inaccuracies in the strength equations. For flexure without axial load, ACI Sec. 9.3.2.1 sets $\phi = 0.90$, and ACI Sec. B9.3.2 sets $\phi = 0.90$ for "tension-controlled sections." Almost all practical beams will be tension-controlled sections and ϕ will be equal to 0.90. The concept of tension-controlled sections will be discussed later in the chapter. The product, ϕM_n, is referred to as the *design moment*, *design strength*, or *factored moment resistance*.

Positive and Negative Moments

A moment that causes compression on the top surface of a beam and tension on the bottom surface will be called a *positive moment*. The compression zones for positive and negative moments are shown shaded in Fig. 4–2. Bending moment diagrams will be plotted on the compression side of the member.

Symbols and Abbreviations

Although symbols are defined as they are first used and are summarized in Appendix B, several *must be understood completely* if one is to understand the theory developed in this book. These include the terms M_u and M_n defined earlier and the cross-sectional dimensions illustrated in Fig. 4–2.

The prime symbol ($'$) generally refers to compressions, as in A_s' and d'.

A_s is the area of reinforcement on the tension face of the beam, tension reinforcement, in.^{2}.

A_s' is the area of reinforcement on the compression side of the beam, compression reinforcement, in.^{2}.

b is the width of the compression face of the beam. This is illustrated in Fig. 4–2 for positive and negative moment regions, in.

b_w is the width of the web of the beam and may or may not be the same as b, in.

d is the distance from the extreme fiber in compression to the centroid of the steel on the tension side of the member. In the positive moment region (Fig. 4–2a) the tension steel is near the bottom of the beam while in the negative moment region it is near the top, in.

d_t is the distance from the extreme compression fiber to the farthest layer of tension steel, in.

f_c' is the specified 28-day compressive strength of the concrete, psi.

f_s is the stress in the tension reinforcement, psi.

f_y is the specified yield strength of the reinforcement, psi.

jd is the *lever arm*, the distance between the resultant compressive force and the resultant tensile force, in.

j is a dimensionless ratio used to define the lever arm, jd. j varies during the life of the beam.

ϵ_{cu} is the assumed concrete strain on the compression face of the beam at flexural failure.

ϵ_s is the strain in the tension reinforcement.

ρ is the longitudinal tension reinforcement ratio, $\rho = A_s/bd$.

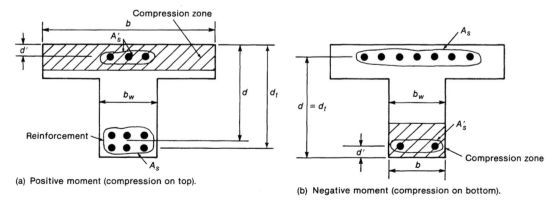

(a) Positive moment (compression on top).

(b) Negative moment (compression on bottom).

Fig. 4–2
Cross-sectional dimensions.

Statics of Beam Action

A *beam* is a structural member that supports applied loads and its own weight primarily by internal moments and shears. Figure 4–3a shows a simple beam that supports its own dead weight, w per unit length, plus an applied load, P. If the axial applied load, N, is equal to zero as shown, the member is referred to as a beam. If N is a compressive force, the member is called a *beam–column*. If it were tensile, the member would be a tension tie. Chapter 4 will be restricted to the very common case where $N = 0$.

The loads, w and P, cause *bending moments* distributed as shown in Fig. 4–3b. The bending moment is a *load effect* determined from the loads using the laws of statics. For a simply supported beam of a given span and a given set of loads, w and P, the moments are independent of the composition and size of the beam.

At any section within the beam, the *internal resisting moment*, M, shown in Fig. 4–3c is necessary to equilibrate the bending moment. An internal resisting shear, V, is also required as shown.

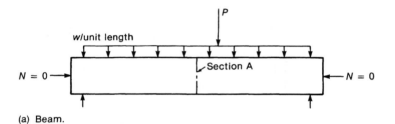

(a) Beam.

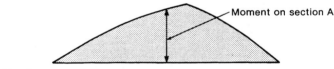

(b) Bending moment diagram.

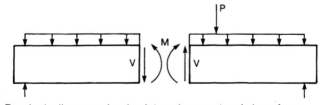

(c) Free body diagrams showing internal moment and shear force.

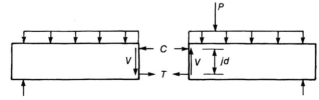

(d) Free body diagrams showing internal moment as a compression–tension force couple.

Fig. 4–3
Internal forces in a beam.

The internal resisting moment, M, results from an internal compressive force, C, and an internal tensile force, T, separated by a lever arm, jd, as shown in Fig. 4–3d. Since there are no external axial loads, N, summation of the horizontal forces gives

$$C - T = 0 \quad \text{or} \quad C = T \tag{4-2}$$

If moments are summed about an axis through the point of application of the compressive force, C, the moment equilibrium of the free body gives

$$M = Tjd \tag{4-3a}$$

Similarly, if moments are summed about the point of application of the tensile force, T,

$$M = Cjd \tag{4-3b}$$

Since $C = T$, these two equations are identical. Equations 4–2 and 4–3 come directly from statics and are equally applicable to beams made of steel, timber, or reinforced concrete.

The conventional *elastic* beam theory results in the equation $\sigma = My/I$, which for an uncracked, homogeneous rectangular beam without reinforcement gives the distribution of stresses shown in Fig. 4–4. The stress diagram shown in Fig. 4–4c and d may be visualized as having a "volume," and hence one frequently refers to the *compressive stress block* and the *tensile stress block*. The resultant of the compressive stresses is the force C given by

$$C = \frac{\sigma_{c\,(\text{max})}}{2}\left(b\frac{h}{2}\right) \tag{4-4}$$

This is equal to the volume of the compressive stress block shown in Fig. 4–4d. In a similar manner one could compute the force T from the tensile stress block. The forces, C and T, act through the centroids of the volumes of the respective stress blocks. In the *elastic* case these forces act at $h/3$ above or below the neutral axis, so that $jd = 2h/3$. From Eqs. 4–3b and 4–4 and Fig. 4–4 we can write

$$M = Cjd \tag{4-5a}$$

$$M = \sigma_{c\,(\text{max})}\frac{bh}{4}\left(\frac{2h}{3}\right) \tag{4-5b}$$

$$M = \sigma_{c\,(\text{max})}\frac{bh^3/12}{h/2} \tag{4-5c}$$

or

$$M = \frac{\sigma I}{y} \tag{4-5d}$$

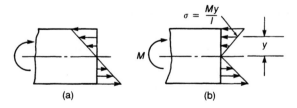

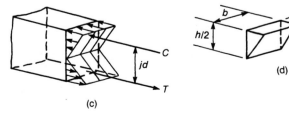

Fig. 4–4
Elastic beam stresses and
stress blocks

Flexure: Basic Concepts, Rectangular Beams

Thus for the elastic case, identical answers are obtained from the traditional beam stress equation 4–5d, and from Eqs. 4–3 using the stress block concept.

The elastic beam theory (Eq. 4–5d) is not used in the design of reinforced concrete beams: first, because the compressive stress–strain curve of concrete is nonlinear, as shown in Fig. 3-15 and, even more important, because the concrete cracks at low tensile stresses, making it necessary to provide steel reinforcement to transfer the tensile force, T. These two factors are easily handled by the stress block concept, combined with Eqs. 4–2 and 4–3.

Flexure Theory for Reinforced Concrete

The theory of flexure for reinforced concrete is based on three basic assumptions, which are sufficient to allow one to calculate the moment resistance of a beam. These are presented first and used to illustrate the behavior of a beam cross section under increasing moment. Following this, four additional simplifying assumptions from the ACI Code are presented to simplify the analysis for practical application.

Flexural Behavior

The cracking pattern and strains measured in a laboratory test of a reinforced concrete beam are shown in Fig. 4–5. The cracks are indicated by the short vertical and inclined lines in Fig. 4–5b and c. A photograph of the beam after failure is shown in Fig. 4–6. The cracks have been marked with black ink so that they show in the photograph.

The strains plotted in Fig. 4–5 were measured on a 16-in. gauge length extending 8 in. on either side of the midspan of the beam. This region is shown shaded in Fig. 4–5a and, as shown in Figs. 4–5c and 4–6, there were two cracks in this region at failure.

The measured strains were used to compute the curvature corresponding to each load level. Curvature, ϕ, is defined as the angle change over a known length and, as shown in the inset of Fig. 4–7, is computed as

$$\phi = \frac{\epsilon}{y} \tag{4-6}$$

where ϵ is the strain at a distance y from the axis of zero strain at the load stage in question. Figure 4–7 relates the bending moment, M, at midspan of the beam to the curvatures at the same location. This is a *moment–curvature diagram*.

Initially, the beam was uncracked, as shown in Fig. 4–5a. The strains at this stage were very small and the stress distribution was essentially linear. The moment and curvature are shown by point A in Fig. 4–7. The moment–curvature diagram for this stage (segment O-B in Fig. 4–7) was linear.

When the stresses at the bottom of the beam reached the tensile strength of the concrete, cracking occurred. After cracking, the tensile force in the concrete was transferred to the steel. As a result, less of the concrete section was effective in resisting moments, as shown by the distribution of stresses in Fig. 4–5b and the stiffness of the beam decreased. Thus the slope of the moment–curvature diagram (shown by B-C-D in Fig. 4–7) also decreased. The crack pattern and strains in Fig. 4–5b correspond to the behavior expected under the loads applied in everyday service (stage C in Fig. 4–5 and point C in Fig. 4–7). The stress distribution in the concrete is still close to linear at this stage. The largest crack had a width of 0.006 in. at this stage. Such a crack is entirely acceptable, as discussed in Chap. 9.

Eventually, the reinforcement reached the yield point shown by point D in Fig. 4–7. The compressive stresses were still close to being linear at this stage. Once yielding had occurred, the curvatures increased rapidly with very little increase in moment, as shown in Fig. 4–7. The increase in curvature can also be seen from the difference between the strain diagrams in Fig. 4–5b and c. The beam failed due to crushing of the concrete at the top of the beam.

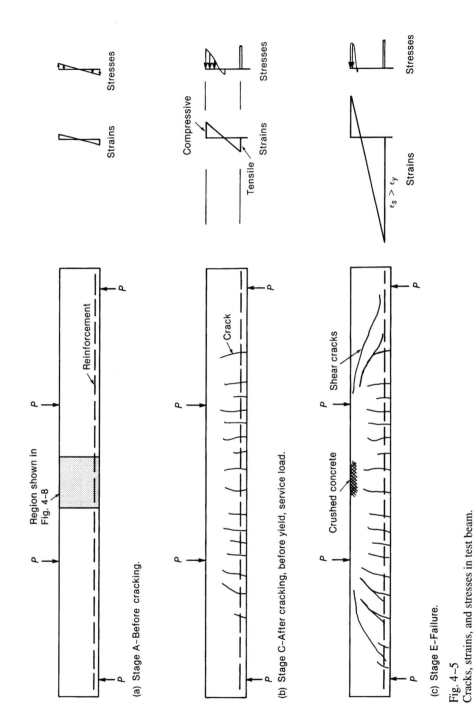

(a) Stage A–Before cracking.

(b) Stage C–After cracking, before yield, service load.

(c) Stage E–Failure.

Fig. 4–5
Cracks, strains, and stresses in test beam.

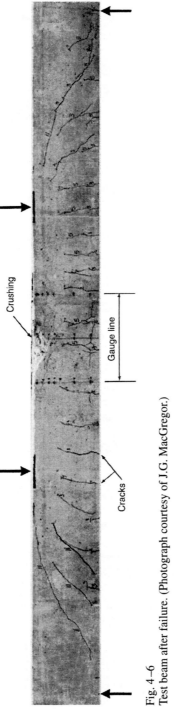

Fig. 4–6
Test beam after failure. (Photograph courtesy of J.G. MacGregor.)

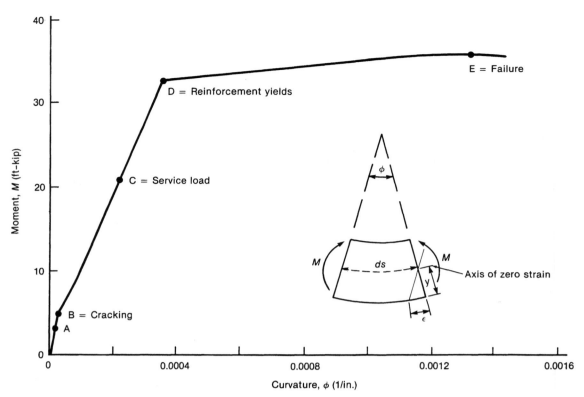

Fig. 4–7
Moment–curvature diagram for test beam.

It is worth noting that although concrete itself is not a very ductile material (see the stress–strain curves in Fig. 3–17) reinforced concrete beams can exhibit large ductilities, as shown by the long, almost flat post yield portion of the moment–curvature diagram. At service load (stage *C*), the midspan deflection of the beam was 0.31 in. or 1/383 of the span. At ultimate, this had increased to 2 in. or 1/60 of the span, showing the great ductility of this beam.

In practice, reinforced concrete design has been carried out in one of two ways. Until the mid–1960s, designers considered the loads expected in service and carried out calculations assuming a linear stress distribution for concrete in compression. This was called *working stress design* and corresponded to the load stage plotted in Fig. 4–5b and point *C* in Fig. 4–7. Since then, calculations have been carried out at the failure state (Fig. 4–5c and point *E* in Fig. 4–7) for loads larger than those expected in service (factored loads), and checks are made of the deflections and cracking at service load levels. This is called *limit states design* or *strength design*, as explained in Sec. 2-3. In this book we concentrate on limit states design. Working stress design principles are used to calculate the deflections and steel stresses at service loads, however, and are explained in Chap. 9.

Basic Assumptions in Flexure Theory

Three basic assumptions are made:

1. Sections perpendicular to the axis of bending which are plane before bending remain plane after bending.

2. The strain in the reinforcement is equal to the strain in the concrete at the same level.

3. The stresses in the concrete and reinforcement can be computed from the strains using stress–strain curves for concrete and steel.

Flexure: Basic Concepts, Rectangular Beams

The first of these is the traditional "plane sections remain plane" assumption made in the development of beam theory. Figure 4–8b is an enlargement of the strain distribution plotted in Fig. 4–5b. The dots represent strains measured on the 16-in. gauge lines shown in Fig. 4–8a. The measured strains are seen to be linear. Figure 4–9 shows a strain distribution measured in tests of two columns subjected to combined axial load and moment.

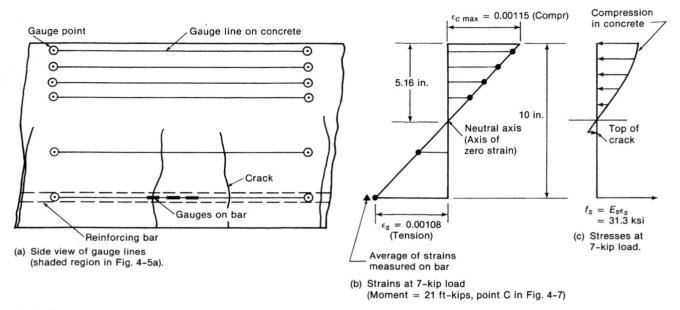

(a) Side view of gauge lines (shaded region in Fig. 4-5a).

(b) Strains at 7–kip load (Moment = 21 ft–kips, point C in Fig. 4–7)

(c) Stresses at 7–kip load.

Fig. 4–8
Distribution of strains and stresses in test beam at service load.

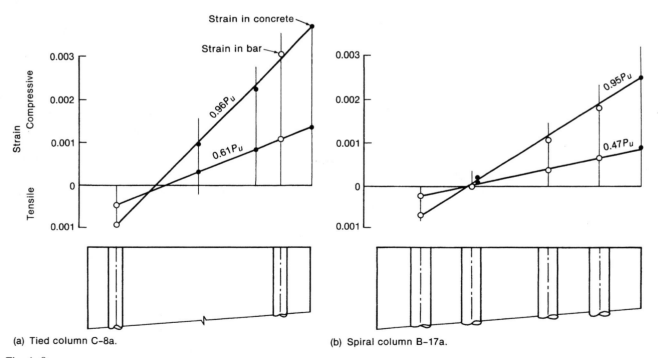

(a) Tied column C-8a.

(b) Spiral column B-17a.

Fig. 4–9
Strains measured in tests of eccentrically loaded columns. (From Ref. 4–1.)

Again the distributions are linear as assumed. This assumption does not hold for so-called "deep beams" with spans shorter than about four times their depths because such members tend to act as tied arches rather than beams (see Chap. 18).

The second assumption is necessary because the concrete and the reinforcement must act together to carry load. This assumption implies perfect bond between the concrete and the steel. The agreement between the strains measured in the steel, plotted with a triangle in Fig. 4–8b, and the line through the dots representing the measured concrete strains show that the steel and concrete do act together as postulated in this assumption. The ACI code combines these first two assumptions as follows:

> 10.2.2–Strain in reinforcement and concrete shall be assumed directly proportional to the distance from the neutral axis. . . .

The stress diagram in Fig. 4–8c was obtained from the strain distribution using the theoretical stress–strain curve for concrete presented in Fig. 3–18b. The moment computed using this stress distribution balances the load on the beam.

Additional Assumptions in Flexure Theory for Design

The three assumptions already made are sufficient to allow calculation of the strength and behavior of reinforced concrete elements. For design purposes, however, the following additional assumptions are introduced to simplify the problem with little loss in accuracy.

4. The tensile strength of concrete is neglected in flexural strength calculations (ACI Sec. 10.2.5).

The strength of concrete in tension is roughly one-tenth of the compressive strength and the tensile force in the concrete below the zero strain axis is small compared to the tensile force in the steel. Hence the contribution of the tensile stresses in the concrete to the flexural capacity of the beam is small and can be neglected. It should be noted that this assumption is made primarily to simplify flexural calculations. In some instances, particularly shear, bond, deflection, and service load calculations for prestressed concrete, the tensile resistance of concrete is utilized.

5. Concrete is assumed to fail when the compressive strain reaches a limiting value.

Strictly speaking, there is no such thing as a limiting compressive strain for concrete. As shown in Fig. 4–7, a simply supported reinforced concrete beam reaches its maximum capacity when the slope, $dM/d\phi$, of the moment–curvature diagram equals zero (point E). Failure occurs when $dM/d\phi$ becomes negative, corresponding to an unstable situation in which further deformations occur at decreasing loads.[4-2] Design calculations are very much simplified, however, if a limiting strain is assumed. Since the moment and curvature at the maximum moment point on the moment–curvature diagram correspond to one particular value of the extreme compressive strain, the moments corresponding to any other strain at the extreme fiber will be smaller. As a result, this assumption will always give conservative estimates of the strength.

The maximum compressive strains, ϵ_{cu}, from tests of reinforced concrete beams, eccentrically loaded columns and eccentrically loaded plain concrete prisms are reproduced in Fig. 4–10. ACI Sec. 10.2.3 specifies a limiting compressive strain equal to 0.003. This compares closely to the lower bound of the test data in Fig. 4–10. In Europe, the CEB Model Code uses a limiting strain of 0.002 for columns under concentric axial load, corresponding to the peak of the stress–strain curve in Fig. 3-17 and 0.0035 for beams and eccentrically loaded columns.[4-4] It should be noted that much higher limiting strains have been measured in members with a significant moment gradient and members in which the concrete is confined by spirals or closely spaced hoops.[4-5, 4-6] Throughout this book the limiting compressive strain will be assumed constant and equal to 0.003.

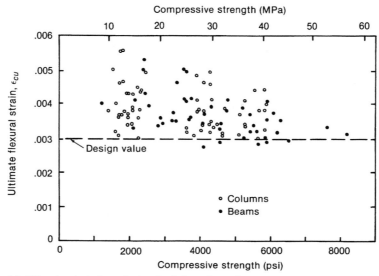

(a) Ultimate strain from tests of reinforced members.

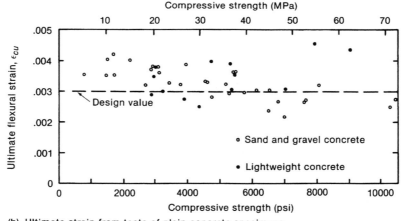

Fig. 4–10
Limiting compressive strain.
(Ref. 4–3.)

(b) Ultimate strain from tests of plain concrete specimens.

6. The compressive stress–strain relationship for concrete may be assumed to be rectangular, trapezoidal, parabolic, or any other shape that results in prediction of strength in substantial agreement with the results of comprehensive tests (ACI Sec. 10.2.6).

Thus rather than using a representative stress–strain curve such as that given in Fig. 3-18b, other diagrams which are easier to use in computations are acceptable, provided that they adequately predict test results. As illustrated in Fig. 4–11, the shape of the stress block in a beam at the ultimate moment can be expressed mathematically in terms of three constants:

k_3 = ratio of the maximum stress f_c'', in the compression zone of a beam to the cylinder strength, f_c'

k_1 = ratio of the average compressive stress to the maximum stress (this is equal to the ratio of the shaded area in Fig. 4–12 to the area of the rectangle ck_3f_c')

4–2 Flexure Theory

93

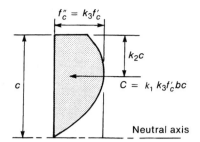

Fig. 4–11
Mathematical description of
compression stress block.

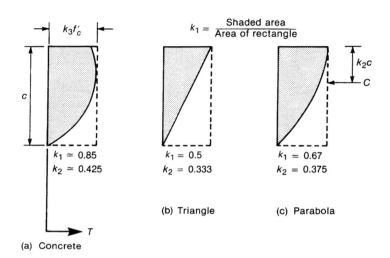

Fig. 4–12
Values of k_1 and k_2 for various stress distributions.

k_2 = ratio of the distance between the extreme compression fiber and the resultant
of the compressive force to the depth of the neutral axis, c

For a rectangular compression zone of width, b, and depth to the neutral axis, c, the resultant compressive force is

$$C = k_1 k_3 f'_c bc \qquad (4\text{-}7)$$

Values of k_1 and k_2 are given in Fig. 4–12 for various assumed compressive stress–strain diagrams or "stress blocks." Tests of plain concrete prisms have yielded the values of $k_1 k_3 f'_c$ shown in Fig. 4–13.

As a further simplification, ACI Sec. 10.2.7 permits the use of the equivalent rectangular concrete stress distribution shown in Fig. 4–14 for ultimate strength calculations. The rectangular distribution is defined by the following:

1. A uniform compressive stress of $\alpha_1 f'_c$ shall be assumed distributed over an equivalent compression zone bounded by the edges of the cross section and a straight line located parallel to the neutral axis at a distance $a = \beta_1 c$ from the fiber of maximum compressive strain, where $\alpha_1 = 0.85$.

2. The distance c from the fiber of maximum strain to the neutral axis is measured perpendicular to that axis.

3. The factor β_1 shall be taken as:

(a) For concrete strengths f'_c up to and including 4000 psi,

$$\beta_1 = 0.85 \qquad (4\text{-}8a)$$

Flexure: Basic Concepts, Rectangular Beams

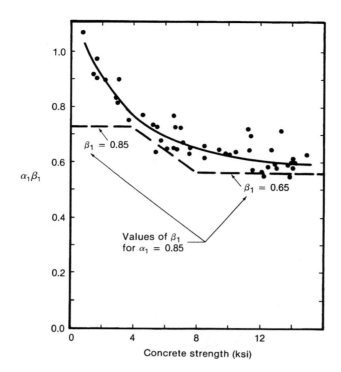

Fig. 4–13
Values of $\alpha_1 \beta_1$ from tests of concrete prisms. (From Ref. 4–7.)

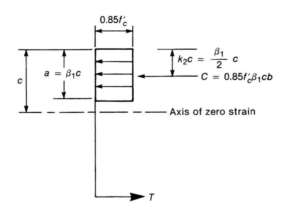

Fig. 4–14
Equivalent rectangular stress block.

(b)   For f_c' between 4000 and 8000 psi,

$$\beta_1 = 1.05 - 0.05\frac{f_c'}{1000} \qquad (4\text{–}8b)$$

(c)   For f_c' greater than 8000 psi,

$$\beta_1 = 0.65 \qquad (4\text{–}8c)$$

The symbols α_1 and β_1 used to describe the rectangular stress block are different from the symbols k_1 k_2, and k_3 used to describe the stress block from tests because the rectangular stress block is described by two symbols, whereas three are needed to describe the stress block from tests.

Studies of the effects of sustained loads on the strength of concrete[4-2] and tests of columns[4-1] suggest that α_1 can be taken equal to 0.85 for commonly occurring concrete strengths. The dashed line in Fig. 4–13 is a lower bound line corresponding to $k_3 = \alpha_1 = 0.85$ and β_1 given by Eq. 4–8.

When working in the SI system of units, the factor β_1 is taken as

$$\beta_1 = 1.09 - 0.008f_c' \qquad (4-8M)$$

but not more than 0.85 nor less than 0.65.

The equivalent rectangular stress block with $\alpha_1 = 0.85$ and β_1 from Eq. 4–8 has been shown[4-3] to give very good agreement with test data for beams (see Fig. 2-3). For columns, the agreement is good up to a concrete strength of about 6000 psi. For columns loaded with small eccentricities and having strengths greater than this, the moment capacity is overestimated by the ACI code stress block. This is because β_1 was chosen as a lower bound on the test data, as suggested by Fig. 4–13. The internal moment arm of the compression force in the concrete about the centroidal axis of a rectangular column is $(h/2 - \beta_1 c/2)$, where c is the depth to the neutral axis (axis of zero strain). If β_1 is too small, the moment arm will be too large and the moment capacity will be overestimated. The following equations have been proposed for α_1 and β_1:

$$\alpha_1 = 0.85 \text{ for } f_c' \leq 8000 \text{ psi} \qquad (4-9a)$$

$$\alpha_1 = 0.85 - \frac{f_c' - 8000}{50,000} \geq 0.73 \text{ for } f_c' > 8000 \text{ psi} \qquad (4-9b)$$

$$\beta_1 = 0.85 \text{ for } f_c' \leq 4000 \text{ psi} \qquad (4-10a)$$

$$\beta_1 = 0.85 - 0.15\left(\frac{f_c' - 4000}{10,000}\right) \geq 0.70 \text{ for } f_c' > 4000 \text{ psi} \qquad (4-10b)$$

4–3 ANALYSIS OF REINFORCED CONCRETE BEAMS

Stress and Strain Compatibility and Equilibrium

Two requirements are satisfied throughout the analysis and design of reinforced concrete beams and columns. These are:

1. *Stress and strain compatibility.* The stress at any point in a member must correspond to the strain at that point. Except for short, deep beams, the distribution of strains over the depth of the member must be linear to satisfy assumptions 1 and 2 presented earlier in this chapter.

2. *Equilibrium.* The internal forces must balance the external load effects, as illustrated in Fig. 4–3 and Eqs. 4–2 and 4–3.

Analysis of the Flexural Capacity of a General Cross Section

The use of equilibrium and strain compatibility in the computation of the capacity of an arbitrary cross section such as the one shown in Fig. 4–15 involves four steps and will be illustrated with an example.

Example 4–1 Calculation of the Moment Capacity of a Beam

The beam shown in Fig. 4–15 is made of concrete with a compressive strength, f_c', of 3000 psi and has three No. 8 bars with a yield strength, f_y, of 60,000 psi.

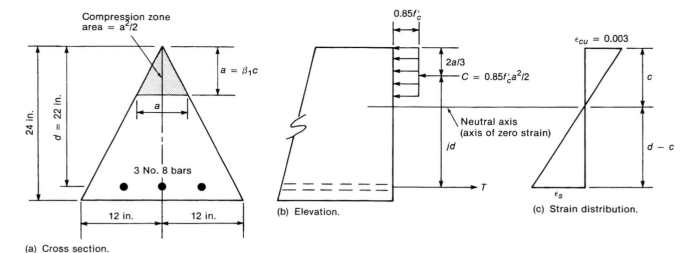

(a) Cross section.

(b) Elevation.

(c) Strain distribution.

Fig. 4–15
Analysis of arbitrary cross section—Example 4–1

1. Initially assume that the stress f_s, in the tension reinforcement equals the yield strength, f_y, and compute the tension force $T = A_s f_y$.

$$A_s = 3 \text{ No. 8 bars} = 3 \times 0.79 \text{ in.}^2 = 2.37 \text{ in.}^2$$

$$f_y = 60,000 \text{ psi for Grade 60 bars}$$

$$T = A_s f_y = 2.37 \text{ in.}^2 \times 60,000 \text{ psi} = 142,200 \text{ lb}$$

The assumption that $f_s = f_y$ will be checked in step 3. If the steel has yielded, a simple solution exists; if not, a more complex solution must be used. This assumption will generally be true since the ACI Code requires that the steel percentage be small enough that the steel will yield.

2. Compute the area of the compression block so that $C = T$. This is done using the equivalent rectangular stress block shown in Fig. 4–15b. The stress block consists of a uniform stress of $0.85 f_c'$ distributed over a depth $a = \beta_1 c$ measured from the extreme compression fiber. The magnitude of the compression force is obtained from equilibrium:

$$C = T = 142,200 \text{ lb}$$

where T was computed in step 1.

By the geometry of this particular triangular beam, shown in Fig. 4–15a, if the depth of the compression zone is a, the width is also a, and the area is $a^2/2$. This is, of course, true only for a beam of this particular shape.

Therefore, $C = (0.85 f_c')(a^2/2)$ and

$$a = \sqrt{\frac{142,200 \text{ lb} \times 2}{0.85 \times 3000}} = 10.56 \text{ in.}$$

3. Check whether $f_s = f_y$.

This is done using strain compatibility. The strain distribution at ultimate is shown in Fig. 4–15c. To plot this it is necessary to know ϵ_{cu} and c:

$$\epsilon_{cu} = 0.003 \qquad \text{(assumption 5, Sec. 4–2)}$$

$$c = \frac{a}{\beta_1}$$

For $f_c' = 3000 \text{ psi}$, $\beta_1 = 0.85$ (from Eqs. 4–8). Therefore, $c = 10.56/0.85 = 12.42 \text{ in.}$

By similar triangles as shown in Fig. 4–15c,

$$\frac{\epsilon_s}{d - c} = \frac{0.003}{c}$$

or

$$\epsilon_s = 0.003\left(\frac{22 - 12.42}{12.42}\right) = 0.00231$$

For Grade 60 reinforcement,

$$\epsilon_y = \frac{60{,}000 \text{ psi}}{29{,}000{,}000 \text{ psi}} = 0.00207$$

Therefore, $\epsilon_s > \epsilon_y$ and $f_s = f_y$. Thus the assumption made in step 1 is satisfied.

4. Compute ϕM_n.

$$M_n = Cjd = Tjd$$

where jd is the lever arm, the distance from the resultant tensile force (at the centroid of the reinforcement) to the resultant compressive force C. Because the area on which the compression stress block acts is triangular in this example, C acts at $2a/3$ from top of the beam. Therefore,

$$jd = d - \frac{2a}{3}$$

and

$$
\begin{aligned}
\phi M_n &= \phi\left[A_s f_y\left(d - \frac{2a}{3}\right)\right] \\
&= 0.9\left[2.37\text{in.}^2 \times 60{,}000 \text{ psi}\left(22 - \frac{2 \times 10.56}{3}\right)\text{in.}\right] \\
&= 1.92 \times 10^6 \text{ in.-lb} \\
&= \frac{1.92 \times 10^6}{12{,}000} \text{ ft-kips} = 160 \text{ ft-kips}
\end{aligned}
$$

Thus the design moment capacity of this beam cross section is 160 ft-kips. ■

This general analysis procedure can be used to compute the moment capacity for beams of any shape. It should be noted that the equations used to compute a, jd, and M_u in Example 4–1 were specifically derived for a beam with a *triangular section* and apply only to such a beam. Since beams are frequently rectangular in cross section, the analysis of rectangular beams is considered later.

Tension, Compression, and Balanced Failures

Depending on the properties of a beam, flexural failures may occur in three different ways:

1. *Tension Failure.* Reinforcement yields before concrete crushes (reaches its limiting compressive strain). Such a beam is said to be *under-reinforced.*

2. *Compression failure.* Concrete crushes before steel yields. Such a beam is said to be *over-reinforced.*

3. *Balanced failure.* Concrete crushes and steel yields simultaneously. Such a beam has *balanced reinforcement.*

In the test specimen shown in Figs. 4–5 to 4–7, the reinforcement yielded before failure occurred and hence the beam developed a tension failure. At failure (point E in Fig. 4–7) the curvature at the section of maximum moment was roughly four times that at yielding (point D). As a result, the beam deflected extensively and developed wide cracks in the final loading stages. This type of behavior is said to be *ductile* since the moment-curvature or the load–deflection diagram has a long plastic region (D–E in Fig. 4–7). If a beam in a

Flexure: Basic Concepts, Rectangular Beams

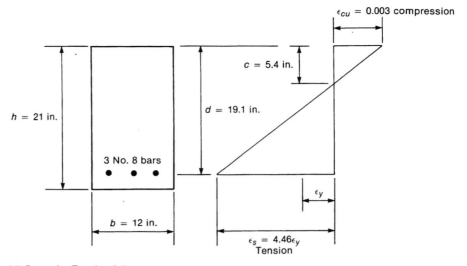

(a) Beam A—Tension failure.

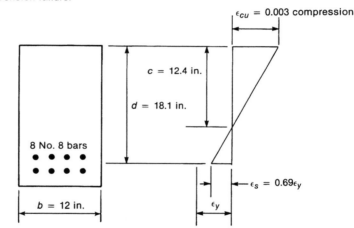

(b) Beam B—Compression failure.

Fig. 4–16
Tension, compression, and
balanced failures.

building fails in a ductile manner, the occupants of the building have warning of the impending failure and hence have an opportunity to leave the building before the final collapse, thus reducing the consequences of collapse.

Moment–curvature diagrams are presented in Fig. 4–16d for three beams shown in Fig. 4–16a, b, and c. The beams differ only in the amount of reinforcement. At failure, the reinforcement in beam A has yielded, as shown by the strain diagram. This beam develops a tension failure and has a ductile moment–curvature response as shown in Fig. 4–16d. As will be seen later in this section, tension failures occur when the *mechanical reinforcement ratio* $\omega = \rho f_y / f_c'$ is small.

In the case of beam B in Fig. 4–16, the concrete in the extreme compression fiber reaches the assumed crushing strain of 0.003 before the steel starts to yield. This is called a *compression failure*. The moment–curvature diagram for such a beam does not have the ductile postyielding response displayed by beam A. If overloaded, this beam may fail suddenly in a *brittle* manner without warning to the occupants of the building, and as a result, such a failure may have serious consequences. Compression failures occur for high values of the mechanical reinforcement ratio ω.

4–3 Analysis of Reinforced Concrete Beams   **99**

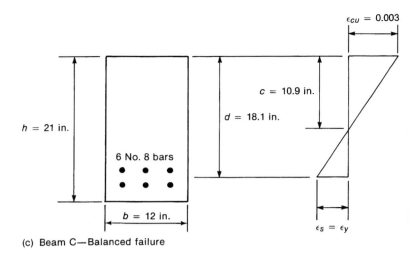

(c) Beam C—Balanced failure

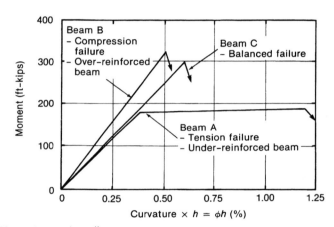

(d) Moment–curvature diagram

Fig. 4–16
(continued)

In the case of beam C, the strain distribution at failure, shown in Fig. 4–16c, involves simultaneous crushing of the concrete and yielding of the steel. This case, which also exhibits a brittle failure as shown in Fig. 4–16d, marks the boundary between ductile tension and brittle compression failures—hence the name *balanced failure*.

Appendix B of ACI 318–95 has introduced the terminology *compression-controlled sections* for sections in which the strain at ultimate in the extreme tension steel layer is less than or equal to the yield strain, $\epsilon_y = f_y/E_s$, in tension. Such sections develop compression failures or balanced failures. Sections having a strain at ultimate in the extreme tension steel layer equal to or greater than 0.005 in tension are referred to as *tension-controlled sections*. Sections falling between these two limits are referred to as *transition sections*.

To reduce the chance that brittle failures will occur, ACI Sec. 10.3.3 requires that beams nominally have properties which ensure that tension failures with $f_s = f_y$ would occur. Alternatively, ACI Sec. B9.3.2 (in ACI Appendix B) allows beams with f_s less than f_y at failure (compression-controlled sections) but for such beams, it requires the lower strength reduction factors, ϕ, used for columns. For beams falling in the transition range, it requires a transition between the ϕ factors for columns and those for beams.

This is discussed more fully later in this chapter.

Flexure: Basic Concepts, Rectangular Beams

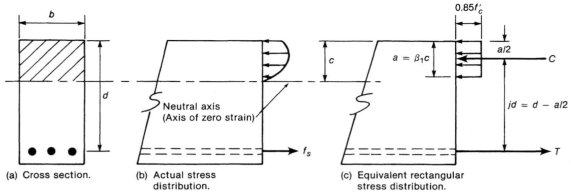

Fig. 4–17
Stresses and forces in a rectangular beam.

Analysis of Rectangular Beams with Tension Reinforcement Only

Equations for M_n and ϕM_n: Tension Steel Yielding

In the preceding section, equilibrium and strain compatibility were used to compute the moment capacity of a particular beam cross section. For the particular case of a rectangular beam the same procedure can be used to derive equations for computing the moment capacity.

Consider the beam shown in Fig. 4–17. The compressive force, C, in the concrete is

$$C = (0.85f_c')ba$$

Then tension force, T, in the steel is

$$T = A_s f_s$$

and for equilibrium, $C = T$. Therefore, the depth, a, of the equivalent rectangular stress block is

$$a = \frac{A_s f_s}{0.85f_c' b}$$

If $f_s = f_y$ as assumed in step 1 of Example 4–1, this becomes

$$a = \frac{A_s f_y}{0.85f_c' b} \qquad (4\text{–}11)$$

It is possible to express the equations of M_n and ϕM_n in several ways based on $M_n = Tjd$, $M_n = Cjd$, or in a nondimensionalized fashion. These three are considered in turn in the following paragraphs.

Equation for M_n Based on $M_n = Tjd$. Summing moments about the line of action of the compressive force, C in Fig. 4–17c gives

$$M_n = Tjd$$

Substituting $T = A_s f_s$ where f_s is equal to f_y and $jd = (d - a/2)$, gives

$$M_n = A_s f_y\left(d - \frac{a}{2}\right) \qquad (4\text{–}12a)$$

and

$$\phi M_n = \phi\left[A_s f_y\left(d - \frac{a}{2}\right)\right] \qquad (4\text{–}12b)$$

This is the basic equation for the flexural capacity of beams.

Equation for M_n Based on $M_n = Cjd$. Alternatively, one could sum moments about the line of action of the tensile force, T. Thus

$$M_n = Cjd$$

Substituting $C = (0.85f_c')ba$ and $jd = (d - a/2)$ gives

$$M_n = (0.85f_c')ba\left(d - \frac{a}{2}\right) \tag{4-13a}$$

and

$$\phi M_n = \phi\left[(0.85f_c')ba\left(d - \frac{a}{2}\right)\right] \tag{4-13b}$$

Nondimensionalized Equations for M_n. If we substitute $A_s = \rho bd$ into Eq. 4-11, we get

$$a = \frac{\rho f_y}{f_c'}\left(\frac{d}{0.85}\right) \tag{4-14a}$$

where $\rho f_y / f_c' = \omega$ and is referred to as the *mechanical reinforcement ratio*. The term ω is frequently used as a measure of the behavior of a beam since it incorporates the three major variables affecting that behavior $(\rho, f_y \text{ and } f_c')$. Thus

$$a = \frac{\omega d}{0.85} \tag{4-14b}$$

Substituting Eq. 4-14b into Eq. 4-13 gives

$$\phi M_n = \phi\left[f_c' bd^2 \omega\left(1 - \frac{\omega}{2 \times 0.85}\right)\right]$$

or

$$\phi M_n = \phi\left[f_c' bd^2 \omega(1 - 0.59\omega)\right] \tag{4-15}$$

This is frequently expressed as

$$\phi M_n = \phi\left(\frac{bd^2}{12,000} k_n\right) \tag{4-16}$$

or, since the most economical design corresponds to $\phi M_n = M_u$,

$$\frac{M_u}{\phi k_n} = \frac{bd^2}{12,000} \tag{4-17}$$

where M_u and M_n are in ft-kips and b and d are in inches. In SI units these equations become

$$\phi M_n = \phi\left(\frac{bd^2}{10^6} k_n\right) \tag{4-16M}$$

or

$$\frac{M_u}{\phi k_n} = \frac{bd^2}{10^6} \tag{4-17M}$$

where M_u and M_n are in kN–m, and b and d are in mm.

In Eqs. 4-16 and 4-17 the term ϕk_n is

$$\phi k_n = \phi\left[f_c' \omega(1 - 0.59\omega)\right] \tag{4-18}$$

with f_c' in psi or MPa. Values of ϕk_n are given in Table A-3 (see Appendix A) for various concrete and steel strengths and reinforcement ratios.

Flexure: Basic Concepts, Rectangular Beams

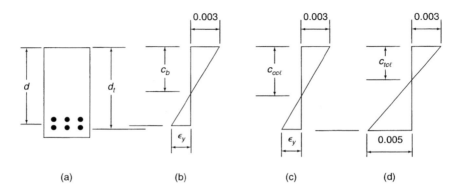

Fig. 4–18
Balanced, compression-
controlled and tension-
controlled sections.

(a)   (b)   (c)   (d)

Determination of Whether $f_s = f_y$

In the derivation of Eqs. 4–11 to 4–18 it was assumed that the steel yielded and $f_s = f_y$. It is necessary to check whether this is true whenever a beam is analyzed or designed. Figure 4–18a shows a beam cross section with two layers of tension reinforcement. The effective depth to the centroid of the reinforcement is d. Consider a beam developing a balanced failure with the concrete starting to crush just as the steel starts to yield. The strain distribution for such a beam is shown in Fig. 4–18b. The depth to the neutral axis at balanced failure is c_b. From similar triangles we can say that

$$\frac{c_b}{d} = \frac{\epsilon_{cu}}{\epsilon_{cu} + \epsilon_y} \tag{4–19}$$

Substituting $\epsilon_{cu} = 0.003$ and multiplying above and below by $E_s = 29,000,000$ psi gives

$$\frac{c_b}{d} = \frac{87,000}{87,000 + f_y} \tag{4–20}$$

where f_y is in psi. If a beam has a neutral axis depth c less than c_b, the steel strains will exceed ϵ_y, and vice versa. Thus, if $c \leq c_b$ at failure, $f_s = f_y$.

Equations 4–12 and 4–13 include $a = \beta_1 c$, rather than c, where a is the depth of the equivalent rectangular stress block. Substituting into Eq. 4–20 gives

$$\frac{a_b}{d} = \beta_1\left(\frac{87,000}{87,000 + f_y}\right) \tag{4–21}$$

where f_y is in psi.

In SI units, $E_S = 200,000$ MPa and Eq. 4–21 becomes

$$\frac{a_b}{d} = \beta_1\left(\frac{600}{600 + f_y}\right) \tag{4–21M}$$

where f_y is in MPa.

To check whether $f_s = f_y$ in design, we shall check whether $a/d \leq a_b/d$. Table A–4 gives values of a_b/d for various concrete and steel strengths.

During design, the exact location of the centroid of the steel is not known until the final reinforcement is chosen. For this reason, it is easier to define strain distributions in terms of depth, d_t, to the layer of steel farthest from the compression face. The net tensile strain in the layer of steel farthest from the compression face is ϵ_t. The words *net tensile strain* refer to the steel strain at nominal strength, exclusive of strains due to effective prestress, creep, shrinkage, or temperature. In other words, for a reinforced concrete beam, the net tensile strain is the strain, ϵ_s, due to the factored live and dead loads on the beam. ACI Sec. B10.3.3 defines a section as being *compression-controlled* if the net tensile strain, ϵ_t, is less than or equal to the yield strain in tension, ϵ_y. The strain distribution corresponding to the *compression-controlled*

4–3 Analysis of Reinforced Concrete Beams

limit is shown in Fig. 4–18c. Here the neutral axis depth is c_{ccl} and the strain ϵ_y occurs in the extreme tension layer at a depth d_t.

Derived in a similar fashion to Eq. 4–21, the depth of the rectangular stress block at failure at the compression-controlled limit, c_{ccl}, is

$$\frac{a_{ccl}}{d_t} = \beta_1 \left(\frac{87,000}{87,000 + f_y} \right) \tag{4-22}$$

where f_y is in psi, and

$$\frac{a_{ccl}}{d_t} = \beta_1 \left(\frac{600}{600 + f_y} \right) \tag{4-22M}$$

where f_y is in MPa.

ACI Sec. B10.3.3 defines a section as being *tension-controlled* if the net tensile strain in the layer of steel farthest from the compression face of the beam equals or exceeds 0.005 in tension. The strain distribution corresponding to the *tension-controlled limit* is shown in Fig. 4–18d. Here the neutral axis depth is c_{tcl}. From Fig. 4–18d using similar triangles,

$$\frac{c_{tcl}}{d_t} = \frac{0.003}{0.003 + 0.005}$$

or

$$\frac{c_{tcl}}{d_t} = 0.375$$

and

$$\frac{a_{tcl}}{d_t} = 0.375 \beta_1 \tag{4-23}$$

The ratios c_{tcl}/d_t and a_{tcl}/d_t are independent of the system of units. Values of a_{ccl}/d_t and a_{tcl}/d_t are given in Table A–4.

Alternative Determination of Whether $f_s = f_y$

Equation 4–20 gives a relationship for c_b/d, where c_b is the neutral axis depth at balanced failure. But $c = a/\beta_1$ and from Eq. 4–14a,

$$a = \frac{\rho_b f_y}{0.85 f_c'} d$$

Therefore, at balanced failure,

$$\frac{c_b}{d} = \frac{\rho_b f_y}{0.85 \beta_1 f_c'} \tag{4-24}$$

where c_b and ρ_b are the neutral-axis depth and steel ratio corresponding to a balanced failure. From Eqs. 4–19 and 4–24,

$$\rho_b = \frac{0.85 \beta_1 f_c'}{f_y} \left(\frac{\epsilon_{cu}}{\epsilon_{cu} + \epsilon_y} \right)$$

If we substitute $\epsilon_{cu} = 0.003$ and multiply above and below by $E_s = 29,000,000$ psi, we obtain

$$\rho_b = \frac{0.85 \beta_1 f_c'}{f_y} \left(\frac{87,000}{87,000 + f_y} \right) \tag{4-25}$$

where f_y and f_c' are in psi.

In SI units, $E_s = 200,000$ MPa and Eq. 4–25 becomes

Flexure: Basic Concepts, Rectangular Beams

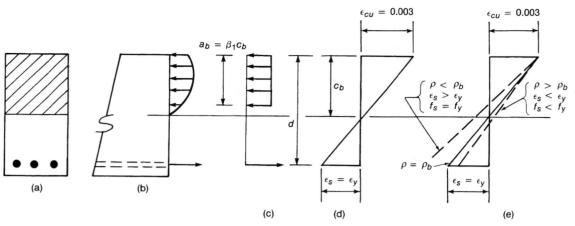

Fig. 4–19
Calulation of balanced steel ratio, ρ_b.

$$\rho_b = \frac{0.85\beta_1 f_c'}{f_y}\left(\frac{600}{600 + f_y}\right) \tag{4–25M}$$

where f_y and f_c' are in MPa.

Equation 4–25 can be used to determine whether a beam will fail in tension or compression. If ρ is less than ρ_b, the tensile force T is smaller than that at balanced, and as a result, the compressive force C is also smaller, requiring less compressive area. As a result, the neutral axis rises in the beam as shown in 4–19e. For such a case, ϵ_s is greater than ϵ_y and the steel will have yielded at failure. This beam is said to be *under-reinforced* since ρ is less than ρ_b. On the other hand, if ρ is greater than ρ_b, ϵ_s will be less than ϵ_y and the depth to the neutral axis must increase as shown in Fig. 4–19e. Such a beam is said to be *over-reinforced*.

Upper Limit on Reinforcement in Beams

In the 1995 ACI Code there are two different approaches to setting an upper limit on the amount of reinforcement which is permitted in a beam. One is given in ACI Sec. 10.3.3, the other in ACI Sec. B10.3.3, located in ACI Appendix B. Appendix B consists of a series of sections that replace sections in the body of the code. If any part of Appendix B is used, all of it must be used in place of the corresponding code sections.

The method presented in ACI Sec. B10.3.3 will be discussed first, followed by a discussion of the method in ACI Sec. 10.3.3. Appendix B will be used in the examples because it is a more universal procedure that is applicable to beams, T beams, columns, and prestressed concrete beams.

Upper Limit on Reinforcement—ACI Sec. B10.3.3. ACI Appendix B does not put an upper limit on the amount of reinforcement in a beam. Instead, ACI Sec. B9.3.2 defines $\phi = 0.9$ for tension-controlled sections, $\phi = 0.70$ for compression-controlled sections with normal stirrups or ties, or $\phi = 0.75$ for members (columns) with the reinforcement enclosed in a closely spaced spiral. Since spirals would not be used in a beam, $\phi = 0.70$ for a compression-controlled beam section. For beams between the limits, ACI Sec. B9.3.2.2 specifies a linear transition from $\phi = 0.90$ to $\phi = 0.70$ given by

$$\phi = 0.356 + \frac{0.204}{c/d_t} \tag{4–26a}$$

or

$$\phi = 0.356 + \frac{0.204}{a/(\beta_1 d_t)} \tag{4–26b}$$

4–3 Analysis of Reinforced Concrete Beams

105

Appendix B was introduced to allow a uniform transition in the strength reduction factor, ϕ, as more and more steel was added to a section.

In this book, Appendix B will be used because it is more rational and more universally applicable[4-8] In beam design, reinforcement will be chosen so that $\phi = 0.90$. This requires that a/d_t not exceed the tension-controlled limit in Eq. 4–23. Table A–4 gives a/d_t values corresponding to the tension- and compression-controlled limits.

Upper Limit on Reinforcement—ACI Sec. 10.3.3. Since an under-reinforced beam fails in a ductile manner and an over-reinforced beam in a brittle manner, ACI Sec. 10.3.3 attempts to prevent nonductile failures by limiting the reinforcement ratio, ρ, to $\leq 0.75\rho_b$. This is roughly equivalent to requiring that ϵ_s be 1.8 to 2.0 times ϵ_y at failure.

Due to the variability of the actual strengths of concrete and steel (understrength concrete and overstrength steel) and the variability of dimensions such as the effective depth, a beam that nominally satisfied $\rho = 0.75\rho_b$ may develop a compression failure. In addition, it is generally difficult to place the reinforcement and the concrete in a beam if ρ exceeds about $0.5\rho_b$, and such beams tend to deflect and crack excessively. For all these reasons it is good practice to limit the maximum steel percentage to $\rho = 0.4$ to $0.5\rho_b$. Values of ρ_b, $0.75\rho_b$, and $0.5\rho_b$ are listed in Table A–5 (see Appendix A).

For rectangular beams it is much easier to base the check of whether $\rho \leq 0.75\rho_b$ on the neutral axis depth ratio c/d or the ratio, a/d, of the depth of the equivalent rectangular stress block, a, to the effective depth of the beam, d. Rearranging Eq. 4–25 and expressing a using Eq. 4–14a gives

$$\frac{a_b}{d} = \beta_1 \left(\frac{87,000}{87,000 + f_y} \right) \tag{4–21}$$

where f_y is in psi and

$$\frac{a_b}{d} = \beta_1 \left(\frac{600}{600 + f_y} \right) \tag{4–21M}$$

where f_y is in MPa.

In Eq. 4–21, a_b is the depth of the equivalent rectangular compressive stress block corresponding to a balanced failure (see Fig. 4–19c). To limit ρ to $0.75\rho_b$ or $0.5\rho_b$, the computed a/d for a given beam should be less than $0.75(a_b/d)$ or $0.5(a_b/d)$, respectively. Table A–4 gives values of a/d ratios corresponding to ρ_b, $0.75\rho_b$, and $0.5\rho_b$.

The tension-controlled limit in ACI Appendix B corresponds to $\rho = 0.563\rho_b$ for a rectangular beam with only tension reinforcement.

EXAMPLE 4–2   Analysis of Singly Reinforced Beams: Tension Steel Yielding

Compute the nominal moment capacities, M_n, of three beams, each with $b = 10$ in., $d = 20$ in. and 3 No. 8 bars giving $A_s = 3 \times 0.79 = 2.37$ in.^{2} and $\rho = A_s/bd = 2.37/(10 \times 20) = 0.0119$. The beam cross section is shown in Fig. 4–20a.

BEAM 1: $f_c' = 3000$ psi **AND** $f_y = 60,000$ psi

1. **Compute a.** Assume that steel stress, f_s, equals f_y (which corresponds to $\rho \leq \rho_b$). This will be checked in step 2. From Eq. 4–11, the depth, a, of the equivalent rectangular stress block is

$$a = \frac{A_s f_y}{0.85 f_c' b}$$

$$= \frac{2.37 \text{ in.}^2 \times 60,000 \text{ psi}}{0.85 \times 3000 \text{ psi} \times 10 \text{ in.}}$$

Therefore, $a = 5.58$ in. (Fig. 4–20a)

2. **Check if $f_s = f_y$ and whether the section is tension-controlled.** If $a/d \leq a_b/d$, f_s will be equal to f_y, where a_b/d is given by Eq. 4–21.

Flexure: Basic Concepts, Rectangular Beams

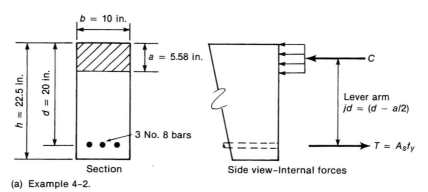

(a) Example 4–2.

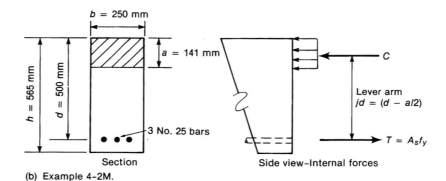

Fig. 4–20
Beams—Examples 4–2 and
4–2M.

(b) Example 4–2M.

$$\frac{a}{d} = \frac{5.58 \text{ in.}}{20 \text{ in.}} = 0.279$$

From Eq. 4–21,

$$\frac{a_b}{d} = \beta_1\left(\frac{87,000}{87,000 + f_y}\right)$$

From Eq. 4–8a, $\beta_1 = 0.85$ for $f'_c = 3000$ psi and

$$\frac{a_b}{d} = 0.85\left(\frac{87,000}{87,000 + 60,000}\right) = 0.503$$

Since the actual $a/d = 0.279$ is less than $a_b/d = 0.503$, $f_s = f_y$.

To check whether the section is tension-controlled, check whether $a/d_t \leq a_{tc\ell}/d_t$. Note that this check is made in terms of a/d_t rather than a/d. For this beam d_t and d are the same. If the steel were in several layers, this would not be true. From Eq. 4–23,

$$a_{tc\ell}/d_t = 0.375\beta_1 = 0.375 \times 0.85 = 0.319$$

Since $a_t/d = 0.279$ is less than 0.319, the section is tension-controlled and $\phi = 0.90$. Note that values of a_b/d and $a_{tc\ell}/d_t$ are given in Table A–4.

3. Compute the nominal moment capacity, M_n. Summing moments about the resultant compressive force, C, the moment capacity is $M_n = Tjd$, where the tension force is $T = A_s f_y$ and the lever arm $jd = (d - a/2)$, giving

$$M_n = A_s f_y\left(d - \frac{a}{2}\right) \tag{4–12a}$$

4–3 Analysis of Reinforced Concrete Beams

Thus

$$M_n = 2.37 \text{ in.}^2 \times 60{,}000 \text{ psi}\left(20 - \frac{5.58 \text{ in.}}{2}\right)$$

$$= 2{,}447{,}000 \text{ in.-lb} = \frac{2{,}447{,}000}{12{,}000} \text{ ft-kips}$$

The nominal moment capacity of beam 1 is $M_n = 204$ ft-kips.

Since the section is tension-controlled, ACI Sec. B9.3.2.1 gives $\phi = 0.90$. The design or factored moment capacity of this beam is

$$\phi M_n = 0.9 \times 204 \text{ ft-kips} = 184 \text{ ft-kips}$$

BEAM 2: SAME AS BEAM 1 EXCEPT THAT $f_c' = 6000$ psi

   1. **Compute a.**

$$a = \frac{2.37 \times 60{,}000}{0.85 \times 6000 \times 10} = 2.79 \text{ in.}$$

   2. **Check if $f_s = f_y$ and whether the section is tension-controlled.** Again we will base the check of $f_s = f_y$ on a_b/d, where

$$\frac{a}{d} = \frac{2.79}{20} = 0.139$$

From Eq. 4–21,

$$\frac{a_b}{d} = \beta_1\left(\frac{87{,}000}{87{,}000 + f_y}\right)$$

where, from Eq. 4–8b,

$$\beta_1 = 1.05 - 0.05\left(\frac{f_c'}{1000}\right) = 0.75$$

$$\frac{a_b}{d} = 0.75\left(\frac{87{,}000}{87{,}000 + 60{,}000}\right) = 0.444$$

Since $a/d = 0.139$ is less than $a_b/d = 0.444$, $f_s = f_y$.

$$\frac{a_{tc\ell}}{d_t} = 0.375\beta_1 = 0.281$$

Again $d_t = d$. Since $a/d_t = 0.139$ is less than $a_{tc\ell}/d_t = 0.281$ the section is tension controlled and $\phi = 0.90$.

   3. **Compute M_n.**

$$M_n = \frac{2.37 \times 60{,}000(20 - 2.79/2)}{12{,}000} = 220 \text{ ft-kips}$$

$$\phi M_n = 0.9 \times 220 \text{ ft-kips} = 198 \text{ ft-kips}$$

Note that doubling the concrete strength increased M_n by only 8%.

BEAM 3: SAME AS BEAM 1 EXCEPT THAT $f_y = 40{,}000$ psi

   1. **Compute a.**

$$a = \frac{2.37 \times 40{,}000}{0.85 \times 3000 \times 10} = 3.72 \text{ in.}$$

   2. **Check if $f_s = f_y$ and whether the section is tension-controlled.**

$$\frac{a}{d} = \frac{3.72}{20} = 0.186$$

$$\frac{a_b}{d} = 0.85\left(\frac{87,000}{87,000 + 40,000}\right) = 0.582$$

$$\frac{a_{tc\ell}}{d_t} = 0.375\beta_1 = 0.319$$

Thus $f_s = f_y$ and the section is tension-controlled. $\phi = 0.90$.

3. **Compute M_n and ϕM_n.**

$$M_n = \frac{2.37 \times 40,000(20 - 3.72/2)}{12,000} = 143 \text{ ft-kips}$$

$$\phi M_n = 0.9 \times 143 \text{ ft-kips} = 129 \text{ ft-kips}$$

Note that reducing f_y by 33% compared to beam 1, reduced M_n by 30%. ∎

EXAMPLE 4–2M Analysis of Singly Reinforced Beams: Tension Steel Yielding—SI Units

Compute the nominal moment capacity, M_n, of a beam with $f'_c = 20$ MPa, $f_y = 400$ MPa, $b = 250$ mm, $d = 500$ mm, and 3 No. 25 bars giving $A_s = 3 \times 500 = 1500$ mm² and $\rho = A_s/bd = 1500/(250 \times 500) = 0.0120$. See Fig. 4–20b.

1. **Compute a.**

$$a = \frac{A_s f_y}{0.85 f'_c b}$$

$$= \frac{1500 \text{ mm}^2 \times 400 \text{ MPa}}{0.85 \times 20 \text{ Mpa} \times 250 \text{ mm}} = 141 \text{ mm}$$

Therefore, $a = 141$ mm.

2. **Check if $f_s = f_y$ and whether the section is tension-controlled.** If $a/d \le a_b/d, f_s = f_y$, where a_b/d is given by Eq. 4–21M.

$$\frac{a}{d} = \frac{141 \text{ mm}}{500 \text{ mm}} = 0.282$$

From Eq. 4–21M,

$$\frac{a_b}{d} = \beta_1\left(\frac{600}{600 + f_y}\right)$$

From Eq. 4–8M, $\beta_1 = 0.85$ for $f'_c = 20$ MPa and

$$\frac{a_b}{d} = 0.85\left(\frac{600}{600 + 400}\right) = 0.510$$

Since the actual $a/d = 0.282$ is less than $a_b/d = 0.510, f_s = f_y$. To check whether the section is tension-controlled, check whether $a_{tc\ell}/d_t \le a/d_t$. Note that this check is in terms of a/d_t. For this beam d_t and d are the same. If the steel were in several layers, this would not be true. From Eq. 4–23,

$$\frac{a_{tc\ell}}{d_t} = 0.375\beta_1 = 0.375 \times 0.85 = 0.319$$

Since $a/d_t = 0.282$ is less than 0.319, the section is tension-controlled and $\phi = 0.90$.

3. **Compute the nominal moment capacity, M_n.** From Eq. 4–12a, M_n is

$$M_n = A_s f_y\left(d - \frac{a}{2}\right)$$

$$= 1500 \text{ mm}^2 \times 400 \text{ N/mm}^2\left(500 - \frac{141}{2}\right) \text{ mm}$$

(where 1 MPa = 1 N/mm²). Therefore,

$$M_n = 258 \times 10^6 \text{ N–mm} = 258 \text{ kN–m}$$

The nominal moment capacity of the beam is $M_n = 258$ kN · m. The design or factored moment capacity, ϕM_n, of this beam is $0.9 \times 258 = 232$ kN − m. ∎

Equations for M_n and ϕM_n: Tension Steel Elastic at Failure

From statics we once again find that

$$C = T$$

and

$$0.85 f_c' ba = A_s f_s$$
$$= \rho E_s \epsilon_s bd$$

From strain compatibility (see Fig. 4–15c),

$$\epsilon_s = \epsilon_{cu}\left(\frac{d - c}{c}\right)$$

Solving these together and observing that $a = \beta_1 c$ gives

$$0.85 f_c' a^2 = \rho E_s \epsilon_{cu} \beta_1 d^2 - \rho E_s \epsilon_{cu} ad$$

or

$$\left(\frac{0.85 f_c'}{\rho E_s \epsilon_{cu}}\right) a^2 + (d)a - \beta_1 d^2 = {}^0 \tag{4–27}$$

This can be solved for a and from Eqs. 4–13, M_n or ϕM_n can be computed.

Beams with $\rho > \rho_b$ are not allowed by ACI Sec. 10.3.3 but are allowed by ACI Sec. B10.3.3. If ρ is found to be greater than $\rho_b (f_s < f_y)$ when checking an existing beam, or when designing a new beam, ACI Sec. B10.3.3 requires that ϕ be taken equal to 0.70. The ACI Code gives no guidance as what value of ϕ to use if an existing beam is found to have $\rho > \rho_b$. It is the author's opinion that the design capacity, ϕM_n, should be calculated using ϕ equal to 0.7 rather than the 0.9 usually used for flexure, due to the brittle nature of a compression failure. (Some authors use $\phi = 0.9$ but limit the moment capacity, ϕM_n, to that corresponding to $\rho = 0.75\rho_b$.)

When ρ is greater than ρ_b, the value of M_n is relatively insensitive to changes in ρ. This is because both f_s and jd decrease as A_s increases. In 1937, Whitney[4-9] used a semiempirical analysis to determine that the moment capacity for compression failures was

$$M_n = 0.333 f_c' bd^2 \tag{4–28}$$

Using the correct solution (Eqs. 4–27 and 4–13a), the constant in Eq. 4–28 is found to range from 0.29 to 0.35 for beams with $\rho = \rho_b$, increasing by roughly 10% for beams with $\rho = 2\rho_b$.

EXAMPLE 4–3 Analysis of Singly Reinforced Beam: Tension Steel Elastic

Compute the nominal moment capacity, M_n, of a beam having $b = 10$ in., $d = 20$ in., $A_s = 4.74$ in.^{2} (6 No. 8 bars), $f_c' = 3000$ psi, and $f_y = 60,000$ psi.

1. Compute a. When analyzing the capacity of a beam one does not know at the start whether the steel will yield or not. Because $f_s = f_y$ in most beams encountered in practice, we will make this assumption and correct it later if necessary. From Eq. 4–11,

$$\text{trial } a = \frac{A_s f_y}{0.85 f_c' b}$$

$$= \frac{4.74 \times 60,000}{0.85 \times 3000 \times 10}$$

Therefore, trial $a = 11.3$ in.

Flexure: Basic Concepts, Rectangular Beams

2. **Check if $f_s = f_y$ and whether the section is tension-controlled.** These checks will be based on Eq. 4–21.

$$\frac{a}{d} = \frac{11.2}{20} = 0.558 \text{ (based on trial a)}$$

$$\frac{a_b}{d} = \beta_1\left(\frac{87,000}{87,000 + f_y}\right) = 0.503$$

Since $a/d = 0.558$ is greater than $a_b/d = 0.503$, this beam will fail in compression with f_s less than f_y. As a result, the value of a computed in step 1 is incorrect and we must start again. Because f_s is less than f_y, the section is compression-controlled.

3. **Recompute a using Eq. 4–27.**

$$\left(\frac{0.85f_c'}{\rho E_s \epsilon_{cu}}\right)a^2 + (d)a - \beta_1 d^2 = 0$$

Where $\rho = A_s/bd = 0.0237$,

$$\left(\frac{0.85 \times 3000 \text{ psi}}{0.0237 \times 29 \times 10^6 \text{ psi} \times 0.003}\right)a^2 + 20a - 0.85 \times 20^2 = 0$$

Therefore, $1.237a^2 + 20a - 340 = 0$ and

$$a = \frac{-20 \pm \sqrt{20^2 - (4 \times -340 \times 1.237)}}{2 \times 1.237} = 10.36 \text{ in.}$$

This computed value of a is less than the 11.2 in. computed in step 1 because the actual steel stress is less than f_y.

4. **Compute M_n using Eq. 4–13a.** Because the steel stress f_s is not known, it is necessary to use Eq. 4–13a to compute M_n rather than Eq. 4–12a.

$$M_n = 0.85f_c' ab\left(d - \frac{a}{2}\right)$$

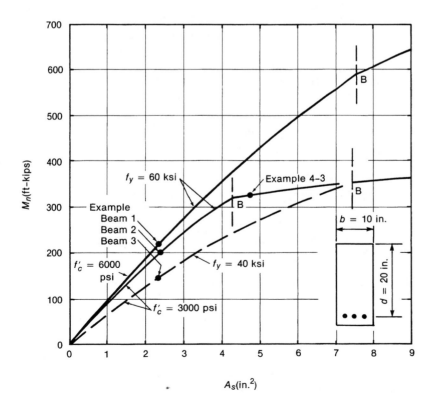

Fig. 4–21
Effect of variables on
strength of beams.

4-3 Analysis of Reinforced Concrete Beams

$$= \frac{0.85 \times 3000 \times 10.36 \times 10(20 - 10.36/2)}{12{,}000}$$

Thus $M_n = 326$ ft-kips. Since the section is compression-controlled and does not contain spiral reinforcement, ACI Sec. B9.3.2.2(b) gives $\phi = 0.70$ and $\phi M_n = 228$ ft-kips.

Whitney's equation for M_n, Eq. 4–28, gives $M_n = 333$ ft-kips. For this example this is essentially the same as the $M_n = 326$ ft-kips computed from the exact equation. ∎

Effect of Variables on M_n for Singly Reinforced Beams

Figure 4–21 compares the moment capacities of a rectangular cross section with varying material strengths and amounts of reinforcement. The upper and lower solid lines are plotted for $f_y = 60{,}000$ psi and concrete strengths, $f_c' = 6000$ and 3000 psi, respectively. The dashed curve is for $f_y = 40{,}000$ psi and $f_c' = 3000$ psi. Each curve consists of two parts, a steep portion to the left of point B for ρ less than ρ_b (tension failures) and a flatter portion to the right of point B for compression failures when ρ is greater than ρ_b.

The major difference between the two solid curves is the bending moment at which the change from tension to compression failures occurs. This is almost directly proportional to the concrete strength. For values A_s less than about 3 in.^{2}, however, the two solid curves in Fig. 4–21 are very close to one another, indicating that the 100% difference between their concrete strengths has relatively little effect on their flexural capacities in this range. This can be seen from Eq. 4–12a, which rearranged slightly is

$$M_n = A_s f_y d\left(1 - \frac{A_s f_y}{1.7 f_c' b d}\right)$$

The three main variables in this equation are A_s, f_y, and d and the moment capacity varies almost linearly with these three. Thus the moment capacities plotted in Fig. 4–21 increase almost linearly with $\rho = A_s/bd$ until ρ_b is reached. On the other hand, the concrete strength, f_c', has a much smaller effect on the strength of under-reinforced beams ($\rho < \rho_b$), acting only to change the depth of the compression zone, a, and hence to change the lever arm, $jd = (d - a/2)$. The points labeled beams 1 and 2 in this figure refer to the strengths computed for these beams in Example 4–2. Here a 100% change in concrete strength, from 3000 psi to 6000 psi, increased the moment capacity by only 8%.

The effect of the yield strength, f_y, can be seen by comparing the dashed curve and the lower solid curve, both of which are plotted for the same concrete strength. The ratio of the ordinates of these two curves is roughly 40/60 for values of ρ less than ρ_b. For example, beam 3 (from Example 4–2) had a capacity of 70% of that of beam 1.

In summary, for steel ratios up to 0.015 or so, the value of M_n is affected almost linearly by A_s, f_y, and although not shown in Fig. 4–21, by d. In this range, M_n is roughly proportional to f_y. The value of M_n at which the behavior changes from tension failures to compression failures and the value of M_n for compression failures are roughly proportional to f_c' and are essentially independent of f_y. Thus the most effective ways to increase the strength of a beam while maintaining a tension failure mode are to increase A_s, f_y, or d. Increasing f_c' is effective only if it allows higher steel percentages to be used.

4–4 DESIGN OF RECTANGULAR BEAMS

General Factors Affecting the Design of Rectangular Beams

Location of Reinforcement

Concrete cracks due to tension and as a result, reinforcement is required where flexure, axial loads, or shrinkage effects cause tensile stresses.

Flexure: Basic Concepts, Rectangular Beams

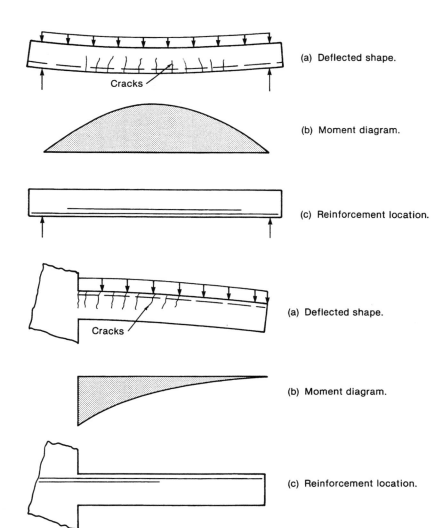

Fig. 4–22
Simply supported beam.

Fig. 4–23
Cantilever beam.

(a) Deflected shape.

Cracks

(b) Moment diagram.

(c) Reinforcement location.

(a) Deflected shape.

Cracks

(b) Moment diagram.

(c) Reinforcement location.

A uniformly loaded, simply supported beam deflects as shown in Fig. 4–22a and has the moment diagram shown in Fig. 4–22b. Because this beam is in positive moment throughout, tensile flexural stresses and cracks are developed along the bottom of the beam. Longitudinal reinforcement is required to resist these tensile stresses and is placed close to the bottom side of the beam as shown in Fig. 4–22c. Since the moments are greatest at midspan, more reinforcement is required at the midspan than at the ends and it may not be necessary to extend all the bars into the supports. In Fig. 4–22c, some of the bars are *cut off* within the span.

A cantilever beam develops negative moment throughout and deflects as shown in Fig. 4–23, the concave surface downward, so that flexural tensions and cracks develop on the top surface. In this case the reinforcement is placed near the top surface as shown in Fig. 4–23c. Since the moments are largest at the fixed end, more reinforcement is required here than at any other point. In some cases some of the bars may be terminated before the free end of the beam. Note that the bars must be anchored into the support.

Commonly, reinforced concrete beams are continuous over several supports, and under gravity loads they develop the moment diagram and deflected shape shown in Fig. 4–24. Again, reinforcement is needed on the tensile face of the beam, which is at the top of the beam in the negative moment regions at the supports, and at the bottom in the positive moment regions at the midspans. Two possible arrangements of this reinforcement are

4–4 Design of Rectangular Beams

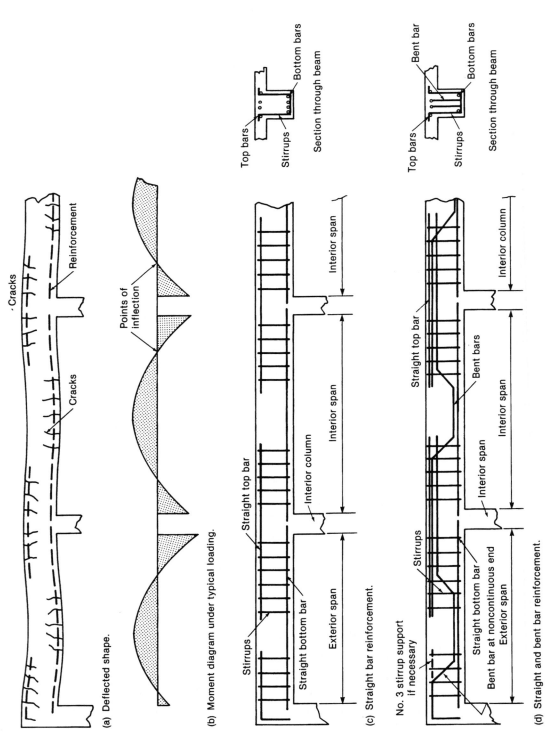

(a) Deflected shape.

(b) Moment diagram under typical loading.

Points of inflection

Cracks

Cracks

Reinforcement

Straight top bar

Stirrups

Straight bottom bar

Interior column

Exterior span

Interior span

Interior span

Interior span

(c) Straight bar reinforcement.

No. 3 stirrup support if necessary

Straight top bar

Bent bars

Stirrups

Straight bottom bar

Bent bar at noncontinuous end

Exterior span

Interior span

Interior span

Interior column

(d) Straight and bent bar reinforcement.

Top bars

Stirrups

Bottom bars

Section through beam

Top bars

Bent bar

Stirrups

Bottom bars

Section through beam

Fig. 4–24
Continuous beam

shown in Fig. 4–24c and d. Today, the straight bar arrangement shown in Fig. 4–24c is used almost exclusively. In some cases a portion of the positive moment or negative moment reinforcement is terminated or cut off when no longer needed. Note, however, that a portion of the steel is extended past the points of inflection, as shown. This is done primarily to account for shifts in the points of inflection due to shear cracking and to allow for changes in loadings and loading patterns. The calculation of bar cutoff points is discussed in Chap. 8.

In addition to longitudinal reinforcement, additional bars, referred to as *stirrups*, are provided to resist shear forces and to hold the various layers of bars in place during construction. These are shown in the cross section in Fig. 4–24. The design of stirrups is discussed in Chap. 6.

Prior to 1965 it was common practice to bend the bottom reinforcement up to the top of the beam when it was no longer required at the bottom. In this way a *bent-up* or *truss bar* could serve as negative and positive reinforcement in the same beam. Such a system is illustrated in Fig. 4–24d.

In conclusion, it is important that designers be able to visualize the deflected shape of a structure. The reinforcing bars for flexure are placed on the tensile face of the member. This is the convex side of the deflected shape.

Construction of Reinforced Concrete Beams and Slabs

The simplest concrete flexural member is the one-way slab shown in Fig. 4–1. The form for such a slab consists of a flat surface, generally built of plywood supported on wooden or steel joists. Whenever possible, the forms are constructed in such a way that they can be reused on several floors. The forms must be strong enough to support the weight of the wet concrete plus construction loads such as workers, concrete buggies, and so on. In addition, the forms must be aligned correctly and *cambered* (arched upward), if necessary, so that the finished floor is flat after the forms are removed.

The reinforcement is supported in the form on wire supports referred to as *bolsters* or *chairs*, which holds the bars at the correct distance above the forms until the concrete has hardened. If the finished slab is expected to be exposed to moisture, wire bolsters may rust, staining the surface. In such a case, small precast concrete blocks or plastic bar chairs may be used instead. Wire bolsters can be seen in the photograph in Fig. 4–25.

Beam forms are most often built of plywood. The size of beam forms is generally chosen to allow maximum reuse of the forms, since the cost of building the forms is a significant part of the total cost of a concrete floor system, as discussed in Sec. 2–8. Some designers prefer to choose 12- or 16-in. beam widths since these widths fit evenly into the width of a standard 4 ft × 8 ft sheet of plywood.

Reinforcement for two beams and some slabs is shown in Fig. 4–25. Here, closed stirrups have been used and the top bars are supported by the top of the closed stirrups. The negative moment bars in the slabs still must be placed. Frequently, the positive moment steel, stirrups, and stirrup support bars for a beam are preassembled into a cage that is dropped into the form.

Relationship between Beam Depth and Deflections

The deflections of a beam can be calculated from equations of the form

$$\Delta_{max} = \frac{C_1 w \ell^4}{EI} \tag{4–29a}$$

Rearranging this and making assumptions concerning steel strains and neutral-axis depth eventually gives an equation of the form

$$\frac{\Delta}{\ell} = C \frac{\ell}{d} \tag{4–29b}$$

Thus for any acceptable ratio of deflection to span lengths, Δ / ℓ, it should be possible to specify span-to-depth ratios, ℓ / d, which, if exceeded, may result in unacceptable deflections. ACI

Fig. 4–25
Intersection of column and
two beams. (Photograph
courtesy of J.G. MacGregor.)

Table 9.5(a) gives minimum thicknesses computed in this way for members *not supporting partitions* or other construction which are liable to be damaged by deflection (see Table A–14 of this book). These minimum thicknesses are frequently used in selecting the overall depths of beams or slabs. Deflections are discussed in Chap. 9.

Concrete Cover and Bar Spacing

It is necessary to have concrete cover between the surface of the slab or beam and the reinforcing bars for four primary reasons:

1. To bond the reinforcement to the concrete so that the two elements act together. The efficiency of the bond increases as the cover increases. A cover of at least one bar diameter is required for this purpose in beams and columns (see Chap. 8).

2. To protect the reinforcement against corrosion. Depending on the environment and the type of member, varying amounts of cover from $\frac{3}{8}$ to 3 in. are required (ACI Sec. 7.7). In highly corrosive environments such as slabs or driveways exposed to deicing salts or ocean spray, the cover should be increased. ACI Commentary Sec. R7.7 allows alternative methods of satisfying the increased cover requirements for elements exposed to the weather. An example of an alternative method might be a waterproof membrane on the exposed surface.

Flexure: Basic Concepts, Rectangular Beams

3. To protect the reinforcement from strength loss due to overheating in the case of fire. The cover for fire protection is specified in the local building code. Generally speaking, $\frac{3}{4}$ in. cover to the reinforcement in a structural slab will provide a 1-hour fire rating, while a $1\frac{1}{2}$-in. cover to the stirrups or ties of beams corresponds to a 2-hour fire rating.

4. Additional cover is sometimes provided on the top of slabs, particularly in garages and factories, so that abrasion and wear due to traffic will not reduce the cover below that required for other purposes.

In this book, the amounts of clear cover will be based on ACI Sec. 7.7.1 unless specified otherwise.

The arrangement of bars within a beam must allow sufficient concrete on all sides of each bar to transfer forces into or out of the bars; sufficient space so that the fresh concrete can be placed or compacted around all the bars; and sufficient space to allow a vibrator to reach through to the bottom of the beam. Figure 4–25 shows the reinforcement at an intersection of two beams and a column. The longitudinal steel in the beams is at the top of the beams because this is a negative moment region. Although this region looks congested, there are adequate openings to place and vibrate the concrete.

ACI Secs. 3.3.2, 7.6.1, and 7.6.2 specify the spacings and arrangements shown in Fig. 4–26. When bars are placed in two or more layers the bars in the top layer must be directly over those in the other layers, to allow the concrete and vibrators to pass through the layers. Conflicts between bars in the columns and other beams should be considered (Figs. 4–25 and 4–27).

Calculation of Effective Depth and Minimum Web Width for a Given Bar Arrangement

The effective depth, d, of a beam is defined as the distance from the extreme compression fiber to the centroid of the longitudinal tensile reinforcement as shown in Fig. 4–2.

EXAMPLE 4–4   Calculation of d and Minimum b

Compute d and the minimum value of b for a beam having bars arranged as shown in Fig. 4–28. The maximum size coarse aggregate is specified as $\frac{3}{4}$ in. The overall depth of the beam is 24 in.

This beam has two different bar sizes. The largest bars are in the bottom layer, to maximize the effective depth and hence the lever arm. Note also that the bars are symmetrically arranged about the centerline of the beam. The bars in the upper layer are directly above those in the lower layer. By placing them on the outside of the section, the top layer of bars can be supported by tying them directly to the stirrups.

From ACI Sec. 7.7.1, the clear cover to the stirrups is 1.5 in. (Fig. 4–26b). From ACI Secs. 7.6.2 and 3.3.2, the minimum distance between layers of bars is the larger of 1 in. or $1\frac{1}{3}$ times the aggregate size, which in this case gives $1\frac{1}{3} \times \frac{3}{4} = 1$ in.

1. Compute the centroid of bars.

Layer	Area, A (in.^{2})	Distance from Bottom, y(in.)	Ay in.^{3}
Bottom	$3 \times 1.00 = 3.00$	$1.5 + \frac{3}{8} + (\frac{1}{2} \times \frac{9}{8}) = 2.44$	7.31
Top	$2 \times 0.79 = \underline{1.58}$	$2.44 + (\frac{1}{2} \times \frac{9}{8}) + 1 + (\frac{1}{2} \times \frac{8}{8}) = 4.50$	$\underline{7.11}$
	Total A = 4.58		Total Ay = 14.42

The centroid is located at $Ay/A = \bar{y} = 14.42/4.58 = 3.15$ in. from the bottom of the beam. The effective depth $d = 24 - 3.15$ in. $= 20.85$ in.—say, $d = 20.8$ in.

2. Compute the minimum web width. This is computed by summing the widths along the most congested layer. The minimum inside radius of a stirrup bend is two times the stirrup diameter,

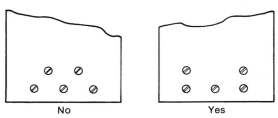

(a) Arrangement of bars in two layers (ACI Sec. 7.6.2).

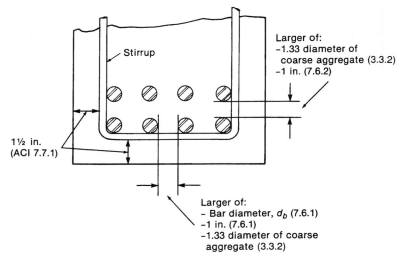

(b) Minimum bar spacing limits in ACI Code.

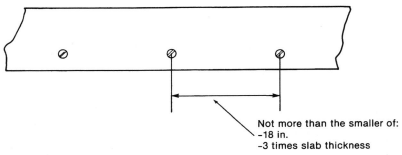

(c) Maximum spacing of flexural reinforcement in slabs (7.6.5)

Fig. 4–26
Bar spacing limits in ACI
Code.

Fig. 4–27
Bar placing problems at the
intersection of two beams.
(From Ref. 4–10)

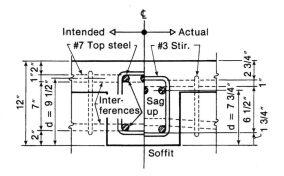

Flexure: Basic Concepts, Rectangular Beams

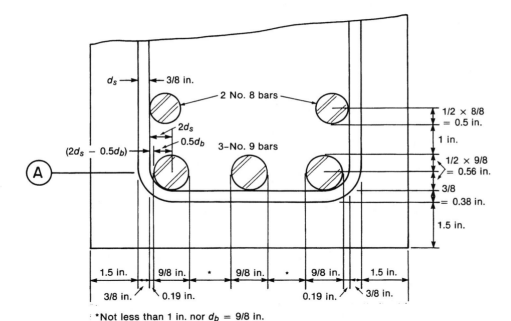

Fig. 4–28
Example 4–4

*Not less than 1 in. nor d_b = 9/8 in.

d_s, which for a No. 3 stirrup is $\frac{3}{4}$ in. (ACI Sec. 7.2.3). For No. 11 or smaller bars there will be a space between the bar and the tie as shown in Fig. 4–28:

$$\text{space} = 2d_s - 0.5d_b$$
$$= 2 \times \frac{3}{8} - 0.5 \times \frac{9}{8} = 0.19 \text{ in.}$$

The minimum horizontal distance between bars is the largest of 1 in., $1\frac{1}{3}$ times the aggregate size, or the bar diameter (see Fig. 4–26b). In this case it is $\frac{9}{8}$ in. Summing the widths along A–A gives

$$b_{min} = 1.5 + \frac{3}{8} + 0.19 + 5\left(\frac{9}{8}\right) + 0.19 + \frac{3}{8} + 1.5$$
$$= 9.76 \text{ in.}$$

Thus the minimum width is 10 in. and design should be based on d = 20.8 in. ■

It is generally satisfactory to estimate the effective depth of a beam using

For beams with one layer of reinforcement,

$$d \simeq h - 2.5 \text{ in.} \tag{4–30a}$$

For beams with two layers of reinforcement,

$$d \simeq h - 3.5 \text{ in.} \tag{4–30b}$$

(The value 3.5 in. given by Eq. 4–30b corresponds to the 3.15 in. computed in Example 4–4.) The value of d should not be overestimated because normal construction practice tends to result in smaller d's than shown on the construction drawings.

For reinforced concrete slabs the minimum clear cover is $\frac{3}{4}$ in. rather than $1\frac{1}{2}$ in., and only one layer of reinforcement is used. This will generally be No. 3, 4, or 5 bars. For this case, Eqs. 4–30a and 4–30b can be rewritten as:

One-way slab spans up to 12 ft:

$$d \simeq h - 1 \text{ in.} \tag{4–30c}$$

4–4 Design of Rectangular Beams   **119**

One-way slab spans over 12 ft:

$$d \simeq h - 1.1 \text{ in.} \tag{4-30d}$$

In SI units, Eqs. 4–30 become:

For beams with one layer of reinforcement,

$$d \simeq h - 65 \text{ mm} \tag{4-30aM}$$

For beams with two layers of reinforcement,

$$d \simeq h - 90 \text{ mm} \tag{4-30bM}$$

For one-way slabs with spans up to 3.5 m,

$$d \simeq h - 25 \text{ mm} \tag{4-30cM}$$

For one-way slabs with spans over 3.5 m,

$$d \simeq h - 30 \text{ mm} \tag{4-30dM}$$

It is important not to overestimate d because normal construction practices lead to smaller values of d than shown on the drawings. Studies of construction accuracy show that on the average, the effective depth of the negative moment reinforcement in slabs is 0.75 in. less than specified.[4-11] In a 5-in.-thick slab, an error of 0.75 in. in the steel placement will reduce the flexural capacity by 19%.

Generally speaking, b should not be less than 10 in. and preferably not less than 12 in. for beams, although with two bars, beam widths as low as 7 in. can be used in extreme cases. The use of a layer of closely spaced bars may lead to a splitting failure along the plane of the bars as explained in Chap. 8. Since such a failure may lead to a loss of bar anchorage as well as corrosion, care should be taken to have at least the minimum bar spacings. (A horizontal crack of this sort can be seen at midspan of the beam shown in Fig. 4–6.) Where there are several layers of bars, a space large enough for the concrete vibrator to pass through should be left between two of the rows. Minimum web widths for various arrangements are given in Table A–6 (see Appendix A). This does not allow extra space for a vibrator to be inserted.

Minimum Reinforcement

If the cracking moment of a beam exceeds the strength of the beam after cracking, a sudden failure could occur with little or no warning when the beam cracks. For this reason ACI Sec. 10.5 requires a minimum amount of flexural reinforcement equal to

$$A_{s,\text{min}} = \frac{3\sqrt{f'_c}}{f_y} b_w d \geq \frac{200 b_w d}{f_y} \tag{4-31}$$

(ACI Eq. 10–3)

where f'_c and f_y are in psi. In SI units this becomes

$$A_{s,\text{min}} = \frac{\sqrt{f'_c}}{4f_y} b_w d \geq \frac{1.4 b_w d}{f_y} \tag{4-31M}$$

(ACI Eq. 10–3M)

where f'_c and f_y are in MPa.

For statically determinate T beams with the flange in tension, the minimum reinforcement is equal to the smaller of

$$A_{s,\min} = \frac{6\sqrt{f_c'}}{f_y}b_w d \qquad (4\text{–}32)$$

(ACI Eq. 10–4)

or the value given by Eq. 4–31 with b_w taken equal to the width of the flange, where f_c' and f_y are in psi. In SI units Eq. 4–32 becomes

$$A_{s,\min} = \frac{\sqrt{f_c'}}{2f_y}b_w d \qquad (4\text{–}32\text{M})$$

(ACI Eq. 10–4M)

The requirements of Eqs. 4–31 and 4–32 need not be applied if the area of reinforcement provided is at least one-third greater than required to provide the required moment capacity.

Design of Rectangular Beams with Tension Reinforcement

In the analysis of beam cross sections, Eqs. 4–1 and 4–15 were presented:

$$\phi M_n \geq M_u \qquad (4\text{–}1\text{b})$$

where M_u represented factored moments due to loads, which for gravity loads equals (ACI Sec. 9.2.1)

$$M_u = 1.4M_D + 1.7M_L$$

where M_D and M_L are the moments due to the unfactored dead and live loads, respectively. The factored resisting moment, ϕM_n, is the couple formed by the internal tensile and compressive forces and can be calculated from

$$\phi M_n = \phi\left[f_c' b d^2 \omega(1 - 0.59\omega)\right] \qquad (4\text{–}15)$$

where $\omega = \rho f_y / f_c'$.

In the design, the problem to be solved involves the selection of a beam cross section to support a given value of the live-load moment, M_L, plus its own dead-load moment and moments due to other loads it may be required to support. In this calculation there are *six* unknowns: b, d, ρ, f_y, f_c' and the beam's own dead-load moment; but only *two* independent equations, Eq. 4–12a (or 4–15) and the relationship between the beam size and the beam's dead-load moment. As a result, it is not possible to design a beam uniquely. The design procedure is thus an iterative process in which four assumptions must be made. Although this requires some intuition and understanding of the construction process, the resulting design freedom makes reinforced concrete the universal and valuable material it is.

The value of f_c' to be used in the design is chosen at the start of the design and used throughout the project. The choice of concrete strength is based on durability considerations if the member is exposed to freezing and thawing, deicing salts, or other aggressive environments. ACI Chapter 4, "Durability Requirements," particularly ACI Table 4.2.2, specifies minimum concrete strengths ranging from 4000 to 5000 psi for various exposures. The strength chosen for durability reasons may be utilized in the design for strength. If durability is not a problem, reinforced concrete beams and slabs are generally constructed of 3000-, 3750-, or 4000-psi concrete, with 3000-psi being the most common. The strength of the concrete in columns may be higher, as discussed in Chap. 11.

The yield strength most commonly used in the United States is 60,000 psi. Steel with a yield strength of 40,000 psi is occasionally used for flexural reinforcement. Only No. 3 to No. 6 bars are available in Grade 40 steel.

Once f_c' and f_y have been chosen for the beam or slab in question, three major independent variables remain: the width, b; the depth, d (or the overall height, h); and the reinforcement ratio, ρ. If b and d are known, it is possible to go directly to the computation of ρ and $A_s = \rho bd$. If not, the following equation is used to arrive at b and d:

$$\frac{M_u}{\phi k_n} = \frac{bd^2}{12{,}000} \qquad (4\text{--}17)$$

where $\phi k_n = \phi[f_c'\,\omega(1 - 0.59\omega)]$ and M_u is in ft-kips. To solve for b and d, it is necessary to assume a value of ω or ρ, compute ϕk_n, and compute $bd^2/12{,}000$ and eventually b and d. Table A–3, which lists ϕk_n, and Table A–7, which lists $bd^2/12{,}000$, may be used to aid in this calculation. A value of ρ equal to 0.01 (corresponding to $\omega \approx 0.12$ to 0.24 or $\rho = 0.30$ to $0.50\rho_b$, depending on the concrete strength) is generally used to get the first value of ϕk_n. Also, $\rho = 0.010$ will correspond to a tension-controlled section that will have $\phi = 0.90$. It is possible to round off the values of b and d to practical sizes at this stage since the values of b and d chosen in this step are then used in a recomputation of ρ and A_s in the next stage.

When designing in metric (SI) units, reinforced concrete beams and slabs are generally constructed of 20, 25, or 30 MPa concrete, with 20 MPa being the most common. In locations where durability is important, ACI Sec. 4.2.2 requires the use of air-entrained concrete having specified maximum water/cementitious materials ratios and minimum strengths ranging from 28 to 35 MPa. The most common yield strength is 400 MPa. Grade 300 reinforcement is only available in sizes 10, 15, and 20.

If b and d are not known, the following equation is used to arrive at b and d when working in SI units:

$$\frac{M_u \times 10^6}{\phi k_{nm}} = bd^2 \qquad (4\text{--}17\text{M})$$

where $\phi k_{nm} = \phi[f_c'\,\omega(1 - 0.59\omega)]$ in metric units, M_u is in kN-m and b and d are in mm. To solve for b and d it is necessary to assume a value of ω or ρ, compute ϕk_{nm}, and compute bd^2 and eventually b and d. A value of ρ equal to 0.01 is generally used to get the first value of ϕk_{nm}.

Two different procedures exist for calculating the area of reinforcement, A_s. If the values of b and d chosen for the beam are close to the calculated values, the steel area can be calculated directly from ρ using

$$A_s = \rho bd$$

The resulting A_s should not be less than A_{smin} (Eq. 4–31) and ρ should not exceed $0.75\rho_b$ (Eq. 4–25 and Table A–5). It is then necessary to check whether this amount of reinforcement is adequate to resist M_u. This is done using Eq. 4–12a as in Example 4–2.

In most cases it is much better, however, to calculate the area of reinforcement using

$$\phi M_n = \phi\left[A_s f_y\left(d - \frac{a}{2}\right)\right] \qquad (4\text{--}12\text{b})$$

where $(d - a/2)$ is referred to as jd. A_s and a are unknown in this equation. It is necessary to assume j, compute A_s, recompute a and $(d - a/2)$ using this value of A_s, and recompute A_s until convergence is obtained. For beams with grade 60 reinforcement, j can range from about 0.95 for minimum reinforcement to about 0.80 for $\rho = 0.75\rho_b$ (see Table A–3). For the most usual steel percentages in beams, j is generally between 0.87 and 0.90. For one-way slabs, which generally have a lower reinforcement ratio than beams, j will generally vary between 0.90 and 0.95. In design problems in this book j will initially be assumed equal to 0.875 for rectangular beams and 0.925 for slabs. Thus, for the initial trial, A_s can be computed from Eq. 4–12b replacing $(d - a/2)$ with jd, where $j \simeq 0.875$ for beams and $j \simeq 0.925$ for slabs, and replacing M_n with M_u/ϕ:

$$A_s = \frac{M_u}{\phi f_y jd} \qquad (4\text{–}33)$$

Since this computed value of A_s is based on an estimate of j, it is necessary to check the accuracy of the estimate by using the A_s selected to compute ϕM_n using Eq. 4–12b. If ϕM_n is not close enough, iterations may be needed. Three general types of design problems exist. These will be discussed in the next three subsections.

Design of Reinforcement When b and h are Known

The first type of design problem is the case in which the dimensions of the concrete section have been established by nonstructural reasons such as architectural appearance, reuse of standard forms, fire resistance, and so on. In this case, b and d (or h) are known and it is only necessary to compute A_s.

EXAMPLE 4–5   Design of Reinforcement When b and h are Known

For architectural reasons it is necessary that the beam shown in Fig. 4–29 be 24 in. wide by 24 in. deep. The strengths of the concrete and steel are 3000 psi and 60,000 psi, respectively. In addition to its own dead load, this beam carries a superimposed service (unfactored) dead load of 1.0 kip/ft and a service live load of 2.45 kips/ft.

Compute the area of reinforcement required at midspan and select the reinforcement.

1.   Compute the factored moment, M_u.

$$\text{Weight/ft of beam} = \frac{(2 \times 2 \times 1)\ \text{ft}^3 \times 150\ \text{lb/ft}^3}{1\ \text{ft of length}} = 600\ \text{lb/ft} = 0.60\ \text{kip/ft}$$

The factored load is

$$U = 1.4D + 1.7L \qquad \text{(ACI Eq. 9.1)}$$

or

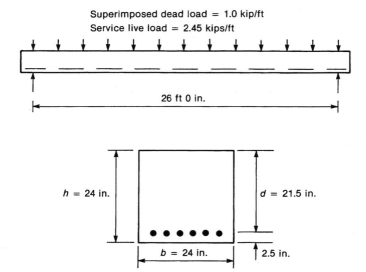

Fig. 4–29
Beam designed in Example 4–5.

$$w_u = 1.4(0.6 + 1.0 \text{ kip/ft}) + 1.7(2.45 \text{ kips/ft})$$
$$= 6.40 \text{ kips/ft}$$

The factored load effect (factored ultimate moment) is

$$M_u = \frac{w_u \ell_n^2}{8}$$

$$= \frac{6.40 \text{ kips/ft} \times 26^2 \text{ ft}^2}{8} = 541 \text{ ft-kips}$$

It is necessary, therefore, to provide $\phi M_n \geq M_u$ or $\phi M_n \geq 541$ ft-kips.

2. Compute the effective depth, d. Because the beam is quite wide, assume that all the bars will be in one layer. From Eq. 4–30a, d can be estimated as

$$d \simeq h - 2.5 \text{ in.}$$
$$\simeq 24 - 2.5 \text{ in.}$$

Therefore, try $d = 21.5$ in.

3. Compute the area of reinforcement, A_s. Assume that

$$jd = d - \frac{a}{2} = 0.875d \text{ (this is equivalent to assuming that } a = 0.25d)$$
$$= 0.875 \times 21.5 \text{ in.} = 18.8 \text{ in.}$$

From Eq. 4–33,

$$A_s = \frac{M_u}{\phi f_y jd}$$

$$= \frac{541 \text{ ft-kips} \times 12 \text{ in./ft}}{0.9 \times 60 \text{ ksi} \times 188 \text{ in.}} = 6.39 \text{ in.}^2$$

Possible choices are (Table A–8):

11 No. 7 bars, $A_s = 6.60$ in.^{2}
7 No. 9 bars, $A_s = 7.00$ in.^{2}
4 No. 9 bars plus 3 No. 8 bars, $A_s = 6.37$ in.^{2}

A check of the required web width as per Example 4–3 (or Table A–6) shows that all of these choices are acceptable. Try 4 No. 9 bars plus 3 No. 8 bars, $A_s = 6.37$ in.^{2}.

4. Check whether $A_s \geq A_{s,\text{min}}$. From Eq. 4–31,

$$A_{s,\text{min}} = \frac{3\sqrt{f_c'}}{f_y} b_w d \geq \frac{200 b_w d}{f_y}$$

$$= \frac{3\sqrt{3000}}{60{,}000 \text{ psi}} \times 24 \times 21.5 \geq \frac{200 \times 24 \times 21.5}{60{,}000}$$

$$= 1.41 \text{ in.}^2 \geq 1.72 \text{ in.}^2.$$

Since 6.37 in.^{2} exceeds 1.72 in.^{2}, $A_s > A_{s,\text{min}}$. (If not OK, increase A_s to $A_{s,\text{min}}$ or satisfy ACI Sec. 10.5.3.)

5. Compute a, check if $f_s = f_y$, and whether the section is tension-controlled. From Eq. 4–9,

$$a = \frac{A_s f_y}{0.85 f_c' b}$$

$$= \frac{6.37 \text{ in.}^2 \times 60{,}000 \text{ psi}}{0.85 \times 3000 \text{ psi} \times 24 \text{ in.}} = 6.25 \text{ in.}$$

$$\frac{a}{d} = \frac{6.25 \text{ in.}}{21.5 \text{ in.}} = 0.290$$

to check if $f_s = f_y$ we shall check whether a/d is less than a_b/d. From Eq. 4–21,

$$\frac{a_b}{d} = \beta_1 \left(\frac{87,000}{87,000 + 60,000} \right) = 0.503 \text{ (see also Table A–4)}$$

Since $a/d = 0.290$ is less than $a_b/d = 0.503$, $f_s = f_y$.

To check whether the section is tension-controlled, we shall check whether a/d_t is less than or equal to $a_{tc\ell}/d_t$. Since all the steel is in one layer, $d = d_t$ and $a/d_t = 0.290$.

$$\frac{a_{tc\ell}}{d_t} = 0.375\beta_1 = 0.375 \times 0.85$$

$$= 0.319 \qquad\qquad (4\text{--}23)$$

Since $a/d_t = 0.290$ is less than $a_{tc\ell}/d_t = 0.319$, the section is tension-controlled and from ACI Sec. B9.3.2.1, $\phi = 0.90$. If the section was compression-controlled or transitional, a lower value of ϕ would be required. Such a beam would be less ductile than a tension-controlled section. If this was not considered acceptable, the section should be enlarged or compression steel added. Compression steel is discussed in Chap. 5.

6. Compute M_n and ϕM_n. Because A_s was calculated using an estimated value of jd, it is necessary to check whether the reinforcement selected provides adequate moment capacity. It may also be desirable to recompute d. From Eq. 4–12a,

$$M_n = A_s f_y \left(d - \frac{a}{2} \right)$$

$$= 6.37 \text{ in.}^2 \times 60 \text{ ksi} \left(21.5 \text{ in.} - \frac{6.25}{2} \right)$$

$$= 7020 \text{ in.-kips} = 585 \text{ ft-kips}$$

$$\phi M_n = 0.9 \times 585 \text{ ft-kips} = 527 \text{ ft-kips}$$

Since this is less than the factored ultimate moment, $M_u = 541$ ft-kips, A_s is too small. This is because the assumed value of $jd = 0.875d$ is greater than the value of $jd = 0.855d$ corresponding to $A_s = 6.37$ in.^{2}. It is therefore necessary to increase the area of steel.

7. Recompute the area of steel required. To do so, recompute the area of steel using the lever arm $(d - a/2)$ based on the value of a computed in Step 5:

$$A_s = \frac{M_u}{\phi f_y(d - a/2)}$$

$$= \frac{541 \times 12}{0.9 \times 60(21.5 - 6.25/2)} = 6.54 \text{ in.}^2$$

Possible choices are:

     11 No. 7 bars, $A_s = 6.60$ in.^{2}
     7 No. 9 bars, $A_s = 7.00$ in.^{2}
     5 No. 9 bars and 2 No. 8 bars, $A_s = 6.58$ in.^{2}

Again, all of these will fit into the web width of 24 in. Try 5 No. 9 bars and 2 No. 8 bars, $A_s = 6.58$ in.^{2}. This exceeds $A_{s,\min} = 1.72$ in.^{2}.

$$a = \frac{A_s f_y}{0.85 f_c' b} = 6.45 \text{ in.}$$

$$\frac{a}{d} = 0.30 \qquad \text{which is less than } \frac{a_b}{d} = 0.503$$

$$\text{and less than } \frac{a_{tc\ell}}{d_t} = 0.319$$

Therefore, $f_s = f_y$ and the section is tension-controlled, $\phi = 0.90$.
Finally,

$$\phi M_n = \frac{0.9[6.58 \times 60(21.5 - 6.45/2)]}{12}$$

$$= 541 \text{ft-kips}$$

since $\phi M_n = M_u$, A_s is OK. **Therefore, use 5 No. 9 bars and 2 No. 8 bars.** ∎

Design of Beams when b and h are Not Known

The second type of design problem involves finding b, d, and A_s. Three sets of decisions not encountered in Example 4–5 must be made here. These are a preliminary estimate of the dead load of the beam, selection of a trial steel percentage, and the final selection of the beam dimensions b and d.

Although no dependable rule of thumb exists for guessing the weight of beams, the weight of a rectangular beam will be roughly 10 to 20% of the loads it must carry. Alternatively, one can estimate h as being roughly 8 to 10% of the span (1 in. deep per foot of span), estimate b as $0.5h$, and use these dimensions to compute a trial weight. These two procedures frequently give quite different values and an intermediate value can be chosen. The dead load guessed at this stage is corrected when the dimensions are finally chosen.

It is then necessary to select a trial steel ratio ρ which is used to calculate ϕk_n to enter Eq. 4–16 to get b and d. This choice is affected by economic considerations; generally, $\rho \simeq 0.01$ is an economical choice; by ductility, generally $\rho \simeq 0.35\rho_b$ to $\rho \simeq 0.4\rho_b$ gives a desirable level of ductility; and by placing considerations since it may be hard to place the reinforcement if ρ exceeds 0.015. For Grade 60 reinforcement, $0.4\rho_b$ is 0.0086 for 3000 psi, 0.0114 for 4000 psi, and 0.0134 for 5000 psi. For this reason we start our design assuming that $\rho = 0.010$ at the point of maximum moment in all cases.

Factors to be considered in choosing the beam dimensions b and h are:

1. A deeper beam requires less reinforcement. The savings here are offset by increased forming costs and either reduced headroom in the story below, or an increase in the overall height of the building.

2. Longer development lengths for closely spaced bars. This is normally not a problem except for top bars in short rectangular beams.

3. It may be possible to avoid deflection calculations if the overall height of the beam exceeds the values given in ACI Table 9.5(a) (see Table A–14 of this book).

4. For rectangular beams it is common practice to select sections with d/b between 1.5 and 2.

The beam size chosen at this stage is rounded off to fit convenient form sizes as discussed earlier. A considerable amount of rounding can be carried out at this stage because the final stage in the design is to compute the required area of steel corresponding to the b and d selected at this stage.

EXAMPLE 4–6 Design of a Beam for which b and d Are Not Known

A beam is to carry its own dead load plus a uniform service live load of 1.75 kips/ft and a uniform superimposed service dead load of 1 kip/ft on a 33-ft span. Select b, d, and A_s if f_c' is 3500 psi and f_y is 60,000 psi.

1. Estimate the dead load of the beam. Estimate the weight of a rectangular beam as 10 to 20% of the loads it must carry. This range corresponds to 0.3 to 0.6 kip/ft. Alternatively, estimate h as being roughly 8 to 10% of the span, and b as $0.5h$, and compute the weight. This gives $h = 2.6$ to 3.3 ft, and following the procedure in step 1 of Example 4–4, these correspond to weights equal to 0.52 to 0.82 kip/ft. Based on these four values estimate the beam weight at 0.5 kip/ft.

2. **Compute the factored moment, M_u.**

$$w_u = 1.4(1 + 0.5) + 1.7(1.75) = 5.08 \text{ kips/ft}$$

$$M_u = \frac{w_u \ell_n^2}{8} = 691 \text{ ft-kips}$$

3. **Compute b and d.** From Eq. 4–17,

$$\frac{M_u}{\phi k_n} = \frac{bd^2}{12,000}$$

where $k_n = f'_c \omega(1 - 0.59\omega)$, $\omega = \rho f_y/f'_c$, and M_u is in ft-kips. To start this calculation assume either ρ or ω. We shall try $\rho \simeq 0.010$. The value of ω for $\rho = 0.01$ is

$$\omega = 0.01 \times \frac{60,000}{3500} = 0.171$$

Therefore,

$$k_n = 3500 \times 0.171(1 - 0.59 \times 0.171) = 538$$

From Eq. 4–17,

$$\frac{bd^2}{12,000} = \frac{M_u}{\phi k_n}$$

$$= \frac{691}{0.9 \times 538} = 1.427$$

or $bd^2 = 17,120$ in.^{3}.

Possible choices (with h calculated using Eqs. 4–30 assuming two layers of reinforcement for the narrower beams and one layer in the 18-in.-wide beam) are

$b = 12$ in. by $d = 37.8$ in. and $h = 37.8 + 3.5 = 41.3$ in.
$b = 16$ in. by $d = 32.7$ in. and $h = 37.2$ in.
$b = 18$ in. by $d = 30.8$ in. and $h = 30.8 + 2.5 = 33.3$ in.

Minimum overall depth to avoid deflection calculations if beam is not supporting brittle partitions [from ACI Table 9.5(a)] for a simple beam is $\ell/16$, which in this case is 24.75 in. All the choices listed exceed this, so deflection should not be a problem.

Try d/b between 1.5 and 2. On this basis we choose $b = 16$ in. and $h = 36$ in. The size chosen has been rounded off to aid in construction of forms, and so on. Assuming two layers of reinforcement, $d = 36 - 3.5 = 32.5$ in. **Use $b = 16$ in., $h = 36$ in., and $d = 32.$ in.**

4. **Check the dead load and revise M_u.** For $b = 16$ in. and $h = 36$ in., the self-weight per foot is

$$(1.33 \times 3 \times 1)\text{ft}^3/\text{ft} \times 0.15 \text{ kip/ft}^3 = 0.600 \text{ kip/ft}$$

Therefore, the total factored moment, M_u becomes 710 ft-kips compared to the original estimate of 691 ft-kips. If the moment, M_u increased by more than about 10%, it may be desirable to repeat steps 3 and 4.

5. **Compute the area of reinforcement A_s.** Assume that

$$jd = \left(d - \frac{a}{2}\right) = 0.875d$$

$$= 28.4 \text{ in.}$$

From Eq. 4–33,

$$A_s = \frac{M_u}{\phi f_y jd}$$

$$= \frac{710 \text{ ft-kips} \times 12 \text{ in./ft}}{0.9 \times 60 \text{ ksi} \times 28.4 \text{ in.}} = 5.56 \text{ in.}^2$$

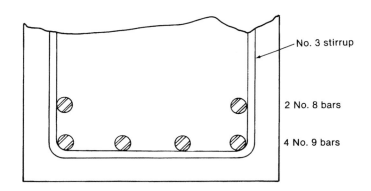

Fig. 4–30
Reinforcement location—
Example 4–6.

From Eq. 4–31

$$A_{s,\min} = \frac{200 \times 16 \times 32.5}{60,000} = 1.73 \text{ in.}^2$$

Since A_s required $> A_{s,\min}$, use A_s required.

Possible choices are:

6 No. 9 bars, $A_s = 6.00$ in.^{2}
7 No. 8 bars, $A_s = 5.53$ in.^{2}
10 No. 7 bars, $A_s = 6.00$ in.^{2}
4 No. 9 bars plus 2 No. 8 bars, $A_s = 5.58$ in.^{2}
4 No. 8 bars plus 4 No. 7 bars, $A_s = 5.56$ in.^{2}

All of these will fit into two layers in $b = 16$ in. if the maximum size aggregate is 3/4 in. (Table A–6). We shall try 4 No. 9 bars plus 2 No. 8 bars arranged as shown in Fig. 4–30, to reduce the number of bars to be placed. The larger bars are placed in the bottom row to give the largest possible lever arm.

If the values of b and d chosen in step 3 are close to the computed ones, A_s could theoretically be computed directly from the value of ρ chosen in step 3 since $A_s = \rho bd$. Generally, however, the rounding of b and d makes it necessary to compute A_s as done in this example.

6. Compute d. For the steel placement in Fig. 4–30, d can be computed using the technique given in Example 4–4. This gives $d = 33$ in. In most cases, this step can be omitted since $d = 32.5$ in. from Eq. 4–30b is close enough.

7. Compute a, check if $f_s = f_y$, and whether the section is tension-controlled. From Eq. 4–11,

$$a = \frac{A_s f_y}{0.85 f_c' b}$$

$$= \frac{5.58 \times 60,000}{0.85 \times 3500 \times 16} = 7.03 \text{ in.}$$

$$\frac{a}{d} = \frac{7.03}{33.0} = 0.213$$

From Eq. 4–21 or Table A–4, a_b/d is 0.503. Since a/d is less than a_b/d, $f_s = f_y$. d is 33 in., $d_t = 36$ in. $- (1.50 + 0.375 + 1.125/2)$in. $= 33.5$ in. Thus

$$\frac{a}{d_t} = \frac{7.03}{33.5} = 0.210 \qquad \frac{a_{tc\ell}}{d_t} = .319$$

since a/d_t is less than $a_{tc\ell}/d_t$, the section is tension-controlled and $\phi = 0.90$.

8. Compute M_n and ϕM_n. Because the area of steel calculated in step 5 was based on an estimate of jd, it is necessary to check whether the reinforcement chosen provides the required moment resistance. From Eqs. 4–12,

$$M_n = A_s f_y \left(d - \frac{a}{2} \right)$$

$$= \frac{5.58 \times 60{,}000(33.0 - 7.03/2)}{12{,}000} = 823 \text{ ft-kips}$$

$$\phi M_n = 0.9 \times 823 \text{ ft-kips} = 740 \text{ ft-kips}$$

Since $\phi M_n \geq M_u$, the design is OK. **Therefore, use b = 16 in., h = 36 in., with f_c'= 3500 psi, f_y = 60,000 psi, and reinforcement as shown in Fig. 4–30.** If ϕM_n were less than M_u or if ϕM_n were much greater than M_u, it would be necessary to recompute the area of steel required (see step 7 of Example 4–5). ∎

EXAMPLE 4–6M Design of a Beam for Which b and h Are Not Known— SI Units

A beam is to carry its own dead load plus a uniform service live load of 25.5 kN/m and a uniform superimposed service dead load of 14.5 kN/m on a 10-m span. Select b, d, and A_s if f_c' is 25 MPa and f_y is 400 MPa.

1. Estimate the dead load of the beam. Estimate the weight of a rectangular beam as 10 to 20% of the loads it must carry. This range corresponds to 4 to 8 kN/m. Alternatively, estimate h as being roughly 8 to 10% of the span, and b as $0.5h$, and compute the weight. This gives h = 0.8 to 1.0 m, and the trial weight as

$$\text{weight of beam} = (0.8 \times 0.4 \times 1)\text{m}^3 \times 2450 \text{ kg/m}^3 \times (9.81/1000)\text{N/kg}$$

$$= 7.69\text{kN/m for } h = 0.8 \text{ m}$$

to 12.0 kN/m for h = 1.0 m. Based on these four values (4, 7.69, 8, and 12 kN/m), estimate the beam weight at 8 kN/m.

2. Compute the factored moment, M_u.

$$w_u = 1.4(14.5 + 8) + 1.7(25.5) = 74.9 \text{ kN/m}$$

$$M_u = \frac{w_u \ell_n^2}{8} = 936 \text{ kN-m}$$

3. Compute b and d. From Eq. 4–17M,

$$\frac{M_u}{\phi k_{nm}} = \frac{bd^2}{10^6}$$

where $\phi k_{nm} = \phi[f_c'\omega(1 - 0.59\omega)]$, $\omega = \rho f_y/f_c'$, and M_u is in kN-m. To start this calculation, assume either ρ or ω. We shall assume that ρ = 0.01. For flexure ϕ = 0.9, except when using the ϕ values from B9.3.2, when it is 0.9 for tension-controlled sections. As we will see later, ρ = 0.01 will always give a tension-controlled beam section. The value for ω for ρ = 0.01 is

$$\omega = 0.01 \times \frac{400}{25} = 0.160$$

Therefore,

$$\phi k_{nm} = 0.9[25 \times 0.160(1 - 0.59 \times 0.160)] = 3.26$$

From Eq. 4–17M,

$$bd^2 = \frac{936 \times 10^6}{3.26} = 287.1 \times 10^6 \text{mm}^2$$

Possible choices (with h calculated using Eq. 4–30M assuming two layers of reinforcement for the narrower beams and one layer in the 450-mm-wide beam) are

b = 300 mm by d = 987 mm and h = 978 + 90 = 1068 mm
b = 400 mm by d = 847 mm and h = 847 + 90 = 937 mm
b = 450 mm by d = 799 mm and h = 799 + 65 = 864 mm

4–4 Design of Rectangular Beams

Minimum overall depth to avoid deflection calculations if the beam is not supporting brittle partitions [from ACI Table 9.5(a)] for a simple beam is $\ell/16$, which in this case is 625 mm. Since all the choices listed above exceed this, deflections should not be a problem.

Try d/b between 1.5 and 2. On this basis we shall choose $b = 400$ mm and $h = 900$ mm. The size chosen was rounded off to aid in the construction of forms, and so on. Assuming two layers of steel, $d = 900 - 90 = 810$ mm. **Use $b = 400$ mm, $h = 900$ mm, and $d = 810$ mm.**

4. Check the dead load and revise M_u.

For $b = 400$ mm and $h = 900$ mm, the self-weight per meter is

$$(0.40 \times 0.90 \times 1.0)\text{m}^3/\text{m} \times 2450 \times (9.81/1000)\text{kN}/\text{m}^3 = 8.65 \text{ kN}/\text{m}$$

Therefore, the total factored moment, M_u, becomes 946 kN-m compared to the original estimate of 936 kN-m. If the moment, M_u, is increased by more than about 10%, it may be desirable to repeat steps 3 and 4.

5. Compute the area of reinforcement, A_s.
Assume that

$$jd = \left(d - \frac{a}{2}\right) = 0.875d$$

$$= 709 \text{ mm}$$

From Eq. 4–33,

$$\text{required } A_s = \frac{M_u}{\phi f_y jd}$$

$$= \frac{947 \times 10^6 \text{ N–mm}}{0.9 \times 400 \text{ MPa} \times 709 \text{ mm}} = 3710 \text{ mm}^2$$

From Eq. 4–31M,

$$A_{s,\text{min}} = \frac{\sqrt{f_c'}}{4f_y}b_w d \geq \frac{1.4 b_w d}{f_y}$$

$$= \frac{\sqrt{25}}{4 \times 400} \times 400 \times 810 = 1012 \text{ mm}^2$$

but not less than

$$\frac{1.4 \times 400 \times 810}{400} = 1134 \text{ mm}^2$$

Since 3710 mm^2 exceeds 1134 mm^2, use the required A_s.
Possible choices are:

6 No. 30 bars, $A_s = 4200$ mm^2
8 No. 25 bars, $A_s = 4000$ mm^2
4 No. 30 bars plus 2 No. 25 bars, $A_s = 3300$ mm^2

All of these will fit into two layers in $b = 400$ mm if the maximum size aggregate is 20 mm. We shall try 4 No. 30 bars plus 2 No. 25 bars arranged as shown in Fig. 4–30 with the 4 No. 30 bars in the bottom layer. The larger bars are placed in the bottom layer to give the largest possible lever arm.

If the values of b and d chosen in step 3 are close to the computed ones, A_s could theoretically be calculated directly from the value of ρ chosen in step 3 since $A_s = \rho bd$. Generally, however, the rounding of b and d makes it necessary to compute A_s as done in this example.

6. Compute d. For steel placed in the manner chosen, d can be computed in the manner given in Example 4–4. This gives $d = 820$ mm. In most cases, this step can be omitted since $d = 810$ mm (from Eq. 4–30bM) is close enough.

7. Compute a and check if the section is tension-controlled.

$$a = \frac{A_s f_y}{0.85 f_c' b}$$

$$= \frac{3800 \times 400}{0.85 \times 25 \times 400} = 179 \text{ mm}$$

$$\frac{a}{d} = \frac{179}{820} = 0.218$$

From Eq. 4–23, the section will be tension-controlled if a/d_t is less than or equal to

$$\frac{a_{tc\ell}}{d_t} = 0.375\beta_1 = 0.319$$

where d_t is the distance from the extreme compression fiber to the centroid of the farthest layer of tension steel $= 900 - (40 + 10 + 30/2) = 835$ mm. The actual a/d_t is

$$\frac{a}{d_t} = \frac{179}{835} = 0.214$$

Since 0.214 is less than 0.319, the section is tension-controlled and $\phi = 0.90$.

8. Compute M_u and ϕM_n. Because the area of steel calculated in step 5 was based on an estimate of jd, it is necessary to check whether the reinforcement chosen provides the required moment resistance. From Eqs. 4–12,

$$M_n = A_s f_y \left(d - \frac{a}{2} \right)$$

$$= \frac{3800 \times 400(820 - 179/2)}{10^6} = 1110 \text{ kN-m}$$

$$\phi M_n = 0.90 \times 1110 \text{ kN-m} = 999 \text{ kN-m}$$

Since $\phi M_n \geq M_u$, the design is OK. **Therefore, use $b = 400$ mm, $h = 900$ mm, with $f_c' = 25$ MPa and $f_y = 400$ MPa, and 4 No. 30 bars and 2 No. 25 bars in two layers in the fashion shown in Fig. 4–30.** If ϕM_n were less than M_u, or if ϕM_n were much greater than M_u, it would be necessary to recompute the area of steel required (see step 7 of Example 4–5). ∎

The third type of rectangular beam design problem occurs when the overall height, h, of the member is predetermined, either to maintain the desired floor to floor clearance or to limit deflections. In this case, the design procedure is identical to Example 4–6 except that in the choice of the concrete section, the value of h, and hence d, is known and bd^2 is solved to find b.

Use of Design Aids in Rectangular Beam Design

The examples solved to date have involved a longhand calculation of all terms. Considerable effort can be saved by using tables available from the American Concrete Institute[4-12] and Concrete Reinforcing Steel Institute,[4-13] among others. Representative tables are presented in Appendix A. To illustrate the use of such tables, Example 4–6 is repeated in Example 4–7.

EXAMPLE 4–7   Use of Design Aids to Design Rectangular Beam When *b* and *d* Are Not Known

The loadings and material strengths are given in Example 4–6.

1. Estimate the dead load of the beam. Choose the self-weight of the beam equal to 0.5 kip/ft, as outlined in Example 4–6.

2. Compute the factored moment, M_u. Using the assumed self-weight, $M_u = 691$ ft-kips.

3. Compute b and d.

$$\frac{M_u}{\phi k_n} = \frac{bd^2}{12,000}$$

Assume that $\rho = 0.4\rho_b$. From Table A–3, $k_n = 538$. Therefore,

$$\frac{bd^2}{12{,}000} = \frac{691}{538} = 1.28$$

From Table A–7 choose b and d to give this value. Choose $b = 16$ in. and $h = 36$ in. to give $d = 32.5$ in.

4. **Recompute the weight and M_u.** As per Example 4–6, $M_u = 710$ ft-kips.

5. **Compute the area of reinforcement, A_s.** From Table A–3, for $\rho = 0.4\rho_b$, $j = 0.899$,,

$$A_s = \frac{M}{\phi f_y j d}$$

$$= \frac{710 \times 12}{0.9 \times 60 \times 0.899 \times 32.5} = 5.40 \text{ in.}^2$$

From Table A–8 we find that 4 No. 9 bars plus 2 No. 8 bars give $A_s = 5.58$ in.^{2}.

$$\rho = \frac{A_s}{bd} = 0.0107$$

From Table A–3 it can be seen that this falls between the ρ corresponding to the tension-controlled limit (bottom of each column) and ρ_{min} (top of table)—therefore OK. At this stage it is optional whether it is necessary to recompute d or use the assumed value. If the bar distribution is markedly different from equal numbers of same-sized bars in each layer, it may be desirable to do so. If not, the values of d from Eq. 4–30 may be used.

6. **Compute a and ϕM_n**

$$a = \frac{5.58 \times 60{,}000}{0.85 \times 3500 \times 16} = 7.03 \text{ in.}$$

ϕM_n(based on original assumed $d = 32.5$ in.)

$$= \frac{0.9 \times 5.58 \times 60{,}000(32.5 - 7.03/2)}{12{,}000} = 728 \text{ ft-kips}$$

This exceeds M_u—therefore OK. **Use $b = 16$ in., $h = 36$ in., and a beam with 4 No. 9 bars and 2 No. 8 bars, as shown in Fig. 4–29.** ■

Direct Solution of Required Area of Steel

An alternative method of solving for the required area of steel can be obtained from Eqs. 4–1, 4–11, and 4–12. From Eq. 4–1 the smallest acceptable value of ϕM_n is

$$M_u = \phi M_n$$

Substituting this and Eq. 4–11 into Eq. 4–12a gives

$$M_u = \phi\left[A_s f_y\left(d - \frac{A_s f_y}{1.7 f_c' b}\right)\right]$$

Rearranging gives

$$\left(\frac{\phi f_y^2}{1.7 f_c' b}\right)A_s^2 - (\phi f_y d)A_s + M_u = 0 \tag{4–34}$$

This is a quadratic equation in x of the type

$$Ax^2 + Bx + C = 0$$

where $x = A_s$. Once f_y, f'_c, b, and d have been chosen, Eq. 4–34 can be solved for a value of A_s. This replaces the iterative calculation used in Examples 4–5 and 4–6 and is useful if programmable calculators are used.

EXAMPLE 4–8 Use Eq. 4–34 to Solve for A_s

The loadings, material strengths, and dimensions are as in Example 4–5. Thus

$$f_y = 60000 \text{ psi} \qquad\qquad f'_c = 3000 \text{ psi}$$

$$b = 24 \text{ in.} \qquad\qquad d = 21.5 \text{ in.}$$

$$M_u = 541 \text{ ft-kips}$$

$$= 6.492 \times 10^6 \text{ in.-kips}$$

1. **Compute the terms in Eq. 4–34.**

$$A = \frac{\phi f_y^2}{1.7 f'_c b}$$

$$= \frac{0.9 \times 60{,}000^2}{1.7 \times 3000 \times 24} = 26.5 \times 10^3$$

$$B = -\phi f_y d$$

$$= -0.9 \times 60{,}000 \times 21.5 = -1.16 \times 10^6$$

$$C = 6.49 \times 10^6$$

2. **Solve Eq. 4–34 for A_s**

$$A_s = \frac{+1.16 \times 10^6 \pm \sqrt{(-1.16 \times 10^6)^2 - 4(26.5 \times 10^3 \times 6.49 \times 10^6)}}{2(26.5 \times 10^3)}$$

$$= 6.58 \text{ in.}^2 \text{ or } 837.28 \text{ in.}^2$$

The higher root is several times the balanced steel ratio and will be discarded. Thus the required $A_s = 6.58 \text{ in.}^2$.

3. **Select the reinforcement and compute ϕM_n as a check.**

Choose 5 No. 9 bars and 2 No. 8 bars, $A_s = 6.58 \text{ in.}^2$.

$$a = \frac{A_s f_y}{0.85 f'_c b} = 6.45 \text{ in.}$$

$a/d = 0.30$ is less than $a_{tc\ell}/d_t = 0.319$ (Table A–4); therefore, $\phi = 0.9$.

Finally, it is good practice to compute ϕM_n to check if the A_s is adequate. This guards against errors in solving Eq. 4–34.

$$\phi M_n = \frac{0.9 \big[6.58 \times 60(21.5 - 6.45/2) \big]}{12} = 541 \text{ ft-kips}$$

Since $\phi M_n = M_u$, A_s is OK. Therefore, use 5 No. 9 and 2 No. 8 bars. ∎

PROBLEMS

4–1 Figure P4–1 shows a simply supported beam and the cross section at midspan. The beam supports a uniform service (unfactored) dead load consisting of its own weight plus 1.4 kips/ft and a uniform service (unfactored) live load of 1.5 kips/ft. the concrete strength is 3000 psi and the yield strength of the reinforcement is 60,000 psi. The concrete is normal-weight concrete.

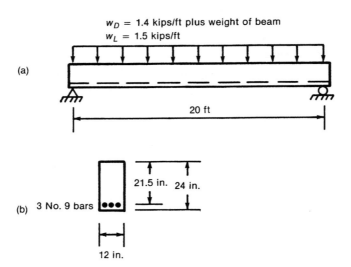

Fig. P4–1

(a) Compute the weight/ft of the beam, the factored load per foot, w_u, and the moment due to the factored loads, M_u, and sketch the bending moment diagram.

(b) Compute ϕM_n for the cross section shown. Is the beam safe?

(c) Draw the cross section at midspan showing

  (1) the location of the compression zone.
  (2) the dimensions of b, d, h, a.

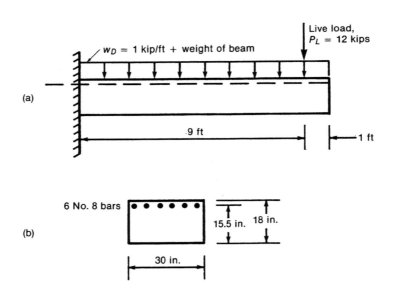

Fig. P4–2.

4–2 Repeat Problem 4–1 for the cantilever beam shown in Fig. P4–2. The beam supports a uniform service (unfactored) dead load of 1 kip/ft plus its own dead load and it supports a concentrated service (unfactored) live load of 12 kips as shown. The concrete is normal-weight concrete with

$f_c' = 4000$ psi and the steel is Grade 60. Draw the section at the support.

4–3 Assuming that the maximum concrete compressive strain is 0.003, compute the steel strain corresponding to the moment M_n for the beam shown in Fig. P4–1. Is $f_s = f_y$ in this beam?

4–4 (a) Compare ϕM_n for singly reinforced rectangular beams having the following properties:

Beam No.	b (in.)	d (in.)	Bars	f_c' (psi)	f_y (psi)
1	12	22	3 No. 7	3,000	60,000
2	12	22	2 No. 9 plus 1 No. 8	3,000	60,000
3	12	22	3 No. 7	3,000	40,000
4	12	22	3 No. 7	4,500	60,000
5	12	33	3 No. 7	3,000	60,000

(b) Taking beam 1 as the reference point, discuss the effects of changing A_s, f_y, f_c', and d on ϕM_n. (Note that each beam has the same properties as beam 1 except for the italicized quantity.)

(c) What is the most effective way of increasing ϕM_n? What is the least effective way?

4–5 For each of the beams shown in Fig. P4–5 and without doing any calculations:

(a) Draw the deflected shape.

(b) Sketch

(1) the bending moment diagram due to the weight of the beam.

(2) the bending moment diagram for the other loads shown.

(3) the sum of the two diagrams.

(c) Show on an elevation view of the beam the location of flexural reinforcement for the final moment diagram from part (b).

4–6 A 16.-ft-span simply supported beam has a rectangular cross section with $b = 14$ in., $d = 19.5$ in., and $h = 22$ in. The beam is made from normal-weight 3500-psi concrete and has 6 No. 6 Grade 40 bars. This beam supports its own dead load plus a uniform service (unfactored) additional dead load of 1.0 kip/ft. Compute the maximum uniform service live load that the beam can support.

4–7 A 12-ft-long cantilever supports its own dead load plus an additional uniform service (unfactored) dead load of 0.5 kip/ft. The beam is made from normal-weight 4000-psi concrete and has $b = 16$ in., $d = 15.5$ in., and $h = 18$ in. It is reinforced with 4 No. 7 grade 60 bars. Compute the maximum service (unfactored) concentrated live load that can be applied at 1 ft from the free end of the cantilever.

4–8 Explain why a rectangular stress block with a maximum stress of $0.85f_c'$ and a depth $a = \beta_1 c$ is used in design.

4–9 Explain the meaning of "over-reinforced beam" and "under-reinforced beam."

4–10 Either explain why ρ is limited to $0.75\rho_b$ (why a is limited to $0.75a_b$) or, explain why a/d_t is limited to a_{tcl}/d_t.

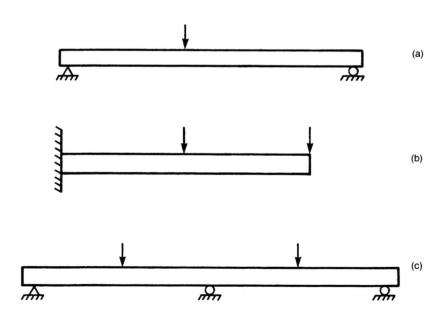

(a)

(b)

(c)

Fig. P4–5

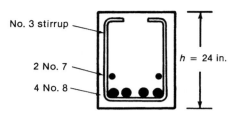

Fig. P4–11

4–11(a) Compute the effective depth, d, and the minimum allowable web width, b_w, of the beam shown in Fig. P4–11. Use the minimum bar spacings allowed by the ACI Code for concrete with 3/4–in. coarse aggregate. Select the cover for concrete not exposed to weather or in contact with the ground.

(b) Compare the computed depth to that given by the approximate Eqs. 4–30.

(c) Check the minimum web width using Table A–6.

4–12 Give three reasons for the minimum cover requirements in the ACI Code. Under what circumstances are greater covers used?

4–13 Give two reasons for the minimum bar spacing requirements in the ACI Code.

4–14 A rectangular beam has $b = 12$ in., $h = 20$ in., and 8 No. 8 bars in two layers of four bars. $f'_c = 3750$ psi and $f_y = 60000$ psi. Compute ϕM_n.

4–15 Select reinforcement for a 20-ft-span rectangular beam with $b = 16$ in. and $h = 21$ in. The beam supports its own weight plus a superimposed service (unfactored) uniform dead load of 0.6 kip/ft and a uniform service live load of 2 kips/ft. Use $f'_c = 3750$ psi and $f_y = 60,000$ psi.

4–16 Select b, d, h, and the reinforcement for a 24–ft-span simply supported rectangular beam that supports its own dead load plus an additional service dead load of 1.5 kips/ft plus a service live load which consists of two concentrated loads of 10 kips, each located at the third points of the span. Use $f'_c = 3000$ psi and $f_y = 60,000$ psi.

4–17 Select b, d, h, and the reinforcement for a 22-ft-span simply supported rectangular beam which supports its own dead load plus a superimposed service dead load of 1.25 kips/ft plus a uniform service load of 2 kips/ft. Use $f'_c = 3000$ psi and $f_y = 60,000$ psi.

4–18 The beam shown in Fig. P4–18 carries its own dead load plus an additional uniform service dead load

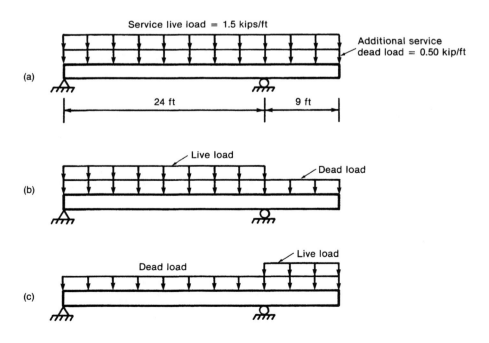

Fig. P4–18

of 0.5 kip/ft and a uniform service live load of 1.5 kips/ft. The dead load acts on the entire beam, of course, but the live load can act on parts of the span. Three possible loading cases are shown in Fig. P4–18.

(a) Draw factored bending moment diagrams for the three loading cases shown and superimpose them to draw a bending moment envelope.

(b) Design the beam, selecting b, d, h, and the reinforcing bars. Use $f_c' = 3750$ psi and $f_y = 60,000$ psi.

(c) Draw an elevation of the beam showing the reinforcement. Estimate the lengths of the top and bottom bars from the bending moment envelope.

(d) Draw cross sections at the points of maximum positive and negative moment.

4–19, 4–20, and 4–21 Use design tables to design the beams described in Problems 4–16, 4–17, and 4–18.

4–22 Write a computer or calculator program to solve for the area of steel, A_s, directly using Eq. 4–34.

4–23 Select the largest possible b and d and the corresponding A_s (based on ACI Sec. 10.5) and the smallest allowable b and d and the corresponding A_s (based on ACI Sec. 10.3.3) to give a resisting moment of $\phi M_n = 250$ ft-kips. In both cases select a section with $b = 0.5d$. Use $f_c' = 3000$ psi and $f_y = 60,000$ psi.

5
Flexure: T Beams, Beams with Compression Reinforcement, and Special Cases

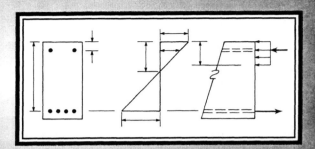

5-1 INTRODUCTION

In Chap. 4 the theory of flexure for reinforced concrete was developed and applied to rectangular beams with flexural reinforcement in the tension zone. Frequently, concrete beams are T or I shaped and sometimes they have reinforcement in both the tension and compression zones. In this chapter the theory of flexure is extended to cover these types of problems.

Beams whose cross sections are not symmetrical about the loading axis, or beams bent about two axes require special treatment because the axis of zero strain (neutral axis) generally is not parallel to the axis about which the resultant moment acts. The analysis of such beams is discussed in Sec. 5–4.

When beams have tension reinforcement in several layers spread over the depth of the beam, or are built of two types of concrete, or contain reinforcement that is not elastic–plastic, strain compatibility must be considered in the calculations. This is discussed in Sec. 5–5.

5-2 T BEAMS

Practical Applications of T Beams

In the floor system shown in Fig. 5–1, the slab is assumed to carry the loads in one direction to beams that carry them in the perpendicular direction. During construction, the concrete in the columns is placed and allowed to harden before the concrete in the floor is placed (ACI Sec. 6.4.5). In the next operation, concrete is placed in the slab and beams in a monolithic pour (ACI Sec. 6.4.6). As a result, the slab serves as the top flange of the beams as indicated by the shading in Fig. 5–1. Such a beam is referred to as a *T beam*. The interior beam, *AB*, has a flange on both sides. The *spandrel beam*, *CD*, with a flange on one side only, is also referred to as a T beam.

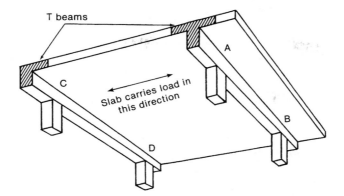

Fig. 5–1
T beams in a one-way beam
and slab floor.

An exaggerated deflected view of the interior beam is shown in Fig. 5–2. This beam develops positive moments at midspan (section *A–A*) and negative moments over the supports (section *B–B*). A photograph of the region between A–A and the left end of such a beam is shown in Fig. 5–3. At midspan the compression zone is in the flange as shown in Fig. 5–2b and d. Generally, it is rectangular as shown in b, although in a few cases, the neutral axis may shift down into the web, giving a T-shaped compression zone as shown in Fig. 5–2d. At the support, the compression zone is at the bottom of the beam and is rectangular, as shown in Fig. 5–2c. For computational purposes, these beams will be classed as a "rectangular beam" if the compression zone is rectangular (Fig. 5–2b and c), and as a "T beam" if T-shaped (Fig. 5–2d).

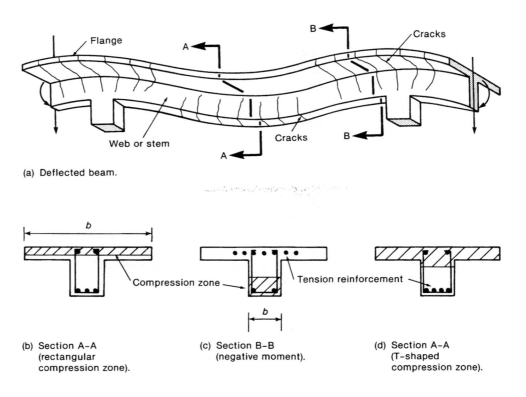

Fig. 5–2
Positive and negative mo-
ment regions in a T beam.

(a) Deflected beam.

(b) Section A–A
(rectangular
compression zone).

(c) Section B–B
(negative moment).

(d) Section A–A
(T-shaped
compression zone).

5–2   T Beams

139

Frequently, a beam-and-slab floor involves slabs supported by beams which, in turn, are supported by other beams referred to as *girders* (Fig. 5–4). Again, all the concrete above the top of the column is placed at one time. Note that the slab acts as a flange for both the beams and girders.

Effective Flange Width and Transverse Reinforcement

The forces acting on the flange of a simply supported T beam are illustrated in Fig. 5–5. At the support there are no compressive stresses in the flange, while at midspan the full width is stressed in compression. The transition requires horizontal shear stresses on the web–flange interface. As a result there is a "shear-lag" effect and the portions of the flange closest to the web are more highly stressed than those portions farther away, as shown in Figs. 5–5 and 5–6.

Figure 5–6 shows the distribution of the flexural compressive stresses in a slab which forms the flanges of a series of parallel beams at a point of maximum positive moment. The compressive stress is a maximum over the web, dropping between the webs. Toward the supports, the variation from maximum to minimum becomes more pronounced.

When proportioning the section for positive moments, an "effective width" is used (Fig. 5–6b). This is the width, b, which when stressed uniformly to $f_{c(max)}$, gives the same compression force as is actually developed in the real compression zone of width b_0.

A number of elastic solutions have been used to estimate the effective flange width.[5-2, 5-3] These solutions suggest that this width is affected by the type of loading (uniform, concentrated), the type of supports, the spacing of the beams, and the relative stiffness of the slabs and beams. However, it must be noted that all such studies have ignored the cracking of the flange observed in tests.

ACI Sec. 8.10 presents rules for estimating this width for design purposes. For an interior beam, ACI Sec. 8.10.2 states that

1. The width of slab effective as a T-beam flange shall not exceed one-fourth the span length of the beam.

2. The effective overhanging slab width on each side of the web shall not exceed the smaller of eight times the slab thickness, or one-half the clear distance to the next beam web.

ACI Secs. 8.10.3 and 8.10.4 give considerably more stringent rules for beams with slabs on one side only and for isolated T beams. In general, the Code rules are a conservative approximation to the elastic solutions for effective width.

The spread of the compression force across the width of the left-hand proportion of the flange in Fig. 5–5 can be idealized by a truss mechanism within the plane of the flange consisting of a series of compression struts, shown by dashed lines in Fig. 5–5b, and transverse tension ties shown by solid lines. The resultant compression force in the overhanging flange at midspan is represented by the force at K. The horizontal shear force applied to the flange at A is transferred out into the flange by the compression strut $A–B$. At B the longitudinal force in this strut is transferred to the strut $B–E$. The transverse component of the force in the strut must be resisted by the transverse tension tie $B–C$, and so on. The ACI Code does not give rules for the design of this transverse steel in the flanges. This is discussed more fully in Sec. 18–9 for both compression and tension flanges. Significant amounts of transverse steel may be needed in tension flanges. The arrangement of this steel may control the effectiveness of the longitudinal tension reinforcement placed in the flanges.

Loads applied to the flange will cause negative moments in the flange where it joins the web. If the slab is continuous and spans perpendicular to the beam as in Fig. 5–1 or perpendicular to the "beams" in Fig. 5–4, the slab reinforcement will be adequate to resist

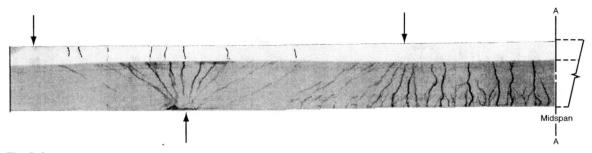

Fig. 5–3
Photograph of a test specimen representing half of the beam shown in Fig. 5–2a. (Photograph courtesy of J. G. MacGregor.)

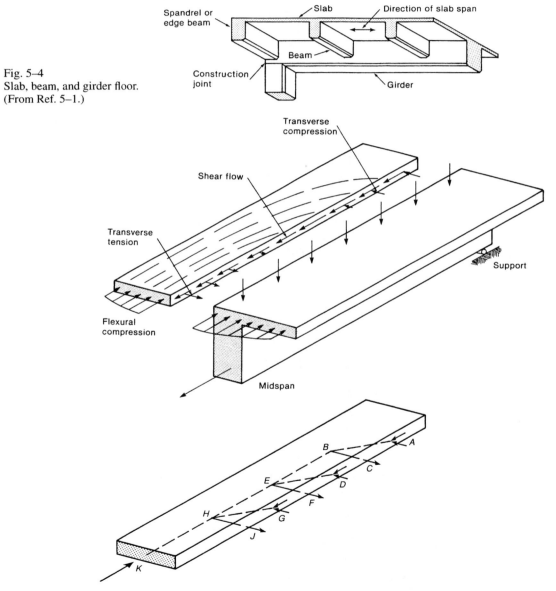

Fig. 5–4
Slab, beam, and girder floor.
(From Ref. 5–1.)

Fig. 5–5
Forces on a T-beam flange.

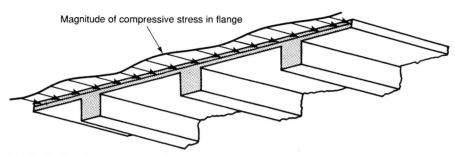

(a) Distribution of maximum flexural compressive stresses

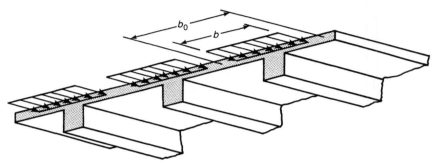

Fig. 5–6
Effective width of T beams.

(b) Flexural compressive stress distribution assumed in design

these moments. If, however, the slab is not continuous (in an isolated T beam) or if the slab reinforcement is parallel to the beam, as is the case of the "girders" in Fig. 5–4, additional reinforcement is required at the top of the slab, perpendicular to the beam (ACI Sec. 8.10.5). This reinforcement is designed assuming that the flange acts as a cantilever loaded with the factored dead and live loads. For an isolated T beam the full overhanging flange width is considered. For a girder in a monolithic floor system (Fig. 5–4), the overhanging part of the effective width is used in this calculation (see Sec. 10–6).

Analysis of T Beams

Generally, the compression zone of a T beam is rectangular, as shown in Fig. 5–2b or c. These may be analyzed as "rectangular beams" with a width, b, as shown. In the unusual case in which the compression zone is T shaped, as shown in Fig. 5–2d, the analysis separated considers the resistance provided by the overhanging flanges and that provided by the remaining rectangular beam.

Consider the beam shown in Fig. 5–7a, with the depth of the stress block, a, greater than the flange thickness, h_f. The internal forces in this beam consist of a compressive force C at the centroid of the compression zone (centroid of shaded area in Fig. 5–7a) and a tensile force $T = A_s f_y$, assuming that the steel yields. These form a resisting moment, $M_n = Cjd$ or $M_n = Tjd$.

To avoid the need for locating the centroid of the shaded area (where a is not yet known), it is convenient to consider two hypothetical beams:

1. *Beam F* (Fig. 5–7c), with a compression zone consisting of the overhanging flanges, area A_f, stressed to $0.85 f_c'$, giving a compressive force, C_f, which acts at the centroid of the area of the overhanging flanges. For equilibrium, beam F has a tensile steel area A_{sf} chosen such that $C_f = T_f$ or $A_{sf} f_y = C_f$. This area of steel, A_{sf}, is a portion of the total A_s and is assumed to have the same centroid as A_s. The moment capacity of this beam, M_{nf}, is the moment of C_f about the tension steel.

Flexure: T Beams, Beams with Compression Reinforcement, and Special Cases

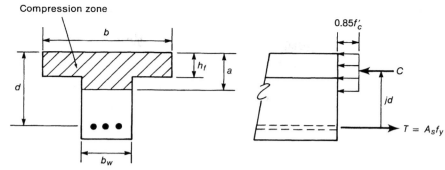

(a) Cross section.

(b) Internal forces.

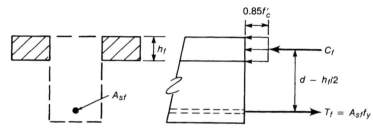

(c) Beam F.

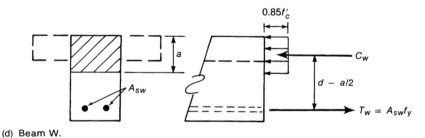

Fig. 5–7
Subdivision of a T beam for
analysis.

(d) Beam W.

2. *Beam W* (Fig. 5–7d), which is a rectangular beam of width b_w, having a compression zone of area $b_w a$ and utilizing the remaining tensile steel $(A_s - A_{sf}) = A_{sw}$. The compressive force in this beam, C_w, acts through the centroid of its compression area. The moment capacity of this beam, M_{nw}, is the moment of C_w about the tensile steel.

The total moment capacity of the T beam is the sum of the moment capacities of the two individual beams, $M_n = M_{nf} + M_{nw}$ calculated as follows:

Beam F

$$\text{Area of compression zone} = (b - b_w)h_f$$

$$\text{Force in compression zone } C_f = (0.85f_c')(b - b_w)h_f$$

To compute the area of reinforcement required in beam F, set $T_f = C_f$, assuming that $f_s = f_y$:

$$A_{sf}f_y = 0.85f_c'(b - b_w)h_f \tag{5–1a}$$

or

5–2 T Beams

143

$$A_{sf} = \frac{0.85f'_c\,(b - b_w)h_f}{f_y} \tag{5–1b}$$

The lever arm is $(d - h_f/2)$.

Summing the moments about the centroid of the tension reinforcement gives

$$M_{nf} = 0.85f'_c\,(b - b_w)h_f\left(d - \frac{h_f}{2}\right) \tag{5–2a}$$

Alternatively, summing the moments about the line of action of C_f gives

$$M_{nf} = A_{sf}f_y\left(d - \frac{h_f}{2}\right) \tag{5–2b}$$

Beam W

Area of tension steel $A_{sw} = A_s - A_{sf}$

Compression force $C_w = 0.85f'_c\,b_w a$

or

$$a = \frac{A_{sw}f_y}{0.85f'_c\,b_w} \tag{5–3}$$

The lever arm is $d - a/2$.

$$M_{nw} = 0.85f'_c\,b_w a\left(d - \frac{a}{2}\right) \tag{5–4a}$$

or

$$M_{nw} = A_{sw}f_y\left(d - \frac{a}{2}\right) \tag{5–4b}$$

T Beam = Beam F + Beam W. The nominal moment capacity of the T beam is the sum of the nominal moment capacities of beam F and beam W:

$$M_n = M_{nf} + M_{nw}$$

giving

$$M_n = \left[0.85f'_c\,(b - b_w)h_f\left(d - \frac{h_f}{2}\right)\right] + \left[0.85f'_c\,b_w a\left(d - \frac{a}{2}\right)\right] \tag{5–5a}$$

or

$$M_n = \left[A_{sf}f_y\left(d - \frac{h_f}{2}\right)\right] + \left[A_{sw}f_y\left(d - \frac{a}{2}\right)\right] \tag{5–5b}$$

Finally, the factored moment capacity is ϕM_n.

Occasionally, a will turn out to be equal to h_f. This may be considered to be a "rectangular beam" for calculation purposes.

Determination of Whether $f_s = f_y$

In the derivation of Eqs. 5–5a and b, it was assumed that $f_s = f_y$. As discussed in Sec. 4–3, this can be checked by comparing the computed c/d or a/d ratios to c_b/d or a_b/d, given by

$$\frac{c_b}{d} = \frac{87,000}{87,000 + f_y} \tag{4–20}$$

and

$$\frac{a_b}{d} = \beta_1\left(\frac{87,000}{87,000 + f_y}\right) \tag{4-21}$$

where f_y is in psi. If f_y is in MPa, 87,000 becomes 600 in both equations. If the computed c/d or a/d ratios are less than those given by Eqs. 4–20 and 4–21, $f_s = f_y$ at failure.

Alternative Determination of Whether $f_s = f_y$

The steel stress f_s will equal f_y if a tension failure or a balanced failure occurs, that is, if $\rho \le \rho_b$. The value of ρ_b corresponds to the strain distribution with $\epsilon_{cu} = 0.003$ at the extreme compression fiber and ϵ_y at the tension steel. For a given f_y, this corresponds to a particular value of the ratio of the depth of the compression zone to the effective depth, a/d. The ratio of a_b/d corresponding to ρ_b is given by Eq. 4–20. This equation is independent of section shape and hence applies to T beams. It should be noted that Eq. 4–19 for ρ_b was derived for a rectangular section and does not apply to a T section.

Upper Limit on Reinforcement in T Beams—ACI Appendix B

No upper limit on the amount of tension reinforcement in beams is given in ACI Appendix B. Instead, the value of ϕ is set at 0.90 for tension-controlled sections and 0.70 for compression-controlled sections without spiral reinforcement in the compression zone. (Spiral reinforcement is described in Sec. 11–2. It is not normally used in beams.) In the transition zone between tension-controlled sections and compression-controlled sections, ϕ varies from 0.90 to 0.70. The resulting penalty in strength due to ϕ being less than 0.90, in effect, limits the practical range of reinforcement to the tension-controlled range. Tension- and compression-controlled sections are defined in ACI Sec. B10.3.3 and Sec. 4–3 of this book.

A *tension-controlled failure* will occur if the ratio, c/d_t, of the neutral axis depth, c, to the depth from the extreme compression fiber to the farthest layer of tension reinforcement, d_t, is less than or equal to the *tension-controlled limit* given by

$$\frac{c_{tc\ell}}{d_t} = 0.375$$

This can also be expressed in terms of the ratio of the depth, a, of the rectangular stress block to d_t at the tension-controlled limit

$$\frac{a_{tc\ell}}{d_t} = 0.375\beta_1 \tag{4-23}$$

A T beam with the flange in compression is almost always a tension-controlled section.

A *compression-controlled failure* with $f_s = f_y$ will occur if the ratio a/d_t is greater than or equal to that at the *compression-controlled limit*

$$\frac{a_{cc\ell}}{d_t} = \beta_1\left(\frac{87,000}{87,000 + f_y}\right) \tag{4-22}$$

ACI Sec. B9.3.2 sets $\phi = 0.70$ for compression-controlled sections without spiral reinforcement in the compression zone.

A *transition failure* will occur if the ratio a/d_t at ultimate is between the compression-controlled limit and the tension-controlled limit. Thus, if a/d_t is between $a_{cc\ell}/d_t$ and $a_{tc\ell}/d_t$ at failure, ϕ is between 0.70 and 0.90 as given by

$$\phi = 0.356 + \frac{0.204}{a/(\beta_1 d_t)} \tag{4-26b}$$

Upper Limit on Reinforcement in T Beams—ACI Sec. 10.3.3

To ensure ductile behavior, ACI Sec. 10.3.3 requires that $\rho \leq 0.75\rho_b$. This can be checked in one of three ways:

1. If the compression zone is rectangular as shown in Fig. 5–2b, $f_s = f_y$ if $a/d < a_b/d$ and $\rho \leq 0.75\rho_b$ if $a/d \leq 0.75a_b/d$. Values of a_b/d and $0.75a_b/d$ are given by Eq. 4–20 or Table A–5. If the compression zone is T-shaped, this calculation cannot be used and it is necessary to make these checks according to method 2 or 3.

2. The area of tension steel corresponding to a balanced failure is

$$A_{sb} = \frac{C_b}{f_y} \tag{5-6}$$

where C_b is the compression force resulting from a compressive stress block with a depth a_b, where a_b is from Eq. 4–21 or Table A–5. The maximum area of steel permitted by ACI Sec. 10.3.3 is then $0.75A_{sb}$.

3. Alternatively, the check of $f_s = f_y$ and $\rho \leq 0.75\rho_b$ can be based on a modified ρ_b for T beams given by

$$\rho_{bT} = \frac{b_w}{b}(\bar{\rho}_b + \rho_f) \tag{5-7}$$

where $\bar{\rho}_b$ is the balanced steel ratio for a rectangular beam given by Eq. 4–25, based on $b = b_w$ and $\rho_f = A_{sf}/(b_w d)$, where A_{sf} is given by Eq. 5–1b. Of these, method 2 is preferred for T-shaped compression zones.

Minimum Reinforcement

It is necessary to have sufficient tension reinforcement so that the moment capacity after cracking exceeds the cracking moment. For a T beam with its flange in compression and for negative moment regions of continuous T beams where the flange is in tension, this is done according to ACI Sec. 10.5.1, by checking whether $A_s = A_{sf} + A_{sw}$ exceeds $A_{s,min}$ given by

$$A_{s,min} = \frac{3\sqrt{f'_c}}{f_y}b_w d \geq \frac{200b_w d}{f_y} \tag{4-31}$$

(ACI Eq. 10–3)

where f'_c and f_y are in psi or

$$A_{s,min} = \frac{\sqrt{f'_c}}{4f_y}b_w d \geq \frac{1.4b_w d}{f_y} \tag{4-31M}$$

where f'_c and f_y are in MPa.
   When a statically determinate T beam, such as a cantilever with a T-shaped cross section, has its flange in tension, ACI Sec. 10.5.2 gives $A_{s,min}$ as

$$A_{s,min} = \frac{6\sqrt{f'_c}}{f_y}b_w d \tag{4-32}$$

(ACI Eq. 10–4)

where f'_c are f_y are in psi or

$$A_{s,min} = \frac{\sqrt{f'_c}}{2f_y}b_w d \tag{4-32M}$$

where f'_c and f_y are in MPa. The amount of minimum reinforcing in a statically determinate T beam with the flange in tension need not exceed the amount given by Eqs. 4–31 and 4–32M, with b_w set equal to the width of the flange.

Flexure: T Beams, Beams with Compression Reinforcement, and Special Cases

Alternatively, ACI Sec. 10.5.3 allows Eqs. 4–31 and 4–32 to be waived at sections where the reinforcement provided is at least 1.33 times that required by analysis.

EXAMPLE 5–1 Analysis of the Positive Moment Capacity of a T Beam: *a* Less Than h_f

An interior T beam in floor system has a clear span, from face to face of the columns, of 18 ft and the cross section shown in Fig. 5–8. The concrete and steel strengths are 3000 psi and 40,000 psi, respectively. Compute the design moment capacity of this beam in the positive moment region.

1. **Compute the effective width of the flange, *b*.** From ACI Sec. 8.10.2,

(a) *b* shall not exceed one-fourth of the span length $= 18/4 = 4.5$ ft $= 54$ in.

(b) The effective overhanging slab width on each side of the web shall not exceed eight times the slab thickness $= 8 \times 5$in. $= 40$in., which gives $b = 40 + 12 + 40 = 92$ in., or

(c) half the clear distance to the next beam web on each side. This gives $b = 108/2 + 12 + 130/2 = 131$ in.

The smallest of these is *b*. Therefore, $b = 54$ in.

2. **Compute *d*.** For two layers of steel

$$d \simeq h - 3.5\text{in.} = 16.5\text{in} \qquad (4\text{--}30b)$$

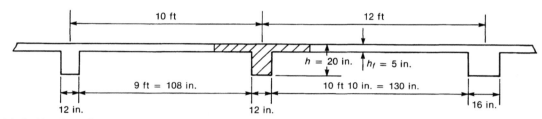

(a) Section through beams and slab.

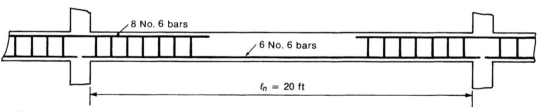

(b) Elevation of beam.

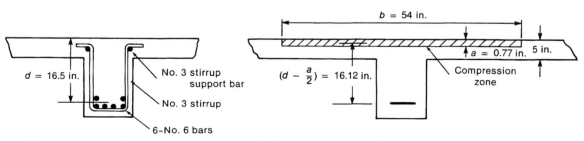

(c) Reinforcement at midspan.

(d) Compression zone and lever arm.

Fig. 5–8
Beam—Examples 5–1 and 5–2.

3. **Compute a.** Since almost all T beams will have the depth of compression stress block, a, less than the flange thickness, h_f, we shall assume "rectangular beam action" as shown in Fig. 5–2b. If the resulting value of a is less than or equal to the flange thickness, h_f, we will continue. If $a > h_f$, it is necessary to analyze the beam for a "T-beam action" as done in Example 5–3.

$$a = \frac{A_s f_y}{0.85 f_c' b}$$

$$= \frac{2.64 \times 40,000}{0.85 \times 3000 \times 54} = 0.767 \text{ in.}$$

Since this is less than $h_f = 5$ in. as shown in Fig. 5–8d, the beam has "rectangular beam action" and ϕM_n can be computed using the basic procedure for rectangular beams (Eqs. 4–12).

4. **Check if $A_s \geq A_{s,\min}$.** Since the flange is in compression, $A_{s,\min}$ is given by

$$A_{s,\min} = \frac{3\sqrt{f_c'}}{f_y} b_w d \geq \frac{200 b_w d}{f_y} \tag{4–31}$$

$$= \frac{3\sqrt{3000}}{40,000} 12 \times 16.5 \geq \frac{200 \times 12 \times 16.5}{40,000}$$

$$= 0.813 \geq 0.990 \text{ in.}^2$$

Since $A_s = 2.64$ in.^{2} exceeds 0.99 in.^{2}, $A_s > A_{s,\min}$.

5. **Check if $f_s = f_y$ and whether the section is tension-controlled.**

$$\frac{a}{d} = \frac{0.767}{16.5} = 0.0432$$

From Eq. 4–21,

$$\frac{a_b}{d} = \beta_1 \left(\frac{87,000}{87,000 + f_y} \right) = 0.582$$

Since 0.0432 is less than 0.582, $f_s = f_y$ at ultimate.

$$d_t = 20 - (1.5 + 0.375 + 0.72/2) = 17.75 \text{ in.}$$

$$\frac{a}{d_t} = \frac{0.767}{17.75} = 0.043$$

This is much less than $a_{tc\ell}/d_t$ at the tension-controlled limit:

$$\frac{a_{tc\ell}}{dt} = 0.375 \beta_1 = 0.319 \tag{4–23}$$

Thus the section is tension-controlled. As a result, $\phi = 0.90$.

Design has been carried out using ACI Appendix B. Alternatively, design could be carried out by ACI Sec. 10.3.3, which limits $\rho < 0.75 \rho_b$. Since $a < h_f$, the compression zone is rectangular and this check can be made by checking if a/d exceeds the value of a/d for $0.75\rho_b$, as was done for rectangular beams.

$$\frac{a}{d} = \frac{0.767}{16.5} = 0.047$$

From Eq. 4–21 (or Table A–5),

$$\frac{a_b}{d} = \beta_1 \left(\frac{87,000}{87,000 + f_y} \right) = 0.582$$

The value of a/d for $0.75\rho_b$ is $0.75 \times 0.582 = 0.437$. Since $0.0470 < 0.437$, $\rho < 0.75\rho_b$—therefore OK.

If $a > h_f$ a different procedure would be needed to check that $\rho \leq 0.75\rho_b$. This is illustrated in Example 5–3. Design would either be carried out using ACI Appendix B or by ACI Sec. 10.3.3. Both have been illustrated for completeness.

Flexure: T Beams, Beams with Compression Reinforcement, and Special Cases

The limits in steps 4 and 5 are very seldom reached in the case of a T beam with the flange in compression. In the event they were, however, it would be necessary to revise the cross section.

6. **Compute ϕM_n.**

$$\phi M_n = \phi\left[A_s f_y\left(d - \frac{a}{2}\right)\right]$$

$$= \frac{0.9\left[2.64 \times 40{,}000(16.5-0.767/2)\right]}{12{,}000}$$

$$= 128 \text{ ft-kips}$$

The positive design moment capacity of the beam shown in Fig. 5–8 is 128 ft-kips. ∎

EXAMPLE 5–2 Analysis of the Negative Moment Capacity of a T Beam

Compute the design negative moment capacity of the T beam shown in Fig. 5–8. The arrangement of the negative moment reinforcement is shown in Fig. 5–9. The concrete and steel strengths are 3000 psi and 40,000 psi, as before.

Because this section is subjected to a negative moment, cracking develops in the top flange (Fig. 5–2) and the compressive zone is at the *bottom* of the beam, as shown in Figs. 5–2c and 5–9. Note that ACI Sec. 10.6.6 requires that a portion of the tension reinforcement be in the flange, allowing all of this reinforcement to be placed in one layer. Two bars are shown in the lower corners of the stirrups. These are there because it is customary to extend some of the positive moment reinforcement into the support. Unless these bars are adequately anchored to develop compressive stresses, which would not be true if the bottom bars were interrupted at the supports as shown in Fig. 5–8b, they should not be included in the calculations.

1. **Compute b.** Since the compression zone is at the bottom of the beam, $b = 12$ in.

2. **Compute d.** Since the tension reinforcement is in one layer,

$$d \simeq h\text{–}2.5\text{in.} = 17.5 \text{ in.} \tag{4–30a}$$

3. **Compute a.**

$$a = \frac{A_s f_y}{0.85 f_c' b} \tag{4–11}$$

$$= \frac{3.52 \times 40{,}000}{0.85 \times 3000 \times 12} = 4.60 \text{ in.}$$

4. **Check if $A_s \geq A_{s,\text{min}}$.** ACI Sec. 10.5.2 applies to statically determinate T beams with the flange in tension. ACI Sec. 10.5.1 applies to "every section of a flexural member where tensile reinforcement is required by analysis except as provided in 10.5.2." Since the beam is continuous, ACI Sec. 10.5.1 applies and

$$A_{s,\text{min}} = \frac{3\sqrt{3000}}{40{,}000}12 \times 17.5 \geq \frac{200 \times 12 \times 17.5}{40{,}000}$$

$$= 0.863 \text{ in.}^2 \geq 1.05 \text{ in.}^2$$

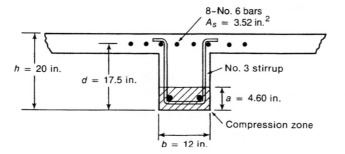

Fig. 5–9
Negative moment section of beam shown in Fig. 5–8—Example 5–2.

Since 3.52 in.^{2} exceeds 1.05 in.^{2}, $A_s \geq A_{s,min}$—therefore OK.

5. **Check if $f_s = f_y$ and if the section is tension-controlled.**

$$\frac{a}{d} = \frac{4.60}{17.5} = 0.263 \qquad (4\text{–}21)$$

$$\frac{a_b}{d} = 0.85 \left(\frac{87,000}{87,000 + 40,000} \right) = 0.582$$

Since a/d is less than a_b/d, $f_s = f_y$.

$$d_t = 17.5 \text{ in.} \qquad \frac{a}{d_t} = 0.263$$

$$\frac{a_{tc\ell}}{d_t} = 0.375\beta_1 = 0.319 \qquad (4\text{–}23)$$

Since 0.263 is less than 0.319, the section is tension-controlled and $\phi = 0.90$.

6. **Compute ϕM_n.**

$$\phi M_n = \phi \left[A_s f_y \left(d - \frac{a}{2} \right) \right]$$

$$= \frac{0.9 \left[3.52 \times 40,000(17.5\text{–}4.60/2) \right]}{12,000} = 161 \text{ ft-kips}$$

Therefore, the design negative moment is $\phi M_n = 161$ ft-kips. ∎

EXAMPLE 5–3 Analysis of a T Beam with the Neutral Axis in the Web

Compute the positive design moment capacity of the beam shown in Fig. 5–10. The concrete and steel strengths are 3000 psi and 60,000 psi, respectively. Although not shown in Fig. 5–10, the beam contains No. 3 stirrups.

1. **Compute b.** This beam is an isolated T beam in which a T-shaped flange is used to increase the area of the compression zone. For such a beam, ACI Sec. 8.10.4 states that the flange thickness shall not be less than one-half the width of web and the effective flange width shall not exceed four times the width of web. By observation, the flange dimensions satisfy this. Thus $b = 18$ in.

2. **Compute d.** $d = 24.5$ in. as shown in Fig. 5–10a.

3. **Compute a.** Assume that the compression zone will be rectangular. Accordingly,

$$a = \frac{A_s f_y}{0.85 f_c' b}$$

$$= \frac{4.74 \times 60,000}{0.85 \times 3000 \times 18} = 6.20 \text{ in.}$$

As shown in Fig. 5–10b, $a = 6.20$ in. is greater than the thickness of the flange, $h_f = 5$ in. This implies that compressive stresses exist in the shaded regions below the flanges. This cannot occur. Because $a > h_f$, our assumption that the compression zone is rectangular is wrong, and our calculated value of a is incorrect. It is therefore necessary to analyze this beam as a "T beam."

4. **Divide the beam into "beam F" and "beam W."** Beam F consists of the overhanging portions of the flanges plus an area of steel, A_{sf}, such that $A_{sf}f_y$ balances the compression in the overhanging flanges (Fig. 5–10c).

(a) **Beam F.** The compression force, C_f, in the overhanging flanges is

$$C_f = (0.85f_c')(b - b_w)h_f$$

$$= (0.85 \times 3000)(18\text{–}10) \times 5 = 102,000 \text{ lb}$$

The area of steel in beam F is

$$A_{sf}f_y = C_f \qquad (5\text{–}1a)$$

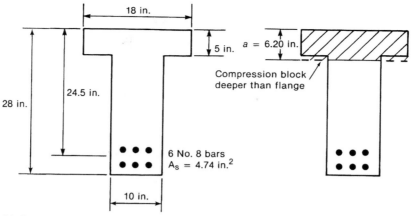

(a) Beam cross section.

(b) Compression zone—Step 3.

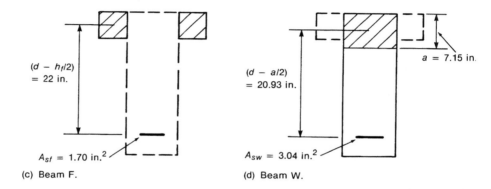

Fig. 5–10
Beam—Example 5–3.

(c) Beam F.

(d) Beam W.

$$A_{sf} = \frac{102{,}000}{60{,}000} = 1.70 \text{ in.}^2$$

Beam F is shown in Fig. 5–10c. Summing the moments about the centroid of the tensile steel gives us

$$M_{nf} = C_f\left(d - \frac{h_f}{2}\right) \tag{5–2a}$$

$$= \frac{102{,}000(24.5 - 5/2)}{12{,}000} = 187 \text{ ft-kips}$$

(b)   Beam W. Beam W consists of the concrete in the web plus the remainder of the tensile reinforcement, as shown in Fig. 5–10d.

$$A_{sw} = A_s - A_{sf}$$

$$= 4.74 - 1.70 = 3.04 \text{ in.}^2$$

For beam W, $b = b_w$ and

$$a = \frac{A_{sw}f_y}{0.85f_c'b_w} \tag{5–3}$$

$$= \frac{3.04 \times 60{,}000}{0.85 \times 3000 \times 10} = 7.15 \text{ in.}$$

Thus the depth of the compression stress block is actually 7.15 in. rather than the 6.20 in. computed in step 3 assuming "rectangular beam action." For beam W, the design moment capacity is

$$M_{nw} = A_{sw}f_y\left(d - \frac{a}{2}\right) \tag{5-4b}$$

$$= \frac{3.04 \times 60{,}000(24.5 - 7.15/2)}{12{,}000} = 318 \text{ ft-kips}$$

The total positive nominal moment capacity is thus

$$M_n = M_{nf} + M_{nw} = 187 + 318 \text{ ft-kips}$$

$$= 505 \text{ ft-kips}$$

5. **Check if $A_s \geq A_{s,min}$.** From Eq. 4–31,

$$A_{s,\,min} = \frac{3\sqrt{3000}}{60{,}000}10 \times 2.45 \geq \frac{200 \times 10 \times 24.5}{60{,}000}$$

$$= 0.671 \text{ in.}^2 \geq 0.817 \text{ in.}^2$$

Since $A_s = 4.74$ in.^{2} exceeds this, $A_s \geq A_{s,min}$—therefore OK.

6. **Check if $f_s = f_y$ and whether the section is tension-controlled.**

$$\frac{a}{d} = \frac{7.17 \text{ in.}}{24.5 \text{ in.}} = 0.293$$

$$\frac{a_b}{d} = 0.85\left(\frac{87{,}000}{87{,}000 + 60{,}000}\right) = 0.503 \tag{4-21}$$

Since 0.293 is less than 0.503, $f_s = f_y$.

$$d_t = 28 - (1.5 + 0.375 + 1.0/2) = 25.62 \text{ in.} \qquad \frac{a}{d_t} = \frac{7.17 \text{ in.}}{25.62 \text{ in.}} = 0.280$$

$$\frac{a_{tc\ell}}{d_t} = 0.85 \times 0.375 = 0.319 \tag{4-23}$$

Since 0.280 is less than 0.319, the section is tension-controlled and $\phi = 0.90$.

7. **Compute ϕM_n.**

$$\phi M_n = 0.90 \times 505 = 455 \text{ ft-kips} \qquad \blacksquare$$

EXAMPLE 5–3M   Analysis of a T Beam with the Neutral Axis in the Web— SI Units

Compute the positive design moment capacity of the beam shown in Fig. 5–11. The concrete and steel strengths are 20 MPa and 400 MPa, respectively. Although they are not shown in Fig. 5–11, the beam contains No. 10 stirrups.

1. **Compute b.** This beam is an isolated T beam in which a T-shaped flange is used to increase the area of the compression zone. For such a beam, ACI Sec. 8.10.4 states that the flange thickness shall not be less than one-half the width of web and the effective flange width shall not exceed four times the width of the web. By observation, the flange dimensions satisfy this. Thus $b = 500$ mm.

2. **Compute d.** $d = 610$ mm, as shown in Fig. 5–11a.

3. **Compute a.** Assume that the compression zone will be rectangular. Accordingly,

$$a = \frac{A_sf_y}{0.85f_c'b}$$

$$= \frac{3000 \times 400}{0.85 \times 20 \times 500} = 141 \text{ mm}$$

Flexure: T Beams, Beams with Compression Reinforcement, and Special Cases

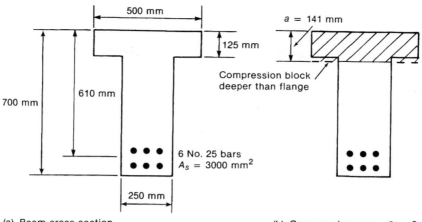

(a) Beam cross section.

(b) Compression zone—Step 3.

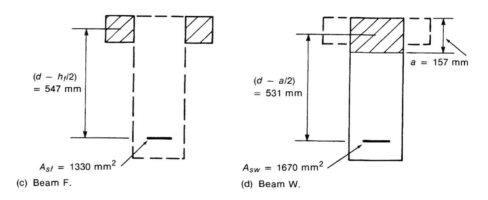

Fig. 5–11
Beam—Example 5–3M.

(c) Beam F.

(d) Beam W.

As shown in Fig. 5–11b, $a = 141$ mm is greater than the thickness of the flange. This implies that compression stresses exist in the shaded regions below the flanges. This cannot occur. Because $a > h_f$, our assumption that the compression zone is rectangular is wrong and our calculated value of a is incorrect. It is therefore necessary to analyze this beam as a "T beam."

4. Divide the beam into "beam F" and "beam W." Beam F is shown in Fig. 5–11c.

  (a) **Beam F.** The compression force, C_f, in overhanging flanges is

$$C_f = (0.85f_c')(b - b_w)h_f$$
$$= 0.85 \times 20(500-250)125$$
$$= 531,000 \text{ N}$$

The area of steel in beam F is

$$A_{sf}f_y = C_f \tag{5-1a}$$

and

$$A_{sf} = \frac{531,000 \text{ N}}{400 \text{ MPa}} = 1330 \text{ mm}^2$$

Thus beam F consists of the overhanging flanges plus tension reinforcement with an area of 1330 mm² as shown in Fig. 5–11c. To calculate the moment capacity of beam F, sum moments about the centroid of the tension steel:

5–2 T Beams

153

$$\text{lever arm} = d - \frac{h_f}{2}$$

$$= 610 - \frac{125}{2} = 547 \text{ mm}$$

$$M_{nf} = \frac{531{,}000 \times 547}{10^6} = 290 \text{ kN-m} \qquad (5\text{--}2a)$$

(b) **Beam W.** Beam W consists of the concrete in the web plus the remainder of the tension reinforcement, A_{sw}, as shown in Fig. 5–11d.

$$A_{sw} = 3000\text{--}1330 = 1670 \text{ mm}^2$$

For beam W, $b = b_w$ and

$$a = \frac{A_{sw}f_y}{0.85f_c'b_w} \qquad (5\text{--}3)$$

$$= \frac{1670 \times 400}{0.85 \times 20 \times 250} = 157 \text{ mm}$$

Thus the depth of the compression stress block is actually 157 mm rather than the 141 mm computed in step 3 assuming "rectangular beam action." For beam W the design moment capacity is

$$M_{nw} = A_{sw}f_y\left(d - \frac{a}{2}\right) \qquad (5\text{--}4b)$$

$$= \frac{1670 \times 400(610\text{--}157/2)}{106} = 355 \text{ kN-m}$$

The total positive nominal moment capacity is thus

$$M_n = M_{nf} + M_{nw} = 290 + 355$$

$$= 645 \text{ kN-m}$$

5. **Check if $A_s > A_{s,\,min}$.** From Eq. 4–31M,

$$A_{s,min} = \frac{\sqrt{f_c'}}{4f_y}b_w d \geq \frac{1.4b_w d}{f_y}$$

$$= \frac{\sqrt{20}}{4 \times 400}250 \times 610 \geq \frac{1.4 \times 250 \times 610}{400}$$

$$= 426 \text{ mm}^2 \geq 534 \text{ mm}^2$$

Since $A_s = 3000$ mm^2 exceeds this, $A_s > A_{s,min}$ — therefore, A_s is OK.

6. **Check if $f_s = f_y$ and whether the section is tension-controlled.**

$$\frac{a}{d} = \frac{157 \text{ mm}}{610 \text{ mm}} = 0.257$$

$$\frac{a_b}{d} = 0.85\left(\frac{600}{600 + 400}\right) = 0.510 \qquad (4\text{--}21M)$$

Since 0.257 is less than 0.510, $f_s = f_y$.

$$d_t = 700 - (40 + 10 + 25/2) = 637.5 \text{ mm} \qquad \frac{a}{d_t} = \frac{157 \text{ mm}}{637.5 \text{ mm}} = 0.246$$

$$\frac{a_{tc\ell}}{d_t} = 0.375\beta_1 = 0.319 \qquad (4\text{--}23)$$

Since 0.246 is less than 0.319, the section is tension-controlled and $\phi = 0.90$.

Flexure: T Beams, Beams with Compression Reinforcement, and Special Cases

7. **Compute ϕM_n.**

$$\phi M_n = 0.90 \times 645 = 581 \ \text{kN-m} \quad \blacksquare$$

Design of T Beams

The design of a T beam involves the choice of the cross section and the reinforcement required. The flange thickness and width are usually established during the design of the floor slab. The size of the beam stem is influenced by the same factors that affect the size of a rectangular beam. In the case of a continuous T beam, the concrete compressive stresses are most critical in the negative moment regions where the compression zone is in the beam stem. Frequently, the stem size is chosen so that $\rho \simeq 0.5\rho_b$ at the point of maximum negative moment.

Once the size of the cross section has been determined, it is possible to compute the area of reinforcement required using Eq. 4–33. This calculation is similar to Example 4–5. In the negative moment region assume that $j = 0.875$ as a first trial value as done in Example 4–5. The large flange width in the positive moment region results in a small value of a (see Fig. 5–8d, for example) and hence a larger value of j. As a first trial value, assume that $j = 0.95$ when designing the positive moment reinforcement. This is equivalent to assuming that $a = 0.1d$.

EXAMPLE 5–4 Design Reinforcement in a T Beam

A T beam with the overall dimensions shown in Fig. 5–8 is subjected to a factored positive moment, M_u, of 230 ft-kips. Using $f_c' = 3000$ psi and $f_y = 60,000$ psi, design the required reinforcement.

1. **Compute the effective flange width.** Following the computations in Example 5–1, b is 60 in.

2. **Compute d.** Assuming that there will be two layers of reinforcement, use $d = h - 3.5$ in. or $d = 16.5$ in.

3. **Compute the area of reinforcement, A_s.** From Eq. 4–33,

$$A_s = \frac{M_u}{\phi f_y jd}$$

Since this is a positive moment region in a T beam, assume that $j = 0.95$.

$$A_s = \frac{230 \times 12,000}{0.9 \times 60,000 \times (0.95 \times 16.5)} = 3.26 \ \text{in.}^2$$

Possible choices from Table A–8:

4 No. 8 bars in one layer, $A_s = 3.16$ in.^{2}
2 No. 9 bars plus 2 No. 8 bars, $A_s = 3.58$ in.^{2}
6 No. 7 bars, 4 in one layer, $A_s = 3.60$ in.^{2}

A check of the web width shows that these will fit into a 12-in. web. Although 4 No. 8 bars give a little less area than required, try this combination since it can go in one layer rather than two. This will increase d from the 16.5 in. estimated in step 2 to 17.5 in. Try 4 No. 8 bars, $A_s = 3.16$ in.^{2}.

4. **Check if $A_s \geq A_{s,\text{min}}$.** From Eq. 4–31,

$$A_{s,\text{min}} = \frac{3\sqrt{f_c'}}{f_y} b_w d \geq \frac{200 b_w d}{f_y}$$

$$= \frac{3\sqrt{3000}}{60,000} \times 12 \times 16.5 \geq \frac{200 \times 12 \times 16.5}{60,000}$$

$$= 0.542 \ \text{in.}^2 \geq 0.66 \ \text{in.}^2$$

Therefore, $A_s > A_{s,\text{min}}$—OK.

5. Compute a and check if $f_s = f_y$ and whether the section is tension-controlled.
Assuming rectangular beam action,

$$a = \frac{A_s f_y}{0.85 f_c' b}$$

$$= \frac{3.16 \times 60,000}{0.85 \times 3000 \times 60} = 1.239 \text{ in.}$$

Since this is less than the flange thickness, rectangular beam action exists.

$$\frac{a}{d} = \frac{1.239}{16.5} = 0.075$$

This is much less than

$$\frac{a_b}{d} = 0.85 \left(\frac{87,000}{87,000 + 60,000} \right) = 0.503$$

Therefore, $f_s = f_y$.

$$d_t = 20 \text{ in.} - (1.5 + 0.375 - 1.00/2) = 17.62 \text{ in.} \qquad \frac{a}{d_t} = \frac{1.239}{17.62} = 0.0703$$

This is very much less than

$$\frac{a_{tc\ell}}{d_t} = 0.375\beta_1 = 0.319 \qquad (4\text{--}23)$$

Therefore, the section is tension controlled and $\phi = 0.90$.

6. Compute ϕM_n.

$$\phi M_n = \phi \left[A_s f_y \left(d - \frac{a}{2} \right) \right]$$

$$= \frac{0.9[3.16 \times 60,000(17.5 - 1.24/2)]}{12,000} = 240 \text{ ft-kips}$$

Since $\phi M_n > M_u$, this is OK. Use 4 No. 8 Grade 60 bars in one layer. The lever arm $(d - a/2) = jd$ was $0.965d$ in this case. The value of $0.95d$ assumed in step 3 normally gives a satisfactory first trial. ■

A complete design of a continuous T beam, including calculation of moment diagrams and proportioning for flexure, shear, and anchorage is carried out in Chap. 10.

5-3 BEAMS WITH COMPRESSION REINFORCEMENT

Occasionally, beams are built with both tension reinforcement and compression reinforcement. The effect of compression reinforcement on the behavior of beams and the reasons why it is used are discussed in the following sections, followed by methods of analyzing such beams.

Effect of Compression Reinforcement on Strength and Behavior

The resultant internal forces at ultimate load, in beams with and without compression reinforcement, are compared in Fig. 5–12. The beam in Fig. 5–12b has compression steel of area A_s' located at d' from the extreme compression fiber. The area of the tension reinforcement, A_s, is the same in both beams. In both beams, the total compressive force $C = T$

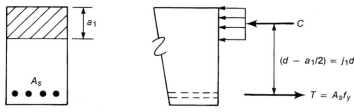

(a) Beam with tension steel only.

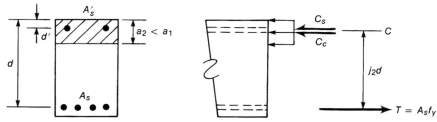

(b) Beam with tension and compression steel.

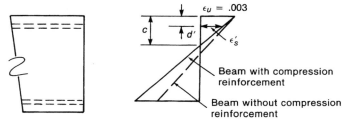

Fig. 5–12
Effect of compression
reinforcement on moment
capacity.

(c) Effect of compression reinforcement on strain distribution in
two beams with the same area of tension reinforcement.

where $T = A_s f_y$. In the beam without compression reinforcement (Fig. 5–12a), this compressive force, C, is entirely resisted by concrete. In the other case, C is the sum of C_c provided by the concrete and C_s provided by the steel. Because some of the compression is resisted by compression reinforcement, C_c will be less than C, with the result that the depth of the compression zone, a_2, in Fig. 5–12b is less than a_1 in Fig. 5–12a.

Summing moments about the resultant compressive force C gives:

Beam without compression steel:

$$M_n = A_s f_y (j_1 d)$$

Beam with compression steel:

$$M_n = A_s f_y (j_2 d)$$

where $j_2 d$ is the distance from the tensile force to the resultant of C_s and C_c.

The only difference between these two expressions is that j_2 is a little larger than j_1 because a_2 is smaller than a_1. Thus for a given amount of tension reinforcement, the addition of compression steel has little effect on the usable ultimate moment capacity, provided

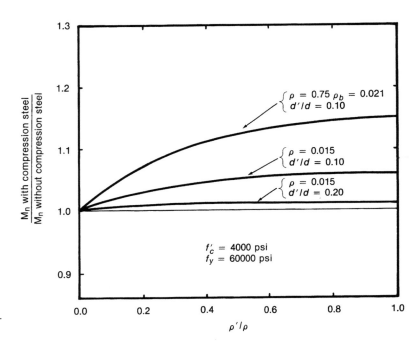

Fig. 5–13
Increase in moment capacity
due to compression reinforcement.

that the tension steel yields in the beam without compression reinforcement. This is illustrated in Fig. 5–13. For normal ratios of tension reinforcement ($\rho \leq 0.015$) the increase in
moment is generally less than 5%.

The effectiveness of compression steel decreases as it is moved away from the compression face. As shown in Fig. 5–12c, if the distance d' from the extreme compression
fiber to the compression steel is increased, the strain ϵ'_s in the compression steel is decreased. As a result, the stress in this steel may be reduced below the yield stress. The effect of increasing d'/d from 0.1 to 0.2 can be seen in Fig. 5–13.

Reasons for Providing Compression Reinforcement

There are four primary reasons for using compression reinforcement in beams.

1. Reduced sustained load deflections. First and most important, the addition of
compression reinforcement reduces the long-term deflections of a beam subjected to sustained loads. Figure 5–14 presents deflection-time diagrams for beams with and without

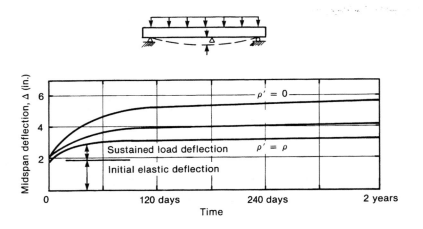

Fig. 5–14
Effect of compression reinforcement on sustained load
deflections. (Adapted from
Ref. 5–4.)

Flexure: T Beams, Beams with Compression Reinforcement, and Special Cases

compression reinforcement. The beams were loaded, in a period of several hours, to the service load level. This load was then maintained for 2 years. At the time of loading (time = 0 in Fig. 5–14), the three beams deflected between 1.6 and 1.9 in., or approximately the same amount. As time passed, the deflections of all three beams increased. The *additional* deflection with time is 195% of the initial deflection for the beam without compression steel ($\rho' = 0$), but only 99% of the initial deflection for the beam with compression steel equal to the tension steel ($\rho' = \rho$). The ACI Code accounts for this in the deflection calculation procedures as outlined in Chap. 9.

Creep of the concrete in the compression zone transfers load from the concrete to the compression steel, reducing the stress in the concrete as occurred in Example 3–4. Because of the lower compression stress in the concrete, it creeps less, leading to a reduction in sustained load deflections.

2. Increased ductility. The addition of compression reinforcement causes a reduction in the depth of the compression stress block, a. Thus, as shown in Fig. 5–12, a_2 is smaller than a_1. As a decreases, the strain in the tension reinforcement at failure increases as shown in Fig. 5–12c, resulting in more ductile behavior. Figure 5–15 compares moment-curvature diagrams for three beams with $\rho < \rho_b$ and varying amounts of compression reinforcement, ρ', where $\rho' = A'_s/bd$. The moment at first yielding of the tension reinforcement is seen to change very little when compression steel is added to these beams. The increase in moment after yielding is largely due to strain hardening of the reinforcement. Because this occurs at very high curvatures and deflections, it is ignored in design. On the other hand, the ductility increases significantly, as shown in Fig. 5–15. This is particularly important in seismic regions or if moment redistribution is desired.

3. Change of mode of failure from compression to tension. When $\rho > \rho_b$, a beam fails in a brittle manner due to crushing of the compression zone before the steel yields. The moment–curvature diagram for such a beam is shown in Fig. 5–16 ($\rho' = 0$). When enough compression steel is added to such a beam, the compression zone is strengthened sufficiently to allow the tension steel to yield before the concrete crushes. The beam then displays a ductile mode of failure as shown in Fig. 5–16. If the compression steel yields, the strain distributions and curvatures at failure in a beam with compression reinforcement will be essentially the same as those in a singly reinforced beam (tension steel only) having a reinforcement ratio of $(\rho - \rho')$. The term $(\rho - \rho')$ is referred to as the *effective reinforcement ratio*. Frequently, designers will add compression steel so that $(\rho - \rho') \leq 0.5\rho_b$.

Two cases where compression steel is frequently used are the negative moment region of continuous T beams and midspan regions of the inverted T beams used to support precast floor panels.

4. Fabrication ease. When assembling the reinforcing cage for a beam, it is customary to provide bars in the corners of the stirrups to hold the stirrups in place in the form

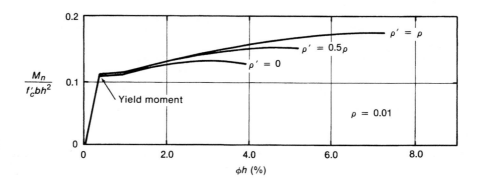

Fig. 5–15
Effect of compression reinforcement on strength and ductility of under-reinforced beams. $\rho < \rho_b$. (From Ref. 5–5.)

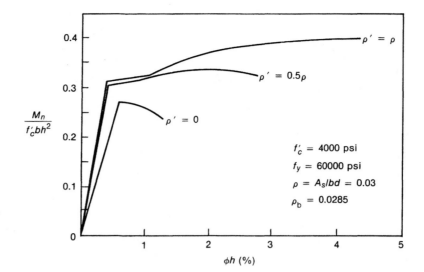

Fig. 5–16
Moment-curvature diagrams for beams with $\rho > \rho_b$, with and without compression reinforcement. (Adapted from Ref. 5–5.)

and also to help anchor the stirrups. If developed properly, these bars in effect are compression reinforcement, although they are generally disregarded in design since they have a small effect on the moment capacity.

Analysis of Beams with Tension and Compression Reinforcement

In the analysis of T beams in the preceding section, the cross section was hypothetically divided into two beams. A similar procedure will be used for a beam with compression reinforcement (Fig. 5–17a). The strain distribution, stresses, and internal forces in this beam are shown in Fig. 5–17b to d. For analysis, we will imagine that this beam is divided into *beam 1*, consisting of the compression reinforcement at the top and sufficient steel at the bottom so that $T_1 = C_s$, and *beam 2*, consisting of the concrete web and the remaining tensile reinforcement, as shown in Fig. 5–17e and f.

The stress in the compression reinforcement has been shown as f_s'. Figure 5–17b shows a strain distribution for a beam with compression steel. From similar triangles,

$$\epsilon_s' = \left(\frac{c - d'}{c}\right)0.003$$

If $\epsilon_s' \geq \epsilon_y$ then $f_s' = f_y$. Replacing c with $c = a/\beta_1$ gives

$$\epsilon_s' = \left(1 - \frac{\beta_1 d'}{a}\right)0.003 \tag{5-8}$$

Setting $\epsilon_s' = \epsilon_y$ and $\epsilon_y = f_y/E_s$, where $E_s = 29 \times 10^6$ psi, we can solve for the limiting value of d'/a for which the compression reinforcement will yield

$$\left(\frac{d'}{a}\right)_{\text{lim}} = \frac{1}{\beta_1}\left(1 - \frac{f_y}{87,000}\right) \tag{5-9}$$

where f_y is in psi, or in SI units with f_y in MPa:

$$\left(\frac{d'}{a}\right)_{\text{lim}} = \frac{1}{\beta_1}\left(1 - \frac{f_y}{600}\right) \tag{5-9M}$$

If the value of d'/a is *greater* than this value, the compression steel will not yield at ultimate [i.e., $f_s' = f_y$ only if $d'/a \leq (d'/a)_{\text{lim}}$]. Values of $(d'/a)_{\text{lim}}$ are given in Table A–10.

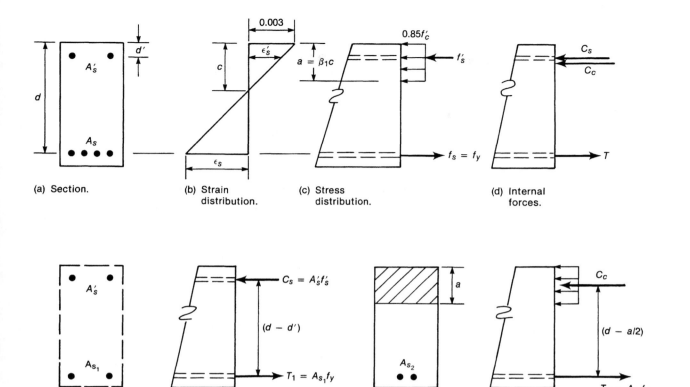

Fig. 5–17
Strains, stresses, and forces in a beam with compression reinforcement.

The procedure followed in computing the moment capacity of a beam with compression reinforcement differs depending on whether the reinforcement yields or not. These two cases are discussed separately.

Case 1: Compression Steel Yields

If the compression steel yields in a particular beam, the analysis is straightforward. It is assumed that the beam can be divided into two imaginary beams, each with $C = T$.

Beam 1 consists of reinforcement in tension and compression and resists moment as a steel force couple. The area of tension steel in this beam is obtained by setting $C_s = T_1$, or

$$A'_s f_y = A_{s1} f_y$$

which gives $A_{s1} = A'_s$. From Fig. 5–17e the nominal moment capacity of this beam is

$$M_{n1} = A'_s f_y (d - d') \tag{5-10}$$

Beam 2 consists of the concrete plus the remaining steel:

$$A_{s2} = A_s - A_{s1}$$

If $f'_s = f_y$, then $A_{s1} = A'_s$. The compression force in the concrete is

$$C_c = 0.85 f'_c b a$$

Since $C = T$ for beam 2, where $T = (A_s - A'_s) f_y$, the depth of the compression stress block, a, is

$$a = \frac{(A_s - A_s')f_y}{0.85f_c'b} \tag{5-11a}$$

From Fig. 5–17f the nominal moment capacity of beam 2 is

$$M_{n2} = (A_s - A_s')f_y\left(d - \frac{a}{2}\right) \tag{5-12}$$

and the total nominal moment capacity of a beam with compression steel is

$$M_n = A_s'f_y(d - d') + (A_s - A_s')f_y\left(d - \frac{a}{2}\right) \tag{5-13a}$$

In the derivation of Eq. 5–13 the compression in the concrete, C_c, was computed using the full rectangular compression zone, ab (Fig. 5–17f), including the area displaced by the compression steel, A_s'. As a result the steel is stressed to $0.85f_c'$, in beam 2 and the additional stress needed to yield it in beam 1 is $(f_y - 0.85f_c')$. To correct for this we have

$$A_{s1} = A_s'\left(1 - \frac{0.85f_c'}{f_y}\right)$$

and

$$M_n = A_s'\left(1 - \frac{0.85f_c'}{f_y}\right)f_y(d - d') \tag{5-13b}$$

$$+ \left[A_s - A_s'\left(1 - \frac{0.85f_c'}{f_y}\right)\right]f_y\left(d - \frac{a}{2}\right)$$

where

$$a = \frac{[A_s - A_s'(1 - 0.85f_c'/f_y)]f_y}{0.85f_c'b} \tag{5-11b}$$

The main effect of $(1 - 0.85f_c'/f_y)$ is to reduce the first term in Eq. 5–13b and increase the second term compared to Eq. 5–13a. These largely offset each other, with the result that for beams with Grade 60 reinforcement, Eq. 5–13a will overestimate M_n by up to 0.3% and normally less. For Grade 40 steel the increase will be about twice as much. Because this increase is so small, the term $(1 - 0.85f_c'/f_y)$ will be ignored for simplicity.

Determination of Whether $f_s = f_y$ in Tension Reinforcement. The derivation of Eq. 5–13a assumed that both the compression steel and the tension steel yielded. It is necessary to check whether this is true. If d'/a is less than or equal to the limiting value given in Eq. 5–9 or 5–9M, the compression steel will yield. The tension steel will yield if a tension failure or a balanced failure occurs. Thus the tension steel will yield if $a/d \leq a_b/d$, where a_b/d is given by Eq. 4–21.

Alternative Determination of Whether $f_s = f_y$ in Tension Reinforcement. The tension steel will yield if a tension or balanced failure occurs. A balanced failure corresponds to a strain distribution with $\epsilon_{cu} = 0.003$ at the extreme compression fiber and ϵ_y at the centroid of the tension steel. Assuming that both the compression steel and the tension steel yield, and substituting $c = a/\beta_1$ and a from Eq. 5–11a into Eq. 4–17, the balanced condition is defined as

$$\frac{(A_s - A_s')f_y}{0.85\beta_1 f_c'bd} = \frac{87{,}000}{87{,}000 + f_y} \tag{5-14a}$$

where f_c' and f_y are in psi. Substituting $\rho = A_s/bd$ and $\rho' = A_s'/bd$ gives

$$(\rho - \rho')_b = \frac{0.85\beta_1 f_c'}{f_y}\left(\frac{87{,}000}{87{,}000 + f_y}\right) \tag{5-15a}$$

In SI units 87,000 becomes 600 and f_c' and f_y are in MPa.

Equation 5–15a is similar to Eq. 4–19 for ρ_b. Note that the effect of compression reinforcement is to reduce the effective value of ρ to $(\rho - \rho')$. Equation 5–15a applies only if $f_s' = f_y$. A similar equation will be derived later for the case when $f_s' < f_y$.

Case 2: Compression Steel Does Not Yield

If the compression reinforcement does not yield, f_s' is not known, and a different solution is required. Assuming that the tensile steel yields, the internal forces in the beam (Fig. 5–17) are

$$T = A_s F_y$$

$$C_c = 0.85 f_c' b a \qquad (5\text{–}16a)$$

Ignoring the correction to the compressive stresses in the steel incorporated into Eq. 5–13b, we have

$$C_s = (E_s \epsilon_s')A_s' \qquad (5\text{–}16b)$$

where ϵ_s' is given by Eq. 5–8. From equilibrium,

$$C_c + C_s = T$$

or

$$0.85 f_c' b a + E_s A_s'\left(1 - \frac{\beta_1 d'}{a}\right)0.003 = A_s f_y$$

This can be reduced to the quadratic equation in a, given by

$$(0.85 f_c' b)a^2 + (0.003 E_s A_s' - A_s f_y)a - (0.003 E_s A_s' \beta_1 d') = 0 \qquad (5\text{–}17)$$

Once the depth of the stress block, a, is known, the nominal moment capacity of the section is

$$M_n = C_c\left(d - \frac{a}{2}\right) + C_s(d - d') \qquad (5\text{–}18)$$

where C_c and C_s are defined by Eqs. 5–16a and b. Note that Eq. 5–17 applies only if $f_s' \le f_y$.

Determination of Whether $f_s = f_y$ in the Tension Reinforcement. The derivation of Eq. 5–17 assumed that the tension steel yielded. It is necessary to check whether this is true. The tension steel will yield if a tension failure or a balanced failure occurs. Thus the tension steel will yield if $a/d \le a_b/d$, where a_b/d is given by Eq. 4–21.

Alternative Determination of Whether $f_s = f_y$ in the Tension Reinforcement. The tension steel will yield if a tension or balanced failure occurs. Equation 5–15a gives the value of $(\rho - \rho')$ corresponding to a balanced failure provided that $f_s' = f_y$. If this is not true, Eq. 5–14a becomes

$$\frac{(A_s - A_s' f_s'/f_y)f_y}{0.85 \beta_1 f_c' bd} = \frac{87,000}{87,000 + f_y} \qquad (5\text{–}14b)$$

and Eq. 5–15a becomes

$$\left(\rho - \frac{\rho' f_s'}{f_y}\right)_b = \frac{0.85 \beta_1 f_c'}{f_y}\left(\frac{87,000}{87,000 + f_y}\right) \qquad (5\text{–}15b)$$

In SI units replace 87,000 with 600 in these equations.

If $(\rho - \rho' f_s'/f_y)$ is less than or equal to $(\rho - \rho' f_s'/f_y)_b$, then $f_s = f_y$ for the tension steel.

Upper Limit on Tension Reinforcement in Beams with Compression Steel—ACI Appendix B. No upper limit on the amount of tension reinforcement is given in ACI Appendix B. Instead, the value of ϕ is set at 0.90 for tension-controlled sections and 0.70 for compression-controlled sections without spiral reinforcement in the compression zone. Enough compression steel would generally be used to ensure a tension-controlled section to take advantage of the higher value of ϕ. A section will be tension-controlled if the a/d ratio at ultimate is less than or equal to

$$\frac{a_{tc\ell}}{d_t} = 0.375\beta_1 \tag{4-23}$$

Upper Limit on Tension Reinforcement in Beams with Compression Steel—ACI Sec. 10.3.3 ACI Sec. 10.3.3 limits the amount of tension steel in beams to 0.75 times that corresponding to a balanced failure, that is, to 0.75 times the amount given by Eqs. 5–15a and 5–15b. The code goes on to say that the portion of ρ_b equalized by compression steel need not be reduced by the 0.75 factor.

Minimum Tension Reinforcement

Minimum tension reinforcement should correspond to Eq. 4–31.

Ties for Compression Reinforcement

As the ultimate load is approached the compression steel in a beam may buckle, causing the surface layer of concrete to spall off, possibly leading to failure. For this reason it is necessary to enclose compression reinforcement with closed stirrups or ties. The design of these ties is covered in ACI Sec. 7.11. The spacing and size of the ties is similar to that of column ties (see Chap. 11).

Examples of the Analysis of Beams with Compression and Tension Reinforcement

Two examples are presented, one for each of the two cases given above. In each example the solution starts by assuming that $f_s' = f_y$ and $f_s = f_y$ since this is the easiest solution. As soon as a has been computed, the assumptions are checked. If this check shows that $f_s' < f_y$, it is necessary to change the solution and base the calculations on Eq. 5–17. If $f_s < f_y$, more compression steel should be added.

EXAMPLE 5–5 Analysis of a Beam with Compression Reinforcement: Compression Reinforcement Yields

The beam shown in Fig. 5–18 has $f_c' = 3000$ psi and $f_y = 60,000$ psi. For this beam, based on the tension steel only, $a = 10.14$ in., $d_t = 23.63$ in., giving $a/d_t = 0.429$, which exceeds the tension-controlled limit $a_{tc\ell}/d_t = 0.319$. As a result, ϕ would be less than 0.90. To allow the use of $\phi = 0.9$ and to give more ductility, 2 No. 7 bars have been added as compression steel. Compute the design moment capacity.

1. Assume that $f_s' = f_y$ and $f_s = f_y$, and divide the beam into two components. The beam is divided into beam 1 and beam 2 (Fig. 5–18b and c). Since the steel is all assumed to yield, $A_{s1} = A_s'$. The area of steel in beam 2 is

$$A_{s2} = A_s - A_{s1}$$

$$= 4.74 - 1.20 = 3.54 \text{ in.}^2$$

2. Compute a for beam 2.

$$a = \frac{(A_s - A_s')f_y}{0.85f_c'b} \tag{5-11a}$$

Flexure: T Beams, Beams with Compression Reinforcement, and Special Cases

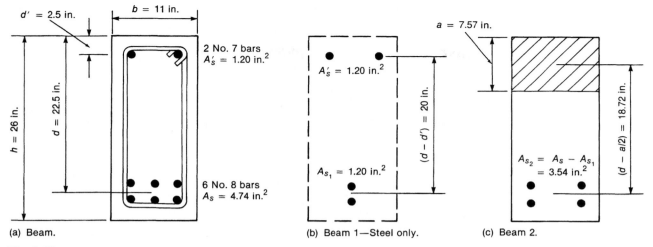

(a) Beam.

Fig. 5–18
Example 5–5.

(b) Beam 1—Steel only.

(c) Beam 2.

$$= \frac{3.54 \times 60{,}000}{0.85 \times 3000 \times 11} = 7.57 \text{ in.}$$

3. Check if compression steel yields. In step 1 we assumed that $f_s' = f_y$. It is necessary to check this assumption.

$$d' = 2.5 \text{ in.}$$

$$\frac{d'}{a} = \frac{2.5}{7.57} = 0.330$$

From Eq. 5–9 or Table A–10,

$$\left(\frac{d'}{a}\right)_{\text{lim}} = \frac{1}{\beta_1}\left(1 - \frac{f_y}{87{,}000}\right) \tag{5-9}$$

$$= \frac{1}{0.85}\left(1 - \frac{60{,}000}{87{,}000}\right) = 0.365$$

Since $d'/a = 0.330$ is *less* than the limiting value of 0.365, the compression steel yields and $f_s' = f_y$.

4. Check if $f_s = f_y$ for the tension steel and whether the section is tension-controlled.

$$a = 7.57 \text{in.}$$

and

$$\frac{a}{d} = \frac{7.57}{22.5} = 0.336$$

From Table A–4, $a_b/d = 0.503$. Since 0.336 is less than 0.503, the tension steel yields.

$$d_t = 26.0 - 1.5 - 0.375 - 1.0/2 = 23.63 \text{ in.}$$

$$\frac{a}{d_t} = \frac{7.57}{23.63} = 0.320$$

The tension-controlled limit is

$$\frac{a_{tc\ell}}{d_t} = 0.375\beta_1 = 0.319$$

Since 0.320 is larger than 0.319, the section is a transition section and ϕ will be less than 0.90. From Eq. 4–26b,

$$\phi = 0.356 + \frac{0.204}{a/(\beta_1 d_t)} = 0.356 + \frac{0.204}{7.57/(0.85 \times 23.63)}$$

$$= 0.897$$

Alternatively, if design was carried out by ACI Sec. 10.3.3, the upper limit on the steel would be such that $\rho - \rho' \leq 0.75\rho_b$. To check this, compute the compression forces C_c and C_s in the concrete and the compression steel for the balanced condition with $a = a_b$. Then compute the area of steel needed to balance these compressions:

$$A_{sb} = \frac{C_c + C_s}{f_y}$$

This comes out to 6.49 in.² and $0.75A_{sb} = 4.87$ in.², which is a little more than $A_s = 4.74$ in.². Thus the beam satisfies the upper limit on steel in ACI Sec. 10.3.3. For design by 10.3.3, $\phi = 0.90$.

Which of these two checks needs to be made depends on whether design is by ACI Appendix B or ACI Sec. 10.3.3. We shall assume that design was being carried out by ACI Appendix B and will use $\phi = 0.897$.

5. **Check if $A_s \geq A_{s,\min}$.** From Eq. 4–31,

$$A_{s,\min} = \frac{3\sqrt{3000}}{60,000} \times 11 \times 22.5 \geq \frac{200 \times 11 \times 22.5}{60,000}$$

$$= 0.68 \geq 0.83 \text{ in.}^2$$

Thus $A_s = 4.74$ in.² exceeds $A_{s,\min}$. This should have been obvious since the reinforcement exceeded the tension-controlled limit.

6. **Compute ϕM_n.**

   (a) **Beam 1:**

   $$\phi M_{n1} = \phi[A_s' f_y(d - d')]$$ (5–10)

   $$= 0.897\left[\frac{1.20 \times 60,000(22.5-2.5)}{12,000}\right] = 107.6 \text{ ft-kips}$$

   (b) **Beam 2:**

   $$\phi M_{n2} = \phi\left[(A_s - A_s')f_y\left(d - \frac{a}{2}\right)\right]$$ 5–12

   $$= 0.897\left[\frac{3.54 \times 60,000(22.5-7.57/2)}{12,000}\right] = 297.1 \text{ ft-kips}$$

The total moment capacity is

$$\phi M_n = \phi M_{n1} + \phi M_{n2} = 405 \text{ ft-kips}$$

The design moment capacity of the beam shown in Fig. 5–18 is 405 ft-kips. ∎

EXAMPLE 5–5M Analysis of a Beam with Compression Reinforcement: Compression Reinforcement Yields–SI Units

The beam shown in Fig. 5–19 has $f_c' = 20$ MPa and $f_y = 400$ MPa. For this beam, based on the tension steel only, $a = 257$ mm, $d_t = 538$ mm, giving $a/d_t = 0.478$, which exceeds the tension-controlled limit $a_{tc\ell}/d_t = 0.319$. As a result, ϕ would be less than 0.90. To allow the use of $\phi = 0.90$ and to increase the ductility, 2 No. 25 bars have been added as compression steel. Compute the design moment capacity.

1. **Assume that $f_s' = f_y$ and $f_s = f_y$, and divide the beam into two components.** The beam is divided into beam 1 and beam 2 (Fig. 5–19b and c). Since the steel is all assumed to yield, $A_{s1} = A_s'$. The area of steel in beam 2 is

Flexure: T Beams, Beams with Compression Reinforcement, and Special Cases

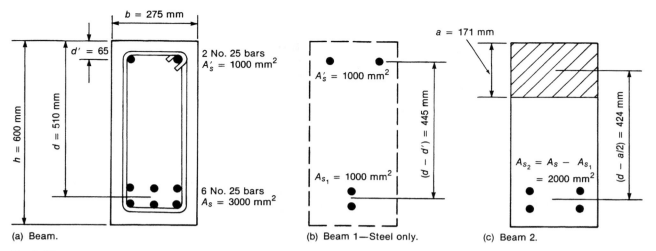

(a) Beam.

(b) Beam 1—Steel only.

(c) Beam 2.

Fig. 5–19
Example 5–5M.

$$A_{s2} = A_s - A_{s1}$$

$$= 3000 - 1000$$

$$= 2000 \text{ mm}^2$$

2. **Compute a for beam 2.**

$$a = \frac{(A_s - A_s')f_y}{0.85f_c'b} \qquad \text{5–11a)}$$

$$= \frac{2000 \times 400}{0.85 \times 20 \times 275} = 171 \text{ mm}$$

3. **Check if compression steel yields.**

$$d' = 65 \text{ mm}$$

$$\frac{d'}{a} = \frac{65}{171} = 0.380$$

From Eq. 5–9M,

$$\left(\frac{d'}{a}\right)_{\text{lim}} = \frac{1}{\beta_1}\left(1 - \frac{f_y}{600}\right) \qquad \text{(5–9M)}$$

$$= \frac{1}{0.85}\left(1 - \frac{400}{600}\right) = 0.392$$

Since $d'/a = 0.380$ is *less* than the limiting value of 0.392, the compression reinforcement yields.

4. **Check if $f_s = f_y$ for the tension steel and whether the section is tension-controlled.**

$$a = 171 \text{ mm} \quad \text{and} \quad \frac{a}{d} = \frac{171}{510} = 0.335$$

From Table A–4M, $a_b/d = 0.510$. Since 0.335 is less than 0.510, the tension steel yields.

$$d_t = 600 - 40 - 10 - \frac{25}{2} = 537 \text{ mm}$$

$$\frac{a}{d_t} = \frac{171}{537} = 0.3184$$

The tension-controlled limit is

$$\frac{a_{tc\ell}}{d_t} = 0.375\beta_1 = 0.3188$$

Since 0.3184 is less than 0.3188, the section is tension-controlled and $\phi = 0.90$.

Alternatively, if design was carried out by ACI Sec. 10.3.3, the upper limit on the steel would be that $\rho - \rho' \leq 0.75\rho_b$. To check this, compute the compression forces C_c and C_s in the concrete and the compression steel for the balanced condition with $a = a_b$. Then compute the area of steel needed to balance these compressions:

$$A_{sb} = \frac{C_c + C_s}{f_y}$$

This comes out to 4040 mm² and $0.75A_{sb} = 3030$ mm², which is a little more than $A_s = 3000$ mm². Thus the beam satisfies the upper limit on steel in 10.3.3. For design by 10.3.3, $\phi = 0.90$.

Which of these two checks needs to be made depends on whether design is by ACI Appendix B or ACI Sec. 10.3.3. We shall assume that design was being carried out by ACI Appendix B.

5. Check if $A_s \geq A_{s,min}$. From Eq. 4–31,

$$A_{s,min} = (392 \geq 491) \text{ mm}^2$$

$$= 491 \text{ mm}^2$$

Thus $A_s = 3000$ mm² exceeds $A_{s,min}$. This should have been obvious since the reinforcement exceeded the tension-controlled limit.

5. Compute ϕM_n.

(a) **Beam 1:**

$$\phi M_{n1} = \phi[A_s' f_y (d - d')] \tag{5-10}$$

$$= \frac{0.9[1000 \times 400(510 - 65)]}{10^6} = 160 \text{ kN-m}$$

(b) **Beam 2:**

$$\phi M_{n2} = \phi\left[(A_s - A_s')f_y\left(d - \frac{a}{2}\right)\right] \tag{5-12}$$

$$= \frac{0.9[(3000 - 1000) \times 400(510 - 171/2)]}{10^6}$$

$$= 306 \text{ kN-m}$$

The total moment capacity is

$$\phi M_n = \phi M_{n1} + \phi M_{n2} = 466 \text{ kN-m}$$

Therefore, the design moment capacity of the beam shown in Fig. 5–19 is 466 kN-m. ∎

EXAMPLE 5–6 Analysis of a Beam with Compression Reinforcement: Compression Reinforcement Does Not Yield

The beam shown in Fig. 5–20 has $f_c' = 3000$ psi and $f_y = 60,000$ psi. It is similar to the beam considered in Example 5–5 except that there is more compression steel. Compute the design moment capacity, ϕM_n. The solution will start in the same way as Example 5–5. If the assumptions made in step 1 prove to be incorrect, a different solution must be used.

1. Assume that $f_s' = f_y$ and $f_s = f_y$ and divide the beam into two components. Beam 1 has 3 No. 8 bars as compression reinforcement and an area of tension reinforcement, A_{s1}, equal to 3 No. 8 bars concentrated at d below the top of the beam.

$$A_{s2} = A_s - A_{s1}$$

$$= 4.74 - 2.37 = 2.37 \text{ in.}$$

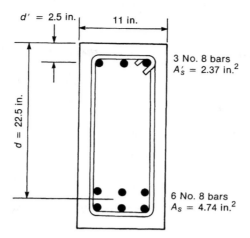

$d' = 2.5$ in. 11 in.

$d = 22.5$ in.

3 No. 8 bars
$A'_s = 2.37$ in.^{2}

6 No. 8 bars
$A_s = 4.74$ in.^{2}

Fig. 5–20
Example 5–6

2. **Compute a for beam 2.**

$$a = \frac{(A_s - A'_s)f_y}{0.85f'_c b} \tag{5–11a}$$

$$= \frac{2.37 \times 60,000}{0.85 \times 3000 \times 11} = 5.07 \text{ in.}$$

3. **Check if compression steel yields.** From Fig. 5–20, $d' = 2.5$ in.

$$\frac{d'}{a} = \frac{2.5}{5.07} = 0.493$$

But this exceeds the limiting value of $d'/a = 0.365$ from Eq. 5–9 or Table A–10. Therefore, the compression steel does not yield. As a result, a from step 2 is incorrect and a must be reevaluated using Eq. 5–19, assuming that the compression steel is elastic.

4. **Solve Eq. 5–19 for a.**

$$(0.85f'_c b)a^2 + (0.003E_s A'_s - A_s f_y)a - (0.003E_s A'_s \beta_1 d') = 0$$

or $28,050a^2 - 78,210a - 438,200 = 0$ and

$$a = \frac{78,210 \pm \sqrt{78,210^2 - (4 \times 28,050 \times -438,200)}}{2 \times 28,050}$$

$$= 5.59 \text{ in.}$$

Note that this is larger than the $a = 5.07$ in. computed in step 2. This is to be expected since the stress in the compression steel is lower than assumed in step 2, and as a result there is a larger compression force in the concrete.

5. **Check if $f_s = f_y$ for the tension steel and whether the section is tension-controlled.**

$$a = 5.59 \text{ in.} \quad \text{and} \quad \frac{a}{d} = \frac{5.59}{22.5} = 0.248$$

From Table A–4, $a_b/d = 0.503$. Since 0.248 is less than 0.503, the tension steel yields.

$$d_t = 26 - 1.5 - 0.375 - \frac{1.0}{2} = 23.63 \text{ in.}$$

$$\frac{a}{d_t} = \frac{5.59}{23.63} = 0.237$$

5–3 Beams with Compression Reinforcement

The tension-controlled limit is

$$\frac{a_{tc\ell}}{d_t} = 0.375\beta_1 = 0.319$$

Thus the section is tension-controlled and $\phi = 0.90$.

6. **Check if $A_s \geq A_{s,min}$.** By inspection, A_s exceeds $A_{s,min}$.

7. **Compute ϕM_n.**

$$\phi M_n = \phi\left[C_c\left(d - \frac{a}{2}\right) + C_s(d - d')\right] \tag{5-18}$$

where

$$C_c = 0.85f_c'ba \tag{5-16a}$$

$$= \frac{0.85 \times 3000 \times 11 \times 5.59}{1000} = 157 \text{ kips}$$

$$C_s = (E_s\epsilon_s')A_s' \tag{5-16b}$$

where

$$\epsilon_s' = \left(1 - \frac{\beta_1 d'}{a}\right)0.003 \tag{5-10}$$

Thus

$$C_s = E_sA_s'\left(1 - \frac{\beta_1 d'}{a}\right)0.003$$

$$= \frac{29 \times 10^6 \times 2.37\left(1 - \dfrac{0.85 \times 2.5}{5.59}\right)0.003}{1000}$$

$$= 128 \text{ kips}$$

and

$$\phi M_n = \frac{0.9\left[157(22.5 - 5.59/2) + 128(22.5 - 2.5)\right]}{12}$$

$$= 423 \text{ ft-kips}$$

Thus the design moment capacity of the beam shown in Fig. 5–20 is 423 ft-kips.

A worthwhile partial check on the calculations can be obtained by comparing $T_s = A_sf_y = 4.74 \times 60 = 284.4$ kips and $C_c + C_s = 156.8 + 127.8 = 284.5$ kips. These should be the same since $C = T$ and are within the accuracy of the calculations. ∎

A comparison of the strengths of the beams considered in Examples 5–5 and 5–6 shows that almost doubling the area of compression steel, resulting in an increase of 20% in the total area of steel in the beam, increased the moment capacity by only 4%. This illustrates the fact that additional compression reinforcement is generally not an effective method of increasing the moment capacity of a beam.

5–4 UNSYMMETRICAL BEAM SECTIONS OR BEAMS BENT ABOUT TWO AXES

Figure 5–21 shows one half of a simply supported beam with an unsymmetrical cross section. The loads lie in a plane referred to as the *plane of loading* and it is assumed that this passes through the shear center of the unsymmetrical section. This beam is free to deflect vertically and laterally between its supports. The applied loads cause moments that must be

Flexure: T Beams, Beams with Compression Reinforcement, and Special Cases

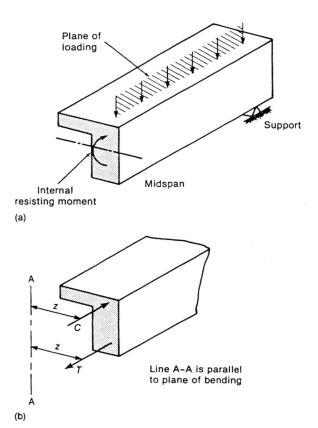

Figure 5–21
location of C and T forces in unsymmetrical beam.

resisted by an internal resisting moment about a horizontal axis, shown by the moment vector in Fig. 5–21a. This internal resisting moment results from compressive and tensile forces C and T as shown in Fig. 5–21b. Because the applied loads do not cause a moment about an axis parallel to the plane of loading (such as A–A), the internal force resultants C and T cannot do so either. As a result, C and T must both lie in the plane of loading or in a plane parallel to it. Both distances z in Fig. 5–21b must be equal.

Figure 5–22 shows a cross section of an inverted L-shaped beam loaded with gravity loads. Because this beam is loaded with vertical loads, leading to moments about a horizontal axis, the line joining the centroids of the compressive and tensile forces must be vertical as shown (both are a distance f from the side of the beam). As a result, the compression zone must be triangular and the neutral axis inclined as shown in Fig. 5–22.

Since $C = T$, and assuming that $f_s = f_y$,

$$\tfrac{1}{2}(3f \times g \times 0.85f_c') = A_s f_y$$

Since the moment is about a horizontal axis, the lever arm must be vertical. Therefore, the case shown in Fig. 5–22,

$$jd = d - \frac{g}{3}$$

and

$$M_n = A_s f_y\left(d - \frac{g}{3}\right)$$

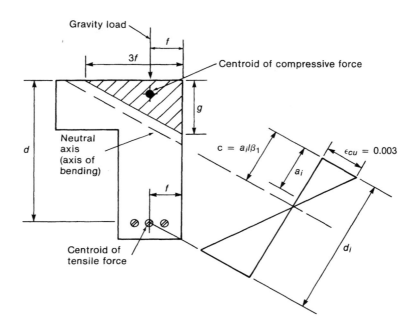

Fig. 5–22
Unsymmetrical beam.

These equations apply only to the beam geometry shown in Fig. 5–22. Different equations or a trial-and-error solution will generally be necessary for other shapes.

The checks of whether $f_s = f_y$ and whether the section is tension-controlled are done by checking whether a_i/d_i and a_i/d_{ti}, measured perpendicular to the neutral axis, are less than the appropriate values from Table A–4. The check of whether $\rho \leq \rho_b$ will require calculation of C_{sb}, A_{sb}, and $0.75\,A_{sb}$.

The discussion to this point has dealt with isolated beams which are free to deflect both vertically and laterally. Such a beam would deflect perpendicular to the axis of bending, that is, both vertically and laterally. If the beam in Fig. 5–22 were the edge beam for a continuous slab that extended to the left to other beams, this slab would prevent lateral deflections. As a result, the neutral axis would be forced to be very close to horizontal and the beam could be designed in the normal fashion.

EXAMPLE 5–7   Analysis of an Unsymmetrical Beam

The beam shown in Fig. 5–23 has an unsymmetrical cross section and an unsymmetrical arrangement of reinforcement. This beam is subjected to vertical loads only. Compute the design moment capacity of this cross section if $f_c' = 3000$ psi and $f_y = 60,000$ psi.

1.   Assume that $f_s = f_y$ and compute the size of the compression zone. The centroid of the three bars is computed to lie at 6.27 in. from the right side of the web. The centroid of the compression zone must also be located this distance from the side of the web. Thus the width of the compression zone is $3 \times 6.27 = 18.8$ in.

Since $C = T$,

$$\tfrac{1}{2}(18.8 \times g \times 0.85f_c') = A_s f_y$$

or

$$g = \frac{2.58 \times 60,000 \times 2}{18.8 \times 0.85 \times 3000}$$

$$= 6.46 \text{ in.}$$

172   Flexure: T Beams, Beams with Compression Reinforcement, and Special Cases

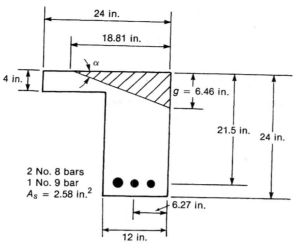

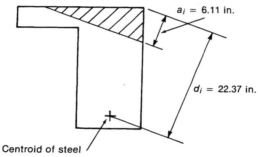

Fig. 5–23
Example 5–7.

(a) Geometry and stress block.

(b) Values of a_i and d_i measured perpendicular to the neutral axis.

The compression zone is shown shaded in Fig. 5–23. The entire area lies within the concrete section, and hence this compression zone can be used for the analysis. If the compression zone were deeper than shown and cut across the reentrant corner, a more complex trial-and-error solution would be required.

2. Check if $f_s = f_y$ and whether the section is tension-controlled. From Fig. 5–23b, $a_i = 6.11$ in. and $d_i = 22.37$ in., giving $a_i/d_i = 0.273$. Since a_i/d_i is less than $a_b/d = 0.503$, $f_s = f_y$.

d_{ti} is the inclined depth from the extreme compression fiber to the extreme tension steel, in this case the No. 9 bar. From the geometry of the section, $d_{ti} = 23.44$ in. and $a_i/d_{ti} = 0.261$. Since this is less than $a_{tc\ell}/d_t = 0.319$, the section is tension-controlled and $\phi = 0.90$.

3. Check if $A_s \geq A_{s,min}$. Since $A_s = 2.58$ in.^{2} exceeds $A_{s,min} = (0.71 \geq 0.86)$ in.^{2} $A_s > A_{s,min}$.

4. Compute ϕM_n.

$$\phi M_n = \phi\left[A_s f_y\left(d - \frac{g}{3}\right)\right]$$

$$= \frac{0.9[2.58 \times 60,000(21.5 - 6.46/3)]}{12,000}$$

$$= 225 \text{ ft-kips}$$

Thus the design moment capacity of the section shown in Fig. 5–23 is 225 ft-kips. Note that the moment calculation is based on the lever arm measured *vertically* (parallel to the plane of loading). ∎

5–4 Unsymmetrical Beam Sections or Beams Bent About Two Axes 173

The analysis procedures presented so far in this chapter and Chap. 4 have been restricted to problems involving:

1. Elastic–plastic reinforcement with a constant yield strength

2. Tension reinforcement and compression reinforcement in two groups of bars that can be represented by compact layers at the centroids of the respective groups

3. All concrete of the same strength

4. A rectangular, T, or other easily definable cross-sectional shape

If any of these restrictions do not apply, a trial-and-error solution based on strain compatibility must be used. The following steps are required in such a solution:

1. Assume a strain distribution defined by a strain, ϵ_{cu}, of 0.003 in the extreme compressive fiber and an assumed value of the depth, c, to the neutral axis.

2. Compute the depth of the rectangular stress block, $a = \beta_1 c$.

3. Compute the strains in each layer of reinforcement from the assumed strain distribution.

4. From the stress–strain curve for the reinforcement and the strains from step 3, determine the stress in each layer of reinforcement.

5. Compute the force in the compression zone and in each layer of reinforcement.

6. Compute $P = C - T$. For a beam without axial force, P equals zero. If the calculated value of P is not equal to zero, adjust the strain distribution and repeat steps 1 to 6 until P is as close to zero as desired. The imbalance should not exceed 0.1 to 0.5% of C.

7. Sum the moments of the internal forces. If $P = 0$, this can be about any convenient axis. We shall sum the moments about the centroid of the cross section. This axis is normally used in columns where P is not zero, as explained in Chap. 11.

EXAMPLE 5-8   Strain Compatibility Analysis of Moment Capacity

Compute the design moment capacity, ϕM_n, of the cross section shown in Fig. 5–24a. The concrete strength is 3500 psi. The reinforcement has the stress–strain curve shown in Fig. 5–24b.

In this solution, *compressive* strains and stresses are taken as *positive*, tensile strains and stresses as negative. As a result, the stress–strain curve for the reinforcement in tension has the equations:

Part O–A, $\epsilon \geq -0.002$:

$$f_s = (29 \times 10^3 \epsilon) \text{ ksi} \tag{5-19a}$$

Part A–B, $\epsilon < -0.002$:

$$f_s = (-55 + 1.5 \times 10^3 \epsilon) \text{ ksi} \tag{5-19b}$$

Similar equations can be derived for the compressive branch.

1. Assume a strain distribution. The first trial strain distribution in Fig. 5–24c is defined by $\epsilon_{cu} = 0.003$ and $c = 8$ in.

2. Compute the depth of the equivalent rectangular stress block.

$$a = \beta_1 c$$

$$= 0.85 \times 8$$

Therefore, $a = 6.8$ in. for first trial.

3. Compute the strains in each layer of reinforcement.

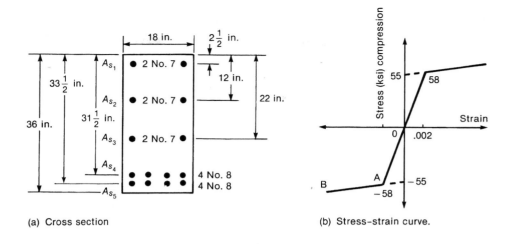

(a) Cross section

(b) Stress-strain curve.

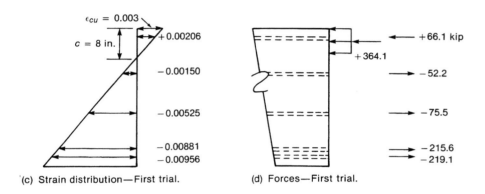

(c) Strain distribution—First trial.

(d) Forces—First trial.

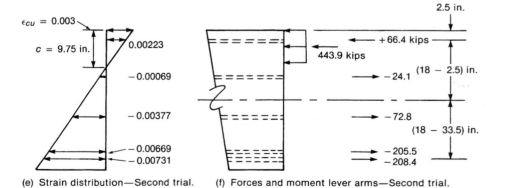

Fig. 5–24
Strain compatibility solution—Example 5–8.

(e) Strain distribution—Second trial.

(f) Forces and moment lever arms—Second trial.

4. **Compute stress in each layer of reinforcement.**

5. **Compute forces in the compression zone and in each layer of reinforcement.**

6. **Compute $P = C - T$.**
   Steps 3 to 6 are carried out in Table 5–1. For a bar located at distance y below the top of the beam, the strain is

TABLE 5–1 Calculation of Internal Forces—Example 5–8[a]

Layer	y (in.)	ϵ	f_s (ksi)	A_s(in.²)	F_s (kips)	C_c (kips)
			First trial: Assume that $c = 8$ in., $a = \beta_1 c = 6.8$ in.			
Compression zone	—	—	—	—	—	$6.8 \times 18 \times 0.85$ $\times 3.5 = +364.1$
A_{s1}	2.5	+0.00206	$58.1 - 0.85$ $\times 3.5 = 55.1$	1.20	+66.1	—
A_{s2}	12	−0.00150	−43.5	1.20	−52.2	—
A_{s3}	22	−0.00525	−62.9	1.20	−75.5	—
A_{s4}	31.5	−0.00881	−68.2	3.16	−215.6	—
A_{s5}	33.5	−0.00956	−69.3	3.16	−219.1	—

$$\Sigma F_s = -496.3 \text{ kips} \quad \Sigma C_c = +364.1 \text{ kips}$$
$$\Sigma F_s + \Sigma C_c = -132.2 \text{ kips}$$

Layer	y (in.)	ϵ	f_s (ksi)	A_s(in.²)	F_s (kips)	C_c (kips)
			Second trial: Assume that $c = 9.75$ in., $a = 8.29$ in.			
Compression zone	—	—	—	—	—	$8.29 \times 18 \times 0.85$ $\times 3.5 = 443.9$
A_{s1}	2.5	+0.00223	$58.3 - 0.85$ $\times 3.5 = 55.4$	1.20	+66.4	—
A_{s2}	12	−0.00069	−20.1	1.20	−24.1	—
A_{s3}	22	−0.00377	−60.7	1.20	−72.8	—
A_{s4}	31.5	−0.00669	−65.0	3.16	−205.5	—
A_{s5}	33.5	−0.00731	−66.0	3.16	−208.4	—

$$\Sigma F_s = -444.4 \text{ kips} \quad \Sigma F_c = +443.9 \text{ kips}$$
$$\Sigma F_s + \Sigma C_c = -0.5 \text{ kips}$$

[a] ϵ, f_s, and F_s are positive in compression.

$$\epsilon = 0.003 - \left(0.003\frac{y}{c}\right) \tag{5–20}$$

For A_{s1}, y is less than a. As a result this layer of steel displaces concrete assumed to be stressed in compression. If $y < a$,

$$f_s = (f_s \text{ from Eq. 5–19}) - 0.85f_c'$$

The strains and forces in the various layers are illustrated in Fig. 5–24c and d. The sum of the forces in the bars and the concrete is 130.2 kips tension. Since there is no axial force in this member, the sum should be zero. Thus the assumed compressed zone is too small.

As a second trial, try $c = 9.75$ in., which results in $a = 8.29$ in. Steps 3 to 6 of the second trial are also given in Table 5–1. At the end of the second trial the forces have converged to within 1 kip. Since this is less than 0.1% of the force in the compression zone, it will be assumed to be close enough. It is sometimes useful to plot $P = C - T$ versus c to help in the choice of c for future trials.

7. Check if $f_s = f_y$ and whether the section is tension-controlled. The stresses in each layer of steel have been computed in Table 5–1 and no further check of f_s is needed. $a/d_t = 9.75/36 = 0.271$. Since this is less than $a_{tc\ell}/d_t = 0.319$, the section is tension-controlled and $\phi = 0.90$.

If the design were carried out according to ACI Sec. 10.3.3, it would be necessary to check if $\rho \leq 0.75 \rho_b$. To do so, compute a_b/d for the balanced case, compute $(C_{cb} + \Sigma F_{si(b)}$ for bars in compression) $= (A_{sb}f_y + \Sigma F_{si(b)}$ for bars in tension), and finally check if $A_s \leq 0.75A_{sb}$.

8. Compute the moments about midheight. Once P has converged to zero, the moments can be computed. The forces from the second trial are shown in Fig. 5–24f. The distances from the midheight are taken positive upward, negative downward. A counterclockwise moment is taken as positive.

$$M_n = 443.9\left(\frac{36}{2} - \frac{8.29}{2}\right) + 66.4(18 - 2.5)$$

$$+ \left[-24.1(18 - 12)\right] + \left[-72.8(18 - 22)\right]$$

$$+ (-205.5(18 - 31.5)) + \left[-208.4(18 - 33.5)\right]$$

$$= 13,330 \text{ in.-kips}$$

Thus the nominal moment capacity of the section shown in Fig. 5–24 is $M_n = 13,330$ in.-kips and the design moment capacity $\phi M_n = 999$ ft-kips. ∎

This type of problem is ideally suited for solution using a spreadsheet.

PROBLEMS

5–1 and 5–2 Compute ϕM_n for the beams shown in Fig. P5–1 and P5–2. Use $f_c' = 3750$ psi for 5–1 and 3000 psi for 5–2 and $f_y = 60,000$ psi.

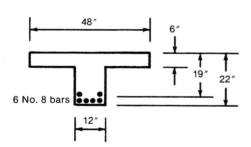

Fig. P5–1

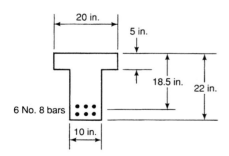

Fig. P5–2

5–3 Compute the negative moment capacity, ϕM_n, for the beam shown in Fig. P5–3. Use $f_c' = 3000$ psi and $f_y = 40,000$ psi.

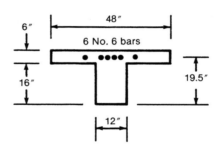

Fig. P5–3

5–4 For the beam shown in Fig. P5–4, $f_c' = 3000$ psi and $f_y = 60,000$ psi.

(a) Compute the effective flange width at midspan.

(b) Compute ϕM_n for the positive and negative moment regions. Check if $f_s = f_y$ at ultimate. At the supports the bars are in one layer, at midspan the No. 8 bars are in the bottom layer, the No. 7 bars in a second layer. (Note that the bottom reinforcement is not adequately anchored in the support to serve as compression reinforcement at the face of the support. Therefore, it can be ignored.)

5–5 The beam shown in Fig. P5–5 carries its own dead load plus an additional service (unfactored) dead load of 1.5 kips/ft plus a service live load of 3.5 kips/ft.

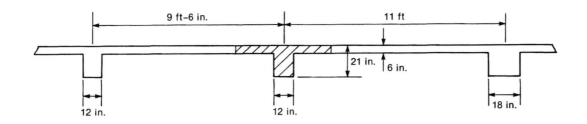

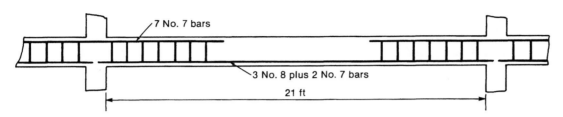

Fig. P5–4

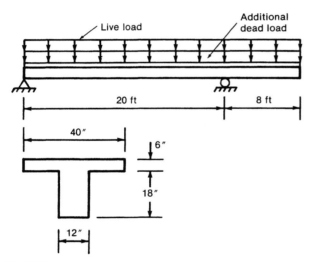

Fig P5–5

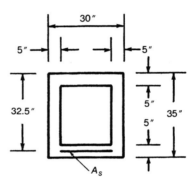

Fig P5–6

5–7 Compute ϕM_n for the beam shown in Fig. P5–7. Use $f_c' = 4000$ psi and $f_y = 60,000$ psi.

(a) Draw bending moment diagrams for the three loading cases shown in Fig. P4–18 and superimpose them to get a bending moment envelope.

(b) Select reinforcement at the negative and positive moment regions. Use $f_c' = 4000$ psi and $f_y = 60,000$ psi.

5–6 Compute ϕM_n for the beam shown in Fig. P5–6. Use $f_c' = 3000$ psi and $f_y = 60,000$ psi and

(a) the reinforcement is 6 No. 8 bars.

(b) the reinforcement is 9 No. 8 bars.

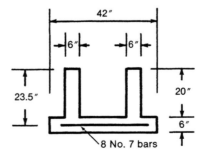

Fig P5–7

5–8 Give three reasons why compression reinforcement is used in beams.

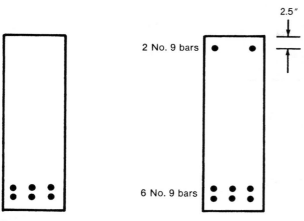
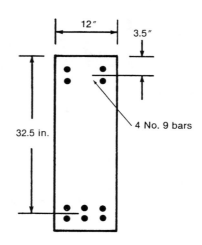

Figure P5–9

5–9 (a) Compute ϕM_n for the three beams shown in Fig. P5–9. In each case $f_c' = 3000$ psi, $f_y = 60,000$ psi, $b = 12$ in., $d = 32.5$ in., and $h = 36$ in.

(b) From the results of part (a) comment on whether adding compression reinforcement is a cost-effective way of increasing the strength, ϕM_n, of a beam.

5–10 Compute ϕM_n for the beam shown in Fig. P5–10. Use $f_c' = 2500$ psi and $f_y = 60,000$ psi. Does the steel yield in this beam at ultimate?

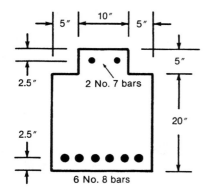

Fig P5–10

5–11 Compute ϕM_n for the beam shown in Fig. P5–11. Use $f_c' = 3000$ psi and $f_y = 60,000$ psi. Check if the steel yields.

5–12 The beam shown in Fig. P5–12 has elastic–plastic reinforcement ($f_s = f_y$ when $\epsilon_s \geq \epsilon_y$) with a yield strength of 60,000 psi. The concrete has $f_c' = 4000$ psi. Compute ϕM_n using a strain compatibility solution.

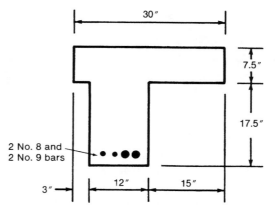

Fig P5–11

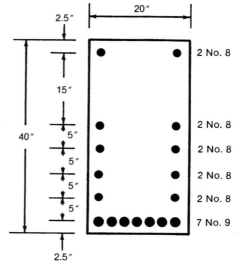

Fig P5–12

6

Shear in Beams

6-1 INTRODUCTION

A beam resists loads primarily by means of internal moments, M, and shears, V, as shown in Fig. 6–1. In the design of a reinforced concrete member, flexure is usually considered first, leading to the size of the section and the arrangement of reinforcement to provide the necessary moment resistance. Limits are placed on the amounts of flexural reinforcement which can be used, to ensure that if failure were ever to occur, it would develop gradually, giving warning to the occupants. The beam is then proportioned for shear. Because a shear failure is frequently sudden and brittle, as suggested by the damage sustained by the building in Fig. 6–2,[6-1] the design for shear must ensure that the shear strength equals or exceeds the flexural strength at all points in the beam.

The manner in which shear failures can occur varies widely depending on the dimensions, geometry, loading, and properties of the members. For this reason there is no unique way to design for shear. In this chapter we deal with the internal shear force, V, in relatively slender beams and the effect of the shear on the behavior and strength of beams. Examples of the design of such beams for shear are given in this chapter. Footings and two-way slabs supported on isolated columns develop shearing stresses on sections around the circumference of the columns, leading to failures in which the column and a conical piece of the slab punch through the slab (Chap. 13). Short deep members such as brackets, corbels, deep beams, and so on, transfer shear to the support by the compressive stresses rather than shear stresses. Such members are considered in Chap. 18.

Chapter 21 of the ACI Code gives special rules for shear reinforcement in members resisting seismic loads. These are reviewed in Chap. 19.

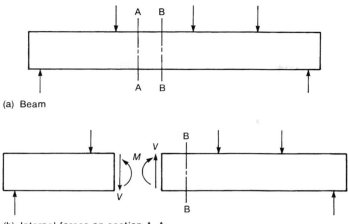

(a) Beam

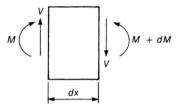

(b) Internal forces on section A–A.

Fig. 6–1
Internal forces in a beam.

(c) Internal forces on portion between sections A–A and B–B.

Fig. 6–2
Shear failure: U.S. Airforce
warehouse. (Photograph
courtesy of C. P. Siess.)

6–1 Introduction

Stresses in an Uncracked Elastic Beam

From the free-body diagram in Fig. 6–1c it can be seen that $dM/dx = V$. Thus shear forces and shear stresses will exist in those parts of a beam where the moment changes from section to section. By the traditional theory for *homogeneous*, *elastic*, *uncracked* beams, we can calculate the shear stresses, v, on elements cut out of a beam (Fig. 6–3a) using the equation

$$v = \frac{VQ}{Ib} \tag{6-1}$$

where

  V = shear force on the cross section

  I = moment of inertia of the cross section

  Q = first moment about the neutral axis of the part of the cross-sectional area lying farther from the neutral axis than the point where the shear stresses are being calculated

  b = width of the member where the stresses are being calculated

It should be noted that equal shearing stresses exist on both the horizontal and vertical planes through an element, as shown in Fig. 6–3a. The horizontal shear stresses are important in the design of construction joints, web-to-flange joints, or regions adjacent to holes in beams. For an *uncracked rectangular* beam, Eq. 6–1 gives the distribution of shear stresses shown in Fig. 6–3b.

The elements in Fig. 6–3a are subjected to combined normal stresses due to flexure, f, and shearing stresses, v. The largest and smallest normal stresses acting on such an element are referred to as *principal stresses*. The principal stresses and the planes they act on are found using a Mohr's circle for stress, as explained in any mechanics of materials textbook. The orientations of the principal stresses on the elements in Fig. 6–3a are shown in Fig. 6–3c.

The surfaces on which principal tension stresses act in the *uncracked* beam are plotted in Fig. 6–4a. These surfaces or *stress trajectories* are steep near the bottom of the beam and flatter near the top. This corresponds with the orientation of the elements shown in Fig. 6–3c. Since concrete cracks when the principal tensile stresses exceed the tensile strength of the concrete, the initial cracking pattern should resemble the network of lines shown in Fig. 6–4a.

The cracking pattern in a test beam is shown in Fig. 6–4b. Two types of cracks can be seen. The vertical cracks occurred first, due to flexural stresses. These start at the bottom of the beam where the flexural stresses are the largest. The inclined cracks at the ends of the beam are due to combined shear and flexure. These are commonly referred to as *inclined cracks*, *shear cracks*, or *diagonal tension cracks*. Such a crack must exist before a beam can fail in shear. Several of the inclined cracks have extended along the reinforcement toward the support, weakening the anchorage of the reinforcement.

Although there is a similarity between the planes of maximum principal tensile stress and the cracking pattern, it is by no means perfect. In reinforced concrete beams, flexural cracks generally occur before the principal tensile stresses at midheight become critical. Once such a crack has occurred, the tensile stress across the crack drops to zero. To maintain equilibrium, a major redistribution of stresses is necessary. As a result, the onset of inclined cracking in a beam cannot be predicted from the principal stresses unless shear cracking precedes flexural cracking. This very rarely happens in reinforced concrete but does occur in some prestressed beams.

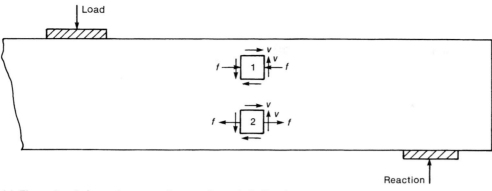

(a) Flexural and shear stresses acting on elements in the shear span.

(b) Distribution of shear stresses.

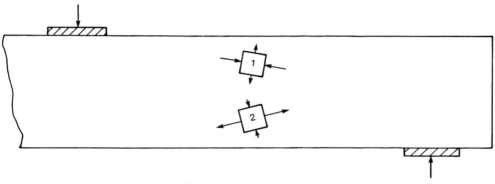

(c) Principal stresses on elements in shear span.

Fig. 6–3
Normal, shear, and principal stresses in homogeneous uncracked beam.

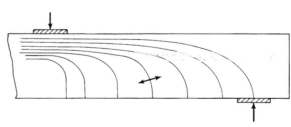

(a) Principal compressive stress trajectories in an uncracked beam.

Fig. 6–4
Principal compressive stress
trajectories and inclined
cracks. (Photograph courtesy
of J. G. MacGregor.)

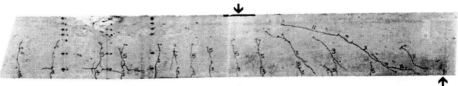

(b) Photograph of half of a cracked reinforced concrete beam.

6-2   Basic Theory

183

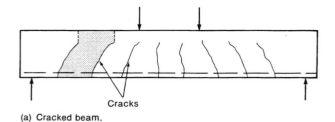

(a) Cracked beam.

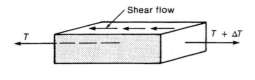

(c) Bottom part of beam.

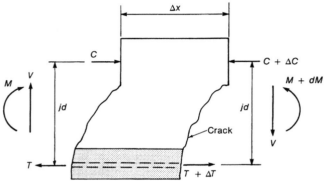

(b) Portion of beam between two cracks.

(d) Average shear stresses.

Fig. 6–5
Calculation of average shear stress between cracks.

Average Shear Stress between Cracks

The initial stage of cracking generally results in vertical cracks which, with increasing load, extend in a diagonal manner, as shown in Fig. 6–4b. The equilibrium of the section of beam between two such cracks (Fig. 6–5b) can be written as

$$T = \frac{M}{jd} \quad \text{and} \quad T + \Delta T = \frac{M + \Delta M}{jd}$$

or

$$\Delta T = \frac{\Delta M}{jd}$$

where jd is the lever arm, which is assumed to be constant. For moment equilibrium of the element,

$$\Delta M = V\Delta x \tag{6–2}$$

and

$$\Delta T = \frac{V\Delta x}{jd} \tag{6–3}$$

If the shaded portion of Fig. 6–5b is isolated as shown in Fig. 6–5c, the force ΔT must be transferred by horizontal shear stresses on the top of the element. The *average* value of these stresses below the top of the crack is

$$v = \frac{\Delta T}{b_w \Delta x}$$

or

$$v = \frac{V}{b_w jd} \qquad (6\text{–}4)$$

where $jd \simeq 0.875d$ and b_w is the thickness of the web. The distribution of *average* horizontal shear stresses is shown in Fig. 6–5d. Since the vertical shear stresses on an element are equal to the horizontal shear stresses on the same element, the distribution of vertical shear stresses will be as shown in Fig. 6–5d. This assumes that about 30% of the shear is transferred in the compression zone. The balance of the shear is transferred across the cracks. In 1970, Taylor[6-2] reported tests of beams without web reinforcement in which he found that about a quarter of the shear was transferred by the compression zone, a quarter by doweling action of the flexural reinforcement, and about half by aggregate interlock along the cracks (see Fig. 6–13, discussed later). Modern shear failure theories assume that a significant amount of the shear is transferred in the web, some of this across inclined cracks.

The ACI design procedure approximates Eq. 6–4 with Eq. 6–5, which does not require the computation of j:

$$v = \frac{V}{b_w d} \qquad (6\text{–}5)$$

Beam Action and Arch Action

In the derivation of Eq. 6–4 it was assumed that the beam was prismatic and the lever arm jd was constant. The relationship between shear and bar force (Eq. 6–3) can be rewritten as[6-3]

$$V = \frac{d}{dx}(Tjd) \qquad (6\text{–}6)$$

which can be expanded as

$$V = \frac{d(T)}{dx}jd + \frac{d(jd)}{dx}T \qquad (6\text{–}7)$$

Two extreme cases can be identified. If the lever arm, jd, remains constant as assumed in normal elastic beam theory,

$$\frac{d(jd)}{dx} = 0 \quad \text{and} \quad V = \frac{d(T)}{dx}jd$$

where $d(T)/dx$ is the *shear flow* across any horizontal plane between the reinforcement and the compression zone as shown in Fig. 6–5c. For beam action to exist, this shear flow must exist.

The other extreme occurs if the shear flow, $d(T)/dx$, equals zero, giving

$$V = T\frac{d(jd)}{dx}$$

6–2   Basic Theory

185

This occurs if the shear flow cannot be transmitted due to the steel being unbonded, or if the transfer of shear flow is prevented by an inclined crack extending from the load to the reactions. In such a case the shear is transferred by *arch action* rather than beam action, as illustrated in Fig. 6–6. In this member the compression force C in the inclined strut and the tension force T in the reinforcement are constant over the length of the shear span.

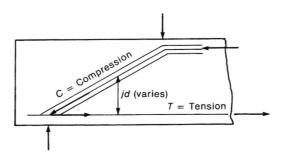

Fig. 6–6
Arch action in a beam.

Shear Reinforcement

In Chap. 4 we saw that horizontal reinforcement was required to restrain the opening of a vertical flexural crack as shown in Fig. 6–7a. An inclined crack opens approximately perpendicular to itself, as shown in Fig. 6–7b, and either a combination of horizontal flexural reinforcement and inclined reinforcement (Fig. 6–7c) or a combination of horizontal and vertical reinforcement (Fig. 6–7d) is required to restrain it from opening. The inclined or vertical reinforcement is referred to as *shear reinforcement* or *web reinforcement* and may be provided by inclined or vertical *stirrups*, as shown in Fig. 6–27 or 6–28. Most often, vertical stirrups are used in North America. The arrangement of stirrups in a beam is illustrated in Fig. 4–24. Inclined stirrups cannot be used in beams resisting shear reversals, such as buildings resisting seismic loads.

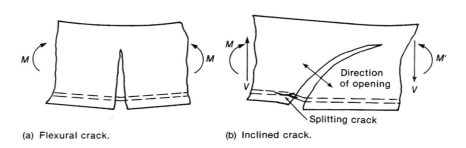

(a) Flexural crack.

(b) Inclined crack.

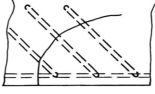

Fig. 6–7
Inclined cracks and shear reinforcement.

(c) Inclined shear reinforcement.

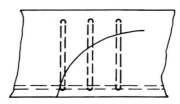

(d) Vertical shear reinforcement.

Shear in Beams

The behavior of beams failing in shear varies widely depending on the relative contributions of beam action and arch action and the amount of web reinforcement.

Behavior of Beams without Web Reinforcement

The moments and shears at inclined cracking and failure of rectangular beams without web reinforcement are plotted in Fig. 6–8b and c as a function of the ratio of the shear span, a, to the depth d (see Fig. 6–8a). The beam cross section remains constant as the span is varied. The maximum moment and shear that can be developed correspond to the nominal moment capacity, M_n, of the cross section plotted as a horizontal line in Fig. 6–8b. The shaded

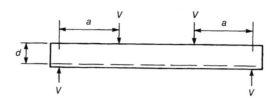

(a) Beam.

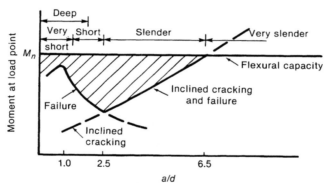

(b) Moments at cracking and failure.

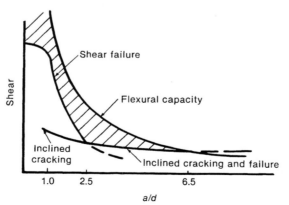

Fig. 6–8
Effect of a/d ratio on shear strength of beams without stirrups.

(c) Shear at cracking and failure.

areas in this figure show the reduction in strength due to shear. Web reinforcement is provided to ensure that the beam reaches the full flexural capacity, M_n.

Figure 6–8b suggests that the shear spans can be divided into four types: very short, short, slender, and very slender. The term *deep beam* is also used to describe beams with very short and short shear spans. Very short shear spans, with a/d from 0 to 1, develop inclined cracks joining the load and the support. These cracks, in effect, destroy the horizontal shear flow from the longitudinal steel to the compression zone and the behavior changes from beam action to arch action, as shown in Fig. 6–6 or 6–9. Here the reinforcement serves as the tension tie of a tied arch and has a uniform tensile force from support to support. The most common mode of failure in such a beam is an anchorage failure at the ends of the tension tie.

Short shear spans, a/d from 1 to 2.5, develop inclined cracks and, after a redistribution of internal forces, are able to carry additional load, in part by arch action. The final failure of such beams will be caused by a bond failure, a splitting failure, or a dowel failure along the tension reinforcement, as shown in Fig. 6–10a, or by crushing of the compression zone over the crack, as shown in Fig. 6–10b. The latter is referred to as a *shear compression failure*. Because the inclined crack generally extends higher into the beam than a flexural crack, failure occurs at less than the flexural moment capacity.

In slender shear spans, a/d from about 2.5 to about 6, the inclined cracks disrupts equilibrium to such an extent that the beam fails at the inclined cracking load as shown in

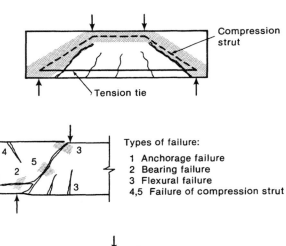

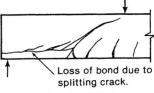

Types of failure:

1 Anchorage failure
2 Bearing failure
3 Flexural failure
4,5 Failure of compression strut

Fig. 6–9
Modes of failure of deep beams, $a/d = 0.5$ to 2.0. (Adapted from Ref. 6–4.)

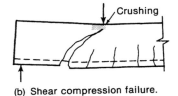

(a) Shear–tension failure.

Fig. 6–10
Modes of failure of short shear spans, $a/d = 1.5$ to 2.5. (Adapted from Ref. 6–4.)

(b) Shear compression failure.

Fig. 6–8b. Very slender beams with a/d greater than about 6 will fail in flexure prior to the formation of inclined cracks.

Figures 6–9 and 6–10 come from Ref. 6–4, which presents an excellent discussion of the behavior of beams failing in shear and the factors affecting their strengths. It is important to note that for short and very short beams, a major portion of the load capacity after inclined cracking is due to load transfer by the compression struts shown in Fig. 6–9. If the beam is not loaded on the top and supported on the bottom in the manner shown in Fig. 6–9, these compression struts are not effective and failure occurs at, or close to, the inclined cracking load.

Because the moment at the point where the load is applied is $M = Va$ for a beam loaded with concentrated loads, as shown in Fig. 6–8a, Fig. 6–8b can be replotted in terms of shear capacity, as shown in Fig. 6–8c. The shear corresponding to a flexural failure is the upper curved line. If stirrups are not provided, the beam will fail at the shear given by the "shear failure" line. This is roughly constant for a/d greater than about 2. Again the shaded area indicates the loss in capacity due to shear. Note that the inclined cracking loads of the short shear spans and slender shear spans are roughly constant. This is recognized in design by ignoring a/d in the equations for the shear at inclined cracking. In the case of slender beams, inclined cracking causes immediate failure if no web reinforcement is provided. For very slender beams, the shear required to form an inclined crack exceeds the shear corresponding to flexural failure and the beam will fail in flexure before inclined cracking occurs.

B-Regions and D-Regions

Figure 6–8 indicates that there is a major change in behavior at a shear span ratio, a/d, of about 2 to 2.5. Longer shear spans carry load by beam action and are referred to as *B-regions*, where the B stands for beam or for Bernoulli, who postulated the linear strain distribution in beams. Shorter shear spans carry load primarily by arch action involving in-plane forces. Such regions are referred to as *D-regions*, where the D stands for discontinuity or disturbed.[6-5]

St. Venant's principle suggests that a local disturbance such as a concentrated load or reaction will dissipate within about one beam depth from the point at which it is applied. Based on this, it is customary to assume that D-regions extend about one member depth each way from concentrated loads, reactions, or abrupt changes in section or direction as shown in Fig. 6–11. The regions between D-regions can be treated as B-regions.

In general, arch action enhances the strength of a section. As a result, B-regions tend to be weaker than corresponding D-regions, as shown by the lower line in Fig. 6–8c for a/d greater than 2 to 2.5. If a shear span consists entirely of D-regions that meet or overlap, as shown by the left end of Fig. 6–11a, its behavior will be governed by arch action. This accounts for the increase in shear strength when a/d is less than 2.

For longer shear spans, such as the right-hand end of the beam in Fig. 6–11a, the shear strength is governed by the B-region and is relatively constant, as shown in Fig. 6–8c. This type of member is discussed in this chapter. D-regions are discussed in Chap. 18.

Inclined Cracking

Inclined cracks must exist before a shear failure can occur. Inclined cracks form in the two different ways shown in Fig. 6–12. In thin-walled I beams in which the a/d ratio is small, the shear stresses in the web are high while the flexural stresses are low. In a few extreme cases and in some prestressed beams, the principal tension stresses at the neutral axis may exceed those at the bottom flange. In such a case a *web-shear crack* occurs (Fig. 6–12a).

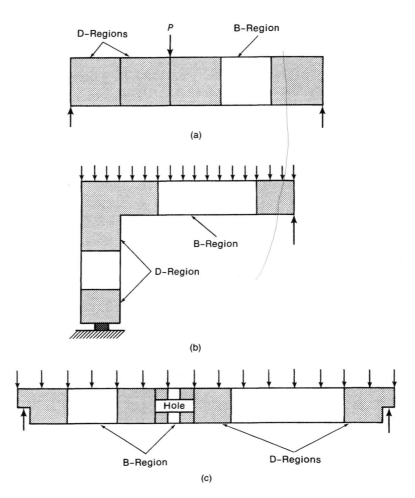

Fig. 6–11
B-regions and D-regions.

The inclined cracking shear can be calculated as the shear necessary to cause a principal tensile stress equal to the tensile strength of the concrete at the centroid of the beam.

In most reinforced concrete beams, however, flexural cracks occur first and extend more or less vertically into the beam, as shown in Fig. 6–4b or 6–12b. These alter the state of stress in the beam causing a stress concentration near the head of the crack. In time, the flexural cracks extend to become *flexure-shear cracks* (Fig. 6–12b), or flexure-shear cracks develop in the uncracked region over the flexural cracks (Fig. 6–4b).

Flexure-shear cracking *cannot* be predicted by calculating the principal stresses in an uncracked beam. For this reason empirical equations have been derived to calculate the flexure-shear cracking load.

The inclined cracks in a T beam loaded to produce positive and negative moments are shown in Fig. 5–3. The slope of the inclined cracks in the negative moment regions changes directions over the support because the shear force changes sign here. All of the inclined cracks in this beam are flexure-shear cracks.

Internal Forces in a Beam without Stirrups

The forces transferring shear across an inclined crack in a beam without stirrups are illustrated in Fig. 6–13. Shear is transferred across line A–B–C by V_{cz}, the shear in the compression zone;

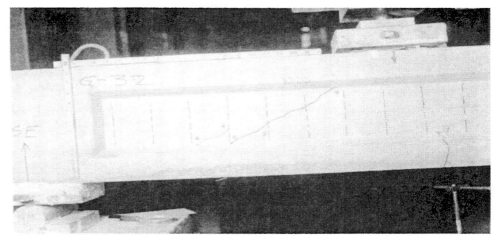

(a) Web-shear cracks.

Fig. 6–12
Types of inclined cracks.
(Photographs courtesy of
J. G. MacGregor.)

(b) Flexure-shear cracks.

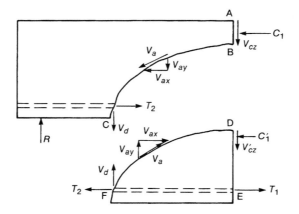

Fig. 6–13
Internal forces in a cracked
beam without stirrups.

V_{ay}, the vertical component of the shear transferred across the crack by interlock of the aggregate particles on the two faces of the crack; and V_d, the dowel action of the longitudinal reinforcement. Immediately after inclined cracking, as much as 40 to 60% of the total shear may be carried by V_d and V_{ay}, together.

Considering the portion D–E–F below the crack, and summing moments about the reinforcement at point E shows that V_d and V_a cause a moment about E which must be equilibrated by a compression force C_1'. Horizontal force equilibrium on section A–B–D–E shows that $T_1 = C_1 + C_1'$, and finally, T_1 and $C_1 + C_1'$ must equilibrate the external moment at this section.

As the crack widens, V_a decreases, increasing the fraction of the shear resisted by V_{cz} and V_d. The dowel shear, V_d, leads to a splitting crack in the concrete along the reinforcement (Fig. 6–10a). When this crack occurs, V_d drops to zero. When V_a and V_d disappear, so do V_{cz}' and C_1', with the result that all the shear and compression are transmitted in the width AB above the crack. This may cause crushing of this region.

It is important to note also that if $C_1' = 0$, $T_2 = T_1$, and as a result, $T_2 = C_1$. In other words, the inclined crack has made the tensile force at point C a function of the moment at section A–B–D–E. This shift in the tensile force must be considered when detailing the bar cutoff points and when anchoring the bars.

The shear failure of a slender beam without stirrups is sudden and dramatic. This is evident from Fig. 6–2. Although this beam had stirrups (which have broken and are hanging down from the upper part of the beam), they were so small as to be useless.

Factors affecting the Shear Strength of Beams without Web Reinforcement

Beams without web reinforcement will fail when inclined cracking occurs or shortly afterwards. For this reason the shear capacity of such members is taken equal to the inclined cracking shear. The inclined cracking load of a beam is affected by five principal variables, some included in design equations and others not.

Tensile Strength of Concrete. The inclined cracking load is a function of the tensile strength of the concrete. The stress state in the web of the beam involves biaxial principal tension and the compression stresses as shown in Fig. 6–30 (see Sec. 3–2). A similar biaxial state of stress exists in a split cylinder tension test (Fig. 3–9), and the inclined cracking load is frequently related to the strength from such a test. As discussed earlier, the flexural cracking which precedes the inclined cracking disrupts the elastic stress field to such an extent that inclined cracking occurs at a principal tensile stress, based on the uncracked section, of roughly a third of f_{ct}.

Longitudinal Reinforcement Ratio, ρ_w. Figure 6–14 presents the shear capacities of simply supported beams without stirrups as a function of the steel ratio, $\rho_w = A_s/b_w d$. The practical range of ρ_w for beams developing shear failures is about 0.0075 to 0.025. In this range, the shear strength is approximately

$$V_c = 2\sqrt{f_c'}\,b_w d \quad \text{lb} \qquad (6\text{–}8)$$
$$\text{(ACI Eq. 11–3)}$$

and in SI units

$$V_c = \frac{\sqrt{f_c'}\,b_w d}{6} \quad \text{N} \qquad (6\text{–}8\text{M})$$

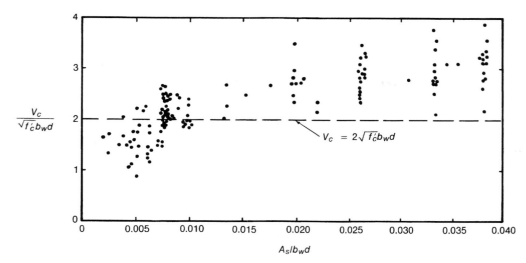

Fig. 6–14
Effect of reinforcement ratio, ρ_w, on shear capacity, V_c, of beams without stirrups.

as indicated by the horizontal dashed line in Fig. 6–14. This equation tends to overestimate V_c for beams with small steel percentages.[6-6]

When the steel ratio, ρ_w, is small, flexural cracks extend higher into the beam and open wider than would be the case for large values of ρ_w. As a result, inclined cracking occurs earlier.

Shear Span-to-Depth Ratio, a/d. The shear span-to-depth ratio, a/d or M/Vd, has some effect on the inclined cracking shears and ultimate shears of shear spans with a/d less than 2, as shown in Fig. 6–8c. Such shear spans are "deep" shear spans (D-regions) and are discussed in Chap. 18. For longer shear spans where B-region behavior dominates, a/d has little effect on the inclined cracking shear (Fig. 6–8c) and can be neglected.

Size of Beam. As the overall depth of a beam increases, the shear stress at inclined cracking tends to decrease for a given f'_c, ρ_w, and a/d.[6-4] As the depth of the beam increases, the crack widths at points above the main reinforcement tend to increase (see Sec. 9–2). This leads to a reduction in aggregate interlock across the crack, resulting in earlier inclined cracking. Collins and Mitchell[6-7] quote tests of geometrically similar uniformly loaded beams *without* web reinforcement made from concrete with constant aggregate size. A beam with $d = 24$ in. failed at a shear of approximately $V_u = 2\sqrt{f'_c}\,b_w d$, while beams with effective depths of 79 and 118 in. failed at V_u less than $1\sqrt{f'_c}\,b_w d$. In beams *with* web reinforcement, on the other hand, the web reinforcement holds the crack faces together so that the aggregate interlock is not lost and the reduction in shear strength due to size is not believed to occur.

Axial Forces. Axial tensile forces tend to decrease the inclined cracking load, while axial compressive forces tend to increase it (Fig. 6–15). As the axial compressive force is increased, the onset of flexural cracking is delayed and the flexural cracks do not penetrate as far into the beam. As a result, a larger shear is required to cause principal tensile stresses equal to the tensile strength of the concrete.

This is only partially true in a prestressed concrete beam. The onset of flexural cracking is delayed by the prestress, but once flexural cracking occurs, the cracks penetrate about the same height as in a comparable reinforced concrete beam. This is because $C = T$ in a prestressed beam just as it is in a reinforced concrete beam. The increase in inclined cracking load for a prestressed concrete beam is largely due to the delay of flexural cracking.

Behavior of Beams with Web Reinforcement

Due to inclined cracking, the strength of beams drops below the flexural capacity as shown in Fig. 6–8b and c. The purpose of web reinforcement is to ensure that the full flexural capacity can be developed.

Prior to inclined cracking, the strain in the stirrups is equal to the corresponding strain of the concrete. Since concrete cracks at a very small strain, the stress in the stirrups prior to inclined cracking will not exceed 3 to 6 ksi. Thus stirrups do not prevent inclined cracks from forming; they come into play only after the cracks have formed.

The forces in a beam with stirrups and an inclined crack are shown in Fig. 6–16. The terminology is the same as in Fig. 6–13. The shear transferred by tension in the stirrups is V_s. Since V_s does not disappear when the crack opens, there will always be a compression force C_1' and a shear force V_{cz}' acting on the part of the beam below the crack. As a result, T_2 will be less than T_1, the difference depending on the amount of web reinforcement. The force T_2 will, however, be larger than $T = M/jd$ based on the moment at C.

The loading history of such a beam is shown qualitatively in Fig. 6–17. The components of the internal shear resistance must equal the applied shear as indicated by the upper 45° line. Prior to flexural cracking, all the shear is carried by the uncracked concrete. Between flexural and inclined cracking, the external shear is resisted by V_{cz}, V_{ay}, and V_d. Eventually, the stirrups crossing the crack yield, and V_s stays constant for higher applied shears. Once the stirrups yield, the inclined crack opens more rapidly. As the inclined crack

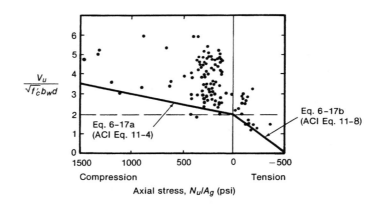

Fig. 6–15
Effect of axial loads on inclined cracking shear.

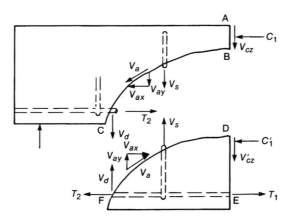

Fig. 6–16
Internal forces in a cracked beam with stirrups.

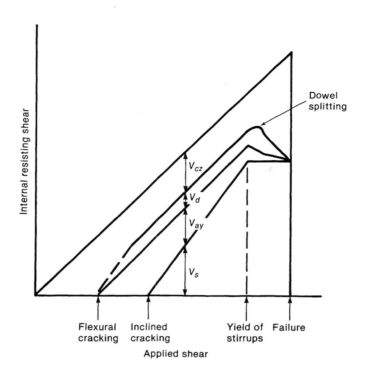

Fig. 6–17
Distribution of internal shears in a beam with web reinforcement. (from Ref. 6–4.)

widens, V_{ay} decreases further, forcing V_d and V_{cz} to increase at an accelerated rate until either a splitting (dowel) failure occurs, or the compression zone crushes due to combined shear and compression.

Each of the components of this process except V_s have a brittle load–deflection response. As a result, it is difficult to quantify the contributions of V_{cz}, V_d, and V_{ay}. In design these are lumped together as V_c, referred to somewhat incorrectly as "the shear carried by the concrete." Thus the nominal shear strength, V_n, is assumed to be

$$V_n = V_c + V_s \qquad (6-9)$$

Traditionally in North America design practice, V_c is taken equal to the failure capacity of a beam without stirrups, which, in turn, is taken equal to the inclined cracking shear as suggested by the line indicating inclined cracking and failure for a/d from 2.5 to 6.5 in Fig. 6–8c. This is discussed more fully in Sec. 6–5.

6–4 TRUSS MODEL OF THE BEHAVIOR OF SLENDER BEAMS FAILING IN SHEAR

The behavior of beams failing in shear must be expressed in terms of a mechanical-mathematical model before designers can make use of this knowledge in design. The best model for beams with web reinforcement is the truss model. This is applied to slender beams in this chapter and to deep beams in Chap. 18.

In 1899 and 1902, respectively, the Swiss engineer Ritter and the German engineer Mörsch, independently, published papers proposing the truss analogy for the design of reinforced concrete beams for shear. These procedures provide an excellent conceptual model to show the forces that exist in a cracked concrete beam.

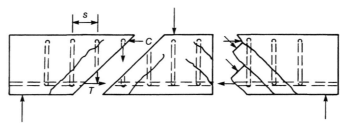

(a) Internal forces in a cracked beam.

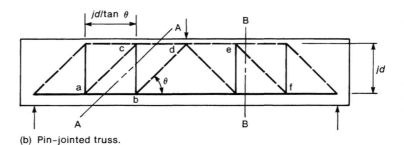

(b) Pin-jointed truss.

Fig. 6–18
Truss analogy.

As shown in Fig. 6–18a, a beam with inclined cracks develops compressive and tensile forces, C and T, in its top and bottom "flanges," vertical tensions in the stirrups and inclined compressive forces in the concrete "diagonals" between the inclined cracks. This highly indeterminate system of forces is replaced by an analogous truss. The simplest truss is shown in Fig. 6–18b; a more complicated truss is shown in Fig. 6–19b.

Several assumptions and simplifications are needed to derive the analogous truss. In Fig. 6–18b the truss has been formed by lumping all of the stirrups cut by section A–A into one vertical member b–c and all the diagonal concrete members cut by section B–B into one diagonal member e–f. This diagonal member is stressed in compression to resist the shear on section B–B. The compression chord along the top of the truss is actually a force in the concrete but is shown as a truss member. The compressive members in the truss are shown with dashed lines to imply that they are really forces in the concrete, not separate truss members. The tensile members are shown with solid lines.

Figure 6–19a shows a beam with inclined cracks. The left end of this beam can be replaced by the truss shown in Fig. 6–19b. In design, the ideal distribution of stirrups would correspond to all stirrups reaching yield by the time the failure load is reached. It will be assumed, therefore, that all the stirrups have yielded and each transmits a force of $A_v f_y$ across the crack, where A_v is the area of the stirrup legs. When this is done, the truss becomes statically determinate. The truss in Fig. 6–19b is referred to as the *plastic truss model* since we are depending on plasticity in the stirrups to make it statically determinate. The beam will be proportioned so that the stirrups yield before the concrete crushes, so that it will not depend on plastic action in the concrete.

This truss model ignores the shear components V_{cz}, V_{ay}, and V_d in Fig. 6–16. Thus it does not assign any shear "to the concrete." A truss analogy that includes such a term will be discussed briefly later.

Construction of a Plastic Truss Model

The construction of a plastic truss model is illustrated by example in Fig. 6–19. In drawing this truss, it is assumed that:

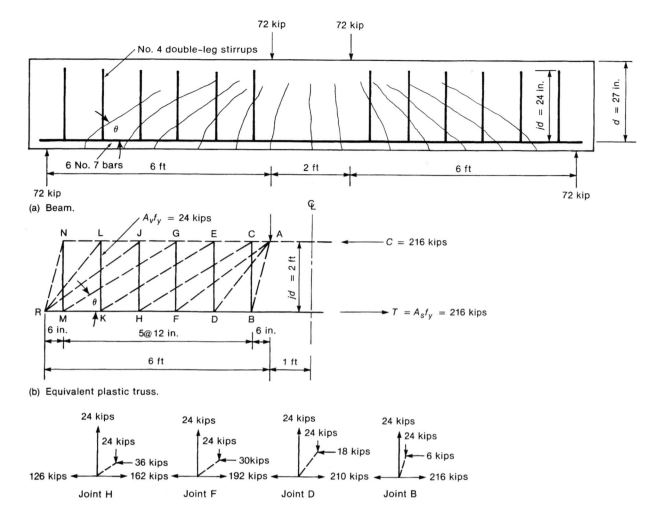

(a) Beam.

(b) Equivalent plastic truss.

(c) Equilibrium of bottom chord joints.

Fig. 6–19
Construction of a plastic truss analogy.

1. The cracks are at an angle θ to the horizontal, where θ is between 25 and 65°.

2. All the shear is resisted by stirrups.

3. The beam is on the verge of a simultaneous shear and flexural failure. Thus it is assumed that all of the stirrups have yielded, each carrying a vertical force of $A_v f_y = 24$ kips.

For the beam in Fig. 6–19 the moment at midspan is 432 ft-kips. Assuming that $jd = (d - a/2)$ is 2 ft, the compression and tension forces C and T at midspan are 216 kips, as shown in Fig. 6–19b.

The vertical applied load of 72 kips must be transmitted by diagonal compression struts (shown by dashed inclined lines) to enough stirrups (shown by solid vertical lines) to equilibrate this force. Since each stirrup can resist a vertical force of 24 kips, three stirrups are required to transmit 72 kips. The vertical applied load of 72 kips will be transmitted by the three diagonals AB, AD, and AF to joints B, D, and F at the bottom of the truss. The right-hand diagram in Fig. 6–17c shows the equilibrium of joint B. The vertical force in the stirrup is $A_v f_y = 24$ kips. The vertical component of the force in the diagonal AB must be 24 kips for equilibrium. From the slope of AB, we find it must have a horizontal component of 6 kips. At midspan, the tension force T in the reinforcement at the bottom of $M/jd = 216$ kips, as shown earlier. Summing the horizontal forces at point B, we find that the tension force between B and D is $216 - 6 = 210$ kips. Similar calculations are carried out at joints D and F.

The stirrup BC transmits the vertical force of $A_v f_y = 24$ kips to the top of the truss at joint C, where it is resisted by the vertical component of the force in diagonal CH, and so on.

The compression diagonals originating at the load (AB, AD, and AF) are referred to as a *compression fan*. The number of such diagonals in the fan must be such that the entire vertical load at A is resisted by the vertical force components in the diagonals meeting at A. A similar compression fan exists at the support R (RN, RL, RJ). Between the compression fans is a *compression field* consisting of the parallel diagonal struts CH, EK, and GM. The angle θ of the compression field is determined by the number of stirrups needed to equilibrate the vertical loads in the fans.

Each of the compression fans occurs in a D-region (discontinuity region). The compression field is a B-region (beam region).

Figure 6–20a shows the crack pattern in a two-span continuous beam. The corresponding truss model is shown in Fig. 6–20b. Figure 6–21 is a close-up of the compression fan over the interior support after failure. The radiating struts in the fan can be clearly seen.

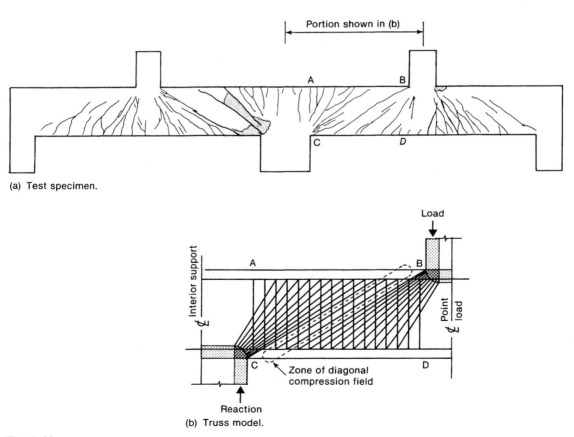

(a) Test specimen.

(b) Truss model.

Fig. 6–20
Crack pattern and truss model for a two-span beam. (From Ref. 6–10.)

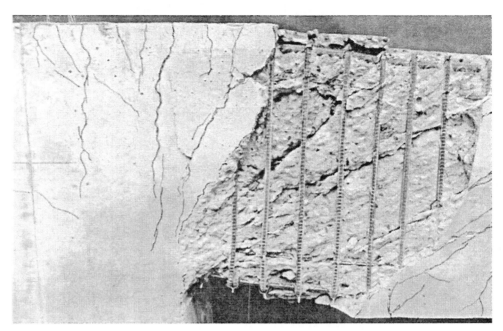

Fig. 6–21
Compression fan at interior support of the beam shown in Fig. 6–20b. (Photograph courtesy of
J. G. MacGregor.)

Simplified Truss Analogy

A statically determinate truss analogy can be derived using the method suggested by
Marti.[6-8, 6-9] Figures 6–22a and b show a uniformly loaded beam with stirrups and a truss
model incorporating all the stirrups and representing the uniform load as a series of concentrated loads at the panel points. The truss in Fig. 6–22b is statically indeterminate but can be
solved if it is assumed that the forces in each stirrup cause that stirrup to just reach yield, as
was done in the preceding paragraphs. For design, it is easier to represent the truss as shown
in Fig. 6–22c, where the tension force in each vertical member represents the force in all the
stirrups within a length $jd/\tan \theta$. Similarly, each inclined compression strut represents a
width of web equal to $jd \cos \theta$. The uniform load has been idealized as concentrated loads
of $w(jd/\tan \theta)$ acting at the panel points. The truss in Fig. 6–22c is statically determinate.
To draw such a truss it is necessary to choose θ. This will be discussed later.

Internal Forces in the Plastic Truss Model

If we consider the free-body diagram cut by section A–A parallel to the diagonals in the
compression field region in Fig. 6–23a, the entire vertical component of the shear force is
resisted by tension forces in the stirrups crossing this section. The horizontal projection of
section A–A is $jd \cot \theta$ and the number of stirrups it cuts is $jd \cot \theta/s$. The force in one stirrup is $A_v f_y$, which can be calculated from

$$A_v f_y = \frac{Vs}{jd \cot \theta} \tag{6–10}$$

The free body shown in Fig. 6–23b is cut by a vertical section between G and J in Fig.
6–19b. Here the vertical force, V, acting on the section must be resisted by an inclined compressive force $D = V/\sin \theta$ in the diagonals (Fig. 6–23c). The width of the diagonals is
$(jd \cos \theta)$ as shown in Fig. 6–23b, and the average compressive stress in the diagonals is

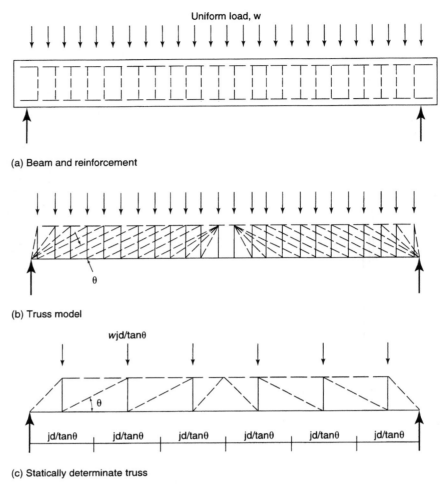

Uniform load, w

(a) Beam and reinforcement

(b) Truss model

$wjd/\tan\theta$

(c) Statically determinate truss

Fig. 6–22
Truss model for design. (From Collins/Mitchell, PRESTRESSED CONCRETE STRUC-
TURES, (c) 1990, p. 339. Reprinted by permission of Prentice Hall, Upper Saddle
River, New Jersey.)

$$f_{cd} = \frac{V}{b_w jd \cos\theta \sin\theta} \qquad (6\text{–}11\text{a})$$

Making use of trigonometric identities, this becomes

$$f_{cd} = \frac{V}{b_w jd}\left(\tan\theta + \frac{1}{\tan\theta}\right) \qquad (6\text{–}11\text{b})$$

where b_w is the thickness of the web. If the web is very thin, this stress may cause the web
to crush as shown in Fig. 6–24.

The shear V on section $B\text{–}B$ has been replaced by the diagonal compression force D
and an axial tension force N_v, as shown in Fig. 6–23c.

$$N_v = \frac{V}{\tan\theta} \qquad (6\text{–}12)$$

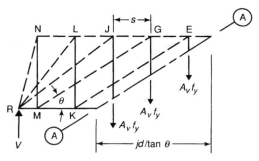

(a) Calculation of forces in stirrups.

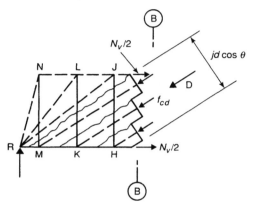

Fig. 6–23
Forces in stirrups and compression diagonals.

(b) Calculation of stress in compression diagonals.

$$D = \frac{V}{\sin \theta}$$

$$N_v = \frac{V}{\tan \theta}$$

(c) Replacement of V with D and N_v.

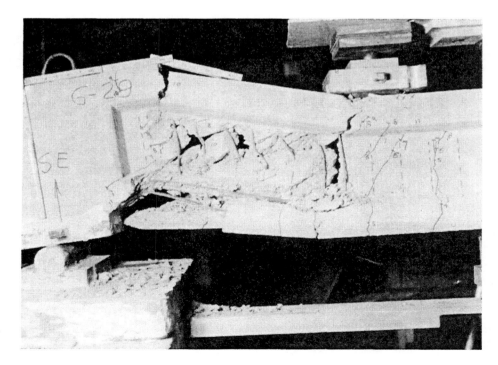

Fig. 6–24
Web crushing failure.
(Photograph courtesy of J. G. MacGregor.)

If it is assumed that the shear stress is constant over the height of the beam, the resultants of D and N_v act at midheight. As a result, a tensile force of $N_v/2$ acts in each of the top and bottom chords as shown in Fig. 6–23b. This reduces the force in the compression chord and increases the force in the tension chord.

In the compression fan regions the angle θ varies and hence N_v varies, approaching zero immediately under the load. The distribution of forces in the tension chord and compression chord of the truss in Fig. 6–19 are shown in Fig. 6–25. The force distribution, C or $T = M/jd$, due to flexure is shown by the dashed lines.

In the compression field region (F to R on the lower chord and A to J on the upper chord), the force in the tension chord halfway between each panel point is larger than $T = M/jd$ by the amount $N_v/2$, and the force in the compression chord is smaller than $C = M/jd$ by the same amount, as shown by the short dashed lines in Fig. 6–25. For this truss, $\cot \theta = 1.5$ and hence $N_v/2 = 0.75V$, which, for $V = 72$ kips, is 54 kips. At the end of the beam it is necessary to anchor the longitudinal bars for a tension force of $N_v/2$ even though the moment is zero.

In the compression fan region under the load, the value of $N_v/2$ in the tension chord gradually reduces to zero as shown in Fig. 6–25a, so that at the point of maximum moment, the force in the reinforcement is $T = M/jd$. The shift in the tension force diagram is equivalent to computing T from a moment diagram which has been shifted away from the points of maximum moment by an amount $(jd \cot \theta)/2$. This is discussed more fully in Chap. 8.

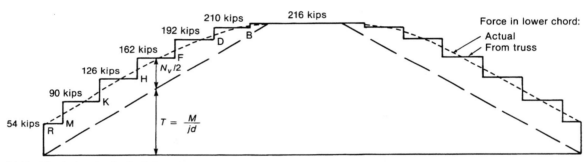

(a) Tension in longitudinal reinforcement.

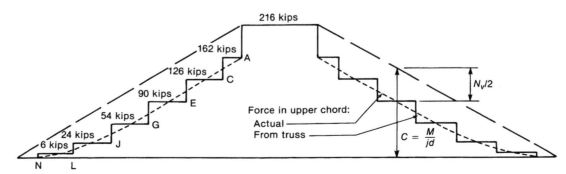

(b) Compression in upper chord.

Fig. 6–25
Forces in lower and upper chords of truss in Fig. 6–19.

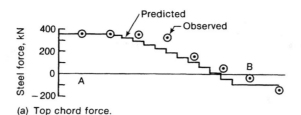

(a) Top chord force.

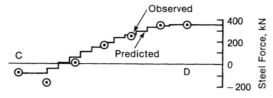

Fig. 6–26
Measured and computed
forces in the top and bottom
reinforcement of the portion
of the beam modeled in Fig.
6–20b. (From Ref. 6–10)

(b) Bottom chord force.

Figure 6–26 compares the measured and computed forces in the top and bottom bars in the beam shown in Fig. 6–20. The diagrams correspond to portion A–B–C–D of the beam modeled in Fig. 6–20b. The truss model accurately predicted the bar forces.

Value of θ in Compression Field Region

When a reinforced concrete beam with stirrups is loaded to failure, inclined cracks initially develop at an angle of 35 to 45° with the horizontal. With further loading, the angle of the compression stresses may cross some of the cracks.[6–7, 6–10] For this to occur, aggregate interlock must exist.

The allowable range of θ is expressed at $0.5 \leq \cot \theta \leq 2.0 (\theta = 26$ to $64°)$ in the Swiss code.[6–11] This range was selected to limit crack widths. A more restricted range, $\frac{3}{5} \leq \cot \theta \leq \frac{5}{3} (\theta = 31$ to $59°)$ is allowed in the European Concrete Committee's Model Code.[6–12] Based on a compatibility analysis, Collins and Mitchell[6–13] proposed limits which can be simplified to

$$\theta_{\min} = 10 + 110 \left(\frac{V_u}{\phi f_c' b_w jd} \right) \quad \text{deg} \tag{6–13a}$$

$$\theta_{\max} = 90 - \theta_{\min} \quad \text{deg} \tag{6–13b}$$

In design, the value of θ should be in the range of $25° \leq \theta \leq 65°$. The choice of a small value of θ reduces the number of stirrups required (Eq. 6–10), but increases the compression stresses in the web (Eq. 6–11) and increases N_v and hence the shift in the moment diagram. The opposite is true for large angles.

In the analysis of a given beam, as done in Figs. 6–19, 6–23, and 6–25, the angle θ is determined by the number of stirrups needed to equilibrate the applied loads and reactions. The angle should be within the limits given except in compression fan regions.

Crushing Strength of Concrete in the Web

The web of the beam will crush if the inclined compressive stress, f_{cd} from Eqs. 6–11, exceeds the strength of the concrete. The compressive strength, f_{ce}, of the concrete in a web which has previously been cracked and which contains stirrups stressed in tension at an angle to the cracks will tend to be less than f_c' as explained in Sec. 3–2. A reasonable limit is $0.25f_c'$ for $\theta = 30°$, increasing to $0.45f_c'$ for $\theta = 45°$. This problem is discussed more fully in Ref. 6–13 and in Sec. 18–2.

Compression Field Theory and Modified Compression Field Theory

The truss analogy represents the beam as a truss with individual stirrups or ties representing groups of stirrups, and individual compression struts, each of which represents a length of web. In the *compression field theory*, originally presented by Collins and Mitchell,[6-13] the web of the beam is represented as a continuum consisting of diagonally cracked concrete crossed by stirrups. The forces in the stirrups are given by Eq. 6–10. The forces in the top and bottom chords are $\pm (M/jd)$ from flexure plus $N_v/2$ from Eq. 6–12. The compressive stress in the web comes from Eq. 6–11. The web is assumed to crush when these stresses reach the compressive strength of cracked concrete. The compressive strength of concrete with cracks parallel to the direction of the compressive forces is discussed in Sec. 3–2, where it is shown to be a function of the width of the cracks, which, in turn, is expressed in terms of the average strain perpendicular to the direction of the cracks, ϵ_1, taken over a gauge length which includes several cracks. When design is carried out using the compression field theory, the crushing strength of the web is expressed using Eq. 3–13, which is a function of ϵ_1. Methods of computing ϵ_1 are given in Refs. 6–7, 6–13, and 6–14.

The *modified compression field theory* is similar to the compression field theory except that a friction or aggregate interlock stress is transferred across the inclined cracks in the web. The magnitude of these stresses is a function of the width of the cracks, which in turn is a function of the crack spacing and reinforcement spacing. The theory is explained in detail in Refs. 6–7 and 6–14. In addition to giving a realistic prediction of the behavior and strength of beams failing in shear, the modified compression field theory is important because it gives a physical explanation for the "shear carried by the concrete," V_c.

6–5 ANALYSIS AND DESIGN OF REINFORCED CONCRETE BEAMS FOR SHEAR—ACI CODE

In the ACI Code, the basic design equation for the shear capacity of slender concrete beams (beams with shear spans containing B-regions) is

$$\phi V_n \geq V_u \tag{6–14}$$
$$\text{(ACI Eq. 11–1)}$$

where V_u is the shear force due to the factored loads, ϕ is a strength reduction factor, taken equal to 0.85 for shear, and V_n, the nominal shear resistance, is

$$V_n = V_c + V_s \tag{6–9}$$
$$\text{(ACI Eq. 11–2)}$$

where V_c is the shear carried by the concrete and V_s is the shear carried by the stirrups.

A shear failure is said to occur when one of several shear limit states is reached. The following paragraphs list the principal limit states and describe how these are accounted for in the ACI Code.

Shear Failure Limit State: Beams without Web Reinforcement

Slender beams without web reinforcement will fail when inclined cracking occurs or shortly afterward. For this reason, the shear strength of such members is taken equal to the inclined cracking shear. The factors affecting the inclined cracking load were discussed in Sec. 6–3.

Design Equations for the Shear Strength of Members without Web Reinforcement

In 1962, the ACI–ASCE Committee on Shear and Diagonal Tension[6-15] presented the following equation for calculating the shear at inclined cracking in beams without web reinforcement:

$$V_c = \left(1.9\sqrt{f_c'} + \frac{2500\rho_w V_u d}{M_u}\right) b_w d \tag{6-15}$$
$$\text{(ACI Eq. 11-5)}$$

The derivation of this equation followed two steps. First, a rudimentary analysis of the stresses at the head of a flexural crack in a shear span was carried out to identify the significant parameters. Then the existing test data were statistically analyzed to establish the constants, 1.9 and 2500, and to drop other terms. The data used in the statistical analysis included "short" and "slender" beams, thereby mixing data from two different behavior types. In addition, most of the beams had high reinforcement ratios, ρ_w. More recent studies have suggested that Eq. 6–15 underestimates the effect of ρ_w for beams without web reinforcement and is not entirely correct in its treatment of the variable a/d, expressed as $V_u d/M_u$ in Eq. 6–15.

For the normal range of variables, the second term in the parentheses in Eq. 6–15 will be equal to about $0.1\sqrt{f_c'}$. If this is substituted into Eq. 6–15, Eq. 6–8 results:

$$V_c = 2\sqrt{f_c'}\, b_w d \tag{6-8}$$
$$\text{(ACI Eq. 11-3)}$$

In 1977, the ACI–ASCE Committee on Shear and Diagonal Tension recommended that Eq. 6–15 no longer be used.[6-6] For this reason it will not be used in this book.

Based on statistical studies of beam data for slender beams without web reinforcement, Zsutty[6-16] derived Eq. 6–16. This equation much more closely models the actual effects of f_c', ρ_w, and a/d than does Eq. 6–15.

$$v_c = 59\left(f_c' \rho_w \frac{d}{a}\right)^{1/3} \text{psi} \tag{6-16}$$

For design, the ACI Code presents both Eqs. 6–8 and 6–15 for computing V_c (ACI Secs. 11.3.1.1 and 11.3.2.1).

For axially loaded members, the ACI Code modifies Eq. 6–8 as follows:

Axial compression (ACI Sec. 11.3.1.2):

$$V_c = 2\left(1 + \frac{N_u}{2000A_g}\right)\sqrt{f_c'}\, b_w d \tag{6-17a}$$
$$\text{(ACI Eq. 11-4)}$$

Axial tension (ACI Sec. 11.3.2.3):

$$V_c = 2\left(1 + \frac{N_u}{500A_g}\right)\sqrt{f_c'}\, b_w d$$

(6–17b)

(ACI Eq. 11–7)

In both of these equations, N_u is positive in compression and $\sqrt{f_c'}$, N_u/A_g, 500, and 2000 all have units of psi. Axially loaded members are discussed more fully in Sec. 6–8.

In SI units, Eqs. 6–8, 6–17a, and 6–17b become

$$V_c = \frac{\sqrt{f_c'}}{6} b_w d$$

(6–8M)

For axial compression:

$$V_c = \left(1 + \frac{N_u}{14A_g}\right)\left(\frac{\sqrt{f_c'}}{6}\right) b_w d$$

(6–17aM)

For axial tension:

$$V_c = \left(1 + \frac{0.3N_u}{A_g}\right)\left(\frac{\sqrt{f_c'}}{6}\right) b_w d$$

(6–17bM)

where N_u is positive in compression and $\sqrt{f_c'}$, N_u/A_g 14, and 0.3 all have units of MPa.

Shear Failure Limit States: Beams with Web Reinforcement

1. Failure due to yielding of the stirrups. In Fig. 6–16 shear was transferred across the surface A–B–C by shear in the compression zone, V_{cz}, the vertical component of the aggregate interlock, V_{ay}, dowel action, V_d, and stirrups, V_s. In the ACI Code V_{cz}, V_{ay}, and V_d are lumped together as V_c, which is referred to as the "shear carried by the concrete." Thus the nominal shear strength, V_n, is assumed to be

$$V_n = V_c + V_s$$

(6–9)

(ACI Eq. 11–2)

The ACI Code further assumes that V_c is equal to the shear strength of a beam without stirrups, which, in turn, is taken equal to the inclined cracking load as given by Eq. 6–8, 6–15, or 6–17. It should be emphasized that taking V_c equal to the shear at inclined cracking is an *empirical* observation from tests, which is *approximately true* if it is assumed that the horizontal projection of the inclined crack is d, as shown in Fig. 6–27. If a flatter crack is used so that (jd $\cot \theta$) is greater than d, a smaller value of V_c must be used. For the values of θ approaching 30° which are used in the plastic truss model, V_c approaches zero, as assumed in that model.

Figure 6–27a shows a free body between the end of a beam and an inclined crack. The horizontal projection of the crack is taken as d, suggesting that the crack is slightly flatter than 45°. If s is the stirrup spacing, the number of stirrups cut by the crack is d/s. Assuming that all the stirrups yield at failure, the shear resisted by the stirrups is

$$V_s = \frac{A_v f_y d}{s}$$

(6–18)

(ACI Eq. 11–15)

This is equivalent to Eq. 6–9 derived from the truss model if $\theta = 45°$ and jd is replaced by d.

If the stirrups are inclined at an angle, α, to the horizontal, as shown in Fig. 6–27b, the number of stirrups crossing the crack is approximately $d(1 + \cot \alpha)/s$, where s is the horizontal spacing of the cracks. The inclined force, F, is

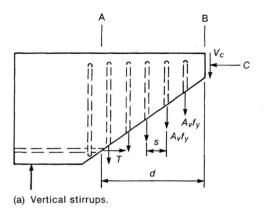

(a) Vertical stirrups.

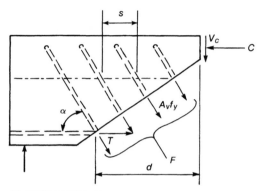

Fig. 6–27
Shear resisted by stirrups.

(b) Inclined stirrups.

$$F = A_v f_y \left[\frac{d(1 + \cot \alpha)}{s} \right] \qquad (6\text{–}19)$$

The shear resisted by the stirrups, V_s, is the vertical component of F, which is $F \sin \alpha$, so that

$$V_s = A_v f_y \left(\sin \alpha + \cos \alpha \right) \frac{d}{s} \qquad (6\text{–}20)$$
$$(\text{ACI Eq. } 11\text{–}16)$$

Figures 6–27 and 6–16 also show that the inclined crack affects the tension force, T, making it larger than the moment diagram would suggest. This is immediately obvious from the truss analogy, but is less obvious in the ACI design method.

If V_u exceeds ϕV_c, stirrups must be provided so that

$$V_u \leq \phi V_n \qquad (6\text{–}14)$$
$$(\text{ACI Eq. } 11\text{–}1)$$

where V_n is given by Eq. 6–9. In design this is generally rearranged to the form

$$\phi V_s \geq V_u - \phi V_c \quad \text{or} \quad V_s \geq \frac{V_u}{\phi} - V_c$$

Introducing Eq. 6–18 and rearranging gives the stirrup spacing:

$$s = \frac{A_v f_y d}{V_u/\phi - V_c} \qquad (6\text{--}21)$$

This equation applies for vertical stirrups.

Stirrups are unable to resist shear unless they are crossed by an inclined crack. For this reason, ACI Sec. 11.5.4.1 sets the maximum spacing of vertical stirrups as the smaller of $d/2$ or 24 in., so that each 45° crack will be intercepted by at least one stirrup (Fig. 6–28a). The maximum spacing of inclined stirrups is such that a 45° crack extending from midheight of the member to the tension reinforcement will intercept at least one stirrup, as shown in Fig. 6–28b.

If $V_u/\phi - V_c = V_s$ exceeds $4\sqrt{f_c'}\,b_w d$, the maximum allowable stirrup spacings are reduced to half those just described. For vertical stirrups, the maximum is the smaller of $d/4$ or 12 in. This is done for two reasons. Closer stirrup spacing leads to narrower inclined cracks and also, the closer stirrup spacing provides better anchorage for the lower ends of the compression diagonals.

In a wide beam with stirrups around the perimeter the diagonal compression in the web tends to be supported by the bars in the corners of the stirrups, as shown in Fig. 6–29a. The situation is improved if there are more than two stirrup legs, as shown in Fig. 6–29b. ACI Commentary Sec. R11.5.6 suggests that the transverse spacing of stirrup legs in wide beams should be limited to a fraction of the width by placing several overlapping stirrups. The *CEB–FIB Model Code 1990*[6-12] suggests that the maximum transverse spacing of stirrup legs should be limited to $2d/3$ or 32 in., whichever is smaller.

2. Shear failure initiated by failure of the stirrup anchorages. Equations 6–18 and 6–21 are based on the assumption that the stirrups will yield at ultimate. This will be true only if the stirrups are well anchored. Generally, the upper end of the inclined crack approaches very close to the compression face of the beam, as shown in Fig. 6–4 or 6–30. At ultimate, the stress in the stirrups approaches or equals the yield strength, f_y, at every point where an inclined crack intercepts a stirrup. Thus the portions of the stirrups shown

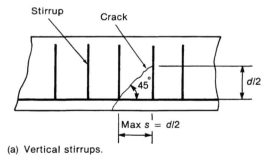

(a) Vertical stirrups.

(b) Inclined stirrups.

Fig. 6–28
Maximum spacing of stirrups.

Shear in Beams

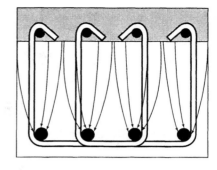

Fig. 6–29
Flow of diagonal compression
force in beams with stirrups.

(a) Widely-spaced stirrup legs   (b) Closely-spaced stirrup legs

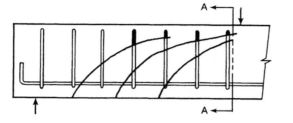

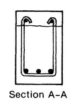

Fig. 6–30
Anchorage of stirrups.

Section A-A

shaded in Fig. 6–30 must be able to anchor f_y. For this reason, ACI Sec. 12.13.1 requires that the stirrups extend as close to the compression and tension faces as cover and bar spacing requirements permit, and in addition, specifies certain types of hooks to anchor the stirrups. The ACI Code requirements for stirrup anchorage are illustrated in Fig. 6–31:

(a) ACI Sec. 12.13.3 requires that each bend away from the ends of a stirrup enclose a longitudinal bar as shown in Fig. 6–31a.

(b) For No. 5 bar or D31 wire stirrups and smaller with any yield strength, ACI Sec. 12.13.2.1 allows the use of a standard hook around longitudinal reinforcement without any specified embedment length. The hooks may be 90°, 135°, or 180°, as shown in Fig. 6–31b. Either 135° or 180° hooks are preferred.

(c) For No. 6, 7, and 8 stirrups with f_y of 40,000 psi, ACI Sec. 12.13.2.1 allows the details shown in Fig. 6–31b.

(d) For No. 6, 7, and 8 stirrups with f_y greater than 40,000 psi, ACI Sec. 12.13.2.2 requires a standard hook around a longitudinal bar plus embedment between midheight of the member and the outside of the hook of at least $0.014 d_b f_y / \sqrt{f_c'}$.

(e) Requirements for welded-wire fabric stirrups formed of sheets bent in U shape or vertical flat sheets are illustrated in ACI Commentary Fig. 12.13.2.3.

(f) In deep members, particularly where the depth varies gradually, it is sometimes advantageous to use lap spliced stirrups described in ACI Sec. 12.13.5 and shown in Fig. 6–31c. This type of stirrup has proven unsuitable in seismic areas.

(g) ACI Sec. 7.11 requires closed stirrups in beams with compression reinforcement, beams subjected to stress reversals, or beams subjected to torsion. ACI Sec. 7.13.2.2 requires closed stirrups in all perimeter or spandrel beams. Closed stirrups may be constructed as shown in Fig. 6–31d. Such a stirrup would not fulfill the force transfer intended in ACI Secs. 7.13.2.2 and 7.13.2.3.

Standard hooks and the development length ℓ_d are discussed in Chap. 8 of this book and ACI Secs. 7.1.3 and 12.2.

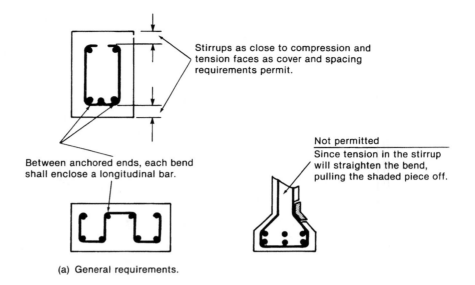

Stirrups as close to compression and tension faces as cover and spacing requirements permit.

Not permitted
Since tension in the stirrup will straighten the bend, pulling the shaded piece off.

Between anchored ends, each bend shall enclose a longitudinal bar.

(a) General requirements.

Standard stirrup hook, ACI Sec. 7.1.3, Must enclose a bar, ACI Sec. 12.13.2.1

(b) Stirrup anchorage requirements for No. 5 and smaller bars as per ACI Secs. 7.1.3 and 12.13.2.1.

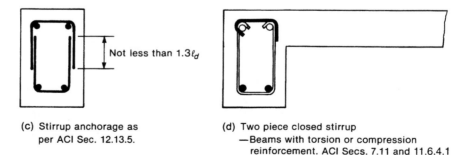

Not less than $1.3\ell_d$

Fig. 6–31
Stirrup detailing requirements.

(c) Stirrup anchorage as per ACI Sec. 12.13.5.

(d) Two piece closed stirrup
—Beams with torsion or compression reinforcement. ACI Secs. 7.11 and 11.6.4.1

Standard stirrup hooks are bent around a smaller-diameter pin than normal bar bends. Very high strength steels may develop small cracks during this bending operation. These cracks may in turn lead to fracture of the bar before the yield strength can be developed. For this reason, ACI Sec. 11.5.2 limits the yield strength used in design calculations to 60,000 psi, except for welded-wire fabric stirrups, for which the limit is 80,000 psi. This is justified because the bend test for the wire used to make welded-wire fabric is more stringent than that for bars. In addition, welded-wire stirrups tend to be more closely spaced than stirrups made from reinforcing bars and give better control of inclined crack widths.

Because the anchorage length available between the inclined crack and the end of the bar is generally very short, the author recommends the use of Grade 40 steel in stirrups except in very large beams. This has the additional advantage that the stirrup spacing may be reduced somewhat, which, in turn, helps prevent excessively wide inclined cracks.

3. Serviceability failure due to excessive crack widths at service loads. Wide inclined cracks in beams are unsightly and may allow water to penetrate the beam, possibly causing corrosion of the stirrups. In tests of three similar beams,[6-17] the maximum service load crack width in a beam with the shear reinforcement provided entirely by bent-up bars was 150% of that in a beam with vertical stirrups. The maximum service load crack width in a beam with inclined stirrups was only 80% of that in the beam with vertical stirrups. In addition, the crack widths were less with closely spaced small-diameter stirrups than with widely spaced large-diameter stirrups.

ACI Sec. 11.5.6.8 attempts to guard against excessive crack widths by limiting the maximum shear that can be transmitted by stirrups to $V_{s(\max)} = 8\sqrt{f_c'}\, b_w d$. In a beam with $V_{s(\max)}$, the stirrup stress will be 34 ksi at service loads, corresponding to a maximum crack width of about 0.014 in.[6-4] Although this limit generally gives satisfactory crack widths, the use of closely spaced stirrups and horizontal steel near the faces of beam webs is also effective in reducing crack widths.

4. Shear failure due to crushing of the web. As indicated in the discussion of the truss analogy, compression stresses exist in the compression diagonals in the web of a beam. In very thin-walled beams, these may lead to crushing of the web. Since the diagonal compression stress is related to the shear stress, v, a number of codes limit the ultimate shear stress to 0.2 to 0.25 times the compression strength of the concrete. The ACI Code limit on V_s for crack control ($V_{s(\max)} = 8\sqrt{f_c'}\, b_w d$) provides adequate safety against web crushing in reinforced concrete beams.

5. Shear failure initiated by failure of the tension chord. The truss analogy shows that the force in the longitudinal tensile reinforcement at a given point in the shear span is a function of the moment at a section located approximately d closer to the nearest section of maximum moment. Partly for this reason, ACI Sec. 12.10.3 requires that flexural reinforcement extend the larger of d or 12 bar diameters past the point where it is no longer needed (except at the supports of simple spans or at the ends of cantilevers).

Minimum Web Reinforcement

Because a shear failure of a beam without web reinforcement is sudden and brittle and because shear failure loads vary widely about the values given by the design equations, ACI Sec. 11.5.5.1 requires a minimum amount of web reinforcement to be provided if the applied shear force, V_u, exceeds half of the factored inclined cracking shear, $\phi(0.5V_c)$, except in:

1. Slabs and footings
2. Concrete joist construction
3. Beams with a total depth not greater than 10 in. (250 mm), $2\frac{1}{2}$ times the thickness of the flange, or one-half the width of the web, whichever is greatest

The exceptions each represent a type of member in which load redistribution can occur across the width of the member, or in the case of joist floors, to adjacent members.

ACI Sec. 7.13.2.2 requires closed stirrups or stirrups with a 135° bend around top steel in all perimeter or spandrel beams. These, acting with the top and bottom reinforcement in the perimeter beams, provide a tension tie around the building to limit the extent of collapse arising from the failure of an interior beam.

Where required, the minimum web reinforcement shall be at least (ACI Sec. 11.5.5.3)

$$A_{v(min)} = \frac{50b_w s}{f_y}$$

(6–22)
(ACI Eq. 11–13)

This is equivalent to providing web reinforcement to transmit a shear stress of 50 psi. For $f_c' = 2500$ psi, 50 psi is half of the shear stress at inclined cracking from Eq. 6–8.

In SI units, Eq. 6–22 becomes

$$A_{v(min)} = \frac{b_w s}{3f_y}$$

(6–22M)

For beams with f_c' greater than 10,000 psi, $\sqrt{f_c'}$ is limited to 100 psi unless the minimum web reinforcement provided satisfies Eq. 6–23 (ACI Sec. 11.1.2.1).

$$A_{v(min)} = \frac{f_c'}{5000}\left(\frac{50b_w s}{f_y}\right) \leq \frac{150b_w s}{f_y}$$

(6–23)

In seismic regions web reinforcement is required in all beams since V_c is taken equal to zero if earthquake-induced shear exceeds half the total shear (see Sec. 19–6 and ACI Sec. 21.7.2.1).

Strength Reduction Factor for Shear

The strength reduction factor, ϕ, for shear and torsion is 0.85 (ACI Sec. 9.3.2.3). This is lower than for flexure because shear failure loads are more variable than flexure failure loads. Special strength reduction factors are required for shear in some members subjected to seismic loads (see Sec. 19–5 and ACI Sec. 9.3.4).

Location of Maximum Shear for the Design of Beams

In a beam loaded on the top flange and supported on the bottom as shown in Fig. 6–32a, the closest inclined cracks that can occur adjacent to the supports will extend outward from the supports at roughly 45°. Loads applied to the beam within a distance, d, from the support in such a beam will be transmitted directly to the support by the compression fan above the 45° cracks and will not affect the stresses in the stirrups crossing the cracks shown in Fig. 6–32. As a result, ACI Sec. 11.1.3.1 states that

> For nonprestressed members sections located less than a distance d from the face of the support may be designed for the same shear, V_u, as that computed at a distance d.

This is permitted only when:

1. The support reaction, in the direction of the applied shear, introduces compression into the end regions of a member.

2. No concentrated load occurs within d from the face of the support.

Thus for the beam shown in Fig. 6–32a, the values of V_u used in design are shown shaded in the shear force diagram of Fig. 6–32b.

This allowance must be applied carefully since it is not applicable in all cases. Figure 6–33 shows five other typical cases that arise in design. If the beam shown in Fig. 6–32 were loaded on the *lower* flange as shown in Fig. 6–33a, the critical section for design

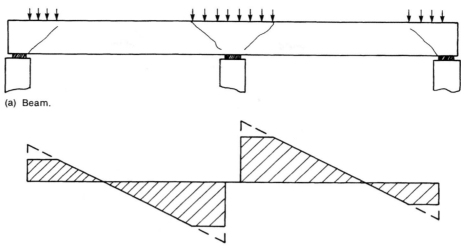

(a) Beam.

(b) Shear force diagram.

Fig. 6–32
Shear force diagram for design.

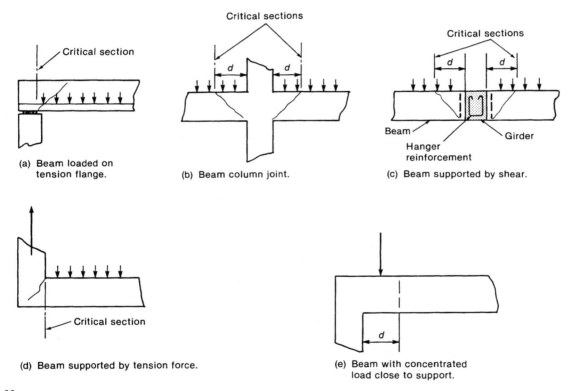

(a) Beam loaded on tension flange.

(b) Beam column joint.

Critical sections

(c) Beam supported by shear.

Beam — Girder
Hanger reinforcement

(d) Beam supported by tension force.

(e) Beam with concentrated load close to support.

Fig. 6–33
Application of ACI Sec. 11.1.3.

would be at the face of the support, since loads applied within d of the support must be transferred across the inclined crack before they reach the support.

A typical beam to column joint is shown in Fig. 6–33b. Here the critical section for design is d away from the section as shown.

If the beam is supported by a girder of essentially the same depth as shown in Fig. 6–33c, the compression fans that form in the supported beams will tend to push the bottom off the supporting beam. The critical sections in the supported beams can be taken at d from the end of the beam *provided* that "hanger reinforcement" is provided to support the reactions from the compression fans. The design of hanger reinforcement is discussed in Sec. 6–6.

Generally, if the beam is supported by a tensile force rather than a compressive force, the critical section will be at the face of the support and the joint must be carefully detailed since the crack will extend into the joint as shown in Fig. 6–33d.

Occasionally, a significant part of the shear at the end of the beam will be caused by a concentrated load acting less than d from the face of the column as shown in Fig. 6–33e. In such a case, the end portion of the beam should be considered to act as a *deep beam* with respect to shear and flexure, and the concentrated load must be considered. The design of deep beams is discussed in Chap. 18.

Shear at Midspan of Uniformly Loaded Beams

In a normal building, the dead and live loads are assumed to be uniform loads. Although the dead load is always present over the full span, the live load may act over the full span, as shown in Fig. 6–34c, or over part of the span, as shown in Fig. 6–34d. Full uniform load over the full span gives the maximum shear at the ends of the beam. Full uniform load over half the span plus dead load on the remaining half gives the maximum shear at midspan. The maximum shears at other points in the span are closely approximated by the *shear force envelope* resulting from these cases (Fig. 6–34e).

The shear at midspan due to a uniform live load on half the span is

$$V_{u(\text{midspan})} = \frac{w_{Lu}\ell}{8} \tag{6-24}$$

This can be positive or negative. Although this has been derived for a simple beam, it is acceptable to apply Eq. 6–24 to continuous beams also.

High-Strength Concrete

Tests suggest that the inclined cracking load of beams increases less rapidly than $\sqrt{f_c'}$ increases for f_c' greater than about 8000 psi. This was offset by an increased effectiveness of stirrups in high-strength concrete beams.[6-18, 6-19] Other tests suggest that the required amount of minimum web reinforcement increases as f_c' increases. For these reasons ACI Sec. 11.1.2 limits $\sqrt{f_c'}$ to 100 psi unless the amount of minimum web reinforcement satisfies Eq. 6–23.

Lightweight Concrete

The inclined cracking load of beams made of lightweight concrete is generally less than that of beams made of normal sand and gravel concrete. ACI Sec. 11.2 requires that the values of V_c be reduced in either one of two ways:

1. Substitute $f_{ct}/6.7$ for $\sqrt{f_c'}$ in Eqs. 6–8, 6–15, and 6–17, where f_{ct} is the split cylinder tensile strength of the concrete and where $f_{ct}/6.7$ shall not exceed $\sqrt{f_c'}$.

2. Multiply V_c from Eqs. 6–8, 6–15, and 6–17 by 0.75 if the concrete is made with lightweight sand and lightweight gravel, or by 0.85 if the concrete is made with normal sand and lightweight gravel.

Comparison of the Truss Model and the ACI Procedure

For beams with large amounts of web reinforcement and failing at high shear stresses, the behavior closely approaches that predicted by the plastic truss model (see Fig. 6–26). On the

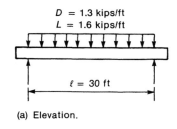

D = 1.3 kips/ft
L = 1.6 kips/ft

ℓ = 30 ft

(a) Elevation.

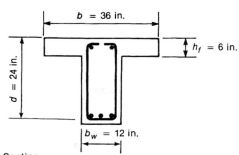

b = 36 in.

h_f = 6 in.

d = 24 in.

b_w = 12 in.

(b) Section.

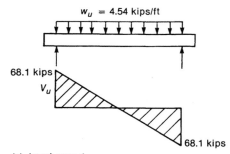

w_u = 4.54 kips/ft

68.1 kips

V_u

68.1 kips

(c) Load case 1.

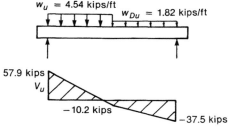

w_u = 4.54 kips/ft

w_{Du} = 1.82 kips/ft

57.9 kips

V_u

−10.2 kips

−37.5 kips

(d) Load case 2.

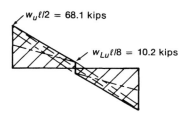

$w_u \ell/2$ = 68.1 kips

$w_{Lu}\ell/8$ = 10.2 kips

(e) Shear force envelope.

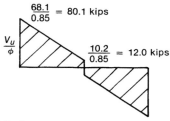

$\dfrac{68.1}{0.85}$ = 80.1 kips

$\dfrac{V_u}{\phi}$

$\dfrac{10.2}{0.85}$ = 12.0 kips

(f) V_u/ϕ diagram.

Fig. 6–34
Beam and shear force diagrams—Example 6.1.

other hand, the truss model predicts zero shear strength for beams without web reinforcement, and tends to underestimate the shear capacity for beams with V_s less than about V_c. Modifications of the plastic truss model that include a V_c term that decreases from $V_c = 2\sqrt{f_c'}$ for $V_u = 2\sqrt{f_c'}$ to zero when $V_u = 6\sqrt{f_c'}$ are incorporated in some codes.[6-11] This empirical fix solves the problem mentioned in the preceding sentence.

The truss model emphasizes two aspects of shear behavior which are frequently overlooked in the ACI procedure. These are the shift in the tensile force diagram, which is of great importance in detailing the longitudinal steel as discussed in Chap. 8, and the need for adequate anchorage of the stirrups in the top and bottom flanges.

Examples of the Design of a Beam for Shear

EXAMPLE 6–1   Design of Vertical Stirrups in a Simply Supported Beam

Figure 6–34b shows the cross section of a simply supported T beam. This beam supports a uniformly distributed service (unfactored) dead load of 1.3 kips/ft, including its own weight and a uniformly distributed service live load of 1.6 kips/ft. Design vertical stirrups for this beam. The concrete strength is 4000 psi, the yield strength of flexural reinforcement is 60,000 psi, and that of the stirrups is 40,000 psi. It is assumed that the longitudinal bars are properly detailed to prevent anchorage and flexural failures. A complete design example including these aspects is presented in Chap. 10.

1. **Compute the factored shear force diagram**

Total factored load:

$$w_u = 1.4 \times 1.3 \text{ kips/ft} + 1.7 \times 1.6 \text{ kips/ft}$$

$$= 4.54 \text{ kips/ft}$$

Factored dead load:

$$w_{Du} = 1.4 \times 1.3 \text{ kips/ft} = 1.82 \text{ kips/ft}$$

Three loading cases should be considered: Fig. 6–34c, Fig. 6–34d, and the mirror opposite of Fig. 6–34d. The three shear force diagrams are superimposed in Fig. 6–34e. For simplicity we shall approximate the shear force envelope with straight lines and design simply supported beams for $V_u = w_u \ell / 2$ at the ends and $V_u = w_{Lu} \ell / 8$ at midspan, where w_u is the total factored live and dead load and w_{Lu} is the factored live load. From Eq. 6–14,

$$V_u \leq \phi V_n$$

Setting these equal, the smallest value of V_n that satisfies Eq. 6–14 is

$$V_n = \frac{V_u}{\phi}$$

This is plotted in Fig. 6–34f.

Since this beam is loaded on the top and supported on the bottom, the critical section for shear is located at $d = 2$ ft from the support. From Fig. 6–34f and similar triangles the shear at d from the support is

$$\frac{V_u}{\phi} \text{ at } d = 80.1 \text{ kips} - \frac{2 \text{ ft}}{15 \text{ ft}} (80.1 - 12.0) \text{ kips}$$

$$= 71.0 \text{ kips}$$

Therefore,

$$\frac{V_u}{\phi} \text{ at } d = 71.0 \text{ kips and min. } V_n = 71.0 \text{ kips}$$

2. **Are stirrups required by ACI Sec. 11.5.5.1?** No stirrups are required if $V_n \leq V_c/2$, where

$$V_c = 2\sqrt{f_c'} b_w d \tag{6-8}$$

$$= 2\sqrt{4000} \text{ psi} \times 12 \text{ in.} \times \frac{24 \text{ in.}}{1000} \tag{ACI Eq. 11-3}$$

$$= 36.4 \text{ kips}$$

Since $V_n = 71.0$ kips exceeds $V_c/2 = 18.2$ kips, stirrups are required.

3. **Check the anchorage of stirrups and maximum spacing.** Try No. 3 double-leg stirrups, $f_y = 40,000$ psi:

$$A_v = 2 \times 0.11 \text{ in.}^2 = 0.22 \text{ in.}^2$$

(a) **Check the anchorage of the stirrups.** Since the diameter of the stirrups is less than No. 6, ACI Sec. 12.13.2.1 states that the stirrups can be anchored by a 90° stirrup hook around a bar. Provide a No. 4. bar in the upper corners of the stirrups to anchor them.

(b) **Determine the maximum spacing.**

Based on the beam depth: ACI Sec. 11.5.4.1 sets the maximum spacing as the smaller of $0.5d = 12$ in. or 24 in. ACI Sec. 11.5.4.3 requires half this spacing if V_n exceeds

$$V_c + 4\sqrt{f_c'} b_w d = 6\sqrt{f_c'} b_w d = 109 \text{ kips}$$

Since the maximum V_n is less than 109 kips, the maximum spacing based on the beam depth is 12 in.

Based on minimum A_v (Eq. 6–22, ACI Eq. 11–14):

$$A_{v(min)} = \frac{50 b_w s}{f_y}$$

Rearranging gives

$$s_{max} = \frac{A_v f_y}{50 b_w}$$

$$= \frac{0.22 \text{ in.}^2 \times 40,000 \text{ psi}}{50 \text{ psi} \times 12 \text{ in.}} = 14.7 \text{ in.}$$

Therefore, the maximum spacing based on the beam depth governs. **Maximum s = 12 in.**

4. Compute the stirrup spacing required to resist the shear forces. For vertical stirrups, from Eq. 6–21,

$$s = \frac{A_v f_y d}{V_u / \phi - V_c}$$

where V_c = 36.4 kips,

At d from the support, V_u / ϕ = 71.0 kips and

$$s = \frac{0.22 \text{ in.}^2 \times 40,000 \text{ psi} \times 24 \text{ in.}}{(71.0 - 36.4) \times 1000 \text{ lb}} = 6.10 \text{ in.}$$

Use s = 6 in. at d from the support. Because this is a reasonable spacing, we can use No. 3 double-leg stirrups as assumed. The stirrup spacing will be changed to 8 in. at the point where this is possible and then to the maximum spacing of 12 in. The intermediate spacings selected are up to the designer. Generally, no more than three different spacings are used, and generally, spacings are varied in multiples of 2 or 3 in.

Compute V_u / ϕ where s = 8 in. Rearranging Eq. 6–21 gives

$$\frac{V_u}{\phi} = \frac{A_v f_y d}{s} + V_c$$

$$= \frac{0.22 \times 40,000 \times 24}{8} + 36,400 = 62.800 \text{ lb} \qquad (6\text{–}25)$$

$$= 62.8 \text{ kips}$$

From Fig. 6–34f and similar triangles, this shear occurs at

$$x = \frac{80.1 - 62.8}{80.1 - 12.0} \times 15 \text{ ft}$$

$$= 3.81 \text{ ft} = 45.7 \text{ in. from the end of the beam}$$

Compute V_u / ϕ where s = 12 in. From Eq. 6–25,

$$\frac{V_u}{\phi} = \frac{0.22 \times 40,000 \times 24}{12} + 36,400 = 54,000 \text{ lb}$$

$$= 54 \text{ kips}$$

This occurs at

$$x = \frac{80.1 - 54.0}{80.1 - 12.0} 15 \times 12$$

$$= 69.0 \text{ in. from the end of the beam}$$

Stirrups must be continued to the point where $V_u/\phi = V_c/2 = 18.2$ kips. This occurs at

$$x = \frac{80.1 - 18.2}{80.1 - 12.0} 15 \times 12$$

$$= 164 \text{ in. from the end of the beam}$$

To summarize, $s = 6$ in. to 45.7 in. from the support, $s = 8$ in. from that point to 69 in. from the support, and $s = 12$ in. from that point to 164 in. from the support. In choosing the numbers of stirrups at each spacing it is assumed that each stirrup reinforces a length of web extending $s/2$ on each side of the stirrup. For this reason the first stirrup is placed at $s/2$ from the support. We shall select the following spacings:

1 at 3 in.
7 at 6 in. (extending to $3 + 42 = 45$ in. from the support)
3 at 8 in. (extending to $45 + 24 = 69$ in. from the support)
8 at 12 in. (extending to $69 + 96 = 165$ in. from the support)

This leaves a 30-in. length at the center without stirrups. Although not required by the Code, we shall place an additional stirrup at each end. The final selection is: **Use No. 3 Grade 40 double-leg stirrups: 1 at 3 in., 7 at 6 in., 3 at 8 in., and 9 at 12 in., each end.** The total number of stirrups in half the beam is 20. Another 20 are required in the other half at similar spacings. The beam is drawn to scale in Fig. 6–35. The cross section is shown in Fig. 6–34b.

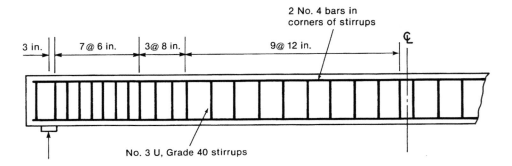

Fig. 6–35
Stirrups in beam—Example 6–1.

EXAMPLE 6–1M   Design of Vertical Stirrups in a Simply Supported Beam— SI Units

Figure 6–36 shows the elevation and cross section of a simply supported T beam. This beam supports a uniformly distributed service (unfactored) dead load of 20 kN/m, including its own weight, and a uniformly distributed service live load of 24 kN/m. Design vertical stirrups for this beam. The concrete strength is 25 MPa, the yield strength of the flexural reinforcement is 400 MPa, and the yield strength of the stirrups is 300 MPa.

1.   **Compute the design factored shear force diagram.** Total factored load

$$w_u = 1.4 \times 20 \text{ kN/m} + 1.7 \times 24 \text{ kN/m}$$

$$= 68.8 \text{ kN/m}$$

Factored dead load:

$$w_{Du} = 1.4 \times 20 \text{ kN/m} = 28.0 \text{ kN/m}$$

Three loading cases should be considered: Fig. 6–36c, Fig. 6–36d, and the mirror opposite of Fig. 6–36d. The three shear force diagrams are superimposed in Fig. 6–36e. For simplicity, we shall approximate the shear force envelope with straight lines and design simply supported beams for $V_u = w_u \ell/2$ at the ends and $V_u = w_{Lu}\ell/8$ at midspan, where w_u is the total factored live and dead load and w_{Lu} is the factored live load, From Eq. 6–14,

$$V_u \le \phi V_n$$

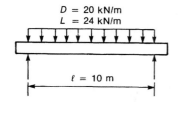

D = 20 kN/m
L = 24 kN/m

ℓ = 10 m

(a) Elevation.

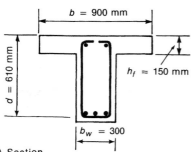

b = 900 mm

$d = 610$ mm

$h_f = 150$ mm

$b_w = 300$

(b) Section.

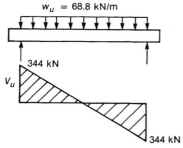

w_u = 68.8 kN/m

344 kN

V_u

344 kN

(c) Load case 1.

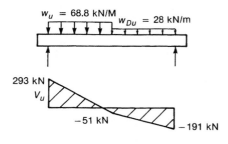

w_u = 68.8 kN/M

w_{Du} = 28 kN/m

293 kN

V_u

−51 kN

−191 kN

(d) Load case 2.

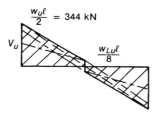

$\frac{w_u \ell}{2}$ = 344 kN

V_u

$\frac{w_{Lu} \ell}{8}$

(e) Shear force envelope.

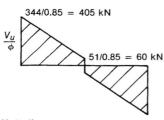

344/0.85 = 405 kN

$\frac{V_u}{\phi}$

51/0.85 = 60 kN

(f) V_u/ϕ diagram.

Fig. 6–36
Beam and shear force diagrams—Example 6–1M.

Setting these equal, the smallest value of V_n that satisfies Eq. 6–14 is

$$V_n = \frac{V_u}{\phi}$$

This is plotted in Fig. 6–36f.

Since this beam is loaded on the top flange and supported on the bottom flange, the critical section is located at $d = 0.61$ m from the support. From Fig. 6–36f and similar triangles, the shear at d from the support is

$$\frac{V_u}{\phi} \text{ at } d = 405 \text{ kN} - \frac{0.61 \text{ m}}{5 \text{ m}}(405 - 60)$$

$$= 363 \text{ kN}$$

Therefore, V_u/ϕ at $d = 363$ kN and min. $V_n = 363$ kN.

2. **Are stirrups required by ACI Sec. 11.5.5.1?** No stirrups are required if $V_n \leq V_c/2$, where

$$V_c = \frac{\sqrt{f_c'}\, b_w d}{6}$$

$$= \frac{\sqrt{25} \text{ MPa} \times 300 \text{ mm} \times 610 \text{ mm}}{6 \times 1000} \qquad (6\text{–}8\text{M})$$

$$= 153 \text{ kN}$$

Since $V_n = 363$ kN exceeds $V_c/2 = 76.3$ kN, stirrups are required.

3. Check anchorage of stirrups and maximum spacing. Try No. 10 double-leg stirrups, $f_y = 300$ MPa.

$$A_v = 2 \times 100 = 200 \text{ mm}^2$$

(a) Check the anchorage of the stirrups. Since the diameter of the stirrups is less than No. 20, ACI Sec. 12.13.2.1 states that the stirrups can be anchored by a 90° stirrup hook around a bar. Provide a No. 10 bar in the upper corners of the stirrups to anchor them.

(b) Determine the maximum spacing.

Based on the beam depth: ACI Sec. 11.5.4.1 requires the smaller of $0.5d = 305$ mm or 600 mm. ACI Sec. 11.5.4.3 requires half the above if V_n exceeds $V_c + \sqrt{f_c'} b_w d/3$:

$$0.5\sqrt{f_c'} b_w d = 458 \text{ kN}$$

Since the maximum V_n is less than 458 kN, the maximum spacing is 305 mm.

Based on minimum A_v (Eq. 6–22M, ACI Eq. 11–14):

$$A_{v(min)} = \frac{b_w s}{3f_y}$$

Therefore,

$$\text{max. } s = \frac{3A_v f_y}{b_w}$$

$$= \frac{3 \times 200 \text{ mm}^2 \times 300 \text{ MPa}}{300 \text{ mm}} = 600 \text{ mm}$$

Therefore, the maximum spacing based on the beam depth governs. **Maximum $s = $ 305 mm.**

4. Compute the stirrup spacing required to resist the shear forces. For vertical stirrups Eq. 6–21 applies:

$$s = \frac{A_v f_y d}{V_u/\phi - V_c} \tag{6–21}$$

where $V_c = 153$ kN. At d from the support, $V_u/\phi = 363$ kN and

$$s = \frac{200 \text{ mm}^2 \times 300 \text{ Mpa} \times 610 \text{ mm}}{(363 - 153) \times 1000 \text{ N}} = 174 \text{ mm}$$

Use $s = 150$ mm at d from the support. Because this is a reasonable spacing, we can use No. 10 double-leg stirrups as assumed. The stirrup spacing will be changed to 200 mm at the point where this is possible and then to the maximum spacing of 300 mm. The intermediate spacings selected are up to the designer. Generally, no more than three different spacings are used, and generally, the spacings are varied in multiples of 50 or 75 mm.

Compute V_u/ϕ where $s = 200$ mm. Rearranging Eq. 6–21 gives

$$\frac{V_u}{\phi} = \frac{A_v f_y d}{s} + V_c \tag{6–25}$$

$$= \frac{200 \times 300 \times 610}{200 \times 1000} + 153 = 336 \text{ kN}$$

From Fig. 6–36f and similar triangles this shear occurs at

$$x = \frac{405 - 336}{405 - 60} \times 5000 \text{ mm}$$

$$= 1000 \text{ mm from the end of the beam}$$

Compute V_u/ϕ where $s = 300$ mm. From Eq. 6–25,

$$\frac{V_u}{\phi} = \frac{200 \times 300 \times 610}{300 \times 1000} + 153 = 275 \text{ kN}$$

This occurs at

$$x = \frac{405 - 275}{405 - 60} \, 5000$$

$$= 1884 \text{ mm from the end of the beam}$$

Stirrups must be continued to the point where $V_u/\phi = V_c/2 = 76.3$ kN. This occurs at

$$x = \frac{405 - 76.3}{405 - 60} \, 5000$$

$$= 4764 \text{ mm from the end of the beam}$$

To summarize, $s = 150$ to 1000 mm from the support, $s = 200$ mm from that point to 1884 mm, and $s = 300$ mm from that point to 4764 mm from the support. In choosing the numbers of stirrups at each spacing it is assumed that each stirrup reinforces a length of web extending $s/2$ on each side of the stirrup. For this reason the first stirrup is placed at $s/2$ from the support. We shall select the following spacings:

1 at 75 mm

6 at 150 mm (extending to $75 + 900 = 975$ mm from the support)

4 at 200 mm (extending to $975 + 800 = 1775$ from the support)

10 at 300 mm (extending to $1775 + 3000 = 4775$ mm from the support)

The final selection is: **Use No. 10 Grade 300 double-leg stirrups: 1 at 75 mm, 6 at 150 mm, 4 at 200 mm, and 10 at 300 mm, each end.** The total number of stirrups in half of the beam is 21. Another 21 stirrups are required in the other half at similar spacings. The elevation and cross section would resemble those in Figs. 6–35 and 6–34b. ■

6–6 HANGER REINFORCEMENT

When a beam is supported by a girder or other beam of essentially the same depth as shown in Fig. 6–33c or 6–37, hanger reinforcement should be provided in the joint. Compression fans form in the supported beams as shown in Fig. 6–37a. The inclined compressive forces will push the bottom off the supporting beam unless they are resisted by hanger reinforcement designed to equilibrate the downward component of the compressive forces in the members of the fan.

No rules are given in the ACI Code for the design of such reinforcement. The following proposals are based on the 1984 Canadian concrete code[6–20] and on Ref. 6–21. In addition to the stirrups provided in the supporting beam for shear, hanger reinforcement with a tensile capacity of $(1 - h'/h_s)$ times the end shear of the supported beam shall be provided in the part of the joint shown shaded in Fig. 6–37b, where h' is the distance from the bottom of the supported beam to the bottom of the supporting beam, and h_s is the depth of the supporting beam. The majority of the hanger reinforcement should be in the supporting beam, although up to a third can be in the supported beam.

These provisions can be waived if the shear at the end of the supported beam is less than $3\sqrt{f_c'} b_w d$, since the inclined cracking is not fully developed at this shear.

The hanger reinforcement should be well anchored top and bottom. The lower layer of reinforcement in the supported beam should be above the reinforcement in the supporting beam.

EXAMPLE 6–2   Design of Hanger Reinforcement in a Beam–Girder Junction

Figure 6–38a shows a beam–girder joint. Each beam transfers a factored end shear of 45 kips to the girder. Design hanger reinforcement assuming the yield strength of the hanger reinforcement is 60 ksi. The shear reinforcement in the beam and girder is No. 3 Grade 60 double–leg stirrups.

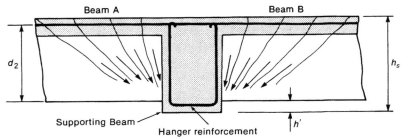

Beam A   Beam B

(a) Compression fan at beam–girder joint.

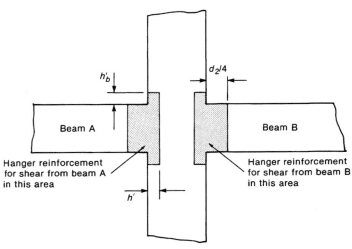

Fig. 6–37
Hanger reinforcement.

(b) Plan of joint area showing location of hanger reinforement.

$$h' = 30 \text{ in.} - 21 \text{ in.} = 9.0 \text{ in.}$$

$$h_s = 30 \text{ in.}$$

The factored tensile force T_h to be transferred by hanger reinforcement is

$$T_h = 2 \times 45 \text{ kips} \left(1 - \frac{9}{30}\right) = 63.0 \text{ kips}$$

The area of hanger reinforcement required is

$$\phi A_h f_y = T_h \tag{6–26}$$

$$A_h = \frac{T_h/\phi}{f_y}$$

$$= \frac{63.0/0.85}{60} = 1.24 \text{ in.}^2$$

Could use

4 No. 4 double-leg stirrups, $A_s = 1.60$ in.^{2}
6 No. 3 double-leg stirrups, $A_s = 1.32$ in.^{2}

Since the shear reinforcement is No. 3 stirrups, we shall select No. 3 stirrups for the hanger steel. It will be placed as shown in Fig. 6–38b. This is in addition to the shear reinforcement already provided. ∎

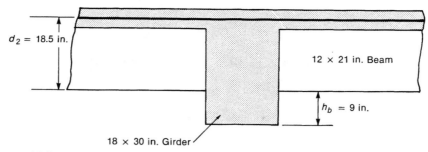

d_2 = 18.5 in.

12 × 21 in. Beam

h_b = 9 in.

18 × 30 in. Girder

(a) Beams.

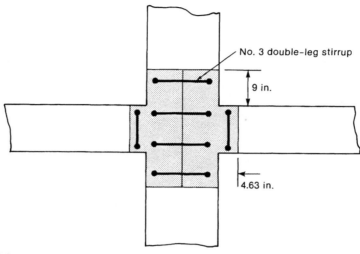

No. 3 double-leg stirrup

9 in.

4.63 in.

Fig. 6–38
Example 6–2.

(b) Joint zone.

6–7 TAPERED BEAMS

In a prismatic beam the average shear stress between two cracks is calculated as

$$v = \frac{V}{b_w jd} \tag{6–4}$$

which is simplified to

$$v = \frac{V}{b_w d} \tag{6–5}$$

In the derivation of Eq. 6–4 it was assumed that jd was constant. If the depth of the beam varies, the compressive and/or tensile forces due to flexure will have vertical components. A segment of a tapered beam is shown in Fig. 6–39. The moment, M, at the left end of the section can be represented by two horizontal force components, C and T, separated by the lever arm jd. The tension force actually acts parallel to the centroid of the reinforcement and hence has a vertical component $T \tan \alpha_T$, where α_T is the angle between the tensile force and horizontal. Similarly, the compressive force acts along a line joining the centroids of the stress blocks at the two sections and hence has a vertical component $C \tan \alpha_c$. The shear force, V, on the left end of the element can be represented as

$$V = V_R + C \tan \alpha_c + T \tan \alpha_T$$

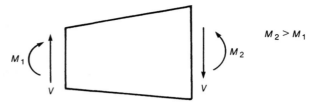

(a) Forces on segment of beam.

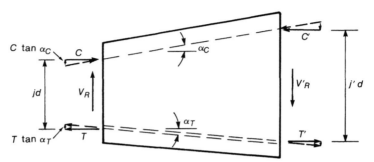

Fig. 6–39
Reduced shear force in non-prismatic beam.

(b) Internal forces and shears.

where V_R is the reduced shear force resisted by the stirrups and the concrete. Substituting $C = T = M/jd$ and letting $\alpha = \alpha_c + \alpha_T$ gives

$$V_R = V - \frac{|M|}{jd} \tan \alpha \qquad (6\text{--}27)$$

where $|M|$ represents the absolute value of the moment and α is positive if the lever arm jd increases in the same direction as $|M|$ increases.

The shear stresses in a tapered beam then become

$$v = \frac{V_R}{b_w d} \qquad (6\text{--}5a)$$

Several examples of the use of Eq. 6–27 are shown in Fig. 6–40.

6–8   SHEAR IN AXIALLY LOADED BEAMS OR COLUMNS

Reinforced concrete beams can be subjected to shear plus axial tensile or compressive forces due to such causes as gravity load effects in inclined members, and stresses resulting from restrained shrinkage or thermal deformations. Similarly, wind or seismic forces cause shear forces in axially loaded columns. Figure 6–41 shows a tied column that failed in shear during an earthquake. The inclined crack in this column resembles Fig. 6–4b rotated through 90°.

Axial forces have three major effects on the shear strength. An axial compressive or tensile force will increase or reduce the load at which flexural and inclined cracks occur. If V_c is assumed to be related to the inclined cracking shear, as is done in the ACI Code, this will directly affect the design. If they have not been considered in the design, axial tensile forces may lead to premature yielding of the longitudinal reinforcement, which in turn will effectively do away with any shear transferred by aggregate interlock. Finally, an exter-

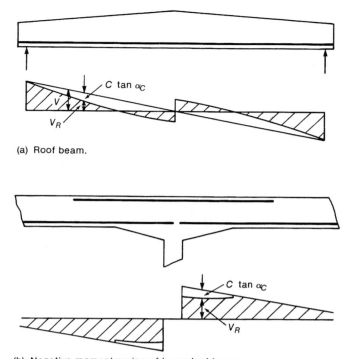

(a) Roof beam.

(b) Negative moment region of haunched beam.

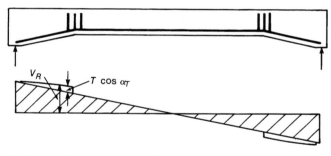

(c) Haunched simply supported beam.

Fig. 6–40
Examples of computation of V_R.

nally applied axial compression, N, will require a larger concrete compressive force and compression zone since $C = T + N$. The shear capacity, V_c, of the enlarged compression zone will tend to be larger than that in a beam without axial compression. The opposite would be true for a beam subjected to axial tensile loads plus shear and bending.

It is important to distinguish between external axial forces which must be included in the calculation of longitudinal force equilibrium, and self-equilibrating force systems such as prestressing forces. In a prestressed beam the tension in the prestressing tendon is equilibrated by compression in the concrete, so that all stages $C = T$. Although the prestressing delays cracking, once cracking occurs the internal forces in a prestressed beam resemble those in a reinforced concrete beam. Hence different design procedures are required for prestressed beams and axially compressed reinforced concrete beams.

6–8   Shear in Axially Loaded Beams or Columns   225

Fig. 6–41
Shear failure in a tied column, 1971 San Fernando earthquake. (Photograph courtesy of U.S. National Bureau of Standards.)

Axial Tension

For axial tensile loadings the nominal shear carried by the concrete is given by

$$V_c = 2\left(1 + \frac{N_u}{500A_g}\right)\sqrt{f_c'}\,b_w d \quad \text{lb} \qquad (6\text{–}17b)$$
$$\text{(ACI Eq. 11–9)}$$

Shear in Beams

where N_u/A_g is expressed in psi and is negative in tension. The term inside the parentheses becomes zero when the average axial stress on the section reaches or exceeds 500 psi in tension, which is roughly the tensile strength of concrete.

In SI units, Eq. 6–17b becomes

$$V_c = \left(1 + \frac{0.3N_u}{A_g}\right)\left(\frac{\sqrt{f_c'}}{6}\right)b_w d \tag{6–17bM}$$

This expression tends to be conservative, especially for high tensions as shown in Fig. 6–15, and although the evidence is ambiguous, tests have shown that beams subjected to tensions large enough to crack them completely through can resist shears approaching those for beams not subjected to axial tensions.[6-4] This shear capacity results largely from aggregate interlock along the tension cracks. It should be noted, however, that if the longitudinal tension reinforcement yields under the action of shear, moment, and axial tension, the shear capacity drops very significantly. This is believed to have affected the failure of the beam shown in Fig. 6–2.

Axial Compression

Axial compression tends to increase the shear strength. The ACI Code provides two expressions to account for this increase. One of these procedures involves an extrapolation of Eq. 6–15. For reasons given when Eq. 6–15 was presented, this equation will not be used in this book. In addition, recent tests indicate that this procedure may be unsafe in some cases.[6-4] The ACI Code also presents the following equation for calculating V_c for members subjected to combined shear, moment, and axial compression:

$$V_c = 2\left(1 + \frac{N_u}{2000A_g}\right)\sqrt{f_c'}\, b_w d \text{ lb} \tag{6–17a}$$
$$\text{(ACI Eq. 11–4)}$$

where N_u/A_g is positive in compression and has units of psi.

In SI units, Eq. 6–17a becomes

$$V_c = \left(1 + \frac{N_u}{14A_g}\right)\left(\frac{\sqrt{f_c'}}{6}\right)b_w d \tag{6–17aM}$$

The design of a beam subjected to axial compression or tension is identical to that for a beam without such forces except that the value of V_c is modified.

EXAMPLE 6–3 Checking the Shear Capacity of a Column Subjected to Axial Compression Plus Shear and Moments

A 12 in. × 12 in. column with $f_c' = 3000$ psi and longitudinal steel and ties having $f_y = 60,000$ psi is subjected to factored axial forces, moments, and shears, as shown in Fig. 6–42.

1. **Compute the nominal shear forces in the column.** Summing moments about the centroid at one end of the column, the factored shear is

$$V_u = \frac{42 + 21 \text{ ft-kips}}{10 \quad \text{ft}} = 6.3 \text{ kips}$$

$$V_n = \frac{V_u}{\phi} = 7.41 \text{ kips}$$

2. **Are stirrups required by ACI Sec. 11.5.5.1?** No stirrups are required if $V_n < V_c/2$, where

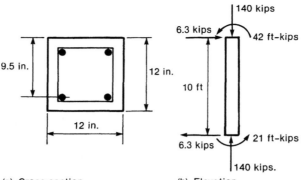

Fig. 6–42
Shear in column—
Example 6–3.

(a) Cross section.　　　　(b) Elevation.

$$V_c = 2\left(1 + \frac{N_u}{2000A_g}\right)\sqrt{f_c'}\,b_w d$$

$$= 2\left(1 + \frac{140,000}{2000 \times 144}\right)\sqrt{3000} \times 12 \times \frac{9.5}{1000} = 18.6 \text{ kips}$$

and $V_c/2 = 9.3$ kips. Since $V_n = 7.41$ kips is less than $V_c/2$, shear reinforcement is not necessary. If it were required, ties at a spacing of not more than $d/2$ would serve as shear reinforcement. These should be used over one-fourth of the column length from each end. ■

6–9 SHEAR IN SEISMIC REGIONS

In seismic regions beams and columns are particularly vulnerable to shear forces. Reversed loading cycles cause crisscrossing inclined cracks which cause V_c to decrease and disappear. As a result, special calculation procedures and special details are required in seismic regions (see Chap. 19).

PROBLEMS

6–1　For the beam shown in Fig. P6–1:
  (a)　Draw a shear force diagram.
  (b)　Show the direction of the principal tensile stresses at middepth at points A, B, and C.

  (c)　On a drawing of the beam sketch the inclined cracks that would develop at A, B, and C.

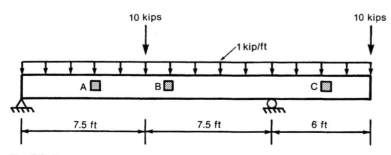

Fig. P6–1

6–2, 6–3, 6–4, and 6–5 Compute ϕV_n for the cross sections shown in Figs. P6–2, P6–3, P6–4, and P6–5. In each case use $f_c' = 3000$ psi and f_y (of the stirrups) = 40,000 psi.

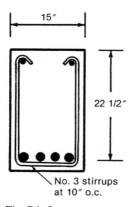

Fig. P6–2

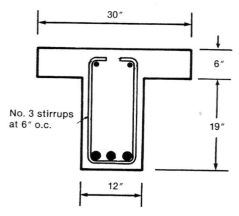

Fig. P6–3

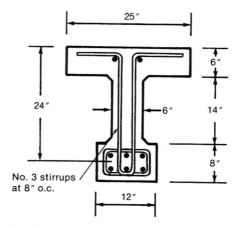

Fig. P6–4

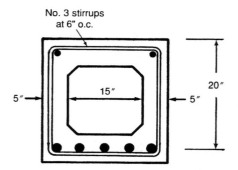

Fig. P6–5

6–6 ACI Sec. 12.13.1 states that "Web reinforcement shall be carried as close to the compression and tension surfaces of a member as cover requirements and proximity of other reinforcement will permit." Explain why.

6–7 ACI Sec. 11.5.4.1 sets the maximum spacing of vertical stirrups at $d/2$. Explain why.

6–8 Figure P6–8 shows a simply supported beam. The beam has No. 3 Grade 40 double-leg stirrups with $A_v f_y = 8.8$ kips and 4 No. 8 Grade 60 longitudinal bars with $A_s f_y = 190$ kips. The plastic truss model for the left end is shown in the figure. Assuming that the stirrups are all loaded to $A_v f_y$:

(a) Use the method of joints to compute the forces in each panel of the compression and tension chords and plot them (see Fig. 6–25). The force in member L_{11}–L_{13} is M_{U12}/jd.

(b) Plot $A_s f_s = M/jd$ on the diagram from part (a) and compare the bar forces from the truss model to those computed from M/jd.

(c) Compute the compression stresses in the diagonal member L_1–U_7 (see Eq. 6–11). The beam width, b_w, is 12 in.

6–9 The beam shown in Fig. P6–9 supports the unfactored loads shown. The dead load includes the weight of the beam.

(a) Draw shearing force diagrams for

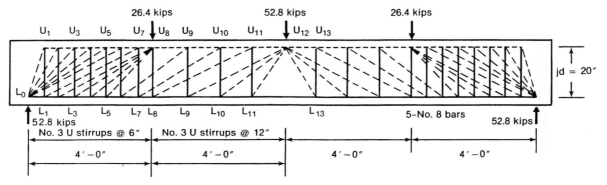

Fig. P6-8

(1) factored dead and live load on the entire beam.

(2) factored dead load on the entire beam plus factored live load on the left half-span.

(3) factored dead load on the entire beam plus factored live load on the right half-span.

(b) Superimpose the diagrams to get a shear force envelope (see Fig. 6–34e). Compare the shear at midspan to that from Eq. 6–24.

(c) Design stirrups. Use $f'_c = 3000$ psi and No. 3 double-leg stirrups with $f_y = 40,000$ psi.

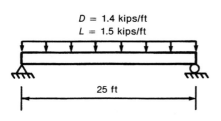
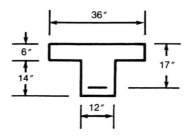

Fig. P6-9

6–10 The beam shown in Fig. P6–10 supports the unfactored loads shown in the figure. The dead load includes the weight of the beam.

(a) Draw shearing force diagrams for

(1) factored dead and live load on the entire length of beam.

(2) factored dead load on the entire beam plus factored live load between B and C.

(3) factored dead load on the entire beam plus factored live load between A and B and between C and D. Loadings (2) and (3) will give the maximum positive and negative shears at B.

(b) Draw the shear force envelope. The shear at B should be the dead load shear plus or minus the shear from Eq. 6–24.

(c) Design stirrups. Use $f'_c = 3500$ psi and f_y of the stirrups of 40,000 psi.

6–11 Figure P6–11 shows an interior span of a continuous beam. The shears at the ends are $\pm w_u \ell_n/2$. The shear at midspan is from Eq. 6–24.

(a) Draw a shear force envelope.

(b) Design stirrups using $f'_c = 2500$ psi and $f_y = 40,000$ psi for the stirrups.

6–12 Figure P6–12 shows a rigid frame and the factored loads acting on the frame. The 7-kip horizontal load can act from the left or the right. $f'_c = 3000$ psi, $f_y = 40,000$ psi.

(a) Design stirrups in the beam.

(b) Are stirrups required in the columns? If so, design the stirrups for the columns.

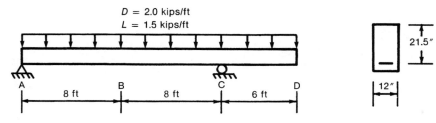

Fig. P6-10

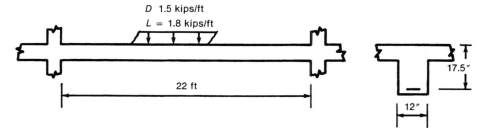

Fig. P6-11

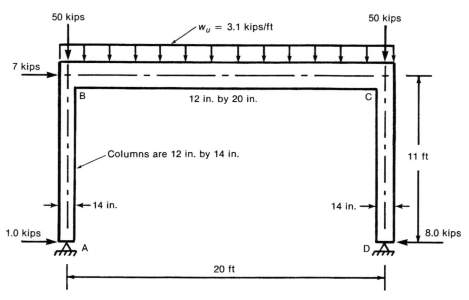

Fig. P6-12

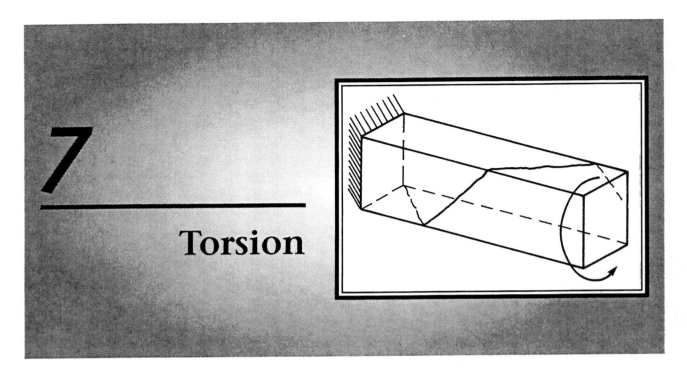

7

Torsion

7–1   INTRODUCTION

A moment acting about the longitudinal axis of a member is called a *twisting moment*, a *torque*, or a *torsional moment, T*. In structures, torsion results from eccentric loading of beams, as shown in Fig. 7–23, which will be discussed later, or from deformations resulting from the continuity of beams or similar members that join at an angle to each other as shown in Figure 7–24, also discussed later.

Shearing Stresses Due to Torsion in Uncracked Members

Solid Members

In a member subjected to torsion, a torsional moment causes shearing stresses on cross-sectional planes and on radial planes extending from the axis of the member to the surface. The element shown in Fig. 7–1 is stressed in shear, τ, due to the applied torque, T. In a circular member, the shearing stresses are zero at the axis of the bar and increase linearly to a maximum stress at the outside of the bar, as shown in Fig. 7–2a. In a rectangular bar, the shearing stresses vary from zero at the center to a maximum at the center of the long sides. Around the perimeter of a square bar, the shearing stresses vary from zero at the corners to a maximum at the center of each side, as shown in Fig. 7–2b.

The distribution of shearing stresses on a cross section can be visualized using the *soap-film analogy*. The equations for the slope of an inflated membrane are analogous to the equations for shearing stress due to torsion. Thus the distribution of shearing stresses can be visualized by cutting an opening in a plate to the shape of the cross section that is loaded in torsion, stretching a membrane or soap film over this opening, and inflating the membrane. Figure 7–3 shows an inflated membrane over a circular opening, representing a circular shaft. The maximum slope at each point in the membrane is proportional to the shearing stress at that point. The shearing stress acts perpendicular to the direction of

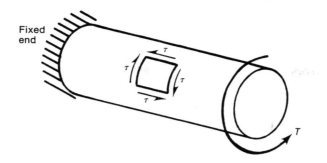

Fig. 7–1
Shear stresses due to torsion.

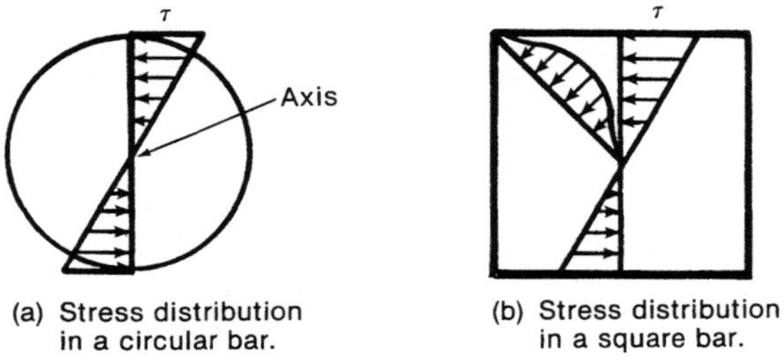

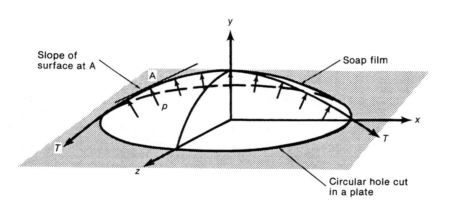

Fig. 7–2
Distribution of torsional
shear stresses in a circular
and a square bar.

(a) Stress distribution
in a circular bar.

(b) Stress distribution
in a square bar.

Fig. 7–3
Soap-film analogy:
circular bar.

the line of maximum slope. Thus the slope of a radial line in the membrane increases
from zero at the middle of the membrane to a maximum at the edge in the same way as
the stress distribution plotted in Fig. 7–2a. Figure 7–4 shows a membrane over a square
opening. Here the slopes of the radial lines correspond to the stress distribution shown
in Fig. 7–2b. A similar membrane for a U-shaped cross section made up from a series
of rectangles is shown in Fig. 7–5a. The corresponding stress distribution is shown
in Fig. 7–5b. For a hollow member with continuous walls, the membrane is similar to
Fig. 7–3 or 7–4 except that the region inside the hollow part is represented by a rigid
plate the shape of the hole.

The torsional moment is proportional to the volume under the membrane. A comparison in Figs. 7–4 and 7–5a shows that for a given maximum slope corresponding to a given
maximum shearing stress, the volume under a solid figure is much greater than that under
an open figure. Thus for a given maximum shearing stress, a solid rectangular cross section
can transmit a much higher torsional moment than an open section. The same is true of a
hollow cross section without slits in the walls.

7–1   Introduction   **233**

The maximum shearing stress in an elastic circular shaft is

$$\tau_{max} = \frac{Tr}{J} \qquad (7\text{–}1)$$

where τ_{max} = maximum shearing stress
T = torsional moment
r = radius of the bar
J = polar moment of inertia, $\pi r^4/2$

In a similar manner, the maximum shearing stress in a rectangular elastic shaft occurs at the center of the long side and can be written as

$$\tau_{max} = \frac{T}{\alpha x^2 y} \qquad (7\text{–}2)$$

where x is the shorter overall dimension of the rectangle, y is the longer overall dimension, and α varies from 0.208 for $y/x = 1.0$ (square bar) to 0.333 for $y/x = \infty$ (an infinitely wide plate).[7-1] An approximation to α is

$$\alpha = \frac{1}{3 + 1.8x/y} \qquad (7\text{–}3)$$

For a cross section made up of a series of thin rectangles, such as that in Fig. 7–5,

$$\tau_{max} \simeq \frac{T}{\Sigma(x^2 y/3)} \qquad (7\text{–}4)$$

where the term $x^2 y/3$ is evaluated for each of the rectangles.

The soap-film analogy and Eqs. 7–1 to 7–4 apply to elastic bodies. For fully plastic bodies, the shearing stress will be constant at all points. Thus the soap-film analogy must be replaced with a figure having constant slopes, producing a cone for a circular shaft or a pyramid for a square member. Such a figure would be formed by pouring sand onto a plate having the same shape as the cross section. This is referred to as the *sand heap analogy*. For a solid rectangular cross section the fully plastic shearing stress is

$$\tau_p = \frac{T}{\alpha_p x^2 y} \qquad (7\text{–}5)$$

where α_p varies from 0.33 for $y/x = 1.0$ to 0.5 for $y/x = \infty$.

The behavior of uncracked concrete members in torsion is neither perfectly elastic nor perfectly plastic as assumed by the soap film and sand heap analogies. However, solutions based on either of these models have been used successfully to predict torsional behavior.

Hollow Members

Figure 7–6a shows a thin-walled tube with continuous walls subjected to a torque about its longitudinal axis. An element *ABCD* cut from the wall is shown in Fig. 7–6b. The thicknesses of the walls along sides *AB* and *CD* are t_1 and t_2, respectively. The applied torque causes shearing forces V_{AB}, V_{BC}, V_{CD}, and V_{DA} on the sides of the element as shown, each equal to the shearing stresses on that side of the element times the area of the side. From

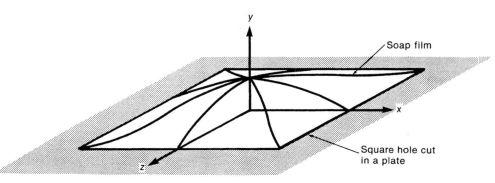

Fig. 7–4
Soap-film analogy: square bar.

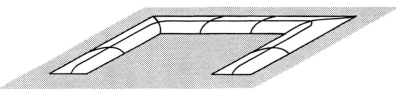

(a) Soap–film analogy.

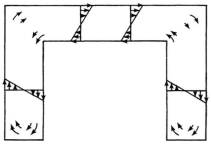

Fig. 7–5
Soap-film analogy: channel
shaped member.

(b) Distribution of shearing stresses.

$\Sigma F_x = 0$, we find that $V_{AB} = V_{CD}$. But $V_{AB} = \tau_1 t_1 dx$ and $V_{CD} = \tau_2 t_2 dx$, which gives $\tau_1 t_1 = \tau_2 t_2$, where τ_1 and τ_2 are the shearing stresses acting on sides AB and CD, respectively. The product τt is referred to as the *shear flow, q*. For equilibrium of a smaller element at corner B of the element in Fig. 7–6b, $\tau_1 = \tau_3$, and similarly at C, $\tau_2 = \tau_4$, as shown in Fig. 7–6c and d, Thus at points B and C on the perimeter of the tube, $\tau_3 t_1 = \tau_4 t_2$. This shows that for a given applied torque, T, the shear flow, q, is constant around the perimeter of the tube. The shear flow has units of (stress $\times$ length) pounds force per inch (N/mm). The name *shear flow* comes from an analogy to water flowing around a circular flume. The amount of water flowing past any given point in the flume is constant at any given period of time.

7–1   Introduction

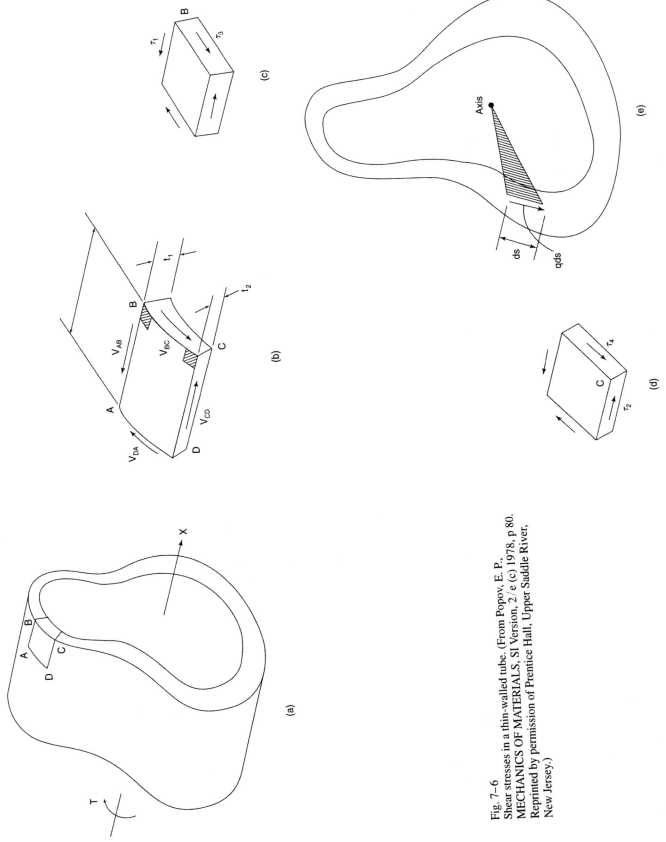

Fig. 7–6
Shear stresses in a thin-walled tube. (From Popov, E. P.,
MECHANICS OF MATERIALS, SI Version, 2/e (c) 1978, p 80.
Reprinted by permission of Prentice Hall, Upper Saddle River,
New Jersey.)

Figure 7–6e shows an end view of the tube. The torsional shear force acting on the length of walls ds is $q\,ds$. The perpendicular distance from this force to the centroidal axis of the tube is r and the moment of this force about the axis is $rq\,ds$, where r is measured from midplane of the wall because that is the line of action of the force $q\,ds$. Integrating this around the perimeter gives the torque in the tube

$$T = \int_p rq\,ds \qquad (7\text{–}6)$$

where $\int_p$ implies integration around the perimeter of the tube. However, q is constant around the perimeter of the tube, and this becomes

$$T = q\int_p r\,ds \qquad (7\text{–}7)$$

From Fig. 7–6e it can be seen that $r\,ds$ is two times the area of the shaded triangle between the elemental length of perimeter, ds, and the axis of the tube. Therefore, $\int_p rds$ is equal to two times the area enclosed by the centerline of the wall thickness. This area is referred to as the *area enclosed by the shear flow path*, A_o. For the cross section shown in Fig. 7–7, A_o is the crosshatched area, including the area of the hole in the center of the tube. Equation 7–7 becomes

$$T = 2qA_o \qquad (7\text{–}8)$$

where $q = \tau t$. Rearranging gives

$$\tau = \frac{T}{2A_o t} \qquad (7\text{–}9)$$

where t is the wall thickness at the point where the shear stress, τ, due to torsion is being computed. The maximum torsional shear stress occurs where the wall thickness is the least. In Fig. 7–7 this would be in the lower flange.

This analysis applies only if the walls of the tube are continuous (no slits parallel to the axis of the tube) or if, in the case of a solid member, the member can be approximated as a tube with continuous walls. Equation 7–9 can be applied to either elastic or inelastic sections. Referring to the shape of the membrane in Fig. 7–3, a tube can be called thin-walled if the change in slope of the membrane is small in the thickness of the wall and can be ignored without serious loss of accuracy.

EXAMPLE 7–1   Compute Torsional Shear Stresses in a Bridge Cross Section Using Thin-Walled Tube Theory

Figure 7–8a shows the cross section of a bridge. Compute the shear stresses, τ, at the top and bottom of the walls and in the lower flange due to an applied torque of 1650 ft-kips.

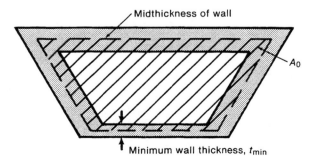

Fig. 7–7
Thin-walled tube.

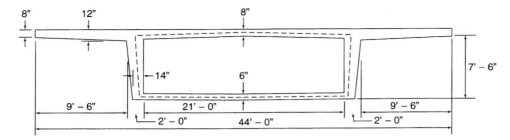

(a) Bridge cross section

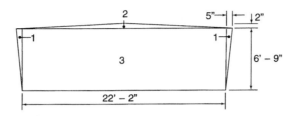

(b) A_o

Fig. 7–8
Cross section of a bridge—Example 7–1.

1. Compute A_o. A_o is the area enclosed by the midplane of the walls of the tube. The dashed line in Fig. 7–8b is the perimeter of A_o. The protruding deck flanges are not part of the tube and are ignored in computing A_o. Divide A_o into triangles and a rectangle as shown in Fig. 7–8b.

$$A_O = (2 \times 6'-9'' \times 5''/2) + (23'-0'' \times 2''/2) + (22'-2'' \times 6'-9'')$$

$$= 405 + 276 + 21{,}546$$

$$= 22{,}227 \text{ in.}^2$$

2. Compute the shear flow, q.

From Eq. 7–7,

$$q = \frac{T}{2A_O} = \frac{1650 \times 12{,}000}{2 \times 22{,}227}$$

$$= 445 \text{ lb/in.}$$

3. Compute the shear stresses. At the top of the wall the thickness, t, is 24 in. The torsional shear stress at the top of the walls is

$$\tau = q/t = 445/24$$

$$= 18.6 \text{ psi}$$

At the bottom of the wall the thickness is 14 in. The torsional shear stress at the bottom of the walls is

$$\tau = q/t = 445/14$$

$$= 31.8 \text{ psi}$$

The thickness of the bottom flange is 6 in. The torsional shear stress in the bottom flange is

$$\tau = q/t = 445/6$$
$$= 74.2 \text{ psi} \quad \blacksquare$$

Principal Stresses Due to Torsion

When the beam shown in Fig. 7–9 is subjected to a torsional moment, T, shearing stresses are developed on the top and front faces as shown by the elements in Fig. 7–9a. The principal stresses on these elements are shown in Fig. 7–9b. The principal tensile stress equals the principal compressive stress and both are equal to the shear stress if T is the only loading. The principal tensile stresses eventually cause cracking which spirals around the body, as shown by the line A–B–C–D–E in Fig. 7–9c.

In a reinforced concrete member, such a crack would cause failure unless it was crossed by reinforcement. This generally takes the form of longitudinal bars in the corners and stirrups. Since the crack spirals around the body, four-sided (closed) stirrups are required.

If a beam is subjected to combined shear and torsion, the two shearing stress components add on one side face and counteract each other on the other, as shown in Fig. 7–10a and b. As a result, inclined cracking starts on the face where the stresses add (crack AB) and extends across the flexural tensile face of the beam (in this case the top since this is a cantilever beam). If the bending moments are sufficiently large, the cracks will extend almost vertically across the back face, as shown by crack CD in Fig. 7–10c. The flexural compression zone near the bottom of the beam prevents the cracks from extending the full height of the front and back faces.

7–2 BEHAVIOR OF REINFORCED CONCRETE MEMBERS SUBJECTED TO TORSION

Pure Torsion

When a concrete member is loaded in pure torsion, shearing stresses, and principal stresses develop as shown in Fig. 7–9a and b. One or more inclined cracks develop when the maximum principal tensile stress reaches the tensile strength of the concrete. The onset of cracking causes failure of an unreinforced member. Furthermore, the addition of longitudinal steel without stirrups has little effect on the strength of a beam loaded in pure torsion.

A rectangular beam with longitudinal bars in the corners and closed stirrups can resist increased load after cracking. Figure 7–11 is a torque-twist curve for such a beam. At the cracking load, point A in Fig. 7–11, the angle of twist increases at constant torque as some of the forces formerly in the uncracked concrete are redistributed to the reinforcement. The cracking extends into the central core of the member, rendering the core ineffective. Figure 7–12 compares the strengths of a series of solid and hollow rectangular beams with the same exterior size and increasing amounts of both longitudinal and stirrup reinforcement.[7-3] Although the cracking torque was lower for the hollow beams, the ultimate strengths were the same for solid and hollow beams having the same reinforcement, indicating that the strength of a cracked reinforced concrete member loaded in pure torsion is governed by the outer skin or tube of concrete containing the reinforcement.

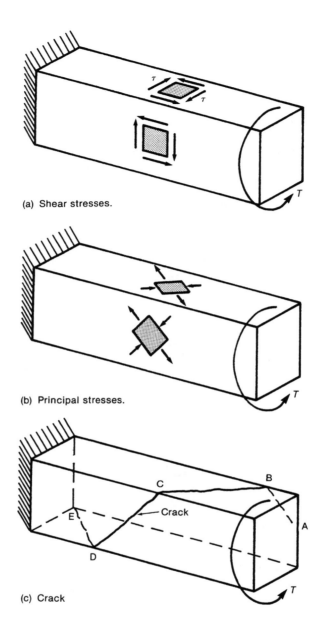

(a) Shear stresses.

(b) Principal stresses.

(c) Crack

Fig. 7–9
Principal tensile stresses and
cracking due to pure torsion.

After cracking of a reinforced beam, failure may occur in several ways. The stirrups, or longitudinal reinforcement, or both, may yield, or for beams that are *overreinforced* in torsion, the concrete between the inclined cracks may crush due to the principal compression stresses prior to yield of the steel. The most ductile behavior results when both reinforcements yield.

Combined Torsion, Moment, and Shear

Torsion seldom occurs by itself. Generally, there are also bending moments and shearing forces. Test results for beams without stirrups, loaded with various ratios of torsion and shear, are plotted in Fig. 7–13. The lower envelope to the data is given by the quarter ellipse

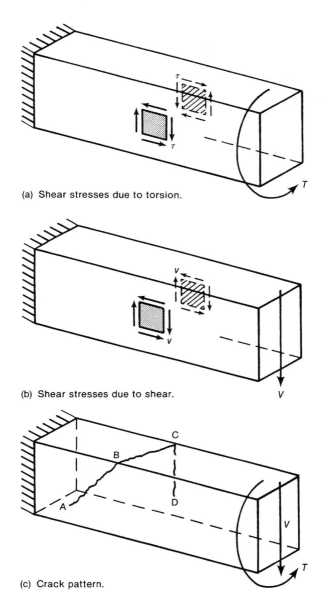

(a) Shear stresses due to torsion.

(b) Shear stresses due to shear.

Fig. 7–10
Combined shear, torsion, and
moment.

(c) Crack pattern.

$$\left(\frac{T_c}{T_{cu}}\right)^2 + \left(\frac{V_c}{V_{cu}}\right)^2 = 1 \tag{7-10}$$

where, in this graph, $T_{cu} = 1.6\sqrt{f_c'}\,x^2y$ and $V_{cu} = 2.68\sqrt{f_c'}\,b_w d$. These beams failed at or soon after inclined cracking.

7–3 DESIGN METHODS FOR TORSION

Two very different theories are used to explain the strength of reinforced concrete members. The first, based on a *skew bending theory* developed by Lessig,[7-5] was presented

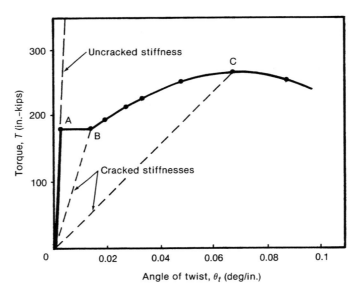

Fig. 7–11
Torque twist curve for a rectangular beam. (From Ref. 7–3.)

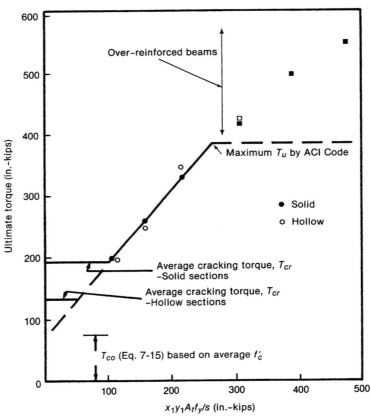

Fig. 7–12
Torsional strength of solid and hollow sections with the same outside dimensions. (From Ref. 7–3.)

in its present form by Hsu[7-3] and was the basis for the torsion design provisions in the 1971 through 1989 ACI Codes. This theory assumes that some shear and torsion is resisted by the concrete, and the rest by shear and torsion reinforcement. The mode of failure is assumed to involve bending on a skew surface resulting from the crack

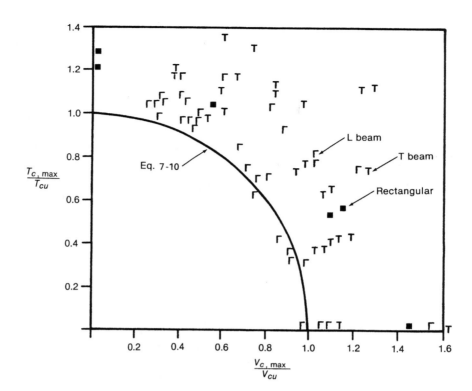

Fig. 7–13
Interaction of torsion and
shear. (From Ref. 7–4.)

The plot shows $\frac{T_{c,max}}{T_{cu}}$ on the vertical axis and $\frac{V_{c,max}}{V_{cu}}$ on the horizontal axis, with a curve labeled Eq. 7-10. Data points are labeled L beam, T beam, and Rectangular.

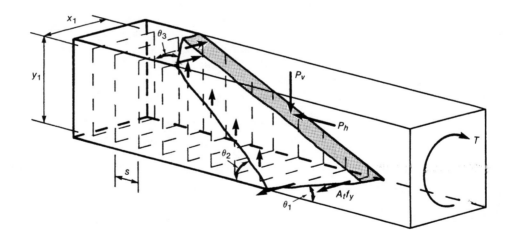

Fig. 7–14
Skew bending failure
surface.

spiraling around three of the four sides of the member, as shown in Figs. 7–10c and 7–14.

    The second design theory is based on a *thin-walled tube/plastic space truss model*, similar to the plastic truss analogy presented in Chap. 6. This theory, presented by Lampert,[7–6] Lampert and Thürlimann,[7–7] and Lampert and Collins,[7–8] forms the basis of the torsion provisions in the Comité Euro-International du Béton Model code,[7-9] the Canadian code,[7-10] and the 1995 ACI Code.

Skew Bending Design Method—1971 Through 1989 ACI Codes

This method will be reviewed briefly for historical interest. For a more complete explanation and examples of its application, refer to earlier editions of this book.

When subjected to pure torsion, a reinforced concrete member cracks when the maximum principal tensile stress reaches the tensile strength of the concrete. Since the principal tensile stress is equal to the shear stress at any given point in a body loaded in pure torsion, the torque at cracking can be obtained by setting τ_{max} equal to f'_t in Eq. 7–2 or 7–5:

$$T_{cr} = f'_t(\alpha x^2 y)$$

Hsu[7-11] selected $\alpha = 0.333$ and $f'_t = 6\sqrt{f'_c}$, giving the cracking torque of a member loaded in pure torsion as

$$T_{cr} = 2\sqrt{f'_c}x^2y \qquad (7–11a)$$

If the section consisted of several rectangles, this was written as

$$T_{cr} = 2\sqrt{f'_c}\,\Sigma\,(x^2y) \qquad (7–11b)$$

The term $\Sigma\,x^2y$ is calculated by dividing the cross section into rectangles, each with short side x and long side y. The arrangement of rectangles that maximizes the sum is used. This calculation tends to underestimate the combined effect since the additional volume of the soap film or sand heap in the junction regions is not considered.

Equation 7–11 applies to uncracked members loaded in pure torsion. Members subjected to combined bending and torsion tend to crack at lower torques, depending on the relative magnitudes of the flexure and the torsion.

Strength of Beams with Web Reinforcement

Figure 7–14 shows a skew-bending failure surface formed by inclined cracks on three sides of a section and a concrete compression zone on the fourth side. This member is loaded with a torque, T. The side on which the compression zone develops and the angles θ_1, θ_2, and θ_3 are functions of the relative magnitudes of the torsional moment and the bending moment and shear force when these exist. To simplify the development, θ_1 will be assumed equal to θ_3. This would be true in a beam subjected to pure torsion. The horizontal forces in the stirrups on the top and bottom faces produce a twisting couple of

$$T_h = n_h(A_t f_s)y_1$$

where n_h = number of horizontal stirrup legs crossed in one face $(x_1 \cot \theta_1/s)$
A_t = area of one leg of a stirrup
f_s = stress in the stirrup
y_1 = vertical distance between the top and bottom surfaces of the stirrup

This can be rewritten as

$$T_h = k_1 A_t f_y \left(\frac{x_1}{s}\right) y_1 \tag{7-12}$$

where k_1 is $(\cot \theta_1 f_s/f_y)$. In a similar manner, the vertical force in the stirrups on the side face and the vertical component P_v, of the compressive force, P_c, form a twisting couple of

$$T_v = n_v(A_t f_s)(k_2 x_1)$$

where n_v is number of vertical stirrups crossed by the crack in the side face $(y_1 \cot \theta_2/s)$, and $k_2 x_1$ is the lever arm between the stirrup forces and the compressive force P_c. This equation can be written in the same form as Eq. 7–12:

$$T_v = k_3 A_t f_y \left(\frac{y_1}{s}\right) x_1 \tag{7-13}$$

where k_3 is $(k_2 \cot \theta_2 f_s/f_y)$. Adding Eqs. 7–12 and 7–13 and replacing $k_1 + k_3$ with α_t gives the torsional resistance provided by the stirrups as

$$T_s = \alpha_t \frac{x_1 y_1}{s} A_t f_y \tag{7-14}$$

Consideration of a number of cross sections led Hsu[7-3] to suggest that T_{co} is roughly 40% of T_{cr}:

$$T_{co} = 0.8 \leq f_c' \Sigma x^2 y \tag{7-15}$$

and that α_t is a function of x_1 and y_1, the distances between the sides of the stirrups:

$$\alpha_t = 0.66 + \frac{0.33 y_1}{x_1} \leq 1.50 \tag{7-16}$$

The vertical component of the stirrup forces is equilibrated by the vertical component, C_v, of the compression force C. The horizontal component, C_h, of the compression force (Fig. 7–14) must be equilibrated by longitudinal steel. The 1989 ACI Code assumes that this will occur if the volume of longitudinal and transverse steel are equal, giving

$$A_\ell = 2A_t \left(\frac{x_1 + y_1}{s}\right) \tag{7-17}$$

This assumption applies only if the cracks are at $45°$ and the beam is subjected to pure torsion. The longitudinal steel is distributed around the perimeter of the closed stirrups and is in addition to the reinforcement provided for flexure.

The torque carried by the concrete after inclined cracking, T_{co}, is less than the cracking torque, T_{cr}, as shown by Fig. 7–12. The ACI Torsion Committee originally proposed that the minimum torsional reinforcement be based on $T_s = T_{cr} - T_{co}$.

Combined Shear, Moment, and Torsion

Torsion seldom occurs be itself. Generally, there are also bending moments and shearing forces. Rearranging Eq. 7–10 and replacing T_{cu} with T_{co} as given by Eq. 7–15, and V_{cu} with V_{co} defined using the usual ACI definition of V_c gives

$$T_c = \frac{T_{co}}{\sqrt{1 + (T_c/V_{co})^2(V_c/T_c)^2}}$$

where

$$\frac{T_{co}}{V_{co}} = \frac{0.8\sqrt{f_c'}\,\Sigma x^2 y}{2\sqrt{f_c'}\,b_w d}$$

$$= \frac{0.4}{C_t}$$

where $C_t = b_w d/\Sigma x^2 y$. Thus T_c becomes

$$T_c = \frac{0.8\sqrt{f_c'}\,\Sigma x^2 y}{\sqrt{1 + (0.4V_c/C_t T_c)^2}}$$

Although this was derived from tests of beams without stirrups, it is assumed to apply equally to the calculation of T_c in beams with web reinforcement, except that the V_c and T_c in the denominator are replaced by V_u and T_u so that the torsion, T_c, resisted by the concrete is

$$T_c = \frac{0.8\sqrt{f_c'}\,\Sigma x^2 y}{\sqrt{1 + (0.4V_u/C_t T_u)^2}} \qquad (7\text{--}18)$$

In a similar manner, Eq. 7–10 can be rearranged to get the shear, V_c, resisted by the concrete in a torsional member:

$$V_c = \frac{2\sqrt{f_c'}\,b_w d}{\sqrt{1 + [2.5C_t(T_u/V_u)]^2}} \qquad (7\text{--}19)$$

When designing for shear and torsion, the stirrups resisting shear are designed for

$$V_u = \phi(V_c + V_s) \qquad (7\text{--}20a)$$

and the stirrups and longitudinal bars resisting torsion are designed to satisfy

$$T_u \le \phi(T_c + T_s) \qquad (7\text{--}20b)$$

The design process is illustrated in Fig. 7–15.

Thin-Walled Tube/Plastic Space Truss Method—1995 ACI Code

A second model for the torsional strength of beams combines the thin-walled tube analogy in Fig. 7–6 and the plastic truss analogy for shear presented in Sec. 6–4. This gives a mechanics-based model of the behavior which is easy to visualize and leads to much simpler calculations than the skew bending theory.

Both solid and hollow members are considered as tubes. The test data for solid and hollow beams in Fig. 7–12 suggest that once torsional cracking has occurred, the concrete in the center of the member has little effect on the torsional strength of the cross section and hence can be ignored. This, in effect, produces an equivalent tubular member.

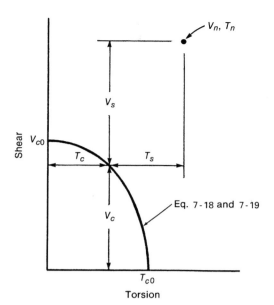

Fig. 7–15
Design for shear and torsion—skew bending theory.

Torsion is assumed to be resisted by shear flow, q, around the perimeter of the member as shown in Fig. 7–16a. The beam is idealized as a thin-walled tube. After cracking, the tube is idealized as a hollow truss consisting of closed stirrups, longitudinal bars in the corners, and compression diagonals approximately centered on the stirrups, as shown in Fig. 7–16b. The diagonals are idealized as being between cracks which are at an angle θ, generally taken as 45° for reinforced concrete.

The derivation of the thin-walled tube/plastic space truss method is presented and compared to tests in Ref. 7–12. In this book, the analogy is referred to as the thin-walled tube analogy because this is the terminology used in the derivation of Eq. 7–9 in mechanics of materials textbooks. The walls of the tube are actually quite thick, being on the order of one-sixth to one-quarter of the smaller side of a rectangular member.

Lower Limit on Consideration of Torsion

Torsional reinforcement is not required if torsional cracks do not occur. In pure torsion the principal tensile stress, σ_1, is equal to the shear stress, τ, at a given location. Thus, from Eq. 7–9 for a thin-walled tube,

$$\sigma_1 = \tau = \frac{T}{2A_0 t} \tag{7–21}$$

To apply this to a solid section it is necessary to define the wall thickness of the equivalent tube prior to cracking. The Canadian Code[7-10] assumes that prior to cracking, the wall thickness, t, is equal to $3A_{cp}/4p_{cp}$, where p_{cp} is the perimeter of the concrete section and A_{cp} is the area enclosed by this perimeter. The area, A_o, enclosed by the centerline of the walls of the tube is taken as $2A_{cp}/3$. Substituting these into Eq. 7–12 gives

$$\sigma_1 = \tau = \frac{T p_{cp}}{A_{cp}{}^2} \tag{7–22}$$

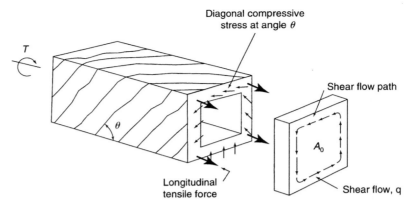

(a) Thin-walled tube analogy

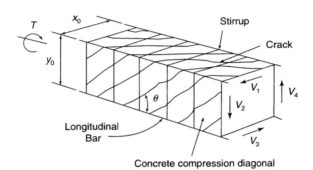

Fig. 7–16
Thin-walled tube/space truss
analogy. (From Ref. 7–12.)

(b) Space truss analogy

Torsional cracking is assumed to occur when the principal tensile stress reaches the tensile strength of the concrete in biaxial tension–compression, taken as $4\sqrt{f'_c}$. Thus the torque at cracking is

$$T_{cr} = 4\sqrt{f'_c}\left(\frac{A_{cp}^2}{p_{cp}}\right) \qquad (7\text{--}23)$$

The tensile strength was taken as $4\sqrt{f'_c}$, less than the $6\sqrt{f'_c}$ used elsewhere because, as shown in Fig. 3–12a, the tensile strength under biaxial compression and tension is less than that in uniaxial tension.

In combined shear and torsion the torsional cracking load follows the circular interaction shown in Fig. 7–13. From Eq. 7–10, a torque of $0.25T_{cr}$ would reduce V_{cr} by about 3%. The ACI Code requires that torsion be considered in design if T_u exceeds $0.25T_{cr}$ given by

$$\phi\sqrt{f'_c}\left(\frac{A_{cp}^2}{p_{cp}}\right) \qquad (7\text{--}24)$$

For an isolated beam, A_{cp} is the area enclosed by the perimeter of the section, including the area of any holes and p_{cp} is the perimeter of the section. For a beam cast monolithically with a floor slab, ACI Sec. 11.6.1 states that the overhanging flange width included is that defined in ACI Sec. 13.2.4, which assumes that the overhanging flange extends (a) the greater of the distances that the beam web projects above or below the flange, (b) but not more than four times the slab thickness.

In SI units, Eq. 7–24 becomes

$$\frac{\phi \sqrt{f_c'}}{12}\left(\frac{A_{cp}^2}{p_{cp}}\right) \tag{7–24M}$$

Area of Stirrups

A beam subjected to pure torsion can be modeled as shown in Fig. 7–16b. A rectangular beam will be considered for simplicity, but a similar derivation could be applied to any cross-sectional shape. The beam is idealized as a space truss consisting of longitudinal bars in the corners, closed stirrups, and diagonal concrete compression members which spiral around the beam between the cracks. The height and width of the truss are y_o and x_o, measured between the centers of the corner bars. The angle of the cracks is θ, which initially is close to 45°, but may become flatter at high torques.

From Eq. 7–8 the shear force per unit length of the perimeter of the tube or truss, referred to as the shear flow, q, is given by

$$q = \frac{T}{2A_o}$$

The total shear force due to torsion along each of the top and bottom sides of the truss is

$$V_1 = V_3 = \frac{T}{2A_o}x_o \tag{7–25a}$$

Similarly, the shear forces due to torsion along each of the two vertical sides

$$V_2 = V_4 = \frac{T}{2A_o}y_o \tag{7–25b}$$

Summing moments about one corner of the truss, we find that the internal torque is

$$T = V_1 y_o + V_2 x_o$$

Substituting for V_1 and V_2 using Eqs. 7–25a and 7–25b gives

$$T = \left(\frac{T}{2A_o}x_o\right)y_o + \left(\frac{T}{2A_o}y_o\right)x_o$$

or

$$T = \frac{2T(x_o y_o)}{2A_o}$$

but, by definition, $x_o y_o = A_o$. Thus we have shown that the internal forces V_1 through V_4 equilibrate the applied torque, T.

A portion of one of the vertical sides is shown in Fig. 7–17. The inclined crack cuts n_2 stirrups, where

$$n_2 = \frac{y_o \cot \theta}{s}$$

where s is the spacing of the stirrups. The force in the stirrups must equilibrate V_2. Assuming that all the stirrups yield at ultimate

$$V_2 = \frac{A_t f_{yv} y_o}{s}\cot \theta \tag{7–26}$$

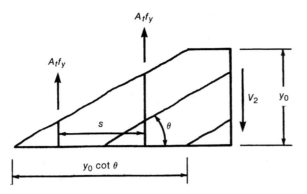

Fig. 7–17
Forces in stirrups.

where f_{yv} is the yield strength of the stirrups. Replacing V_2 with Eq. 7–25b and taking T equal to the nominal torsion capacity, T_n, gives

$$T_n = \frac{2A_o A_t f_{yv}}{s} \cot \theta \qquad (7\text{–}27)$$

$$(\text{ACI Eq. 11–21})$$

where θ may be taken as any angle between 30 and 60°. For nonprestressed concrete, ACI Sec. 11.6.3.6 suggests that θ be taken as 45° because this corresponds to the angle assumed in the derivation of the equation for designing stirrups for shear. The factors affecting the choice of θ are discussed later in this section.

After cracking, the torsional resistance is provided by the tube formed by the walls of the space truss. ACI Sec. 11.6.3.6 allows the area A_o to be taken as $0.85A_{oh}$, where A_{oh} is the area enclosed by the outermost closed stirrups.

Hsu[7–13] has given more accurate values of t and A_o as

$$t = \frac{4T_n}{A_c f_c'} \qquad (7\text{–}28)$$

and

$$A_o = A_{cp} - \frac{2T_n p_{cp}}{A_{cp} f_c'} \qquad (7\text{–}29)$$

where p_{cp} and A_{cp} are the perimeter of the concrete section and the area enclosed by that perimeter, respectively. For very large members, the use of Eqs. 7–28 and 7–29 may lead to smaller amounts of torsional reinforcement than Eq. 7–27. A_o will be taken as $0.85A_{oh}$ in this book.

Area of Longitudinal Reinforcement

As shown by the force triangle in Fig. 7–18, the shear force V_2 can be resolved into a diagonal compression force, D_2, parallel to the concrete struts and an axial tension force, N_2, where D_2 and N_2 are given by

$$D_2 = \frac{V_2}{\sin \theta} \qquad (7\text{–}30)$$

$$N_2 = V_2 \cot \theta \qquad (7\text{–}31)$$

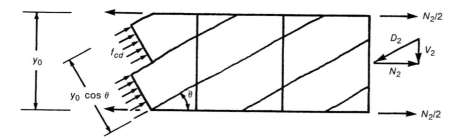

Fig. 7–18
Side of space truss—resolution of shear force V_2.

Because the shear flow, q, is constant from point to point along side 2, the force N_2 acts along the centroidal axis of side 2. For a beam with longitudinal bars in the top and bottom corners of side 2, half of N_2 will be resisted by each corner bar. A similar resolution of forces occurs on each side of the truss. For a rectangular member as shown in Fig. 7–16b, the total longitudinal force is

$$N = 2(N_1 + N_2)$$

Substituting Eqs. 7–25a and b and 7–31, and taking T equal to T_n gives

$$N = \frac{T_n}{2A_o}2(x_o + y_o)\cot\theta \tag{7–32}$$

where $2(x_o + y_o)$ is the perimeter of the closed stirrup, p_h. Longitudinal reinforcement with a total area of A_ℓ must be provided for the longitudinal force, N. Assuming this reinforcement yields at ultimate, with a yield strength of $f_{y\ell}$, gives

$$A_\ell f_{y\ell} = N$$

or

$$A_\ell = \frac{T_n p_h}{2A_o f_{y\ell}}\cot\theta \tag{7–33}$$

For convenience in design, A_ℓ can be expressed in terms of the area of the torsional stirrups. Substituting Eq. 7–27 into Eq. 7–33 gives

$$A_\ell = \left(\frac{A_t}{s}\right)p_h\left(\frac{f_{yv}}{f_{y\ell}}\right)\cot^2\theta \tag{7–34}$$
(ACI Eq. 11–22)

Because the individual wall tension forces N_1, N_2, N_3, and N_4 act along the centroidal axes of the sides, the total force, N, acts along the centroidal axis of the member. For this reason, the longitudinal torsional reinforcement must be distributed evenly around the perimeter of the cross section so that the centroid of the bar areas coincides approximately with the centroid of the member. One bar is placed in each corner of the stirrups to anchor the compression struts where the compressive forces change direction around the corner.

Combined Shear and Torsion

In ACI Codes prior to 1995, a portion, T_c of the torsion was carried by concrete, and a portion, V_c, of the shear was carried by concrete. When both shear and torsion acted, an elliptical interaction diagram was assumed between T_c and V_c and stirrups were provided for the rest of the torsion and shear as shown in Fig. 7–15. The derivation of Eqs. 7–27 and 7–34 for the space truss analogy assumed that all the torsion was carried by reinforcement, T_s, without any "torsion carried by concrete," T_c. When shear and torsion act together, the 1995 ACI Code assumes that V_c remains constant and T_c remains equal to zero

$$V_n = V_c + V_s \qquad\qquad (6-9)$$
$$\text{(ACI Eq. 11–2)}$$

$$T_n = T_s \qquad\qquad (7-35)$$

where V_c is given by Eq. 6–8 (ACI Eq. 11–3). The application of Eqs. 6–9 and 7–35 is illustrated in Fig. 7–19, which can be compared to the previous design method, illustrated in Fig. 7–15. The assumption that there is no interaction between V_c and T_c greatly simplifies the calculations. Design comparisons carried out by ACI Committee 318 showed that for combinations of low V_u and high T_u, with v_u less than about $0.8(\phi 2\sqrt{f'_c})$ psi, the 1995 code method requires more stirrups than are required by previous ACI Codes. For v_u greater than this value, the 1995 method requires the same or marginally fewer stirrups than the skew bending method.

Maximum Shear and Torsion

A member loaded by torsion or by combined shear and torsion may fail by yielding of the stirrups and longitudinal reinforcement as assumed in the derivation of Eqs. 7–27 and 7–34, or by crushing of the concrete due to the diagonal compressive forces, D_2, shown in Fig. 7–18. A serviceability failure may occur if the inclined cracks are too wide at service loads. The limit on combined shear and torsion in ACI Sec. 11.6.3.1 was derived to limit service load crack widths, but as shown later, it also gives a lower bound on the web crushing capacity.

Crack Width Limit. As explained in Sec. 6–5, ACI Sec. 11.5.6.8 attempts to guard against excessive crack widths by limiting the maximum shear, V_s, that can be trans-

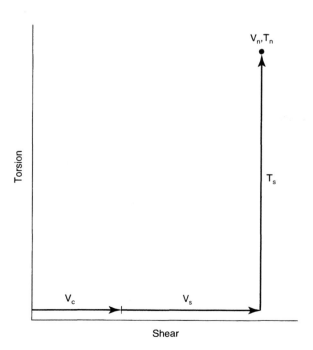

Fig. 7–19
Design for shear and tor-
sion–1995 ACI Code.

Torsion

ferred by stirrups to $8\sqrt{f_c'}\,b_w d$. In ACI Sec. 11.6.3.1 the same concept is used, expressed in terms of stresses. The shear stress, v, due to direct shear is $V_u/b_w d$. From Eq. 7–9 with A_o after torsional cracking taken as $0.85A_{oh}$ and $t = A_{oh}/p_h$, the shear stress, τ, due to torsion is $T_u p_h/(1.7A_{oh}^2)$. In a hollow section these two shear stresses are additive on one side, at point A in Fig. 7–20a, and the limit is given by

$$\frac{V_u}{b_w d} + \frac{T_u p_h}{1.7A_{oh}^2} \le \phi\left(\frac{V_c}{b_w d} + 8\sqrt{f_c'}\right) \tag{7–36a}$$
$$\text{(ACI Eq. 11–19)}$$

If the wall thickness varies around the cross section, as for example in Fig. 7–8, ACI Sec. 11.6.3.2 states that Eq. 7–36a is evaluated at the location where the left-hand side is the greatest.

If a hollow section has a wall thickness, t, less than A_{oh}/p_h, ACI Sec. 11.6.3.3 requires that the actual wall thickness be used. Thus the second term of Eq. 7–36a becomes $T_u/1.7A_{oh}t$. Alternatively, the second term on the left-hand side of Eq. 7–36a can be taken as $T_u/A_o t$ where A_o and t are computed as in Example 7–1.

In a solid section, the shear stresses due to direct shear are assumed to be distributed uniformly across the width of the web while the torsional shear stresses exist only in the walls of the thin-walled tube as shown in Fig. 7–20b. In this case, a direct addition of the two terms tends to be conservative and a root-square summation is used.

$$\sqrt{\left(\frac{V_u}{b_w d}\right)^2 + \left(\frac{T_u p_h}{1.7A_{oh}^2}\right)^2} \le \phi\left(\frac{V_c}{b_w d} + 8\sqrt{f_c'}\right) \tag{7–36b}$$
$$\text{(ACI Eq. 11–18)}$$

The right-hand sides of Eqs. 7–36a and 7–36b include the term $V_c/b_w d$ so that the same equation can be used for prestressed concrete members or members with axial tension or compression that have different values of V_c.

In SI units, the right-hand sides of Eqs. 7–36a and 7–36b become

$$\phi\left(\frac{V_c}{b_w d} + \frac{8\sqrt{f_c'}}{12}\right) \tag{7–36M}$$

Web Crushing Limit.   Failure can also occur due to crushing of the concrete in the walls of the tube due to the inclined compressive forces in the struts between cracks. As will now be shown, this sets a higher limit on the stresses than Eqs. 7–36a and 7–36b.

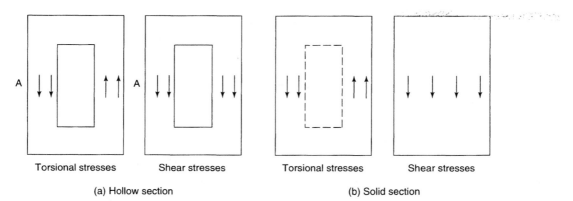

| Torsional stresses | Shear stresses | Torsional stresses | Shear stresses |

(a) Hollow section   (b) Solid section

Fig. 7–20
Addition of shear stresses due to torsion and shear. (From Ref. 7–17.)

The diagonal compressive force in a vertical side of the member shown in Fig. 7–18 is given by Eq. 7–30. This force acts on a width $y_o \cos \theta$, as shown in Fig. 7–18. The resulting compressive stress is

$$f_{cd} = \frac{V_2}{t y_o \cos \theta \sin \theta} \tag{7-37}$$

Substituting Eq. 7–9, again taking A_o equal to $0.85 A_{oh}$ and approximating t as A_{oh}/p_h, gives f_{cd} due to torsion as

$$f_{cd} = \frac{T_u p_h}{1.7 A_{oh}^2 \cos \theta \sin \theta} \tag{7-38}$$

The compressive stresses due to shear may be calculated in a similar manner as

$$f_{cd} = \frac{V_u}{b_w d \cos \theta \sin \theta} \tag{7-39}$$

For a solid section, these will be added using a root-square as explained in the derivation of Eq. 7–36b giving

$$f_{cd} = \sqrt{\left(\frac{V_u}{b_w d \cos \theta \sin \theta}\right)^2 + \left(\frac{T_u p_h}{1.7 A_{oh}^2 \cos \theta \sin \theta}\right)^2} \tag{7-40}$$

The value of f_{cd} from Eq. 7–40 should not exceed the crushing strength of the cracked concrete in the tube, f_{ce}. As discussed in Sec. 3–2, Collins and Mitchell[7-14] have related f_{ce} to the strains in the longitudinal and transverse reinforcement in the tube. For $\theta = 45°$ and longitudinal and transverse strains, $\epsilon = 0.002$, equal to the yield strain of Grade 60 steel, Collins and Mitchell predict $f_{ce} = 0.549 f_c'$. Setting f_{cd} in Eq. 7–40 equal to $0.549 f_c'$ and evaluating $\cos \theta \sin \theta$ for $\theta = 45°$, the upper limit on the shears and torques as determined by crushing of the concrete in the walls of the tube becomes

$$\sqrt{\left(\frac{V_u}{b_w d}\right)^2 + \left(\frac{T_u p_h}{1.7 A_{oh}^2}\right)^2} \le \phi(0.275 f_c') \tag{7-41}$$

The limit in Eqs. 7–36a and 7–36b has been set at $\phi(v_c + 8\sqrt{f_c'})$ to limit crack widths where, for reinforced concrete, v_c can be assumed to be $2\sqrt{f_c'}$, giving a limit of $\phi 10\sqrt{f_c'}$. The limit of $\phi(0.275 f_c')$ in Eq. 7–41 will always exceed $\phi 10\sqrt{f_c'}$ for f_c' greater than 1324 psi. Since reinforced concrete members will always have f_c' greater than 1324 psi, only the crack width limits, Eqs. 7–36a and 7–36b, are included in the ACI Code. Two simplifications were made in the derivation of Eq. 7–41. First, the calculation of f_{cd} in Eq. 7–39 involved the effective depth, d, while the calculation of f_{cd} in Eq. 7–37 used the wall height, y_o, which is about $0.9d$. Second, all the shear was assumed to be carried by truss action without a V_c term. These were considered to be reasonable approximations in view of the level of accuracy of the right-hand sides of Eqs. 7–36 and 7–41.

In Ref. 7–12 the code limit, Eq. 7–36b, is compared to tests of reinforced concrete beams in pure torsion which failed due to crushing of the concrete in the tube. The limit gave an acceptable lower bound on the test results.

Value of θ

ACI Sec. 11.6.3.6 allows the value of θ to be taken as any value between 30° and 60°, inclusive. ACI Sec. 11.6.3.7 requires that the value of θ used in calculating the area of longitudinal steel, A_ℓ, be the same as used to calculate A_t. This is because a reduction in θ leads to (a) a reduction in the required area of stirrups, A_t, as shown by Eq. 7–27; (b) an increase in the required area of longitudinal steel, A_ℓ, as shown by Eq. 7–34; and (c) an increase in f_{cd} as shown by Eq. 7–40.

ACI Sec. 11.6.3.6 suggests a default value of $\theta = 45°$ for reinforced concrete members. This will be used in the examples.

Combined Moment and Torsion

Torsion causes an axial tensile force N, given by Eq. 7–32. Half of this, $N/2$, is assumed to act in the top chord of the space truss, and half in the bottom chord, as shown in Fig. 7–21a. Flexure causes a compression–tension couple, $C = T = M_u/jd$, shown in Fig. 7–21b, where $j \approx 0.9$. For combined moment and torsion these internal forces add together, as shown in Fig. 7–21c. The reinforcement provided for the flexural tension force, T, and that provided for the tension force in the lower chord due to torsion, $N/2$, must be added together as required by ACI Sec. 11.6.3.8.

In the flexural compression zone, the force C tends to cancel out some, or all, of $N/2$. ACI Sec. 11.6.3.9 allows the area of the longitudinal torsion reinforcement in the compression zone to be reduced by an amount equal to $M_u/(0.9df_{y\ell})$, where M_u is the moment that acts in conjunction with the torsion at the section being designed. It is necessary to determine this reduction at a number of sections because the bending moment varies along the length of the member. If several loading cases must be considered in design, M_u and T_u must be from the same loading case. Normally, the reduction in the area of the compression steel is not significant, as will be shown in Example 7–2.

Torsional Stiffness

The torsional stiffness, K_t, of a member of length ℓ is defined as the torsional moment, T, required to cause a unit twist in the length ℓ:

$$K_t = \frac{T}{\phi_t \ell} \tag{7–42}$$

or

$$K_t = \frac{T}{\theta_t} \tag{7–43}$$

where ϕ_t is the angle of twist per unit length and $\theta_t = \phi_t \ell$ is the total twist in the length ℓ. For a thin-walled tube of length ℓ, the total twist can be found by virtual work by equating the external work done when the torque, T, acts through a virtual angle change θ_t, to the internal work done when the shearing stresses due to torsion, τ, act through a shear strain, $\gamma = \tau/G$.

$$T\theta_t = \int_V \tau \gamma \, dV$$

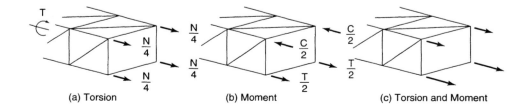

Fig. 7–21
Internal forces due to combined torsion and moment.

(a) Torsion   (b) Moment   (c) Torsion and Moment

7–3   Design Methods for Torsion

where $\int_V$ implies integration over the volume. Replacing γ with τ/G, τ with Eq. 7–9, and $d_V = \ell t\, ds$ gives

$$T\theta_t = \ell \int_p \frac{T}{2A_o t} \left(\frac{T}{2A_o t G} \right) t\, ds$$

where $\int_p$ implies integration around the perimeter of the tube. This reduces to

$$\theta_t = \ell \frac{T}{4A_o^2 G} \int_p \frac{ds}{t} \tag{7–44}$$

If the wall thickness t is constant,

$$\theta_t = \frac{T p_o}{4A_o^2 t G} \tag{7–45}$$

where p_o is the perimeter of A_o. Substituting Eq. 7–45 into Eq. 7–43 gives the torsional stiffness as

$$K_t = \left(\frac{4A_o^2 t}{p_o} \right) \frac{G}{\ell} = \frac{CG}{\ell} \tag{7–46}$$

where C, the torsional constant, refers to the term in parentheses. Equation 7–46 is similar to the equation for the flexural stiffness of a beam that is fixed at the far end:

$$K = \frac{4EI}{\ell}$$

In this equation, EI is the *flexural rigidity* of the section. The term CG in Eq. 7–46 is the *torsional rigidity* of the section.

Figure 7–11 shows a measured torque-twist curve for pure torsion for a member having a moderate amount of torsional reinforcement (longitudinal steel, plus stirrups at roughly $d/3$). From Eqs. 7–43 and 7–46, the slope, T/ϕ_t, of a radial line through the origin and any point on the torque-twist curve gives the effective value of the term CG corresponding to that torque.

Prior to torsional cracking, the value of CG corresponds closely to the uncracked value, $\beta_t x^2 y G$. At torsional cracking there is a sudden increase in ϕ_t and hence a sudden drop in the effective value of CG. In this test, the value of CG immediately after cracking (line $0-B$ in Fig. 7–11) was one-fifth of the value before cracking. At failure (line $0-C$) the effective CG was roughly one-sixteenth of the uncracked value. This drastic drop in torsional stiffness allows a significant redistribution of torsion in certain indeterminate beam systems. Methods of estimating the postcracking torsional stiffness of beams are given by Collins and Lampert.[7–15]

Equilibrium and Compatibility Torsion

Torsional loadings can be separated into two basic categories: *equilibrium torsion,* where the torsional moment is required for the equilibrium of the structure, and *compatibility torsion,* where the torsional moment results from the compatibility of deformations between members meeting at a joint. Figure 7–22a shows a cantilever beam supporting an eccentrically applied load P which causes torsion. Figure 7–22b shows the cross section of a beam supporting precast floor slabs. Torsion will result if the dead loads of slabs A and B differ, or if one supports a live load and the other does not. The torsion in the beams in Fig. 7–22a and b must be resisted by the structural system if the beam is to remain in equilibrium. If the applied torsion is not resisted, the beam will rotate about its axis until the structure collapses. Similarly, the canopy shown in Fig. 7–22c applies a tor-

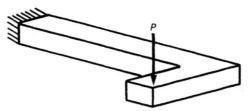

(a) Cantilever beam with eccentrically applied load.

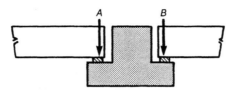

(b) Section through a beam supporting precast floor slabs.

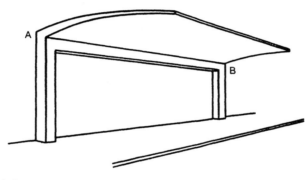

Fig. 7–22
Examples of equilibrium torsion. (From Ref. 7–17.)

(c) Canopy.

sional moment to the beam. For this structure to stand, the beam must resist the torsional moment and the columns must resist the resulting bending moments.

The beam $A-B$ in Fig. 7–23a develops a slope at each end when loaded and develops the bending moment diagram shown in Fig. 7–23b. If, however, end A is built monolithically with a cross beam $C-D$, as shown in Fig. 7–23a, beam $A-B$ can develop an end slope at A only if beam $C-D$ twists about its own axis. If ends C and D are restrained against rotation, a torsional moment, T, will be applied to beam $C-D$ at A, as shown in Fig. 7–23d. An equal and opposite moment, M_A, acts on $A-B$. The magnitude of these moments depends on the relative magnitudes of the torsional stiffness of C–D and the flexural stiffness of $A-B$. If C and D were free to rotate about the axis $C-D$, T would be zero. On the other hand, if C and D could not rotate, and if the torsional stiffness of C–D was very much greater than the flexural stiffness of $A-B$, the moment M_A would approach a maximum equal to the moment that would be developed if A was a fixed end. Thus the moment M_A and the twisting moments T result from the need for the end slope of beam $A-B$ at A to be *compatible* with the

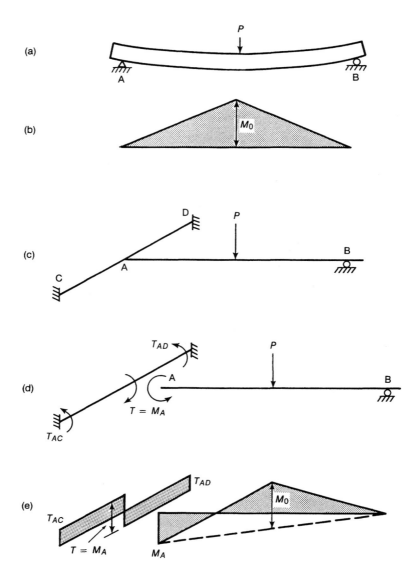

Fig. 7–23
Compatibility torsion.

angle of twist of beam $C-D$ at point A. Note that the moment M_A causes a reduction in the moment at midspan of beam $A-B$, as shown in Fig. 7–23e.

When a beam cracks in torsion, its torsional stiffness drops significantly, as discussed in the preceding section. As load is applied to beam $A-B$, torsional moments build up in member $C-D$ until it cracks due to torsion. With the onset of cracking, the torsional stiffness of $C-D$ decreases and the torque, T, and the moment M_A drop. When this happens, the moment at midspan of $A-B$ must increase. This phenomenon is discussed by Collins and Lampert[7-15] and is the basis of ACI Secs. 11.6.2 and 11.6.3.

If the torsional moment, T_u, is required to maintain equilibrium, ACI Sec. 11.6.2.1 requires that the members involved be designed for T_u. On the other hand, in those cases where compatibility torsion exists and a reduction of the torsional moment can occur due to redistribution of moments, ACI Sec. 11.6.2.2 permits T_u to be reduced to

$$\phi 4\sqrt{f_c'}\left(\frac{A_{cp}^2}{p_{cp}}\right) \tag{7-47}$$

at the sections located d away from the faces of the supports. This is roughly the cracking torque of a member loaded in pure torsion. The resulting torsional reinforcement will help limit crack widths to acceptable values at service loads. If the torsional moments are reduced, it is necessary to redistribute these moments to adjoining members.

In SI units, Eq. 7–47 becomes

$$\frac{\phi\sqrt{f_c'}}{3}\left(\frac{A_{cp}^2}{p_{cp}}\right) \tag{7–47M}$$

Calculation of Torsional Moments

Equilibrium Torsion: Statically Determinate Case

In torsionally statically determinate beams, such as shown in Fig. 7-22a, the torsional moment at any section can be calculated by cutting a free-body diagram at that section.

Equilibrium Torsion: Statically Indeterminate Case

In the case shown in Fig. 7-22c, the torsional moment, t, transmitted to the beam per foot of length of beam $A-B$, is the moment of the weight of a 1-ft strip of the projecting canopy about the line of action of the vertical reactions. The distribution of the torque along beam $A-B$ will be such that

$$\text{change in slope between the columns at } A \text{ and } B \ = \int_A^B \frac{t\,dx}{CG}$$

If this change in angle is zero and the distribution of CG along the member is symmetrical about the midspan, the torque at A and B will be $\pm t\ell/2$ (Fig. 7–24a). If the flexural stiffness of column B were less than column A, end B would rotate more than end A, and the torque diagram will not be symmetrical since more torque would go to the stiffer end, end A.

On the other hand, the increased torsion at end A will lead to earlier torsional cracking at that end. When this occurs, the effective torsional rigidity, CG, at end A will decrease, reducing the stiffness at A. This will cause a redistribution of the torsional moments along the beam so that the final distribution approaches the symmetrical distribution.

The interaction between the torque, T, the column stiffness, K_c, and the torsional stiffness, K_t, along the beams makes it extremely difficult to estimate the torque diagram in statically indeterminate cases. At the same time, however, it is necessary to design the beam for the full torque necessary to equilibrate the loads on the overhang. Provided that the beam is detailed to have adequate ductility, a safe design will result for any reasonable distribution of torque, T, which is in equilibrium with the loads. This can vary from $T = t\ell$ at one end and $T = 0$ at the other (Fig. 7–24b), to the reverse. In such cases, however, wide torsional cracks would develop at the end where T was assumed to be zero, since this end has to twist through the angle necessary to reduce T from its initial value to zero. In most cases it is sufficiently accurate to design stirrups in this class of beams for $T = \pm t\ell/2$ at each end using the torque diagram shown in Fig. 7–24a.

Compatibility Torsion

Based on some assumed set of torsional and flexural stiffnesses, an elastic grid or plate analysis or an approximation to such an analysis leads to torsional moments in the edge members. These can then be redistributed to account for the effects of torsional stiffness. This process was illustrated in Fig. 7–23 and will be considered in Example 7–3.

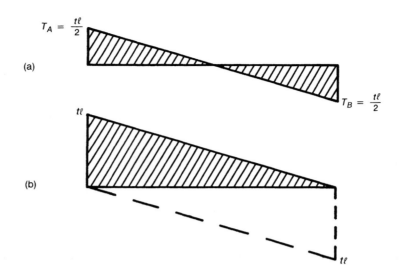

(a)

$T_A = \dfrac{t\ell}{2}$

$T_B = \dfrac{t\ell}{2}$

$t\ell$

(b)

$t\ell$

Fig. 7–24
Moments in beam $A-B$ of
the structure shown in Fig.
7–22c.

7–4   DESIGN FOR TORSION, SHEAR, AND MOMENT—1995 ACI CODE

The design procedure for combined torsion, shear, and moment involves designing for the moment, ignoring the torsion and shear, and then providing stirrups and longitudinal reinforcement to give adequate shear and torsional strength. The basic design equations are

$$\phi V_n \geq V_u$$

(6–14)
(ACI Eq. 11–1)

$$V_n = V_c + V_s$$

(6–9)
(ACI Eq. 11–2)

and

$$\phi T_n \geq T_u$$

(7–48)
(ACI Eq. 11–20)

where ϕ is the strength reduction factor for shear and torsion, taken equal to 0.85 (ACI Sec. 9.3), and T_n is given by Eq. 7–27 (ACI Eq. 11–21).

Selection of Cross Section for Torsion

A torsional moment is resisted by shearing stresses in the uncracked member (Figs. 7–2 and 7–5b), and by the shear flow forces (V_1 and V_2 in Fig. 7–16b) in the member after cracking. For greatest efficiency, the shearing stresses and shear flow forces should flow around the member in the same circular direction and should be located as far from the axis of the member as possible. Thus the solid square member in Figs. 7–2b and 7–4 is more efficient than the U-shaped member in Fig. 7–5. For equal volumes of material, a closed tube will be more efficient than a solid section. For building members, solid rectangular sections are generally used for practical reasons. For bridges, box sections are frequently used. Open U-sections like Fig. 7–5b, made up of beams and a deck, are common but they are weak and lack stiffness in torsion.

Although the 1995 ACI Code does not require fillets at the inside corners of a hollow section, it is good practice to provide them. Fillets reduce the stress concentrations where the inclined compressive forces flow around inside corners and also aid in the removal of the formwork. Section 11.6.1.2 of the 1989 ACI Commentary suggested that each side of a fillet should by $x/6$ in length if there are fewer than 8 longitudinal bars, where x is the smaller dimension of a rectangular cross section, and $x/12$ if the section has 8 or more longitudinal bars, but not necessarily more than 4 in.

Location of Critical Section for Torsion

In Sec. 6–5 and Figs. 6–29 and 6–30, the critical section for shear was found to be located at a distance d away from the face of the support. For an analogous reason, ACI Sec. 11.6.2.4 allows sections located at less than d from the support to be designed for the same torque, T_u, that exists at a distance d from the support. This would not apply if a large torque was applied within a distance d from the support.

Definitions of A_{cp} and p_{cp}

A_{cp} is the total area enclosed by the perimeter of the cross section and includes the area of any holes, provided that there are no gaps or slits in the perimeter. In the case of a torsional member cast monolithically with a slab, A_{cp} includes the area of the portion of the slab defined in ACI Sec. 13.2.4 on each side of the beam where there is a flange. The perimeter p_{cp} is the perimeter of A_{cp}, including the flange portions.

Definition of A_{oh}

ACI Sec. 11.6.3 states that the area enclosed by the shear flow path, A_o, shall be determined by analysis except that it is permissible to take A_o as $0.85A_{oh}$, where A_{oh} is the area enclosed by the centerline of the outermost closed stirrups as shown in Fig. 7–25. In most cases it is adequate to use this definition, although it may be on the conservative side for very large box girders. Equations 7–28 and 7–29, derived by Hsu[7-13] satisfy the requirement for analysis. Figure 7–25 shows A_{oh} for several cross sections and illustrates the importance of having a square or wide rectangular section to resist torsion.

Torsional Reinforcement: Amounts and Details

Torsional reinforcement consists of closed stirrups satisfying Eq. 7–27 (ACI Eq. 11–21) and longitudinal bars satisfying Eq. 7–34 (ACI Eq. 11–22). According to ACI Sec. 11.6.3.8, these are added to the longitudinal bars and stirrups provided for flexure and shear. It is possible to reduce the area of longitudinal torsional reinforcement in the flexural compression zone as discussed earlier in this section (ACI Sec. 11.6.3.9).

When designing for shear, a given size stirrup, with area of both legs A_v, is chosen and the required spacing, s, is computed. When considering combined shear and torsion, it is necessary to add the stirrups required for shear to those required for torsion. The area of stirrups required for shear and torsion will be computed in terms of A_v/s and A_t/s, both with units of in.^{2}/in. of length of beam. These can be added. Since A_v refers to both legs of a stirrup, while A_t refers to only one leg, the total A_{v+t}/s is

$$\frac{A_{v+t}}{s} = \frac{A_v}{s} + \frac{2A_t}{s} \tag{7–49}$$

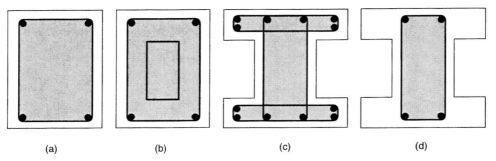

Fig. 7–25
Examples of A_{oh}. (From Ref. 7–17.)

where A_{v+t} refers to the cross-sectional area of both legs of a stirrup. It is now possible to select A_{v+t} and compute a spacing s. If a stirrup in a wide beam had more than two legs for shear, only the outer legs would be included in the summation in Eq. 7–49.

Types of Torsional Reinforcement and Its Anchorage

Because the inclined cracks can spiral around the beam, as shown in Fig. 7–9c, 7–10, or 7–16b, stirrups are required in all four faces of the beam. For this reason, ACI Sec. 11.6.4.1 requires the use of longitudinal bars plus either (a) closed stirrups, perpendicular to the axis of the member, (b) closed cages of welded-wire fabric with wires transverse to the axis of the member, or (c) spirals. These should extend as close to the perimeter of the member as cover requirements will allow so as to make A_{oh} as large as possible.

Tests by Mitchell and Collins[7–16] have examined the types of stirrup anchorages required. Figure 7–26a shows one corner of the space truss model shown in Fig. 7–16b. The inclined compressive stresses in the concrete, f_{cd}, have components parallel to the top and side surfaces as shown in Fig. 7–26b. The components acting toward the corner are balanced by tensions in the stirrups. The concrete outside the reinforcing cage is only poorly anchored at best, however, and the shaded region will spall off if the compression in the outer shell is large. For this reason, ACI Sec. 11.6.4.2(a) requires that stirrups be anchored with 135° hooks around a longitudinal bar if the corner can spall. If the concrete around the stirrup anchorage is restrained against spalling by a flange or slab or similar member, ACI Sec. 11.6.4.2(b) allows the use of the anchorage details shown in Fig. 7–27a.

ACI Sec. 11.6.4.3 requires that longitudinal reinforcement for torsion be developed at both ends. Since the maximum torsions generally act at the ends of a beam, it is generally necessary to anchor the longitudinal torsional reinforcement for its yield strength at the face of the support. This may require hooks or horizontal U-shaped bars lap spliced with the longitudinal torsion reinforcement. A common error is to extend the bottom reinforcement in spandrel beams loaded in torsion, 6 in. into the support as allowed in ACI Sec. 12.11.1. Generally, this is not adequate to develop the longitudinal bars needed to resist torsion. Figure 7–28 shows a spandrel beam in a test slab which failed in torsion due, in part, to inadequate anchorage of the bottom reinforcement in the support.

Minimum Torsional Reinforcement

When the factored torsional moment exceeds

$$\phi\sqrt{f_c'}\left(\frac{A_{cp}^{2}}{p_{cp}}\right) \tag{7–24}$$

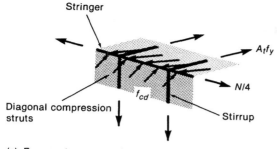

Stringer

Diagonal compression
struts

f_{cd}

Stirrup

$A_t f_y$

$N/4$

(a) Forces at a corner of the space truss.

Spalls off

Diagonal compressive
stresses

$A_t f_y$

(b) Spalling at corner of space truss.

Fig. 7–26
Compressive strut forces at a corner of a torsional member.

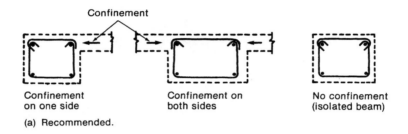

Confinement

Confinement
on one side

Confinement on
both sides

No confinement
(isolated beam)

(a) Recommended.

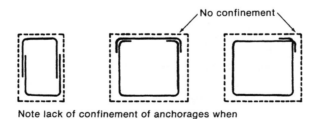

No confinement

Fig. 7–27
Anchorage of closed stirrups.
(From Ref. 7–18.)

Note lack of confinement of anchorages when
compared to similar members in (a).

(b) Not recommended.

or, in SI units

$$\frac{\phi\sqrt{f_c'}}{12}\left(\frac{A_{cp}^2}{p_{cp}}\right) \qquad (7\text{–}24\text{M})$$

the larger of (a) the torsional reinforcement satisfying the strength requirements of ACI Sec. 11.6.3, or (b) the minimum reinforcement required by ACI Sec. 11.6.5 must be provided. ACI Sec. 11.6.5.2 specifies that the minimum area of closed stirrups shall be

$$A_v + 2A_t \geq \frac{50b_w s}{f_{yv}} \qquad (7\text{–}50)$$

(ACI Eq. 11–23)

Fig. 7–28
Torsional failure of a spandrel beam. (Photograph courtesy of J. G. MacGregor.)

In SI units this becomes

$$A_v + 2A_t \geq \frac{b_w s}{3f_y} \qquad (7\text{--}50M)$$

In Hsu's tests [7-3] of rectangular reinforced concrete members subjected to pure torsion, two beams failed at the torsional cracking load. In these beams, the total ratio of the volume of the stirrups and longitudinal reinforcement to the volume of the concrete was 0.802 and 0.88%. A third beam with a volumetric ratio of 1.07% failed at 1.08 times the torsional cracking torque. All the other beams tested by Hsu had volumetric ratios of 1.07% or greater, and failed at torques in excess of 1.2 times the cracking torque. This suggests that beams with similar concrete and steel strengths loaded in pure torsion should have a volumetric ratio of torsional reinforcement in the order of 0.9 to 1.0%. Thus the minimum volumetric ratio should be set at about 1%:

$$\frac{A_{\ell, \min} s}{A_{cp} s} + \frac{A_t p_h}{A_{cp} s} \geq 0.01$$

or

$$A_{\ell, \min} = 0.01 A_{cp} - \frac{A_t p_h}{s}$$

If the constant 0.01 is assumed to be a function of the material strengths in the test specimens, the first term on the right-hand side of this equation can be rewritten as $7.5\sqrt{f_c'}/f_{y\ell}$. In the 1971 to 1989 ACI Codes, a transition was provided between the total volume of reinforcement required by the equation for $A_{\ell, \min}$ for pure torsion, and the much smaller amount of minimum reinforcement required in beams subjected to shear without torsion. This was accomplished by multiplying the same term by $\tau/(\tau + v)$, giving

$$A_{\ell, \min} = \frac{7.5\sqrt{f_c'}}{f_{y\ell}} A_{cp}\left(\frac{\tau}{\tau + v}\right) - \left(\frac{A_t}{s}\right) p_h \left(\frac{f_{yv}}{f_{y\ell}}\right) \qquad (7\text{--}51)$$

During the development of the 1995 torsion provisions, it was assumed that a practical limit on $\tau/(\tau + v)$ was $2/3$ for beams that satisfied Eq. 7–34. When this was introduced, Eq. 7–51 became

$$A_{\ell,\,min} = \frac{5\sqrt{f'_c}}{f_{y\ell}} A_{cp} - \left(\frac{A_t}{s}\right) p_h \left(\frac{f_{yv}}{f_{y\ell}}\right) \qquad\qquad (7\text{–}52)$$

(ACI Eq. 11–24)

This equation was derived for the case of pure torsion. When it is applied to combined shear, moment, and torsion, it is not clear how much of the area of the stirrups should be included in A_t/s. In this book we shall assume that A_t/s in Eq. 7–52 is the actual amount of transverse reinforcement provided for shear and torsion, where A_t is for one leg of a closed stirrup. The value of A_t/s should not be taken less than $25b_w/f_{yv}$ the amount corresponding to Eq. 7–50.

In SI units, Eq. 7–52 becomes

$$A_{\ell,\,min} = \frac{5\sqrt{f'_c}}{12f_{y\ell}} A_{cp} - \left(\frac{A_t}{s}\right) p_h \left(\frac{f_{yv}}{f_{y\ell}}\right) \qquad\qquad (7\text{–}52\text{M})$$

Spacing of Torsional Reinforcement

Figure 7–26 shows that the corner bars in a beam help to anchor the compressive forces in the struts between cracks. If the stirrups are too far apart, or if the longitudinal bars in the corners are too small in diameter, the compressive forces will tend to bend the longitudinal bars outward, weakening the beam. ACI Sec. 11.6.6.1 limits the stirrup spacing to the smaller of $p_h/8$ or 12 in., where p_h is the perimeter of the outermost closed stirrups.

Because the axial force due to torsion, N, acts along the axis of the beam, ACI Sec. 11.6.6.2 specifies that the longitudinal torsional reinforcement be distributed around the perimeter of the closed stirrups with the centroid of the steel approximately at the centroid of the cross section. The longitudinal steel has a maximum spacing of 12 in. The longitudinal reinforcement should be inside the stirrups with a bar inside each corner of the stirrups. The diameter of the longitudinal bars should be at least $1/24$ of the stirrup spacing but not less than 0.375 in. In tests,[7–16] corner bars with a diameter of $1/31$ of the stirrup spacing bent outward at failure.

ACI Sec. 11.6.6.3 requires that torsional reinforcement continue a distance $(b_t + d)$ past the point where the torque is less than

$$\phi\sqrt{f'_c}\left(\frac{A_{cp}{}^2}{p_{cp}}\right) \qquad\qquad (7\text{–}24)$$

where b_t is the width of that part of the cross section containing the closed stirrups. This length takes into account the fact that torsional cracks spiral around the beam. In SI units, Eq. 7–24 is divided by 12.

Maximum Yield Strength of Torsional Reinforcement

ACI Sec. 11.6.3.4 limits the yield strength used in design calculations to 60 ksi. This is done to limit crack widths at service loads.

High-Strength Concrete and Lightweight Concrete

In the absence of tests of high-strength concrete beams loaded in torsion, ACI Sec. 11.1.2 limits the value of $\sqrt{f'_c}$ to 100 psi in all torsional calculations. This affects only ACI Secs. 11.6.1, 11.6.2.2, 11.6.3.1, and 11.6.5.3.

For the reasons presented in Sec. 6–5, the quantity $\sqrt{f_c'}$, which is related to the tensile strength, is reduced by ACI Sec. 11.2 for beams constructed of lightweight concrete. Unfortunately, this section was not updated when the torsion section was revised. We shall assume that it applies to the code sections mentioned in the preceding paragraph. The reduction is applied in the manner presented in Sec. 6–5.

7–5 APPLICATION OF 1995 ACI CODE DESIGN METHOD FOR TORSION

Review of the Steps in the Design Method

1. Calculate the bending moment, M_u, diagram, or envelope for the member.

2. Select b, d, and h based on the ultimate flexural moment. For problems involving torsion, square cross sections are preferable.

3. Based on b and h, draw final M_u, V_u, and T_u diagrams or envelopes. Calculate the area of reinforcement required for flexure.

4. Determine whether torsion must be considered. Torsion must be considered if T_u exceeds the torque given by Eq. 7–24. Otherwise, it can be neglected and the design is carried out according to Chap. 6 of this book.

5. Determine whether the case involves equilibrium or compatibility torsion. If it is the latter, the torque may be reduced to the value given by Eq. 7–47 at the sections d from the faces of the supports. If this is done, the moments and shears in the other members must be adjusted accordingly.

6. Check if the section is large enough for torsion. If T_u exceeds the values given by Eq. 7–36a or 7–36b, enlarge the section.

7. Compute the area of stirrups required for shear. This is done using Eqs. 6–8, 6–9, 6–14, and 6–18 (ACI Eqs. 11–1, 11–2, 11–3, and 11–17). To facilitate the addition of stirrups for shear and torsion calculate

$$\frac{A_v}{s} = \frac{V_s}{f_{yv}d}$$

If V_s exceeds $8\sqrt{f_c'}\,b_w d$, the cross section is too small and must be enlarged. This is satisfied automatically by step 6.

8. Compute the area of stirrups required for torsion using Eqs. 7–48 and 7–27 (ACI Eqs. 11–20 and 11–21). Again, these will be computed in terms of A_t/s.

9. Add the stirrup amounts together using Eq. 7–49 and select the stirrups. The area of stirrups must exceed the minimum given by Eq. 7–50 (ACI Eq. 11–23). The spacing must satisfy ACI Secs. 11.6.4.4, 11.6.6.1, and 11.6.6.3. Stirrups must be closed.

10. Design the longitudinal reinforcement for torsion using Eq. 7–34 (ACI Eq. 11–22) and add to that provided for flexure. The longitudinal reinforcement for torsion must exceed the minimum given by Eq. 7–52 (ACI Eq. 11–24) and must satisfy ACI Secs. 11.6.4.3, 11.6.6.2, and 11.6.6.3.

EXAMPLE 7–2 Design for Torsion, Shear, and Moment: Equilibrium Torsion

The cantilever beam shown in Fig. 7–29a supports its own dead load plus a concentrated load as shown. The beam is 54 in. long and the concentrated load acts at a point 6 in. from the end of the beam and 6 in. away from the vertical axis of the member. The unfactored concentrated load consists of a 20-kip dead load and a 20-kip live load. Use $f'_c = 3000$ psi and $f_y = 60,000$ psi.

1. Compute the bending moment diagram. Estimate the size of the member. The minimum depth of control flexural deflections is $\ell/8 = 6.75$ in. (ACI Table 9.5a or Table A–14 of this book). This seems too small in view of the loads involved. As a first trial, use a 16 in. wide by 24 in. section with $d = 21.5$ in.

$$w = \frac{16 \times 24}{144} \times 0.15 = 0.40 \text{ kip/ft}$$

$$\text{factored uniform load} = 0.56 \text{ kip/ft}$$

$$\text{factored concentrated load} = 1.4 \times 20 + 1.7 \times 20 = 62 \text{ kips}$$

The bending moment diagram is as shown in Fig. 7–29b, with the maximum $M_u = 254$ ft-kips.

2. Select b, d, and h for flexure (see Example 4–6).

From Eqs. 4–17,

$$\frac{M_u}{\phi k_n} = \frac{bd^2}{12,000}$$

where $\phi k_n = \phi[f'_c \omega(1 - 0.59\omega)]$, $\omega = \rho f_y/f'_c$, and M_u is in ft-kips. Try $\rho \approx 0.01$:

$$\omega = \frac{0.01 \times 60,000}{3000} = 0.20$$

$$\phi k_n = 0.9 \times 3000 \times 0.20(1 - 0.59 \times 0.20)$$

$$= 476 \text{ (see also Table A–3)}$$

Therefore,

$$\frac{bd^2}{12,000} = \frac{254}{476} = 0.533$$

$$bd^2 = 6390 \text{ in.}^3$$

Possible choices are

$b = 12$ in., $d = 23.1$, and $h = 25.6$ in.
$b = 14$ in., $d = 21.4$, and $h = 23.9$ in.
$b = 16$ in., $d = 20.0$, and $h = 22.5$ in.

Use $b = 14$ in., $d = 21.5$ in., and $h = 24$ in. Since this is smaller than the section originally chosen, it is necessary to recompute w and M_u.

$$w = 0.35 \text{ kip/ft} \quad \text{and} \quad w_u = 0.49 \text{ kip/ft}$$

$$M_u = 253 \text{ ft-kips at root of cantilever}$$

$$A_{s(req'd)} = \frac{M_u \times 12,000}{\phi f_y jd}$$

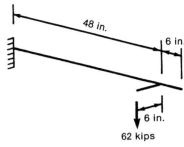

(a) Beam.

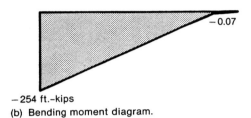

(b) Bending moment diagram.

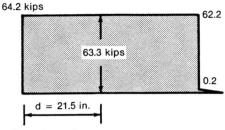

(c) Shear force diagram.

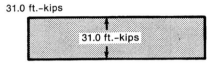

(d) Torque diagram.

Fig. 7–29
Cantilever beam—Example 7–2.

Assume that $j = 0.875$.

$$A_{s(req'd)} = \frac{253 \times 12{,}000}{0.9 \times 60{,}000(0.875 \times 21.5)}$$

$$= 2.99 \text{ in.}^2$$

Check M_n for $A_s = 2.99$ in.^{2}. From Eq. 4–11,

$$a = 5.02 \text{ in.}$$

$$\phi M_n = \frac{0.9 \times 2.99 \times 60{,}000(21.5 - 5.02/2)}{12{,}000}$$

$$= 255 \text{ ft-kips}$$

Therefore, use a 14 in. by 24 in. section with $d = 21.5$ in. $A_s = 2.99$ in.^{2} is required at the top for flexure.

3. Compute the final $M_u, V_u,$ and T_u diagrams.

The shear force and torque diagrams are shown in Fig. 7–29. The shear and torque at d from the face of the support are shown.

4. Should torsion be considered?

For the cross section: $A_{cp} = 14 \times 24 = 336$ in.^{2} and $p_{cp} = 2(14 + 24) = 76$ in. From ACI Sec. 11.6.1, torsion can be neglected if T_u is less than

$$\phi \sqrt{f_c'} \left(\frac{A_{cp}^2}{p_{cp}} \right) = 0.85\sqrt{3000} \left(\frac{336^2}{76} \right) = 69{,}160 \text{ in.-lb} = 5.76 \text{ ft-kips} \qquad (7\text{–}24)$$

Since $T_u = 31.0$ ft-kips, torsion must be considered.

Torsion

5. Equilibrium or compatibility torsion?

The torsion is needed for equilibrium; therefore, design for $T_u = 31.0$ ft-kips.

6. Is the section big enough to resist the torsion?

For a solid cross section, ACI Sec. 11.6.3.1(a) requires the section to satisfy

$$\sqrt{\left(\frac{V_u}{b_w d}\right)^2 + \left(\frac{T_u p_h}{1.7 A_{oh}^2}\right)^2} \leq \phi\left(\frac{V_c}{b_w d} + 8\sqrt{f_c'}\right) \tag{7-36b}$$

(ACI Eq. 11–18)

From ACI Sec. 11.3.1.1, take $V_c = 2\sqrt{f_c'}b_w d$. A_{oh} = area within centerline of closed stirrups. Assume 1.5 in. of cover and No. 4 stirrups as shown in Fig. 7–30.

$$A_{oh} = (14 - 2 \times 1.5 - 0.5)(24 - 2 \times 1.5 - 0.5) = 215 \text{ in.}^2$$

$$p_h = 2(10.5 + 20.5) = 62 \text{ in.}$$

$$\sqrt{\left(\frac{63,300}{14 \times 21.5}\right)^2 + \left(\frac{31.0 \times 12,000 \times 62}{1.7 \times 215^2}\right)^2} \leq 0.85(2\sqrt{3000} + 8\sqrt{3000})$$

$$\sqrt{43,810 + 86,100} = 361 \text{ psi} \qquad 0.85 \times 10\sqrt{3000} = 466 \text{ psi}$$

Since 361 psi is less than 466 psi, the cross section is large enough.

7. Compute the stirrup area required for shear.

From Eqs. 6–9 and 6–14

$$V_u \leq \phi(V_c + V_s)$$

$$V_c = 2\sqrt{f_c'}b_w d = 2 \times \sqrt{3000} \times 14 \times 21.5$$

$$= 32.97 \text{ kips}$$

$$V_s = \frac{63.3}{0.85} - 32.97 = 41.5 \text{ kips}$$

From Eq. 6–18 (ACI Eq. 11–15),

$$V_s = \frac{A_v f_y d}{s} \qquad \text{or} \qquad \frac{A_v}{s} = \frac{V_s}{f_y d}$$

$$\frac{A_v}{s} = \frac{41,500}{60,000 \times 21.5} = 0.0322$$

For shear we require stirrups with $A_v/s = 0.0322$.

8. Compute the stirrup area required for torsion.

From Eq. 7–48 (ACI Eq. 11–20), $\phi T_n \geq T_u$. Therefore,

$$T_n = \frac{31.0 \times 12,000}{0.85} = 437,650 \text{ in.-lb}$$

From Eq. 7–26 (ACI Eq. 11–22),

$$T_n = \frac{2A_o A_t f_{yv}}{s} \cot\theta \qquad \text{or} \qquad \frac{A_t}{s} = \frac{T_n}{2A_o f_{yv}} \cot\theta$$

From ACI Sec. 11.6.3.6:

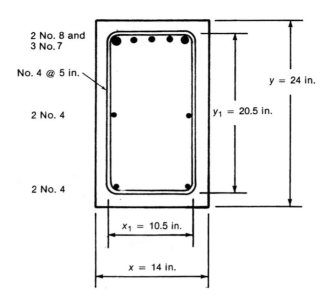

Fig. 7–30
Cantilever beam—Example 7–2.

$$A_o = 0.85 A_{oh} = 0.85 \times 215$$

$$= 182.8 \text{ in.}^2$$

$$\theta = 45°$$

$$\frac{A_t}{s} = \frac{437,650}{2 \times 182.8 \times 60,000} = 0.020$$

For torsion we require stirrups with $A_t/s = 0.020$.

9. Add the stirrup areas and select stirrups.

From Eq. 7–49,

$$\frac{A_{v+t}}{s} = \frac{A_v}{s} + \frac{2A_t}{s}$$

$$= 0.0322 + 2 \times 0.020 = 0.0722 \text{ in.}^2/\text{in.}$$

Check minimum stirrups: From Eq. 7–50 (ACI Eq. 11–23),

$$\frac{A_v + 2A_t}{s} \geq \frac{50 b_w}{f_{yv}}$$

$$\text{minimum } \frac{A_{v+t}}{s} = \frac{50 \times 14}{60,000} = 0.0117$$

Since 0.0722 exceeds 0.0117, the minimum does not govern.

For No. 3 stirrups, A_{v+t} (two legs) $= 0.22$ in.^{2} and the required $s = 3.05$ in. For No. 4 stirrups, A_{v+t} (two legs) $= 0.40$ in.^{2} and the required $s = 5.54$ in. The minimum stirrup spacing (ACI Sec. 11.6.6.1) is the smaller of $p_h/8 = 62/8 = 7.75$ in., or 12 in.

Use No. 4 closed stirrups at 5 in. on centers.

10. Design the longitudinal reinforcement for torsion. From Eq. 7–34 (ACI Eq. 11–22),

$$A_\ell = \left(\frac{A_t}{s}\right) p_h \left(\frac{f_{yv}}{f_{y\ell}}\right) \cot^2\theta$$

where A_t/s is the amount computed in step 8.

$$A_\ell = (0.020) \times 62 \times \frac{60,000}{60,000} \cot^2 45°$$

$$= 1.24 \text{ in.}^2$$

From Eq. 7–52 (ACI Eq. 11–24), the minimum A_ℓ is

$$A_{\ell \text{ min.}} = \frac{5\sqrt{f_c'}A_{cp}}{f_{y\ell}} - \left(\frac{A_t}{s}\right)p_h\left(\frac{f_{yv}}{f_{y\ell}}\right)$$

where A_t/s will be based on the stirrup area actually provided in step 9, No. 4 double-leg stirrups at 5 in. on centers. Since A_t is one leg of a stirrup, $A_t/s = 0.20/5 = 0.040$.

$$A_\ell = \frac{5\sqrt{3000} \times 336}{60,000} - (0.040) \times 62 \times \frac{60,000}{60,000}$$

$$= -0.95 \text{ in.}^2$$

Since this is negative, it does not apply and the minimum A_ℓ does not govern. Provide $A_\ell = 1.24$ in.^{2}.

To satisfy the 12-in. maximum spacing specified in ACI Sec. 11.6.6.2, we need at least 6 bars each of area 0.207 in.^{2}. There must be a longitudinal bar in each corner of the stirrups. The minimum bar diameter is $1/24$ of the stirrup spacing $= 5/25 = 0.21$ in.

Provide 4 No. 4 bars in the bottom half of the beam and add $1.24 - 4 \times 0.20 = 0.44$ in.^{2} to the flexural steel. The total area of steel required in the top face of the beam $= 2.99 + 0.44 = 3.43$ in.^{2}.

6 No 7 bars: $A_s = 3.60$ in.^{2}, will not fit in one layer.
3 No. 8 and 2 No. 7 bars: $A_s = 3.57$ in.^{2}, will fit in one layer.
2 No. 8 and 3 No. 7 bars: $A_s = 3.38$ in.^{2}, will fit in one layer.

Provide 2 No. 8 and 3 No. 7 bars at the top of the beam.

ACI Sec. 11.6.3.9 allows one to subtract $M_u/(0.9df_{y\ell})$ from the area of longitudinal steel in the flexural compression zone. We shall not do this for two reasons: First, M_u varies along the length of the beam, and second, we need two bars of minimum diameter 0.21 in. in the corners of the stirrups.

The final beam design is shown in Fig. 7–30. It is interesting to note that the cross section is identical to the one chosen using the 1989 ACI Code in the preceding edition of this book. ■

EXAMPLE 7–3   Compatibility Torsion

The one-way joist system shown in Fig. 7–31 supports a total factored dead load of 157 psf and a factored live load of 170 psf, totaling 327 psf. Design the end span, *AB*, of the exterior spandrel beam on grid line 1. The factored dead load of the beam and the factored loads applied directly to it total 1.1 kips/ft. The spans and loadings are such that the moments and shears can be calculated using the moment coefficients from ACI Sec. 8.3.3 (see Sec. 10–2 of this book). Use $f_y = 60,000$ psi and $f_c' = 4000$ psi.

1. Compute the bending moments for the beam. In laying out the floor, it was found that joists with an overall depth of 18.5 in. would be required. The slab thickness is 4.5 in. The spandrel beam was made the same depth to save forming costs. The columns supporting the beam are 24 in. square. For simplicity in forming the joists, the beam overhangs the inside face of the columns by $1\frac{1}{2}$ in. Thus the initial choice of beam size is $h = 18.5$ in., $b = 25.5$ in., and $d = 16$ in.

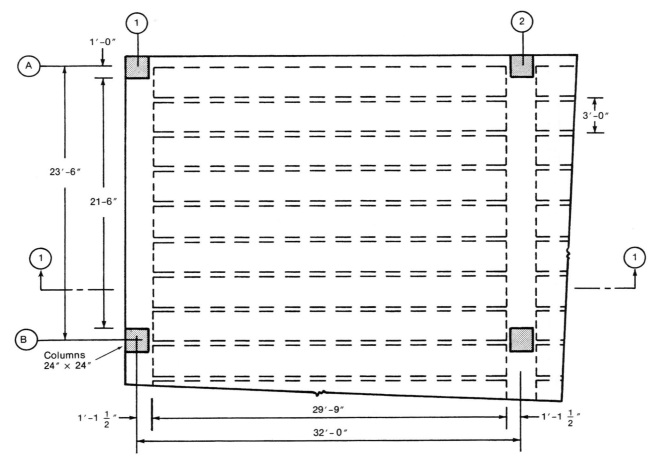

(a) Plan.

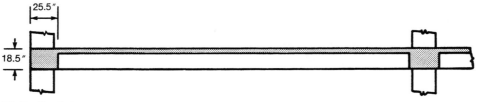

(b) Section 1–1.

Fig. 7–31
Joist floor—Example 7–3.

Although the joist loads are transferred to the beam by the joist webs, we shall assume a uniform load for simplicity. Very little error is introduced by this assumption. The joist reaction per foot of length of beam is

$$\frac{w\ell}{2} = \frac{0.327 \text{ ksf} \times 29.75 \text{ ft}}{2} = 4.86 \text{ kips/ft}$$

The total load on the beam is

$$w = 4.86 + 1.1 = 5.96 \text{ kips/ft}$$

The moments in the edge beam are:

Exterior end negative: $-M_u = \dfrac{w\ell_n^2}{16} = -172$ ft-kips

Midspan positive: $+M_u = \dfrac{w\ell_n^2}{14} = +197$ ft-kips

First interior negative: $-M_u = \dfrac{w\ell_n^2}{10} = -276$ ft-kips

2. Compute *b*, *d*, and *h*. Since *b* and *h* have already been selected, we shall check whether they are sufficiently large to ensure a ductile flexural behavior. Going through such a check, we find that $\rho \simeq 0.4\rho_b$ at the first interior negative moment point and the ratio, ρ, is smaller at other points. Thus the section has adequate size for flexure. The areas of steel required for flexure are:

Exterior end negative: $A_s = 2.64$ in.^{2}
Midspan positive: $A_s = 3.02$ in.^{2}
First interior negative: $A_s = 4.23$ in.^{2}

The actual steel will be chosen when the longitudinal torsion reinforcement has been calculated.

3. Compute the final M_u, V_u, and T_u diagrams. The moment and shear diagrams for the edge beam, computed using the ACI moment coefficients (ACI Sec. 8.3.3; Sec. 10–2 of this book), are plotted in Fig. 7–32a and b.

The joists are designed as having a clear span of 29.75 ft from the face of one beam to the face of the other beam. Because the exterior ends of the joists are "built integrally with" a "spandrel beam," ACI Sec. 8.3.3 gives the exterior negative moment in the joists as

$$-M_u = \frac{w\ell_n^2}{24}$$

Rather than consider the moments in each individual joist, we shall compute an average moment per foot of width of support:

$$-M_u = \frac{0.327 \text{ ksf} \times 29.75 \text{ ft}^2}{24} = -12.1 \text{ ft-kips/ft}$$

Although this is a bending moment in the joist, it acts as a twisting moment on the edge beam. As shown in Fig. 7–33a, this moment and the end shear of 4.86 kips/ft act at the face of the edge beam. Summing moments about the center of the columns (point *A* in Fig. 7–33a) gives the moment transferred to the column as 18.4 ft-kips/ft.

For the design of the edge beam for torsion, we need the torque about the axis of the beam. Summing moments about the centroid of the edge beam (Fig. 7–33b) gives the torque *t* as

$$t = 18.4 \text{ ft-kips/ft} - 5.96 \text{ kips/ft} \times \frac{0.75}{12} \text{ ft}$$

$$= 18.0 \text{ ft-kips/ft}$$

The forces and torque acting on the edge beam per foot of length are shown in Fig. 7–33b.

If the two ends of the beam *A–B* are fixed against rotation by the columns, the total torque at each end will be

$$T = \frac{t\ell_n}{2}$$

If this is not true, the torque diagram can vary within the range illustrated in Fig. 7–24. For the reasons given earlier, we shall assume that $T = t\ell_n/2$ at each end of member *A–B*. This gives the torque diagram shown in Fig. 7–32c.

4. Should torsion be considered? If T_u exceeds the following, it must be considered

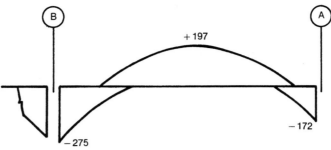

+ 197

B   A

− 172

− 275

(a) Moments, M_u (ft–kips).

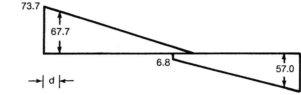

73.7

67.7

6.8

57.0

d

(b) Shears, V_u (kips).

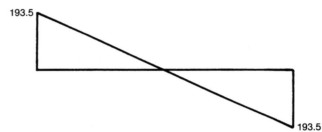

193.5

193.5

(c) Torque (ft–kips).

50.3

50.3

Fig. 7–32
Moments, shears, and torques
in end span of edge beam—
Example 7–3

(d) Reduced Torque, T_u (ft–kips).

$$\phi\sqrt{f_c'}\left(\frac{A_{cp}^{2}}{p_{cp}}\right)$$

The effective cross section for torsion is shown in Fig. 7–34. ACI Sec. 11.6.1 states that the overhanging flange shall be as defined in ACI Sec. 13.2.4. The projection of the flange is the smaller of the height of the web below the flange (14 in.), or four times the thickness of the flange (18 in.).

$$A_{cp} = 18.5 \times 25.5 + 4.5 \times 14 = 535 \text{ in.}^2$$

$$p_{cp} = 18.5 + 25.5 + 14 + 4.5 + 39.5 = 102 \text{ in.}$$

$$0.85\sqrt{4000}\left(\frac{535^2}{102}\right) = 150{,}900 \text{ in.-lb}$$

$$= 12.57 \text{ ft-kips}$$

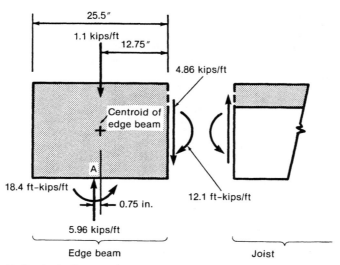

(a) Freebody diagram of edge beam.

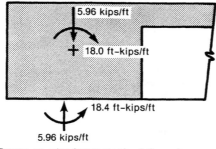

(b) Forces on edge beam resolved through centroid of edge beam.

Fig. 7–33
Forces on edge beam—
Example 7–3.

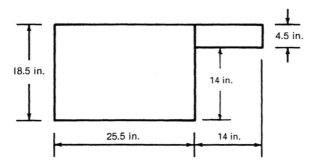

Fig. 7–34
Effective section for torsion—Example 7–3.

Since the maximum torque of 193.5 ft-kips exceeds this, torsion must be considered.

5. (a). Equilibrium or compatibility torsion? The torque resulting from the 0.75 in. offset of the axes of the beam and column (see Fig. 7–33) is necessary for the equilibrium of the structure and hence is equilibrium torque. The torque at the ends of the beam due to this is

$$5.96 \times \frac{0.75}{12} \times \frac{21.5}{2} = 4.00 \text{ ft-kips}$$

On the other hand, the torque resulting from the moments at the ends of the joists exists only because the edge beam is assumed to have a torsional stiffness. If the torsional stiffness were to decrease to zero, this torque would disappear. This part of the torque is therefore compatibility torsion.

Since the loading involves compatibility torsion, we can reduce the maximum torsional moments at d from the faces of the columns to

$$\phi 4\sqrt{f_c'}\left(\frac{A_{cp}^2}{p_{cp}}\right) = 0.85 \times 4\sqrt{4000} \times \frac{535^2}{102} = 603,400 \text{ in.-kips} = 50.3 \text{ ft-kips} \qquad (7\text{--}47)$$

but not less than the equilibrium torque of 4.00 ft-kips. The reduced torque diagram is plotted in Fig. 7–32d. The reduced torques are about one-fourth of the original torques. The reduced distributed torque, t, due to moments at the ends of the joists has decreased to

$$\frac{50.3 - 4.00}{18.83/2} = 4.92 \text{ ft-kips/ft}$$

5 (b). Adjust the moments in the joists. The moment diagram for the joists, assuming that the exterior negative moment is $-w\ell_n^2/24$, is plotted in Fig. 7–35a. This is given per foot of width of floor. If the negative moment at the center of the joint between the joist and beam is decreased from -17.3 ft-kips/ft to -4.92 ft-kips/ft, a moment of $-0.5(17.3 - 4.92) = -6.19$ ft-kips/ft is carried over to the other end of the joist, as shown in Fig. 7–35b. At the faces of the beams, the changes in the joist end moments are $+11.7$ ft-kips/ft and -5.54 ft-kips/ft. At midspan the change is $+3.1$ ft-kips/ft. The resulting moment diagram per foot of width is shown in Fig. 7–35c. Each joist supports a 3-ft-wide strip and hence supports three times these moments. The exterior negative moment steel in the joist should be designed for a negative moment since it is necessary to develop torsional cracks in the spandrel beam before the redistribution can occur. A good rule of thumb is to design the exterior negative steel for the moment computed from $w\ell_n^2/24$, as shown by the dashed line in Fig. 7–35c.

6. Is the section big enough for the torsion? For a solid section, the limit on shear and torsion is given by

$$\sqrt{\left(\frac{V_u}{b_w d}\right)^2 + \left(\frac{T_u p_h}{1.7 A_{oh}^2}\right)^2} \leq \phi\left(\frac{V_c}{b_w d} + 8\sqrt{f_c'}\right) \qquad (7\text{--}36b)$$
$$\text{(ACI Eq. 11--18)}$$

From Fig. 7–36,

$$A_{oh} = (18.5 - 2 \times 1.5 - 0.5)(25.5 - 2 \times 1.5 - 0.5)$$

$$= 330 \text{ in.}^2$$

$$p_h = 2(15.0 + 22.0)$$

$$= 74 \text{ in.}$$

$$\sqrt{\left(\frac{67,700}{25.5 \times 16}\right)^2 + \left(\frac{50.3 \times 12,000 \times 74}{330^2}\right)^2} = \sqrt{27,530 + 168,200}$$

$$= 442 \text{ psi}$$

From Eq. 6–8 (ACI Eq. 11–3),

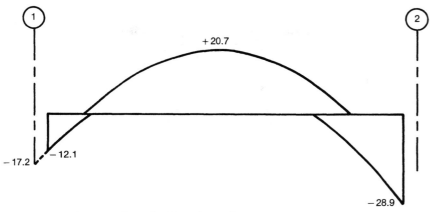

(a) Moment in joists per foot of width before adjustment.

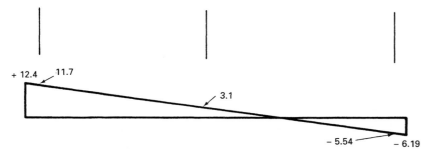

(b) Redistribution of moment per foot of width.

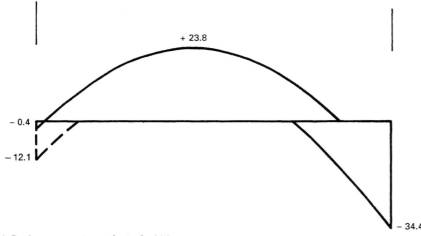

(c) Design moments per foot of width.

Fig. 7–35
Moments in end span of
joist—Example 7–3.

$$V_c = 2\sqrt{f_c'}\,b_w d$$

$$\phi(2\sqrt{f_c'} + 8\sqrt{f_c'}) = 0.85(10\sqrt{4000})$$

$$= 538 \text{ psi}$$

Since 442 psi is less than 538 psi, the section is large enough.

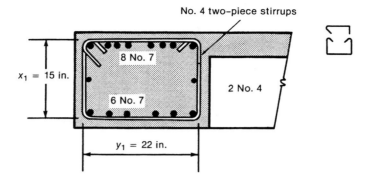

No. 4 two-piece stirrups

8 No. 7

$x_1 = 15$ in.

2 No. 4

6 No. 7

$y_1 = 22$ in.

Fig. 7–36
Section *A–A* through edge
beam. Joist reinforcement
omitted for clarity.

7. Compute the stirrup area required for shear. From Eqs. 6–9 and 6–14 (ACI Eqs. 11–1 and 11–2),

$$V_s = \frac{V_u}{\phi} - V_c$$

and from Eq. 6–18 (ACI Eq. 11–15):

$$\frac{A_v}{s} = \frac{V_u/\phi - V_c}{f_{yv} d}$$

where from Eq. 6–8 (ACI Eq. 11–3):

$$V_c = 2\sqrt{4000} \times 25.5 \times \frac{16}{1000} = 51.6 \text{ kips}$$

Thus:

$$\frac{A_v}{s} = \frac{V_u/0.85 - 51.6}{60 \times 16}$$

where V_u is in kips.

Figure 7–37a illustrates the calculation of $V_u/\phi - V_c$. Figure 7–37b is a plot of the A_v/s required for shear along the length of the beam. The values of A_v/s for shear and A_t/s for torsion (step 8) will be superimposed in step 9.

8. Compute the stirrups required for torsion. From Eqs. 7–27 and 7–48 (ACI Eqs. 11–21 and 11–20) and taking $\theta = 45°$ and $A_o = 0.85 A_{oh}$ gives

$$\frac{A_t}{s} = \frac{T_u/\phi}{2 \times 0.85 A_{oh} f_{yv}}$$

$$= \frac{T_u/\phi \times 12,000}{2 \times 0.85 \times 330 \times 60,000}$$

$$= 0.000357 \frac{T_u}{\phi}$$

where T_u is in ft-kips. This is plotted in Fig. 7–37d.

9. Add the stirrup areas and select the stirrups.

$$\frac{A_{v+t}}{s} = \frac{A_v}{s} + \frac{2A_t}{s} \tag{7–49}$$

278   Torsion

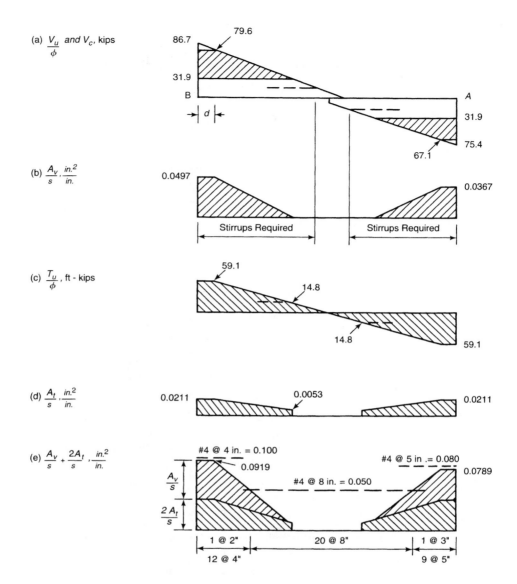

(a) $\dfrac{V_u}{\phi}$ and V_c, kips

79.6

86.7

31.9

B

$\rightarrow | d | \leftarrow$

A

31.9

75.4

67.1

(b) $\dfrac{A_v}{s}, \dfrac{in.^2}{in.}$

0.0497

0.0367

Stirrups Required   Stirrups Required

(c) $\dfrac{T_u}{\phi}$, ft - kips

59.1

14.8

14.8

59.1

(d) $\dfrac{A_t}{s}, \dfrac{in.^2}{in.}$

0.0211

0.0053

0.0211

(e) $\dfrac{A_v}{s} + \dfrac{2A_t}{s}, \dfrac{in.^2}{in.}$

#4 @ 4 in. = 0.100

#4 @ 5 in .= 0.080

0.0919

0.0789

$\dfrac{A_v}{s}$

#4 @ 8 in. = 0.050

$\dfrac{2A_t}{s}$

1 @ 2"   20 @ 8"   1 @ 3"

12 @ 4"   9 @ 5"

Fig. 7–37
Calculation of stirrups for
shear and torsion.

A_{v+t}/s is plotted in Fig. 7–37e. The maximum allowable spacings are:

For shear (ACI Sec. 11.5.4.1): $d/2 = 8$ in.

For torsion (ACI Sec. 11.6.6.1): The smaller of $p_h/8 = 74/8 = 9.25$ in., or 12 in.

 The horizontal lines in Fig. 7–37e are the values of A_{v+t}/s for No. 4 closed stirrups at spacings of 4 in. $= 2 \times 0.20/4.0 = 0.100$, 5 in., and 8 in. Stirrups must extend to points where $V_u/\phi = V_c/2$, or to $(d + b_t)$, where b_t is the width of the portion of the edge beam with closed stirrups, $= 16 + 25.5 = 41.5$ in. past the point where torsional reinforcement is no longer needed, that is, past the points where $T_u/\phi = $ (the torque given by Eq. 7–23) $/\phi = 12.57/0.85 = 14.8$ ft-kips. These points are located in Fig. 7–37b and d. Since they are closer than 41.5 in. to midspan, stirrups are required over the entire span.

Provide No. 4 closed stirrups:

End A: 1 @ 3 in., 9@ 5 in.

End B: 1 @ 2 in., 12 @ 4 in., then @ 8 in. on centers throughout the rest of the span.

10. **Design the longitudinal reinforcement for torsion.**

$$A_\ell = \left(\frac{A_t}{s}\right) p_h \left(\frac{f_{yv}}{f_{y\ell}}\right) \cot^2 \theta \tag{7-34}$$

where A_t/s is the amount computed in step 8. This varies along the length of the beam. For simplicity we shall keep the longitudinal steel constant along the length of the span and shall base it on the maximum $A_t/s = 0.0211$ in.^{2}/in. Again, $\theta = 45°$.

$$A_\ell = 0.0211 \times 74 \times 1.0 \times 1.0 = 1.56 \text{ in.}^2$$

The minimum A_ℓ is given by Eq. 7–52 (ACI Eq. 11–24):

$$A_{\ell, \min} = \frac{5\sqrt{f'_c} A_{cp}}{f_{y\ell}} - \left(\frac{A_t}{s}\right) p_h \left(\frac{f_{yv}}{f_{y\ell}}\right)$$

where A_t/s shall not be less than $25b_w/f_{yv} = 25 \times 25.5/60,000 = 0.0106$. Again, A_t/s varies along the span. The maximum A_ℓ will correspond to the minimum A_t/s. In the center region of the beam No. 4 stirrups at 8 in. have been chosen (see Fig. 7–37e). As a result, we shall take $A_t/s = 0.20/8 = 0.025$ in Eq. 7–52.

$$A_{\ell, \min} = \frac{5\sqrt{4000} \times 535}{60,000} - 0.025 \times 74 \times 1.0$$

$$= 2.82 - 1.85 = 0.97 \text{ in.}^2$$

Since $A_\ell = 1.56$ in.^{2} exceeds $A_{\ell, \min} = 0.97$ in.^{2}, use 1.56 in.^{2}.

From ACI Sec. 11.6.6.2, the longitudinal steel is distributed around the perimeter of the stirrups with a maximum spacing of 12 in. There must be a bar in each corner of the stirrups, and these bars have a minimum diameter of 1/24 of the stirrup spacing, but not less than a No. 3 bar.

Minimum bar diameter corresponds to maximum stirrup spacing = 8 in., $8/24 = 0.33$ in.

To satisfy the 12 in. maximum spacing, we need **3 bars at the top and bottom and one halfway up each side.** A_s per bar = $1.56/8 = 0.195$ in.^{2}. **Use No. 4 bars for longitudinal steel.**

The longitudinal torsion steel required at the top of the beam is provided by increasing the area of flexural steel provided at each end, and by lap splicing 3 No. 4 bars with the negative moment steel. The lap splices should be at least a Class B tension lap for a No. 4 top bar (see Table 8–4), since all the bars are spliced at the same point.

Exterior end negative moment: $A_s = 2.64 + 3 \times 0.195 = 3.23$ in.^{2}
   Use 4 No. 7 and 2 No. 6 = 3.28 in.^{2}. These fit in one layer.

First interior negative moment: $A_s = 4.23 + 0.58 = 4.81$ in.^{2}.
   Use 8 No. 7 = 4.80 in.^{2}. These fit in one layer.

The longitudinal torsional steel required at the bottom is obtained by increasing the area of steel at midspan. The increased area of steel will be extended from support to support.

Midspan positive moment: $A_s = 3.02 + 0.58 = 3.60$ in.^{2}
   Use 6 No. 7 = 3.60 in.^{2}. These fit in one layer.

The steel finally chosen is shown in Fig. 7–38. A section through the beam at the first interior support is shown in Fig. 7–36. The cutoff points for the flexural steel were based on Fig. A–5b, except that the area of positive moment steel anchored in the supports by hooks and lap splices was taken equal to the larger of the amounts given in Fig. A–5b and the bottom layer of $A_\ell = 3 \times 0.195 = 0.58$ in.^{2}. This was rounded up arbitrarily to 2 No. 7 bars. ■

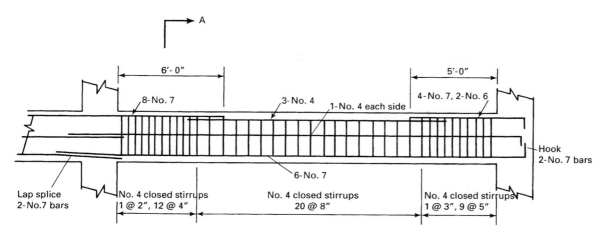

Fig. 7–38
Reinforcement in edge beam—Example 7–3.

PROBLEMS

7–1 A cantilever beam 8 ft long and 18 in. wide supports its own dead load plus a concentrated load located 6 in. from the end of the beam and 4.5 in. away from the vertical axis of the beam. The concentrated load is 15 kips dead load and 20 kips live load. Design reinforcement for flexure, shear, and torsion. Use $f_y = 60,000$ psi for all steel and $f_c' = 3750$ psi.

7–2 Explain why the torsion in the edge beam A–B in Fig. 7–22c is called "equilibrium torsion," while the torsion in the edge beam A1–B1 in Fig. 7–31 is "compatibility torsion."

8

Development, Anchorage, and Splicing of Reinforcement

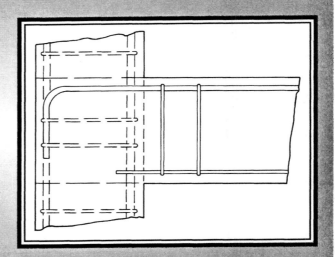

8-1   INTRODUCTION

In a reinforced concrete beam the flexural compressive forces are resisted by concrete, while the flexural tensile forces are provided by reinforcement as shown in Fig. 8–1. For this process to exist, there must be a force transfer, or *bond*, between the two materials. The forces acting on the bar are shown in Fig. 8–1b. For the bar to be in equilibrium, bond stresses must exist. If these disappear, the bar will pull out of the concrete and the tensile force, T, will drop to zero, causing the beam to fail.

Bond stresses must be present whenever the stress or force in a reinforcing bar changes from point to point along the length of the bar. This is illustrated by the free-body diagram in Fig. 8–2. If f_{s2} is greater than f_{s1}, bond stresses, μ, must act on the surface of the bar to maintain equilibrium. Summing forces parallel to the bar, one finds that the average bond stress, μ_{avg}, is

$$\Delta f_s \frac{\pi d_b^2}{4} = \mu_{\text{avg}}(\pi d_b)\ell$$

and

$$\mu_{\text{avg}} = \frac{\Delta f_s d_b}{4\ell} \tag{8-1}$$

If ℓ is taken as a very short length, dx, this equation can be written as

$$\frac{df_s}{dx} = \frac{4\mu}{d_b} \tag{8-2}$$

where μ is the *true bond stress* acting in the length dx.

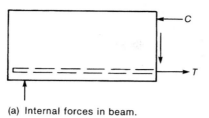

(a) Internal forces in beam.

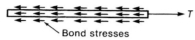

Bond stresses

(b) Forces on reinforcing bar.

Fig. 8–1
Need for bond stresses.

Average Bond Stress in a Beam

In a beam, the force in the steel at a crack can be expressed as

$$T = \frac{M}{jd} \tag{8–3}$$

where jd is the internal lever arm and M is the moment acting at the section. If we consider a length of beam between two cracks as shown in Fig. 8–3, the moments acting at the two cracks are M_1 and M_2. If the beam is reinforced with one bar of diameter d_b, the forces on the bar are as shown in Fig. 8–3c. Summing horizontal forces gives

$$\Delta T = (\pi d_b)\mu_{avg}\,\Delta x \tag{8–4}$$

where d_b is the diameter of the bar, or

$$\frac{\Delta T}{\Delta x} = (\pi d_b)\mu_{avg}$$

But

$$\Delta T = \frac{\Delta M}{jd}$$

giving

$$\frac{\Delta M}{\Delta x} = (\pi d_b)\mu_{avg}\,jd$$

From the free-body diagram in Fig. 8–3d, we can see that $\Delta M = V\Delta x$ or $\Delta M / \Delta x = V$. Therefore,

$$\mu_{avg} = \frac{V}{(\pi d_b)jd} \tag{8–5}$$

Fig. 8–2
Relationship between change
in bar stress and bond stress.

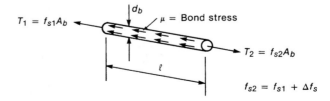

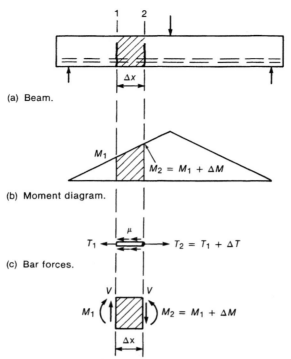

(a) Beam.

(b) Moment diagram.

(c) Bar forces.

M_1

$M_2 = M_1 + \Delta M$

T_1 μ $T_2 = T_1 + \Delta T$

V V

$M_1 \left(\uparrow \right)$ $\left(\downarrow \right) M_2 = M_1 + \Delta M$

Δx

Fig. 8–3
Average flexural bond stress.

(d) Part of beam between sections 1 and 2.

If there is more than one bar, the bar perimeter (πd_b) is replaced with the sum of the perimeters, ΣO, giving

$$\mu_{\text{avg}} = \frac{V}{\Sigma O \, jd} \tag{8–6}$$

Equations 8–5 and 8–6 give the *average bond stress* between two cracks in a beam. As shown later, the actual bond stresses vary from point to point between the cracks.

Bond Stresses in an Axially Loaded Prism

Figure 8–4a shows a prism of concrete containing one reinforcing bar, which is loaded in tension. At the cracks the stress in the bar is $f_s = P/A_s$. Between the cracks, a portion of the load is transferred to the concrete by bond, and the resulting distributions of steel and concrete stresses are shown in Fig. 8–4b and c. From Eq. 8–2 we see that the bond stress at any point is proportional to the slope of the steel stress diagram at that same point. Thus the bond stress distribution is shown in Fig. 8–4d. Since the stress in the steel is equal at each of the cracks, the force is also equal, so that $\Delta T = 0$ at the two cracks, and from Eq. 8–4 we see that the average bond stress, μ_{avg}, is also equal to zero. Thus for the average bond stress to equal zero, the total area under the bond stress diagram between any two cracks in Fig. 8–4d must equal zero when $\Delta T = 0$.

    The bond stresses given by Eq. 8–2 and plotted in Fig. 8–4d are referred to as *true bond stresses* or *in-and-out bond stresses* (they transfer stress into the bar and back out again) to distinguish them from the *average bond stresses* calculated from Eq. 8–1.

     Development, Anchorage, and Splicing of Reinforcement

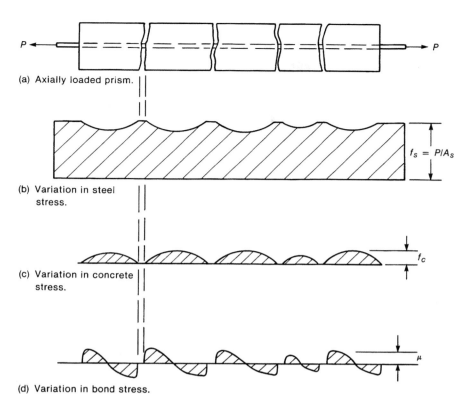

Fig. 8–4
Steel, concrete, and bond
stress in a cracked prism.

(a) Axially loaded prism.

(b) Variation in steel stress.

$f_s = P/A_s$

(c) Variation in concrete stress.

f_c

(d) Variation in bond stress.

μ

True Bond Stresses in a Beam

At the cracks in a beam, the bar force can be computed from Eq. 8–3. If the concrete and the bar are bonded together, a portion of the tensile force will be resisted by the concrete at points between the cracks. As a result, the tensile stresses in the steel and the concrete at the level of the steel will vary as shown in Fig. 8–5c and d. This gives rise to the bond stress distribution plotted in Fig. 8–5e. Once again there are in-and-out bond stresses, but now the total area under the bond stress diagram is not zero. The average bond stress in Fig. 8–5e must equal the value given by Eq. 8–5.

Bond Stresses in a Pull-Out Test

The easiest way to test the bond strength of bars in a laboratory is by means of the *pull-out test*. Here a concrete cylinder containing the bar is mounted on a stiff plate and a jack is used to pull the bar out of the cylinder, as shown in Fig. 8–6a. In such a test, the concrete is compressed and hence does not crack. The stress in the bar varies as shown in Fig. 8–6b and the bond stress (Eq. 8–2) varies as shown in *c*. This test does not give values representative of the bond strength of beams because the concrete is not cracked and hence there is no in-and-out bond stress distribution. Also, the bearing stresses of the concrete against the plate cause a frictional component that resists the transverse expansion which would result from Poisson's ratio. Prior to 1950, pull-out tests were used extensively to determine the bond strength of bars. Since then various types of beam tests have been used to study bond strength.[8-1]

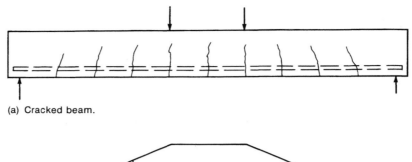

(a) Cracked beam.

M_1

(b) Moment diagram.

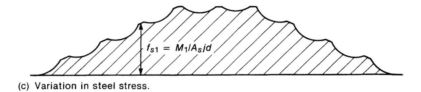

$f_{s1} = M_1/A_s jd$

(c) Variation in steel stress.

(d) Tensile stress in concrete.

μ_{avg}

(e) Bond stresses.

Fig. 8–5
Steel, concrete, and bond
stresses in a cracked beam.

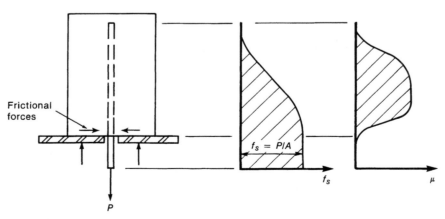

Frictional
forces

$f_s = P/A$

f_s

μ

P

Fig. 8–6
Stress distribution in a pull-
out test.

(a) Test method.

(b) Bar stress.

(c) Bond stress.

Development, Anchorage, and Splicing of Reinforcement

A smooth bar embedded in concrete develops bond by adhesion between the concrete and the bar, and by a small amount of friction. Both of these effects are quickly lost when the bar is loaded in tension, particularly because the diameter of the bar decreases slightly due to Poisson's ratio. For this reason, smooth bars are generally not used as reinforcement. In cases where smooth bars must be embedded in concrete (anchor bolts, stirrups made of small diameter bars, etc.), mechanical anchorage in the form of hooks, nuts and washers on the embedded end, or similar devices are used.

Although adhesion and friction are present when a deformed bar is loaded for the first time, these bond transfer mechanisms are quickly lost, leaving the bond to be transferred by bearing on the deformations of the bar as shown in Fig. 8–7a. Equal and opposite bearing stresses act on the concrete as shown in Fig. 8–7b. The forces on the concrete have a longitudinal and a radial component (Fig. 8–7c and d). The latter causes circumferential tensile stresses in the concrete around the bar. Eventually, the concrete will split parallel to the bar and the resulting crack will propagate out to the surface of the beam. The splitting cracks follow the reinforcing bars along the bottom or side surfaces of the beam as shown

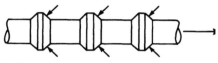

(a) Forces on bar.

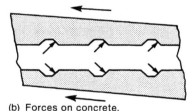

(b) Forces on concrete.

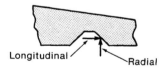

Longitudinal   Radial

(c) Components of force
    on concrete.

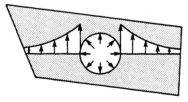

(d) Radial forces on concrete
    and splitting stresses shown
    on a section through the bar.

Fig. 8–7
Bond transfer mechanism.

in Fig. 9–3e. Once these cracks develop, the bond transfer drops rapidly unless reinforcement is provided to restrain the opening of the splitting crack.

The load at which splitting failure develops is a function of:

1. The minimum distance from the bar to the surface of the concrete or to the next bar. The smaller this distance, the smaller the splitting load.

2. The tensile strength of the concrete.

3. The average bond stress. As this increases, the wedging forces increase, leading to a splitting failure.

These factors are discussed more fully in Refs. 8–2, 8–3, and 8–4. Typical splitting failure surfaces are shown in Fig. 8–8. The splitting cracks tend to develop along the shortest distance between a bar and the surface or between two bars. In Fig. 8–8 the circles touch the edges of the beam where the distances are shortest.

If the cover and bar spacings are large compared to the bar diameter, a *pull out failure* can occur, where the bar and the annulus of concrete between successive deformations pull out along a cylindrical failure surface joining the tips of the deformations.

8–3   DEVELOPMENT LENGTH

Because the actual bond stress varies along the length of a bar anchored in a zone of tension, the ACI Code uses the concept of *development length* rather than bond stress. The development length, ℓ_d, is the shortest length of bar in which the bar stress can increase from zero to the yield strength, f_y. If the distance from a point where the bar stress equals f_y to the end of the bar is less than the development length, the bar will pull out of the concrete. The development lengths are different in tension and compression because a bar loaded in ten-

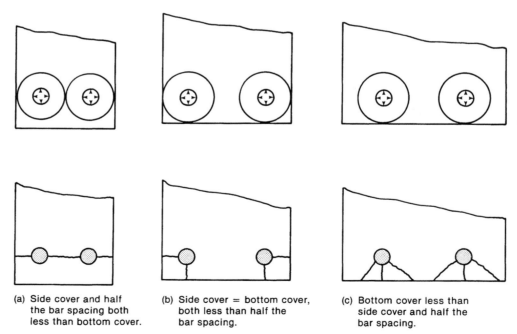

(a) Side cover and half the bar spacing both less than bottom cover.

(b) Side cover = bottom cover, both less than half the bar spacing.

(c) Bottom cover less than side cover and half the bar spacing.

Fig. 8–8
Typical splitting failure surfaces.

sion is subject to in-and-out bond stresses and hence requires a considerably longer development length.

The development length can be expressed in terms of the ultimate value of the average bond stress by setting Δf_s in Eq. 8–1 equal to f_y:

$$\ell_d = \frac{f_y d_b}{4\mu_{\text{avg},u}} \tag{8–7}$$

where $\mu_{\text{avg},u}$ is the value of μ_{avg} at bond failure in a beam test.

Tension Development Lengths

Basic Tension Development Equation

In 1977, Orangun et al.[8–2] fitted a regression equation through the results of a large number of bond and splice tests. The resulting equation for bar development length, ℓ_d, included terms for bar diameter d_b, the bar stress to be developed, $\sqrt{f_c'}$, the cover and/or bar spacing, and the transverse steel ratio. It served as the basis of the development length provisions in the 1989 ACI Code. These provisions proved difficult to use, however, and between 1989 and 1995, ACI Committee 318 and the ACI bond committee simplified the design expressions. This took two stages. First, a basic expression was developed for the development length, ℓ_d, expressing it as a multiple of the bar diameter, d_b, to give a ready point of comparison. This is given in ACI Sec. 12.2.3 as

$$\frac{\ell_d}{d_b} = \frac{3}{40} \frac{f_y}{\sqrt{f_c'}} \frac{\alpha\beta\gamma\lambda}{\dfrac{c + K_{tr}}{d_b}} \tag{8–8}$$

(ACI Eq. 12–1)

where the term $(c + K_{tr})/d_b$ is limited to not larger than 2.5 to prevent pull-out bond failures, and the length ℓ_d is not taken less than 12 in.

ℓ_d is the development length, in.

d_b is the bar diameter, in.

α is a bar location factor given in ACI Sec. 12.2.4.

β is an epoxy coating factor given in ACI Sec. 12.2.4.

γ is a bar diameter factor given in ACI Sec. 12.2.4.

λ is a lightweight concrete factor given in ACI Sec. 12.2.4.

c is the smaller of (a) the smallest distance measured from the surface of the concrete to the *center* of a bar being developed, or

(b) one-half of the *center-to-center* spacing of the bars being developed.

K_{tr} is a transverse reinforcement factor given in ACI Sec. 12.2.4.

The factors $\alpha, \beta, \gamma, \lambda$, and K_{tr} will be presented later.

The second stage in the simplification of Eq. 8–8 was to substitute common values of c and K_{tr} into Eq. 8–8 as described below.

Simplified Tension Development Length Equations

In most cases, Eq. 8–8 would be difficult to use in design because c and K_{tr} vary along the length of a member. This equation was simplified by substituting lower limit values of c and K_{tr} for the most common design cases to get more widely applicable equations which

TABLE 8–1 Equations for Development Length Ratios[a]

	No. 6 and Smaller Bars and Deformed Wires	No. 7 and Larger Bars
Case 1: Clear spacing of bars being developed or spliced not less than d_b, **and** stirrups or ties throughout ℓ_d not less than the code minimum	$$\frac{\ell_d}{d_b} = \frac{f_y\alpha\beta\lambda}{25\sqrt{f_c'}}$$ (8–9)	$$\frac{\ell_d}{d_b} = \frac{f_y\alpha\beta\lambda}{20\sqrt{f_c'}}$$ (8–10)
or		
Case 2: Clear spacing of bars being developed or spliced not less than $2d_b$ and clear cover not less than d_b		
Other cases	$$\frac{\ell_d}{d_b} = \frac{3f_y\alpha\beta\lambda}{50\sqrt{f_c'}}$$ (8–11)	$$\frac{\ell_d}{d_b} = \frac{3f_y\alpha\beta\lambda}{40\sqrt{f_c'}}$$ (8–12)

[a]The length ℓ_d computed using Eqs. 8–9 to 8–12 shall not be taken less than 12 in.

TABLE 8–1M Equations for Development Length Ratios—SI Units[a]

	No. 20 and Smaller Bars and Deformed Wires	No. 25 and Larger Bars
Case 1: Clear spacing of bars being developed or spliced not less than d_b, and stirrups or ties throughout ℓ_d not less than the code minimum	$$\frac{\ell_d}{d_b} = \frac{12f_y\alpha\beta\lambda}{25\sqrt{f_c'}}$$ (8–9M)	$$\frac{\ell_d}{d_b} = \frac{12f_y\alpha\beta\lambda}{20\sqrt{f_c'}}$$ (8–10M)
or		
Case 2: Clear spacing of bars being developed or spliced not less than $2d_b$ and clear cover not less than d_b		
Other cases	$$\frac{\ell_d}{d_b} = \frac{18f_y\alpha\beta\lambda}{25\sqrt{f_c'}}$$ (8–11M)	$$\frac{\ell_d}{d_b} = \frac{18f_y\alpha\beta\lambda}{20\sqrt{f_c'}}$$ (8–12M)

[a]The length ℓ_d computed using Eqs. 8–9M to 8–12M shall not be taken less than 300 mm.

did not explicitly include these factors. For deformed bars or deformed wire, ACI Sec 12.2.2 defines the development length as given in Tables 8–1 and 8–1M.

The cases 1 and 2 described in the top row of Tables 8–1 and 8–1M are illustrated in Fig. 8–9a and b. The "code minimum" stirrups and ties mentioned in case 1 correspond to the minimum amounts and maximum spacings specified in ACI Secs. 11.5.4.1, 11.5.5.3, and 7.10.5. Values of ℓ_d/d_b computed from Equations 8–9 to 8–12 are tabulated in Tables A–11 and A–11M in Appendix A. Table A–6 can be used to get the minimum web widths corresponding to a $1.0d_b$ spacing (case 1). If the actual web width exceeds the value from this table, it is not necessary to further check the spacings when determining whether the beam is case 1.

A clear spacing of bars being developed or spliced of not less than d_b corresponds to $c = (\text{space of } d_b/2) + d_b/2 = d_b$. Minimum stirrups correspond to K_{tr} in the order of $0.5d_b$. Thus, for this combination $(c + K_{tr})/d_b \approx 1.5$. Similarly, a clear spacing of bars being developed or spliced of not less than $2d_b$ with clear cover not less than d_b and no stir-

Development, Anchorage, and Splicing of Reinforcement

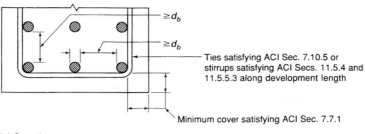

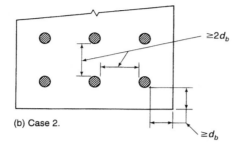

(a) Case 1.

(b) Case 2.

Fig. 8–9
Explanation of Cases 1 and 2
in ACI Sec.12.2.2.

rups or ties corresponds to $c = 1.5d_b$ and $(c + K_{tr})/d_b = 1.5$. The bar size factor γ is 0.8 for No. 6 and smaller bars and 1.0 for No. 7 and larger bars. Substituting these values of $(c + K_{tr})/d_b$ and γ into Eq. 8–8 gives Eqs. 8–9 and 8–10.

The minimum bar spacings and covers in ACI Secs. 7.6.1 and 7.7 correspond to $c = 1.0d_b$ and $(c + K_{tr})/d_b = 1.0$. Substituting this and γ into Eq. 8–8 gives Eqs. 8–11 and 8–12.

Factors in Eqs. 8–8 through 8–12

The Greek letter factors in Eqs. 8–8 through 8–12 are defined in ACI Sec. 12.2.4 as follows:

α = bar location factor

Horizontal reinforcement so placed that more than 12 in. of fresh concrete is cast in the member below the development length or splice 1.3

Other reinforcement 1.0

Horizontal reinforcement with more than 12 in. of fresh concrete below it at the time the bar is embedded in concrete is referred to as *top reinforcement.* During the placement of the concrete, water and mortar migrate vertically upward through the concrete, collecting on the underside of reinforcing bars. If the depth below the bar exceeds 12 in., sufficient mortar will collect to weaken the bond significantly. This applies to the top reinforcement in beams with depths greater than 12 in. and to horizontal steel in walls cast in lifts greater than 12 in. The factor was reduced from 1.4 to 1.3 in 1989 based on tests in Ref. 8–5.

β = coating factor

Epoxy-coated bars or wires with cover less than $3d_b$, or clear spacing less than $6d_b$ 1.5

All other epoxy-coated bars or wires 1.2

Uncoated reinforcement 1.0

The product of $\alpha\beta$ need not be taken greater than 1.7.

Tests of epoxy-coated bars have indicated that there is negligible friction between concrete and the epoxy-coated bar deformations. As a result, the forces acting on the deformations and the concrete in Fig. 8–7a and b act perpendicular to the surface of the deformations. In a bar without an epoxy coating, friction between the deformation and the concrete allows the forces on the deformation and the concrete to act at a flatter angle than shown in Fig. 8–7. Because of this, the radial force components are larger in an epoxy-coated bar than in a normal bar for a given longitudinal force component, and hence splitting occurs at a lower longitudinal force.[8–6] The 1.5 value of β corresponds to cases where splitting failures occur. For larger covers and spacings pull out failures tend to occur and the effect of epoxy coating is smaller.

γ = **bar size factor**

No. 6 and smaller bars and deformed wires 0.8

No. 7 and larger bars. 1.0

Comparison of Eq. 8–8 with a large collection of bond and splice tests showed that a shorter development length was possible for smaller bars.

λ = **lightweight aggregate concrete factor**

When lightweight aggregate concrete is used 1.3
However, when f_{ct} is specified, λ shall be permitted to be taken as
$6.7\sqrt{f_c'}/f_{ct}$ but not less than 1.0

When normal-weight concrete is used 1.0

The tensile strength of lightweight concrete is generally less than for normal-weight concrete and hence the splitting load will be less. In addition, in some lightweight concretes the wedging forces that the bar deformations exert on the concrete may cause localized crushing, which allows bar slip to occur. The factor λ does not differentiate between sand-lightweight and all-lightweight concretes as is done in shear calculations. This is based on the results of tests of hooked bar anchorages, which did not show differences between the two types of lightweight concrete.

c = **spacing or cover dimension, in.**

c is the smaller of:

(a) the smallest distance measured from the surface of the concrete to the *center* of the bar, or

(b) one-half of the *center-to-center* spacing of the bars.

It is important to note that c is defined relative to the center of the bars in both cases.

K_{tr} = **transverse reinforcement index**

$$= \frac{A_{tr}f_{yt}}{1500sn}$$

where
the factor 1500 has units of lb/in.^{2}

A_{tr} = total cross-sectional area of all transverse reinforcement within the spacing s, which crosses the potential plane of splitting along the reinforcement being developed within the development length, in.^{2} (illustrated in Fig. 8–10).

f_{yt} = specified yield strength of the transverse reinforcement, psi

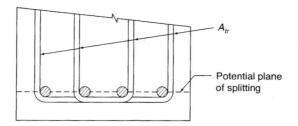

Fig. 8–10
Definition of A_{tr}

s = maximum center-to-center spacing of transverse reinforcement within ℓ_d, in.

n = number of bars or wires being developed along the plane of splitting

ACI Sec. 12.2.4 allows K_{tr} to be taken equal to zero to simplify the calculations, even if there is transverse reinforcement.

Excess Flexural Reinforcement

If excess flexural reinforcement is provided, the bar stress that must be developed is less than f_y. In such a case, ACI Sec. 12.2.5 allows ℓ_d to be multiplied by (A_s required/A_s provided). This multiplier could also be expressed as f_s/f_y. If room is available, it is good practice to ignore this factor, thus ensuring that the steel is fully anchored if a change in the use of the structure requires the bars to be fully stressed. In statically indeterminate structures, the increased stiffness resulting from the additional reinforcement may lead to higher moments at the over-reinforced section. In such a case the steel would be more highly stressed than expected from the ratio of areas. This multiplier is not applied in the design of members resisting seismic loads.

The development length, ℓ_d, calculated as the product of ℓ_d from ACI Sec. 12.2.2 or 12.2.3 and the factors given above, shall not be taken less than 12 in. except when computing the length of lap slices.

Approximate Derivation of Eq. 8–9

Although Eq. 8–9 was derived from test results, the following analysis illustrates the factors affecting the splitting load. Consider a cylindrical prism of concrete of diameter $2c$, containing a bar of diameter d_b, as shown in Fig. 8–11a. The radial components of the forces on the concrete, shown in Fig. 8–7b, c, and d, cause a pressure p on a portion of the cross section of the prism, as shown in Fig. 8–11b. This is equilibrated by tensile stresses in the concrete on either side of the bar. In Fig. 8–11c, the distribution of these stresses has arbitrarily been assumed to be triangular. Splitting is assumed to occur when the maximum tensile stress is equal to the tensile strength of the concrete, f_{ct}. For equilibrium in the vertical direction in a length of prism equal to ℓ,

$$pd_b\ell = \frac{2}{K}\left(c - \frac{d_b}{2}\right)f_{ct}\ell$$

where $K = 2$ for the triangular distribution of concrete stresses shown in Fig. 8–11c. Rearranging gives

$$p = \left(\frac{c}{d_b} - \frac{1}{2}\right)f_{ct}$$

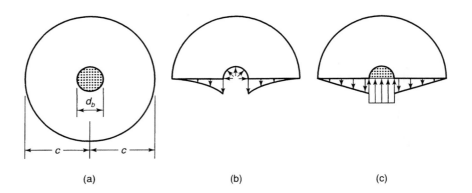

Fig. 8–11
Stresses in a concrete prism
subjected to bond stresses.

(a) (b) (c)

If the forces shown in Fig. 8–7b are assumed to act at 45°, the bond stress, $\mu_{avg,u}$, at the onset of splitting is equal to p. Taking $f_{ct} = 6\sqrt{f_c'}$ gives

$$\mu_{avg,u} = 6\sqrt{f_c'}\left(\frac{c}{d_b} - \frac{1}{2}\right) \qquad (8\text{--}13)$$

The length of bar required to raise the bar stress from zero to f_y is called the development length, ℓ_d. From Eq. 8–7,

$$\frac{\ell_d}{d_b} = \frac{f_y}{4\mu_{avg,u}}$$

Substituting Eq. 8–13 for $\mu_{avg,u}$ gives

$$\frac{\ell_d}{d_b} = \frac{f_y}{24\sqrt{f_c'}\left(\dfrac{c}{d_b} - \dfrac{1}{2}\right)}$$

Taking $c = 1.5d_b$ as done in the derivation of Eqs. 8–9 and 8–10 gives

$$\frac{\ell_d}{d_b} = \frac{f_y}{24\sqrt{f_c'}}$$

This is very similar to Eq. 8–9 if α, β, and λ are taken equal to 1.0. Three major assumptions were made in this derivation: (a) the distribution of the tensile stresses in the concrete, (b) the angle of inclination of the forces on the deformations, and (c) the replacement of the concentrated forces on the deformations with a force that is uniformly distributed along the length of the bar.

The similarity of the final result and Eq. 8–9 reinforces the validity of the splitting model.

Compression Development Lengths

Compression development lengths are considerably shorter than tension development lengths because some force is transferred to the concrete by bearing at the end of the bar, and because there are no cracks in such an anchorage region, and hence no in-an-out bond. The basic compression development length is (ACI Sec. 12.3)

Development, Anchorage, and Splicing of Reinforcement

$$\ell_{dbc} = \frac{0.02d_b f_y}{\sqrt{f_c'}} \text{ but not less than } 0.0003d_b f_y \qquad (8\text{--}14)$$

Values of ℓ_{db} are given in Tables A–12 and A–12M. The development length in compression is found as the product of ℓ_{db} and the applicable modification factors given in ACI Sec. 12.3.3 for excess reinforcement and enclosure by spirals or ties. The resulting development length shall not be less than 8 in. (ACI Sec. 12.3.1).

Development Lengths for Bundled Bars

Where a large number of bars are required in a beam or column, the bars are sometimes placed in bundles of 2, 3, or 4 bars (ACI Sec. 7.6.6). The effective perimeter for bond failure of bundles is less than the total perimeter of the individual bars in the bundle. The ACI Sec. 12.4 accounts for this by requiring an increased development length of 1.2 times the individual bar development lengths for bars in a 3-bar bundle and 1.33 times for bars in a 4-bar bundle. The value of d_b used in ACI Sec. 12.2 shall be taken as the diameter of a hypothetical single bar having the same area as the bundle.

Development Lengths for Coated Bars

In bridge decks and parking garages it is becoming common practice to use epoxy-coated or galvanized reinforcing bars to reduce corrosion problems. Epoxy-coated bars are covered by the factor β in ACI Sec. 12.2.4. The zinc coating on galvanized bars can affect the bond properties due to chemical reaction with the alkali in the concrete. This can be prevented by treating the bars with a solution of chromate after galvanizing. If this is done, the bond is essentially the same as for normal reinforcement.

Development Lengths for Welded Wire Fabric

ACI Sec. 12.8 provides rules for development of welded plain wire fabric in tension. The development of *plain wire fabric* depends on the mechanical anchorage from at least two cross wires. The nearest of these cross wires must be located 2 in. or farther from the critical section. Wires that are closer to the critical section cannot be counted for anchorage. The second cross wire must not be located closer to the critical section than

$$\ell_d = 0.27 \frac{A_w}{s_w} \frac{f_y}{\sqrt{f_c'}} \lambda \qquad (8\text{--}15)$$

where A_w and s_w are the cross-sectional area, and spacing of the wire being developed, respectively. The development length may be reduced by multiplying by the factor in ACI Sec. 12.2.5 for excess reinforcement, but may not be taken as less than 6 in. except when computing the length of splices according to ACI Sec. 12.19.

    Deformed fabric derives anchorage from bond stresses along the deformed wires and from mechanical anchorage from the cross wires. The ASTM specification for welded deformed wire fabric does not require as strong welds for deformed fabric as for plain wire fabric. ACI Sec. 12.7 gives the basic development length of deformed fabric with at least one cross wire in the development length, but not closer than 2 in. to the critical section, as ω times the development length computed from ACI Sec. 12.2.2, 12.2.4, and 12.2.5, where

$$\omega = \frac{f_y - 35,000}{f_y} \tag{8-16}$$

or

$$\omega = \frac{5d_b}{s_w} \tag{8-17}$$

The development length computed shall not be taken less than 8 in., except when computing the length of splices according to ACI Sec. 12.18 or stirrup anchorages according to ACI Sec. 12.13.

Tests have shown that the development length of deformed wire fabric is not affected by epoxy coating, and for this reason, the epoxy coating factor, β, is taken equal to 1.0 for epoxy-coated deformed wire fabric.

Deformed welded wire fabric may have some plain wires in one or both directions. For the purpose of determining the development length, such a fabric shall be considered to be a plain wire fabric if any of the wires in the direction that development is being considered are plain wires (ACI Sec. 12.7.4).

8–4 HOOKED ANCHORAGES

Behavior of Hooked Anchorages

Hooks are used to provide additional anchorage when there is insufficient length available to develop a bar. Unless otherwise specified, the so-called *standard hook* described in ACI Sec. 7.1 would be used. Details of 90° and 180° standard hooks and standard stirrup and tie hooks are given in Fig. 8–12. It is important to note that a standard hook on a large bar takes up a lot of room, and the actual size of the hook is frequently quite critical in detailing a structure.

A 90° hook loaded in tension develops forces in the manner shown in Fig. 8–13a. The stress in the bar is resisted by bond on the surface of the bar and by bearing on the concrete inside the hook.[8-7] The hook moves inward, leaving a gap between it and the concrete outside the bend. Because the compressive force inside the bend is not colinear with the applied tensile force, the bar tends to straighten out, producing compressive stresses on the outside of the tail. Failure of a hook almost always involves crushing of the concrete inside the hook. If the hook is close to a side face, the crushing will extend to the surface of the concrete, removing the side cover. Occasionally, the concrete outside the tail will crack allowing the tail to straighten.

The stresses and slip measured at points along a hook at a bar stress of $1.25f_y$ (75 ksi) in tests of No. 7 hooks are plotted in Fig. 8–13b and c.[8-7] The axial stresses in the bar decrease due to bond on the lead-in length and bond and friction on the inside of the bar. The magnitude and direction of slip at A, B, and C are shown by the arrows. The slip at point A is 75% larger for the 180° hook than for the 90° hook.

The amount of slip depends, among other things, on the angle of the bend and the orientation of the hook relative to the direction of concrete placing. The slip of hooks displays a top-bar effect. In tests, top cast hooks, oriented so that weaker mortar was trapped

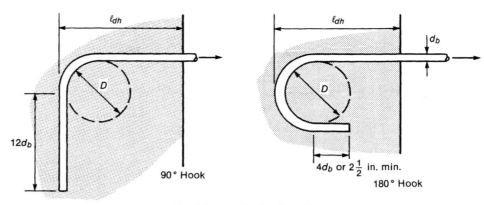

No. 3 through No. 8: $D = 6d_b$
No. 9, 10, 11 : $D = 8d_b$
No. 14 and 18 : $D = 10d_b$

(a) Standard hooks—ACI Sec. 7.1 and 7.2.1.

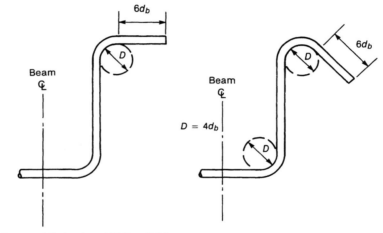

Fig. 8–12
Standard hooks.

(b) Stirrup and tie hooks—ACI Sec. 7.1.3
For No. 3

inside the bend during casting, slipped 50 to 100% more than bottom cast bars at a given bar stress.[8-8]

Tests on bars hooked around a corner bar show that 10 to 30% larger tensile stresses can be developed at a given end slip.

Design of Hooked Anchorages

ACI Sec. 12.5.1 does not distinguish between 90° and 180° hooks or top or bottom bar hooks. The development length of a hook, ℓ_{dh} (illustrated in Fig. 8–12a), is computed as the product of a basic development length, ℓ_{hb}, and a series of multipliers. The final development length shall not be less than $8d_b$ or 6 in., whichever is smaller. The basic development length for a hooked bar with f_y equal to 60,000 psi shall be

$$\ell_{hb} = \frac{1200d_b}{\sqrt{f_c'}} \tag{8–18}$$

8–4   Hooked Anchorages

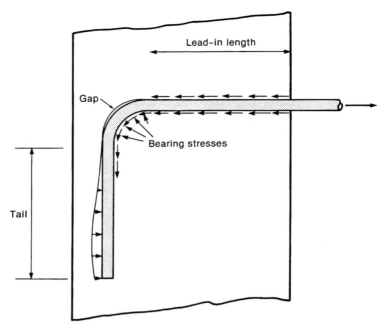

Fig. 8–13
Behavior of hooks.

(a) Forces acting on bar.

Values of ℓ_{hb} are given in Tables A–13 and A–13M.

The length ℓ_{hb} is multiplied by whichever of the following are applicable:

12.5.3.1   bars with f_y other than 60,000 psi $\times f_y/60{,}000$

For clarity in presentation here, ACI Sec. 12.5.3.2 has been divided into two subsections:

12.5.3.2(a)   For 180° hooks on No. 11 and smaller bars with side cover normal to the plane of the hook not less than $2\frac{1}{2}$ in. $\times 0.7$

12.5.3.2(b)   For 90° hooks on No. 11 and smaller bars with side cover normal to the plane of the hook not less than $2\frac{1}{2}$ in. *and* cover on the bar extension beyond the hook (tail) not less than 2 in. $\times 0.7$

12.5.3.3   For No. 11 and smaller bars, if the hook is enclosed vertically or horizontally by stirrups or ties spaced no farther apart than $3d_b$, where d_b is the diameter of the hooked bar $\times 0.8$

In calculating the development lengths of hooks in No. 14 and No. 18 bars, the reductions in 12.5.3.2 and 12.5.3.3 above are not permitted.

12.5.3.4   Where anchorage or development for f_y is not specifically required (as, for example, by ACI Sec. 12.11.2), the required hook length can be multiplied by the ratio A_s required$/A_s$ provided

Again, this is equivalent to multiplying by the ratio f_s/f_y. Unless space is limited, it is good practice to take this ratio equal to 1.0 so that the bars can be fully utilized if this is ever required.

12.5.3.5   For hooks embedded in all-lightweight or sand-lightweight concrete $\times 1.3$

Finally, the length of the hook, ℓ_{dh}, shall not be less than 8 bar diameters or 6 in., whichever is greater.

Development, Anchorage, and Splicing of Reinforcement

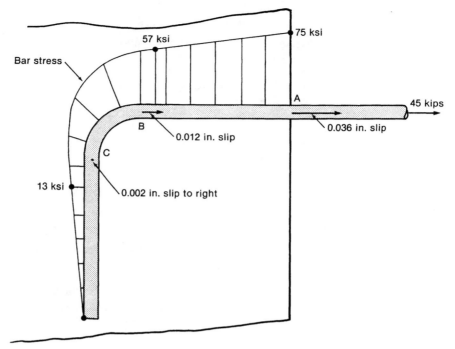

(b) Stresses and slip—90° standard hook.

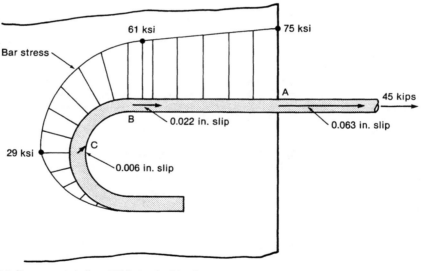

(c) Stresses and slip—180° standard hook

Fig. 8–13
Continued.

At the discontinuous ends of members where *both* the side cover to the hook, and the top (or bottom) cover to the lead-in length are less than $2\frac{1}{2}$ in., ACI Sec. 12.5.4 requires that the hook be enclosed over the length, ℓ_{dh}, by ties or stirrups spaced no farther apart than $3d_b$, where d_b is the diameter of the hooked bar. In such a region the reduction in hook length due to ties in ACI Sec. 12.5.3.3 above is not permitted. ACI Sec. 12.5.4 applies if the bar stresses are such that hooks are required at such points as the ends of simply supported beams (particularly if these are deep beams), at the free ends of cantilevers, and at the ends of members that terminate in a joint. Hooked bars at discontinuous ends of slabs are assumed to have confinement from the slab on each side of the hook, and hence ACI Sec. 12.5.4 is not applied. Although ACI Sec. 12.5.4 requires confinement of the full length ℓ_{dh}, the confinement is effective only in the region that will spall off if crushing occurs inside the hook. Since this region extends 6 to 10 bar diameters from the inside of the hook, 3 or 4 stirrups will be adequate. Hooks may not be used to develop bars in compression because bearing on the outside of the hook is not efficient.

Special requirements in ACI Sec. 21.5.4.1 govern hooks in joints in frames resisting seismic loads. Because such hooks must be inside the column, these rules implicitly allow for confinement of the hooks.

8–5 DESIGN FOR ANCHORAGE

The basic rule governing the development and anchorage of bars is that:

> The calculated tension or compression in reinforcing bars at every section of reinforced concrete members shall be developed on each side of that section by embedment length, hooks, and anchorage, or a combination of these. (ACI Sec. 12.1)

This requirement is satisfied in various ways for different types of members. Three examples of bar anchorage, two for straight bars and one for hooks, are presented in this section. The question of anchorage of bars in beams is discussed in Sec. 8–6.

EXAMPLE 8–1   Anchorage of a Straight Bar

A 16-in.-wide cantilever beam frames into the edge of a 16-in.-thick wall as shown in Fig. 8–14. At ultimate, the No. 8 bars at the top of the cantilever are stressed to their yield strength at point A at the face of the wall. Compute the minimum embedment of the bars into the wall. The concrete is sand-lightweight concrete with a strength of 3000 psi. The yield strength of the flexural reinforcement is 60,000 psi. Construction joints are located at the bottom and top of the beam as shown in Fig. 8–14. The beam has closed No. 3 stirrups with $f_y = 40,000$ psi at a spacing of 7.5 in. throughout its length. The cover is 1.5 in. to the stirrups. The 3 No. 8 bars are inside the No. 4 at 12 in. vertical steel in each face of the wall. The wall steel is Grade 60.

We shall do this problem twice, first using ACI Sec. 12.2.2 (Table 8–1) and again using ACI Sec. 12.2.3 (Eq. 8–8).

1.   Determine the spacing and confinement case. The clear side cover to the No. 8 bars in the wall is $1.5 + 0.5 = 2$ in. $= 2d_b$. The clear spacing of the bars is

$$\frac{16 - 2(1.5 + 0.5) - 3 \times 1.0}{2} = 4.5 \text{ in.} = 4.5d_b$$

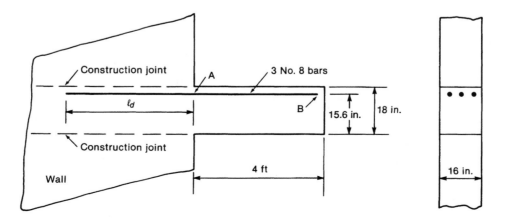

Fig. 8–14
Cantilever beam—Examples
8–1 and 8–2.

There are no stirrups or ties in the wall. Since the clear spacing between the bars is not less than $2d_b$ and the clear cover to the No. 8 bars exceeds d_b, this is case 2, and for No. 8 bars, Eq. 8–10 applies.

2. **Compute the development length.** From Eq. 8–10,

$$\frac{\ell_d}{d_b} = \frac{f_y \alpha \beta \lambda}{20\sqrt{f'_c}}$$

where

$\alpha = 1.3$ because there will be more than 12 in. of fresh concrete under the bar when the concrete is placed in the beam

$\beta = 1.0$ because the bars are not epoxy coated

$\lambda = 1.3$ because the concrete is sand lightweight

$$\frac{\ell_d}{d_b} = \frac{60,000 \times 1.3 \times 1.0 \times 1.3}{20\sqrt{3000}} = 92.6$$

Thus the development length is $\ell_d = 92.6 \times 1.0 = 92.6$ in.

The bars must extend 92.6 in. into the wall to develop the full yield strength. **Extend the bars 8 ft into the wall.**

Alternatively, this could be solved using ACI Sec. 12.2.3 and Eq. 8–8. The necessary steps will be repeated.

1. **Compute the development length.** From Eq. 8–8

$$\frac{\ell_d}{d_b} = \frac{3}{40}\frac{f_y}{\sqrt{f'_c}}\frac{\alpha\beta\gamma\lambda}{\dfrac{c + K_{tr}}{d_b}}$$

where α, β, and λ are as before and $\gamma = 1.0$ because the bars are No. 8.

$c =$ the smaller of:

(a) the distance from the center of the bar to the nearest concrete surface

The side cover $= 1.50 + 0.50 + 1.0/2 = 2.5$ in.

The top cover at $A = 1.5 + 1.0/2 = 2.0$ in.

(b) half the center to center spacing of the bars $= 0.5\left(\dfrac{16 - 2 \times 2.5}{2}\right) = 2.75$ in.

Therefore, $c = 2.0$ in.

$$K_{tr} = \frac{A_{tr}f_{yt}}{1500sn}$$

where

$s =$ spacing of the transverse reinforcement within the development length, ℓ_d
$= 12$ in.

$A_{tr} =$ total cross-sectional area of reinforcement crossing the plane of splitting within the spacing s (one No. 4 bar each face)

$= 2 \times 0.20 = 0.40$ in.^{2}

$f_{yt} = 60,000$ psi for the wall steel
$n =$ number of bars being anchored $= 3$

Thus

$$K_{tr} = \frac{0.40 \text{ in.}^2 \times 60,000 \text{ psi}}{1500 \text{ lb/in.}^2 \times 12 \text{ in.} \times 3} = 0.444 \text{ in.}$$

The term

$$\frac{c + K_{tr}}{d_b} = \frac{2.0 \text{ in. } + 0.444 \text{ in.}}{1.0 \text{ in.}} = 2.44$$

but not more than 2.5. Substituting into Eq. 8–8 gives

8–5 Design for Anchorage

$$\frac{\ell_d}{d_b} = \frac{3}{40} \frac{60,000}{\sqrt{3000}} \frac{1.3 \times 1.0 \times 1.0 \times 1.3}{2.44} = 56.8$$

Thus $\ell_d = 56.8 \times 1.0$ in. $= 56.8$ in. Based on this, the bars should be extended 56.8 in. into the wall, say 5 ft. **Extend the bars 5 ft into the wall.**

In this case, there is a large difference between ℓ_d computed using Eq. 8–10 and ℓ_d computed using Eq. 8–8. This is because 8–10 was derived using $(c + K_{tr})/d_b$ equal to 1.5. In this case, it is equal to 2.44. ∎

EXAMPLE 8–1M   Anchorage of a Straight Bar—SI Units

A 400-mm-wide cantilever beam frames into the edge of a 400-mm-thick wall similar to Fig. 8–14. At ultimate, the three No. 25 bars at the top of the wall are stressed to their yield strength at point A at the face of the wall. Compute the minimum embedment of the bars into the wall. The concrete is sand/low-density concrete with a strength of 20 MPa. The yield strength of the flexural reinforcement is 400 MPa. Construction joints are located at the bottom and top of the beam as shown in Fig. 8–14. The beam has closed No. 10 stirrups with $f_y = 300$ MPa at a spacing of 180 mm throughout its length. The cover is 40 mm to the stirrups. The three No. 25 bars are inside the No. 15 vertical steel in each face of the wall.

1. **Determine the spacing and confinement case.** The clear side cover to the No. 25 bars in the wall is $40 + 15 = 55$ mm $= 2.2d_b$. The clear spacing of the bars is

$$\frac{400 - 2(40 + 15) - 3 \times 25}{2} = 107.5 \text{ mm} = 4.3 \; d_b$$

There are no stirrups or ties in the wall.

Since the clear spacing between the bars is not less than $2d_b$ and the clear cover to the No. 8 bars exceeds d_b, this is case 2 and, for No. 25 bars, Eq. 8–10M applies.

2. **Compute the development length.** From Eq. 8–10M,

$$\frac{\ell_d}{d_b} = \frac{12f_y\alpha\beta\lambda}{20\sqrt{f_c'}}$$

where

   $\alpha = 1.3$ because there will be more than 300 mm of fresh concrete under the bar when the concrete is placed in the beam

   $\beta = 1.0$ because the bars are not epoxy coated

   $\lambda = 1.3$ because the concrete is sand lightweight

$$\frac{\ell_d}{d_b} = \frac{12 \times 400 \times 1.3 \times 1.0 \times 1.3}{20\sqrt{20}} = 90.7$$

Thus the development length is $\ell_d = 90.7 \times 25 = 2270$ mm.

The bars must extend 2270 mm into the wall to develop the full yield strength. **Extend the bars 2.3 m into the wall.** ∎

Example 8–1 was also solved using Eq. 8–8. This will not be done in the SI example since the process is the same.

EXAMPLE 8–2   Development of a Bar in a Cantilever

The cantilever shown in Fig. 8–14 extends 48 in. from the face of the wall. The bars shown are stressed to their yield strength at the face of the wall. Is there adequate development length for No. 8

bars in the span? If not, what is the largest-size bar that can be used? The beam has No. 3 Grade 40 closed stirrups at 7.5 in. on centers along its full length.

The point of maximum bar force occurs at the face of the wall (point A in Fig. 8–14). The bar must be developed on each side of this point. To accomplish this, the bar must extend a minimum of ℓ_d into the support (Example 8–1) and a minimum of ℓ_d into the span.

1. **Determine the spacing and confinement case.** From Example 8–1, the clear spacing between the No. 8 bars in the beam is $4.5d_b$. For the 7.5-in. stirrup spacing, the minimum area of stirrups by ACI Sec. 11.5.5.3 is

$$A_v = 50 \frac{b_w s}{f_y} \qquad\qquad (6\text{–}22)$$

$$\text{(ACI Eq. 11–13)}$$

$$= 50 \frac{16 \times 7.5}{40,000} = 0.15 \text{ in.}^2$$

Stirrups provided are No. 3 double leg stirrups, $A_v = 0.22$ in.². The spacing does not exceed the minimum of $d/2 = 7.80$ in.

Since the clear spacing between the bars is at least d_b and the stirrups exceeds the minimum amount required, this is case 1 and for No. 8 bars Eq. 8–10 applies.

2. **Compute the development length.** From Example 8–1, the development length for a No. 8 top bar is 92.6 in.

Since the bars extend $60 - 1.5 = 58.5$ in. into the beam from the face of the wall, there is insufficient length to develop a No. 8 bar. We must use smaller bars with shorter ℓ_d or hook the No. 8 bars at B. Try 6 No. 6 bars ($A_s = 2.64$ in.² compared to 2.37 in.² for 3 No. 8 bars). For the new bar arrangement we must start over.

1. **(Repeated) Determine the spacing and confinement case.** The clear side cover to the bars in the wall portion is $1.5 + 0.5 = 2$ in. $= 2d_b$. The clear spacing between the 6 No. 6 bars is

$$\frac{16 - 2 \times 2 - 6 \times 0.75}{5} = 1.50 \text{ in.} = 2d_b$$

Since the stirrups exceed the Code minimum and the bar spacing is not less than d_b, this is case 1 and Eq. 8–9 applies.

2. **(Repeated) Compute the development length.** From Eq. 8–9,

$$\frac{\ell_d}{d_b} = \frac{f_y \alpha \beta \lambda}{25\sqrt{f_c'}} = \frac{60,000 \times 1.3 \times 1.0 \times 1.3}{25\sqrt{3000}} = 74.0$$

and $\ell_d = 74.0 \times 0.75$ in. $= 55.5$ in. Since 58.5 in. > 55.5 in., No. 6 bars can be developed without hooks at the free end. **Use 6 No. 6 bars.** ∎

ACI Sec. 12.2.5 allows the development length to be reduced by multiplying ℓ_d by the ratio $(A_s \text{ required})/(A_s \text{ provided}) = 2.37 \text{ in.}^2/2.64 \text{ in.}^2 = 0.898$. We will not take advantage of this because the reduced anchorage length will not allow the bars to be used for their full capacity in the event of a change of use of the structure.

EXAMPLE 8–2M Development of a Bar in a Cantilever—SI Units

A cantilever similar to the one shown in Fig. 8–14 extends 1600 mm from the face of the wall. The 3 No. 25 Grade 400 bars are stressed to their yield strength at the face of the wall. Is there adequate development length for No. 25 bars in the span? If not, what is the largest size bar that can be used? The beam has No. 10 Grade 300 closed stirrups at 180 mm on centers along its full length.

The point of maximum bar force occurs at the face of the wall (point A in Fig. 8–14). The bar must be developed on each side of this point. To accomplish this, the bar must extend a minimum of ℓ_d into the support (Example 8–1M) and a minimum of ℓ_d into the span.

1. Determine the spacing and confinement case. From Example 8–1M, the clear spacing between the No. 25 bars in the beam is $4.3d_b$. For the 180-mm stirrup spacing, the minimum area of stirrups by ACI Sec. 11.5.5.3 is

$$A_v = \frac{b_w s}{3f_y}$$ (6–22M)

(ACI Eq. 11–13)

$$= \frac{400 \times 180}{3 \times 300} = 80 \text{ mm}^2$$

Stirrups provided are No. 10 double-leg stirrups, $A_v = 200 \text{ mm}^2$. The spacing does not exceed the minimum of $d/2 = 192$ mm.

Since the clear spacing between the bars is at least d_b and the stirrups exceed the minimum amount required, this is case 1, and for No. 8 bars, Eq. 8–10M applies.

2. Compute the development length. From Example 8–1, the development length for a No. 25 top bar is 2270 mm. Since the bars extend $1600 - 40 = 1560$ mm into the beam from the face of the wall, there is insufficient length to develop a No. 25 bar. We must use smaller bars with shorter ℓ_d, or hook the No. 25 bars at B. Try 5 No. 20 bars ($A_s = 1500 \text{ mm}^2$ for both bar combinations). For the new bar arrangement we must start over.

1. (Repeated) Determine the spacing and confinement case. The clear side cover to the bars in the wall portion is $40 + 15 = 55$ mm $= 2.75d_b$. The clear spacing between the 5 No. 20 bars is

$$\frac{400 - 2 \times 55 - 5 \times 20}{4} = 47.5 \text{ mm} = 2.37d_b$$

Since the stirrups exceed the Code minimum and the bar spacing is not less than d_b, this is case 1 and Eq. 8–9M applies.

2. (Repeated) Compute the development length. From Eq. 8–9M,

$$\frac{\ell_d}{d_b} = \frac{12f_y\alpha\beta\lambda}{25\sqrt{f_c'}} = \frac{12 \times 400 \times 1.3 \times 1.0 \times 1.3}{25\sqrt{20}} = 72.6$$

and $\ell_d = 72.6 \times 20 = 1451$ mm. Since 1560 mm $>$ 1451 mm, No. 20 bars can be developed without hooks at the free end. **Use 5 No. 20 bars.** ∎

EXAMPLE 8–3   Hooked Bar Anchorage into a Column

The exterior end of a 16-in.-wide by 24-in.-deep continuous beam frames into a 24-in. square column, as shown in Fig. 8–15. The column has 4 No. 11 longitudinal bars. The negative moment reinforcement at the exterior end of the beam consists of 4 No. 8 bars. The concrete is 3000-psi normal-weight concrete. The steel strength is 60,000 psi. Design the anchorage of the 4 No. 8 bars into the column.

1. Compute development length for beam bars. If the No. 8 bars extended straight into the column, they would be confined by the vertical steel, not the column ties. As a result, this is either an "other case" in Table 8–1, or it must be solved by Eq. 8–8. We shall use Eq. 8–8 because the column steel provides considerable confinement.

$$\frac{\ell_d}{d_b} = \frac{3}{40} \frac{f_y}{\sqrt{f_c'}} \frac{\alpha\beta\gamma\lambda}{\left(\dfrac{c + K_{tr}}{d_b}\right)}$$

where

$\alpha = 1.3$ (top bars)
$\beta = 1.0$ (no coating)
$\gamma = 1.0$ (No. 8 bars)
$\lambda = 1.0$ (normal-weight concrete)
$c = $ the smaller of:

Development, Anchorage, and Splicing of Reinforcement

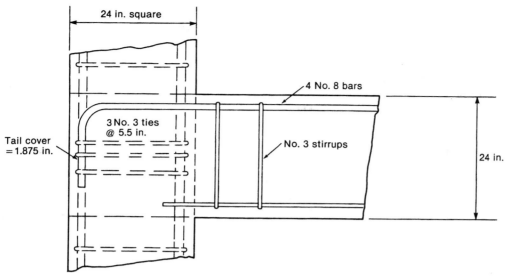

(a) Section A–A.

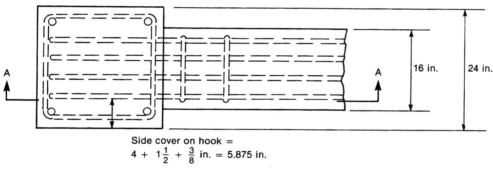

Side cover on hook =
$4 + 1\frac{1}{2} + \frac{3}{8}$ in. = 5.875 in.

Fig. 8–15
Column-beam joint—Example 8–3.

(a) the distance from the center of the bar to the nearest edge of the concrete

Side and top distance = $1.50 + 0.375 + 1.0/2 = 2.375$ in.

(b) half the center to center spacing of the bars $\dfrac{1}{2}\left(\dfrac{16 - 2 \times 2.375}{3}\right) = 1.875$ in.

Therefore, $c = 1.875$ in.

$$K_{tr} = \frac{A_{tr}f_{yt}}{1500sn}$$

A_{tr} = the column reinforcement crossing the splitting plane in the plane of the layer of bars

= 4 No. 11 bars = $4 \times 1.56 = 6.24$ in.^{2}

s = spacing of transverse reinforcement (we shall assume that the column steel confines the entire width of the column = 24 in.)

$$K_{tr} = \frac{6.24 \times 60,000}{1500 \times 24 \times 4} = 2.60 \text{ in.}$$

$$\frac{c + K_{tr}}{d_b} = \frac{1.875 + 2.60}{1.0} = 4.475 \text{ but not more than } 2.5$$

Substituting into Eq. 8–8 gives

$$\frac{\ell_d}{d_b} = \frac{3}{40} \frac{60,000}{\sqrt{3000}} \frac{1.3 \times 1.0 \times 1.0 \times 1.0}{2.5} = 42.7$$

Thus the development length is $42.7 \times 1.0 = 42.7$ in. Since this greatly exceeds the width of the column, it is necessary to use hooks to anchor the bars.

2. Compute the development length for hooked beam bars. The basic development length for a hooked bar is

$$\ell_{hb} = \frac{1200 d_b}{\sqrt{f_c'}}$$

The constant 1200 has units of psi, as does $\sqrt{f_c'}$. Therefore,

$$\ell_{hb} = \frac{1200 \times 1.0}{\sqrt{3000}} = 21.9 \text{ in.}$$

Assume that the 4 No. 8 bars will extend into the column inside the vertical column bars as shown in Fig. 8–15. ACI Sec. 11.11.2 requires minimum ties in the joint area. Assume that these are No. 3 ties. The required spacing of No. 3 closed ties by ACI Sec. 11.11.2 is computed using ACI Eq. 11–13:

$$A_v = \frac{50 b_w s}{f_y}$$

or

$$s = \frac{0.11 \text{ in.}^2 \times 60,000 \text{ psi}}{50 \text{ psi} \times 24 \text{ in.}}$$

$$= 5.5 \text{ in.}$$

The side cover to hooked bars is

4 in. offset in the side of the beam + 1.5-in. cover + 0.375-in. ties = 5.875 in.

This exceeds $2\frac{1}{2}$ in. and is therefore OK. The top cover to the lead-in length in joint exceeds $2\frac{1}{2}$ in. since the joint is in the column.

The cover on bar extension beyond hook (tail of hook) is

1.5-in. cover to ties + 0.375-in. ties = 1.875 in.

$$\ell_{dh} = \ell_{hb} \times \text{multipliers in ACI Sec. 12.5.3}$$

12.5.3.2(b): The side cover exceeds 2.5 in., but the cover on the bar extension is less than 2 in.; therefore, the multiplier = 1.0. ACI Sec. 12.5.4 does not apply since the side cover and top cover both exceed 2.5 in. Therefore, only the minimum ties required by ACI Sec. 11.11.2 are required: No. 3 ties at 5.5 in. These are spaced farther apart than $3d_b = 3 \times 1.0$ in. and therefore ACI Sec. 12.5.3.3 does not apply; therefore, the multiplier = 1.0.

$$\ell_{dh} = \ell_{hb} \times 1.0 \times 1.0$$

$$= 21.9 \text{ in.} \geq 8d_b \text{ or 6 in.} \qquad \text{therefore OK}$$

The hook development length available is

24 in. − cover on bar extension = 22.1 in.

Since 22.1 in. exceeds 21.9 in., the hook extension is OK.

Check the vertical height of a standard hook on a No. 8 bar. From Fig. 8–12a, the vertical height of a 90° standard hook is $4d_b + 12d_b = 16$ in. This will fit into the joint. Therefore, anchor the 4 No. 8 bars into the joint as shown in Fig. 8–15. ∎

Why Bars Are Cut Off

In reinforced concrete, reinforcement is provided near the tensile face of beams to provide the tension component of the internal resisting couple. A continuous beam and its moment diagram are shown in Fig. 8–16. At midspan, the moments are positive, and reinforcement is required near the bottom face of the member as shown in Fig. 8–16a. The opposite is true at the supports. For economy, some of the bars can be *terminated* or *cut off* where they are no longer needed. The location of the cutoff points is discussed in this section.

Four major factors affect the location of bar cutoffs:

1. Bars can be cut off where they are no longer needed to resist tensile forces or where the remaining bars are adequate to do so. The location of points where bars are no longer needed is a function of the flexural tensions resulting from the bending moments *and* the effects of shear on these tensile forces.

2. There must be sufficient extension of each bar, on each side of every section, to develop the force in that bar at that section. This is the basic rule governing the development of reinforcement, presented in Sec. 8–5 (ACI Sec. 12.1).

3. Tension bars, cut off in a region of moderately high shear force, cause a major stress concentration, which can lead to major inclined cracks at the bar cutoff.

4. Certain constructional requirements are specified in the Code as good practice.

Generally speaking, bar cutoffs should be kept to a minimum, particularly in zones of tension to simplify design and construction.

In the following sections, the location of theoretical cutoff points for flexure, referred to as *flexural cutoff points*, is discussed. This is followed by a discussion of how these flexural cutoff locations must be modified to account for shear, development, and constructional requirements to get the *actual cutoff points* used in construction.

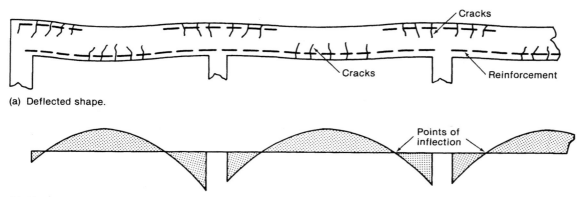

(a) Deflected shape.

(b) Moment diagram under typical loading.

Fig. 8–16
Moments and reinforcement in a continuous beam.

Location of Flexural Cutoff Points

The calculation of the flexural cutoff points will be illustrated using the simply supported beam shown in Fig. 8–17a. At midspan this beam has 5 No. 8 reinforcing bars shown in section in Fig. 8–17c. At points C and C', two of these bars are cut off, leaving 3 No. 8 bars in the end portions of the beam, as shown in Fig. 8–17b.

The beam is loaded with a uniform load of 6.6 kips/ft, including its self-weight, which gives the diagram of ultimate moments, M_u, shown in Fig. 8–17d. This is referred to as the *required moment diagram*, since at each section the beam must have a capacity, ϕM_n, at least equal to M_u. The maximum required moment at midspan:

$$M_u = \frac{w\ell_n^2}{8} = 330 \text{ ft-kips}$$

Assuming 3000-psi concrete and Grade 60 reinforcement, the moment capacity, ϕM_n, of the section with 5 No. 8 bars is 331 ft-kips, which is adequate at midspan. At points away from midspan the required M_u is less than 330 ft-kips, as shown by the moment diagram in Fig. 8–17d. Since $\phi M_n = \phi A_s f_y jd$ (Eqs. 4–1b and 4–12a), less reinforcement (less A_s) is required at points away from midspan. This is accomplished by "cutting off" some of the bars where they are no longer needed. In the example illustrated in Fig. 8–17, it has been arbitrarily decided that 2 No. 8 bars will be cut off where they are no longer needed. The remaining 3 No. 8 bars give a moment capacity, $\phi M_n = 211$ ft-kips. Thus the two bars *theoretically* can be cut off when $M_u \leq 211$ ft-kips since the remaining 3 bars will then be strong enough to resist M_u. From the equation for the required moment diagram (Fig. 8–17d) we find the $M_u = 211$ ft-kips at 3.99 ft from each support. Thus the two bars that are to be cut off are no longer needed for flexure in the outer 3.99 ft of each end of the beam and *theoretically* can be cut off at those points as shown in Fig. 8–17e.

Figure 8–17f is a plot of the moment capacity, ϕM_n, at each point in the beam and is referred to as a *moment capacity diagram*. At midspan (point E in Fig. 8–17e), the beam has 5 bars and hence has a capacity of 331 ft-kips. To the left of point C the beam contains 3 bars, giving it a capacity of 211 ft-kips. The distance CD represents the development length, ℓ_d, for the two bars cut off at C. At the ends of the bars at point C, these 2 bars are undeveloped and hence cannot resist stresses. As a result, they do not add to the moment capacity at C. On the other hand, the bars are fully developed at D and in the region from D to D' they could be stressed to f_y, if required. In this region, the moment capacity is 331 ft-kips.

The three bars that extend into the supports are cut off at points A and A'. At A and A' these bars are undeveloped and as a result the moment capacity is $\phi M_n = 0$ at A and A'. At points B and B' the bars are fully developed and the moment capacity $\phi M_n = 211$ ft-kips.

In Fig. 8–17g the moment capacity diagram from Fig. 8–17f and the required moment diagram from Fig. 8–17d are superimposed. Because the moment capacity is greater than or equal to the required moment at all points, the beam has adequate capacity *for flexure, neglecting the effects of shear*.

In the calculation of the moment capacity and required moment diagrams in Fig. 8–17, only flexure was considered. Shear has a significant effect on the stresses in the longitudinal tensile reinforcement and must be considered in determining the cutoff points. This is discussed in the next section.

Effect of Shear on Bar Forces and Location of Bar Cutoff Points

In Sec. 6–4 the truss analogy was presented to model the shear strength of beams. It was shown in Figs. 6–19, 6–20, and 6–25a that inclined cracking increased the tension force in the flexural reinforcement. This will be examined more deeply here.

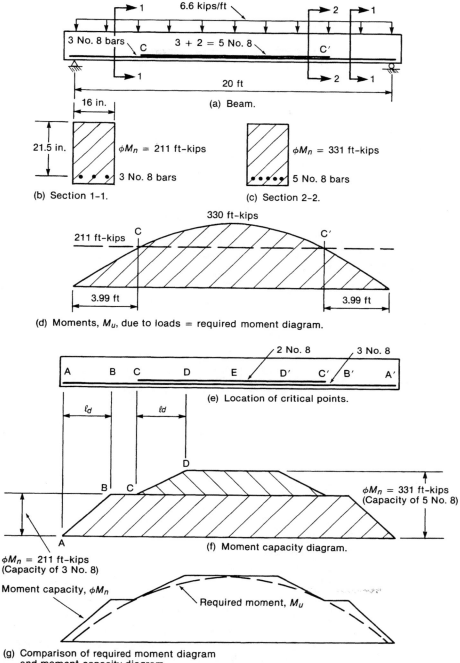

Fig. 8–17
Required moment and
moment capacity.

(a) Beam.

(b) Section 1-1.

(c) Section 2-2.

(d) Moments, M_u, due to loads = required moment diagram.

(e) Location of critical points.

(f) Moment capacity diagram.

(g) Comparison of required moment diagram
and moment capacity diagram.

Figure 8–18a shows a beam with inclined and flexural cracks. From flexure theory the tensile force is $T = M/jd$. If jd is assumed to be constant, the distribution of T is the same as the distribution of moment, as shown in Fig. 8–18b. The maximum value of T is 216 kips between the loads.

In Fig. 6–19b the same beam is idealized as a truss. The distribution of the tensile force in the longitudinal reinforcement from the truss model is shown by the solid stepped

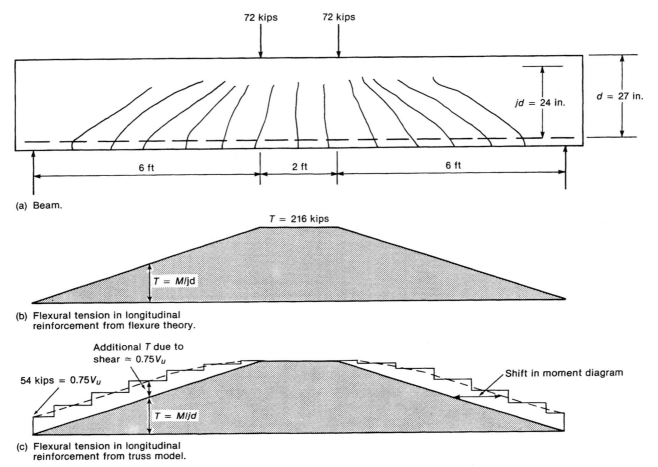

(a) Beam.

T = 216 kips

T = M/jd

(b) Flexural tension in longitudinal
reinforcement from flexure theory.

Additional T due to
shear ≈ 0.75V_u

54 kips = 0.75V_u

Shift in moment diagram

T = M/jd

(c) Flexural tension in longitudinal
reinforcement from truss model.

Fig. 8–18
Flexural tension in longitudinal reinforcement.

line in Fig. 8–18c. For comparison, the steel force diagram due to flexure is shown by the line labeled $T = M/jd$.

The presence of inclined cracks has increased the force in the tension reinforcement at all points in the shear span except in the region of maximum moment, where the tensile force of 216 kips equals that computed from flexure. The increase in tensile force gets larger as one moves away from the point of maximum moment as the slope of the compression diagonals decreases. For the 34° struts in the truss in Fig. 6–19b (1.5 horizontal to 1 vertical), the force in the tensile reinforcement has been increased by $0.75V_u$ in the end portion of the shear span. Another way of looking at this is to assume that the force in the tension steel corresponds to a moment diagram which has been shifted $0.75jd$ toward the support.

For beams having struts at other angles, θ, the increase in the tensile force is $V_u/(2 \tan \theta)$, corresponding to a shift in the moment diagram of $jd/(2 \tan \theta)$.

The ACI Code does not explicitly treat the effect of shear on the tensile force. Instead, ACI Sec. 12.10.3 arbitrarily requires that longitudinal tension bars be extended a minimum distance equal to the greater of d or 12 bar diameters past the theoretical cutoff point for flexure. This accounts for the shift due to shear plus:

contingencies arising from unexpected loads, yielding of supports, shifting of points of inflection or other lack of agreement with assumed conditions governing the design of elastic structures. . . [8-9]

Development, Anchorage, and Splicing of Reinforcement

In the beam in Fig. 8–18 there is a tensile force of $0.75V_u$ in the tensile reinforcement at the face of the support. If the shear stresses are large enough to cause significant inclined cracking, say greater than $v_u = 4\sqrt{f_c'}$ it is good practice to anchor these bars for this force. The actual force depends on the angle θ, but $0.75V_u$ is a reasonable value. This is especially important for short, deep beams, as pointed out in ACI Sec. 12.10.6 and illustrated in Chap. 18 of this book.

Development of Bars at Points of Maximum Bar Force

For reinforcement and concrete to act together, each bar must have adequate embedment on both sides of each section to develop the force in the bar at that section. In beams this is critical at:

1. Points of maximum positive and negative moment, which are points of maximum bar stress

2. Points where adjacent reinforcement is cut off or bent (ACI Sec. 12.10.2)

Thus bars must extend at least a development length, ℓ_d, each way from such points or be anchored with hooks or mechanical anchorages.

It is clear why this applies at points of maximum bar stress, such as point E in Fig. 8–17e, but the situation of bar cutoffs needs more explanation. In Fig. 8–17 the selection of bar cutoffs for flexure alone was discussed. To account for bar forces resulting from shear effects, cutoff bars are then extended away from the point of maximum moment a distance d or 12 bar diameters past the flexural cutoff point. This is equivalent to using the modified bending moment, M_u', shown in dashed lines in Fig. 8–19a to select the cutoffs. The beam in Fig. 8–17 required 5 bars at midspan. If all 5 bars extended the full length of the beam, the bar stress diagram would resemble the modified moment diagram as shown in Fig. 8–19b. Instead, 2 bars will be cut off at points C and C' where the moment $M_u' = 211$ ft-kips, which is equal to the moment capacity, ϕM_n, of the cross section with 3 No. 8 bars (Fig. 8–19c). The stresses in the cutoff bars and the full-length bars are plotted in Fig. 8–19d and e, respectively. Points D and D' are located at a development length, ℓ_d, away from the ends of the cutoff bars. By this point in the beam the cutoff bars are fully effective. As a result, all 5 bars act to resist the applied moments between D and D' and the stresses in both sets of bars are the same (Fig. 8–19d and e), and are the same as if all bars extended the full length of the beam (Fig. 8–19b). Between D and C, the stress in the cutoff bars reduces to zero, while the stress in the remaining 3 bars increases. At point C, the stress in the remaining 3 bars reaches the yield strength, f_y, as assumed in selecting the cutoff points. For the bars to reach their yield strength at point C, the distance AC must not be less than the development length, ℓ_d. If AC is less than ℓ_d, the required anchorage can be obtained by hooking the bars at A, by using smaller bars, or by extending all 5 bars into the support.

Development of Bars in Positive Moment Regions

Figure 8–20 shows a uniformly loaded, simple beam and its bending moment diagram. As a first trial, the designer has selected 2 No. 14 bars as reinforcement. These run the full length of the beam and are enclosed in minimum stirrups. The development length of a No. 14 grade 60 bar in 3000-psi concrete is 93 in. The point of maximum bar stress is at midspan and since the bars extend 9 ft 6 in. = 144 in. each way from midspan, they are developed at midspan.

Because the bending moment diagram for a uniformly loaded beam is a parabola, it is possible for the bar stress to be developed at midspan and not be developed at, for example, the quarter points of the span where the moment is three-fourths of the maximum. This is illustrated in Fig. 8–20b, where the moment capacity and the required moment diagrams are compared. The moment capacity is assumed to increase linearly from zero at the

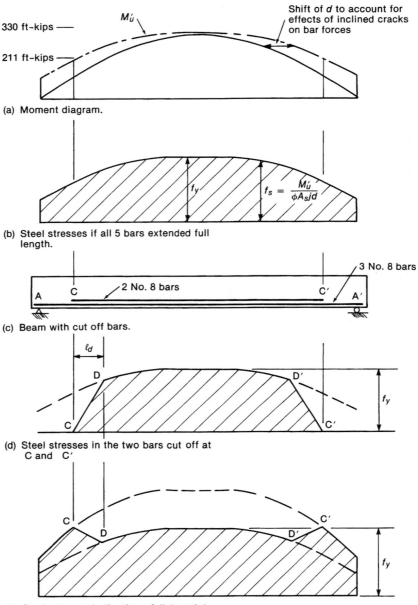

330 ft–kips ——

M_u'

Shift of d to account for effects of inclined cracks on bar forces

211 ft–kips ——

(a) Moment diagram.

f_y $f_s = \dfrac{M_u'}{\phi A_s jd}$

(b) Steel stresses if all 5 bars extended full length.

3 No. 8 bars

A   C   2 No. 8 bars   C′   A′

(c) Beam with cut off bars.

ℓ_d

D   D′ f_y

C   C′

(d) Steel stresses in the two bars cut off at C and C′

C   C′

D   D′ f_y

(e) Steel stresses in the three full–length bars.

Fig. 8–19
Steel stresses in vicinity
of bar cutoff points.

ends of the bars, to $\phi M_n = 363$ ft-kips at a distance $\ell_d = 93$ in. from the ends of the bars. Between points A and B, the required moment exceeds the moment capacity. Stated in a different way, the bar stresses required at points between A and B are larger than those which can be developed in the bar.

Ignoring the extension of the bar into the support for simplicity, it can be seen from Fig. 8–21 that the slope of the rising portion of the moment capacity diagram cannot be less than indicated by line OA. If the moment capacity diagram had the slope OB, the bars would have insufficient development for the required stresses in the shaded region of Fig. 8–21. Thus the slope of the moment capacity diagram, $d(\phi M_n)/dx$, cannot be less than that of the tangent to the required moment diagram dM_u/dx at $x = 0$. The slope of

Development, Anchorage, and Splicing of Reinforcement

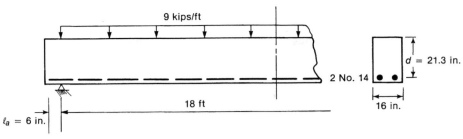

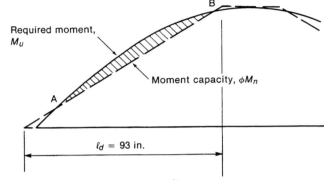

(a) Beam and section.

(b) Required moment and moment capacity diagrams.

Fig. 8–20
Anchorage of positive
moment reinforcement.

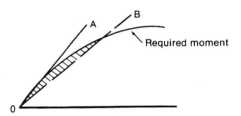

Fig. 8–21
Anchorage at point of
zero moment.

the moment capacity diagram is $\phi M_n / \ell_d$. The slope of the required moment diagram is $dM_u / dx = V_u$. Thus the least slope the moment capacity diagram can have is

$$\frac{\phi M_n}{\ell_d} = V_u$$

Thus the longest development length that can be tolerated is

$$\ell_d = \frac{\phi M_n}{V_u}$$

where M_n is the moment capacity based on the bars in the beam at 0 and V_u is the shear at 0.

ACI Sec. 12.11.3 requires that *at simple supports* where the reaction induces compressive confining stresses in the bars (as would be the case in Fig. 8–20), the size of the positive moment reinforcement should be small enough that the development length ℓ_d, satisfies

$$\ell_d \leq \frac{1.3M_n}{V_u} + \ell_a \qquad (8\text{--}19)$$

where ℓ_a is the embedment length past the centerline of the support. The factor 1.3 accounts for the fact that transverse compression tends to increase the bond strength by offsetting some of the splitting stresses. When the beam is supported in such a way that there are no bearing stresses above the support, the factor 1.3 becomes 1.0 (giving Eq. 8–20, ACI Eq. 12–2).

When the bars are hooked with the point of tangency of the hook outside the centerline of the support, or if mechanical anchors are provided, ACI Sec. 12.11.3 does not require that Eq. 8–19 be satisfied. It should be noted that hooked bars at a support can lead to bearing failures unless they are carefully detailed. Figure 8–22a shows, to scale, a support of a simple beam. The potential crack illustrated does not encounter any reinforcement. In precast beams the end of the beam is often reinforced as shown in Fig. 8–22b.

At positive moment *points of inflection* (points where the positive moment envelope passes through zero), a similar situation exists, except that there are not transverse bearing stresses. Here the Code requires that the diameter of the positive moment reinforcement should be small enough that the development length, ℓ_d, satisfies

$$\ell_d \leq \frac{M_n}{V_u} + \ell_a \tag{8–20}$$

where ℓ_a is the longer of the effective depth, d, or 12 bar diameters, but not more than the actual embedment of the bar in the negative moment region past the point of inflection. The ACI Code does not specify this last condition, which is added here for completeness.

Equations 8–19 and 8–20 are written in terms of M_n, rather than ϕM_n, since M_n leads to a slightly more conservative value. It should be noted that the derivation of Eqs. 8–19 and 8–20 did not consider the shift in the bar force due to shear. As a result, these equations do not provide a sufficient check of the end anchorage of bars at simple supports of short, deep beams or beams supporting shear forces larger than about $V_u = 4\sqrt{f_c'}b_w d$. In such cases the bars should be anchored in the support for a force of at least $V_u/2$ and preferably $0.75V_u$, as discussed in the preceding section.

Equations 8–19 and 8–20 are not applied in negative moment regions, because the shape of the moment diagram is concave downward such that the only critical point for anchorage is the point of maximum bar stress.

EXAMPLE 8–4   Checking the Development of a Bar in a Positive Moment Region

The beam in Fig. 8–20 has 2 No. 14 bars and No. 3 U stirrups at 10 in. o.c. The material strengths are $f_c' = 3000$ psi and $f_y = 60,000$ psi. The beam supports a total factored load of 9.0 kips/ft. Check if ACI Sec. 12.11.3 is satisfied.

$$\ell_d \leq 1.3 \frac{M_n}{V_u} + \ell_a \tag{8–19}$$

1. Determine the spacing and confinement case for No. 14 bars. Bar spacing $= 16 - 2 (1.5 + 0.375) - 2 \times 1.69 = 8.87$ in.

The beam has code minimum stirrups. Therefore, the beam is case 1.

2. Compute development length for No. 14 bars. From Eq. 8–10

$$\frac{\ell_d}{d_b} = \frac{f_y \alpha \beta \lambda}{20\sqrt{f_c'}}$$

$$= \frac{60,000 \times 1.0 \times 1.0 \times 1.0}{20 \times \sqrt{3000}} = 54.8$$

Thus the development length is 54.8×1.69 in. $= 92.7$ in.

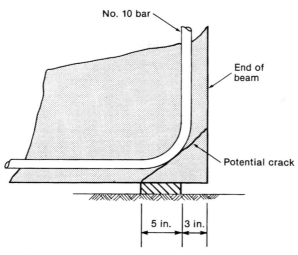

No. 10 bar

End of
beam

Potential crack

5 in. 3 in.

(a) Bearing failure.

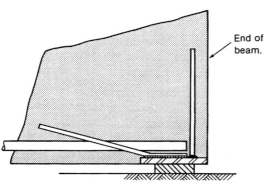

End of
beam.

Fig. 8–22
Simple beam supports.

(b) Precast bearing detail.

3. **Solve Eq. 8–19 for the required** ℓ_d. At the support there are 2 No. 14 bars:

$$M_n = 2 \times 2.25 \times 60,000\left(21.3 - \frac{2 \times 2.25 \times 60,000}{1.7 \times 3000 \times 16}\right)$$

$$= 4.86 \times 10^6 \text{ in.–lb} = 4860 \text{ in.–kips}$$

At the support,

$$V_u = \frac{w\ell}{2} = \frac{9.0 \times 18}{2}$$

$$= 81 \text{ kips}$$

ℓ_a = extension of bar past centerline of support = 6 in.

Thus

$$1.3\frac{M_n}{V_u} + \ell_a = \frac{1.3 \times 4860 \text{ in.-kips}}{81 \text{ kips}} + 6 \text{ in.}$$

$$= 84 \text{ in.}$$

8–6   Bar Cutoffs and Development of Bars in Flexural Members   **315**

But $\ell_d = 92.7$ in. is greater than 84 in.; therefore, No. 14 bars cannot be used. Try 6 No. 8 bars. It is necessary to recompute ℓ_d.

    4. Determine the spacing and confinement case for No. 8 bars. Either compute the spacing, which works out to 1.25 in. $= 1.25d_b$, or use Table A–6 to find the minimum web width for 6 No. 8 bars is 15.5 in. Since 16 in. exceeds 15.5 in. the bar spacing exceeds d_b. The beam has code minimum stirrups. Therefore, this is case 1.

    5. Compute development length for No. 8 bars. From Eq. 8–10,

$$\frac{\ell_d}{d_b} = \frac{f_y \alpha \beta \lambda}{20\sqrt{f_c'}} = \frac{60{,}000 \times 1.0 \times 1.0 \times 1.0}{20 \times \sqrt{3000}}$$

$$= 54.8$$

Thus $\ell_d = 54.8 \times 1.0 = 54.8$ in. (See also Table A–9.)

    6. Solve Eq. 8–19 for the required ℓ_d. $M_n = 5067$ in.-kips.

$$1.3\,\frac{M_n}{V_u} + \ell_a = \frac{1.3 \times 5067}{81} + 6 = 87.3 \text{ in.}$$

Since 54.8 in. < 87.3 in., this is acceptable. **Use 6 No. 8 bars.** ■

Effect of Discontinuities at Bar Cutoff Points

The bar stress diagrams in Fig. 8–19d and e suggest that a severe discontinuity in bar stresses exists in the vicinity of points where bars are cut off in a region of flexural tension. One effect of this discontinuity is a reduction in the inclined cracking shear in this vicinity.[8-10] The resulting inclined crack starts at, or near, the end of the cutoff bars. ACI Sec. 12.10.5 prohibits bar cutoffs in a zone of flexural tension unless *one* of the following is satisfied:

    1. The shear, V_u, at the cutoff point is not greater than $2/3\phi(V_c + V_s)$ (ACI Sec. 12.10.5.1).

    2. Extra stirrups are provided over a length of $0.75d$ starting at the end of the cutoff bar and extending along it. The maximum spacing of the extra stirrups is $s = d/8\beta_b$, where β_b is the ratio of the area of the bars that are cut off to the area of the bars immediately before the cutoff. The area of the stirrups, A_v, is not to be less than $60b_w s/f_y$ (ACI Sec. 12.10.5.2).

    3. For No. 11 bars and smaller, the continuing reinforcement provides twice the area required for flexure at the cutoff point and V_u is not greater than $0.75\phi(V_c + V_s)$ (ACI Sec. 12.10.5.3).

    As a result of the difficulties in satisfying these sections, designers frequently will extend all bars into the supports in simple beams, or past the points of inflection in continuous beams.

Requirements for Structural Integrity

A structure is said to have *structural integrity* if localized damage does not spread progressively to other parts of the structure. The 1989 ACI Code introduced Sec. 7-13, which provides details to improve the integrity of joist construction, beams without stirrups, perimeter beams, and precast buildings.

    There are two underlying concepts, *catenary action* and providing *ties* around buildings to anchor the catenary forces. If a support is damaged or removed from a continuous beam, the beam should be able to bridge the gap by catenary action, admittedly at large deflections. In continuous beams without stirrups the top steel over the damaged support will tear out through the top of the beam and catenary action will not develop. ACI Sec. 7.13.2.1, which deals with joist construction (joists do not usually have stirrups), and ACI Sec. 7.13.2.3, which applies to beams without closed stirrups, require a portion of the bottom reinforcement to be continuous over the support or lap spliced at the support to provide catenary action. At the outside of the building the catenary must be anchored into the supporting members with hooks.

ACI Sec. 7.13.2.2 requires that some of the negative and positive moment reinforcement in perimeter beams be continuous around the structure and be tied with closed stirrups or stirrups with 135° hooks around top bars so that these beams will serve to anchor catenary forces from interior beams.

8–7 CALCULATION OF BAR CUTOFF POINTS

General Procedure

Bar cutoff points will be calculated in a three-stage procedure. First, *flexural cutoff points* (points where bars are no longer required for flexure) will be determined from bending moment diagrams. Second, the bars will be extended to satisfy various detailing rules. The resulting cutoff points will be referred to as the *actual cutoff points*. Finally, extra stirrups will be designed for those points where bars are cut off in a zone of flexural tension. Two examples will be presented, one based on the calculation of flexural cutoff points from the equations of the moment diagram and the second based on the use of graphs of these diagrams.

The ACI Code sections governing bar cutoff locations are 7.13, 12.1, 12.10, 12.11, and 12.12. These sections can be summarized into six "rules" governing detailing requirements:

All Bars

Rule 1.   Bars must extend the longer of d or $12d_b$ past the flexural cutoff points except at supports or the ends of cantilevers.

Rule 2.   Bars must extend at least ℓ_d from the point of maximum bar stress or from the flexural cutoff points of adjacent bars (ACI Secs. 12.10.2, 12.10.4, and 12.12.2)

Positive Moment Bars

Rule 3.   **(a)**   *Simple supports.* At least one-third of the positive moment reinforcement must extend 6 in. into the support (ACI Sec. 12.11.1).

(b)   *Continuous interior beams with closed stirrups.* At least one-fourth of the positive moment reinforcement must extend 6 in. into the support (ACI Secs. 12.11.1 and 7.13.2.3).

(c)   *Continuous interior beams without closed stirrups.* At least one-fourth of the positive moment reinforcement must be continuous or shall be spliced near the support with a class A tension splice and at noncontinuous supports be terminated with a standard hook (ACI Sec. 7.13.2.3).

(d)   *Continuous perimeter beams.* At least one-fourth of the positive moment reinforcement required at midspan shall be made continuous around the perimeter of the building and must be enclosed within closed stirrups or stirrups with 135° hooks around top bars. The required continuity of reinforcement may be provided by splicing the bottom reinforcement at or near the supports with class A tension splices (ACI Sec. 7.13.2.2).

(e) *Beams forming part of a frame that is the primary lateral load resisting system for the building.* This reinforcement must be anchored to develop the specified yield strength, f_y, at the face of the support (ACI Sec. 12.11.2).

Rule 4. At the positive moment point of inflection, and at simple supports, the positive moment reinforcement must satisfy Eqs. 8–19 and 8–20 (ACI Sec. 12.11.3).

Negative Moment Bars

Rule 5. Negative moment reinforcement must be anchored into or through supporting columns or members (ACI Sec. 12.12.1).

Rule 6. **(a)** *Interior beams.* At least one-third of the negative moment reinforcement must be extended by the greatest of d, $12d_b$, or $\ell_n/16$ past the negative moment point of inflection (ACI Sec. 12.12.3).

(b) *Perimeter beams.* In addition to satisfying rule 6a, one-sixth of the negative reinforcement required at the support must be made continuous at midspan. This can be achieved by means of a class A tension splice at midspan (ACI Sec. 7.13.2.2).

EXAMPLE 8–5   Calculation of Bar Cutoff Points Based on Equations of Moment Diagrams.

The beam shown in Fig. 8–23a is constructed of 3000-psi concrete and Grade 60 reinforcement. It supports a factored dead load of 0.42 kip/ft and a factored live load of 3.4 kips/ft. The cross sections at the points of maximum positive and negative moment are shown in Fig. 8–23b. Two loading cases must be considered as shown in Figs. 8–24a and 8–26a.

1. Locate flexural cutoffs for positive moment reinforcement. The positive moment in span *AB* is governed by the loading case in Fig. 8–24a. From a free body of a part of span *AB* (Fig. 8–24d) the equation for M_u at a distance x from *A* is

$$M_u = 46.5x - \frac{3.82x^2}{2} \text{ ft-kips}$$

At midspan, the beam has 2 No. 9 plus 2 No. 8 bars. The 2 No. 8 bars will be cut off. The capacity of the remaining bars is

$$\phi M_n = \frac{0.9 \times (2 \times 1.0) \times 60,000\left(21.5 - \dfrac{2.0 \times 60,000}{1.7 \times 3000 \times 12}\right)}{12,000}$$

$$= 176 \text{ ft-kips}$$

Therefore, flexural cutoff points occur where $M_u = 176$ ft-kips. Setting $M_u = 176$ ft-kips, the equation for M_u can be rearranged to

$$1.91x^2 - 46.5x + 176 = 0$$

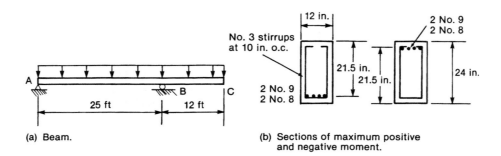

Fig. 8–23
Beam—Example 8–5.

(a) Beam.

(b) Sections of maximum positive and negative moment.

Development, Anchorage, and Splicing of Reinforcement

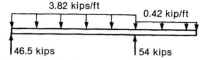

(a) Beam and loads.

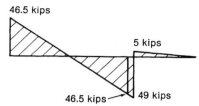

(b) Shear force diagram.

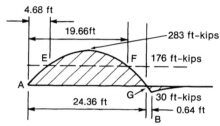

(c) Bending moment diagram.

(d) Free body diagram.

Fig. 8–24
Calculation of flexural
cutoff points for positive
moment—Example 8–5.

This is a quadratic equation of the form

$$Ax^2 + Bx + C = 0$$

and has the solution

$$x = \frac{-B \pm \sqrt{B^2 - 4AC}}{2A}$$

Thus $x = 4.68$ ft. and 19.66 ft from A. These flexural cutoff points are shown in Fig. 8–24c and 8–25a. They will be referred to as *flexural cutoff points E and F*. In a similar fashion, flexural cutoff point G is found to be 24.4 ft from A or 0.64 ft from B.

2. Compute the development lengths for the bottom bars.

$$\text{Bar spacing} = \frac{12 - 2(1.5 + 0.375) - 2 \times 1.128 - 2 \times 1.0}{3} = 1.33 \text{ in.}$$

Since the bar spacing exceeds d_b for both the No. 8 and 9 bars, and since the beam has minimum stirrups, the bars satisfy case 1 in Table 8–1. From Eq. 8–10 (or from Table A–11),

$$\frac{\ell_d}{d_b} = \frac{f_y \alpha \beta \lambda}{20 \sqrt{f_c'}} = \frac{60{,}000 \times 1.0 \times 1.0 \times 1.0}{20 \times \sqrt{3000}} = 54.8$$

Thus for the No. 8 bars, $\ell_d = 54.8$ in., and for the No. 9 bars, $\ell_d = 61.8$ in.

8–7 Calculation of Bar Cutoff Points

319

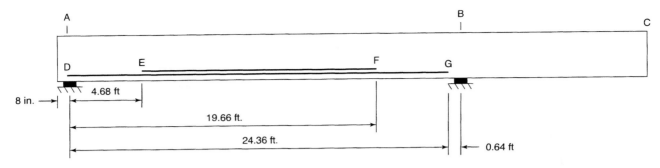

(a) Flexural cutoff points for positive moment steel.

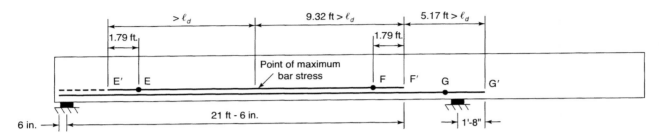

(b) Actual cutoff points for positive moment steel.

Fig. 8–25
Location of positive moment cutoff points—Example 8–5.

3. Locate actual cutoff points for positive moment reinforcement. The actual cutoff points are determined from the flexural cutoff points using the six "rules" stated earlier. Since the location of cutoffs G and D are affected by the locations of cutoffs E and F, the latter are established first, starting with F.

(a) Cutoff F. 2 No. 8 bars are cut off; must satisfy rules 1 and 2.

Rule 1. Extend the bars by the larger of $d = 21.5$ in. $= 1.79$ ft and $12d_b = 1$ ft. Therefore, the first trial position of the actual cutoff is at $19.66 + 1.79 = 21.45$ ft from the center of the support at A, say, 21 ft 6 in. (see point F' in Fig. 8–25b).

Rule 2. Bars must extend at least ℓ_d past the points of maximum bar stress. For the bars cut off at F', the maximum bar stress occurs near midspan, at 12.18 ft from A. The distance from the point of maximum bar stress to the actual bar cutoff is $21.5 - 12.18 = 9.32$ ft. ℓ_d for the No. 8 bars is 54.8 in. The distance available is more than ℓ_d —therefore OK. **Cut off 2 No. 8 bars at 21 ft 6 in. from A** (shown as point F' in Fig. 8–25b).

(b) Cutoff G. 2 No. 9 bars are cut off; must satisfy rules 2 and 3 at G, and rule 4 at the positive moment point of inflection.

Rule 3. At least one-third of the bars must extend 6 in. into the support. As a first trial we will extend the 2 No. 9 bars 6 in. into the support at B (25 ft 6 in. from A).

Rule 2. Bars must extend at least ℓ_d past actual cutoffs of adjacent bars. ℓ_d for No. 9 bottom bar $= 61.8$ in. $= 5.15$ ft. Distance from F' to $G' = (25$ ft 6 in.$) - (21$ ft 6 in.$) = 4$ ft. Bar does not extend ℓ_d, therefore, extend the bars to 21.45 ft $+ 5.15$ ft $= 26.6$ ft, say, 26 ft 8 in.

Rule 4. Must satisfy Eq. 8–20 at point of inflection (point where the moment is zero). Therefore, at G, $\ell_d \leq M_n/V_u + \ell_a$. The point of inflection is 0.64 ft from the support (Fig. 8–24c). At this point V_u is 46.5 kips (Fig. 8–24b) and the moment capacity M_n for the bars in the beam at the point of inflection (2 No. 9 bars) is

$$M_n = 176 \text{ ft-kips} \times \frac{12}{0.9} = 2345 \text{ in.-kips}$$

ℓ_a = larger of d (21.5 in.) or $12d_b$ (13.5 in.) but not more than the actual extension of the bar past the point of inflection (26.67 − 24.36 = 2.31 ft = 27.7 in.). Therefore, ℓ_a = 21.5 in., and

$$\frac{M_n}{V_u} + \ell_a = \frac{2345}{46.5} + 21.5$$

$$= 71.9 \text{ in.}$$

Since this exceeds ℓ_d = 61.8 in., OK. **Cut off 2 No. 9 bars 1 ft 8 in. from B** (shown as point G' in Fig. 8–25b).

(c) **Cutoff E.** 2 No. 8 bars are cut off; must satisfy rules 1 and 2.

Rule 1. Extend the bars d = 1.79 ft. past the flexural cutoff point. Therefore, the actual cutoff E' is at 4.68 − 1.79 = 2.89 ft (2 ft 10 in.) from A.

Rule 2. The distance from the point of maximum moment to the actual cutoff exceeds ℓ_d = 54.8 in.—therefore OK. **Cut off 2 No. 8 bars at 2 ft 10 in. from A** (point E' in Fig. 8–25b; note that this is changed later).

(d) **Cutoff D.** 2 No. 9 bars are cut off; must satisfy rules 2, 3, and 4.

Rule 3. Extend 2 No. 9 bars 6 in. past A.

Rule 2. Bars must extend ℓ_d from actual cutoff E', where ℓ_d = 61.8 in. (No. 9 bars). The maximum possible length available is 2 ft 10 in. + 6 in. = 40 in. Since this is less than ℓ_d, we must either extend the end of the beam, hook the ends of the bars, use smaller bars, or eliminate the cutoff E'. We shall do the latter. Therefore, **extend all 4 bars 6 in. past point A**.

Rule 4. Must satisfy Eq. 8–19 at the support.

$$\ell_d \le \frac{1.3M_n}{V_u} + \ell_a$$

$$V_u = 46.4 \text{ kips}$$

$$M_n = \frac{3.58 \times 60{,}000\left(21.5 - \dfrac{3.58 \times 60{,}000}{1.7 \times 3000 \times 12}\right)}{1000}$$

$$= 3860 \text{ in.-kips}$$

$$\ell_a = 6 \text{ in.}$$

$$1.3\frac{M_n}{V_n} + \ell_a = 1.3\frac{3860}{46.5} + 6 = 113.9 \text{ in.}$$

Since this exceeds ℓ_d, rule 4 is satisfied. The actual cutoff points are illustrated in Fig. 8–25b.

4. Locate flexural cutoffs for negative moment reinforcement. The negative moment is governed by the loading case in Fig. 8–26a. The equations for the negative bending moments are:

Between A and B, x measured from A:

$$M_u = -5.8x - \frac{0.42\,x^2}{2} \text{ ft-kips}$$

Between C and B, x_1 measured from C:

$$M_u = \frac{-3.82x_1^2}{2} \text{ ft-kips}$$

Over the support at B, the reinforcement is 2 No. 9 bars plus 2 No. 8 bars. The 2 No. 8 bars are no longer required when the moment is less than ϕM_n = 176 ft-kips (capacity of the beam with 2 No. 9 bars).

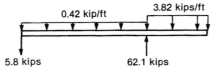

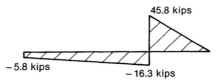

(a) Beam and loads.

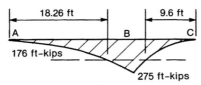

(b) Shear force diagram.

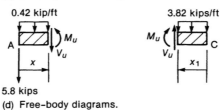

(c) Bending moment diagram.

Fig. 8–26
Calculation of flexural cutoff
points for negative mo-
ment—Example 8–5.

(d) Free-body diagrams.

Between A and B:

$$-176 = -5.8x - 0.21x^2$$

$$x = -45.9 \text{ ft or } 18.26 \text{ ft from } A$$

Therefore, the flexural cutoff point for 2 No. 8 top bars in span AB is at 18.28 ft from A.

Between B and C:

$$-176 = -1.91x_1^2$$

$$x_1 = 9.60 \text{ ft from } C$$

Therefore, the flexural cutoff point for 2 No. 8 top bars in span BC is at 9.60 ft from C. The flex-ural cutoff points for the negative moment steel are shown in Fig. 8–27a, and are lettered H, J, K, and L.

5. **Compute development lengths for the top bars.** Since there is more than 12 in. of concrete below the top bars, $\alpha = 1.3$. For the No. 8 bars, $\ell_d = 71.2$ in., and for the No. 9 bars, $\ell_d = 80.3$ in.

6. **Locate the actual cutoff points for the negative moment reinforcement.** Again, the inner cutoffs will be considered first since their location affects the design of the outer cutoffs. The choice of actual cutoff points is illustrated in Fig. 8–27b.

(a) **Cutoff J.** 2 No. 8 bars cut off; must satisfy rules 1, 2, and 5.

Rule 1. Extend bars by $d = 1.79$ ft past the flexural cutoff. Cut off at $18.26 - 1.79 = 16.47$ ft from $A = 8.53$ ft from B, say 8 ft 6 in.

Development, Anchorage, and Splicing of Reinforcement

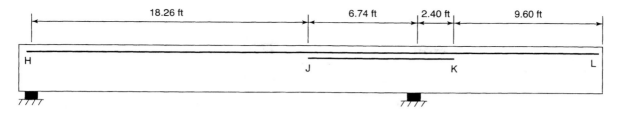

(a) Flexural cutoff points for negative moment steel.

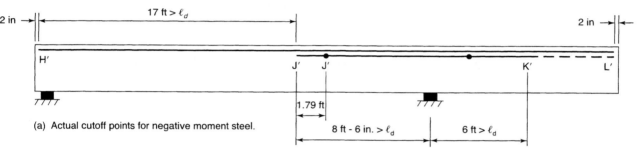

(a) Actual cutoff points for negative moment steel.

Fig. 8–27
Location of negative moment cutoff points—Example 8–5.

Rule 2. The bars must extend ℓ_d from the point of maximum bar stress. For the 2 No. 8 top bars, the maximum bar stress is at B. The actual bar extension is 8.5 ft = 102 in. This exceeds ℓ_d = 71.2 in.—therefore OK.

Rule 5. Bar must be anchored at support. This will be satisfied when cutoff K is selected. Therefore, **cut off 2 No. 8 bars at 8 ft 6 in. from B** (point J' in Fig. 8–26b).

(b) Cutoff H. 2 No. 9 bars cut off; must satisfy rule 2.

Rule 2. Bar must extend ℓ_d past J', where ℓ_d = 80.3 in. Length available = 17 ft—therefore OK. **Extend 2 No. 9 bars to 2 in. from the end of the beam** (point H' in Fig. 8–27b).

(c) Cutoff K. 2 No. 8 bars are cut off; must satisfy rules 1, 2, and 5. The flexural cutoff is at 9.60 ft from C (2.40 ft from B).

Rule 1. Extend bars d = 1.79 ft. The end of the bars is at 2.40 + 1.79 = 4.19 ft from B, say, 4 ft 3 in.

Rule 2. Extend ℓ_d past B. ℓ_d for a No. 8 top bar = 71.2 in. The extension of 4 ft 3 in. is thus not enough. Try extending the No. 8 top bars 6 ft past B to point K'.

Rule 5. The bars must be anchored at the support. Since the No. 8 bars extend more than ℓ_d on either side of the support, they are adequately anchored. Therefore, cut off 2 No. 8 bars at 6 ft from B (point K' in Fig. 8–27b). Note that this is changed in the next step.

(d) Cutoff L. 2 No. 9 bars are cut off. Rule 2 applies.

Rule 2. The bars must extend ℓ_d past K'. ℓ_d for a No. 9 top bar = 80.3 in. = 6.69 ft. The actual extension is 11.83 − 6 = 5.83 ft, which is less than ℓ_d—therefore not OK. Two solutions are available, either extend all the bars to the end of the beam, or change the bars to 6 No. 7 bars in two layers. We shall do the former. The final actual cutoff points are shown in Fig. 8–28.

7. Check whether extra stirrups are required at cutoffs. ACI Sec. 12.10.5 prohibits bar cutoffs in a tension zone unless:

12.10.5.1: V_u at actual cutoff $\leq \frac{2}{3}\phi \left(V_c + V_s\right)$ at that point, or

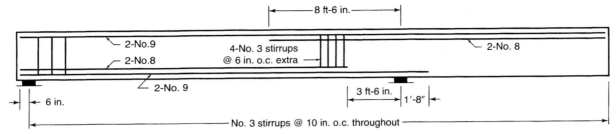

Fig. 8–28
Reinforcement details—Example 8–5.

12.10.5.2: Extra stirrups are provided at actual cutoff point, or

12.10.5.3: The continuing flexural reinforcement at the flexural cutoff has twice the required A_s and $V_u \leq 0.75 \, \phi(V_c + V_s)$.

Since we have selected flexural cutoff points on the basis of the continuing steel having 1.0 times the required A_s, ACI Sec. 12.10.5.3 will not govern. Therefore, we will provide extra stirrups unless ACI Sec. 12.10.5.1 is satisfied.

As shown in Fig. 8–23b, the beam has No. 3 double-leg stirrups at 10 in. o.c. throughout its length. From Eqs. 6–9, 6–14 and 6–18,

$$\phi V_n = \phi(V_c + V_s)$$

$$= \frac{0.85\left(2\sqrt{3000} \times 12 \times 21.5 + \dfrac{0.22 \times 60,000 \times 21.5}{10}\right)}{1000}$$

$$= 48.1 \text{ kips}$$

and $\frac{2}{3}\phi V_n = 32.1$ kips. Thus if V_u at the *actual* cutoff point exceeds 32.1 kips, extra stirrups will be required.

(a) Cutoff F'. Located at 21 ft 6 in. from A or 3 ft 6 in. from B. Flexural tensions occur at this point due to load case 1 (Fig. 8–24c). The shear at F' for load case 1 is

$$V_u = -49.0 \text{ kips } + 3.5 \text{ ft} \times 3.82 \text{ kips } = -35.6 \text{ kips}$$

where the sign simply indicates the direction of the shear force. Since $V_u = 35.6$ kips exceeds $\frac{2}{3}\phi V_n = 32.1$ kips, extra stirrups are required. Extra stirrups must extend $0.75d = 0.75 \times 21.5 = 16.1$ in. along the No. 8 bars from cutoff point F'.

$$\text{Spacing} \leq \frac{d}{8\beta_b}$$

where

$$\beta_b = \frac{\text{area of bars cut off}}{\text{total area immediately prior to cutoff}}$$

$$= \frac{2 \times 0.79 \text{ in.}^2}{2 \times 1.0 \text{ in.}^2 + 2 \times 0.79 \text{ in.}^2} = 0.44$$

Therefore, the stirrup spacing is

$$s \leq \frac{21.5 \text{ in.}}{8 \times 0.44} = 6.09 \text{ in.}$$

Try 4 extra No. 3 double-leg stirrups at 6 in. o.c. Check:

$$\text{required } A_v = \frac{60b_w s}{f_y}$$

$$= \frac{60 \times 12 \times 6}{60,000} = 0.072 \text{ in.}^2$$

A_v provided $= 0.22$ in.^{2}—therefore OK. **Use 3 extra No. 3 double-leg stirrups at 6 in. o.c., the first one at 2 in. from end of cutoff bars at F$'$.**

(b) **Cutoff E.** All bars were extended into the support at A, so supplementary shear reinforcement is not required here.

(c) **Cut off J$'$.** The cutoff is located at 8 ft 6 in. from B. Flexural tension occurs in these bars due to loading case 2 (Fig. 8–26). By inspection V_u is considerably less than $\frac{2}{3}\phi V_u = 32.1$ kips. Therefore, no extra stirrups are required at this cutoff.

The final reinforcement details are shown in Fig. 8–28. For nonstandard beams such as this, a detail of this sort should be shown in the contract drawings. ■

The calculations just carried out are tedious and if the underlying concepts are not understood, the detailing provisions become a meaningless mumbo jumbo. The reader is advised to reread Sec. 8–6 to clarify the concepts involved.

Several things can be done to simplify these calculations. One is to extend all of the bars past their respective points of inflection so that no bars are cut off in zones of flexural tension. This reduces the number of cutoffs required and eliminates the need for extra stirrups, on one hand, while requiring more flexural reinforcement on the other. A second method is to determine the flexural cutoff points graphically. This is discussed in the next section.

Graphical Calculation of Flexural Cutoff Points

The flexural capacity of a beam is $\phi M_n = \phi A_s f_y jd$, where jd is the internal level arm and is relatively insensitive to the amount of reinforcement. If it is assumed that jd is constant, then ϕM_n is directly proportional to A_s. Since in design, ϕM_n is set equal to M_u, we can then say that the amount of steel, A_s, required at any section is directly proportional to M_u at that section. If it is desired to cut off a third of the bars at a particular cutoff point, the remaining two-thirds of the bars would have a capacity of two-thirds of the maximum ϕM_n, and hence this cutoff would be located where M_u was two-thirds of the maximum M_u.

Figure A–1 in Appendix A is a schematic graph of the bending moment envelope for a typical interior span of a multispan continuous beam designed for maximum negative moments of $w\ell_n^2/11$ and a maximum positive moment of $w\ell_n^2/16$ (as per ACI Sec. 8.3.3). Similar graphs for end spans are given in Figs. A–2 to A–4.

Figure A–1 can be used to locate the flexural cutoff points and points of inflection for typical interior uniformly loaded beams, *satisfying the limitations of ACI Sec. 8.3.3.* Thus the extreme points of inflection for positive moments (points where positive moment diagram equals zero) are at $0.146\ell_n$ from the faces of the two supports, while the corresponding negative moment points of inflection are at $0.24\ell_n$ from the supports. This means that positive moment steel must extend from midspan to at least $0.146\ell_n$ from the supports, while negative moment steel must extend at least $0.24\ell_n$ from the supports. The use of Figs. A–1 to A–4 is illustrated in Example 8–6. A more complete example is given in Chap. 10.

EXAMPLE 8–6   Use of Bending Moment Diagrams to Select Bar Cutoffs

A continuous T beam having the section shown in Fig. 8–29a carries a factored load of 3.26 kips/ft and spans 22 ft, center to center of 16-in.-square columns. The design has been carried out using the moment coefficients in ACI Sec. 8.3.3. Locate the bar cutoff points. For simplicity, all the bars will be carried past the points of inflection before cutting them off, except in the case of the positive

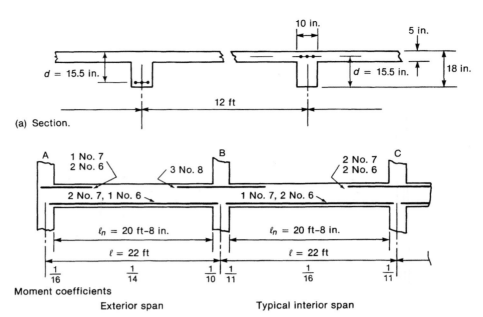

(a) Section.

(b) Elevation.

A 1 No. 7
2 No. 6
3 No. 8
B
2 No. 7
2 No. 6
C

2 No. 7, 1 No. 6
1 No. 7, 2 No. 6

ℓ_n = 20 ft-8 in.
ℓ_n = 20 ft-8 in.

ℓ = 22 ft
ℓ = 22 ft

Moment coefficients

$\frac{1}{16}$ $\frac{1}{14}$ $\frac{1}{10}$ $\frac{1}{11}$ $\frac{1}{16}$ $\frac{1}{11}$

Exterior span
Typical interior span

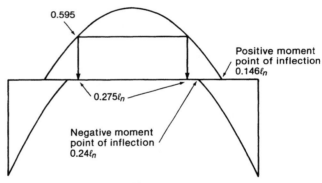

0.595

Positive moment
point of inflection
$0.146\ell_n$

$0.275\ell_n$

Negative moment
point of inflection
$0.24\ell_n$

(c) Use of Fig. A–1 to locate flexural cutoff points.

Fig. 8–29
Example 8–6.

moment steel in span *BC*, where the No. 7 bar will be cut off earlier to illustrate the use of the bar cut-off graph. The concrete strength is 3000 psi and the steel yield strength is 60,000 psi. Double-leg No. 3 stirrups are provided at 7.5 in. throughout.

1. Determine the positive moment steel cutoff points for a typical interior span—span *BC*.

(a) Development lengths of bottom bars. From Table A–6, the minimum web width for 2 No. 7 bars and 1 No. 6 is 9 in. Since the beam width exceeds this, the bar spacing is at least d_b, and since there are at least code minimum stirrups, this is case 1 development. From Table A–11 for $\beta = 1.0$ and $\lambda = 1.0$:

No. 6 bar, $\ell_d = 43.8\ d_b = 32.9$ in. $= 2.74$ ft.
No. 7 bar, $\ell_d = 54.8\ d_b = 48$ in. $= 4$ ft.

(b) Cutoff point for one No. 7 bar. Cut off 1 No. 7 bar when it is no longer needed, and extend the remaining 2 No. 6 bars into the supports.

Development, Anchorage, and Splicing of Reinforcement

After the No. 7 bar is cut off, the A_s remaining $= 0.88$ in.^{2} or $0.88/1.48 = 0.595$ times A_s at midspan. Therefore, the No. 7 bar can be cut off where M_u is 0.595 times the maximum moment. From Fig. A–1, this occurs at $0.275\ell_n$ from each end for the positive moment in a typical interior span. This is illustrated in Fig. 8–29c.

Therefore, the flexural cutoff point for the No. 7 bar is at

$$0.275\ell_n = 0.275 \times 20.67 \text{ ft}$$

$$= 5.68 \text{ ft from the faces of the columns}$$

To compute the actual cutoff points, we must satisfy detailing rules 1 and 2.

Rule 1. The bar must extend by the longer of $d = 15.5$ in. or $12d_b = 12 \times 0.875$ in. $= 10.5$ in. past the flexural cutoff. Therefore, extend the No. 7 bar 15.5 in. $= 1.29$ ft. The end of bar will be at $5.68 - 1.29$ ft $= 4.39$ ft, say 4 ft 4 in. from face of column as shown in Fig. 8.30b.

Rule 2. Bars must extended ℓ_d from the point of maximum bar stress. ℓ_d for No. 7 bottom bar $= 48$ in. Clearly, the distance from the point of maximum bar stress at midspan to the end of bar exceeds this. Therefore, **cut off the No. 7 bar at 4 ft 4 in. from the column faces in the interior span.**

(c) Cutoff point for remaining positive moment steel. The flexural cutoff points for the 2 No. 6 bars are at the positive moment points of inflection. From Fig. A–1, these are at $0.146\ell_n = 3.02$ ft from the face of the supports. At these points we must satisfy detailing rules 1, 2, 3, and 4.

Rule 3. The application of rule 3 depends on whether the beam is an interior beam or perimeter beam and whether it has closed stirrups, on one hand, and no stirrups or open

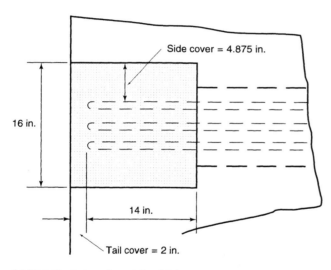

(a) Plan of exterior column-beam joint.

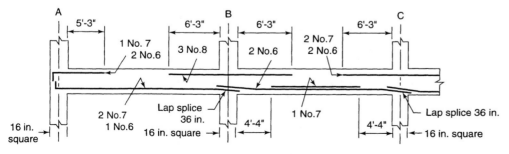

(b) Final bar details

Fig. 8–30
Reinforcement details—Example 8–6.

stirrups, on the other. This is an interior beam. Rule 3b applies if the beam has closed stirrups. Otherwise, rule 3c applies. We shall assume that the stirrups are open stirrups since this is the normal case. Thus rule 3c applies: at least one-fourth of the positive moment steel must be lap spliced in the support with a class A splice which is lapped $1.0\ell_d$ (see Sec. 8–8). We will extend the 2 No. 6 bars into the support so as to have a bar in each lower corner of the stirrups and for simplicity we will lap splice both of them $\ell_d = 32.9$ in., say 36 in.

Rule 1. The bars must extend a distance d past the flexural cutoff. This is satisfied by running the bars into the support.

Rule 2. The bars must extend by ℓ_d past the cutoff point of the No. 7 bar, ℓ_d for a No. 6 bottom bar $= 32.9$ in. More than ℓ_d is available—therefore OK.

Rule 4. At the positive moment point of inflection Eq. 8–20 must be satisfied: $\ell_d \le M_n / V_u + \ell_a$. At the point of inflection there are 2 No. 6 bars. M_n for these bars is

$$M_n = 2 \times 0.44 \times \frac{60{,}000\left(15.5 - \dfrac{2 \times 0.44 \times 60{,}000}{1.7 \times 3000 \times 62} \right)}{1000}$$

$$= 810 \text{ in.-kips}$$

The points of inflection are at $0.146\ell_n = 3.02$ ft from the faces of the columns.

$$V_u \text{ at point of inflection} = w\left(\frac{\ell_n}{2} - 0.146\ell_n \right)$$

$$V_u = 3.26\left(\frac{20.67}{2} - 3.02 \right)$$

$$= 23.8 \text{ kips}$$

ℓ_a is the greater of $d = 15.5$ in. or $12d_b = 12 \times 0.75 = 9$ in. but not more than the actual extension of the bar past the point of inflection, which is 3.02 ft plus half of the lap splice length (3.02 ft $+$ 0.5 $\times$ 36 in.) Therefore, $\ell_a = 15.5$ in.

$$\frac{M_n}{V_u} + \ell_a = \frac{810 \text{ in.-kips}}{23.8 \text{ kips}} + 15.5 \text{ in.}$$

$$= 49.5 \text{ in.}$$

Since $\ell_d = 32.9$ in. is less than 49.5 in., the anchorage at the point of inflection is OK. Therefore, **extend 2 No. 6 bars 16.5 in., say 18 in. past the center of the support.**

Check the shear at the points where the No. 7 bars are cut off (ACI Sec. 12.10.5):

$$\phi(V_c + V_s) = 0.85\left(2 \times \sqrt{3000} \times 10 \times 15.5 + \frac{0.22 \times 60{,}000 \times 15.5}{7.5} \right)$$

$$= 37.6 \text{ kips}$$

The shear at the cutoff point is

$$V_u = 3.26\left(\frac{20.67}{2} - 4.33 \right)$$

$$= 19.6 \text{ kips}$$

Since $V_u = 19.6$ kips is less than $\frac{2}{3} \times 37.6 = 25.1$ kips, no extra stirrups are required at the points where the No. 7 bar is cut off.

2. Determine the negative moment steel cutoffs for end B, span BC. To simplify the calculations, detailing, and construction, all the bars will be extended past the negative moment point of inflection and cutoff. From Fig. A–1, the negative moment point of inflection is at $0.24\ell_n = 0.24 \times 20.67$ ft $= 4.96$ ft from the face of the column. Therefore, the flexural cutoff point is at 4.96 ft from face of column.

Development, Anchorage, and Splicing of Reinforcement

Development lengths of top bars. Again this is case 1. From Table A–11 for $\beta = 1.0$ and $\lambda = 1.0$:

$$\ell_d \text{ for No. 6} = 57.0 d_b = 42.8 \text{ in}$$
$$\ell_d \text{ for No. 7} = 71.2 d_b = 62.3 \text{ in.}$$
$$\ell_d \text{ for No. 8} = 71.2 d_b = 71.2 \text{ in.}$$

The actual cutoff point is determined by applying rules 1, 2, 5, and 6.

Rule 6. At least one-third of the negative moment reinforcement must extend $d = 15.5$ in. $= 1.29$ ft, $12d_b = 12$ in. (at B, less at C), or $\ell_n/16 = 20.67/16 = 1.29$ ft past the point of inflection. Therefore, the bars must be extended by $4.96 + 1.29 = 6.25$ ft.

Therefore, try a cutoff point at 6 ft 3 in. from the face of the columns.

Rule 1. Already checked by checking rule 6.

Rule 2. Bars must extend ℓ_d past the point of maximum bar stress. For the No. 8 bars at support B, $\ell_d = 71.2$ in. The actual bar extension $= 6$ ft. 3 in. is adequate.

Rule 5. Negative moment bars must be anchored into the support. This will be checked by checking the anchorage of the No. 8 bars on the opposite side of columns B and C.

Therefore, **cut off top bars at 6 ft 3 in. from the face of column B** (see Fig. 8–30b).

Repeating the calculations at end C of span BC, the development length of the No. 7 top bars is 5 ft 3 in. Since this is less than the 6 ft 3 in. chosen earlier using rule 6, we shall **cut off top bars at 6 ft 3 in. from the face of column C**.

3. Determine the positive moment cutoffs for the end span—span AB. Since the beam AB frames into a column at the exterior end, it must be designed using moment coefficients of $\frac{1}{16}, \frac{1}{14}$, and $\frac{1}{10}$ (ACI Sec. 8.3.3). Therefore, use Fig. A–2 to select the cutoff points. For simplicity, extend all the positive moment bars 6 in. into the support. This will satisfy rules 1 and 2.

Rule 3. To satisfy rule 3c we shall hook the 2 No. 7 bars into the support at A and will lap splice these bars $1.0\ell_d$ of a No. 6 bar $= 32.9$ in., say 36 in. with the 2 No. 6 bars from span BC.

Rule 4. At the positive moment point of inflection $\ell_d \leq M_n/V_u + \ell_a$. For 2 No. 7 plus 1 No. 6, $M_n = 1355$ in.-kips. From Fig. A–2, the positive moment point of inflection is at $0.1\ell_n = 2.07$ ft from the face of the exterior column.

$$V_u = w\left(\frac{\ell_n}{2} - 2.07\right) = 27.0 \text{ kips}$$

$$\ell_a = d = 15.5$$

Therefore,

$$\frac{M_n}{V_u} + \ell_a = \frac{1335}{27.0} + 15.5$$

$$= 65 \text{ in.}$$

This exceeds ℓ_d, which is 48 in. for No. 7 bottom bar—therefore OK.

Therefore, **extend the No. 6 bar 6 in. into the supports, hook the 2 No. 7 bars at support A, and lap splice them 36 in. at support B.**

4. Determine the negative moment cutoffs for the interior end of the end span. At the interior end of the end span (at end B of span AB), the negative moment point of inflection is at $0.24\ell_n$ from the face of the interior column (Fig. A–2). Following the calculations for the negative moment bars in the interior span, we can **cut off the top bars at end B or span AB at 6 ft 3 in. from the face of the column** (see Fig. 8–30b).

5. Determine the negative moment cutoffs for the exterior end of the end span. It is frequently difficult to anchor the top bars at the exterior end of a beam. These bars develop a stress of f_y at the face of the exterior column and hence must be anchored into the column for this stress (rule 5).

The exterior column is 16 in. square. Since the development length for a No. 7 top bar is 36.8 in., we clearly cannot extend the bars straight into the column. Try hooks:

$$\text{basic development length } \ell_{hb} \text{ (Eq. 8–18)} = \frac{1200 d_b}{\sqrt{f'_c}}$$

$$= \frac{1200 \times 0.875}{\sqrt{3000}} = 19.2 \text{ in.}$$

This is multiplied by 0.7 if the hooks have adequate cover (ACI Sec. 12.5.4.2). Figure 8–30a shows a plan view of the exterior column-beam joint. The side cover to the hooked bars exceeds 2.5 in. and the bars can be placed with 2 in. of cover to the tail of the hook. Therefore,

$$\ell_{dh} = 19.2 \times 0.7 = 13.4 \text{ in.}$$

There is a space of 14 in. available in the column; therefore, we can use the top steel selected earlier with standard hooks in the column. Note that in some cases ACI Sec. 12.5.4 will require that the hook be enclosed within ties at $3d_b$, where d_d is the diameter of hooked bar. Note also that the need for 2 in. of cover on the tail of the hook should be shown on the contract drawings.

Since the bars can be anchored into column, continue. (If the bars could not be anchored, repeat step 5 using the next-smaller-size bars.)

From Fig. A–2, the negative moment point of inflection is at $0.164\ell_n = 3.39$ ft from the face of the exterior column. Here anchorage rules 2, 5, and 6 must be satisfied (rule 1 is included in rule 6).

> **Rule 6.** Extend the bars by $d = 1.29$ ft, $12d_b = 0.875$ ft, of $\ell_n/16 = 1.29$ ft beyond the point of inflection to $3.39 + 1.29 = 4.68$ ft or 4 ft 9 in. from the support.
>
> **Rule 2.** Bars must extend ℓ_d from the face of the column. ℓ_d for a No. 7 top bar is 62.3 in. Therefore, extend the top bars 5 ft 3 in. into the span.
>
> **Rule 5.** Already satisfied by hooking the bars into the column.

Therefore, extend the top bars at the exterior end 5 ft 3 in. into the span and provide standard hooks into the column. The final reinforcement layout is shown in Fig. 8–30b. ■

Frequently, bending moments cannot be calculated using the coefficients in ACI Sec. 8.3.3. This occurs, for example, in beams supporting concentrated loads, such as reactions from other beams; for beams of widely varying span lengths, such as the case that may occur in schools or similar buildings where two wide rooms are separated by a corridor, or for beams with a change in the uniform loads along the span. For such cases, the bending moment graphs in Fig. A–1 to A–4 cannot be used. Here the designer must plot his or her own required moment envelope from structural analyses and ensure that the moment capacity diagram falls outside the required moment envelope.

Standard Cutoff Points

For beams and one-way slabs which satisfy the limitations on span lengths and loadings in ACI Sec. 8.3.3 (two or more spans, uniform loads, roughly equal spans, factored live load not greater than 3 times factored dead load), the bar cutoff points shown in Fig. A–5 can be used. These are based on Figs. A–1 to A–4 assuming the span-to-depth ratios are not less than 10 for beams and 18 for slabs.

These cutoff points do not apply to beams forming part of the primary lateral load-resisting system of a building. It is necessary to check whether the top bar extensions equal or exceed ℓ_d for the bar spacing actually used. Figure A–5 can be used as a guide for beams and slabs reinforced with epoxy-coated bars, but it is necessary to check the cutoff points selected.

For the beam in Example 8–6, Fig. A–5 would give negative moment reinforcement cutoffs at 5 ft 2 in. from the face of column A and 6 ft 11 in. from the face of column B. These are close to the computed extensions.

Reinforcement Details for Beams Resisting Seismic Loads.

Special cutoff requirements are given in ACI Secs. 21.3.2 and 21.8.2 for longitudinal reinforcement in beams in frames resisting seismic loads (see Sec. 19–6).

8–8 SPLICES

Tension Lap Splices

In a lapped splice, the force in one bar is transferred to the concrete, which transfers it to the adjacent bar. The force transfer mechanism shown in Fig. 8–31a is clearly visible from the crack pattern sketched in Fig. 8–31b.[8-11] The transfer of forces out of the bar into the concrete causes radially outward pressures on the concrete as shown in Fig. 8–31c, which, in turn, cause splitting cracks along the bars similar to those shown in Fig. 8–8a. Once these

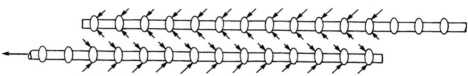

(a) Forces on bars at splice.

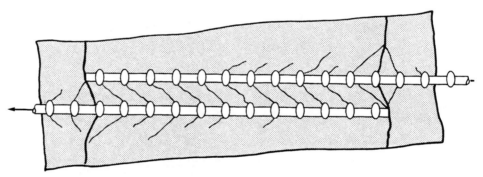

(b) Internal cracks at splice.

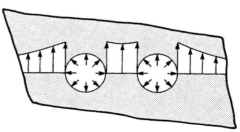

Fig. 8–31
Tension lap splices.

(c) Radial forces on concrete and splitting stresses shown on a section through the splice.

occur, the splice fails as shown in Fig. 8–32. The splitting cracks generally initiate at the ends of the splice, where the splitting pressures tend to be larger than at the middle. As shown in Fig. 8–31b, large transverse cracks occur at the discontinuities at the ends of the spliced bars. Transverse reinforcement in the splice region delays the opening of the splitting cracks and hence improves the splice capacity.

ACI Sec. 12.15 distinguishes between two types of tension lap splices depending on the fraction of the bars spliced in a given length and on the reinforcement stress at the splice. Table R12.15.2 of the ACI Commentary is reproduced as Table 8–2. The equivalence between (A_s provided$/A_s$ required) and f_s/f_y given in Table 8–2 does not appear in the 1995 ACI Code, although it was expressed in terms of f_s/f_y in the 1971 Code and the intent has not changed. The splice lengths for each class of splice are as follows:

Class A splice: $1.0\ell_d$

Class B splice: $1.3\ell_d$

Because the stress level in the bar is accounted for in Table 8–2, the reduction in the development length for excess reinforcement allowed in ACI Sec. 12.2.5 is *not* applied in computing ℓ_d for this purpose.

Fig. 8–32
Failure of a tension lap splice. (Photograph courtesy of J. G. MacGregor.)

TABLE 8–2   Type of Tension Lap Splices Required

$\dfrac{A_s \text{ Provided}}{A_s \text{ Required}}$	or	$\dfrac{f_s}{f_y}$	Maximum Percentage of A_s Spliced within Required Lap Length	
			50%	100%
2 or more		0.5 or less	Class A	Class B
Less than 2		More than 0.5	Class B	Class B

Source: ACI Sec. 12.15.2.

The center-to-center distance between two bars in a lap splice cannot be greater than one-fifth of the splice length with a maximum of 6 in. (ACI Sec. 12.14.2.3) Bars larger than No. 11 cannot be lap spliced, except at footing-to-column joints (ACI Sec. 15.2.8.4). Lap splices should always be enclosed within stirrups, ties, or spirals to delay or prevent the complete loss of capacity indicated in Fig. 8–32. As indicated in ACI Secs. 12.2.2 and 12.2.3, the presence of transverse steel may lead to shorter ℓ_d and hence shorter splices. ACI Sec. 21.3.2.3 requires that tension lap splices of flexural reinforcement in beams resisting seismic loads be enclosed by hoops or spirals.

Compression Lap Splices

In a compression lap splice, a portion of the force transfer is through bearing of the end of the bar on the concrete.[8-12, 8-13] This, and the fact that no transverse tension cracks exist in the splice length, allow compression lap splices to be much shorter than tension lap splices (ACI Sec. 12.16). Frequently, a compression lap splice will fail by spalling of the concrete under the ends of the bars. The design of column splices is discussed in Chap. 11.

Welded, Mechanical, and Butt Splices

In addition to lap splices, bars stressed in tension or compression may be spliced by welding, or by various mechanical devices, such as sleeves filled with molten cadmium metal (Fig. 11–25) or threaded sleeves. The use of such splices is governed by ACI Secs. 12.14.3 and 12.16.3. Descriptions of some commercially available splices are given in Ref. 8–14.

PROBLEMS

8–1 Figure P8–1 shows a cantilever beam with $b = 12$ in. containing 3 No. 7 bars which are anchored in the column by standard 90 hooks. $f'_c = 5000$ psi and $f_y = 60,000$ psi. Assuming that the steel is stressed to f_y at the face of the column, can these bars

   (a) be anchored by hooks into the column. The clear cover to the side of the hook is $2\frac{3}{4}$ in. The clear cover to the bar extension beyond the bend is $1\frac{7}{8}$ in. The joint is enclosed by ties at 6 in. o.c.

   (b) be developed in the beam. The bar ends 2 in. from the end of the beam. The beam has No. 3 double leg stirrups at 7.5 in.

8–2 Give two reasons why the tension development length is longer than the compression development length.

8–3 Why do bar spacing and cover to the surface of the bar affect bond strength?

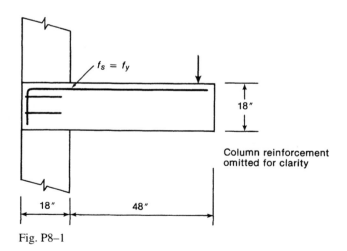

Fig. P8–1

8–4 A simply supported rectangular beam with $b = 14$ in. and $d = 17.5$ in. and No. 3 minimum stirrups spans 14 ft and supports a total factored uniform load of 6.5 kips/ft, including its own dead load. It is built of 3000-psi concrete and contains 2 No. 10

Grade 60 bars which extend 5 in. past the centers of the supports at each end and No. 3 minimum stirrups. Does this beam satisfy ACI Sec. 12.11.3? If not, what is the largest size bars which can be used?

8–5 Why do ACI Secs. 12.10.3 and 12.12.3 require that bars extend d past their flexural cutoff points?

8–6 Why does ACI Sec. 12.10.2 define "points within the span where adjacent reinforcement terminates" as critical sections for development of reinforcement in flexural members?

8–7 A rectangular beam with cross section $b = 14$ $in.$, $h = 24$ in., and $d = 21.7$ in. supports a total factored load of 3.9 kips/ft, including its own dead load. The beam is simply supported with a 22-ft span. It is reinforced with 6 No. 6 Grade 60 bars, two of which are cut off between midspan and the support and four of which extend 12 in. past the centers of the supports. $f'_c = 4000$ psi. The beam has No. 3 stirrups satisfying ACI Sec. 11.5.4 and 11.5.5.3.

    (a) Plot to scale the factored moment diagram. $M = w\ell x/2 - wx^2/2$, where x is the distance from the support and ℓ is the span.

    (b) Redraw this diagram shifted a distance d toward the supports as shown in Fig. 8–19a.

    (c) Plot a resisting moment diagram and locate the cutoff points for the two cutoff bars.

8–8 Why does ACI Sec. 12.10.5 require extra stirrups at bar cutoff points in some cases?

The beam shown in Fig. P8–9 is built of 3000-psi concrete and Grade 60 steel. The effective depth $d = 18.63$ in. The beam supports a total factored uniform load of 5.25 kips/ft, including its own dead load. The frame is not part of the lateral load-resisting system for the building. Use Figs. A–1 to A–4 to select cutoff points in Problems 8–9 to 8–11.

8–9 Select cutoff points for span AB based on the following requirements:

    (a) Extend all positive moment bars into the columns before cutting them off.

    (b) Extend all negative moment bars past the negative moment point of inflection before cutting them off.

    (c) Check the anchorage of the negative moment bars at A and modify the bar size if necessary.

    (d) Compare the answer to Fig. A–5b.

8–10 Repeat Problem 8–9(a) and (b) for span BC.

8–11 Select cutoff points for span AB based on the following requirements:

    (a) Extend all negative moment bars at A past the negative moment point of inflection.

    (b) Cut off the 2 No. 6 positive moment bars when no longer needed at each end. Extend the remaining bars into the columns.

    (c) Cut off two of the negative moment bars at B when no longer needed. Extend the remaining bars past the point of inflection.

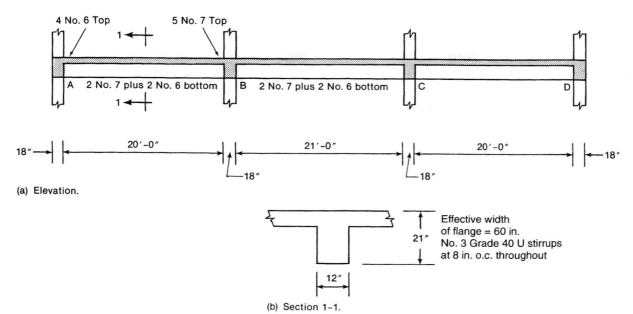

(a) Elevation.

(b) Section 1-1.

Effective width
of flange = 60 in.
No. 3 Grade 40 U stirrups
at 8 in. o.c. throughout

Fig. P8–9

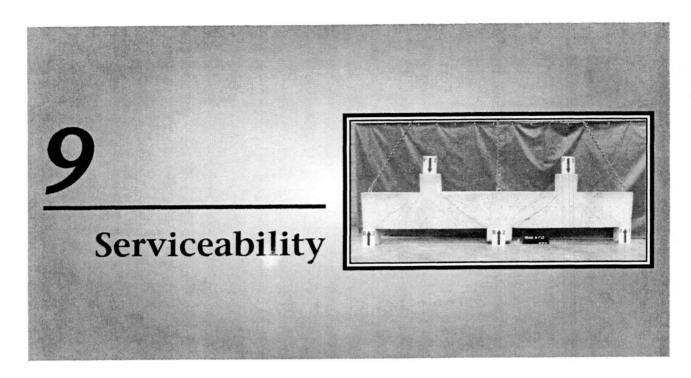

9
Serviceability

9–1 INTRODUCTION

In Chap. 2, limit states design was discussed. The limit states (states at which the structure becomes unfit for its intended function) were divided into two groups: those leading to collapse and those that disrupt the use of structures but do not cause collapse. These were referred to as *ultimate limit states* and *serviceability limit states,* respectively. The major serviceability limit states for reinforced concrete structures are: excessive crack widths, excessive deflections, and undesirable vibrations. These are discussed in this chapter. Although fatigue is an ultimate limit state, it occurs at service loads and is considered here.

Historically, deflections and crack widths have not been a problem for reinforced concrete building structures. With the advent of strength design and Grade 60 reinforcement, however, the reinforcement stresses at service loads have increased by about 50%. Since crack widths, deflections, and fatigue are all related to steel stress, each of these has become more critical.

9–2 ELASTIC ANALYSIS OF BEAM SECTIONS

At service loads the distribution of stresses in the compression zone of a cracked beam is close to being linear, as shown in Fig. 4–8, and the steel is elastic. As a result, an elastic calculation gives a good estimate of the concrete and steel stresses at service loads. This type of analysis is also referred to as a *straight-line theory* analysis since a linear stress distribution is assumed. The straight-line theory is used in calculating the stiffness EI at service loads for deflection calculations and the steel stresses for use in crack width or fatigue calculations.

Calculation of *EI*

Modulus of Elasticity and Modular Ratio

The ACI Code assumes that concrete has a modulus of elasticity of (ACI Sec. 8.5.1)

$$E_c = w_c^{1.5} \, (33 \sqrt{f_c'}) \text{ psi} \qquad (9\text{--}1)$$

where w_c is the unit weight of concrete, lb/ft^3. For normal-weight concrete this reduces to $57,000 \sqrt{f_c'}$ psi. This is the modulus of elasticity of concrete stressed to working stress levels. As discussed in Sec. 3–3, E_c is also affected strongly by the modulus of elasticity of the aggregate. In critical situations it may be desirable to allow for this. Reinforcing steel has a modulus of elasticity of $E_s = 29 \times 10^6$ psi.

The ratio E_s/E_c is referred to as the *modular ratio, n,* and has values ranging from $n = 9.3$ for 3000-psi concrete to 6.6 for 6000-psi concrete. This means that for a given strain less than the yield strain, the stress in steel will be six to nine times that in concrete subjected to the same strain.

Transformed Section

At service loads, the beam is assumed to act elastically. The basic assumptions in elastic bending are that strains are linearly distributed over the depth of the member and that the stresses can be calculated from the strains using the relationship $\sigma = E\epsilon$. This leads to the elastic bending equation, $\sigma = My/I$. When a beam made of two materials is loaded, the different values of E for the two materials lead to a different stress distribution since one material is stiffer and accepts more stress for a given strain than the other. However, the elastic beam theory can be used if the beam is hypothetically *transformed* to either an all-steel beam or an all-concrete beam, customarily the latter. This is done by replacing the area of steel with an area of concrete having the same axial stiffness AE. Since $E_s/E_c = n$, the resulting area of concrete will be nA_s. This transformed area is assumed to be concentrated at the same point in the cross section as the real steel area.

When the steel is in a compression zone or in an uncracked tension zone, its transformed area is nA_s, but it displaces an area of concrete equal to A_s. As a result, compression steel is transformed to an equivalent concrete area of $(n - 1)A_s'$. [In the days of working stress design, compression steel was transformed to an equivalent concrete area of $(2n - 1)A_s'$ to reflect the effect of creep on the stresses.]

Prior to flexural cracking, the beam shown in Fig. 9–1a has the transformed section in Fig. 9–1b. Since the steel is displacing concrete which could take stress, the transformed area is $(n - 1)A_s$ for both layers of steel. The cracked transformed section is shown in Fig. 9–1c. Here the steel in the compression zone displaces stressed concrete and has the transformed area of $(n - 1) A_s'$, while that in the tension zone does not and hence has an area of nA_s.

The neutral axis of the cracked section occurs at a distance $c = kd$ below the top of the section. For an elastic section, the neutral axis occurs at the centroid of the area, which is defined as that point where

$$\Sigma A_i \bar{y}_i = 0 \qquad (9\text{--}2)$$

where $\bar{y}_i$ is the distance from the centroidal axis to the centroid of the ith area. The solution of Eq. 9–2 is illustrated in Example 9–1.

EXAMPLE 9–1 Calculation of the Transformed Section Properties

The beam shown in Fig. 9–1 is built of 4000-psi concrete. Compute the location of the centroid and the moment of inertia for both the uncracked section and the cracked section.

UNCRACKED TRANSFORMED SECTION

The modulus of elasticity of the concrete is

$$E_c = 57,000 \sqrt{f_c'} = 3.605 \times 10^6 \text{ psi}$$

The modular ratio is

$$n = \frac{E_s}{E_c}$$

$$= \frac{29 \times 10^6}{3.605 \times 10^6} = 8.04$$

Since all the steel is in uncracked parts of the beam, the transformed areas of the two layers of steel are

$$\text{Top steel:} \quad (8.04 - 1) \times 1.20 = 8.45 \text{ in.}^2$$

$$\text{Bottom steel:} \quad (8.04 - 1) \times 2.40 = 16.91 \text{ in.}^2$$

See Fig. 9–1b.

The centroid of the transformed section is located at:

Part	Area (in.²)	y_{top} (in.)	Ay_t (in.³)
Concrete	$12 \times 24 = 288$	12	3456
Top steel	8.45	2.5	21.1
Bottom steel	16.91	21.5	363.6
	313.4		$\Sigma Ay_t = 3840.7$
		$\bar{y}_{top} = \dfrac{3840.7}{313.4} = 12.26 \text{ in.}$	

Therefore, the centroid is 12.26 in. below the top of the section. The moment of inertia is:

Part	Area (in.²)	$\bar{y}$ (in.)	$I_{\text{own axis}}$ (in.⁴)	$A\bar{y}^2$ (in.⁴)
Concrete	288	−0.26	13,824	19.5
Top steel	8.45	9.76	—	805
Bottom steel	16.91	−9.24	—	1440
			$I_{gt} = 16{,}090 \text{ in.}^4$	

The uncracked transformed moment of inertia $I_{gt} = 16,090$ in.⁴. This is 16% larger than the gross moment of inertia of the concrete alone, referred to as I_g. The moments of inertia of the steel areas about their own centroidal axes are small and were neglected.

CRACKED TRANSFORMED SECTION

Assume that the neutral axis is lower than the top steel. The transformed areas are

$$\text{Top steel:} \quad (8.04 - 1) \times 1.20 = 8.45 \text{ in.}^2$$

$$\text{Bottom steel:} \quad 8.04 \times 2.40 = 19.30 \text{ in.}^2$$

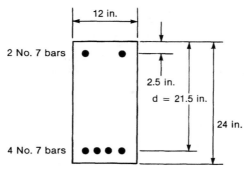

2 No. 7 bars

4 No. 7 bars

d = 21.5 in.

2.5 in.

12 in.

24 in.

(a) Cross section.

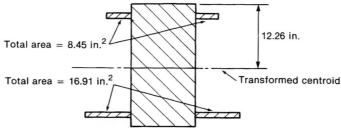

Total area = 8.45 in.^{2}

Total area = 16.91 in.^{2}

Transformed centroid

12.26 in.

(b) Uncracked transformed section.

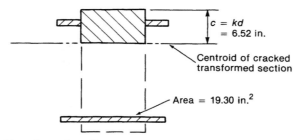

$c = kd$
$= 6.52$ in.

Centroid of cracked transformed section

Area = 19.30 in.^{2}

Fig. 9–1
Transformed sections—
Example 9–1.

(c) Cracked transformed section.

Let the depth to the neutral axis be c, as shown in Fig. 9–1c and sum the moments of the areas about the neutral axis to compute c:

Part	Area (in.^{2})	$\bar{y}$ (in.)	$A\bar{y}$ (in.^{3})
Compression zone	$12c$	$c/2$	$12c^2/2 = 6c^2$
Top steel	8.45	$c - 2.5$	$8.45c - 21.13$
Bottom steel	19.30	$c - 21.5$	$19.30c - 414.95$

But by definition c is the distance to the centroid when $\Sigma A\bar{y} = 0$. Therefore, $6c^2 + 27.75c - 436.08 = 0$ and

$$c = \frac{-27.75 \pm \sqrt{27.75^2 + 4 \times 6 \times 436.08}}{2 \times 6}$$

$$= 6.52 \text{ in. or } -11.5 \text{ in.}$$

9–2 Elastic Analysis of Beam Sections

The positive value lies within the section. Since the top steel is in the compression zone, the initial assumption is OK. Therefore, **the centroidal (neutral) axis is at 6.52 in. below the top of the section.**

Compute the moment of inertia:

Part	Area (in.^{2})	y (in.)	$I_{own\ axis}$ (in.^{4})	Ay^2 (in.^{4})
Compression zone	$12 \times 6.52 = 78.25$	6.52/2	277	832
Top steel	8.45	$(6.52 - 2.5) = 4.02$	—	137
Bottom steel	19.30	$(6.52 - 21.5) = -14.98$	—	4330
			$I = 5580$ in.^{4}	

Thus the cracked transformed moment of inertia is $I_{cr} = 5580$ in.^{4} ∎

In this example, I_{cr} is only 35% of that of the uncracked transformed section and 40% of that of the concrete section alone. This illustrates the great reduction in stiffness due to cracking.

The procedure followed in Example 9–1 can be used to locate the neutral axis for any shape of cross section bent in uniaxial bending. For the special case of rectangular beams without compression reinforcement, Eq. 9–2 gives

$$\frac{bc^2}{2} - nA_s(d - c) = 0$$

Substituting $c = kd$ and $\rho = A_s/bd$ gives

$$\frac{b(kd)^2}{2} - \rho nbd(d - kd) = 0$$

Dividing by bd^2 and solving for k gives

$$k = \sqrt{2\rho n + (\rho n)^2} - \rho n \qquad (9\text{--}3)$$

where $\rho = A_s/bd$ and $n = E_s/E_c$. Equation 9–3 can be used to locate the neutral-axis position directly, thus simplifying the calculation of the moment of inertia. This equation applies only for rectangular beams without compressive reinforcement.

Service Load Stresses in a Cracked Beam

The compression stresses in the beam shown in Fig. 9–2 vary linearly from zero at the neutral axis to a maximum stress of f_c at the extreme compression fiber. The total compressive force C is

$$C = \frac{f_c bkd}{2}$$

This force acts at the centroid of the triangular stress block or $kd/3$ from the top. For the straight-line stress distribution the lever arm jd is

$$jd = d - \frac{kd}{3} = d\left(1 - \frac{k}{3}\right)$$

If the moment at service loads is M_s, we can write

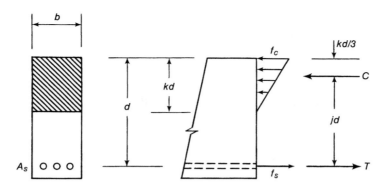

Fig. 9–2
Stress distribution in straight-
line theory.

$$M_s = Cjd = \frac{f_c bkd}{2} jd$$

and

$$f_c = \frac{2M_s}{jkbd^2} \qquad (9\text{–}4)$$

Similarly, taking moments about C yields

$$M_s = Tjd = f_s A_s jd$$

and

$$f_s = \frac{M_s}{A_s jd} \qquad (9\text{–}5)$$

This analysis ignores the effects of creep which will tend to increase the stress in the tension steel a small amount.

EXAMPLE 9–2   Calculation of the Service Load Steel Stress in a Rectangular Beam

A rectangular beam similar to the one shown in Fig. 9–2 has $b = 10$ in.., $d = 20$ in., 3 No. 8 Grade 60 bars, and $f_c' = 3000$ psi. Compute f_s at service loads if the service live-load moment is 50 ft-kips and the service dead load moment is 70 ft-kips.

    1.   Compute k and j.

$$E_c = 57{,}000 \sqrt{3000} = 3.12 \times 10^6 \text{ psi} \qquad (9\text{–}3)$$

$$n = \frac{E_s}{E_c}$$

$$= \frac{29 \times 10^6}{3.12 \times 10^6} = 9.29$$

$$\rho = \frac{A_s}{bd}$$

$$= \frac{2.37}{10 \times 20} = 0.0019$$

$$\rho n = 0.110$$

$$k = \sqrt{2\rho n + (\rho n)^2} - \rho n$$

$$= \sqrt{2 \times 0.110 + 0.110^2} - 0.110$$

$$= 0.372$$

$$j = 1 - \frac{k}{3} = 0.876$$

2. **Compute f_s at M_s.**

$$M_s = 50 + 70 = 120 \text{ ft-kips} \qquad (9\text{-}5)$$

$$= 1440 \text{ in.-kips}$$

$$f_s = \frac{M_s}{A_s jd}$$

$$= \frac{1440}{2.37 \times 0.876 \times 20} = 34.7 \text{ ksi}$$

The steel stress at service loads is 34.7 ksi. ∎

Age-Adjusted Transformed Section

Creep causes an increase in the compressive strains in a beam and a resulting drop in the neutral axis and an increase in the strains in the tension steel. The concept of age-adjusted transformed sections, introduced in Sec. 3–4 for axially loaded members, can be used to estimate stresses and deflections in beams subjected to sustained loads.[9-1 to 9-3]

9–3   CRACKING

Types of Cracks

Tensile stresses induced by loads, moments, and shears cause distinctive crack patterns, as shown in Fig. 9–3. Members loaded in direct tension crack right through the entire cross section, with a crack spacing ranging from 0.75 to 2 times the minimum thickness of the member. In the case of a very thick tension member with reinforcement in each face, small surface cracks develop in the layer containing the reinforcement (Fig. 9–3a). These join in the center of the member. As a result, for a given total change in length, the crack width at B is greater than at A.

Members subjected to bending moments develop flexural cracks as shown in Fig. 9–3b. These vertical cracks extend almost to the zero-strain axis (neutral axis) of the member. In a beam with a web that is more than 3 to 4 ft high, the cracking is relatively closely spaced at the level of the reinforcement, with several cracks joining or disappearing above the reinforcement as shown in Fig. 9–3b. Again, the crack width at B will frequently exceed that at A. The cracks in the middle third of the beam shown in Figs. 4–5 and 4–6 are flexural cracks.

Cracks due to shear have a characteristic inclined shape as shown in Fig. 9–3c and Figs. 6–4 and 6–12. Such cracks extend upward as high as the neutral axis and sometimes into the compression zone. Torsion cracks are similar. In pure torsion, they spiral around the beam. In a normal beam where shear and moment also act, they tend to be pronounced on the face where the direct shear stresses and the shear stresses due to torsion add, and less pronounced or absent on the opposite face, where the stresses counteract (Fig. 9–3d).

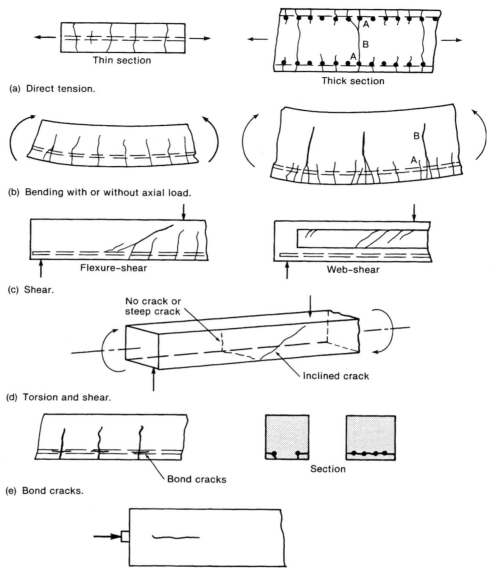

(a) Direct tension.

(b) Bending with or without axial load.

(c) Shear.

(d) Torsion and shear.

(e) Bond cracks.

(f) Concentrated load.

Fig. 9–3
Load-induced cracks.
(From Ref. 9–4.)

Bond stresses lead to splitting along the reinforcement as shown in Fig. 9–3e. Concentrated loads will sometimes cause splitting cracks or "bursting cracks" of the type shown in Fig. 9–3f.

At service loads, the final cracking pattern has generally not developed completely, with the result that there are normally only a few cracks at points of maximum stress at this load level.

Cracks also develop due to imposed deformations such as differential settlements, shrinkage, and temperature differentials. If shrinkage is restrained, as in the case of a thin floor slab attached at each end to a massive stiff structure, shrinkage cracks may occur. Generally, however, shrinkage simply increases the width of load-induced cracks.

A frequent cause of cracking in structures is restrained contraction resulting from the cooling down to ambient temperatures of very young members which have expanded due to the heat of hydration which developed as the concrete was setting. This most typically occurs where a length of wall is cast on a foundation which was cast some time before. As the wall

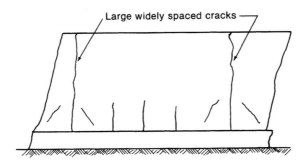

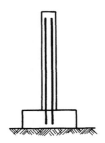

Fig. 9–4
Heat of hydration cracking.

cools, its contraction is restrained by the foundation. A typical *heat of hydration cracking* pattern is shown in Fig. 9–4. Such cracking can be controlled by controlling the heat rise due to the heat of hydration and the rate of cooling, or both; by placing the wall in short lengths; or by reinforcement considerably in excess of normal shrinkage reinforcement.[9-5, 9-6]

Plastic shrinkage and slumping of the concrete, which occurs as newly placed concrete bleeds and the surface dries, result in settlement cracks along the reinforcement, Fig. 9–5a, or a random cracking pattern, referred to as *map cracking,* Fig. 9–5b. These types of cracks can be avoided by proper mix design and by preventing rapid drying of the surface during the first hour or so after placing. Map cracking can also occur due to alkali-aggregate reaction.

Rust occupies two to three times the volume of the metal from which it is formed. As a result, if rusting occurs, a bursting force is generated at the bar location which leads to splitting cracks and an eventual loss of cover (Fig. 9–5c). Such cracking looks similar to bond cracking (Fig. 9–3e) and may accompany bond cracking. The various mechanisms of cracking are discussed in detail in Refs. 9–5 to 9–7.

Development of Cracks Due to Loads

Figure 9–6 shows an axially loaded prism. Cracking starts when the tensile stress in the concrete (shown by the shaded area in Fig. 9–6b) reaches the tensile strength of the concrete (shown by the outer envelope) at some point in the bar. When this occurs, the prism cracks. At the crack, the entire force in the prism is carried by the reinforcement. Bond gradually builds up the stress in the concrete on either side of the crack until, with further loading, the stress reaches the tensile strength at some other section, which then cracks (Fig. 9–6c). With increasing load, this process continues until the distance between the cracks is not big enough for the tensile stress in the concrete to increase

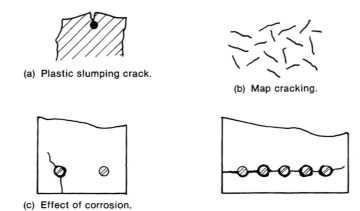

(a) Plastic slumping crack.

(b) Map cracking.

(c) Effect of corrosion.

Fig. 9–5
Other types of cracks.

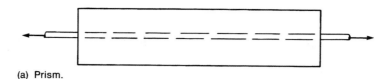

(a) Prism.

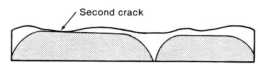

(b) Variation of tensile strength and stress along prism.

(c) Tensile stresses after first crack.

(d) Tensile stresses after three cracks.

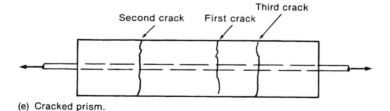

(e) Cracked prism.

Fig. 9–6
Cracking of an axially loaded prism.

enough to cause cracking. Once this stage is reached, the crack pattern has stabilized and further loading merely widens the existing cracks. The distance between stabilized cracks is a function of the overall member thickness, the cover, the efficiency of the bond, and several other things. Roughly, however, it is two to three times the bar cover. Cracks that extend completely through the member generally occur at roughly one member thickness apart.

Figure 9–7b and c show the variation in the steel and concrete stresses along an axially loaded prism with a stabilized crack pattern. At the cracks, the steel stress and strain are at a maximum and can be computed from a cracked section analysis. Between the cracks, there is stress in the concrete. This reaches a maximum midway between two cracks. The total width, w, of a given crack is the difference in the elongation of the steel and the concrete over a length A–B equal to the crack spacing:

$$w = \int_A^B (\epsilon_s - \epsilon_c)\, dx \qquad (9\text{–}6)$$

9–3 Cracking

345

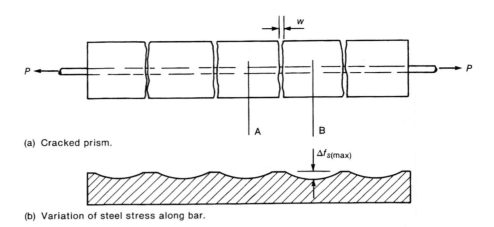

(a) Cracked prism.

(b) Variation of steel stress along bar.

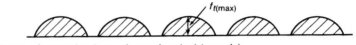

Fig. 9–7
Stresses in concrete and steel
in a cracked prism.

(c) Variation of concrete stress along prism (not to scale).

where ϵ_s and ϵ_c are the strains in the steel and concrete at a given location between A and B and x is measured along the axis of the prism.

The crack spacing, s, and the variation in ϵ_s and ϵ_c are difficult to determine in practice and empirical equations are generally used to compute the crack width. The best known of these are the Euro-International Concrete Committee (CEB) procedure[9–7] based on Eq. 9–6 and the Gergely–Lutz equation,[9–8] derived statistically from a number of test series.

As discussed in Chap. 8, bond stresses are transferred from steel to concrete by means of forces acting on the deformation lugs on the surface of the bar. These lead to cracks in the concrete adjacent to the ribs as shown in Fig. 9–8. In addition, the tensile stress in the concrete decreases as one moves away from the bar, leading to less elongation of the surface of the concrete than of the concrete at the bar. As a result, the crack width at the surface of the concrete exceeds that at the bar.

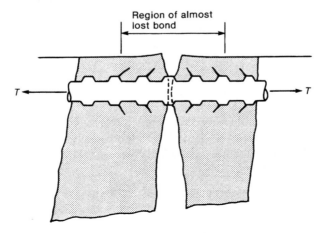

Fig. 9–8
Cracking at bar deformations.
(From Ref. 9–4.)

Serviceability

In a deep flexural member, the distribution of crack widths over the depth shows a similar effect, particularly if several cracks combine as shown in Fig. 9–3b. The crack width at B frequently exceeds that at the level of the reinforcement.

Reasons for Controlling Crack Widths

Crack widths are of concern for three main reasons: appearance, leakage, and corrosion. Wide cracks are unsightly and sometimes lead to concern by owners and occupants. Crack surveys reported in Ref. 9–7 suggest that on easily observed, clean, smooth surfaces, cracks wider than 0.01 to 0.013 in. can lead to public concern. Wider cracks are tolerable if the surfaces are less easy to observe or are not smooth. On the other hand, cracks in exposed surfaces may be accentuated by streaks of dirt or leached materials. Sandblasting the surface will accentuate the apparent width of cracks, making small cracks much more noticeable.

Crack control is important in the design of liquid-retaining structures. Leakage is a function of crack width. Corrosion of reinforcement has traditionally been related to crack width. Modern studies[9-9] suggest that the factors governing the eventual development of corrosion are independent of the crack width, although the period of time required for corrosion to start is a function of crack width. Reinforcement surrounded by concrete will not corrode until an electrolytic cell can be established. This will occur when carbonization of the concrete reaches the steel, or when chlorides penetrate through the concrete to the bar surface. The time taken for this to occur will depend on whether or not the concrete is cracked, the environment, the thickness of the cover, and the permeability of the concrete. If the concrete is cracked, the time required for a corrosion cell to be established is a function of the crack width. After a period of 5 to 10 years, however, the amount of corrosion is essentially independent of crack width.

Generally speaking, corrosion is most apt to occur if:

1. Chlorides or other corrosive substances are present.
2. The relative humidity exceeds 60%. Below this value corrosion rarely occurs except in the presence of chlorides.
3. Ambient temperatures are high, accelerating the chemical reaction.
4. Wetting and drying cycles occur such that the concrete at the level of the steel is alternately wet and dry. Corrosion does not occur in permanently saturated concrete because the water prevents oxygen flow to the steel.
5. Stray electrical currents occur in the bars.

Corrosion of steel embedded in concrete is discussed in Ref. 9–10. ACI Sec. 4.4 deals with corrosion.

Limits on Crack Width

There are no universally accepted rules for maximum crack widths. The ACI Code crack control limits are based on a maximum crack width of 0.016 in. for interior exposure and 0.013 in. for exterior exposure. What constitutes interior and exterior exposure is not defined. In addition to crack control provisions, there are special requirements in ACI Chap. 4 for the composition of concrete subjected to special exposure conditions.

The Euro-International Concrete Committee (CEB)[9-7, 9-11] limits the mean crack width (about 60% of the maximum crack width) as a function of exposure condition, sensitivity of reinforcement to corrosion, and duration of the loading condition.

ACI Code Crack Control Provisions

Crack control is handled indirectly in the ACI Code by defining rules for the distribution of reinforcement in beams and one-way slabs. These are based on the Gergely–Lutz equation,

$$w = 0.76 \, \beta f_s \sqrt[3]{d_c A} \qquad (9\text{--}7)$$

where

w = crack width in units of 0.001 in.

β = distance from the neutral axis to the bottom fiber, divided by the distance to the reinforcement

f_s = service load stress in the reinforcement, ksi

d_c = distance from the extreme tension fiber to the center of the reinforcing bar located closest to it, in.

A = effective tension area of concrete surrounding the tension reinforcement, and having the same centroid as that reinforcement, divided by the number of bars or wires, in.^{2}

The definition of d_c and A is illustrated in Example 9–3.

Equation 9–7 was derived statistically from the maximum crack widths at the flexural tension surface of members as reported by a number of investigators[9-8] and is intended to predict the "maximum crack width." Because the scatter in crack widths is large, there will always be some cracks wider than predicted by Eq. 9–7. Ten percent of the measured crack widths used to derive Eq. 9–7 exceeded 1.5 times the value given by the equation, while 2% were less than 0.5 times the calculated width.

Because of the scatter in crack widths, the ACI Code does not limit crack width per se, but rather limits the magnitude of the term:

$$z = f_s \sqrt[3]{d_c A} \qquad (9\text{--}8)$$

Rewriting Eq. 9–7 gives $w = 0.076\beta z$. For interior and exterior exposures, the critical crack widths were taken equal to 0.016 in ($w = 16$) and 0.013 in. ($w = 13$), respectively, and β was taken equal to 1.2. This gave values of z of 175 kips/in. and 145 kips/in., respectively, for these two classes of exposure (ACI Sec. 10.6.4). For thin one-way slabs, the ratio β will be larger than 1.2 and for such members, the ACI Commentary suggests a value of β of 1.35, leading to values of z that are 1.2/1.35 times the code values. Thus for slabs, the values of z should become roughly 155 and 130 kips/in. for interior and exterior exposure, respectively.

The steel stress f_s can be computed using Eq. 9–5. Alternatively, ACI Sec. 10.6.4 allows the value of f_s at service loads to be taken as $0.60f_y \approx f_y/(1.55/0.9)$, where $(1.55/0.9)$ represents the average load factor divided by the strength reduction factor, ϕ, for flexure. For the beam considered in Example 9–2, this is close to the computed value.

EXAMPLE 9–3   Checking the Distribution of Reinforcement in a Beam

At the point of maximum positive moment, a beam contains the reinforcement shown in Fig. 9–9. The reinforcement has a yield strength of 60,000 psi. Is this distribution satisfactory for exterior exposure?

For exterior exposure, $z = f_s \sqrt[3]{d_c A}$ must not exceed 145 kips/in.

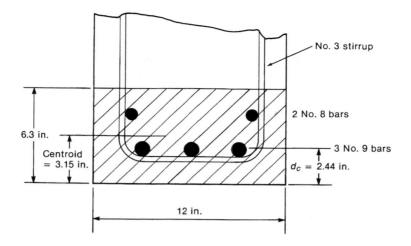

Fig. 9–9
Beam—Example 9–3.

$$f_s = 0.6f_y \text{ ksi} = 36 \text{ ksi}$$

$$d_c = 1.5 \text{ in.} + 0.375 \text{ in.} + (0.5 \times \tfrac{9}{8} \text{ in.})$$

$$= 2.44 \text{ in.}$$

The centroid of the bar group is located at 3.15 in. above the extreme tension fiber (bottom fiber) as calculated in Example 4-4. Thus the area of concrete having the same centroid as the bars is the area shown shaded in Fig. 9–9.

$$A = \frac{\text{shaded area}}{\text{number of bars}}$$

Since the bars are of two sizes, it is necessary to express the number of bars as an equivalent number of the larger size. Thus

$$N = \frac{A_s}{\text{Area of No. 9 bar}}$$

$$= \frac{4.58}{1.0}$$

$$= 4.58$$

$$A = \frac{6.3 \text{ in.} \times 12 \text{ in.}}{4.58}$$

$$= 16.51 \text{ in.}^2$$

and

$$z = 36 \text{ ksi} \sqrt[3]{2.44 \text{ in.} \times 16.51 \text{ in.}^2}$$

$$= 123 \text{ kips}/\text{in.}$$

Thus this steel distribution is acceptable. ∎

In theory this should be checked at every section of maximum positive and negative moment. In practice, however, it is normally necessary to check z only at those positive and negative moment sections having the smallest numbers of bars, N, since A will be a maximum at these locations.

9–3 Cracking

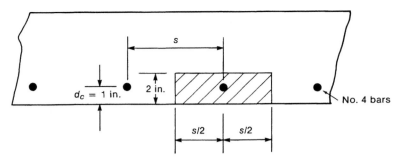

Fig. 9–10
Slab—Example 9–4.

In one-way slabs, the flexural reinforcement distribution is frequently determined by checking whether the bar spacing is less than that corresponding to the maximum value of z. The maximum bar spacing is calculated in Example 9–4. It is not necessary to compute z for two-way slabs (ACI Secs. 10.6.1 and 10.6.2).

EXAMPLE 9–4 Calculation of the Maximum Bar Spacing for Bars in a One-Way Slab

Figure 9–10 shows a section through a slab with No. 4 bars at a spacing, s. The bars have $f_y = 60,000$ psi and a minimum cover of $\frac{3}{4}$ in. Compute the maximum value of s for interior exposure.

For interior exposure, ACI Sec. 10.6.4 gives $z \leq 175$ kips/in. The ACI Commentary suggests that this value be reduced to 155 kips/in. for slabs as noted earlier. Thus $z = f_s \sqrt[3]{d_c A}$ should not exceed 155, where

$$d_c = 1.0 \text{ in.}$$

$$A = 2 \times 1.0 \times s$$

$$f_s = 36 \text{ ksi}$$

and

$$155 = 36\sqrt[3]{1.0 \times 2.0s}$$

The maximum spacing $s = 39.9$ in. But maximum spacing is governed by ACI Sec. 7.6.5, which limits s to three times the slab thickness, or 18 in. ∎

In general, ACI Sec. 7.6.5 will govern the bar spacing in slabs for interior exposure. For exterior exposure additional cover is required and for this exposure case, the crack control provisions govern. For very thick cover as in footings, ACI Sec. 10.6.4 breaks down. In such cases it is customary to disregard ACI Sec. 10.6.4.

In the negative moment regions of T beams, the flanges will be stressed in tension and will crack as shown in Fig. 5–2. To restrict the width of the cracks in the flanges, ACI Sec. 10.6.6 requires that "part" of the flexural tension reinforcement be distributed over a width equal to the smaller of the effective flange width or $\ell_n / 10$, and requires that "some" longitudinal reinforcement be provided in the outer portions of the flange. The terms "part" and "some" are not defined. This can be accomplished by placing roughly one-fourth to one-half the reinforcement in the overhanging portions of the flange at, or near, the maximum spacing for crack control, and placing the balance over the web of the beam.

Web Face Reinforcement

In beams deeper than about 3 ft, the widths of flexural cracks may be as large, or larger, at points above the reinforcement than they are at the level of the steel as shown in Fig. 9–3b.

To control the width of these cracks, ACI Sec. 10.6.7 requires additional reinforcement near the side faces of the beam. The area of skin reinforcement, A_{sk}, per foot of height on each side face shall be

$$A_{sk} \geq 0.012(d - 30) \text{ in.}^2 \qquad (9\text{--}9)$$

The maximum spacing of the skin reinforcement shall not exceed the smaller of $d/6$ or 12 in. Where other code provisions require more steel those provisions and their spacings shall govern. These requirements grew out of research reported in Ref. 9–12.

For large beams this can be a significant amount of reinforcement. If strain compatibility is used to determine the stresses in the web face steel, the effect of this steel on ϕM_n can be included in design (see Sec. 5–5).

9–4 DEFLECTIONS: RESPONSE OF CONCRETE BEAMS

In this section we deal with the flexural deflections of beams or slabs. Deflections of frames are considered in Sec. 9–5.

Load–Deflection Behavior of a Concrete Beam

Figure 9–11a traces the load–deflection history of the fixed-ended, reinforced concrete beam shown in Fig. 9–11b. Initially, the beam is uncracked and is stiff (0–A). With further

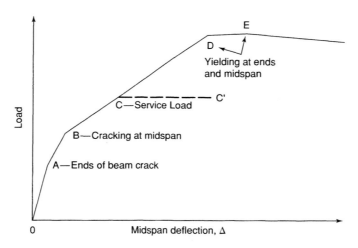

(a) Load deflection diagram.

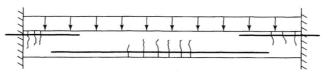

(b) Beam and loading.

Fig. 9–11
Load—deflection behavior of a concrete beam.

load, cracking occurs when the moment at the ends exceeds the cracking moment. When a section cracks, its moment of inertia decreases leading to a decrease in the stiffness of the beam. This causes a reduction in stiffness (A–B) in the load–deflection diagram in Fig. 9–11a. Cracking in the midspan region causes a further reduction of stiffness (point B). Eventually, the reinforcement would yield at the ends, or at midspan, leading to large increases in deflection with little change in load (points D and E). The service load level is represented by point C. The beam is essentially elastic at point C, the nonlinear load deflection being caused by a progressive reduction of flexural stiffness due to increased cracking as the loads are increased.

With time, the service load deflection would increase from C to C', due to creep of the concrete. The short-time, or instantaneous, deflection under service loads (point C) and the long-time deflection under service loads (point C') are both of interest in design.

Flexural Stiffness and Moment of Inertia

The deflection of a beam is calculated by integrating the curvatures along the length of the beam. For an elastic beam, the curvature, ϕ, is calculated as $\phi = M/EI$, where EI is the flexural stiffness of the cross section. If EI is constant, this is a relatively routine process. For reinforced concrete, however, three different EI values must be considered. These can be illustrated using a moment curvature diagram for a length of beam, including several cracks, shown in Fig. 9–12d. The slope of any radial line through the origin in such a diagram is $M/\phi = EI$.

Before cracking, the entire cross section shown in Fig. 9–12b is stressed due to loads. The moment of inertia of this section is called the uncracked moment of inertia and the corresponding EI can be represented by the radial line O–A in Fig. 9–12d. The cross section at a crack is shown in Fig. 9–12c. This has a much lower moment of inertia than the uncracked section. The EI calculated using the cracked moment of inertia is less than the uncracked EI, and corresponds relatively well to the curvatures at loads approaching yield, as shown by the radial line O–B in Fig. 9–12d. At service loads, points C_1 and C_2 in Fig. 9–12d, the EI values are between these two extremes. The actual EI at service load levels varies considerably, as shown by the difference in the slope of the lines O–C_1 and O–C_2, depending on the relative magnitudes of the cracking moment, M_{cr}, the service load moment, M_a, and the yield moment, M_y. The variation in EI with moment is shown in Fig. 9–12e, obtained from Fig. 9–12d.

The transition from uncracked to cracked moment of inertia reflects two different phenomena. Figure 9–7b and c show the tensile stresses in the reinforcement and the concrete in a prism. At loads only slightly above the cracking load, a significant fraction of tensile force between cracks is in the concrete and hence the member behaves more like an uncracked section than a cracked section. As the loads are increased, internal cracking of the type shown in Fig. 9–8 occurs, with the result that the steel strains increase with no significant change in the tensile force in the concrete. At very high loads, the tensile force in the concrete is insignificant compared to that in the steel, and hence the member approaches a completely cracked section. The effect of the tensile forces in the concrete on EI is referred to as *tension stiffening*.

Figure 9–13 shows the distribution of EI along the beam shown in Fig. 9–11b. The EI varies from the uncracked value at points where the moment is less than the cracking moment, to a partially cracked value at points of high moment. Since the use of such a distribution of EI values would make the deflection calculations tedious, an overall average or *effective EI* value is used. The effective moment of inertia must account for both the tension stiffening and the variation of EI along the member.

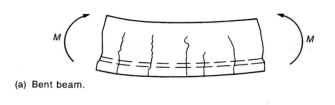

(a) Bent beam.

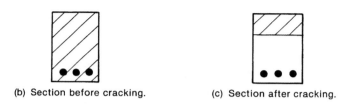

(b) Section before cracking.　　　(c) Section after cracking.

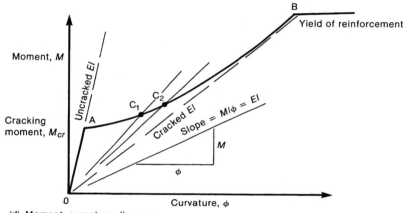

(d) Moment–curvature diagram.

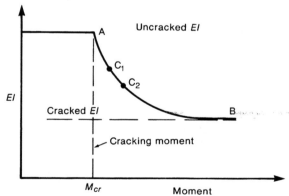

(e) Variation in EI with moment.

Fig. 9–12
Moment—curvature diagram
and variation in EI.

Effective Moment of Inertia

The slope of line OA in Fig. 9–12d is approximately EI_{gt} while that of line OB is approximately EI_{cr}. At points between cracking (point A) and yielding of the steel (point B), intermediate values of EI exist. Branson[9-13] derived the following equation to express the transition from I_{gt} to I_{cr} that is observed in experimental data:

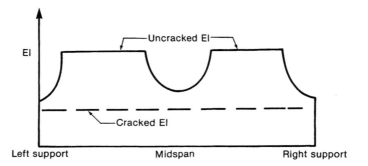

Fig. 9–13
Variation of *EI* along length of beam shown in Fig. 9–11b.

$$I_e = \left(\frac{M_{cr}}{M_a}\right)^a I_{gt} + \left[1 - \left(\frac{M_{cr}}{M_a}\right)^a\right] I_{cr}$$

where

M_{cr} = cracking moment = $f_r I_g / y_t$ (ACI Eq. 9–8)

I_g = moment of inertia of the concrete section

f_r = modulus of rupture = $7.5\sqrt{f_c'}$ (ACI Eq. 9–9)

y_t = distance from centroid to extreme tension fiber

The ACI Code defines M_a as the maximum moment in the member at the stage at which deflection is being computed. A better definition would be

M_a = maximum moment in the member at the loading stage for which the moment of inertia is being computed or at any previous loading stage

In some structures, such as two-way slabs, construction loads may exceed service loads. If the construction loads cause cracking, the effective moment of inertia will be reduced.

For a region of constant moment, Branson found the exponent a to be 4. This accounted for the tension stiffening action. For a simply supported beam, Branson suggested that both the tension stiffening and the variation in *EI* along the length of the member could be accounted for by using $a = 3$. For simplicity, the ACI Code equation is written in terms of the moment of inertia of the gross concrete section, I_g, ignoring the small increase in the moment of inertia due to the reinforcement:

$$I_e = \left(\frac{M_{cr}}{M_a}\right)^3 I_g + \left[1 - \left(\frac{M_{cr}}{M_a}\right)^3\right] I_{cr} \qquad (9–10\text{a})$$
$$(\text{ACI Eq. 9–7})$$

This can be rearranged to

$$I_e = I_{cr} + (I_g - I_{cr})\left(\frac{M_{cr}}{M_a}\right)^3 \qquad (9–10\text{b})$$

For a continuous beam, the I_e values may be quite different in the negative and positive moment regions. In such a case, the positive moment value may be assumed to apply between the points of contraflexure and the negative moment values in the end regions. ACI Sec. 9.5.2.4 suggests the use of the average I_e value. A better suggestion is given in Ref. 9–13:

Beams with two ends continuous:

$$\text{average } I_e = 0.70 I_{em} + 0.15(I_{e1} + I_{e2}) \qquad (9–11\text{a})$$

Beams with one end continuous:

$$\text{average } I_e = 0.85 I_{em} + 0.15(I_{e\,\text{continuous end}}) \qquad (9\text{–}11b)$$

where I_{em}, I_{e1}, and I_{e2} are the values of I_e at midspan and the two ends of the beam. Moment envelopes or moment coefficients should be used in computing M_a and I_e at the positive and negative moment sections.

Calculation of Deflections

When a concrete beam is loaded, it undergoes a deflection referred to as an *immediate deflection, Δ_i*. If the load remains on the beam, additional *sustained load deflections* occur due to creep. These two types of deflections are discussed separately.

Instantaneous Deflections

Equations for calculating the instantaneous deflections of beams for common cases are summarized in Fig. 9–14. In this figure M_{pos} and M_{neg} refer to the maximum positive and negative moments, respectively. The deflections listed are the maximum deflections along the beam except for the second beam and the last beam shown in Fig. 9–14.

The centerline deflection for a continuous beam with uniform loads and unequal end moments can be computed using superposition as shown in Fig. 9–15. Thus

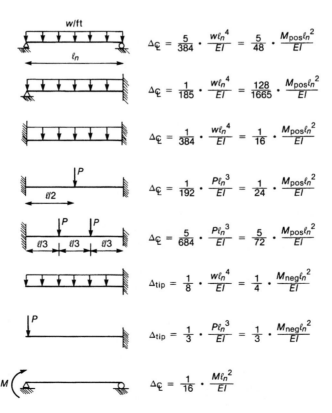

Fig. 9–14
Deflection equations.

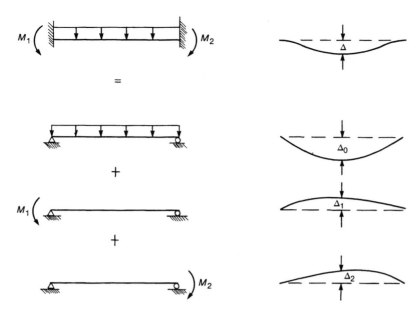

Fig. 9–15
Calculation of deflection for beam with unequal end moments.

$$\Delta = \Delta_0 + \Delta_1 + \Delta_2$$

From Fig. 9–14, assuming M_0 (the moment due to uniform loads on a simple beam), M_1, and M_2 are all positive so that the signs will be compatible:

$$\Delta = \frac{5}{48} \frac{M_0 \ell_n^2}{EI} + \frac{3}{48} \frac{M_1 \ell_n^2}{EI} + \frac{3}{48} \frac{M_2 \ell_n^2}{EI}$$

The midspan moment M_m is

$$M_m = M_0 + \frac{M_1}{2} + \frac{M_2}{2}$$

Therefore,

$$\Delta = \frac{\ell_n^2}{48\,EI}\left[5M_m - \frac{5}{2}(M_1 + M_2) + 3(M_1 + M_2)\right]$$

$$= \frac{5}{48} \frac{\ell_n^2}{EI}\big[M_m + 0.1(M_1 + M_2)\big] \qquad (9\text{–}12\text{a})$$

where Δ is the midspan deflection, M_m is the midspan moment, and M_1 and M_2 are the moments at the two ends which must be given the correct algebraic sign. Generally, M_1 and M_2 will be negative. Alternatively, Eq. 9–12a can be written

$$\Delta = \frac{5\ell_n^2}{48\,EI}(1.2M_m - 0.2M_0) \qquad (9\text{–}12\text{b})$$

The moments M_m, M_1 and M_2 in Eq. 9–12 should all result from the same loading. Equation 9–12a suggests that the loading pattern which produces maximum positive moment will give the largest deflections. In beam design, however, moment envelopes or moment coefficients are used to compute the critical moments at each section (see Sec. 10–3). These moments represent the worst effects of several different loading cases and when used in Eqs. 9–12a and 9–12b can give excessively high computed deflections. A prismatic fixed-end beam loaded with uniform load has moments of $-\frac{1}{12}w\ell^2$ and $\frac{1}{24}w\ell^2$ at the ends and

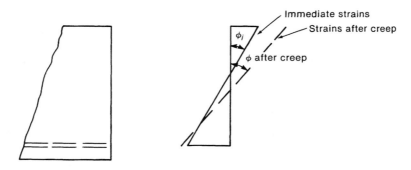

Fig. 9–16
Effect of creep on strains and curvature.

midspan. For a typical interior span ACI Sec. 8.3.3 gives moments of $-\frac{1}{11}$ and $\frac{1}{16}w\ell^2$. Using these in Eq. 9–12a gives 1.77 times the deflection computed using $-\frac{1}{12}$ and $\frac{1}{24}$, and using $\frac{1}{16}$ in Eq. 9–12b gives two times the deflection from $-\frac{1}{12}$ and $\frac{1}{24}$. Frequently, the critical deflections are sustained load deflections due to the dead load. Since dead load is not a pattern load, the appropriate moment coefficients are closer to $-\frac{1}{12}$ and $\frac{1}{24}$ than are the ACI coefficients. In conclusion, the use of the ACI moment coefficients will generally over estimate the deflections, particularly when Eq. 9–12b is used. If the ACI moment coefficients are used and the resulting computed deflections are excessive, it may be desirable to recompute the deflections using more realistic moments.

Sustained Load Deflections

Under sustained loads, concrete undergoes creep strains and the curvature of a cross section increases, as shown in Fig. 9–16. Because the lever arm is reduced, there is a small increase in the steel force, but for normally reinforced sections this will be minimal. At the same time, the compressive stress in the concrete decreases slightly because the compression zone is larger.

    If compression steel is present, the increased compressive strains will cause an increase in stress in the compression reinforcement which will shift some of the compressive force from the concrete to the compression steel. As a result, the compressive stress in the concrete decreases, resulting in reduced creep strains. The larger the ratio of compression reinforcement, $\rho' = A_s'/bd$, the greater the reduction in creep, as shown in Fig. 5–14. Based on test data, Branson[9–14] derived Eq. 9–13, which gives the ratio, λ, of the *additional* sustained load deflection to the immediate deflection. The total immediate and sustained load deflection is $(1 + \lambda)\Delta_i$:

$$\lambda = \frac{\xi}{1 + 50\rho'}$$

(9–13)

(ACI Eq. 9–10)

where $\rho' = A_s'/bd$ at midspan for simple and continuous beams and at the support for cantilever beams, and ξ is a factor between 0 and 2, depending on the time period over which sustained load deflections are of interest. Values of ξ are given in ACI Sec. 9.5.2.5 and Fig. 9–17.

Deflection Calculation by Integration of Curvatures

It is frequently more accurate to compute deflections by integrating the curvatures along the beam. This is particularly true for lightly reinforced sections with service load moments close to the cracking moment. If the curvatures, Ψ, due to loads, creep, and shrinkage are known at

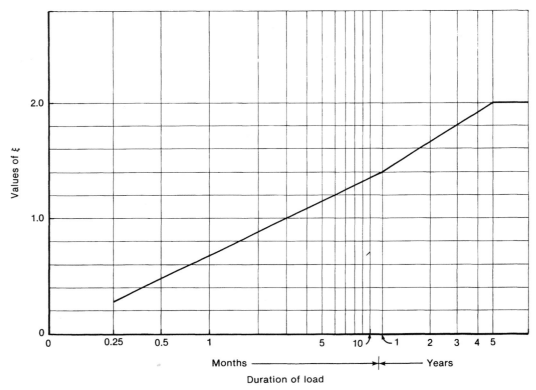

Fig. 9–17
Value of ξ.

the ends and midspan, and if they are assumed to be parabolically distributed along the length of the span, the midspan deflection, Δ, of a simple or continuous beam can be computed from

$$\Delta = \frac{\ell^2}{96}(\Psi_1 + 10\Psi_2 + \Psi_3) \qquad (9\text{–}14)$$

where ℓ is the span length, and Ψ_1, Ψ_2, and Ψ_3 are the curvatures at one end, midspan, and the other end, respectively. Guidance in the use of Eq. 9–14 is given in Ref. 9–15.

9–5   CONSIDERATION OF DEFLECTIONS IN DESIGN

Limits on Deflections

Visual Appearance

Deflections greater than about 1/250 of the span $(\ell/250)$ are generally visible for simple or continuous beams, depending on a number of factors, such as floor finish.[9-16] Deflections will be most noticeable if it is possible to compare the deflected member with an undeflected member or a horizontal line. Uneven deflections of the tips of cantilevers can be particularly noticeable. The critical deflection is the total immediate and sustained deflection minus any camber.

Damage to Nonstructural Elements

Excessive deflections cause cracking of partitions or malfunctioning of doors and windows. Such problems can be handled by limiting the deflections that occur after installation of the nonstructural elements or by designing the nonstructural elements to accommodate the required amount of movement.

Surveys of partition damage have shown that damage to brittle partitions can occur with deflections as small as $\ell/1000$. A frequent limit on deflections causing damage is $\ell/480$ deflection occurring after attachment of the nonstructural element (ACI Table 9.5b). This is the sum of the immediate deflection due to live loads plus the sustained portion of the deflections due to dead load and any sustained live load:

$$\Delta = \Delta_{iL} + \lambda_{(t_0,\infty)}\Delta_{iD} + \lambda_\infty \Delta_{iLS} \qquad (9-15)$$

where Δ_{iD}, Δ_{iL}, and Δ_{iLS} are the instantaneous deflections due to dead load, live load, and the sustained portion of the live load, respectively; $\lambda_{(t_0,\infty)}$ is the value of λ (Eq. 9–13) based on the value of ξ for 5 years or more, minus the value of ξ at the time t_0 when the partitions, and so on, are installed (see Fig. 9–17). Thus if the nonstructural elements were installed 3 months after the shores were removed, the value of ξ used in calculating $\lambda_{(t_0,\infty)}$ would be $2.0 - 1.0 = 1.0$. Finally, λ_∞ is the value of λ based on $\xi = 2.0$ since it is assumed that all the sustained live-load deflection will occur after the partitions are installed.

As is evident from Eq. 9–10, the effective moment of inertia, I_e, decreases as the loads increase, once cracking has occurred. As a result, the value of I_e is larger when only dead load acts than when dead and live loads act together. There are two schools of thought about the value of I_e to be used in the calculation of the Δ_{iD} in Eq. 9–15. Some designers use the value of I_e which is effective when only dead load is supported. Others reason that the portion of the sustained dead-load deflection which is significant is that occurring after the partitions are in place (i.e., that occurring during the live-loading period). This second line of reasoning implies that Δ_{iD} in Eq. 9–15 should be based on the value of I_e corresponding to dead plus live loads. The author supports the second procedure, particularly in view of the fact that construction loads or shoring loads may lead to premature cracking of the structure. The solution of Eq. 9–15 is illustrated in Example 9–6.

Disruption of Function

Excessive deflections may interfere with the use of the structure. This is particularly true if the members support machinery that must be carefully aligned. The corresponding deflection limits will be prescribed by the manufacturer of the machinery or by the owner of the building.

Excessive deflections may also cause problems with drainage of floors or roofs. The deflection due to the weight of water on a roof will cause additional deflections, allowing it to hold more water. In extreme cases, this may lead to progressively deeper and deeper water, resulting in a *ponding failure* of the roof.[9-17] Ponding failures are most apt to occur with long-span roofs in regions where roof design loads are small.

Damage to Structural Elements

The deflections corresponding to structural damage are several times those causing serviceability problems. If a member deflects so much that it comes in contact with other members, the load paths may change, causing cracking.

Allowable Deflections

In principle, the designer should select the deflection limit that applies to his structure, based on the characteristics and function of the structure. In this respect, allowable deflections given in codes are only guides.

The ACI Code gives values of maximum permissible computed deflections in ACI Table 9.5(b). The ACI Deflection Committee have also proposed a set of allowable deflections.[9-18]

Accuracy of Deflection Calculations

The ACI Code deflection calculation procedure gives results that are within $\pm 20\%$ of the true values, provided that M_{cr}/M_a is less than 0.7 or greater than 1.25.[9-19] Since a small amount of cracking can make a major difference in the value of I_e as indicated by the big decrease in slope from O-A to O-C in Fig. 9–12d, the accuracy is considerably poorer when the service load moment is close to the cracking moment. This is because random variations in the tensile strength of the concrete will change M_{cr}, and hence I_e.

For these reasons it is frequently satisfactory to use a simplified deflection calculation as a first trial, and then refine the calculations if deflections appear to be a problem.

Deflection Control by Span-to-Depth Limits

Equation 4–29b shows the relationship between beam depth and deflections:

$$\frac{\Delta}{\ell} = C\frac{\ell}{d} \tag{4–29b}$$

where the term C allows for the support conditions, grade and amount of steel. For a given desired Δ/ℓ a table of ℓ/d values can be derived. ACI Table 9.5(a) is such a table and gives minimum thicknesses of beams and one-way slabs unless deflections are computed. This table is frequently used to select member depths. It should be noted, however, that this table applies only to *members not supporting, or attached to, partitions or other construction likely to be damaged by large deflections*. If the member supports such partitions, and so on, the ACI Code requires calculation of deflections.

Table A–14 in Appendix A summarizes the ACI limits and presents other more stringent limits for the case of beams supporting brittle partitions.[9-20]

EXAMPLE 9–5   Calculation of Immediate Deflection

The T beam shown in Fig. 9–18 supports unfactored dead and live loads of 0.87 kip/ft and 1.2 kips/ft, respectively. It is built of 3000-psi concrete and Grade 60 reinforcement. The moments used in the design of the beam were calculated using the moment coefficients in ACI Sec. 8.3.3. Calculate the immediate midspan deflection. Assume that the construction loads did not exceed the dead load.

1. **Is the beam cracked at service loads?** The cracking moment is

$$M_{cr} = \frac{f_r I_g}{y_t}$$

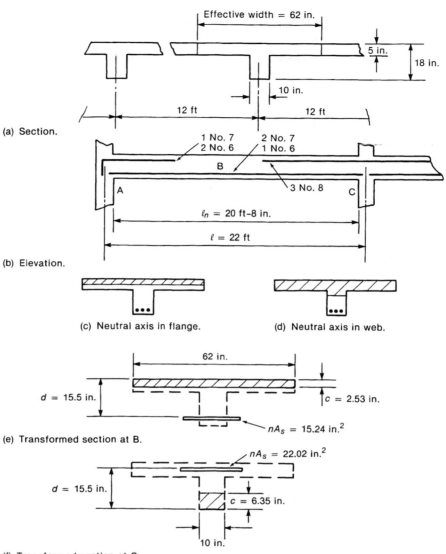

(a) Section.

(b) Elevation.

(c) Neutral axis in flange.　　　(d) Neutral axis in web.

(e) Transformed section at B.

(f) Transformed section at C.

Fig. 9–18
End span of beam—
Examples 9–5 and 9–6.

Compute I_g for the uncracked T section (ignore the effect of the reinforcement for simplicity):

$$\text{flange width} = \text{effective flange width from ACI Sec. 8.10.2}$$

$$= 62 \text{ in.}$$

(a) **Compute the centroid of the cross section.**

Part	Area (in.²)	y_{top} (in.)	Ay_{top} (in.³)
Flange	62 × 5 = 310	2.5	775
Web	13 × 10 = 130	11.5	1495
	Area = 440 in.²		ΣAy_t = 2270 in.³

9–5　Consideration of Deflections in Design　　　**361**

The centroid is located at $2270/440 = 5.16$ in. below the top of the beam.

$$\bar{y}_{top} = 5.16 \text{ in.} \qquad \bar{y}_{bottom} = 12.84 \text{ in.}$$

(b)   Compute the moment of inertia, I_g.

Part	Area (in.^{2})	$\bar{y}$, (in.)	$I_{own\ axis}$ (in.^{4})	$A\bar{y}^2$ (in.^{4})
Flange	310	$5.16 - 2.5 = 2.66$	646	2192
Web	130	$5.16 - 11.5 = -6.34$	1831	5227
				$I_y = 9896$ in.^{4}

(c)   Determine the flexural cracking moment.

$$M_{cr} = \frac{f_r I_g}{y_t}$$

where $f_r = 7.5\sqrt{f_c'} = 411$ psi.
In negative moment regions at A and C:

$$M_{cr} = \frac{411 \text{ psi} \times 9896 \text{ in.}^4}{5.16 \text{ in.}} = 788,200 \text{ in.-lb}$$

$$= -65.7 \text{ ft-kips}$$

$$\text{negative moment at } A = -\frac{w\ell_n^2}{16}$$

$$\text{dead-load moment} = -\frac{0.87 \times 20.67^2}{16} = -23.2 \text{ ft-kips}$$

$$\text{dead-plus-live-load moment} = \frac{(0.87 + 1.20) \times 20.67^2}{16}$$

$$= -55.3 \text{ ft-kips}$$

Since both of these are less than M_{cr}, section A will not be cracked at service loads.

$$\text{negative moment at } C = \frac{w\ell_n^2}{10}$$

$$\text{dead-load moment} = -\frac{0.86 \times 20.67^2}{10}$$

$$= -36.7 \text{ ft-kips (not cracked)}$$

$$\text{dead-plus-live-load moment} = -\frac{2.07 \times 20.67^2}{10}$$

$$= -88.4 \text{ ft-kips (cracked)}$$

In the positive moment region:

$$M_{cr} = \frac{411 \times 9896}{12.84} = 26.4 \text{ ft-kips}$$

$$\text{positive moment at } B = \frac{w\ell_n^2}{14}$$

$$\text{dead-load moment} = \frac{0.87 \times 20.67^2}{14}$$

$$= 26.6 \text{ ft-kips (just cracked)}$$

$$\text{dead-plus-live-load moment} = \frac{2.07 \times 20.67^2}{14}$$

$$= 63.2 \text{ ft-kips (cracked)}$$

Therefore, it will be necessary to compute I_{cr} and I_e at midspan and at support C.

2. Compute I_{cr} at midspan. It is not known whether this section will have a rectangular compression zone (Fig. 9–18c) or a T-shaped compression zone (Fig. 9–18d). We shall assume that the compression zone is rectangular for the first trial.

To locate the centroid, either follow the calculation in Example 9–1, or since the compression zone is assumed to be rectangular and there is no compression steel, use Eq. 9–3 to compute k where $c = kd$.

$$k = \sqrt{2\rho n + (\rho n)^2} - \rho n$$

where

$$n = \frac{E_s}{E_c}$$

$$= \frac{29 \times 10^6}{57{,}000\sqrt{3000}} = 9.29$$

$$\rho \text{ (based on } b = 62 \text{ in.)} = \frac{1.64}{62 \times 15.5} = 0.00171$$

$$\rho n = 0.0159$$

$$k = \sqrt{2 \times 0.0159 + 0.0159^2} - 0.0159$$

$$= 0.163$$

Therefore,

$$c = kd = 0.163 \times 15.5$$

$$= 2.53 \text{ in.}$$

Since this is less than the thickness of the flange, the compression zone is rectangular as assumed. If it was not, use the general method as in Example 9–1.

(a) Compute the cracked moment of inertia at B. Transform the reinforcement as shown in Fig. 9–18e.

$$nA_s = 9.29 \times 1.64 = 15.24 \text{ in.}^2$$

Part	Area (in.²)	$\bar{y}$ (in.)	$I_{\text{own axis}}$ (in.⁴)	$A\bar{y}^2$ (in.⁴)
Compression zone	$2.53 \times 62 = 156.86$	$2.53/2$	84	251
Reinforcement	15.24	$2.53 - 15.5 = -12.97$	—	2564
			$I_{cr} = $	2899 in.⁴

Therefore, I_{cr} at $B = 2899$ in.⁴

3. Compute I_{cr} at support C. The cross section at support C is shown in Fig. 9–18f. The transformed area of the reinforcement is

$$nA_s = 9.29 \times 2.37 = 22.02 \text{ in.}^2$$

$$\rho = \frac{2.37}{10 \times 15.5} = 0.0153$$

From Eq. 9–3, $k = 0.410$ and

$$c = kd = 6.35 \text{ in.}$$

The positive moment reinforcement is not developed for compression at the support and therefore will be neglected.

Part	Area (in.^{2})	$\bar{y}$ (in.)	$I_{\text{own axis}}$ (in.^{4})	$A\bar{y}^2$ (in.^{4})
Compression zone	$6.35 \times 10 = 63.5$	6.35/2	213	640
Reinforcement	22.02	6.35 − 15.5	—	1844
				$I_{cr} = 2697$ in.^{4}

In summary, $I_g = 9896$ in.^{4}, I_{cr} at $B = 2899$ in.^{4}, and I_{cr} at $C = 2697$ in.^{4}

4. **Compute immediate dead-load deflection.**

(a) **Compute I_e at A.** Since $M_a = -23.2$ ft-kips is less than $M_{cr} = -65.7$ ft-kips, $I_e = 9896$ in.^{4} at A for dead loads.

(b) **Compute I_e at B.** $M_a = 26.6$ ft-kips and $M_{cr} = 26.4$ ft-kips. Therefore, $M_{cr}/M_a = 0.992$ and $(M_{cr}/M_a)^3 = 0.978$.

$$I_e = \left(\frac{M_{cr}}{M_a}\right)^3 I_g + \left[1 - \left(\frac{M_{cr}}{M_a}\right)^3\right] I_{cr}$$

$$= 0.978 \times 9896 + (1 - 0.978) \times 2899$$

$$= 9739 \text{ in.}^4 \text{ at } B \text{ for dead loads}$$

(c) **Compute I_e at C.** $M_a = -36.7$ ft-kips and $M_{cr} = -65.7$ ft-kips. Therefore, the section is not cracked, and $I_e = 9896$ in.^{4}

(d) **Compute the weighted average, I_e.**

$$I_e = 0.7I_m + 0.15(I_{e1} + I_{e2})$$

$$= 0.7 \times 9739 + 0.15(9896 + 9896) \tag{9–11a}$$

$$= 9790 \text{ in.}^4$$

(e) **Immediate dead-load deflection from Eq. 9–12.**

$$\Delta = \frac{5}{48}\frac{\ell_n^2}{EI}[M_m + 0.1(M_1 + M_2)]$$

where

$\ell_n = 20.67$ ft
$E = 57{,}000\sqrt{3000} = 3.122 \times 10^6$ psi
$M_m =$ midspan moment due to dead loads $= 26.6$ ft-kips
$M_1 =$ moment at A due to dead loads $= -23.2$ ft-kips
$M_2 =$ moment at C due to dead loads $= -36.7$ ft-kips

$$\Delta = \frac{5}{48} \times \frac{(20.67 \times 12)^2[26.6 + 0.1(-23.2 - 36.2)] \times 12{,}000}{3.122 \times 10^6 \times 9790}$$

$$= 0.052 \text{ in.}$$

Therefore, the immediate dead-load deflection is 0.052 in.

5. Compute the immediate live-load deflection.

When the live load is applied to the beam, the moments will increase leading to more cracking. As a result, I_e decreases. The deflection that occurs when the live load is applied is

$$\Delta_{iL} = \Delta_{i,L+D} - \Delta_{iD}$$

(a) Compute I_e at A. Since $M_a = -55.3$ ft-kips is less than $M_{cr} = -65.7$ ft-kips, the section is uncracked and $I_e = 9896$ in.⁴ at A for dead + live loads.

(b) Compute I_e at B. $M_a = 63.2$ ft-kips, $M_{cr} = 26.4$ ft-kips; therefore, $M_{cr}/M_a = 0.418$ and $(M_{cr}/M_a)^3 = 0.0729$.

$$I_e = 0.0729 \times 9896 + (1 - 0.0729)2899$$

$$= 3409 \text{ in.}^4$$

(c) Compute I_e at C. $M_a = -88.4$ ft-kips and $M_{cr} = -65.7$ ft-kips; therefore, $M_{cr}/M_a = 0.743$ and $(M_{cr}/M_a)^3 = 0.411$.

$$I_e = 0.411 \times 9896 + (1 - 0.411)2697$$

$$= 5652 \text{ in.}^4$$

(d) Compute the weighted average, I_e.

$$I_e = 0.7 \times 3409 + 0.15(9896 + 5652)$$

$$= 4720 \text{ in.}^4$$

(e) Compute the immediate dead + live load deflection.

$$\Delta_{i,D+L} = \frac{5}{48} \frac{\ell_n^2}{EI} [M_m + 0.1(M_1 + M_2)]$$

$$= \frac{5(20.67 \times 12)^2 [63.2 + 0.1(-55.3 - 88.4) \times 12,000]}{48 \times 3.122 \times 10^6 \times 4720}$$

$$= 0.255 \text{ in.}$$

Therefore, the immediate dead + live load deflection is 0.255 in. The deflection increased by $0.255 - 0.052 = 0.203$ in. when the live load was applied to the beam.

When the live load is removed, the beam will not return to its dead-load deflection of 0.052 in. because it is now cracked with $I_e = 4720$ in.⁴, rather than the $I_e = 9790$ in.⁴ that governed the dead-load deflections. The new dead-load deflection will be $M_D/M_{D+L} \times 0.255$ in. $= 0.107$ in. When the live load is reapplied, the deflection will increase from 0.107 in. to 0.255 in., an increase of 0.148 in.

In the deflection calculations M_m, M_1, and M_2 were based on the moment coefficients from ACI Sec. 8.3.3. These give a total moment of $M = w\ell^2/6.6$ rather than $w\ell^2/8$. As a result, the deflections tend to be overestimated. The exact amount depends on which loading case governed. ∎

ACI Table 9.5(b) limits the computed immediate live-load deflections to $\ell/180$ and $\ell/360$, respectively, for flat roofs and floors that do not support, or are not attached to elements likely to be damaged by large deflections. Conservatively, this should be taken as the initial value of 0.203 in. For this beam $\ell/360$ is 22×12 in.$/360 = 0.733$ in. where ℓ is the span from center to center of supports.

If this beam does not support partitions that can be damaged by deflections, ACI Table 9.5(a) would have indicated that it was not necessary to compute deflections since the overall depth of the beam ($h = 18$ in.) exceeds the minimum value of $\ell/21 = 22 \times 12/21 = 12.6$ in. given in the table.

If the structure just barely passes the allowable deflection calculations, consideration should be given to means of reducing the deflections since deflection calculations are of dubious accuracy. The easiest way to do this is to add compression steel or deepen the section.

EXAMPLE 9–6   Calculation of Deflections Occurring after Attachment of Nonstructural Elements

The beam considered in Example 9–5 will be assumed to support partitions that would be damaged by excessive deflections. Again, it supports a dead load of 0.87 kip/ft and a live load of 1.2 kips/ft, 25% of which is sustained. The partitions are installed at least 3 months after the shoring is removed. Will the computed deflections exceed the allowable in the end span?

The controlling deflection is (Eq. 9–15)

$$\Delta = \Delta_{iL} + \lambda_{(t_0, \infty)}\Delta_{iD} + \lambda_\infty \Delta_{iLS}$$

In Example 9–5 the following quantities were calculated:

$$\Delta_{iD+L} = 0.255 \text{ in.}$$

$$M_D \text{ (at midspan)} = 26.6 \text{ ft-kips}$$

$$M_{D+L} \text{ (at midspan)} = 63.2 \text{ ft-kips}$$

Therefore, $M_L = 63.2 - 26.6$ ft-kips $= 36.6$ ft-kips.

The immediate live-load deflection is

$$\Delta_{iL} = \Delta_{iD+L} \times \frac{M_L}{M_{D+L}}$$

$$= 0.255 \times \frac{36.6}{63.2} = 0.148 \text{ in.}$$

Twenty-five percent of this results from the sustained live loads, $\Delta_{iLS} = 0.037$ in.

The immediate dead-load deflection based on I_e which is effective when both dead and live loads are acting:

$$\Delta_{iD} = \Delta_{iD+L} \times \frac{M_D}{M_{D+L}}$$

$$= 0.255 \times \frac{26.6}{63.2} = 0.107 \text{ in.}$$

Creep multipliers (Eq. 9–13):

$$\lambda = \frac{\xi}{1 + 50\rho'}$$

The beam being checked does not have compression reinforcement, and hence $\rho' = 0$. If it had such reinforcement the value of ρ' in Eq. 9–13 is the value at midspan. Note that the bottom steel entering the supports at A and B is not developed for compression in the support and hence does not qualify as compression steel in this calculation.

$$\lambda_\infty = \frac{2.0}{1 + 0} = 2.0$$

$$\lambda_{(t_0, \infty)} = \frac{2.0 - 1.0}{1 + 0} = 1.0$$

The deflection occurring after the partitions are in place is

$$\Delta = 0.148 + 1.0 \times 0.107 + 2.0 \times 0.037$$

$$= 0.329 \text{ in.}$$

The maximum permissible computed deflection is $\ell/480$ [ACI Table 9.5(b)], where ℓ is the span from center to center of the supports:

$$\frac{22 \times 12 \text{ in.}}{480} = 0.55 \text{ in.}$$

Therefore, this floor is satisfactory. ∎

9–6 FRAME DEFLECTIONS

Section 9–4 dealt with vertical deflections of floors and roofs. Several other types of deflection must be considered in the design of concrete frame structures. The most obvious of these are lateral deflections of the frame itself. These must be controlled to prevent discomfort to occupants, or damage to partitions, and so on, in high frames, and because lateral deflections lead to second-order or $P-\Delta$ effects which are inversely proportional to the lateral stiffness. Guidance in the selection of acceptable lateral sway indices, Δ/H, is given in Ref. 9–21.

If the frame of a building is exposed to atmospheric temperature changes and the interior is maintained at a constant temperature, the thermal contraction and expansion of the exterior gives rise to relative vertical deflections of interior and exterior columns. This problem is discussed in Ref. 9–22.

In tall concrete buildings with columns and shear walls, the columns will generally be more highly stressed than the walls and will shorten more due to elastic shortening and creep. The relative shortening may exceed 1 in. after several years. As a result, the floors may slope from the walls to the columns. Procedures for estimating and dealing with these deflections are discussed in Ref. 9–23.

Concrete buildings with brick exteriors occasionally encounter difficulties due to the incompatible long-term deformations of concrete and brick. With time, the concrete frame shortens due to creep, while the bricks expand with time due to an increase in moisture content above their kiln-dried condition. This, plus thermal deformations of the brick, leads to crushing and cracking of the brick if the joints in the brickwork close so that bricks begin to carry load. The expansion of bricks due to moisture gain is in the order of 0.0002 in./in., while the shortening of a concrete column due to creep strains may be in the order of 0.0005 to 0.0010 in./in., giving a total relative strain between the brick and the concrete of as much as 0.0012 in./in. In a 12-ft-high story, this amounts to a shortening of the frame of 0.17 in. relative to the brick cladding. This and temperature effects must be considered in designing cladding.

9–7 VIBRATIONS

Reinforced concrete buildings are not, as a rule, subject to vibration problems. This is because of their mass and stiffness. Occasionally, long-span floors or assembly structures may be susceptible to vibrations induced by persons walking, dancing, or exercising. These activities induce vibrations of approximately 2 to 4 cycles per second (hertz). If a floor system or supporting structure has a natural frequency of less than 5 cycles per second, its vibrational properties should be examined more closely to ensure that floor vibrations will not be a problem. Guidance on this problem may be found in Refs. 9–24 and 9–25.

Fortunately, it is relatively easy to estimate the natural frequency of a floor if its deflection can be calculated:

$$f = 0.18 \sqrt{\frac{g}{\Delta_{is}}}$$

(9–16)

$$= 3.5 \sqrt{\frac{1}{\Delta_{is}(\text{in.})}} \qquad \text{cycles per second}$$

where Δ_{is} is the immediate static deflection at the center of the floor due to the loads expected to be on the floor when it is vibrating.[9-25] These would include the dead load and that portion of the live load on the floor when vibrating. The factor 3.5 in Eq. 9–16 actually varies from 3.1 to 4.0 for different types of floors (simple beams, fixed slabs, etc.), but the resulting error in f is small compared to inaccuracies in the deflection calculations themselves.

For continuous beams the first vibration mode involves alternate spans deflecting up and down at the same time. The resulting modal deflections will be larger than those due to downward loads on each span. To use Eq. 9–16 in such a case the deflections from a special frame analysis with dead plus some fraction of live load applied upward and downward on alternate spans are needed.

9–8   FATIGUE

Structures subjected to cyclic loads may fail in fatigue. This type of failure requires (1) generally in excess of 1 million load cycles, and (2) a change of reinforcement stress in each cycle in excess of about 20 ksi. Since dead-load stresses account for a significant portion of the service load stresses in most concrete structures, the latter case is infrequent. As a result, fatigue failures of reinforced concrete structures are rare. An overview of the fatigue strength of reinforced concrete structures is given in Ref. 9–26 (see also Sec. 3–9).

Fatigue failure of the concrete itself occurs due to a progressive growth of microcracking. The fatigue strength of plain concrete in compression or tension for a life of 10 million cycles is roughly 55% of the static strength. It is affected by the minimum stress in the cycle and by the stress gradient. Concrete loaded in flexural compression has a 15 to 20% higher fatigue strength than axially loaded concrete, because the development of microcracking in highly strained areas is inhibited by the restraint provided by less strained areas. The fatigue strength of concrete is not sensitive to stress concentrations.

The ACI fatigue committee[9-26] recommends that the compressive stress range, f_{cr}, should not exceed

$$f_{cr} = 0.4f_c' + 0.47f_{\min}$$

(9–17)

where $f_{\min}$ is the minimum compressive stress in the cycle (positive in compression). The ACI bridge committee[9-27] limits the compressive stress at service load to $0.5f_c'$.

The fatigue strength of deformed reinforcing bars is affected by the stress range and the minimum stress in the stress cycle, but is almost independent of the yield strength of the bar.

In tests, bends with the minimum radius of bend permitted by the ACI Code reduced the fatigue strength in the region of the bend by a half, while tack welding of stirrups to longitudinal bars reduced the strength of the latter by more than a third.[9-26] The fatigue strength of bars is also affected by the radius at the bottom of the deformation lugs.

For bars of normal lug geometry, the ACI bridge committee[9-27] recommends that the stress range not exceed

$$f_r = 23.4 - 0.33f_{\min} \qquad \text{ksi}$$

(9–18)

where f_r is the algebraic difference between the maximum and minimum stresses at service loads and $f_{\min}$ is the minimum stress at service loads. In both cases tension is positive. At bends or tack welds the allowable f_r should be taken as half the value given by Eq. 9–18.

The fatigue strength of stirrups is less than that of longitudinal bars[9-26] due in part to a slight kinking of the stirrup where it crosses an inclined crack, and possibly due to an increase in the stirrup stress resulting from a reduction in the shear carried by the concrete in cyclically loaded beams. No firm design recommendations are currently available. A reasonable design procedure would be to limit the stirrup stress range to 0.75 times f_r from Eq. 9–17, where the stirrup stress range would be calculated on the basis of the ACI shear design procedures, but using $V_c/2$.

The effect of cyclic loading on bond strength is approximately the same as that on the compressive strength of concrete.[9-28] Hence a relationship similar to Eq. 9–17 would adequately estimate the allowable bond stress range. During the initial stages of the cyclic load, the loaded end of the bar will slip relative to the concrete due to internal cracking around the bar. The slip ceases once the cracking has stabilized.

PROBLEMS

9–1   Explain the differences in appearance of flexural cracks, shear cracks, and torsional cracks.

9–2   Why is it necessary to limit the widths of cracks?

9–3   Does the beam shown in Fig. P9–3 satisfy the ACI Code crack control provisions for interior exposure? $f_y = 60$ ksi.

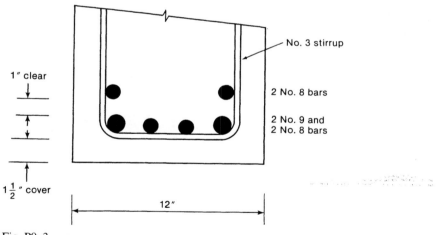

Fig. P9–3

9–4   Compute the maximum spacing of No. 5 bars in a one-way slab with 1 in. of clear cover that will satisfy the ACI Code crack control provisions for exterior exposure. $f_y = 60$ ksi.

9–5 and 9–6   For the cross sections shown in Figs. P9–5 and P9–6, compute

(a)   the gross moment of inertia, I_g.

(b)   the location of the neutral axis of the cracked section and I_{cr}.

(c) I_{eff} for $M_a = 0.6\phi M_n$.

The beams have 1.5 in. of clear cover and No. 3 stirrups. The concrete strength is 3000 psi.

9–7   Why are deflections limited in design?

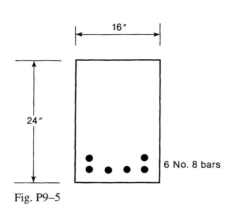

Fig. P9–5

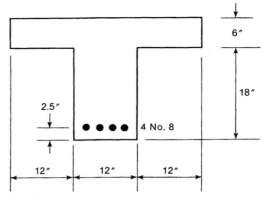

Fig. P9–6

9–8 A simply supported beam with the cross section shown in Fig. P9–5 has a span of 25 ft and supports an unfactored dead load of 1.5 kips/ft, including its own self-weight plus an unfactored live load of 1.5 kips/ft. The concrete strength is 3000 psi. Compute

(a) the immediate dead load deflection.

(b) the immediate dead-plus-live-load deflection.

(c) the deflection occurring after partitions are installed. Assume that the partitions are installed 2 months after the shoring for the beam is removed and assume that 20% of the live load is sustained.

9–9 Repeat Problem 9–8 for a beam having the same dimensions and tension reinforcement, but with 2 No. 8 bars as compression reinforcement.

9–10 The beam shown in Fig. P9–10 is made of 4000-psi concrete and supports unfactored dead and live loads of 1 kip/ft and 1.1 kips/ft. Compute

(a) the immediate dead load deflection.

(b) the immediate dead-plus-live-load deflection.

(c) the deflection occurring after partitions are installed. Assume that the partitions are installed 4 months after the shoring is removed and assume that 10% of the live load is sustained.

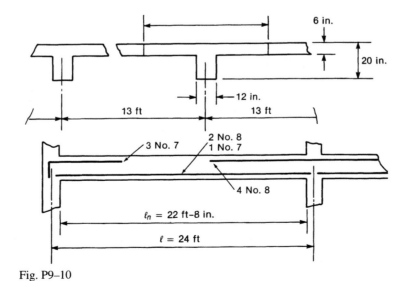

Fig. P9–10

10

Continuous Beams and One-Way Slabs

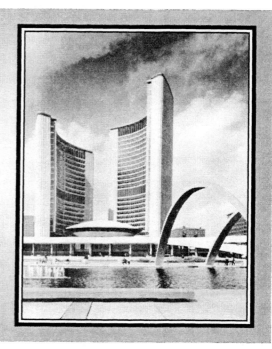

10-1 INTRODUCTION

In the design of a continuous beam or slab it is necessary to consider several ultimate limit states: failure by flexure, shear, bond, and possibly torsion; and several serviceability limit states: excessive deflections, crack widths, and vibrations. Each of these topics has been covered in a preceding chapter. In this chapter the distribution of moments in continuous beams is discussed, followed by a series of design examples showing the overall design of continuous one-way slabs and beams.

10-2 CONTINUITY IN REINFORCED CONCRETE STRUCTURES

The construction of two cast-in-place reinforced concrete structures is illustrated in Figs. 10-1 to 10-3. In Fig. 10-1 formwork is under construction for the beams and slabs supporting a floor. The raised platform-like areas are the bottom forms for the slabs. The rectangular openings between them are forms for the beam concrete. The beams meet at right angles at the columns.

Once the formwork is complete and the reinforcement has been placed in the forms (Fig. 10-2), the concrete for the slabs and beams will be placed in one monolithic pour (see ACI Sec. 6.4.6). Following this, the columns for the next story are erected, as shown in Fig. 10-3. ACI Sec. 6.4.5 requires that the concrete in the columns or walls have set before the concrete in the floor supported by those columns is placed. This sequence is required because the column concrete will tend to settle in the forms while in the plastic state. If the floor concrete had been placed, this would leave a gap between the column concrete and the beam. By placing the floor concrete after the column concrete is no longer plastic, any gap that formed as the concrete settled will be filled. The resulting construction joints can be clearly seen at the bottom and top of each column in Fig. 10-3.

Fig. 10–1
Formwork for a beam-and-slab floor. (Photograph courtesy of J. G. MacGregor.)

Fig. 10–2
Beam and slab reinforcement in the forms. (Photograph courtesy of J. G. MacGregor.)

Continuous Beams and One-Way Slabs

Fig. 10–3
Construction of columns in a tall building. (Photograph courtesy of J. G. MacGregor.)

As a result of this placing sequence, each floor acts as a continuous unit. Because the column reinforcement extends through the floor, the columns act with the floors to form a continuous frame.

Braced and Unbraced Frames

A frame is said to be *unbraced* if it relies on frame action to resist lateral loads or lateral deformations. Thus the frame in Fig. 10–4a is unbraced, while that in Fig. 10–4b is braced by the shear wall. If the lateral stiffness of the bracing element in a story exceeds 6 to 10 times the sum of the lateral stiffnesses of all the columns in that story, a story can be considered to be braced (ACI Sec. 10.11.4). Most concrete buildings are braced by walls, elevator shafts or stairwells. The floor members in an unbraced frame must resist moments induced by lateral loads as well as gravity loads, while in most cases, the lateral load moments can be ignored in the design of beams in a braced frame. The design examples in this chapter are limited to beams in braced frames.

One-Way Slab and Beam Floors

Figure 10–5 illustrates the simplest type of one-way slab and beam floor. Here the load P applied at A is carried by the shaded strip of slab to the beams at B and C. The beams, in turn, carry the load to the columns at $D, E, F,$ and G. Frequently, such a floor will have beams in two directions, as shown by the plan layouts in Fig. 10–6. Here the load from the slab is transmitted to the beams, which transfer the load to *girders,* which in turn are supported by the

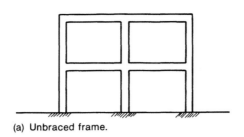

(a) Unbraced frame.

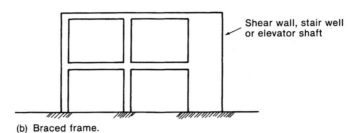

Fig. 10–4
Braced and unbraced frames.

Shear wall, stair well
or elevator shaft

(b) Braced frame.

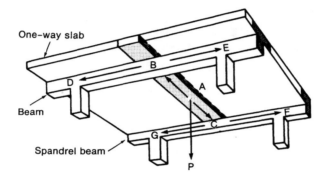

One-way slab

Beam

Spandrel beam

Fig. 10–5
One-way slab-and-beam
floor.

columns. (A girder is a main supporting beam, generally one that supports other beams.) When designing such a floor, the slabs are designed first, then the beams, and finally the girders. In this way, the dead load of the slab is known when the beam is designed, and so on.

Slabs without beams are referred to as *flat slabs* or *flat plates*. Slabs with beams on an approximately square grid with columns at each corner of each square are referred to as *two-way beam and slab system*. Flat slabs and two-way slabs are discussed more fully in Chaps. 13 to 15.

Occasionally, the beams are made wide and shallow. Such a structure is referred to as a *slab-band* floor. These are designed as one-way or two-way slab systems, depending on the geometry of the slab panels.

10–3 MOMENTS AND SHEARS IN CONTINUOUS BEAMS

Continuous slabs, beams, and frames are statically indeterminate structures. Three families of procedures are available for computing moments and shears in such members:

1. Elastic analyses, such as slope-deflection, moment distribution, and matrix methods

2. Plastic analyses

3. Approximate analyses, such as the use of moment coefficients or such procedures as the portal and cantilever methods

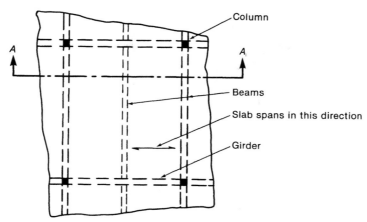

(a) Plan-beams at middle
    of panel.

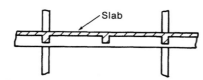

(b) Section A-A.

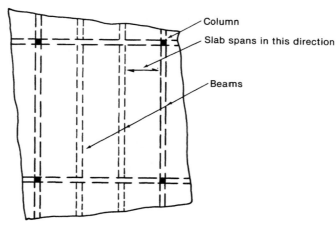

Fig. 10-6
One-way slab-and-beam
floors.

(c) Plan-beams at third points
    of panel.

The elastic analysis of reinforced concrete structures is discussed in textbooks on structural analysis. The design examples carried out in this chapter are based on the approximate moment coefficients presented in the ACI Code. The validity of these coefficients is discussed in the following paragraphs.

Pattern Loadings

The largest moments in a continuous beam or a frame occur when some spans are loaded and others are not. Diagrams, referred to as *influence lines*, are used to determine which spans should and should not be loaded. An influence line is a graph of the variation in the

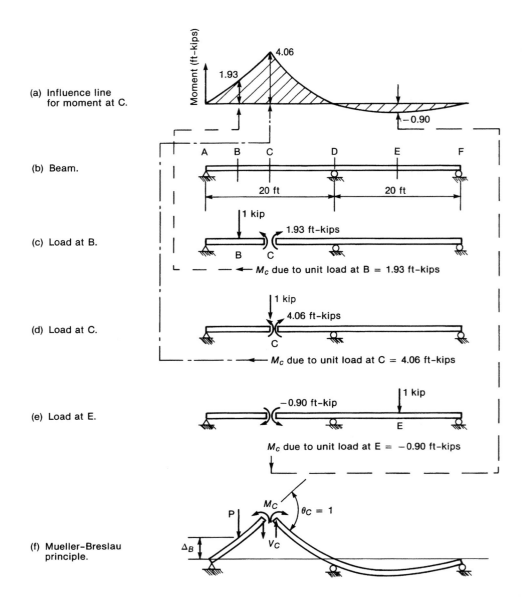

(a) Influence line
    for moment at C.

(b) Beam.

(c) Load at B.

M_C due to unit load at B = 1.93 ft-kips

(d) Load at C.

M_C due to unit load at C = 4.06 ft-kips

(e) Load at E.

M_C due to unit load at E = −0.90 ft-kips

(f) Mueller–Breslau
    principle.

Fig. 10–7
Concept of influence lines.

moment, shear, or other effect at *one particular point* in a structure due to a unit load that moves across the beam.

Figure 10–7a is an influence line for moment at point C in the two-span beam shown in Fig. 10–7b. The horizontal axis refers to the *position* of a unit (1 kip) load on the beam, and the vertical ordinates are the *moment at C* due to a 1-kip load at the point in question. The derivation of the ordinates at B, C, and E is illustrated in Fig. 10–7c to e. When a unit load acts at B, it causes a moment of 1.93 ft-kips at C (Fig. 10–7c). Thus the ordinate at B in Fig. 10–7a is 1.93 ft-kips. Figure 10–7d and e show that the moments at C due to loads at C and E are 4.06 and − 0.90 ft-kip, respectively. These are the ordinates at C and E in Fig. 10–7a and are referred to as *influence ordinates*. If a concentrated load of P kips acted at point E, the moment at C would be P times the influence ordinate at E, $M = -0.90P$. If a uniform load of w acted on the span A-D the moment at C would be w times the area of the influence diagram from A to D.

Figure 10–7a shows that a load placed anywhere between A and D will cause positive moment at point C, while a load anywhere between D and F will cause a negative moment at C. Thus to get the maximum positive moment at C, we must load span AD only.

Two principal methods are used to calculate influence lines. In the first, a 1-kip load is placed successively at evenly spaced points across the span and the moment, or shear, at the point for which the influence line is being drawn, is calculated as was done in Fig. 10–7c to e. The second procedure, known as the Mueller-Breslau principle, is based on the principle of virtual work, which states that the total work done during a virtual displacement of a structure is zero if the structure is in equilibrium. The use of the Mueller-Breslau principle to compute an influence line for moment at C is illustrated in Fig. 10–7f. The beam is broken at point C and displaced so that M_c does work by acting through an angle change θ_c. Note that there was no shearing displacement at C, so V_c did not do work. The load, P, at B was displaced upward by an amount Δ_B and hence did negative work. The total work done during this imaginary displacement was

$$M_c\theta_c - P\Delta_B = 0$$

and

$$M_c = P\left(\frac{\Delta_B}{\theta_c}\right) \tag{10–1}$$

where Δ_B / θ_c is the influence ordinate at B. Thus the deflected shape of the structure for such a displacement has the same shape as the influence line for moment at C (see Fig. 10–7a and f).

The Mueller-Breslau principle is presented here as a qualitative guide to the shape of the influence line to determine where to load a structure to cause maximum moments or shears at various points.

Figure 10–8 illustrates influence lines, drawn using the Mueller-Breslau principle. Figure 10–8a shows the influence line for moment at B. The loading pattern that will give the largest positive moment at B consists of loads on all spans having positive influence ordinates. Such a loading is shown in Fig. 10–8b and is referred to as an *alternate span* loading or a *checkerboard* loading.

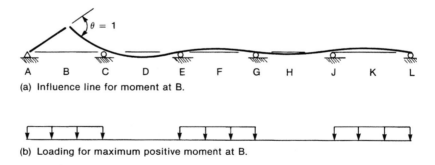

(a) Influence line for moment at B.

(b) Loading for maximum positive moment at B.

(c) Influence line for moment at C.

Fig. 10–8
Influence lines for moment and loading patterns.

(d) Loading for maximum negative moment at C.

The influence line for moment at the support C is found by breaking the structure at C and allowing a positive moment, M_c, to act through an angle change θ_c. The resulting deflected shape (Fig. 10–8c) is the influence line for M_c. The maximum *negative* moment at C will result from loading all spans having negative influence ordinates. (Fig. 10–8d.) This is referred to as an *adjacent span* loading.

Influence lines for shear can be drawn by breaking the structure at the point in question and allowing the shear at that point to act through a unit shearing displacement, Δ, as shown in Fig. 10–9. During this displacement, the parts of the beam on the two sides of the break must remain parallel so that the moment at the section does not do work. The loadings required to cause maximum positive shear at sections A and B in Fig. 10–9 are shown in Fig. 10–9b and d.

Based on this sort of reasoning ACI Sec. 8.9.2 requires design for the following loading patterns:

1. Factored dead load on all spans with factored live load on two adjacent spans and no live load on any other spans. This will give the maximum negative moment and maximum shear at the support between the two loaded spans. This loading case is repeated for each interior support.

2. Factored dead load on all spans with factored live load on alternate spans. This will give the maximum positive moments at the middle of the loaded spans, minimum positive moments (which may be negative) at the middle of the unloaded spans, and maximum negative moment at the exterior support.

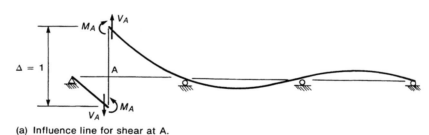

(a) Influence line for shear at A.

(b) Loading for maximum positive shear at A.

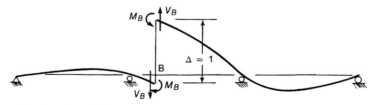

(c) Influence line for shear at B.

Fig. 10–9
Influence lines for shear.

(d) Loading for maximum positive shear at B.

Continuous Beams and One-Way Slabs

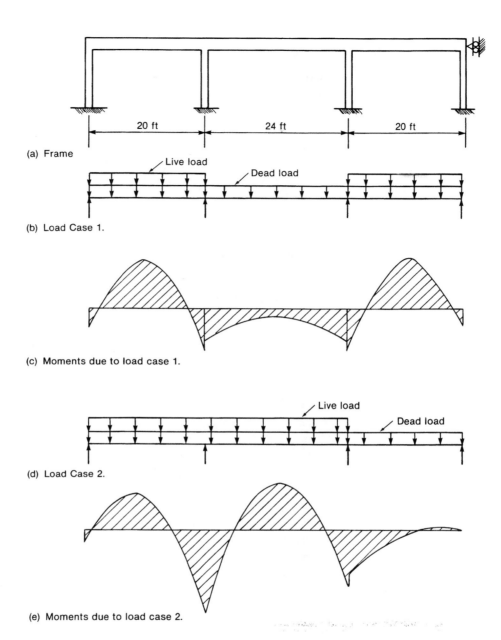

(a) Frame

(b) Load Case 1.

(c) Moments due to load case 1.

(d) Load Case 2.

(e) Moments due to load case 2.

Fig. 10–10
Moments in a three-span
beam.

Following this procedure for the three-bay frame shown in Fig. 10–10 gives the three loading cases and three moment diagrams shown. A fourth loading case is obtained by loading the middle and right spans. Because the factored live load is considerably larger than the factored dead load, the negative live-load moments at the middle of the unloaded spans will be larger than the positive dead-load moments at these points, giving a total moment that is negative.

When the four moment diagrams are superimposed, the *moment envelope* shown in Fig. 10–10h results. At every section the beam should be designed to have positive and negative moment capacity, ϕM_n, at least equal to the envelope values.

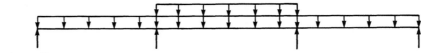

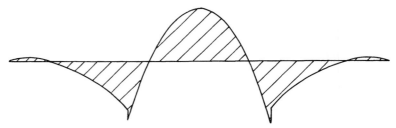

(f) Load case 3.

(g) Moments due to load case 3.

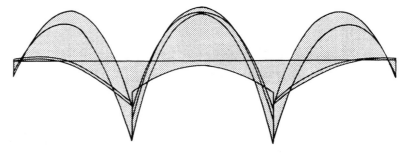

Fig. 10–10
(Continued)

(h) Moment envelope.

ACI Moment Coefficients

Because the calculations necessary to derive the moment envelope in Fig. 10–10 are tedious, approximate moment coefficients are presented in ACI Sec. 8.3.3 which can be used to calculate the moment and shear envelopes for nonprestressed continuous beams or one-way slabs that meet the following criteria:

1. There are two or more spans.

2. The spans are approximately equal, with the longer of two adjacent spans not more than 1.2 times the length of the shorter one.

3. The loads are uniformly distributed.

4. The unit live load does not exceed three times the unit dead load. The word "unit" means the unfactored load per foot.

5. The beams must be prismatic.

6. The beam must be in a braced frame without significant moments due to lateral loads (not stated in the ACI Code, but necessary for the coefficients to apply).

The maximum positive and negative moments and shears are computed from the following expressions:

$$M_u = C_m(w_u \ell_n^2) \qquad (10\text{–}2)$$

Continuous Beams and One-Way Slabs

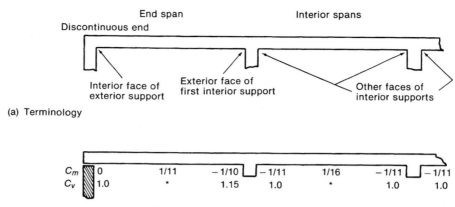

(a) Terminology

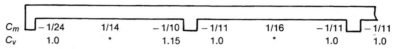

C_m	0	1/11	−1/10	−1/11	1/16	−1/11	−1/11
C_v	1.0	*	1.15	1.0	*	1.0	1.0

(b) Moment coefficients–Discontinuous end unrestrained, more than 2 spans

C_m	−1/24	1/14	−1/10	−1/11	1/16	−1/11	−1/11
C_v	1.0	*	1.15	1.0	*	1.0	1.0

(c) Moment coefficients–Discontinuous end integral with support where support is a spandrel girder

C_m	−1/16	1/14	−1/10	−1/11	1/16	−1/11	−1/11
C_v	1.0	*	1.15	1.0	*	1.0	1.0

Fig. 10–11
ACI moment and shear coefficients.

(d) Moment coefficients–Discontinuous end integral with support where support is a column

* not given in ACI Code; See Eq. 10–4.

$$V_u = C_v \left(\frac{w_u \ell_n}{2} \right) \tag{10–3}$$

where w_u is the total factored dead and live load per unit length, C_m and C_v are moment and shear coefficients from Fig. 10–12 (ACI Sec. 8.3.3), and ℓ_n is:

1. For negative moment at the interior face of the exterior support, for positive moment, and for shear, ℓ_n = clear span of the span in question.

2. For negative moment at interior supports, ℓ_n = average of the clear spans of the adjacent spans.

The terminology used to identify the critical sections is illustrated in Fig. 10–11a.

For the interior span of the structure shown in Fig. 10–10, the moment envelope plotted in Fig. 10–10h has a maximum negative moment of $(1/11.1)w_u\ell_n^2$ and a maximum positive moment of $(1/17.1)w_u\ell_n^2$. The corresponding ACI moment coefficients are $\frac{1}{11}$ and $\frac{1}{16}$. These total 123% of $w\ell^2/8$, the extra 23% being the result of the different loading cases considered.

The shears at the ends of the beams are taken as the simple beam shear $w_u \ell_n / 2$, except at the exterior face of the first interior support, where it is 1.15 times the simple beam shear. Because there is more fixity at this end of the exterior span than at the exterior end, the moments are greater here, as shown in Figs. 10–10c and e. This leads to a shear greater than $w_u \ell_n / 2$ at the interior end, and a shear less than $w_u \ell_n / 2$ at the exterior end. The decrease of the exterior end is not recognized by the ACI shear coefficients.

Although the ACI Code says nothing about shear at midspan, the critical shear at this point results from factored live load on half of the span and is approximately equal to $w_{Lu} \ell_n / 8$. Thus $C_v = 0.25 w_{Lu} / w_u$.

$$V_u = \frac{0.25 w_{Lu}}{w_u} \left(\frac{w_u \ell_n}{2} \right) \tag{10-4a}$$

This is explained in Sec. 6–5 and illustrated in Example 6–1.

At the middle of the exterior span the ACI shear coefficients give a shear of $1.15(w_u \ell_n / 2)$ minus $w_u \ell_n / 2$, which gives $0.15(w_u \ell_n / 2)$. In design calculations we shall use the larger of the shear given by Eq. 10–4a and

$$V_u = 0.15 \left(\frac{w_u \ell_n}{2} \right) \tag{10-4b}$$

Frequently, the girders in a one-way slab, beam, and girder system are loaded by concentrated loads from beams framing in at midpoint, or at the third points as shown in Fig. 10–6. In such cases, the moment coefficients presented in Figs. 10–11b to d do not apply and a structural analysis is required. It is commonplace, however, to approximate the girder moments as follows:

1. Compute the total factored simple beam moments, M_0, due to the uniform and concentrated loads. If the beam were loaded only with uniform loads, for example, these would be $M_0 = w_u \ell_n^2 / 8$.

2. Calculate the approximate moment envelope values at the end and midspan as $8C_m$ times M_0, where C_m is the appropriate moment coefficient from Fig. 10–11.

10–4   ONE-WAY SLABS

One-way slab and beam systems having plans similar to those shown in Fig. 10–6 are commonly used, especially for spans of greater than 20 ft and for heavy live loads. Generally, the ratio of the long side to the short side of the slab panels exceeds 1.5.

Structural Action

For design purposes, a one-way slab is assumed to act as a series of parallel, independent 1-ft-wide strips of slab, continuous over the supporting beams. The slab strips span the short direction of the panel, as strips A and B in Fig. 10–12 do. Near the ends of the panel adjacent to the girders, some load is resisted by bending of the longitudinal strips (strip C) and less by the transverse strips (strip A). Thus near the girders the load is supported by two-way slab action, which is discussed more fully in Chap. 13. In one-way slab design, this is ignored when designing the one-way slab strips, but is accounted for by providing top reinforcement extending into the top of the slabs on each side of the girders across the ends of the panel. If this reinforcement is omitted, wide cracks may develop in the top of the slab along $D-E$.

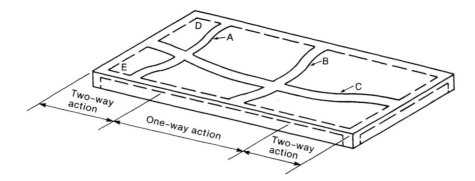

Fig. 10–12
One-way and two-way slab action.

Thickness of One-Way Slabs

Except for very heavily loaded slabs, such as slabs supporting several feet of earth, the thickness is chosen so that deflections will not be a problem. Occasionally, the thickness will be governed by shear or flexure and these will be checked in each design. Table 9–5(a) of the ACI Code gives minimum thicknesses of slabs *not supporting or attached to* partitions or other construction liable to be damaged by large deflections. No guidance is given for other cases. Table A–14 gives minimum thicknesses for one-way slabs which do and do not support such partitions.

Sometimes slab thicknesses are governed by heat transmission during a fire. Thus, by this criterion, the fire rating of a floor is the number of hours necessary for the temperature of the unexposed surface to rise by a given amount, generally 250°F. Based on a 250°F temperature rise, a $3\frac{1}{2}$-in.-thick slab will give a 1-hour fire rating, a 5-in. slab, a 2-hour fire rating, and a $6\frac{1}{4}$-in. slab, a 3-hour fire rating.[10–1] Generally, slab thicknesses are selected in $\frac{1}{4}$-in. increments up to 6 in., and $\frac{1}{2}$-in. increments for thicker slabs.

Cover

Concrete cover to the reinforcement provides corrosion resistance, fire resistance, provides a wearing surface, and is necessary to develop bond between steel and concrete. ACI Sec. 7.7.1 gives minimum covers for corrosion protection in slabs as:

1. Concrete *not exposed* to weather or in contact with the ground: No. 11 bars and smaller, $\frac{3}{4}$ in.

2. Concrete *exposed* to weather or in contact with the ground: No. 5 bars and smaller, $1\frac{1}{2}$ in.; No. 6 bars and larger, 2 in.

The words "exposed to weather" imply direct exposure to moisture changes. The undersides of exterior slabs are not considered exposed to weather unless subject to alternate wetting and drying, including that due to condensation, leakage from the exposed top surface, or runoff. ACI Commentary Sec. 7.7.5 recommends a minimum cover of 2 in. for slabs exposed to chlorides, such as deicing salts (as in parking garages). Alternatively, epoxy-coated bars can be used in such cases. ACI Secs. 4.2 and 4.4 require special concretes in structures exposed to chlorides.

The structural endurance of a slab exposed to fire depends, among other things, on the cover to the reinforcement. Building codes give differing fire ratings for various covers. Reference 10–1 suggests that for normal ratios of service load moment to ultimate moment, $\frac{3}{4}$ in. cover will give a $1\frac{1}{4}$-hour fire rating, 1 in. cover, about $1\frac{1}{2}$ hours, and $1\frac{1}{2}$ in. cover, about 3 hours.

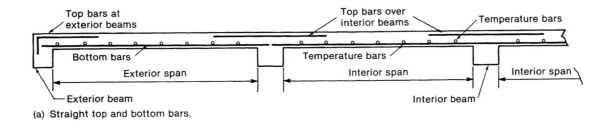

Top bars at exterior beams

Top bars over interior beams

Temperature bars

Bottom bars

Temperature bars

Exterior span

Interior span

Interior span

Exterior beam

Interior beam

(a) Straight top and bottom bars.

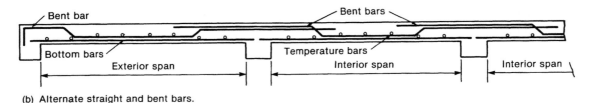

Bent bar

Bent bars

Bottom bars

Temperature bars

Exterior span

Interior span

Interior span

(b) Alternate straight and bent bars.

Fig. 10–13
Sections through one-way slabs showing reinforcement.

Reinforcement

Reinforcement details for one-way slabs are shown in Fig. 10–13. The straight bar arrangement in Fig. 10–13a is almost always used in buildings. Prior to the 1960s slab reinforcement was arranged using the bent-bar straight-bar arrangement shown in Fig. 10–13b. The cutoff points shown in Fig. A–5 can be used if the slab satisfies the requirements for use of the moment coefficients in ACI Sec. 8.3.3. Cutoff points in slabs not satisfying ACI 8.3.3 are obtained using the procedure given in Examples 8–5 and 8–6. One-way slabs are designed assuming a 1-ft-wide strip. The area of reinforcement is then computed as A_s/ft of width. The area of steel is the product of the area of a bar times the number of bars per foot.

$$A_s/\text{ft} = A_b\left(\frac{12 \text{ in.}}{\text{bar spacing in inches}}\right) \tag{10–5}$$

where A_b is the area of one bar (see Table A–9 in Appendix A). In SI units, Eq. 10–5 becomes (see Table A–9M)

$$A_s/\text{m} = A_b\left(\frac{1000 \text{ mm}}{\text{bar spacing in mm}}\right) \tag{10–5M}$$

The maximum spacing of bars in slabs is three times the slab thickness or 18 in., whichever is smaller (ACI Sec. 7.6.5). The maximum bar spacing is also governed by crack control provisions, as discussed in Sec. 9–3 (ACI Sec. 10.6.4). For covers of $\frac{3}{4}$ in. or 1 in., ACI Sec. 7.6.5 governs. For exterior exposure, which requires $1\frac{1}{2}$- to 2-in. minimum covers, ACI Sec. 10.6.4 will govern.

Because a slab is thinner than the beams supporting it, the concrete in the slab shrinks more rapidly than the concrete in the beams. This may lead to shrinkage cracks in the slab. Shrinkage cracks perpendicular to the span will be crossed by flexural reinforcement which will limit the crack widths. To limit the width of shrinkage cracks parallel to the span, additional *shrinkage and temperature reinforcement* is placed perpendicular to the flexural reinforcement. The amount required is specified in ACI Sec. 7.12, which requires the following ratios of reinforcement area to gross concrete area:

1. Slabs with Grade 40 or 50 deformed bars: 0.0020

2. Slabs with Grade 60 deformed bars or welded-wire fabric (smooth or deformed): 0.0018

3. Slabs with reinforcement with a yield strength, f_y, in excess of 60,000 psi at a yield strain of 0.35%: $\dfrac{0.0018 \times 60,000}{f_y}$ but not less than 0.0014

Shrinkage and temperature reinforcement is spaced not farther apart than five times the slab thickness, or 18 in. Splices of such reinforcement must be designed to develop the full yield strength of the bars in tension.

It should be noted that shrinkage cracks may be wide even when this amount of shrinkage reinforcement is provided.[10-2] In buildings, this may occur when shear walls, large columns, or other stiff elements provide restraint to shrinkage and temperature movements. ACI Sec. 7.12.1.2 states that if shrinkage and temperature movements are restrained significantly, the requirements of ACI Secs. 8.2.4 and 9.2.7 shall be considered. These sections ask the designer to make a realistic assessment of the shrinkage deformations and to estimate the stresses resulting from these movements. If the shrinkage movements are restrained completely, the shrinkage steel will yield at the cracks, resulting in a few wide cracks. About three times the minimum shrinkage and temperature reinforcement specified in ACI Sec. 7.12 is required to limit the shrinkage cracks to reasonable widths. Alternatively, unconcreted control strips may be left during construction, to be filled in with concrete after the initial shrinkage has occurred. Methods of limiting shrinkage and temperature cracking in concrete structures are reviewed in Ref. 10–3.

ACI Sec. 10.5.4 specifies that the minimum flexural reinforcement shall be the same as the amount required in ACI Sec. 7.12 for shrinkage and temperature except that maximum spacing of flexural reinforcement is three times the slab thickness (ACI Sec. 7.6.5).

Generally, No. 4 and larger bars are used for flexural reinforcement in slabs, since No. 3 bars tend to be bent out of position by workers walking on the reinforcement during construction. This is more critical for top reinforcement than bottom reinforcement, since the effective depth, d, of the top steel is reduced if it is pushed down, while that of the bottom steel is increased. Sometimes No. 3 bars are used for bottom steel, and for shrinkage and temperature reinforcement in conjunction with No. 4 top bars.

Examples

The examples of slabs, beams, and girders designed in this chapter are based on the floor design for a building similar to the building shown in Fig. 10–3 designed by B. James Wensley Architect Ltd. and Read Jones Christoffersen Ltd. Engineers. In the examples, the floor plan has been modified to simplify the presentation, and the design loads have been changed to correspond to the ASCE 7–95[10-4] design load standard.

A typical floor plan is shown in Fig. 10–14. The exterior column spacing was selected to give a uniform window width around the building. Because the exterior column spacing was greater than the width of the elevator shaft, the interior girders G3 were placed at an angle. This required more complicated formwork than a rectangular grid, but the formwork was reused floor after floor and hence these costs were minimized. Example 10–1 considers the design of the eight-span, one-way slabs along section A-A. A partial sec-

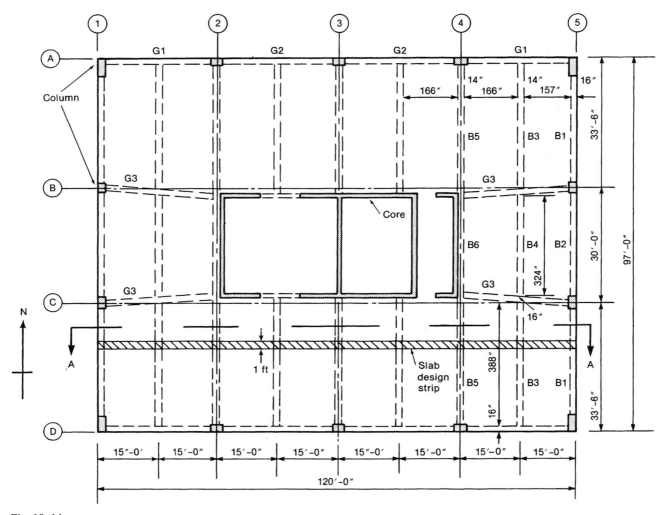

Fig. 10–14
Typical floor plan.

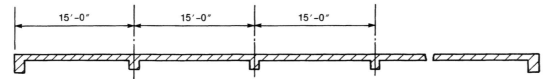

Fig. 10–15
Section A-A.

tion of these slabs is shown in Fig. 10–15. Example 10–2 considers the design of beams
B3–B4–B3. The underside of a typical floor is shown in Fig. 10–16. The photograph is
taken looking along beams B3–B4–B3.

The design loads are selected using ASCE 7–95 *Minimum Design Loads for
Buildings and Other Structures*.[10-4] The dead loads include the weight of the floor, 0.5 psf
for asphalt tile or carpet, 4 psf for mechanical equipment and lighting fixtures hung below
the floor, and 2 psf for a suspended ceiling.

ASCE 7–95 specifies the minimum live loads for offices at 50 psf and for lobbies,
100 psf (see Table 2–1). The live loads for file storage rooms and computer rooms are based
on the actual usage of the room. An allowance must be made for the weight of partitions

(considered to be live load in ASCE 7–95) if the floor loading is 80 psf or less. The design live loads can be reduced as a function of the floor area, as discussed in Sec. 2–7. No live-load reduction is allowed for one-way slabs. In consultation with the architect and owner, it has been decided that all floors will be designed for a live load of 100 psf, including par-titions. This has been done to allow flexibility in office layouts.

A second loading case required in ASCE 7–95 for office buildings is a concentrated load of 2000 lb acting on a 30 in. by 30 in. area placed on the floor to produce the maxi-mum moments or shears in the structural members. The uniform live load is not on the structure when this load acts. For the slab spans in Fig. 10–14, the uniform live loads will always give greater moments and shears in the beams than the concentrated load. If it is as-sumed that the concentrated load is supported by a strip of slab extending one slab thick-ness on each side of the 30-in. area, it can be shown that for this example, the uniform load governs the slab design also. For this reason, the concentrated loads are not specifically considered in Examples 10–1 and 10–2.

The columns are built using 5000-psi concrete; the floors will be 3750-psi con-crete. These strengths have been chosen on the basis of ACI Sec. 10–15, which allows the floor to have a lower concrete strength than the columns. The reinforcement strength is 60 ksi for flexural reinforcement and 40 ksi for shear reinforcement, throughout.

EXAMPLE 10–1   Design of a One-Way Slab

Design the eight-span floor slab spanning east-west in Fig. 10.14. A typical 1-ft-wide design strip is shown shaded. A partial section through this strip is shown in Fig. 10–15. The beams are assumed to be 14 in. wide. The concrete strength is 3750 psi and the reinforcement strength is 60,000 psi. The live load is 100 psf.

1. Estimate the thickness of the floor. The initial selection of the floor thickness will be based on ACI Table 9.5(a) (Table A–14) which gives the minimum thicknesses unless deflections are computed, for members not supporting partitions likely to be damaged by large deflections. In consultation with the architect and owner, it has been decided that the partitions will be movable metal partitions that can accommodate floor deflections.

$$\text{End bay:} \quad \text{min. } h = \frac{\ell}{24}$$

$$= \frac{172 \text{ in.}}{24} = 7.2 \text{ in.}$$

$$\text{Interior bays:} \quad \text{min. } h = \frac{\ell}{28}$$

$$= \frac{180}{28} = 6.43 \text{ in.}$$

Therefore, try a 7.25-in. slab. Assuming $\frac{3}{4}$ in. clear cover and No. 4 bars,

$$d = 7.25 - \left(0.75 + \frac{0.5}{2}\right) = 6.25 \text{ in.}$$

Before the thickness is finalized, it will be necessary to check whether it is adequate for moment and shear.

2. Compute the trial factored loads. Based on the thickness selected in step 1, it is now possible to compute a first estimate of the factored uniform loads. The dead load is:

Slab:

$$w_D = \frac{7.25 \text{ in.}}{12 \text{ in./ft.}} \times 150 \text{ lb/ft}^3$$

$$= 90.6 \text{ lb/ft}^2 \text{ of floor surface}$$

Other dead loads:

Floor cover	0.5 psf
Mechanical equipment	4 psf
Ceiling	2 psf

Total dead load $= 97.1$ psf. The live load is

$$w_L = 100 \text{ psf}$$

The factored load is

$$w_u = 1.4 \times 97.1 + 1.7 \times 100 = 306 \text{ psf}$$

The load per foot on the strip shown shaded in Fig. 10–14 is 1 ft $\times$ 306 psf $= 306$ lb/ft.

Since $w_L < 3w_D$ and the other requirements of ACI Sec. 8.3.3 are met, use the ACI moment coefficients to calculate the moments.

3. Check the thickness required for moment. In Chap. 4 the calculation of the required d was discussed. Equation 4–17 gave

$$\frac{bd^2}{12,000} = \frac{M_u}{\phi k_n} \text{ ft-kips}$$

where $\phi k_n = \phi[f_c'\omega(1 - 0.59)\omega]$, $\omega = \rho f_y/f_c'$. The maximum reinforcement which would normally be used would correspond to the tension-controlled limit for the neutral axis depth given by Eq. 4–23. However, slabs seldom have more than $\rho = 0.01$. Check whether the slab thickness chosen is adequate if $\rho \leq 0.01$. For $f_c' = 3750$ psi and $f_y = 60,000$ psi:

$$\omega = \frac{0.01 \times 60,000}{3750} = 0.160$$

$$\phi k_n = 0.9[3750 \times 0.160(1 - 0.59 \times 0.160)] = 489$$

(see also Table A–3). This can be used to compute bd^2 for a given M_u. The maximum moment M_u will occur at the first or second interior support (ACI Sec. 8.3.3)

First interior support:

$$M_u = \frac{w_u \ell_n^2}{10}$$

where

$$\ell_n = [(157 \text{ in.} + 166 \text{ in.})/2]/12 \text{ in./ft}$$
$$= 13.46 \text{ ft.}$$

$$M_u = \frac{306 \text{ lb/ft} \times (13.46 \text{ ft})^2}{10}$$

$$= 5544 \text{ ft-lb/ft of width} = 5.54 \text{ ft-kips/ft}$$

Second interior support:

$$M_u = \frac{w_u \ell_n^2}{11}$$

where

$$\ell_n = 13.83 \text{ ft.}$$
$$M_u = 5.32 \text{ ft-kips/ft}$$

Therefore, maximum $M_u = 5.54$ ft-kips/ft

$$bd^2 = \frac{5.54 \times 12,000}{489} = 136 \text{ in.}^3$$

But we are designing a 1-ft-wide strip, $b = 12$ in. and

$$d = \sqrt{\frac{136}{12}} = 3.37 \text{ in.}$$

The minimum $d = 3.37$ in. to keep $\rho \leq 0.01$. Since $d = 6.25$ in. computed in step 1 exceeds this, the slab will be OK for flexure.

4. Check if thickness is adequate for shear. Shear reinforcement is required if $V_u > \phi V_c$ (ACI Sec. 11.5.5.1a). Since it is difficult to place shear reinforcement in a slab, the upper limit on V_u will be taken equal to ϕV_c.

If all spans were the same, the maximum shear would occur at the exterior face of the first interior support where $V_u = 1.15 w_u \ell_n/2$. Since the spans are unequal, we will check that location plus a typical interior support:

First interior support:

$$V_u = \frac{1.15 \times 306 \ \text{lb/ft} \times (157/12) \ \text{ft}}{2}$$

$$= 2302 \ \text{lb/ft of width}$$

Typical interior support:

$$V_u = \frac{306(166/12)}{2} = 2117 \ \text{lb/ft}$$

$$\phi V_c = 0.85(2\sqrt{f_c'} \ b_w d)$$

$$= 0.85(2\sqrt{3750} \times 12 \times 6.25) = 7808 \ \text{lb/ft}$$

Since $\phi V_c > V_u$ the slab chosen is adequate for shear. Therefore, **use h = 7.25 in., d = 6.25 in., and w_u = 306 lb/ft.** When slab thicknesses are selected on the basis of deflection control (ACI Table 9–5a), flexure and shear seldom govern.

5. Design the reinforcement. The calculations are done in Table 10–1. First several constants used in that table must be calculated. The maximum moment has been calculated in step 2 as 5.54 ft-kips. We will calculate the reinforcement required here and the corresponding lever arm, jd, and then use that value of jd at all other critical sections. Since the moment is smaller at all other sections, this value of jd will be on the safe (too small) side and will lead to values of A_s that are slightly too large, again on the safe side.

$$A_s = \frac{M_u \times 12,000}{\phi f_y jd} \tag{4–33}$$

For a first trial assume that $jd = 0.925d$ for a slab. Therefore,

$$A_s = \frac{5.54 \times 12,000}{0.9 \times 60,000(0.925 \times 6.25)}$$

$$= 0.213 \ \text{in.}^2/\text{ft}$$

If this is provided exactly,

TABLE 10–1 Calculations for One-Way Slab—Example 10–1

Line						
1. ℓ_n, ft	13.08	13.08	13.46	13.83	13.83	13.83
2. $w_u \ell_n^2$	52.35	52.35	55.44	58.53	58.53	
3. M coefficients	1/24	1/14	1/10 1/11	1/16	1/11	1/16
4. M_u, ft-kip/ft	2.18	3.74	5.54 ~~5.04~~	3.66	5.32	3.66
5. $A_{s(req'd)}$, in.²/ft	~~0.080~~	~~0.137~~	0.202	~~0.134~~	0.194	
6. $A_{s(min)}$, in.²/ft	0.157	0.157	~~0.157~~	0.157	~~0.157~~	
7. Choose steel	#4 @ 15	#4 @ 15	#4 @ 12	#4 @ 15	#4 @ 12	
8. A_s provided	0.16	0.16	0.20	0.16	0.20	

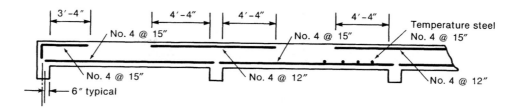

Fig. 10–17
Reinforcement—Example
10.1.

$$a = \frac{0.213 \times 60,000}{0.85 \times 3750 \times 12} = 0.334 \text{ in.}$$

and

$$jd = d - \frac{a}{2}$$

$$= 6.25 - \frac{0.334}{2} = 6.08 \text{ in.}$$

(Note that $j = 0.973$ in this case.)

Since the first estimate of A_s was based on a guess of jd, we will recompute the required A_s:

$$A_s(\text{in.}^2/\text{ft}) = \frac{M_u \times 12,000}{0.9 \times 60,000 \times 6.08}$$

$$= 0.0365 \times M_u \text{ (ft-kips/ft)}$$

The area of steel required at any section will be computed as $A_s(\text{in.}^2/\text{ft}) = 0.0365 \times M_u$ (ft-kips/ft). At the first interior support, this gives $A_s = 0.202$ in.²/ft.

Determine the minimum flexural reinforcement using ACI Sec. 10.5.4, which refers to ACI Sec. 7.12 for amount and spacing:

$$A_{s(\text{min})} = 0.0018bh$$

$$= 0.0018 \times 12 \text{ in.} \times 7.25 \text{ in.}$$

$$= 0.157 \text{ in.}^2/\text{ft}$$

For maximum spacing, ACI Sec. 7.6.5 gives $3h = 21.75$ in., but not more than 18 in. Therefore, maximum spacing $= 18$ in.

The remaining flexural calculations are given in Table 10–1. The choice of the reinforcement in line 7 of this table is made using Eq. 10–5 or Table A–9. The resulting steel arrangement is shown in Fig. 10–17. The cutoff points have been computed using Fig. A–5c since the slab geometry allowed the use of the ACI moment coefficients.

6.   Determine the shrinkage and temperature reinforcement. ACI Sec. 7.12 requires shrinkage and temperature reinforcement perpendicular to the span of the slab:

$$A_s = 0.0018bh = 0.157 \text{ in.}^2/\text{ft}$$

maximum spacing $= 18$ in.

Therefore, provide **No. 4 bars at 15 in. o.c., as shrinkage and temperature reinforcement.** These are placed on top of the lower layer of steel, as shown in Fig. 10–17.

7.   Design the transverse top steel at girders. Due to localized two-way action adjacent to the girders (G1, G2, G3, etc., in Fig. 10–14), top reinforcement is required in the slab perpendicular to the girders. This will be designed when the girders are designed.   ■

This completes the design of this slab. The slab thickness was selected in step 1 using ACI Table 9–5(a) to limit deflections. A 6.5-in. thickness was acceptable in the six interior spans, but a 7.25-in. thickness was required in the end spans. If the entire floor

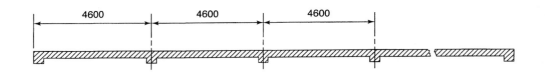

Fig. 10–18
Section A-A—Example
10–1M.

were made 6 in. thick instead of 7.25 in., about 45 cubic yards of concrete could be saved per floor, with a resultant saving of 180 kips of dead load per floor. In a 20-story building this represents a considerable saving. With this in mind the floor was redesigned 6 in. thick and the computed deflections were found to be acceptable. The calculations will not be repeated here, but the slab thickness used in the design of the beams in Example 10–2 will be taken as 6 in.

EXAMPLE 10–1M   Design of a One-Way Slab—SI Units

Design an eight-span floor slab similar to the strip shown shaded in Fig. 10–14. A 1-m-wide strip will be considered. A partial section through this strip is shown in Fig. 10–18. The beams are assumed to be 350 mm wide. The concrete strength is 25 MPa and the reinforcement strength is 400 MPa. The live load is 4.8 kPa.

1. Estimate the thickness of the floor. The initial selection of the floor thickness will be based on ACI Table 9.5(a) (Table A–14), which gives the minimum thickness unless deflections are computed, for members not supporting partitions likely to be damaged by large deflections. In consultation with the architect and owner, it has been decided that the partitions will be movable metal partitions that can accommodate floor deflections.

End bay: min. $h = \dfrac{\ell}{24}$

$$= \frac{4600 \text{ mm} - \dfrac{350 \text{ mm}}{2}}{24} = 184 \text{ mm}$$

Interior bays: min. $h = \dfrac{\ell}{28}$

$$= \frac{4600 \text{ mm}}{28} = 164 \text{ mm.}$$

Therefore, try a 190-mm slab. Assuming 20 mm of clear cover and No. 15 bars,

$$d = 190 - \left(20 + \frac{15}{2}\right) = 162.5 \text{ mm}$$

Before the thickness is finalized, it will be necessary to check whether it is adequate for moment and shear.

2. Compute the trial factored loads.

Based on the thickness selected in step 1, it is now possible to compute a first estimate of the factored uniform loads. The dead load per square meter of slab is:

Slab:

$$w_{Ds} = (0.190 \times 1.0 \times 1.0)\text{m}^3 \times 2450 \text{ kg/m}^3 \times (9.81/1000)\text{kN/kg}$$

$$= 4.57 \text{ kN per m}^2 \text{ of floor surface} = 4.57 \text{ kPa.}$$

Other dead loads:

Floor cover   0.025 kPa

$$\text{Mechanical equipment} \qquad 0.20 \text{ kPa}$$
$$\text{Ceiling} \qquad 0.10 \text{ kPa}$$

Total dead load $w_D = 4.90$ kPa. The live load $= 4.8$ kPa. The factored load is

$$w_u = 1.4 \times 4.9 + 1.7 \times 4.8 = 15.02 \text{ kPa (kN/m}^2)$$

The load per meter on the strip of slab being designed is $1\text{m} \times 15.02 \text{ kN/m}^2 = 15.02$ kN/m.

Since $w_L < 3w_D$ and the other requirements of ACI Sec. 8.3.3 are met, use the ACI moment coefficients to calculate the moments.

3. **Check the thickness required for moment.** In Chap. 4, calculation of the required d was discussed. Equation 4–17M gives

$$\frac{bd^2}{10^6} = \frac{M_u}{\phi k_n} \qquad \text{kN·m}$$

where $\phi k_n = \phi[f_c' \omega (1 - 0.59\omega)]$ and $\omega = \rho f_y / f_c'$. The maximum reinforcement that would normally be used would be that corresponding to the tension controlled limit for the neutral axis depth given by Eq. 4–23. However, slabs seldom have more than $\rho = 0.01$. Check whether the slab thickness chosen is adequate if $\rho \leq 0.01$. For $f_c' = 25$ MPa and $f_y = 400$ MPa,

$$\omega = \frac{0.01 \times 400}{25} = 0.160$$

$$\phi k_n = 0.9[25 \times 0.160(1 - 0.59 \times 0.160)] = 3.26$$

This can be used to compute bd^2 for a given M_u. The maximum moment M_u will occur at the first or second interior support (ACI Sec. 8.3.3).

First interior support:

$$M_u = \frac{w_u \ell_n^2}{10}$$

where

$$\ell_n = \frac{(4600 - 350 - 350/2) + (4600 - 350)}{2} = 4162 \text{ mm}$$

$$M_u = \frac{15.02 \text{ kN/m} \times 4.162^2 \text{m}^2}{10} = 26.0 \text{ kN-m per meter width of slab}$$

Second interior support:

$$M_u = \frac{w_u \ell_n^2}{11}$$

where $\ell_n = 4600 - 350 = 4250$ mm and

$$M_u = \frac{15.02 \times 4.250^2}{11} = 24.7 \text{ kN-m/m}$$

Therefore, the maximum $M_u = 26.0$ kN-m/m and

$$bd^2 = \frac{26.0 \times 10^6}{3.26} = 7,975,000 \text{ mm}^3$$

But we are designing a 1-m-wide strip. Therefore, $b = 1000$ mm and

$$d = \sqrt{\frac{7{,}975{,}000}{1000}} = 89.3 \text{ mm}$$

The minimum d is 89.3 mm to keep $\rho \leq 0.01$. Since $d = 162.5$ mm computed in step 1 exceeds this, the slab will be OK for flexure.

4. Check if thickness is adequate for shear. Shear reinforcement is required in slabs if $V_u > \phi V_c$ (ACI Secs. 11.5.5.1a and 11.5.6.1). Since it is difficult to place shear reinforcement in a slab, the upper limit on V_u will be taken as ϕV_c.

If all the spans were the same, the maximum shear would occur at the exterior face of the first interior support, where $V_u = 1.15 w_u \ell_n / 2$. Since the spans are unequal, we will check that location plus a typical interior support:

First interior support:

$$\ell_n \text{ for shear} = 4600 - 350 - \frac{350}{2} = 4075 \text{ mm}$$

$$V_u = \frac{1.15 \times 15.02 \times 4.075}{2} = 35.2 \text{ kN/m of width}$$

Typical interior support:

$$\ell_n = 4600 - 350 = 4250 \text{ mm}$$

$$V_u = \frac{15.02 \times 4.250}{2} = 31.9 \text{ kN/m}$$

$$\phi V_c = 0.85\left(\sqrt{\frac{f_c'}{6}} b_w d \right)$$

$$= 0.85\left(\frac{\sqrt{25}}{6} \times 1000 \times 162.5 \right) = 115{,}100 \text{ N}$$

$$= 115.1 \text{ kN}$$

Since $\phi V_c > V_u$, the slab chosen is adequate for shear. Therefore, **use $h = 190$ mm, $d = 162.5$ mm, and $w_u = 15.02$ kN/m^2.** When slab thicknesses are selected on the basis of deflection control (ACI Table 9.5a), flexure and shear seldom govern.

5. Design the reinforcement. The calculations are done in Table 10–1M. First, several constants used in that table must be calculated. The maximum moment has been calculated in step 2 as 26.0 kN-m/m. We will calculate the reinforcement required at that point and the corresponding lever

TABLE 10–1M Calculations for One-Way Slab—Example 10–1M

Line							
1. ℓ_n, m	4.075	4.075	4.163		4.250	4.250	4.250
2. $w_u \ell_n^2$	249.4	249.4	260.3		271.3	271.3	271.3
3. M coefficients	1/24	1/14	1/10	1/11	1/16	1/11	1/16
4. M_u, kN-m/m	10.39	17.81	26.03	~~23.66~~	16.96	24.66	16.96
5. $A_{s(reqd)}$, mm^2/m	~~182.7~~	~~313.1~~	457.6		~~298.2~~	433.5	~~298.2~~
6. $A_{s,min}$, mm^2/m	342	342	~~342~~		342	~~342~~	342
7. Choose steel	#10 @ 280	#10 @ 280	@10 @ 220		#10 @ 280	#10 @ 220	#10 @ 280
8. A_s provided	357	357	455		357	455	357

arm, jd, and use that value of jd at all other critical sections. Since the moment is smaller at all other sections, this value of jd will be on the safe side (too small) and will lead to values of A_s that are slightly too large, again on the safe side.

$$A_s = \frac{M_u \times 10^6}{\phi f_y jd} \tag{4–33}$$

For a first trial, assume that $jd = 0.925d$ for a slab. Therefore,

$$A_s = \frac{26.0 \times 10^6}{0.90 \times 400 \times (0.925 \times 162.5)}$$

$$= 480 \text{ mm}^2/\text{m}$$

If this is provided exactly,

$$a = \frac{480 \times 400}{0.85 \times 25 \times 1000} = 9.04 \text{ mm}$$

and

$$jd = d - \frac{a}{2}$$

$$= 162.5 - \frac{9.04}{2} = 158.0 \text{ mm}$$

(Note that $j = 0.972$ in this case.)

Since the first estimate of A_s was based on a guess of jd, will recompute the required A_s:

$$A_s (\text{mm}^2/\text{m}) = \frac{M_u \times 10^6}{0.90 \times 400 \times 158}$$

$$= 17.58 \times M_u (\text{kN}-\text{m}/\text{m})$$

The area of steel required at each critical section will be computed as $A_s(\text{mm}^2/\text{m}) = 17.58 \times M_u$ ($\text{kN}-\text{m}/\text{m}$). At the first interior support this gives $A_s = 457 \text{ mm}^2/\text{m}$.

Determine the minimum flexural reinforcement using ACI Sec. 10.5.4, which refers to ACI Sec. 7.12 for the amount and gives the maximum spacing as the lesser of three times the slab thickness or 500 mm.

$$A_{s, min} = 0.0018bh$$

$$= 0.0018 \times 1000 \text{ mm} \times 190 \text{ mm}$$

$$= 342 \text{ mm}^2/\text{m}$$

The maximum spacing is $3h = 3 \times 190 = 570$ mm but not more than 500 mm. Therefore, the maximum spacing is 500 mm.

The remaining flexural calculations are given in Table 10–1M. The choice of the reinforcement in line 7 of this table is made using Eq. 10–5M or Table A–9M. The resulting steel arrangement is shown in Fig. 10–19. The cutoff points have been computed using Fig A–5c since the slab geometry allowed the use of the ACI moment coefficients.

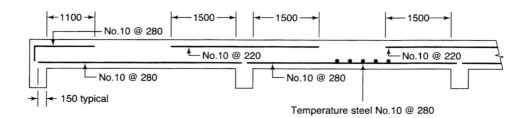

Fig. 10–19
Reinforcement—Example
10–1M

6. Select the shrinkage and temperature reinforcement. ACI Sec. 7.12 requires shrinkage and temperature reinforcement perpendicular to the span of the slab:

$$A_s = 0.0018bh = 342 \text{ mm}^2/\text{m}$$

$$\text{maximum spacing} = 500 \text{ mm}$$

Therefore, **provide No. 10 bars at 280 mm on centers as shrinkage and temperature reinforcement.** These are placed on top of the lower layer of steel, as shown in Fig. 10–19.

7. Design the transverse top steel at girders. Due to localized two-way action adjacent to the girders (G1, G2, G3, etc., in Fig. 10–14), top reinforcement is required in the slab perpendicular to the girders. This will be designed when the girders are designed. ■

10–5 CONTINUOUS BEAMS

The three major stages in the design of a continuous beam are: design for flexure, design for shear, and design of bar details. In addition, it is necessary to consider deflections and crack control and in some cases, torsion.

When the area supported by a beam exceeds 400 ft², it is frequently possible to use a reduced live load in calculating the moments and shears in the beam. The concept of live-load reduction factors has been discussed in Sec. 2–7. Depending on the applicable building code, the live-load reduction factor will be a function of the tributary area of the beam or its influence area. The following example uses the ASCE 7–95 loading code, which is based on influence areas (see Sec. 2–7).

EXAMPLE 10–2 Design of a Continuous T Beam

Design the beam B3–B4–B3 in Fig. 10–14. This beam supports its own dead load plus load from a 6-in. slab. This is thinner than the slab designed in Example 10–1, as explained at the end of that example. The beam is supported on girders at lines A, B, C, and D as shown in Fig. 10–14 and is symmetrical about the centerline of the building. Figure 10–16 is a photograph looking along the beam. Use $f'_c = 3750$ psi, $f_y = 60,000$ psi for flexural reinforcement, and $f_y = 40,000$ psi for stirrups. Base the loadings on the ASCE 7–95 design loading code.

1. Compute the trial factored loads on the beam. Because the size of the beam stem is unknown at this stage, it is not possible to compute the final loads for use in the design. For preliminary purposes estimate the size of the beam stem. Once the size has been established, the factored load will be corrected and used in subsequent calculations.

The ASCE 7–95 design loading code allows live-load reductions based on *influence areas*. For a beam, this is the area between the beam in question and the faces of the beams on either side of it over the loaded length of the beam as shown in Fig. 2–10. Three influence areas will be considered in the design of beam B3–B4–B3, as shown in Fig. 10–20.

(a) To compute the positive moment at midspan of beam B3, and negative moment at the exterior end of beam B3, it is necessary to load spans AB and CD. Since the majority of the moment in the left span results from loads on that span, we shall consider the influence area to be that shown in Fig. 10–20a:

$$A_I = \frac{388(157 + 14 + 166)}{12 \times 12} = 908 \text{ ft}^2$$

From Eq. 2–9 the reduced live load, L, is

$$L = L_0\left(0.25 + \frac{15}{\sqrt{A_I}}\right)$$

where L_0 is the unreduced live load=100 psf and

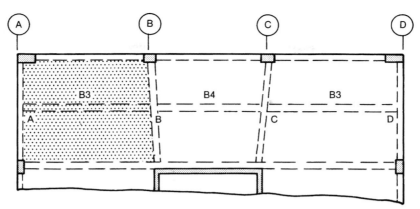

(a) Influence area for positive moment–B3.

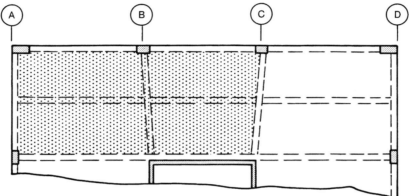

(b) Influence area for interior negative moment.

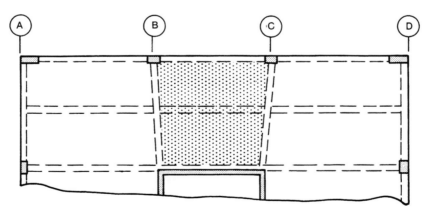

Fig. 10–20
Influence areas—Example
10–2.

(c) Influence area for positive moment–B4.

$$L = 100 \left(0.25 + \frac{15}{\sqrt{908}} \right) = 0.748 \times 100$$

$$= 74.8 \text{ psf}$$

(b) To compute the maximum negative moment at the interior support, B, spans AB and BC would be loaded. The corresponding influence area is shown in Fig. 10–20b. Here

$$A_I = \frac{(157 + 14 + 166)(388 + 16 + 324)}{12 \times 12}$$

$$= 1704 \text{ ft}^2$$

and

$$L = 100\left(0.25 + \frac{15}{\sqrt{1704}}\right) = 61.3 \text{ psf}$$

(c) To compute the positive moment in the center span (beam $B4$), span BC would be loaded. The influence area is shown in Fig. 10–20c. Here

$$A_I = \frac{324(157 + 14 + 166)}{12 \times 12} = 758 \text{ ft}^2$$

and

$$L = 100\left(0.25 + \frac{15}{\sqrt{758}}\right) = 79.5 \text{ psf}$$

If this beam was being analyzed using a computer analysis throughout, one would probably use a single reduced live load throughout. This would be based on the smallest influence area.

The size of the stem beam $B3$–$B4$–$B3$ will be chosen on the basis of negative moment at the first interior support. For this location the factored load on the beam is as follows:

(a) Dead load (not including the portion of the beam stem below the slab):

Slab	75 psf
Floor, ceiling, mechanical	6.5 psf
	81.5 psf

Live load (reduced): 61.3 psf
Total factored load from slab:

$$1.4 \times 81.5 + 1.7 \times 61.3 = 218.3 \text{ psf}$$

Assume that the tributary width which loads beam $B3$–$B4$–$B3$ extends half-way from this beam to the two adjacent beams, or

$$\left(\frac{157}{2} + 14 + \frac{166}{2}\right) \text{in.} = 175.5 \text{ in.} = 14.63 \text{ ft}$$

Therefore, the factored load per foot from the slab is

$$14.63 \text{ ft} \times 218.3 \text{ psf} = 3.19 \text{ kips per foot}$$

It is now necessary to estimate the weight of the beam stem. Two very approximate methods of doing this were given in Example 4–6. These are:

(1) The factored dead load of the stem $\simeq$ 10 to 20% of the other factored loads on the beam. This gives 0.32 to 0.64 kip/ft.
(2) The overall depth of beam $h \simeq$ 8 to 10% of ℓ_n and $b_w \simeq 0.5h$. This gives the overall $h = 31$ to 39 in., with the stem extending 25 to 33 in. below the slab, and gives $b_w = 15$ to 20 in. The factored load from stems of such sizes ranges from 0.55 to 0.94 kip/ft.

As a first trial, assume the factored weight of the stem to be 0.70 kip/ft.

$$\text{total trial load per foot} = 3.19 + 0.70 \text{ kip/ft.}$$

$$= 3.89 \text{ kips/ft}$$

say, 3.9 kips/ft.

 2. **Choose the actual size of the beam stem.** The size of the stem is governed by three factors: deflections, moment capacity at the point of maximum negative moment, and shear capacity. In addition, the overall depth of the floor should be minimized to reduce the overall height of the building.

 (a) **Determine the minimum depth based on deflections.** ACI Table 9.5(a) (Table A–14) gives minimum depths unless deflections are checked. Since the anticipated partitions will be flexible enough to undergo some deflections, this table can be used.

Beam $B3$ is one-end continuous, $f_y = 60{,}000$ psi, and normal-density concrete. Therefore,

$$\text{min. } h = \frac{\ell}{18.5}$$

where ℓ shall be taken as the span center to center of supports. Therefore,

$$\text{min. } h \text{ based on deflections} = \frac{388 + 16 \text{ in.}}{18.5} = 21.8 \text{ in.}$$

 (b) **Determine the minimum depth based on the moment at the first interior support.**

$$\text{Moment} \simeq \frac{w_u \ell_n^2}{10} \quad \text{where } \ell_n \text{ is the average of the spans}$$

$$= \frac{(388 + 324) \text{ in.}}{2} = 356 \text{ in.} = 29.67 \text{ ft}$$

Therefore,

$$M_u \simeq \frac{3.9 \times 29.67^2}{10} = 343 \text{ ft-kips}$$

 In Chap. 4 beam sizes were chosen starting with the assumption that a desirable steel ratio was $\rho = 0.01$. In a continuous beam a slightly higher steel ratio is often used at the point of maximum negative moment. Assume that ρ is limited to 0.013 and that there is no compression reinforcement. (From Table A–5, the ρ corresponding to the tension-controlled limit is 0.0169.) From Table A–3, $\phi k_n = 616$. Thus

$$\frac{bd^2}{12{,}000} = \frac{M_u}{\phi k_n}$$

$$= \frac{343}{616} = 0.557$$

and $bd^2 = 6682$. Possible choices are

$$b = 10 \text{ in.}, d = 25.8 \text{ in.}$$

$$b = 12 \text{ in.}, d = 23.6 \text{ in.}$$

$$b = 14 \text{ in.}, d = 21.8 \text{ in.}$$

Select the last of these. With one layer of reinforcement at the supports $h \simeq 21.8 + 2.5 = 24.3$ in. This exceeds the minimum h based on deflections. **Try a 14 in. by 24 in. stem, with $d = 21.5$ in.**

 (c) **Check the shear capacity of the stem.**

$$V_u = \phi(V_c + V_s)$$

From ACI Sec. 11.3.1.1., $V_c = 2\sqrt{f_c'}\, b_w d$, and from ACI Sec. 11.5.6.8, the maximum V_s is $8\sqrt{f_c'}\, b_w d$. Therefore, the absolute maximum V_u allowed by the code $= \phi(2 + 8)\sqrt{f_c'}\, b_w d$ where

$$\phi = 0.85$$

$$\text{maximum } V_u = \phi(V_c + V_s)$$

The maximum V_u due to loads on the beam $= 1.15\, w_u \ell_n / 2$ where ℓ_n for beam $B3$ is 388 in.

$$V_u = 72.5 \text{ kips}$$

Therefore,

$$\text{minimum } b_w d = \frac{72.5 \times 1000\, lb}{8.5\sqrt{3750}} = 139 \text{ in.}$$

Thus, for $b_w = 14$ in., $d = 10$ in. Therefore, shear does not govern.

(d) Summary. Use

$$\mathbf{b = 14\ in.}$$

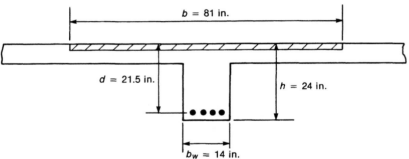

(a) Positive moment region.

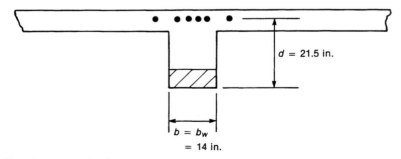

Fig. 10–21
Beam sections.

(b) Negative moment region.

$$h = 24 \text{ in. } (18 \text{ in. below slab})$$

$$d = 21.5 \text{ in.,} \text{ assuming one layer of steel at all sections}$$

This section is shown in Fig. 10–21.

3. Compute the dead load of the stem and recompute the total load per foot.

$$\text{weight per foot of the stem below the slab} = \frac{14 \times 18 \times 12}{1728} \times 0.15 = 0.263 \text{ kip/ft}$$

$$\text{total factored load} = 3.19 + 1.4 \times 0.263$$

$$= 3.56 \text{ kips/ft (at interior supports)}$$

Note that although the weight of the beam stem was only $0.263/0.7 = 0.38$ times the original guess, the error in the original estimate of the total factored load was only 9%. If the final factored load per foot increased by more than about 10%, it would be desirable to recalculate the beam size and load per foot.

4. Calculate the flange width for the positive moment regions. The beam acts as a T beam in the positive moment regions. The effective width of the flange is the smallest of (ACI Sec. 8.10.2):

$$0.25 \, \ell_n \text{ (based on the shorter span for simplicity)} = 0.25 \times 324 \text{ in.} = 81 \text{ in.}$$

$$b_w + 2(8 \times 6) = 14 \text{ in.} + 2 \times 48 \text{ in.} = 110 \text{ in.}$$

$$b_w + \frac{157 \text{ in.}}{2} + \frac{166 \text{ in.}}{2} = 175.5 \text{ in.}$$

Therefore, the effective flange width is 81 in., as shown in Fig. 10–21.

5. Compute the moments. The moments can be computed using frame analysis, or if the structure fits the limitations of ACI Sec. 8.3.3, the moment coefficients can be used. The structure is a non-prestressed continuous beam (ACI Sec. 8.3.2) with more than two spans. The ratio of clear spans = 388 in./324 in. = 1.198. This is just less than the upper limit of 1.20 for use of the moment coefficients. The loads are uniformly distributed.

$$\text{factored live load per foot} = 1.7 \times 100 \times 14.63$$

$$= 2.49 \text{ kips/ft}$$

$$\text{factored dead load per foot} = 1.4(81.5 \times 14.63 + 0.263)$$

$$= 2.04 \text{ kips/ft}$$

Therefore, the live load is less than three times the dead load.

TABLE 10–2 Calculation of Moments for Beam B3–B4–B3—Example 10–2

Line					
1. ℓ_n, ft	32.33	32.33	29.67	27.0	29.67
2. w_u, kip/ft	3.90	3.90	3.56	4.02	3.56
3. $w_u \ell_n^2$	4074	4074	3135	2927	3135
4. C_m	1/24	1/14	1/10 1/11	1/16	1/11 1/10
5. $M_u = C_m w_u \ell_n^2$, ft-kip	−170	291	−314	+183	−314

TABLE 10–3 Calculation of Reinforcement Required—Example 10–2

Line				
5. M_u, ft-kips	−170	291	−314	183
6. A_s coefficient	0.0117	0.0105	0.0117	0.0105
7. $A_{s(req'd)}$, in.^{2}	1.99	3.06	3.67	1.92
8. $A_s > A_{s(min)}$	Yes	Yes	Yes	Yes
9. Bars selected	2 No. 7	4 No. 8	6 No. 7	1 No. 8
	2 No. 6			2 No. 7
10. A_s provided, in.^{2}	2.08	3.16	3.60	1.99
11. b_w OK	—	Yes	—	Yes

Therefore, use the ACI moment coefficients in this design. The moment calculations are presented in Table 10–2. Since the structure is symmetrical about the center of the building, only a part of the beam is shown in this table. The values of ℓ_n in line 1 are the clear span lengths except that when computing the negative moment at supports B and C, ℓ_n is the average of the two adjacent spans. The factored uniform loads in line 2 differ for each span, due to the different live-load reduction factors for the various sections. Because the exterior end of span AB is supported on a spandrel girder, the moment coefficient at A is 1/24. The resulting moment diagram is shown in Table 10–3.

6.   Design the reinforcement. The reinforcement will be designed in Table 10–3. This is a continuation of Table 10–2. Prior to entering Table 10–3, however, it is necessary to compute some constants for use in subsequent calculations of area of reinforcement.

(a)   Compute the area of steel required at the point of maximum negative moment (first interior support).

$$A_s = \frac{M_u \text{ ft-kips} \times 12,000}{\phi f_y jd} \tag{4–33}$$

Because there is negative moment at the support, the beam acts as a rectangular beam with compression in the web, as shown in Fig. 10–21b. Assume that $j = 0.875$:

$$A_s = \frac{314 \times 12,000}{0.9 \times 60,000(0.875 \times 21.5)}$$

$$= 3.70 \text{ in.}^2$$

If exactly this much is used,

$$a = \frac{3.70 \times 60,000}{0.85 \times 3750 \times 14}$$

$$= 4.98 \text{ in.}$$

$$\frac{a}{d} = \frac{a}{d_t} = \frac{4.98}{21.5} = 0.232$$

Since this is less than $a_b/d = 0.503$ from Table A-4, $f_s = f_y$ at ultimate. Since $a/d_t = 0.232$ is less than $a_{tcl}/d_t = 0.319$, the section is tension-controlled and $\phi = 0.90$.

Because the original calculation of A_s was based on a guess of jd, recompute A_s as

$$A_s = \frac{M_u(\textit{ft-kips}) \times 12,000}{0.9 \times 60,000(21.5 - 4.98/2)}$$

$$= 0.0117M_u$$

In lines 6 and 7 of Table 10–3 the area of steel required in negative moment regions is calculated as $A_s = 0.0117M_u$. The "constant" 0.0117 was evaluated at the point of maximum negative moment and will be conservative at all other points of negative moment.

(b) **Determine the area of steel required at the point of maximum positive moment (point B near the middle of the exterior span).** In the positive moment regions, the beam acts as a T-shaped beam with compression in the top flange. Assume that the compression zone is rectangular, as shown in Fig. 10–21a. Take $j = 0.95$ for the first calculation of A_s:

$$A_s = \frac{291 \times 12,000}{0.9 \times 60,000(0.95 \times 21.5)}$$

$$= 3.17 \text{ in.}^2$$

Assume that a is less than h_f; then

$$a = \frac{A_s f_y}{0.85 f_c' b}$$

where b is the effective flange width of 81 in.

$$a = \frac{3.17 \times 60,000}{0.85 \times 3750 \times 81} = 0.74 \text{ in.}$$

$$\frac{a}{d} = \frac{a}{d_t} = \frac{0.74}{21.5} = 0.0344$$

From Table A–4 it is seen that this is much less than the $a/d = 0.503$ for balanced failure; therefore, $f_s = f_y$. Similarly, $a/d_t = 0.0344$ is much less than $a/d_t = 0.319$ at the tension-controlled limit; therefore, $\phi = 0.90$.

To compute A_s, use

$$A_s = \frac{M_u \text{ (ft-kips)} \times 12,000}{0.9 \times 60,000(21.5 - 0.74/2)}$$

$$= 0.0105M_u$$

The area of steel required at positive moment regions is calculated in lines 6 and 7 of Table 10–3 using $A_s = 0.0105M_u$.

(c) **Determine the minimum reinforcement.** The minimum reinforcement required is (ACI Sec. 10.5.1)

$$A_{s,\,min} = \frac{3\sqrt{f_c'}}{f_y} b_w d \geq \frac{200 b_w d}{f_y} \qquad (4\text{–}31)$$

$$\text{(ACI Eq. 10–3)}$$

$$= \frac{3\sqrt{3750}}{60,000} \times 14 \times 21.5 \geq \frac{200 \times 14 \times 21.5}{60,000}$$

$$= 0.922 \geq 1.00 \text{ in.}^2$$

A_s exceeds $A_{s,min}$ in both the positive and negative moment regions.

(d) **Calculate the area of steel and select the bars.** The remaining calculations are done in Table 10–3. The areas computed in line 7 exceed $A_{s(min)}$ in all cases. If they did not, $A_{s(min)}$ would be used. The bars selected at each location and their areas are given in lines 9 and 10. The bars were selected using Table A–10. Small bars were selected at the exterior support since they have to be hooked into the

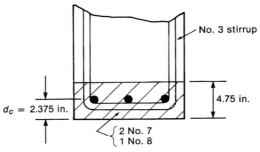

(a) Calculation of A in positive moment region.

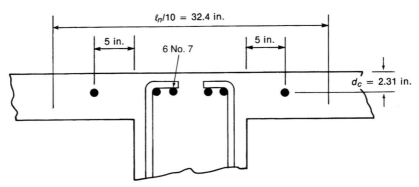

Fig. 10–22
Distribution of reinforcement.

(b) Bar spacing at interior support.

support and there may not be enough room for a standard hook on larger bars. The area selected at supports B and C is slightly less (2% less) than the required area. This is acceptable since the areas selected at the positive moment regions are greater than required. Check whether the bars will fit into the beam (line 11, Table 10–3) using Table A–6. In the negative moment regions, some of the bars can be placed in the slab beside the beams and hence it is not necessary to check whether they will fit into the web width.

7. Check the distribution of the reinforcement.

(a) Positive moment region. Since f_y exceeds 40,000 psi it is necessary to satisfy ACI Secs. 10.6.3 and 10.6.4. This will be checked at the section with the smallest number of positive moment bars, the middle of span B-C. Here there are 2 No. 7 bars and 1 No. 8 bar, with $A_s = 1.99$ in.^{2}, as shown in Fig. 10–22a.

$$\text{equivalent number of No. 8 bars} = \frac{1.99 \text{ in.}^2}{0.79 \text{ in.}^2} = 2.52$$

The centroid of the bar is located d_c above the bottom of the beam:

$$d_c = 1.5 \text{ in. cover} + 0.375 \text{ in. stirrups} + \frac{1.0}{2} \text{ in.}$$

$$= 2.375 \text{ in.}$$

The shaded area in Fig. 10–22 is $2d_c$ in height and has an area of $4.75 \times 14 = 66.5$ in.^{2} Therefore,

$$A = \frac{\text{shaded area in Fig. 10–20a}}{\text{equiv. number of bars}}$$

$$= \frac{66.5 \text{ in.}^2}{2.52} = 26.39 \text{ in.}^2 \text{ per bar}$$

$$z = f_s \sqrt[3]{d_c A}$$

where $f_s = 0.6 f_y$ where f_s is in ksi $= 36$ ksi, and

$$z = 36 \text{ ksi}(\sqrt[3]{2.375 \times 26.39}) \text{ in.}$$

$$= 143 \text{ kips/in.}$$

For interior exposure, z cannot exceed 175 kips/in. The bar spacing is adequate in the positive moment regions.

(b) Negative moment region. ACI Sec. 10.6.6 says "part" of the negative moment steel shall be distributed over a width equal to the smaller of the effective flange width = 81 in., or $\ell_n / 10 = 324/10 = 32.4$ in. At each of the interior negative moment regions there are six top bars. Two of these will be placed in the corners of the stirrups, as shown in Fig. 10–22b, two over the beam web, and the other two will be placed in the slab. (Note that for bars placed in the slab to be completely effective, there must be reinforcement perpendicular to the beam in the slab. In this case the slab reinforcement will serve this purpose.) The two bars placed in the slab will be placed to give a value of z approaching the maximum allowed. Here

$$d_c = 1.5 + 0.375 + \frac{0.875}{2} = 2.31 \text{ in}$$

and $f_s = 36$ ksi. For $z = 175$ kips/in. the maximum allowable value of A is

$$175 = 36\sqrt[3]{2.31A}$$

$$\text{maximum } A = \frac{(175/36)^3}{2.31}$$

$$= 49.67 \text{ in.}^2$$

For $d_c = 2.31$ the maximum bar spacing (Fig. 10–22b)

TABLE 10–4 Calculation of Shear Forces—Example 10–3

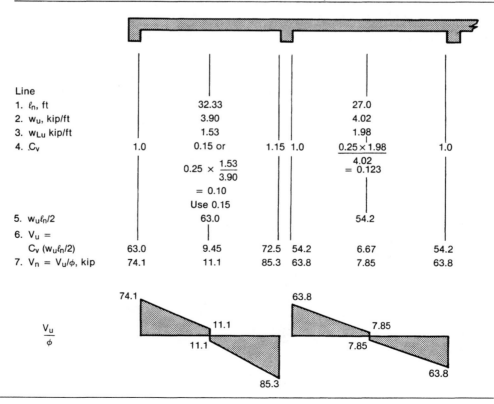

Line							
1. ℓ_n, ft		32.33			27.0		
2. w_u, kip/ft		3.90			4.02		
3. w_{Lu} kip/ft		1.53			1.98		
4. C_v	1.0	0.15 or $0.25 \times \dfrac{1.53}{3.90}$ $= 0.10$ Use 0.15	1.15	1.0	$\dfrac{0.25 \times 1.98}{4.02}$ $= 0.123$	1.0
5. $w_u\ell_n/2$		63.0			54.2		
6. $V_u =$ $C_v (w_u\ell_n/2)$	63.0	9.45	72.5	54.2	6.67	54.2
7. $V_n = V_u/\phi$, kip	74.1	11.1	85.3	63.8	7.85	63.8

$\dfrac{V_u}{\phi}$

74.1

11.1

11.1

85.3

63.8

7.85

7.85

63.8

$$s = \frac{A}{2d_c}$$

$$= \frac{49.67 \text{ in}^2}{2 \times 2.31 \text{ in.}} = 10.75 \text{ in.}$$

Thus within a width of 32.4 in. we must place 6 bars. These cannot be farther apart than 10.75 in. ACI Sec. 10.6.6 requires that "part" of the longitudinal reinforcement be distributed over the effective flange width or over a width of $\frac{1}{10}$ of the span (324 in./10 = 32.4 in.), whichever is smaller. We shall arbitrarily place two bars in the slab at 5 in. outside the web of the beam.

ACI Sec. 10.6.6 requires "some" longitudinal reinforcement in the slab outside this band. We shall assume that the shrinkage and temperature steel already in the slab will satisfy this requirement. The final distribution of negative moment steel is as shown in Fig. 10–22b.

8. Design the shear reinforcement. The shear force diagrams are calculated in Table 10–4 and shown at the bottom of that table. The shear coefficients at midspan of the beams (line 4) are based on Eq. 10–4.

(a) Exterior end of B3. The critical section for shear is located at $d = 21.5$ in. from the support. From Table 10–4 and similar triangles, the shear $V_n = V_u/\phi$ at d from the support is

$$\frac{V_u}{\phi} \text{ at } d = 74.1 - \frac{21.5}{194}(74.1{-}11.1)$$

$$= 67.1 \text{ kips}$$

ACI Sec. 11.5.5.1 requires stirrups if $V_n \geq V_c/2$, where

$$V_c = 2\sqrt{f_c'}\, b_w d$$

$$= 2\sqrt{3750} \times 14 \times \frac{21.5}{1000} = 36.8 \text{ kips}$$

$$\frac{V_c}{2} = 18.4 \text{ kips}$$

Since $V_u/\phi = 67.1$ kips exceeds $V_c/2 = 18.4$ kips, stirrups are required.

Try No. 3 Grade 40 double-leg stirrups with a 90° hook enclosing a No. 4 stirrup support bar. The maximum stirrup spacing is the smaller of

$$\frac{d}{2} = 10.75 \text{ in.} \qquad \text{(ACI Sec. 11.5.4.1)}$$

and

$$s = \frac{A_v f_y}{50 b_w} = 12.6 \text{ in.} \qquad \text{(ACI Sec. 11.5.5.3)}$$

Use 10 in. as the maximum spacing.

The spacing required to support the shear forces is

$$s = \frac{A_v f_y d}{V_u/\phi - V_c} \qquad (6{-}21)$$

At d from end A, $V_u/\phi = 67.1$ kips and

$$s = \frac{0.22 \times 40{,}000 \times 21.5}{(67.1{-}36.8) \times 1000}$$

$$= 6.24 \text{ in.} \qquad \text{say, 6 in. on centers}$$

The stirrup spacing will be changed to 8 in. and then to 10 in. Compute V_u/ϕ where $s = 8$ in. From Eq. 6–25,

$$\frac{V_u}{\phi} = \frac{A_v f_y d}{s} + V_c$$

$$= \frac{0.22 \times 40 \times 21.5}{8} + \frac{36,800}{1000}$$

$$= 60.5 \text{ kips}$$

This occurs at

$$x = \frac{74.1 - 60.5}{74.1 - 11.1} \times 194 \text{ in.} = 41.9 \text{ in. from end } A$$

Compute V_u / ϕ and x where $s = 10$ in.:

$$\frac{V_u}{\phi} = 55.7 \text{ kips} \qquad x = 56.7 \text{ in.}$$

At the exterior end of $B3$, use No. 3 Grade 40 double-leg stirrups: 1 at 3 in., 7 at 6 in., 2 at 8 in., 13 at 10 in. on centers.

(b) **Interior end of $B3$.** The shear at d from the support at B is

$$\frac{V_u}{\phi} = 85.3 - \frac{21.5}{194}(85.3 - 11.1)$$

$$= 77.1 \text{ kips}$$

The spacing required at this point

$$s = \frac{0.22 \times 40,000 \times 21.5}{(77.1 - 36.8)1000}$$

$$= 4.7 \text{ in.}, \qquad \text{use } s = 4.5 \text{ in.}$$

Change the stirrup spacing to 6 in. and then to 10 in. V_u / ϕ where $s = 6$ in.:

$$\frac{V_u}{\phi} = \frac{0.22 \times 40 \times 21.5}{6} + \frac{36,800}{1000}$$

$$= 68.3 \text{ kips}$$

This occurs at

$$x = \frac{85.3 - 68.3}{85.3 - 11.1} \times 194 = 44.4 \text{ in. from end } B$$

$V_u / \phi = 55.7$ kips at point where spacing can change from 6 in. to 10 in. this occurs at

$$x = 77.4 \text{ in. from end } B$$

At the interior end of $B3$ use No. 3 Grade 40 double-leg stirrups: 1 at 2 in., 10 at 4.5 in., 5 at 6 in. and 11 at 10 in. on centers.

(c) **Ends of beam $B4$.** The calculations for beam $B4$ are carried out in the same manner as for beam $B3$. **At each end of beam $B3$ use No. 3 Grade 40 double-leg stirrups: 1 at 4 in., 3 at 8 in., and 11 at 10 in.**

9. **Check the development lengths and design bar cutoffs.**

(a) **Perform the preliminary calculations.**

Detailing requirements:

$$d = 21.5 \text{ in.}$$

$12d_b = 10.5$ in. for No. 7 and 12 in. for No. 8.

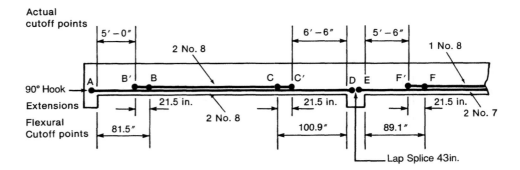

Fig. 10–23
Calculation of bar cutoff
points for positive moment
steel.

$$\frac{\ell_n}{16} = \frac{388}{16} = 24.25 \text{ in. for } B3$$

$$= \frac{324}{16} = 20.25 \text{ in. for } B4$$

Therefore, d always exceeds $12d_b$, but $\ell_n / 16$ exceeds d in beam $B3$.

(b) Select cutoffs for positive moment steel in $B3$. The reinforcement at midspan is 4 No. 8 bars (Table 10–3). Extend two of these into the supports and cut off two where they are no longer required. The remaining bars have 50% of the original area; therefore, the flexural cutoff is at the point where $M_u = 50\%$ of the maximum M_u at midspan. From Fig. A–3, this occurs at $0.21\ell_n = 0.21 \times 388 = 81.5$ in. from the exterior end, and $0.26\ell_n = 100.9$ in. from the interior end. These are points B and C in Fig. 10–23. To compute the actual cutoff points we must satisfy the detailing rules in ACI Secs. 12.1, 12.10, 12.11, and 12.12. These are summarized in the form of six "rules" in Sec. 8–7 of this book. The following discussion will refer to the rules given in that section.

Before doing so we shall calculate ℓ_d for the bottom bars.

Spacing and confinement case. The bottom bars have clear spacing and cover of at least d_b and are enclosed by at least minimum stirrups. Therefore, this is case 1 in Table 8–1 (ACI Sec. 12.2.2). The bars are No. 8.

$$\frac{\ell_d}{d_b} = \frac{f_y \alpha \beta \lambda}{20\sqrt{f_c'}} \tag{8–10}$$

$$= \frac{60,000 \times 1.0 \times 1.0 \times 1.0}{20 \times \sqrt{3750}} = 49$$

For the No. 8 bars, $d_b = 1.0$ in. and thus $\ell_d = 49$ in.
At B and C we must satisfy rules 1 and 2:

> **Rule 1.** Bars must extend d or $12d_b$ past flexural cutoff points. Extend bars $d = 21.5$ in. to points B' and C' (Fig. 10–23).
>
> **Rule 2.** Bars must extend ℓ_d past the point of maximum bar stress. The maximum bar stress occurs at midspan. By inspection, the distances from midspan to B' and C' exceeds $\ell_d = 49$ in. Therefore, cut off 2 No. 8 bars at 81.5 in. minus 21.5 in., equals 5 ft, from the interior fact of the exterior column, and 79.4 in., say 6 ft 6 in., from the exterior face of the interior column, as determined using rule 1.

At A and D we must satisfy rules 2, 3, and 4.

> **Rule 3.** This is an interior beam with U stirrups. Therefore, rule 3c applies. At least one-fourth of the bars must be hooked into the support at A and lapped spliced at the support at D. Two (half) of the bars extend in each case. We will hook and lap splice both of these. We shall base the lap splice length on ℓ_d for the No. 7 bars from span $B4$.

Rule 2. Bars must extend ℓ_d past the actual cutoff points of adjacent bars. The actual extensions past B' and C' exceed 60 in. and 78 in., respectively, both of which are greater than $\ell_d = 49$ in.

Rule 4. At positive moment points of inflection the bars must satisfy

$$\ell_d \leq \frac{M_n}{V_u} + \ell_a \qquad (8\text{–}20)$$

From Fig. A–3 the positive moment points of inflection are at $0.098\,\ell_n = 38$ in. from the exterior end and $0.146\,\ell_n = 56.6$ in. from the interior end. At each of these points the remaining steel is 2 No. 8 bars.

$$M_n = 2 \times 0.79 \times 60{,}000 \left(21.5 - \frac{2 \times 0.79 \times 60{,}000}{1.7 \times 3750 \times 81} \right)$$

$$= 2020 \text{ in.-kips}$$

Check at the exterior end. From the shear force diagram in Table 10–4, the V_u at 38 in. from exterior end is

$$\frac{V_u}{\phi} = 74.1 - \frac{38}{194}(74.1\text{–}11.1) = 61.8 \text{ kips}$$

$$V_u = 52.5 \text{ kips}$$

ℓ_a = the smaller of the actual extension, which exceeds 38 in., or d or $12d_b = 21.5$ in.

$$= 21.5 \text{ in.}$$

Therefore,

$$\frac{M_n}{V_u} + \ell_a = \frac{2020}{52.5} + 21.5 = 60.0 \text{ in.}$$

Since this exceeds $\ell_d = 49$ in., OK.

Check at interior end:

$$V_u \text{ at 56.6 in. from interior end} = 54.1 \text{ kips}$$

$$\frac{M_n}{V_u} + \ell_a = \frac{2020}{54.1} + 21.5$$

$$= 58.9 \text{ in.} - \text{therefore OK}$$

Summary of the bar cutoff points for the positive moment steel in beam B3: Hook 2 No. 8 bars in the exterior support and lap splice these bars in the interior support. Cut off 2 No. 8 bars at 5 ft from the face of the exterior column and 6 ft 6 in. from the face of the interior column. These locations are shown as points, A, B', C', and D in Fig. 10–23.

(c) Select cutoffs for positive moment steel in B4. At midspan we have 1 No. 8 and 2 No. 7 bars. Run the 2 No. 7 bars into the supports and cut off the No. 8 bar. A_s remaining $= 1.20/1.99 = 0.60$ times the original amount. Therefore, the flexural cutoff is located at the point where the moment is 0.60 times the midspan moment. From Fig. A–1, this occurs at $0.275\ell_n = 0.275 \times 324 = 89.1$ in. from the face of the support (Fig. 10–23). Applying rules 1 and 2 we that find the bar must extend 21.5 in. past this point to 67.6 in. from the support. Therefore, cut off 1 No. 8 bar at 5 ft 6 in. from the support.

The bars have a spacing and cover of at least d_b and are enclosed by at least minimum stirrups. Therefore, for No. 8 bars $\ell_d = 49.0$ in., and for the No. 7 bars, $\ell_d = 49 \times 0.875 = 42.9$ in.

The remaining bars must satisfy rules 2, 3, and 4. Inspection shows that rule 2 is satisfied.

Rule 3. Again rule 3c applies. Lap splice the 2 No. 7 bars with the bars from beam *B3.* The length of a Class A lap splice for a No. 7 bar is $1.0\ell_d = 42.9$ in. Therefore, lap splice the bars 43 in.

Rule 4. The positive moment points of inflection are $0.146\ \ell_n = 47.3$ in. from supports. At these sections the reinforcement is 2 No. 7 bars.

$$M_n = 2 \times 0.60 \times 60,000 \left(21.5 - \frac{2 \times 0.60 \times 60,000}{1.7 \times 3750 \times 81} \right)$$

$$= 1540 \text{ in.-kips}$$

$$V_u = 47.5 \text{ kips}$$

$$\frac{M_n}{V_u} + \ell_a = \frac{1540}{47.5} + 21.5 = 53.9 \text{ in.}$$

This exceeds ℓ_d —therefore OK. The cutoff points for the positive moment steel are as indicated in Fig. 10–23.

(d) Select cutoffs for the negative moment steel at the exterior end of *B3.* At the exterior end there are 2 No. 7 bars and 2 No. 6 bars. These must be anchored in the spandrel beam, which is 16 in. wide. The basic development lengths of standard hooks are (ACI Sec. 12–5)

$$\ell_{hb} = \frac{1200 d_b}{\sqrt{f_c'}}$$

For

$$\text{No. 7 bar: } \ell_{hb} = 17.2 \text{ in.}$$

$$\text{No. 6 bar: } \ell_{hb} = 14.7 \text{ in.}$$

$$\text{No. 5 bar: } \ell_{hb} = 12.25 \text{ in.}$$

These can be multiplied by 0.7 if the side cover (perpendicular to the plane of the hook) exceeds 2.5 in. and for 90° hooks, if the tail cover is not less than 2 in. The hooks in question enter the spandrel beam perpendicular to its axis and hence satisfy the side cover requirement. The tail cover can be set at 2 in. Therefore,

$$\ell_{dh} \text{ for No. 7 bar} = 17.2 \times 0.7 = 12.0 \text{ in.}$$

This can be accommodated in the spandrel beam. **Therefore, anchor the top bars into the spandrel beam with 90° standard hooks.**

Extend all the top bars past the negative moment point of inflection before cutting them off. From Fig. A–3 the negative moment point of inflection is at $0.108\ell_n = 41.9$ in. from the face of the support (point *G* in Fig. 10–24). At this cutoff point we must satisfy rules 1, 2, and 6.

Rule 6 requires at least one-third of the bars to extend the longer of *d*, $12d_b$, or $\ell_n / 16$ past the point of inflection. $\ell_n / 16 = 24.3$ in. governs here. Therefore, cut off all bars at 5 ft 6 in. from the face of the support.

Rule 6 automatically satisfies rule 1. Rule 2 requires that the bars extend at least ℓ_d from the face of the support.

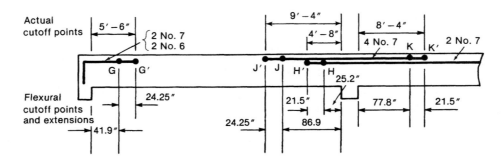

Fig. 10–24
Calculation of bar cutoff points for negative moment steel.

Continuous Beams and One-Way Slabs

ℓ_d for No. 6 top bar $= 50.9d_b = 38.2$ in., say 39 in.
ℓ_d for a No. 7 top bar $= 63.7d_b = 55.7$, say 56 in.

Since 5 ft 6 in. exceeds 56 in., rule 2 is satisfied. **Therefore, cut off all top bars at the exterior support at 5 ft 6 in. from the face of the support.**

(e) **Select cutoffs for the negative moment steel at the interior end of B3.** The negative moment steel at the first interior support consists of 6 No. 7 bars. We will cut off 2 of these bars when they are no longer needed for flexure and extend the remaining bars ($0.67A_s$) past the negative moment point of inflection. From Fig. A–3, $M_u = 0.67M_{u(max)}$ at $0.065\ell_n = 25.2$ in. from the face of the support (point H in Fig. 10–24). Here detailing rules 1 and 2 apply.

Rule 1. Bars must extend $d = 21.5$ in. past H to 46.7 in., say 4 ft from the face of the support.

Rule 2. Bars must extend a distance ℓ_d from the point of maximum bar stress (at the face of the support). 4 ft does not exceed $\ell_d = 56$ in. Therefore, extend bars so that cutoff H' is 56 in. from the face of the support.

Cut off 2 No. 7 top bars at 4 ft 8 in. from the exterior face of the first interior support. The remaining 4 bars will be extended past the negative moment point of inflection. From Fig. A–3, this is $0.224\ell_n = 86.9$ in. from the face of the support (point J in Fig. 10–24). Here detailing rules 2 and 6 apply.

Rule 6. At least one-third of bars must extend $\ell_n/16 = 24.25$ in. past the point of inflection to 111.2 in. from the face of the support, say 9 ft 4 in. from the face of the support.

Rule 2. Bars must extend $\ell_d = 56$ in. past the actual cutoff H'. Therefore, the bars must extend to 112 in. from the face of the support. This is exactly satisfied.

Cut off 4 No. 7 top bars at 9 ft 4 in. from the exterior face of the first interior support.

(f) **Select cutoffs for the negative moment steel in beam B4.** When the interior span of a three-span continuous beam is less than about 90% as long as the end spans, it can have negative moment at midspan under some loads, even though this is not apparent from the ACI moment coefficients. For this reason we will extend 2 No. 7 top bars the full length of the beam. The remaining 4 bars will be extended past the negative moment point of inflection and cut off. From Fig. A–1 the negative moment point of inflection is $0.24\ell_n = 77.8$ in. from the face of the column (point K in Fig. 10–24).

Rule 6 says that these bars must extend by the larger of d, $12d_b$, or $\ell_n/16$ past this point to $77.8 + 21.5 = 99.3$ in. from the face of the support.

Rule 2 says that these bars must extend by $\ell_d = 56$ in. past the face of the support—therefore OK.

Cut off 4 No. 7 top bars at 8 ft 4 in. from the interior face of the interior support, and extend the remaining 2 No. 7 top bars the full length of the interior span (see Fig. 10–24).

(g) **Check shear at points where bars are cut off in a zone of flexural tension.** Cutoffs B', C', F', and H' occur in zones of flexural tension. ACI Sec. 12.10.5 requires special consideration of these regions if V_u exceeds two-thirds of $\phi(V_c + V_s)$.

Cutoff B': 5 ft from support, $V_u = 46.4$ kips. At this point there are No. 3 double-leg stirrups at 8 in. on centers. $V_c = 36.8$ kips.

$$V_s = \frac{0.22 \times 40 \times 21.5}{8} = 23.6 \text{ kips}$$

$$\phi(V_c + V_s) = 0.85(36.8 + 23.6) = 51.4 \text{ kips}$$

Thus $V_u = 0.90$ of the shear permitted and ACI Sec. 12.10.5.1 and 12.10.5.3 do not apply. Therefore, we must add extra stirrups. From ACI Sec. 12.10.5.2,

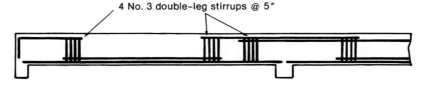

4 No. 3 double–leg stirrups @ 5″

Fig. 10–25
Additional stirrups at bar cutoff points.

$$\text{maximum spacing} = \frac{d}{8\beta_b}$$

where

$$\beta_b = \frac{\text{area of bars cut off}}{\text{area immediately before cutoff}} = 0.5$$

Therefore, the maximum spacing $= 21.5/(8 \times 0.5) = 5.38$ in.

The added stirrups must provide $A_v = 60b_w s/f_y$. This corresponds to a maximum spacing of

$$s = \frac{A_v f_y}{60b_w}$$

$$= \frac{0.22 \times 40{,}000}{60 \times 14}$$

$$= 10.5 \text{ in.} \qquad (5.38 \text{ in. governs})$$

Extra stirrups are required for $0.75d = 16.1$ in. **Provide 4 No. 3 double-leg Grade 40 stirrups at 5 in. o.c., the first stirrup at 2.5 in. from the end of the bar at B'** (see Fig. 10–25). These extend along the bar which is cut off at B'.

Cutoff C': Similar calculations show that extra stirrups are required at cutoff C'. These will be the same as at B'.

Cutoff F': At 5 ft 6 in. from the support, $V_u = 34.9$ kips. At this point there are No. 3 double-leg stirrups at 10 in. o.c. and $\phi(V_c + V_s) = 47.35$ kips. Thus $V_u = 0.74$ of the shear permitted and again extra stirrups are required. We shall use the same detail as before.

Cutoff H': At 4 ft from the support, $V_u = 56.9$ kips. At this point there are No. 3 stirrups at 6 in. o.c. and $\phi(V_c + V_s) = 58.1$ kips. Again stirrups are required and the same detail will be used as at B' to avoid confusion. The extra stirrups are shown in Fig. 10–25.

10. Design web face steel.

Since d does not exceed 3 ft, no side face steel is required (ACI Sec. 10.6.7).

This concludes the design. The bar detailing in step 9 of this example is quite lengthy. The process could be simplified considerably if all bars were extended past the points of inflection before being cut off or if the standard bar details in Fig. A–5 were used. ∎

In Example 10–2, the moments were computed using the ACI moment coefficients. Figure 10–26 shows the elastically computed moment envelope for this beam computed using the reduced live load of 79.5 psf computed in step 1c for positive moment in beam $B4$ (the center span). The beam was modeled for analysis with springs at the two discon-

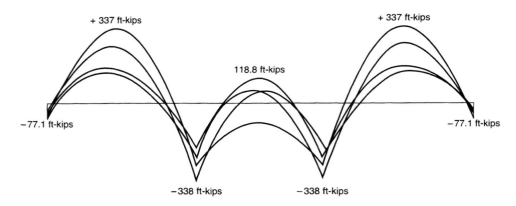

Fig. 10–26
Elastic moment envelopes—Example 10–2.

Continuous Beams and One-Way Slabs

tinuous ends to represent torsional stiffness of the spandrel beams. Their spring stiffness was derived using Eq. 7–46.

The moments plotted in Fig. 10–26 can be compared to the moments from the ACI moment coefficients, shown in the moment diagram in the heading of Table 10–3. Moments in the end spans and at the interior supports are relatively close from the two procedures. Moment redistribution, discussed in Sec. 10–8, brings them even closer. The negative moments at the exterior ends are directly a function of the assumed torsional stiffness of the spandrel beams. The interior span positive moment is overestimated by the ACI moment coefficients, which, in addition, fail to predict the negative moment at midspan of the interior span.

10–6   DESIGN OF GIRDERS

Girders support their own weight plus the concentrated loads from the beams they support. It is customary to compute the moments and shears in the girder assuming that the girder supports concentrated loads equal to the beam reactions, plus a uniform load equal to the self-weight of the girder plus the live load applied directly over the girder. This neglects two-way action in the slab adjacent to the girder. The use of moment coefficients in such a case was discussed in Sec. 10–3.

When the flexural reinforcement in the slab runs parallel to the girder, the two-way action shown in Fig. 10–12 causes negative moments in the slab along the slab-girder connection, as indicated by the curvature of slab strip *B* in Fig. 10–12. To reinforce for these moments, ACI Sec. 8.10.5 requires slab reinforcement transverse to the girder as shown in Fig. 10–27. The reinforcement is designed assuming that the slab acts as a cantilever, projecting a distance equal to the effective overhanging slab width and carrying the factored dead load and live load supported by this portion of the slab.

Frequently, edge girders are loaded in torsion due to beams framing into them between the ends of the girder. This is *compatibility torsion* since it exists only because the free end rotation of the beam is restrained by the torsional stiffness of the girder. In such a case, ACI Sec. 11.6.2.2 allows a reduction in the design torque as discussed in Chap. 7. If torque is present, the stirrups must be closed stirrups (ACI Secs. 7.11.2 and 11.6.4.1).

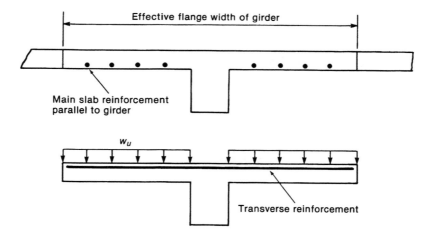

Fig. 10–27
Transverse reinforcement over girders.

Long-span floors for relatively light live loads can be constructed as a series of closely spaced, cast-in-place T beams or joists with a cross section as shown in Fig. 10–28. The joists span one way between beams. Most often, removable metal forms referred to as *fillers* or *pans* are used to form the joists. Occasionally, joist floors are built using clay tile fillers which serve as forms for the concrete in the ribs and which are left in place to serve as the ceiling (ACI Sec. 8.11.5).

When the dimensions of the joists conform to ACI Secs. 8.11.1 to 8.11.3, they are eligible for less cover to the reinforcement than for beams (ACI Sec. 7.7.1c) and for a 10% in-

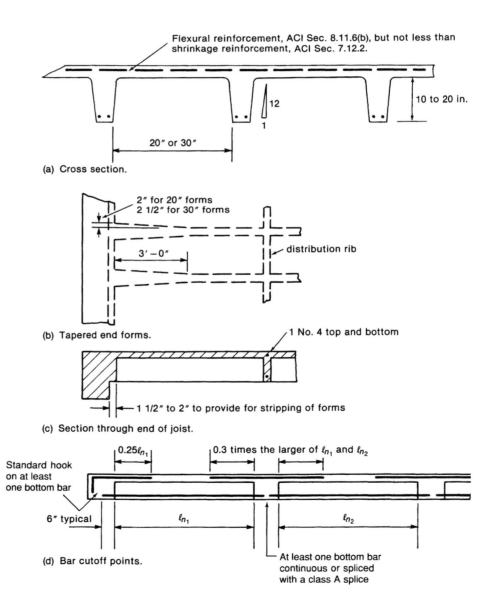

Fig. 10–28
One-way joist construction.

crease in the shear, V_c, carried by the concrete (ACI Sec. 8.11.8). The principal requirements are that the floor be a monolithic combination of regularly spaced ribs and a top slab with:

1. Ribs not less than 4 in. in width
2. Depth of ribs not more than $3\frac{1}{2}$ times the minimum web width
3. Clear spacing between ribs not greater than 30 in.

Ribbed slabs not meeting these requirements are designed as slabs and beams.

The slab thickness is governed by ACI Sec. 8.11.6.1, which requires a thickness of not less than 2 in. for joints formed with 20-in.-wide pans and $2\frac{1}{2}$ in. for 30-in. pans. The slab thickness and cover to the reinforcement are also a function of the fire-resistance rating required by the local building code. For a 1-hour fire rating, a $\frac{3}{4}$-in. cover and a 3- to $3\frac{1}{2}$-in. slab are required. For a 2-hour rating, a 1-in. cover and a $4\frac{1}{2}$-in. slab are required. These vary from code to code and as a function of the type of ceiling used. Joist floor systems are also used for parking garages. Here the top cover of the slab reinforcement and the quality of concrete used must be chosen to reduce the possibility of corrosion (ACI Chap. 4). Epoxy-coated or galvanized bars are often used in such applications. Wheel loads on the slab and abrasion of the concrete in traffic paths should also be considered.

The dimensions of standard removable pan forms are shown in Fig. 10–28. The forms are available in widths of 20 and 30 in., measured at the bottom of the ribs. Standard depths are 6, 8, 10, 12, 14, 16, or 20 in. End forms have one end filled in to serve as the side form for the supporting beams. Tapered end forms allow the width of the ends of the joists to increase as shown in Fig. 10–28b, giving additional shear capacity at the end of the joists. The 20-in. tapered end pans reduce in width from 20 in. to 16 in. over a 3-ft length. The 30-in. pans reduce to 25 in., also in a 3-ft length. Pan forms wider than 30 in. are also available, but the resulting floors or roofs must be designed as conventional slabs and beams (ACI Sec. 8.11.4).

When laying out such a floor, the rib and slab thicknesses are governed by strength, fire rating, and the available space. The overall depth and rib thickness are governed by deflections and shear. No stirrups are used in most cases. Generally, it is most economical to use a constant depth of form for all joists in a given floor. The most economical forming results if the joists and supporting beams have the same depth. Frequently, this will involve beams which are considerably wider than the columns. Such a system is referred to as a *joist-band* system. Frequently, it is not possible to have beams and joists of the same depth, however, and the beams will be deeper than the joists. Generally, it is most economical to use untapered end forms whenever possible, using tapered forms only at supports having higher-than-average shears.

Although not required by the ACI Code, load-distributing ribs perpendicular to the joists are provided at midspan or the third points of long spans. These have at least one continuous No. 4 bar at the top and the bottom. The CRSI Handbook suggests no load distributing ribs in spans of up to 20 ft. one at midspan for spans of 20 to 30 ft, and two at the third points for spans over 30 ft.

For joist floors meeting the requirements of ACI Sec. 8.3.3, the moment and shear coefficients can be used in design, taking ℓ_n as the clear span of the joists themselves. For uneven spans, it is necessary to analyze the floor. If the joists are supported on wide beams or wide columns, the width of the supports should be modeled in the analysis. The negative moments in the ends of the joists will be underestimated if this is not done.

The bar cutoffs given in Fig. 10–28d apply if the joists fit the requirements of ACI Sec. 8.3.3. The CRSI Handbook recommends cutting off half of the top and bottom bars at shorter lengths. If this is done, the shear at the points where bars are cut off in a zone of flexural tension should satisfy ACI Sec. 12.10.5. ACI Sec. 7.13.2.1 requires at least one bottom bar to be continuous or spliced over supports with a Class A tension splice. At edge beams at least one of the bottom bars in each joist must be anchored into the beam with a standard hook.

10–8   MOMENT REDISTRIBUTION

As a continuous beam is loaded beyond the service loads, the tension reinforcement will eventually yield at some section. With further loading, this section will deform as a plastic hinge. However, increased loads cause an increase in the overall moments in the member, and since the moment cannot increase at the plastic hinge, the shape of the moment diagram must shift, more moment going to those sections which are still elastic. Eventually, a mechanism forms and the plastic capacity is reached. Procedures for carrying out inelastic analyses of concrete structures are given in Refs. 10–5 to 10–7.

For a uniformly loaded fixed-end beam, the maximum elastic positive moments ($w\ell^2/24$), are half the maximum negative moment ($w\ell^2/12$). As a result, approximately twice as much reinforcement is required at the supports compared to midspan. Sometimes this leads to congestion of the steel at the supports. The plastic moment distribution is much more variable, depending on the actual plastic moment capacities in the positive and negative moment regions and moments can be redistributed from the elastic distribution, allowing reductions in the peak moments with corresponding increases in the lower moments. The amount of redistribution that can be tolerated is governed by two aspects. First, the hinging section must be able to undergo the necessary inelastic deformations. Since the inelastic rotational capacity is a function of the reinforcement ratio, as shown in Fig. 4–16, this implies an upper limit on the reinforcement ratio. Second, hinges should not have occurred at service loads, since wide cracks develop at hinge locations.

The ACI Code provides two methods for determining how much redistribution can be allowed. ACI Sec. 8.4 allows the negative moments at the supports to be increased or decreased by not more than

$$20\left(1 - \frac{\rho - \rho'}{\rho_b}\right) \quad \text{percent} \tag{10–6}$$

This can only be done at sections where $\rho - \rho'$ is less than $0.5\rho_b$, where ρ_b is as defined in Eq. 4–25.

Alternatively, ACI Sec. B8.4 allows the negative moments at the supports to be increased or decreased by not more than $1000\epsilon_t$ percent, with a maximum of 20%, provided that $\epsilon_t \geq 0.0075$ at the section where moments are being reduced, where ϵ_t is the tensile strain in the layer of steel closest to the tension face of the member.

Regardless of the procedure employed, moment redistribution can be used only if the moments have been computed using an elastic analysis and the modified negative moments must be used in calculating the moments at other sections, the shears, and the bar cutoff points.

EXAMPLE 10–3   Moment Redistribution

Compute the design moments for the three-span beam from Example 10–2 using an elastic analysis and moment redistribution. Because Appendix B was used in the design of Example 10–2, we shall use it here. Step 5 of the example will be repeated.

$$\epsilon_t = \frac{21.5 - 5.69}{5.69} \times 0.003$$

$$= 0.0083$$

Thus the maximum redistribution that is allowed is $1000\epsilon_t$ percent = 8.3%. The reduced moment is thus $(1 - 0.083) \times 338 = 310$ ft-kips. Check whether $A_s = 3.60$ in.^{2} is adequate.

$$\phi M_n = \frac{0.90 \times 3.60 \times 60,000 \left(21.5 - \dfrac{4.84}{2} \right)}{12,000}$$

$$= 309.1 \text{ ft-kips}$$

We shall assume that this is close enough.

It is now necessary to compute the positive moments corresponding to this value of the negative moment. These moments correspond to a factored load of 3.56 kips/ft on two adjacent spans and an elastically computed moment of -61.8 ft-kips at the exterior end. For this loading the maximum positive moment is 289 ft-kips. The maximum positive moment in the end spans comes from a load case with uniform loads of 3.90 kips/ft on the two end spans. For this case, the interior negative moments were -278 ft-kips, which is less than the negative moment after redistribution (-309 ft-kips). Hence this moment diagram cannot be redistributed.

The final moments after redistribution are:

Exterior end negative moment	-77.1 ft-kips
Exterior span positive moment	337 ft-kips
Interior support negative moment	309 ft-kips
Midspan interior span: positive	118.8 ft-kips
negative	-92.1 ft-kips

In this case, very little was accomplished by redistributing the moments. ■

10–9   SUMMARY

The two lengthy examples presented in this chapter appear somewhat cumbersome the first time through. With practice and understanding of the principles, the design of beams and slabs becomes straightforward. The beams and slabs designed for one floor can frequently be used many times throughout the building.

PROBLEMS

10–1 A five-span one-way slab is supported on 12-in.-wide beams with a center-to-center spacing of beams of 16 ft. It carries a superimposed dead load of 10 psf and a live load of 100 psf. Using $f_c' = 3000$ psi and $f_y = 60,000$ psi, design the slab. Draw a cross section showing the reinforcement. Use Fig. A–5 to locate bar cut-off points.

10–2 A four-span one-way slab is supported on 12-in.-wide beams with the center-to-center spacing of

beams equal to 14, 16, 16, and 14 ft. The slab carries a superimposed dead load of 20 psf and a live load of 150 psf. Design the slab using $f_c' = 3000$ psi and $f_y = 60,000$ psi. Select bar cutoff points using Fig. A–5 and draw a cross section showing the reinforcement.

10–3 A three-span continuous beam supports 6-in.-thick one-way slabs which span 15 ft center to center of beams. The beams have clear spans, face to face of 16-in.-square columns, of 27, 30, and 27 ft. The floor supports ceiling, duct-work, and lighting fixtures weighing a total of 8 psf, ceramic floor tile weighing 16 psf, partitions equivalent to a uniform dead load of 20 psf, and a live load of 100 psf. Design the beam using $f_c' = 3000$ psi. Use $f_y = 60,000$ psi for flexural reinforcement and $f_y = 40,000$ psi for shear reinforcement. Calculate cutoff points, extending all reinforcement past points of inflection. Draw an elevation view of the beam and enough cross sections to summarize the design.

10–4 Repeat Problem 10–3, but use standard bar cutoff points from Fig. A–5.

10–5 Repeat Problem 10–3, but cut off up to 50% of the negative and positive moment bars in each span where they are no longer needed.

10–6 Explain the reason for the two live-load patterns specified in ACI Sec. 8.9.2.

11

Columns: Combined Axial Load and Bending

11-1 INTRODUCTION

A column is a vertical structural member transmitting axial compressive loads, with or without moments. The cross-sectional dimensions of a column are generally considerably less than its height. Columns support vertical loads from the floors and roof and transmit these loads to the foundations.

In construction, the reinforcement and concrete for the beams and slabs in a floor are placed. Once this concrete has hardened, the reinforcement and concrete for the columns over that floor are placed, followed by the next-higher floor. This process is illustrated in Figs. 10–3, 11–1, and 11–2. Figure 11–1 shows a completed column prior to construction of the formwork for the next floor. This is a *tied column*, so called because the longitudinal bars are tied together with smaller bars at intervals up the column. One set of ties is visible just above the concrete. The longitudinal (vertical) bars protruding from the column will extend through the floor into the next-higher column and will be lap spliced with the bars in that column. The longitudinal bars are bent inward to fit inside the cage of bars for the next-higher column. (Other splice details are sometimes used; see Fig. 11–25.) A reinforcement cage in place ready for the column forms is shown in Fig. 11–2. The lap splice at the bottom of the column and the ties can be seen in this photograph. Typically, a column cage is assembled on sawhorses prior to erection and is then lifted into place by a crane.

The more general terms *compression members* or *members subjected to combined axial load and bending* are sometimes used to refer to columns, walls, and members in concrete trusses or frames. These may be vertical, inclined, or horizontal. A column is a special case of a compression member that is vertical.

Stability effects must be considered in the design of compression members. If the moments induced by slenderness effects weaken a column appreciably, it is referred to as

Fig. 11–1
Tied column under construction. (Photograph courtesy of J. G. MacGregor)

Fig. 11–2
Reinforcement cage for a tied column. (Photograph courtesy of J. G. MacGregor)

Columns: Combined Axial Load and Bending

a *slender column* or a *long column*. The great majority of concrete columns are sufficiently stocky that slenderness can be ignored. Such columns are referred to as *short columns*. Slenderness effects are discussed in Chap. 12.

Although the theory developed in this chapter applies to columns in seismic regions, such columns require special detailing to resist the shear forces and repeated cyclic loads from earthquakes. This is discussed in Chap. 19.

11-2 TIED AND SPIRAL COLUMNS

Over 95% of all columns in buildings in nonseismic regions are tied columns similar to those shown in Fig. 11-1 and 11-2. Tied columns may be square, rectangular, L shaped, circular, or any other required shape. Occasionally when high strength and/or high ductility are required, the bars are placed in a circle and the ties are replaced by a bar bent into a helix or spiral, with a pitch of $1\frac{3}{8}$ to $3\frac{3}{8}$ in. Such a column, called a *spiral column*, is illustrated in Fig. 11-3. Spiral columns are generally circular, although square or polygonal shapes are sometimes used. The spiral acts to restrain the lateral expansion of the column core under axial loads causing crushing, and in doing so, delays the failure of the core, making the column more ductile, as discussed in the next section.

In seismic regions the ties are heavier and much more closely spaced than in Figs. 11-1 and 11-2. Spiral columns are used more extensively in such regions.

Behavior of Tied and Spiral Columns

Figure 11-4a shows a portion of the core of a spiral column enclosed by one and a half turns of the spiral. Under compressive loads the concrete in this column shortens longitudinally under the stress f_1, and due to Poisson's ratio it expands laterally. This lateral expansion is especially pronounced at stresses in excess of 70% of the cylinder strength, as discussed in Chap. 3. In a spiral column, the lateral expansion of the concrete inside the spiral (referred to as the *core*) is restrained by the spiral. This stresses the spiral in tension as shown in Fig. 11-4b. For equilibrium the concrete is subjected to lateral compressive stresses f_2. An element taken out of the core (Fig. 11-4c) is subjected to triaxial compression. In Chap. 3, triaxial compression was shown to increase the strength of concrete:

$$f_1 = f_c' + 4.1f_2 \tag{3-15}$$

Later in this chapter this equation is used to derive an equation for the amount of spiral reinforcement needed in a column.

In a tied column in a nonseismic region, the ties are spaced roughly the width of the column apart, and as a result, provide relatively little lateral restraint to the core. Outward pressure on the sides of the ties due to lateral expansion of the core merely bends them outward, developing a negligible hoop stress effect. Hence normal ties have little effect on the strength of the core in a tied column. They do, however, act to reduce the unsupported length of the longitudinal bars, thus reducing the danger of buckling of those bars as the bar stress approaches yield.

Figure 11-5 presents load deflection diagrams for a tied column and a spiral column subjected to axial loads. The initial parts of these diagrams are similar. As the maximum load is reached, vertical cracks and crushing develop in the concrete shell outside the ties or spiral, and this concrete spalls off. When this occurs in a tied column, the capacity of the core that remains is less than the load and the concrete core crushes and the reinforcement buckles outward between ties. This occurs suddenly, without warning, in a brittle manner.

When the shell spalls off the spiral column, the column does not fail immediately because the strength of the core has been enhanced by the triaxial stresses resulting from the

Fig. 11–3
Spiral column. (Photograph
courtesy of J. G. MacGregor)

effect of the spiral reinforcement. As a result, the column can undergo large deformations, eventually reaching a *second maximum load* when the spirals yield and the column finally collapses. Such a failure is much more ductile and gives warning of the impending failure together with possible load redistribution to other members. It should be noted, however, that this is accomplished only at very high strains. For example, the strains necessary to reach the second maximum load correspond to a shortening of about 1 in. in an 8-ft-high column, as shown in Fig. 11–5a. When spiral columns are eccentrically loaded, the second maximum load may be less than the initial maximum, but the deformations at failure are still large, allowing load redistribution (Fig. 11–5b). Because of their greater ductility, spiral columns are assigned a capacity reduction factor, ϕ, of 0.75, rather than the value of 0.70 used for tied columns.

Spiral columns are used when ductility is important or where high loads make it economical to utilize the extra strength resulting from the higher ϕ factor. Figures 11–6 and 11–7 show a tied and a spiral column, respectively, after an earthquake. Both columns are in the same building and have undergone the same deformations. The tied column has failed completely, while the spiral column, although badly damaged, is still supporting a load. The very minimal ties in Fig. 11–6 were inadequate to confine the core concrete. Had the column contained ties detailed according to ACI Sec. 21.4, it would have performed much better.

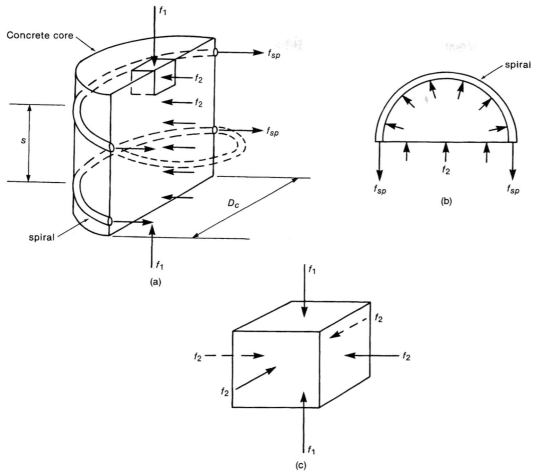

Fig. 11–4
Triaxial stresses in core of spiral column

Strength of Axially Loaded Columns

When a symmetrical column is subjected to a concentric axial load, P, longitudinal strains, ϵ, develop uniformly across the section as shown in Fig. 11–8a. Because the steel and concrete are bonded together, the strains in the concrete and steel are equal. For any given strain it is possible to compute the stresses in the concrete and steel using the stress–strain curves for the two materials. The forces, P_c and P_s, in the concrete and steel are equal to the stresses multiplied by the corresponding areas. The total load on the column, P_0, is the sum of these two quantities. Failure occurs when P_0 reaches a maximum. For a steel with a well-defined yield strength (Fig. 11–8c), this occurs when $P_c = f_c'' A_c$ and $P_s = f_y A_{st}$, where $f_c'' = C f_c'$ is the strength of the concrete loaded as a column. Based on tests of 564 columns carried out at the University of Illinois and Lehigh University from 1927 to 1933,[11–2] the ACI Code takes C equal to 0.85. Thus for a column with a well-defined yield strength the axial load capacity is

$$P_0 = 0.85 f_c'(A_g - A_{st}) + f_y A_{st} \qquad (11\text{–}1)$$

where A_g is the gross area and A_{st} is the area of the steel. This equation represents the summation of the fully plastic strength of the steel and the concrete. If the reinforcement is not

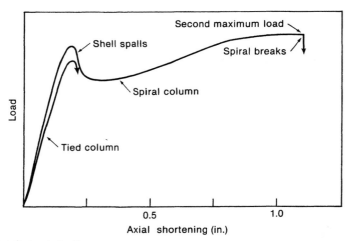

(a) Axially loaded columns.

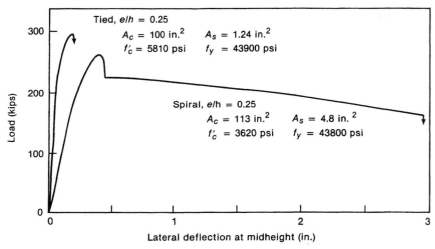

Fig. 11–5
Load-deflection behavior of
tied and spiral columns.
(Adapted from Ref. 11–1)

(b) Eccentrically loaded columns.

elastic–perfectly plastic, failure occurs when P_0 reaches a maximum, but this may not coincide with the strain at which the maximum P_c occurs.

11–3   INTERACTION DIAGRAMS

Almost all compression members in concrete structures are subjected to moments in addition to axial loads. These may be due to the load not being centered on the column, as shown in Fig. 11–9b, or may result from the columns resisting a portion of the unbalanced moments at the ends of the beams supported by the columns (Fig. 11–9c). The distance e is referred to as the *eccentricity* of the load. These two cases are the same since the eccentric load P in Fig. 11–9b can be replaced by a load P acting along the centroidal axis, plus a moment, $M = Pe$, about the centroid. The load P and moment M are calculated with respect

Columns: Combined Axial Load and Bending

Fig. 11-6
Tied column destroyed in 1971 San Fernando earthquake. (Photograph courtesy of National Bureau of Standards.)

to the geometric centroidal axis because the moments and forces obtained from structural analysis are referred to this axis.

To illustrate conceptually the interaction between moment and axial load in a column, an idealized homogeneous and elastic column with a compressive strength, f_{cu}, equal to its tensile strength, f_{tu}, will be considered. For such a column failure would occur in compression when the maximum stresses reached f_{cu} as given by

$$\frac{P}{A} + \frac{My}{I} = f_{cu} \qquad (11\text{--}2)$$

where

    A, I = area and moment of inertia of the cross section, respectively

        y = distance from the centroidal axis to the most highly compressed surface (surface A-A in Fig. 11–9a), positive to the right

        P = axial load, positive in compression

      M = moment, positive as shown in Fig. 11–9c

Dividing both sides of Eq. 11–2 by f_{cu} gives

Fig. 11–7
Spiral column damaged by
1971 San Fernando earth-
quake. Although this column
has been deflected sideways
20 in., it is still carrying load.
(Photograph courtesy of
National Bureau of
Standards.)

$$\frac{P}{f_{cu}A} + \frac{My}{f_{cu}I} = 1$$

The maximum axial load the column can support occurs when $M = 0$, and is $P_{max} = f_{cu}A$. Similarly, the maximum moment that can be supported occurs when $P = 0$, and is $M_{max} = f_{cu}I/y$. Substituting P_{max} and M_{max} gives

$$\frac{P}{P_{max}} + \frac{M}{M_{max}} = 1 \tag{11–3}$$

This equation is known as an *interaction equation* because it shows the interaction of, or relationship between, P and M at failure. It is plotted as the line AB in Fig. 11–10. A simi-

Columns: Combined Axial Load and Bending

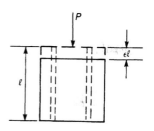

(a) Strains in column.

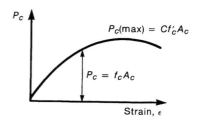

(b) Load resisted by concrete.

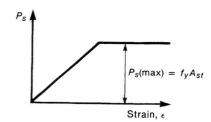

(c) Load resisted by steel.

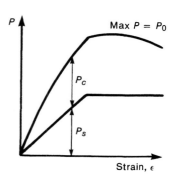

Fig. 11–8
Resistance of an axially
loaded column

(d) Total load resisted by column.

lar equation for a tensile load, P, governed by f_{tu}, gives the line BC in this figure, and the lines AD and DC result if the moments have the opposite sign.

Figure 11–10 is referred to as an *interaction diagram*. Points on the lines plotted in this figure represent combinations of P and M corresponding to the resistance of the section. A point inside the diagram, such as E, represents a combination of P and M that will not cause failure. Load combinations falling on the line or outside the line, such as point F, will equal or exceed the resistance of the section and hence will cause failure.

Figure 11–10 is plotted for an elastic material with $f_{tu} = -f_{cu}$. Figure 11–11a shows an interaction diagram for an elastic material with a definite value of f_{cu} but with the tensile strength, f_{tu}, equal to zero, and Fig. 11–11b shows a diagram for a material with $f_{tu} = -f_{cu}/2$. Lines AB and AD indicate load combinations corresponding to failure initiated by compression (governed by f_{cu}), while lines BC and DC indicate failures initiated by

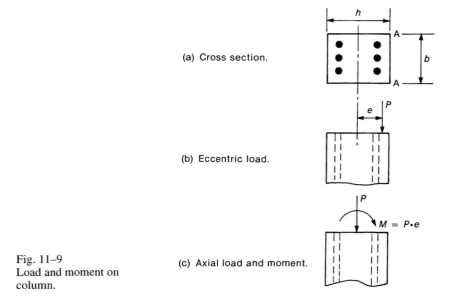

(a) Cross section.

(b) Eccentric load.

$M = P \cdot e$

(c) Axial load and moment.

Fig. 11–9
Load and moment on
column.

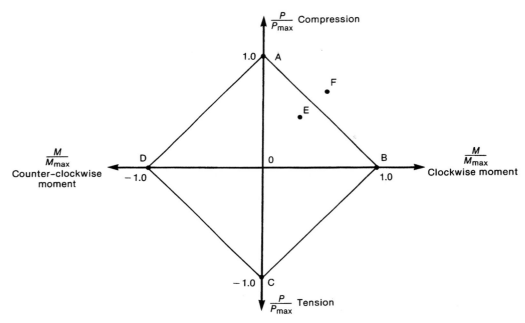

Fig. 11–10
Interaction diagram for an elastic column, $|f_{cu}| = |f_{tu}|$.

tension. In each case, the points B and D in Figs. 11–10 and 11–11 represent *balanced failures* in which the tensile and compressive resistances of the material are reached simultaneously.

Reinforced concrete is not elastic and has a tensile strength that is lower than its compressive strength. An effective tensile strength is developed, however, by reinforcing bars on the tension face of the member. For these reasons the calculation of an interaction diagram for reinforced concrete is more complex than for an elastic material. The general shape of the diagram resembles Fig. 11–11b, however.

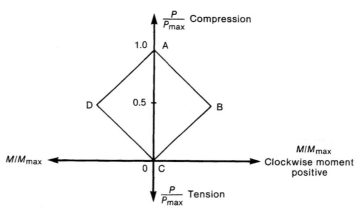

(a) Material with $f_{tu} = 0$.

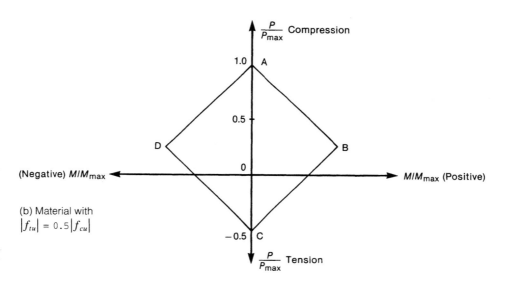

(b) Material with $|f_{tu}| = 0.5|f_{cu}|$

11–4   INTERACTION DIAGRAMS FOR CONCRETE COLUMNS

Strain Compatibility Solution

Concept and Assumptions

Although it is possible to derive a family of equations to evaluate the strength of columns subjected to combined bending and axial loads (see Ref. 11–3 and Sec. 11–7), these equations are tedious to use. For this reason, interaction diagrams for columns are generally computed by assuming a series of strain distributions, each corresponding to a particular point on the interaction diagram, and computing the corresponding values of P and M. Once enough such points have been computed, the results are summarized in an interaction diagram.

The calculation process is illustrated in Fig. 11–12 for one particular strain distribution. The cross section is illustrated in Fig. 11–12a, and one assumed strain distribution is shown in Fig. 11–12b. The maximum compressive strain is set at 0.003, corresponding to

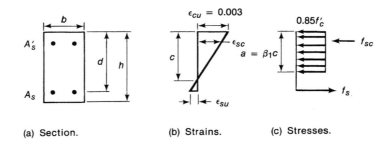

(a) Section. (b) Strains. (c) Stresses.

Fig. 11–12
Calculation of P_n and M_n for
a given strain distribution.

(d) Internal forces. (e) Resultant forces.

failure of the section. The location of the neutral axis and the strain in each level of reinforcement are computed from the strain distribution. This information is then used to
compute the size of the compression stress block and the stress in each layer of reinforcement, as shown in Fig. 11–12c. The forces in the concrete and the steel layers, shown in
Fig. 11–12d, are computed by multiplying the stresses by the areas on which they act.
Finally, the axial force P_n is computed by summing the individual forces in the concrete
and steel, and the moment M_n is computed by summing the moments of these forces about
the geometric centroid of the cross section. These values of P_n and M_n represent one point
on the interaction diagram.

Figure 11–13 illustrates a series of strain distributions and the resulting points on the
interaction diagram. Strain distribution A and point A represent pure axial compression.
Point B corresponds to crushing at one face and zero tension at the other. If the tensile
strength of concrete is ignored in the calculations, this represents the onset of cracking on
the bottom face of the section. All points lower than this in the interaction diagram represent cases in which the section is partially cracked. Point C corresponds to a strain distribution with a maximum compression strain of 0.003 on one side of the section and a tensile
strain of ϵ_y, the yielding strain of the reinforcement, at the level of the tension steel. This
represents a *balanced failure* in which crushing of the concrete and yielding of the tension
steel develop simultaneously. Point C, the farthest right point on the interaction diagram,
represents the change from *compression failures* for higher loads and *tension failures* for
lower loads. This point corresponds to point B in Fig. 11–11b. Point D in Fig. 11–13 corresponds to the strain distribution shown. Here the reinforcement has been strained to several
times the yield strain before the concrete reaches its crushing strain. This implies ductile
behavior. In contrast, for the strain distribution B, the column fails as soon as the maximum
compressive strain reaches 0.003. Since the tension steel has not yielded, there are no large
deformations prior to failure and this column fails in a brittle manner. The actual calculation procedure is discussed more fully later in this section.

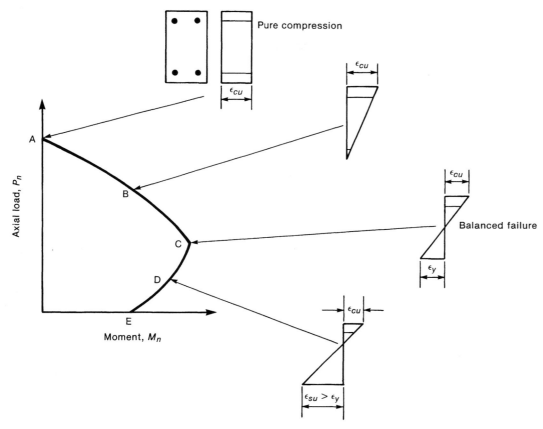

Fig. 11–13
Strain distributions corresponding to points on interaction diagram.

Maximum Axial Load

As seen earlier, the strength of a column under truly concentric axial loading can be written as

$$P_0 = (0.85f_c')(A_g - A_{st}) + f_y(A_{st}) \qquad (11\text{–}1)$$

where

$$0.85f_c' = \text{maximum concrete stress}$$
$$A_g = \text{gross area of the section (concrete and steel)}$$
$$f_y = \text{yield strength of the reinforcement}$$
$$A_{st} = \text{total area of reinforcement in the cross section}$$

The value of $0.85f_c'$ was derived from tests.[11–1, 11–2]

The strength given by Eq. 11–1 cannot normally be attained in a structure because almost always there will be moments present, and as shown by Figs. 11–10, 11–11, and 11–13, any moment leads to a reduction in the axial load capacity. Such moments or eccentricities arise from unbalanced moments in the beams, misalignments of columns from floor to floor, uneven compaction of the concrete across the width of the section, or misalignment of the reinforcement. An examination of Fig. 11–1 will show that the reinforcement has been displaced to the left in this column. Hence in this case, the centroid of the theoretical column resistance does not coincide with the axis of the column as built. The misalignment of the reinforcement in Fig. 11–1 is considerably greater than the allowable tolerances for reinforcement location (ACI Sec. 7.5.2.1), and such a column would not be acceptable.

To account for the effect of accidental moments, ACI Secs. 10.3.5.1 and 10.3.5.2 specify that the maximum load on a column may not exceed 0.85 times the value from Eq. 11–1 for spiral columns and 0.8 times Eq. 11–1 for tied columns.

Spiral columns:

$$\phi P_{n(\text{max})} = 0.85\phi\left[0.85f_c'(A_g - A_{st}) + f_y(A_{st})\right] \qquad (11\text{–}4a)$$
$$(\text{ACI Eq. } 10\text{–}1)$$

Tied columns:

$$\phi P_{n(\text{max})} = 0.80\phi\left[0.85f_c'(A_g - A_{st}) + f_y(A_{st})\right] \qquad (11\text{–}4b)$$
$$(\text{ACI Eq. } 10\text{–}2)$$

This limit will be included in the calculations of the interaction diagram. The difference between the allowable values for spiral and tied columns reflects the more ductile behavior of spiral columns.

Strength Reduction Factor for Columns

In the design of columns, the axial load and moment capacities must satisfy

$$\phi P_n \geq P_u \quad \phi M_n \geq M_u \qquad (11\text{–}5)$$

where

P_u and M_u = factored load and moment applied to the column, computed from a frame analysis

P_n and M_n = nominal strengths of the column cross section

ϕ = strength reduction factor; the value of ϕ is the same in both relationships in Eq. 11–5

Values of ϕ for columns are given in ACI Sec. 9.3.2.2(b) and in ACI Appendix B.9.3.2. For high values of axial load, both procedures specify ϕ equal to 0.70 for tied columns and 0.75 for spiral columns. The value for spiral columns is lower than the 0.9 used for flexure because the strength of a column is affected more strongly by variations in concrete strength than that of a beam, and because the failure of a column is likely to have more serious consequences than that of a beam. The value for tied columns is lower still because the failure of a tied column is much more brittle than that of a spiral column.

The horizontal axes of the interaction diagrams in Figs. 11–10, 11–11, and 11–13 correspond to pure bending or flexure. Here the strength reduction factor, ϕ, is equal to or approaches 0.9. As a result a transition is needed between $\phi = 0.7$ or 0.75 for high axial loads to $\phi = 0.9$ for pure bending. The transition is made in the tension failure region and reflects the increase in ductility in this region. The transition is made differently in ACI Secs. 9.3.2.2(b) and B.9.3.2. Although both procedures will be explained, B.9.3.2 will be used in calculations as done in earlier chapters because the procedures in ACI Appendix B can be used in a wider range of situations and are more rational than ACI Sec. 9.3.2.2(b).

ACI Section 9.3.2.2(b). The change in ϕ begins at an axial load capacity, ϕP_a, which is equal to the smaller of the balanced load, ϕP_b, or $0.1f_c'A_g$ Generally, ϕP_b exceeds $0.1f_c'A_g$ except for a few nonrectangular columns. The value of ϕ varies linearly as ϕP_n decreases to zero.

ACI Section B.9.3.2. In ACI Appendix B the transition in ϕ is a function of the strain, ϵ_t, in the layer of reinforcement farthest from the compressive face. When this strain is between 0.003 in compression and $-\epsilon_y$ (where tensile strains are negative), the section is said to be *compression-controlled*. When ϵ_t is smaller (more negative and hence more tensile) than -0.005, the section is said to be *tension-controlled*. The value of ϕ varies linearly as a function of ϵ_t as ϵ_t varies from $-\epsilon_y$ to -0.005.

Both methods will be discussed more fully in the development of the calculation procedure for interaction diagrams.

Derivation of Computation Method

In this section the relationships needed to compute the various points on an interaction diagram are derived using strain compatibility and mechanics. The calculation of an interaction diagram involves the basic assumptions and simplifying assumptions stated in Sec. 4–2 of this book and ACI Sec. 10.2. For simplicity, the derivation and the computational example in the next section is limited to rectangular tied columns, as shown in Fig. 11–14a. The extension of this procedure to other cross sections is discussed later in this section.

Throughout the computations it is necessary to rigorously observe a sign convention for stresses, strains, and forces. Compression has been taken as positive in all cases.

Concentric Compressive Axial Load Capacity and Maximum Axial Load Capacity

The theoretical top point on the interaction diagram is calculated using Eq. 11–1. The maximum usable axial loads are computed using Eqs. 11–4. For a symmetrical section, the corresponding moment will be zero. Unsymmetrical sections are discussed briefly later in this chapter. For such a section Eq. 11–12 is used to compute the moment.

General Case

The general case involves the calculation of P_n acting at the centroid, and M_n about the centroid, for an assumed strain distribution with $\epsilon_{cu} = 0.003$. The column cross section and the assumed strain distribution are shown in Fig. 11–14a and b. Four layers of reinforcement are shown, layer 1 having strain ϵ_{s1} and area A_{s1}, and so on. Layer 1 is closest to the "least compressed" surface and is at a distance d_1 from the "most compressed" surface.

The strain distribution will be defined by setting $\epsilon_{cu} = 0.003$ and assuming a value of ϵ_{s1}. Because an iterative calculation will be necessary to consider a series of cases, as

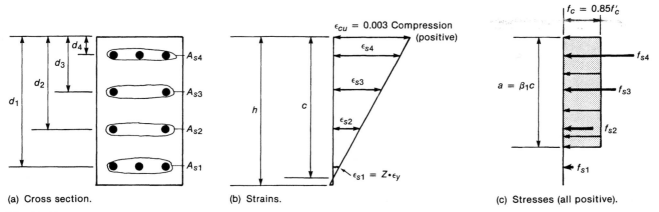

(a) Cross section. (b) Strains. (c) Stresses (all positive).

Fig. 11–14
Notation and sign convention.

shown in Fig. 11–13, this will be done by setting $\epsilon_{s1} = Z\epsilon_y$, where Z is an arbitrarily chosen value. Positive values of Z correspond to positive (compressive) strains (as shown in Fig. 11–14b). For example, $Z = -1$ corresponds to $\epsilon_{s1} = -1\epsilon_y$, the yield strain in tension. Such a strain distribution would correspond to the balanced failure condition.

From Fig. 11–14b using similar triangles,

$$c = \left(\frac{0.003}{0.003 - Z\epsilon_y}\right)d_1 \tag{11-6}$$

and

$$\epsilon_{si} = \left(\frac{c - d_i}{c}\right)0.003 \tag{11-7}$$

where ϵ_{si} and d_i are the strain in the ith layer of steel and the depth to that layer.

Once the values of c and ϵ_{s1}, ϵ_{s2}, and so on, are known, the stresses in the concrete and in each layer of steel can be computed. For elastic–plastic reinforcement with the stress–strain curve illustrated in Fig. 11–15,

$$f_{si} = \epsilon_{si}E_s \quad \text{but} \quad -f_y \le f_{si} \le f_y \tag{11-8}$$

The stresses in the concrete are represented by the equivalent rectangular stress block described in assumption 6 in Sec. 4–2 of this book (ACI Sec. 10.2.7). The depth of this stress block is $a = \beta_1 c$ where a, shown in Fig. 11–14c, obviously cannot exceed the overall height of the section, h. The factor β_1 is given by

$$\beta_1 = 1.05 - 0.05\left(\frac{f_c'}{1000}\right) \tag{4-8b}$$

but not more than 0.85 nor less than 0.65.

The next step is to compute the compressive force in the concrete, C_c, and the forces in each layer of reinforcement, F_{s1}, F_{s2}, and so on. This is done by multiplying the stresses by corresponding areas. Thus

$$C_c = (0.85f_c')(ab) \tag{11-9}$$

For a nonrectangular section, the area (ab), would be replaced by the area of the compression zone having a depth, a, measured perpendicular to the neutral axis.

If a is less than d_i,

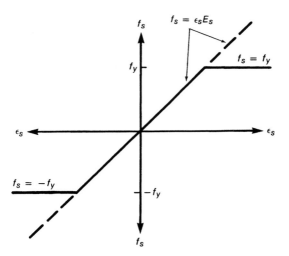

Fig. 11–15
Calculation of stress in steel
(Eq. 11–8).

Columns: Combined Axial Load and Bending

$$F_{si} = f_{si}A_{si} \text{ (positive in compression)} \qquad (11\text{--}10a)$$

If a is greater than d_i for a particular layer of steel, the area of the reinforcement in that layer has been included in the area (ab) used to compute C_c. As a result, it is necessary to subtract $0.85f_c'$ from f_{si} before computing F_{si}:

$$F_{si} = (f_{si} - 0.85f_c')A_{si} \qquad (11\text{--}10b)$$

The resulting forces C_c and F_{s1} to F_{s4} are shown in Fig. 11–16b.

The axial load capacity, P_n, for the assumed strain distribution is the summation of the axial forces:

$$P_n = C_c + \sum_{i=1}^{n} F_{si} \qquad (11\text{--}11)$$

The moment capacity M_n for the assumed strain distribution is found by summing the moments of all the internal forces about the *centroid* of the column. The moments are summed about the centroid of the section since this is the axis about which moments are computed in a conventional structural analysis. In the 1950s and 1960s moments were sometimes calculated about the *plastic centroid*, the location of the resultant force in a column strained uniformly in compression, case A in Fig. 11–13. The centroid and plastic centroid are the same point in a symmetrical column with symmetrical reinforcement.

All the forces are shown positive (compressive) in Figs. 11–14 and 11–16. A positive internal moment corresponds to a compression at the top face, and

$$M_n = C_c\left(\frac{h}{2} - \frac{a}{2}\right) + \sum_{i=1}^{n} F_{si}\left(\frac{h}{2} - d_i\right) \qquad (11\text{--}12)$$

Pure Axial Tension Case

The strength under pure axial tension is computed assuming that the section is completely cracked through and subjected to a uniform strain equal to, or less (more tensile) than, $-\epsilon_y$. The stress in all the layers of reinforcement is therefore $-f_y$ (yielding in tension), and

$$P_{nt} = \sum_{i=1}^{n} -f_y A_{si} \qquad (11\text{--}13)$$

The axial tensile capacity of the concrete is, of course, ignored. For a symmetrical section the corresponding moment will be zero. For an unsymmetrical section Eq. 11–12 is used to compute the moment.

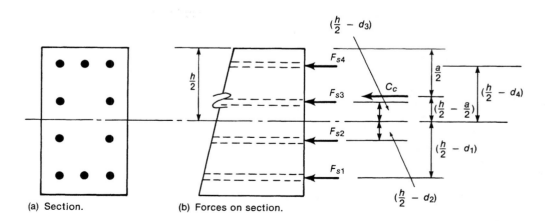

Fig. 11–16
Internal forces and moment arms.

(a) Section.

(b) Forces on section.

Calculation of Strength Reduction Factor, ϕ

As discussed earlier, ϕ varies from 0.7 (or 0.75 for spiral columns) to 0.9 for pure bending. This is done differently in ACI Secs. 9.3.2.2(b) and B.9.3.2. Both procedures will be presented here for completeness, although only ACI Sec. B.9.3.2 will be used in the examples.

ACI Section B.9.3.2 ϕ varies linearly with ϵ_t, the strain in the layer of reinforcement farthest from the extreme compression fiber, as ϵ_t varies from $-\epsilon_y$ to -0.005 (where tension is negative). Thus, for a *tied column*:

If ϵ_t is greater (more compressive) than or equal to $-\epsilon_y$:

$$\phi = 0.7 \tag{11-14a}$$

If ϵ_t is between $-\epsilon_y$ (yield in tension) and -0.005 for Grade 60 steel:

$$\phi = 0.56 - 68\epsilon_t \tag{11-14b}$$

If ϵ_t is less (more tensile) than -0.005:

$$\phi = 0.90 \tag{11-14c}$$

Note that the minus sign in Eq. 11–14b arises from the sign convention, compression positive.

For *spiral columns*, the coefficients 0.7, 0.56, and 68 become 0.75, 0.65, and 50, respectively for Grade 60 steel.

ACI Section 9.3.2.2(b). For *tied columns* ϕ varies linearly as the nominal axial load capacity, P_n, varies from P_a to zero, where P_a is equal to the smaller of P_b or $(0.10/0.70)f'_c A_g = 0.143 f'_c A_g$.

If P_n is greater than or equal to P_a:

$$\phi = 0.7 \tag{11-15a}$$

If P_n is between P_a and zero:

$$\phi = 0.9 - \frac{0.2P_n}{P_a} \tag{11-15b}$$

If P_n is less than or equal to zero:

$$\phi = 0.9 \tag{11-15c}$$

In the computational procedure given above, the balanced failure load, P_b, corresponds to $\epsilon_{s1} = -1\epsilon_y$ (i.e., $Z = -1$).

For *spiral columns*, the coefficients 0.7, 0.2, and 0.143 become 0.75, 0.15, and 0.133, respectively.

EXAMPLE 11-1   Calculation of an Interaction Diagram

Compute four points on the interaction diagram for the column shown in Fig. 11–17a. Use $f'_c = 5000$ psi and $f_y = 60,000$ psi. $A_g = bh$ is 256 in.², $A_{s1} = 4$ in.², $A_{s2} = 4$ in.², $A_{st} = \Sigma A_{si} = 8$ in.², and $\rho_t = A_{st}/A_g = 0.031$. The yield strain, $\epsilon_y = f_y/E_s$, is 60,000 psi/29,000,000 psi $= 0.00207$.

1. **Compute the concentric axial load capacity and maximum axial load capacity.**
From Eq. 11–1

$$P_0 = (0.85f'_c)(A_g - A_{st}) + f_y(A_{st})$$

$$= (0.85 \times 5 \text{ ksi})(256 - 8) \text{ in.}^2 + 60 \text{ ksi} \times 8 \text{ in.}^2$$

$$= 1054 \text{ kips} + 480 \text{ kips} = 1534 \text{ kips}$$

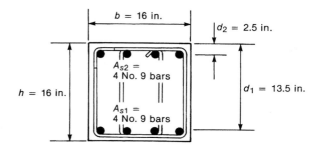

(a) Section.

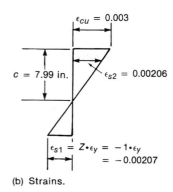

(b) Strains.

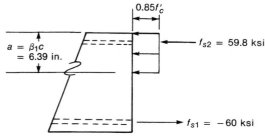

(c) Stresses.

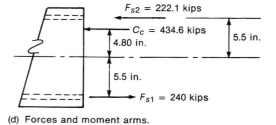

(d) Forces and moment arms.

Fig. 11–17
Calculations—Example 11–1, $\epsilon_{s1} = -1 \cdot \epsilon_y$, $(Z = -1)$.

This is the nominal concentric axial load capacity. The value used in drawing a design interaction diagram would be ϕP_0, where $\phi = 0.70$, because the concentric axial load capacity is obviously greater than either P_b or $0.143 f_c' A_g$ and this is a tied column. Thus

$$\phi P_0 = 1074 \text{ kips}$$

P_0 and ϕP_0 are plotted as points A and A' in Fig. 11–18.

For this column with $\rho_t = 0.031$ (or 3.1%), the 480 kips carried by the reinforcement is roughly 30% of the 1534-kip nominal capacity of the column. For axially loaded columns, reinforcement will generally carry between 10 and 35% of the total capacity of the column.

The maximum load allowed on this column (ACI Sec. 10.3.5.2) is given by Eq. 11–4b (ACI Eq. 10.2):

$$\phi P_{n(\text{max})} = 0.80 \phi \left[(0.85 f_c')(A_g - A_{st}) + f_y(A_{st}) \right]$$

$$= 0.80 \phi P_0 = 859 \text{ kips}$$

This load is plotted as a horizontal solid line in Fig. 11–18. The portion of the ϕP_n, ϕM_n interaction diagram above this line is shown with a dashed line because this capacity cannot be used in design.

2. **Compute ϕP_n and ϕM_n for the general case.** To get a complete interaction diagram, a number of strain distributions must be considered and the corresponding values of P_n, M_n, ϕP_n, and ϕM_n calculated. Ideally, these would include a sequence of strain gradients corresponding to values of $\epsilon_{s1} = +0.5\epsilon_y, +0.375\epsilon_y$, and so on, corresponding, in turn, to values of $Z = +0.5, +0.375$, $+0.125, 0, -0.25, -0.5, -0.75, -1, -1.5, -2, -2.5, -3, -4, -6$, and so on. These intervals are made successively larger because the points get closer and closer together as Z gets larger. Points corresponding to $Z = 0 (f_{s1} = 0)$, -0.5, $(f_{s1} = 0.5 f_y$ in tension), and -1 $(f_{s1} = f_y$ in tension, giving a balanced failure) are important in detailing column splices and should be shown on the interaction diagram. In this example we compute the values of ϕP_n and ϕM_n for $Z = -1, -2$, and -4.

3. **Compute ϕ and ϕM_n for balanced failure ($\epsilon_{s1} = -\epsilon_y$).**

    (a) **Determine c and the strains in the reinforcement.** The column cross section and the strain distribution corresponding to $\epsilon_{s1} = -1\epsilon_y (Z = -1)$ are shown in Fig. 11–17a and b.

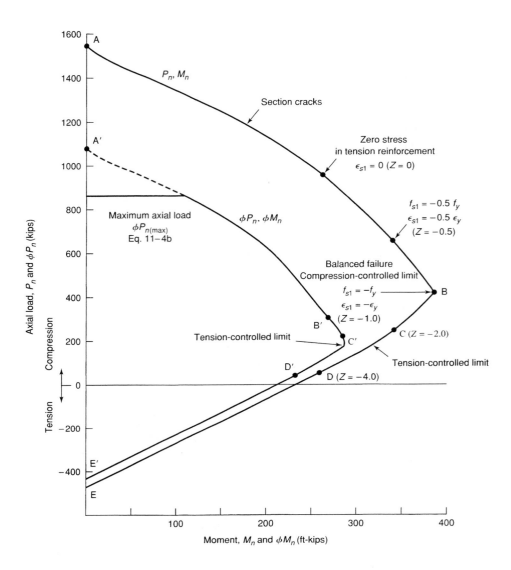

Fig. 11–18
Interaction diagram—
Example 11–1.

The strain in the bottom layer of steel is $-1\epsilon_y = -0.00207$. From similar triangles the depth to the neutral axis is

$$c = \frac{0.003}{0.003 - (-1 \times 0.00207)} d_1 \qquad (11\text{–}6)$$

$$= \frac{0.003}{0.003 + 0.00207} \times 13.5 \text{ in.} = 7.99 \text{ in.}$$

Using similar triangles, the strain in the compression steel is

$$\epsilon_{s2} = \left(\frac{c - d_2}{c}\right) 0.003 \qquad (11\text{–}7)$$

$$= \left(\frac{7.99 - 2.5}{7.99}\right) 0.003 = 0.00206$$

The strain in layer 1 is

Columns: Combined Axial Load and Bending

$$\epsilon_{s1} = Z\epsilon_y = -0.00207$$

(b)   Compute the stresses in reinforcement layers. The stress in reinforcement layer 2 is (Eq. 11–8)

$$f_{s2} = \epsilon_{s2}E_s \qquad \text{but} \qquad -f_y \le f_{s2} \le f_y$$

$$\epsilon_{s2}E_s = 0.00206 \times 29{,}000 \text{ ksi} = 59.8 \text{ ksi}$$

but $-60 \le f_{s2} \le +60$ ksi. Therefore, $f_{s2} = 59.8$ ksi. Because this is positive, it is compressive. The stress in layer 1 is found to be -60 ksi.

(c)   Compute a. The depth of the equivalent rectangular stress block is $a = \beta_1 c$, where a cannot exceed h. For $f_c' = 5000$ psi,

$$\beta_1 = 1.05 - 0.05\left(\frac{f_c'}{1000 \text{ psi}}\right) = 0.80$$

and

$$a = \beta_1 c$$

$$= 0.80 \times 7.99 \text{ in.} = 6.39 \text{ in.}$$

This is less than h; therefore, this value can be used. If a exceeded h, $a = h$ would be used. The stresses computed in steps 2 and 3 are shown in Fig. 11–17c.

(d)   Compute the forces in the concrete and steel. The force in the concrete, C_c, is equal to the average stress, $0.85f_c'$, times the area of the rectangular stress block, ab:

$$C_c = (0.85f_c')(ab) \qquad\qquad (11\text{–}9)$$

$$= 0.85 \times 5 \text{ ksi} \times 6.39 \text{ in.} \times 16 \text{ in.} = 434.6 \text{ kips}$$

The distance $d_1 = 13.5$ in. to reinforcement layer 1 exceeds $a = 6.39$ in. Hence this layer of steel lies outside the compression stress block and does not displace concrete included in the area (ab) when computing C_c. Thus

$$F_{s1} = f_{s1}A_{s1}$$

$$= -60 \text{ ksi} \times 4 \text{ in.}^2 = -240 \text{ kips (negative} = \text{tension)}$$

Reinforcement layer 2 lies in the compression zone, since $a = 6.39$ in. exceeds $d_2 = 2.5$ in. Hence we must allow for the stress in the concrete displaced by the steel when we compute F_{s2}. From Eq. 11–10b,

$$F_{s2} = (f_{cs2} - 0.85f_c')A_{s2}$$

$$= (59.8 - 0.85 \times 5) \text{ ksi} \times 4 \text{ in.}^2 = 222.1 \text{ kips}$$

The forces in the concrete and steel are shown in Fig. 11–17d.

(e)   Compute P_n. The nominal axial load capacity, P_n, is found by summing the axial force components (Eq. 11–11):

$$P_n = C_c + \Sigma F_{si}$$

$$= 434.6 - 240 + 222.1 = 417 \text{ kips}$$

Since $\epsilon_{s1} = -\epsilon_y$ (yield in tension), this is the balanced failure condition and $P_n = P_b$.

(f)   Compute M_n. From Fig. 11–17d the moment of C_c, F_{s1}, and F_{s2} about the centroid of the section is (Eq. 11–12)

$$M_n = C_c\left(\frac{h}{2} - \frac{a}{2}\right) + F_{s1}\left(\frac{h}{2} - d_1\right) + F_{s2}\left(\frac{h}{2} - d_2\right)$$

$$= 434.6 \text{ kips}\left(\frac{16}{2} - \frac{6.39}{2}\right) \text{ in.} + \left[-240(8 - 13.5)\right] + 222.1(8 - 2.5)$$

$$= 2088 + 1320 + 1222 \text{ in.-kips} = 4630 \text{ in.-kips}$$

Therefore, $M_n = M_b = 386$ ft-kips.

11–4   Interaction Diagrams for Concrete Columns

(g) **Compute ϕ, ϕP_n, and ϕM_n.** ϕ will be computed according to ACI Sec. B.9.3.2. The strain ϵ_t, in the layer of reinforcement farthest from the compression face is $\epsilon_{s1} = -0.00207 = -\epsilon_y$. Thus Eq. 11–14a applies:

$$\phi = 0.70$$

$$\phi P_n = 0.70 \times 417 = 292 \text{ kips}$$

$$\phi M_n = 0.70 \times 386 = 270 \text{ ft-kips}$$

This completes the calculations for one value of $\epsilon_{s1} = Z(\epsilon_y)$ and gives the points B and B' in Fig. 11–18. Other values of Z are now assumed and the calculations are repeated until one has enough points to complete the diagram.

4. **Compute ϕP_n and ϕM_n for $Z = -2$.** To illustrate the calculation of ϕ for cases falling between the compression-controlled limit and the tension-controlled limit, the computations will be repeated for the strain distribution corresponding to $Z = -2$.

(a) **Determine c and the strains in the reinforcement.** From similar triangles with $\epsilon_1 = -2\epsilon_y$ (substituting $Z = -2$ into Eq. 11–6) gives

$$c = 5.67 \text{ in.}$$

From Eq. 11–7, the strain in the compression steel is

$$\epsilon_{s2} = 0.00168$$

The strain in the tension reinforcement is

$$\epsilon_{s1} = -2 \times 0.00207 = -0.00414$$

(b) **Compute the stress in the reinforcement layers.**

$$f_{s2} = 0.00168 \times 29{,}000 \text{ ksi} = 48.7 \text{ ksi}$$

This is within the range $\pm f_y$—therefore OK.

$$f_{s1} = -60 \text{ ksi since } |\epsilon_{s1}| > |\epsilon_y|$$

(c) **Compute a.**

$$a = 0.80 \times 5.67 = 4.54 \text{ in.}$$

(d) **Compute the forces in the concrete and steel.** The compression force in the concrete is

$$C_c = 0.85 \times 5 \text{ ksi} \times 4.54 \text{ in.} \times 16 \text{ in.} = 308.7 \text{ kips}$$

The force in the tension reinforcement is

$$F_{s1} = -60 \text{ ksi} \times 4 \text{ in.}^2 = -240 \text{ kips}$$

Since $a = 4.54$ in. is greater than $d_2 = 2.5$ in., Eq. 11–10b is used to compute F_{s2}.

$$F_{s2} = (48.7 - 0.85 \times 5) \times 4 = 177.8 \text{ kips}$$

(e) **Compute P_n.** Summing forces perpendicular to the section (Eq. 11–11) gives

$$P_n = 308.7 - 240 + 177.8 = 246.5 \text{ kips}$$

(f) **Compute M_n.** Summing moments about the centroid of the section (Eq. 11–12) gives

$$M_n = 308.7\left(8 - \frac{4.54}{2}\right) + (-240 \times -5.50) + (177.8 \times 5.50)$$

$$= 4067 \text{ in.-kips} = 338.9 \text{ ft-kips}$$

(g) **Compute ϕ, ϕP_n, and ϕM_n.** The strain $\epsilon_t = \epsilon_{s1} = -0.00414$ is between $-\epsilon_y$ and -0.005. Therefore, Eq. 11–14b applies and

$$\phi = 0.56 - 68\epsilon_t$$

where ϵ_t is the strain in the extreme tensile layer of steel, ϵ_{s1}

$$\phi = 0.56 - (68 \times -0.00414) = 0.842$$

$$\phi P_n = 0.842 \times 246.5 = 207.6 \text{ kips}$$

$$\phi M_n = 0.842 \times 338.9 \text{ ft-kips} = 285.4 \text{ ft-kips}$$

P_n, M_n, ϕP_n, and ϕM_n calculated for this strain distribution are plotted as points C and C' in Fig. 11–18. The peculiar shape of this portion of the ϕP_n, ϕM_n interaction diagram is due to the transition from $\phi = 0.70$ to $\phi = 0.90$.

5. Compute ϕP_n and ϕM_n for $Z = -4$. Repeating the calculations for $Z = -4$ gives

$$P_n = 44.2 \text{ kips} \qquad M_n = 257.6 \text{ ft-kips}$$

Since $\epsilon_t = \epsilon_{s1} = -4 \times 0.00207$ is less than -0.005, $\phi = 0.90$ and

$$\phi P_n = 39.8 \text{ kips} \qquad \phi M_n = 231.8 \text{ ft-kips}$$

These are plotted as points D and D' in Fig. 11–18.

6. Compute the capacity in axial tension. The final loading case to be considered in this example is concentric axial tension. The strength under such a loading is equal to the yield strength of the reinforcement in tension as given by Eq. 11–13:

$$P_{nt} = \sum_{i=1}^{n} (-f_y A_{si})$$

$$= -60 \text{ ksi } (4 + 4) \text{ in.}^2 = -480 \text{ kips}$$

This is an axial tension. Because the section is symmetrical, $M = 0$.

The design capacity in pure tension is ϕP_{nt}, where $\phi = 0.9$. Thus

$$\phi P_{nt} = 0.9 \times -480 \text{ kips} = -432 \text{ kips}$$

Points E and E' in Fig. 11–18 represent the pure tension case. ∎

Interaction Diagrams for Circular Columns

The strain compatibility solution described in the preceding section can also be used to calculate the points on an interaction diagram for a circular column. As shown in Fig. 11–19b, the depth to the neutral axis, c, is calculated from the assumed strain diagram using similar triangles (or from Eq. 11–6). The depth of the equivalent rectangular stress block, a, is again $\beta_1 c$.

The resulting compression zone is a segment of a circle, of depth a, as shown in Fig. 11–19d. To compute the compressive force and its moment about the centroid of the column, it is necessary to be able to compute the area and centroid of the segment. These terms

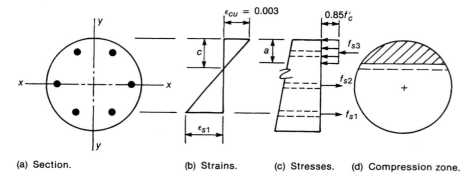

Fig. 11–19
Circular column.

(a) Section.　　　　　(b) Strains.　　　(c) Stresses.　　(d) Compression zone.

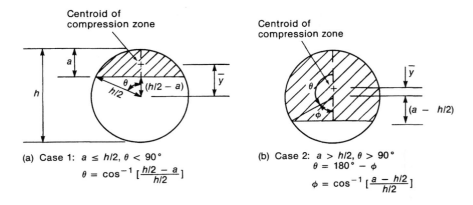

Fig. 11–20
Circular segments.

Centroid of compression zone

Centroid of compression zone

(a) Case 1: $a \leq h/2$, $\theta < 90°$

$$\theta = \cos^{-1}\left[\frac{h/2 - a}{h/2}\right]$$

(b) Case 2: $a > h/2$, $\theta > 90°$
$$\theta = 180° - \phi$$

$$\phi = \cos^{-1}\left[\frac{a - h/2}{h/2}\right]$$

can be expressed as a function of the angle θ shown in Fig. 11–20. The area of the segment is

$$A = h^2\left(\frac{\theta - \sin\theta\cos\theta}{4}\right) \tag{11–16}$$

where θ is expressed in radians (1 radian $= 180°/\pi$). The moment of this area about the center of the column is

$$A\bar{y} = h^3\left(\frac{\sin^3\theta}{12}\right) \tag{11–17}$$

The shape of the interaction diagram of a circular column is affected by the number of bars and their orientation relative to the direction of the neutral axis. Thus the moment capacity about axis x–x in Fig. 11–19a is less than that about axis y–y. Since the designer has little control over the arrangement of the bars in a circular column, the interaction diagram should be computed using the least favorable bar orientation. For circular columns with more than 8 bars, this problem vanishes since the bar placement approaches a continuous ring.

If the circular column is a *spiral* column, Eqs. 11–14a to 11–14c (ACI Sec. B9.3.2) become:

If ϵ_t is greater (more compressive) than $-\epsilon_y$:

$$\phi = 0.75 \tag{11–14d}$$

If ϵ_t is between $-\epsilon_y$ and -0.005 for Grade 60 steel:

$$\phi = 0.65 - 50\epsilon_t \tag{11–14e}$$

If ϵ_t is less (more tensile) than -0.005:

$$\phi = 0.90 \tag{11–14f}$$

or alternatively for a spiral column, Eqs. 11–15a to 11–15c (ACI Sec. 9.3.2.2) become:

If P_n is greater than or equal to P_a:

$$\phi = 0.75 \tag{11–15d}$$

If P_n is between P_a and zero:

$$\phi = 0.9 - \frac{0.15P_n}{P_a} \tag{11–15e}$$

If P_n is less than or equal to zero:

$$\phi = 0.9 \qquad (11\text{–}15\mathrm{f})$$

where P_a is the smaller of P_b and $0.133 f'_c A_g$.

For a *circular tied* column, ϕ is given by Eqs. 11–14a to 11–14c or 11–15a to 11–15c. It should be noted that the nondimensional interaction diagrams given in Figs. A–12 to A–14 include the ϕ factors for spiral columns. They cannot be used to design circular tied columns unless an adjustment is made to ϕ.

Properties of Interaction Diagrams for Reinforced Concrete Columns

Nondimensional Interaction Diagrams

Frequently, it is useful to express interaction diagrams independently of column dimensions. This can be done by dividing the axial load values, P_n or ϕP_n, by the column area, A_g, or by $f'_c A_g$ (1 /0.85 times the axial load capacity of the concrete alone) and dividing the moment values, M_n or ϕM_n, by $A_g h$ or by $f'_c A_g h$ (which has the units of moments). A family of such curves is plotted in Figs. A–6 to A–14 in Appendix A. Design aids, including sets of similar curves, are published by the American Concrete Institute[11-4] and others.[11-5] Use of these diagrams is illustrated in several of the examples given later in this chapter.

Eccentricity of Load

In Fig. 11–9 it was shown that a load P, applied to a column at an eccentricity e, was equivalent to a load P acting through the centroid, plus a moment $M = Pe$ about the centroid. A radial line through the origin in an interaction diagram has the slope P/M, or $P/Pe = 1/e$. For example, the balanced load and moment computed in Example 11–1 correspond to an eccentricity of 386 ft-kips /417 kips = 0.93 ft, and a radial line through this point (point B in Fig. 11–18) would have a slope of $1/0.93$ ft. The pure moment case may be considered to have an eccentricity $M/P = \infty$, since $P = 0$.

In a nondimensional interaction diagram such as Fig. A–6, a radial line has a slope equal to $(P/A_g)/(M/A_g h)$. Substituting $M = Pe$ shows that the line has the slope h/e or $1/(e/h)$, where e/h represents the ratio of the eccentricity to the column thickness. Radial lines corresponding to several eccentricity ratios are plotted in Fig. A–6. Cutting across these radial lines in Fig. A–6 are dashed lines corresponding to tensile stresses in the bars on the tensile face of the column of $f_s = 0$, $f_s = -0.5 f_y$ and $f_s = -f_y$. These are used when designing column splices, as discussed later in this chapter.

Unsymmetrical Columns

Up to this point, only interaction diagrams for symmetrical columns have been shown. If the column cross section is symmetrical about the axis of bending, the interaction diagram is symmetrical about the vertical $M = 0$ axis, as shown in Fig. 11–21a. For unsymmetrical columns, the diagram is tilted as shown in Fig. 11–21b provided the moments are taken about the geometric centroid. The calculation of an interaction diagram for such a member follows the same procedure as Example 11–1 except that for the cases of uniform compressive or tensile strains (P_0 and P_t), the unsymmetrical bar placement gives rise to a moment of the steel forces about the centroid. Figure 11–21a is drawn for the column shown in Fig. 11–17a and is the same as the nominal interaction diagram in Fig. 11–18. Figure 11–21b is drawn for a similar cross section with 4 No. 9 bars in one face and 2 No. 9 bars in the other. For positive moment the face with 4 bars is in tension. As a result, the balanced load for positive moment is less than that for negative moment.

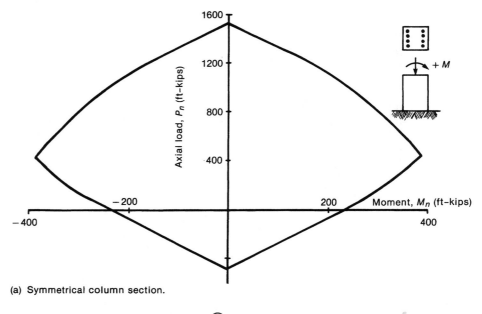

(a) Symmetrical column section.

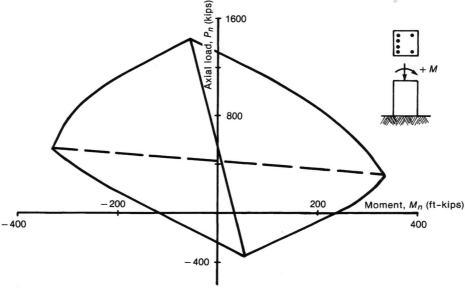

Fig. 11–21
Interaction diagrams for symmetrical and unsymmetrical columns.

(b) Unsymmetrical column section.

In a similar manner, a uniform compressive strain of 0.003 across the section, corresponding to the maximum axial load capacity, leads to a moment, since the forces in the two layers of steel are unequal.

Simplified Interaction Diagrams for Columns

Generally, designers have access to published interaction diagrams or computer programs to compute interaction diagrams for use in design. Occasionally, this is not true, as for example, in the design of hollow bridge piers, elevator shafts, or unusually shaped members. Interaction diagrams for such members can be calculated using the strain compatibility so-

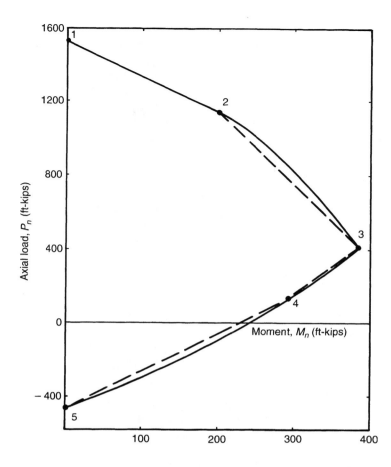

Fig. 11–22
Simplified interaction
diagram.

lution presented earlier. In most cases it is adequate to represent the interaction diagram by a series of straight lines joining the load and moment values corresponding to the following five strain distributions:

1. A uniform compressive strain of 0.003. This gives point 1 in Fig. 11–22.

2. A strain diagram corresponding to incipient cracking, passing through a compressive strain of 0.003 on one face and zero strain on the other (point 2 in Fig. 11–22).

3. The balanced strain distribution and limiting compression-controlled strain distribution having a compressive strain of 0.003 on one face and a tensile strain of $-\epsilon_y$ in the reinforcement layer nearest to the tensile face (point 3).

4. The limiting tension-controlled strain distribution having a compressive strain of 0.003 on one face and a tensile strain of -0.005 in the reinforcement layer nearest to the tensile face (point 4).

5. A uniform tensile strain of $-\epsilon_y$ in the steel with the concrete cracked (point 5).

Figure 11–22 compares the interaction diagram for Example 11–1 to an interaction diagram drawn by joining the five points described above. The five-point diagram is sufficiently accurate for design.

When computing ϕ by ACI Sec. B.9.3.2, $\phi = 0.7$ for points 1, 2, and 3, and is equal to 0.9 for points 4 and 5.

It is difficult to calculate the pure moment case directly. If this value is required for a symmetrical section, it can be estimated as the larger of (1) the flexural capacity ignoring

11–4   Interaction Diagrams for Concrete Columns   **445**

the reinforcement in the compressive zone, or (2) the moment computed ignoring the concrete and assuming a strain of $5\epsilon_y$ in the reinforcement adjacent to each face. For the column in Example 11–1 the pure moment capacity, M_{0n}, from a strain compatibility solution was 236 ft-kips, compared to 227 from the dashed line in Fig. 11–22, 235 ft-kips computed as a beam ignoring the compression reinforcement, and 220 ft-kips as a steel couple ignoring the concrete. The column in question has a high steel percentage. The accuracy of these approximations decreases as ρ approaches the minimum allowed.

11–5   DESIGN OF SHORT COLUMNS

Types of Calculations—Analysis and Design

In Chap. 4 two types of computations were discussed. If the cross section is known and it is necessary to compute the capacity, an *analysis* is carried out. On the other hand, if the loads and moments are known and it is necessary to select a cross section to resist them, the procedure is referred to as *design* or *proportioning*. A design problem is solved by guessing a section, analyzing whether it will be satisfactory, revising the section, and reanalyzing it. In each case the analysis portion of the problem is most easily carried out using interaction diagrams.

Factors Affecting the Choice of Column

Choice of Column Type

Figure 11–23 compares the interaction diagrams for three columns, each with the same f_c' and f_y, the same total area of longitudinal steel, A_{st}, and the same gross area, A_g. The columns differ in the arrangement of the reinforcement, as shown in the figure. To obtain the same gross area, the spiral column had a diameter of 18 in. while the tied columns were 16 in. square.

For eccentricity ratios, e/h, less than 0.1, the spiral column is more efficient in terms of load capacity. This is due to ϕ being 0.75 for spiral columns compared to 0.7 for tied columns and is also due to the higher axial load allowed by ACI Eq. 10–1. This economy tends to be offset by more expensive forming and by the cost of the spiral, which may exceed that of the ties.

For eccentricity ratios, e/h, greater than 0.2, a tied column with bars in the faces farthest from the axis of bending is most efficient. Even more efficiency can be obtained by using a rectangular column to increase the depth perpendicular to the axis of bending.

Tied columns with bars in four faces are used for e/h ratios of less than about 0.2 and also when moments exist about both axes. Spiral columns are relatively infrequent in non-seismic areas. In seismic areas or other situations where ductility is important, spiral columns are used more extensively.

Choice of Material Properties and Reinforcement Ratios

In small buildings the concrete strength in the columns is selected to be equal to that in the floors so that one grade of concrete can be used throughout. Frequently, this will be 3000 or 3500 psi.

In tall buildings, the concrete strength in the columns is often higher than that in the floors, to reduce the column size. When this occurs, the designer must consider the transfer of the column loads through the weaker floor concrete. Tests of column–floor junctions subjected to axial column loads have shown that the lateral restraint provided by the sur-

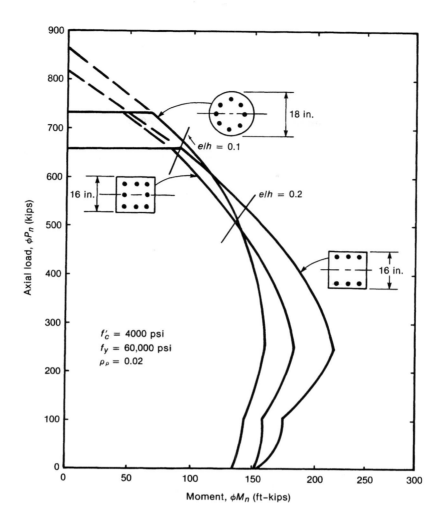

Fig. 11–23
Effect of column type on
shape of interaction diagram.

rounding floor members enables a floor slab to transmit the column loads, provided that the strength of column concrete does not exceed 1.4 times that of the floor concrete.[11-6] ACI Sec. 10.15 allows this strength differential and provides several ways of dealing with higher strength differentials. In tall buildings it is not uncommon to use 3750-psi concrete in the floors and 5000 psi $\simeq$ 1.4 × 3750 concrete in the columns.

In the vast majority of columns built, Grade 60 reinforcement is used. Unless there is a good reason for using Grade 40 reinforcement, the yield strength should be taken as 60,000 psi except for stirrups and ties, which frequently are Grade 40 steel.

ACI Sec. 10.9.1 limits the area, A_{st}, of longitudinal reinforcement in tied and spiral columns to not less than 0.01 times the gross area, A_g (i.e., $\rho_t = A_{st}/A_g$ not less than 0.01, or 1%), and not more than $0.08A_g$ ($0.06A_g$ in the seismic regions). Under sustained loads, creep of the concrete gradually transfers load from the concrete to the reinforcement. In tests of axially loaded columns, column reinforcement yielded under sustained service loads if the ratio of longitudinal steel was less than roughly 0.01.[11-2] (See Example 3–4.)

Although the code allows a maximum steel ratio of 0.08, it is generally very difficult to place this amount of steel in a column, particularly if lapped splices are used. Tables A–16 and A–17 give maximum steel percentages for various column sizes. These range from roughly 3 to 5 or 6%. In addition, the most economical tied column section generally involves ρ_t of 1 to 2%. As a result, tied columns seldom have ρ_t greater than 3%.

Exceptions to this are the lower columns in a tall building, where the column size must be limited for architectural reasons. In such a case, the bars may be tied in bundles of 2 to 4 bars. Design requirements for bundled bars are given in ACI Secs. 7.6.6 and 12.14.2.2. Because they are used for high axial loads, spiral columns generally have steel ratios between 2.5 and 5%.

The minimum number of bars in a rectangular column is 4 and in a circular column or spiral column is 6 (ACI Sec. 10.9.2). Almost universally, an even number of bars is used in a rectangular column so that the column is symmetrical about the axis of bending, and almost universally, all the bars are the same size. Table A–18 gives the areas of various combinations of bars.

Estimating the Column Size

The initial stage in column design involves estimating the required size of the column. There is no simple rule for doing this, since the axial load capacity of a given cross section varies depending on the moment acting on the section. For very small values of M, the column size is governed by the maximum axial load capacity given by Eq. 11–4. Rearranging Eq. 11–4, simplifying, and rounding down the coefficients gives the approximate relationship

$$A_{g(\text{trial})} \geq \frac{P_u}{0.45(f_c' + f_y \rho_t)} \tag{11–18a}$$

where $\rho_t = A_{st}/A_g$.

Equation 11–18a was derived for tied columns. For spiral columns, the coefficient 0.8 in Eq. 11–4b becomes 0.85 and ϕ is 0.75. Hence, for such columns,

$$A_{g(\text{trial})} \geq \frac{P_u}{0.55(f_c' + f_y \rho_t)} \tag{11–18b}$$

Both of these equations will tend to underestimate the column size if there are moments present, since they correspond roughly to the horizontal line portion of the ϕP_n, ϕM_n interaction diagram in Fig. 11–18.

Column widths and depths are generally varied in increments of 2 in. Although the ACI Code does not specify a minimum column size, the minimum dimension of a cast-in-place tied column should not be less than 8 in. and preferably not less than 10 in. The diameter of a spiral column should not be less than about 12 in.

Slender Columns

A slender column deflects laterally under load. This increases the moments in the column and hence weakens the column (see Sec. 12–1). Slenderness effects are discussed in Chapter 12. Chapter 11 deals only with short columns in braced frames, the most commonly occurring case. ACI Sec. 10.12.2 states that it is permissible to neglect slenderness effects if

$$\frac{k\ell_u}{r} \leq 34 - 12\left(\frac{M_1}{M_2}\right) \tag{11–19}$$

where

k = effective length factor which, for a braced frame, will be less than or equal to 1.0

ℓ_u = unsupported height of column from top of floor to the bottom of the beams or slab in the floor above

r = radius of gyration, equal to 0.3 and 0.25 times the overall depth of rectangular and circular columns, respectively

M_1/M_2 = ratio of the moments at the two ends of the column which, for a braced frame, will generally be between +0.5 and −0.5

This limit is discussed more fully in Sec. 12–2. In this chapter we shall assume that $k = 1.0$ and $M_1/M_2 = +0.5$. This will almost always be conservative. For this combination, columns are short if $k\ell_u/r \leq 28$. For a square column, this corresponds to $\ell_u/h \leq 8.4$.

Bar Spacing Requirements

ACI Sec. 7.7.1 requires a clear concrete cover of not less than $1\frac{1}{2}$ in. to the ties or spirals in columns. More cover may be required for fire protection in some cases. The concrete for a column is placed in the core inside the bars and must be able to flow out between the bars and the form. To facilitate this, the ACI Code requires that the minimum clear distance between longitudinal bars shall not be less than the larger of 1.5 times the longitudinal bar diameter, 1.5 in. (ACI Sec. 7.6.3), or $1\frac{1}{3}$ times the size of the coarse aggregate (ACI Sec. 3.3.2). These clear distance limitations also apply to the clear distance between lap spliced bars and adjacent bars or lap splices (ACI Sec. 7.6.4). Because the maximum number of bars occurs at the splices, the spacing of bars at this location generally governs. The spacing limitations at splices in tied and spiral columns are illustrated in Fig. 11–24. Tables A–16 and A–17 give the maximum number of bars that can be used in rectangular tied columns and circular spiral columns, respectively, assuming that the bars are lapped as

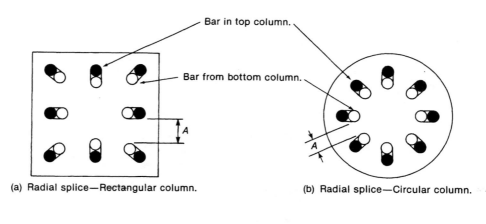

(a) Radial splice—Rectangular column. (b) Radial splice—Circular column.

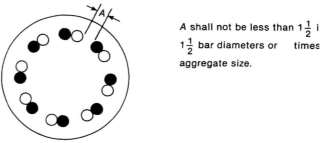

(c) Side-by-side lap splice

A shall not be less than $1\frac{1}{2}$ i $1\frac{1}{2}$ bar diameters or   times aggregate size.

Fig. 11–24
Arrangement of bars at lap splices in columns.

shown in Fig. 11–24a or b. These tables also give the area, A_{st}, of these combinations of bars and the ratio $\rho_t = A_{st}/A_g$.

Reinforcement Splices

In most buildings in nonseismic zones, the longitudinal bars in the columns are spliced just above each floor. Lap splices as shown in Fig. 11–29 are most widely used, although in large columns with large bars, mechanical splices or butt splices may be used (see Figs. 11–25 and 11–26). Lap splices, welded or mechanical splices, and end-bearing splices are covered in ACI Secs. 12.17.2, 12.17.3, and 12.17.4, respectively.

The requirements for lap splices vary depending on the state of stress in the bar at the ultimate load. In columns subjected to combined axial load and bending, tensile stresses may occur on one face of the column, as seen in Example 11–1 (see Fig. 11–17c, for example). Design interaction charts, such as Fig. A–6, frequently include lines indicating the eccentricities for which various tensile stresses occur in the reinforcement closest to the tensile face of the column. In Fig. A–6 these are labeled $f_s = 0, f_s = 0.5f_y$, and $f_s = f_y$ (in tension). The range of eccentricities for which various types of splices are required is shown schematically in Fig. 11–27.

Column splice details are important to the designer for two major reasons. First, a compression lap splice will automatically be provided by the reinforcement detailer unless a different lap length is specified by the designer. Hence if the bar stress at ultimate is tensile, compression laps may be inadequate and the designer should compute and specify the laps required. Second, if Class B lap splices are required (see Fig. 11–27) the required splice lengths may be excessive. For closely spaced bars larger than No. 8 or 9 the length of such a splice may exceed 5 ft and thus may be half or more of the height of the average story. The splice lengths will be minimized by choosing the smallest practical bar sizes and the highest practical concrete strength. Alternatively, welded or mechanical splices should be used. Generally, all the bars in a column will have the same length of splice regardless of whether they are on the tension or compression face. This is done to reduce the chance of field errors. The required splice lengths are computed using the following steps:

Fig. 11–25
Metal-filled bar splice: tension or compression.
(Photograph courtesy of Erico Products Inc.)

Columns: Combined Axial Load and Bending

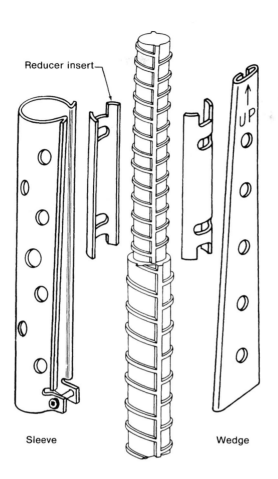

Fig. 11–26
Wedged sleeve bar splice: compression only. (Drawing courtesy of Gateway Building Products.)

Reducer insert

Sleeve

Wedge

1. Establish whether compression lap splices or tension lap splices are required (see Fig. 11–27 or Fig. A–6, etc.)

2. If *compression* lap splices are required, compute the basic compression splice length according to ACI Sec. 12.17.2.1 and, if desired, modify for the presence of ties or spirals according to ACI Sec. 12.17.2.4 or 12.17.2.5 (see Table A–19).

3. If *tension* lap splices are required, compute ℓ_d from ACI Sec. 12.2.2 or 12.2.3. Since ACI Sec. 7.6.3 requires that longitudinal column bars have a minimum clear spacing of $1.5d_b$ and since they will always be enclosed by ties or spirals, ℓ_d is computed for case 1 in Table 8–1. ℓ_d is then multiplied by 1.0 or 1.3 for Class A or Class B splices.

Bars loaded in compression only can be spliced by end bearing splices (ACI Secs. 12.16.4 and 12.17.4) provided that the splices are staggered or extra bars are provided at the splice locations so that the continuing bars in each face of the column at the splice location have a tensile strength at least 25% of that of all the bars in that face. In an end bearing splice the ends of the two bars to be spliced are cut as squarely as possible and held in contact with a wedged sleeve device (Fig. 11–26). Butt splices are used only on vertical, or almost vertical, bars enclosed by stirrups or ties. Tests show that the force transfer in end bearing splices is superior to that in lapped splices, even if the ends of the bars are slightly off square (up to 3° total angle between the ends of the bars).

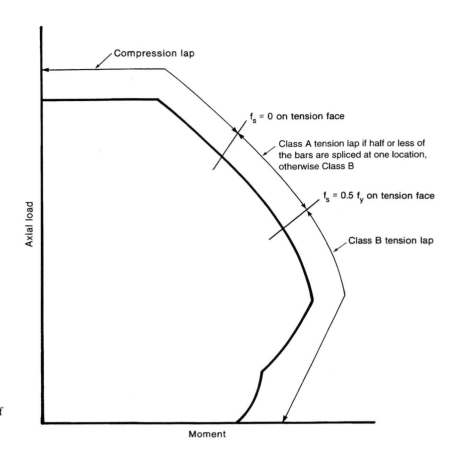

Fig. 11–27
Types of lap splices required if bars are lap spliced at every floor. (From Ref. 11–12).

Spacing and Construction Requirements for Ties

Ties are provided in reinforced concrete columns for four reasons:

1. Ties restrain the longitudinal bars from buckling out through the surface of the column

ACI Secs. 7.10.5.1, 7.10.5.2, and 7.10.5.3 give limits on the size, spacing, and the arrangement of the ties so that they are adequate to restrain the bars. The minimum tie size is a No. 3 bar for longitudinal bars up to No. 10, and a No. 4 bar for larger longitudinal bars, or for bundled bars. The vertical spacing of ties shall not exceed 16 longitudinal bar diameters to limit the unsupported length of these bars, and shall not exceed 48 tie diameters to ensure that the cross-sectional area of the ties is adequate to develop the forces needed to restrain buckling of the longitudinal bars. The maximum spacing is also limited to the least dimension of the column. In seismic regions much closer spacings are required (ACI Sec. 21.4.4).

ACI Sec. 7.10.5.3 outlines the arrangement of ties in a cross section. These are illustrated in Fig. 11–28. A bar is adequately supported against lateral movement if it is located at a corner of a tie or if the dimension *x* in Fig. 11–28 is 6 in. or less. Diamond-shaped and octagonal-shaped ties are not uncommon and keep the center of the column open so that placing the vibration of the concrete is not impeded by the cross-ties. The ends of the ties are anchored by a 90° or 135° bend around a bar, plus an extension of at least 6 tie bar di-

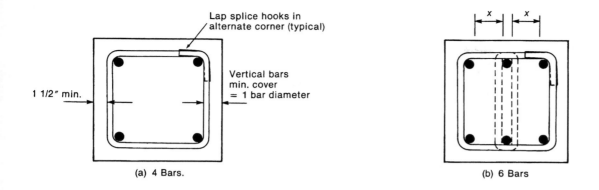

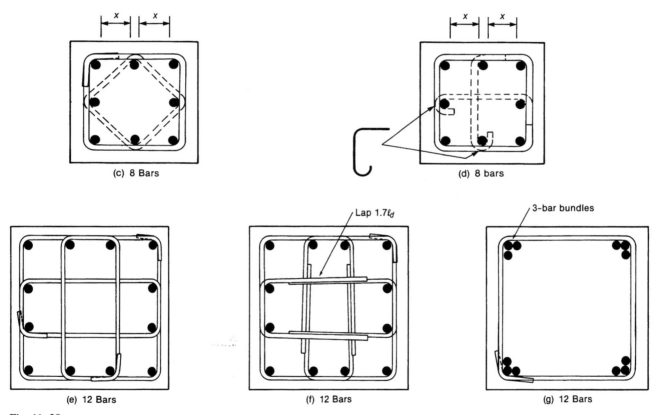

(a) 4 Bars.

(b) 6 Bars

Note: ties shown dashed in (b) (c) and (d)
may be omitted if $x < 6$ in.

(c) 8 Bars

(d) 8 bars

(e) 12 Bars

(f) 12 Bars

(g) 12 Bars

Fig. 11–28
Typical tie arrangements.

ameters but not less than $2\frac{1}{2}$ in. In seismic areas, a 135° bend plus a 6-tie-diameter extension is required.

2. Ties hold the reinforcement cage together during the construction process, as shown in Figs. 11–2 and 11–3.

3. Properly detailed ties confine the concrete core, providing increased ductility.

4. Ties serve as shear reinforcement for columns.

If the shear V_u/ϕ exceeds $0.5V_c$, shear reinforcement is required (ACI Sec. 11.5.5.1). Ties can serve as shear reinforcement but to be effective cannot be spaced farther apart than $d/2$ (ACI Sec. 11.5.4.1). This is generally less than the minimum tie spacing allowed in ACI Sec. 7.10.5.2. If shear governs, the smaller spacing ($d/2$) must be used and the area of all legs parallel to the direction of the shear force must satisfy ACI Eqs. 11–2 and 11–15 in addition to ACI Sec. 7.10.5.

Specially fabricated welded wire reinforcement cages incorporating the longitudinal reinforcement and ties are sometimes an economical solution to constructing column cages. Each cage consists of several interlocking sheets bent to form one to three sides of the cage. The bars or wires forming the ties are hooked around longitudinal bars to make the ties continuous. The layout of the cages must be planned carefully so that bars from one part of the cage do not interfere with those from another when they are assembled in the field.

Ties may be formed from continuously wound wires with a pitch and area conforming to the tie requirements.

ACI Secs. 7.10.5.4 and 7.10.5.5 require that the bottom and top ties be placed as shown in Fig. 11–29. ACI Sec. 7.9.1. requires that bar anchorages in connections of beams and columns be enclosed by ties, spirals, or stirrups. Generally, ties are most suitable for this purpose and should be arranged as shown in Fig. 11–29b.

Finally, extra ties are required at the outside (lower) end of offset bends at column splices (see Fig. 11–29) to resist the horizontal force component in the sloping portion of the bar. The design of these ties is described in ACI Sec. 7.8.1.

Amount of Spirals and Spacing Requirements

The minimum spiral reinforcement required by the ACI Code was chosen so that the second maximum load of the core and longitudinal reinforcement would roughly equal the initial maximum load of the entire column before the shell spalled off (see Fig. 11–5a). Figure 11–4 shows that the core of a spiral column is stressed in triaxial compression and Eq. 3–15 indicates that the strength of concrete is increased by such a loading.

The amount of spiral reinforcement is defined using a spiral reinforcement ratio, ρ_s, equal to the ratio of the volume of the spiral reinforcement to the volume of the core measured out to out of the spirals, enclosed by the spiral. For one turn of the spiral shown in Fig. 11–4a:

$$\rho_s = \frac{A_{sp}L_{sp}}{A_cL_c}$$

where

A_{sp} = area of the spiral bar = $\pi d_{sp}^2/4$
d_{sp} = diameter of the spiral bar
l_{sp} = length of one turn of the spiral = πD_c
D_c = diameter of the core, out to out of the spirals
A_c = area of the core = $\pi D_c^2/4$
l_c = spiral pitch = s

Thus

$$\rho_s = \frac{(A_{sp})(\pi D_c)}{(\pi D_c^2/4)s}$$

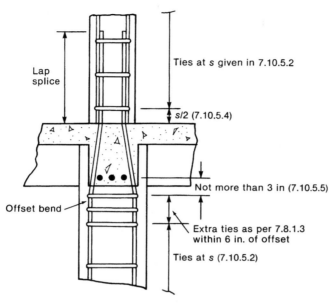

Lap
splice

Ties at s given in 7.10.5.2

s/2 (7.10.5.4)

Not more than 3 in (7.10.5.5)

Offset bend

Extra ties as per 7.8.1.3
within 6 in. of offset

Ties at s (7.10.5.2)

(a) Tie spacing at interior column-beam joint.

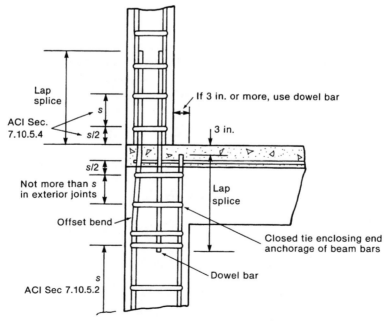

Lap
splice

ACI Sec.
7.10.5.4

s

s/2

If 3 in. or more, use dowel bar

3 in.

s/2

Not more than s
in exterior joints

Lap
splice

Offset bend

Closed tie enclosing end
anchorage of beam bars

s

Dowel bar

ACI Sec 7.10.5.2

Fig. 11–29
Tie spacing at column-beam
joints

(b) Ties at exterior column-beam joint.

or

$$\rho_s = \frac{4A_{sp}}{sD_c} \qquad (11\text{--}20)$$

From the horizontal force equilibrium of the free body in Fig. 11–4a,

$$2f_{sp}A_{sp} = f_2 D_c s \qquad (11\text{--}21)$$

From Eqs. 11–20 and 11–21,

$$f_2 = \frac{f_{sp}\rho_s}{2} \qquad (11\text{--}22)$$

From Eq. 11–1 the strength of a column at the first maximum load before the shell spalls off is

$$P_0 = 0.85 f_c'(A_g - A_{st}) + f_y A_{st} \qquad (11\text{--}1)$$

and the strength after the shell spalls is

$$P_2 = 0.85 f_1(A_c - A_{st}) + f_y A_{st}$$

Thus if P_2 is to equal P_0, $0.85 f_1(A_c - A_{st})$ must equal $0.85 f_c'(A_g - A_{st})$. Since A_{st} is small compared to A_g or A_c we can disregard it, giving

$$f_1 = \frac{A_g f_c'}{A_c} \qquad (11\text{--}23)$$

Substituting Eqs. 3–15 and 11–22 into Eq. 11–23, taking f_{sp} equal to the yield strength of the spiral bar, f_y, rearranging, and rounding down the coefficient gives

$$\rho_s = 0.45\left(\frac{A_g}{A_c} - 1\right)\frac{f_c'}{f_y} \qquad (11\text{--}24)$$
$$(\text{ACI Eq. 10--6})$$

There is experimental evidence that more spiral reinforcement may be needed in high-strength concrete spiral columns than given by Eq. 11–24[11-7] to ensure a ductile failure. To date this has not been considered by the ACI Code committee.

Design requirements for column spirals are presented in ACI Secs. 10.9.3 and 7.10.4.1 to 7.10.4.9. ACI Sec. 7.10.4.2 requires that in cast-in-place construction spirals be at least No. 3. in diameter. The spacing is determined by three rules:

1. Equation 11–24 (ACI Eq. 10–6) can be solved for the maximum spacing that can be used if the second maximum load is to equal or exceed the initial maximum load. Combining Eqs. 11–20 and 11–24 and rearranging gives the maximum *center-to-center* spacing of the spirals:

$$s \le \frac{\pi d_{sp}^2 f_y}{0.45 D_c f_c'\left[(A_g/A_c) - 1\right]} \qquad (11\text{--}25)$$

2. For the spirals to bind the core effectively they must be spaced relatively close together. ACI Sec. 7.10.4.3 limits the *clear* spacing between spirals to not more than 3 in.

3. To avoid problems in placing the concrete, spirals should be as far apart as possible. The *clear* spacing between spirals should not be less than $1\frac{1}{3}$ times the size of the coarse aggregate (ACI Sec. 3.3.3) and never less than 1 in. (ACI Sec. 7.10.4.3).

Table A–15 in Appendix A gives spiral spacings that satisfy all three rules given above. The termination of spirals at the top and bottom of columns is governed by ACI

Secs. 7.10.4.6 to 7.10.4.8. Again, joint reinforcement satisfying ACI Sec. 7.9.1 is also required, and generally ties are provided for this purpose.

Design Examples

Three examples will be considered to illustrate the design of column cross sections. In the second case the strength will be governed by the maximum axial load capacity, and in the third, the axial load will fall in the range where ϕ varies from the value for a column to that for a beam. Although no examples of analysis per se are presented, the analysis of a given cross section is identical to the three examples, except that it is not necessary to select the material properties and column size (step 1 in the examples).

EXAMPLE 11–2 Design of a Tied Column for a Given P_u and M_u

Design a tied column cross section to support P_u = 350 kips, M_u = 110 ft-kips, and V_u = 16 kips. The column is in a braced frame and has an unsupported length of 10 ft 6 in.

1. Select the material properties, trial size, and trial reinforcement ratio. Select f_y = 60 ksi and f_c' = 3 ksi. The most economical range of ρ_t is 1 to 2%. Assume that ρ_t = 0.015 for the first trial value. From Eq. 11–18a,

$$A_{g(\text{trial})} \geq \frac{P_u}{0.45(f_c' + f_y \rho_t)}$$

$$\geq \frac{350}{0.45(3 + 60 \times 0.015)}$$

$$\geq 199 \text{ in.}^2 \text{ or } 14.1 \text{ in. square}$$

Since moments act on this column Eq. 11–18a will underestimate the column size. Choose a 16-in-square column.

To determine the preferable bar arrangement, compute the e/h ratio:

$$e = \frac{M_u}{P_u}$$

$$= \frac{110 \text{ ft-kips}}{350 \text{ kips}} = 0.314 \text{ ft}$$

$$\tfrac{e}{h} = 0.236$$

For this range of e/h, Fig. 11–23 indicates that a column with bars in two faces will be most efficient. Use a tied column with bars in two faces.

Slenderness can be neglected if

$$\frac{k\ell_u}{r} \leq 34 - 12\left(\frac{M_1}{M_2}\right) \tag{11–19}$$

Since this is a braced frame, $k \leq 1.0$, and M_1/M_2 will normally be between +0.5 and −0.5. We shall assume that k = 1.0, and M_1/M_2 = +0.5. The left-hand side of Eq. 11–19 is

$$\frac{k\ell_u}{r} = \frac{1.0 \times 126 \text{ in.}}{0.3 \times 16 \text{ in.}} = 26.3$$

The right-hand side is

$$34 - 12\left(\frac{M_1}{M_2}\right) = 34 - 12(+0.5) = 28$$

Since 26.3 is less than 28, slenderness can be neglected.

Summary for the trial column. A 16 in. × 16 in. tied column with bars in two faces, $f_c' = 3{,}000$ psi, and $f_y = 60{,}000$ psi.

2. **Compute** γ. The interaction diagrams in Appendix A are each drawn for a particular value of the ratio, γ, of the distance between the centers of the outside layers of bars to the overall depth of the column. To estimate γ, assume that the longitudinal bars are No. 8 and the ties are No. 3 bars. For $1\frac{1}{2}$ in. of clear cover to the ties (see Fig. 11–30),

$$\gamma = \frac{16 - 2(1.5 + 0.375 + 0.5)}{16} = 0.703$$

Since the interaction diagrams in Appendix A are given for $\gamma = 0.60$ and $\gamma = 0.75$, it will be necessary to interpolate. If the bars chosen are not No. 8, it will be necessary to recompute γ.

3. **Use interaction diagrams to determine** ρ_t. The interaction diagrams are entered using

$$\frac{\phi P_n}{A_g} = \frac{P_u}{A_g} \qquad \text{ksi}$$

$$= \frac{350}{16 \times 16} = 1.367$$

$$\frac{\phi M_n}{A_g h} = \frac{M_u}{A_g h} \qquad \text{ksi}$$

$$= \frac{110 \times 12}{16 \times 16 \times 16} = 0.322$$

From Fig. A–6 (interaction diagram for $\gamma = 0.6$),

$$\rho_t = 0.014$$

From Fig. A–7 (for $\gamma = 0.75$),

$$\rho_t = 0.013$$

Use linear interpolation to compute value for $\gamma = 0.703$:

$$\rho_t = 0.014 - 0.001 \times \frac{0.103}{0.15} = 0.0133$$

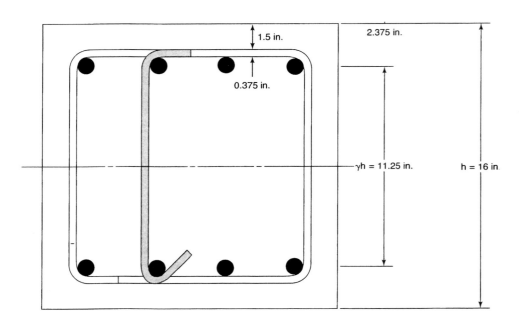

Fig. 11–30
Computation of γ—
Example 11–2.

Columns: Combined Axial Load and Bending

If the value of ρ_t computed here exceeds 0.03 to 0.04, a larger section should be chosen. If ρ_t is less than 0.01, either use 0.01 (the minimum allowed by ACI Sec. 10.9.1) or recompute using a smaller cross section.

4. **Select the reinforcement.**

$$A_{st} = \rho_t A_g$$

$$= 0.0133 \times 16 \times 16 \text{ in.}^2 = 3.41 \text{ in.}^2$$

Possible combinations are (Table A–22):

4 No. 9 bars, $A_{st} = 4.00$ in.^{2}, 2 in each face
8 No. 6 bars, $A_{st} = 3.52$ in.^{2}, 4 in each face
6 No. 7 bars, $A_{st} = 3.60$ in.^{2}, 3 in each face

An even number of bars will be chosen so that the reinforcement is symmetrical about the bending axis. From Table A–18 it is seen that none of these violates the minimum bar spacing rules. **Try a 16-in.-square column with 6 No. 7 bars.**

5. **Check the maximum load capacity.** P_u should not exceed $\phi P_{n(\max)}$ given by Eq. 11–4. The upper horizontal lines in the interaction diagrams represent $\phi P_{n(\max)}$ and the section chosen falls below the upper limit. Actually, this check is necessary only if one were using interaction diagrams that did not show this cutoff.

6. **Design the lap splices.** From Figs. A–6 and A–7, the stress in the bars adjacent to the tensile face for $P_u/bh = 1.367$ and $M_u/bh^2 = 0.322$ is about $0.2f_y$ in tension. From ACI Secs. 12.17.2.2 and 12.17.2.3, the splice must be a Class B splice if more than half of the bars are spliced at any section or a Class A splice if half or less are spliced at one location. Normally, all the bars would be spliced at the same location. We shall assume that this is done. The splice length is $1.3\ell_d$. From ACI Sec. 12.2.2, ℓ_d for No. 7 bars is

$$\ell_d = \left(\frac{f_y \alpha \beta \lambda}{20\sqrt{f_c'}}\right) d_b$$

$$= \left(\frac{60{,}000 \times 1.0 \times 1.0 \times 1.0}{20 \times \sqrt{3000}}\right) \times 0.875$$

$$= 47.9 \text{ in.}$$

The splice length is

$$1.3\ell_d = 1.3 \times 47.9$$

$$= 62.3 \text{ in.}$$

This is a very long splice. It would approach half of the story height. For this reason, consider using 8 No. 6 bars, four in each face. The clear spacing between the bars would be 3.08 in. which exceeds $1.5d_b$. For No. 6 bars, ℓ_d is

$$\ell_d = \left(\frac{f_y \alpha \beta \lambda}{25\sqrt{f_c'}}\right) d_b = 38.3 \text{ in.}$$

The splice length is $1.3 \times 38.3 = 49.8$ in., say 50 in.

7. **Select the ties.** From ACI Sec. 7.10.5.1, No. 3 ties are the smallest allowed. The required spacing (ACI Sec. 7.10.5.2) is the smallest of:

$$16 \text{ longitudinal bar diameters} = 16 \times \frac{3}{4} = 12 \text{ in.}$$

$$48 \text{ tie diameters} = 48 \times \frac{3}{8} = 18 \text{ in.}$$

$$\text{least dimension of column} = 16 \text{ in.}$$

If $V_u > 0.5\phi V_c$, the ties must satisfy ACI Chap. 11. as well as ACI Sec. 7.10.5.

$$V_c = 2\left(\frac{1 + N_u}{2000A_g}\right)\sqrt{f_c'}\, b_w d$$

$$= 2\left(1 + \frac{350{,}000}{2000 \times 16 \times 16}\right)\sqrt{3000} \times 16 \times 13.6$$

$$= 40.1 \text{ kips}$$

$V_u = 16$ kips is less than $0.5\phi V_c = 0.5 \times 0.85 \times 40.1 = 17.1$ kips. Therefore, ACI Sec. 7.10.5 governs. (If $0.5\phi V_c < V_u \leq \phi V_c$, it would be necessary to satisfy ACI Secs. 7.10.5, 11.5.4.1, and 11.5.5.3.) **Use No. 3. ties at 12 in. o.c.** The tie arrangement is shown in Fig. 11–30. ACI Sec. 7.10.5.3 requires that all corner bars and alternate other bars be at the corner of a tie. To accomplish this, we shall add a cross-tie.

Summary of the design. Use a 16 in. $\times$ 16 in. tied column with 8 No. 6 bars, $f_y = 60{,}000$ psi, and $f_c' = 3000$ psi. Use No. 3 ties and cross-ties at 12 in. on centers as shown in Fig. 11–30. Alternate the end with the 90° hook. Lap splice all the longitudinal bars 50 in. ∎

EXAMPLE 11–3 Design of a Spiral Column for a Large Axial Load and Small Moment

Design a spiral column cross section to support factored forces and end moments of $P_u = 1600$ kips and $M_u = 150$ ft-kips.

1. Select the material properties, trial size, and trial reinforcement ratio. Use $f_c' = 4$ ksi and $f_y = 60$ ksi. Try $\rho_t = 0.04$. From Eq. 11–18b,

$$A_{g(\text{trial})} = \frac{P_u}{0.55(f_c' + f_y\rho_t)} = 455 \text{ in.}^2$$

This corresponds to a diameter of 24.1 in. We shall try 24 in.

2. Compute γ. Assume that the spiral is made from a No. 3 bar and the longitudinal bars are No. 10:

$$\gamma = \frac{24 - 2(1.5 + 0.375 + 1.27/2)}{24} = 0.791$$

3. Use interaction diagrams to determine ρ_t.

$$\frac{P_u}{A_g} = \frac{1600}{\pi \times 12^2} = 3.54$$

$$\frac{M_u}{A_g h} = \frac{150 \times 12}{\pi \times 12^2 \times 24} = 0.166$$

Interpolating between Figs. A–13 and A–14 (interaction diagrams for spiral columns with $\gamma = 0.75$ and 0.90 and the correct material strengths) gives $\rho_t = 0.038$. Because the values of P_u/A_g and $M_u/A_g h$ fall in the upper horizontal part of the diagram, Eq. 11–4a could be used to solve for A_{st} directly.

4. Select the reinforcement.

$$A_{st} = \rho_t A_g$$

$$= 0.038(12^2\pi) = 17.19 \text{ in.}^2$$

From Table A–18, 11 No. 11 bars give 17.16 in.^{2} and Table A–21 shows that these will fit into a 24-in.-diameter column. Try 11 No. 11 bars.

5. Check the maximum load capacity. From Eq. 11–4a,

$$\phi P_{n,\text{max}} = 0.85 \times 0.75\left[0.85 \times 4(\pi \times 12^2 - 17.16) + 60 \times 17.16\right]$$

$$= 1600 \text{ kips} \quad \text{OK}$$

6. Select the spiral. The minimum-size spiral is No. 3 (ACI Sec. 7.10.4.2). The center-to-center pitch, s, to ensure the second maximum load is equal to the initial maximum load is given by Eq. 11–25:

$$s < \frac{\pi d_{sp}^2 f_y}{0.45 D_c f_c' \left[(A_g / A_c) - 1 \right]}$$

where

d_{sp} = diameter of spiral = 0.375 in.
f_y = yield strength of spiral bar = 60,000 psi
D_c = diameter of core, outside to outside of spirals = 24 in. $-$ 2 × 1.5 in. = 21 in.
A_g = gross area = $12^2 \pi$ = 452 in.^{2}
A_c = area of core = $10.5^2 \pi$ = 346 in.^{2}

Thus

$$s \leq \frac{\pi \times 0.375^2 \times 60,000}{0.45 \times 21 \times 4000 \left[(452/346) - 1 \right]}$$

$$\leq 2.29 \text{ in.}$$

Thus the center-to-center pitch cannot exceed 2.29 in. but must also satisfy detailing requirements. The maximum clear spacing (ACI Sec. 7.10.4.3) is 3 in. (maximum pitch = 3.0 + 0.375 = 3.375 in.). The minimum clear spacing (ACI Sec. 7.10.4.3) is 1 in. or $1\frac{1}{3}$ times the size of the coarse aggregate (ACI Sec. 3.3.3). For $\frac{3}{4}$-in. aggregate the minimum pitch is 1.375 in.

Use a No. 3 spiral with $2\frac{1}{4}$-in. pitch. Table A–15 could be used to select the spiral directly.

7. Design the lap splices. From the interaction diagrams, f_s is compression. Therefore, use compression lap splices (ACI Sec. 12.16.1). From Table A–19 these must be 42.3 in. long. Since the bars are enclosed in a spiral this can be multiplied by 0.75, giving a splice length of 31.7 in. (ACI Sec. 12.17.2.5).

Summary of design. Use a 24-in.-diameter column with 11 No. 11 bars, f_y = 60,000 psi, and f_c' = 4000 psi. Use a No. 3 spiral with a pitch of $2\frac{1}{4}$ in. and f_y = 60,000 psi. Lap splice all bars 32 in. just above every floor. ∎

EXAMPLE 11–4 Design of a Tied Column for a Small Axial Load

If the interaction diagrams used in designing the column are plotted for ϕP_n and ϕM_n and have the transition range of ϕ included, as do all the interaction diagrams in Appendix A, the solution proceeds exactly as in Example 11–2. Occasionally, however, one may obtain design tables or interaction diagrams that do not include ϕ, or do not show the correct transition for ϕ. In the tension failure region such diagrams look like the line B–C–D in Fig. 11–18. Diagrams that correctly include ϕ have a discontinuity in the tension failure region and resemble line B'–C'–D' in this figure. When working with diagrams that do not include ϕ, follow the procedure given in Sec. 11–4 and Example 11–1. ∎

11–6 CONTRIBUTIONS OF STEEL AND CONCRETE TO COLUMN STRENGTH

The interaction diagram for a column with reinforcement in two faces can be considered to be the sum of three components: (1) the load and moment resistance of the concrete, and (2) and (3) the load and moment resistance provided by each of the two layers of reinforcement.

In Fig. 11–31, the P_n, M_n interaction diagram from Example 11–1 and Fig. 11–18 is replotted showing the load and moment resistance contributed by the concrete and the two layers of steel. The center curved portion, O–J–A–G–D–L, represents the compressive force in the concrete, C_c, and the moment of this force about the centroid of the column. The shaded band represents the force in the "tensile" reinforcement adjacent to the less compressed face, force F_{s1} in Example 11–1, and the moment of this force about the centroid of the column. The outer band is the force F_{s2} in the "compressive" reinforcement adjacent to the most compressed face, and its moment about the centroid. A reexamination of Example 11–1 shows that at the balanced failure condition the force C_c was 434.6 kips and the moment of C_c about the centroid was 174 ft-kips. These are plotted in Fig. 11–31 as the vector OA. The force F_{s1} was -240 kips and its moment about the centroid was 110 ft-kips, plotted as vector AB. Finally, the force F_{s2} was 222.1 kips and its moment was 101.8 ft-kips, plotted as vector BC. An examination of line $OABC$ shows that the majority of the axial load capacity at the balanced load comes from the concrete, with the forces in the tension and compression steel layers essentially canceling each other out. All three constituents contribute to the moment capacity.

A different case is represented by line $ODEF$ in Fig. 11–31. Here the force F_{s1} is in compression and hence adds to the axial load capacity and subtracts from the moment capacity, as shown by the vector DE. An intermediate case, in which $F_{s1} = 0$, is shown by line OGH. A similar case, in which $F_{s2} = 0$, is shown by line OJK.

The portion of the interaction diagram due to the concrete is a continuous curve. The discontinuity in the overall interaction diagram at the balanced point is due to the reinforcement. The two branches of the interaction diagram in Figs. 11–18 and 11–31 are relatively straight. This is characteristic of a column with a high steel ratio. In the case of low steel ratios the curve is much more curved because the concrete portion dominates.

The effects of the individual reinforcement layers can be seen from Fig. 11–31. From the top of the diagram down to about F both layers of steel are effective in increasing the axial capacity. At H, only the steel on the compression face is effective. At C both layers are effective in increasing the moment capacity. At K only the reinforcement on the tension face is effective and finally, at M both layers add to the tensile capacity. To optimize the design of a member subjected to axial load and bending, one may wish to provide different amounts of reinforcement in the two faces. Figure 11–31 can be used as a guide to where the reinforcement should go. Generally speaking, however, columns are built with the same reinforcement in both faces to minimize the chance of putting the bars in the wrong face and because moments frequently change sign along the length of a column, or due to winds from alternate sides of the building. In the case of culverts, arch sections, or rigid frame legs, however, it may be desirable to have unsymmetrical reinforcement.

11–7 APPROXIMATE SOLUTION FOR TIED COLUMNS FAILING IN COMPRESSION

The interaction diagram calculation for the capacity of column cross sections, although theoretically correct, does not give a direct solution for use in sizing a column. In 1942, Whitney[11–8] developed an approximate equation for the compression failure branch of the interaction diagram. His analysis assumed that $f_y = 50,000$ psi. Today most columns are built using 60,000 psi. As a result, a modified Whitney equation based on $f_y = 60,000$ psi is presented here. The following assumptions are necessary:

> **1.** The column has a rectangular cross section with reinforcement in two layers parallel to the axis of bending and equal distances from that axis.

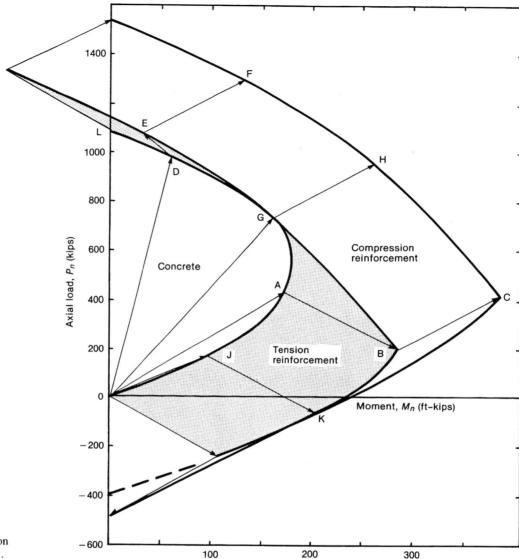

Fig. 11-31
Components of interaction
diagram—Example 11-1.

2. The compression reinforcement has yielded. This is generally true for compression failures, particularly for small eccentricities.

3. The area of concrete displaced by compression steel may be neglected.

4. The interaction diagram for compression failures can be represented by a straight line from the pure axial load capacity P_{n0} to the load and moment corresponding to a balanced failure.

5. The depth of the compression stress block for a balanced failure is $a = \beta_1 c$. From strain compatibility assuming that $f_c' \leq 4000$ psi and $f_y = 60,000$ psi, $a = 0.51d$.

Summing moments about the tension reinforcement gives

$$P_n\left(e + \frac{d - d'}{2}\right) = C\left(d - \frac{a}{2}\right) + C_s(d - d')$$

11-7   Approximate Solution for Tied Columns Failing in Compression   **463**

or

$$P_n = \frac{C(d - a/2)}{e + [(d - d')/2]} + \frac{C_s}{[e/(d - d')] + \frac{1}{2}}$$

Substituting yields

$$C\left(d - \frac{a}{2}\right) = [0.85f_c'(0.51d)b]\left(d - \frac{0.51d}{2}\right)$$

$$= 0.323f_c'bd^2$$

and $C_s = A_s'f_y$ gives

$$P_n = \frac{0.323f_cbd^2}{e + \frac{1}{2}(d - d')} + \frac{A_s'f_y}{[e/(d - d')] + \frac{1}{2}}$$

which can be rewritten as

$$P_n = \frac{0.323f_c'bh}{(eh/d^2) + [(d - d')h]/2d^2} + \frac{A_s'f_y}{[e/(d - d')] + \frac{1}{2}} \tag{11-26}$$

If the eccentricity corresponding to balanced failure is known, this equation can be used to solve for the balanced failure load, P_{nb}. For this equation to fit the range from P_{n0} to P_{nb} it is necessary to satisfy the boundary condition for $e = 0$. Here

$$P_n = 0.85f_c'bh + 2f_yA_s'$$

Substituting $e = 0$ and this value of P_n into Eq. 11–26 gives

$$\frac{(d - d')h}{2d^2} = 0.38 \tag{11-27}$$

Substituting this into Eq. 11–26 and recognizing that $A_{st} = 2A_s'$ gives the *modified Whitney equation*:

$$P_n = \frac{0.323f_c'bh}{eh/d^2 + 0.38} + \frac{A_{st}f_y}{1 + 2e/(d - d')} \tag{11-28}$$

EXAMPLE 11–5 Calculation of P_n Using Modified Whitney Equation

In step 3 of Example 11–1 the balanced load and moment for the column shown in Fig. 11–17 were computed. They were $P_{nb} = 417$ kips and $M_{nb} = 4630$ in.-kips. This corresponds to an eccentricity $e = 4630/417 = 11.1$ in. Compute P_n for this column for $e = 11.1$ in.
    Substituting into Eq. 11–28 gives

$$P_n = \frac{0.323 \times 5 \times 16 \times 16}{(11.1 \times 16)/13.5^2 + 0.38} + \frac{8 \times 60}{1 + (2 \times 11.1)/(13.5 - 2.5)}$$

$$= 305 + 159 = 464 \text{ kips}$$

$$M_n = 464 \times 11.1 \text{ in.} = 5153 \text{ in.-kips} \qquad \blacksquare$$

These exceed the values computed by the theoretically correct strain compatibility solution by about 10%. This results primarily from the approximation introduced in

Eq. 11–27. Assumption 3 also introduced some error, as did the use of f_c' higher than 4000 psi. Studies of interaction diagrams suggest that Eq. 11–28 will be unconservative if $P_n < 0.6f_c'bh$. Finally, Eq. 11–28 does not apply if the axial load is greater than that given by Eq. 11–4b.

EXAMPLE 11–6   Using the Modified Whitney Equation to Select Reinforcement for a Column

Select a column cross section for $P_u = 560$ kips and $M_u = 210$ ft-kips. Use $f_c' = 5000$ psi and $f_y = 60,000$ psi.

1. Select the column size. Based on Eq. 11–18a with $\rho_t = 0.02$, a 14.2-in.-square column is required. Try a 16-in.-square column. For this column,

$$b = 16 \text{ in.} \qquad h = 16 \text{ in.}$$

$$d = 13.5 \text{ in.} \qquad d - d' = 11.0 \text{ in.}$$

From the load and moment,

$$e = \frac{210 \times 12 \text{ in.-kips}}{560 \text{ kips}} = 4.5 \text{ in.}$$

2. Is the modified Whitney equation applicable? Equation 11–28 can be used if $P_n \geq 0.6f_c'bh = 768$ kips. For this problem

$$P_n = \frac{P_u}{\phi} = \frac{560}{0.7}$$

$$= 800 \text{ kips}$$

Therefore, Eq. 11–28 can be used.

3. Solve for A_{st}. Substituting into Eq. 11–28 gives

$$800 \text{ kips} = \frac{0.323 \times 5 \text{ ksi} \times 16 \times 16 \text{ in.}^2}{[(4.5 \text{ in.} \times 16 \text{ in.})/(13.5 \text{ in.})^2] + 0.38} + \frac{A_{st} \text{ in.}^2 \times 60 \text{ ksi}}{1 + (2 \times 4.5 \text{ in.})/(11.0 \text{ in.})}$$

$$= 533 + 33A_{st}$$

$$A_{st} = 8.08 \text{ in.}^2$$

Use a 16-in.-square column with $A_{st} = 8.08$ in.², $f_c' = 5$ ksi, and $f_y = 60$ ksi. ■

From an interaction diagram solution for this case the required A_{st} is 8.00 in.² In this case the modified Whitney Equation was quite accurate.

Although it is possible to derive similar equations for the strength of columns failing in tension, these are less accurate and since most practical columns fail in compression, are less necessary. In the tension failure region interaction diagrams should be used.

11–8 BIAXIALLY LOADED COLUMNS

Up to this point in the chapter we have dealt with columns subjected to axial loads accompanied by bending about one axis. It is not unusual for columns to support axial forces and

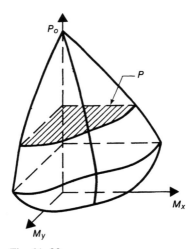

Fig. 11–32
Interaction surface for axial
load and biaxial bending.

bending about two perpendicular axes. One common example is a corner column in a frame.

For a given cross section and reinforcing pattern, one can draw an interaction diagram for axial load and bending about either axis. As shown in Fig. 11–32, these interaction diagrams form the two edges of an interaction surface for axial load and bending about two axes. The calculation of *each* point on such a surface involves a double iteration: (1) the strain gradient across the section is varied, and (2) the angle of the neutral axis is varied. For the same reasons discussed in Sec. 5–3, the neutral axis will generally not be parallel to the resultant moment vector. The calculation of interaction diagrams for biaxially loaded columns is discussed in Ref. 11–9.

A horizontal section through such a diagram resembles a quadrant of a circle or an ellipse at high axial loads, and depending on the arrangement of bars, it becomes considerably less circular near the balanced load, as shown in Fig. 11–32.

Three procedures are commonly used to design rectangular columns subjected to biaxial loads:

1. The biaxial eccentricities, e_x and e_y, can be replaced by an equivalent uniaxial eccentricity, e_{0x}, and the column designed for uniaxial bending and axial load.[11–10] We shall define e_x as the component of the eccentricity parallel to the side x and the x axis, as shown in Fig. 11–33, such that the moment, M_y, about the y axis is

$$M_{uy} = P_u e_x \qquad M_{ux} = P_u e_y \qquad (11\text{–}29\text{a,b})$$

If

$$\frac{e_x}{x} \geq \frac{e_y}{y} \qquad (11\text{–}30)$$

then the column can be designed for P_u and a factored moment $M_{0y} = P_u e_{0x}$, where

$$e_{0x} = e_x + \frac{\alpha e_y}{y} x \qquad (11\text{–}31)$$

where for $P_u / f_c' A_g \leq 0.4$,

$$\alpha = \left(0.5 + \frac{P_u}{f_c' A_g}\right) \frac{f_y + 40{,}000}{100{,}000} \geq 0.6 \qquad (11\text{–}32\text{a})$$

and for $P_u / f_c' A_g > 0.4$,

$$\alpha = \left(1.3 - \frac{P_u}{f_c' A_g}\right) \frac{f_y + 40{,}000}{100{,}000} \geq 0.5 \qquad (11\text{–}32\text{b})$$

In Eq. 11–32, f_y is in psi. If the inequality in Eq. 11–30 is not satisfied, the x's and y's are interchanged in Eq. 11–31.

This procedure is limited in application to columns that are symmetrical about two axes with a ratio of side lengths, x/y, between 0.5 and 2.0. The reinforcement should be in all four faces of the column. The use of Eqs. 11–29 to 11–32 is illustrated in Example 11–7.

2. Charts[11–4] or relationships[11–11] are available for the 45° section through the interaction surface (M_x and M_y at point A in Fig. 11–34). The design is then based on straight-line approximations to horizontal slices through the interaction surface as shown by the dashed lines in Fig. 11–34.

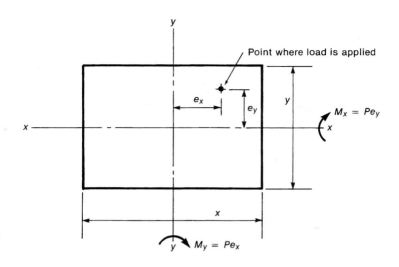

Fig. 11–33
Definition of terms: biaxially
loaded columns.

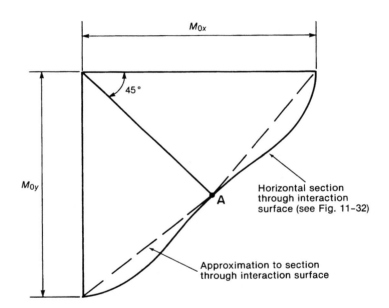

Fig. 11–34
Approximation of section
through intersection surface.

3. ACI Commentary Secs. 10.3.5 and 10.3.6 present Eq. 11–33 originally presented by Bresler,[11-12] for calculating the capacity under biaxial bending:

$$\frac{1}{P_u} = \frac{1}{\phi P_{nx}} + \frac{1}{\phi P_{ny}} - \frac{1}{\phi P_{n0}} \qquad (11\text{–}33)$$

where

P_u = factored axial load

ϕP_{nx} = factored axial load capacity corresponding to the eccentricity e_x and the steel provided, with $e_y = 0$

ϕP_{ny} = factored axial load capacity corresponding to the eccentricity e_y and the steel provided, with $e_x = 0$

ϕP_{n0} = factored axial load capacity for $e_x = 0$ and $e_y = 0$ (Eq. 11–1)

where, as shown in Fig. 11–33 and Eq. 11–29,

M_{ux} = moment about x axis, $P_u e_y$

e_x = eccentricity measured parallel to the x axis

$$= \frac{M_{uy}}{P_u} = \frac{P_u e_x}{P_u}$$

e_y = eccentricity measured parallel to the y axis

x = length of side of column parallel to the x axis

y = length of side of column parallel to the y axis

This procedure is widely used and is illustrated in Example 11–8.

EXAMPLE 11–7 Design of a Biaxially Loaded Column: Equivalent Eccentricity Method

Select a tied column cross section to resist factored loads and moments of $P_u = 360$ kips, $M_{ux} = 70$ ft-kips, and $M_{uy} = 80$ ft-kips. Use $f_y = 60$ ksi and $f'_c = 3$ ksi. The first procedure, based on Eqs. 11–29 to 11–32, will be used.

1. Select a trial section. Assume that $\rho_t = 0.015$. Use a section with bars in the four faces since the column is loaded biaxially.

$$A_{g(trial)} \geq \frac{P_u}{0.45(f'_c + f_y \rho_t)}$$

$$\geq \frac{360}{0.45(3 + 60 \times 0.015)}$$

$$\geq 205 \text{ in.}^2 \text{ or } 14.3 \text{ in. square}$$

Try a 16-in.-square column with $f'_c = 3$ ksi and $f_y = 60$ ksi.

2. Compute γ.

$$\gamma = 0.703 \text{ (step 2, Example 11–2)}$$

3. Compute e_x, e_y, and e_{0x}, or e_{0y}. From the definition of the moments and eccentricities in Fig. 11–33 and Eq. 11–29,

$$e_x = \frac{M_{uy}}{P_u}$$

$$= \frac{80 \times 12}{360} = 2.67 \text{ in.}$$

$$e_y = \frac{M_{ux}}{P_u}$$

$$= \frac{70 \times 12}{360} = 2.33 \text{ in.}$$

By inspection, $e_x/x \geq e_y/y$ therefore, use Eq. 11–31 as given. If this was not true, transpose the x and y terms and subscripts in Eq. 11–31 before using.

$$\frac{P_u}{f'_c A_g} = \frac{360}{3 \times 256} = 0.469 > 0.4$$

Therefore, use Eq. 11–32b to compute α:

$$\alpha = \left(1.3 - \frac{P_u}{f'_c A_g}\right)\frac{f_y + 40,000}{100,000} \text{ but not less than } 0.5$$

$$= (1.3 - 0.469)\frac{60,000 + 40,000}{100,000} = 0.831$$

From Eq. 11–31,

$$e_{0x} = e_x + \frac{\alpha e_y x}{y}$$

$$= 2.67 + 0.831 \times 2.33 \times \frac{16}{16} = 4.61 \text{ in.}$$

Thus the equivalent uniaxial moment is

$$M_{0y} = P_u e_{0x}$$

$$= 360 \times 4.61 = 1658 \text{ in.-kips}$$

The column is designed for $P_u = 360$ kips and $M_{0y} = 1658$ in.-kips.

4. Use interaction diagrams to determine ρ_t. Since the column has biaxial bending, we will select a section with bars in four faces. The interaction diagrams are entered with

$$\frac{P_u}{A_g} = \frac{360}{256} = 1.41 \text{ ksi}$$

and

$$\frac{M_{0y}}{A_g h} = \frac{1658}{16^3} = 0.405 \text{ ksi}$$

From Figs. A–9 and A–10

For $\gamma = 0.60$: $\rho_t = 0.033$
For $\gamma = 0.75$: $\rho_t = 0.026$

Using linear interpolation, $\rho_t = 0.031$ for $\gamma = 0.703$.

5. Compute A_{st} and select the reinforcement.

$$A_{st} = \rho_t A_g = 7.94 \text{ in.}^2$$

From Table A–22, select **8 No. 9 bars, three in each face, A_{st} = 8.00 in.²**. Design ties and lap splices as in earlier examples. ■

EXAMPLE 11–8   Design of a Biaxially Loaded Column: Bresler Reciprocal Load Method

Repeat Example 11–7 using Eq. 11–33.

1. **Select a trial section.** Select A_g as in Example 11–7. To use Eq. 11–33 it is necessary also to estimate the reinforcement required. Try a 16-in.-square column, $f'_c = 3$ ksi, $f_y = 60$ ksi, and 8 No. 8 bars, three in each face, and No. 3 ties.

2. **Compute γ.**

$$\gamma = 0.703$$

3. **Compute ϕP_{nx}.** ϕP_{nx} is the factored axial load capacity corresponding to e_x and ρ_t.

$$\rho_t = \frac{8 \times 0.79}{16 \times 16} = 0.0247$$

$$\frac{e_x}{x} = \frac{M_{uy}}{P_u x} = \frac{80 \text{ ft-kips} \times 12}{360 \text{ kips} \times 16 \text{ in.}}$$

$$= 0.167$$

From Fig. A–9, for $e_x/x = 0.167$ and $\rho_t = 0.0247$ gives $\phi P_{nx}/bh = 1.76$ for $\gamma = 0.60$. From Fig. A–10, $\phi P_{nx}/bh = 1.85$ for $\gamma = 0.75$. Interpolating gives $\phi P_{nx}/bh = 1.82$ and $\phi P_{nx} = 466$ kips.

4. **Compute ϕP_{ny}.**

$$\frac{e_y}{y} = \frac{M_{ux}}{P_u y} = \frac{70 \times 12}{360 \times 16} = 0.146$$

From Fig. A–9, $\phi P_{ny}/bh = 1.88$ for $\gamma = 0.60$. From Fig. A–10, $\phi P_{ny}/bh = 1.95$ for $\gamma = 0.75$. Interpolating gives $\phi P_{ny}/bh = 1.93$ and $\phi P_{ny} = 494$ kips.

5. **Compute ϕP_{n0}.** From Figs. A–9 and A–10, ϕP_{n0} is the point where the interaction curve for $\rho_t = 0.0247$ would intersect the vertical axis, $\phi P_{n0}/bh = 2.75$, and $\phi P_{n0} = 704$ kips.

6. **Solve for P_u.**

$$\frac{1}{P_u} = \frac{1}{\phi P_{nx}} + \frac{1}{\phi P_{ny}} - \frac{1}{\phi P_{n0}}$$

$$= \frac{1}{466} + \frac{1}{494} - \frac{1}{704}$$

$$= 364 \text{ kips}$$

The required capacity is 360 kips; therefore, the column design is adequate. **Use 8 No. 8 bars, three in each face, $A_{st} = 6.32$ in.^{2}.** ∎

The amount of steel required by the two methods differs by 25%. Since both methods are empirical, it is not possible to state which design is closer to the truth.

PROBLEMS

11–1 The column shown in Fig. P11–1 is made of 4000-psi concrete and Grade 60 steel.

(a) Compute the theoretical capacity of the column for pure axial load.

(b) Compute the maximum permissible ϕP_n for the column.

11–2 Why does a spiral improve the behavior of a column?

11–3 Compute the balanced axial load and moment capacity of the column shown in Fig. P11–1. Use $f'_c = 4000$ psi and $f_y = 60,000$ psi.

11–4 For the column shown in Fig. P11–4 use a strain compatibility solution to compute five points on the

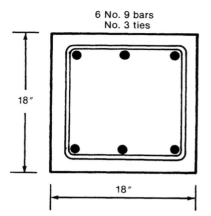

6 No. 9 bars
No. 3 ties

18″

18″

Fig. P11–1

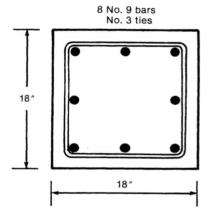

8 No. 9 bars
No. 3 ties

18″

18″

Fig. P11–4

interaction diagram corresponding to points 1 to 5 in Fig. 11–22. Plot the interaction diagram. Use $f_c' = 3000$ psi and $f_y = 60,000$ psi.

11–5 Write a program for a calculator or computer to solve for points on an interaction diagram for rectangular columns with up to 10 layers of steel. (On a Hewlett-Packard calculator such a program involves about 250 steps.)

11–6 Use the interaction diagrams in Appendix A to compute the maximum moment, M_u, which can be supported by the column shown in Fig. P11–1 if

(a) $P_u = 583$ kips.

(b) $P_u = 130$ kips.

(c) $e = 4$ in.

Use psi $f_c' = 3000$ and $f_y = 60,000$ psi.

11–7 Use the interaction diagrams in Appendix A to select tied column cross sections to support the loads given below. In each case, use $f_c' = 3000$ psi and $f_y = 60,000$ psi. Design ties. Calculate the required splice lengths assuming that the bars extending up from the column below are the same diameter as in the column you have designed, and draw a typical cross section of the column showing the bars and ties.

(a) $P_u = 390$ kips, $M_u = 220$ ft-kips, square column with bars in two faces.

(b) $P_u = 710$ kips, $M_u = 50$ ft-kips, square column with bars in four faces.

(c) $P_u = 130$ kips, $M_u = 240$ ft-kips, square column with bars in four faces.

11–8 Use the interaction diagrams in Appendix A to select spiral column cross sections to support the loads given below. In each case, use $f_c' = 4000$ psi and $f_y = 60,000$ psi. Design spirals. Calculate the required splice lengths and draw a typical cross section of the column showing the bars and spiral.

(a) $P_u = 600$ kips, $M_u = 65$ ft-kips.

(b) $P_u = 500$ kips, $M_u = 150$ ft-kips.

11–9 Why are tension splices required in some columns?

11–10 Use the modified Whitney equation to select a cross section and reinforcement to support $P_u = 390$ kips and $M_u = 220$ ft-kips. Use $f_c' = 3000$ psi and $f_y = 60,000$ psi.

11–11 Select a cross section and reinforcement to support $P_u = 450$ kips, $M_{ux} = 100$ ft-kips, and $M_{uy} = 130$ ft-kips. Use $f_c' = 3000$ psi and $f_y = 60,000$ psi.

12

Slender Columns

12–1 INTRODUCTION

Definition of Slender Columns

An eccentrically loaded, pin-ended column is shown in Fig. 12–1a. The moments at the ends of the column are

$$M_e = Pe \qquad (12\text{–}1)$$

When the loads P are applied, the column deflects laterally by an amount Δ as shown. For equilibrium, the internal moment at midheight must be (Fig. 12–1b)

$$M_c = P(e + \Delta) \qquad (12\text{–}2)$$

The deflection increases the moments for which the column must be designed. In the symmetrical column shown here, the maximum moment occurs at midheight where the maximum deflection occurs.

Figure 12–2 shows an interaction diagram for a reinforced concrete column. This diagram gives the combinations of axial load and moment which are required to cause failure of a column cross section or a very short length of column. The dashed radial line $O\text{–}A$ is a plot of the end moment on the column in Fig. 12–1. Since this load is applied at a constant eccentricity, e, the end moment, M_e, is a linear function of P, as given by Eq. 12–1. The curved, solid line $O\text{–}B$ is the moment M_c at midheight of the column, given by Eq. 12–2. At any given load P, the moment at midheight is the sum of the end moment, Pe, and the moment due to the deflections, $P\Delta$. The line $O\text{–}A$ is referred to as a *load-moment curve* for the end moment, while the line $O\text{–}B$ is the load–moment curve for the maximum column moment.

Failure occurs when the load–moment curve $O\text{–}B$ for the point of maximum moment intersects the interaction diagram for the cross section. Thus the load and moment at failure

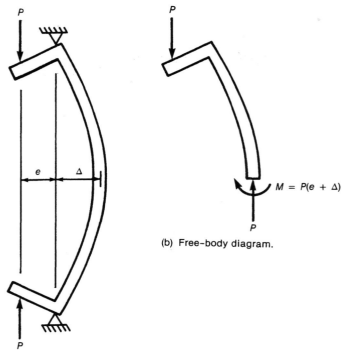

(b) Free-body diagram.

$$M = P(e + \Delta)$$

Fig. 12–1
Forces in a deflected column.

(a) Column.

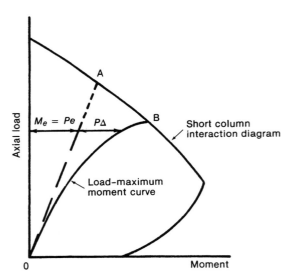

Fig. 12–2
Load and moment in a column.

are denoted by point *B* in Fig. 12–2. Because of the increase in maximum moment due to deflections, the axial load capacity is reduced from *A* to *B*. This reduction in axial load capacity results from what are referred to as *slenderness effects*.

A *slender column* is defined as a column that has a significant reduction in its axial load capacity due to moments resulting from lateral deflections of the column. In the derivation of the ACI Code "a significant reduction" was arbitrarily taken as anything greater than about 5%.[12-1]

Buckling of Axially Loaded Elastic Columns

Figure 12–3 illustrates three states of equilibrium. If the ball in Fig. 12–3a is displaced laterally and released, it will return to its original position. This is *stable equilibrium*. If the ball in Fig. 12–3c is displaced laterally and released, it will roll off the hill. This is *unstable equilibrium*. The transition between stable and unstable equilibrium is *neutral equilibrium*, illustrated in Fig. 12–3b. Here the ball will remain in the displaced position. Similar states of equilibrium exist for the axially loaded column in Fig. 12–4a. If the column returns to its original position when it is pushed laterally at midheight and released, it is in stable equilibrium; and so on.

Figure 12–4b shows a portion of a column that is in a state of neutral equilibrium. The differential equation for this column is

$$EI\frac{d^2y}{dx^2} = -Py \tag{12–3}$$

In 1744, Leonhard Euler derived Eq. 12–3 and its solution:

$$P_c = \frac{n^2\pi^2EI}{\ell^2} \tag{12–4}$$

where

EI = flexural stiffness
ℓ = length of the column
n = number of half-sine waves in the length of the column

Cases with $n = 1$, 2, and 3 are illustrated in Fig. 12–4c. The lowest value of P_c will occur with $n = 1.0$. This gives what is referred to as the *Euler buckling load*:

$$P_E = \frac{\pi^2EI}{\ell^2} \tag{12–5}$$

Such a column is shown in Fig. 12–5a. If this column were unable to move sideways at midheight, as shown in Fig. 12–5b, it would buckle with $n = 2$ and the buckling load would be

$$P_c = \frac{2^2\pi^2EI}{\ell^2}$$

which is four times the critical load of the same column without the midheight brace.

Another way of looking at this involves the concept of the *effective length* of the column. The effective length is the length of a pin-ended column having the same buckling load. Thus the column in Fig. 12–5c has the same buckling load as that in Fig. 12–5b. The effective length of the column is $\ell/2$ in this case, where $\ell/2$ is the length of each of the half-sine waves in the deflected shape of the column in Fig. 12–5b. The effective length, $k\ell$, is equal to ℓ/n. The *effective length factor* is $k = 1/n$. Equation 12–4 is generally written as

$$P_c = \frac{\pi^2EI}{(k\ell)^2} \tag{12–6}$$

Four idealized cases are shown in Fig. 12–6 together with the corresponding values of the effective length, $k\ell$. Frames a and b are prevented against deflecting laterally. They are said to be *braced against sidesway*. Frames c and d are free to sway laterally when they buckle. They are called *unbraced* or *sway* frames. The critical loads of the columns shown in Fig. 12–6 are in the ratio $1 : 4 : 1 : \frac{1}{4}$.

Thus it is seen that the restraints against end rotation and lateral translation have a major effect on the buckling load of axially loaded elastic columns. In actual structures,

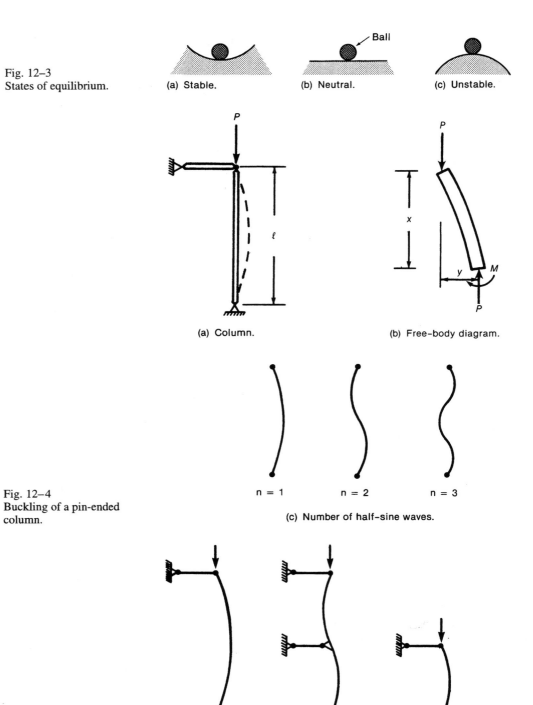

Fig. 12–3
States of equilibrium.

(a) Stable.　(b) Neutral.　(c) Unstable.

(a) Column.　(b) Free-body diagram.

n = 1　n = 2　n = 3

(c) Number of half-sine waves.

Fig. 12–4
Buckling of a pin-ended
column.

Fig. 12–5
Effective length of columns.

(a)　(b)　(c)

fully fixed ends, such as those assumed in Fig. 12–6b to d, rarely, if ever, occur. This is discussed later in the chapter.

In the balance of this chapter we consider, in order, the behavior and design of pinended columns, as in Fig. 12–6a; restrained columns in frames that are braced against lat-

Slender Columns

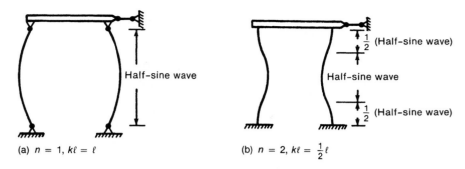

(a) $n = 1, k\ell = \ell$ Half-sine wave

(b) $n = 2, k\ell = \frac{1}{2}\ell$ $\frac{1}{2}$ (Half-sine wave)   Half-sine wave $\frac{1}{2}$ (Half-sine wave)

Frames braced against sidesway.

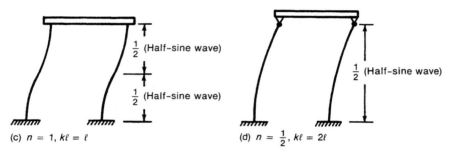

(c) $n = 1, k\ell = \ell$ $\frac{1}{2}$ (Half-sine wave) $\frac{1}{2}$ (Half-sine wave)

(d) $n = \frac{1}{2}, k\ell = 2\ell$ $\frac{1}{2}$ (Half-sine wave)

Fig. 12–6
Effective lengths of idealized columns.

Frames free to sway laterally.

eral displacement (*braced frames*), Fig. 12–6b; and restrained columns in frames which are free to translate sideways (*unbraced frames* or *sway frames*), Fig. 12–6c and d.

Slender Columns in Structures

Pin-ended columns are rare in cast-in-place concrete construction, but do occur in precast construction. Occasionally, these will be slender, as, for example, the columns supporting the back of a precast grandstand.

Most concrete building structures are braced frames with the bracing provided by shear walls, stairwells, or elevator shafts which are considerably stiffer than the columns themselves (Figs. 10–4 and 10–15). Occasionally, unbraced frames are encountered near the tops of tall buildings, where the stiff elevator core may be discontinued before the top of the building, or in industrial buildings where an open crane bay exists.

Most building columns fall in the short-column category.[12-1] Exceptions occur in industrial buildings or in buildings that have a high main-floor story for architectural reasons. An extreme example is shown in Fig. 12–7. The left corner column has a height of 50 times its least thickness. Some bridge piers fall into the slender-column category.

Organization of Chapter 12

To aid in understanding, the presentation of slender columns is divided into three progressively more complex parts. Slender pin-ended columns are discussed in Sec. 12–2. Restrained columns in nonsway frames are discussed in Secs. 12–3 and 12–4. These sections build on the material in Sec. 12–2. Finally, restrained columns in sway frames are discussed in Secs. 12–5 to 12–7.

Fig. 12–7
Bank of Brazil building,
Porto Alegre, Brazil. Each
floor cantilevers out over the
floor below it. (Photograph
courtesy of J. G. MacGregor.)

12–2 BEHAVIOR AND ANALYSIS OF PIN-ENDED COLUMNS

Lateral deflections of a slender column cause an increase in the column moments, as illustrated in Fig. 12–1 and Eq. 12–2. These increased moments cause an increase in the deflections, which in turn lead to an increase in the moments. As a result, the load–moment line $O–B$ in Fig. 12–2 is nonlinear. If the axial load is below the critical load, the process will converge to a stable situation. If the load is greater than the critical load, it will not. This is referred to as a *second-order* process, since it is described by a second-order differential equation (Eq. 12–3).

Material Failures and Stability Failures

Load–moment curves are plotted in Fig. 12–8 for columns of three different lengths, all loaded as shown in Fig. 12–1 with the same end eccentricity, e. The load–moment curve $O–A$ for a relatively short column is practically the same as the line Pe. For a column of moderate length, line $O–B$, the deflections become significant, reducing the failure load. This column fails when the load–moment curve intersects the interaction diagram at point B. This is called a *material failure* and is the type of failure expected in most practical columns in braced frames. If the column is very slender, it may reach a deflection Δ at which the value of the $\partial M / \partial P$ approaches infinity or becomes negative. When this occurs, the column becomes unstable, since with further deflections, the moment capacity will drop. This type of failure is known as a *stability failure* and occurs only with very slender braced columns or slender columns in sway frames.[12–2]

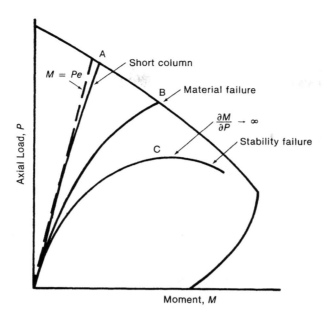

Fig. 12–8
Material and stability
failures.

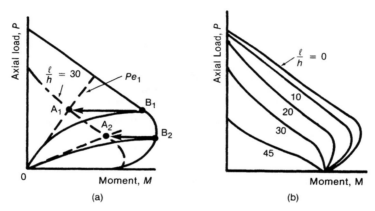

Fig. 12–9
Slender column interaction
curves. (From Ref. 12–1.)

Slender Column Interaction Curves

In discussing the effects of variables on column strength, it is sometimes convenient to use slender column interaction curves. Line $O–B_1$ in Fig. 12–9a shows the load–maximum moment curve for a column with slenderness $\ell/h = 30$ and a given end eccentricity, e_1. This column fails when the load–moment curve intersects the interaction diagram at point B_1. At the time of failure, the load and moment at the *end* of the column are given by point A_1. If this process is repeated a number of times, we get the *slender column interaction curve* shown by the broken line passing through A_1 and A_2, and so on. Such curves show the loads and maximum *end* moments causing failure of a given slender column. A family of slender column interaction diagrams is given in Fig. 12–9b for columns with the same cross section but different slenderness ratios.

Moment Magnifier for Symmetrically Loaded Pin-Ended Beam Column

The column from Fig. 12–1 is shown in Fig. 12–10a. Under the action of the end moments, M_0, it deflects an amount Δ_0. This will be referred to as the *first-order* deflection. When the

12–2   Behavior and Analysis of Pin-Ended Columns   **479**

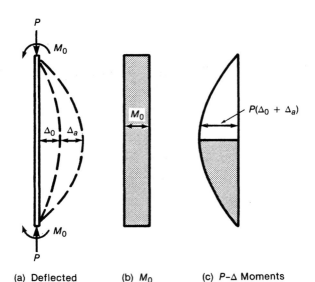

Fig. 12–10
Moments in deflected
column.

(a) Deflected   (b) M_0 (c) P-Δ Moments
    column

axial loads P are applied, the deflection increases by the amount Δ_a. The final deflection at midspan is $\Delta = \Delta_0 + \Delta_a$. This *total* deflection will be referred to as the *second-order deflection*. It will be assumed that the final deflected shape approaches a half-sine wave. The primary moment diagram, M_0, is shown in Fig. 12–10b, and the secondary moments, $P\Delta$, are shown in Fig. 12–10c. Since the deflected shape is assumed to be a sine wave, the $P - \Delta$ moment diagram is also a sine wave. Using the moment area method and observing that the deflected shape is symmetrical, the deflection Δ_a is the moment about the support of the portion of the M/EI diagram between the support and midspan, shown shaded in Fig. 12–10c. The area of this portion is

$$\text{area} = \left[\frac{P}{EI}(\Delta_0 + \Delta_a) \right] \frac{\ell}{2} \times \frac{2}{\pi}$$

and its centroid is ℓ/π from the support. Thus

$$\Delta_a = \left[\frac{P}{EI}(\Delta_0 + \Delta_a) \frac{\ell}{2} \times \frac{2}{\pi} \right]\left(\frac{\ell}{\pi} \right)$$

$$= \frac{P\ell^2}{\pi^2 EI}(\Delta_0 + \Delta_a)$$

where $\pi^2 EI/\ell^2 = P_E$, the Euler buckling load of a pin-ended column. Thus

$$\Delta_a = (\Delta_0 + \Delta_a)\frac{P}{P_E}$$

Rearranging gives

$$\Delta_a = \Delta_0 \left(\frac{P/P_E}{1 - P/P_E} \right) \tag{12–7}$$

Since the final deflection Δ is the sum of Δ_0 and Δ_a:

$$\Delta_a = \Delta + \Delta_0 \left(\frac{P/P_E}{1 - P/P_E} \right)$$

or

$$\Delta = \frac{\Delta_0}{1 - P/P_E} \tag{12-8}$$

This equation shows that the second-order deflection, Δ, increases as P/P_E increases, reaching infinity when $P = P_E$.

The maximum bending moment is

$$M_c = M_0 + P\Delta$$

Here M_c is referred to as the *second-order moment*, and M_0 is referred to as the *first-order moment*. Substituting Eq. 12–8 gives

$$M_c = M_0 + \frac{P\Delta_0}{1 - P/P_E} \tag{12-9}$$

For the moment diagram shown in Fig. 12–10b,

$$\Delta_0 = \frac{M_0 \ell^2}{8EI} \tag{12-10}$$

Substituting this and $P = (P/P_E)\pi^2 EI/\ell^2$ into Eq. 12–9 gives

$$M_c = \frac{M_0(1 + 0.23P/P_E)}{1 - P/P_E} \tag{12-11}$$

The coefficient 0.23 is a function of the shape of the M_0 diagram.[12-3] It becomes –0.38, for example, for a triangular moment diagram with M_0 at one end of the column and zero moment at the other.

In the ACI Code the $(1 + 0.23P/P_E)$ term is omitted because the factor 0.23 varies as a function of the moment diagram and Eq. 12–11 is given essentially as

$$M_c = \delta M_0 \tag{12-12}$$

where δ is called the *moment magnifier* and is given by

$$\delta = \frac{1}{1 - P/P_c} \tag{12-13}$$

where P_c is given by Eq. 12–6 and is equal to P_E for a pin-ended column. Equation 12–13 underestimates the moment magnifier for the column loaded with equal end moments but approaches the truth when the end moments are not equal.

Effect of Unequal End Moments on the Slender Column Strength

Up to now, we have only considered pin-ended columns subjected to equal moments at the two ends. This is a very special case for which the maximum deflection moment, $P\Delta$, occurs at a section where the applied load moment, Pe, is also a maximum. As a result, these quantities can be added directly, as done in Fig. 12–1 and Eq. 12–2.

In the usual case, the end eccentricities, $e_1 = M_1/P$ and $e_2 = M_2/P$, are not equal, giving applied moment diagrams as shown shaded in Fig. 12–11b and c. The maximum value of Δ occurs between the ends of the column while the maximum e occurs at one end of the column. As a result, e_{max} and Δ_{max} cannot be added directly. Two different cases exist. For a slender column with small end eccentricities, the maximum sum of $e + \Delta$ may occur between the ends of the column as shown in Fig. 12–11b. For a shorter column, or a column with large end eccentricities, the maximum sum of $e + \Delta$ will occur at one end of the column as shown in Fig. 12–11c.

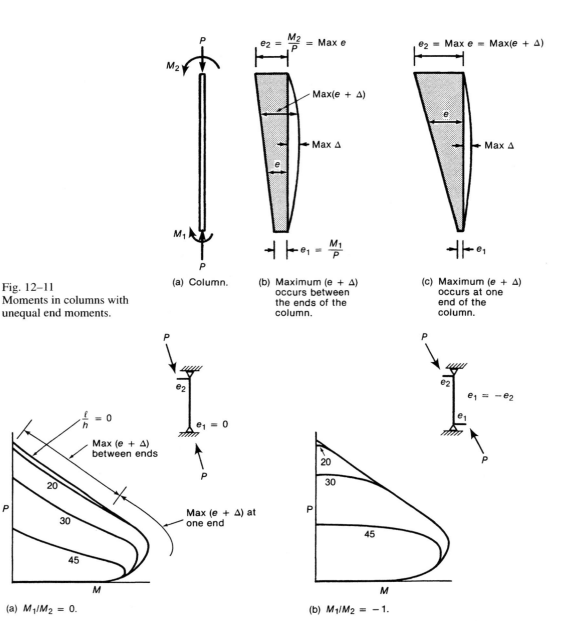

Fig. 12–11
Moments in columns with
unequal end moments.

(a) Column.

(b) Maximum $(e + \Delta)$
occurs between
the ends of the
column.

(c) Maximum $(e + \Delta)$
occurs at one
end of the
column.

(a) $M_1/M_2 = 0$.

(b) $M_1/M_2 = -1$.

Fig. 12–12
Effect of M_1/M_2 ratio on slender column
interaction curves for hinged columns.
(From Ref. 12–1.)

These two types of behavior can be identified in the slender column interaction diagrams shown in Figs. 12–9b and 12–12. For $e_1 = e_2$ (Fig. 12–9b), the interaction diagram
for $\ell/h = 20$, for example, shows a reduction in strength throughout the range of eccentricities. For moment applied at one end only ($e_1/e_2 = 0$, Fig. 12–12a), the maximum
$e + \Delta$ occurs between the ends of the column for small eccentricities, and at one end for
large eccentricities. In the latter case there is no slenderness effect and the column can be
considered as a "short column" by the definition given in Sec. 12–1.

In the case of reversed curvature with $e_1/e_2 = -1$ the slender column range is even
smaller, so that a column with $\ell/h = 20$ subjected to reversed curvature has no slenderness
effects for most eccentricities, as shown in Fig. 12–12b. At low loads, the deflected

Slender Columns

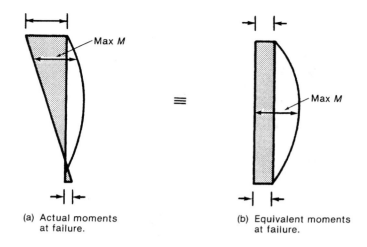

(a) Actual moments
at failure.

$\equiv$

(b) Equivalent moments
at failure.

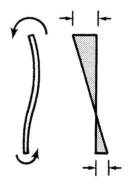

Fig. 12–13
Equivalent moment factor,
C_m.

(c) Single curvature column.

(d) Double curvature column.

shape of such a column is a symmetrical S-shape. As failure approaches, the column tends to *unwrap*, moving from the initial antisymmetrical deflected shape toward a single curvature shape. This results from the inevitable lack of uniformity along the length of the column.

In the moment magnifier design procedure, the column subjected to unequal end moments shown in Fig. 12–13a is replaced with a similar column subjected to equal moments of $C_m M_2$ at both ends, as shown in Fig. 12–13b. The moments, $C_m M_2$, are chosen so that the maximum magnified moment is the same in both columns. The expression for the *equivalent moment factor C_m* was originally derived for use in the design of steel beam columns[12–4] and was adopted without change for concrete design[12–1] (ACI Sec. 10.12.3.1):

$$C_m = 0.6 + 0.4 \frac{M_1}{M_2} \geq 0.4 \qquad (12–14)$$

$$(\text{ACI Eq. } 10–14)$$

In this equation, M_1 and M_2 are the smaller and larger end moments, respectively, calculated using a conventional first-order elastic analysis. The sign convention for the ratio of M_1/M_2 is illustrated in Fig. 12–13c. If the moments M_1 and M_2 act to bend in single curvature without a point of contraflexure between the ends, as shown in Fig. 12–13c, M_1/M_2 is positive. If the moments M_1 and M_2 bend the column in double curvature with a point of zero moment between the two ends, as shown in Fig. 12–13d, M_1/M_2 is negative.

Equation 12–14 applies only to hinged columns or columns in braced frames, loaded with axial loads and end moments. In all other cases, including columns subjected to transverse loads between their ends and concentrically loaded columns (no end moment), C_m is taken equal to 1.0 (ACI Sec. 10.12.3.1). The term C_m is not included in the equation for the moment magnifier for unbraced frames.

Column Stiffness, *EI*

The calculation of the critical load, P_c, using Eq. 12–6 involves the use of the flexural stiffness, *EI*, of the column. The value of *EI* chosen for a given column section, axial load level, and slenderness must approximate the *EI* of the column *at the time of failure*, taking into account the type of failure (material failure or stability failure) and the effects of cracking, creep, and nonlinearity of the stress–strain curves at the time of failure. Figure 12–14 shows moment–curvature diagrams for three different load levels for a typical column cross section. (P_b is the balanced failure load.) A radial line in such a diagram has a slope $M/\phi = EI$. The value of *EI* depends on the particular radial line selected. In a material failure, failure occurs when the most highly stressed section fails (point *B* in Fig. 12–8). For such a case, the appropriate radial line should intercept the end of the moment–curvature diagram, as shown for the $P = P_b$ (balanced load) case in Fig. 12–14. On the other hand, a stability failure occurs before the cross section fails (point *C* in Fig. 12–8). This corresponds to a steeper line in Fig. 12–14 and thus a higher value of *EI*. The multitude of radial lines that can be drawn in Fig. 12–14 suggests that there is no all-encompassing value of *EI* for slender concrete columns. The Australian and Swiss Codes base their *EI* values on the moment and curvature at balanced failure, given by the dashed line in Fig. 12–14.

References 12–1 and 12–5 describe empirical attempts to derive values of *EI*. The expressions adopted by the ACI Code Committee in 1971 were

$$EI = \frac{0.2E_cI_g + E_sI_{se}}{1 + \beta_d}$$

(12–15)

(ACI Eq. 10–12)

or

$$EI = \frac{0.40E_cI_g}{1 + \beta_d}$$

(12–16)

(ACI Eq. 10–13)

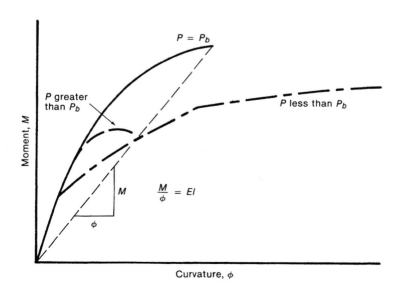

Fig. 12–14
Moment–curvature diagrams
for a column cross section.

Slender Columns

where

E_c, E_s = moduli of elasticity of concrete (ACI Sec. 8.5.1) and steel, respectively

I_g = gross moment of inertia of the concrete section about its centroidal axis ignoring the reinforcement

I_{se} = moment of inertia of the reinforcement about the centroidal axis of the concrete section

The term $(1 + \beta_d)$ reflects the effect of creep on the column deflections and is discussed later.

Equation 12–15 is more accurate than Eq. 12–16 but is more difficult to use since I_{se} is not known until the steel is chosen. It can be rewritten in a more usable form, however. The term I_{se} in Eq. 12–15 can be rewritten as

$$I_{se} = C\rho_t\gamma^2 I_g \qquad (12\text{–}17)$$

where

C = constant depending on the steel arrangement
ρ_t = total longitudinal reinforcement ratio
γ = ratio of the distance between the centers of the outermost bars to the column thickness (illustrated in Table 12–1)

Values of C are given in Table 12–1. Substituting Eq. 12–17 into Eq. 12–15 and rearranging gives

$$EI = \frac{E_c I_g}{1 + \beta_d}\left(0.2 + \frac{C\rho_t\gamma^2 E_s}{E_c}\right) \qquad (12\text{–}18)$$

It is then possible to estimate EI without knowing the exact steel arrangement by choosing ρ_t, estimating γ from the column dimensions, and using the appropriate value of C from Table 12–1. For the common case of bars in four faces and $\gamma \simeq 0.75$, this reduces to

$$EI = \frac{E_c I_g}{1 + \beta_d}\left(0.2 + \frac{1.2\rho_t E_s}{E_c}\right) \qquad (12\text{–}19)$$

This equation can be used for the preliminary design of columns.

Two different sets of EI values are given in the slender column sections of the ACI Code. ACI Sec. 10.12.3 gives Eqs. 12–15 and 12–16 for use in ACI Eq. 10–11 to compute P_c when using the moment magnifier method. These represent the behavior of a single, highly loaded column.

ACI Sec. 10.11.1 gives a different set of values of the moment of inertia, I, for use:

(a) In elastic frame analyses used to compute the moments in columns and beams, and the lateral deflections of frames, and

(b) To compute the Ψ used in computing the effective length factor, k.

The lateral deflection of a frame is affected by the stiffnesses of all the beams and columns in the frame. For this reason, the moments of inertia in ACI Sec. 10.11.1 are intended to represent an overall average of the I values for each type of member in a frame. In a similar manner, the effective length of a column in a frame is affected by the flexural stiffnesses of a number of beams and columns. It is incorrect to use the I values from ACI Sec. 10.11.1 when computing the critical load using ACI Eq. 10–11.

Effect of Sustained Loads on Pin-Ended Columns

Up to this point, the discussion has been limited to columns failing under short-time loadings. Columns in structures, on the other hand, are subjected to sustained dead loads and sometimes to sustained live loads. The creep of the concrete under sustained loads increases the column deflections, increasing the moment $M = P(e + \Delta)$, and thus weaken-

TABLE 12–1   Calculation of I_{se}[a]

Type of Column	Number of Bars	I_{se}	$\dfrac{I_{se}}{I_g} = C\rho_t\gamma^2$
(square column, $b \times h$, γh)	—	$0.25A_{st}\gamma^2h^2$	$3\rho_t\gamma^2$
(column, γh)	3 per face	$0.167A_{st}\gamma^2h^2$	$2\rho_t\gamma^2$
	6 per face	$0.117A_{st}\gamma^2h^2$	$1.4\rho_t\gamma^2$
(column, γh)	8 bars (3 per face)	$0.187A_{st}\gamma^2h^2$	$2.2\rho_t\gamma^2$
	12 bars (4 per face)	$0.176A_{st}\gamma^2h^2$	$2.10\rho_t\gamma^2$
	16 bars (5 per face)	$0.172A_{st}\gamma^2h^2$	$2.06\rho_t\gamma^2$
(tall column, γh, h)	$h = 2b$ 16 bars as shown About strong axis	$0.128A_{st}\gamma^2h^2$	$1.54\rho_t\gamma^2$
(wide column, γh)	$b = 2h$ About weak axis	$0.219A_{st}\gamma^2h^2$	$2.63\rho_t\gamma^2$
(circular column, γh)	—	$0.125A_{st}\gamma^2h^2$	$2\rho_t\gamma^2$
(square column with circular bar pattern)	—	$0.125A_{st}\gamma^2h^2$	$1.5\rho_t\gamma^2$

[a]Total area of steel $= A_{st} = \rho_t A_c$.

Source: Ref. 12–5

ing the column. The load–moment curve of Fig. 12–2 can be replotted, as shown in Fig. 12–15, for columns subjected to sustained loads. In Fig. 12–15a, the column is loaded rapidly to the service load (line O–A). The service load acts for a number of years and during this time the creep deflections and resulting second-order effects increase the moment, as shown by line A–B. Finally, the column is rapidly loaded to failure, as shown by line BC. The failure load corresponds to point C. Had the column been rapidly loaded to failure without the period of sustained service load, the load–moment curve would resemble line O–A–D with failure corresponding to point D. The effect of the sustained loads has been to increase the midheight deflections and moments, causing a reduction in the failure load from D to C. On the reloading (line BC), the column deflections are governed by the EI corresponding to rapidly applied loads.

The second type of column behavior under sustained loads is referred to as *creep buckling*. Here, as shown in Fig. 12–15b, the column deflections continue to increase under

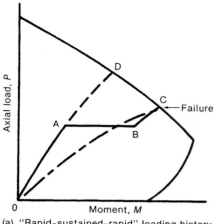

(a) "Rapid-sustained-rapid" loading history.

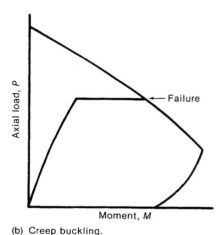

Fig. 12–15
Load-moment behavior for
columns subjected to sus-
tained loads.

(b) Creep buckling.

the sustained load, causing failure under the sustained load itself.[12–6,12–7] This occurs only under high sustained loads greater than about 70% of the short-time capacity represented by point D. Since the sustained load will rarely exceed the strength reduction factor, ϕ, divided by the dead-load factor, $0.70/1.4 = 0.5$, times the column capacity, this type of failure is not specifically addressed in the ACI Code design procedures.

Two different design procedures are widely used to account for creep effects. In the *reduced modulus procedure*,[12–1,12–7,12–8] the value of E used to compute P_c is reduced to give the correct failure load. This procedure is illustrated by the broken line, O–C, in Fig. 12–15a.

The second procedure replaces the column loaded at eccentricity e, with one loaded at an eccentricity $e + \Delta_{0,cr}$ where $\Delta_{0,cr}$ is the creep deflection that would remain in the column in Fig. 12–15a if it were unloaded after reaching point B.[12–9]

The ACI Code moment magnifier procedure uses the reduced modulus procedure. The value of EI is reduced by dividing by $(1 + \beta_d)$, as shown in Eqs. 12–15 and 12–16, where for hinged columns and columns in restrained frames, β_d is defined as the ratio of the factored axial load due to dead load, to the total factored axial load. In a lightly reinforced column, creep of the concrete frequently causes a significant increase in the steel stress so that the compression reinforcement yields at a lower load than it would under a rapidly applied load. This effect is empirically accounted for in Eq. 12–15 by dividing both $E_c I_g$ and $E_s I_{se}$ by $(1 + \beta_d)$.

12–2 Behavior and Analysis of Pin-Ended Columns

ACI Sec. 10.0 gives three definitions of β_d, depending primarily on whether the frame is nonsway or sway. To be stable, a pin-ended column must be in a structure that restricts sway of the ends of the column. In addition, it does not develop end moments if the structure sways sideways. In effect, a pin-ended column is always a nonsway column. For columns in a nonsway frame, ACI Sec. 10.0 defines β_d as the ratio of the maximum factored axial dead load to the total factored axial load. This definition applies to pin-ended columns.

Limiting Slenderness Ratios for Slender Columns

Most columns in structures are sufficiently short and stocky to be unaffected by slenderness effects. To avoid checking slenderness effects in all cases, ACI Sec. 10.12.2 allows slenderness effects to be ignored in the case of hinged columns or columns in braced frames if

$$\frac{k\ell_u}{r} < 34 - 12\frac{M_1}{M_2} \qquad (12\text{--}20)$$

in Eq. 12–20, k refers to the effective length factor, which is 1.0 for a pin-ended column (ACI Sec. 10.12.1), ℓ_u is the unsupported height (ACI Sec. 10.11.3.1), and r is the radius of gyration, taken as $0.3h$ for rectangular sections and $0.25h$ for circular sections (ACI Sec. 10.11.2). For other shapes, the value of r can be calculated from the area and moment of inertia of the cross section. By definition $r = \sqrt{I/A}$. The sign convention for M_1/M_2 is given in Fig. 12–13.

Pin-ended columns having slenderness ratios less than the right-hand side of Eq. 12–20 should have a slender column strength of 95% or more of the short column strength.

Arrangement of ACI Code Sections 10.10 to 10.13

In the 1995 ACI Code, the sections dealing with the design of slender columns were completely rewritten and rearranged. ACI Sec. 10.10 is an overall umbrella section. It sets requirements for a rigorous slenderness analysis (ACI Sec. 10.10.1) and allows the use of the more approximate moment magnifier analysis described in ACI Sec. 10.11, 10.12, and 10.13.

ACI Sec. 10.11 "Magnified Moments—General" gives general requirements for the design of slender columns in both non-sway and sway frames. ACI Sec. 10.11.4 gives methods of distinguishing between non-sway and sway columns. If a column is in a non-sway frame, design involves ACI Secs. 10.11 and 10.12, "Magnified Moments—Non-sway Frames." If a column is in a sway frame, design involves ACI Secs. 10.11 and 10.13 "Magnified Moments—Sway Frames." In some relatively unusual cases involving very slender columns in sway frames, ACI Sec. 10.13.5 requires a further moment magnification using ACI Sec. 10.12.3.

Summary of ACI Moment Magnifier Design Procedure for Slender Pin-Ended Columns

A pin-ended column must be braced by a frame or other structure to remain stable. Hence it will be designed using ACI Sec. 10.11 "Magnified Moments—General" and ACI Sec. 10.12 "Magnified Moments—Non-sway Frames."

1. Length of column. The unsupported length ℓ_u is the clear distance between members capable of giving lateral support (ACI Sec. 10.11.3.1). In the case of a pin-ended column ℓ_u is the distance between the hinges.

2. Effective length. For a pin-ended column the effective length $k = 1.0$ (ACI Sec. 10.11.2).

3. **Radius of gyration.** For a rectangular section $r = 0.3h$, and for a circular section $r = 0.25h$. (See ACI Sec. 10.11.2.) For other sections, r can be calculated from the area and moment of inertia of the concrete section as $r = \sqrt{I_g/A_g}$.

4. **Consideration of slenderness effects.** For a pin-ended column, ACI Sec. 10.12.2 allows slenderness to be neglected if $k\ell_u/r$ satisfies Eq. 12–20. The sign convention for M_1/M_2 is given in Fig. 12–13. ACI Sec. 10.11.5 gives an upper limit of $k\ell_u/r = 100$ for columns designed according to ACI Sec. 10.11 to 10.13.

5. **Minimum moment.** ACI Sec. 10.12.3.2 requires that the maximum end moment on the column, M_2, not be taken less than

$$M_{2,\,min} = P_u(0.6 + 0.03h) \qquad (12\text{–}21)$$
$$(\text{ACI Eq. 10–15})$$

where the 0.6 and h are in inches. When M_2 is less than $M_{2,min}$, C_m shall either be taken equal to 1.0, or evaluated using the actual end moments.

6. **Moment magnifier equation.** ACI Sec. 10.12.3 states the columns shall be designed for the factored axial load, P_u, and the magnified moment, M_c, defined by

$$M_c = \delta_{ns}M_2 \qquad (12\text{–}22)$$
$$(\text{ACI Eq. 10–9})$$

The subscript ns refers to nonsway. The moment M_2 is defined as the larger end moment acting on the column. ACI Sec. 10.12.3 goes on to define δ_{ns} as

$$\delta_{ns} = \frac{C_m}{1 - P_u/(0.75P_c)} \geq 1.0 \qquad (12\text{–}23)$$
$$(\text{ACI Eq. 10–10})$$

where

$$C_m = 0.6 + 0.4\frac{M_1}{M_2} \geq 0.4 \qquad (12\text{–}14)$$
$$(\text{ACI Eq. 10–14})$$

$$P_c = \frac{\pi^2 EI}{(k\ell_u)^2} \qquad (12\text{–}24)$$
$$(\text{ACI Eq. 10–11})$$

and

$$EI = \frac{0.2E_cI_g + E_sI_{se}}{1 + \beta_d} \qquad (12\text{–}15)$$
$$(\text{ACI Eq. 10–12})$$

or

$$EI = \frac{0.40E_cI_g}{1 + \beta_d} \qquad (12\text{–}16)$$
$$(\text{ACI Eq. 10–13})$$

Equation 12–23 is Eq. 12–13 rewritten to include the equivalent moment factor, C_m, and to include a strength reduction factor, ϕ, taken equal to 0.75 for all slender columns.[12–10] The number 0.75 has been used in Eq. 12–23 rather than the symbol ϕ, to avoid confusion with the ϕ factor used in design of the column cross section, which is 0.70 for tied columns and 0.75 for spiral columns.

If the computed value of δ_{ns} is less than 1.0, the maximum moment occurs at one end of the column. In this case δ_{ns} is set equal to 1.0.

12–2 Behavior and Analysis of Pin-Ended Columns   **489**

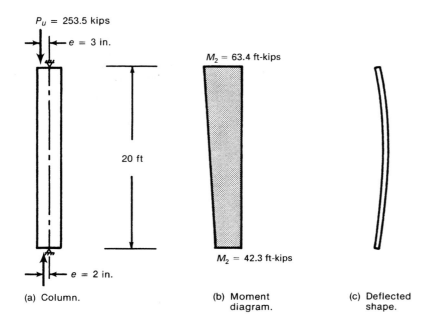

$P_u = 253.5$ kips

$e = 3$ in.

$M_2 = 63.4$ ft-kips

20 ft

$M_2 = 42.3$ ft-kips

$e = 2$ in.

Fig. 12–16
Column—Example 12–1.

(a) Column.

(b) Moment
diagram.

(c) Deflected
shape.

EXAMPLE 12–1   Design of a Slender Pin-Ended Column

Design a 20-ft-tall column to support an unfactored dead load of 90 kips and an unfactored live load of 75 kips. The loads act at an eccentricity of 3 in. at the top and 2 in. at the bottom, as shown in Fig. 12–16. Use $f'_c = 3000$ psi and $f_y = 60,000$ psi.

1.   Compute the factored loads and moments and M_1/M_2.

$$P_u = 1.4D + 1.7L$$

$$= 1.4 \times 90 \text{ kips} + 1.7 \times 75 \text{ kips} = 253.5 \text{ kips}$$

The moment at the top is

$$M = P_u \times e$$

$$= 253.5 \text{ kips} \times \frac{3 \text{ in.}}{12 \text{ in.}} = 63.4 \text{ ft-kips}$$

The moment at the bottom is

$$M = 253.5 \text{ kips} \times \frac{2 \text{ in.}}{12 \text{ in.}}$$

$$= 42.3 \text{ ft-kips}$$

By definition, M_2 is the larger end moment in the column. Therefore, $M_2 = 63.4$ ft-kips and $M_1 = 42.3$ ft-kips. The ratio M_1/M_2 is taken to be positive, since the column is bent in single curvature (see Figs. 12–16c and 12–13c). Thus $M_1/M_2 = 0.667$.

2.   Estimate the column size. From Eq. 11–18a, assuming that $\rho_t = 0.015$,

$$A_{g(\text{trial})} \geq \frac{P_u}{0.45(f'_c + f_y\rho_t)}$$

$$A_{g(\text{trial})} \geq \frac{253.5 \times 1000}{0.45(3000 + 60,000 \times 0.015)}$$

$$= 144 \text{ in.}^2$$

This suggests that a 12 in. $\times$ 12 in. column would be satisfactory. It should be noted that Eq. 11–18a was derived for short columns and will underestimate the size of slender columns.

3. **Is the column slender?** From Eq. 12–20, a column is short if

$$\frac{k\ell_u}{r} < 34 - 12\frac{M_1}{M_2}$$

For the 12 in. $\times$ 12 in. section from step 2, where $k = 1.0$, since the column is pin-ended, and $r = 0.3h = 0.3 \times 12$ in. $= 3.6$ in.:

$$\frac{k\ell_u}{r} = \frac{1.0 \times 240 \text{ in.}}{3.6 \text{ in.}} = 66.7$$

For $M_1/M_2 = 0.667$,

$$34 - 12\frac{M_1}{M_2} = 34 - 12 \times 0.667 = 26$$

Since $k\ell_u/r = 66.7$ exceeds 26, the column is quite slender. This suggests that the 12 in. $\times$ 12 in. section may be inadequate. We shall select a 16 in. x 16 in. section for the first trial.

ACI Sec. 10.11.5 requires a special analysis if $k\ell_u/r$ exceeds 100. This is not required.

4. **Check if the moments are less than the minimum.** ACI Sec. 10.12.3.2 requires that a braced column be designed for a minimum eccentricity of $0.6 + 0.03h = 1.08$ in. Since the maximum end eccentricity exceeds this, design for the moments from step 1.

5. **Compute EI.** At this stage, the area of reinforcement is not known. It is thus not possible to use Eq. 12–15 to compute EI at this stage, but either Eq. 12–16 or 12–19 can be used. From Eq. 12–16,

$$EI = \frac{0.40\,E_c I_g}{1 + \beta_d}$$

where

$$E_c = 57{,}000\sqrt{f_c'} = 3.122 \times 10^6 \text{ psi} \quad (\text{ACI Sec. 8.5.1})$$
$$I_g = bh^3/12 = 5461 \text{ in.}^4$$

The term β_d is the ratio of the factored dead load to the total factored axial load:

$$\beta_d = \frac{1.4 \times 90}{253.5} = 0.497$$

Thus

$$EI = \frac{3.122 \times 10^6 \text{ psi} \times 5461 \text{ in.}^4}{2.5(1 + 0.497)}$$
$$= 4.56 \times 10^9 \text{ in.}^2\text{-lb}$$

Alternatively, using Eq. 12–19, assuming that $\rho_t = 0.015$, gives

$$EI = \frac{3.122 \times 10^6 \times 5461}{1 + 0.497}\left(0.2 + 1.2 \times 0.015 \times \frac{29 \times 10^6}{3.122 \times 10^6}\right)$$
$$= 4.18 \times 10^9 \text{ in.}^2\text{-lb}$$

We shall use $EI = 4.56 \times 10^9$ in.^{2}-lb. Generally, one would use Eq. 12–15 or 12–19 if ρ_t exceeded about 0.02, since they give higher values of EI.

6. **Compute the magnified moment.** From Eq. 12–22,

$$M_c = \delta_{ns}M_2$$

where

$$\delta_{ns} = \frac{C_m}{1 - P_u/0.75P_c} \geq 1.0 \qquad (12\text{--}23)$$
$$(\text{ACI Eq. } 10\text{--}10)$$

$$C_m = 0.6 + 0.4 \frac{M_1}{M_2} \geq 0.4 \qquad (12\text{--}14)$$

$$= 0.6 + 0.4 \times 0.667 = 0.867$$
$$P_c = \frac{\pi^2 EI}{(k\ell_u)^2} \qquad (12\text{--}24)$$

where $k = 1.0$ since the column is pin-ended.

$$P_c = \frac{\pi^2 \times 4.56 \times 10^9 \text{ in.}^2\text{-lb}}{(1.0 \times 240 \text{ in.})^2} = 781,300 \text{ lb}$$

$$= 781.3 \text{ kips}$$

and

$$\delta_{ns} = \frac{0.867}{1 - 253.5/(0.75 \times 781.3)} \geq 1.0$$
$$= 1.528$$

Normally, if δ_{ns} exceeds 2.0, one should select a larger cross section. Thus the magnified moment is

$$M_c = 1.528 \times 63.4 \text{ ft-kips} = 96.9 \text{ ft-kips}$$

7. Select the column reinforcement. Interaction diagrams for a 16 in. × 16 in. column with 4 No. 7 bars, 4 No. 8 bars, and 4 No. 9 bars are given in Fig. 12–17. The column reinforcement must be designed to resist $P_u = 253.5$ kips and $M_c = 96.9$ ft-kips. The interaction diagrams show that 4 No. 7 bars would be adequate. This gives a steel ratio of 0.0094, which is less than the minimum steel ratio of 0.01 required in ACI Sec. 10.9.1. For this reason, we will **use a 16 in. × 16 in. column with 4 No. 8 bars.** Alternatively, the calculations could be repeated using a 15 in. × 15 in. column. A 15 in. × 15 in. column will not be used since δ_{ns} is 1.97 for such a column, which the author considers to be too high. ■

12–3 BEHAVIOR OF RESTRAINED COLUMNS IN NONSWAY FRAMES

Effect of End Restraints on Braced Frames

A simple indeterminate frame is shown in Fig. 12–18a. A load P and an unbalanced moment M_{ext} are applied to the joint at each end of the column. The moment M_{ext} is equilibrated by the moment M_c in the column and the moment M_r in the beam as shown in Fig. 12–18b. By moment distribution

$$M_c = \left(\frac{K_c}{K_c + K_b} \right) M_{\text{ext}} \qquad (12\text{--}25)$$

where K_c and K_b are the flexural stiffnesses of the column and the beam, respectively, at the upper joint. Thus K_c represents the moment required to bend the end of the column through a unit angle. The term in parentheses in Eq. 12–25 is the distribution factor for the column.

The total moment, M_{max}, in the column at midheight is

$$M_{\text{max}} = M_c + P\Delta \qquad (12\text{--}26)$$

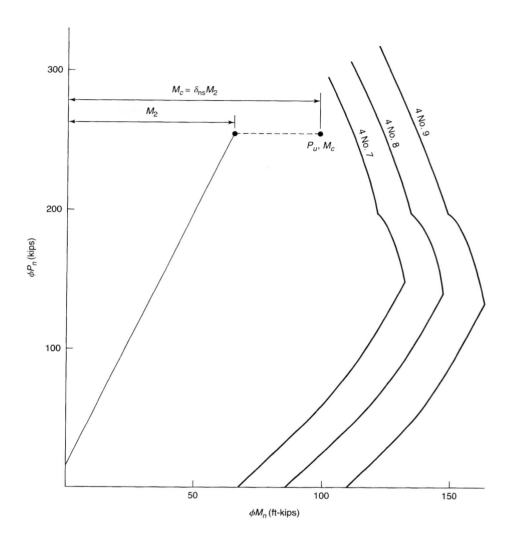

Fig. 12–17
Selection of reinforcement—
Example 12–1

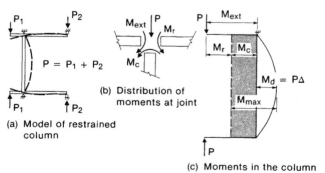

Fig. 12–18
Moments in a restrained column. (From Ref. 12–1.)

(a) Model of restrained column

(b) Distribution of moments at joint

(c) Moments in the column

As discussed earlier, the combination of the $P\Delta$ moments and M_c give rise to a larger total deflection and hence a larger rotation at the ends of the column than would be the case if just M_c acted. As a result, one effect of the axial force is to reduce the column stiffness, K_c. When this occurs, Eq. 12–25 shows that the fraction of M_{ext} assigned to the column drops, thus reducing M_c. Inelastic action in the column tends to hasten this reduction in column stiffness, again reducing the moment developed at the ends of the column. On the

12–3   Behavior of Restrained Columns in Nonsway Frames   **493** other hand, a reduction in the beam stiffness, K_b, due to cracking or inelastic action in the beam will throw moment back to the column.

This is illustrated in Fig. 12–19, which shows frame F2 tested by Furlong and Ferguson.[12–11] The columns in this frame had $\ell/h = 20$ $(k\ell_u/r = 57)$, and an initial eccentricity ratio $e/h = 0.106$. The loads bent the column in symmetrical single curvature. Failure occurred at section A at midheight of one of the columns. In Fig. 12–19b, load–moment curves are presented for section A and for section B, located at one end of the column. The moment at section B corresponds to the moment M_c in Eq. 12–26 and Fig. 12–18. Although the loads P and βP were proportionally applied, the variation in moment at B is not linear due to a decrease in K_c as the axial load is increased. As the moment at the ends of the columns decreased, the moments at midspan of the beams had to increase to maintain equilibrium. The moment at A, the failure section, is equal to the sum of the moment M_c at section B, plus the moment due to the column deflection, $P\Delta$.

Figures 12–20 to 12–22 trace the deflections and moments in slender columns in braced frames under increasing loads. These are based on inelastic analyses by Cranston[12–12] of reinforced concrete columns with elastic end restraints.

Figure 12–20 illustrates the behavior of a tied column with a slenderness ratio $\ell/h = 15$ $(k\ell/r = 33)$ with equal end restraints, loaded with an axial load P and an ex-

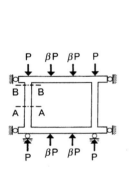
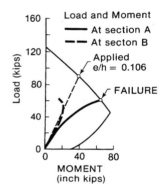

Fig. 12–19
Load–moment behavior of a
column in a braced frame.
(From Ref. 12–1.)

(a) Test specimen.

(b) Measured load—
moment response.

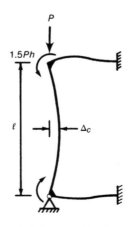
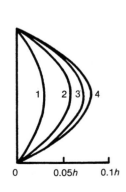
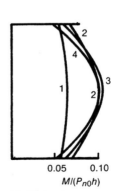
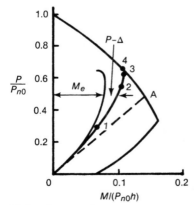

(a) Column and loads.

(b) Deflections.

(c) Moments.

(d) Load–moment curves.

Fig. 12–20
Behavior of a short restrained column bent in symmetrical single curvature. $\ell/h = 15$, $\rho_t = 0.01$,
$K_b = 2.5K_c$ at both ends (From Ref. 12–12.)

ternal moment applied to the joint of $1.5hP$. A first-order analysis indicates that the end moments on the column itself are $0.25hP$, as shown by the dashed line O–A in Fig. 12–20d. As the loads are increased, the column is deflected as shown in Fig. 12–20b. The moment diagrams in the column at the same four stages are shown in part c. The maximum midheight moment occurred at load stage 3. The increase in deflections from 3 to 4 (Fig. 12–20b) was more than offset by the decrease in end moments (Fig. 12–20c). Figure 12–20d traces the load–moment curves at midheight (centerline) and at the ends (line labeled M_e). The total moment at midheight is the sum of M_e and $P\Delta$. Because the end moments decreased more rapidly than the $P\Delta$ moments increased, the load–moment line for the midheight section curls upward. The failure load, point 4 in Fig. 12–20d, is higher than the failure load ignoring slenderness effects, point A. This column was *strengthened* by slenderness effects.

Figure 12–21 is a similar plot for a tied column with a slenderness ratio of $\ell/h = 40$ $(k\ell/r = 88)$. Such a column would resemble the columns in Fig. 12–7. At failure the column deflection approached *60% of the overall depth* of the column, as shown in

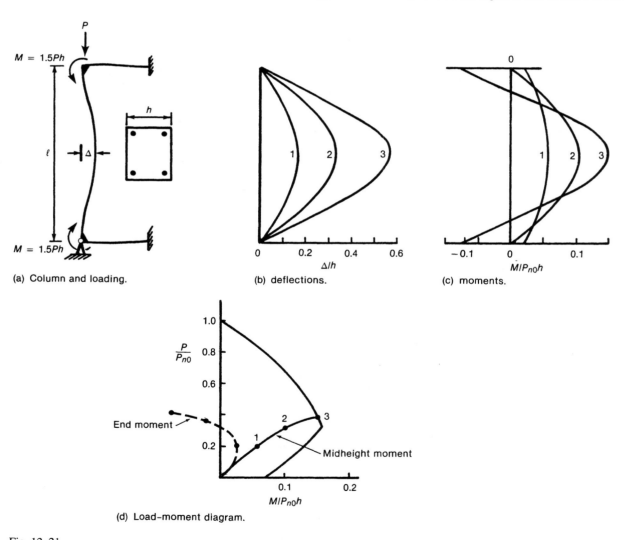

(a) Column and loading.

(b) deflections.

(c) moments.

(d) Load–moment diagram.

Fig. 12–21
Behavior of a very slender restrained column bent in symmetrical single curvature. $\ell/h = 40$, $\rho_t = 0.01$, $K_b = 2.5K_c$ at both ends. (From Ref. 12–12.)

Fig. 12–21b. The moments at the ends of the columns decreased, reaching zero at load stage 2 and becoming negative. This reduction in end moments was more than offset by the $P\Delta$ moments due to the deflections. The load–moment curves for the ends of the column and midheight are shown in Fig. 12–21d.

The behavior shown in Figs. 12–20 and 12–21 is typical for reinforced concrete columns bent in single curvature ($M_1/M_2 \leq 0$). In such columns both end moments decrease as P increases, possibly changing sign. The maximum moments in the column may or may not increase, depending on the relative magnitudes of the decrease in end moments compared to the $P\Delta$ moments.

For columns loaded in double curvature ($M_1/M_2 < 0$), the behavior is different, as illustrated in Fig. 12–22. Assuming that the larger end moment, M_2, is positive and the smaller, M_1, is negative, it can be seen that both end moments become more negative, just as they did in Fig. 12–21. The difference, however, is that M_2 decreases and eventually be-

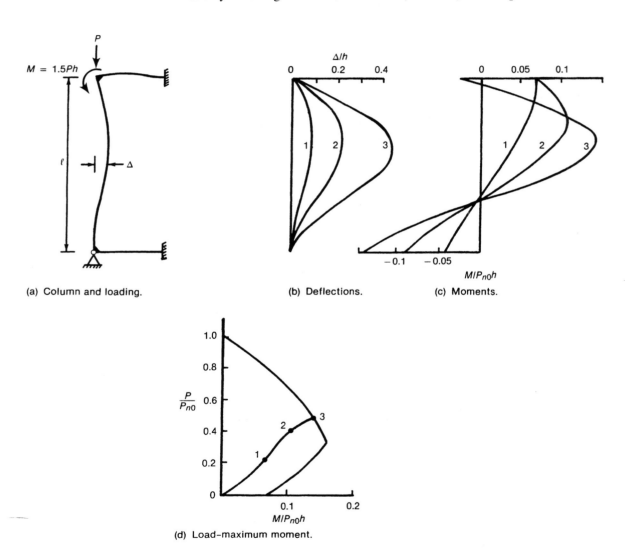

(a) Column and loading.

(b) Deflections.

(c) Moments.

(d) Load-maximum moment.

Fig. 12–22
Behavior of a very slender restrained column bent in double curvature. $\ell/h = 40$, $\rho_t = 0.01$, $K_b = 2.5K_c$ at top and $6K_c$ at bottom. (From Ref. 12–12.)

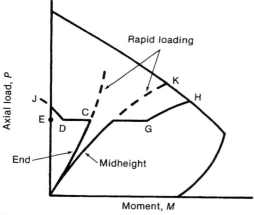

(a) Column weakened by creep.

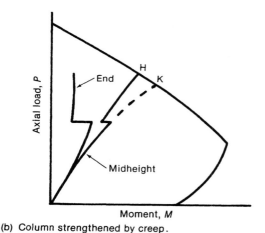

Fig. 12–23
Effect of sustained loads on moments in columns in braced frames.

(b) Column strengthened by creep.

comes negative, while M_1 becomes larger (more negative). At failure of the column in Fig. 12–22, the negative moment at the bottom of the column is almost as big as the maximum positive moment.

Effect of Sustained Loads on Columns in Braced Frames

Consider a frame similar to frame $F2$ shown in Fig. 12–19 which is loaded rapidly to service load level, is held at this load level for several years, and then is loaded rapidly to failure. If the columns are slender, the behavior plotted in Fig. 12–23a would be expected.[12-13] During the sustained load period, the creep deflections cause a reduction in the column stiffness K_c, which in turn leads to a reduction in the column end moments (at section B), as shown by the horizontal line C–D in Fig. 12–23a and a corresponding increase in the midspan moments in the beams. At the same time, however, the $P\Delta$ moment increases due to the increase in deflections. At the end of the sustained load period, the end moment is indicated by the distance E–D in Fig. 12–23a, while the total $P\Delta$ moment at midheight is shown by D–G. Failure of such a column occurs when the load–moment line intersects the interaction diagram at H. Failure may also result from the reversal of sign of the end moments, shown by J in Fig. 12–23a, if the end restraints are unable to resist the reversed moment. The dashed lines indicate the load–moment curve for the end and midheight sections in a column loaded to failure in a short time. The decrease in load from K to H is due to the creep effect.

For a short column in a similar frame, the reduction in end moment due to creep may be larger than the increase in the $P\Delta$ moment, resulting in a strengthening of the column[12-13] as illustrated in Fig. 12–23b.

The reduction of column end moments due to creep greatly reduces the risk of creep buckling (Fig. 12–15b) of columns of braced frames.

12–4 DESIGN OF COLUMNS IN NONSWAY FRAMES

Design Approximation for the Effect of End Restraints in Nonsway Frames

Figure 12–24a shows a restrained column in a frame. The solid line in Fig. 12–24b is the moment diagram (including slenderness effects) for this column at failure (similar to Fig. 12–20c or 12–21c). Superimposed on this is the corresponding first-order moment diagram for the same load level. In design it is convenient to replace the restrained column with an equivalent hinged end column of length ℓ_i, the distance between the points on the second-order moment diagram where the moments are equal to the end moments in the first-order diagram (Fig. 12–24c). This equivalent hinged column is then designed for the axial load, P, and the end moments, M_2, from the first-order analysis.

Unfortunately, the length ℓ_i is difficult to compute. In all modern concrete and steel design codes the empirical assumption is made that ℓ_i can be taken equal to the effective length for elastic buckling, $k\ell$. The accuracy of this assumption is discussed in Ref. 12–14, which concludes that $k\ell$ slightly underestimates ℓ_i for an elastically restrained, elastic column.

The concept of effective lengths was discussed earlier in this chapter for the four idealized cases shown in Fig. 12–6. The effective length of a column, $k\ell_u$, is defined as the

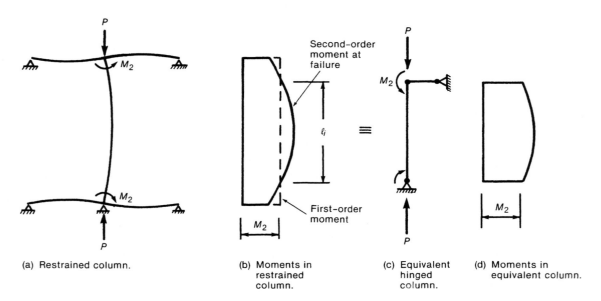

(a) Restrained column.

(b) Moments in restrained column.

(c) Equivalent hinged column.

(d) Moments in equivalent column.

Fig. 12–24
Replacement of restrained column with an equivalent hinged column for design.

length of an equivalent pin-ended column having the same buckling load. When a pin-ended column buckles, its deflected shape is a half-sine wave, as shown in Fig. 12–6a. The effective length of a restrained column is taken equal to the length of a complete half-sine wave in the deflected shape.

Figure 12–6b to d are drawn assuming truly fixed ends. This condition seldom, if ever, actually exists. In buildings, columns are restrained by beams or footings which always allow some rotation of the ends of the column. Thus the three cases considered in Fig. 12–6 will actually deflect as shown in Fig. 12–25, and the effective lengths will be greater than the values for completely fixed ends. The actual value of k for an elastic column is a function of the relative stiffnesses, ψ, of the beams and columns at each end of the column, where ψ is

$$\psi = \frac{\Sigma \, (E_c I_c / \ell_c)}{\Sigma \, (E_b I_b / \ell_b)} \tag{12–27}$$

where the subscripts b and c refer to beams and columns, respectively, and the lengths ℓ_b and ℓ_c are measured center to center of the joints. The summation signs refer to all the compression members meeting at a joint in one case, and all the beams or other restraining members at the joint in the other.

If $\psi = 0$ at one end of the column, the column is fully fixed at that end. Similarly, $\psi = \infty$ denotes a perfect hinge. Thus, as ψ approaches zero at the two ends of a column in a braced frame, k approaches 0.5, the value for a fixed-ended column. Similarly, when ψ approaches infinity at the two ends of a braced column, k approaches 1.0, the value for a pin-ended column. This is illustrated in Table 12–2. The value for columns that are fully fixed at both ends is 0.5, found in the lower left corner of the table. The value for columns that are pinned at both ends is 1.0, found in the upper right corner.

In practical structures there is no such thing as a truly fixed end or a truly hinged end. Reasonable upper and lower limits on ψ are 20 and 0.2. For columns in braced frames, k should never be taken less than 0.6. In sway frames k should never be taken less than 1.2 for columns restrained at both ends.

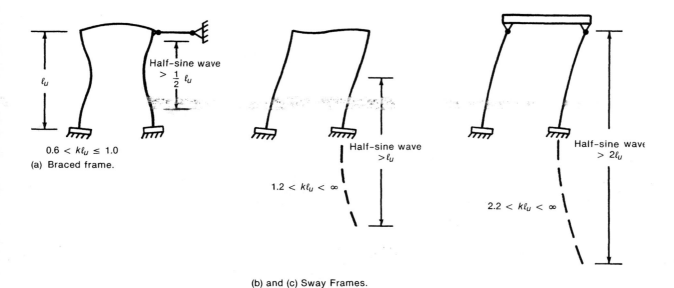

(a) Braced frame.

(b) and (c) Sway Frames.

Fig. 12–25
Effective length of column in frames.

TABLE 12–2 Effective Length Factors for Braced Frames

Top		k				
Hinged		0.70	0.81	0.91	0.95	1.00
Elastic $\psi = 3.1$		0.67	0.77	0.86	0.90	0.95
Elastic $\psi = 1.6$		0.65	0.74	0.83	0.86	0.91
Stiff $\psi = 0.4$		0.58	0.67	0.74	0.77	0.81
Fixed		0.50	0.58	0.65	0.67	0.70
		Fixed	Stiff	Elastic	Elastic	Hinged
				Bottom		

Calculation of k Using Tables

Table 12–2 can be used to select values of k for the design of braced frames. The shaded areas correspond to one or both ends truly fixed. Since such a case rarely, if ever, occurs in practice, this part of the table should not be used. The column and row labeled "stiff" represents a practical degree of fixity. Because k values for sway frames can vary widely, no similar table is given for such frames.

Calculation of k Using Nomographs

The nomographs given in Fig. 12–26 are also used to compute k. To use these nomographs, ψ is calculated at both ends of the column using Eq. 12–27 and the appropriate value of k is found as the intersection of the line labeled k and a line joining the values of ψ at the two ends of the column. The calculation of ψ is discussed in a later section.

The nomographs in Fig. 12–26 were derived[12–15,12–16] considering a typical interior column in an infinitely high and infinitely wide frame, in which all of the columns have the same cross section and length, as do all beams. Equal loads are applied at the tops of each of the columns, while the beams remain unloaded. All columns are assumed to buckle at the same moment. As a result of these very idealized and quite unrealistic assumptions, the

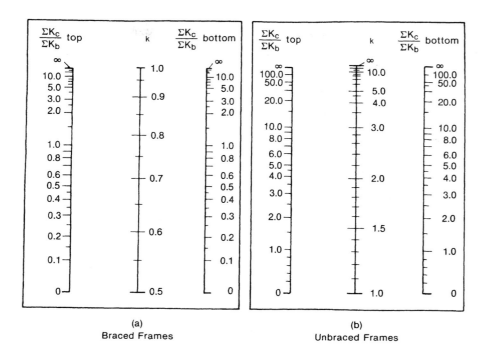

Fig. 12–26
Nomograph for effective
length factors.

(a)
Braced Frames

(b)
Unbraced Frames

nomographs tend to underestimate the value of k for elastic frames of practical dimensions by up to 15%.[12–14] This then leads to an underestimate of the magnified moments, M_c.

The lowest practical value for k in an unbraced frame is about 1.2. When smaller values are obtained from the nomographs, it is good practice to use 1.2.

Equations for k

Approximate equations for k are presented in Sec. R10.12.1 of the ACI Commentary. These have been derived to give a conservative approximation of the effective length factor. Equations 12–28, 12–29, and 12–32 are from the 1972 British Standard Code of Practice,[12–12,12–17] while Eqs. 12–30 and 12–31 were derived for use in the ACI Strength Design Handbook.[12–18]

For braced columns, an upper bound to the effective length factor may be taken as the smaller of the following two expressions:

$$k = 0.7 + 0.05(\psi_A + \psi_B) \leq 1.0 \tag{12–28}$$

$$k = 0.85 + 0.05\psi_{min} \leq 1.0 \tag{12–29}$$

where ψ_A and ψ_B are the values of ψ at the two ends of the column and ψ_{min} is the smaller of the two values. Values of ψ are calculated using Eq. 12–27 as explained in the following section.

For unbraced compression members restrained at both ends, the effective length may be taken as:

For $\psi_m < 2$:

$$k = \frac{20 - \psi_m}{20}\sqrt{1 + \psi_m} \tag{12–30}$$

For $\psi_m \geq 2$:

$$k = 0.9\sqrt{1 + \psi_m} \qquad (12\text{–}31)$$

where ψ_m is the average of the ψ values at the two ends of the column.

For unbraced compression members that are hinged or free at one end, the effective length factor may be taken as

$$k = 2.0 + 0.3\psi \qquad (12\text{–}32)$$

where ψ is the value at the restrained end.

Calculation of ψ

The stiffness ratio, ψ, is calculated using Eq. 12–27. The values of $E_c I_c$ and $E_b I_b$ should be realistic for the state of loading immediately prior to failure of the columns. Generally at this stage of loading, the beams are extensively cracked and the columns are uncracked or slightly cracked. Ideally, the values of EI should reflect the degree of cracking and the actual reinforcement present. This is not practical, however, since this information is not known at this stage of design. ACI Secs. 10.12.1 and 10.13.1 state that the calculation of k shall be based on ψ based on the E and I values given in ACI Sec. 10.11.1. When computing ψ, β_d can be taken as zero.

When calculating I_b for a T beam, the flange width can be taken as defined in ACI Sec. 8.10.2 or 8.10.3. For common ratios of flange thickness to overall depth, h, and flange width to web width, b_w, the gross moment of inertia, I_g, is approximately twice the moment of inertia of a rectangular section with dimensions b_w and h.

The value of ψ at the lower end of a column supported on a footing can be calculated from relationships presented in the *PCI Design Handbook*.[12-19] Equation 12–27 can be rewritten as

$$\psi = \frac{\Sigma K_c}{\Sigma K_b} \qquad (12\text{–}33)$$

where ΣK_c and ΣK_b are the sums of the flexural stiffnesses of the columns and the restraining members (beams) at a joint. At a column-to-footing joint, $\Sigma K_c = 4E_c I_c / \ell_c$ for a braced column restrained at its upper end and ΣK_b is replaced by the rotational stiffness of the footing and soil, taken equal to

$$K_f = \frac{M}{\theta_f} \qquad (12\text{–}34)$$

where M is the moment applied to the footing and θ_f is the rotation of the footing. The stress under the footing is the sum of $\sigma = P/A$, which causes a uniform downward settlement, and $\sigma = My/I$, which causes a rotation. The rotation θ_f is

$$\theta_f = \frac{\Delta}{y} \qquad (12\text{–}35)$$

where y is the distance from the centroid of the footing area. If k_s is the subgrade modulus, defined as the stress required to compress the soil by a unit amount ($k_s = \sigma/\Delta$), then θ_f is

$$\theta_f = \frac{\sigma}{k_s y} = \frac{My}{I_f} \times \frac{1}{k_s y}$$

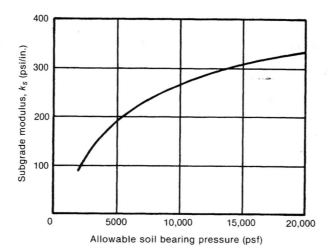

Fig. 12–27
Approximate relationship between allowable soil bearing pressure and subgrade modulus, k_s. (From Ref. 12–19.)

Substituting this into Eq. 12–34 gives

$$K_f = I_f k_s \qquad (12\text{–}36)$$

where I_f is the moment of inertia of the contact area between the bottom of the footing and the soil and k_s is the subgrade modulus, which can be taken from Fig. 12–27. Thus the value of ψ at a footing-to-column joint for a column restrained at its upper end is

$$\psi = \frac{4E_c I_c / \ell_c}{I_f k_s} \qquad (12\text{–}37)$$

Since hinges are never completely frictionless, a value of $\psi = 10$ is frequently used for hinged ends, rather than $\psi = \infty$.

Definition of Braced and Unbraced Frames

The preceding discussions of column behavior and effective length factors were based on the assumption that frames could be separated into "completely braced" frames or "completely unbraced" frames. A column may be considered to be "braced" in a given direction if the lateral stability of the structure as a whole is provided by walls, bracing, or buttresses designed to resist all lateral forces in that direction. A column is completely "unbraced" in a given plane if all resistance to lateral loads comes from bending of the columns.

In actual fact there is no such thing as a "completely braced" frame, and no clear-cut boundary exists between braced and unbraced frames. Some frames are clearly unbraced, as, for example, the frames shown in Fig. 12–25b and c. Other frames are connected to shear walls, elevator shafts, and so on, which restrict the lateral movements of the frame as shown in Fig. 12–25a. Since no such wall is completely rigid, however, there will always be some lateral movement of a braced frame, and hence some $P\Delta$ moments will result from the lateral deflections.

For the purposes of design, a story or a frame can be considered "braced" if horizontal displacements do not significantly reduce the vertical load capacity of the structure. Since the ACI Code moment magnifier design procedure accounts for slenderness by magnifying moments, this criterion could be restated as: A frame can be considered "braced" if the $P\Delta$ moments due to lateral deflections are small compared to the first-order moments due to lateral loads. ACI Sec. 10.11.4.1 allows designers to assume that a frame is nonsway if the increase in column end moments due to second-order effects does not exceed 5% of the first-order moments. This test should be carried out at the end of the column where the magnified end moment is the largest.

Alternatively, ACI Sec. 10.11.4.2 allows designers to assume that a story in a frame is nonsway if

$$Q = \frac{\Sigma P_u \Delta_0}{V_u \ell_c} \qquad (12\text{–}38)$$
$$(\text{ACI Eq. }10\text{–}7)$$

is less than or equal to 0.05, where Q is the *stability index*, ΣP_u is the total vertical load in all the columns and walls in the story in question, V_u is the shear in the story due lateral loads, Δ_0 is the first-order relative deflection between the top and bottom of that story *due to V_u*, and ℓ_c is the height of the story measured from center to center of the joints above and below the story. This concept, which is explained and developed more fully in Sec. 12–6, results in a similar limit as ACI Sec. 10.11.4.1. It was originally presented in Ref. 12–20.

ACI Commentary Sec. R10.11.4 suggests that, frequently, the test of whether a story is sway or nonsway can be done by inspection by comparing the total lateral stiffness of all the columns in a story to that of the bracing elements in the story, such as walls or shear trusses. The Commentaries to the 1971 to 1989 ACI Codes suggested that a story would be nonsway (braced) if the sum of the lateral stiffnesses, ΣK_ℓ, for the bracing elements exceeded six times ΣK_ℓ for the columns in the direction under consideration. The lateral stiffness of a column or bracing element is $K_\ell = V/\Delta$, where V is the shear in the member and Δ is the relative lateral displacement of the ends of the column due to that shear.

Summary of Moment Magnifier Design Procedure for Slender Columns in Braced Frames

If a column is in a non-sway frame, design involves ACI Secs. 10.11, "Magnified Moments—General," and 10.12, "Magnified Moments—Non-sway Frames."

1. Length of column. The unsupported length, ℓ_u, is defined in ACI Sec. 10.11.3.1 as the clear height between slabs or beams capable of giving lateral support to the column.

2. Effective length. ACI Sec. 10.12.1 states that the effective length factors, k, of columns in nonsway frames shall be 1.0 or less. The effective length factors can be estimated using Table 12–2, or Fig. 12–26, or Eqs. 12–28 to 12–32. The last two of these procedures require that the ratio, ψ, of EI/ℓ of the columns and beams be known. This factor is given by Eq. 12–27. ACI Sec. 10.12.1 says that ψ should be based on the E and I values in ACI Sec. 10.11.1.

3. Determination of whether the frame is braced. Frequently, this can be done by inspection by seeing if the bracing elements, such as walls, are considerably stiffer than the columns. Alternatively, the frame can be assumed to be nonsway if Q from Eq. 12–38 is not greater than 0.05.

4. Radius of gyration. For a rectangular cross section, $r = 0.3h$, and for a circular cross section, $r = 0.25h$. For other sections, r can be calculated from the area and moment of inertia of the concrete section as $r = \sqrt{I_g/A_g}$ (ACI Sec. 10.11.2).

5. Consideration of slenderness effects. For columns in braced frames, ACI Sec. 10.12.2 allows slenderness to be neglected if

$$\frac{k\ell_u}{r} < 34 - 12\frac{M_1}{M_2} \qquad (12\text{--}20)$$

For columns in unbraced frames, ACI Sec. 10.12.2 allows slenderness to be neglected if $k\ell_u/r$ is less than 22. If $k\ell_u/r$ exceeds 100, design shall be based on a second-order analysis. The sign convention for M_1/M_2 is illustrated in Fig. 12–13c and d.

6. Minimum moment. For columns in braced frames, the larger end moment, M_2, shall not be taken less than

$$M_{2,\min} = P_u(0.6 + 0.03h) \qquad (12\text{--}21)$$
$$(\text{ACI Eq. } 10\text{--}15)$$

about each axis separately, where 0.6 and h are in inches.

7. Moment magnifier equation. ACI Sec. 10.12.3 states that columns in nonsway frames shall be designed for the factored axial load, P_u, and a magnified factored moment, M_c, given by

$$M_c = \delta_{ns}M_2 \qquad (12\text{--}22)$$
$$(\text{ACI Eq. } 10\text{--}9)$$

where M_2 is the larger end moment, and δ_{ns} is given by

$$\delta_{ns} = \frac{C_m}{1 - P_u/0.75P_c} \geq 1.0 \qquad (12\text{--}23)$$
$$(\text{ACI Eq. } 10\text{--}10)$$

$$C_m = 0.6 + 0.4\left(\frac{M_1}{M_2}\right) \geq 0.4 \qquad (12\text{--}14)$$
$$(\text{ACI Eq. } 10\text{--}14)$$

where the sign convention for M_1/M_2 is as illustrated in Fig. 12–13c and d.

$$P_c = \frac{\pi^2 EI}{(k\ell_u)^2} \qquad (12\text{--}24)$$
$$(\text{ACI Eq. } 10\text{--}11)$$

and

$$EI = \frac{0.2E_cI_g + E_sI_{se}}{1 + \beta_d} \qquad (12\text{--}15)$$
$$(\text{ACI Eq. } 10\text{--}12)$$

or

$$EI = \frac{0.40E_cI_g}{1 + \beta_d} \qquad (12\text{--}16)$$
$$(\text{ACI Eq. } 10\text{--}13)$$

The term β_d has three definitions, only one of which applies to columns in nonsway frames. For such columns,

$$\beta_d = \frac{\text{maximum factored axial dead load in the column}}{\text{total factored axial load in the column}} \qquad (12\text{--}39a)$$

Equations 12–18 and 12–19 may also be used to compute EI for use in Eq. 12–24. The EI values given in ACI Sec. 10.11.1 cannot be used to compute EI for use in Eq. 12–24. The EI values in ACI Sec. 10.11.1 approach the average values for an entire story and are intended for use in first- and second-order frame analyses.

If P_u exceeds $0.75P_c$ in Eq. 12–23, δ_{ns} will be negative. If the stiffness were lower than expected, such a column would be unstable. Hence, if P_u exceeds $0.75P_c$, the column

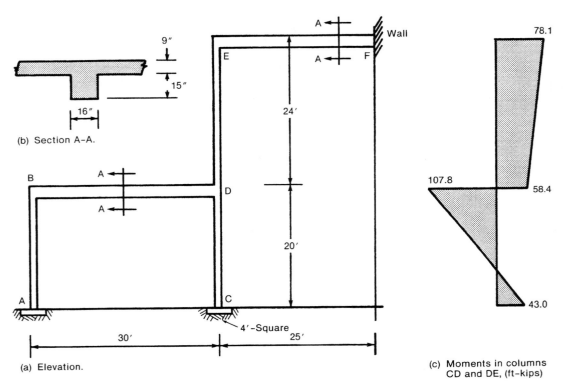

9″

15″

16″

(b) Section A-A.

Wall

78.1

24′

107.8

58.4

B

A

D

20′

A

C

43.0

4′-Square

30′

25′

(a) Elevation.

(c) Moments in columns
CD and DE, (ft–kips)

Fig. 12–28
Braced frame—Example 12–2

cross section should be enlarged. Indeed, if δ_{ns} exceeds 2.0, strong consideration should be given to enlarging the column cross section because the calculations become very sensitive to the assumptions made.

EXAMPLE 12–2 Design of the Columns in a Braced Frame

Figure 12–28 shows part of a typical frame in an industrial building. The frames are spaced 20 ft apart. The columns rest on 4-ft-square footings. The soil bearing capacity is 4000 psf. Design columns C–D and D–E. Use $f'_c = 3000$ psi and $f_y = 60,000$ psi for beams and columns.

1. **Determine the column loads from a frame analysis.**

	Column *CD*	Column *DE*
Service loads, P	Dead = 80 kips	Dead = 50 kips
	Live = 24 kips	Live = 14 kips
Service moments at	Dead = −60 ft-kips	Dead = 42.4 ft-kips
tops of columns	Live = −14 ft-kips	Live = 11.0 ft-kips
Service moments at	Dead = −21 ft-kips	Dead = −32.0 ft-kips
bottoms of columns	Live = −8 ft-kips	Live = −8 ft-kips

Clockwise moments on the ends of members are positive. All wind forces are assumed to be resisted by the end walls of the building.

2. **Determine the factored loads.**

(a) **Column CD:**

$$P_u = 1.4 \times 80 + 1.7 \times 24 = 152.8 \text{ kips}$$

$$\text{moment at top} = 1.4 \times -60 + 1.7 \times -14 = -107.8 \text{ ft-kips}$$

$$\text{moment at bottom} = 1.4 \times -21 + 1.7 \times -8 = -43.0 \text{ ft-kips}$$

The factored moment diagram is shown in Fig. 12–28c. By definition (ACI Sec. 10.0), M_2 is always positive and M_1 is positive if the column is bent in single curvature (Fig. 12–13c and d). Since column CD is bent in double curvature (Fig. 12–28c), M_{1b} is negative. Thus, for slender column design, $M_2 = +107.8$ ft-kips and $M_1 = -43.0$ ft-kips.

(b) **Column DE:**

$$P_u = 93.8 \text{ kips}$$

$$\text{moment at top} = +78.1 \text{ ft-kips}$$

$$\text{moment at bottom} = -58.4 \text{ ft-kips}$$

Thus $M_2 = +78.1$ ft-kips and $M_1 = +58.4$ ft-kips. M_1 is positive since the column is in single curvature.

3. **Make a preliminary selection of the column size.** From Eq. 11–18a for $\rho_t = 0.015$,

$$A_{g(\text{trial})} \geq \frac{P_u}{0.45(f_c' + f_y\rho_t)}$$

$$= \frac{152.8}{0.45(3 + 0.015 \times 60)}$$

$$= 87.1 \text{ in.}^2 \text{ or } 9.3 \text{ in. square}$$

Because of the slenderness and because of the large moments, we shall take a larger column. Try 14 in. × 14 in. columns throughout.

4. **Are the columns slender?** From ACI Sec. 10.12.2, a column in a braced frame is short if $k\ell_u/r$ is less than $34 - 12M_1/M_2$.

(a) **Column CD:**

$$\ell_u = 20 \text{ ft} - 2 \text{ ft} = 18 \text{ ft} \qquad \text{(ACI Sec. 10.11.1)}$$

$$= 216 \text{ in.}$$

$$r = 0.3 \times 14 \text{ in.} = 4.2 \text{ in.} \qquad \text{(ACI Sec. 10.11.3)}$$

From Table 12–2, $k = 0.77$. Thus

$$\frac{k\ell_u}{r} = \frac{0.77 \times 216}{4.2} = 39.6$$

$$34 - 12\left(\frac{M_1}{M_2}\right) = 34 - 12\left(-\frac{43.0}{107.8}\right) = 38.8$$

Since $39.6 > 38.8$, column CD is just slender.

(b) **Column DE:**

$$\ell_u = 24 \text{ ft} - 2 \text{ ft} = 264 \text{ in.}$$

$$k = 0.86$$

$$\frac{k\ell_u}{r} = \frac{0.86 \times 264}{4.2} = 54.1$$

$$34 - 12\left(\frac{M_1}{M_2}\right) = 34 - 12\left(\frac{58.4}{78.1}\right) = 25.0$$

Thus column DE is also slender. Neither column exceeds the $k\ell_u/r = 100$ limit in ACI Sec. 10.11.5.

5. Check if the moments are less than the minimum. ACI Sec. 10.12.3.2 requires that braced slender columns be designed for a minimum eccentricity of $(0.6 + 0.03h)$ in. For 14-in. columns, this is 1.02 in. Thus column CD must be designed for a moment M_2 at least:

$$P_u e_{\min} = 152.8 \text{ kips} \times 1.02 \text{ in.} = 13.0 \text{ ft-kips}$$

and column DE for a moment of at least 8.0 ft-kips. Since the actual moments exceed these values, the columns shall be designed for the actual moments.

6. Compute EI. Since the reinforcement is not known at this stage of the design we can use either Eq. 12–16 or 12–19 to compute EI. From Eq. 12–16

$$EI = \frac{0.40 \, E_c I_g}{1 + \beta_d}$$

where

$$E_c = 57,000\sqrt{f_c'} = 3.12 \times 10^6 \text{ psi} \qquad \text{(ACI Sec. 8.5.1)}$$
$$I_g = 14^4/12 = 3201 \text{ in.}^4$$
$$0.40 \, E_c I_g = 4.00 \times 10^9 \text{ in.}^2\text{-lb}$$

  (a) Column CD:

$$\beta_d = \frac{1.4 \times 80}{152.8} = 0.733 \qquad (12\text{–}39a)$$

$$EI = \frac{4.00 \times 10^9}{1 + 0.733} = 2.31 \times 10^9 \text{ lb-in}^2$$

  (b) Column DE:

$$\beta_d = \frac{1.4 \times 50}{93.8} = 0.746$$

$$EI = \frac{4.00 \times 10^9}{1.746} = 2.29 \times 10^9 \text{ lb-in}^2$$

7. Compute the effective length factors. Three methods of estimating the effective length factors, k, have been presented. In this example we calculate k using all three methods to illustrate their use. In practice, only one of these procedures would be used.

$$\psi = \frac{\Sigma \, E_c I_c/\ell_c}{\Sigma \, E_b I_b/\ell_b} \qquad (12\text{–}27)$$

where ACI Sec. 10.12.1 says that E and I shall be as in ACI Sec. 10.11.1. Thus $I_c = 0.70 I_g$ and $I_b = 0.35 I_g$, where I_g is the gross moment of inertia of the cross section. For the beam section shown in Fig. 12–28b, ACI Sec. 8.10.2 gives the effective flange width as 90 in. Using this width gives $I_g = 36,600$ in.^{4}, so $I_b = 0.35 \times 36,600 = 12,810$ in.^{4}. Similarly, $I_c = 0.70 \times 14^4/12 = 2240$ in.^{4}. In Eq. 12–27, ℓ_c and ℓ_b are the spans of the column and beam, respectively, measured center to center of the joints in the frame.

  (a) Column DE: The value of ψ at E is

$$\psi E = \frac{E_c \times 2240/(24 \times 12)}{E_b \times 12,810/300}$$

where $E_c = E_b$. Thus $\psi_E = 0.182$. The value of ψ at D is

$$\psi_D = \frac{E_c \times 2240/288 + E_c \times 2240/240}{E_b \times 12,810/360} = 0.481$$

The value of k from Fig. 12–26 is $k = 0.63$. The value of k from Eq. 12–28 is

$$k = 0.70 + 0.05(0.182 + 0.481) = 0.733$$

The value of k from Table 12–2 is $k = 0.86$.

As pointed out in the discussion of Fig. 12–26, the effective-length nomographs tend to underestimate the values of k for beam columns in practical frames.[12-14] Analyses of elastic beam columns suggest that Eqs. 12–28 to 12–32 tend to give a better value of k. Since Table 12–2 gives reasonable values without the need to calculate ψ, it has been used to compute k in this example. Thus we shall use $k = 0.86$ for column DE.

(b)   Column CD: The value of ψ at D is

$$\psi_D = 0.481$$

The column is restrained at C by the rotational resistance of the soil under the footing and is continuous at D. From Eq. 12–27,

$$\psi = \frac{4E_c I_c / \ell_c}{I_f k_s}$$

where I_f is the moment of inertia of the contact area between the footing and the soil and k_s is the subgrade modulus obtained from Fig. 12–27.

$$I_f = \frac{48^4}{12} = 442{,}400 \text{ in}^4$$

$$\psi_c = \frac{4 \times 3.122 \times 10^6 \text{ lb/in.}^2 \times 3201 \text{ in.}^4 / 240 \text{ in}}{48^4 / 12 \text{ in.}^4 \times 160 \text{ lb/in.}^3}$$

$$= 2.35$$

From Fig. 12–26: $k = 0.76$
From Eq. 12–28: $k = 0.84$
From Table 12–2: $k = 0.77$

Use $k = 0.77$ for column CD.

8.   Compute the magnified moments. From Eq. 12–22,

$$M_c = \delta_{ns} M_2$$

where

$$\delta_{ns} = \frac{C_m}{1 - (P_u / 0.75 P_c)} \geq 1.0 \qquad (12\text{–}23)$$
$$(\text{ACI Eq. }10\text{–}10)$$

(a)   Column CD:

$$C_m = 0.6 + 0.4\left(-\frac{43.0}{107.8}\right) \geq 0.4 \qquad (12\text{–}14)$$
$$(\text{ACI Eq. }10\text{–}14)$$

$$= 0.440$$

$$P_c = \frac{\pi^2 EI}{(k\ell_u)^2} \qquad (12\text{–}24)$$
$$(\text{ACI Eq. }10\text{–}11)$$

$$\ell_u = 18 \text{ ft} = 216 \text{ in}$$

$$P_c = \frac{\pi^2 \times 2.31 \times 10^9 \text{ lb}-\text{in.}^2}{(0.77 \times 216 \text{ in.})^2}$$

$$= 823{,}500 = 824 \text{ kips}$$

$$\delta_{ns} = \frac{0.440}{1 - 152.8/(0.75 \times 824)}$$

$$= 0.585 \geq 1.0$$

Therefore, $\delta_{ns} = 1.0$. This means that the section of maximum moment remains at the end of the column.

$$M_c = 1.0 \times 107.8 = 107.8 \text{ ft-kips}$$

Column CD is designed for $P_u = 152.8$ kips and $M_u = M_c = 107.8$ ft-kips.

(b) Column DE:

$$C_m = 0.6 + 0.4\left(\frac{58.4}{78.1}\right) = 0.899$$

$$P_c = \frac{\pi^2 \times 2.29 \times 10^9}{(0.86 \times 264)^2} = 439 \text{ kips}$$

$$\delta_{ns} = \frac{0.899}{1 - \dfrac{93.8}{0.75 \times 439}} \geq 1.0$$

$$= 1.257$$

This column is affected by slenderness.

$$M_c = 1.257 \times 78.1 = 98.2 \text{ ft-kips}$$

Column DE is designed for $P_u = 93.8$ kips and $M_u = M_c = 98.2$ ft-kips.

9. Select the reinforcement. Figure 12–29 gives interaction diagrams for 14 in. × 14 in. columns with 4 No. 8 bars, 4 No. 9 bars, and 4 No. 10 bars.

Column CD: Use 14 in. × 14 in. column with 4 No. 8 bars.
Column DE: Use 14 in. × 14 in. column with 4 No. 10 bars. ∎

12–5 BEHAVIOR OF RESTRAINED COLUMNS IN SWAY FRAMES

Statics of Sway Frames

An unbraced frame is one that depends on moments in the columns to resist lateral loads and lateral deflections. Such a frame is shown in Fig. 12–30a. The sum of the moments at the tops and bottoms of all the columns must equilibrate the applied lateral load moment, $H\ell$, plus the moment due to the vertical loads, $\Sigma\, P\Delta$. Thus

$$\Sigma\, (M_{\text{top}} + M_{btm}) = H\ell + \Sigma\, P\Delta \qquad (12\text{--}40)$$

It should be noted that both columns have deflected laterally by the same amount Δ. For this reason it is not possible to consider columns independently in an unbraced frame.

If a sway frame includes some pin-ended columns, as might be the case in a precast building, the vertical loads in the pin-ended columns are included in $\Sigma\, P$ in Eqs. 12–38 and 12–40. Such columns are referred to as *leaning columns* because they depend on the frame for stability.

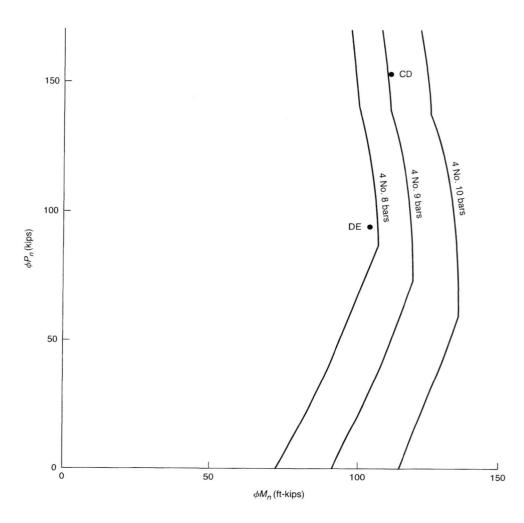

Fig. 12–29
Interaction diagrams—
Example 12–2

The moment diagram due to the lateral loads is shown in Fig. 12–30b, and that due to the $P\Delta$ moments in Fig. 12–30c. It can be seen that these are directly additive, because the maximum lateral load moments and the maximum $P\Delta$ moments occur at the ends of the column. For this reason, the equivalent moment factor, C_m, given by Eq. 12–14 does not apply to sway columns since the maximum lateral load moments and maximum $P\Delta$ moments add directly. On the other hand, Eq. 12–11 becomes

$$M_c = \frac{M_0(1 - 0.18P/P_E)}{1 - P/P_E} \qquad (12\text{–}41)$$

The term $(1 - 0.18P/P_E)$ reflects the shape of the moment diagram in Fig. 12–30b, which differs from the rectangular diagram (Fig. 12–10b) used to derive Eq. 12–11. Again this term has been neglected in the ACI Code. This is conservative.

It is also important to note that if hinges were to form at the ends of the beams in the frame as shown in Fig. 12–30d, the frame would be unstable. Thus the beams must resist the *full magnified end moment* from the columns for the frame to remain stable (ACI Sec. 10.13.7).

Loads causing sway are seldom sustained, although such cases can be visualized, such as a frame that resists the horizontal reaction from an arch roof or a frame resisting lat-

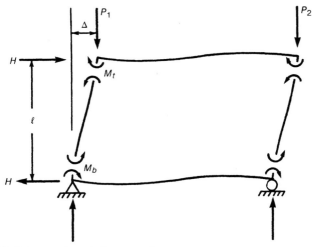

(a) Column moments in a sway frame.

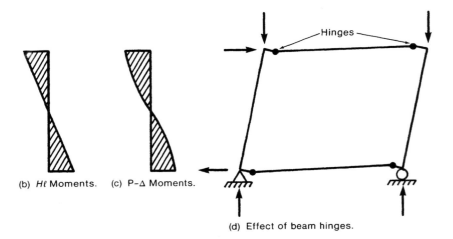

(b) $H\ell$ Moments. (c) P-Δ Moments.

(d) Effect of beam hinges.

Fig. 12–30
Column moments in a sway frame.

eral earth loads. If a sustained load acts on an unbraced frame, the deflections increase with time, leading directly to an increase in the $P\Delta$ moment. This process is very sensitive to small variations in material properties and loadings. As a result, structures subjected to sustained lateral loads should always be braced. Indeed, braced frames should be used wherever possible, regardless of whether the loads are short time or sustained.

M_{ns} and M_s Moments

Two different types of moments occur in frames:

1. Moments due to loads not causing sway, M_{ns}
2. Moments due to loads causing appreciable sway, M_s

These moments are considered separately in the ACI design procedure because they are magnified differently, as the frame deflects. This is illustrated in Fig. 12–31, which is based on a similar figure by Nathan.[12–21]

A conventional moment distribution analysis of the frame shown in Fig. 12–31a would consist of two stages. The frame would be held against sway at, for example, corner

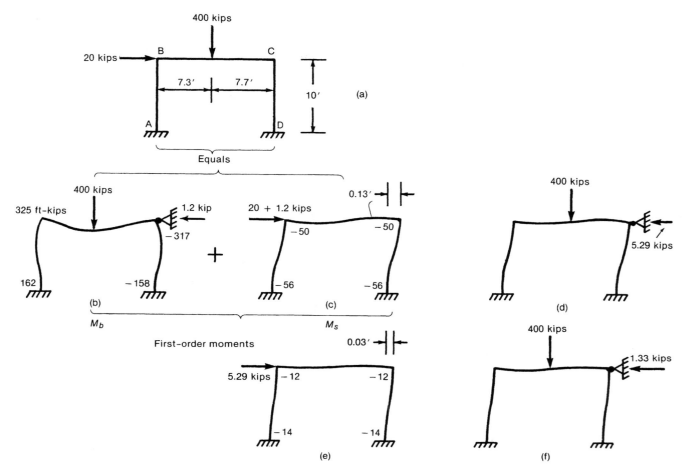

Fig. 12–31
M_{ns} and M_s moments in a portal frame.

C and the gravity loads would be applied, giving the moments shown in Fig. 12–31b. Since the 400-kip load does not act symmetrically on the frame, a horizontal reaction of 1.2 kips is required at C to prevent sway. In the second stage, the applied lateral load of 20 kips and the holding force of 1.2 kips are applied to the frame, which is now free to sway. This gives the moments and deflections shown in Fig. 12–31c. The moments obtained from a conventional moment distribution analysis are the sum of those given in Fig. 12–31b and c.

When the 400-kip load acts on the laterally displaced frame, a new holding force of 5.29 kips is required to prevent any further lateral deflections, as shown in Fig. 12–31d. This force can be canceled out by adding the loading case shown in Fig. 12–31e. The 400-kip load interacts with this new deflection to give the holding force in f, and so on. The magnified sway moments $\delta_s M_s$, are the sum of the moments in Figs. 12–31c, e, and so on. The moments in Figs. 12–31e and g (not shown), and so on, are proportional to those in Fig. 12–31c and bear no relationship to the moments in Fig. 12–31b, except that the force system in part b gave rise to the 1.2-kip holding force in Figs. 12–31b and c. For this reason the ACI Code differentiates between the M_{ns} and M_s moments. This was proposed initially in Ref. 12–22.

Today, most frames are analyzed using direct stiffness analyses which do not involve holding forces. To simplify the subdivision of moments, the ACI Code allows one to disregard the holding forces for loads "that result in no appreciable sidesway." When this is done, the M_{ns} moments are obtained directly from a first-order direct stiffness analysis of the frame under such loads. For the frame in Fig. 12–31, such an analysis would give moments within 2% of those obtained in Fig. 12–31b.

The term "no appreciable sidesway" is not defined in the code or commentary. The 1977 through 1989 ACI Commentaries defined "no appreciable sidesway" as being a lateral deflection of $\Delta \leq 1/1500$ of the story height at factored loads.

12–6 CALCULATION OF MOMENTS IN SWAY FRAMES USING SECOND-ORDER ANALYSES

First-Order and Second-Order Analysis

A *first-order frame analysis* is one in which the effect of lateral deflections on bending moments, axial forces, and lateral deflections is ignored. The resulting moments and deflections are linearly related to the loads. In a *second-order frame analysis*, the effects of deflections on moments, and so on, is considered. The resulting moments and deflections include the effects of slenderness and hence are nonlinear with respect to the load. Because the moments are directly affected by the lateral deflections as shown in Eq. 12–40, it is important that the stiffnesses, EI, used in the analysis be representative of the stage immediately prior to ultimate.

Second-Order Analysis

Load Level for the Analysis

In a second-order analysis, column moments and lateral frame deflections increase more rapidly than the loads. Thus it is necessary to calculate the second-order effects at the factored load level. Using the ACI Code load factors and assuming that the sway deflections result from wind, the analysis would be carried out for the load levels:

$$U = 0.75(1.4D + 1.7L + 1.7W) \tag{12–42a}$$
$$= 1.05D + 1.275L + 1.275W$$
$$U = 0.9D + 1.3W \tag{12–42b}$$

As discussed later, it is also necessary to check for sidesway buckling under the load level:

$$U = 1.4D + 1.7L \tag{12–42c}$$

Stiffness Reduction Factor

The term 0.75 in the denominator of Eq. 12–23 is a stiffness reduction factor, ϕ_K, which accounts for variability in the critical load, P_c, and variability introduced by the assumptions in the moment magnifier calculation. This factor leads to an increase in the magnified moments. A similar stiffness reduction factor is needed when a second-order analysis is carried out. Two things combine to allow the use of a value of ϕ_K larger than 0.75 when a second-order analysis is carried out. First, the modulus of elasticity, E_c, used in the frame analysis is based on the specified strength, f_c', while the deflections are a function of the E_c based on the average concrete strength which is 600 to 1400 psi higher than f_c'. Second, the second-order analysis is a better representation of the behavior of the frame than the sway magnifier given by Eq. 12–48. The moments of inertia given in ACI Sec. 10.11.1 have been multiplied by 0.875, which, when combined with the underestimate of E_c, lead to an overestimation of the second-order deflections in the order of 20 to 25%, corresponding to an implicit value of ϕ_K of 0.80 to 0.85.

Stiffness of the Members

Ultimate Limit State. The appropriate stiffness to be used in strength calculations must be chosen to estimate the lateral deflections accurately at the factored load level. They must be simple to apply, because a frame consists of many cross sections with differing reinforcement ratios and differing degrees of cracking. Furthermore, the reinforcement amounts and distributions are not known at the time the analysis is carried out. Based on studies of the flexural stiffness of beams with cracked and uncracked regions, MacGregor and Hage[12-20] recommend that the beam stiffnesses be taken as $0.4E_cI_g$ when carrying out a second-order analysis. In ACI Sec. 10.11.1 this value has been multiplied by a stiffness reduction factor of 0.875, giving $I = 0.35I_g$.

Two levels of behavior must be distinguished in selecting the EI of columns. The lateral deflections of the frame are influenced by the stiffness of all the members in the frame and by the variable degree of cracking of these members. Thus the EI used in the frame analysis should be an average value. On the other hand, when designing an individual column in a frame using Eq. 12–23, the EI used in calculating δ_{ns} must be for that column. This EI must reflect the greater chance that a particular column will be more cracked, or be weaker than the overall average, and hence this EI will tend to be smaller than the average EI for all the columns acting together. Reference 12–20 recommends the use of $EI = 0.8E_cI_g$ when carrying out second-order analyses of frames. ACI Sec. 10.11.1 gives this value multiplied by 0.875, or $EI = 0.70E_cI_g$ for this purpose. On the other hand, when calculating the moment magnifiers δ_{ns} and δ_s using Eqs. 12–23 and 12–48 (ACI Eqs. 10–10 and 10–19), EI must be taken as given by Eq. 12–15 or 12–16.

The value of EI for shear walls may be taken equal to the value for beams in regions where the wall is cracked due to flexure or shear, and the value for columns where the wall is uncracked. If the factored moments and shears from an analysis based on $EI = 0.70E_cI_g$ for the walls, indicate that a portion of the wall will crack based on the modulus of rupture, the analysis should be repeated with $EI = 0.35E_cI_g$ for the cracked parts of the wall.

Servicability Limit State. The moments of inertia given in ACI Sec. 10.11.1 are for the ultimate limit state. At service loads, the members are cracked less than they are at ultimate. When computing deflections or vibrations at service loads, the values of I should be representative of the degree of cracking at service loads. The Commentary R10.11.1 suggests that I at service loads may be taken as $1/0.70 = 1.43$ times the values given in ACI Sec. 10.11.1.

Foundation Rotations

The rotations of foundations subjected to column end moments reduce the fixity at the foundations and lead to larger sway deflections. These are particularly significant in the case of shear walls or large columns, which resist a major portion of the lateral loads. The effects of foundation rotations can be included in the analysis by modeling each foundation as an equivalent beam having the stiffness given by Eq. 12–36.

Effect of Sustained Loads

The loads causing appreciable sidesway are generally short duration loads, such as wind or earthquake, and, as a result, do not cause creep deflections. In the unlikely event that sustained lateral loads act on the structure, the *EI* values used in the frame analysis should be reduced. ACI Sec. 10.11.1 states that in such a case, I shall be divided by $(1 + \beta_d)$, where, for this case, β_d is defined in ACI Sec. 10.0, definition (b) as

$$\beta_d = \frac{\text{maximum factored sustained shear within a story}}{\text{total factored shear in that story}} \tag{12–39b}$$

Methods of Second-Order Analysis

Computer programs that carry out second-order analyses are widely available. The principles of such an analysis are presented in the following paragraphs. Methods of second-order analysis are reviewed in Refs. 12–20 and 12–23.

Iterative P–Δ Analysis

When a frame is displaced sideways under the action of lateral and vertical loads, as shown in Figs. 12–30 and 12–32, the column end moments must equilibrate the lateral loads and a moment equal to $(\Sigma P)\Delta$:

$$\Sigma (M_{\text{top}} + M_{btm}) = H\ell_c + \Sigma P\Delta \tag{12–40}$$

where Δ is the lateral deflection of the top of the story relative to the bottom of the story. The moment $\Sigma P\Delta$ in a given story can be represented by shear forces, $(\Sigma P)\Delta/\ell_c$, where ℓ_c is the story height, as shown in Fig. 12–32b. These shears give an overturning moment of $(\Sigma P\Delta/\ell_c)/(\ell_c) = (\Sigma P)\Delta$. Figure 12–32c shows the story shears in two different stories. The algebraic sum of the story shears from the columns above and below a given floor gives rise to a *sway force* acting on that floor. At the jth floor the sway force is

$$\text{sway force}_j = \frac{(\Sigma P_i)\Delta_i}{\ell_i} - \frac{(\Sigma P_j)\Delta_j}{\ell_j} \tag{12–43}$$

The sway forces are added to the applied lateral loads at each floor level, and the structure is reanalyzed giving new lateral deflections and larger column moments. If the deflections increase by more than about 5%, new $\Sigma P\Delta/\ell$ forces and sway forces are computed and the structure is reanalyzed for the sum of the applied lateral loads and the new sway forces. This process is continued until convergence is obtained.

At discontinuities in the stiffness of the building or discontinuities in the applied loads, the sway force may be negative. In such a case it acts in the opposite direction to that shown in Fig. 12–32c.

Ideally one correction must be made to this process. The P–Δ moment diagram for a given column is the same shape as the deflected column, as shown in Fig. 12–30c, while the moment diagram due to the $P\Delta/\ell$ shears is a straight-line diagram, similar to the $H\ell$ moments shown in Fig. 12–30b. As a result, the area of the real $P\Delta$ moment diagram is larger than that of the straight-line moment diagram. It can be shown by the moment–area theorems that the deflections due to the real diagram will be larger than those due to the $P\Delta/\ell$ shears. The increase in deflection varies from zero for a very stiff column with very flexible restraining beams, to 22% for a column that is fully fixed against rotation at each end. A reasonable average value is about 15%. The increased deflection can be accounted for by taking the story shears as $\gamma\Sigma P\Delta/\ell$, where γ is a *flexibility factor*[12–23] which ranges from 1.0 to 1.22 and can be taken equal to 1.15 for practical frames. Unfortunately, most commercially available second-order analysis programs do not include this correction. For this reason, we shall omit γ also.

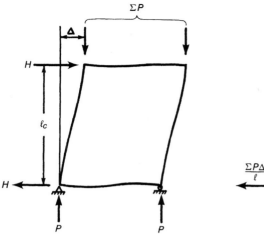

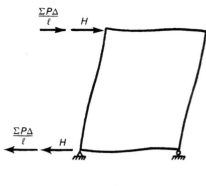

(a) Loads on a sway frame.

(b) $\Sigma P\Delta/\ell$ Shears in a story.

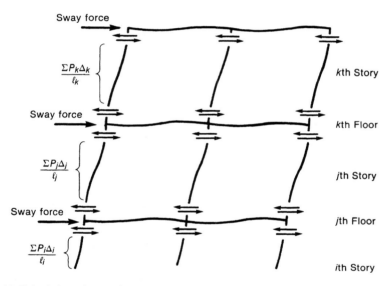

Fig. 12–32
Iterative P–Δ analyses.

(c) Calculation of sway forces.

Direct P–Δ Analysis for Sway Frames

The iterative calculation procedure described in the preceding section can be described mathematically by using an infinite series. The sum of this series gives the second-order deflection, Δ:

$$\Delta = \frac{\Delta_0}{1 - \gamma(\Sigma P_u)\Delta_0/(V_u\ell_c)} \qquad (12\text{--}44)$$

where

V_u = shear in the story due to wind loads acting on the frame above the story in question

ℓ_c = story height

12–6 Calculation of Moments in Sway Frames   **517**

$\Sigma P_u =$ the total axial load in all the columns in the story

$\gamma \simeq 1.15$

$\Delta_0 =$ first-order deflection due to the story shear, V

$\Delta =$ second-order deflection

Both Δ_0 and Δ refer to the lateral deflection of the top of the story relative to the bottom of the story.

Since the moments in the frame are directly proportional to the deflections, the second-order moments are

$$M = \delta_s M_s = \frac{M_0}{1 - \gamma(\Sigma P_u)\Delta_0/(V_u \ell_c)} \qquad (12-45)$$

where M_0 and M are the first- and second-order moments, respectively.

ACI Sec. 10.11.4.2 defines the stability index for a story as

$$Q = \frac{\Sigma P_u \Delta_0}{V_u \ell_c} \qquad (12-38)$$
$$\text{(ACI Eq. 10-7)}$$

Substituting this into Eq. 12-45 and omitting the flexibility factor γ gives

$$\delta_s M_s = \frac{1}{1 - Q} \geq M_s \qquad (12-46)$$
$$\text{(ACI Eq. 10-18)}$$

Direct P–Δ Analysis for Braced Structures

In a braced structure, the relative deflections of the top and bottom of a story are largely controlled by the slope of the shear wall or bracing element in that story. In such a case the relative deflection of a given story is not independent of that in adjacent stories and Eqs. 12–44 and 12–45 cannot be used. In Ref. 12–23 it is shown that the moments and deflections of the entire frame can be magnified using the frame magnifier, δ_f, given by

$$\delta_f = \frac{1}{1 - \sum_{i=1}^{n} (\Sigma \gamma P_u / \ell_c)_i \Delta_{0i}^2 \left[\sum_{i=1}^{n} (\Sigma V_u)_i \Delta_{0i} \right]} \qquad (12-47)$$

The procedure for calculating the second-order moments and deflections is the same as for the unbraced frame, except that the same magnifier δ_f is used in all braced stories. This procedure is applicable to braced frames and to the braced portions of partially braced structures.

12-7   DESIGN OF COLUMNS IN SWAY FRAMES

Overview

In the 1995 ACI Code the sections dealing with the design of slender columns were completely rewritten and rearranged. ACI Sec. 10.10 is the overall umbrella section. It sets requirements for an accurate slenderness analysis (ACI Sec. 10.10.1) and allows the use of the more approximate moment magnifier analysis described in ACI Secs. 10.11, 10.12, and 10.13.

ACI Sec. 10.11 "Magnified Moments—General" gives general requirements for the design of slender columns in both nonsway and sway frames. ACI Sec. 10.11.4 gives methods of distinguishing between nonsway and sway columns. If a column is in a sway frame, design involves ACI Secs. 10.11 and 10.13 "Magnified Moments—Sway Frames." The 1995 ACI Code design procedure for slender columns in sway frames consists of five steps:

1. **The unmagnified moments, M_{ns}, due to loads not causing appreciable sway are computed.** This is done using a regular first-order elastic frame analysis using the member stiffnesses given in ACI Sec. 10.11.1. For the load combination

$$U = 0.75(1.4D + 1.7L + 1.7W) = 1.05D + 1.275L + 1.275W \quad (12-42a)$$

the M_{ns} moments would result from $1.05D + 1.275L$.

2. **The magnified sway moments, $\delta_s M_s$, are computed.** Three alternative methods of carrying out this computation are given in ACI Sec. 10.13.4. For the load combination quoted in step 1, the M_s moments result from $1.275W$.

3. **The magnified sway moments, $\delta_s M_s$ are added to the unmagnified nonsway moments, M_{ns}.** This is done at each end of each column (ACI Sec. 10.13.3).

4. **Check whether the maximum moment occurs between the ends of the column.** Normally, the maximum moment in the column will be at one end and the column is designed for this moment. However, if the axial loads on the column are high and the slenderness exceeds the limit given in ACI Sec. 11.13.5, it is necessary to check if the moment at some section between the ends of the column exceeds the maximum end moment. This is done using the braced frame magnifier from ACI Sec. 10.12.3.

5. **Check whether sidesway buckling can occur under gravity loads alone.** This is done using ACI Sec. 11.13.6.

Each of these steps will be discussed in the following sections, followed by examples. The design process is developed and discussed in Ref. 12–24.

Computation of $\delta_s M_s$

ACI Sec. 10.13.4 allows designers to compute the magnified sway moments, $\delta_s M_s$, in frames by one of three ways. In order of decreasing accuracy these are: second-order analyses, a direct solution of the iterative P–Δ analyses, and the sway moment magnifier, δ_s, used in ACI Codes since 1971.

Computation of $\delta_s M_s$ Using Second-Order Analyses

ACI Sec. 10.13.4.1 allows the use of second-order analyses to compute $\delta_s M_s$. Second-order analyses were discussed in Sec. 12–6. If torsional displacements of the frame are significant, a three-dimensional second-order analysis should be used. There is no easy way to incorporate torsional effects into the calculation of $\delta_s M_s$ by ACI Sec. 10.13.4.2 or 10.13.4.3.

Computation of $\delta_s M_s$ Using Direct P–Δ Analysis

ACI Sec. 10.13.4.2 permits the use of a direct calculation of P–$\overset{\backprime}{\Delta}$ moments using an equation similar to Eq. 12–45. In ACI Sec. 10.13.4.2 this is written as

$$\delta_s M_s = \frac{M_s}{1 - Q} \geq M_s \quad (12-46)$$

$$(\text{ACI Eq. } 10-18)$$

where

$$Q = \frac{\Sigma P_u \Delta_0}{V_u \ell_c}$$

(12–38)

(ACI Eq. 10–7)

Computation of $\delta_s M_s$ Using Sway Frame Moment Magnifier

ACI Sec. 10.13.4.3 allows the use of the traditional sway frame moment magnifier to compute the magnified sway moments, $\delta_s M_s$, where

$$\delta_s M_s = \frac{M_s}{1 - \Sigma P_u / (0.75 \, \Sigma P_c)} \geq M_s$$

(12–48)

(ACI Eq. 10–19)

Here ΣP_u and ΣP_c refer to the sums of the axial loads and critical loads, respectively, for all the columns in the story being analyzed. In this case, the values of P_c are calculated using the effective lengths, $k\ell_u$, evaluated for columns in a sway frame, and β_d defined as

$$\beta_d = \frac{\text{maximum factored sustained shear in the story}}{\text{total factored shear in the story}}$$

(12–39b)

In most sway frames the story shear is due to wind or seismic loads and is not sustained, giving $\beta_d = 0$. The use of the summation terms in Eq. 12–48 accounts for the fact that sway instability involves all the columns in the story (see Fig. 12–30).

Moments at the Ends of the Columns

The unmagnified nonsway moments, M_{ns}, are added to the magnified sway moments, $\delta_s M_s$, at each end of the column using

$$M_1 = M_{1ns} + \delta_s M_{1s}$$

(12–49a)

(ACI Eq. 10–16)

$$M_2 = M_{2ns} + \delta_s M_{2s}$$

(12–49b)

(ACI Eq. 10–17)

The addition is carried out for the moments at the top and bottom. The larger absolute sum of the resulting end moments is called M_2 and the smaller is called M_1.

Maximum Moment between the Ends of the Column

In most columns in sway frames, the maximum moments will occur at the ends of the column and will have the values given by Eqs. 12–49a and 12–49b. Occasionally, for very slender, highly loaded columns the deflections of the column may cause the maximum column moment to exceed the moment at one end in a fashion analogous to the moments in the braced frame shown in Fig. 12–22c. At load stage 1 in this figure, the maximum moment is at the top of the column. At load stage 2 the moment at the upper third point of the column exceeds that at the top. If

$$\frac{\ell_u}{r} > \frac{35}{\sqrt{\dfrac{P_u}{f_c' A_g}}}$$

(12–50)

(ACI Eq. 10–20)

the moment at some point between the ends of the column may exceed the larger end moment.[12–24]

If ℓ_u / r exceeds the value given by Eq. 12–50 the maximum moment along the length of the column must be computed. This is done using the nonsway moment magnifier, δ_{ns}, given by Eq. 12–23 (ACI Eq. 10–10). This moment magnifier computes the maximum moment in a column having the end moments M_1 and M_2. If $\delta_{ns} \leq 1.0$, the maximum moment is at the end of the column. If $\delta_{ns} > 1.0$, it will be between the ends. Equation 12–23 includes the critical load, P_c. This is computed using the nonsway effective length, k, from ACI Sec. 10.12.1. Since the end moments include sway moments, β_d is the value for a sway frame given by Eq. 12–39b.

Sidesway Buckling Under Gravity Loads

As shown in Fig. 12–6, a frame may buckle in a sidesway manner under gravity loads alone. ACI Sec. 10.13.6 requires a check of this possibility. This is done using the load combination that gives the largest gravity load, $U = 1.4D + 1.7L$. Since there are three methods to calculate $\delta_s M_s$, three corresponding methods are given to check sidesway buckling.

(a) If $\delta_s M_s$ has been computed using second-order analyses, ACI Sec. 10.13.6a limits the ratio of second-order deflections to first-order deflections to 2.5. This requires a special second-order analysis based on the EI values from ACI Sec. 10.11.1 divided by $(1 + \beta_d)$ with β_d from Eq. 12–39a [ACI Sec. 10.0 definition (c)], with the gravity load from $1.4D + 1.7L$, and any arbitrarily chosen lateral load. Generally, one would choose a set of lateral loads already considered in the design such as the wind loads.

(b) If $\delta_s M_s$ has been computed using the direct P–Δ analysis, ACI Sec. 10.13.6b limits Q to 0.60, where Q is based on ΣP_u from $1.4D + 1.7L$ and Δ_0 computed for any arbitrary story shear, V_u, using the EI values from ACI Sec. 10.11.1 divided by β_d from Eq. 12–39a [ACI Sec. 10.0 definition (c)]. This is a relatively straightforward calculation and may be used as an initial check of the tendency for sidesway buckling when second-order analyses have been used. From Eq. 12–46 it can be seen that $Q = 0.60$ corresponds to $\delta_s = 2.5$.

(c) If $\delta_s M_s$ has been computed using the sway moment magnifier, ACI Sec. 10.13.6c limits δ_s from Eq. 12–48 (ACI Eq. 10–19) to not more than 2.5, where δ_s is based on ΣP_u from $1.4D + 1.7L$ and ΣP_c is based on effective lengths from ACI Sec. 10.13.1 and the EI values from Eqs. 12–15 and 12–16 (ACI Eqs. 10–12 and 10–13) with β_d from Eq. 12–39a [ACI Sec. 10.0 definition (c)].

Minimum Moment

The ACI Code specifies a minimum moment $M_{2,min}$ to be considered in the design of columns in nonsway frames but does not do so for columns in sway frames. This will only be a problem for the load combination $U = 1.4D + 1.7L$ acting on a sway frame, since this load combination does not involve $\delta_s M_s$. For this load combination we shall design for the larger of M_2 and $M_{2,min}$.

Example 12–3 Design of the Columns in an Unbraced Frame

Figure 12–33 shows the plan of the main floor and a section through a five-story building. The building is clad with nonstructural precast panels. There are no structural walls or other bracing. The beams in the north–south direction are all 18 in. wide with an overall depth of 30 in. The floor slabs are 7 in. thick. Design an interior and exterior column in the ground-floor level of the frame along column line 3 for dead load, live load, and north-south wind forces. Use $f_c' = 4000$ psi and $f_y = 60,000$ psi.

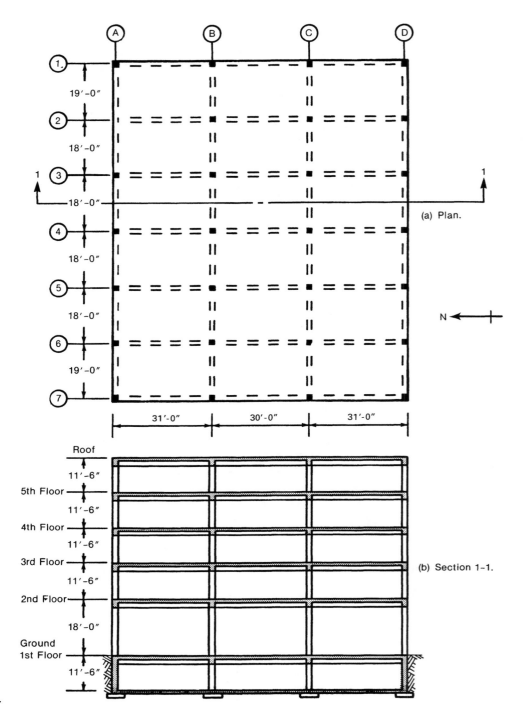

Fig. 12–33
Sway frame—Example 12–3.

1. Make a preliminary selection of the column size. A preliminary calculation of the gravity loads in the first-story columns based on the respective tributary areas on the roof and each floor above the main floor gives the unfactored dead and unreduced live loads in the columns between the ground floor and the second floor as:

Exterior column: Dead load = 176 kips
                   Live load = 10.5 kips from the roof and 104 kips of floor load

Interior column: Dead load = 381 kips

Live load = 22.1 kips from the roof and 223 kips floor load

ASCE 7–95[12–25] allows the floor live loads to be reduced as a function of the influence area, A_I, of floor supported by the columns. Snow loads are not reduced. For a column, the influence area is four times the tributary area of the column (see Fig. 2–10). Thus $A_I = 4(18 \text{ ft} \times 30.25/2 \text{ ft}) = 1089 \text{ ft}^2$ per floor × 4 floors = 4356 ft². The live load due to use and occupancy can be multiplied by

$$\frac{L}{L_0} = 0.25 + \frac{15}{\sqrt{A_I}} \tag{2–9}$$

$$= 0.477$$

but not less than 0.50 for columns supporting one floor, nor less than 0.40 for columns supporting two or more floors. For the interior column the influence area is $A_I = 4(18 \text{ ft} \times 30.12 \text{ ft}) \times 4 \text{ floors} = 8676 \text{ ft}^2$ and $L/L_0 = 0.411$. Thus the reduced live loads and factored axial loads in the first-story columns are:

Exterior column: Reduced live load = 10.5 kips + 0.477 × 104 kips = 60.1 kips

Factored load = 1.4 × 176 kips + 1.7 × 60.1 kips = 349 kips

Interior column: Reduced live load = 22.1 kips + 0.411 × 223 kips = 113.8 kips

Factored load = 1.4 × 381 + 1.7 × 113.8 = 727 kips

For the exterior column from Eq. 11–17a for $\rho_t = 0.015$,

$$A_{g(\text{trial})} \geq \frac{P_u}{0.45(f_c' + f_y\rho_t)}$$

$$= \frac{349 \text{ kips}}{0.45(4 \text{ ksi} + 60 \text{ ksi} \times 0.015)}$$

$$= 158 \text{ in.}^2 \text{ or } 12.6 \text{ in. by } 12.6 \text{ in.}$$

We shall try an 18 in. × 18 in. column. This is larger than $A_{g(\text{trial})}$ to allow for slenderness and to allow for the moments in the column in load cases involving wind.

For the interior column, $A_{g(\text{trial})} = 330 \text{ in.}^2$ or 18.2 in. x 18.2 in. We will try 18 in. x 18 in. interior columns so that all columns will be the same size. This may necessitate an increase in the steel ratio, ρ_t. Assume that the columns are 18 in. x 18 in. in the basement and second floor also.

FACTORED LOAD COMBINATIONS

Three different load cases will be considered:

Case 1: gravity and wind loads, $U = 0.75(1.4D + 1.7L \pm 1.7W)$

                       $= 1.05D + 1.275L \pm 1.275W$, where the wind can blow from the north or the south

Case 2: gravity and wind loads, $U = 0.9D \pm 1.3W$

Case 3: gravity loads only, $U = 1.4D + 1.7L$

Since this is a symmetrical frame, the gravity loads will not cause appreciable sidesway. Thus the dead and live loads give rise to the M_{ns} moments, while the wind loads cause the M_s moments.

For clarity, all the calculations for case 1 will be presented before starting case 2, and so on.

LOAD CASE 1: $U = 0.75(1.4D + 1.7L + 1.7W) = 1.05D + 1.275L \pm 1.275W$

2. Is the frame sway or nonsway? ACI Sec. 10.11.4 defines a story in a frame as being a nonsway story if

$$Q = \frac{\Sigma P_u \Delta_0}{V_u \ell_c} \leq 0.05 \tag{12–38}$$

$$\text{(ACI Eq. 10–7)}$$

where for the first story,

$$\Sigma P_u = \text{total reduced factored axial load in all 28 columns in the floor} = 10{,}565 \text{ kips}$$

$\Delta_0 = 0.429$ in. (from a first-order elastic analysis)

$V_u = $ total factored shear in the first story in all seven frames $= 188$ kips

$\ell_c = $ story height center to center of joints $= 18 \text{ ft} \times 12 = 216$ in.

$$Q = \frac{10{,}565 \text{ kips} \times 0.429 \text{ in.}}{188 \text{ kips} \times 216 \text{ in.}} = 0.112$$

Thus the first story of the frame is clearly a sway story. We shall treat the entire frame as a sway frame.

3. Are the columns slender? From ACI Sec. 10.11.4.2 a column in an unbraced frame is slender if $k\ell_u/r \geq 22$. However, k is not known at this stage. Since k will not normally be less than 1.2, we shall use this value in our check. For the main floor,

$$\ell_u = 18.0 \text{ ft} - 2.5 \text{ ft}$$

$$= 15.5 \text{ ft} \times 12 = 186 \text{ in.}$$

From ACI Sec. 10.11.3, $r = 0.3h$. For 18-in.-square columns,

$$\frac{k\ell_u}{r} = \frac{1.2 \times 186}{0.3 \times 18} = 41.3$$

Thus the columns are slender. If this calculation gave $k\ell_u/r$ values close to 22, it should be repeated with a better estimate of k.

4. Compute the factored axial loads P_u and the M_{ns} moments from a first-order frame analysis. For the main floor columns the values are:

	Exterior Columns	Interior Columns
Factored axial force, P_u (kips)		
1.05 Dead load	184.8	399.6
1.275 reduced live load	88.5	144.8
Factored M_{ns} moments (ft-kips)		
At top	−103.1	12.61
At bottom	−108.5	16.49

The M_s moments will be computed in the second-order analysis in step 5. The axial loads due to live load have been reduced as a function of influence area, but the live load M_{ns} moments have not. These moments result primarily from loading on the beams and slabs in one floor and L/L_0 for the beams is much closer to one than for the columns.

5. Compute $\delta_s M_s$ and M_c using a second-order analysis. The $\delta_s M_s$ moments come directly from a second-order analysis for the loads, causing appreciable sway. A commercial second-order analysis program would normally be used. To illustrate the process, we will use an iterative P–Δ analysis.

(a) Compute ΣP_u, $\Sigma P_u \Delta_0 / \ell_c$, and sway forces. The total load in the columns supporting the second floor is 10,565 kips. For the columns supporting the third floor, $\Sigma P_u = 8{,}589$ kips.

The horizontal deflections of the ground, second, and third floors, respectively, are 0 in., 0.429 in., and 0.556 in.

The $P_u \Delta_0 / \ell_c$ shears are:

First story:
$$\frac{\Sigma P_u \Delta_0}{\ell_c} = \frac{10{,}565 \text{ kips} \times (0.429 - 0) \text{ in.}}{(18 \times 12) \text{ in.}}$$

$$= 20.98 \text{ kips}$$

Second story:
$$\frac{\Sigma P_u \Delta_0}{\ell_c} = \frac{8589 \text{ kips} \times (0.556 - 0.429) \text{ in.}}{11.5 \times 12 \text{ in.}}$$

$$= 7.90 \text{ kips}$$

Sway force at the second floor level = 22.64 kips −7.90 kips = 13.08 kips.
The calculation of the sway forces for the entire building can be tabulated as follows:

Floor	Story	ΣP_u kips	Floor Δ (in.)	Story Δ (in.)	$\dfrac{\Sigma P_u \Delta_0}{\ell_c}$	Sway Force
Roof			0.721			0.356
	5th	2,047		0.024	0.356	
5th			0.697			1.422
	4th	4,461		0.055	1.778	
4th			0.642			2.300
	3rd	6,544		0.086	4.078	
3rd			0.556			3.826
	2nd	8,589		0.127	7.904	
2nd			0.429			13.076
	1st	10,565		0.429	20.98	
Ground						

This frame is now reanalyzed for 1.275W plus the sway forces. The unfactored wind loads at each floor are: roof, 25.6 kips; third, fourth, and fifth floors, 37.9 kips each; and second floor, 48.7 kips. The total factored wind forces plus sway forces at each floor are:

Roof: $1.275 \times 25.6 + 0.356 = 33.0$ kips
5th floor: $1.275 \times 37.9 + 1.422 = 49.7$ kips
4th floor: $1.275 \times 37.9 + 2.300 = 50.6$ kips
3rd floor: $1.275 \times 37.9 + 3.826 = 52.1$ kips
2nd floor: $1.275 \times 48.7 + 13.08 = 75.2$ kips

These loads are resisted by the seven frames in the north–south direction.

(b) Check convergence for the second cycle. The deflections from the two cycles of iteration are:

	Δ-First Cycle (in.)	Δ-Second Cycle (in.)	2nd/1st
Roof	0.721	0.769	1.07
2nd floor	0.429	0.466	1.09

This clearly has not converged.

(c) Calculate sway forces and loads based on the deflections from the second cycle.

Floor	Story	ΣP_u (kips)	Floor Δ (in.)	Story Δ (in.)	$\dfrac{\Sigma P_u \Delta_0}{\ell_c}$	Sway Force
Roof			0.769			0.371
	5th	2,047		0.025	0.371	
5th			0.744			1.407
	4th	4,461		0.055	1.778	
4th			0.689			2.442
	3rd	6,544		0.089	4.220	
3rd			0.600			4.120
	2nd	8,589		0.134	8.340	
2nd			0.466			14.45
	1st	10,565		0.466	22.79	

Reanalyze the structure for the following horizontal loads:

Roof: $1.275 \times 25.6 + 0.371 = 33.0$ kips
5th floor: $1.275 \times 37.9 + 1.407 = 49.7$ kips
4th floor $1.275 \times 37.9 + 2.442 = 50.8$ kips
3rd floor $1.275 \times 37.9 + 4.120 = 52.4$ kips
2nd floor $1.275 \times 48.7 + 14.45 = 76.5$ kips

(d) Check convergence for the third cycle.

	Δ–2nd Cycle (in.)	Δ–3rd Cycle (in.)	3rd/2nd
Roof	0.769	0.773	1.005
2nd floor	0.466	0.469	1.006

This has converged adequately.

(e)   Compute M_1 and M_2. The column end moments from the third cycle of iteration are:
Exterior columns–1st story: From the third cycle of the second-order analysis:

Top: $\delta_s M_s = 73.9$ ft-kips
Bottom: $= 79.5$ ft-kips

From Eq. 12–49 (ACI Eqs. 10–16 and 10–17), with M_{ns} from step 4:

$$M_{end} = M_{ns} + \delta_s M_s$$

Because the wind can blow from either the north or the south, we shall take this as

$$M_{end} = M_{ns} \pm \delta_s M_s$$

Top: $M_{top} = -103.1 - 73.9 = -177$ ft-kips. This is the smaller end moment; therefore, this is M_1.
Bottom: $M_{bottom} = -108.5 - 79.5 = -188$ ft-kips. This is M_2.

Interior columns—1st story:

Top: $\delta_s M_s = 91.1$ ft-kips
M_{top} $= 12.6 + 91.1 = 103.7$ ft-kips. This is the smaller end moment; therefore, this is M_1.
Bottom: $\delta_s M_s = 94.1$ ft-kips
     $M_{bottom} = 16.5 + 94.1 = 110.6$ ft-kips. This is M_2.

6.   Check if the maximum moment is between the ends of the column. ACI Sec. 10.13.5 requires a check of whether the maximum moment occurs away from the ends of the column if

$$\frac{\ell_u}{r} > \frac{35}{\sqrt{\dfrac{P_u}{f'_c A_g}}} \qquad (12\text{–}50)$$

$$(\text{ACI Eq. }10\text{–}20)$$

where $\ell_u = 18 \times 12$ in. $- 30$ in. $= 186$ in. and $r = 0.30 \times 18$ in. $= 5.4$ in.

$$\frac{\ell_u}{r} = \frac{186 \text{ in.}}{5.4 \text{ in.}} = 34.4$$

$$f'_c = 4 \text{ ksi}$$

$$A_g = 324 \text{ in.}^2$$

For the exterior column, $P_u = 273.3$ kips and

$$\frac{35}{\sqrt{\dfrac{P_u}{f'_c A_g}}} = \frac{35}{\sqrt{\dfrac{273.3 \text{ kips}}{4 \text{ ksi} \times 324 \text{ in.}^2}}}$$

$$= 76.2$$

Since $\ell_u/r = 34.4$ is less than 76.2, the maximum moment in the exterior column is at one end of the column and is $M_2 = 188$ ft-kips.

For the interior column, $P_u = 544.4$ kips and

$$\frac{35}{\sqrt{\dfrac{544.4}{4 \times 324}}} = 54.0$$

Since $\ell_u/r = 34.4$ is less than 54.0, the maximum moment in the interior column is $M_2 = 110.6$ ft-kips.

If ℓ_u/r was greater than either of these limits, it would be necessary to compute the maximum moment along the length of that column. According to ACI Sec. 10.3.5, this is done using Eq. 12–23 (ACI Eq. 10–10) with M_1 and M_2 as computed above, β_d as defined for the load combination under consideration, and k as defined in ACI Sec. 10.12.1. Although both columns passed the test, we shall go through this calculation for the interior column to illustrate the procedure. These calculations are italicized because they are not necessary when ℓ_u/r is less than the limit from Eq. 12–50.

The maximum moment in the column is given by

$$M_c = \delta_{ns} M_2 \qquad\qquad\qquad (12\text{–}22)$$
$$(\text{ACI Eq. 10–9})$$

where

$$\delta_{ns} = \frac{C_m}{1 - P_u/0.75 P_c} \geq 1.0 \qquad\qquad (12\text{–}23)$$
$$(\text{ACI Eq. 10–10})$$

$$C_m = 0.6 + 0.4\left(\frac{M_1}{M_2}\right) \geq 0.4 \qquad\qquad (12\text{–}14)$$

$$= 0.6 + 0.4\left(\frac{-91.1}{103.7}\right) \geq 0.4$$
$$(\text{ACI Eq. 10–14})$$

$$= 0.249 \geq 0.4$$

$$= 0.4$$

$$P_c = \frac{\pi^2 EI}{(k\ell_u)^2} \qquad\qquad\qquad (12\text{–}24)$$
$$(\text{ACI Eq. 10–11})$$

$$EI = \frac{0.40 E_c I_g}{1 + \beta_d} \qquad\qquad\qquad (12\text{–}16)$$
$$(\text{ACI Eq. 10–13})$$

This is a sway frame and definition (b) for β_d in ACI Sec. 10.0 applies:

$$\beta_d = \frac{\text{maximum factored sustained shear within a story}}{\text{total factored shear in the story}} \qquad (12\text{–}39\text{b})$$

where the shear is due to rapidly applied wind loads and the sustained shear is thus zero, giving

$$\beta_d = 0.0$$

and

$$EI = \frac{0.40 \times (57{,}000\sqrt{4000}) \times 18^4/12}{1 + 0.0}$$

$$= 12.61 \times 10^9 \ lb\text{-}in^2$$

ACI Sec. 10.13.5 says that k shall be from ACI Sec. 10.12.1 which gives $k \leq 1.0$. We shall assume $k = 1.0$ as a first trial. If $\delta_{ns} > 1.0$ we shall compute k and repeat the calculation.

$$P_c = \frac{\pi^2 \times 12.61 \times 10^6 \ lb\text{-}in.^2}{(1.0 \times 186 \ in.)^2}$$

$$= 3597 \; kips$$

$$\delta_{ns} = \frac{0.40}{1 - 544.4/0.75 \times 3597} \geq 1.0$$

$$= 0.501 \geq 1.0$$

$$= 1.0$$

Since $\delta_{ns} = 1.0$, the maximum moment in the column is $M_2 = 103.7$ ft-kips.

This bears out the results of the test carried out in step 6, which showed that the maximum moment would be at one end of the column. As stated earlier, all calculations in italicized print need not be done if the columns satisfy Eq. 12–50 (ACI Eq. 10–20).

Summary for Load Case 1

Exterior columns: $P_u = 273$ kips, $M_c = 188$ ft-kips
Interior columns: $P_u = 544$ kips, $M_c = 103.7$ ft-kips

The column cross section will be chosen when all load combinations have been completed.

LOAD CASE 2: $U = 0.9D \pm 1.3W$

Steps 4, 5, and 6 will be repeated.

4. Compute the factored axial loads P_u and the M_{ns} moments from a first-order frame analysis. For the main floor columns the values are:

	Exterior Columns	Interior Columns
Factored axial force, P_u (kips)		
0.9 Dead load	158.4	342.5
Factored M_{ns} moments (ft-kips)		
At top	–47.0	5.57
At bottom	–49.4	7.37

5. Compute $\delta_s M_s$ and M_c using a second-order analysis.
   (a) Compute ΣP_u, $\Sigma P_u \Delta_0 / \ell_c$, and sway forces.

Floor	Story	ΣP_u (kips)	Floor Δ (in.)	Story Δ (in.)	$\dfrac{\Sigma P_u \Delta_0}{\ell_c}$	Sway Force
Roof			0.735			0.251
	5th	1384		0.025	0.251	
5th			0.710			0.780
	4th	2588		0.055	1.031	
4th			0.655			1.446
	3rd	3884		0.088	2.477	
3rd			0.567			2.363
	2nd	5178		0.129	4.840	
2nd			0.438			8.280
	1st	6472		0.438	13.12	
Ground			0.00			

The total factored wind forces plus sway forces at each floor are:

Roof: $1.3 \times 25.6 + 0.251 = 33.5$ kips
5th floor $1.3 \times 37.9 + 0.780 = 50.1$ kips
4th floor $1.3 \times 37.9 + 1.446 = 50.7$ kips
3rd floor $1.3 \times 37.9 + 2.363 = 51.6$ kips
2nd floor $1.3 \times 48.7 + 8.280 = 71.6$ kips.

These loads are resisted by the seven frames in the north–south direction. The frame is now reanalyzed for these forces.

   (b)   Check convergence for the second cycle. The deflections from the two cycles of iteration are:

	Δ-First Cycle (in.)	Δ-Second Cycle (in.)	2nd/1st
Roof	0.735	0.765	1.04
2nd floor	0.438	0.461	1.05

This has not converged.

   (c)   Carry out a third cycle of iteration. The deflections converged after the third cycle.

   (d)   Compute M_1 and M_2. The column end moments from the third cycle of iteration are:

Exterior columns—1st story:

Top: $\delta_s M_s = 72.6$ ft-kips
Bottom: $= 78.2$ ft-kips

$$M_{end} = M_{ns} \pm \delta_s M_s$$

Top: $M_{top} = -47.0 - 72.6 = -120$ ft-kips. This is M_1.
Bottom: $M_{bottom} = -49.4 - 78.2 = -128$ ft kips. This is M_2.

Interior columns—1st story:

Top: $\delta_s M_s = 89.7$ ft-kips
    $M_{top} = 5.57 + 89.7 = 95.3$ ft-kips. This is M_1.

Bottom: $\delta_s M_s = 92.6$ ft-kips
    $M_{bottom} = 7.37 + 92.6 = 100$ ft-kips. This is M_2.

   6.   Check if the maximum moment is between the ends of the column. Since this check was satisfied in load case 1, which has a larger P_u, the maximum moment will be at one end of the column.

Summary for Load Case 2

Exterior columns: $P_u = 158.4$ kips, $M_c = 128$ ft-kips
Interior columns: $P_u = 342.5$ kips, $M_c = 100$ ft-kips

LOAD CASE 3: $U = 1.4D + 1.7L$

Although this load case does not include lateral loads, the columns must be designed for P_u and M_2 from this loading. In addition, this load case will be used to check whether the frame is subject to sidesway buckling under gravity loads since this case involves the largest gravity loads considered in design. ACI Sec. 10.13.6 presents three ways to check the tendency for sidesway buckling, depending on the method used to compute $\delta_s M_s$. When $\delta_s M_s$ has been computed using a second-order analysis, ACI Sec. 10.13.6(a) requires that

$$\frac{second - order\ lateral\ deflections}{first - order\ lateral\ deflections} \leq 2.5$$

when computed for a gravity load of $1.4D + 1.7L$ plus any arbitrarily chosen lateral load. In this calculation we shall take the lateral load as $1.275W$ since we already have deflections calculated for this loading. ACI Sec. 10.13.6 further requires the member stiffnesses be taken as $EI/(1 + \beta_d)$, where EI is from ACI Sec. 10.11.1 and where, for the gravity load sidesway buckling check,

$$\beta_d = \frac{\text{maximum factored sustained axial load}}{\text{total factored axial load}} \qquad (12\text{--}39a)$$

[See ACI Sec. 10.13.6 or definition (c) in ACI Sec. 10.0.] Because the live-load reduction factor changes from story to story, β_d also changes from story to story. The most slender columns are in the first story and we will use the value of β_d from this story throughout.

Total factored sustained axial load in all the columns in the first story = 10,400 kips

Total factored axial load in all the columns in the first story = 14,420 kips

$$\beta_d = \frac{10,400}{14,420} = 0.721$$

Thus all member stiffnesses will be divided by $(1 + \beta d) = 1 + 0.721$. This will give lateral deflections which are 1.721 times those already computed for the load of 1.275W. As a result, the second-order effects will be larger.

4. Compute the factored axial loads P_u and the M_{ns} moments from a first-order frame analysis. For the main floor columns the values are:

	Exterior Columns	Interior Columns
Factored axial force, P_u (kips)		
1.4D	246.4	533
1.7 Reduced live load	118.0	193
Factored M_{ns} moments (ft-kips)		
At top	−137.5	16.81
At bottom	−144.6	21.99

5. Compute $\delta_s M_s$ and M_c using a second-order analysis.

(a) Compute ΣP_u, $\Sigma P_u \Delta_0 / \ell_c$, and sway forces.

Floor	Story	ΣP_u (kips)	Floor Δ (in.)	Story Δ (in.)	$\dfrac{\Sigma P_u \Delta_0}{\ell_c}$	Sway Force
Roof			1.241			0.81
	5th	2,729		0.041	0.811	
5th			1.200			3.28
	4th	5948		0.095	4.095	
4th			1.105			5.26
	3rd	8725		0.148	9.357	
3rd			0.957			8.78
	2nd	11,426		0.219	18.133	
2nd			0.738			31.14
	1st	14,420		0.738	49.268	
Ground			0.00			

The total factored wind forces plus sway forces at each floor are:

Roof: $1.275 \times 25.6 + 0.81 = 33.5$ kips
5th floor: $1.275 \times 37.9 + 3.28 = 51.6$ kips
4th floor: $1.275 \times 37.9 + 5.26 = 53.6$ kips
3rd floor: $1.275 \times 37.9 + 8.78 = 57.1$ kips
2nd floor: $1.275 \times 48.7 + 31.14 = 93.2$ kips.

These loads are resisted by the seven frames in the north–south direction. The frame is analyzed for these loads using the reduced EI values.

(b) Check convergence for the second cycle. The deflections from the two cycles of iteration are:

	Δ-First Cycle (in.)	Δ-Second Cycle (in.)	2nd/1st
Roof	1.241	1.435	1.16
2nd floor	0.738	0.887	1.20

This clearly has not converged.

(c) Carry out a third iteration. The deflections from the second and third iterations are:

	Δ-Second Cycle (in.)	Δ-Third Cycle (in.)	3rd/2nd
Roof	1.435	1.469	1.02
2nd floor	0.887	0.917	1.03

We shall assume that this is adequate convergence.

(d) Check if sidesway buckling will occur under gravity loads. The ratio of the first-order lateral deflections to the second-order lateral deflections is 0.917/0.738 = 1.24 at the second floor. Since this is less than 2.5, the frame is not in danger of sidesway buckling under gravity loads.

It is not necessary to consider the moments from the second-order analysis because this was not a real loading case. However, the columns must be designed for the axial loads and first-order M_{ns} moments due to $U = 1.4D + 1.7L$ given in step 4.

6. Check if the maximum moment is between the ends of the column. From ACI Sec. 10.13.5, the maximum moments will exceed those at the ends of the column if

$$\frac{\ell_u}{r} > \frac{35}{\sqrt{\dfrac{P_u}{f'_c A_g}}} \qquad\qquad \begin{array}{c}(12\text{--}50)\\(\text{ACI Eq. 10--20})\end{array}$$

$$\text{First story: } \frac{\ell_u}{r} = \frac{216 \text{ in. } -30 \text{ in.}}{0.3 \times 18 \text{ in.}} = 34.4$$

$$\text{Interior column: } \frac{35}{\sqrt{\dfrac{P_u}{f'_c A_g}}} = \frac{35}{\sqrt{\dfrac{726}{4 \times 324}}} = 46.8$$

Since 34.4 is less than 46.8, the maximum moment in the column is at one end.

7. Check minimum moment. The code does not require columns in sway frames to be designed for a minimum moment. However, we shall conservatively design the column for larger of the computed moments and $M_{2,min}$ given by Eq. 12–21 (ACI Eq. 10–15).

Exterior columns:

$$M_{2,min} = P_u(0.6 + 0.03h) = 364 \text{ kips}(0.6 + 0.03 \times 18) \text{ in.}$$
$$= 19.8 \text{ ft-kips—does not govern}$$

Interior columns:

$$M_{2,\,min} = 726 \text{ kips } (0.6 + 0.03 \times 18) \text{ in.}$$
$$= 69.0 \text{ ft-kips—governs}$$

Summary for Load Case 3

   Exterior columns: $P_u = 364$ kips, $M_c = 144.6$ ft-kips
   Interior columns: $P_u = 726$ kips, $M_c = 69.0$ ft-kips

12–7 Design of Columns in Sway Frames   **531**

SUMMARY OF P_U AND M_C FOR LOAD CASES 1 TO 3

The columns must be designed for the following combinations of axial load and moment:

Exterior columns:

Load case 1: P_u = 273 kips, M_c = 188 ft-kips

Load case 2: P_u = 158.4 kips, M_c = 128 ft-kips

Load case 3: P_u = 364 kips, M_c = 144.6 ft-kips

Interior columns:

Load case 1: P_u = 544 kips, M_c = 103.7 ft-kips

Load case 2: P_u = 342 kips, M_c = 100 ft-kips

Load case 3: P_u = 726 kips, M_c = 69.0 ft-kips

8. Select the reinforcement. Figure 12–34 is interaction diagrams for an 18-in.-square column with 4 No. 8 and 4 No. 9 bars. These columns satisfy the minimum reinforcement requirements of ACI Sec. 10.9.1. It can be seen that an 18-in.-square column with 4 No. 8 bars satisfies all three load cases for the exterior columns. 4 No. 9 bars are needed in an interior column.

The sections selected must also satisfy ACI Sec. 10.3.5.2, which gives the horizontal top part of the interaction diagrams. They are therefore adequate.

Exterior columns: Use 18 in. × 18 in. columns with 4 No. 8 bars.

Interior columns: Use 18 in. × 18 in. columns with 4 No. 9 bars.

9.   Check the beam capacity. ACI Sec. 10.11.6 requires that beams in sway frames have adequate flexural capacity to resist the total magnified moments at each joint. In a complete design such a check would have to be made for load cases 2 and 3.   ■

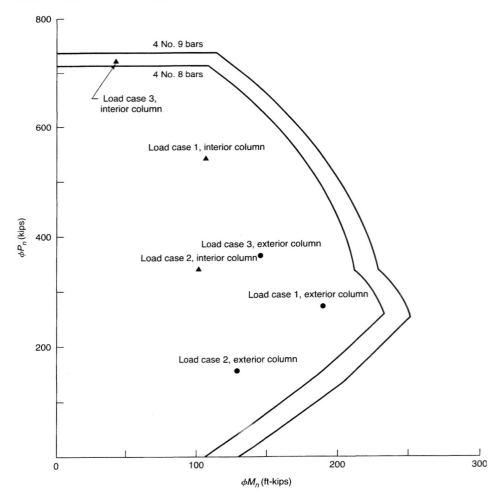

Fig. 12–34
Interaction diagrams—
Example 12–3.

EXAMPLE 12–4   Sway Frame: Using Direct P–Δ Analysis—ACI Sec. 10.13.4.2

Consider the frame from Example 12–3. For brevity we shall only consider load cases 1 and 3. The structural analysis has been carried out using the E and I values given in ACI Sec. 10.11.1. **Steps 1, 2, and 3** are as in Example 12–3.

LOAD CASE 1: $U = 0.75(1.4D + 1.7L + 1.7W) = 1.05D + 1.275L + 1.275W$

4. **Compute the factored axial loads P_u, the M_{ns} moments, the M_s moments, and deflections Δ_0 from a first-order frame analysis.** For the main floor columns the values are:

	Exterior Columns	Interior Columns
Factored axial force, P_u (kips)		
1.05 Dead load	184.8	399.6
1.275 Reduced live load	88.5	144.8
Factored M_{ns} moments (ft-kips)		
At top	−103.1	12.61
At bottom	−108.5	16.49
Factored M_s moments (ft-kips)		
At top	67.3	83.2
At bottom	72.6	86.0

The first-order lateral deflections are: roof 0.721 in., 5th floor 0.697 in., 4th floor 0.642 in., 3rd floor 0.556 in., and 2nd floor 0.429 in.

5. **Compute $\delta_s M_s$ and M_c according to ACI Secs. 10.13.4.2 and 10.13.3.**

(a) **Compute the story magnifiers.**
From ACI Sec. 10.13.4.2,

$$\delta_s M_s = \frac{M_s}{1 - Q} \geq M_s \qquad (12\text{–}46)$$
$$(\text{ACI Sec. } 10\text{–}18)$$

where

$$Q = \frac{\Sigma P_u \Delta_0}{V_u \ell_c} \qquad (12\text{–}38)$$
$$(\text{ACI Eq. } 10\text{–}7)$$

Story	ΣP_u (kips)	V_u (kips)	Story Δ_0 (in.)	$\dfrac{\Sigma P_u \Delta_0}{V_u \ell_c}$	δ_s
5th	2,047	32.6	0.024	0.0104	1.011
4th	4,461	80.9	0.055	0.022	1.023
3rd	6,544	129.3	0.086	0.032	1.033
2nd	8,589	177.6	0.127	0.045	1.047
1st	10,565	240	0.429	0.087	1.096

(b) **Compute $\delta_s M_s$, M_1, and M_2.**
Exterior columns: For the exterior columns in the first story, the column end moments are:

$$M_{s,\text{top}} = \pm\, 67.3 \text{ ft-kips, and } \delta_s M_{s,\text{top}} = 1.096 \times \pm\, 67.3 \text{ ft-kips} = \pm\, 73.7 \text{ ft-kips}$$

$M_{s,\text{bottom}} = \pm 72.6$ ft-kips, $\delta_s M_{s,\text{bottom}} = 1.096 \times \pm 72.6$ ft-kips $= \pm 79.6$ ft-kips

Total moment at the ends: $\quad M_{\text{top}} = -103.1 - 73.7 = -176.8$ ft-kips—this is M_1.

$\qquad\qquad\qquad\qquad M_{\text{bottom}} = -108.5 - 79.6 = 188.1$ ft-kips—this is M_2.

Interior columns:

$M_{s,\text{top}} = \pm 83.2$ ft-kips, $\delta_s M_{s,\text{top}} = 1.096 \times \pm 83.2$ ft-kips $= \pm 91.2$ ft-kips

$M_{s,\text{bottom}} = \pm 86.0$ ft-kips, $\delta_s M_{s,\text{bottom}} = 1.096 \times \pm 86.0$ ft-kips $= \pm 94.3$ ft-kips

Total moment at the ends: $\quad M_{\text{top}} = 12.61 + 91.2 = 103.8$ ft-kips—this is M_1.

$\qquad\qquad\qquad\qquad M_{\text{bottom}} = 16.49 + 94.3 = 110.8$ ft-kips—this is M_2.

Step 6 is the same as in Example 12–3

Summary for Load Case 1

Exterior columns: $\quad P_u = 273$ kips, $M_c = 188.1$ ft-kips

Interior columns: $\quad P_u = 544$ kips, $M_c = 110.8$ ft-kips

LOAD CASE 3: $U = 1.4D + 1.7L$

Although this load case does not include lateral loads, it must be considered because the columns must be designed for the axial loads and moments from this case. In addition, it is necessary to check whether the frame is subject to sidesway buckling under high gravity loads.

4. Compute the factored axial loads P_u and the M_{ns} moments from a first-order frame analysis. For the main floor columns the values are:

	Exterior Columns	Interior Columns
Factored axial force, P_u (kips)		
1.4D	246	533
1.7 Reduced L	118	193
Factored M_{ns} moments (ft-kips)		
At top	−137.5	16.81
At bottom	−144.6	21.99

5. Compute $\delta_s M_s$ and M_c according to ACI Secs. 10.13.4.2 and 10.13.3. Since there are no lateral loads, $M_s = 0$ in all columns. Therefore, $\delta_s M_s = 0$.

Exterior columns:

$\qquad M_{\text{top}} = M_{ns} + \delta_s M_s = -137.5$ ft-kips $+ 0 = -137.5$ ft-kips. This is M_1.

$\qquad M_{\text{bottom}} = -144.6$ ft-kips. This is M_2.

Interior columns:

$\qquad M_{\text{top}} = 16.81$ ft-kips. This is M_1.

$\qquad M_{\text{bottom}} = 21.99$ ft-kips. This is M_2.

6. Check if maximum moment is at the end of the column. This is the same as step 6 in load case 3 in Example 12–3. The maximum moment is at one end of the column.

7. Check if sidesway buckling can occur under gravity loads. When $\delta_s M_s$ is computed using ACI Sec. 10.13.4.2, this check is done by checking whether Q computed using ΣP_u for $1.4D + 1.7L$ exceeds 0.60. In the calculations EI is taken as the value corresponding to the gravity load condition [i.e., $EI/(1 + \beta_d)$ with β_d equal to the ratio of the sustained axial load to the total axial load]. Since the most slender columns are in the first story, we shall compute β_d for this story. Thus:

sustained axial load in all columns in the first story $= 10{,}400$ kips

total axial load in all the columns in the first story $= 14{,}420$ kips

$$\beta_d = \frac{10{,}400}{14{,}420} = 0.721$$

Thus, all member stiffnesses will be divided by $(1 + \beta_d) = 1 + 0.721$.

To compute Q we need V_u and Δ_0 for the story, where Δ_0 is defined as the lateral story deflection due to V_u. In load case 3, $V_u = 0$ and hence $\Delta_0 = 0$. To compute Q we need some shear and the corresponding deflections. Since we have already computed the story deflections for $1.275W$, we will use V_u and Δ_0 for this case. In that load case, the deflections were computed using EI computed with $\beta_d = 0$. For the check of sidesway buckling, we must use $\beta_d = 0.721$. The reduced stiffness will give lateral deflections which are $1 + \beta_d = 1.721$ times those already computed for the load of $1.275W$.

V_u due to $1.275W = 188$ kips

Δ_0 in first story from step 4 of load case 1 = 0.429 in.

Δ_0 based on reduced stiffness $= 0.429 \times 1.721 = 0.738$ in.

$$Q = \frac{\Sigma P_u \Delta_0}{V_u \ell_c} = \frac{14{,}420 \text{ kips} \times 0.738 \text{ in.}}{188 \text{ kips} \times 216 \text{ in.}}$$

$$= 0.262$$

Since $Q = 0.262$ is less than 0.60, sidesway buckling will not be a problem.

8. Check minimum moment. This is the same as step 7, load case 3, in Example 12–3.

Summary for Load Case 3

Exterior columns: $P_u = 364$ kips, $M_c = 144.6$ ft-kips

Interior columns: $P_u = 726$ kips, $M_c = 69.0$ ft-kips

This concludes Example 12–4. ∎

EXAMPLE 12–5   Sway Frame: Using Sway Moment Magnifier—ACI Sec. 10.13.4.3

Repeat Example 12–3 using ACI Sec. 10.13.4.3 to compute the magnified moments. For brevity we will only consider load cases 1 and 3.

Steps 1, 2, and 3 are as in Example 12–3. The frame is a sway frame. The column size will be taken as 18 in. square for all columns. The same three load cases will be considered.

LOAD CASE 1: $U = 0.75(1.4D + 1.7L \pm 1.7W) = 1.05D + 1.275L + 1.275W$

4. Compute the factored axial loads P_u, the M_{ns} moments, and the M_s moments from a first-order frame analysis. For the main floor columns the values are:

	Exterior Columns	Interior Columns
Factored axial force, P_u (kips)		
1.05 Dead load	184.8	399.6
1.275 Reduced live load	88.5	144.8
Factored M_{ns} moments (ft-kips)		
At top	−103.1	12.61
At bottom	−108.5	16.49
Factored M_s moments (ft-kips)		
At top	67.3	83.2
At bottom	72.6	86.0

5. Compute the effective length factors. Effective length factors can be computed using the nomographs in Fig. 12–26 or Eqs. 12–28 to 12–32. We shall use the nomographs with ψ based on the EI values given in ACI Sec. 10.11.1 with $\beta_d = 0$. The span lengths, ℓ_c and ℓ_b, are center to center of joints.

(a) **Compute $E_c I_c / \ell_c$ for the columns.**

$$E_c = 57{,}000 \sqrt{f_c'} = 3.60 \times 10^6 \text{ psi} \qquad \text{(ACI Sec. 8.5.1)}$$

$$I_c = 0.70 I_g = 0.70 \left(\frac{18^4}{12} \right) = 6120 \text{ in.}^4$$

Columns over the second floor and below the first floor:

$$\ell_c = 11 \text{ ft } 6 \text{ in.} = 138 \text{ in.}$$

$$\frac{E_c I_c}{\ell_c} = 160 \times 10^6 \text{ in.-lb}$$

Columns between the first and second floors:

$$\ell_c = 18 \text{ ft} = 216 \text{ in.}$$

$$\frac{E_c I_c}{\ell_c} = 102 \times 10^6 \text{ in.-lb}$$

(b) **Compute $E_b I_b / \ell_b$ for the beams.** Assume that the flanges of the beams are as given in ACI Secs. 8.10.2 and 8.10.3, $0.25\ell = 7.5$ ft. The gross moment of inertia of the T beam is $I_g = 80{,}100$ in.^{4} (Normally for an interior T beam, it is sufficiently accurate to take $I_g = 2(b_w h^3/12) = 81{,}000$ in.^{4})

$$I_b = 0.35 I_g = 0.35 \times 80{,}100 = 28{,}000 \text{ in.}^4$$

Beam between lines A and B, line 3:

$$\ell_b = 31 \text{ ft minus } 9 \text{ in.} = 363 \text{ in.}$$

$$\frac{E_b I_b}{\ell_b} = 278 \times 10^6 \text{ in.-lb}$$

Beam between lines B and C, line 3:

$$\ell_b = 360 \text{ in.}$$

$$\frac{E_b I_b}{\ell_b} = 280 \times 10^6 \text{ in. lb}$$

(c) **Compute ψ and k.** *Exterior columns, north and south walls:*

$$\psi = \frac{\Sigma(E_c I_c / \ell_c)}{\Sigma(E_b I_b / \ell_b)}$$

Two columns and one beam meet at the joint at the top of the exterior column in the first story.

$$\psi_{top} = \frac{160 \times 10^6 + 102 \times 10^6}{278 \times 10^6} = 0.942$$

The bottom of the exterior columns at A–2 to A–6 and D–2 to D–6 are supported on a pilaster built into the basement wall. When the frame is being deflected in the north–south direction the stiffness of the pilaster and wall for bending about an axis parallel to the wall will be taken the same as for a column of the same size. As a result,

$$\psi_{bottom} = 0.942$$

From Fig. 12–26,

$$k = 1.28 \text{ for sway case}$$

$$k = 0.77 \text{ for braced case}$$

Interior column:

$$\Psi_{\text{top}} = \Psi_{\text{bottom}} = \frac{160 \times 10^6 + 102 \times 10^6}{278 \times 10^6 + 280 \times 10^6}$$

$$= 0.470$$

$k = 1.15$ for the sway case. Use $k = 1.20$ as a minimum practical value for a sway column. $k = 0.68$ for the braced case.

Corner columns and side wall columns: These columns ($A–1$, $B–1$, $C–1$, $D–1$, and $A–7$ through $D–7$) are nearly fixed at their bottom ends by the basement wall, which is perpendicular to the axis of bending being considered. Use $k = 1.2$ for the sway case, $k = 0.65$ for the braced case.

6. Compute the magnified moments for load case 1.

(a) Compute EI. Since the reinforcement is not known at this stage, use either Eq. 12–16 or 12–19 to calculate *EI.* From Eq. 12–16,

$$EI = \frac{E_c I_g / 2.5}{1 + \beta_d}$$

$$E_c = 3.60 \times 10^6 \text{ psi} \quad I_g = 8750 \text{ in.}^4$$

$$\frac{E_c I_g}{2.5} = 12.61 \times 10^9 \text{ in.}^2/\text{lb}$$

Since this is a sway frame, definition (b) for β_d in ACI Sec. 10.0 applies. Since there is no sustained load shear in the story, $\beta_d = 0$. Therefore,

$$EI = \frac{12.61 \times 10^9}{1 + 0} = 12.61 \times 10^9 \text{ in.}^2\text{-lb}$$

From ACI Sec. 10.13.4.3,

$$\delta_s M_s = \frac{M_s}{1 - \Sigma P_u / (0.75 \Sigma P_c)} \geq M_s \qquad (12\text{–}48)$$
$$\text{(ACI Eq. 10–19)}$$

where

ΣP_u = the sum of the factored axial loads in all the columns in the story for load case 1

$$= 10{,}565 \text{ kips}$$

$$P_c = \frac{\pi^2 EI}{(k\ell_u)^2} \qquad (12\text{–}24)$$
$$\text{(ACI Eq. 10–11)}$$

For the columns in the north and south walls

$$P_c = \frac{\pi^2 \times 12.61 \times 10^9 \text{ in.}^2\text{-lb}}{(1.28 \times 186 \text{ in.})^2} = 2196 \text{ kips}$$

For interior columns:

$$P_c = \frac{\pi^2 \times 12.61 \times 10^9}{(1.20 \times 186)^2} = 2498 \text{ kips}$$

P_c for corner and sidewall columns is also 2498 kips.

$$\Sigma P_c = 10 \times 2196 + 10 \times 2498 + 8 \times 2498 = 66{,}924 \text{ kips}$$

and

$$\delta_s M_s = \frac{M_s}{1 - 10{,}565 \text{ kips} / (0.75 \times 66{,}924)} \geq M_s$$

$$= 1.27 M_s$$

12–7 Design of Columns in Sway Frames   537

Exterior columns in first story:

Top of column: $M_{ns} = -103.1$ ft-kips

$M_s = \pm 67.3$ ft-kips, $\delta_s M_s = 1.27 \times \pm 67.3 = \pm 85.2$ ft-kips.

From ACI Sec. 10.13.3: $M_{top} = -103.1 - 85.2 = 188.3$ ft-kips. This is M_1.

Bottom of column: $M_{ns} = -108.5$ ft-kips

$M_s = \pm 72.6$ ft-kips, $\delta_s M_s = 1.27 \times \pm 72.6 = \pm 92.2$ ft-kips

$M_{bottom} = -108.5 - 92.2 = -200.7$ ft-kips. This is M_2.

Interior columns in first story:

Top of column: $M_{ns} = 12.61$ ft-kips

$M_s = \pm 83.2$ ft-kips, $\delta_s M_s = 1.27 \times \pm 83.2 = \pm 105.7$ ft-kips

$M_{top} = 12.61 + 105.7 = 118.3$ ft-kips. This is M_1.

Bottom of column: $M_{ns} = 16.49$ ft-kips

$M_s = \pm 86.0$ ft-kips, $\delta_s M_s = 1.27 \times \pm 86.0 = \pm 109.2$ ft-kips

$M_{bottom} = 16.49 + 109.2 = 125.7$ ft-kips. This is M_2.

7. Check whether the maximum column moment occurs between the ends of the column. If

$$\frac{\ell_u}{r} > \frac{35}{\sqrt{\dfrac{P_u}{f'_c A_g}}} \qquad (12\text{–}50)$$

$$\text{(ACI Eq. 10–20)}$$

it is necessary to check whether the moment at a point between the ends of the column exceeds that at the end of the column. We will check this first for an interior column since they have the largest axial loads, P_u.

$$\frac{\ell_u}{r} = \frac{186 \text{ in.}}{0.30 \times 18 \text{ in.}} = 34.4$$

$$\frac{35}{\sqrt{\dfrac{P_u}{f'_c A_g}}} = \frac{35}{\sqrt{\dfrac{544 \text{ kips}}{4 \text{ ksi} \times 324 \text{ in.}^2}}} = 54.0$$

Since $34.4 < 54.0$, the maximum moment is at the end of the column. The same is true for the exterior column.

Summary for Load Case 1

Exterior column: $P_u = 273$ kips, $M_c = 200.7$ ft-kips

Interior column: $P_u = 544$ kips, $M_c = 125.7$ ft-kips

LOAD CASE 3: $U = 1.4D + 1.7L$

4. Compute the factored axial loads P_u and the M_{ns} moments from a first-order frame analysis. For the main floor columns the values are:

	Exterior Columns	Interior Columns
Factored axial force, P_u (kips)		
1.4D	246	533
1.7 Reduced L	118	193
Factored M_{ns} moments (ft-kips)		
At top	−137.5	16.81
At bottom	−144.6	21.99

5. **Compute the effective length factors.** These will be the same as in load case 1.

6. **Compute $\delta_s M_s$ and M_c according to ACI Secs. 10.13.4.2 and 10.13.3.** Since there are no lateral loads, $M_s = 0$ in all columns. Therefore, $\delta_s M_s = 0$ also.

Exterior columns:

$M_{top} = M_{ns} + \delta_s M_s = -137.5 \text{ ft-kips} + 0 = -137.5 \text{ ft-kips}$. This is M_1.
$M_{bottom} = -144.6 \text{ ft-kips}$. This is M_2.

Interior columns:

$M_{top} = 16.81 \text{ ft-kips}$. This is M_1.
$M_{bottom} = 21.99 \text{ ft-kips}$. This is M_2.

7. **Check if the maximum moment occurs between the ends of the column.** This check is the same as step 7 of load case 3 in Example 12–4.

8. **Check if frame can undergo sidesway buckling under gravity loads.** When $\delta_s M_s$ is calculated using ACI Sec. 10.13.4.3, ACI Sec. 10.13.6(c) says that sidesway buckling will not be a problem if δ_s computed using ΣP_u for $1.4D + 1.7L$, and ΣP_c based on $0.40 EI/(1 + \beta_d)$ with β_d computed as the ratio of the total sustained axial loads to total axial loads, is positive and does not exceed 2.5.

From Eq. 12–48 (ACI Eq. 10–19),

$$\delta_s = \frac{1}{1 - \Sigma P_u / (0.75 \Sigma P_c)} \geq M_s$$

where, for the first story, total sustained loads = 10,400 kips, total axial loads $\Sigma P_u = 14{,}420$ kips and

$$\beta_d = \frac{10{,}400}{14{,}420} = 0.721$$

$$P_c = \frac{\pi^2 EI}{(k\ell_u)^2}$$

$$EI = \frac{0.40 E_c I_g}{1 + \beta_d} = \frac{12.61 \times 10^9}{1 + 0.721}$$

$$= 7.327 \times 10^9 \text{ in.}^2\text{-lb}$$

For the columns in the north and south walls

$$P_c = \frac{\pi^2 \times 7.327 \times 10^9}{(1.28 \times 186)^2}$$

$$= 1276 \text{ kips}$$

For the interior columns and the columns in the end walls

$$P_c = \frac{\pi^2 \times 7.327 \times 10^9}{(1.20 \times 186)^2}$$

$$= 1452 \text{ kips}$$

$$\Sigma P_c = 10 \times 1276 + 10 \times 1452 + 8 \times 1452 = 38{,}896 \text{ kips}$$

$$\delta_s = \frac{1}{1 - 14{,}420 / 0.75 \times 38{,}896} = 1.98$$

Since $\delta_s = 1.98$ is less than 2.5, sidesway buckling will not occur.

9. **Check minimum moment.** This is the same as step 7, load case 3, Example 12–3.

Summary for Load Case 3:

Exterior columns: $P_u = 364 \text{ kips}$, $M_c = 144.6 \text{ ft-kips}$
Interior columns: $P_u = 726 \text{ kips}$, $M_c = 69.0 \text{ ft-kips}$

This concludes Example 12–5. ∎

12–7 Design of Columns in Sway Frames

539

Comparison of Magnified Moments by the Three Methods for Load Case 1

For the exterior columns:

ACI Sec. 10.13.4.1: second-order analysis, $M_c = 188$ ft-kips

ACI Sec. 10.13.4.2: Q factor analysis, $M_c = 188$ ft-kips

ACI Sec. 10.13.4.3: sway moment magnifier, $M_c = 188$ ft-kips

For the interior columns:

ACI Sec. 10.13.4.1: second-order analysis, $M_c = 103.7$ ft-kips

ACI Sec. 10.13.4.2: Q factor analysis, $M_c = 111$ ft-kips

ACI Sec. 10.13.4.3: sway moment magnifier, $M_c = 118$ ft-kips

The three methods gave very similar design moments. If a second-order analysis is available, the easiest solution is using such an analysis. If not, the direct $P–\Delta$ (Q magnifier) method is easiest, and is recommended. The sway magnifier method involves a lot of extra calculations to compute effective length factors and critical loads for all the columns.

12–8 GENERAL ANALYSIS OF SLENDERNESS EFFECTS

ACI Sec. 10.10.1 presents requirements for a general analysis of column and frame stability. It lists a number of things the analysis must include. Such an analysis must be compared to a wide range of test results and must predict failure loads within 15% of the measured failure loads. When using the analysis, the structure must be reanalyzed if the final member sizes are more than 10% different from those assumed in the analysis.

The ACI moment magnifier method for designing columns is limited by its derivation to prismatic columns with the same reinforcement from end to end. Tall bridge piers or other tall compression members may vary in cross section along their lengths. In such a case the design can proceed as follows:

1. Derive moment–curvature diagrams for the various sections along the column for the factored axial loads at each section.

2. Use the Vianello method combined with Newmark`s numerical method[12–26] to solve for the deflected shape and the magnified moments at each section, taking into account the loadings and end restraints. A computer program to carry out such a solution is described in Ref. 12–27.

These procedures tend to slightly underestimate the magnified moments because the solution is discretized and shear deformations are ignored.

PROBLEMS

12–1 A hinged end column 18 ft tall supports unfactored loads of 100 kips dead load and 60 kips live load. These loads are applied at an eccentricity of 2 in. at the bottom and 4 in. at the top. Both eccentricities are on the same side of the centerline of the column. Use a tied column with $f_c' = 3000$ psi and $f_y = 60,000$ psi.

12–2 Repeat Problem 12–1 with the top eccentricity to the right of the centerline and the bottom eccentricity to the left.

12–3 Figure P12–3 shows an exterior column in a multistory frame. The dimensions are center to center of joints. The beams are 12 in. wide by 18 in. in overall depth. The floor slab is 6 in. thick. The building includes a service core which resists the majority of the lateral loads. Use $f_c' = 3000$ psi and $f_y = 60,000$ psi. The loads and moments on column AB are:

Factored dead load:

$$\text{axial force} = 176 \text{ kips}$$
$$\text{moment at top} = 55 \text{ ft-kips}$$
$$\text{moment at bottom} = 55 \text{ ft-kips}$$

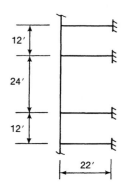

Fig. P12–3

Factored live load:

$$\text{axial force} = 210 \text{ kips}$$
$$\text{moment at top} = 72 \text{ ft-kips}$$
$$\text{moment at bottom} = 50 \text{ ft-kips}$$

Select a cross section and reinforcement.

12–4 Redesign the columns in the main floor of Example 12–3 assuming that the floor-to-floor height of the first story is 16 ft 0 in. rather than 18 ft 0 in.

13

Two-Way Slabs: Behavior, Analysis, and Direct Design Method

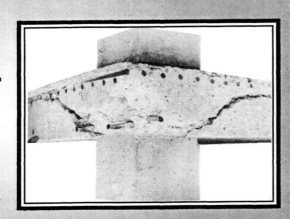

13-1   INTRODUCTION

The structural systems designed in Chapter 10 involved one-way slabs which carried loads to beams, which, in turn, transmitted the loads to columns, as shown in Fig. 4–1. If the beams were incorporated within the depth of the slab itself, the system shown in Fig. 13–1 results. Here the slab carries load in two directions. The load at *A* may be thought of as being carried from *A* to *B* and *C* by one strip of slab, and from *B* to *D* and *E*, and so on, by other slab strips. Since the slab must transmit loads in two directions, it is referred to as a *two-way slab*.

Two-way slabs are a form of construction unique to reinforced concrete, among the major structural materials. It is an efficient, economical, and widely used structural system. In practice, two-way slabs take various forms. For relatively light loads, as experienced in apartments or similar buildings, *flat plates* are used. As shown in Fig. 13–2a, such a slab is simply a slab of uniform thickness supported on columns. In an apartment building, the top of the slab would be carpeted and the bottom of the slab would be finished as the ceiling for the story below. Flat plates are most economical for spans from 15 to 20 ft (4.5 to 6 m).

For larger spans, the thickness required to transmit the vertical loads to the columns exceeds that required for bending. As a result, the concrete at the middle of the panel is not efficiently used. To lighten the slab, reduce the slab moments, and save material, the slab at midspan can be replaced by intersecting ribs, as shown in Fig. 13–2b. Note that near the columns, the full depth is retained to transmit loads from the slab to the columns. This type of slab is known as a *waffle slab* or a two-way joist system, and is formed with fiberglass or metal "dome" forms. Waffle slabs are used for spans from 25 to 40 ft (7.5 to 12 m).

For heavy industrial loads, the *flat slab* system shown in Fig. 13–2c may be used. Here the load transfer to the column is accomplished by thickening the slab near the column, using *drop panels* and/or by flaring the top of the column to form a *column capital*. The drop panel commonly extends about one-sixth of the span each way from each column, giving extra strength is the column region while minimizing the amount of concrete at midspan. *Flat slabs* are used for loads in excess of 100 psf (5 kPa) and for spans of

542

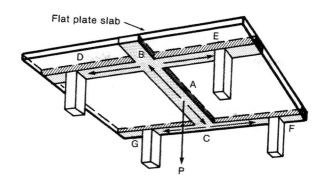

Fig. 13–1
Two-way flexure.

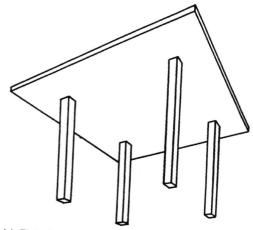

(a) Flat plate.

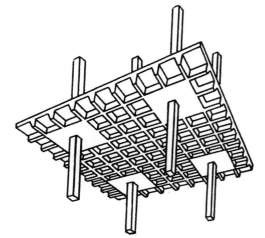

(b) Waffle slab.

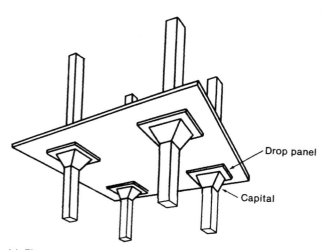

Drop panel

Capital

(c) Flat slab.

Fig. 13–2
Types of two-way slabs.

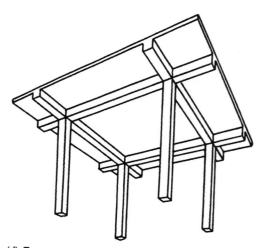

(d) Two-way slab with beams.

13–1   Introduction

20 to 30 ft (6 to 9 m). Capitals of the type shown in Fig. 13–2c are less common today than they were in the 1920s and 1930s due to the costs of forming the capitals.

Sometimes a slab system will incorporate beams between some or all of the columns. If the resulting panels are roughly square, the structure is referred to as a *two-way slab with beams* (Fig. 13–2d).

13–2   HISTORY OF TWO-WAY SLABS

One of the most interesting chapters in the development of reinforced concrete structures concerns the two-way slab. Because the mechanics of slab action were not understood when the first slabs were built, a number of patented systems developed together with a number of semi-empirical design methods. The early American papers on slabs attracted copious and very colorful discussion, each patent holder attempting to prove that his theories were right and all others were wrong.

It is not clear who built the first flat slabs. In their excellent review of the history of slabs, Sozen and Siess claim that the first American true flat slab was built by C. A. P. Turner in 1906 in Minneapolis.[13–1] In the same year Maillart built a flat slab in Switzerland. Turner's slabs, known as mushroom slabs because the columns flared out to join the slab, had steel running in bands in four directions, the two orthogonal directions and the diagonals. These bands draped down from the top of the slab over the columns to the bottom at midspan. Some of the steel was bent down into the columns, and bars bent in a circle were placed around the columns (Fig. 13–3).

The early slab buildings were built at the risk of the designer, who frequently had to put up a bond for several years and often had to load test the slabs before the owners would accept them. Turner based his designs on analyses carried out by H. T. Eddy, which were based on an incomplete plate analysis theory. During this period, the use of the crossing beam analogy in design led to a mistaken feeling that only part of the load had to be carried in each direction, so that statics somehow did not apply to slab construction.

In 1914, J. R. Nichols[13–2] used statics to compute the total moment in a slab panel. This analysis forms the basis of slab design in the current ACI Code and is presented later in this chapter. The first sentence of his paper stated: "Although statics will not suffice to determine the stresses in a flat slab floor of reinforced concrete, it does impose certain lower limits on these stresses." Eddy[13–3] attacked this concept, saying: "The fundamental erroneous assumption of this paper appears in the first sentence. . . . " Turner[13–3] thought the paper "to involve the most unique combination of multifarious absurdities imaginable from either a logical, practical or theoretical standpoint." A. W. Buel[13–3] stated that he was "unable to find a single fact in the paper nor even an explanation of facts." Rather, he felt it was "contradicted by facts." Nichols' analysis suggested that the then current slab designs underestimated the moments by 30 to 50%. The emotions expressed by the discussers appear to be inversely proportional to the amount of underdesign in their favorite slab design system.

Although Nichols' analysis was generally accepted as being correct by the mid–1920s, it was not until 1971 that the ACI Code fully recognized it and required flat slabs to be designed for 100% of statics.

13–3   BEHAVIOR OF A SLAB LOADED TO FAILURE IN FLEXURE

There are four or more stages in the behavior of a slab loaded to failure:

1. Before cracking, the slab acts as an elastic plate and, for short-time loads the deformations, stresses, and strains can be predicted from an elastic analysis.

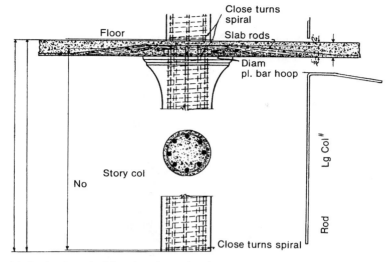

(a) Section through slab and mushroom head

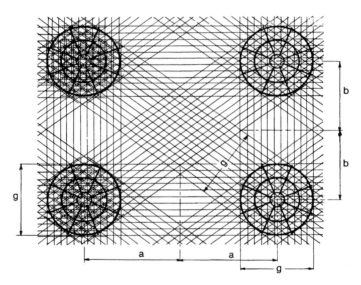

Fig. 13–3
C. A. P. Turner mushroom
slab.

(b) Plan of reinforcement

2. After cracking and before yielding of the reinforcement, the slab is no longer of constant stiffness, since the cracked regions have a lower flexural stiffness, *EI,* than the uncracked regions; and the slab is no longer isotropic since the crack pattern may differ in the two directions. Although this violates the assumptions in the elastic theory, tests indicate that the elastic theory still predicts the moments adequately. Normal building slabs are generally partially cracked at service loads.

3. Yielding of the reinforcement eventually starts in one or more regions of high moment and spreads through the slab as moments are redistributed from yielded regions to areas that are still elastic. The progression of yielding through a slab fixed on four edges is illustrated in Fig. 13–4. In this case the initial yielding occurs due to negative moments which form localized plastic hinges at the centers of the long sides (Fig. 13–4b). These hinges spread along the long sides and eventually, new hinges form at the ends of the slab

(Fig. 13–4c). Meanwhile, the positive moments increase in strips across the center of the slab in the short direction because of the moment redistribution caused by the plastic hinges at the ends of these strips. Eventually, yielding occurs due to positive moments in these strips, as shown in Fig. 13–4c. With further load, the regions of yielding, known as *yield lines,* divide the slab into a series of trapezoidal or triangular elastic plates, as shown in Fig. 13–4d. The loads corresponding to this stage of behavior can be estimated using a *yield line analysis.*

4. Although the yield lines divide the plate to form a plastic mechanism, the hinges jam with increased deflection and the slab forms a very flat compression arch, as shown, in Fig. 13–5. This assumes that the surrounding structure is stiff enough to provide reactions for the arch. This stage of behavior is not counted on in design at present.

This review of behavior has been presented to point out first, that elastic analyses of slabs begin to lose their accuracy as the loads exceed the service loads; and second, that a great deal of redistribution of moments occurs after yielding first starts. A slab supported on and continuous with stiff beams or walls has been considered here. In the case of a slab supported on isolated columns as shown in Fig. 13–2a, similar behavior would be observed except that the first cracking would be on the top of the slab around the column, followed by cracking of the bottom of the slab midway between the columns.

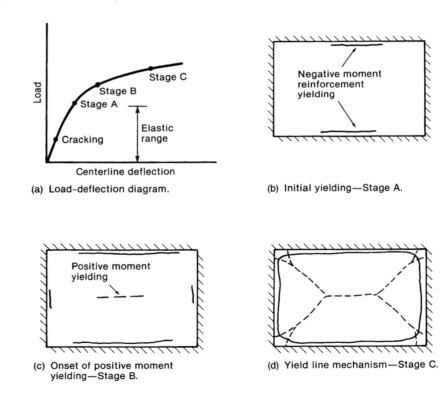

Fig. 13–4
Inelastic action in a slab fixed on four sides.

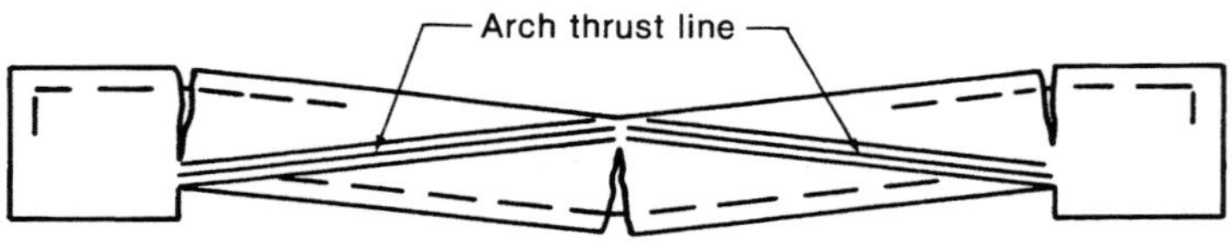

(a) Load–deflection diagram.

(b) Initial yielding—Stage A.

(c) Onset of positive moment yielding—Stage B.

(d) Yield line mechanism—Stage C.

Fig. 13–5
Arch action in slab.

Slabs that fail in flexure are extremely ductile. Slabs, particularly flat plate slabs, may also fail in shear. Shear failures tend to be brittle. Shear in slabs is discussed in Sec. 13–7.

13–4 STATICAL EQUILIBRIUM OF TWO-WAY SLABS

Figure 13–6 shows a floor made up of simply supported planks supported by simply supported beams. The floor carries a load of w lb/ft^2 or w kPa. The moment per foot of width in the planks at section A–A is

$$m = \frac{w\ell_1^2}{8} \text{ ft-kips/ft} \qquad (\text{kN} \cdot \text{m/m})$$

The total moment in the entire width of the floor is

$$M_f = (w\ell_2)\frac{\ell_1^2}{8} \text{ ft-kips} \qquad (\text{kN} \cdot \text{m}) \qquad (13\text{–}1)$$

This is the familiar equation for the maximum moment in a simply supported floor of width ℓ_2 and span ℓ_1.

The planks apply a uniform load of $w\ell_1/2/$ft (or /m) on each beam. The moment at section B–B in one beam is

$$M_{1b} = \left(\frac{w\ell_1}{2}\right)\frac{\ell_2^2}{8} \text{ ft-kips} \qquad (\text{kN} \cdot \text{m})$$

The total moment in both beams is

$$M = (w\ell_1)\frac{\ell_2^2}{8} \text{ ft-kips} \qquad (\text{kN} \cdot \text{m}) \qquad (13\text{–}2)$$

It is important to note that the full load was transferred east and west by the planks, causing a moment equivalent to $w\ell^2/8$ in the planks and then was transferred north and south by the beams, causing a similar moment in the beams. Exactly the same thing happens in the two-way slab shown in Fig. 13–7. The total moment required along sections A–A and B–B, respectively, are

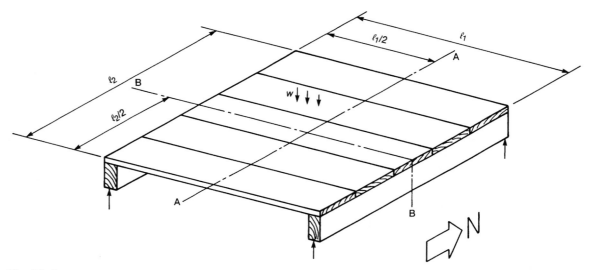

Fig. 13–6
Moments in a plank-and-beam floor.

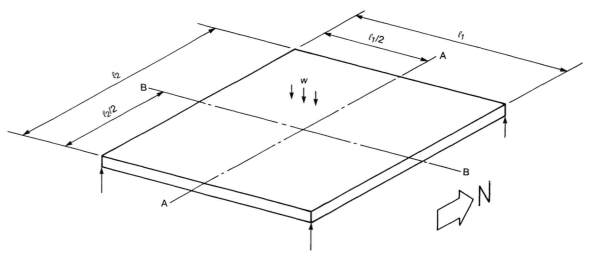

Fig. 13–7
Moments in a two-way slab

$$M = (w\ell_2)\frac{\ell_1^2}{8} \qquad M = (w\ell_1)\frac{\ell_2^2}{8} \qquad\qquad (13\text{–}1, 13\text{–}2)$$

Again, the full load was transferred east and west and then the full load was transferred north and south, this time by the slab in both cases. This, of course, must always be true regardless of whether the structure has one-way slabs and beams, two-way slabs, or some other system.

Nichols' Analysis of Slabs

The analysis used to derive Eqs. 13–1 and 13–2 was first published in 1914 by Nichols.[13–2] Nichols' original analysis was presented for a slab on round columns rather than the point supports assumed in deriving Eqs. 13–1 and 13–2. Because rectangular columns are more common today, the following derivation considers that case. Assume:

1. A rectangular, typical interior panel in a large structure
2. All the panels in the structure are uniformly loaded with the same load

These two assumptions are made to ensure that the lines of maximum moment, and hence the lines on which the shears and twisting moments are equal to zero, will be lines of symmetry in the structure. This allows one to isolate the portion of the slab shown shaded in Fig. 13–8a. This portion is bounded by lines of symmetry.

The reaction to the vertical loads is transmitted to the slab by shear around the face of the columns. It is necessary to know, or assume, the distribution of this shear to compute the moments in this slab panel. The maximum shear transfer occurs at the corners of the column, with lesser amounts transferred in the middle of the sides of the column. For this reason we shall assume:

3. Column reaction is concentrated at the four corners of the column

Figure 13–8b shows a side view of the slab element with the forces and moments acting on it. The applied load is $(w\ell_1\ell_2/2)$ at the center of the shaded panel minus the load on the area occupied by the column $(wc_1c_2/2)$. This is equilibrated by the upward reaction at the corners of the column.

The total *statical moment*, M_0, is the sum of the negative moment M_1, and the positive moment, M_2, as computed by summing moments about line A–A:

Two-Way Slabs: Behavior, Analysis, and Direct Design Method

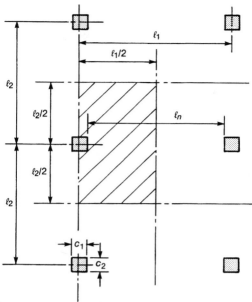

(a) Plan of slab element.

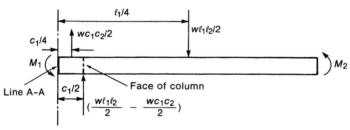

(b) Side view of slab element.

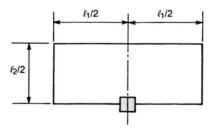

Fig. 13–8
Slab considered in Nichols'
analysis.

(c) Plan of second slab element.

$$M_0 = M_1 + M_2 = \left(\frac{w\ell_1\ell_2}{2}\right)\frac{\ell_1}{4} - \left(\frac{wc_1c_2}{2}\right)\frac{c_1}{4} - \left(\frac{w\ell_1\ell_2}{2} - \frac{wc_1c_2}{2}\right)\frac{c_1}{2}$$

and

$$M_0 = \frac{w\ell_2}{8}\left[\ell_1^2\left(1 - \frac{2c_1}{\ell_1} + \frac{c_2c_1^2}{\ell_2\ell_1^2}\right)\right] \tag{13-3}$$

The ACI Code has simplified this expression slightly by replacing the term in the square brackets with ℓ_n^2, where ℓ_n is the clear span between the faces of the columns, given by

$$\ell_n = \ell_1 - c_1$$

13–4   Statical Equilibrium of Two-Way Slabs   **549** and where

$$\ell_n^2 = \ell_1^2 \left(1 - \frac{2c_1}{\ell_1} + \frac{c_1^2}{\ell_1^2}\right) \tag{13-4}$$

A comparison of Eqs. 13–3 and 13–4 shows that ℓ_n^2 differs only slightly from the term in brackets in Eq. 13–3, and the equation for the statical moment can be written as

$$M_0 = \frac{w\ell_2\ell_n^2}{8} \tag{13-5}$$

(ACI Eq. 13–3)

For circular columns, Nichols assumed the shear to be uniformly distributed around the face of the column, leading to

$$M_0 = \frac{w\ell_2\ell_1^2}{8}\left[1 - \frac{4d_c}{\pi\ell_1} + \frac{1}{3}\left(\frac{d_c}{\ell_1}\right)^3\right] \tag{13-6}$$

where d_c is the diameter of the column or the column capital. Nichols approximated this as

$$M_0 = \frac{w\ell_2\ell_1^2}{8}\left(1 - \frac{2d_c}{3\ell_1}\right)^2 \tag{13-7}$$

ACI Sec. 13.6.2.2 expresses this using Eq. 13–5, where ℓ_n is based on the span between equivalent square columns having the same area as the circular columns. In this case, $c_1 = d_c\sqrt{\pi}/2 = 0.886d_c$.

For square columns the practical range of c_1/ℓ_1 is roughly 0.05 to 0.15. For $c_1/\ell_1 = 0.05$ and $c_1 = c_2$, Eqs. 13–3 and 13–5 give $M_0 = Kw\ell_2\ell_1^2/8$, where $K = 0.900$ and 0.903, respectively. For $c_1/\ell_1 = 0.15$ the respective values of K are 0.703 and 0.723. Thus Eq. 13–5 closely represents the moments in a slab supported on square columns, becoming more conservative as c_1/ℓ_1 increases.

For circular columns the practical range of d_c/ℓ_1 is roughly 0.05 to 0.20. For $d_c/\ell_1 = 0.05$, Eq. 13–6 gives $K = 0.936$, while Eq. 13–5 with ℓ_n defined using $c_1 = d_c\sqrt{\pi}/2$ gives $K = 0.913$. For $d_c/\ell_1 = 0.2$, the corresponding values of K from Eqs. 13–6 and 13–5 are 0.748 and 0.677, respectively. Thus for circular columns Eq. 13–5 tends to underestimate M_0 by up to 10%, compared to Eq. 13–6.

If the equilibrium of the element shown in Fig. 13–8c were studied, a similar equation for M_0 would result, with ℓ_1 and ℓ_2, and c_1 and c_2, interchanged. This indicates once again that the total load must satisfy moment equilibrium in both the ℓ_1 and ℓ_2 directions.

13–5 DISTRIBUTION OF MOMENTS IN SLABS

Relationship between Slab Curvatures and Moments

The principles of elastic analysis of two-way slabs are presented briefly in Sec. 15–1. The basic equation for moments is Eq. 15–6. Frequently in studies of concrete plates, Poisson's ratio, ν, is taken equal to zero. When this is done, Eq. 15–6 reduces to

$$m_x = -\frac{Et^3}{12}\left(\frac{\partial^2 z}{\partial x^2}\right)$$

$$m_y = -\frac{Et^3}{12}\left(\frac{\partial^2 z}{\partial y^2}\right) \tag{13-8}$$

$$m_{xy} = -\frac{Et^3}{12}\left(\frac{\partial^2 z}{\partial x\,\partial y}\right)$$

In these equations, $\partial^2 z / \partial x^2$ represents the curvature in a slab strip in the x direction, and $\partial^2 z / \partial y^2$ represents the curvature in a strip in the y direction. Thus, by visualizing the deflected shape of a slab, one can qualitatively estimate the distribution of moments.

Figure 13–9a shows a rectangular slab that is fixed on all sides on stiff beams. Three transverse strips are shown. The deflected shape of these strips and the corresponding moment diagrams are shown in Fig. 13–9b to d. Where the deflected shape is concave downward, the moment causes compression on the bottom: that is, the moment is negative. This may be seen also from Eq. 13–8. Since z was taken as positive downward, a positive curvature, $\partial^2 z / \partial x^2$, corresponds to a curve that is concave downward. From Eq. 13–8, a positive curvature corresponds to a negative moment. The magnitude of the moment is proportional to the curvature.

The largest deflection, Δ_2, occurs at the center of the panel. As a result, the curvatures and hence the moments in strip B are larger than those in strip A. The center portion of strip

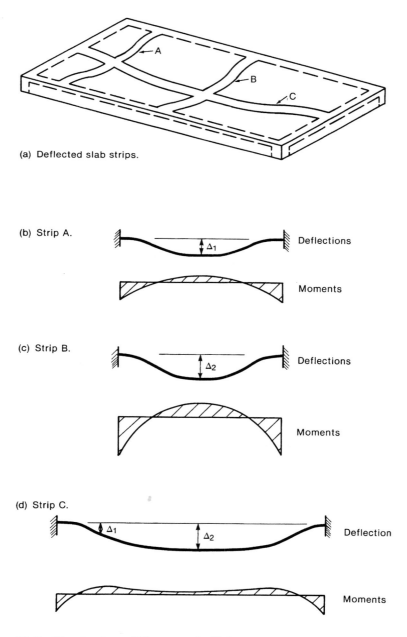

(a) Deflected slab strips.

(b) Strip A.

Deflections

Moments

(c) Strip B.

Deflections

Moments

(d) Strip C.

Deflection

Moments

Fig. 13–9
Relationship between slab
curvatures and moments.

13–5   Distribution of Moments in Slabs

551

C is essentially straight, indicating that most of the loads in this region are being transmitted by one-way action across the short direction of the slab.

The existence of the twisting moments, m_{xy}, in a slab can be illustrated by the crossing-strip analogy. Figure 13–10 shows section *B–B* through the slab shown in Fig. 13–9. Here the slab is represented by a series of crossing beams, some parallel to *B–B* and the others shown in cross section, parallel to *C–C*. The slab strips perpendicular to the section (shown in cross section) must twist as shown. This is due to the m_{xy} twisting moments.

Moments in Slabs Supported on Stiff Beams or Walls

The distributions of moments in a series of square and rectangular slabs will be presented in one of two graphical treatments. The distribution of the negative moments, M_A, or the positive moments, M_B, along lines across the slab will be depicted as shown in Fig. 13–11b.

Fig. 13–10
Deflection of strip B of Fig. 13–9. Note twisting in lower layer.

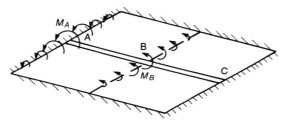

(a) Moments at edge and middle of slab.

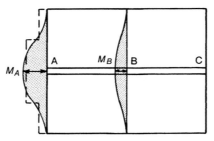

(b) Distribution of moments at edge and middle.

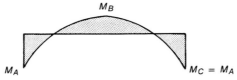

Fig. 13–11
Types of moment diagrams: four-edged fixed slab.

(c) Moments in strip ABC.

Two-Way Slabs: Behavior, Analysis, and Direct Design Method

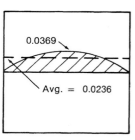

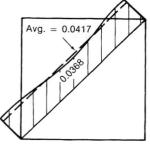

Fig 13–12
Moments in a square slab
hinged on four edges. (From
Ref. 13–4.)

(a) Moments across center of slab.

(b) Moments across diagonal.

These distributions may be shown as continuous curves, as shown by the solid lines and shaded areas, or as a series of steps, as shown by the dashed line. The height of the curve at any point indicates the magnitude of the moment at that point. Occasionally, the distribution of bending moments in a strip A–B–C across the slab will be plotted as shown in Fig. 13–11c or 13–9.

The moments will be expressed in terms of Cwb^2, where b is the short dimension of the panel. The value of C would be 0.125 in a square simply supported one-way slab. The units will be ft-lb/ft of width, or kN · m/m. In all cases, the moment diagrams are for uniformly loaded slabs.

Figure 13–12 shows moment diagrams for a simply supported square slab. The moments act about the lines shown (similar to Fig. 13–11a). The largest moments in the slab are about an axis along the diagonal, as shown in Fig. 13–12b.

It is important to remember that the total load must be transferred from support to support by moments in the slab or beams. Figure 13–13 shows a slab which is simply supported at two ends and supported by stiff beams along the other two sides. The moments in

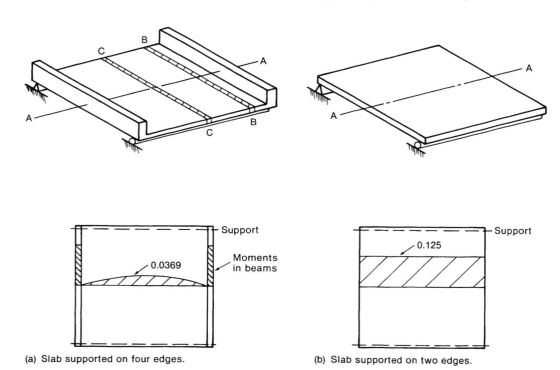

(a) Slab supported on four edges.

(b) Slab supported on two edges.

Fig. 13–13
Effect of beam stiffness on moments in slab.

the slab (which are the same as in Fig. 13–12a) account for only 19% of the total moments. The balance of the moments are divided between the two beams as shown.

Figure 13–13b shows the effect of reducing the stiffness of the two edge beams. Here the stiffness of the edge beams per unit width has been reduced to be equal to that of the slab. Now the entire moment is resisted by one-way slab action. The fact that the moments about line A–A are constant indicates that the midspan curvatures of all the strips spanning from support to support are equal. On the other hand, the curvatures in a strip along B–B in Fig. 13–13a are much smaller than those in strip C–C since B–B is next to the stiff beam. This explains why the slab moments in Fig. 13–13a decrease as one approaches the edge beams.

The corners of a simply supported slab tend to curl up off their supports as discussed later in this section. The moment diagrams presented here assume that the corners are held down by downward point loads at the corners.

The distribution of moments in several other slab configurations are plotted in Figs. 13–14 to 13–16. The moments in the square fixed-edge slab shown in Fig. 13–15a ac-

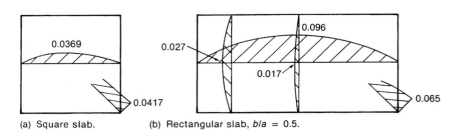

Fig. 13–14
Bending moments per unit width in rectangular slabs with simply supported edges. (From Ref. 13–4.)

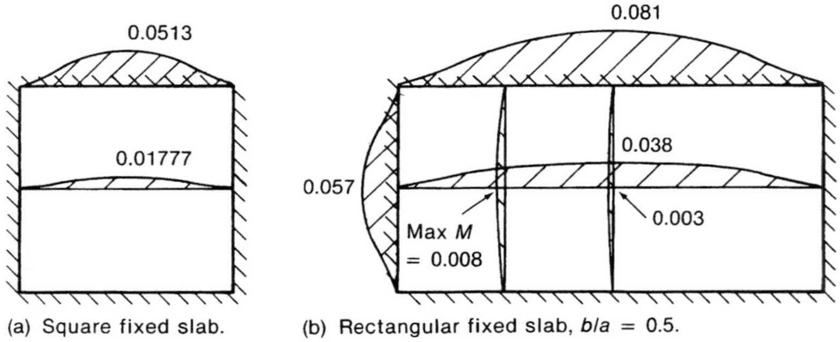

(a) Square slab. (b) Rectangular slab, $b/a = 0.5$.

Fig. 13–15
Bending moments per unit width in rectangular slabs with fixed edges. (From Ref. 13–4.)

(a) Square fixed slab. (b) Rectangular fixed slab, $b/a = 0.5$.

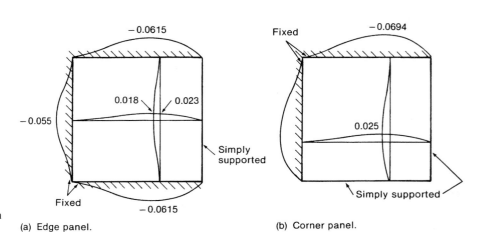

Fig. 13–16
Bending moments in edge and corner slabs supported on stiff beams.

(a) Edge panel. (b) Corner panel.

Two-Way Slabs: Behavior, Analysis, and Direct Design Method

count for 36% of the statical moments, the balance is in the beams. As the ratio of length to width increases, the center portion of the slab approaches one-way action.

Moments in Slabs Supported on Isolated Columns

In a flat plate or flat slab, the slab is supported directly on the columns without any beams. Here the stiffest portions of the slab are those running from column to column along the four sides of a panel. As a result, the moments are largest in these parts of the slab.

Figure 13–17a illustrates the moments in a typical interior panel of a very large slab in which all panels are uniformly loaded with equal loads. The slab is supported on circular columns with a diameter $c = 0.1\,\ell$. The largest negative and positive moments occur in the strips spanning from column to column. In Fig. 13–17b and c, the curvatures and moment diagrams are shown for strips along lines $A–A$ and $B–B$. Both strips have negative moment adjacent to the columns and positive moment at midspan. In Fig. 13–17d, the moment diagram from Fig. 13–17a is replotted to show the average moments over a *column strip* of width $\ell_2/2$ and a *middle strip* between the two column strips. The ACI Code design procedures consider the average moments over the width of the middle and column strips. A

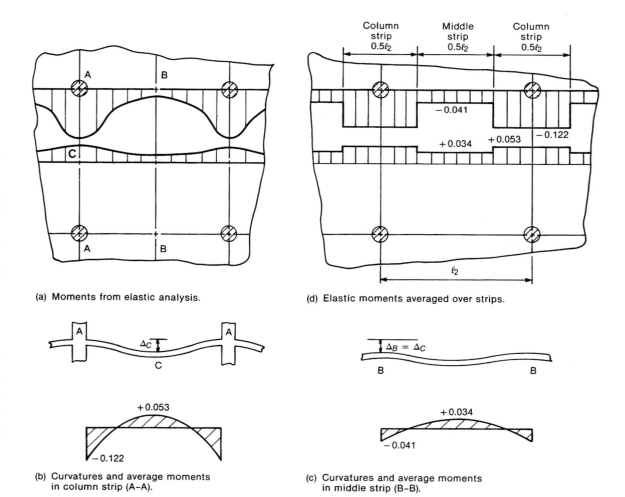

(a) Moments from elastic analysis.

(d) Elastic moments averaged over strips.

(b) Curvatures and average moments in column strip (A–A).

(c) Curvatures and average moments in middle strip (B–B).

Fig. 13–17
Moments in a slab supported on isolated columns, $\ell_2/\ell_1 = 1.0$, $c/\ell = 0.1$

13–5 Distribution of Moments in Slabs

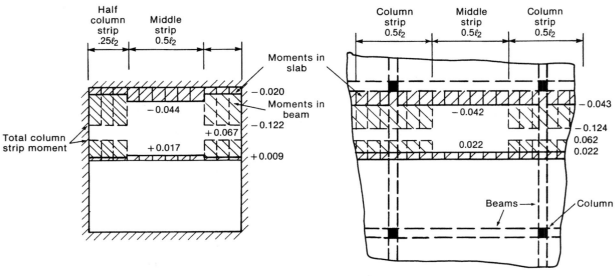

Fig. 13–18
Moments in slabs, $\ell_2 / \ell_1 = 1.0$.

comparison of Fig. 13–17a and d shows that immediately adjacent to the columns, the theoretical elastic moments may be considerably larger than indicated by the average values.

The total moment accounted for here is

$$w\ell_n^2 \big[(0.122 \times 0.5\ell_2) + (0.041 \times 0.5\ell_2) \big.$$
$$\big. + (0.053 \times 0.5\ell_2) + (0.034 \times 0.5\ell_2) \big] = 0.125 w\ell_2 \ell_n^2$$

The distribution of moments given in Fig. 13–15 for a slab fixed on four edges and supported on rigid beams is replotted in Fig. 13–18a with the moments averaged over column strip and middle strip bands in the same way as the flat-plate moments were in Fig. 13–17. In addition, the sum of the beam moments and the column strip slab moments has been divided by the width of the column strip and plotted as the *total column strip moment*. The distribution of moments in Fig. 13–17d closely resembles the distribution of middle strip and total column strip moments in Fig. 13–18a.

An intermediate case in which the beam stiffness, I_b, equals the stiffness, I_s, of a slab of width ℓ_2, is shown in Fig. 13–18b. Although the division of moment between slab and beams differs, the distribution of total moments is again similar to that shown in Fig. 13–17d or 13–18a.

The slab design procedures in the ACI Code take advantage of this similarity in the distributions of the total moments in presenting a unified design procedure for the whole spectrum, from slabs supported on isolated columns to slabs supported on beams in two directions.

13–6   DIRECT DESIGN METHOD: INTRODUCTION

ACI Sec. 13.5.1 allows slabs to be designed by any procedure that satisfies equilibrium and geometric compatibility, provided that every section has a strength at least equal to the required strength, and provided that serviceability conditions are satisfied. Two slab design procedures are presented in detail in the ACI Code. These are the Direct Design

Method, considered in this chapter, and the Equivalent Frame Design Method, presented in Chap. 14. These two methods differ primarily in the way in which the slab moments are computed. The design examples in both Chaps. 13 and 14 are similar to allow comparison of the two sets of design moments and other aspects. Other methods, such as the yield line method or the strip method, both presented in Chap. 15, are allowed under ACI Sec. 13.5.1.

Steps in Slab Design

The steps in the design of a two-way slab include:

1. Choose the layout and type of slab to be used. The various types of two-way slabs and their uses have been discussed briefly in Sec. 13–1. The choice of type of slab is strongly affected by architectural or construction considerations.

2. Choose the slab thickness. Generally, the slab thickness is chosen to prevent excessive deflection in service. Equally important, the slab thickness chosen must be adequate for shear at both interior and exterior columns.

3. Choose the design method. The *Equivalent Frame Method* uses an elastic frame analysis to compute the positive and negative moments in the various panels in the slab. The *Direct Design Method* uses coefficients to compute these moments.

4. Compute the positive and negative moments in the slab.

5. Determine the distribution of the moments across the width of the slab. The lateral distribution of moments within a panel depends on the geometry of the slab and the stiffness of the beams (if any). This procedure is the same in both design methods.

6. If there are beams, a portion of the moment must be assigned to the beams.

7. Reinforcement is designed for the moments from steps 5 and 6.

8. The shear strengths at the columns are checked.

Several of the parameters used in this process will be defined prior to carrying out slab designs.

Limitations on the Use of the Direct Design Method

The Direct Design Method is easier to use than the Equivalent Frame Method, but can only be applied to fairly regular multipanel slabs. The limitations are given in ACI Sec. 13.6.1 and include:

1. There must be a minimum of three continuous spans in each direction. Thus a nine-panel structure (3 by 3) is the smallest that can be considered. If there are fewer than three panels, the interior negative moments from the direct design method tend to be too small.

2. Rectangular panels must have a long span/short span ratio not greater than 2. One-way action predominates as the span ratio reaches and exceeds 2.

3. Successive span lengths in each direction shall not differ by more than one-third of the longer span. This limit is imposed so that certain standard reinforcement cutoff details can be used.

4. Columns may be offset from the basic rectangular grid of the building by up to 0.1 times the span parallel to the offset. In a building laid out in this way, the actual column locations are used in determining the spans of the slab to be used in calculating the design moments.

5. All loads must be due to gravity only. The direct design method cannot be used for unbraced laterally loaded frames, foundation mats, or prestressed slabs.

6. The service (unfactored) live load shall not exceed two times the service dead load. Strip or checkerboard loadings with large live/dead load ratios may lead to moments larger than those assumed in this method of analysis.

7. For a panel with beams between supports on all sides the relative stiffness of the beams in the two perpendicular directions given by $(\alpha_1 \ell_2^2)/(\alpha_2 \ell_1^2)$ shall not be less than 0.2 or greater than 5. The term α is defined in the next section and ℓ_1 and ℓ_2 are the spans in the two directions.

Limitations 2 and 7 do not allow use of the direct design method for slab panels that transmit load as one-way slabs. Limitation 6 was changed from "three times" to "two times" in the 1995 code to reduce the effect of pattern loads and hence to eliminate the need to check for the effect of such loads.

Beam-to-Slab Stiffness Ratio, α

Slabs are frequently built with beams spanning from column to column around the perimeter of the building. These beams act to stiffen the edge of the slab and help to reduce the deflections of the exterior panels of the slab. Very heavily loaded slabs and long-span waffle slabs sometimes have beams joining all columns in the structure.

In the ACI Code, the effects of beam stiffness on deflections and the distribution of moments are expressed as a function of α, defined as the flexural stiffness, $4EI/\ell$, of the beam divided by the flexural stiffness of a width of slab bounded laterally by the centerlines of the adjacent panels on each side of the beam.

$$\alpha = \frac{4E_{cb}I_b/\ell}{4E_{cs}I_s/\ell}$$

Since the lengths, ℓ, of the beam and slab are equal, this is simplified and expressed in the code as

$$\alpha = \frac{E_{cb}I_b}{E_{cs}I_s} \qquad (13\text{–}9)$$

where E_{cb} and E_{cs} are the moduli of elasticity of the beam concrete and slab concrete, respectively, and I_b and I_s are the moments of inertia of the uncracked beams and slabs. The sections considered in computing I_b and I_s are shown shaded in Fig. 13–19. The span perpendicular to the direction being designed is ℓ_2. In Fig. 13–19c the panels adjacent to the beam under consideration have different transverse spans. The calculation of ℓ_2 in such a case is illustrated in Example 13–4. If there is no beam, $\alpha = 0$.

ACI Sec. 13.2.4 defines a beam in monolithic or fully composite construction as the beam stem plus a portion of the slab on each side of the beam extending a distance equal to the projection of the beam above or below the slab, whichever is greater, but not greater than four times the slab thickness. This is illustrated in Fig. 13–20.

Once the size of the slab and beam have been chosen, values of α can be computed from first principles or from Fig. 13–21. The first kinks in these curves are due to a change in scale; the second represent the point at which the flange width limits are reached.

EXAMPLE 13–1 Calculation of α for an Edge Beam

An 8-in.-thick slab is provided with an edge beam which has a total depth of 16 in. and a width of 12 in., as shown in Fig. 13–22a. The slab and beam were cast monolithically and have the same concrete strength and the same E_c. Compute α. Since $f'_{cs} = f'_{cb}$, $E_{cb} = E_{cs}$, and Eq. 13–9 reduces to $\alpha = I_b/I_s$.

1. Compute I_b. The cross section of the beam is as shown in Fig. 13–22b. The centroid of this beam is located 7.00 in. from the top of the slab. The moment of inertia of the beam is

Two-Way Slabs: Behavior, Analysis, and Direct Design Method

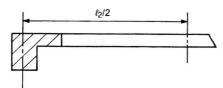

(a) Section for I_b—Edge beam.

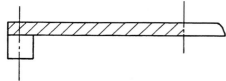

(b) Section for I_s—Edge beam.

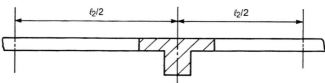

(c) Section for I_b—Interior beam.

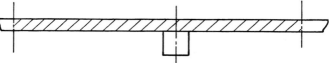

Fig. 13–19
Beam and slab sections for
calculation of α.

(d) Section for I_s—Interior beam.

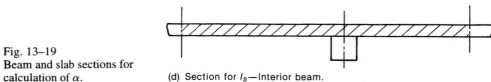

Fig. 13–20
Cross section of beams.
(From Ref. 13–5)

$$I_b = \left(12 \times \frac{16^3}{12}\right) + (12 \times 16) \times 1^2 + \left(8 \times \frac{8^3}{12}\right) + (8 \times 8) \times 3^2$$

$$= 5205 \text{ in.}^4$$

2. **Compute I_s.** I_s is computed for the shaded portion of the slab in Fig. 13–22c.

$$I_s = 126 \times \frac{8^3}{12} = 5376 \text{ in.}^4$$

3. **Compute α.**

$$\alpha = \frac{I_b}{I_s} = \frac{5205}{5376}$$

$$= 0.968$$

13–6   Direct Design Method: Introduction

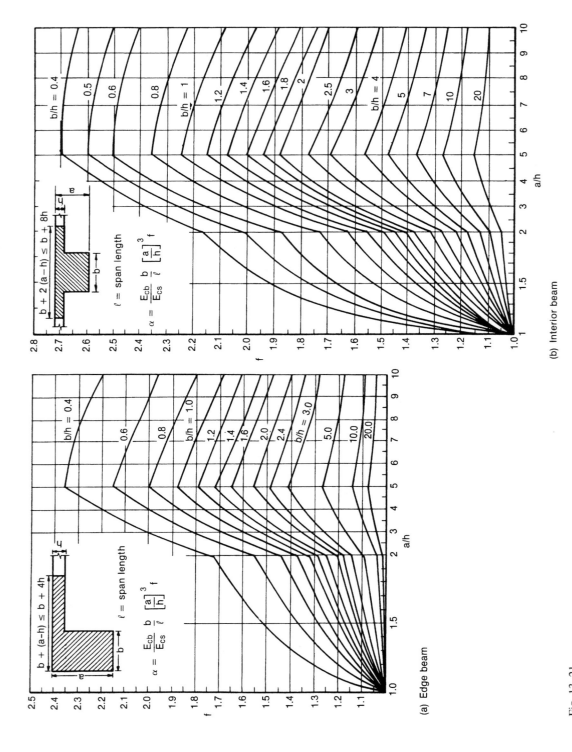

(a) Edge beam

(b) Interior beam

Fig. 13–21
Charts for computing α. (From Ref. 13–6. Courtesy of the Portland Cement Association.)

Two-Way Slabs: Behavior, Analysis, and Direct Design Method

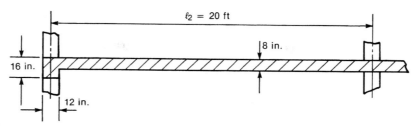

(a) Section through edge of slab.

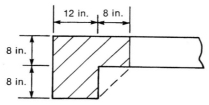

(b) Edge beam.

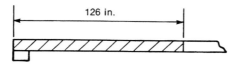

Fig. 13–22
Slab—Example 13–1.

(c) Section of slab.

Alternatively, using Fig. 13–21a,

$$\alpha = \frac{E_{cb}}{E_{cs}} \frac{b}{\ell} \left(\frac{a}{h}\right)^3 f$$

where $E_{cb} = E_{cs}$. From the graph for $a/h = 2$ and $b/h = 12/8 = 1.5, f = 1.27$ and

$$\alpha = \frac{12}{126} \left(\frac{16}{8}\right)^3 \times 1.27 = 0.968 \qquad \blacksquare$$

Minimum Thickness of Two-Way Slabs

ACI Sec. 9.5.3 defines minimum thicknesses which are intended to limit slab deflections to acceptable values. Thinner slabs can be used if it can be shown by computation that the slab deflections will not be excessive. The computation of slab deflections is discussed in Sec. 13–12.

Slabs without Beams between Interior Columns

The minimum thicknesses of slabs without beams between interior columns and having a ratio of long to short spans of 2 or less, the minimum thickness is as given in Table 13–1 (ACI Table 9.5c), but not less than 5 in. in slabs without drop panels, or 4 in. in slabs with drop panels having the dimensions defined in ACI Secs. 13.3.7.1 and 13.3.7.2.

ACI Sec. 9.5.3.4 allows thinner slabs to be used if calculated deflections satisfy limits given in ACI Table 9.5(b). See Sec. 13–12.

As noted in the footnote to Table 13–1, a beam must have a stiffness ratio α of 0.8 or more to be called an edge beam. It can be shown that an edge beam which has a depth, $a,$ at least twice the slab thickness, $h,$ and an area, $ab_w,$ at least $4h^2$, will always

TABLE 13-1 Minimum Thickness of Slabs without Interior Beams

Yield Strength, f_y^b, (psi)	Without Drop Panels[a]			With Drop Panels[a]		
	Exterior Panels		Interior Panels	Exterior Panels		Interior Panels
	Without Edge Beams	With Edge Beams[c]		Without Edge Beams[c]	With Edge Beams	
40,000	$\ell_n/33$	$\ell_n/36$	$\ell_n/36$	$\ell_n/36$	$\ell_n/40$	$\ell_n/40$
60,000	$\ell_n/30$	$\ell_n/33$	$\ell_n/33$	$\ell_n/33$	$\ell_n/36$	$\ell_n/36$
75,000	$\ell_n/28$	$\ell_n/31$	$\ell_n/31$	$\ell_n/31$	$\ell_n/34$	$\ell_n/34$

Source: Ref. 13–5.

[a]The required geometry of a drop panel is defined in ACI Secs. 13.3.7.1 and 13.3.7.2.

[b]For yield strengths between the values given, use linear interpolation.

[c]Slabs with beams between columns along exterior edges. The value of α for the edge beam shall not be less than 0.8.

have α greater than 0.8. This rule of thumb can be used to simplify the selection of slab thicknesses.

Excessive slab deflections are a serious problem in many parts of North America. The thickness calculated using Table 13–1 should be rounded up to the next 1/4 or even 1/2 in. Rounding the thickness up will give a slightly stiffer slab and hence smaller deflections. Studies of slab deflections presented in Ref. 13–7 suggest that slabs without interior beams should be about 10% thicker than the ACI minimum values to avoid excessive deflections.

Slabs with Beams between the Interior Supports

For slabs with beams between interior supports, ACI Sec. 9.5.3.3 gives the following minimum thicknesses:

(a) For $\alpha_m \leq 0.2$, the minimum thicknesses in Table 13–1 shall apply.

(b) For $0.2 < \alpha_m < 2.0$, the thickness shall not be less than

$$h = \frac{\ell_n[0.8 + (f_y/200,000)]}{36 + 5\beta(\alpha_m - 0.2)} \quad \text{but not less than 5 in.} \quad (13\text{--}10)$$
$$\text{(ACI Eq. 9–11)}$$

(c) For $\alpha_m > 2.0$, the thickness shall not be less than

$$h = \frac{\ell_n[0.8 + (f_y/200,000)]}{36 + 9\beta} \quad \text{but not less than 3.5 in.} \quad (13\text{--}11)$$
$$\text{(ACI Eq. 9–12)}$$

(d) At discontinuous edges, an edge beam with a stiffness ratio α not less than 0.8 shall be provided or the slab thickness shall be increased by at least 10% in the edge panel.

where

$h = $ overall thickness

$\ell_n = $ longer clear span of the slab panel under consideration

α_m = the average of the values of α for the four sides of the panel

β = longer clear span/shorter clear span of the panel

Minimum thicknesses are computed as the first step of Examples 13–7 and 13–9.

The thickness of slabs may also be governed by shear. This is particularly serious if moments are transferred to the columns at edge columns and may be serious at interior columns between two spans that are greatly different in length. The selection of slab thicknesses to satisfy shear requirements is discussed in Sec. 13–8. Briefly, that section suggests that the trial slab thickness be chosen such that $V_u \simeq 0.5$ to $0.55(\phi V_c)$ at edge columns and $V_u \simeq 0.85$ to $1.0(\phi V_c)$ at interior columns.

Distribution of Moments within Panels

Statical Moment, M_0

For design, the slab is considered to be a series of frames in the two directions, as shown in Fig. 13–23. These frames extend to the middle of the panels on each side of the column lines. In each span of each of these frames it is necessary to compute the total statical moment, M_0:

$$M_0 = \frac{w_u \ell_2 \ell_n^2}{8}$$

(13–5)

(ACI Eq. 13–3)

where

w_u = factored load per unit area

ℓ_2 = transverse width of the strip

ℓ_n = clear span between columns

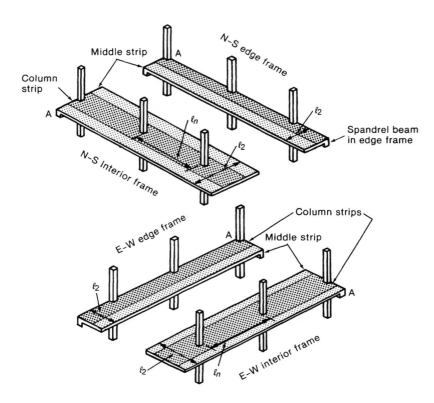

Fig. 13–23
Division of slab into frames for design.

In computing ℓ_n, circular columns or column capitals of diameter d_c are replaced by equivalent square columns with side lengths $0.886d_c$. Values of ℓ_2 and ℓ_n are shown in Fig. 13–23 for panels in each direction. Example 13–2, illustrates the calculation of M_0 in a typical slab panel.

EXAMPLE 13–2   Computation of Statical Moment, M_0

Compute the statical moment, M_0, in the slab panels shown in Fig. 13–24. The slab is 8 in. thick and supports a live load of 100 psf.

1.   Compute the factored uniform loads.

$$w_u = 1.4\left(\frac{8}{12} \times 0.15\right) \text{ksf} + 1.7(0.100) \text{ksf}$$

$$= 0.310 \text{ ksf}$$

Note that if the local building code allows a live-load reduction, the live load should be multiplied by the appropriate live-load reduction factor before computing w_u.

2.   Consider panel A spanning from column 1 to column 2. Slab panel A is shown shaded in Fig. 13–24a. The moments computed in this part of the example would be used to design the reinforcement running parallel to line 1–2 in this panel. From Eq. 13–5 (ACI Eq. 13–3);

$$M_0 = \frac{w\ell_2\ell_n^2}{8}$$

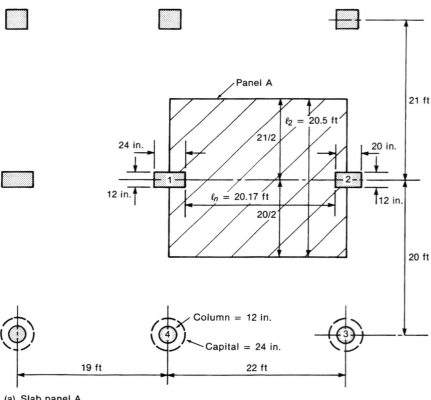

Fig. 13–24
Slab–Example 13–2.   (a) Slab panel A.

where

$$\ell_n = \text{clear span of slab panel}$$

$$= 22 \text{ ft } - \frac{1}{2}\left(\frac{20}{12}\right) \text{ft} - \frac{1}{2}\left(\frac{24}{12}\right) \text{ft} = 20.17 \text{ ft}$$

$$\ell_2 = \text{width of slab panel}$$

$$= (21/2) \text{ ft } + (20/2) \text{ ft} = 20.5 \text{ ft}$$

Therefore,

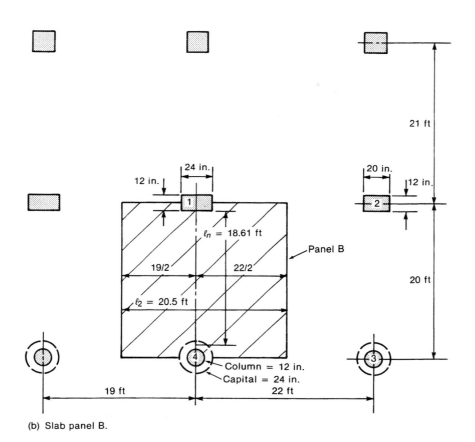

(b) Slab panel B.

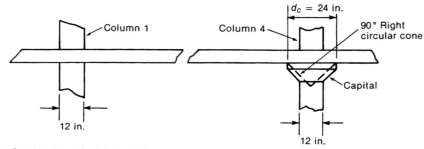

(c) Section through slab panel B.

Fig. 13–24
(Continued)

$$M_0 = \frac{0.310 \text{ ksf} \times 20.5 \text{ ft} \times 20.17^2 \text{ ft}^2}{8}$$

$$= 323 \text{ ft-kips}$$

3. **Consider panel B, spanning from column 1 to column 4.** Slab panel B is shown shaded in Fig. 13–24b. The moments computed here would be used to design the reinforcement running parallel to line 1–4 in this panel.

A section through the slab showing columns 1 and 4 is shown in Fig. 13–24c. Column 4 has a *column capital*. ACI Sec. 13.1.2 defines the effective diameter of this capital as the diameter, measured at the bottom of the slab or drop panel, of the largest right circular cone or pyramid with a $90°$ vertex that can be included within the column and capital. The outline of such a cone is shown with dashed lines in Fig. 13–24c and the diameter, d_c, is 24 in. For the purposes of computing ℓ_n, the circular supports are replaced by equivalent square columns having a side length $c_1 = d_c \sqrt{\pi}/2$ or equal to $0.886 d_c$. Thus

$$\ell_n = 20 \text{ ft} - \frac{1}{2}\left(\frac{12}{12}\right) \text{ ft} - \frac{1}{2}\left(0.886 \times \frac{24}{12}\right) \text{ ft} = 18.61 \text{ ft}$$

$$\ell_2 = \frac{19}{2} \text{ ft} + \frac{22}{2} \text{ ft} = 20.5 \text{ ft}$$

$$M_0 = \frac{0.310 \text{ ksf} \times 20.5 \times 18.61^2 \text{ ft}^2}{8} = 275 \text{ ft-kips} \quad ■$$

Definition of Column Strips and Middle Strips

As seen in Sec. 13–5 and Fig. 13–17a, the moments vary continuously across the width of the slab panel. To aid in steel placement, the design moments are averaged over the width of *column strips* over the columns and *middle strips* between the column strips as shown in Fig. 13–17d. The widths of these strips are defined in ACI Secs. 13.2.1 and 13.2.2 and are illustrated in Fig. 13–25. The column strips in both directions extend one-fourth of the smaller span, ℓ_{min}, each way from the column line.

Positive and Negative Moments in Panels

In the Direct Design Method, the total statical moment M_0 is divided into positive and negative moments according to rules given in ACI Sec. 13.6.3. These are illustrated in Fig. 13–26. In interior spans, 65% of M_0 is assigned to the negative moment region and 35% to the positive moment regions. This is approximately the same as a uniformly loaded, fixed-ended beam where the negative moment is two-thirds of $w\ell^2/8$ and the positive moment is one-third.

The exterior end of the span has considerably less fixity than on the interior support. The division of M_0 in an end span into positive and negative moment regions is given in Table 13–2. In this table "exterior edge unrestrained" refers to a slab whose exterior edge rested on, but was not attached to, a masonry wall, for example, while "exterior edge fully restrained" refers to a slab whose exterior edge was supported by, and was continuous with, a concrete wall with a flexural stiffness as large or larger than that of the slab.

If the computed negative moments on two sides of a support are different, the negative moment section of the slab is designed for the larger of the two unless a moment distribution is carried out to divide the moment between the members meeting at the joint. The use of such a moment distribution is illustrated in Example 13–9.

Provision for Pattern Loadings

In the design of continuous reinforced concrete beams, analyses are carried out for several distributions of live load to get the largest values of positive and negative moment in each

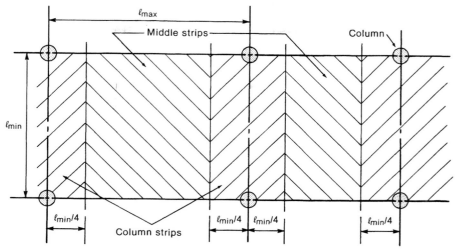

(a) Short direction of panel.

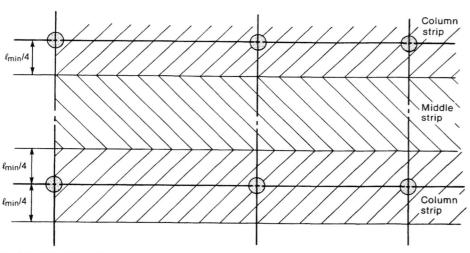

Fig. 13–25
Definition of column and
middle strips.

(b) Long direction of panel.

span. In the case of a slab designed by the Direct Design Method, no such analysis is done. The Direct Design Method can only be used when the live load does not exceed two times the dead load. The effects of pattern loadings are not large in such a case.[13-8]

Provision for Concentrated Loads

Some building codes require that office floors be designed for the largest effects of either a uniform load, or a concentrated load placed anywhere on the floor distributed over a 30 in. by 30 in. area. The ACI slab design provisions apply only to the uniform case. Woodring and Siess[13-9] studied the moments due to concentrated loads acting on square interior slab panels. For 20-ft-, 25-ft-, and 30-ft-square flat plates the largest moments due to a 2000-lb concentrated load on a 30-in.-square area were equivalent to uniform loads of 39 or -23 psf, 27 or -16 psf, and 20 or -13 psf, respectively, for the three sizes of slab. The positive (downward) equivalent uniform loads are smaller than the 50- or 100-psf loads used in office floor design and hence do not govern. The negative equivalent uniform loads indicate that at some point in the slab the worst effect of the concentrated load is equivalent to that

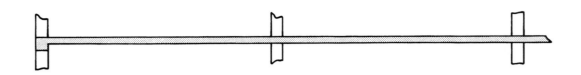

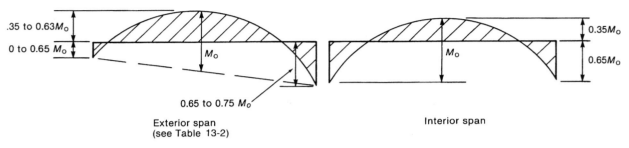

Fig. 13–26
Assignment of M_0 to positive and negative moment regions.

TABLE 13–2 Distribution of Total Factored Static Moment, M_0, in an End Span

	(1) Exterior Edge Unrestrained	(2) Slab *with* Beams between *All* Supports	(3) Slab *without* Beams between *Interior* Supports Without Edge Beam	(4) With Edge Beam	(5) Exterior Edge Fully Restrained
Interior negative factored moment	0.75	0.70	0.70	0.70	0.65
Positive factored moment	0.63	0.57	0.52	0.50	0.35
Exterior negative factored moment	0	0.16	0.26	0.30	0.65

Source: ACI Sec. 13.6.3.3.

of an upward uniform load. Since these are all less than the dead load of a concrete floor slab, the concentrated load case does not govern here either. In summary, therefore, design for the uniform load case will satisfy the 2000 lb concentrated load case for 20-ft-square and larger slabs. For larger concentrated loads the strip method can be used (see Sec. 15–4).

Distribution of Moments between Middle Strips and Column Strips

ACI Sec. 13.6.4 defines the fraction of the negative and positive moments assigned to the column strips. The remaining negative and positive moment is assigned to the adjacent half middle strips. The division is a function of $\alpha_1 \ell_2 / \ell_1$, which depends on the aspect ratio of

Two-Way Slabs: Behavior, Analysis, and Direct Design Method

the panel, ℓ_2/ℓ_1, and the relative stiffness, α_1, of the beams (if any) spanning in the direction of the panel. The calculation of α was discussed earlier in this section.

For a flat plate, $\alpha_1\ell_2/\ell_1$ is taken equal to zero since $\alpha = 0$ if there are no beams. In this case, 75% of the negative moment is in the column strip and the remaining 25% is divided equally between the two adjacent half middle strips, 12.5% to each (ACI Sec. 13.6.4.1). Similarly, 60% of the positive moment is assigned to the column strip and the remaining 40% is divided, with 20% going to each adjacent half middle strip (ACI Sec. 13.6.4.4).

At an exterior edge, the division of the exterior end negative moment in the strip spanning perpendicular to the edge also depends on the relative torsional stiffness of the edge beam, β_t, where β_t is calculated as the shear modulus, G, times the torsional constant of the edge beam, C, divided by the EI of the slab spanning perpendicular to the edge beam (i.e., EI for a slab having a width equal to the length of the edge beam from the center of one span to the center of the other span, as shown in Fig. 13–19d). Assuming that $v = 0$ gives $G = E/2$, and β_t is defined as

$$\beta_t = \frac{E_{cb}C}{2E_{cs}I_s} \tag{13–12}$$

where the cross section of the edge beam is as defined in ACI Sec. 13.7.5.1 and Fig. 13–27. Note that the cross section defined to compute β_t differs in some cases from that used to compute the flexural stiffness of the beam (Fig. 13–19). Conditions a, b, and c in

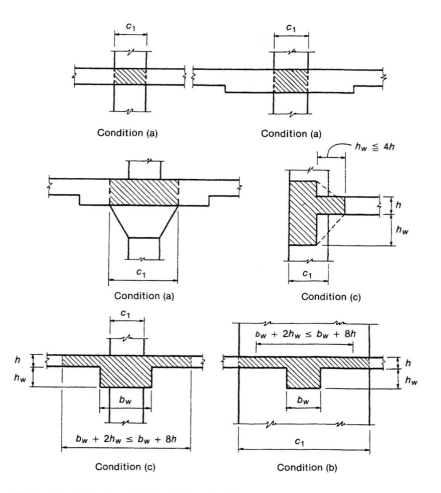

Fig. 13–27
Torsional members. (From Ref. 13–6. Courtesy of the Portland Cement Association.)

13–6 Direct Design Method: Introduction

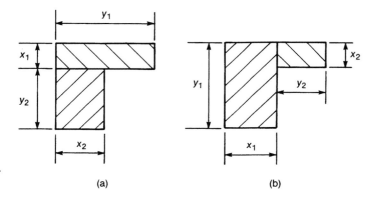

Fig. 13–28
Division of edge members for
calculation of C.

(a)   (b)

Fig. 13–27 refer to ACI Sec. 13.7.5.1 a, b, and c, respectively. If there are no edge beams, β_t can be taken equal to zero.

The term C in Eq. 13–12 refers to the torsional constant of the edge beam. This is roughly equivalent to a polar moment of inertia. It is calculated by subdividing the cross section into rectangles and carrying out the following summation:

$$C = \sum \left[\left(1 - 0.63\frac{x}{y} \right) \frac{x^3 y}{3} \right] \tag{13–13}$$

where x is the shorter side of a rectangle and y is the longer side. The subdivision of the cross section of the torsional members is illustrated in Fig. 13–28. Several possible combinations of rectangles may have to be tried to get the *maximum* value of C. To do so, the wide rectangles should be made as large as possible. Thus the rectangles chosen in Fig. 13–28b will give larger values of C than those in Fig. 13–28a.

ACI Sec. 13.6.4.2 assigns from 45 to 100% of the negative moment to the column strip at the exterior end of a panel (at points A in Fig. 13–23). The exact distribution depends on ℓ_1/ℓ_2, $\alpha_1\ell_2/\ell_1$, and β_t.

If a beam is present in the column strip (parallel to it), a portion of the column strip moment is assigned to the beam as specified in ACI Sec. 13.6.5. If the beam has $\alpha_1\ell_2/\ell_1$ greater than 1.0, 85% of the column strip moments are assigned to the beam and 15% to the slab. This is discussed more fully in Sec. 13–13 and Example 13–9.

EXAMPLE 13–3   Calculation of Moments in an Interior Panel of a Flat Plate

Figure 13–29 shows an interior panel of a flat-plate floor in an apartment building. The slab thickness is 5.5 in. The slab supports a design live load of 50 psf and a superimposed dead load of 25 psf for partitions. The columns and slab have the same strength concrete. The story height is 9 ft. Compute the column strip and middle strip moments in the short direction of the panel.

1.   Compute the factored loads.

$$w_u = 1.4\left(\frac{5.5}{12} \times 150 + 25\right) + 1.7(50) = 216 \text{ psf}$$

Note that if the local building code allows a live-load reduction factor, the 50-psf live load should be multiplied by the appropriate factor. The reduction should be based on the area $\ell_1 \times \ell_2$.

2.   Compute the moments in the short span of the slab.

(a)   Compute ℓ_n, ℓ_2 and divide the slab into column and middle strips.

$$\ell_n = 13.17 - \frac{10}{12} = 12.33 \text{ ft}$$

$$\ell_2 = 14.5 \text{ ft}$$

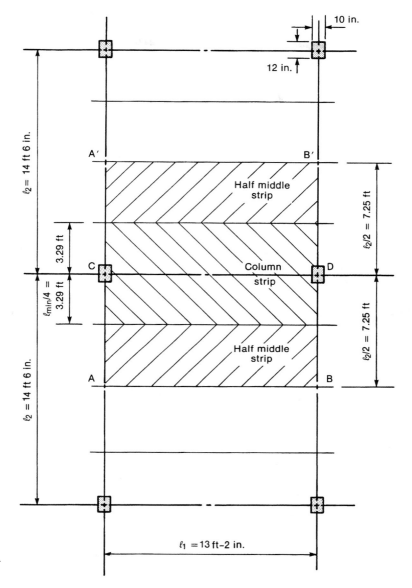

Fig. 13–29
Interior panel of a flat plate—
Example 13–3.

The column strip extends the smaller of $\ell_2/4$ or $\ell_1/4$ on each side of the column centerline, as shown in Fig. 13–25 (ACI Sec. 13.2.1). Thus the column strip extends $13.17/4 = 3.29$ ft on each side of column centerline. The total width of the column strip is 6.58 ft. Each half middle strip extends from the edge of the column strip to the centerline of the panel. The total width of two half middle strips is $14.5 - 6.58 = 7.92$ ft.

    (b) Compute M_0.

$$M_0 = \frac{w_u \ell_2 \ell_n^2}{8} = \frac{0.216 \times 14.5 \times 12.33^2}{8}$$

$$= 59.5 \text{ ft-kips} \tag{13–5}$$

$$\text{(ACI Eq. 13–3)}$$

    (c) Divide M_0 into negative and positive moments. From Code Sec. 13.6.3.2:

$$\text{negative moment} = 0.65 M_0 = -38.7 \text{ ft-kips}$$

$$\text{positive moment} = 0.35 M_0 = 20.8 \text{ ft-kips}$$

This process is illustrated in Fig. 13–30a and the resulting distribution of total moments is shown in Fig. 13–30b.

(d) Divide the moments between the column and middle strips.

Negative moments: From ACI Sec. 13.6.4.1 for $\alpha_1 \ell_2 / \ell_1 = 0$ (taken equal to zero since $\alpha_1 = 0$ because there are no beams between columns A and B in this panel):

$$\text{column strip negative moment} = 0.75 \times -38.7 \text{ ft-kips}$$
$$= -29.0 \text{ ft-kips}$$
$$\text{middle strip negative moment} = 0.25 \times -38.7 \text{ ft-kips}$$
$$= -9.7 \text{ ft-kips}$$

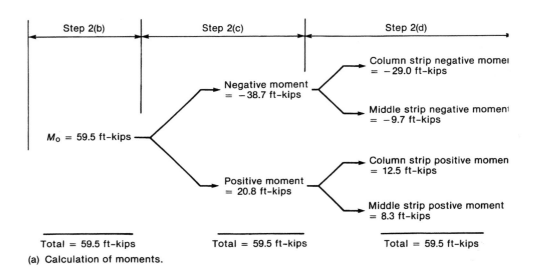

(a) Calculation of moments.

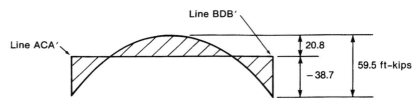

(b) Division of M_o into positve and negative moments.
— Step 2(c).

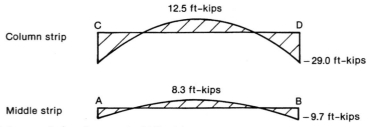

(c) Total moments in column and middle strips.

Fig. 13–30
Calculation of moments—Example 13–3.

Half of this, -4.85 ft-kips, goes to each adjacent half middle strip. Since the adjacent bays have the same width, ℓ_2, a similar moment is assigned to the other half of each middle strip so that the total middle strip negative moment is 9.7 ft-kips.

Positive moments: From ACI Sec. 13.6.4.4, where $a_1\ell_2/\ell_1 = 0$:

$$\text{column strip positive moment} = 0.6 \times 20.8 = 12.5 \text{ ft-kips}$$

$$\text{middle strip positive moment} = 0.4 \times 20.8 = 8.3 \text{ ft-kips}$$

This calculation is illustrated in Fig. 13–30a. The resulting distributions of moments in the column strip and middle strip are summarized in Fig. 13–30c. In Fig. 13–31, the moments in each strip have been divided by the width of that strip. This diagram is very similar to the theoretical elastic distribution of moments given in Fig. 13–17d.

3. Compute the moments in the long span of the slab. Although not asked for in this example, in a slab design it would now be necessary to repeat steps 2(a) to 2(e) for the long span. ∎

EXAMPLE 13–4 Calculation of Moments in an Exterior Panel of a Flat Plate

Compute the positive and negative moments in the column and middle strips of the exterior panel of the slab between columns B and E in Fig. 13–32. The slab is 8 in. thick and supports a superimposed service dead load of 25 psf and a service live load of 60 psf. The beam is 12 in. wide by 16 in. in overall depth and is cast monolithically with the slab.

1. Compute the factored loads.

$$w_u = 1.4\left(\frac{8}{12} \times 150 + 25\right) + 1.7(60) = 277 \text{ psf}$$

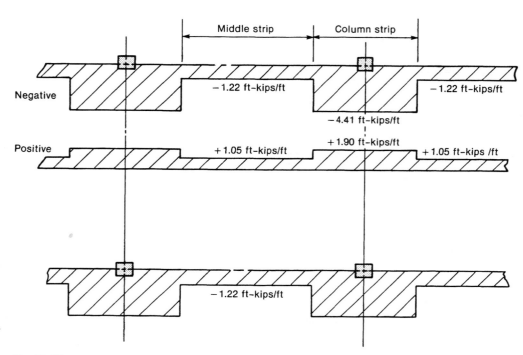

Fig. 13–31
Distribution of moments in an interior panel of a flat slab—Example 13–3.

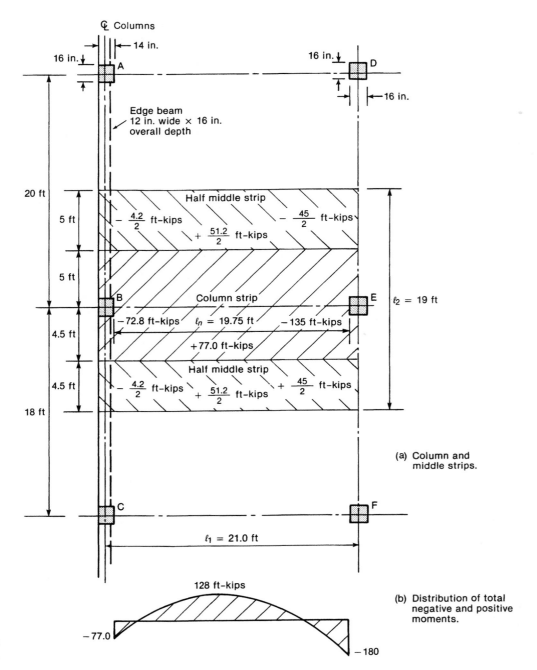

Fig. 13–32
Calculation of moments in
end span—Example 13–4.

2. **Compute the moments in span *BE*.**

   (a) **Compute ℓ_n, ℓ_2 and divide the slab into middle and column strips.**

   $$\ell_n = 21.0 - \frac{1}{2}\left(\frac{14}{12}\right) - \frac{1}{2}\left(\frac{16}{12}\right) = 19.75 \text{ ft}$$

   $$\ell_2 = 19 \text{ ft}$$

The column strip extends smaller of $\ell_2/4$ or $\ell_1/4$ on each side of the column centerline. Since ℓ_1 is greater than either value of ℓ_2, base this on ℓ_2. The column strip extends $20/4 = 5$ ft toward AD and $18/4 = 4.5$ ft toward CF from line BE, as shown in Fig. 13–32a. The total width of the column strip is 9.5 ft. The half middle strip between BE and CF has a width of 4.5 ft, and the other one is 5 ft, as shown.

(b) Compute M_0.

$$M_0 = \frac{w_u \ell_2 \ell_n^2}{8} = \frac{0.277 \times 19 \times 19.75^2}{8}$$

$$= 257 \text{ ft-kips}$$

(13–5)
(ACI Eq. 13–3)

(c) Divide M_0 into positive and negative moments. The distribution of the moment to the negative and positive regions is as given in Table 13–2 (ACI Sec. 13.6.3.3). In the terminology of Table 13–2 this is a "slab without beams between interior supports with edge beam." From Table 13–2, the total moment is divided as:

interior negative $M_u = 0.70 M_0 = -180$ ft-kips

positive $M_u = 0.50 M_0 = 128$ ft-kips

exterior negative $M_0 = 0.30 M_0 = -77$ ft-kips

This calculation is illustrated in Fig. 13–33. The resulting negative and positive moments are shown in Fig. 13–32b.

(d) Divide the moments between the column and middle strips (see Fig. 13–33).
Interior negative moments: This division is a function of $\alpha_1 \ell_2/\ell_1$, which again equals 0 since there are no beams parallel to BE. From ACI Sec. 13.6.4.1,

interior column strip negative moment $= 0.75 \times -180 = -135$ ft-kips

$= -14.2$ ft-kips/ft of width of column strip

interior middle strip negative moment $= -45$ ft-kips

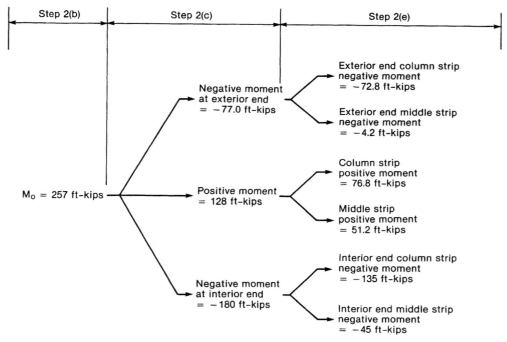

Fig. 13–33
Calculation of moments in end span—Example 13–4.

Half of this goes to each of the half middle strips beside column strip *BE*.

Positive moments: From ACI Sec. 13.6.4.4,

$$\text{column strip positive moment} = 0.60 \times 128 = 76.8 \text{ ft-kips}$$

$$= 8.1 \text{ ft-kips/ft}$$

$$\text{middle strip positive moment} = 51.2 \text{ ft-kips}$$

Half of this goes to each half middle strip.

Exterior negative moment: From ACI Sec. 13.6.4.2 the exterior negative moment is divided as a function of $\alpha_1 \ell_2 / \ell_1$ (again equal to zero since there is no beam parallel to ℓ_1) and β_t, where

$$\beta_t = \frac{E_{cb}C}{2E_{cs}I_s} \tag{13–12}$$

where E_{cb} and C refer to the attached torsional member shown in Fig. 13–34 and E_{cs} and I_s refer to the strip of slab being designed (the column strip and the two half middle strips shown shaded in Fig. 13–32a). To compute C, divide the edge beam into rectangles. The two possibilities shown in Fig. 13–34 will be considered. For Fig. 13–34a, Eq. 13–13 (ACI Eq. 13–7) gives

$$C = \frac{(1 - 0.63 \times 12/16)12^3 \times 16}{3} + \frac{(1 - 0.63 \times 8/8)8^3 \times 8}{3}$$

$$= 5367 \text{ in.}^4$$

For Fig. 13–34b, $C = 3741 \text{ in.}^4$. The larger of these two values is used; therefore, $C = 5367 \text{ in.}^4$.

I_s is the moment of inertia of the strip of slab being designed. It has $b = 19$ ft and $h = 8$ in.

$$I_s = \frac{(19 \times 12) \times 8^3}{12} = 9728 \text{ in.}^4$$

Since f_c' is the same in the slab and beam, $E_{cb} = E_{cs}$ and

$$\beta_t = \frac{5367}{9728} = 0.552$$

Interpolating in the table given in ACI Sec. 13.6.4.2:

For $\beta_t = 0$: 100% to column strip
For $\beta_t = 2.5$: 75% to column strip

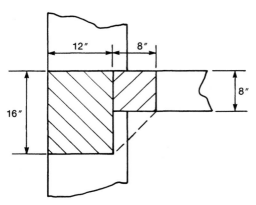

(a) Attached torsional member.

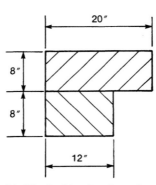

(b) Attached torsional member.

Fig. 13–34
Slab, column, and edge
beam—Example 13–4.

Therefore, for $\beta_t = 0.552$, 94.5% to column strip. Therefore,

$$\text{exterior column strip negative moment} = 0.945(-77.0) = -72.8 \text{ ft-kips}$$

$$= -7.66 \text{ ft-kips/ft}$$

$$\text{exterior middle strip negative moment} = -4.2 \text{ ft-kips} \qquad \blacksquare$$

Moments in Columns and Transfer of Moments to Columns

Exterior Columns

When design is carried out using the Direct Design Method, ACI Sec. 13.6.3.6 specifies that the moment that is transferred from a slab to an edge column is $0.3M_0$. This moment is used to compute the shear stresses due to moment transfer to the edge column, as shown in Sec. 13–8. Although the ACI Code does not specifically state, this moment can be assumed to be about the centroid of the shear perimeter. The exterior negative moment from the Direct Design Method calculation is divided between the columns above and below the slab in proportion to the column stiffnesses, $4EI/\ell$. The resulting column moments are used in the design of the columns.

Interior Columns

At interior columns, the moment transfer calculations and the total moment used in the design of the columns above and below the floor are based on an unbalanced moment resulting from an uneven distribution of live load. The unbalanced moment is computed assuming that the longer span adjacent to the column is loaded with the factored dead load and half the factored live load, while the shorter span carries only the factored dead load. The total unbalanced negative moment at the joint is thus

$$M = 0.65\left[\frac{(w_d + 0.5w_\ell)\ell_2\ell_n^2}{8} - \frac{w_d'\,\ell_2'\,(\ell_n')^2}{8}\right]$$

where w_d and w_ℓ refer to the factored dead and live loads on the longer span and w_d', ℓ_2', and ℓ_n' refer to the shorter span adjacent to the column. The factor 0.65 is the fraction of the static moment assigned to the negative moment at an interior support. The factors 0.65 and $\frac{1}{8}$ combine to give 0.081. A portion of the unbalanced moment is distributed to the slabs, and the rest goes to the columns. Since slab stiffnesses have not been calculated, it is assumed that most of the moment is transferred to the columns, giving

$$M_{col} = 0.07\left[(w_d + 0.5w_\ell)\ell_2\ell_n^2 - w_d'\,\ell_2'\,(\ell_n')^2\right] \qquad (13\text{–}14)$$
$$\text{(ACI Eq. 13–4)}$$

The moment, M_{col}, is used to design the slab-to-column joint. It is distributed between the columns above and below the joint in the ratio of their stiffnesses to determine the moments used to design the columns.

Design of Edge Beams for Shear and Moment

When a slab panel contains a beam, either an edge beam or an interior beam between the columns, the moments in the panel are divided between the slab and the beam, as discussed in Sec. 13–13 and Example 13–9. The design of edge beams for flat-plate floors proceeds exactly as in that example.

A shear failure in a beam results from an inclined crack caused by flexural and shearing stresses. This crack starts at the tensile face of the beam and extends diagonally to the compression zone near a concentrated load as explained in Chap. 6. In the case of a two-way slab or footing, the two shear failure mechanisms shown in Fig. 13–35 are possible. *One-way shear* or *beam action shear* (Fig. 13–35a) involves an inclined crack extending across the entire width of the structure. *Two-way shear* or *punching shear* involves a truncated cone or pyramid-shaped surface around the column as shown in Fig. 13–35b. Generally, the punching shear capacity of a slab or footing will be considerably less than the one-way shear capacity. In design it is necessary to consider both failure mechanisms, however.

This section is limited to footings and slabs without beams. The shear strength of slabs with beams is discussed in Sec. 13–13.

Behavior of Slabs Failing in Two-Way Shear

As shown in Sec. 13–5, the maximum moments in a uniformly loaded flat plate occur around the columns and lead to a circular crack around each column. After additional loading, the cracks necessary to form a fan yield line mechanism develop and at about the same time, inclined or shear cracks form on the truncated conical surface shown in Fig. 13–35b. These cracks can be seen in Fig. 13–36, which shows a slab that has been sawn through along two sides of the column after the slab had failed in two-way shear.

In Chap. 6, truss models are used to explain the behavior of beams failing in shear. Alexander and Simmonds[13–10] have explained punching shear failures using the truss model shown in Fig. 13–37. Prior to the formation of the inclined cracks shown in Fig. 13–36, the shear is transferred by shear stresses. Once the cracks have formed, shear cannot be transferred across them. Now the shear is transferred by inclined struts *A–B* and *C–D* extending from the bottom of the slab at the column to the reinforcement at the top of the slab at *A* and *D*. Similar struts exist on all four sides of the column. The horizontal component of the

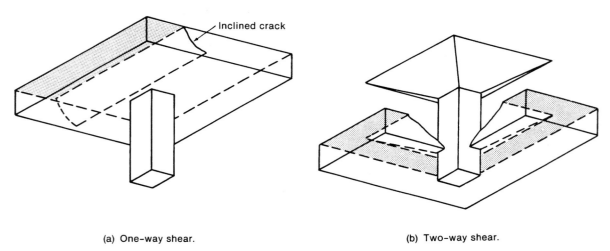

(a) One-way shear.　　　　　　　　　　(b) Two-way shear.

Fig. 13–35
Shear failures in a slab.

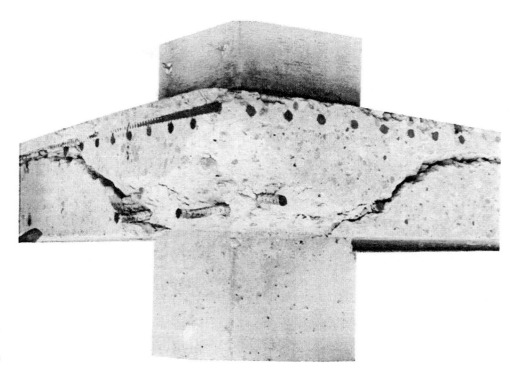

Fig. 13–36
Inclined cracks in a slab after a shear failure. (Photograph courtesy of J. G. MacGregor.)

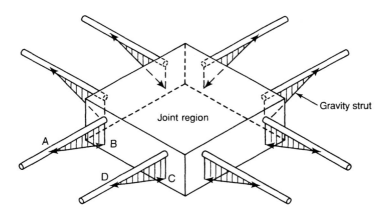

Fig. 13–37
Truss model for shear transfer at an interior column.

force in the struts causes a change in the force in the reinforcement at *A* and *D,* the vertical component pushes up on the bar and is resisted by tensile stresses in the concrete between the bars. Eventually, this concrete cracks in the plane of the bars and a punching failure results. Such a failure occurs suddenly with little, if any, warning. Once a punching shear failure has occurred, the shear capacity of that particular joint is completely lost. In the case of a two-way slab, as the slab slides down the column, the negative moment steel near the top of the slab rips out of the top of the slab, leaving no physical connection between the slab and the column. Thus although a two-way slab possesses great ductility if it fails in flexure, it has very little if it fails in shear. Excellent reviews of the factors affecting the shear strength of slabs are presented in Refs. 13–11 and 13–12.

13–7   Shear Strength of Two-Way Slabs

579

Design for Two-Way Shear

Initially, this discussion will consider the case of shear transfer without appreciable moment transfer. The case when both shear and moment are transferred from the slab to the column is discussed in Sec. 13–8.

Based on extensive tests, Moe[13-13] concluded that the critical section for shear was located at the surface of the column. ACI-ASCE Committee 326 (now 445)[13-14] accepted Moe's conclusions, but showed that a much simpler design equation could be derived by considering a critical section located at $d/2$ away from the face of the column, where d is the average effective depth of the slab. This simplification has been incorporated in the ACI Code.

Location of the Critical Perimeter

Two-way shear is assumed to be critical on a vertical section through the slab or footing and extending around the column. According to ACI Sec. 11.12.1.2, this section is chosen so that it is never less than $d/2$ from the face of the column and so that its length b_0, is a minimum. Although this would imply that the corners of the shear perimeter should be rounded, the intent of the Code is that the critical section for rectangular columns should be rectangular. Several examples are given in Fig. 13–38. In slabs with drop panels around the column two critical sections should be considered as shown in Fig. 13–39. When openings are located at less than 10 times the slab thickness from a column, ACI Sec. 11.12.5 requires that the critical perimeter be reduced as shown in Fig. 13–40a.

The critical perimeter for edge or corner columns is not clearly defined in the ACI Code. In 1978, ACI-ASCE Committee 426[13-15] recommended that the critical sections in Fig. 13–40b and c be considered.

Design Equations: Two-Way Shear with Negligible Moment Transfer

Unbalanced floor loads, or lateral loads, on a flat-plate building require that both moments and shears be transferred from the slab to the columns. In the case of interior columns in a braced flat-plate building, the worst loading case for shear generally corresponds to a negligible moment transfer from the slab to the column. Similarly, columns generally transfer little or no moment to footings.

Design for two-way shear without moment transfer is carried out using Eqs. 6–9, 6–14, and 13–15 to 13–17 (ACI Eqs. 11–1, 11–2, and 11–35 to 11–37). The basic equation for shear design states

$$V_u \le \phi V_n \qquad \text{(6–14)}$$
$$\text{(ACI Eq. 11–1)}$$

where V_u is the factored shear force due to the loads and V_n is the nominal shear resistance of the slab or footing. For shear, the strength reduction factor, ϕ, equals 0.85.

For uniformly loaded two-way slabs the tributary areas used to calculate V_u are bounded by lines of zero shear. For interior panels these lines can be assumed to pass through the center of the panel. For edge panels the moment coefficients in ACI Sec. 13.6.3.3 correspond to lines of zero shear at $0.45\ell_n$ and $0.44\ell_n$ from the exterior support in flat plates with and without edge beams, respectively (see Fig. 13–41). For simplicity the lines of zero shear are frequently assumed to occur at midspan. This is conservative for shear at the exterior columns, where V_u will be overestimated but is unconservative for the shear at the first interior columns.

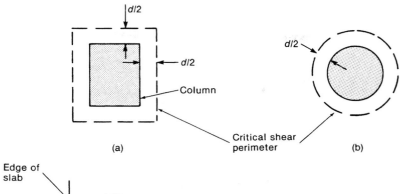

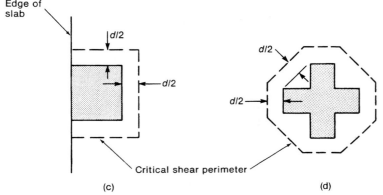

Fig. 13–38
Location of critical shear
perimeters.

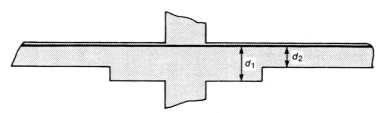

(a) Section through drop panel.

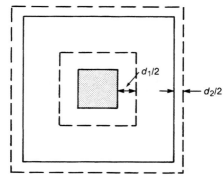

Fig. 13–39
Critical sections in slab with
drop panels.

(b) Critical sections.

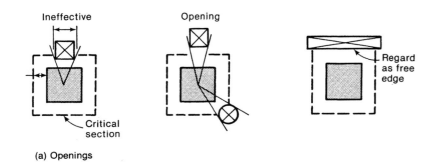

(a) Openings

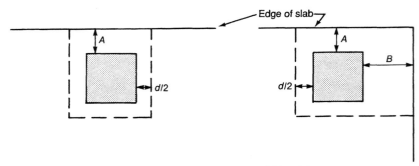

(b) Perimeters if A and B do not exceed the greater of 4h or 2ℓ$_d$.

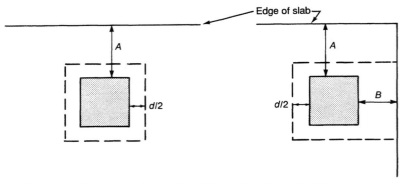

Fig. 13–40
Effect of openings and edges
on critical perimeter.

(c) Perimeters if A exceeds the greater of 4h or 2ℓ$_d$, but B does not.

ACI Eq. 11–2 defines V_n as follows:

$$V_n = V_c + V_s \qquad\qquad (6\text{--}9)$$
$$\text{(ACI Eq. 11--2)}$$

where V_c and V_s are the shear resistances attributed to the concrete and the shear reinforcement, respectively. In most slabs V_s is zero. For two-way shear V_c is taken as the smallest of:

$$\text{(a)} \qquad V_c = \left(2 + \frac{4}{\beta_c}\right)\sqrt{f_c'}\,b_o d \qquad\qquad (13\text{--}15)$$
$$\text{ACI Eq. 11--35)}$$

where β_c is the ratio of long side to short side of the column, concentrated load, or reaction area. For nonrectangular columns this is defined as shown in Fig. 13–42.

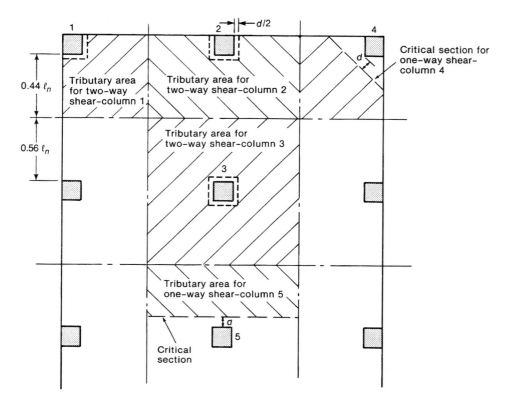

Fig. 13–41
Critical sections and tributary
areas for shear in a flat plate.

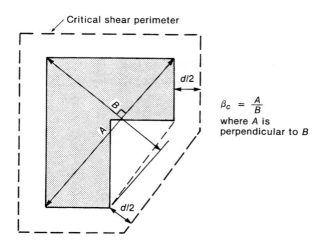

Fig. 13–42
Definition of β_c for irregular
shaped column.

$$\text{(b)} \quad V_c = \left(\frac{\alpha_s d}{b_o} + 2\right)\sqrt{f_c'}\,b_o d \qquad (13\text{–}16)$$

$$\text{(ACI Eq. 11–36)}$$

where α_s is 40 for interior columns, 30 for edge columns, and 20 for corner columns; and

$$\text{(c)} \quad V_c = 4\sqrt{f_c'}\,b_o d \qquad (13\text{–}17)$$

$$\text{(ACI Eq. 11–37)}$$

The distribution of shear stresses around the column is approximately as shown in Fig. 13–43 with higher shear stresses transferred in the vicinity of the corners.[13-12] For very large columns or rectangular columns with two long sides the shear stress between the corners decreases, approaching the value for one-way shear, $2\sqrt{f_c'}$.

Equation 13–15 applies to the case of rectangular columns. It is compared to test data in Fig. 13–44a.[13-16] Equation 13–16 applies to very large columns. It is compared to test data in Fig. 13–44b.[13-17] Equation 13–17 applies if β_c is less than 2 or if b_o/d is less than 20 in the case of an interior column.

One-Way Shear in Slabs

In the case of a uniformly loaded slab, the critical section for one-way shear is located at d from the face of the support or at d from the face of a drop panel or other change in thickness (Fig. 13–35b, ACI Sec. 11.12.1.1). The tributary areas for one-way shear in a slab are illustrated in Fig. 13–41 (for columns 4 and 5). The shear strength on the critical section is computed as for beams (see Chap. 6) using Eqs. 6–14, 6–9, and 6–8 (ACI Eqs. 11–1, 11–2, and 11–3), where

$$V_c = 2\sqrt{f_c'}b_w d \quad \text{lb} \tag{6–8}$$

One-way shear is seldom critical in flat plates or flat slabs, as will be seen from Example 13–5.

EXAMPLE 13–5 Checking One-Way and Two-Way Shear at an Interior Column in a Flat Plate

Figure 13–45 shows an interior column in a large uniform flat-plate slab. The slab is 6 in. thick and has $d = 5$ in. for reinforcement perpendicular to the long side of the column and $d = 4.5$ in. in the other direction. The slab supports a uniform superimposed dead load of 20 psf and a uniform superimposed live load of 60 psf. The concrete strength is 3000 psi. The moments transferred from the slab to the column (or vice versa) are negligible. Check whether the shear capacity is adequate.

1. Determine the factored uniform load.

$$w_u = 1.4\left(\frac{6}{12} \times 150 + 20\right) + 1.7 \times 60 = 235 \text{ psf}$$

$$= 0.235 \text{ ksf}$$

2. Check the one-way shear. One-way shear is critical at a distance d from the face of the column. Thus the critical sections for one-way shear are A–A and B–B in Fig. 13–45. The loaded areas causing shear on these sections are cross hatched. Their outer boundaries are lines of symmetry on which $V_u = 0$. Since the tributary area for section A–A is bigger, this section will be more critical.

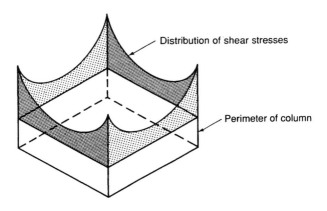

Distribution of shear stresses

Perimeter of column

Fig. 13–43
Distribution of shear around the perimeter of a square column.

Two-Way Slabs: Behavior, Analysis, and Direct Design Method

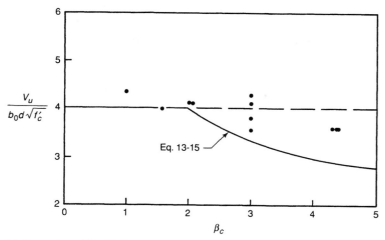

(a) Comparison of Eq. 13-15 to tests of rectangular columns

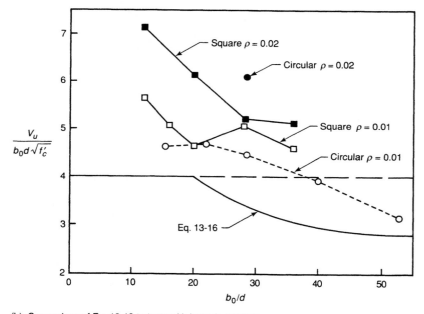

Fig. 13–44
Comparison of design equations to test data.

(b) Comparison of Eq. 13-16 to tests with large b_0/d ratios

(a) **Compute V_u at section A–A.**

$$V_u = 0.235 \text{ ksf} \times 8.08 \text{ ft} \times 18 \text{ ft} = 34.2 \text{ kips}$$

(b) **Compute ϕV_n for one-way shear.** Since there are no stirrups or other shear reinforcement (from Eqs. 6–9 and 6–14),

$$\phi V_n = \phi V_c$$

where V_c for one way shear is given by Eq. 6–8:

$$\phi V_c = 0.85(2\sqrt{f_c'}bd)$$

$$= 0.85\left(2\sqrt{3000} \times (18 \text{ ft} \times 12 \text{ in.}) \times \frac{5 \text{ in.}}{1000}\right)$$

$$= 100.6 \text{ kips}$$

13–7 Shear Strength of Two-Way Slabs

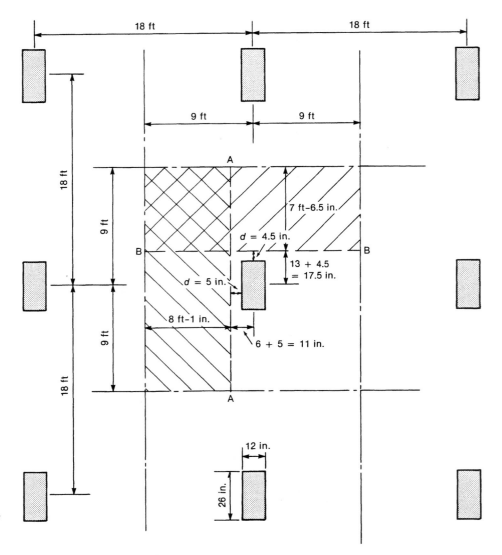

Fig. 13–45
Critical sections for one-way
shear at interior column—
Example 13–5.

Since $\phi V_c > V_u$, the slab is OK in one-way shear.

3. Check the two-way shear. Two-way shear is critical on a rectangular section located at $d/2$ away from the face of the column, as shown in Fig. 13–46. The load on the cross hatched area causes shear on the critical perimeter. Once again, the outer boundaries of this area are lines of symmetry where V_u is assumed to be zero. The average depth, $d = 4.75$ in., will be used in these computations.

(a) **Compute V_u on the critical perimeter.**

$$V_u = 0.235 \text{ ksf} \left[(18 \text{ ft} \times 18 \text{ ft}) - \left(\frac{16.75 \text{ in.}}{12} \times \frac{30.75 \text{ in.}}{12} \right) \text{ft}^2 \right]$$
$$= 75.3 \text{ kips}$$

(b) **Compute ϕV_c for the critical section.** V_c is the smallest of:

$$\text{(a)} \quad V_c = \left(2 + \frac{4}{\beta_c} \right) \sqrt{f_c'} b_o d \qquad \qquad \text{(13–15)}$$
$$\text{(ACI Eq. 11–35)}$$

Two-Way Slabs: Behavior, Analysis, and Direct Design Method

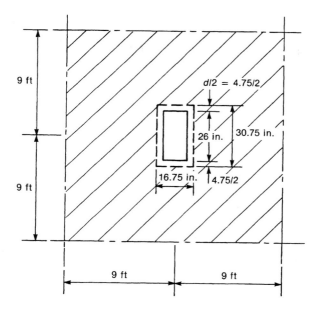

Fig. 13–46
Loaded area and critical section for two-way shear—Example 13–5.

where

$$\beta_c = \frac{26}{12} = 2.167$$

$$b_0 = 2(16.75 + 30.75) = 95 \text{ in.}$$

Therefore,

$$V_c = \left(2 + \frac{4}{2.167}\right)\sqrt{3000} \times 95 \times \frac{4.75}{1000}$$

$$= 95.1 \text{ kips}$$

(b) $\quad V_c = \left(\dfrac{\alpha_s d}{b_o} + 2\right)\sqrt{f_c'}b_o d$ (13–16)
(ACI Eq. 11–36)

where $\alpha_s = 40$ for an interior column. Therefore,

$$V_c = \left(\frac{40 \times 4.75}{95} + 2\right)\sqrt{3000} \times 95 \times \frac{4.75}{1000}$$

$$= 98.9 \text{ kips}$$

(c) $\quad V_c = 4\sqrt{f_c'}b_o d$ (13–17)
(ACI Eq. 11–37)

$$= 4\sqrt{3000} \times 95 \times \frac{4.75}{1000} = 98.9 \text{ kips}$$

Therefore, $V_c = 95.1$ kips and $\phi V_c = 0.85 \times 95.1 = 80.8$ kips. **Since ϕV_c exceeds V_u, the slab is OK in two-way shear.** ∎

Shear Reinforcement

If ϕV_c is less than V_u, the shear capacity can be increased by:

    1. Thickening the slab over the entire panel
    2. Using a drop panel to thicken the slab adjacent to the column

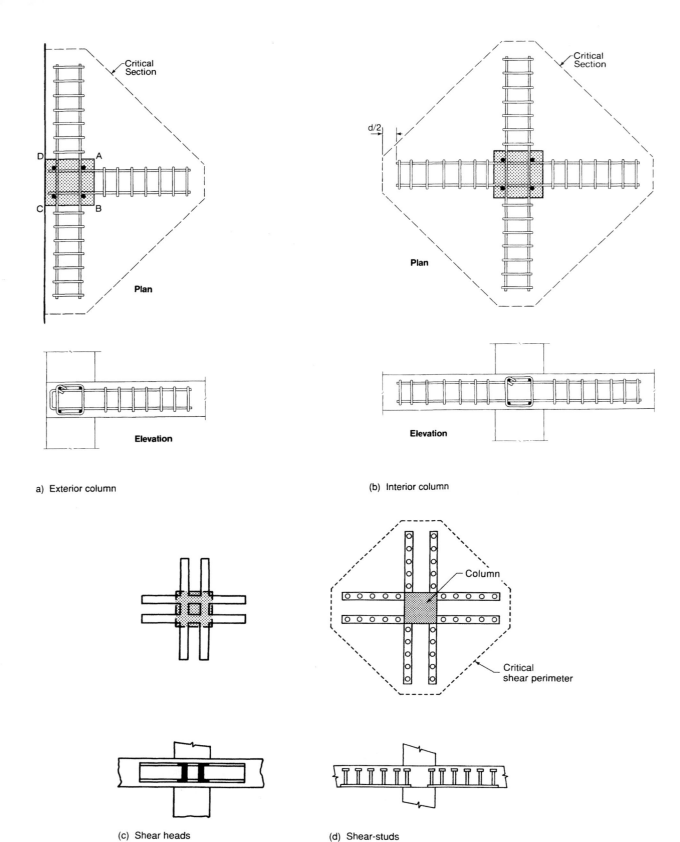

a) Exterior column

b) Interior column

c) Shear heads

d) Shear-studs

Fig. 13–47
Shear reinforcement in slabs.

3. Increasing b_o by increasing the column size or adding a fillet or a capital around the column

4. Adding shear reinforcement

Shear reinforcement, although not widely used in slabs, is generally one the three types shown in Fig. 13–47.

Closed stirrups with bars in all four corners are designed as stirrups would be in a beam. The stirrup cages must extend out at least as far as the critical section on which $\phi V_c = \phi(2\sqrt{f_c'})b_o d$ first exceeds V_u (ACI Sec. 11.12.3.1). Generally speaking, this type of stirrup is not practical in slabs less than about 12 in. thick. Figure 13–47a and b are drawn to scale for a 10-in. slab. Because the stirrup cage must fit between the layers of flexural steel in the two directions, two different sizes of stirrups are required as shown. The stirrups must be spaced at $d/2$, which requires a large number of stirrups. The truss theory used in Chap. 6 to explain the action of stirrups in beams requires that stirrups be anchored to the tension and compression chords. Anchorage to the tension chord is accomplished by bending the stirrups around bars near the tension face of the slab. Anchorage to the compression chord requires the stirrup to extend into the compression zone. This is difficult to ensure in a thin slab where the compression zone may be less than the bar cover and as a result, stirrups may not be fully effective in a thin slab. Finally, great care must be taken to anchor the forces developed in the stirrups within the shallow depth of the slab. For this reason closed stirrups with 135° hooks are recommended. If 90° hooks are used on the top or bottom surfaces, the tails will tend to straighten, pushing the cover off. Because of all of this ACI Sec. 11.12.3.2 limits V_n to $6\sqrt{f_c'}\,b_o d$ at the face of the column.

Structural steel *shearheads,* shown in Fig. 13–47c, may be used at interior columns. Shearheads at exterior columns require special provisions to transfer to the column the unbalanced moment due to loads on the arm perpendicular to the edge of the slab. The design of shearheads is discussed in ACI Sec. 11.12.4 and Ref. 13–18.

A third type of shear reinforcement that is gaining acceptance is fabricated from flat steel bars with resistance welded vertical rods, each capped with a circular plate as shown in Fig. 13–47d. Extensive testing has shown these to be very effective.[13-19, 13-20] These *shear studs* are easy to place and have a minimal effect on the placement of other reinforcement. Guidance for the design of this type of shear reinforcement is given in Ref. 13–21.

13–8 COMBINED SHEAR AND MOMENT TRANSFER IN TWO-WAY SLABS

Behavior of Slab–Column Connections Loaded with Shear and Moment

When lateral loads or unbalanced gravity loads cause a transfer of moment between the slab and column, or vice versa, the behavior is complex, involving flexure, shear, and torsion in the portion of the slab attached to the column as shown in Figs. 13–48 and 13–49, which will be discussed more fully later. Depending on the relative strengths in these three modes, failures can take various forms. Research on moment and shear transfer in slabs is reviewed in Refs. 13–11, 13–12, and 13–22. The truss model discussed in Sec. 13–7 and Fig. 13–37 can be extended to the case of moment and shear transfer.[13-23] Figure 13–50 shows an interior slab–column connection subjected to a large moment and a small shear. As a result, the struts on the front side correspond to those in Fig. 13–37 and resist downward forces, while those on the back resist uplift.

A truss model for an edge column is shown in Fig. 13–51. It is similar to the model for the interior column except that the back face cannot be utilized to transfer shear or

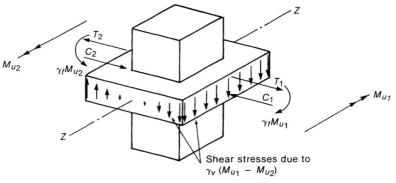

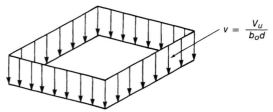

(a) Transfer of unbalanced moments to column.

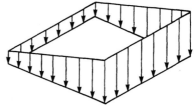

(b) Shear stresses due to V_u.

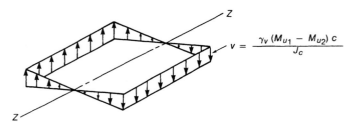

$$v = \frac{\gamma_v (M_{u1} - M_{u2}) c}{J_c}$$

(c) Shear due to unbalanced moment.

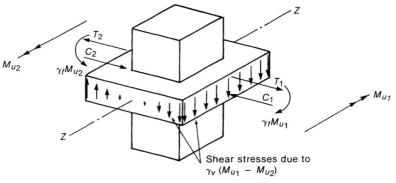

Fig. 13–48
Shear stresses due to shear
and moment transfer at an
interior column.

(d) Total shear stresses.

moment. If the top reinforcement outside the column (bar A) is used to transfer moment to the column, the bar forces must be transferred to the column by horizontal anchoring struts. The components of the anchoring strut force acting perpendicular to the side face of the column must be resisted by bars such as bar B in Fig. 13–51.[13–23] This action is relatively ineffective, with the result that bar A is not fully stressed. Figure 13–52 shows the steel strains measured at ultimate in the top bars perpendicular to the edge at the inner face of the edge column in a test specimen without an edge beam.[13–24] Only those bars that enter the column have yielded.

Two-Way Slabs: Behavior, Analysis, and Direct Design Method

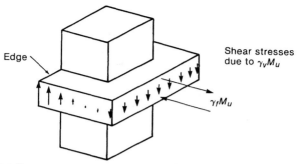

Edge

Shear stresses
due to $\gamma_v M_u$

$\gamma_f M_u$

(a) Transfer of moment at edge column.

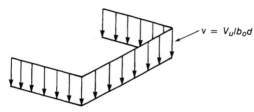

$v = V_u/b_o d$

(b) Shear stresses due to V_u.

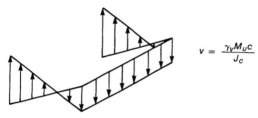

$v = \dfrac{\gamma_v M_u c}{J_c}$

(c) Shear stresses due to M_u.

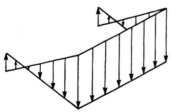

(d) Total shear stresses.

Fig. 13–49
Shear stresses due to shear
and moment transfer at an
edge column.

Figure 13–53 is a plot of the moments and shears transferred at failure in a number of test specimens of edge columns in slabs without edge beams.[13–25] The vertical axis represents the ratio of the ultimate moment in a test, M_s, taken about the centroid of the critical shear perimeter, to the nominal moment capacity, M_n, of the flexural reinforcement within a width $c_2 + 3h$, centered on the column, where c_2 is the width of the column and h is the slab thickness. (It should be noted that M_s was erroneously called M_u in Ref. 13–25 as pointed out in the closure to the discussion of Ref. 13–25.) The horizontal axis is the ratio of the ultimate shear transferred in a test, V_u, to the shear capacity in the absence of moment transfer, $V_c = 4\sqrt{f_c'}b_o d$. A safe lower envelope to the strengths is given by the

13–8 Combined Shear and Moment Transfer in Two-Way Slabs

591

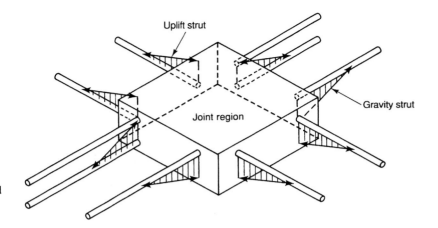

Fig. 13–50
Truss model for moment and shear transfer at an interior column. (From Ref. 13–10.)

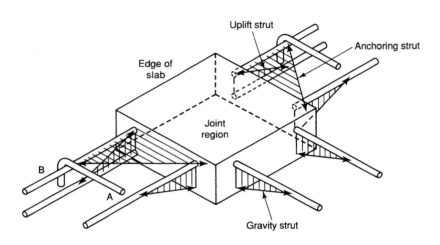

Fig. 13–51
Truss model for an edge column. (From Ref. 13–23.)

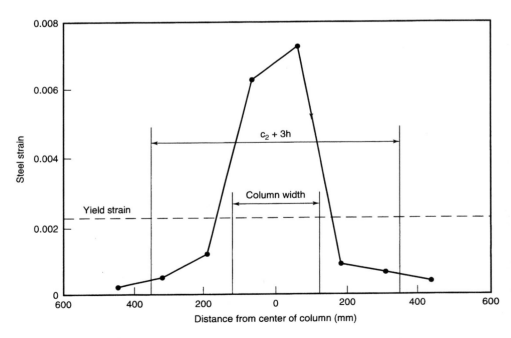

Fig. 13–52
Measured strains in bars adjacent to an edge column at failure. (From Ref. 13–24.)

         Two-Way Slabs: Behavior, Analysis, and Direct Design Method

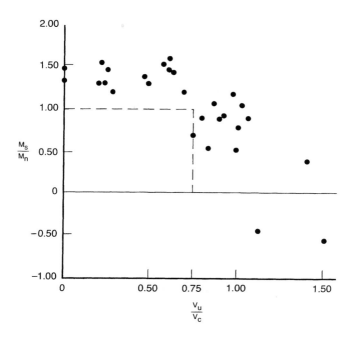

Fig. 13–53
Interaction between shear and
moment at edge columns.

bilinear interaction diagram shown by the dashed lines in Fig. 13–53. This suggests that if V_u is less than or equal to $0.75V_c$, the connection can transfer a moment at least equal to the nominal moment capacity of the slab steel within the width $c_2 + 3h$. This does not imply that this moment is all transferred by flexure, some will be transferred by the eccentricity of the shear force on the front face of the critical section and some by torsion on the side faces. The strains plotted in Fig. 13–52 indicate that the actual moment transferred by flexure in the width $c_2 + 3h$ will be considerably less than M_n in some tests.

Moment and Shear Transfer at Slab–Column Connections—Design Methods

Two design methods have been proposed for designing slab-column connections transferring shear and moment. The most fundamental of these represents the connection as the joint region plus beams projecting out from each face of the column where there is an adjacent slab.[13–22] The shear capacity of the connection is the sum of the shear capacities of the individual beams. The moment capacity is the sum of the moment capacities of the beams on the front and back faces plus the torsional capacities of the beams on the side faces. This model has never been developed enough for design office use and is mentioned only as a guide to understanding the response of such connections.

The ACI Code presents an empirical method[13–26, 13–27] for designing for shear and moment transfer which assumes that the shear stresses on a critical section located at $d/2$ away from the face of the column due to the direct shear, V_u, shown in Figs. 13–48b and 13–49b, can be added to the shear stresses on the same section due to moment transfer, shown in Figs. 13–48c and 13–49c. Failure is assumed to occur when the maximum sum of the shear stresses reaches a limiting value. This analysis is highly idealized and ignores such things as the effect of cracking of the slab as shown in Fig. 13–36, the uneven distribution of the shear stresses around a column as shown in Fig. 13–43, and the strut-and-tie action after cracking shown in Figs. 13–50 and 13–51.

Figure 13–48 illustrates the moment and shear transfer at an interior column where a shear, V_u, and an unbalanced moment, $M_{u1} - M_{u2}$, are transferred from the slab to the column. A portion of the unbalanced moment, equal to $\gamma_f(M_{u1} - M_{u2})$, is transferred by flexural stresses (shown as T_1, T_2, C_1, and C_2 in Fig. 13–48a) in the slab adjacent to the column. ACI Sec. 13.5.3.2 requires that this moment be transferred by providing sufficient flexural reinforcement within a width extending 1.5 times the slab, or drop panel, thickness on each side of the column. The reinforcement already designed for flexure in this region can be used for this purpose.

Based on test results on interior slab column joints with square columns, Hanson and Hanson[13-27] arbitrarily set $\gamma_f = 0.6$. Thus they assumed that $0.6M_u$ was transferred by flexure. The rest of the moment, $\gamma_v M_u = (1 - \gamma_f)M_u$, or $0.4M_u$, was assumed to be transferred by shearing stresses on the critical section as shown in Figs. 13–48c and 13–49c.

The shearing stresses resulting from the shear, V_u, and the unbalanced moment, $\gamma_v(M_{u1} - M_{u2})$, are shown in Fig. 13–48b and c. The combined stress shown in Fig. 13–48d is given by

$$v_u = \frac{V_u}{b_o d} \pm \frac{\gamma_v M_u c}{J_c} \tag{13–18}$$

where M_u is the unbalanced moment, $M_{u1} - M_{u2}$, c is the distance from the centroidal axis (Z–Z) of the shear perimeter in Fig. 13–48c to the point where the shear stresses are being computed, and J_c is analogous to the polar moment of inertia of the shear perimeter about the axis Z–Z.

The maximum value of v_u from Eq. 13–18 must satisfy

$$v_u \leq \phi v_n \tag{13–19}$$

where

$$\phi v_n = \frac{\phi V_c}{b_o d} \tag{13–20}$$
$$\text{(ACI Eq. 11–43)}$$

for slabs without shear reinforcement where V_c is from Eqs. 13–15 to 13–17 or for slabs with shear reinforcement

$$\phi v_n = \frac{\phi(V_c + V_s)}{b_o d} \tag{13–21}$$
$$\text{(ACI Eq. 11–44)}$$

The distribution of shear stresses at an edge column is also calculated using Eq. 13–18. Here the shear perimeter is generally three-sided, as shown in Fig. 13–49, and the centroid of the shear perimeter does not coincide with the centroid of the column. In this case, M_u in Eq. 13–18 is the unbalanced moment acting about the centroid of the shear perimeter.

In the case of an edge column, the centroid of the critical perimeter lies closer to the inside face of the column than to the outside. As a result, the shear stresses due to the moment shown in Fig. 13–49c are largest at the outside corners of the shear perimeter. If M_u is large and V_u is small, a negative shear stress may occur at these points. If M_u due to the combination of lateral loads and gravity loads is positive on this joint, rather than negative as shown in Fig. 13–49, the largest shear stress will occur at the outside corners.

Fraction of Unbalanced Moment Transferred by Flexure, γ_f

ACI Secs. 13.5.3.2 and 13.5.3.3 define the fraction of the moment transferred by flexure, γ_f, using Eq. 13–22 (ACI Eq. 13–1):

$$\gamma_f = \frac{1}{1 + \left(\frac{2}{3}\right)\sqrt{b_1/b_2}} \tag{13–22}$$

where b_1 is the total width of the critical section measured perpendicular to the axis about which the moment acts, and b_2 is the total width parallel to the axis. In Fig. 13–48a, b_1 is perpendicular to the axis Z–Z.

For a square critical section with $b_1 = b_2$, γ_f is 0.60 and $\gamma_v = 1 - \gamma_f$ is 0.40, meaning that 60% of the unbalanced moment is transferred to the column by flexure and 40% by eccentric shear stresses. Equation 13–22 was derived to give $\gamma_f = 0.6$ for $b_1 = b_2$, as proposed by Hanson and Hanson,[13-27] and to provide a transition to $\gamma_f = 1.0$ for a slab attached to the side of a wall, and to γ_f approaching zero for a slab attached to the end of a long wall.

ACI Sec. 13.5.3.3 allows the designer to modify the value of γ_f in certain circumstances, provided that the value of the steel ratio ρ required to resist the moment $\gamma_f M_u$ within the width of the column, c_2, plus 1.5h on each side of the column, $c_2 + 3h$), does not exceed $0.375\rho_b$, based on the adjusted γ_f. For interior, edge and corner columns, ACI Sec. 13.5.3.3 can be paraphrased as:

1. Interior column. The value of γ_f from Eq. 13–22 may be increased by up to 25%, provided that V_u does not exceed $0.4\phi V_c$.

2. Edge column, moments about an axis parallel to the edge. The value of γ_f from Eq. 13–22 may be increased up to 1.0, provided that V_u does not exceed $0.75\phi V_c$.

3. Edge column, moments about an axis perpendicular to the edge. The value of γ_f from Eq. 13–22 may be increased by up to 25%, provided that V_u does not exceed $0.4\phi V_c$.

4. Corner column. The value of γ_f from Eq. 13–22 may be increased up to 1.0, provided that V_u does not exceed $0.5\phi V_c$.

Fraction of Unbalanced Moment Transferred by Shear, γ_v

The moment transferred to the column by eccentric shear stresses is $\gamma_v M_u$, where

$$\gamma_v = 1 - \gamma_f \qquad (13-23)$$

Properties of the Shear Perimeter

To solve Eq. 13–18 it is necessary to compute the length of the perimeter, b_o, the location of the centroidal axis Z–Z, and the polar moment of inertia, J_c. The latter is computed as the polar moment of inertia of the faces of the shear perimeter which are perpendicular to the axis Z–Z, plus the area of the parallel faces times the square of the distance from those faces to the centroid. The polar moment of inertia of a rectangle about an axis Z–Z, perpendicular to the plane of the rectangle and displaced a distance, $\bar{x}$, from the centroid of the rectangle (Fig. 13–54), is given by

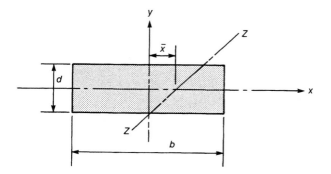

Figure 13–54
Rectangle considered in computing polar inertia of side.

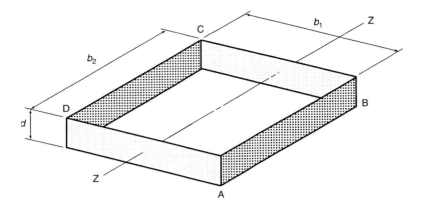

(a) Critical perimeter of an interior column

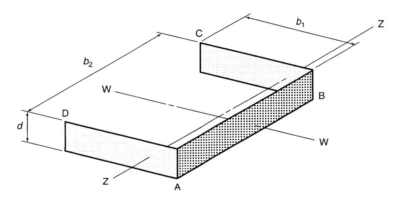

(b) Critical perimeter of edge column

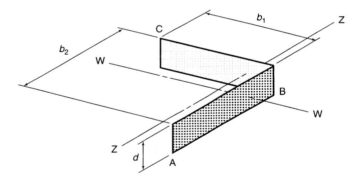

Fig. 13–55
Critical shear perimeters.

(c) Critical shear perimeter of corner column

$$J = I_x + I_y + A\bar{x}^2$$

$$= \frac{bd^3}{12} + \frac{db^3}{12} + (bd)\bar{x}^2$$

(13–24)

Interior Column

The centroid of the shear perimeter of the interior columns shown in Fig. 13–55a passes through the centroid of the sides *DA* and *CB*. The calculation of the polar moment of inertia involves

Two-Way Slabs: Behavior, Analysis, and Direct Design Method

$$J_c = \frac{2(b_1 d^3)}{12} + \frac{2(db_1^3)}{12} + 2(b_2 d)\left(\frac{b_1}{2}\right)^2$$

$$\underbrace{\quad\quad}_{I_x} \quad \underbrace{\quad\quad}_{I_y} \quad \underbrace{\quad\quad\quad}_{\substack{A\bar{x}^2 - \text{faces } AB \\ \text{and } CD}}$$

$$\underbrace{\hspace{6cm}}_{\substack{J - \text{faces } BC \\ \text{and } DA}}$$

where

$$b_1 = c_1 + 2(d/2)$$
= length of shear perimeter perpendicular to the axis of bending

$$b_2 = c_2 + 2(d/2)$$
= length of shear perimeter parallel to the axis of bending

c_1 = width of column perpendicular to the axis of bending

c_2 = width of column parallel to the axis of bending

If large openings are present adjacent to the column, the shear perimeter will be discontinuous, as shown in Fig. 13–40. If this occurs, the calculation of the location of the centroid and J_c should include the effect of the holes.

Edge Columns

For the three-sided perimeter shown in Fig. 13–55b and taking b_1 as the length of the side perpendicular to the edge, as shown, the location of the centroidal axis Z–Z is calculated as follows:

Moments about an axis parallel to the edge.

$$c_{AB} = \frac{\text{moment of area of the sides about } AB}{\text{area of the sides}}$$

$$= \frac{2(b_1 d)b_1/2}{2(b_1 d) + b_2 d} \tag{13–26}$$

The polar moment J_c of the shear perimeter is given by

$$J_c = 2\left(\frac{b_1 d^3}{12}\right) + 2\left(\frac{db_1^3}{12}\right) + 2(b_1 d)\left(\frac{b_1}{2} - c_{AB}\right)^2 + (b_2 d)c^2{}_{AB}$$

$$\underbrace{\quad}_{I_x} \quad \underbrace{\quad\quad}_{I_y} \quad \underbrace{\quad}_{A\bar{x}^2} \quad \underbrace{\quad}_{\substack{A\bar{x}^2 - \text{face} \\ AB}} \tag{13–27}$$

$$\underbrace{\hspace{5cm}}_{J \text{ about } Z\text{–}Z \text{ of faces } DA \text{ and } BC}$$

Moments about an axis perpendicular to the edge. Frequently, moments about an axis perpendicular to the edge are also transferred from the slab to the column. In this case Eq. 13–18 becomes

$$v_u = \frac{V_u}{b_o d} \pm \frac{\gamma_v M_{u1} c}{J_{c1}} \pm \frac{\gamma_v M_{u2} c}{J_{c2}} \tag{13–18a}$$

where M_{u1} and J_{c1} refer to the moments from the span perpendicular to the edge and M_{u2} and J_{c2} refer to the moments from the span parallel to the edge. J_{c1} is given by Eq. 13–27. A new equation is required for J_{c2} about axis W–W in Fig. 13–55b:

$$c_{CB} = c_{AD} = \frac{b_2}{2}$$

$$J_{c2} = \underbrace{2(b_1 d)c_{CB}^2}_{\substack{A\bar{x}^2 - \text{faces} \\ CB \text{ and } AD}} + \underbrace{\frac{b_2 d^3}{12}}_{I_x} + \underbrace{\frac{b_2^3 d}{12}}_{I_y} \qquad 13\text{–}28$$

$$\underbrace{}_{J \text{ about } W\text{—}W \text{ of face } AB}$$

Corner Columns

For the two-sided perimeter shown in Fig. 13–55c with sides b_1 and b_2, the location of the centroidal axis Z–Z is

$$c_{AB} = \frac{(b_1 d)b_1/2}{b_1 d + b_2 d} \qquad (13\text{–}29)$$

The polar moment of inertia, J_c, of the shear perimeter is

$$J_c = \underbrace{\frac{b_1 d^3}{12}}_{I_x} + \underbrace{\frac{b_1^3 d}{12}}_{I_y} + \underbrace{b_1 d\left(\frac{b_1}{2} - c_{AB}\right)^2}_{A\bar{x}^2} + \underbrace{(b_2 d)c_{AB}^2}_{A\bar{x}^2 - \text{face } AB} \qquad (13\text{–}30)$$

$$\underbrace{}_{J \text{ about } Z\text{–}Z \text{ of face } BC}$$

Circular Columns

For combined shear and moment calculations, ACI-ASCE Committee 426[13–15] recommends that the shear perimeter of circular columns be based on that of a square column with the same centroid and the same length of perimeter. In this case the equivalent square column would have sides of length $c = \sqrt{\pi} \, d_c/2 = 0.886 d_c$, where d_c is the diameter of the column. Two cases are illustrated in Fig. 13–56.

Use of Principal Axes and Principal Moments of Inertia

Equation 13–30 for J_c of a corner column is written relative to the orthogonal x–x and y–y axes. For a square corner column, the largest or *major principal moment of inertia* of the shear perimeter is about an axis on the 45° diagonal of the column through the corner, and the smallest or *minor principal moment of inertia* is about a perpendicular axis. In a discussion of the 1995 ACI Code revisions, Ghali[13–28] proposed two changes:

1. He pointed out that J_c was not a standard term in mechanics and hence might be difficult to evaluate in some cases. He suggested that J_c be replaced by I_x and I_y, where I_x and I_y are d times (the moments of inertia about the x and y axes, respectively, of the line formed by the intersection of the critical shear perimeter and the top of the slab). It can be shown that I_x and I_y are about 3% smaller than J_{cx} and J_{cy} because the term labeled as I_x in Eq. 13–24 is omitted.

2. He also pointed out that the stresses computed using moments of inertia relative to the orthogonal x and y axes do not equilibrate the applied moments for critical sections where the x and y axes are not principal axes. He suggested that the shear stresses arising from moment transfer be computed using the combined stress equation:

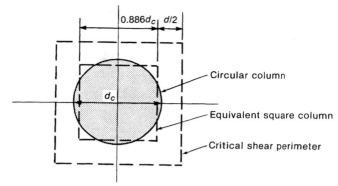

(a) Interior column.

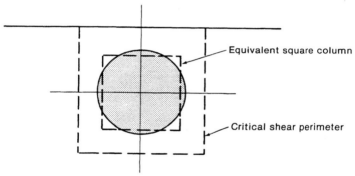

Figure 13–56
Critical shear perimeters for
moment and shear transfer at
circular columns.

(b) Exterior column.

$$v_u = \frac{V_u}{b_o d} + \left(\frac{\gamma_{vx} M_{ux} I_y - \gamma_{vy} M_{uy} I_{xy}}{I_x I_y - I_{xy}^2} \right) y + \left(\frac{\gamma_{vy} M_{uy} I_x - \gamma_{vx} M_{ux} I_{xy}}{I_x I_y - I_{xy}^2} \right) x \quad (13\text{--}31)$$

where

γ_{vx} and γ_{vy} are the values of γ_v relative to the x and y axes, respectively

I_x and I_y are defined above

I_{xy} is d times (the product of inertia about the x and y axes of the line representing the critical section)

When one or both of the x and y axes are axes of symmetry, the x and y axes are the principal axes and $I_{xy} = 0$. Substituting this into Eq. 13–31 gives Eq. 13–18a. The principal axis problem primarily affects corner columns, especially when shear reinforcement is used. Ghali presents special values of γ_v for use with Eq. 13–31.

ACI Committee 318 did not accept this proposal because comparison of the two procedures to test data for corner columns without shear reinforcement showed that the orthogonal axes method was adequately safe while the principal axes procedure was quite conservative.

Calculation of Moment about Centroid of Shear Perimeter

The distribution of stresses calculated using Eq. 13–18, and illustrated in Fig. 13–49, assumes that V_u acts through the centroid of the shear perimeter and that M_u is about the centroidal axis of this perimeter. In structural analyses the calculated V_u and M_u may correspond to a different set of axes, and if so, it is necessary to compute the value of M_u corresponding to V_u acting through the centroid of the shear perimeter.

Interior Column

If the shear perimeter is symmetrical about two axes and has no gaps in it of the type shown in Fig. 13–40, the axis used to compute moments and shears will coincide with that of the shear perimeter and no corrections are needed.

Edge Column: Moments Calculated by Direct Design Method

ACI Sec. 13.6.3.6 sets the gravity load moment to be transferred between slab and column equal to $0.3M_0$. This moment is assumed to act about the centroid of the shear perimeter.

Edge Column: Moments Calculated by Equivalent Frame Analysis

In the Equivalent Frame Method given in Chap. 14, moments are calculated at the point where the members meet at the center of the slab—column joint. The method used to transfer these moments to the centroidal axis of the shear perimeter is described in Sec. 14–2.

EXAMPLE 13–6   Checking Combined Shear and Moment Transfer at an Edge Column

A 12 in. by 16 in. column is located 4 in. from the edge of a flat slab without edge beams as shown in Fig. 13–57. The slab is 6.5 in. thick, with an average effective depth of 5.5 in. The concrete and reinforcement strengths are 3500 psi and 60,000 psi, respectively. The Direct Design Method gives the statical moment, M_0, in the edge panel as 152 ft-kips. The shear from the edge panel is 31.3 kips. The portion of the slab outside the centerline of the column produces a factored shear force of 4.0 kips acting at 6 in. outside the center of the column. There is no moment about the axis perpendicular to the edge.

    1. Locate the critical shear perimeter. The critical shear perimeter is located at $d/2$ from the sides of the column. Because the shear perimeter that intercepts the edge of the slab at D and C is shorter than one located at $d/2$ away from the outside face of the column, the perimeter shown in Fig. 13–57 is assumed to be critical.

    2. Compute the centroid of the shear perimeter.

$$c_{AB} = \frac{\Sigma Ay}{A} \qquad \text{where } y \text{ is measured from } AB$$

$$= \frac{2(18.75 \times 5.5)18.75/2}{2(18.75 \times 5.5) + 21.5 \times 5.5} = 5.96 \text{ in.}$$

Therefore, $c_{AB} = 5.96$ in. and $c_{CD} = 12.79$ in.

    3. Compute the moment about the centroid of the shear perimeter. For the portion of the slab between the centerline of the edge columns and the centerline of the first interior columns, $M_0 = 152$ ft-kips and $V_u = 31.3$ kips. ACI Sec. 13.6.3.6 defines the moment to be transferred as $0.3M_0 = 45.6$ ft-kips. We shall assume that this is about the centroid of the shear perimeter and that V_u from the edge panel acts through this point. The portion of the slab outside the column centerline

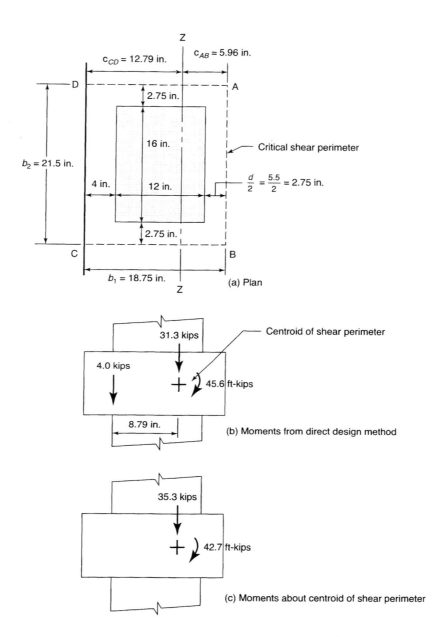

Figure 13-57
Slab—column joint—
Example 13-6.

has a shear of $V_{uc} = 4$ kips acting at 6 in. $+$ 2.79 in. from the centroid of the shear perimeter. The total moment about the centroid of the shear perimeter is

$$M_u = 45.6 \text{ ft-kips} - 4 \text{ kips} \times 8.79/12 \text{ ft} = 42.7 \text{ ft-kips}$$

The total shear to be transferred is

$$V_u = 31.3 \text{ kips} + 4 \text{ kips} = 35.3 \text{ kips}$$

   4. Compute ϕV_c and $V_u/\phi V_c$. V_c is the smallest value given by Eqs. 13–15 to 13–17 (ACI Eqs. 11–35 to 11–37).

   (a) $$V_c = \left(2 + \frac{4}{\beta_c}\right)\sqrt{f_c}\, b_0 d \qquad (13\text{–}15)$$

where

$$\beta_c = \frac{\text{long side of the column}}{\text{short side of the column}} = \frac{16}{12}$$

$$= 1.33$$

$$b_0 = 2 \times 18.75 \text{ in.} + 21.5 \text{ in.} = 59 \text{ in.}$$

$$\phi V_c = 0.85\left(2 + \frac{4}{1.33}\right)\sqrt{3500} \times 59 \times 5.5 = 81.7 \text{ kips}$$

(b) $$V_c = \left(\frac{\alpha_s d}{b_0} + 2\right)\sqrt{f_c'} \, b_o d$$ (13–16)

where $\alpha_s = 30$ for an edge column.

$$\phi V_c = 0.85\left(\frac{30 \times 5.5}{59} + 2\right)\sqrt{3500} \times 59 \times 5.5 = 78.3 \text{ kips}$$

(c) $$V_c = 4\sqrt{f_c'} \, b_o d$$ (13–17)

$$\phi V_c = 0.85 \times 4\sqrt{3500} \times 59 \times 5.5 = 65.3 \text{ kips}$$

Therefore $\phi V_c = 65.3$ kips and $V_u/\phi V_c = 35.3 \text{ kips}/65.3 \text{ kips} = 0.541$.

5. **Determine the fraction of the moment transferred by flexure, γ_f.**

$$\gamma_f = \frac{1}{1 + \frac{2}{3}\sqrt{\frac{b_1}{b_2}}}$$ (13–22)

$$= \frac{1}{1 + \frac{2}{3}\sqrt{\frac{18.75}{21.5}}}$$ (ACI Eq. 13–1)

$$= 0.616$$

ACI Sec. 13.5.3.3 allows γ_f to be increased up to 1.0 if $V_u/\phi V_c$ does not exceed 0.75 and the resulting $\rho \leq 0.375\rho_b$ within a width $c_2 + 3h$ centered on the column.. From step 4, $V_u/\phi V_c = 0.541$. Therefore, try γ_f equal to 1.0 and check the reinforcement needed.

6. **Design the reinforcement for moment transfer by flexure.**

$$\text{Width effective for flexure} = c_2 + 3h = 16 \text{ in.} + 3 \times 6.5 \text{ in.}$$

$$= 35.3 \text{ in.}$$

$$\text{Moment} = 1.0 \times 42.7 \text{ ft-kips} = 42.7 \text{ ft-kips}$$

Assume that $jd = 0.925d$.

$$A_s = \frac{M_u}{\phi f_y jd}$$

$$= \frac{42.7 \times 12,000}{0.9 \times 60,000 \times 0.925 \times 5.5}$$

$$= 1.87 \text{ in.}^2$$

Try 10 No. 4 bars $= 2.00 \text{ in.}^2$. Since this is based on a guess of jd, compute a for $A_s = 2.00 \text{ in.}^2$ and recompute A_s using that value of a.

$$a = \frac{A_s f_y}{0.85 f_c' b}$$

$$= \frac{2.00 \times 60,000}{0.85 \times 3500 \times 35.5}$$

$$= 1.136 \text{ in.}$$

$$A_S = \frac{M_u}{\phi f_y(d - a/2)} = \frac{42.7 \times 12,000}{0.9 \times 60,000(5.5 - 1.136/2)}$$

$$= 1.92 \text{ in.}^2$$

Use 10 No. 4 bars $= 2.00 \text{ in.}^2$. For this A_s,

$$\frac{a}{d} = \frac{1.14}{5.5} = 0.207$$

From Eq. 4–21,

$$\frac{a_b}{d} = \beta_1\left(\frac{87,000}{87,000 + f_y}\right)$$

$$= 0.85\left(\frac{87,000}{87,000 + 60,000}\right)$$

$$= 0.503$$

and

$$0.375\frac{a_b}{d} = 0.189$$

Since a/d exceeds $0.375a_b/d$, ρ exceeds $0.375\rho_b$ and γ_f cannot be increased to 1.0. It can, however, be increased to some value between 0.616 and 1.0, provided that the value of ρ for the resulting moment does not exceed $0.375\rho_b$. We shall arbitrarily provide 8 No. 4 bars with $A_s = 1.60 \text{ in.}^2$. These will give $\rho < 0.375\rho_b$ and will transfer $\phi M_n = 36.3$ ft-kips $= \gamma_f M_u$. The amount transferred by shear is $\gamma_v M_u = 42.7$ ft-kips $- 36.3$ ft-kips $= 6.4$ ft-kips.

7. **Compute the torsional moment of inertia, J_c.** From Eq. 13–27,

$$J_c = 2\left(\frac{b_1 d^3}{12}\right) + 2\left(\frac{db_1^3}{12}\right) + 2(b_1 d)\left(\frac{b_1}{2} - c_{AB}\right)^2 + (b_2 d)c_{AB}^2$$

$$= \frac{2 \times 18.75 \times 5.5^3}{12} + \frac{2 \times 5.5 \times 18.75^3}{12}$$

$$+ 2(18.75 \times 5.5)\left(\frac{18.75}{2} - 5.96\right)^2 + (21.5 \times 5.5)5.96^2$$

$$= 13,170 \text{ in.}^4$$

8. **Compute the shear stresses.**

$$v_u = \frac{V_u}{b_0 d} \pm \frac{\gamma_v M_u c}{J_c} \tag{13–18}$$

$$v_u = \frac{35,300}{59 \times 5.5} \pm \left(\frac{6.4 \times 12,000}{13,170}\right)c$$

$$= 108.7 \pm 5.83c$$

The shear stress at AB is

$$v_{u,AB} = 108.7 + 5.83 \times 5.96$$

$$= 143.4 \text{ psi}$$

The shear stress at CD is

$$v_{u,CD} = 108.7 - 5.83 \times 12.79$$

$$= 34.1 \text{ psi}$$

From step 4, $\phi V_c = 65.3$ kips,

$$\phi v_c = \frac{\phi V_c}{b_0 d} = \frac{65,300}{59 \times 5.5}$$

$$= 201 \, \text{psi}$$

Since ϕv_c exceeds v_u, the shear is OK. **Use a 12 in. by 16 in. column as shown in Fig. 13–57 with 8 No. 4 bars at 5 in. on centers, centered on the column.** ∎

Shear and Moment Transfer to a Spandrel Beam

The negative moments at the exterior end of the panel cause torsional moments in the spandrel beam or in the strip of slab between the exterior columns. These are compatibility torsional moments since they exist due to the fact that the spandrel beam restrains the rotation of the edge of the panel. In such a case, ACI Sec. 11.6.2.2 allows the torsional moment in the spandrel beam at a section d away from the face of the columns to be reduced to

$$\phi 4 \sqrt{f_c' \left(\frac{A_{cp}^2}{p_{cp}} \right)} \tag{7–47}$$

Consider the slab in Example 13–4, A_{cp} and p_{cp} are the area and perimeter of the beam cross section defined in ACI Sec. 13.2.4 (see Fig. 13–20). For Example 13–4 the spandrel beam has the section shown in Fig. 13–34, and

$$A_{cp} = 12 \times 16 + 8 \times 8 = 256 \, \text{in.}^2$$
$$p_{cp} = 16 + 12 + 8 + 8 + 8 + 20 = 72 \, \text{in.}$$

Assuming that $f_c' = 3000$ psi,

$$\phi 4 \sqrt{f_c'} \left(\frac{A_{cp}^2}{p_{cp}} \right) = 0.85 \times 4\sqrt{3000} \left(\frac{256^2}{72} \right)$$
$$= 14.1 \, \text{ft-kips}$$

Thus the torsional moment at $d = 13.5$ in. on each side of the column can be reduced to 14.1 ft-kips. The remaining torque of 77.0 ft-kips $- 2 \times 14.1$ ft-kips $= 48.8$ ft-kips is either redistributed as was done in Example 7–2, or is transferred to the column and the spandrel beam between the column and the sections d away from the column. We shall assume that it is transferred to the column and edge beam and will take the effective width for this transfer as the smaller of the distance between the critical sections for torsion $= 16 + 2 \times 13.5 = 43$ in. or the width $c_2 + 3h = 16 + 3 \times 8 = 40$ in., the width considered effective for flexure in a slab without edge beams. The moment transferred would require 1.68 in.², or 9 No. 4 bars in this width.

Normally, the torsional moment in the spandrel beam would not be redistributed in the design of a two-way slab, especially if the slab serves as the spandrel beam. These calculations have been carried out to illustrate the process. The spandrel beam would need No. 4 closed stirrups at 6 in. on centers if designed for the shear and the full torsion, and No. 3 closed stirrups at 6.25 in. on centers if designed for the shear and the reduced torsion. This reduction in the stirrups required in the spandrel beam is accompanied by an increase in the widths of the torsional cracks in the spandrel beam.

Shear in Slabs: Design

When designing two-way slabs or footings with little or no transfer of moments from the slab to the column, it is customary to select the slab thickness on the basis of $V_u \simeq 0.85$ to $1.0(\phi V_c)$, unless there are holes adjacent to the column. The presence of holes adjacent to the column reduces the shear perimeter and, unless the holes are symmetrically placed around the column, introduces an eccentricity between the line of action of the shear and the centroid of the shear perimeter.

At an edge column in a two-way slab subjected to gravity loads, the shear stresses resulting from moment transfer may be of the same order of magnitude as those from direct shear transfer, and hence the initial trial slab thickness should be chosen on the basis of $V_u \simeq 0.5 \text{ to } 0.55(\phi V_c)$. The unbalanced moment to be transferred can be reduced by cantilevering the slab past the column centerline. The moment from the cantilever will offset some of the unbalanced moment to be transferred.

In unbraced, laterally loaded flat-plate frames, the shear induced by the lateral load moments will increase the shears at either the inside or outside face of an edge column. Either case may be serious.

As pointed out in Sec. 13–7, if ϕv_c is less than v_u when both moment and shear effects are calculated, the shear capacity can be increased by:

1. Thickening the slab over the entire panel
2. Adding a drop panel
3. Adding a fillet or capital around the column
4. Adding shear reinforcement
5. Increasing the column size

Alternatives 2, 3, and 5 are generally the most economical. Since the volume of concrete in columns is small compared to that in slabs, it is frequently economical to select the column size to alleviate any possible shear problems. ACI Sec. 11.12.6.2 allows shear reinforcement to be used in connections transferring moment and shear.

13–9 DETAILS AND REINFORCEMENT REQUIREMENTS

Drop Panels

Drop panels are thicker portions of the slab adjacent to the columns, as shown in Fig. 13–58 or Fig. 13–2c. They are provided for three main reasons:

1. The minimum thickness of slab required to limit deflections (see Sec. 13–6) may be reduced by 10% if the slab has drop panels conforming to ACI Sec. 13.3.7. The drop panel stiffens the slab in the region of highest moments and hence reduces the deflection.

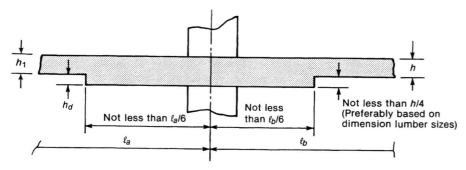

Figure 13–58
Drop panels.

Minimum size of drop panels—ACI Sec. 9.5.3.2 and 13.3.7.

2. A drop panel with dimensions conforming to ACI Sec. 13.3.7 can be used to reduce the amount of negative moment reinforcement required over a column in a flat slab. By increasing the overall depth of the slab, the lever arm, jd, used in computing the area of steel is increased, resulting in less required reinforcement in this region.

3. A drop panel gives additional depth at the column, thereby increasing the area of the critical shear perimeter.

ACI Secs. 9.5.3.2 and 13.3.7 give essentially the same requirements for the minimum size of a drop panel. These are illustrated in Fig. 13–58. The only difference between the two sections is that ACI Sec. 13.3.7.3 states that when computing the negative moment flexural reinforcement, the thickness of the drop panel below the slab used in the calculations shall not be taken greater than one-fourth of the distance from the edge of the drop panel to the edge of the column or column capital. If the drop panel were deeper than this it is assumed that the maximum compression stresses would occur at some point above the bottom of the drop panel so that the full depth is not effective.

For economy in form construction, the thickness of the drop shown as h_d in Fig. 13–58 should be related to actual lumber dimensions, such as $\frac{3}{4}$ in., $1\frac{1}{2}$ in., $3\frac{1}{2}$ in., $5\frac{1}{2}$ in. (nominal 1 in., 2 in., 4 in., or 6 in. lumber sizes) or some combination of these, plus the form plywood thickness. The drop should always be underneath the slab (in a slab loaded with gravity loads), so that the negative moment steel is straight over its entire length.

Column Capitals

Occasionally, the top of the columns will be flared outward, as shown in Fig. 13–59 or 13–2c. This is done to provide a larger shear perimeter at the column and to reduce the clear span, ℓ_n, used in computing moments.

ACI Sec. 6.4.6 requires that the capital concrete be placed at the same time as the slab concrete. As a result, the floor forming becomes considerably more complicated and expensive. For this reason, other alternatives, such as drop panels or shear reinforcement in the slab, should be considered before capitals are selected. If capitals must be used, it is desirable to use the same size throughout the project.

The diameter or effective dimension of the capital, d_c or c, is defined in ACI Sec. 13.1.2 as that part of the capital lying within the largest right circular cone or pyramid with a 90° vertex that can be included within the outlines of the supporting column. The diameter is measured at the bottom of the slab or drop panel, as illustrated in Fig. 13–59. This diameter c is used to define the effective width for moment transfer, $c_2 + 3h$, and to define the clear span, ℓ_n. Concrete in the capital outside the 45° lines (see Fig. 13–59) can be used to increase the shear strength.

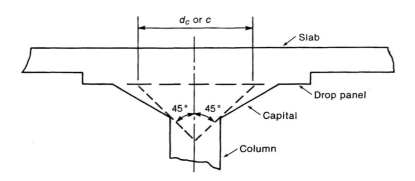

Fig. 13–59
Effective diameter of capital.

Two-Way Slabs: Behavior, Analysis, and Direct Design Method

Reinforcement

Placement Sequence

In a flat plate or flat slab, the moments are largest in the slab strips spanning the long direction of the panels. As a result, the reinforcement for the long span is generally placed closer to the top and bottom of the slab than the short-span reinforcement. This gives the largest effective depth for the largest moment. For slabs supported on beams having α greater than about 1.0, the opposite is true and the reinforcing pattern should be reversed. If a particular placing sequence has been assumed in the reinforcement design, it should be shown or noted on the drawings. It is important, however, that the same arrangements of layers should be maintained throughout the entire floor to avoid confusion in the field. Thus if the east–west reinforcement is nearest the surface in one area, this arrangement should be maintained over the entire slab if at all possible.

Cover and Effective Depth

ACI Sec. 7.7.1 specifies, the minimum clear cover to the surface of the reinforcement as $\frac{3}{4}$ in. for No. 11 and smaller bars provided that the slab is not exposed to earth or to weather. For concrete exposed to weather, the minimum clear cover is $1\frac{1}{2}$ in. for No. 5 and smaller bars and 2 in. for larger bars. Concrete parking decks exposed to deicing salts should have greater cover and epoxy-coated bars. ACI Commentary Sec. 7.7.5 suggests 2 in. of cover in such a case. It may be necessary to increase the cover for fire resistance. This will be specified in the local building code.

The reinforcement in a two-way slab with spans of up to 20 ft will generally be No. 4 bars; up to 25 ft, No. 5 bars; and over 25 ft, No. 5 or No. 6 bars. For the long span of a flat plate or flat slab, $d = h - 3/4 - 0.5d_b$, and for the short span $d = h - 3/4 - 1.5d_b$. For preliminary design these can be taken as:

For flat-plate or flat slab spans up to 25 ft (7 m):

$$\text{long span } d \simeq h - 1.1 \text{ in. (30 mm)} \qquad (13\text{–}32\text{a})$$
$$\text{short span } d \simeq h - 1.7 \text{ in. (45 mm)} \qquad (13\text{–}32\text{b})$$

For flat-plate or flat slab spans over 25 ft (7 m):

$$\text{long span } d \simeq h - 1.15 \text{ in. (30 mm)} \qquad (13\text{–}32\text{c})$$
$$\text{short span } d \simeq h - 1.9 \text{ in. (50 mm)} \qquad (13\text{–}32\text{d})$$

It is important not to overestimate d in slabs because normal construction inaccuracies tend to result in smaller values of d than shown on the drawings.

Spacing Requirements, Minimum Reinforcement, and Minimum Bar Size

ACI Sec. 13.4.1 requires that the minimum area of reinforcement provided for flexure should not be less than:

$0.0020bh$ if Grade 40 or 50 deformed bars are used

$0.0018bh$ if Grade 60 deformed bars or welded-wire fabric is used

The maximum spacing of reinforcement at points of maximum positive and negative moments in the middle and column strips shall not exceed two times the slab thickness (ACI Sec. 13.3.2) or 18 in. (ACI Sec. 7.12.3).

Although there is no code limit on bar size, the Concrete Reinforcing Steel Institute recommends that top steel in slab should not be less than No. 4 bars at 12 in. on centers, to give adequate rigidity to prevent displacement of the bars under ordinary foot traffic before the concrete is placed.

Calculation of the Required Area of Steel

The calculation of the steel required is based on Eq. 4–33 as illustrated in Examples 4–4 and 10–1.

$$A_s = \frac{M_u}{\phi f_y jd} \tag{4–33}$$

where $jd = 0.90$ to $0.95d$ for slabs or normal proportions. As recommended in Sec. 4–2 and Example 10–1, j can be assumed to be 0.925 for the first trial. Once a trial value of A_s has been computed for the section of maximum moment, the depth of the compression zone, a, will be computed and used to compute a more correct value of $jd = d - a/2$. This will be used to compute A_s at all sections in the slab. It is also necessary to check whether A_s exceeds $A_{s(min)}$ at all sections and whether $\rho \leq 0.75\rho_b$.

Bar Cutoffs and Anchorages

For slabs without beams ACI Sec. 13.3.8.1 allows the bars to be cut off as shown in Fig. 13–60). Where adjacent spans have unequal lengths, the extension of the negative moment bars past the face of the support is based on the length of the longer span.

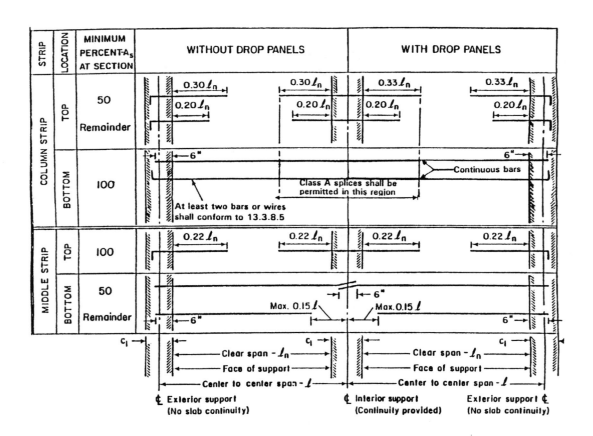

Fig. 13–60
Minimum cutoff points for slabs without beams (ACI Sec. 13.3.8.1).

Two-Way Slabs: Behavior, Analysis, and Direct Design Method

Prior to 1989, ACI Fig. 13.3.8 also showed details for slabs with alternate straight and bent bars. While such a bar arrangement is still allowed, the details have been deleted from ACI Fig. 13.3.8 in the Code because the bent–straight bar arrangement is rarely used now. If such a slab were designed or checked, the bar details should be based on Fig. 13.4.8 of the 1983 Code modified in accordance with 1995 ACI Sec. 13.3.8.5.

ACI Sec. 13.3.4 requires that all negative moment steel perpendicular to an edge be bent, hooked, or otherwise anchored in spandrel beams, columns, and walls along the edge to develop f_y in tension. If there is no edge beam this steel should still be hooked to act as torsional reinforcement.

Detailing Slab Reinforcement at an Edge Column

The shear and moment transfer from the slab to an exterior or corner column assumes that the edge of the slab will act as a torsional member (see the truss in Fig. 13–51). ACI-ASCE Committee 352[13-29] has recommended details for edge column to slab connections. Based in part on these recommendations, the author recommends the following reinforcing details:

(a) The top steel required to transfer the moment $\gamma_f M_u$ according to ACI Sec. 13.5.3.2 should be placed in a width equal to the smaller of $2(1.5h) + c_2$ or $2c_e + c_2$ centered on the column (Fig. 13–61), where c_e is the distance from the inner face of the column to the edge of the slab but not more than the depth of the column, c_1, and c_2 is the width of the column.

(b) Within the width defined in (a), edge reinforcement should be provided at a spacing of $1.5d$ or less. This can be provided by hooking the top steel perpendicular to the edge with 180° bends or with hairpin bars at least No. 3 in size. Each leg of a hairpin should have an extension of at least ℓ_d of the top bars.

Structural Integrity Reinforcement

When a punching shear failure occurs, it completely removes the shear capacity at a column and the slab drops, pulling the top reinforcement out through the top of the slab as shown in Fig. 13–36. If the slab falls on the slab below, that slab will probably fail also, causing a progressive type of failure. Research reported by Mitchell and Cook[13-30] suggests that this can be prevented by providing reinforcement through the column at the bottom of the slab. ACI Sec. 13.3.8.5 requires that all bottom steel in the column strip in each direction be continuous or lap spliced with a Class A lap splice (see Fig. 13–60). At least two of the bottom bars in each direction must pass through the column core. The words "column core" mean these bars should be between the corner bars of the column. At exterior and corner columns at least two bars perpendicular to the edge should be bent, hooked, or

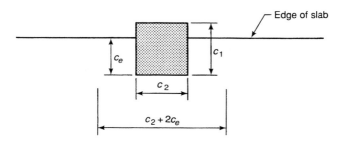

Fig. 13–61
Width effective for moment transfer at exterior column.

otherwise anchored within the column core. The bars passing through the column are referred to as *integrity steel*.

Corner Reinforcement

At discontinuous corners where the slab is supported on relatively stiff beams, as shown in Fig. 13–62a, there is a tendency for the corner to lift off its support unless a downward reaction is provided at the corner. In Fig. 13–62a, point *B* has deflected downward relative to the simply supported edges *A–D* and *C–D*. If corner *D* is held down, strips *A–C* and *B–D* develop the curvatures shown. Strip *A–C* develops positive moments and ideally should be reinforced with bars parallel to *A–C* at the bottom of the slab as shown in Fig. 13–62b. Strip *B–D* develops negative moments and should be reinforced with reinforcement parallel to *B–D* at the top of the slab as shown in Fig. 13–62c. Generally, the diagonal bars are replaced with orthogonal mats of bars parallel to the two sides of the slab. Two such mats are required, one at the top of the slab and one at the bottom.

ACI Sec. 13.3.6 requires special corner reinforcement in the exterior corners of all slabs with edge beams having α greater than 1.0. The reinforcement in each direction in each face is designed for a moment per unit width equal to the largest positive moment per unit width in the panel in question and extends one-fifth of the longer span each way from the corner.

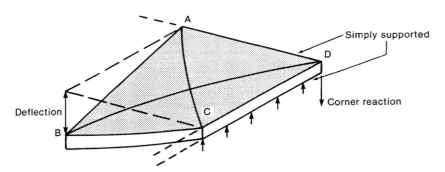

(a) Deformations of corner of simply supported slab.

Fig. 13–62
Moments in corner of slab supported on beams.

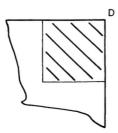

(b) Reinforcement at bottom of slab.

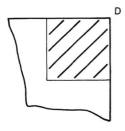

(c) Reinforcement at top of slab.

Two-Way Slabs: Behavior, Analysis, and Direct Design Method

**EXAMPLE 13–7   Design of a Flat-Plate Floor without Spandrel Beams:
Direct Design Method**

Figure 13–63 shows a plan of a part of a flat-plate floor. There are no spandrel beams. The floor supports its own dead load plus 25 psf for partitions and finishes, and a live load of 40 psf. The slab extends 4 in. past the exterior face of the column to support an exterior wall that weighs 300 lb/ft of length of wall. The story-to-story height is 8 ft 10 in. Use 4000-psi normal-weight concrete and Grade 40 reinforcement. The columns are 12 in. × 12 in., or 12 in. × 20 in. and are oriented as shown. Select the thickness, compute the design moments, and select the reinforcement in the slab.

To fully design the portion of slab shown, it is necessary to consider two east–west strips (Fig. 13–64a), and two north–south strips (Fig. 13–64b). In steps 3 and 4 the computations are carried out for the east–west strip along column line 2 and the edge strip running east–west along line 1. Steps 5 and 6 deal similarly with the north–south strips.

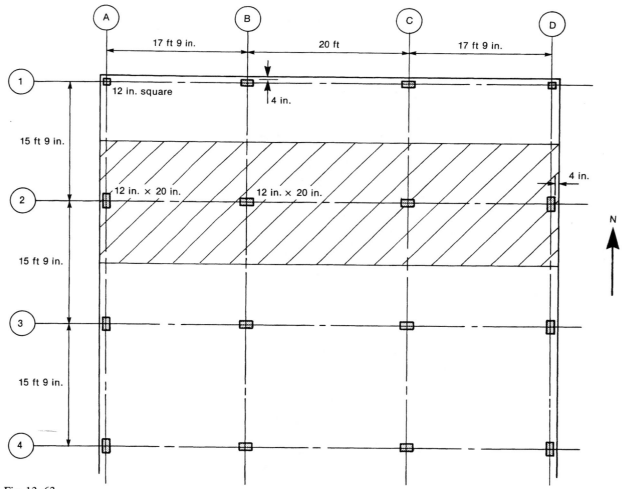

Fig. 13–63
Plan of flat-plate floor—Example 13–7.

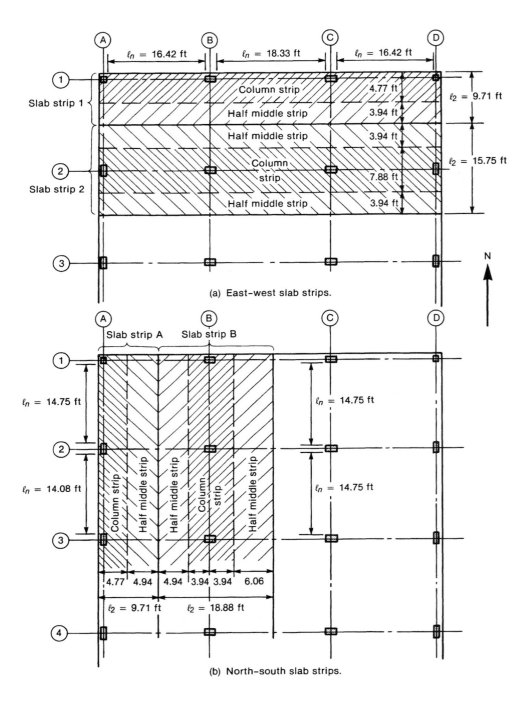

Fig. 13–64
Division of slab into strips
for design—Example 13–7.

(b) North-south slab strips.

1. Select the design method. ACI Sec. 13.6.1 allows use of the Direct Design Method if:

   (a) Minimum of three consecutive spans each way—therefore OK.

   (b) Longer span/shorter span ≤ 2: $20.0/17.5 < 2$—therefore OK.

   (c) Successive span lengths differ by not more than one-third of the longer span. Thus short span/long span $\geq 0.667 : 17.75/20 = 0.89$—therefore OK.

   (d) Columns offset up to 10%—OK.

   (e) All loads are uniformly distributed gravity loads. Strictly speaking, the wall load is not uniformly distributed, but use DDM.

(f) Unfactored live load not greater than two times the unfactored dead load. Using Table 13–1, estimate the slab thickness as $\ell/36 = 20 \times 12/36 \simeq 6.5$ in.. The approximate dead load $= 6.5/12 \times 150 + 25 = 106$ psf. This exceeds one-half of the live load—therefore OK.

(g) No beams—therefore, ACI Sec. 13.6.1.6 does not apply and it is not necessary to make this check.

Therefore, use the Direct Design Method.

2. **Select the thickness.**

(a) **Determine the thickness to limit deflections.** From Table 13–1 the minimum thicknesses of panels 1 to 4 are:

Panel 1–2–A–B (corner):

$$\text{max. } \ell_n = (17 \text{ ft } 9 \text{ in.}) - (6 + 10) \text{ in.} = 197 \text{ in.}$$

$$\text{min. } h = \frac{\ell_n}{33} = 5.97 \text{ in.}$$

Panel 1–2–B–C (edge):

$$\text{max. } \ell_n = 220 \text{ in.}$$

$$\text{min. } h = \frac{\ell_n}{33} = 6.67 \text{ in.}$$

Panel 2–3–A–B (edge):

$$\text{max. } \ell_n = 197 \text{ in.}$$

$$\text{min. } h = 5.97 \text{ in.}$$

Panel 2–3–B–C (interior):

$$\text{max. } \ell_n = 200$$

$$\text{min. } h = \frac{\ell_n}{36} = 6.11 \text{ in.}$$

Try $h = 6.75$ in. Because slab deflections are frequently a problem, it is best to round the thickness up rather than down.

(b) **Check the thickness for shear.** Check at columns B2 and B1. Figure 13–65 gives the critical shear perimeters and tributary areas for these two columns, based on $h = 6.75$ in.

$$w_u = 1.4\left(\frac{6.75}{12} \times 150 + 25\right) + 1.7(40) = 221 \text{ psf}$$

$$\text{Average } d \simeq 6.75 \text{ in.} - 1.4 \text{ in.} \simeq 5.35 \text{ in.}$$

Column B2:

$$b_o = 2(17.35 + 25.35) = 85.4 \text{ in.}$$

$$V_u = 221\left[\left(\frac{20 + 17.75}{2}\right) \times 15.75 - \frac{17.35 \times 25.35}{144}\right]$$
$$= 65,000 \text{ lb}$$

From Eq. 13–15,

$$\phi V_c = 0.85\left(2 + \frac{4}{\beta_c}\right)\sqrt{f_c'}\, b_0 d$$

$$= 0.85\left(2 + \frac{4}{20/12}\right)\sqrt{4000} \times 85.4 \times 5.35$$

$$= 108,000 \text{ lb}$$

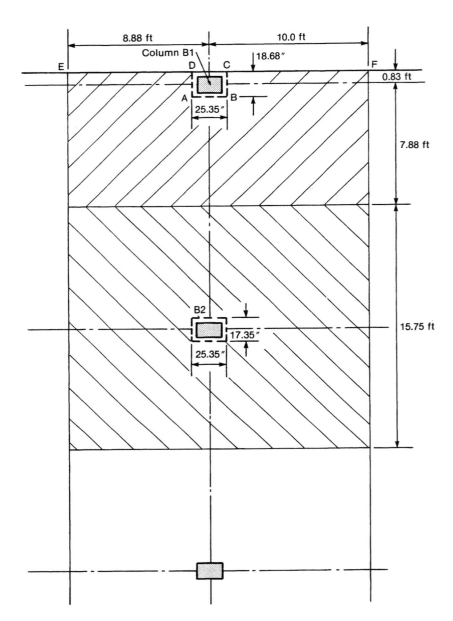

Fig. 13–65
Initial critical shear perimeters and tributary areas for columns B1 and B2.

From Eq. 13–16

$$\phi V_c = 0.85 \left(\frac{\alpha_s d}{b_0} + 2 \right) \sqrt{f_c'} \, b_0 d$$

$$= 0.85 \left(\frac{40 \times 5.35}{85.4} + 2 \right) \sqrt{4000} \times 85.4 \times 5.35$$

$$= 110,700 \text{ lb}$$

From Eq. 13–17

$$\phi V_c = \phi 4 \sqrt{f_c'} b_0 d = 0.85 \times 4 \sqrt{4000} \times 85.4 \times 5.35$$

$$= 98,250 \text{ lb}$$

Therefore, $\phi V_c = 98,250$ lb. Thickness is OK since $\phi V_c > V_u$.

Because this calculation ignores the shear stresses due to moment, it is wise to consider thickening the slab if ϕV_c is less than about $1.2 V_u$ at an interior column.

Column B1:

$$b_0 = 18.68 + 25.35 + 18.68 = 62.7 \text{ in.}$$

$$V_u = 221\left[18.88 \times \left(\frac{7.88}{2} + 0.833\right) - \left(\frac{18.68 \times 25.35}{144}\right)\right] + (300 \times 18.88)$$

(the last term is the weight of the wall)

$$= 35,620 + 5660 = 41,300 \text{ lb}$$

ϕV_c is the smallest of:

$$\phi V_c = 0.85\left(2 + \frac{4}{20/12}\right)\sqrt{4000} \times 62.7 \times 5.35$$

$$= 79,350 \text{ lb}$$

$$\phi V_c = 0.85\left(\frac{30 \times 5.35}{62.7} + 2\right)\sqrt{4000} \times 62.7 \times 5.35$$

$$= 82,200 \text{ lb}$$

$$\phi V_c = 0.85(4\sqrt{4000} \times 62.7 \times 5.35) = 72,100 \text{ lb}$$

$$\phi V_c = 1.75 \text{ times } V_u$$

Although this seems OK, this calculation ignores the portion of the moment that is transferred by shear stresses. Generally, if ϕV_c is less than 1.8 to 2 times V_u at an exterior column, the slab may have inadequate shear capacity for combined shear and moment. This will be checked in step 9 of this example. We shall increase the thickness to 7 in. **Use a 7-in. slab throughout.**

(c) Compute the final value of w_u.

$$w_u = 1.4\left(\frac{7}{12} \times 150 + 25\right) + 1.7(40)$$

$$= 225.5 \text{ psf, say } 226 \text{ psf}$$

If the area of any of the panels had exceeded 400 ft^2, it would be possible to reduce the live load before factoring it in this calculation. In such a case, w_u may differ from panel to panel. Live-load reduction factors are used in step 3 of Example 13–9.

(d) Compute α for the exterior beams. Since there are no edge beams, $\alpha = 0$.

3. Compute the moments in slab strip along column line 2 (Fig. 13–64a). This piece of slab acts as a rigid frame spanning between columns *A2, B2, C2,* and *D2.* In this slab strip, slab panels *A2–B2* and *C2–D2* are "end panels" and *B2–C2* is an "interior panel." Columns *A2* and *D2* are "exterior columns" and *B2* and *C2* are "interior columns."

These calculations are generally tabulated as shown in Tables 13–3 to 13–6. The individual calculations will be followed in detail for slab strips 2 and 1 shown in Fig. 13–64a.

Line 1 (Table 13–3). ℓ_1 is the center-to-center span in the direction of the strip being designed.

Line 2. ℓ_n is the clear span in the direction of the strip (see also Example 13–3).

Line 3. ℓ_2 is the width perpendicular to ℓ_1. The lengths ℓ_2 and ℓ_n are shown in Fig. 13–64a.

Line 4. w_u may differ from panel to panel, especially if panel areas are large enough to be affected by live-load reductions.

Line 5. See also Example 13–3.

Line 6. The moment in the exterior panels is divided using Table 13–2 (ACI Sec. 13.6.3.3) as done in Example 13–4:

$$\text{negative moment at exterior end of end span} = 0.26 M_0$$

$$\text{positive moment in end span} = 0.52 M_0$$

$$\text{negative moment at interior end of end span} = 0.70 M_0$$

TABLE 13-3 Calculation of Negative and Positive Moments for Slab Strip 2— Example 13-7

	A2	B2	C2	D2
1. ℓ_1 (ft)	17.75	20	17.75	
2. ℓ_n (ft)	16.42	18.33	16.42	
3. ℓ_2 (ft)	15.75	15.75	15.75	
4. w_u (ksf) from step 2(c)	0.226	0.226	0.226	
5. $M_0 = w_u\ell_2\ell_n^2/8$ (ft-kips)	120.0	149.5	120.0	
6. Moment coefficients	-0.26 0.52 -0.70	-0.65 0.35 -0.65	-0.70 0.52 -0.26	
7. Negative and positive moments (ft-kips)	-31.2 62.4 -84.0	-97.2 52.3 -97.2	-84.0 62.4 -31.2	
8. Sum of column moments	31.2	24.1	24.1	31.2

TABLE 13-4 Calculation of Negative and Positive Moments for Slab Strip 1— Example 13-7

	A1	B1	C1	D1
1. ℓ_1 (ft)	17.75	20	17.75	
2. ℓ_n (ft)	16.42	18.33	16.42	
3. ℓ_{2edge} (ft)	9.71	9.71	9.71	
4. w_u (ksf)	0.226	0.226	0.226	
5. $M_0 = w_u\ell_{2edge}\ell_n^2/8$ (ft-kips)	74.0	92.2	74.0	
6. Moment coefficients	-0.26 0.52 -0.70	-0.65 0.35 -0.65	-0.70 0.52 -0.26	
7. Negative and positive moments	-19.2 38.5 -51.8	-59.9 32.3 -59.9	-51.8 38.5 -19.2	
8. Column moments from slab load (ft-kips)	19.2	14.9	14.9	19.2
9. Wall load/ft (kip/ft)	0.42	0.42	0.42	
10. Wall M_0 (ft/kips)	14.2	17.6	14.2	
11. Negative and positive moments from wall load (ft-kips)	-3.7 7.4 -9.9	-11.5 6.2 -11.5	-9.9 7.4 -3.7	
12. Column moments from wall load (ft-kips)	3.7	1.6	1.6	3.7
13. Total column moments (ft-kips)	22.9	16.5	16.5	22.9

For the interior panel, from ACI Sec. 13.6.3.2,

$$\text{negative} = 0.65M_0, \quad \text{positive} = 0.35M_0$$

Line 7. This is the product of lines 5 and 6. If the negative moments on the two sides of an interior column differ by less than about 15 to 20% of the larger, design the slab at the joint for the larger negative moment. If the difference exceeds this, carry out a moment distribution at the joint (see step 3 of Example 13-9).

TABLE 13–5 Calculation of Negative and Positive Moments for Slab Strip B— Example 13–7

	B1			B2		B3
1. ℓ_1 (ft)		15.75			15.75	
2. ℓ_n (ft)		14.75			14.75	
3. ℓ_2 (ft)		18.88			18.88	
4. w_u (ksf)		0.226			0.226	
5. $M_0 = w_u \ell_2 \ell_n^2 / 8$ (ft-kips)		116.0			116.0	
6. Moment coefficients	-0.26	0.52	-0.70	-0.65	0.35	-0.65
7. Negative and positive moments (ft-kips)	-30.2	60.3	-81.2	-75.4	40.6	-75.4
8. Σ Column moments (ft-kips)	30.2			9.8		9.8

TABLE 13–6 Calculation of Negative and Positive Moments for Slab Strip A— Example 13–7

	A1			A2		A3
1. ℓ_1 (ft)		15.75			15.75	
2. ℓ_n (ft)		14.42			14.08	
3. ℓ_2 (ft)		9.71			9.71	
4. w_u (ksf)		0.226			0.226	
5. $M_0 = w_u \ell_2 \ell_n^2 / 8$ (ft-kips)		57.0			54.4	
6. Moment coefficients	-0.26	0.52	-0.70	-0.65	0.35	-0.65
7. Negative and positive moments (ft-kips)	-14.8	29.7	-39.9	-35.4	19.0	-35.4
8. Column moments from slab load (ft-kips)	14.8			5.8		5.8
9. Wall load/ft (kip/ft)		0.42			0.42	
10. Wall M_0 (ft-kips)		10.9			10.4	
11. Negative and positive moments) from wall load (ft-kips)	-2.8	5.7	-7.6	-6.7	3.6	-6.7
12. Column moments from wall load (ft-kips)	2.8			0.9		0.0
13. Total column moments (ft-kips)	17.6			6.7		5.8

Line 8. ACI Sec. 13.6.9.2 requires that interior columns be designed for the moment given by Eq. 13–17:

$$M_{col} = 0.07\big[(1.4 \times 112.5 + 0.5 \times 1.7 \times 40) \times 15.75 \times 18.33^2$$
$$- (1.4 \times 112.5)15.75 \times 16.42^2\big] = 24.1 \text{ ft-kips}$$

Therefore, design the interior columns for a total moment of 24.1 ft-kips divided between the columns above and below the joint in the ratio of their stiffnesses.

The unbalanced moment at the exterior edge (-31.2 ft-kips) is divided between the columns above and below the joint in the ratio of their stiffnesses.

4. Compute the moments in the slab strip along column line 1 (Fig. 13–64a). This strip of slab acts as a rigid frame spanning between columns $A1$, $B1$, $C1$, and $D1$. Spans $A1$ to $B1$, and $C1$

to $D1$, are "end panels" and span $B1$ to $C1$ is an "interior panel" in this strip. Similarly, columns $A1$ and $D1$ are "exterior columns" and $B1$ and $C1$ are "interior columns" in this slab strip.

The calculations are carried out in Table 13–4. These are similar to those for slab strip 2 except that the panel is not as wide, and there is a uniformly distributed wall load that causes additional moments in the slab. Since the wall load occurs along the edge of the slab, these moments will be assumed to be resisted entirely by the exterior column strip. The wall load moments will be calculated separately (lines 9–12 in Table 13–4) and added in at a later step (line 4 of Table 13–7).

Line 3. ℓ_2 is the transverse span measured center to center of supports. ACI Sec. 13.6.2.4 states that for edge panels, ℓ_2 in the equation for M_0 should be replaced by $\ell_{2\,edge}$, which is the width from the edge of the slab to a line halfway between column lines 1 and 2.

Line 8. The interior columns ($B1$ and $C1$) are designed for the moment from the slab given by Eq. 13–14:

$$M_{col} = 0.07\big[(1.4 \times 112.5 + 0.5 \times 1.7 \times 40)9.71 \times 18.33^2$$
$$- (1.4 \times 112.5)9.71 \times 16.42^2\big] = 14.9 \text{ ft-kips}$$

Line 9. The factored dead load of the wall is $1.4 \times 0.3 = 0.42 \text{ kip/ft}$

Line 10. The statical moment of the wall load is $w\ell_n^2/8$. Therefore, in span $A1$–$B1$:

$$M_{0\,wall} = 0.42 \times \frac{16.42^2}{8} = 14.2 \text{ ft-kips}$$

Line 11. The negative and positive moments due to the wall load are found by multiplying the moments in line 10. by the coefficients in line 6.

Line 12. The column moments at the interior columns are assumed to be equal to the unbalanced moments at the joint. Equations 13–14 is not applied here since there is no unbalanced live load from the wall.

Line 13. The total column moments are the sum of lines 8 and 12.

5.   Compute the moments in the slab strip along column line B (Fig. 13–64b). This strip of slab acts as a rigid frame spanning between columns $B1$, $B2$, $B3$, and so on. The calculations are carried out in Table 13–5 and follow step 3 of this example.

6.   Compute the moments in the slab strip along column line A (Fig. 13–64b). This strip of slab acts as a rigid frame spanning between columns $A1$, $A2$, $A3$, and so on. In this strip, span $A1$–$A2$ is an "end span," $A2$–$A3$, and so on, are "interior spans." Column $A1$ is an "exterior column" and columns $A2$ and $A3$ are "interior columns" in this slab strip. The calculations are carried out in Table 13–6 and follow step 4 of this example.

7.   Distribute the negative and positive moments to the column and middle strips and design the reinforcement: strips spanning east and west (strips 1 and 2).

In steps 3 and 4, the moments in the strips along column lines 1 and 2 were computed. These moments must be distributed to the various middle and column strips to enable the east–west reinforcement to be designed.

   (a)   Divide the slab strips into middle and column strips. In each panel, the column strip extends 0.25 times the smaller of ℓ_1 and ℓ_2 from the line joining the columns, as shown in Fig. 13–25 and, for this example, in Fig. 13–64a. Thus the column strips extend 15 ft 9 in./4 = 47.25 in. on each side of the column lines. The total width of the column strip is 2×47.25 in. = 7.88 ft. The width of the middle strip is 7.87 ft. The edge–strip has a width of 47.25 + 10 in. = 4.77 ft.

   (b)   Divide the moments between the column and middle strip and design the reinforcement. The calculations are carried out in Table 13–7 for the east–west strips. This table is laid out to resemble a plan of the part of the slab shown shaded in Fig. 13–64a. The columns are shown by square dots and are numbered. The heavy line represents the edge of the slab. The computations leading up to the entries in each line of this table are summarized below. The calculations are repeated for each of the negative and positive moment regions. Since this slab has only three spans, it is not necessary to consider the typical interior negative moment.

Line 1. These moments are taken from line 7 in Tables 13–3 and 13–4. At the first interior support, the larger of the two adjacent negative moments is used (ACI Sec. 13.6.3.4).

Line 2. these are the moment coefficients.

TABLE 13–7   Division of Moment to Column and Middle Strips: East-West Strips–Example 13–7

		Column Strip	Middle Strip	Column Strip	Middle Strip	Edge Column Strip
		7.87	7.88	7.87	7.88	4.77

Exterior Negative Moments

			A3		A2		A1
1. Slab Moment (ft–kip)			−31.2		−31.2		−19.2
2. Moment Coefficients	0.0		1.00	0.0 0.0	1.00	0.0 0.0	1.00
3. Moment to Column and Middle Strips (ft–kip)			−31.2	0	−31.2	0	−19.2
4. Wall Moment (ft–kip)							−3.7
5. Total Moment in Strip (ft–kip)			−31.2	0	−31.2	0	−22.9
6. A_s required (in.^{2})			1.84	0	1.84	0	1.35
7. Min A_s (in.^{2})			1.32	1.32	1.32	1.32	0.80
8. Choose Steel			10 #4 bars	7 #4 bars	10 #4 bars	7 #4 bars	7 #4 bars
9. A_s provided (in.^{2})			2.00	1.40	2.00	1.40	1.40

End Span Positive Moments

1. Slab Moment (ft–kip)			62.4		62.4		38.5
2. Moment Coefficients	0.2		0.6	0.2 0.2	0.6	0.2 0.4	0.6
3. Moment to Column and Middle Strips (ft–kip)	12.5		37.4	12.5 12.5	37.4	12.5 15.4	23.1
4. Wall Moment (ft–kip)							7.4
5. Total Moment in Strip (ft–kip)			37.4	25.0	37.4	27.9	30.5
6. A_s required (in.^{2})			2.20	1.47	2.20	1.64	1.79
7. Min A_s (in.^{2})			1.32	1.32	1.32	1.32	0.80
8. Choose Steel			11 #4 bars	8 #4 bars	11 #4 bars	9 #4 bars	9 #4 bars
9. A_s provided (in.^{2})			2.20	1.60	2.20	1.80	1.80

First Interior Negative Moments

			B3		B2		B1
1. Slab Moment (ft–kip)			−97.2		−97.2		−59.9
2. Moment Coefficients			0.75	0.125 0.125	0.75	0.125 0.25	0.75
3. Moment to Column and Middle Strips (ft–kip)			−72.9	−12.15 −12.15	−72.9	−12.15 15.0	−44.9
4. Wall Moment (ft–kip)							−11.5
5. Total Moment in Strip (ft–kip)			−72.9	−24.3	−72.9	−27.2	−56.4
6. A_s required (in.^{2})			4.29	1.43	4.29	1.60	3.32
7. Min A_s (in.^{2})			1.32	1.32	1.32	1.32	0.80
8. Choose Steel			14 #5 bars	8 #4 bars	14 #5 bars	8 #4 bars	11 #5 bars
9. A_s provided (in.^{2})			4.34	1.60	4.34	1.60	3.41

Interior Positive Moments

1. Slab Moment (ft–kip)			52.3		52.3		32.3
2. Moment Coefficients			0.60	0.20 0.20	0.60	0.20 0.40	0.60
3. Moment to Column and Middle Strips (ft–kip)			31.4	10.45 10.45	31.4	10.45 12.9	19.4
4. Wall Moment (ft–kip)							6.1
5. Total Moment in Strip (ft–kip)			31.4	20.9	31.4	23.4	25.5
6. A_s required (in.^{2})			1.85	1.23	1.85	1.38	1.50
7. Min A_s (in.^{2})			1.32	1.32	1.32	1.32	0.8
8. Choose Steel			10 #4 bars	7 #4 bars	10 #4 bars	7 #4 bars	8 #4 bars
9. A_s provided (in.^{2})			2.00	1.40	2.00	1.40	1.60

(a) *Exterior negative moments.* ACI Sec. 13.6.4.2 gives the fraction of the exterior negative moment resisted by the column strip. This section is entered using β_t, $\alpha_1 \ell_2/\ell_1$, and ℓ_2/ℓ_1. For a slab without edge beams, $\beta_t = 0$, and for a slab without beams parallel to ℓ_1 for the span being designed, $\alpha_1 = 0$. Thus, from the table in ACI Sec. 13.6.4.2, 100% of the exterior negative moment is assigned to the column strips.

(b) *Positive moment regions.* ACI Sec. 13.6.4.4 gives the fraction of positive moments assigned to the column strips. Since $\alpha_1 = 0$, $\alpha_1 \ell_2/\ell_1 = 0$, and 60% is assigned to the column strip. The remaining 40% is assigned to the adjacent half middle strips. At the edge, there is only one adjacent half middle strip, so 40% of the panel moment goes to it. At an interior column strip, there are two adjacent half middle strips and $0.5 \times 40 = 20\%$ goes to each.

13–10   Design of Slabs without Beams

(c) *Interior negative moment regions.* ACI Sec. 13.6.4.1 specifies the fraction of interior negative moments assigned to the column strips. Since $\alpha_1 = 0$, 75% goes to the column strip and the remaining 25% to the middle strips.

Line 3. This is the product of lines 1 and 2. The arrows in Table 13–7 illustrate the distribution of the moments between the various parts of the slab.

Line 4. The weight of the north–south wall along column line 1 causes moments in the edge column strip. The moments given here come from line 10 of Table 13–3.

Line 5. The calculation of this term is indicated by the arrows in Table 13–7. Line 5 is the sum of lines 3 and 4.

Line 6. The area of steel will be calculated using Eq. 4–33 as done in Example 10–1.

$$A_s = \frac{M_u}{\phi f_y jd} \tag{4–33}$$

(a) Compute d. Since the largest moments in the panel occur at support $B2$ in the slab strip 2 (see Table 13–3) place reinforcement as shown in Fig. 13–66. Thus

$$d = 7 \text{ in.} - \tfrac{3}{4}\text{in.} - \tfrac{1}{2} \text{ bar diameter}$$

$$= 5.94 \text{ in. (assumming No. 5 bars) (or use Eq. 13-32a)}$$

(b) Compute trial A_s required at the section of maximum moment (first interior negative column strip). Largest $M_u = -72.9$ ft-kips. Assume that $j = 0.925$.

$$A_{s(\text{req'd})} = \frac{72.9 \times 12,000}{0.9 \times 40,000 \times 0.925 \times 5.94} \tag{4–33}$$

$$= 4.42 \text{ in.}^2$$

(c) Compute a and a/d and check whether section is tension-controlled.

$$a = \frac{A_s f_y}{0.85 f_c' b} = \frac{4.42 \times 40,000}{0.85 \times 4000(7.87 \times 12)}$$

$$= 0.55 \text{ in.}$$

$$\frac{a}{d} = \frac{0.55}{5.94} = 0.093$$

This is much less than the a/d in Table A–4 for the tension-controlled limit—therefore, $\phi = 0.9$.

(d) Compute jd and the constant for computing A_s.

$$jd = d - \frac{a}{2} = 5.94 - \frac{0.55}{2}$$

$$= 5.66 \text{ in.}$$

$$A_s(\text{in.}^2) = \frac{M_u(\text{ft-kips}) \times 12,000}{0.9 \times 40,000 \times 5.66}$$

Therefore,

$$A_s(\text{in.}^2) = 0.0588 M_u(\text{ft-kips}) \tag{A}$$

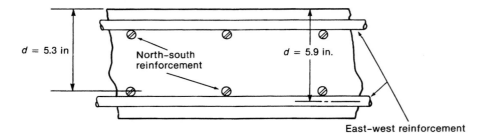

Fig. 13–66
Arrangement of bars in
slab—Example 13–7.

Two-Way Slabs: Behavior, Analysis, and Direct Design Method

TABLE 13–8 Division of Moment to Column and Middle Strips: North-South Strips–Example 13–7

	Edge Column Strip 4.77	Middle Strip 9.88	Column Strip 7.88	Middle Strip 12.12	Column Strip 7.88
	A1		B1		C1
Exterior Negative Moments					
1. Slab Moment (ft-kip)	−14.8		−30.2		−30.2
2. Moment Coefficients	1.00	0	1.00	0	1.00	0
3. Moment to Column and Middle Strips (ft-kip)	−14.8	0	−30.2	0	−30.2
4. Wall Moment (ft-kip)	−2.8				
5. Total Moment in Strip (ft-kip)	−17.6	0	−30.2	0	−30.2
6. A_s required (in.^{2})	1.15	0	2.00	0	2.00
7. Min A_s (in.^{2})	0.80	1.66	1.32	2.04	1.32
8. Choose Steel	6 #4 bars	9 #4 bars	10 #4 bars	11 #4 bars	10 #4 bars
9. A_s provided (in.^{2})	1.20	1.80	2.00	2.20	2.00
End Span Positive Moments					
1. Slab Moment (ft-kip)	29.7		60.3		60.3
2. Moment Coefficients	0.6	0.4 / 0.2	0.6	0.2 / 0.2	0.6 / 0.2
3. Moment to Column and Middle Strips (ft-kip)	17.8	11.9 / 12.1	36.2	12.1 / 12.1	36.2 / 12.1
4. Wall Moment (ft-kip)	5.7				
5. Total Moment in Strip (ft-kip)	23.5	23.9	36.2	24.1	36.2
6. A_s required (in.^{2})	1.55	1.58	2.39	1.59	2.39
7. Min A_s (in.^{2})	0.80	1.66	1.32	2.04	1.32
8. Choose Steel	8 #4 bars	9 #4 bars	12 #4 bars	11 #4 bars	12 #4 bars
9. A_s provided (in.^{2})	1.60	1.80	2.40	2.20	
	A2		B2		C2
First Interior Negative Moments					
1. Slab Moment (ft-kip)	−39.9		−81.2		−81.2
2. Moment Coefficients	0.75	0.25 / 0.125	0.75	0.125 / 0.125	0.75 / 0.125
3. Moment to Column and Middle Strips (ft-kip)	−29.9	−10.0 / −10.2	−60.9	−10.2 / −10.2	−60.9 / −10.2
4. Wall Moment (ft-kip)	−7.6				
5. Total Moment in Strip (ft-kip)	−37.5	−20.2	−60.9	−20.4	−60.9
6. A_s required (in.^{2})	2.48	1.33	4.03	1.34	4.03
7. Min A_s (in.^{2})	0.80	1.66	1.32	2.04	1.32
8. Choose Steel	8 #5 bars	9 #4 bars	13 #5 bars	11 #4 bars	13 #5 bars
9. A_s provided (in.^{2})	2.48	1.80	4.03	2.20	4.03
Interior Positive Moments					
1. Slab Moment (ft-kip)	19.0		40.6		40.6
2. Moment Coefficients	0.6	0.4 / 0.2	0.6	0.2 / 0.2	0.6 / 0.2
3. Moment to Column and Middle Strips (ft-kip)	11.4	7.6 / 8.1	24.4	8.1 / 8.1	24.4 / 8.1
4. Wall Moment (ft-kip)	3.6				
5. Total Moment in Strip (ft-kip)	15.1	15.7	24.4	16.2	24.4
6. A_s required (in.^{2})	1.00	1.04	1.61	1.07	1.61
7. Min A_s (in.^{2})	0.80	1.66	1.32	2.04	1.32
8. Choose Steel	6 #4 bars	9 #4 bars	8 #4 bars	11 #4 bars	8 #4 bars
9. A_s provided (in.^{2})	1.20	1.80	1.60	2.20	1.60
	A3		B3		C3
Typical Interior Negative Moment					
1. Slab Moment (ft-kip)	−35.4		−75.4		−75.4
2. Moment Coefficients	0.75	0.25 / 0.125	0.75	0.125 / 0.125	0.75 / 0.125
3. Moment to Column and Middle Strips (ft-kip)	−26.6	−8.9 / −9.4	−56.6	−9.4 / −9.4	−56.6 / −9.4
4. Wall Moment (ft-kip)	−6.7				
5. Total Moment in Strip (ft-kip)	−33.3	−18.3	−56.6	−18.8	−56.6
6. A_s required (in.^{2})	2.20	1.20	3.74	1.24	3.74
7. Min A_s (in.^{2})	0.80	1.66	1.32	2.04	1.32
8. Choose Steel	8 #5 bars	9 #4 bars	12 #5 bars	11 #4 bars	12 #5 bars
9. A_s provided (in.^{2})	2.48	1.80	3.72	2.00	3.72

13–10   Design of Slabs without Beams

The values of A_s in lines 6 of Table 13–7 are computed using Eq. A.

Line 7. The minimum A_s is specified in ACI Sec. 13.3.1 (see Sec. 13–9). $A_{s(\min)} = 0.002bh$ for Grade 40 reinforcement. Maximum bar spacing $= 2h$ (ACI Sec. 13.3.2) but not more than 18 in. (ACI Sec. 7.12.2.2). Therefore, the maximum spacing $= 14$ in.

Edge column strip:

$$A_{s(\min)} = 0.002bh = 0.002(4.77 \times 12) \times 7$$
$$= 0.80 \text{ in.}^2$$

$$\text{minimum number of bar spaces} = \frac{4.77 \times 12}{14} = 4.09$$

Therefore, the minimum number of bars $= 5$.

Other strips:

$$A_{s(\min)} = 0.0020(7.88 \times 12) \times 7 = 1.32 \text{ in.}^2$$

Minimum number of bars $= 7$.

Line 8. The final bar choices are given in line 8.

Line 9. The actual areas of reinforcement provided are given in line 9. The reinforcement chosen for each east–west strip is shown in the section in Fig. 13–67. In the calculations for line 6 it was assumed that No. 5 bars would be used, giving $d = 5.94$ in. In most cases the bars chosen were No. 4 bars. As a result, $d = 6.0$ in. could be used in some parts of the slab. This will not make enough difference to repeat the calculation.

8. Distribute the negative and positive moments to the column and middle strips and design the reinforcement: strips spanning north and south (strips *A* and *B*). In steps 5 and 6, the

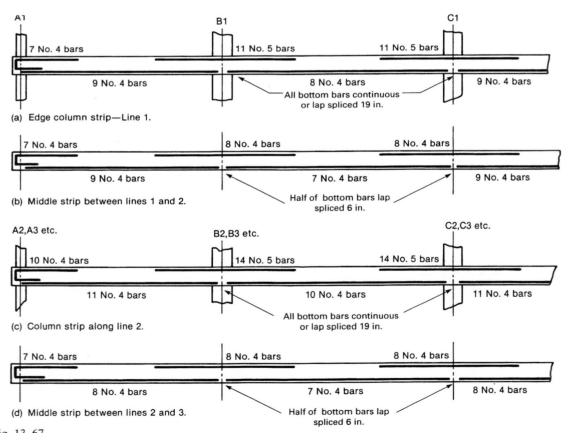

(a) Edge column strip—Line 1.

(b) Middle strip between lines 1 and 2.

(c) Column strip along line 2.

(d) Middle strip between lines 2 and 3.

Fig. 13–67

Schematic diagram of reinforcement in east–west—Example 13–7 moments were computed for the strips along column lines A and B. These moments must be assigned to the middle and column strips so that the north–south reinforcement can be designed. This proceeds in the same manner as step 7, using the moments from lines 7 and 10 of Tables 13–5 and 13–6.

(a) **Divide the slab strips into middle and column strips.** See Fig. 13–64b.

(b) **Divide the moments between the column and middle strips and design the reinforcement.** For the north–south strips, the calculations are carried out in Table 13–8. This table is laid out to represent a plan view of the shaded part of Fig. 13–64b.

Line 1. These moments are from line 7 of Tables 13–5 and 13–6.

Line 6. The area of steel is computed as in Table 13–7 except that d is smaller, as shown in Fig. 13–66. Assuming No 5 bars, $d \simeq 7 - 0.75 - 1.5(0.625) = 5.31$ in..

(a) Compute a trial A_s.

$$\text{largest } M_u = -60.9 \text{ ft-kips}$$

$$A_{s(\text{req'd})} = \frac{60.9 \times 12{,}000}{0.9 \times 40{,}000 \times 0.925 \times 5.31} = 4.13 \text{ in.}^2$$

(b) Compute a and a/d and check whether section is tension-controlled.

$$a = \frac{A_s f_y}{0.85 f_c' b} = \frac{4.13 \times 40{,}000}{0.85 \times 4000(7.88 \times 12)} = 0.51 \text{ in.}$$

$$\frac{a}{d} = 0.097$$

This is much less than the a/d in Table A–4 for the tension-controlled limit—therefore, $\phi = 0.9$.

(c) Compute jd and the constant for computing A_s.

$$jd = d - \frac{a}{2} = 5.31 - \frac{0.51}{2} = 5.05 \text{ in.}$$

$$A_s(\text{in.}^2) = \frac{M_u(\text{ft-kips}) \times 12{,}000}{0.9 \times 40{,}000 \times 5.05}$$

Therefore,

$$A_s(\text{in.}^2) \simeq 0.0660 M_u(\text{ft-kips}) \tag{B}$$

The values of A_s in lines 6 of Table 13–8 are computed using Eq. B.

Line 7. For the column strips the minimum A_s and minimum number of bars are as given in Table 13–7, because the column strip width is the same in both directions. In the middle strips, different amounts of steel are required.

Middle strip between lines A and B:

$$A_{s(\text{min})} = 0.0020(9.88 \times 12) \times 7 = 1.66 \text{ in.}^2$$

$$\text{minimum number of bar spaces} = 8.5$$

$$\text{minimum number of bars} = 9$$

Middle strip between lines B and C:

$$A_{s(\text{min})} = 0.0020(12.12 \times 12) \times 7 = 2.04 \text{ in.}^2$$

$$\text{minimum number of bars} = 11$$

Line 8. The final bar choices are given in line 8.

Line 9. The reinforcement chosen for each north–south strip is given in line 9 and shown in the sections in Fig. 13–68.

9. Check the shear at the exterior columns for combined shear and moment transfer. Either column A2 or B1 will be the most critical. Since the tributary area is largest for B1, we shall limit our check to that column, although both joints should be checked.

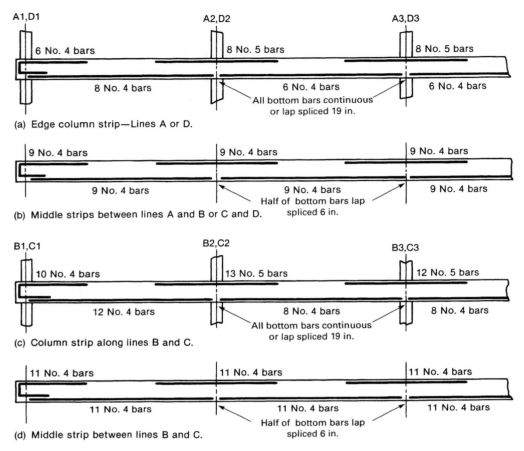

(a) Edge column strip—Lines A or D.

(b) Middle strips between lines A and B or C and D.

(c) Column strip along lines B and C.

(d) Middle strip between lines B and C.

Fig. 13–68
Schematic diagram of reinforcement in north–south strips—Example 13–7.

Since the reinforcement is No. 4 bars at all exterior ends, the effective depths will be a little bigger than shown in Fig. 13–66. Thus $d_{EW} = 6.0$ in., $d_{NS} = 5.5$ in., and $d_{avg} = 5.75$ in.. Following the calculations in Example 13–6 gives:

(a) Locate the critical shear perimeter. The critical shear perimeter is at $d/2$ from the face of the column, where d is the average depth. Since the shortest perimeter results from the section shown in Fig. 13–69a, this perimeter will be used.

$$d_{avg} = 5.75 \text{ in.} \qquad \frac{d}{2} = 2.88 \text{ in.}$$

Thus $b_1 = 18.88$ in. and $b_2 = 25.75$ in.

(b) Locate the centroid of the shear perimeter. For moments about the $Z–Z$ axis:

$$y_{AB} = \frac{2 \times (18.88 \times 5.75) \times 18.88/2}{2(18.88 \times 5.75) + (25.75 \times 5.75)}$$

$$= 5.61 \text{ in.}$$

Therefore, $c_{AB} = 5.61$ in. and $c_{CD} = 13.27$ in. For moments about the $W–W$ axis:

$$c_{CB} = c_{AD} = \frac{25.75}{2} = 12.88 \text{ in.}$$

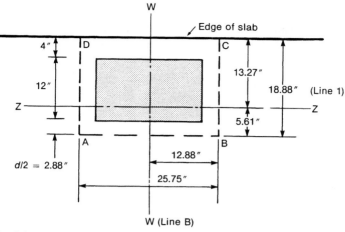

(a) Critical section—Column B1.

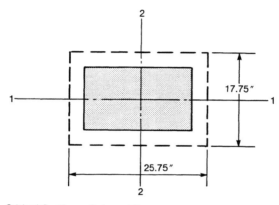

(b) Critical Section—Column B2

Fig. 13–69
Final critical shear perimeters: columns $B1$ and $B2$—Example 13–7.

(c) **Compute the moment about the centroid of the shear perimeter.** The tributary area of column $B1$ is shown in Fig. 13–65 and the critical shear perimeter is shown in Fig. 13–69a.

$$V_u = 0.226 \text{ ksf} \left[18.88 \text{ ft} \times (7.88 + 0.83) \text{ ft} - \frac{18.88 \text{ in.} \times 25.75 \text{ in.}}{144} \text{ ft}^2 \right]$$

$$= 36.4 \text{ kips}$$

The shear due to the wall load outside the critical perimeter (between E and D, and C and F in Fig. 13–65)

$$V_{uw} = \left(18.88 \text{ ft} - \frac{25.75}{12} \text{ ft} \right) \times 1.4 \times 0.3 \text{ kips/ft} = 7.0 \text{ kips}$$

total $V_u = 43.4$ kips

For slabs designed by the Direct Design Method, the moment transferred from the slab to the column about axis Z–Z in Fig. 13–69a is $0.3M_0$, where, from line 5 in Table 13–5, $M_0 = 116.0$ ft-kips. Therefore, the moment to be transferred to the column is $0.3 \times 116.0 = 34.8$ ft-kips. We shall ignore the moment from the cantilever portion of the slab. If the slab cantilevered several feet outside the column, or if a heavy wall load existed on the cantilever and it could be shown that this heavy wall load would be in place *before* significant construction or live loads were placed on the slab itself, the moment about Z–Z could be reduced.

13–10 Design of Slabs without Beams

Moments about axis W–W in Fig. 13–69a come from the slab strips parallel to the edge. Line 7 of Table 13–4 gives the moments for all panels loaded. Line 8 gives the moments transferred to the columns when one adjacent panel is loaded and the other panel is not. Since the shear check involves the shear due to all panels loaded, we shall use the moments from line 7. From line 7 in Table 13–4, we find that the moment of the edge panel loads about axis W–W of column B1 is $59.9 - 51.8 = 8.1$ ft-kips. From line 13 we find that the unbalanced moment due to the wall load is 1.6 ft-kips. The total moment to be transferred is 9.7 ft-kips.

In summary, the connection at B1 must be designed for $V_u = 43.4$ kips, $M_{ZZ} = 34.8$ ft-kips, and $M_{WW} = 9.7$ ft-kips. These moments act about the centroids of the shear perimeter in the two directions.

(d)   Compute ϕV_c and $V_u / \phi V_c$. V_c is the smallest value given by Eqs. 13–15 to 13–17 (ACI Eqs. 11–35 to 11–37).

$$\phi V_c = \phi\left(2 + \frac{4}{\beta_c}\right)\sqrt{f'_c}\, b_0 d \qquad (13\text{–}15)$$

where $b_0 = 2 \times 18.88$ in. $+ 25.75$ in. $= 63.51$ in.

$$\phi V_c = 0.85\left(2 + \frac{4}{20/12}\right)\sqrt{4000} \times 63.51 \times 5.75$$

$$= 86.4 \text{ kips} \qquad (13\text{–}16)$$

$$\phi V_c = \phi\left(\frac{\alpha_s d}{b_0} + 2\right)\sqrt{f'_c}\, b_0{}_d$$

$$= 0.85\left(\frac{30 \times 5.75}{63.51} + 2\right)\sqrt{4000} \times 63.51 \times 5.75$$

$$= 92.6 \text{ kips}$$

$$\phi V_c = \phi 4\sqrt{f'_c}\, b_0 d \qquad (13\text{–}17)$$

$$= 0.85 \times 4 \times \sqrt{4000} \times 63.51 \times 5.75$$

$$= 78.5 \text{ kips}$$

Therefore, $\phi V_c = 78.5$ kips and $V_u / \phi V_c = 43.4/78.5 = 0.553$.

(e)   Determine the fraction of the moment transferred by flexure, γ_f.
Moments about the Z–Z axis:

$$\gamma_f = \frac{1}{1 + \frac{2}{3}\sqrt{b_1/b_2}} \qquad (13\text{–}22)$$

$$\text{(ACI Eq. 13–1)}$$

where, for moment about the Z–Z axis, $b_1 = 18.88$ in. and $b_2 = 25.75$ in.

$$\gamma_f = \frac{1}{1 + \frac{2}{3}\sqrt{\frac{18.88}{25.75}}}$$

$$= 0.637$$

ACI Sec. 13.5.3.3 allows γ_f to be increased up to 1.0 if $V_u / \phi V_c$ does not exceed 0.75 and the resulting $\rho < 0.375\rho_b$ within a width of $c_2 + 3h$ centered on the column. From step (d), $V_u / \phi V_c = 0.553$. Therefore, take $\gamma_f = 1.0$ and check the reinforcement required.

Moments about the W–W axis: Exchanging b_1 and b_2 in Eq. 13–22 gives $\gamma_{f2} = 0.562$. ACI Sec. 13.5.3.3 allows γ_f to be increased by up to 25% if $V_u / \phi V_c \leq 0.4$. Since $V_u / \phi V_c = 0.553$, no increase can be made in γ_{f2}. The moment transferred by flexure is $\gamma_{f2} M_u = 0.562 \times 9.7$ ft-kips $= 5.45$ ft-kips.

(f)   Design the reinforcement required for moment transfer by flexure.
Moments about the Z–Z axis:

Width effective for flexure $= c_2 + 3h = 20$ in. $+ 3 \times 7$ in. $= 41$ in.

Moment $= 1.0 \times 34.8$ ft-kips $= 34.8$ ft-kips

Assume that $jd = 0.925d$.

$$A_s = \frac{M_u}{\phi f_y jd}$$

$$= \frac{34.8 \times 12{,}000}{0.9 \times 60{,}000 \times 0.925 \times 5.75}$$

$$= 1.45 \text{ in.}^2$$

The steel provided in step 8 is 10 No. 4 bars (see Fig. 13–68c) in a column strip width of 7.88 ft = 94.5 in., or roughly 9.5 in on centers. The bars within the 41-in. effective width can be used for the moment transfer. Add 4 additional bars in this region, giving $A_s = 1.60$ in.^{2} in the effective width.

Since the computed A_s was based on a guess for jd, we shall compute a for the A_s chosen and recompute A_s.

$$a = \frac{A_s f_y}{0.85 f_c' b} = \frac{1.60 \times 60{,}000}{0.85 \times 4000 \times 41}$$

$$= 0.689 \text{ in.}$$

$$A_s = \frac{M_u}{f_y(d-a/2)} = \frac{34.8 \times 12{,}000}{0.9 \times 60{,}000(5.75-0.689/2)}$$

$$= 1.43 \text{ in.}^2 \qquad \text{—therefore, the steel chosen has adequate capacity}$$

$$\frac{a}{d} = \frac{0.689}{5.75} = 0.120$$

From Eq. 4–21

$$\frac{a_b}{d} = \beta_1\left(\frac{87{,}000}{87{,}000 + f_y}\right) = 0.85\left(\frac{87{,}000}{87{,}000 + 40{,}000}\right)$$

$$= 0.582$$

and $0.375 a_b/d = 0.375 \times 0.582 = 0.218$. Since $a/d = 0.120$ is less than $0.375 a_b/d = 0.218$, $\rho < 0.375\rho_b$ and we can use $\gamma_f = 1.0$. As a result, it is not necessary to transfer any of the moment about axis Z–Z by shear.

Moments about the W–W axis: Effective width for moment transfer = 12 in. + 4 in. + 1.5 × 7 in. = 26.5 in. For this moment, 2 No. 5 bars are required in this width. This will be accomplished if the bars are uniformly distributed in this region.

(g) **Compute the shear stresses.**

$$v_u = \frac{V_u}{b_0 d} \pm \frac{\gamma_{v1} M_{u1} c}{J_{c1}} \pm \frac{\gamma_{v2} M_{u2} c}{J_{c2}} \tag{13–18a}$$

where M_{u1}, and so on, refer to axis Z–Z, and M_{u2}, and so on, refer to axis W–W.

$$b_0 = 63.51 \text{ in.}$$

$$\gamma_{v1} = 1 - \gamma_{f1} = 0.0$$

$$\gamma_{v2} = 1 - \gamma_{f2} = 0.438$$

Since $\gamma_{v1} = 0.0$, the second term in the equation for v_u drops out. As a result, it is not necessary to compute J_{c1}.

$$J_{c2} = 2(b_1 d)c_{CB}^2 + \frac{b_2 d^3}{12} + \frac{b_2^3 d}{12}$$

$$= 2(18.88 \times 5.75)12.882 + \frac{25.75 \times 5.75^3}{12} + \frac{25.75^3 \times 5.75}{12}$$

$$= 44{,}600 \text{ in.}^4$$

The maximum stresses along sides CB and AD are

$$v_u = \frac{43.4 \times 1000}{63.51 \times 5.75} \pm 0.0 \pm \frac{0.438 \times 9.7 \times 12{,}000 \times 12.88}{44{,}600}$$

$$= 119 + 0.0 \pm 15$$

$$= 134 \text{ psi at } B \text{ and } 104 \text{ psi at } A$$

Since there is no shear reinforcement in the slab, $\phi v_n = \phi V_c/b_0 d$, where from step (d), $\phi V_c = 78.5$ kips. Thus

$$\phi v_n = \frac{78.5 \times 1000}{63.51 \times 5.75} = 215 \text{ psi}$$

Since $v_u < \phi v_n$, the shear is OK in this column–slab connection.

If $v_u > \phi v_c$, it would be necessary to modify the connection. Solutions would be to thicken the slab, use stronger concrete, enlarge the column, or use shear reinforcement. The choice of solution should be based on a study of the extra costs involved.

10. Check the shear at an interior column for combined shear and moment transfer.
Tables 13–3 and 13–5 indicate that moments of 24.1 ft-kips and 9.8 ft-kips are transferred from the slab to column B_2 from strip 2 and strip B, respectively. Check the shear at this column.

(a) **Locate the critical shear perimeter.** See Fig. 13–69b.

(b) **Locate the centroid of the shear perimeter.** Since the perimeter is continuous, the centroids pass through the centers of the sides.

(c) **Compute the forces to be transferred.** The tributary area for column $B2$ is shown in Fig. 13–65.

$$V_u = 0.226\left(18.88 \times 15.75 - \frac{25.75 \times 17.75}{144}\right)$$

$$= 66.5 \text{ kips}$$

The moments to be transferred come from Tables 13–3 and 13–5. Line 7 of these tables gives the moments for all panels loaded. Line 8 of these tables gives moments transferred to the columns when one adjacent panel is loaded and the other panel is not. Since the shear check involves the shear due to all panels loaded, we shall use the moments from line 7. In line 7 of Table 13–3, the difference between the negative moments on the two sides of column $B2$ is $97.2 - 84.0 = 13.2$ ft-kips. From Table 13–5 it is $81.2 - 75.4 = 5.8$ ft-kips.

The forces transferred to the column are $V_u = 66.5$ kips, $M_{u1} = 5.8$ ft-kips, and $M_{u2} = 13.2$ ft-kips.

(d) **Compute the fraction of the moment transferred by flexure.**

$$\gamma_{f1} = 0.644 \qquad \gamma_{f2} = 0.554$$

ACI Sec. 13.5.3.3 allows γ_f to be increased by up to 25% if $V_u/\phi V_c \le 0.4$. Again, Eq. 13–17 governs and

$$\phi V_c = \phi 4\sqrt{f_c'}\,b_0 d = 0.85 \times 4\sqrt{4000} \times 87 \times 5.75$$

$$= 107.6 \text{ kips}$$

$$\frac{V_u}{\phi V_c} = \frac{66.5}{107.6} = 0.618$$

Since $V_u/\phi V_c$ exceeds 0.4, γ_f cannot be increased about either axis.

(e) **Compute the torsional moment of inertia, J_c.** Bending about axis 1–1 (from Eq. 13–25):

$$J_{c1} = 2\left(\frac{b_1 d^3}{12}\right) + 2\left(\frac{d b_1^3}{12}\right) + 2(b_2 d)\left(\frac{b_1}{2}\right)^2$$

Two-Way Slabs: Behavior, Analysis, and Direct Design Method

where $b_1 = 17.75$ in. and $b_2 = 25.75$ in. (see Fig. 13–69b).

$$J_{c1} = 29,200 \text{ in.}^4$$

Bending about axis 2–2: here $b_1 = 25.75$ in., $b_2 = 17.75$ in., and $J_{c2} = 51,000$ in.^{4}

(f) **Design the reinforcement for moment transfer.** By inspection, the reinforcement already in the slab is adequate.

(g) **Compute the shear stresses.** Maximum stresses occur at the corner where all three terms are additive.

$$b_0 = 2(17.75 + 25.75) = 87.0 \text{ in.}$$

$$v_u = \frac{66.5}{87.0 \times 5.75} + \frac{0.356 \times 5.8 \times 12 \times 8.88}{29,200} + \frac{0.445 \times 13.2 \times 12 \times 12.88}{51,000}$$

$$= 0.133 + 0.008 + 0.018 = 0.159 \text{ ksi}$$

$$v_u = 159 \text{ psi} \quad \phi v_c = 215 \text{ psi}$$

Since $\phi v_c > v_u$, the shear is OK at this column. The moment transfer increased the shear stress by 20%, from 133 psi to 159 psi.

11. Check the shear at the corner column.

Tables 13–4 and 13–6 indicate that moments of 22.9 ft-kips and 17.6 ft-kips are transferred from the slab to the corner column, $A1$, from strip 1 and strip A, respectively. Two shear perimeters will be considered. Two-way shear may be critical on the perimeter shown in Fig. 13–70a, while one-way shear may be critical on the section shown in Fig. 13–70b.

Two-Way Shear

(a) **Locate the critical perimeter.** The critical perimeter is as shown in Fig. 13–70a. For two-way shear, we shall base the calculations of J_c and the stresses on the orthogonal x- and y-axes.

(b) **Locate the centroid of the perimeter.**

$$\bar{x} = \frac{(18.88 \times 5.75)(18.88/2)}{2 \times 18.88 \times 5.75}$$
$$= 4.72 \text{ in. from inside corner}$$

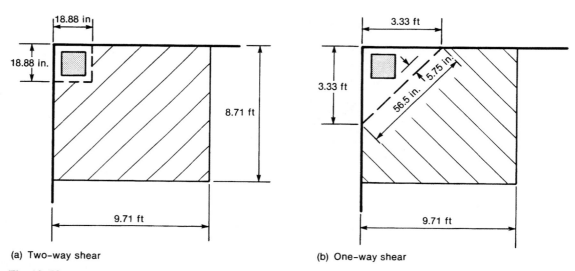

(a) Two-way shear   (b) One-way shear

Fig. 13–70
Critical shear perimeters and tributary areas: column $A1$—Example 13–7.

(c) **Compute the forces to be transferred.**

$$V_u = 0.226\left(9.71 \times 8.71 - \frac{18.88^2}{144}\right) = 18.6 \text{ kips}$$

$$\text{Shear due to the wall load} = \left(9.71 + 8.71 - 2 \times \frac{18.88}{12}\right) \times 1.4 \times 0.3$$

$$= 6.42 \text{ kips}$$

$$\text{total shear} = 25.0 \text{ kips}$$

ACI Sec. 13.6.3.6 states that the amount of moment to be transferred from a slab to an edge column is $0.3M_0$. Although no similar statement is made for a corner column, we shall assume that the same moment should be transferred.

For strip 1: $M_0 = 74.0$ ft-kips, $0.3M_0 = 22.2$ ft-kips
For strip A: $M_0 = 57.0$ ft-kips, $0.3M_0 = 17.1$ ft-kips

Again, it will be assumed that these moments and the shears all act through the centroid of the shear perimeter.

(d) **Determine the fraction of the moment transferred by flexure.** Since the critical section is square, Eq. 13–22 gives $\gamma_f = 0.6$ about each axis. ACI Sec. 13.5.3.3 allows γ_f to be increased to 1.0, provided that $V_u/\phi V_c \leq 0.5$ and $\rho \leq 0.375\rho_b$ in the strip of slab which is effective for moment transfer, width $= 4 + 12 + 1.5 \times 7 = 26.5$ in.

$$\phi V_c = \phi 4\sqrt{f_c'}\, b_0 d = 41.1 \text{ kips}$$

$$\frac{V_u}{\phi V_c} = \frac{25.0}{41.1} = 0.608$$

Therefore, no change can be made in γ_f.

(e) **Design the reinforcement for moment transfer by flexure.** Since $\gamma_f = 0.6$, 0.6 times 7 No. 4 bars must be placed within 26.5 in. of the corner in strip 1 and 0.6 times 6 No. 4 bars within 26.5 in. of the corner in strip A.

(f) **Compute the torsional moment of inertia, J_c.** Bending about the Z–Z axis

$$J_c = \frac{18.88 \times 5.75^3}{12} + \frac{5.75 \times 18.88^3}{12}$$

$$+ (18.88 \times 5.75)\left(\frac{18.88}{2} - 4.72\right)^2 + (5.75 \times 18.88)4.72^2$$

$$= 8360 \text{ in.}^4$$

(g) **Compute the shear stresses.**

$$v_u = \frac{V_u}{b_0 d} \pm \frac{\gamma_{v1}M_{u1}c}{J_{c1}} \pm \frac{\gamma_{v2}M_{u2}c}{J_{c2}}$$

where $b_0 = 2 \times 18.88 = 37.75$ in. and $\gamma_{v1} = 1 - \gamma_{f1} = 0.4$. All of these terms add at the inside corner to give the maximum v_u:

$$v_u = \frac{25.0}{37.75 \times 5.75} + \frac{0.4 \times 22.2 \times 12 \times 4.72}{8360} + \frac{0.4 \times 17.1 \times 12 \times 4.72}{8360}$$

$$= 0.115 + 0.060 + 0.046 = 0.221 \text{ ksi}$$

$$= 221 \text{ psi}$$

$$\phi v_c = \phi 4\sqrt{f_c'} = 215 \text{ psi}$$

Therefore, the corner column is *not adequate* in two-way shear. A check using a 15-in.-square column with the exterior faces 4 in. from the outside shows that such a column would have adequate shear capacity.

One-Way Shear

(a) Locate the critical section. This check will be made using the original column size, although it will be necessary to enlarge the column. The critical section is as shown in Fig. 13–70b. This section is located $d_{avg} = 5.75$ from the corner of the column.

(b) Compute the shear on the critical section.

$$V_u = 0.226\left(9.71 \times 8.71 - \frac{3.33^2}{2}\right) = 17.86 \text{ kips}$$

$$\text{shear due to wall load} = (9.71 - 3.33) + (8.71 - 3.33) \times 1.4 \times 0.3$$

$$= 4.94 \text{ kips}$$

$$\text{total shear} = 22.8 \text{ kips}$$

(c) ϕV_c for the critical section is

$$\phi V_c = 2\sqrt{f_c'} \, bd = 2\sqrt{4000} \times 56.5 \times \frac{5.75}{1000}$$

$$= 41.1 \text{ kips}$$

Therefore, the slab is OK in one-way shear. Since it is not OK in two-way shear, **use a 15-in.-square corner column.** This change will reduce the moments and reinforcement in the corner panel a small amount. The slab is OK as designed and we will not repeat the calculations.

This completes the design of the slab. The bar cutoffs are calculated using Fig. 13–60. ∎

13–11 CONSTRUCTION LOADS ON SLABS

Most two-way slab buildings are built using *flying forms,* which can be removed sideways out of the building and are then lifted or *flown* up to form a higher floor. When the flying form is removed from under a slab the weight of the slab is taken by posts or *shores* which are wedged into place to take the load. Sets of flying forms and shores can be seen in Fig. 10–3. To save on the number of shores needed, it is customary to have only three to six floors of shoring below a slab at the time the concrete is placed. As a result, the weight of the fresh concrete is supported by the three to six floors below it. Because these floors are of different ages, they each take a different fraction of the load of the new slab. The calculation of the construction loads on slabs is presented in Ref. 13.31. Depending on the number of floors that are shored and the sequence of casting and form removal, the maximum construction load on a given slab may reach 1.8 to 2.2 times the dead load of the slab. This can approach the capacity of the slab, particularly if, as is the usual case, the slab has not reached its full strength when the construction loads occur. These high loads cause cracking of the green slabs and lead to larger short- and long-time deflections than would otherwise be expected.[13-32]

13–12 SLAB DEFLECTIONS

Two-way slabs frequently deflect excessively causing sagging floors and damage to partitions, doors, and windows. ACI Sec. 9.5.3.2 gives minimum thicknesses of two-way slabs to avoid excessive deflections. ACI Sec. 9.5.3.4 allows thinner slabs to be used if calculated deflections are less than the allowable deflections given in ACI Table 9.5(b). Generally, it is good practice to round up the thicknesses given in ACI Sec. 9.5.3.2 to the next-larger half-inch.

The calculation of deflections of two-way slabs is generally based on a crossing beam analogy in which the average deflection of the midspans of the column strips in one direction are added to the midspan deflection of the perpendicular middle strip as shown in Fig. 13–71. The calculation follows the procedures given in Chap. 9.

The effective moment of inertia is calculated using Eqs. 9–10 and 9–11 (ACI Eq. 9–7). To use this equation, one needs M_a, defined in the code as the maximum moment in the member at the stage at which deflection is calculated. A better definition would be the maximum moment acting on the member at the stage in question or at any previous stage. In Sec. 13–11 it was pointed out that construction loads frequently reach two times the dead load of the slab. In computing slab deflections, M_a should be based on the larger of 1.0 (dead plus live load) and 2.0 (dead load). Furthermore, M_{cr} should properly be representative of the age at which M_a acted.

ACI Table 9.5(b) gives values of maximum permissible computed deflections as a function of the span length ℓ. Unfortunately, the code does not define what ℓ is in the case of a two-way slab. The author believes that the column strip and middle strip deflections should be limited using the length of the appropriate strip and the midpanel deflection should be limited using the length of the diagonal of the panel.

EXAMPLE 13–8   Calculation of Deflections of an Interior Panel of a Flat-Plate Floor

Compute the deflections of the panel B–C–2–3 of the slab shown in Fig. 13–63.

1.   Compute the deflections of the column strips between $B2$ and $C2$ and between $B3$ and $C3$.

(a)   **Compute M_a.** Factored load on slab = 226 psf [from Example 13–7, step 2(c)]. Loads for deflection calculation:

Dead load: 1.0 (dead) = 87.5 + 25 = 112.5
Service load: 1.0 (dead + live) = 87.5 + 25 + 40 = 152.5 psf
Construction load: 2.0 (dead load of slab) = 2 × 87.5 = 175 psf

Therefore, cracking would be governed by construction loads. Take M_a as 175/226 = 0.774 times the column strip moment from Table 13–7.

Negative at $B2$ and $C2$ = 0.774 × −72.9 = 56.4 ft-kips

Positive at midspan = 0.774 × 31.4 = 24.3 ft-kips

(b)   **Compute M_{cr}.** Assume that the maximum construction load on a given slab occurs when it is 14 days old. From Eq. 3–5, a 14-day-old slab has a strength about $0.88f'_c$. From ACI Eq. 9–8,

$$M_{cr} = \frac{f_r I_g}{y_t}$$

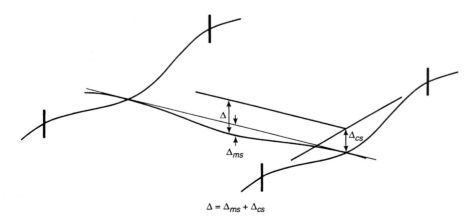

Fig. 13–71
Superposition of column strip and middle strip deflections.

$\Delta = \Delta_{ms} + \Delta_{cs}$

Two-Way Slabs: Behavior, Analysis, and Direct Design Method

where

$$f_r = 7.5\sqrt{f'_c} = 7.5\sqrt{0.88 \times 4000}$$

$$= 445 \text{ psi}$$

$$I_g = \frac{(7.87 \times 12) \times 7^3}{12} = 2699 \text{ in.}^4$$

$$y_t = 3.5 \text{ in.}$$

$$M_{cr} = \frac{445 \times 2699}{3.5 \times 12,000} = 28.6 \text{ ft-kips}$$

(c) Compute I_{cr} and I_e.

Negative moment region: From Fig. 13–67, $A_s = 14$ No. 5 bars.

$$\rho = \frac{14 \times 0.31}{7.87 \times 12 \times 5.66} = 0.0081$$

$$n \text{ at time of cracking} = \frac{29,000,000}{57,000\sqrt{0.88 \times 4000}} = 8.58$$

$n\rho = 0.0696$ and from Eq. 9–3, $k = 0.31$.

$$I_{cr} = \frac{(7.87 \times 12) \times (0.31 \times 5.66)^3}{3}$$

$$+ 14 \times 0.31 \times 8.58(5.66 - 0.31 \times 5.66)^2$$

$$= 170 + 568 = 738 \text{ in.}^4$$

From Eq. 9–10b,

$$I_e = I_{cr} + (I_g - I_{cr})\left(\frac{M_{cr}}{M_a}\right)^3$$

$$= 738 + (2699 - 738) \times \left(\frac{28.6}{56.4}\right)^3 = 994 \text{ in.}^4$$

Positive moment region: Since $M_a < M_{cr}$, $I_e = I_g = 2699 \text{ in.}^4$.
Weighted average value of I_e: From Eq. 9–11a,

$$\text{average } I_e = 0.70I_{em} + 0.15(I_{e1} + I_{e2})$$

$$= 0.7 \times 2699 + 0.15(994 + 994)$$

$$= 2187 \text{ in.}^4$$

(d) Compute the dead-load deflections. From Eq. 9–12, the dead-load moments are:

Negative: $\dfrac{112.5}{226} \times -72.9 = -36.3 \text{ ft-kips}$

Positive: $\dfrac{112.5}{226} \times 31.4 = 15.6 \text{ ft-kips}$

$$\Delta = \frac{5}{48}\frac{\ell_n^2}{EI}\left[M_m + 0.1(M_1 + M_2)\right]$$

$$= \frac{5}{48}\frac{(14.75 \times 12)^2}{57,000\sqrt{4000} \times 2187}\left[15.6 + 0.1(-36.3 - 36.3)\right] \times 12,000$$

$$= 0.041 \text{ in.}$$

13–12 Slab Deflections   **633**

The midspan deflection of the column strips is 0.041 in. under unfactored dead load.

2. **Compute the deflection of the middle strip between lines 2 and 3.**
    (a) **Compute M_a and M_{cr}.** Again M_a is due to construction loads.

$$\text{Negative at line 2} = 0.774 \times (-20.4) = -15.8 \text{ ft-kips}$$

$$\text{Positive at midspan} = 0.774 \times 16.2 = 12.5 \text{ ft-kips}$$

$$\text{Negative at line 3} = 0.774 \times (-18.8) = -14.6 \text{ ft-kips}$$

$$I_g = \frac{12.12 \times 12 \times 7^3}{12} = 4157 \text{ in.}^4$$

$$M_{cr} = 44.0 \text{ ft-kips}$$

   (b) **Compute I_e.** Since $M_a < M_{cr}$ at all points, $I_e = I_g = 4157$ in.^{4}.
   (c) **Compute the dead-load deflections.**

$$\text{Negative moment at line 2} = 0.498 \times (-20.4) = -10.2 \text{ ft-kips}$$

$$\text{Positive moment} = 0.498 \times 16.2 = 8.1$$

$$\text{Negative moment at line 3} = 0.498 \times (-18.8) = -9.4 \text{ ft-kips}$$

$$\Delta = \frac{5}{48} \frac{(14.75 \times 12)^2}{57,000\sqrt{4000} \times 4157} \left[8.1 + 0.1(-10.2 - 9.4) \right] \times 12,000$$

$$= 0.016 \text{ in.}$$

The midspan deflection of the middle strip is 0.016 under unfactored dead load.

3. **Compute the dead-load deflection at the middle of the panel.**

$$\Delta = \text{average column strip } \Delta + \text{ middle strip } \Delta$$

$$= 0.041 + 0.016 = 0.057 \text{ in.}$$

To study the deflections of the panel completely, it would be necessary to consider several load cases. See Examples 9–5 and 9–6. ∎

13–13   DESIGN OF SLABS WITH BEAMS IN TWO DIRECTIONS

Because of its additional depth, a beam is stiffer than the adjacent slab and as a result, moments are attracted to the beam. This was discussed in Sec. 13–5 and illustrated in Figs. 13–13 and 13–18. The average moments in the column strip are almost the same in a flat plate (Fig. 13–17d) and in a slab with beams between all columns (Fig. 13–18b). In the latter case, the column strip moment is divided between the slab and the beam. This reduces the reinforcement required in the slab in the column strip, although the beam must now be reinforced.

The greater stiffness of the beams reduces the overall deflections, allowing a thinner slab to be used than in the case of a flat plate. The advantage of slabs with beams in two directions lies in their reduced weight. Also, two-way shear does not govern for most two-way slabs with beams, again allowing thinner slabs. This is offset by the increased overall depth of the floor system and increased forming and reinforcement placing costs.

The Direct Design Method for computing moments in the slab and beams is the same as the procedure used in slabs without beams, with one additional step. Thus the designer will

1. Compute M_0.
2. Divide M_0 between the positive and negative moment regions.
3. Divide the positive and negative moments between the column and middle strips.

The additional step needed is:

4. Divide the column strip moments between the beam and the slab.

The amount of moment assigned to the column and middle strips in step 3 and the division of moments between the beam and slab in step 4 are a function of $\alpha_1 \ell_2/\ell_1$, where α_1 is the beam-to-slab stiffness ratio in the direction in which the reinforcement is being designed (see Sec. 13–16 and Example 13–1).

When slabs are supported on beams having $\alpha_1 \ell_2/\ell_1 \geq 1.0$, the beams must be designed for shear forces computed assuming tributary areas bounded by 45° lines at the corners of the panels, and the centerlines of the panels, as shown in Fig. 13–72. If the beams have $\alpha_1 \ell_2/\ell_1$ between 0 and 1.0, the shear forces computed from these tributary areas are multiplied by $\alpha_1 \ell_2/\ell_1$. In such a case the remainder of the shear must be transmitted to the column by shear in the slab. The ACI Code is silent on how this is to be done. The most common interpretation involves using two-way shear in the slab between the beams and one-way shear in the beams, as shown in Fig. 13–73. Frequently, problems are encountered when $\alpha_1 \ell_2/\ell_1$ is less than 1.0 because the two-way shear perimeter is inadequate to transfer the portion of the shear not transferred by the beams.

The size of the beams is also governed by their shear strength and flexural strength. The cross section should be large enough so that $V_u \leq \phi(V_c + V_s)$, where an upper practical limit on $V_c + V_s$ would be about $(6\sqrt{f_c'}b_w d)$. The critical location for flexure is the of maximum negative moment, at which point the reinforcement ratio, ρ, should not exceed about $0.5\rho_b$ and must not exceed $0.75\rho_b$.

The ACI Code allows any value of $\alpha_1 \ell_2/\ell_1$ from zero (no beams) to large values of $\alpha_1 \ell_2/\ell_1$ (very stiff beams). As mentioned earlier, practical difficulties arise in design, particularly with respect to shear, if $\alpha_1 \ell_2/\ell_1$ is between 0 and 1.0 and it is recommended that this range of beam stiffnesses be avoided.

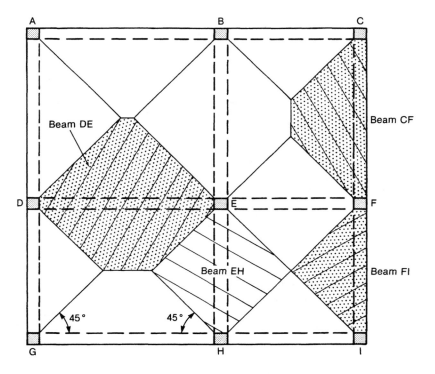

Fig. 13–72
Tributary areas for computing shear in beams supporting two-way slabs.

13–13 Design of Slabs with Beams in Two Directions

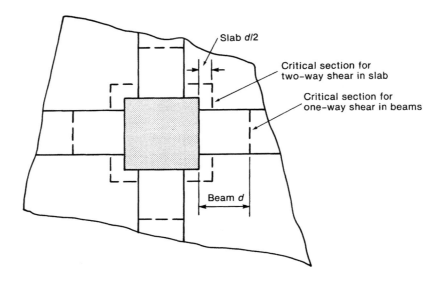

Fig. 13–73
Shear perimeters in slabs
with beams.

EXAMPLE 13–9 Design of a Two-Way Slab with Beams in Both Directions: Direct Design Method

Figure 13–74 is a plan of a part of a floor with beams between all columns. For simplicity in forming, the beams have been made the same width as the columns. The floor supports its own weight, superimposed dead loads of 5 psf for ceiling and mechanical fixtures and 25 psf for future partitions, plus a live load of 80 psf. The exterior wall weighs 300 lb/ft and is supported by the edge beam. The heights of the stories above and below the floor in question are 12 ft and 14 ft, respectively. Lateral loads are resisted by an elevator shaft not shown in the plan. Design the east–west strips of slab along column lines A and B using normal-weight 3000-psi concrete and Grade 60 reinforcement.

This example illustrates several things not included in Example 13–7. These include:

(a) The effect of edge beams on the required thickness

(b) The use of live-load reduction factors

(c) The effect of beams on the division of moments between the slab and beam in the column strip

(d) The distribution of negative moments where the direct design method gives large differences between the moments on each side of a support

(e) The calculation of shear in the beam and slab

For live loads of 100 psf or less, on panels having an area greater than 400 ft^2, ASCE 7–95 allows a live-load reduction based on Eq. 2–9:

$$w = w_0\left(0.25 + \frac{15}{\sqrt{A_I}}\right) \tag{2–9}$$

where w and w_0 are the reduced and specified live loads and A_I is the influence area. In the case of a two-way slab, A_I is the panel area measured from center to center of the columns.

1. Select the design method. ACI Code Section 13.6.1 sets limits on the use of the Direct Design Method. These are checked as done in step 1 of Example 13–7. One additional check is necessary (ACI Sec. 13.6.1.6): The beams will be selected so that all are the same width and depth. As a result, the ratio of $\alpha_1 \ell_2^2 / \alpha_2 \ell_1^2$ should fall within the bounds given. Therefore, use the Direct Design Method.

2. Select the slab thickness and beam size. The slab thickness is chosen to satisfy deflection requirements once the beam size is known. If $\alpha_1 \ell_2 / \ell_1$ exceeds 1.0 for all beams, all the shear is transferred to the columns by the beams, making it unnecessary to check shear while selecting the slab thickness. If there were only edge beams, the minimum slab thickness for deflection would be governed by Table 13–1 (ACI Table 9.5c) and would be 8.18 in., based on $\ell_n = 22$ ft $- 6$ in. (See Table 13–1). To

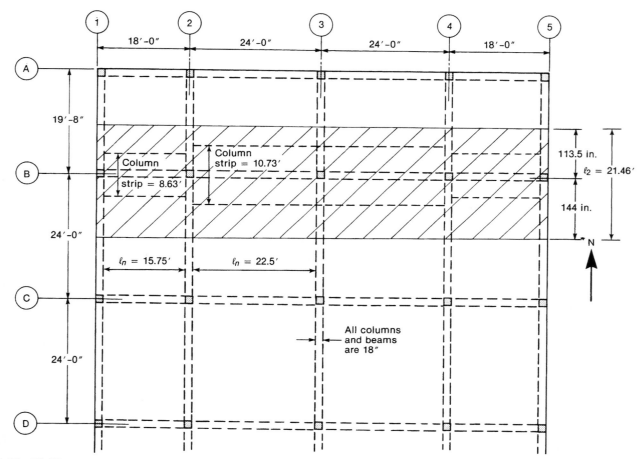

Fig. 13–74
Plan of two-way slab with beams—Example 13–9.

select a thickness for a slab with beams between interior columns, the thickness will be arbitrarily reduced by 15%, due to the stiffening effect of the beams, giving a trial thickness of 7 in. Assume a beam with an overall depth of about 2.5 times that of the slab to give a value of α a little greater than 1.0.

For the first trial select a slab thickness of 7 in. and a beam 18 in. wide by 18 in. deep. Check the thickness using Eqs. 13–10 and 13–11. The cross sections of the beams are shown in Fig. 13–75. First compute α.

$$\alpha = \frac{E_{cb}I_b}{E_{cs}I_s} \tag{13–9}$$

For the edge beam, the centroid is 7.94 in. below the top of the section, giving $I_b = 10,944$ in.[4] The width of slab working with the beam along line A is 122.5 in., giving $I_s = 3501$ and $\alpha = 3.13$.

For the beam along line 1, the width of slab is 112.5 in., giving $\alpha = 3.40$. For the interior beams $I_b = 12,534$ in.[4], along line B, slab width = 275.5 in., giving $\alpha = 1.70$; along lines C and 3, slab width = 288 in., giving $\alpha = 1.52$; and along line 2, the slab width = 247.5 in., giving $\alpha = 1.77$.

The thickness computations are given in Table 13–9. A 6.92-in. slab is required in the largest panel, so that 7-in. thickness chosen is satisfactory.

Before continuing, check whether the shears at the column with the largest tributary area exceed the shear capacity of the beams at that column.

$$w = 1.4\left(\frac{7}{12} \times 0.15 + 0.005 + 0.025\right) + 1.7(0.080)$$
$$= 0.301 \text{ ksf}$$

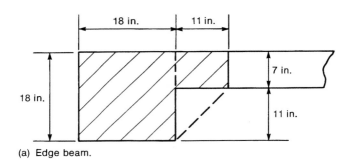

(a) Edge beam.

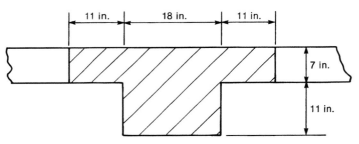

Fig. 13–75
Cross sections of beams—
Example 13–9.

(b) Interior beam.

TABLE 13–9   Computation of Minimum Thickness—Example 13–9

	Panel			
	A–B–1–2 (Corner)	A–B–2–3	B–C–1–2	B–C–2–3 (Interior)
Maximum ℓ_n	(19ft 8 in.) − 18 in. − 9in.	24 ft − 18 in.		
	209 in.	270 in.	270 in.	270 in.
Minimum ℓ_n	(18 ft 0 in.) − 18 in. − 9 in.			
	189 in.	209 in.	189 in.	270 in.
β	$\dfrac{209}{189} = 1.106$	1.292	1.429	1.00
α_m	$\dfrac{3.13 + 1.77 + 1.70 + 3.40}{4}$ $= 2.50$	2.03	2.10	1.63
Applicable equation	13–11, ACI 9–12	13–11, 9–12	13–11, 9–12	13–10, 9–11
h from equation	5.0	6.24	6.08	**6.88**
Minimum h	3.5	3.5	3.5	5.0

Assume that the value of d is $17 - 2.5 = 14.5$ in.

$$V_u = 0.301\left[24 \text{ ft} \times 24 \text{ ft} - \left(\frac{18 + 2 \text{ beam } d}{12}\right)^2\right]$$
$$= 168.8 \text{ kips}$$

ϕV_c for the four beams is

$$\Sigma \phi V_c = 4\left[0.85 \times 2\sqrt{3000} \times 18 \times \frac{14.5}{1000}\right]$$
$$= 97.2 \text{ kips}$$

Thus $V_u \simeq 1.7\Sigma\phi V_c$, which will not be excessive if stirrups are used. Therefore, use the beams selected earlier.

3. Compute the moments in the slab strip along column line B (Fig. 13–74). This strip of slab acts as a rigid frame spanning between columns $B1$, $B2$, $B3$, and so on. In this slab strip the ℓ_1 direction is parallel to line B, the ℓ_2 direction is perpendicular. Slab panels $B1$-$B2$ and $B4$-$B5$ are "end panels"; the other two are "interior panels."

Values of ℓ_n, and ℓ_2, and so on, are shown in Fig. 13–74. Since the structure is symmetrical about line 3, only half the structure needs to be considered.

The calculations are carried out in Table 13–10. The calculations are the same as in Table 13–3, except as noted below.

Line 4(a). The influence area of the panel is $\ell_n\ell_2$.

Line 4(b). The reduced live loads are based on Eq. 2–9.

Line 4(c). In computing the dead load, the weight of the beam stem has been omitted. It will be considered in line 8.

$$w_u = 1.4\left(\frac{7}{12} \times 0.150 + 0.005 + 0.025\right) + 1.7w_\ell$$

Line 7(b). ACI Sec. 13.6.3.4 requires that the negative moment sections either be designed for the larger negative moment at that support, or that the unbalanced negative moment be distributed in accordance with the stiffnesses of the adjoining elements. The two moments at $B2$ differ by more than 15% and hence a distribution will be carried out. Because of the two dimensional nature of slabs, the carryover of moments is not strictly the same as in beams. In the direct design method it is sufficiently accurate to distribute the moments without any carryovers. Figure 13–76 shows the joint at $B2$.

$$\text{Unbalanced moment at joint } B2 = 257.2 - 140.2$$
$$= 117 \text{ ft-kips}$$

The distribution factor to span $B1$-$B2$ is

$$\text{DF}_{B1-B2} = \frac{(K_s + K_b)_{B1-B2}}{(K_s + K_b)_{B1-B2} + (K_s + K_b)_{B2-B3} + \Sigma K_c}$$

where K_s, K_b, and K_c refer to the stiffnesses of the slabs, beams, and columns. In computing the stiffnesses for this purpose the variation in the cross sections along the lengths of slabs and columns will be ignored, giving $K = 4EI/\ell$. For slab $B1$-$B2$:

$$I_s = \frac{21.46 \times 12 \times 7^3}{12}$$

$$= 7361 \text{ in.}^4$$

ℓ (center to center of supports) $= 17.25$ ft

$$K_{s,B1-B2} = \frac{4 \times 7361E_c}{17.25 \times 12}$$

$$= 142 E_c$$

TABLE 13–10 Calculation of Negative and Positive Moments for Slab Strip B— Example 13–9

	B1			B2		B3
1. ℓ_1 (ft)		17.25			24	
2. ℓ_n (ft)		15.75			22.5	
3. ℓ_2 (ft)		21.46			21.46	
4. (a) Area A_l (ft^2)		15.75 × 21.46 = 338			483	
(b) Reduced live load (ksf)		0.080			0.0746	
(c) w_u (ksf)		0.301			0.291	
5. Slab moment, M_0 (ft-kips)		200			396	
6. Moment coefficients	−0.16	0.57	−0.70	−0.65	0.35	−0.65
7. Negative and positive moments (ft-kips)						
(a) Line 5 × line 6	−32.0	114.2	−140.2	−257.2	138.5	−257.2
(b) From distribution of negative moments at B2	0	−19.2	−38.5	+27.7	+13.9	0
(c) Sum (ft-kips)	−32.0	95.0	−178.7	−229.5	152.4	−257.2
8. Factored load on beam (kips/ft)		0.263			0.263	
9. Beam M_0 (ft-kips)		9.0			18.3	
10. Negative and positive moments from beam load						
(a) Line 6 × line 9	−1.4	5.1	−6.3	−11.9	6.4	−11.9
(b) Distribution of negative moments at B2	0	−0.9	−1.8	+1.3	0.7	
(c) Sum (ft-kips)	−1.4	4.2	−8.1	−10.6	7.1	−11.9
11. Column moments						
(a) From Eq. 13–14					112.0	48.5
(b) From slab and beam	33.4				53	

For beam B1–B2, $I_b = 10{,}340$ in.^{4} (step 2) and $K_b = 200E_c$. Similar calculations for B2–B3 give $K_s = 102E_c$ and $K_b = 144E_c$. For column B_2 above the slab, $L = 12$ ft, $I_c = 8748$ in.^{4}, and $K_c = 243E_c$, while for column B2 below the slab, $K_c = 208E_c$.

$$\mathrm{DF}_{B1-B2} = \frac{142E_c + 200E_c}{(142E_c + 200E_c) + (102E_c + 144E_c) + (243E_c + 208E_c)}$$

$$= 0.329$$

$$\mathrm{DF}_{B2-B3} = 0.237$$

$$\mathrm{DF}_{\mathrm{columns}} = 0.434 \text{ (see Fig. 13–76a)}$$

unbalanced moment to slab B1−B2 = 0.329 × 117 = 38.5 ft-kips

$$\text{change in midspan moment} = \frac{38.5}{2} = 19.2 \text{ ft} - \text{kips}$$

Line 8. This is the weight of the beam stem below the slab plus any loads applied directly to the beam. (The live loads on the beam have been included in M_0 for the slab.)

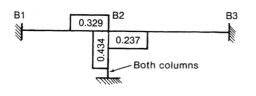

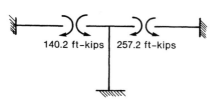

(a) Slab and equivalent column.

(b) Unbalanced moment at joint.

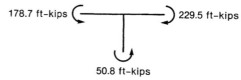

(c) Moments at joint after distribution.

Fig. 13–76
Moment distribution at joint
B2.

$$w = 1.4\left(\frac{11 \times 18}{144} \times 0.15\right) = 0.289 \text{ kip/ft}$$

Line 9. $M_0 = w\ell_n^2/8$.

Line 10. This is calculated using the distribution factors used in line 7(b).

Line 11(b). These moments are the unbalanced moments assigned to the column in the moment distribution.

4. Compute the moments in the slab strip along column line A (Fig. 13–74). This strip of slab acts as a rigid frame spanning between columns $A1$, $A2$, and $A3$. The slab strip includes an edge beam ($A1$–$A2$, $A2$–$A3$, etc.) parallel to the slab spans. In this slab strip, panels $A1$–$A2$ and $A4$–$A5$ are "end panels" and the other two are "interior panels." The calculations are carried out in Table 13–11 and proceed in the same way as the calculations in Table 13–10, except as noted below:

Line 5. For the purpose of calculating M_0 in line 5 of this table, ACI Sec. 13.6.2.4 requires that for edge panels, ℓ_2 be taken as the distance from the edge of the slab to the panel centerline.

Line 7. The unbalanced moment at $A2$ is 59.7 ft-kips. The distribution factors for spans $A1$–$A2$, $A2$–$A3$ and the columns are 0.371, 0.259, and 0.37, respectively. Thus the moment distributed to span $A1$–$A2$ is $0.371 \times 59.7 = 22.1$ ft-kips.

Line 8. This is the weight of the beam stem and the wall supported by the beam.

$$w = 1.4\left(\frac{11 \times 18}{144} \times 0.15 + 0.300\right) = 0.709 \text{ kip/ft}$$

5 and 6. Compute the moments in the north–south slab strips. Although a complete solution would include computation of moments in north–south strips as well as east–west strips, these will be omitted here.

7. Distribute the negative and positive moments to the column strips and middle strips and to beams: strips spanning east–west. In steps 3 and 4, the moments along column lines A and B were computed. These moments must now be distributed to column and middle strips and the column strip moment must be divided between the slab and the beam.

(a) Divide the slabs into column and middle strips. See Fig. 13–74. The widths of the column strips as defined in ACI Sec. 13.2.1 vary along their lengths, as shown by the dashed lines in Fig. 13–74. Thus along line B, the column strip width is 8.63 ft in span $B1$–$B2$ and 10.73 ft in span $B2$–$B3$. Assume a constant-width column strip of width 9 ft for simplicity in detailing the slab.

(b) Divide the moments between the column strips and middle strips. These calculations are carried out in Table 13–12, which is laid out to resemble a plan of the slab. Arrows illustrate the flow of moments to the various parts of the slab. The division of moments is a function of the beam stiffness ratio for the beam parallel to the strip being designed, and the aspect ratio of the

TABLE 13–11 Calculation of Negative and Positive Moments for Slab Strip A— Example 13–9

		A1			A2		A3
1. ℓ_1(ft)			17.25			24.0	
2. ℓ_n(ft)			15.75			22.5	
3. ℓ_2(ft)			10.21			10.21	
4. (a) Area A_I (ft^2)			176			245	
(b) Reduced live load (ksf)			0.080			0.080	
(c) w_u (ksf)			0.301			0.301	
5. Slab moment, M_0 (ft-kips)			95.3			194.5	
6. Moment coefficients		−0.16	0.57	−0.70	−0.65	0.35	−0.65
7. Negative and positive moments from slab M_0 (ft-kips) (a) Line 5 × line 6		−15.2	54.3	−66.7	−126.4	68.1	−126.4
(b) From distribution of moments at A2		0	−11.0	−22.1	+15.5	+7.7	0
(c) Sum (ft-kips)		−15.2	43.3	−88.8	−110.9	75.8	−129.7
8. Factored load on beam (kips/ft)			0.709			0.709	
9. Beam M_0 (ft-kips)			22.0			44.9	
10. Negative and positive moments from beam M_0 (a) Line 6 × line 9		−3.5	12.5	−15.4	−29.2	15.7	−29.2
(b) Distribution of moments at A2		0	−2.5	−5.1	+3.5	1.7	0
(c) Sum (ft-kips)		−3.5	10.0	−20.5	−25.7	17.4	−29.2
11. Column moments (a) From Eq. 13–14		—			55.0		23.1
(b) From slab and beam	19.1			27.3			

panel. For the east–west strips these terms are summarized below. Values of α were calculated in step 2 and values of ℓ_1 and ℓ_2 are given in Tables 13–10 and 13–11. For the interior strips, ℓ_2 is taken equal to the value used in calculating the moments in the strips.

Panel B1–B2: $\ell_1 = 17.25$ ft, $\ell_2 = 21.46$ ft, $\alpha_1 = 1.70$, $\dfrac{\ell_2}{\ell_1} = 1.24$, $\dfrac{\alpha_1 \ell_2}{\ell_1} = 2.11$

Panel B2–B3: $\ell_1 = 24.0$ ft, $\ell_2 = 21.46$ ft, $\alpha_1 = 1.70$, $\dfrac{\ell_2}{\ell_1} = 0.89$, $\dfrac{\alpha_1 \ell_2}{\ell_1} = 1.51$

Panel C1–C2: $\ell_1 = 17.25$ ft, $\ell_2 = 24.0$ ft, $\alpha_1 = 1.52$, $\dfrac{\ell_2}{\ell_1} = 1.39$, $\dfrac{\alpha_1 \ell_2}{\ell_1} = 2.11$

Panel C2–C3: $\ell_1 = 24.0$ ft, $\ell_2 = 24.0$ ft, $\alpha_1 = 1.52$, $\dfrac{\ell_2}{\ell_1} = 1.0$, $\dfrac{\alpha_1 \ell_2}{\ell_1} = 1.52$

The ACI Code is not clear about which ℓ_2 should be used when considering an edge panel. Current practice involves taking ℓ_2 equal to the total width of the edge panel.

TABLE 13–12 Division of Moment to Column and Middle Strips: East–West strips—Example 13–9

	Column Strip	Middle Strip	Column Strip	Middle Column Strip	Edge Column
	9.0	15.0	9.0	10.17	5.0

Exterior Negative Moments — C1, B1, A1

	Column Strip (9.0)	Middle Strip (15.0)	Column Strip (9.0)	Middle Column Strip (10.17)	Edge Column (5.0)
	C1 ▦		B1 ▦		A1 ▦
1. Slab Moment (ft–kip)	−35.8		−32.0		−15.2
2. Moment Coefficients	0.89	0.055 0.055	0.89	0.055 0.20	0.80
3. Moment to Column and Middle Strips (ft–kip)	−31.9	−2.0 −1.8	−28.5	−1.8 −3.0	−12.2
4. (a) Column Strip Moments to Slabs, Beams (ft–kip)	Slab −4.8 Beam −27.1		Slab −4.3 Beam −24.2		Slab −1.8 Beam −10.4
(b) Moments due to Loads on Beams (ft–kip)	−1.0		−1.0		−3.4
5. Total Moment in Slabs, Beams (ft–kip)	−4.8 −28.1	−3.8	−4.3 −25.2	−4.8	−1.8 −13.8

End Span Positive Moments

	Column Strip (9.0)	Middle Strip (15.0)	Column Strip (9.0)	Middle Column Strip (10.17)	Edge Column (5.0)
1. Slab Moment (ft–kip)	127.7		114.2		54.3
2. Moment Coefficients	0.633	0.184 0.161	0.678	0.161 0.28	0.72
3. Moment to Column and Middle Strips (ft–kip)	80.8	23.4 18.4	77.4	18.4 15.2	39.1
4. (a) Column Strip Moments to Slabs, Beams (ft–kip)	Slab 12.1 Beam 68.7		Slab 11.6 Beam 65.8		Slab 5.9 Beam 33.2
(b) Moments due to Loads on Beams (ft–kip)	3.6		3.6		9.6
5. Total Moment in Slabs, Beams (ft–kip)	12.1 72.3	41.8	11.6 69.4	33.6	5.9 42.8

First Interior Negative Moments — C2, B2, A2

	Column Strip (9.0)	Middle Strip (15.0)	Column Strip (9.0)	Middle Column Strip (10.17)	Edge Column (5.0)
	C2 ▦		B2 ▦		A2 ▦
1. Slab Moment (ft–kip)	−256.7		−229.5		−110.9
2. Moment Coefficients	0.633	0.184 0.161	0.678	0.161 0.28	0.72
3. Moment to Column and Middle Strips (ft–kip)	−162.5	−47.1 −36.9	−155.7	−36.9 −31.1	−79.8
4. (a) Column Strip Moments to Slabs, Beams (ft–kip)	Slab −24.4 Beam −138.1		Slab −23.4 Beam −132.3		Slab −12.0 Beam −67.8
(b) Moments due to Loads on Beams (ft–kip)	−9.6		−9.6		−24.7
5. Total Moment in Slabs, Beams (ft–kip)	−24.4 −147.7	−84.0	−23.4 −141.9	−68.0	−12.0 −92.5

Interior Positive Moments

	Column Strip (9.0)	Middle Strip (15.0)	Column Strip (9.0)	Middle Column Strip (10.17)	Edge Column (5.0)
1. Slab Moment (ft–kip)	170.4		152.4		75.8
2. Moment Coefficients	0.633	0.184 0.161	0.678	0.161 0.28	0.72
3. Moment to Column and Middle Strips (ft–kip)	107.9	31.3 24.5	103.4	24.5 21.2	54.6
4. (a) Column Strip Moments to Slabs, Beams (ft–kip)	Slab 16.2 Beam 91.7		Slab 15.5 Beam 87.9		Slab 8.2 Beam 46.4
(b) Moments due to Loads on Beams (ft–kip)	6.4		6.4		16.8
5. Total Moment in Slabs, Beams (ft–kip)	16.2 98.1	55.8	15.5 94.3	45.7	8.2 63.2

Interior Negative Moments — C3, B3, A3

	Column Strip (9.0)	Middle Strip (15.0)	Column Strip (9.0)	Middle Column Strip (10.17)	Edge Column (5.0)
	C3 ▦		B3 ▦		A3 ▦
1. Slab Moment (ft–kip)	−287.6		−257.2		−129.7
2. Moment Coefficients	0.633	0.184 0.161	0.678	0.161 0.28	0.72
3. Moment to Column and Middle Strips (ft–kip)	−182.0	−52.8 −41.4	−174.4	−41.4 −36.3	−93.4
4. (a) Column Strip Moments to Slabs, Beams (ft–kip)	Slab −27.3 Beam −154.7		Slab −26.2 Beam −148.2		Slab −14.0 Beam −79.4
(b) Moments due to Loads on Beams (ft–kip)	−8.1		−8.1		−28.1
5. Total Moment in Slabs, Beams (ft–kip)	−27.3 −162.8	−94.2	−26.2 −156.3	−77.7	−14.0 −107.5

Panel $A1$–$A2$: $\ell_1 = 17.25$ ft, $\ell_2 = 18.92$ ft, $\alpha_1 = 3.13$, $\dfrac{\ell_2}{\ell_1} = 1.10$, $\dfrac{\alpha_1 \ell_2}{\ell_1} = 3.44$

Panel $A2$–$A3$: $\ell_1 = 24.0$ ft, $\ell_2 = 18.92$ ft, $\alpha_1 = 3.13$, $\dfrac{\ell_2}{\ell_1} = 0.79$, $\dfrac{\alpha_1 \ell_2}{\ell_1} = 2.47$

Line 1. These moments come from lines 7(d) of Tables 13–10 and 13–11. The moments in Strip C are from a similar set of calculations. Note that the larger of the two slab moments is used at the first interior support.

Line 2. Exterior negative moments: ACI Sec. 13.6.4.2 gives the fraction of the exterior negative moment resisted by the column strip. This is determined using ℓ_2/ℓ_1 and $\alpha_1 \ell_2/\ell_1$ calculated above plus β_t, where

$$\beta_t = \frac{E_{cb}C}{2E_{cs}I_s} \qquad (13\text{–}12)$$

where C is the torsional constant for the beam computed using Eq. 13–16. The edge beam cross section effective for torsion is defined in ACI Sec. 13.7.5. The effective cross section is shown in Fig. 13–75. To compute C, the beam is divided into rectangles. The maximum value of C corresponds to the rectangles shown in Fig. 13–75a. $C = 13,700$ in.^{4}

The I_s in Eq. 13–12 is the moment of inertia of the slab span framing into the edge beam. Thus $I_s = \ell_2 h^3/12$. Since the slab and beam are cast at the same time, $E_{cb} = E_{cs}$.

(a) *Slab strip A:* $I_s = (10.21 \times 12)\dfrac{7^3}{12} = 3502$ in.^{4}, $\beta_t = \dfrac{13,700}{2 \times 3502} = 1.96$

Interpolating in the table in ACI Sec. 13.6.4.2 for $\ell_2/\ell_1 = 1.10$, $\alpha_1\ell_2/\ell_1 = 3.44$, and $\beta_t = 1.96$ gives 0.78 of the exterior negative moment to the column strip (at column $A1$).

(b) *Slab strip B:* $I_s = 7361$ in.^{4}, $\beta_t = 0.93$, $\ell_2/\ell_1 = 1.24$, and $\alpha_1\ell_2/\ell_1 = 2.11$. Interpolating in ACI Sec. 13.6.4.2 gives 0.88 times the exterior negative moment to the column strip (at column $B1$).

(c) *Slab strip C:* $I_s = 8232$ in.^{4}, $\beta_t = 0.83$, $\ell_2/\ell_1 = 1.39$, and $\alpha_1\ell_2/\ell_1 = 2.11$ gives 0.88 times the exterior moment to the column strip.

Positive moments:

(a) *Slab strip A:* $\ell_2/\ell_1 = 1.10$, $\alpha_1\ell_2/\ell_1 > 1.0$. From ACI Sec. 13.6.4.4, 72% of the positive moments go to the column strips.

(b) *Slab strip B:* $\ell_2/\ell_1 = 1.24$, $\alpha_1\ell_2/\ell_1 > 1.0$. From ACI Sec. 13.6.4.4, 67.8% of the positive moments go to the column strips.

(c) *Slab strip C:* 63.3% of positive moments go to the column strip.

Interior negative moments: Because $\alpha_1\ell_2/\ell_1 > 1.0$, ACI Sec. 13.6.4.1 gives the same division of interior negative moment between column and middle strips as for positive moments. Thus interpolating in ACI Sec. 13.6.4.1, 72, 67.8, and 63.3% of negative moments in slab strips, A, B, and C go to the column strip.

Line 4(a). The column strip moments are divided between the slab and the beam following the rules given in ACI Secs. 13.6.5.1 and 13.6.5.2. If $\alpha_1\ell_2/\ell_1 \geq 1.0$, 85% of the column strip moment is assigned to the beam. For this slab, this is true in all cases.

Line 4(b). These moments come from line 10(c) of Tables 13–10 and 13–11. They are assumed to be carried entirely by the edge beams.

Line 5. The final moments in the middle strip between lines A and B, in the slab in the column strip along line B, and in the beam along line B are plotted in Fig. 13–77.

8. Design the slab reinforcement. The division of moments is similar in the other direction. The design of slab reinforcement is carried out in the same way as in previous examples and will not be repeated here. ACI Sec. 13.3.6 requires special corner reinforcement at exterior corners. At corners $A1$ and $A5$ provide one mat each way parallel to the sides of the slab at the bottom and top of the slab. These are designed for a moment equal to the maximum positive moment per foot of width occurring in the corner panels. From Table 13–12 this is 33.6 ft-kips/10.17 ft = 3.30 ft-kips/ft. The steel should extend $\ell_n/5 = 209/5 = 42$ in. each way from the corner.

9. Design the beams. The beams must be designed for moment, shear, and bar anchorage, as in the examples in Chap. 10. The edge beams are subjected to a torque which must be considered in the design.

(a) Flexure. The flexural design is similar to that of any continuous beam, except that the moments used are those given in line 5 of Table 13–12. Figure 13–77 shows the moments used in the design of the beam along line *B*. When selecting the reinforcement, the cover to the reinforcement was taken as that for a beam rather than that for a slab.

(b) Torsion. The distribution of negative moments along the edge *A1–B1–C1*, was calculated in Table 13–12. A similar analysis of the perpendicular slab strips gave the distribution of exterior negative moments in the slab along *A1–A2–A3*, shown in Fig. 13–78a. The distribution of these moments per foot of length of the edge members is shown in Fig. 13–78b. A torque diagram has been computed in the same way that a shear force diagram would be for simply supported beams loaded with distributed loads equivalent to the distributed torques. This, in effect, assumes that the edge beams are rigidly fixed against rotation at the columns so that the total angle change $\theta = TL/GJ$ equals zero in each span.

According to ACI Sec. 11.6.1 and 11.6.2.4, torsional reinforcement is required if

$$T_u \geq \phi\sqrt{f_c'}\left(\frac{A_{cp}^2}{p_{cp}}\right)$$

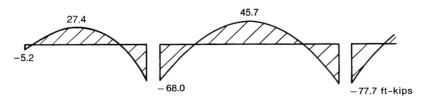

(a) Moments in middle strip between lines A and B.

(b) Moments in slab in column strip along line B.

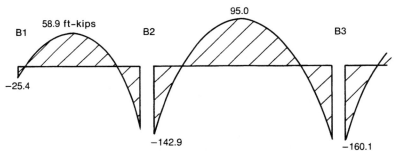

(c) Moments in beam along line B.

Fig. 13–77
Moments in slab strips.

13–13   Design of Slabs with Beams in Two Directions

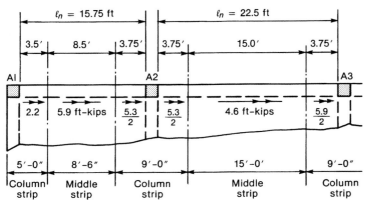

(a) Moments at edge of slab (ft-kip).

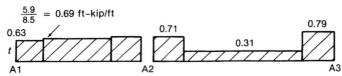

$\frac{5.9}{8.5} = 0.69$ ft-kip/ft

(b) Distributed torques, t, on edge beams (ft-kips/ft).

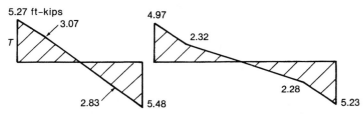

(c) Torque diagram, T, for edge beams (ft-kips).

Fig. 13–78
Torque on edge beam.

at sections located d away from the face of the columns. ACI Sec. 11.6.1 defines the cross section of the edge beam as given in ACI Sec. 13.2.4 and as illustrated in Fig. 13–75a. For this section,

$$A_{cp} = 18 \times 18 + 11 \times 7 = 401 \text{ in.}^2 \qquad p_{cp} = 3 \times 18 + 3 \times 11 + 7 = 94 \text{ in.}$$

$$\phi\sqrt{f_c'}\left(\frac{A_{cp}^2}{p_{cp}}\right) = 0.85\sqrt{3000}\left(\frac{401^2}{94}\right)$$

$$= 79{,}600 \text{ in.-lb} = 6.64 \text{ ft-kips}$$

Figure 13–78c shows that T_u does not exceed this at any point and hence torsion can be ignored, except that closed stirrups will be used in the spandrel beams. ACI Sec. 11.6.2.3 allows the designer to assume a uniform distribution of t along the beam rather than the more elaborate analysis used here.

(c) Shear. Since $\alpha_1 \ell_2 / \ell_1 \geq 1.0$ for all beams, ACI Sec. 13.6.8.1 requires that the beams be designed for the shear caused by loads on the tributary areas shown in Fig. 13–79a. For the beams along line A the corresponding beam loads and shear force diagrams are shown in Figs. 13–79b and c. Design of stirrups is in accordance with ACI Secs. 11.1 to 11.5. ∎

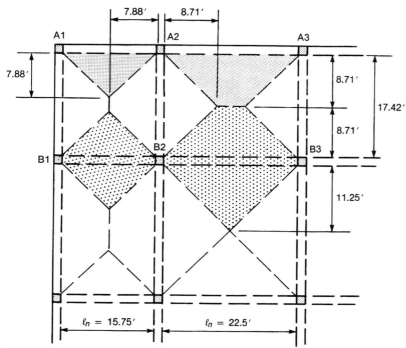

(a) Tributary areas for beams along lines A and B.

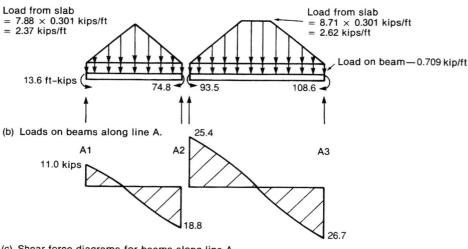

Load from slab
= 7.88 × 0.301 kips/ft
= 2.37 kips/ft

Load from slab
= 8.71 × 0.301 kips/ft
= 2.62 kips/ft

Load on beam—0.709 kip/ft

13.6 ft-kips

74.8   93.5   108.6

(b) Loads on beams along line A.

A1   A2   25.4   A3

11.0 kips

18.8   26.7

(c) Shear force diagrams for beams along line A.

Fig. 13–79
Shear on edge beam.

PROBLEMS

13–1 Compute α for the beam shown in Fig. P13–1. The concrete for the slab and beam was placed in one pour.

13–2 Compute the column strip and middle strip moments in the long-span direction for an interior panel of the flat slab floor shown in Fig. 13–29. The slab is

6 in. thick, the design live load is 40 psf, and the superimposed dead load is 5 psf for ceiling, flooring, and so on, and 25 psf for partitions. The columns are 10 in. × 12 in. as shown in Fig. 13–29.

13–3 Compute the column strip and middle strip moments in an exterior bay of the flat-plate slab shown in Fig. P13–3. The slab is $7\frac{1}{2}$ in. thick and supports a superimposed service dead load of 25 psf and a service live load of 50 psf. There is no edge beam. The columns are all 18 in. square.

13–4 A 7 in.-thick flat-plate slab with spans of 20 ft in each direction is supported on 16 in. × 16 in. columns. The effective depths are 5.9 and 5.3 in. in the two directions. The slab supports its own dead load plus 20 psf unfactored superimposed dead load and 40 psf unfactored live load. The concrete strength is 4000 psi. Check one-way and two-way shear at a typical interior support.

13–5 The slab described in Problem 13–4 is supported in 10 in. × 24 in. columns. Check one-way and two-way shear at a typical interior support.

13–6 The slab shown in Fig. P13–6 supports a superimposed dead load of 25 psf and a live load of 60 psf. The slab extends 4 in. past the exterior face of the column to support an exterior wall which weighs 400 lb/ft of length of wall. The story-to-story height is 9 ft. Use 4500-psi concrete and Grade 60 reinforcement.

(a) Select thickness.

(b) Compute moments and design the reinforcement for slab strips running north and south.

(c) Check shear and moment transfer at columns A2 and B2. Neglect unbalanced moments about column line 2.

13–7 For the slab shown in Fig. P13–6 and the loadings and material strengths given in Problem 13–6:

(a) Select thickness.

(b) Compute moments and design the reinforcement for slab strips running east and west.

(c) Check shear and moment transfer at columns C1 and C2. Neglect unbalanced moments about column line C.

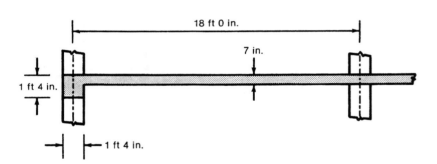

Fig. P13–1

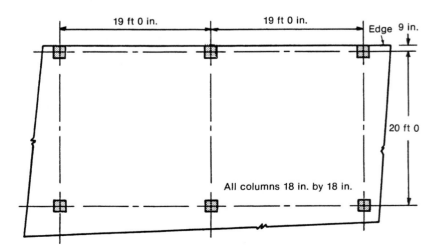

Fig. P13–3

Two-Way Slabs: Behavior, Analysis, and Direct Design Method

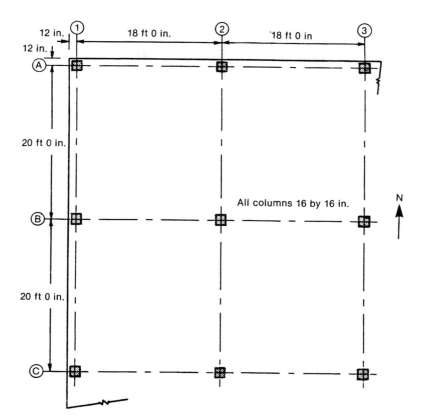

Fig. P13–6

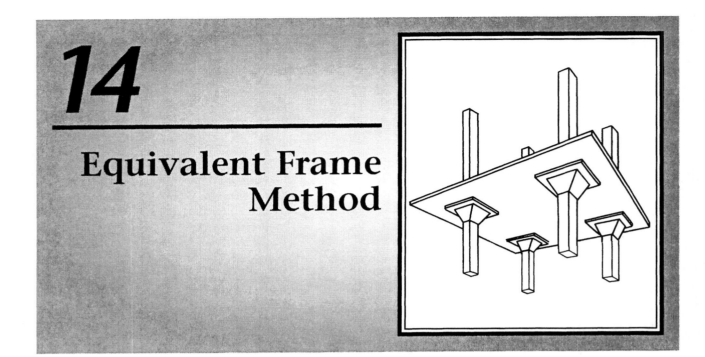

14

Equivalent Frame Method

14-1 INTRODUCTION

The ACI Code presents two parallel methods for calculating moments in two-way slab systems. These are the Direct Design Method presented in Chap. 13 and the Equivalent Frame Method presented in this chapter. In addition, ACI Sec. 13.5.1 allows other methods if they meet certain requirements. The relationship between the Direct Design Method and the Equivalent Frame Method is explained in Sec. 13–6 of this book. In the Direct Design Method, the statical moment, M_0, is calculated for each panel. This moment is then divided between positive and negative moment regions using arbitrary moment coefficients and the positive moments are adjusted to reflect pattern loadings. In the Equivalent Frame Method, all of this is accomplished by frame analyses.

The use of frame analyses to analyze slabs was first proposed by Peabody[14-1] in 1948 and a method of slab analysis referred to as "design by elastic analysis" was incorporated in the 1956 and 1963 editions of the ACI Code. In the late 1940s Siess and Newmark[14-2,14-3] studied the application of moment distribution analyses to two-way slabs on stiff beams. Following extensive research on two-way slabs carried out at the University of Illinois, Corley and Jirsa[14-4] presented a more refined method of frame analysis for slabs. This has been incorporated in the 1971 and subsequent ACI Codes. Corley and Jirsa considered only gravity loads. Studies of the use of frame analyses for laterally loaded column-slab structures[14-5] led to treatment of this problem in the 1983 and subsequent ACI Codes.

The Equivalent Frame Method is intended for use in analyzing moments in any practical building frame. Its scope is thus wider than the Direct Design Method, which is subject to the limitations presented in Sec. 13–6 (ACI Sec. 13.6.1). It works best for frames that can also be designed by the Direct Design Method, however.

This chapter builds on the basic knowledge of the behavior and design of slabs in flexure and shear presented in Chap. 13.

The slab is divided into a series of equivalent frames running in the two directions of the building, as shown in Fig. 13–23 (ACI Sec. 13.7.2). These frames consist of the slab, any beams that are present, and the columns above and below the slab. For gravity load analysis, the code allows analysis of an entire equivalent frame extending over the height of the building, or each floor can be considered separately with the far ends of the columns fixed. These frames are, in turn, divided up into column and middle strips as described in Sec. 13–6 and Fig. 13–25.

The original derivation of the Equivalent Frame Method assumed that moment distribution would be the procedure used to analyze the slabs, and some of the concepts in the method are awkward to adopt to other methods of analysis. In this chapter the Equivalent Frame Method is presented for use with moment distribution. This is followed by a brief discussion of how to use direct stiffness computer programs in carrying out an Equivalent Frame analysis.

Calculation of Stiffness, Carryover, and Fixed-End Moments

In the moment distribution method, it is necessary to compute *flexural stiffnesses, K, carryover factors*, COF, *distribution factors*, DF, and *fixed-end moments*, FEM, for each of the members in the structure. For a prismatic member, fixed at the far end, with negligible axial loads, the flexural stiffness is

$$K = \frac{kEI}{L} \tag{14-1}$$

where $k = 4$ and the carryover factor is ± 0.5, the sign depending on the sign convention used for moments. For a prismatic uniformly loaded beam the fixed-end moments are $w\ell^2/12$.

In the Equivalent Frame Method, the increased stiffness of members within the column-slab joint region is accounted for, as is the variation in cross section at drop panels. As a result all members have a stiffer section at each end, as shown in Fig. 14–1. If the *EI* used in Eq. 14–1 is that at midspan of the slab strip, k will be greater than 4. Similarly, the carryover factor will be greater than 0.5 and the fixed-end moments will be greater than $w\ell^2/12$.

Several methods are available for computing values of k, COF, and fixed-end moments. Frequently, these are computed using the *column analogy* developed by Hardy Cross. Cross observed an analogy between the equations used to compute stresses in an unsymmetrical column loaded with axial loads and moments, and the equations used to compute moments in a fixed-end beam. For more details, see Ref. 14–6.

Tables and charts for computing k, COF, and fixed-end moments are given in Appendix A as Tables A–20 and A–23.

Properties of Slab-Beams

The horizontal members in the equivalent frame are referred to as *slab-beams*. These may consist of a slab, a slab and a drop panel, or a slab with a beam running parallel to the equivalent frame. ACI Sec. 13.7.3 explains how these are to be modeled for analysis:

1. At points outside of joints or column capitals the moment of inertia may be based on the gross area of the concrete. Variations in the moment of inertia along the length shall be taken into account.

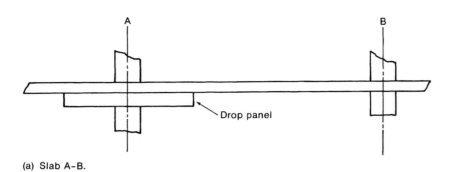

(a) Slab A–B.

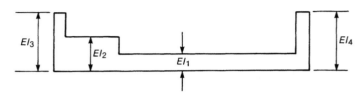

Fig. 14–1
Variation in stiffness along
span.

(b) Distribution of EI along slab.

Thus for the slab with a drop panel shown in Fig. 14–2a, the moment of inertia at section A–A is that for a slab of width ℓ_2 (Fig. 14–2c). At section B–B through the drop panel, the moment of inertia is for a slab having the cross section shown in Fig. 14–2d. Similarly, for a slab system with a beam parallel to ℓ_1 as shown in Fig. 14–3a, the moment of inertia for section C–C is that for a slab and beam section as shown in Fig. 14–3c. Section D–D is cut through a beam running perpendicular to the page.

> **2.** The moment of inertia of the slab-beams from the center of the column to the face of the column, bracket, or capital (as defined in ACI Sec. 13.1.2 and Fig. 13–59) shall be taken as the moment of inertia of the slab-beam at the face of the column, bracket, or capital divided by the quantity $(1 - c_2/\ell_2)^2$, where ℓ_2 is the transverse width of the equivalent frame (Fig. 13–23) and c_2 is the width of the support parallel to ℓ_2.

The application of this is illustrated in Figs. 14–2 and 14–3. Tables A–20 and A–22[14–7,14–8] present moment distribution constants for flat plates and slabs with drop panels. For most practical cases these eliminate the need to use the column analogy solution.

EXAMPLE 14–1   Calculation of the Moment Distribution Constants for a Flat-Plate Floor

Figure 14–4 shows a plan of a flat-plate floor without spandrel beams. The floor is 7 in. thick. Compute the moment distribution constants for the slab-beams in the equivalent frame along line 2, shown shaded in Fig. 14–4.

SPAN A2–B2

At end A, $c_1 = 12$ in., $\ell_1 = 213$ in., $c_2 = 20$ in., and $\ell_2 = 189$ in.. Thus $c_1/\ell_1 = 0.056$ and $c_2/\ell_2 = 0.106$. Interpolating in Table A–20, the fixed-end moment is

$$M = 0.084w\,\ell_2\ell_1^2$$

The stiffness is

$$K = \frac{4.11EI_1}{\ell_1}$$

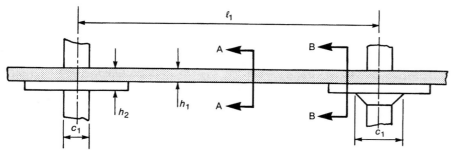

(a) Slab with drop panels.

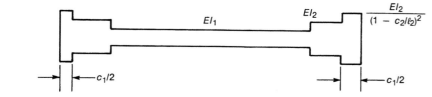

(b) Variation in EI along slab-beam.

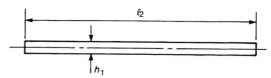

(c) Cross section used in compute I_1—Section A-A.

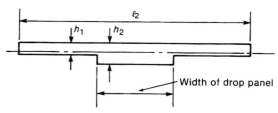

(d) Cross section used to compute I_2—Section B-B.

Fig. 14–2
EI values for slab with drop panel.

and the carryover factor is -0.508.

SPAN *B2–C2*

For span *B2–C2*, $c_1 = 20$ in., $\ell_1 = 240$ in., $c_2 = 12$ in., and $\ell_2 = 189$ in. Thus $c_1/\ell_1 = 0.083$, $c_2\ell_2 = 0.064$. From Table A–20, the fixed-end moment is

$$M = 0.084 w \ell_2 \ell_1{}^2$$

The stiffness is

$$K = \frac{4.10 EI_1}{\ell_1}$$

and the carryover factor is -0.507. ∎

EXAMPLE 14–2   Calculation of the Moment Distribution Constants for a Two-Way Slab with Beams

Figure 14–5 shows a two-way slab with beams between all columns. The slab is 7 in. thick and all the beams are 18 in. wide by 18 in. in overall depth. Compute the moment distribution constants for the slab-beams in the equivalent frame along line *B* (shown shaded in Fig. 14–5).

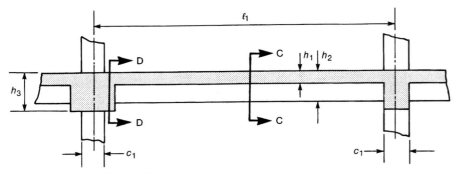

(a) Slab with beams in two directions.

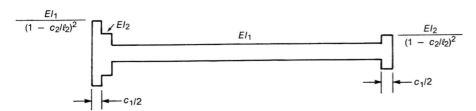

(b) Variation in EI along slab beam.

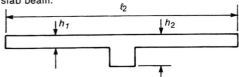

(c) Cross section used to compute I_1—Section C–C.

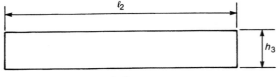

Fig. 14–3
EI values for slab and beam.

(d) Cross section used to compute I_2—Section D–D.

SPAN B1–B2

Span $B1$–$B2$ is shown in Fig. 14–6. A cross section at midspan is shown in Fig. 14–6b. The centroid of this section lies 4.27 in. below the top of the slab and its moment of inertia is

$$I_1 = 20{,}700 \text{ in.}^4$$

The columns at both ends are 18 in. square, giving $c_1/\ell_1 = 18/207 = 0.087$ and $c_2/\ell_2 = 0.070$. Since this member has uniform stiffness between the joint regions, we can use Table A–20 to get the moment distribution constants.

$$\text{Fixed-end moment: } M = 0.084w\,\ell_1^2$$

$$\text{Stiffness: } K = \frac{4.10EI_1}{\ell_1}$$

$$\text{Carryover factor} = -0.507$$

SPAN B2–B3

Here $c_1/\ell_1 = 18/288 = 0.0625$ and $c_2/\ell_2 = 0.070$. From Table A–20

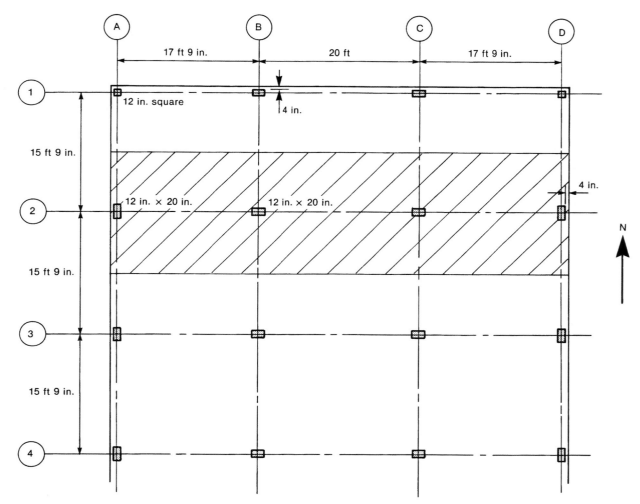

Fig 14-4
Plan of flat plate floor—Examples 14-1, 14-3, and 14-5.

Fixed-end moment: $M = 0.084w\ell_1^2$

Stiffness: $K = \dfrac{4.07EI_1}{\ell_1}$

Carryover factor $= -0.505$ ∎

Properties of Columns

In computing the stiffnesses and carryover factors for columns, ACI Sec. 13.7.4 states:

1. The moment of inertia of columns at any cross section outside of the joints or column capitals may be based on the gross area of the concrete, allowing for variations in the actual moment of inertia due to changes in the column cross section along the length of the column.

2. The moment of inertia of columns shall be assumed to be infinite within the depth of the slab-beam at a joint.

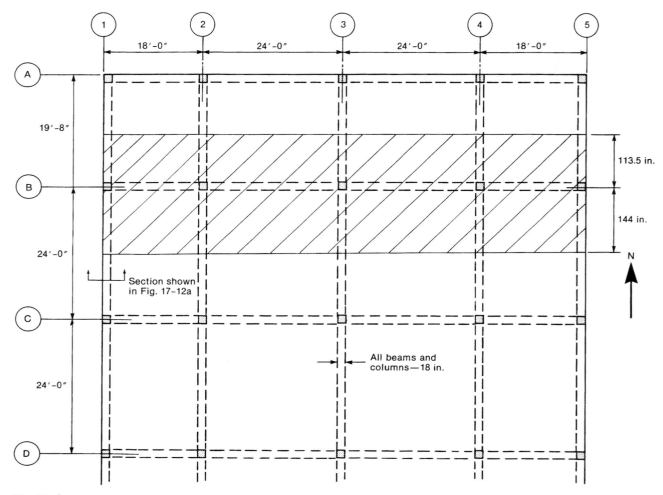

Fig. 14–5
Two-way slabs with beams—Examples 14–2 and 14–4.

Figure 14–7 illustrates this for four common cases. Again, the column analogy can be used to solve for the moment distribution constants or the values given in Table A–23[14–7] can be used.

Torsional Members and Equivalent Columns

When the beam and column frame shown in Fig. 14–8a is loaded, the ends of the column and beam undergo equal rotations where they meet at the joint. If the flexural stiffness, $K = M/\theta$, is known for the two members, it is possible to calculate the joint rotations and the end moments in the members. Similarly, in the case shown in Fig. 14–8b, the ends of the slab and the wall both undergo equal end rotations when the slab is loaded. When a flat plate is connected to a column as shown in Fig. 14–8c, the end rotation of the column is equal to the end rotation of the strip of slab C–D, which is attached to the column. The rotation at A of strip A–B is greater than the rotation at point C, however, because there is less restraint to the rotation of the slab at this point. In effect, the edge of the slab has twisted, as shown in Fig. 14–8d. As a result, the *average* rotation of the edge of the slab is greater than the rotation of the end of the column.

Equivalent Frame Method

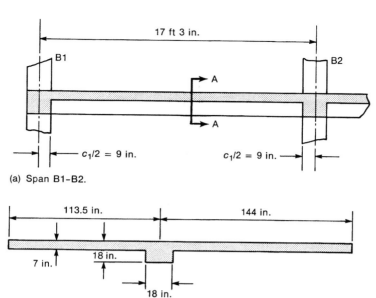

(a) Span B1–B2.

113.5 in.   144 in.

7 in.   18 in.

18 in.

Fig. 14–6
Span $B1$–$B2$—Example 14–2.   (b) Section A-A.

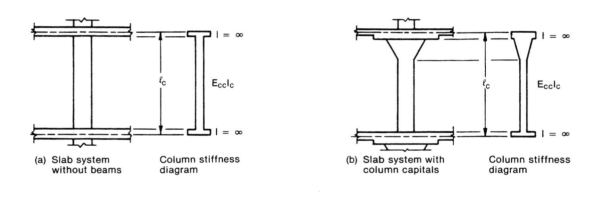

(a) Slab system   Column stiffness
without beams   diagram

(b) Slab system with   Column stiffness
column capitals   diagram

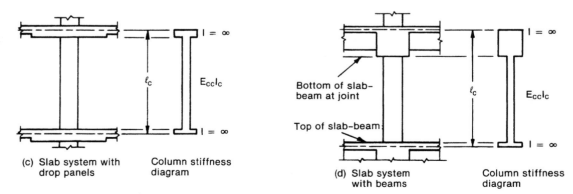

(c) Slab system with   Column stiffness
drop panels   diagram

Bottom of slab-
beam at joint

Top of slab-beam

(d) Slab system   Column stiffness
with beams   diagram

Fig. 14–7
Sections for the calculation of column stiffness, K_c. (From Ref. 14–10.)

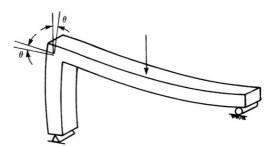

(a) Beam and column frame.

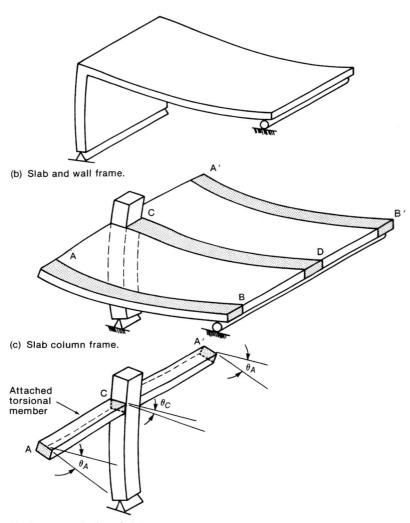

(b) Slab and wall frame.

(c) Slab column frame.

Attached torsional member

(d) Column and edge of slab.

Fig. 14–8
Frame action and twisting of edge member.

To account for this in slab analysis, the column is assumed to be attached to the slab-beam by the transverse torsional members $A-C$ and $C-A'$. One way of including these members in the analysis is the use of the concept of an *equivalent column,* which is a single element consisting of the *columns* above and below the floor and *attached torsional members,* as shown in Fig. 14–8d. The stiffness of the equivalent column, K_{ec}, represents the combined stiffnesses of the columns and attached torsional members:

$$K_{ec} = \frac{M}{\text{average rotation of the edge beam}} \qquad (14\text{--}2)$$

The inverse of a stiffness, $1/K$, is called the *flexibility*. The flexibility of the equivalent column, $1/K_{ec}$, is equal to the average rotation of the joint between the "edge beam" and the rest of the slab when a unit moment is transferred from the slab to the equivalent column. This average rotation is the rotation of the end of the columns, θ_c, plus the average twist of the beam, $\theta_{t,\text{avg}}$, both computed for a unit moment.

$$\theta_{ec} = \theta_c + \theta_{t,\text{avg}} \qquad (14\text{--}3)$$

The value of θ_c for a unit moment is $1/\Sigma K_c$, where ΣK_c refers to the sum of the flexural stiffnesses of the columns above and below the slab. Similarly, the value of $\theta_{t,\text{avg}}$ for a unit moment is $1/K_t$, where K_t is the torsional stiffness of the attached torsional members. Substituting into Eq. 14–3 gives

$$\frac{1}{K_{ec}} = \frac{1}{\Sigma K_c} + \frac{1}{K_t} \qquad (14\text{--}4)$$

If the torsional stiffness of the attached torsional members is small, K_{ec} will be much smaller than ΣK_c.

The derivation of the torsional stiffness of the torsional members or "edge beams" is illustrated in Fig. 14–9. Figure 14–9a shows an equivalent column with attached torsional members that extend halfway to the next columns in each direction. A unit torque $T = 1$ is applied to the equivalent column with half going to each arm. Since the assembly is stiffest adjacent to the column, the moment, t, per unit length of the edge beam is arbitrarily assumed to be as shown in Fig. 14–9b. The height of this diagram at the middle of the column has been chosen to give a total area equal to 1.0, the value of the applied moment, T.

The applied torques give rise to the twisting moment diagram shown in Fig. 14–9c. Since half of the torque is applied to each arm, the maximum twisting moment is $\frac{1}{2}$. The twist angle per unit length of torsional member is shown in Fig. 14–9d. This is calculated by dividing the twisting moment at any point by CG, the product of the torsional constant, C (similar to a polar moment of inertia) and the modulus of rigidity, G. The total twist of the end of an arm relative to the column is the summation of the twists per unit length and is equal to the area of the diagram of twist angle per unit length in Fig. 14–9d. Since this is a parabolic diagram, the angle of twist at the outer end of the arm is one-third of the height times the length of the diagram:

$$\theta_{t,\text{end}} = \frac{1}{3} \frac{(1 - c_2/\ell_2)^2}{2CG} \left[\frac{\ell_2}{2} \left(1 - \frac{c_2}{\ell_2} \right) \right]$$

Replacing G with $E/2$ gives

$$\theta_{t,\text{end}} = \frac{\ell_2 (1 - c_2/\ell_2)^3}{6CE}$$

This is the rotation of the end of the arm. The rotation required for use in Eq. 14–3 is the average rotation of the arm, which is assumed to be a third of the end rotation:

$$\theta_{t,\text{avg}} = \frac{\ell_2 (1 - c_2/\ell_2)^3}{18CE} \qquad (14\text{--}5)$$

Finally, the torsional stiffness of one arm is calculated as $K_t = M/\theta_{t,\text{avg}}$, where the moment resisted by one arm is taken as $\frac{1}{2}$, giving

$$K_t(\text{one-arm}) = \frac{9EC}{\ell_2 (1 - c_2/\ell_2)^3}$$

ACI Commentary Sec. R13.7.5 expresses the torsional stiffness of the two arms as

$$K_t = \sum \frac{9E_{cs}C}{\ell_2(1 - c_2/\ell_2)^3} \tag{14-6}$$

where ℓ_2 refers to the transverse spans on each side of the column. For a corner column there is only one term in the summation.

If a beam parallel to the ℓ_1 direction (a beam along C–D in Fig. 14–8) frames into the column, a major fraction of the exterior negative moment is transferred directly to the column without involving the attached torsional member. In such a case K_{ec} underestimates the stiffness of the column. This is empirically allowed for by multiplying K_t by the ratio I_{sb}/I_s, where I_{sb} is the moment of inertia of the slab and beam together, while I_s is the moment of inertia of the slab neglecting the beam stem (ACI Sec. 13.7.5.4).

The cross section of the torsional members is defined in ACI Sec. 13.7.5.1(a) to (c) and is illustrated in Fig. 13–27. Note that this cross section may differ from that used to compute the flexural stiffness of the beam (ACI Sec. 13.2.4 and Fig. 13–20). It also differs from the definition of the beam section for design for torsion in ACI Sec. 11.6.1.

The constant C in Eq. 14–6 is calculated by subdividing the cross section into rectangles and carrying out the following summation:

$$C = \sum \left[\left(1 - 0.63\frac{x}{y}\right)\frac{x^3 y}{3} \right] \tag{14-7}$$

where x is the shorter side of a rectangle and y is the longer side. The subdivision of the cross section of the torsional members is illustrated in Fig. 13–28 and explained in Sec. 13–6.

When using a moment distribution analysis, the frame analysis is carried out for a frame with slabs having stiffnesses, K_s, equivalent columns having stiffnesses, K_{ec}, and possibly beams parallel to the slabs with stiffnesses, K_b.

EXAMPLE 14–3 Calculation of K_t, ΣK_c, and K_{ec} for an Edge Column and an Interior Column

The 7-in.-thick flat plate shown in Fig. 14–4 is attached to 12 in. × 20 in. columns oriented with the 12 in. dimension perpendicular to the edge, as shown in Fig. 14–4. The story-to-story height is 8 ft 10 in. The slab and columns are 4000-psi concrete. Compute K_t, ΣK_c, and K_{ec} for the connections between the slab strip along line 2 and columns $A2$ and $B2$.

1. Compute the values for the exterior column, $A2$.

(a) Define the cross section of the torsional members. According to ACI Sec. 13.7.5.1, the attached torsional member at the exterior column corresponds to condition (a) in Fig. 13–27 as shown in Fig. 14–10a. Here $x = 7$ in. and $y = 12$ in.

(b) Compute C.

$$C = \sum \left(1 - 0.63\frac{x}{y}\right)\frac{x^3 y}{3} = \left(1 - 0.63\frac{7}{12}\right)\frac{7^3 \times 12}{3} \tag{14-7}$$
$$= 868 \text{ in.}^4$$

(c) Compute K_t.

$$K_t = \sum \frac{9E_{cs}C}{\ell_2(1 - c_2/\ell_2)^3} \tag{14-6}$$

where the summation refers to the beams on either side of line 2 and ℓ_2 refers to the length ℓ_2 of the beams on each side of line 2. Since both beams are similar:

$$K_t = 2\left[\frac{9E_{cs} \times 868}{15.75 \times 12\left(1 - \dfrac{20}{15.75 \times 12}\right)^3}\right] = 116E_{cs}$$

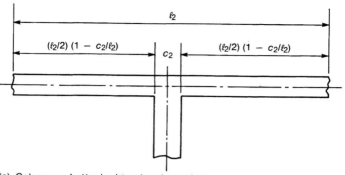

(a) Column and attached torsional member.

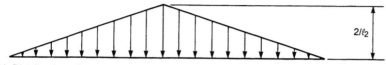

(b) Distribution of torque per unit length along column center line.

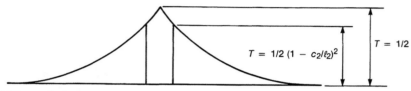

(c) Torque diagram.

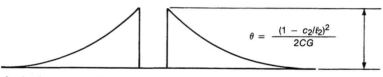

(d) Angle change per unit length.

Fig 14–9
Calculation of K_t. (From Ref. 14–10.)

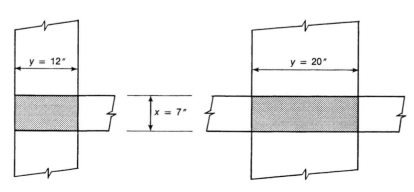

Fig 14–10
Attached torsional members—Example 14–3.

(a) Column A2.

(b) Column B2.

(d) Compute ΣK_c for the edge columns. The height center to center of the floor slabs is 8 ft 10 in. = 106 in. The distribution of stiffnesses along the column is similar to Fig. 14–7a. The edge columns are bent about an axis parallel to the edge of the slab.

$$I_c = 20 \times \frac{12^3}{12}$$

$$= 2880 \text{ in.}^4$$

For this column the overall height $\ell = 106$ in., the unsupported or clear height $\ell_u = 99$ in., and $\ell/\ell_u = 1.071$. The distance from the centerline of the slab to the top of the column proper, t_a, is 3.5 in., as is the corresponding distance, t_b, at the bottom of the bottom of the column. Interpolating in Table A–23 for $\ell/\ell_u = 1.071$ and $t_a/t_b = 1.0$ gives

$$K_c = \frac{4.76EI_c}{\ell_c}$$

and the carryover factor is -0.55. Since there are two columns (one above the floor and one under) each with the same stiffness,

$$\Sigma K_c = 2 \left(\frac{4.76E_{cc} \times 2880}{106} \right) = 259E_{cc}$$

(e) Compute the equivalent column stiffness K_{ec} for the edge column connection.

$$\frac{1}{K_{ec}} = \frac{1}{\Sigma K_c} + \frac{1}{K_t} = \frac{1}{259E_{cc}} + \frac{1}{116E_{cs}}$$

Since the slab and the columns have the same strength concrete, $E_{cc} = E_{cs} = E_c$. Therefore, $K_{ec} = 79.9E_c$.

Note that K_{ec} is only 30% of ΣK_c. This illustrates the large reduction in effective stiffness due to the lack of a stiff torsional member at the edge.

2. Compute the values at the interior column, *B2*. The torsional member at column B2 also has a section corresponding to condition (a) in Fig. 13–27a with $x = 7$ in. and $y = 20$ in. as shown in Fig. 14–10b. Thus $C = 1782$ in.^{4} and $K_t = 237.3E_c$. In the slab-strip along line 2, these columns are bent around their strong axes and have $I_c = 8000$ in.^{4}. Again

$$\Sigma K_c = 2 \left(\frac{4.76E_c I}{\ell} \right) = 718E_c$$

$$K_{ec} = 178E_c$$

It is important to note that unless K_t is very large, K_{ec} will be much smaller than ΣK_c. ∎

EXAMPLE 14–4 Calculation of K_{ec} for an Edge Column in a Two-Way Slab with Beams

The two-way slab with beams between all columns in Fig. 14–5 is supported on 18-in.-square columns. The floor-to-floor height is 12 ft 0 in. above the floor in question and 14 ft 0 in. below the floor in question. All beams are 18 in. wide by 18 in. in overall depth. The concrete strength is 3000 psi in the slab and beams and 4000 psi in the columns. Compute K_{ec} for the connections between the slab strip along line *B* and columns *B1* and *B2*. A vertical cross section through the structure is shown in Fig. 14–11a.

1. Compute the values at the exterior column, *B1*.

(a) Define the cross section of the torsional member. The torsional member at the exterior edge has the cross section shown in Fig. 14–12a (ACI Secs. 13.7.5.1 and 13.2.4).

(b) Compute *C*. To compute *C*, divide the torsional member into rectangles to maximize *C* as shown in Fig. 14–12a:

$$C = \left(1 - 0.63 \frac{18}{18} \right) \frac{18^3 \times 18}{3} + \left(1 - 0.63 \frac{7}{11} \right) \frac{7^3 \times 11}{3}$$

$$= 13,700 \text{ in.}^4$$

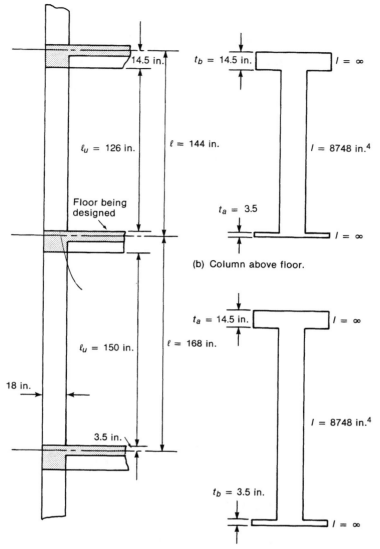

Fig. 14–11
Exterior column to slab joint
at $B1$—Example 14–4.

(a) Columns, beams, and slabs.

(b) Column above floor.

(c) Column below floor.

(c) Compute K_t.

$$K_t = \sum \frac{9E_{cs}C}{\ell_2(1 - c_2/\ell_2)^3} \tag{14–6}$$

For span $A1$–$B1$, $\ell_2 = 18$ ft 11 in. $= 227$ in., while for span $B1$–$C1$, $\ell_2 = 24$ ft 0 in. $= 288$ in. Thus

$$K_t = \frac{9E_{cs} \times 13{,}700}{227(1 - 18/227)^3} + \frac{9E_{cs} \times 13{,}700}{288(1 - 18/288)^3}$$

$$= 1216E_{cs}$$

Because the span for which moments are being determined contains a beam parallel to the span, K_t is multiplied by the ratio of the moment of inertia, I_{sb}, of a cross section including the beam stem (Fig. 14–6b) to the moment of inertia, I_s, of the slab alone. From Example 14–2, $I_{sb} = 20{,}700$ in.[4]

$$I_s = \frac{257.5 \times 7^3}{12} = 7360 \text{ in.}^4$$

Therefore, $I_{sb}/I_s = 2.81$, and $K_t = 2.81 \times 1216E_{cs} = 3420E_{cs}$.

(d) Compute ΣK_c for the edge columns at $B1$. The columns and slabs at $B1$ are shown in Fig. 14–11a. The distributions of moments of inertia along the columns are shown in Fig. 14–11b and c.

$$I_c = \frac{18^4}{12} = 8748 \text{ in.}^4$$

For the bottom column, $\ell = 168$ in., $\ell_u = 150$ in., and $\ell/\ell_u = 1.11$. The ratio of $t_a/t_b = 14.5 \text{ in.}/3.5 \text{ in.} = 4.14$. From Table A–23,

$$K_{c,\text{lower}} = \frac{5.73EI}{\ell} = \frac{5.73E_{cc} \times 8748}{168}$$

$$= 298E_{cc}$$

For the column over the floor, $\ell = 144$ in., $\ell_u = 126$ in., and $\ell/\ell_u = 1.14$. Since moment distribution is being carried out at the lower end, $t_a = 3.5$ in., $t_b = 14.5$ in., and $t_a/t_b = 0.24$. From Table A–23,

$$K_{c,\text{upper}} = \frac{4.93E_{cc} \times 8748}{144} = 300E_c$$

$$\Sigma K_c = 598E_{cc}$$

(e) Compute K_{ec}.

$$\frac{1}{K_{ec}} = \frac{1}{\Sigma K_c} + \frac{1}{K_t}$$

$$\frac{1}{K_{ec}} = \frac{1}{598E_{cc}} + \frac{1}{3420E_{cs}}$$

where

$$E_{cc} = 57,000 \sqrt{4000} = 3.61 \times 10^6 \text{ psi (ACI Sec. 8.5.1)}$$
$$E_{cs} = 57,000 \sqrt{3000} = 3.12 \times 10^6 \text{ psi}$$

Thus $E_{cc} = 1.15E_{cs}$ and $K_{ec} = 573E_{cs}$ for the edge column-to-slab connection.

2. Compute the values at the interior column, $B2$. The cross section of the torsional member is shown in Fig. 14–12b. For this section $C = 14,450$ in.^{4} and K_t of the torsional members is $K_t = 1279E_{cs}$. This is multiplied by the ratio $I_{sb}/I_s = 2.81$, giving $K_t = 3590E_{cs}$. Again $\Sigma K_c = 598E_{cc}$ and $E_{cc} = 1.15E_{cs}$. Thus $K_{ec} = 565E_{cs}$ at the interior column $B2$. ∎

Arrangement of Live Loads for Structural Analysis

The placement of live loads to produce maximum moments in a continuous beam or one-way slab was discussed in Sec. 10–2 and illustrated in Fig. 10–8. Similar loading patterns are specified in ACI Sec. 13.7.6 for the analysis of two-way slabs. If the unfactored live load does not exceed 0.75 times the unfactored dead load, it is not necessary to consider pattern loadings, and only the case of full factored live and dead load on all spans need be analyzed (ACI Sec. 13.7.6.2). This is based on the assumption that the increase in live-load moments due to pattern loadings compared to uniform live loads will be small compared to the dead-load moments. It also recognizes the fact that a slab is sufficiently ductile in flexure to allow moment redistribution.

If the unfactored live load exceeds 0.75 times the unfactored dead load, pattern loadings described in ACI Sec. 13.7.6.3 need to be considered:

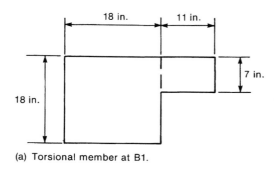

(a) Torsional member at B1.

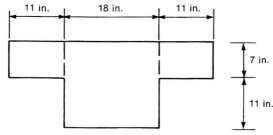

Fig. 14–12
Torsional members at $B1$ and
$B2$—Example 14–3.

(b) Torsional member at B2.

1. For maximum positive moment, factored dead load on all spans and 0.75 times the full factored live load on the panel in question and on alternate panels.

2. For maximum negative moment at an interior support, factored dead load on all panels and 0.75 times the full factored live load on the two adjacent panels.

The final design moments shall not be less than for the case of full factored dead and live load on all panels. (ACI Sec. 13.7.6.4).

ACI Sec. 13.7.6.3 is an empirical attempt to recognize the small probability of full live load in a pattern loading situation. Again, the possibility of moment redistribution is recognized.

Moments at the Face of Supports

The Equivalent Frame Analysis gives moments at the point where ends of the members meet in the center of the joint. ACI Sec. 13.7.7 permits the moments at the face of rectilinear supports to be used in the design of the slab reinforcement. The critical sections for negative moment are illustrated in Fig. 14–13. For columns extending more than $0.175\ell_1$ from the face of the support, the moments can be reduced to the values existing at $0.175\ell_1$ from the center of the joint. This limit is necessary since the representation of the slab-beam stiffness in the joint is not strictly applicable for very long narrow columns.[14–9]

If the slab meets the requirements for slabs designed by the Direct Design Method (see Sec. 13–6 and ACI Sec. 13.6.1) the total design moments in a panel can be reduced so that the absolute sum of the positive moment and the average negative moment does not exceed the statical moment, M_0, for that panel, where M_0 is as given by Eq. 13–5 (ACI Eq. 13–3). Thus for the case illustrated in Fig. 14–14, the computed moments, M_1, M_2, and M_3 would be multiplied by the ratio

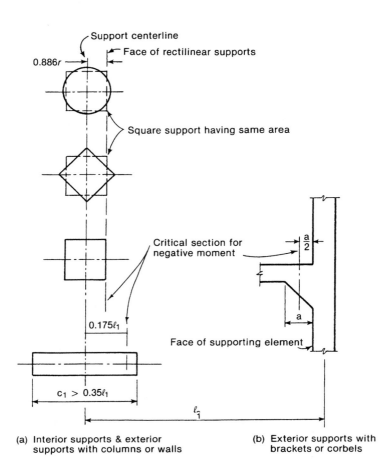

Fig. 14–13
Critical sections for negative moment. (From Ref. 14–7. Courtesy of the Portland Cement Association.)

(a) Interior supports & exterior supports with columns or walls

(b) Exterior supports with brackets or corbels

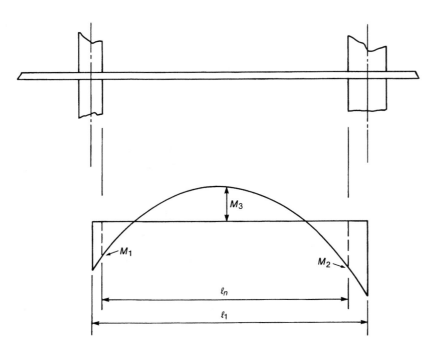

Fig. 14–14
Negative and positive moment in slab beam.

$$\frac{w_u \ell_2 \ell_n^2}{8} \bigg/ \left(\frac{M_1 + M_2}{2} + M_3 \right)$$

This adjustment is carried out only if this ratio is less than 1.

Division of Moments to Column Strips, Middle Strips, and Beams

Once the negative and positive moments have been determined for each equivalent frame, these are distributed to column and middle strips in accordance with ACI Secs. 13.6.4 and 13.6.6 in exactly the same way as in the Direct Design Method. This is discussed more fully in Sec. 13–6 and is illustrated in Examples 13–3, 13–4, 13–7, and 13–9.

For panels with beams between the columns on all sides, the distribution of moments to the column and middle strips according to ACI Secs. 13.6.4 and 13.6.6 is valid only if $\alpha_1 \ell_2^2 / \alpha_2 \ell_1^2$ falls between 0.2 and 5.0. Cases falling outside this range tend to approach one-way action and other methods of slab analysis are required.

Selection of Slab Thickness

Slab thicknesses are generally chosen to satisfy ACI Sec. 9.5.3 (see Sec. 13–6 and Table 13–1). Thinner slabs may be used if calculations show that deflections will not be excessive. Deflection estimates should include an allowance for long-time deflections and should allow for cracking of the beams and slabs. Frequently, construction loads will exceed the service loads. The moment M_a used in evaluating the equivalent moment of inertia of the slab should be based on the largest unfactored moments encountered prior to the time at which deflections are checked.

Calculation of Moment about Centroid of Shear Perimeter

An equivalent frame analysis gives moments and shears at the ends of the members where they meet at the center of the column-slab joint. In Fig. 14–15b, M_{sj} is the moment in the slab at the center of the joint computed in the frame analysis, V_{slab} is the shear from the slab to the right of the column centerline, and V_{cant} is the shear from the cantilever, if one exists. The moment from the cantilever is

$$M_c = V_{cant} e_c$$

For the joint to be in equilibrium, a moment of $M_{sj} - M_c$ and a vertical force, $V_{slab} + V_{cant}$, must be transferred to the column. Figure 14–15c shows a new free body of the joint region, with the vertical reaction, $V_{cant} + V_{slab}$, acting through the centroid of the shear perimeter which is located at a distance e_{sc} from the center of the joint. Summing moments about the centroid of the shear perimeter in Fig. 14–15c gives the moment to be transferred as

$$M_{sj} - V_{cant}(e_c + e_{sc}) - V_{slab} e_{sc} - M_u = 0$$

Replacing $V_{cant} e_c$ by M_c gives

$$M_u = M_{sj} - M_c - (V_{slab} + V_{cant}) e_{sc} \tag{14–8}$$

The connection is then designed for $(V_{slab} + V_{cant})$ and M_u.

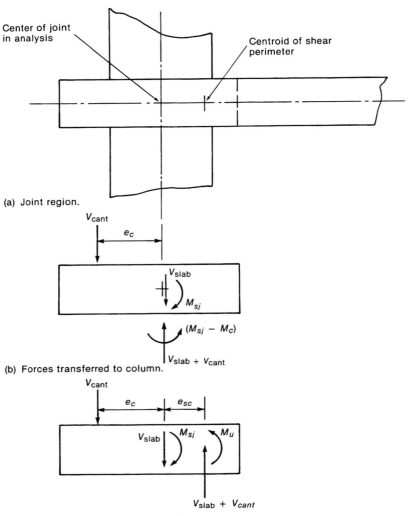

(a) Joint region.

(b) Forces transferred to column.

(c) Column reactions resolved through centroid of shear

Fig. 14–15
Moment about centroid of
shear perimeter.

EXAMPLE 14–5   Design of a Flat-Plate Floor Using the Equivalent Frame Method

Figure 14–4 shows a plan of a flat-plate floor without spandrel beams. Design this floor using the Equivalent Frame Method. Use 4000-psi concrete for the columns and slab and Grade 40 reinforcement. The story-to-story height is 8 ft 10 in. The floor supports its own dead load plus 25 psf for partitions and finishes and a live load of 40 psf.

This is the same slab that was designed in Example 13–7. Only those parts of the design that differ from Example 13–7 are discussed here. The steps in the design process are numbered the same as in Example 13–7.

1. Select the design method. Although the slab satisfies the requirements for the use of the Direct Design Method, it has been decided to base the design on moments calculated using the Equivalent Frame Method.

2. Select the thickness. The selection of thickness is based on Table 13–1 and also on providing adequate shear strength. Based on the calculations in Example 13–7, a 7-in. slab will be used.

3. Compute the moments in the equivalent frame along column line 2. (Examples 13–7 and 14–5 differ in this step.) The strip of slab along column line 2 acts as a rigid frame spanning between columns A2, B2, C2, and D2. For the purposes of analysis, the columns above and below the slabs will be assumed fixed at their far ends.

(a) Determine moment distribution coefficients for slab-beams. From Example 14–1:

Span A2–B2:

$$K_{A2-B2} = \frac{4.11EI_1}{\ell_1}$$

$$= \frac{411E_c \times 5402}{213} = 104E_c$$

$$\text{COF}_{A2-B2} = -0.508$$

$$K_{B2-A2} = 104E_c \qquad \text{COF}_{B2-A2} = -0.508$$

$$\text{fixed-end moments} = 0.084w\ell_2\ell_1^2$$

Span B2–C2:

$$K_{B2-C2} = \frac{4.10EI_1}{\ell_1}$$

$$= \frac{4.10E_c \times 5402}{288} = 76.9E_c$$

$$\text{COF} = -0.507$$

$$\text{fixed-end moments} = 0.084w\ell_2\ell_1^2$$

Span C2–D2: Same as *A2–B2*.

(b) Determine the moment distribution coefficients for the equivalent columns. From Example 14–3:

Column *A2*: $K_{ec} = 79.9E_c$, COF $= -0.55$
Column *B2*: $K_{ec} = 178E_c$, COF $= -0.55$

(c) Compute the distribution factors. The distribution factors are computed in the usual manner, thus:

$$DF_{A2-B2} = \frac{K_{A2-B2}}{K_{A2-B2} + K_{ecA2}}$$

$$= \frac{104E_c}{104E_c + 79.9E_c} = 0.566$$

$$DF_{\text{column } A2} = 0.434$$

The distribution factors and carryover factors are shown in Fig. 14–16. The cantilever members projecting outward at joints *A2* and *D2* refer to the slab that extends outside the column to support the wall.

(d) Select the loading cases and compute the fixed-end moments. Since $w_L = 40$ psf is less than three-fourths of $w_D = 112.5$ psf, only the case of uniform live load on each panel need be considered (ACI Sec. 13.7.6.2). Since the influence area of each of the panels is less than 400 ft^2, live-load reductions do not apply.

$$w_u = 1.4\left(\frac{7}{12} \times 0.15 + 0.025\right) + 1.7(0.040) = 0.226 \text{ ksf}$$

For the slab-beam in question, $\ell_2 = 15.75$ ft.
Span A2–B2:

$$M = 0.084w_u\ell_2\ell_1^2$$

$$= 0.084 \times 0.226 \times 15.75 \times 17.75^2 = 94.0 \text{ ft-kips}$$

(Note that the moment, *M,* is based on the center-to-center span, ℓ_1, rather than on the clear span, ℓ_n, used in the Direct Design Method.)

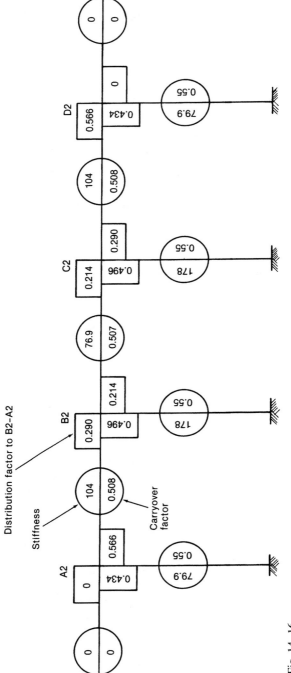

Fig. 14-16
Stiffness, carryover, and distribution factors—Example 14-5

Span B2–C2:

$$M = 0.084 \times 0.226 \times 15.75 \times 20^2 = 119.6 \text{ ft-kips}$$

The weight of the slab outside line *A* and the wall load cause a small cantilever moment at joint *A2*. It is assumed that the wall load acts 2 in. outside the exterior face of the column, or 8 in. from the center of the column.

$$M = (1.4 \times 15.75 \times 0.300) \times \frac{8}{12} + \left(1.4 \times 15.75 \times \frac{7}{12} \times 0.150\right)\frac{(10/12)^2}{2}$$

$$= 5.08 \text{ ft-kips}$$

Span A2–B2:

(e) **Carry out a moment distribution analysis.** The moment distribution analysis is carried out in Table 14–1. The sign convention used takes clockwise moments *on the joints* as positive. Thus anticlockwise moments *on the ends of members* are positive. When this sign convention is used, the carryover factors are positive rather than negative. The resulting moment and shear diagrams are plotted in Fig. 14–17a and b. The moment diagram is plotted on the compression face of the member and positive and negative moments in the slab-beam correspond to the normal sign convention for continuous beams. The moment diagram calculated in Example 13–7 is plotted in Fig. 14–17c for comparison.

(f) **Calculate the moments at the faces of the supports.** The moments at the face of the supports are calculated by subtracting the areas of the shaded parts of Fig. 14–17b from the moments at the ends of the members.

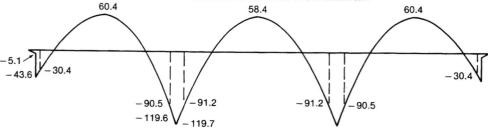

(a) Moment diagram from equivalent frame analysis (ft–kips).

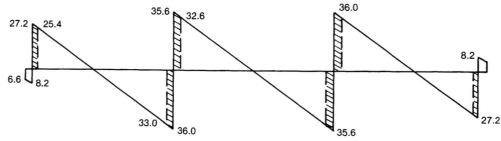

(b) Shear force diagram (kips).

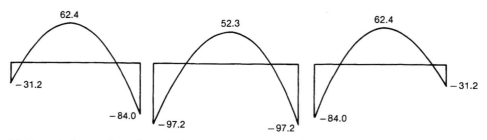

(c) Moment diagram from direct design method (ft–kips).

Fig. 14–17
Moments and shears in frame along line 2—Example 14–5.

TABLE 14–1 Moment Distribution—Example 14–5

	A2			B2			C2			D2		
		COF = 0.508			COF = 0.508	COF = 0.507				COF = 0.508		
	0	0.434	0.566	0.290	0.496	0.214	0.214	0.496	0.290	0.566	0.434	0
	Cant.	Coln.	Slab	Slab	Coln.	Slab	Slab	Coln.	Slab	Slab	Coln.	Cant.
FEM	−5.1	0	+94.0	−94.0	0	+119.6	−119.6	0	+94.0	−94.0	0	+5.1
B1		−38.6	−50.3	−7.4	−12.7	−5.5	+5.5	+12.7	+7.4	+50.3	+38.6	
C1			−3.8	−25.6		+2.8	−2.8		+25.6	+3.8		
B2		+1.7	+2.1	+6.6	+11.3	+4.9	−4.9	−11.3	−6.6	−2.1	−1.7	
C2			+3.4	+1.1		−2.5	+2.5		−1.1	−3.4		
B3		−1.5	−1.9	+0.4	+0.7	+0.3	−0.3	−0.7	−0.4	+1.9	+1.5	
C3			+0.2	−1.0		−0.2	+0.2		+1.0	−0.2		
B4		−0.1	−0.1	+0.3	+0.6	+0.3	−0.3	−0.6	−0.3	+0.1	−0.1	
Sum	−5.1	−38.5	43.6	−119.6	−0.10	+119.7	−119.7	+0.10	+119.6	−43.6	+38.5	+5.1
Sum at joint		0			0			0			0	

(g) Reduce the moments to M_0. For slabs that fall within the limitation for the use of the Direct Design Method, the total moments between faces of columns can be reduced to M_0. This slab satisfies ACI Sec. 13.6.1, so this reduction can be used.
Span A2–B2:

$$\text{Total moment} \simeq \frac{30.4 + 90.5}{2} + 60.4 = 120.9 \text{ ft-kips}$$

$$M_0 = \frac{w\ell_2\ell_n^2}{8} = 120 \text{ ft-kips}$$

Thus the computed moments in span A2–B2 could be multiplied by

$$\frac{M_0}{\text{total moment}} = \frac{120}{120.9}$$

$$= 0.993$$

and the resulting values used for design.
Span B2–C2:

$$\text{Total moment} = \frac{91.2 + 91.2}{2} + 58.4 = 149.6 \text{ ft-kips}$$

$$M_0 = 149.5 \text{ ft-kips}$$

The changes are so small in this example that no adjustments will be made.

(h) Determine the column moments. The total moment transferred to the columns at A2 is 38.5 ft-kips. This moment is distributed between the two columns in the ratio of the stiffnesses, K_c. Since the columns above and below have equal stiffnesses, half goes to each column, giving the bending moment diagrams shown in Fig. 14–18. The columns are designed for the moments at the edge of the floor slabs. At B2 a very small moment is transferred to the columns.

4, 5, and 6. Compute the moments in the equivalent frames along column lines 1, B, and A. To complete the moment analysis it is necessary to compute the moments in equivalent frames along the edge and interior column lines in both directions, as done in Example 13–7.

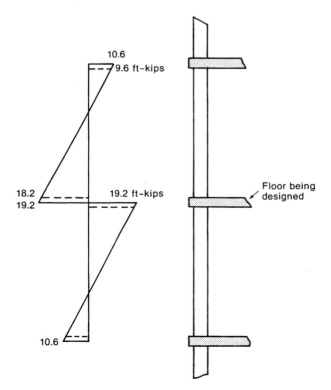

Fig. 14–18
Moments in column A2—
Example 14–5.

7 and 8. Distribute the negative and positive moments to the column and middle strips and design the reinforcement. These steps are carried out in exactly the same way as in Example 13–7.

9. Check the shear at the exterior columns. The check of moment and shear transfer at the exterior columns differs from that carried out in the Direct Design Method calculation because the moment and shear have been calculated at the center of the joint. These must be corrected using Eq. 14–8 to get the moment about the centroid of the shear perimeter when the shear acts through this point.

Since we have analyzed the equivalent frame along line 2, we shall check the shear and moment transfer at column $A2$. The connections with columns $B2$ and $C2$ may be more critical, however.

(a) Locate the critical shear perimeter. The critical perimeter is located at $d/2$ from the face of the column, where d is the average depth, 5.75 in. (see Fig. 14–19a). The centroid of this section is at 5.61 in. from line A–B and the torsional moment of inertia is $J_c = 14,900$ in.[4] about the Z–Z axis and 44,600 in.[4] about the W–W axis (Example 13–7).

(b) Compute the shear on the section and the moment about the centroid of the section. A free-body diagram of the joint is shown in Fig. 14–19b. When the shear forces act through the centroid of the shear perimeter the corresponding moment is (Eq. 14–8)

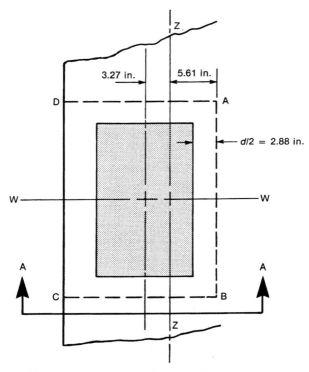

(a) Critical shear perimeter—Column A-2.

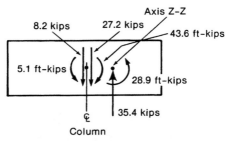

(b) Freebody of critical shear perimeter (Section A-A).

Fig. 14–19
Slab–column joint $A2$—
Example 14–5.

674   Equivalent Frame Method

$$M_u = 43.6 - 5.1 - (27.2 + 8.2)\frac{3.27}{12}$$

$$= 28.9 \text{ ft-kips}$$

The rest of the combined moment and shear transfer check proceeds as in Example 13–7.

10, 11, and 12.   Check the shear at the interior and corner columns and design the torsional reinforcement. These steps are similar to the corresponding steps in Example 13–7.   ∎

14–3   USE OF COMPUTERS IN THE EQUIVALENT FRAME METHOD

The Equivalent Frame Method was derived assuming that the structural analysis would be carried out using the moment distribution method and fixed-end moments, stiffnesses, and equivalent column stiffnesses are computed for use in such an analysis. If a standard frame analysis program based on the stiffness method is to be used, the frame must be specially modeled to get answers that agree with those obtained using the Equivalent Frame Method.

The slab column frame shown in Fig. 14–20a can be modeled for a computer analysis as shown in Fig. 14–20b. For this very simple frame, a total of 27 joints and 26 members

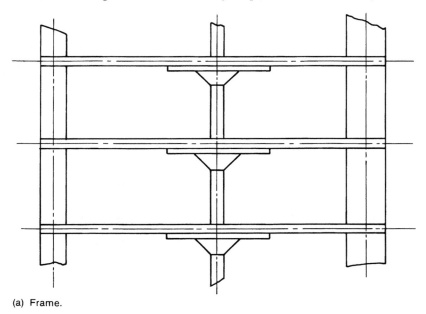

(a) Frame.

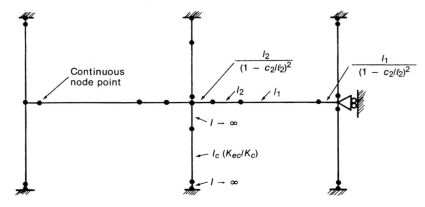

Fig. 14–20
Computer model of braced frame.

(b) Computer model.

are needed, compared to 9 joints and 8 members if the variation in stiffness is ignored. In addition, it is necessary for the analyst to evaluate K_{ec} for each column line, before computing the equivalent values of the moment of inertia of the columns.

Alternatively, specially written computer programs can be used. The most common of these is "Analysis and Design of Slab Systems" developed and marketed by the Portland Cement Association. Another such program is EFRAME, described in Ref. 14–5.

14–4 EQUIVALENT FRAME ANALYSIS OF LATERALLY LOADED UNBRACED FRAMES

A frame consisting of columns and either flat plates or slabs with drop panels, which does not have shear walls or other bracing elements, is inefficient in resisting lateral loads and is subject to significant lateral drift deflections. These deflections are amplified by the $P-\Delta$ moments resulting from gravity loads. As a result, flat-plate structures generally are braced by shear walls.

Unbraced slab-column frames are sometimes used for low buildings or for the top few floors in a tall building where the story-to-story drift in the upper stories may be reduced by terminating the shear wall before the top of the building. For such cases it is necessary to analyze equivalent frame structures for both gravity and lateral loads and add the results. For the gravity load analysis, the Equivalent Frame analysis described in Sec. 14–1 can be used without modification. For lateral load analysis, however, this Equivalent Frame underestimates the lateral deflections and hence the $P-\Delta$ effects, because it is based on uncracked EI values.

For both gravity and lateral load analyses, ACI Sec. 13.7.2.3 requires that the slab-beam strips be attached to columns by torsional members. For laterally loaded frames, ACI Sec. 13.5.1.2 requires that the effects of cracking and reinforcement be taken into account when computing the stiffness of the frame members. The ACI Commentary Sec. R13.5.1.2 suggests that this can be done using a reduced ℓ_2 of 0.25 to $0.50\ell_2$ when computing the moment of inertia of the slab-beam. Reference 14–5 suggests using $0.33I_{sb}$ instead. The procedure for analyzing an unbraced slab and column structure according to the ACI Code is as follows:

1. *For gravity loads:*
   (a) Compute K_{sb} for the slab-beams using the full width ℓ_2 and assuming that the slab is uncracked.
   (b) Compute K_c for each of the columns at a joint assuming the columns are uncracked.
   (c) Compute K_t for uncracked torsional members.
   (d) Compute K_{ec}, using ΣK_c and K_t.
   (e) Analyze the frame for gravity load effects using K_{sb} and K_{ec}, as in Sec. 14–1 and Example 14–5.

2. *For lateral loads:*
   (a) Compute the cracked slab-beam stiffness, $K_{sb,cr} = 0.33K_{sb}$, where K_{sb} is from step 1(a).
   (b) Compute an effective stiffness of each column at a joint as $(K_{ec}/\Sigma K_c)$ at that joint times the stiffness of the column in question, where K_{ec} and K_c are the values computed in steps 1(b) and 1(e).
   (c) Analyze the frame for lateral effects using $K_{sb,cr}$ and K_c.

3. *For combined gravity and lateral loads:* Superimpose steps 1 and 2.

This is a complex series of steps. It is hoped that by the next code revision, the Equivalent Frame Method will be restated in a clearer and a more computer-compatible form.

PROBLEMS

14–1 Design the north-south strips in slab shown in Fig. P13–4 using the Equivalent Frame Method. Loadings, dimensions, and material strengths are as given in Problem 13–4.

14–2 Repeat Problem 14–1 for the east-west strips in Fig. P13–4.

14–3 Compute the moments in east-west strips along lines A and B in the slab shown in Fig. 14–5. The floor supports its own weight, superimposed dead loads of 5 psf for ceiling and mechanical fixtures and 25 psf for future partitions, plus a live load of 80 psf. The exterior wall weighs 300 lb/ft and is supported by the edge beam. The stories above and below the floor in question are 12 ft and 14 ft high, respectively. Lateral loads are resisted by an elevator shaft. Use 3000-psi concrete and Grade 60 reinforcement.

15

Two-Way Slabs: Elastic, Yield Line, and Strip Method Analyses

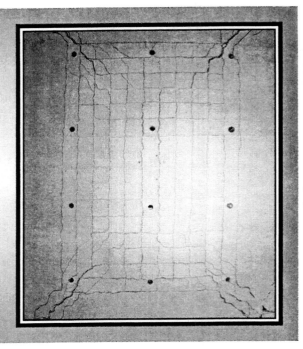

15–1 ELASTIC ANALYSIS OF SLABS

The concepts involved in the elastic analysis of slabs are reviewed very briefly here to show the relationship between the internal moments in the slab and the loads. In addition, and even more important, this will show the relationship between moments and slab curvatures.

Slabs may be subdivided into *thick* slabs with a thickness greater than about one-tenth of the span, *thin* slabs with a thickness less than about one-fortieth of the span, and *medium-thick* slabs. Thick slabs transmit a portion of the loads as a flat arch and have significant in-plane compressive forces, with the result that the internal resisting compressive force C is larger than the internal tensile force T. Thin slabs transmit a portion of the loads acting as a tension membrane, and hence T is larger than C. A medium-thick slab does not exhibit either arch action or membrane action and hence has $T = C$.

Figure 15–1 shows an element cut from a medium-thick, two-way slab. This element is acted on by the moments shown in Fig. 15–1a and the shears and loads shown in Fig. 15–1b. (The figures are separated for clarity.)

Two types of moments exist on each edge: bending moments m_x and m_y about axes parallel to the edges and twisting moments m_{xy} and m_{yx} about axes perpendicular to the edges. The moments are shown by moment vectors represented by double-headed arrows. The moment in question acts about the arrow according to the right-hand-screw rule. The moments m_x, and so on, are defined for a unit width of the edge they act on, as are the shears V_x, and so on. The m_x and m_y moments are positive, corresponding to compression on the top surface. The twisting moments on adjacent edges both act to cause compression on the same surface of the slab at the corner between the two edges, as shown in Fig. 15–1a.

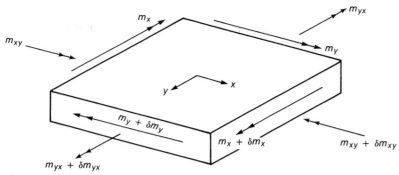

(a) Bending and twisting moments on a slab element.

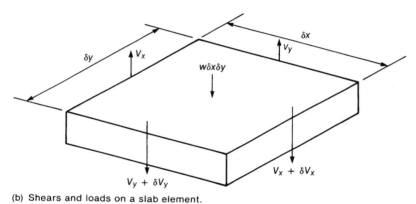

Fig. 15–1
Moments and forces in a
medium-thick plate.

(b) Shears and loads on a slab element.

Summing vertical forces gives:

$$\frac{\partial V_x}{\partial x} + \frac{\partial V_y}{\partial y} = -w \tag{15–1}$$

Summing moments about lines parallel to the x and y axes and neglecting higher-order terms gives, respectively:

$$\frac{\partial m_y}{\partial y} + \frac{\partial m_{xy}}{\partial x} = V_y \quad \text{and} \quad \frac{\partial m_x}{\partial x} + \frac{\partial m_{yx}}{\partial y} = V_x \tag{15–2}$$

It can be shown that $m_{xy} = m_{yx}$. Differentiating Eq. 15–2 and substituting into Eq. 15–1 gives the basic equilibrium equation for medium-thick slabs:

$$\frac{\partial^2 m_x}{\partial x^2} + \frac{2\partial^2 m_{xy}}{\partial x\,\partial y} + \frac{\partial^2 m_y}{\partial y^2} = -w \tag{15–3}$$

This is purely an equation of statics and applies regardless of the behavior of the plate material. For an elastic plate the deflection, z, can be related to the applied load by means of

$$\frac{\partial^4 z}{\partial x^4} + 2\frac{\partial^4 z}{\partial x^2 \partial y^2} + \frac{\partial^4 z}{\partial y^4} = -\frac{w}{D} \tag{15–4}$$

where the *plate rigidity, D,* is

$$D = \frac{Et^3}{12(1 - \nu^2)} \tag{15-5}$$

where ν is Poisson's ratio. The term D is comparable to the *EI* value of a unit width of slab.

In an elastic plate analysis, Eq. 15–4 is solved to determine the deflections, z, and the moments are calculated from

$$m_x = -D\left(\frac{\partial^2 z}{\partial x^2} + \frac{\nu \partial^2 z}{\partial y^2}\right)$$

$$m_y = -D\left(\frac{\partial^2 z}{\partial y^2} + \frac{\nu \partial^2 z}{\partial x^2}\right) \tag{15-6}$$

$$m_{xy} = -D(1 - \nu)\frac{\partial^2 z}{\partial x \, \theta y}$$

where z is the positive downward.

Solutions to Eq. 15–6 may be found in books on elastic plate theory.

15–2 DESIGN OF REINFORCEMENT FOR MOMENTS FROM A FINITE ELEMENT ANALYSIS

The most common way of solving Eq. 15–6 is to use a finite element analysis. Such an analysis gives values of m_x, m_y, and m_{xy} in each element where m_x, m_y, and m_{xy} are moments per unit width. A portion of an element bounded by a diagonal crack is shown in Fig. 15–2. The moments on the x and y faces from the finite element analysis are shown in Fig. 15–2b. The moment about an axis parallel to the crack is m_c given by

$$m_c \, ds = (m_x \, dy + m_{xy} k \, dy) \cos \theta + (m_y k \, dy + m_{xy} \, dy) \sin \theta$$

or

$$m_c = \left(\frac{dy}{dx}\right)^2 (m_x + k^2 m_y + 2k m_{xy}) \tag{15-7}$$

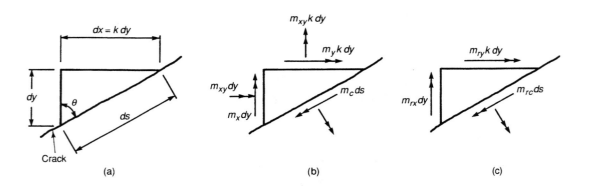

Fig. 15–2
Resolution of moments. (From Ref. 15–1 by Hillerborg, "Strip Method of Design," published by E&F, N Spon, London, 1975.)

Two-Way Slabs: Elastic, Yield Line, and Strip Method Analyses

This slab is to be reinforced with bars in the x and y direction with positive moment capacities m_{rx} and m_{ry} per unit width. The corresponding moment capacity at the assumed crack is

$$m_{rc} = \left(\frac{dy}{dx}\right)^2 (m_{rx} + k^2 m_{ry}) \qquad (15\text{-}8)$$

where m_{rc} must equal or exceed m_c to provide adequate strength. Equating these and solving for the minimum we get

$$m_{ry} = m_y + \frac{1}{k} m_{xy}$$

Since m_{ry} must equal or exceed m_y to account for the effects of m_{xy}, $(1/k)m_{xy} \geq 0$, which gives

$$m_{ry} = m_y + \frac{1}{k}\left|m_{xy}\right|$$
$$m_{rx} = m_x + k\left|m_{xy}\right| \qquad (15\text{-}9)$$

where k is a positive number. This must be true for all crack orientations (i.e., for all values of k). As k is increased, m_{ry} goes down and m_{rx} goes up. The smallest sum of the two (i.e., the smallest total reinforcement) depends on the slab in question, but $k = 1$ is the best choice for a wide range of moment values.[15-1,15-2]

The reinforcement at the bottom of the slab in each direction is designed to provide positive moment resistances of

$$m_{ry} = m_y + \left|m_{xy}\right|$$
$$m_{rx} = m_x + \left|m_{xy}\right| \qquad (15\text{-}10a)$$

If either of these is negative, it is set equal to zero. Similarly, the steel at the top of the slab is designed to provide negative moment resistances of

$$m_{ry} = m_y - \left|m_{xy}\right|$$
$$m_{rx} = m_x - \left|m_{xy}\right| \qquad (15\text{-}10b)$$

If either of these is positive, it is set equal to zero.

EXAMPLE 15–1   Computation of Design Moments at a Point

A finite element analysis gives the moments in an element as $m_x = 5$ ft-kips/ft, $m_y = -1$ ft-kips/ft, and $m_{xy} = -2$ ft-kips/ft. Compute the moments to be used to design reinforcement.

    (a)   Steel at bottom of slab

$$m_{ry} = -1 + \left|-2\right| = +1 \text{ ft-kips/ft}$$
$$m_{rx} = 5 + \left|-2\right| = +7 \text{ ft-kips/ft}$$

    (b)   Steel at top of slab

$$m_{ry} = -1 - \left|-2\right| = -3 \text{ ft-kips/ft}$$
$$m_{rx} = 5 - \left|-2\right| = +3 \text{ ft-kips/ft}$$

Since m_{rx} is positive, it is set equal to zero when designing steel at the top of the slab. ∎

Under overload conditions in a slab failing in flexure the reinforcement will yield first in a region of high moment. When this occurs, this portion of the slab acts as a plastic hinge, able to resist its hinging moment, but no more. When the load is increased further, the hinging region rotates plastically and the moments due to additional loads are redistributed to adjacent sections, causing them to yield, as shown in Fig. 13–4. The bands in which yielding has occurred are referred to as *yield lines* and divide the slab into a series of elastic plates. Eventually, enough yield lines exist to form a plastic mechanism in which the slab can deform plastically without an increase in the applied load.

A *yield line analysis* uses rigid plastic theory to compute the failure loads corresponding to given plastic moment resistances in various parts of the slab. It does not give any information about deflections or the loads at which yielding first starts. Although the concept of yield line analysis was first presented by A. Ingerslev in 1921–1923, K. W. Johansen developed modern yield line theory.[15–3] This type of analysis is widely used for slab design in the Scandinavian countries. The yield line concept is presented here to aid in the understanding of slab behavior between service loads and failure. Further details are given in Refs. 15–4 to 15–6.

Yield Criterion

It is assumed that the moment–curvature relationship for a slab section is elastic–plastic, with a plastic moment capacity equal to ϕM_n, the design flexural capacity of the section. To limit deflections, floor slabs are generally considerably thicker than required for flexure, and as a result, they seldom have reinforcement ratios exceeding 0.3 to $0.4\rho_b$. In this range of reinforcement ratios, the moment–curvature response is essentially elastic–plastic.

If yielding occurs along a line at an angle α to the reinforcement, as shown in Fig. 15–3, the bending and twisting moments are assumed to be distributed uniformly along the yield line and are the maximum values provided by the flexural capacities of the reinforcement layers crossed by the yield line. It is further assumed that there is no kinking of the reinforcement as it crosses the yield line. In Fig. 15–3 the reinforcement in the x and y directions provide moment capacities of m_x and m_y per unit width (ft-lb/ft). The bending moment, m_b, and twisting moment, m_t, per unit length of the yield line can be calculated from the moment equilibrium of the element. In this calculation, α will be measured counterclockwise from the x axis; the bending moments m_x, m_y, and m_b will be positive if they cause tension in the bottom of the slab; and the twisting moment, m_t, is positive if the moment vector points away from the section, as shown.

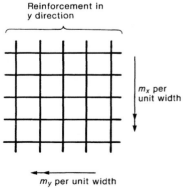

(a) Reinforcement pattern.

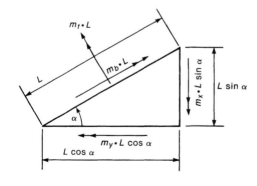

(b) Moments on an element.

Fig. 15–3
Yield criterion.

Considering the equilibrium of the element in Fig. 15–3b:

$$m_b L = m_x (L \sin \alpha) \sin \alpha + m_y (L \cos \alpha) \cos \alpha$$

giving

$$m_b = m_x \sin^2 \alpha + m_y \cos^2 \alpha \qquad (15\text{–}11)$$

and

$$m_t = \left(\frac{m_x - m_y}{2} \right) \sin 2\alpha \qquad (15\text{–}12)$$

These equations apply only for orthogonal reinforcement. If $m_x = m_y$, these two equations reduce to $m_b = m_x = m_y$ and $m_t = 0$, regardless of the angle of the yield line.

Locations of Axes and Yield Lines

Yield lines form in regions of maximum moment and divide the plate into a series of elastic plate segments. When the yield lines have formed, all further deformations are concentrated at the yield lines and the slab deflects as a series of stiff plates joined together by long hinges, as shown in Fig. 15–4. The pattern of deformation is controlled by *axes* that pass along line supports and over columns, as shown in Fig. 15–5, and by the yield lines. Since the individual plates rotate about the axes and/or yield lines, these axes and lines must be straight. To satisfy compatibility of deformations at points such as A and B in Fig. 15–4, the yield line dividing two plates must intersect the intersection of the axes about which those plates are rotating. Figure 15–5 shows the locations of axes and yield lines in a number of slabs subjected to uniform loads.

Figure 15–6 shows the underside of a reinforced concrete slab similar to the slab shown in Fig. 15–4. The wide cracks extending in from the corners and the band of cracks along the middle of the plate mark the locations of the yield lines.

Methods of Solution

Once the yield lines have been chosen, it is possible to compute the values of m necessary to support a given set of loads, or vice versa. The solution can be carried out by the *equilibrium method,* in which equilibrium equations are written for each plate segment, or by the *virtual work method,* in which some part of the slab is given a virtual displacement and the resulting work is considered.

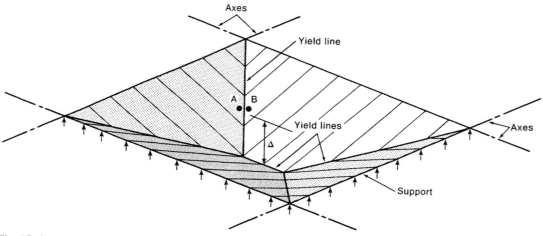

Fig. 15–4
Deformations of a slab with yield lines.

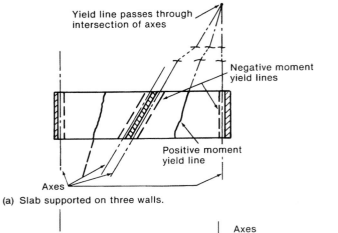

(a) Slab supported on three walls.

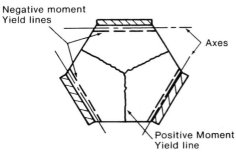

(c) Slab supported on three walls.

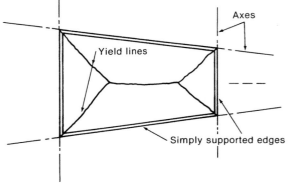

(b) Trapezoidal simply supported slab.

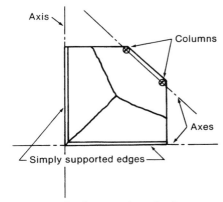

(d) Slab with simple supports and columns.

Fig. 15–5
Examples of yield line patterns.

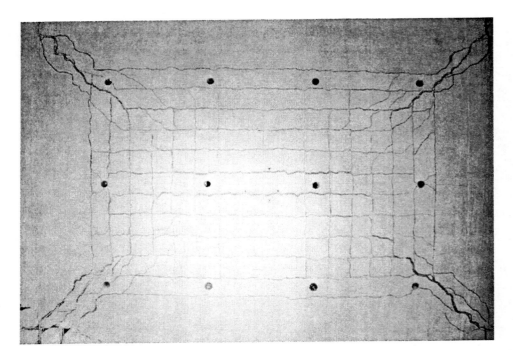

Fig. 15–6
Yield lines in a slab that is
simply supported on four
edges. (Photograph courtesy
of J.G. MacGregor)

When the equilibrium method is used, considerable care must be taken to show all of the forces acting on each element, including the twisting moments, especially when several yield lines intersect or when yield lines intersect free edges. For this reason, some building codes require that yield line calculations be done by the virtual work method. The introductory presentation here will be limited to the virtual work method of solution. Further information on the equilibrium method can be obtained in Refs. 15–3 to 15–6.

Virtual Work Method

Once the yield lines have been chosen, some point on the slab is given a virtual displacement, δ, as shown as in Fig. 15–7c. The external work done by the loads when displaced this amount is

$$\text{external work} = \sum \iint w\delta \, dx \, dy$$

$$= \sum (W\Delta_c) \tag{15–13}$$

where w = load on an element of area

δ = deflection of that element

W = total load on a plate segment

Δ_c = deflection of the centroid of that segment

The total external work is the sum of the work for each plate.

The internal work done by rotating the yield lines is

$$\text{internal work} = \sum (m_b\ell\theta) \tag{15-14}$$

where m_b = bending moment per unit length of yield line

ℓ = length of the yield line

θ = angle change at that yield line

The total internal work done during the virtual displacement is the sum of the internal work done on each yield line. Since the yield lines are assumed to have formed before the virtual displacement is imposed, no elastic deformations occur during the virtual displacement.

The principle of virtual work states that for equilibrium,

$$\text{external work} = \text{internal work}$$

$$\sum(W\Delta_c) = \sum(m_b\ell\theta) \tag{15–15}$$

The virtual work solution is an *upper bound solution*. That is, the load W is higher than or equal to the true failure load. If an incorrect set of yield lines is chosen, W will be too large for a given m or conversely, the value of m for a given W will be too small.

EXAMPLE 15–2   Yield Line Analysis of One-Way Slab by Virtual Work

The one-way slab shown in Fig. 15–7 has reinforcement at the top at the ends with a moment capacity m_{x1} per unit width and reinforcement at the bottom at midspan with a capacity m_{x2} per unit width. It is loaded with a uniform load of w per unit area. Compute the moment capacity required to support w.

1.   Select the axes and yield lines. Axes and negative moment yield lines will form along the faces of the supports and a positive moment yield line will form at midspan as shown by the dashed and wavy lines in Fig. 15–7b.

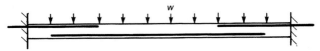

(a) Cross section.

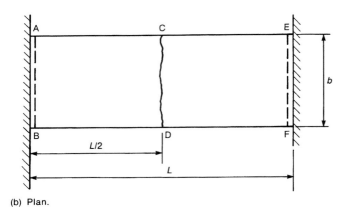

(b) Plan.

(c) Deformed shape.

Fig. 15–7
Slab–Example 15–2.

2. Give the slab a virtual displacement. Some point in the slab is displaced downward by an amount δ. In Fig. 15–7c, line *C–D* has been displaced δ.

3. Compute the external work. The total load on plate segment *A–B–C–D* is $W = w(bL/2)$ The displacement of the resultant load W: $\Delta_c = \delta/2$. Therefore, the external work done on plate *A–B–C–D* is

$$W\Delta_c = w\left(\frac{bL}{2}\right)\frac{\delta}{2}$$

The total external work is

$$\text{external work} = 2\left(\frac{wbL\delta}{4}\right)$$

4. Compute the internal work. The negative moment yield line at *A–B* rotates through an angle θ, where $\theta = \delta/(L/2)$. The internal work done in rotating through this angle is $m_{x1}b(2\delta/L)$. The total internal work done at all three yield lines is

$$\text{internal work} = 2\left[m_{x1}b\left(\frac{2\delta}{L}\right)\right] + m_{x2}b\left(\frac{4\delta}{L}\right)$$

5. Equate the external and internal work.

$$2\left(\frac{wbL\delta}{4}\right) = 4(m_{x1} + m_{x2})b\frac{\delta}{L}$$

and

$$m_{x1} + m_{x2} = \frac{wL^2}{8}$$

686   Two-Way Slabs: Elastic, Yield Line, and Strip Method Analyses

Thus any combination of m_{x1} and m_{x2} that equal $wL^2/8$ will satisfy equilibrium. If the distance A–C was taken as $0.4L$ rather than $L/2$, the values of $(m_{x1} + m_{x2})$ for the two plates would differ, indicating that an incorrect central yield line had been chosen. ∎

EXAMPLE 15–3 Yield Line Analysis of a Square Slab by Virtual Work

The simply supported square slab shown in Fig. 15–8 has a positive moment capacity, $m_x = m_y$. Compute the value of m required to support a uniform load w.

1. Select the axes and yield lines. Yield lines will form along the diagonals which, as shown in Fig. 13–12, are lines of maximum moment. The segments will rotate about axes along the four supports.

2. Give the slab a virtual displacement. Point E is displaced downward by an amount δ.

3. Compute the external work.

Load on plate segment A–B–E: $W = w\left(\dfrac{L^2}{4}\right)$

Deflection of centroid of plate: $\Delta_c = \dfrac{\delta}{3}$

External work done on plate A–B–E: $W\Delta_c = \left(\dfrac{wL^2}{4}\right)\dfrac{\delta}{3}$

Total external work: $\sum W\Delta_c = 4\left(\dfrac{wL^2}{4}\right)\dfrac{\delta}{3}$

4. Compute the internal work. Consider the yield line A–E. Length, $\ell = L/\sqrt{2}$. As shown in Fig. 15–8b, the rotation of yield line A–E is

$$\theta = \theta_1 + \theta_2$$

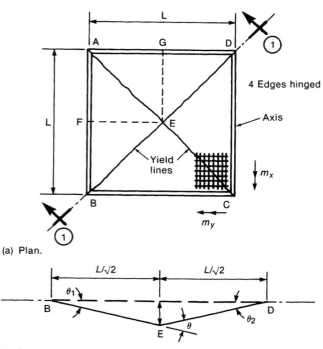

(a) Plan.

(b) Section 1–1.

Fig. 15–8
Slab–Example 15–3.

where

$$\theta_1 = \sqrt{2}\delta/L = \theta_2$$
$$\theta = 2\sqrt{2}\delta/L$$

The internal work on line A–$E = m_b \ell \theta$. From Eq. 15–11, m_b is equal to m_x and m_y. Thus $m_b \ell \theta = m(L/\sqrt{2})(2\sqrt{2}\delta/L) = 2m\delta$. The total work on all four yield lines is

$$\sum (m_b \ell \theta) = 4(2m\delta)$$

5. **Equate the external and internal work.**

$$\frac{wL^2 \delta}{3} = 8m\delta \quad \text{and} \quad m = \frac{wL^2}{24}$$

Thus the reinforcement in both directions should be designed for

$$\phi M_n = \frac{w_u L^2}{24}$$

where w_u is the factored uniform load. ∎

In Example 15–3 it was necessary to compute the value of m_b along the yield lines and the rotation of the yield lines. Generally, it is easier to consider the rotations about the axes which the slab segments rotate, especially if these are parallel and perpendicular to the reinforcement. Figure 15–9 shows a portion of the slab considered in Example 15–3. The rotation of yield line A–E is made up of θ_1 caused by the rotation θ_y of plate A–F–E about axis A–F and θ_2 caused by the rotation θ_x of plate A–G–E about axis A–G, where

$$\theta_1 = \theta_y \cos \alpha_1$$
$$\theta_2 = \theta_x \cos \alpha_2$$

The internal work done on yield line A–E is

$$m_b \ell \theta = m_b \ell (\theta_y \cos \alpha_1 + \theta_x \cos \alpha_2)$$
$$= m_b [\theta_y (\ell \cos \alpha_1) + \theta_x (\ell \cos \alpha_2)]$$

where $\ell \cos \alpha_1$ and $\ell \cos \alpha_2$ are the projections of ℓ on the y and x axes and are equal to L_y and L_x, respectively. Since only the component of m_b about an axis parallel to A–F, m_x, can do work when rotated about the y axis, and similarly only m_y can do work when rotated about the x axis, replace m_b with these components, giving

$$m_b \ell \theta = m_x L_y \theta_y + m_y L_x \theta_x$$

and

$$\text{total internal work} = \sum (m_x L_y \theta_y + m_y L_x \theta_x) \qquad (15\text{–}16)$$

This is equivalent to considering the yield line as a series of infinitesimal steps parallel to the x and y axes with moment capacities m_x and m_y, respectively. As rotation of A–F–E occurs about the x axis (A–F), only the component m_x on the steps does work. Similarly, when A–G–E rotates about A–G, only the component m_y does work.

EXAMPLE 15–4 Yield Line Analysis of a Square Slab Using Virtual Work

Redo Example 15–3 using Eq. 15–16 to compute the internal work. Steps 1, 2, and 3 are unchanged.

4. **Compute the internal work.** Consider plate A–B–E (see Fig. 15–8):

$$\theta_x = 0 \qquad \theta_y = \frac{\delta}{L/2}$$

Two-Way Slabs: Elastic, Yield Line, and Strip Method Analyses

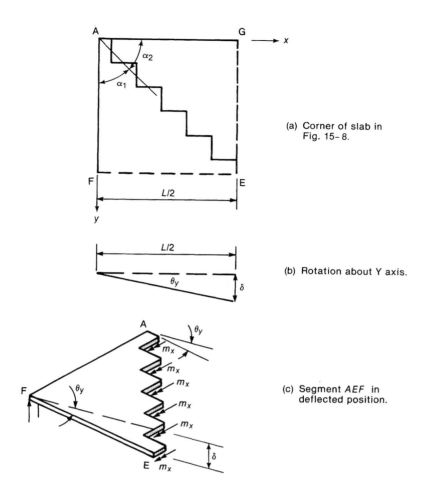

(a) Corner of slab in Fig. 15-8.

(b) Rotation about Y axis.

(c) Segment *AEF* in deflected position.

Fig.15–9
Rotation of segment *AEF*
about *Y* axis.

$$\text{internal work} = m_x L_y \theta_y + m_y L_x \theta_x$$

$$= m_x L \left(\frac{2\delta}{L}\right) + 0$$

$$= 2m_x \delta$$

Similarly for plate *A–D–E:*

$$\theta_x = \frac{\delta}{L/2} \qquad \theta_y = 0$$

$$\text{internal work} = 0 + m_y L \left(\frac{2\delta}{L}\right)$$

$$\text{total internal work} = 2(2m_x \delta) + 2(2m_y \delta)$$

Since $m_x = m_y = m$, the total internal work is $8m\delta$, which is the same value calculated in step 4 of Example 15–3. Thus $m = wL^2/24$. ∎

EXAMPLE 15–5 Calculation of the Load Capacity of a Reinforced Concrete Slab

Figure 15–10 shows a rectangular slab that is fixed on four sides and has negative moment (top) reinforcement with the capacities shown in Fig. 15–10a, and positive moment (bottom) reinforcement as shown in Fig. 15–10b. Compute the load w corresponding to failure of this slab.

15-3 Yield Line Analysis of Slabs   689

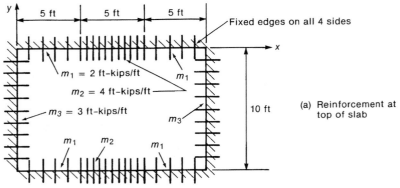

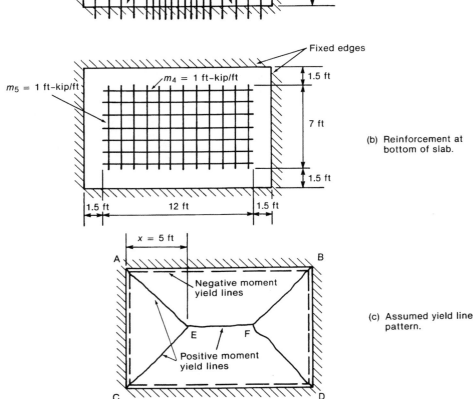

Fig. 15–10
Slab–Example 15–5.

1. Select the axes and yield lines. Axes will form along the four sides of the slab. Negative moment yield lines will form along the axes and positive moment yield lines are assumed as shown in Fig. 15–10c. The distance x to the intersection of the yield lines will be assumed to be 5 ft for the first trial.

2. Give the slab a virtual displacement. Line E–F is given a virtual displacement of δ.

3. Compute the external work. Panel A–C–E:

$$W\Delta_c = w\left(10 \times \frac{5}{2}\right)\frac{\delta}{3} = 8.33 w\delta \text{ ft-kips}$$

Panel $ABFE$:

$$W\Delta_c = w\left(10 \times \frac{5}{2}\right)\frac{\delta}{3} + (w \times 5 \times 5)\frac{\delta}{2} = 20.83 w\delta \text{ ft-kips}$$

Total external work $= (2 \times 8.33 + 2 \times 20.83)w$

$$= 58.3 w\delta \text{ ft-kips}$$

Two-Way Slabs: Elastic, Yield Line, and Strip Method Analyses

4. Compute the internal work. Panel *ACE:*

$$m_x L_y \theta_y = \left(3 \text{ ft-kips/ft} \times 10 \text{ ft} \times \frac{\delta}{5}\right) + \left(1 \text{ ft-kip/ft} \times 7 \text{ ft} \times \frac{\delta}{5}\right)$$

$$= 7.4\delta \text{ ft-kips}$$

Panel *ABFE:*

$$m_y L_x \theta_x = \left(2 \times 10 \times \frac{\delta}{5}\right) + \left(4 \times 5 \times \frac{\delta}{5}\right) + \left(1 \times 12 \times \frac{\delta}{5}\right)$$

$$= 10.4\delta \text{ ft-kips}$$

Total internal work:

$$\sum m\ell\theta = 2 \times 7.4\delta + 2 \times 10.4\delta$$

$$= 35.6\delta \text{ ft-kips}$$

5. Equate the external and internal work.

$$58.3w\delta = 35.6\delta \quad \text{and} \quad w = 0.610 \text{ ksf}$$

For this assumed yield line pattern, the load this slab can support is 610 psf.

After trying several values of *x,* a value of $x = 6.1$ ft is found to give the lowest load, $w = 602$ psf. This represents the total factored dead- and live-load capacity. ∎

Yield Line Patterns at Discontinuous Corners

In Examples 15–3 to 15–5 the yield lines were assumed to extend along the diagonals into the corners of the slabs. In Sec. 13–9 and Fig. 13–62, the localized bending in the corners of a simply supported slab was discussed. As a result of this bending, the yield line patterns in the corners of such a slab fork out to the sides of the slab as shown in Fig. 15–11a. If the corner of the slab is free to lift, it will do so. If the corner is held down, the slab will form a crack or form a yield line across the corner as shown in Fig. 15–11b. References 15–3 and 15–4 show that inclusion of the corner segments or *corner levers* will reduce the uniform load capacity of a simply supported slab by up to 9% compared

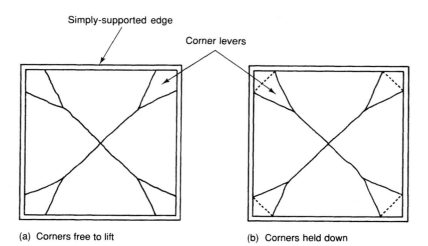

Fig. 15–11
Corner levers in simply-supported slabs.

(a) Corners free to lift

(b) Corners held down

to an analysis that ignores them. In actual slabs this reduction is offset by increases in capacity due to strain hardening of the reinforcement and membrane action in the slab.

Yield Line Patterns at Columns or Concentrated Loads

Figure 15–12 shows a slab panel supported on circular columns. The element at A is subjected to bending and twisting moments. If this element is rotated, however, it is possible to orient it so that the element is acted on by bending moments only, as shown in Fig. 15–12c. The moments determined in this way are known as the principal moments, and represent the maximum and minimum moments on any element at this point. On the element shown near the face of the column, the m_1 moment is the largest moment. As a result, a circular crack develops around the column, and eventually the reinforcement crossing the circumference of the column yields. The moments about lines radiating out from the center of the column lead to radial cracks, and as a result, a circular *fan-shaped* yield pattern develops as shown in Fig. 15–13a. Under a concentrated load, a similar fan pattern also develops, as shown in Fig. 15–13b. Note that these fans involve both positive and negative moment yield lines, shown by solid lines and dashed lines, respectively.

EXAMPLE 15–6 Calculation of the Capacity of a Fan Yield Pattern

Compute the moments required to resist a concentrated load P if a fan mechanism develops. Assume that the negative moment capacity is m_1 and the positive moment capacity is m_2 per unit width in all directions. Figure 15–13c shows a triangular segment from the fan in Fig. 15–13b. This segment subtends an angle α and has an outside radius r.

Consider segment A–B–C.

1. Give point A a virtual displacement of δ.

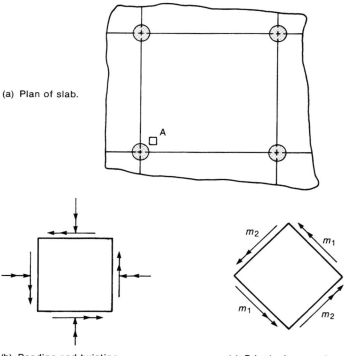

(a) Plan of slab.

Fig. 15–12
Principal moments adjacent
to column in flat plate.

(b) Bending and twisting
     moments on element at A.

(c) Principal moments on
     element at A.

Two-Way Slabs: Elastic, Yield Line, and Strip Method Analyses

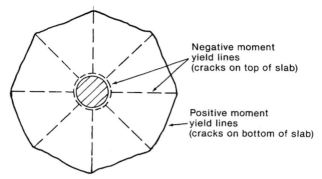

(a) Fan yield line at column in a flat plate.

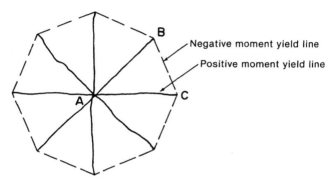

(b) Fan yield line around a downward
concentrated load at A.

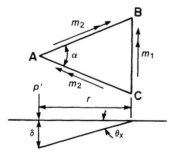

(c) Segment ABC.

Fig. 15–13
Fan yield lines.

2. **Compute the external work done.**

$$\text{External work} = P'\delta$$

where P' is the fraction of P carried by segment ABC.

3. **Compute the internal work done.** Consider one typical triangle. From Eq. 15–16,

$$\text{internal work} = \sum(m_x L_x \theta_x + m_y L_y \theta_y)$$

where θ_x is the rotation about line BC:

$$\theta_x = \frac{\delta}{r} \qquad \theta_y = 0$$

15-3   Yield Line Analysis of Slabs   **693**

For yield line A–B, the internal work is: $m_2(\alpha r/2)(\delta/r)$; for yield line B–C, the internal work is: $m_1(\alpha r)(\delta/r)$. The total internal work on this triangle is

$$m_1\alpha\delta + m_2\alpha\delta.$$

The total number of triangles is $2\pi/\alpha$, so the total internal work is

$$2\pi\delta(m_1 + m_2)$$

Setting this equal to the external work gives

$$m_1 + m_2 = \frac{P}{2\pi} \qquad \blacksquare$$

From a similar analysis, Johnson[15-5] has computed the yield line moments for a circular fan around a column of diameter, d_c, in a uniformly loaded square plate with spans ℓ_1, to be

$$m_1 + m_2 = w\ell_1^2\left[\frac{1}{2\pi} - 0.192\left(\frac{d_c}{\ell_1}\right)^{2/3}\right] \qquad (15\text{-}17)$$

If $d_c = 0$, this reduces to the answer obtained in Example 15–6.

If the slab fails due to two-way or punching shear (Chap. 13) the fan mechanism may not fully form. On the other hand, the attainment of a fan mechanism may bring on a punching shear failure.

15–4 STRIP METHOD

Simple Strip Method: Edge-Supported Slabs

The strip method developed by Hillerborg[15-1,15-7] is a slab design method that gives a lower-bound equilibrium solution to the moments in a slab. As such, it gives a lower bound (safe) estimate of the *capacity* of the slab in flexure. The equilibrium condition for an elastic slab is given by Eq. 15–3. If reinforcement is provided in the x and y directions, one possible solution that satisfies equilibrium is obtained by assuming that $m_{xy} = 0$ and dividing the total load w between the load carried in the x direction and that carried in the y direction:

$$\frac{\partial^2 m_x}{\partial x^2} = -w_x$$
$$\frac{\partial^2 m_y}{\partial y^2} = -w_y \qquad (15\text{-}18)$$

The physical significance of this is that the load w is divided into two parts, one carried by strips in the x direction, the other by strips in the y direction. The twisting of the strips due to m_{xy} (see Fig. 13–10) is ignored. By ignoring m_{xy}, m_x and m_y will tend to be overestimated.

Choice of Strips and w_x and w_y: Edge-Supported Slabs

The choice of strips and the division of the load w between those strips is entirely arbitrary. Three possible choices for a simply supported square slab are shown in Figs. 15–14 to 15–16. In Fig. 15–14 the slab is assumed to carry half the load in the x direction and half in the y direction. This gives the maximum $m_x = m_y = w\ell^2/16 = 0.0625w\ell^2$. By comparison Fig. 13–12 shows that the average maximum moment (on the diagonal) is $0.0417w\ell^2$.

A second possible choice is given in Fig. 15–15a. Here the load is carried directly to the nearest edge, with w_x and w_y either w or 0. Thus a narrow strip along A–A carries the

Two-Way Slabs: Elastic, Yield Line, and Strip Method Analyses

load shown in Fig. 15–15b and has the moment diagram shown in Fig. 15–15c. This gives the moments about the middle of the plate as shown in Fig. 15–15d. The average moment m_x about the center of the plate is $0.0417w\ell^2$. Since each narrow strip would require different reinforcement, this choice of strips would not be a practical solution.

A third possible set of strips and distribution of w is shown in Fig. 15–16. Here two-way action is assumed in the corners and in the middle of the slab, and one-way action is assumed along the edges. A full arrowhead indicates that all the load w is carried in the direction of the arrow. Half arrowheads indicate the load is divided between the two directions. The average moment m_x about the center of the plate is $0.0469w\ell^2$.

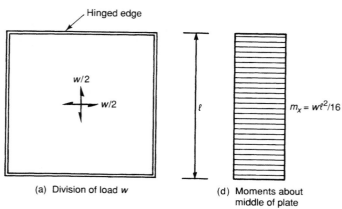

(a) Division of load w

(d) Moments about middle of plate

Fig. 15–14
Choice of strips and division of load w – First trial. (From Ref. 15–1, by Hillerborg, "Strip Method of Design," published by E&F, N Spon, London, 1975.)

(b) Loads on strip

(c) Moments in strip

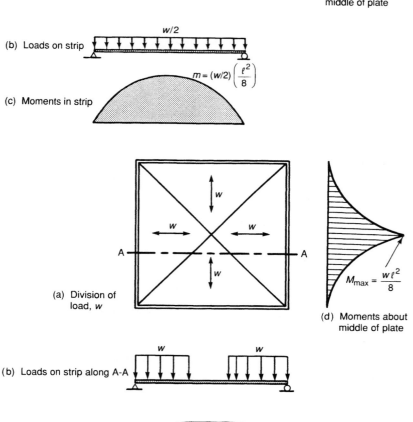

(a) Division of load, w

(d) Moments about middle of plate

Fig. 15–15
Choice of strips and division of load w – Second trial. (From Ref. 15–1, by Hillerborg, "Strip Method of Design," published by E&F, N Spon, London, 1975.)

(b) Loads on strip along A-A

(c) Moments in strip along A-A

15-4   Strip Method

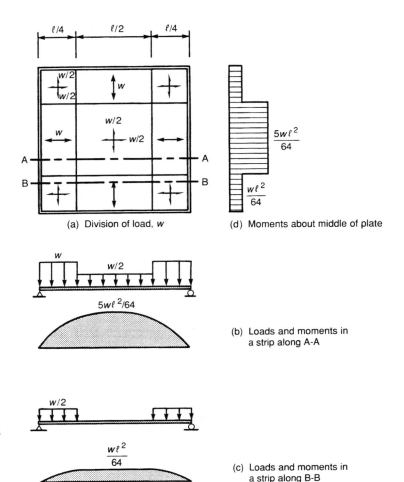

(a) Division of load, w

(d) Moments about middle of plate

$$\frac{5w\ell^2}{64}$$

$$\frac{w\ell^2}{64}$$

(b) Loads and moments in a strip along A-A

$$5w\ell^2/64$$

w
$w/2$

Fig. 15–16
Choice of strips and division
of load w – Third trial. (From
Ref. 15–1, by Hillerborg,
"Strip Method of Design,"
published by E&F, N Spon,
London, 1975.)

$w/2$

$$\frac{w\ell^2}{64}$$

(c) Loads and moments in a strip along B-B

All three of these patterns will give values of m_x and m_y that are on the safe side (too high), but when the combination of m_x, m_y, and m_{xy} is considered (see Sec. 15–2), a safe amount of reinforcement results. The pattern shown in Fig. 15–16 has the advantage that only two types of strips need to be used in each direction. As a result, it is widely used in practice.

EXAMPLE 15–7   Calculation of Moment Field in an Edge-Supported Slab

Compute the distribution of moments in the slab shown in Fig. 15–17. The total load on the slab is 150 psf.

1. Divide the slab into strips. If all four edges were hinged or all were fixed, a good choice of edge strip width would be one-fourth of the narrow width of the panel. This will be used for the north-south edge strips. Since the ends of the panel are fixed and hence will attract more load than the hinged ends, the east-west edge strips will arbitrarily be taken 50% wider. The load will be assumed to be distributed in the directions shown by the arrows.

2. Compute the moments in the strips.
Strip 1–2: The loads on a 1-ft wide strip in strip 1–2 are shown in Fig. 15–17b. This gives rise to the moments in Fig. 15–17c. The maximum moment is

$$m = 375 \times 5 - \frac{75 \times 5^2}{2} = 937 \text{ ft-lb/ft}$$

Strip 3–4: The loads on strip 3–4 are shown in Fig. 15–17d. The maximum moment is

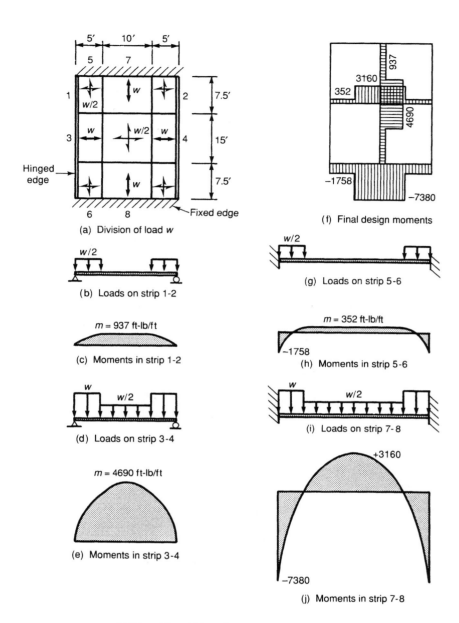

(a) Division of load *w*

(b) Loads on strip 1-2

m = 937 ft-lb/ft

(c) Moments in strip 1-2

(d) Loads on strip 3-4

m = 4690 ft-lb/ft

(e) Moments in strip 3-4

(f) Final design moments

(g) Loads on strip 5-6

m = 352 ft-lb/ft

(h) Moments in strip 5-6

(i) Loads on strip 7-8

(j) Moments in strip 7-8

Fig. 15–17
Slab–Example 15–7

$$m = 1125 \times 10 - 150 \times 5 \times 7.5 - 75 \times 5 \times 2.5 = 4690 \text{ ft-lb/ft}$$

Strip 5–6: The loads are shown in Fig. 15–17g. The maximum positive and negative moments for this case are

$$\text{fixed-end moments} = \frac{wa^2}{6}\left(3 - \frac{2a}{\ell}\right)$$

$$= \frac{75 \times 7.5^2}{6}\left(3 - \frac{2 \times 7.5}{30}\right) = 1758 \text{ ft-lb/ft}$$

$$\text{midspan moment} = \frac{wa^2}{2} - \text{fixed-end moment}$$

$$= \frac{75 \times 7.5^2}{2} - 1758 = 352 \text{ ft-lb/ft}$$

Strip 7–8: This can be considered as the sum of a strip uniformly loaded with 75 lb/ft and a strip loaded the same way as strip 5–6.

15-4   Strip Method   **697**

$$\text{fixed-end moments} = \frac{75 \times 30^2}{12} + 1758 = -7380 \text{ ft-lb/ft}$$

$$\text{midspan moment} = \frac{75 \times 30^2}{24} + 352 = 3160 \text{ ft-lb/ft}$$

The final moments are shown in Fig. 15-17f.   ■

Advanced Strip Method: Corner-Supported Elements

The simple strip method presented in the preceding section relies on the slab and the rigid supports to transmit the load in a given direction (see Fig. 13–13). As a result, it is possible to divide the load w between the two directions. For a slab on point supports, the slab must transmit the full load in both directions, as shown in Fig. 13–7. For such a case Hillerborg has presented the *advanced strip method.*[15-1] In its simplest form[15-8] this involves dividing the slab up into *simple strips* spanning one way to beam or wall supports and *corner-supported elements* spanning two directions. The latter, shown shaded in Fig. 15–18, are denoted by crossed arrows with full arrowheads which indicate that all the load is transmitted in each direction.

The corner-supported element proposed by Hillerborg satisfies the following requirements.

1. The edges are parallel to the reinforcement directions.
2. It carries a uniform load w per unit area.
3. It is supported on only one corner.
4. There are no shear forces on the sides of the element.
5. There are no twisting moments on the sides of the element.
6. All bending moments along one edge have the same sign or are zero.

Requirements 4 and 5 locate the edges of the corner-supported element at the point of maximum positive moment in the panel. It will further be assumed that

7. The element is square or rectangular.
8. The bending moment is constant along each half of each edge of the corner-supported element.

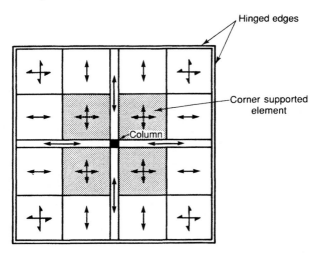

Fig. 15–18
Slab with simple strips and corner-supported elements.

Two-Way Slabs: Elastic, Yield Line, and Strip Method Analyses

In effect, the last assumption divides the elements into column strips and middle strips and allows for a constant moment over the width of each strip as assumed in the ACI Code.

Figure 15–19 shows the forces and moments acting on a corner-supported element. The widths of the element are b_x and b_y. The vertical reaction at the corner is

$$R = wb_xb_y \qquad (15\text{-}19)$$

Moment equilibrium about the y axis gives

$$m_{xsc}\frac{b_y}{2} + m_{xsm}\frac{b_y}{2} + m_{xmc}\frac{b_y}{2} + m_{xmm}\frac{b_y}{2} = \frac{wb_yb_x{}^2}{2} \qquad (15\text{-}20)$$

where m_{xsc} is the moment about the x axis through the support (s) in the column strip (c), m_{xmm} is the moment about midspan of the panel (first m), in the middle strip (second m), and so on. A similar equation can be written taking moments about the x axis. These are essentially the same as Nichols' equation 13–3.

The division of moment between support and midspan (between m_{xs} and m_{xm}) can be made (a) based on the moment coefficients in ACI Sec. 13.6.3.3, (b) based on the elastically computed moments from a strip as in Fig. 15–20b loaded as shown there, or (c) arbitrarily, provided that equilibrium is satisfied. The division of moments between column and middle strips along the sides of the element could be based on ACI Sec. 13.6.4 as done in Example 15–8. Hillerborg assumes that the midspan moments m_{xmc} and m_{xmm} are equal, and the support moments $m_{xsm} = 0$ and $m_{xsc} = 2$ times the average m_{xs} value.

EXAMPLE 15–8   Calculation of Moment Field Using Advanced Strip Method

The slab shown in Fig. 15–20a is supported on three edges by walls and by two columns as shown. Compute the distribution of moments for $w = 150$ psf.

1.   Divide the slab into strips and corner-supported elements. The arrangement chosen is shown in Fig. 15–20a. There are bands of simple strip elements along each wall and simple strip elements between each column. There are six corner supported elements. The boundaries of the various elements is chosen from the locations of the maximum positive moments in the strips. Assume that strip 2B–3B–3G–2G is loaded as shown in Fig. 15–20b and has the moment diagram shown. Lines C and F are located at the points of maximum positive moment. Similarly, lines 2 and 5 are at the points of maximum positive moment in strip 1C–6C–6D–1D.

2.   Compute the moments in strip 1C–6C–6D–1D and distribute these to column and middle strip. This strip is loaded as shown in Fig. 15–20c. The width of the strip is 8.8 ft. The statical moment considering the 8.8 ft width is

$$M_0 = \frac{150 \times 8.8 \times 15^2}{8 \times 1000} = 37.1 \text{ ft-kips}$$

(a)   *At lines 1 and 6:* $M = -0.65 \times 37.1 = -24.1$ ft-kips. From ACI Sec. 13.6.4.3, this is uniformly distributed across the width of the strip.

(b)   *At lines 2 and 5:* $M = 0.35 \times 37.1 = 13.0$ ft-kips. From ACI Sec. 13.6.4.4, $60\% = 7.80$ ft-kips is assigned to the column strip, 5.2 ft-kips to the middle strip.

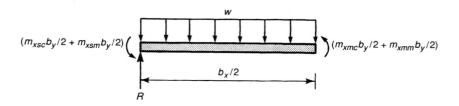

Fig. 15–19
Edge view of a corner-supported element.

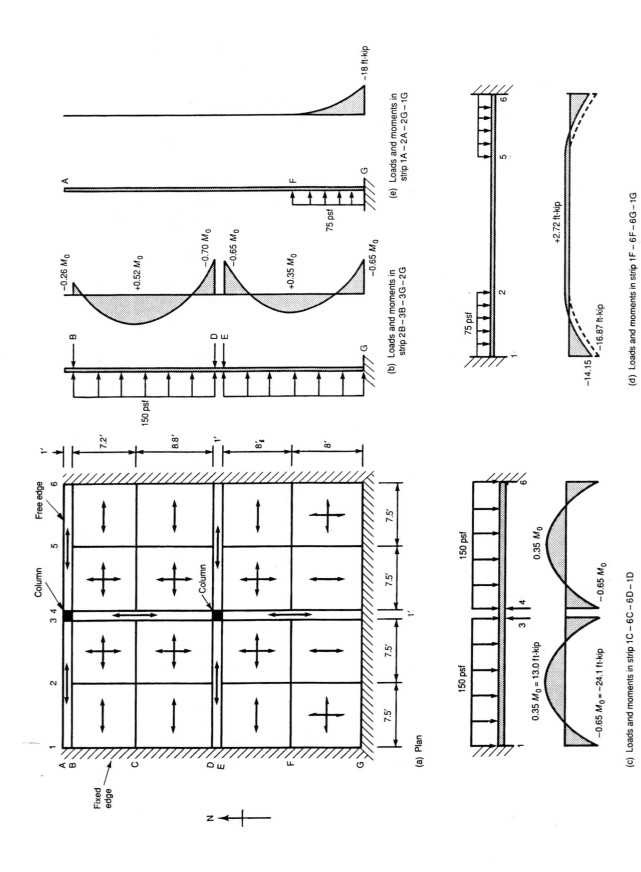

(e) Loads and moments in strip 1A − 2A − 2G − 1G

(b) Loads and moments in strip 2B − 3B − 3G − 2G

(d) Loads and moments in strip 1F − 6F − 6G − 1G

(a) Plan

(c) Loads and moments in strip 1C − 6C − 6D − 1D

Fig. 15–20
Slab–Example 15–8

(c) *At lines 3 and 4:* $M = -0.65 \times 37.1 = -24.1$ ft-kips. From ACI Sec. 13.6.4.1, 75% = -18.1 ft-kips is assigned to the column strip, -6.0 ft-kips to the middle strip.

3. Compute the moments in other east–west strips. For the strips between lines *A* and *B*, *B* and *C*, *D* and *E,* and *E* and *F* these are computed in the same way, allowing for the differing widths of the strips.

The strip between lines *F* and *G* is a simple strip loaded as shown in Fig. 15–20d and having the moment diagram shown.

4. Combine the moments from adjacent east–west middle and column strips. The column strip along edge *A* extends 1 ft plus half of 7.2 ft, totaling 4.6 ft from the edge of the slab. At the column, this strip has a total negative moment of $-24.1/8.8 = -2.74$ ft-kips in strip *A–B* and $-18.1 \times 7.2/8.8 = -14.81$ ft-kips from strip *B–C* for a total negative moment of -17.55 ft-kips in the 4.6-ft width or -3.82 ft-kips/ft (see Fig. 15–21a).

In the middle strip along *C* the negative moment at lines 3 and 4 is -6 ft-kips from strip *B–C* for a total of -10.91 ft-kips in the 8-ft width or -1.36 ft-kips/ft.

The middle strip along line *F* consists partly of strip *E–F* and partly of the simple strip *F–G*. As shown in Fig. 15–20d, the simple strip has positive moment at lines 3 and 4. These will be disregarded. However, to satisfy equilibrium in strip *E–F* the moment diagram will be shifted downward by 2.72 ft-kips, giving the dashed moment diagram in Fig. 15–20d. The moments in the middle strip along *F* will be taken as twice those in the middle strip portion of strip *E–F*.

The positive moments about line 2 are computed in the same manner. From step 2(a) the negative moments along the wall (about line 1) between *C* and *D* are -24.1 ft-kips in the 8.8-ft strip. From ACI Sec. 13.6.4.3 the moment is uniformly distributed across the width of the strip giving -2.74 ft-kips/ft. This moment exists from *A* to *F*. Between *F* and *G* the moment per unit width is (from Fig. 15–20d) $-16.87/8$ ft $= -2.11$ ft-kips/ft.

The final moments about lines 1, 2, and 3 are shown in Fig. 15–21a.

5. Compute the moments in the north–south strips.

The calculations proceed in the same manner as for the east–west strips. The moments in the strips between 2 and 3, 3 and 4, and 4 and 5 are computed from Fig. 15–20b. The strips along the walls (between 1 and 2 and between 5 and 6) are assumed to be loaded as shown in Fig. 15–20e.

The moments are distributed across the width of the slab in the same way as in steps 2 and 4. The final design moments about lines *B, C, D, F* and *G* are shown in Fig. 15–21b. ∎

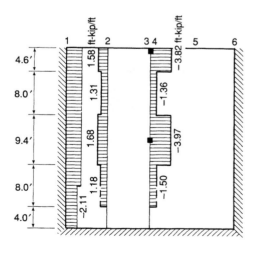

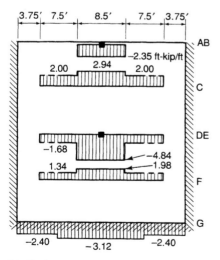

Fig. 15–21
Design Moments–Example
15–8.

(a) Design moments about lines 1, 2, 3

(b) Design moments about lines AB, C, DE, F, G

PROBLEMS

Using a yield line analysis, compute the moment capacities required in the slabs shown in the following figures.

15–1 Fig. P15–1: $m_x = m_y$ and uniform load $= w$.

15–2 Fig. P15–2: positive and negative moment capacities in both directions are m and load $= w$. Try several yield line patterns to get the maximum m.

15–3 Using a yield line analysis, compute the maximum uniform ultimate load, w, which the slab in Fig. P15–3 can support. The slab is 6 in. thick. The reinforcement in the north–south direction has $\frac{3}{4}$ in. cover to the top and bottom surfaces. The east–west reinforcement is inside the north–south steel. Use $f'_c = 3000$ psi and $f_y = 40,000$ psi.

15–4 Compute the total negative and positive moment required to support an ultimate load w of 250 psf. Assume a circular fan yield line at a column supporting an interior panel of a slab with spans $\ell_1 = 20$ ft, $d_c = 14$ in. If the overall thickness of the slab is 7 in., $f'_c = 3000$ psi, $f_y = 60,000$ psi, and the bottom steel is No. 4 at 15 in. o.c., what top reinforcement is required? Use the average d in your calculations.

15–5 Compute the moments in the slab shown in Fig. P15–5 using the strip method.

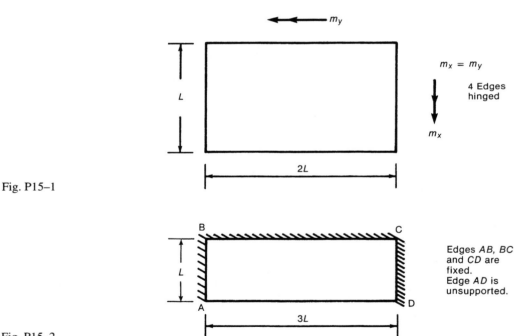

Fig. P15–1

Fig. P15–2

Two-Way Slabs: Elastic, Yield Line, and Strip Method Analyses

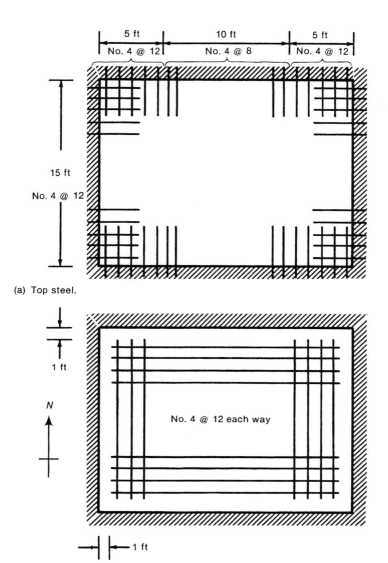

(a) Top steel.

(b) Bottom steel.

Fig. P15–3

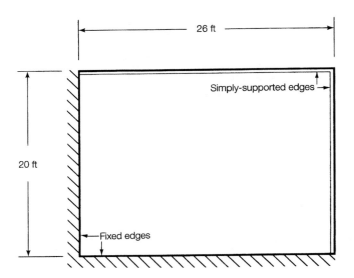

Fig. P15–5

Two-Way Slabs: Elastic, Yield Line, and Strip Method Analyses

16

Footings

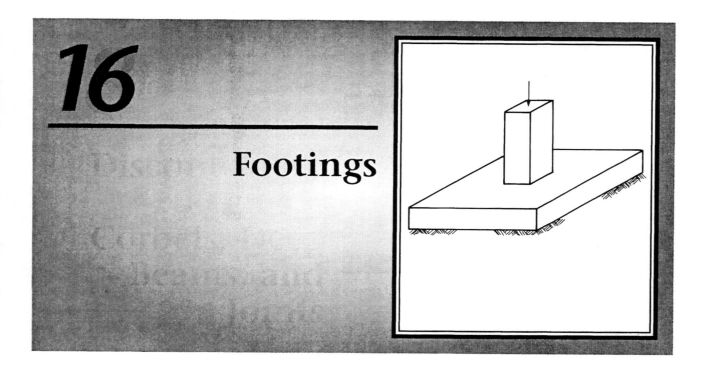

16–1 INTRODUCTION

Footings and other foundation units transfer the loads from the structure to the soil or rock supporting the structure. Because the soil is generally much weaker than the concrete columns and walls that must be supported, the contact area between the soil and the footing is much larger than that of the supported member.

The more common types of footings are illustrated in Fig. 16–1. *Strip footings* or *wall footings* display essentially one-dimensional action, cantilevering out on each side of the wall. *Spread footings* are pads that distribute the column load to an area of soil around the column. These distribute the load in two directions. Sometimes spread footings have pedestals, are stepped, or are tapered to save materials. A *pile cap* transmits the column load to a series of *piles*, which, in turn, transmit the load to a strong soil layer at some depth below the surface. *Combined footings* transmit the loads from two or more columns to the soil. Such a footing is often used when one column is close to a property line. A *mat or raft foundation* transfers the loads from all the columns in a building to the underlying soil. Mat foundations are used when very weak soils are encountered. *Caissons* 2 to 5 ft in diameter are sometimes used instead of piles to transmit heavy column loads to deep foundation layers. Frequently, these are enlarged at the bottom (*belled*) to apply load to a larger area.

The choice of foundation type is selected in consultation with the geotechnical engineer. Factors to be considered are the soil strength, the soil type, the variability of the soil type over the area and with increasing depth, and the susceptibility of the soil and the building to deflections.

Strip, spread, and combined footings are considered in this chapter since these are the most basic and most common types.

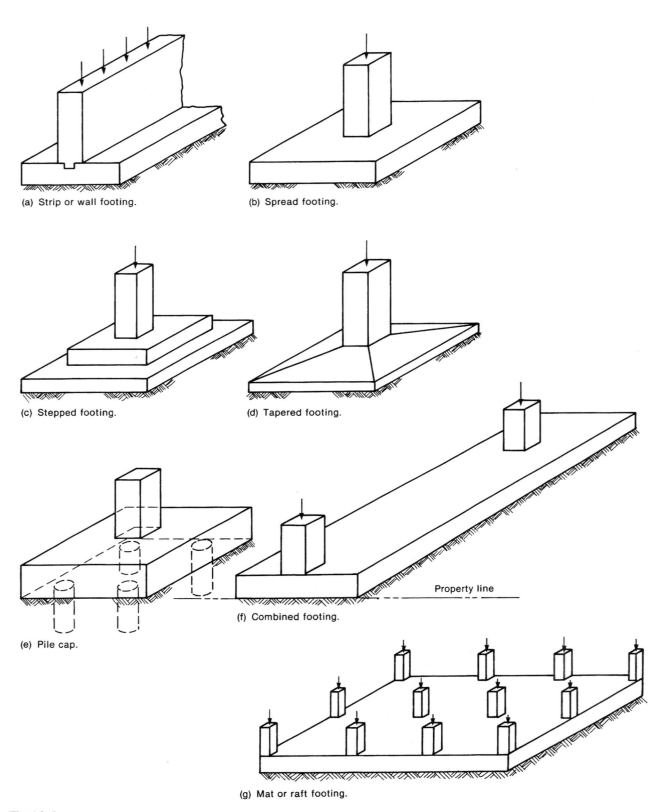

(a) Strip or wall footing.

(b) Spread footing.

(c) Stepped footing.

(d) Tapered footing.

(e) Pile cap.

Property line

(f) Combined footing.

(g) Mat or raft footing.

Fig. 16–1
Types of footings.

The distribution of soil pressure under a footing is a function of the type of soil and the relative rigidity of the soil and the foundation pad. A concrete footing on sand will have a pressure distribution similar to Fig. 16–2a. The sand near the edges of the footing tends to displace laterally when the footing is loaded, tending to a decrease in soil pressure near the edges. On the other hand, the pressure distribution under a footing on clay is similar to Fig. 16–2b. As the footing is loaded, the soil under the footing deflects in a bowl-shaped depression, relieving the pressure under the middle of the footing. For design purposes, it is customary to assume the soil pressures are linearly distributed, such that the resultant vertical soil force is collinear with the resultant downward force.

Limit States Governed by the Soil

There are three primary modes of failure of isolated foundations.[16-1] These involve (1) a bearing failure of the footing (Fig. 16–3) in which the soil under the footing moves downwards and out from under the footing, (2) a serviceability failure in which excessive differential settlements of adjacent footings causes structural and architectural damage, or (3) excessive total settlement.

The first type of failure is controlled by limiting the service load stress under the footing to less than an allowable stress:

$$q_a = \frac{q_{ult}}{FS} \qquad (16\text{–}1)$$

where q_{ult} is the stress corresponding to the failure of the footing and FS is a factor of safety on the order of 2.5 to 3. Values of q_a are obtained from the principles of geotechnical engineering and depend on the shape of the footing, the depth of the footing, the overburden or surcharge on top of the footing, the position of the water table, and the type of soil. When using a value of q_a provided by a geotechnical engineer, it is necessary to know what assumptions have been made in arriving at this allowable soil pressure, particularly with respect to overburden and depth to the footing.

It should be noted that q_a is a service load stress, whereas the rest of the structure has been designed using factored loads corresponding to the ultimate limit states. The method of accounting for this discrepancy is explained later.

The second major limit state affecting footing design is differential settlement. Settlement occurs in two stages: (1) immediate settlement as loads are applied, and (2) for fine-grained soils such as clays, a long-term settlement known as consolidation. Procedures for minimizing differential settlements involve a degree of geotechnical theory outside the realm of this book.

Elastic Distribution of Soil Pressure under a Footing

The soil pressure under a footing is calculated assuming linearly elastic action in compression but no tensile strength across the contact between the footing and the soil. If the column load is applied at, or near, the middle of the footing, as shown in Fig. 16–4, the stress, q, under the footing is

$$q = \frac{P}{A} \pm \frac{My}{I} \qquad (16\text{–}2)$$

where

P = vertical load, positive in compression

A = area of the contact surface between the soil and the footing

Fig. 16–2
Pressure distribution under
footings.

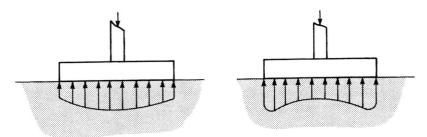

(a) Footing on sand.

(b) Footing on clay.

Fig. 16–3
Bearing failure of footing.

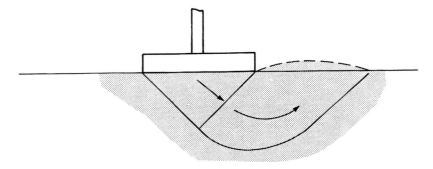

Fig. 16–4
Soil pressure under a footing:
loads within kern.

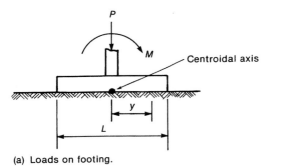

(a) Loads on footing.

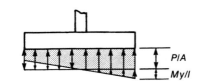

(b) Soil pressure distribution.

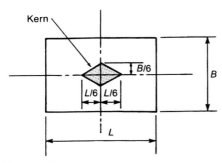

(c) Kern dimensions.

Footings

I = moment of inertia of this area

M = moment about the centroidal axis of the area

y = distance from the centroidal axis to the point where the stresses are being calculated

The moment, M, can be expressed as Pe, where e is the eccentricity of the load relative to the centroidal axis of the area A. The maximum eccentricity, e, for which Eq. 16–2 applies is that which first causes $q = 0$ at some point. Larger eccentricities will cause a portion of the footing to lift off the soil, since the soil–footing interface cannot resist tension. For a rectangular footing this occurs when the eccentricity exceeds

$$e_k = \frac{L}{6} \tag{16–3}$$

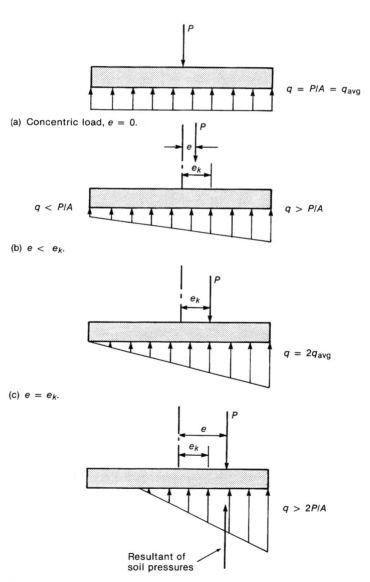

Fig. 16–5
Pressures under eccentrically loaded footing.

16–2 Soil Pressure Under Footings

709

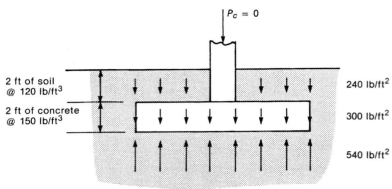

$P_c = 0$

2 ft of soil @ 120 lb/ft³
2 ft of concrete @ 150 lb/ft³

240 lb/ft²

300 lb/ft²

540 lb/ft²

(a) Self weight and soil surcharge.

P_c

240 lb/ft²

300 lb/ft²

540 lb/ft²

$q_n = P_c/A$

Gross soil pressure = $540 + q_n$ lb/ft²

(b) Gross soil pressure.

P_c

Net soil pressure. $q_n = P_c/A$

(c) Net soil pressure.

Fig. 16–6
Gross and net soil pressures.

This is referred to as the *kern distance*. Loads applied within the *kern*, the shaded area in Fig. 16–4c, will cause compression over the entire area of the footing and Eq. 16–2 can be used to compute q.

The pressures under a series of rectangular footings are shown in Fig. 16–5. If the load is axially applied, the soil pressure q equals $q_{avg} = P/A$. If the load acts through the kern point (Fig. 16–5c), $q = 0$ at one side and $q = 2q_{avg}$ at the other. If the load falls outside the kern point, the resultant upward load is equal and opposite to the resultant downward load as shown in Fig. 16–5d. Generally, such a pressure distribution would not be acceptable, since it makes inefficient use of the footing concrete, tends to overload the soil and may tilt.

Gross and Net Soil Pressures

Figure 16–6a shows a 2-ft-thick spread footing with a column at its center and with its top surface located 2 ft below the ground surface. There is no column load at this stage. The total downward load from the weights of the soil and the footing is 540 psf. This is balanced by an equal and opposite upward pressure of 540 psf. As a result, the net effect on the concrete footing is zero. There are no moments or shears in the footing due to this loading.

When the column load P_c is added, the pressure under the footing increases by $q_n = P_c/A$, as shown in Fig. 16–6b. The total soil pressure is $q = 540 + q_n$. This is referred to as the *gross soil pressure* and must not exceed the allowable soil pressure, q_a. When moments and shears in the concrete footing are calculated, the upward and downward pressures of 540 psf cancel out, leaving only the *net soil pressure*, q_n, to cause internal forces in the footing, as shown in Fig. 16–6c.

In design, the area of the footing is selected so that the *gross soil pressure* does not exceed the allowable soil pressure. The flexural reinforcement and the shear strength of the footing are then calculated using the *net soil pressure*. Thus the area of the footing is selected as

$$A = \frac{D(\text{structure, footing, surcharge}) + L}{q_a} \tag{16–4}$$

where D and L refer to dead and live loads.

For load combinations including wind, W, most codes allow a 33% increase in q_a. For such a load combination, the required area would be

$$A = \frac{D(\text{structure, footing, surcharge}) + L + W}{1.33 q_a} \tag{16–5}$$

but not less than the value given by Eq. 16–4. In Eqs. 16–4 and 16–5 the loads are the unfactored service loads.

Once the area of the footing is known, it is necessary to base the rest of the design of the footing on soil stresses due to the factored loads. These stresses are referred to as *factored net soil stresses*, q_{nu}, and for concentrically loaded footings, are the larger of

$$q_{nu} = \frac{1.4D(\text{structure}) + 1.7L}{A} \tag{16–6}$$

or

$$q_{nu} = \frac{0.75[1.4D(\text{structure}) + 1.7L + 1.7W]}{A} \tag{16–7}$$

The factored net soil pressure, q_{nu}, is based on the factored loads and will exceed q_a in most cases. This is acceptable because the factored loads are roughly 1.6 times the service loads, while the factor of safety implicit in q_a is 2.5 to 3. Hence the factored net soil pressure will be less than the pressure causing failure of the soil.

If both load and moment are transmitted to the footing, it is necessary to use Eq. 16–2 if the load is within the kern, or other relationships as illustrated in Fig. 16–5d to compute q_{nu}. In such calculations the factored loads would be used.

16–3 STRUCTURAL ACTION OF STRIP AND SPREAD FOOTINGS

The behavior of footings has been studied experimentally at various times. Our current design procedures have been strongly affected by the tests reported in Refs. 16–2 and 16–3.

The design of a footing must consider bending, development of reinforcement, shear, and the transfer of load from the column or wall to the footing. Each of these is considered separately here, followed by a series of examples in the ensuing sections. In this section only axially loaded footings with uniformly distributed soil pressures, q_{nu}, are considered.

Flexure

A spread footing is shown in Fig. 16–7. Soil pressures acting under the cross-hatched portion of the footing in Fig. 16–7b cause the moments about axis A–A at the face of the column. From Fig. 16–7c we see that these moments are:

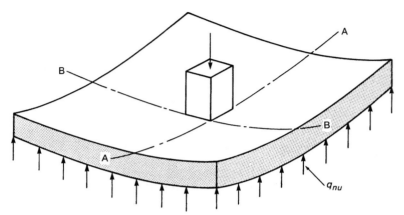

(a) Footing under load.

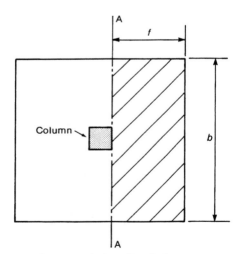

(b) Tributary area for moment at section A–A.

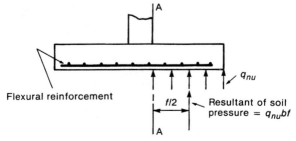

(c) Moment About Section A–A.

Fig. 16–7
Flexural action of a spread
footing.

712   Footings

$$M_u = (q_{nu}bf)\frac{f}{2} \qquad (16\text{-}8)$$

where $q_{nu}bf$ is the resultant of the soil pressure on the crosshatched area and $f/2$ is the distance from the resultant to section A–A. This moment must be resisted by reinforcement placed as shown in Fig. 16–7c. The maximum moment will occur adjacent to the face of the column on section A–A, or a similar section on the other side of the column.

In a similar manner, the soil pressures under the portion outside of section B–B in Fig. 16–7a will cause a moment about section B–B. Again, this must be resisted by flexural reinforcement perpendicular to B–B at the bottom of the footing, resulting in two layers of steel, one each way.

The critical sections for moment are taken as (ACI Secs. 15.3 and 15.4.2):

1. For footings supporting square or rectangular concrete columns or walls, at the face of the column or wall

2. For footings supporting circular or regular polygonal columns, at the face of an imaginary square column with the same area

3. For footings supporting masonry walls, halfway between the middle and the edge of the wall

4. For footings supporting a column with steel base plates, halfway between the face of the column and the edge of the base plate

The moments per unit length vary along lines A–A and B–B, with the maximum occurring adjacent to the column. To simplify reinforcement placing, however, ACI Sec. 15.4.3 states that the reinforcement shall be distributed uniformly across the entire width of the footing. A banded arrangement is used in rectangular footings, as will be illustrated in Example 16–3.

ACI Sec. 10.5.4 states that for footings of uniform thickness, the minimum area of flexural tensile reinforcement shall be the same as that required for shrinkage and temperature reinforcement in ACI Sec. 7.12. For Grade 40 steel this $A_{s,min} = 0.0020bh$, and for Grade 60 steel it is $A_{s,min} = 0.0018bh$. This amount of steel should provide a moment capacity between 1.1 and 1.5 times the flexural cracking moment and hence should be enough to prevent sudden failures at the onset of cracking. ACI Sec. 10.5.4 gives the maximum spacing of the reinforcement in a footing as the lesser of three times the thickness or 18 in.

Development of Reinforcement

The footing reinforcement is chosen assuming that the reinforcement stress reaches f_y along the maximum moment section at the face of the column. The reinforcement must extend far enough on each side of the points of maximum bar stress to develop this stress. In other words, the bars must extend ℓ_d from these points or be hooked. The maximum bar stresses are assumed to occur at the critical sections for moment.

Shear

Footings may fail in shear as a wide beam, as shown in Fig. 16–8a or due to punching, as shown in Fig. 16–8b. These are referred to as one-way shear and two-way shear and are discussed more fully in Sec. 13–7.

One-Way Shear

A footing failing due to one-way shear is designed as a beam with (ACI Sec. 11.12.1.1):

$$V_u \le \phi(V_c + V_s) \qquad (6\text{-}9 \text{ and } 6\text{-}14)$$

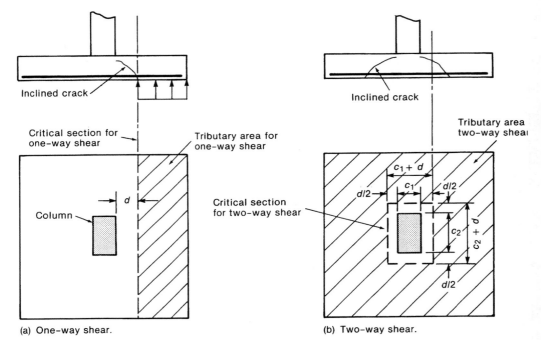

Fig. 16–8
Critical sections and tributary areas for shear in a spread footing.

(a) One-way shear.

(b) Two-way shear.

where

$$V_c = 2\sqrt{f'_c}\,b_w d \qquad (6\text{–}8)$$

Web reinforcement is very seldom used in strip footings or spread footings, due to the difficulty in placing it and the fact that it is cheaper and easier to deepen the footing than it is to provide stirrups. Hence $V_s = 0$ in most cases. The inclined crack shown in Fig. 16–8a intercepts the bottom of the member about d from the face of the column. As a result, the critical section for one-way shear is located at d away from the face of the column or wall as shown in plan in Fig. 16–8a. For footings supporting columns with steel base plates, the critical section is d away from a line halfway between the face of the column and the edge of the base plate. The shear V_u is q_{nu} times the tributary area shown shaded in Fig. 16–8a.

Two-Way Shear

Research[16–4] has shown that the critical section for punching shear is at the face of the column, while the critical loaded area is that lying outside the area of the portion punched through the slab. To simplify the design equations, the critical shear perimeter for design purposes has been defined as lying $d/2$ from the face of the column as shown by the dashed line in Fig. 16–8b (ACI Secs. 11.12.1.2 and 11.12.1.3). For the column shown in Fig. 16–8b, the length, b_o, of this perimeter is

$$b_o = 2(c_1 + d) + 2(c_2 + d) \qquad (16\text{–}9)$$

where c_1 and c_2 are the lengths of the sides of the column and d is the average effective depth in the two directions. The tributary area assumed critical for design purposes is shown shaded in Fig. 16–8b.

Since web reinforcement is rarely used in a footing, $V_u \leq \phi V_c$, where from ACI Sec. 11.12.2.1 V_c shall be the smallest of:

$$\text{(a)} \quad V_c = \left(2 + \frac{4}{\beta_c}\right)\sqrt{f'_c}\,b_o d \qquad (13\text{–}15)$$
$$\text{(ACI Eq. 11–35)}$$

where β_c is the ratio of the long side to the short side of the column (c_2/c_1 in Fig. 16–8b) and b_o is the perimeter of the critical section.

$$\text{(b)} \quad V_c = \left(\frac{\alpha_s d}{\beta_o} + 2\right)\sqrt{f_c'}\,b_o d \qquad \qquad (13\text{–}16)$$
$$\text{(ACI Eq. 11–36)}$$

where α_s is 40 for columns in the center of footing, 30 for columns at an edge of a footing and 20 for columns at a corner of a footing, and

$$\text{(c)} \quad V_c = 4\sqrt{f_c'}\,b_o d \qquad \qquad (13\text{–}17)$$
$$\text{(ACI Eq. 11–37)}$$

Transfer of Load from Column to Footing

The column applies a concentrated load on the footing. This load is transmitted by bearing stresses in the concrete and by stresses in the dowels or column bars that cross the joint. The design of such a joint is considered in ACI Sec. 15.8. The area of the dowels can be less than that of the bars in the column above, provided that the area of the dowels is at least 0.005 times the column area (ACI Sec. 15.8.2.1) and is adequate to transmit the necessary forces. Such a joint is shown in Fig. 16–9. An 18 in. × 18 in. column with $f_c' = 5000$ psi and 8 No. 8 Grade 60 bars is supported by a footing made of 3000-psi concrete and has 4 No. 7 bars as dowels. The dowels extend into the footing a distance equal to the compression development length of the No. 7 bars in 3000-psi concrete (19 in.) and into the column a distance equal to the greater of the compression splice length for No. 7 bars in 5000 psi concrete (26 in.) and the compression development length of the No. 8 bars (18. in.).

The total capacity of the column for pure axial load is 968 kips, of which 212 kips is carried by the steel and the rest by the concrete, as shown in Fig. 16–9b. At the joint, the area of the dowels is less than that of the column bars and the force transmitted by them is $\phi A_{sd} f_y$, where A_{sd} is the area of the dowels and ϕ is that for tied columns. As a result, the load carried by the concrete has increased. In Fig. 16–9 the dowels are hooked so that they can be supported on and tied to the footing reinforcement. The hooks cannot be used to develop compressive force in the bars (ACI Sec. 12.5.5).

This joint could fail by (limit states) crushing of the concrete at the bottom of the column, where the column bars are no longer effective, by crushing of the concrete in the footing under the column, by bond failure of the dowels in the footing, or by a failure of the lap splice between the dowels and the column bars. Each of these potential failure modes must be considered in the design.

The maximum bearing load on the concrete is defined in ACI Sec. 10–17 as $\phi(0.85 f_c' A_1)$, where $\phi = 0.70$, and A_1 is the area of the contact surface. When the supporting surface is wider on all sides than the loaded area, the maximum bearing load may be taken as

$$\phi(0.85 f_c' A_1)\sqrt{\frac{A_2}{A_1}} \qquad \qquad (16\text{–}10)$$

but not more than $\phi(1.7 f_c' A_1)$, where A_2 is the area of the lower base of a right pyramid or cone formed by extending lines out from the sides of the bearing area at a slope of 2 horizontal to 1 vertical to the point where the first such line intersects an edge. This is illustrated in Fig. 16–10. The first intersection with an edge occurs at point B, resulting in the area A_2 shown crosshatched in Fig. 16–10b. See also Sec. 18–8.

Two distinct cases must be considered: (1) joints that do not transmit computed moments to the footing, and (2) those that do. These will be discussed separately. If no moments are transmitted, or if the eccentricity falls within the kern of the column, there will

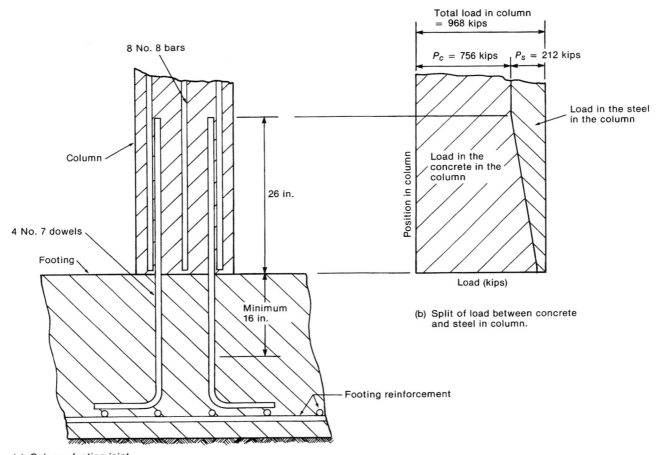

8 No. 8 bars

Column

4 No. 7 dowels

Footing

26 in.

Minimum
16 in.

Footing reinforcement

(a) Column-footing joint.

Fig. 16–9
Column–footing joint.

Total load in column
= 968 kips

P_c = 756 kips P_s = 212 kips

Load in the steel
in the column

Position in column

Load in the
concrete in the
column

Load (kips)

(b) Split of load between concrete
and steel in column.

be compression over the full section. The total force transferred by bearing is then calculated as $(A_g - A_{sd})$ times the smaller of the bearing stresses allowed on the column or the footing, where A_g is total area of the column and A_{sd} is the area of the bars or dowels crossing the joint. Any additional load must be transferred by dowels.

If moments are transmitted to the footing, bearing stresses will exist over part, but not all of the column cross section. The number of dowels required can be obtained by considering the area of the joint as an eccentrically loaded column with a maximum concrete stress equal to the smaller of the bearing stresses allowed on the column or the footing. Sufficient reinforcement must cross the interface to provide the necessary axial load and moment capacity. Generally, this requires that all the column reinforcement must cross the interface. This steel must be spliced in accordance with the requirements for column splices.

Practical Aspects

Three other aspects warrant discussion prior to doing examples. The minimum cover to the reinforcement in footings cast against the soil is 3 in. (ACI Sec. 7.7.1). This allows for small irregularities in the surface of the excavation and for potential contamination of the bottom layer of concrete with soil. Sometimes the bottom of the excavation for the footing

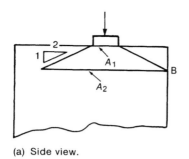

(a) Side view.

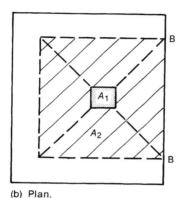

(b) Plan.

Fig. 16–10
Definition of A_1 and A_2.

is covered with a lean concrete seal coat to prevent the bottom becoming uneven after rain-storms and to give a level surface for placing reinforcement.

The minimum depth of the footing above the bottom reinforcement is 6 in. for footings on soil and 12 in. for footings on piles (ACI Sec. 15.7). ACI Sec. 10.6.4, covering the distribution of flexural reinforcement in beams and one-way slabs, does not apply to footings.

16–4 STRIP OR WALL FOOTINGS

A wall footing cantilevers out on both sides of the wall as shown in Figs. 16–1a and 16–11. The soil pressure causes the cantilevers to bend upward and as a result, reinforcement is re-quired at the bottom of the footing, as shown in Fig. 16–11. The critical sections for design for flexure and anchorage are at the face of the wall (Section A–A in Fig. 16–11. One-way shear is critical at a section a distance d from the face of the wall (section B–B in Fig. 16–11). Thicknesses of wall footings are chosen in 1-in. increments, and widths in 2- or 3-in. increments.

EXAMPLE 16–1 Design of a Wall Footing

A 12-in.-thick concrete wall carries a service (unfactored) dead load of 10 kips per foot and a service live load of 12.5 kips per foot. The allowable soil pressure, q_a, is 5000 psf at the level of the base of the footing, which is 5 ft below the final ground surface. Design a wall footing using $f'_c = 3000$ psi and $f_y = 60,000$ psi. The density of the soil is 120 lb/ft³.

   1. Estimate the size of the footing and the factored net pressure. Consider a 1-ft strip of footing and wall. Allowable soil pressure = 5 ksf; allowable net soil pressure

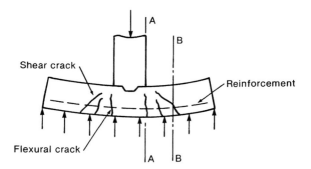

Fig. 16–11
Structural action of a strip
footing.

$= 5$ ksf $-$ weight / ft² of the footing and the soil over the footing. Since the thickness of the foot-ing is not known at this stage, it is necessary to guess a thickness for a first trial. Generally, the thick-ness will be 1 to 1.5 times the wall thickness. We shall try a 12-in.-thick footing. Therefore, $q_n = 5 - (1 \times 0.15 + 4 \times 0.12) = 4.37$ ksf.

$$\text{Area required} = \frac{10 \text{ kips} + 12.5 \text{ kips}}{4.37 \text{ ksf}}$$

$$= 5.15 \text{ ft}^2 \text{ per foot of length}$$

Try a footing 5 ft 2 in. wide.

$$\text{Factored net pressure, } q_{nu} = \frac{1.4 \times 10 + 1.7 \times 12.5}{5.167} = 6.82 \text{ ksf}$$

In the design of the concrete and reinforcement, we shall use $q_{nu} = 6.82$ ksf.

    2. Check the shear. Shear usually governs the thickness of footings. Only one-way shear is significant in a wall footing. Check it at d away from the face of the wall (section B–B in Fig. 16–11)

$$d = 12 \text{ in. } - 3 \text{ in cover } - \tfrac{1}{2} \text{ bar diameter } \simeq 8.5 \text{ in.}$$

The tributary area for shear is shown shaded in Fig. 16–12a.

$$V_u = 6.82 \text{ ksf} \left(\frac{16.5}{12} \times 1\right) \text{ft}^2 = 9.38 \text{ kips/ft}$$

$$\phi V_c = \phi 2 \sqrt{f_c'} b_w d = \frac{0.85 \times 2 \times \sqrt{3000} \times 12 \times 8.5}{1000}$$

$$= 9.50 \text{ kips/ft}$$

Since $V_u < \phi V_c$ the footing depth is OK. If V_u is larger or considerably smaller than ϕV_c, choose a new thickness and repeat steps 1 and 2. **Use a 12-in.-thick footing, 5 ft 2 in. wide.**

    3. Design the reinforcement. The critical section for moment is at the face of the wall (sec-tion A–A in Fig. 16–11). The tributary area for moment is shown in shaded Fig. 16–12b.

$$M_u = 6.82 \frac{(25/12)^2 \times 1}{2} \text{ ft-kips/ft} = 14.8 \text{ ft-kips/ft of length}$$

$$M_u = \phi M_n = \phi A_s f_y j d$$

Footings are generally very lightly reinforced. Therefore, assume that $j = 0.925$. Therefore,

$$A_s = \frac{14.8 \times 12,000}{0.9 \times 60,000(0.925 \times 8.5)} = 0.418 \text{ in.}^2/\text{ft}$$

From ACI Secs. 10.5.4 and 7.12.2

           Footings

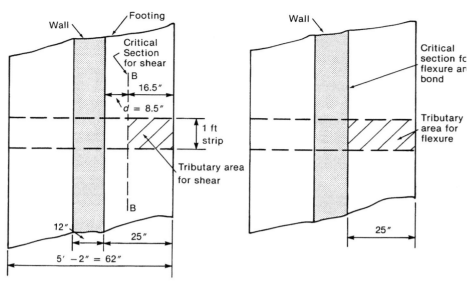

(a) Plan view of footing showing tributary area for shear.

(b) Plan view showing tributary area for moment.

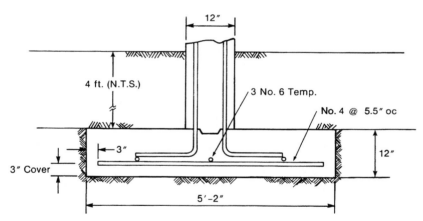

Fig. 16–12
Strip footing—Example 16–1.

(c) Reinforcement details.

$$\text{Minimum } A_s = 0.0018bh$$

$$= 0.0018 \times 12 \times 12$$

$$= 0.26 \text{ in.}^2/\text{ft (does not govern)}$$

Maximum spacing of bars (ACI Sec. 7.6.5) = $3h$ or 18 in. Therefore, maximum = 18 in. **Try No. 5 bars at 9 in. o.c.** $A_s = 0.41$ in.^{2}/ft. Since the calculation of A_s was based on a guess of the value j, re-compute ϕM_n:

$$a = \frac{0.41 \times 60,000}{0.85 \times 3000 \times 12} = 0.804 \text{ in}$$

Since $a/d = 0.804/8.5 = 0.095$ is much less than a/d for the tension-controlled limit from Table A–4, the section is tension-controlled and $\phi = 0.90$.

$$\phi M_n = \frac{0.9 \times 0.41 \times 60,000(8.5 - 0.804/2)}{12,000} = 14.94 \text{ ft-kips}$$

Required $M_u = 14.8$ ft-kips/ft—therefore OK.

4. Check the development. The clear spacing of the bars being developed exceeds $2d_b$ and the clear cover exceeds d_b. Therefore, this is case 2 development in Tables 8–1 and A–11. From Table A–11 ℓ_{db}/d_b for a No. 5 bottom bar in 3000-psi concrete is 43.8 in. The development length is

$$\ell_d = \frac{\ell_{db}}{d_b} d_b \beta \lambda$$

where $\beta = 1.0$ for uncoated reinforcement and $\lambda = 1.0$ for normal-weight concrete.

$$\ell_d = 43.8 \times 0.625 \times 1.0 \times 1.0$$
$$= 27.4 \text{ in.}$$

The distance from the point of maximum bar stress (at the face of the wall) to the end of the bar is 25 in. $-$ 3 in. cover on the ends of the bars = 22 in. This is less than $\ell_d = 27.4$ in. Therefore, we either must hook the bars or use a smaller diameter. For a No. 4 bar, $\ell_d = 21.9$ in. **Use No. 4 bars at 5.5 in. on centers.** This is still case 2 development and 22 in. just satisfies ℓ_d.

5. Select the temperature reinforcement. By ACI Sec. 7.12.2 we require

$$A_s = 0.0018bh - 0.0018 \times 62 \times 12$$
$$= 1.34 \text{ in.}^2$$

The maximum spacing is $5 \times 12 = 60$ in. or 18 in. **Provide 3 No. 6 bars for shrinkage reinforcement.**

6. Design the connection between the wall and the footing. ACI Sec. 15.8.2.2 requires that reinforcement equivalent to the minimum vertical wall reinforcement extend from the wall into the footing. A cross section through the wall footing designed in this example is shown in Fig. 16–12c. ∎

16–5 SPREAD FOOTINGS

Spread footings are square or rectangular pads which spread a column load over an area of soil that is large enough to support the column load. The soil pressure causes the footing to deflect upward as shown in Fig. 16–7a, causing tension in two directions at the bottom. As a result, reinforcement is placed in two directions at the bottom, as shown in Fig. 16–7c. Two examples will be presented: a square axially loaded footing and a rectangular axially loaded footing.

EXAMPLE 16–2 Design of a Square Spread Footing

A square spread footing supports an 18-in.-square column supporting a service dead load of 400 kips and a service live load of 270 kips. The column is built of 5000-psi concrete and has 8 No. 9 longitudinal bars with $f_y = 60,000$ psi. Design a spread footing to be constructed using 3000-psi concrete and Grade 60 bars. The top of the footing will be covered with 6 in. of fill with a density of 120 lb/ft^3 and a 6-in. basement floor (Fig. 16–13). The basement floor loading is 100 psf. The allowable bearing pressure on the soil is 6000 psf.

1. Estimate the footing size and the factored net soil pressure. Allowable net soil pressure $q_n = 6$ ksf $-$ (weight/ft^2 of the footing and the soil and floor over the footing and the floor loading). Estimate the overall thickness of the footing between one and two times the width of the column, say 27 in.

$$q_n = 6.0 - \left(\frac{27}{12} \times 0.15 + 0.5 \times 0.12 + 0.5 \times 0.15 + 0.100\right)$$
$$= 5.43 \text{ ksf}$$

$$\text{Area required} = \frac{400 \text{ kips} + 270 \text{ kips}}{5.43 \text{ ksf}}$$

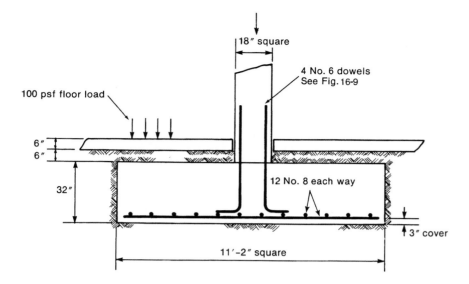

Fig. 16–13
Spread footing—Example
16–2.

$$= 123.4 \text{ ft}^2 = 11.11 \text{ ft square}$$

Try a footing 11 ft 2 in. square by 27 in. thick

$$\text{Factored net soil pressure} = \frac{1.4 \times 400 + 1.7 \times 270}{11.17^2}$$
$$= 8.17 \text{ ksf}$$

2. Check the thickness for two-way shear. Generally, the thickness of a spread footing is governed by two-way shear. The shear will be checked on the critical perimeter at $d/2$ from the face of the column and, if necessary, the thickness will be increased or decreased. Since there is reinforcement in both directions, the average d will be used.

$$\text{Average } d = 27 \text{ in.} - (3 \text{ in. cover}) - (1 \text{ bar diameter})$$
$$= 23 \text{ in.}$$

The critical shear perimeter (ACI Sec. 11.11.1.2) is shown dashed in Fig. 16–14a. The tributary area for two-way shear is shown crosshatched.

$$V_u = 8.17 \text{ ksf} \left[11.17^2 - \left(\frac{41}{12} \right)^2 \right] \text{ ft}^2 = 923 \text{ kips}$$

Length of critical shear perimeter:

$$b_o = 4 \times 41 \text{ in.} = 164 \text{ in.}$$

ϕV_c is the smallest of

(a) $\quad \phi V_c = \phi \left(2 + \frac{4}{\beta_c} \right) \sqrt{f_c'} b_o d$

$$\beta_c = \frac{\text{long side of column}}{\text{short side of column}} = 1.0$$
$$\phi V_c = \frac{0.85(2 + 4/1)\sqrt{3000} \times 164 \times 23}{1000} = 1053 \text{ kips}$$

(b) $\quad \phi V_c = \phi \left(\frac{\alpha_s d}{b_o} + 2 \right) \sqrt{f_c'} b_o d$

$$= 0.85 \left(\frac{40 \times 23}{164} + 2 \right) \frac{\sqrt{3000} \times 164 \times 23}{1000} = 1336 \text{ kips}$$

16–5 Spread Footings

721

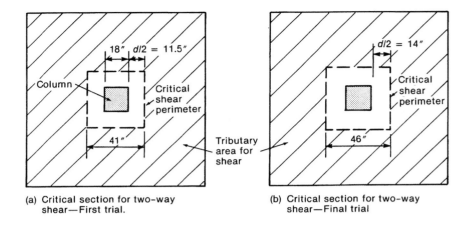

(a) Critical section for two-way
   shear—First trial.

(b) Critical section for two-way
   shear—Final trial

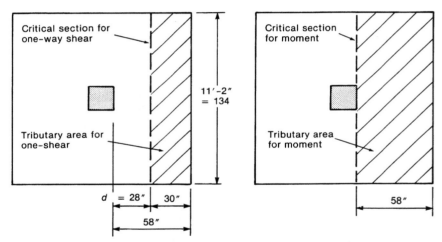

(c) Critical section for one-way shear

(d) Critical section for moment

Fig. 16–14
Critical sections—Example
16–2.

(c) $\phi V_c = \phi 4\sqrt{f_c'}b_o d = 702$ kips

Since ϕV_c is less than V_u, the footing is not thick enough. Try $h = 32$ in. Because the footing is thicker, it weighs more. Hence a larger area may be required.

$$q_n = 6.0 - \left(\frac{32}{12} \times 0.15 + 0.5 \times 0.12 + 0.5 \times 0.15 + 0.100\right)$$

$$= 5.37 \text{ksf}$$

$$\text{Area required} = \frac{400 + 270}{5.37}$$

$$= 124.8 \text{ft}^2$$

$$= 11.17 \text{ft square}$$

Try an 11 ft 2 in. square footing, 32 in. thick.

$$\text{Factored net soil pressure, } q_{nu} = \frac{1.4 \times 400 + 1.7 \times 270}{11.17^2}$$

$$= 8.17 \text{ ksf}$$

$$\text{Average } d = 32 - 3 - 1$$

$$= 28 \text{ in.}$$

The new critical shear perimeter and tributary area for shear are shown in Fig. 16–14b.

$$V_u = 8.17\left[11.17^2 - \left(\frac{46}{12}\right)^2\right] = 899 \text{ kips}$$

Again ACI Eq. 11–38 governs:

$$\phi V_c = 0.85 \times 4\sqrt{3000} \times (4 \times 46) \times \frac{28}{1000} = 959 \text{ kips}$$

This is more than adequate. A check using $h = 30$ in. shows that a 30-in.-thick footing is not adequate. **Use an 11 ft 2 in. square footing, 32 in. thick.**

3. **Check the one-way shear.** Although one-way shear is seldom critical, we shall check it. The critical section for one-way shear is located at d away from the face of the column (ACI Sec. 11.12.1.1) as shown in Fig. 16–14 c.

$$V_u = 8.17 \text{ ksf}\left(11.17 \text{ ft} \times \frac{30}{12} \text{ ft}\right) = 228 \text{ kips}$$

$$\phi V_c = \phi 2\sqrt{f_c'}b_w d = 0.85 \times 2\sqrt{3000} \times 134 \times \frac{28}{1000}$$

$$= 349 \text{ kips}$$

Therefore, **OK in one-way shear.**

4. **Design the reinforcement.** The critical section for moment and reinforcement anchorage is shown in Fig. 16–14 d. The ultimate moment is

$$M_u = 8.17\left[11.17 \times \frac{(58/12)^2}{2}\right] = 1066 \text{ ft-kips}$$

Assuming that $j = 0.9$, the area of steel required is

$$A_s = \frac{1066 \times 12,000}{0.9 \times 60,000(0.9 \times 28)} = 9.40 \text{ in.}^2$$

The average value of d was used in this calculation for simplicity. The same steel will be used in both directions.

$$\text{Minimum } A_s \text{ (ACI Secs. 10.5.3 and 7.12.2)} = 0.0018bh$$

$$= 0.0018 \times 134 \times 32 = 7.72 \text{ in.}^2$$

$$\text{(does not govern)}$$

$$\text{Maximum spacing (ACI Sec. 7.6.5)} = 18 \text{ in.}$$

Try 12 No. 8 bars each way; $A_s = 9.48$ in.^{2}. Recompute ϕM_n as a check.

$$a = \frac{9.48 \times 60,000}{0.85 \times 3000 \times 134} = 1.66 \text{ in.}$$

Since $a/d = 1.66/28 = 0.06$ is much less than a/d for the tension-controlled limit from Table A–4, $\phi = 0.90$.

$$\phi M_n = \phi A_s f_y\left(d - \frac{a}{2}\right)$$

$$= \frac{0.9 \times 9.48 \times 60,000(28 - 1.66/2)}{12,000} = 1159 \text{ ft-kips}$$

Since this exceeds M_u, we can reduce the amount of steel required. **Try 9 No. 9 bars, A_s = 9.0 in.^{2}.** $\phi M_n = 1102$ ft-kips—therefore OK.

   5.   Check the development. The clear spacing of the bars being developed exceeds $2d_b$ and the clear cover exceeds d_b. Therefore, this is case 2 development in Tables 8–1 and A–11. From Table A–11 ℓ_{db}/d_b for a No. 9 bottom bar in 3000-psi concrete is 54.8. The development length is

$$\ell_d = \frac{\ell_{db}}{d_b} d_b \beta \lambda$$

where $\beta = 1.0$ for uncoated reinforcement and $\lambda = 1.0$ for normal-weight concrete.

$$\ell_d = 54.8 \times 1.128 \times 1.0 \times 1.0$$

$$= 61.8 \text{ in.}$$

Bar extension past point of maximum moment = 58 in. − 3 in. = 55 in. This is less than 61.8 in.— therefore not OK. For a No. 8 bar, $\ell_d = 54.8$ in., which is just OK. **Use 12 No. 8 bars each way, A_s = 9.48 in.^{2}.**

   6.   Design the column–footing joint. The column–footing joint is shown in Fig. 16–9. The factored load at the base of the column is

$$1.4 \times 400 + 1.7 \times 270 = 1019 \text{ kips}$$

The allowable bearing on the footing (ACI Sec. 10.17.1) $= \phi(0.85 f'_c A_1)\sqrt{A_2/A_1}$, but not more than $\phi 1.7 f'_c A_1$, where A_1 is the contact area between the column and the footing and A_2 is the area determined as shown in Fig. 16–10. By inspection, $\sqrt{A_2/A_1}$ exceeds 2, and hence the maximum bearing load on the footing is

$$0.7 \times 0.85 \times 3 \times 18^2 \times 2 = 1157 \text{ kips}$$

The allowable bearing on the base of the column

$$\phi(0.85 f'_c A_1) = 0.7 \times 0.85 \times 5 \times 18^2$$

$$= 964 \text{ kips}$$

Thus the maximum load that can be transferred by bearing is 964 kips and dowels are needed to transfer the excess load

$$\text{Area of dowels required} = \frac{1019 - 964}{\phi f_y}$$

where $\phi = 0.7$ will be used. This is the ϕ value for columns and for bearing. Thus

$$\text{area of dowels} = \frac{1019 - 964}{0.7 \times 60}$$

$$= 1.31 \text{ in.}^2$$

The area of dowels must also satisfy ACI Sec. 15.8.2.1:

$$\text{Area of dowels} \geq 0.005 A_g = 1.62 \text{ in.}^2$$

Try 4 No. 6 dowels, dowel each corner bar. The dowels must extend into the footing a compression development length for a No. 6 bar in 3000-psi concrete. This is 16 in. The bars will be extended down to the level of the main footing steel and hooked 90°. The hooks will be tied to the main steel to hold the dowels in place. The dowels must extend into the column a distance equal to the greater of a compression splice for the dowels (23 in.) or the compression development length of the column bars (25 in.). **Use 4 No. 6 dowels, dowel each corner bar. Extend dowels 25 in. into column** (see Fig. 16–13).   ∎

Rectangular Footings

Rectangular footings may be used when there is inadequate clearance for a square footing. In such a footing, the reinforcement in the short direction is placed in the three bands shown in Fig. 16–15, with a closer bar spacing in the band under the column than in the two end bands. The band under the column has a width equal to the length of the short side of the footing, but not less than the width of the column if that is greater, and is centered on the column. Under long narrow columns it should not be less than the width of the column. The reinforcement in the band shall be $2/(\beta + 1)$ times the total reinforcement in the short direction, where β is the ratio of the long side of the footing to the short side (ACI Sec. 15.4.4). The reinforcement within each band is distributed evenly, as is the reinforcement in the long direction.

EXAMPLE 16–3   Design of a Rectangular Spread Footing

Redesign the footing from Example 16–2 assuming that the maximum width of the footing is limited to 9 ft. Steps 1 and 2 of Example 16–3 would proceed in the same sequence as in Example 16–2, leading to a footing 9 ft wide by 13 ft 8 in. long by 32 in. thick. The factored net soil pressure is 8.29 ksf.

1. Check the one-way shear. One-way shear may be critical in a rectangular footing and must be checked. The critical section and tributary area for one-way shear are shown in Fig. 16–16a.

$$V_u = 8.29\left(\frac{45}{12} \times 9\right) = 280 \text{ kips}$$

$$\phi V_c = 0.85 \times 2\sqrt{3000} \times 108 \times \frac{28}{1000} = 282 \text{ kips}$$

This is just OK in one-way shear.

2. Design the reinforcement in the long direction. The critical section for moment and reinforcement anchorage is shown in Fig. 16–16b. The ultimate moment is

$$M_u = 8.29\left(9 \times \frac{(73/12)^2}{2}\right) = 1380 \text{ ft-kips}$$

Assuming that $j = 0.9$ the area of steel required is

$$A_s = \frac{1380 \times 12,000}{0.9 \times 60,000(0.9 \times 28)} = 12.17 \text{ in.}^2$$

$$A_{s(\text{min})} = 0.0018 \times 108 \times 32 = 6.22 \text{ in.}^2 \text{ (does not govern)}$$

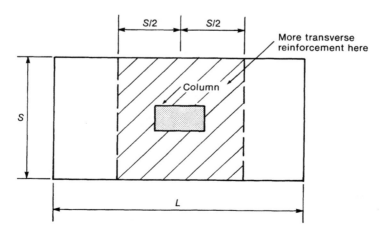

Fig. 16–15
Rectangular footing.

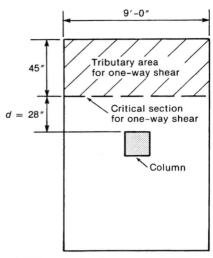

(a) Critical section for one-way shear.

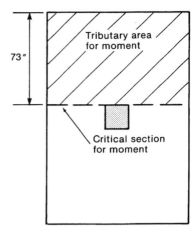

(b) Critical section for moment—
Long direction.

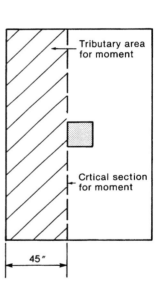

(c) Critical section for moment—
Short direction.

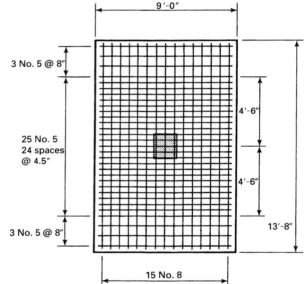

(d) Bar placement.

Fig. 16–16
Example 16–3.

Could use

$$16 \text{ No. 8 bars, } A_s = 12.64 \text{ in.}^2$$
$$13 \text{ No. 9 bars, } A_s = 13.00 \text{ in.}^2$$
$$10 \text{ No. 10 bars, } A_s = 12.70 \text{ in.}^2$$

Try 16 No. 8 bars $\phi M_n = 1514$ ft-kips; try 15 No. 8 bars, $\phi M_n = 1424$ ft-kips. Check development,

$$\ell_d = 54.8 \times 1.0 \times 1.0 \times 1.0 = 54.8 \text{ in.}$$

Length available = 70 in.—therefore OK. **Use 15 No. 8 bars in the long direction.**

3. Design the reinforcement in the short direction. The critical section for moment and reinforcement anchorage is shown in Fig. 16–16c.

$$M_u = 8.29\left(13.67 \times \frac{(45/12)^2}{2}\right) = 797 \text{ ft-kips}$$

Assuming that $j = 0.9$, $A_s = 7.03$ in.^{2},

$$A_{s(min)} = 0.0018 \times 164 \times 32 = 9.45 \text{ in.}^2 \text{ (this governs)}$$

Try 12 No. 8 bars, $A_s = 9.48$ in.^{2}.

Check development. $\ell_d = 54.8$ in. and the length available $= 28.5$ in.—therefore not OK. We must consider smaller bars. Try 31 No. 5 bars, $A_s = 9.61$ in.^{2}, $\ell_d = 27.4$ in.—therefore OK. Use 31 No. 5 bars in the short direction of the footing. The arrangement of the bars in the transverse direction (ACI Sec. 15.4.4.2):

$$\beta = \frac{\text{long side}}{\text{short side}} = 1.519$$

In the middle strip of width 9 ft, provide

$$\frac{2}{1.519 + 1} \times 31 \text{ bars} = 24.6 \text{ bars}$$

Provide 25 No. 5 bars in the middle strip and provide 3 No. 5 bars in each end strip. The final design is shown in Fig. 16–16d. ■

Footings Transferring Vertical Load and Moment

Very occasionally, footings must transmit both axial load and moment to the soil. The design of such a footing proceeds in the same manner as that for a square or rectangular footing, except for three things. First, a deeper slab will be necessary since there will be shear stresses developed by both direct shear and moment. The calculations concerning this are discussed in Chap. 13. Second, the soil pressures will be uneven, as discussed in Sec. 16–2 and shown in Fig. 16–5b. Third, the design for two-way shear must consider moment and shear (see Sec. 13–8).

The uneven soil pressures will lead to a tilting settlement of the footing, which will relieve some of the moment if the moment results from compatibility at the fixed end. The tilting can be reduced by offsetting the footing so that the column load acts through the center of the footing base area. If the moment is necessary for equilibrium, it will not be relieved by rotation of the foundation.

16–6 COMBINED FOOTINGS

Combined footings are used when it is necessary to support two columns on one footing, as shown in Fig. 16–1 or 16–17. When an exterior column is so close to a property line that a spread footing cannot be used, a combined footing is often used to support the edge column and an interior column.

The shape of the footing is chosen such that the centroid of the area in contact with soil coincides with the resultant of the column loads supported by the footing. Common shapes are shown in Fig. 16–17. For the rectangular footing in Fig. 16–17a, the distance from the exterior end to the resultant of the loads is half the length of the footing. If the interior column load is much larger than the exterior column load, a tapered footing may be used, Fig. 16–17b. The location of the centroid can be adjusted to agree with the resultant of the loads by dividing the area into a parallelogram or rectangle of area A_1, plus a triangle of area A_2, such that $A_1 + A_2 = $ required area, and $\bar{y}_1 A_1 + \bar{y}_2 A_2 = y_R A$.

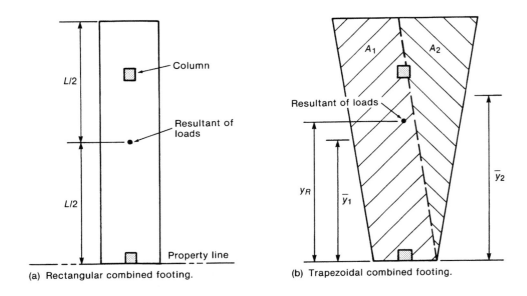

(a) Rectangular combined footing.

(b) Trapezoidal combined footing.

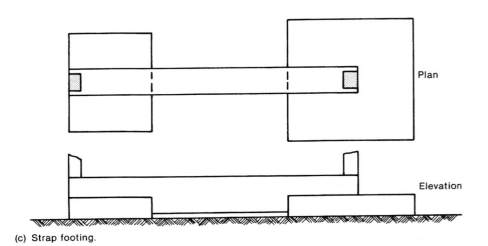

Fig. 16-17
Types of combined footings.

(c) Strap footing.

Sometimes a combined footing will be designed as two isolated pads joined by a *strap* or stiff beam as shown in Fig. 16–17c. Here the exterior footing acts as a wall footing, cantilevering out on the two sides of the strap. The interior footing can be designed as a two-way footing. The strap is designed as a beam and may require shear reinforcement in it.

For design, the structural action of a combined footing is idealized as shown in Fig. 16–18a. The soil pressure is assumed to act on longitudinal beam strips, *A–B–C* in Fig. 16–18a. These transmit the load to hypothetical cross beams, *A–D* and *B–E*, which transmit the upward soil reactions to the columns. For the column placement shown, the longitudinal beam strips would deflect as shown in Fig. 16–18b, requiring the reinforcement shown. The deflected shape and reinforcement of the cross beams are shown in Fig. 16–18c. The cross beams are generally assumed to extend a distance *d*/2 on each side of the columns.

Footings

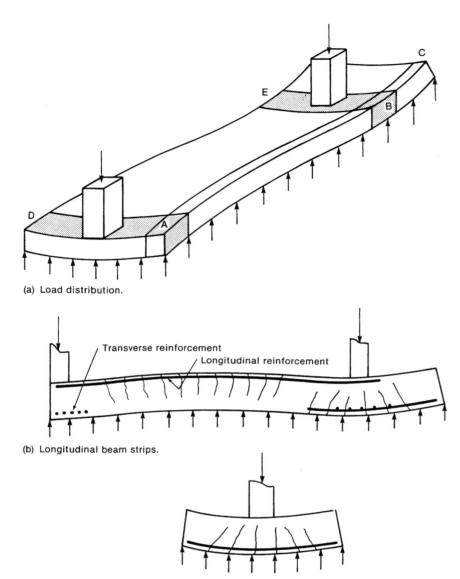

(a) Load distribution.

Transverse reinforcement

Longitudinal reinforcement

(b) Longitudinal beam strips.

Fig. 16–18
Structural action of combined
footing.

(c) Transverse beam strips.

EXAMPLE 16–4 Design of a Combined Footing

A combined footing supports a 24 in. × 16 in. exterior column carrying a service dead load of 200 kips and a service live load of 150 kips, and a 24-in.-square interior column carrying service loads of 300 kips dead load and 225 kips live load. The distance between the columns is 20 ft, center to center. The allowable soil bearing pressure is 5000 psf at a depth of 4 ft below the finished basement floor. The basement floor is 5 in. thick and supports a live load of 100 psf. The density of the fill above the footing is 120 lb/ft^3. Design the footing, assuming that $f_c' = 3000$ psi and $f_y = 60,000$ psi.

1. Estimate the size and the factored net pressure. Allowable net soil pressure, $q_n = 5$ ksf − (weight/ft^2 of the footing and the soil, the floor over the footing, and the basement floor loading). This can be calculated, as in the previous examples, using the actual densities and thicknesses of the soil and concrete, or can be approximated using an average density for the soil and concrete. This will be taken as 140 lb/ft^3. Therefore,

$$q_n = 5.0 - (4 \times 0.14 + 0.10) = 4.34 \text{ ksf}$$

$$\text{area required} = \frac{200 + 150 + 300 + 225}{4.34} = 201.6 \text{ ft}^2$$

The resultant of the column loads is located at

$$\frac{8 \text{ in.} \times 350 \text{ kips} + 248 \text{ in.} \times 525 \text{ kips}}{350 \text{ kips} + 525 \text{ kips}} = 152 \text{ in.}$$

from the exterior face of the exterior column (Fig. 16–19a). To achieve uniform soil pressures, the centroid of the footing area will be located at 152 in. from exterior edge. Thus the footing will be 304 in. = 25 ft 4 in. long. The width of the footing will be 7.96 ft, say 8 ft.

The factored net pressure, q_{nu}, is

$$q_{nu} = \frac{1.4(200 + 300) + 1.7(150 + 225)}{25.33 \times 8} = 6.6 \text{ ksf}$$

2. Determine the bending moment and shear force diagrams for the longitudinal action. The factored loads on the footing and the corresponding bending moment and shearing force diagrams for the footing are shown in Fig. 16–19. These are plotted for the full 8-ft width of the footing. The design must satisfy flexure and one-way shear in the longitudinal direction, flexure in the transverse direction, and punching shear at each of the columns.

3. Determine the thickness required for maximum positive moment. Because the cross section is so massive (8 ft wide by 2 to 3 ft deep), using more than about 0.5% reinforcement will lead to very large numbers of large bars. As a first trial we shall select the depth, assuming 0.5% reinforcement. Since this member acts as a beam, the minimum flexural reinforcement ratio from ACI Sec. 10.5.1 is $3\sqrt{f_c'}/f_y \geq 200/f_y = 0.0033$. From Eq. 4–17,

$$\frac{M_u}{\phi k_n} = \frac{bd^2}{12,000}$$

where $k_n = f_c'\omega(1 - 0.59\omega)$, $\omega = \rho f_y/f_c'$, and M_u is in ft-kips. For $\rho = 0.005$, $\phi k_n = 254$ (Table A–3):

$$\frac{bd^2}{12,000} = \frac{2354}{254} = 9.27$$

For $b = 96$ in., $d = 34$ in. **As a first trial, we shall assume an overall thickness of 36 in. with $d = 32.5$ in.**

4. Check the two-way shear at the interior column. The critical perimeter is a square with sides $24 + 32.5 = 56.5$ in. long, giving $b_o = 4 \times 56.5 = 226$ in. The shear, V_u, is the column load minus the force due to soil pressure on the area within the critical perimeter.

$$V_u = 802.5 - 6.6\left(\frac{56.5}{12}\right)^2 = 656 \text{ kips}$$

ϕV_c is the smallest of

(a) $\quad \phi V_c = \phi\left(2 + \dfrac{4}{\beta_c}\right)\sqrt{f_c'}b_o d$ where $\beta_c = 1.0$

$$= 0.85\left(2 + \frac{4}{1}\right)\sqrt{3000} \times 226 \times \frac{32.5}{1000} = 2052 \text{ kips}$$

(b) $\quad \phi V_c = \phi\left(\dfrac{\alpha_s d}{b_o} + 2\right)\sqrt{f_c'}b_o d$

$$= 0.85\left(\frac{40 \times 32.5}{226} + 2\right)\sqrt{3000} \times 226 \times \frac{32.5}{1000} = 2651 \text{ kips}$$

(c) $\quad \phi V_c = 4\sqrt{f_c'}b_o d$

$$= 0.85 \times 4\sqrt{3000} \times 226 \times \frac{32.5}{1000} = 1368 \text{ kips}$$

Therefore, the depth of the footing is more than adequate for two-way shear at this column. (Further calculations would show that $d = 22$ in. would be adequate.)

Footings

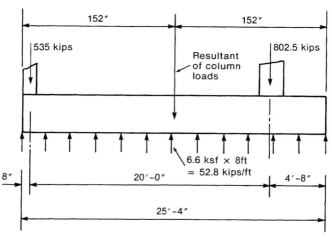

(a) Freebody diagram.

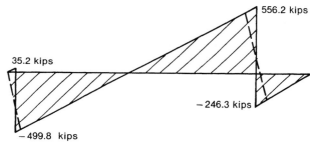

(b) Shear force diagram.

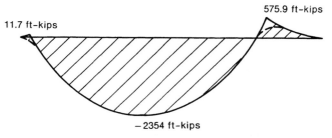

Fig. 16–19
Bending moment and shearing force diagrams—
Example 16–4.

(c) Bending moment diagram.

5. Check the two-way shear at the exterior column. The shear perimeter around the exterior column is three-sided, as shown in Fig. 16–20a. The distance from line A–B to the centroid of the shear perimeter is

$$\frac{2(32.5 \times 32.25) \times 32.25/2}{32.5(2 \times 32.25 + 56.5)} = 8.60 \text{ in.} \tag{13–26}$$

The force due to soil pressure on the area within the critical perimeter is

$$6.6\left(32.25 \times \frac{56.5}{144}\right) = 83.5 \text{ kips}$$

A free-body diagram of the critical perimeter is shown in Fig. 16–20b. Summing moments about the centroid of the shear perimeter gives

16–6 Combined Footings

731

$$M_u = 535 \text{ kips} \times 15.65 \text{ in.} - 83.5 \text{ kips} \times 7.53 \text{ in.}$$

$$= 7744 \text{ in.-kips}$$

This moment must be transferred to the footing by shear stresses and flexure, as explained in Chap. 13. The moment of inertia of the shear perimeter is

$$J_c = 2\left[\left(32.25 \times \frac{32.5^3}{12}\right) + \left(32.5 \times \frac{32.25^3}{12}\right)\right.$$

$$\left. + (32.25 \times 32.5)(16.125 - 8.60)^2\right] + (56.5 \times 32.5)8.60^2$$

$$= 620,700 \text{ in.}^4$$

The fraction of moment transferred by flexure is

$$\gamma_f = \frac{1}{1 + 2\sqrt{b_1/b_2}/3} \tag{13-22}$$

$$\frac{1}{1 + 2\sqrt{32.25/56.5}/3} = 0.665$$

ACI Sec. 13.5.3.3 allows an adjustment in γ_f at edge columns in two-way slab structures. We shall not make this adjustment since the structural action of this footing is quite different from that of a two-way slab.

The fraction transferred by shear is $\gamma_v = 1 - \gamma_f = 0.335$.

The shear stresses due to the direct shear and the shear due to moment transfer will add at points C and D in Fig. 16–20, giving the largest shear stresses on the critical shear perimeter.

$$v_u = \frac{V_u}{b_o d} + \frac{\gamma_v M_u c}{J_c}$$

$$= \frac{535 - 83.5}{(2 \times 32.25 + 56.5) \times 32.5} + \frac{0.335 \times 7744(32.25 - 8.60)}{620,700}$$

$$= 0.115 + 0.099$$

$$= 0.214 \text{ ksi}$$

ϕv_c is computed as $\phi V_c / b_o d$, where ϕV_c is the smallest value from ACI Eqs. 11–35 to 11–37.

(a) $\quad \phi v_c = 0.85\left(2 + \frac{4}{(24/16)}\right)\sqrt{3000} = 217 \text{ psi}$

(b) $\quad \phi v_c = 0.85 \dfrac{30 \times 32.5}{121} = 468 \text{ psi}$

(c) $\quad \phi v_c = 0.85 \times 4\sqrt{3000} = 186 \text{ psi}$

Thus $\phi v_c = 186$ psi or 0.186 ksi and the thickness is not adequate for shear at the exterior column. In this case the direct shear stress and the maximum shear stress due to moment are approximately equal. Additional calculations show that an overall depth of 40 in. with $d = 36.5$ in. is required for punching shear at the exterior column. For the final choice, the total moment transferred to the footing is 8417 in.-kips of which 5202 in.-kips are transferred by flexure. **Use a combined footing 25 ft 4 in. by 8 ft in plan, 3 ft 4 in. thick, with effective depth 36.5 in.**

6. **Check the one-way shear.** One-way shear is critical at d from the face of the interior column:

$$V_u = 556.2 \text{ kips} - \left(\frac{12 + 36.5}{12}\right)\text{ft} \times 52.8 \text{ kips/ft} = 343 \text{ kips}$$

$$\phi V_c = 0.85 \times 2\sqrt{3000} \times 96 \times \frac{36.5}{1000} = 326 \text{ kips}$$

Since $\phi V_c < V_u$, we have two alternatives. These are to increase the effective depth to 38.5 in. so that $\phi V_c = V_u$, or to provide stirrups. Since this member is vital to the integrity of the entire structure, we shall do the latter and provide stirrups to the point in the structure where $V_u = \phi V_c/2$.

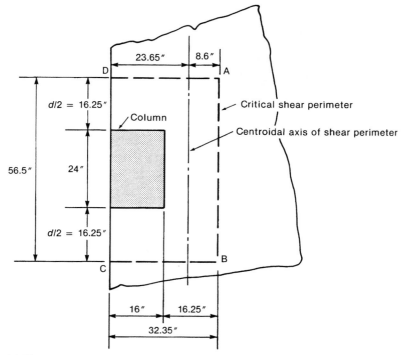

(a) Plan.

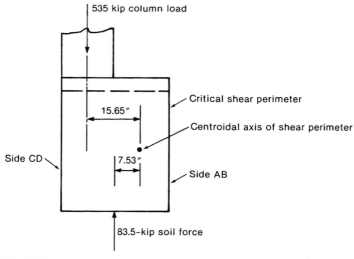

(b) Freebody of joint area

Fig. 16–20
Shear calculations at exterior
column—Example 16–4.

$$\phi V_s = V_u - \phi V_c = 16.5 \text{ kips}$$

$$\phi V_s = \frac{\phi A_v f_y d}{s}$$

The maximum $s = d/2$, which gives 18 in. as a reasonable spacing. For $s = 18$ in., the required A_v is

$$A_v = \frac{\phi V_s s}{\phi f_y d} = \frac{16.5 \times 18}{0.85 \times 60 \times 36.5}$$

$$= 0.16 \text{ in.}^2$$

16–6 Combined Footings   733

Because of the extreme width of this element, multileg stirrups are required to provide anchorage for the compression diagonals in the truss model. Use No. 3 Grade 60, six-leg stirrups, giving $A_v = 0.66$ in.^{2}. Extend these until $V_u < \phi 0.5 V_c$ (ACI Sec. 11.5.5.1). This occurs at 7.4 ft from the face of the column. **Use 5 sets of No. 3 six-leg stirrups, 1 at 9 in. and the rest at 18 in. at each end of the span between the columns.**

7. **Design the flexural reinforcement.**

(a) **Midspan (negative moment):**

$$A_s = \frac{M_u \times 12,000}{\phi f_y j d}$$

Estimate $j = 0.95$ since ρ will be very small.

$$A_s = \frac{2354 \times 12,000}{0.9 \times 60,000 \times 0.95 \times 36.5} = 15.09 \text{ in.}^2$$

Min. $A_s = \dfrac{200bd}{f_y}$

$$= \frac{200 \times 96 \times 36.5}{60,000}$$

$$= 11.68 \text{ in.}^2 \text{ (does not govern)}$$

Try 19 No. 8 bars, $A_s = 15.01$ in.^{2}.

$$\phi M_n = \frac{0.9 \times 15.01 \times 60,000(36.5 - 3.68/2)}{12,000} = 2341 \text{ ft-kips}$$

Since this is less than 1% under M_u, we shall accept it. **Use 19 No. 8 top bars at midspan.**

(b) **Interior column (positive moment).** The positive moment at the interior support requires that $A_s = 3.7$ in.^{2}, which is less than $A_{s,min}$. **Use 15 No. 8 bottom bars, $A_s = 11.85$ in.^{2} at the interior column.**

8. **Check the development of the top bars.** ℓ_d for a No. 8 top bar in 3000-psi concrete (Table A–11):

$$\ell_d = 71.2 \times 1.00 \times 1.0 \times 1.0 = 71.2 \text{ in.} = 5.93 \text{ ft}$$

We shall extend all the bars into the column regions at both ends. At points of inflection (1.09 ft from the center of the interior column and under the exterior column), $V_u = 499$ kips, and by ACI Sec. 12.11.3,

$$\frac{M_n}{V_u} + \ell_a \geq \ell_d \qquad\qquad (8\text{--}20)$$

$$(\text{ACI Eq. } 12\text{--}2)$$

$$\frac{M_n}{V_u} = \frac{2340/0.9}{499} = 5.21 \text{ ft}$$

where ℓ_a must be at least $5.93 - 5.21 = 0.72$ ft to satisfy Eq. 8–20.

At the exterior support, ℓ_a is the extension beyond the center of the support. This will not exceed 5 in., which is not enough to satisfy Eq. 8–20. Therefore, the top bars will all have to be hooked at the exterior end. This is also necessary to anchor the bars transferring the unbalanced moment from the column to the footing.

At the interior column, extend the bars to the interior face of the column. This will give ℓ_a in excess of 0.72 ft.

9. **Check the development of the bottom bars.** ℓ_d for a No. 8 bottom bar in 3000-psi concrete is

$$\ell_d = 54.8 \times 1.00 \times 1.0 \times 1.0 = 54.8 \text{ in.}$$

Extend the bottom bars to 3 in. from the exterior end and d past the point of inflection. Cut off the bottom bars at 4 ft 5 in. from the centerline of the interior column toward the exterior column.

10. Design the transverse "beams." Transverse strips under each column will be assumed to transmit the load from the longitudinal beam strips into the column as shown in Fig. 16–18a. The width of the beam strips will be assumed to extend $d/2$ on each side of the column. The actual width is unimportant since the moments to be transferred are independent of the width of the transverse beams. Figure 16–21 shows a section through the transverse beam under the interior column. The factored load on this column is 802.5 kips. This is balanced by an upward net force of 802.5 kips/8 ft = 100.3 kips/ft. The maximum moment in this transverse beam is

$$M_u = \frac{100.3 \times 3^2}{2} = 451 \text{ ft-kips}$$

Assuming that $j = 0.95$ and $d = 35.5$ in., the required A_s is

$$A_s = \frac{451 \times 12,000}{0.9 \times 60,000 \times 0.95 \times 35.5} = 2.97 \text{ in.}^2$$

$$A_{s(min)} = \frac{200}{60,000} \times 60.5 \times 35.5 = 7.16 \text{ in.}^2 \text{ (this governs)}$$

Use 9 No. 8 transverse bottom bars at the interior column, $A_s = 7.11$ in.². $A_{s(min)}$ also controls at the exterior column. **Use 6 No. 8 transverse bottom bars at the exterior column.** Because two-way shear cracks would extend roughly the entire width of the footing, we shall hook the transverse bars at both ends for adequate anchorage outside the inclined cracks.

11. Design the column to footing dowels. This is similar to step 6 in Example 16–2 and will not be repeated here. The complete design is detailed in Fig. 16–22. ∎

16–7 MAT FOUNDATIONS

A mat foundation supports all the columns in a building, as shown in Fig. 16–1g. They are used when buildings are founded on soft or irregular soils in locations where pile foundations cannot be used. Design is carried out assuming that the foundation acts as an inverted slab. The distribution of soil pressure is affected by the relative stiffness of the soil and foundation, with more pressure being developed under the columns than at points between columns. Detailed recommendations for the design of such foundations are given in Refs. 16–5 and 16–6.

16–8 PILE CAPS

Piles may be used when the surface soil layers are too soft to properly support the loads from the structure. The pile loads are either transmitted to a stiff bearing layer some distance below the surface, or are transmitted to the soil by friction along the length of the pile. Treated timber piles have capacities of up to about 30 tons. Precast concrete and steel piles

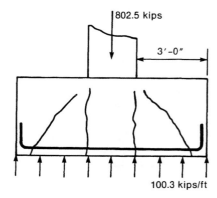

Fig. 16–21
Transverse beam at interior
column—Example 16–4.

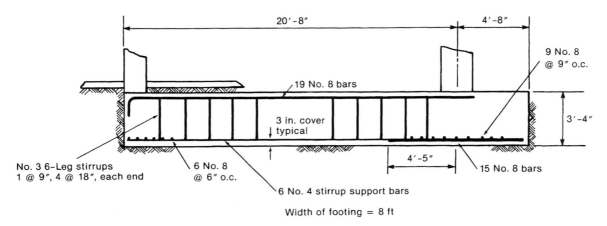

Fig. 16–22
Combined footing—Example 16–4.

may have capacities from 40 to 200 tons. The most common pile caps for high-strength piles comprise groups of 2 to 6 piles, although groups of 25 to 30 piles have been used. The center-to-center pile spacing is a function of the type and capacity of the piles, but frequently is on the order of 3 ft. It is not uncommon for a pile to be several inches to a foot away from its intended location due to problems during pile driving.

The structural action of a four-pile group is shown schematically in Fig. 16–23. The pile cap is a special case of a "deep beam" and can be idealized as a three-dimensional truss or strut-and-tie model, with four compression struts transferring load from the column to the tops of the piles, and four tension ties equilibrating the outward components of the compression thrusts (Fig. 16–23b). The tension ties have constant force in them and must be anchored for the full horizontal tie force outside the intersection of the pile and the compression strut (outside points A and B in Fig. 16–23c). Hence the bars must either extend ℓ_d past the centerlines of the piles, or they must be hooked outside this point. Tests of model pile caps are reported in Ref. 16–7. Strut-and-tie models are discussed in Chap. 18.

The modes of failure (limit states) for such a pile cap include: (1) crushing under the column or over the pile, (2) bursting of the side cover where the pile transfers its load to the pile cap, (3) yielding of the tension tie, (4) anchorage failure of the tension tie, (5) a two-way shear failure where the cone of material inside the piles punches downward, or (6) failure of the compression struts. Least understood of these is the two-way shear failure mode. ACI Sec. 11.12 is not strictly applicable, since the failure cone differs from the one assumed in the Code. Guidance in selecting allowable shear stresses is given in Ref. 16–8. This reference is based on analyses rather than tests, however. It should be noted that Ref. 16–8 gives inadequate attention to the anchorage of the horizontal bars. ACI Sec. 15.5.3 applies to pile caps. The minimum thickness is sometimes governed by the development length of the dowels from the pile cap into the column.

For the pile cap shown in Fig. 16–23, the total horizontal tie force in one direction can be calculated from the force triangle shown in Fig. 16–23b. The factored downward force is 1240 kips equilibrated by vertical forces of 310 kips in each pile. Considering the joint at A shown in Fig. 16–23c we find that the horizontal steel force in tie A–B required to equilibrate the force system is 207 kips, corresponding to $\phi A_s f_y$, which taking $\phi = 0.85$ for shear, requires 7 No. 7 bars for tie A–B, and a similar number for each of the ties B–C, C–D, and A–D. The use of trusses in design is discussed in Chap. 18.

Reference 16–8 recommends an edge distance of 15 in. measured from the center of the pile for piles up to 100-ton capacity, and 21 in. for higher-capacity piles, to prevent bursting of the side cover over the piles.

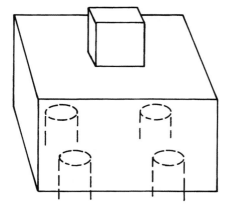

(a) Pile cap.

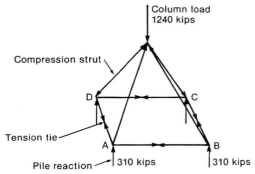

Column load
1240 kips

Compression strut

D   C

Tension tie

A   B

Pile reaction   310 kips   310 kips

(b) Internal forces in pile cap.

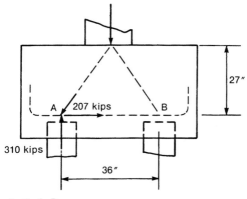

27"

A   207 kips   B

310 kips

36"

(c) Force in tie A–B.

Fig. 16–23
Forces in a pile cap.

PROBLEMS

Design wall footings for the following conditions.

16–1 Service dead load = 6 kips/ft, service live load = 8 kips/ft. Wall is 12 in. thick. Allowable soil pressure, q_a = 4000 psf at 3 ft below the final grade. f_c' = 2500 psi and f_y = 60,000 psi.

16–2 Service dead load = 18 kips/ft, service live load = 8 kips/ft. Wall is 16 in. thick. Allowable soil pressure, q_a = 6000 psf at ground surface. f_c' = 3000 psi and f_y = 60,000 psi.

Design square spread footings for the following conditions.

16–3 Service dead load = 350 kips, service live load = 275 kips. Soil density = 130 lb/ft^3.

Allowable soil pressure = 4500 psf at 5 ft below the basement floor. Column is 18 in. square. f_c' = 3000 psi and f_y = 60,000 psi. Place bottom of footing at 5 ft below floor level.

16–4 Service dead load = 400 kips, service live load = 250 kips. Soil density 120 lb/ft^3. Allowable soil pressure = 6000 psf. Column cross section is 30 in. × 12 in. f_c' = 3000 psi and f_y = 60,000 psi. Select the elevation of the top of the footings so that there is 6 in. of soil and a 6 in. slab on grade above the footing.

17

Shear Friction, Horizontal Shear Transfer, and Composite Concrete Beams

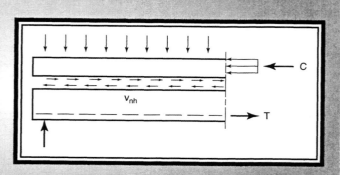

17-1   INTRODUCTION

This chapter considers the shear strength of interfaces between members or parts of members that can slip relative to one another. Among other cases, this includes the interface between a beam and a slab cast later than the beam, but expected to act in a composite manner.

17-2   SHEAR FRICTION

From time to time shear must be transferred across an interface between two members that can slip relative to one another. The shear carrying mechanism is known variously as *aggregate interlock, interface shear transfer,* or *shear friction.* The latter term is used here.

Behavior in Tests

Extensive tests have been carried out by Mattock [17-1,17-2] and others, [17-3] using specimens of the general configuration shown in Fig. 17–1. The specimens were either intentionally cracked along the dashed line in Fig. 17–1 prior to testing, or were uncracked. Typical test results are shown in Fig. 17–2, where the shear stress at failure, v_u, is plotted against ρf_y, where ρ is the ratio of the area of the transverse reinforcement across the shear surface to the area of the shear surface. Considerably higher strenghs were attained for uncracked sections than for cracked sections. For high values of ρf_y both curves reached an upper limit on v_u.

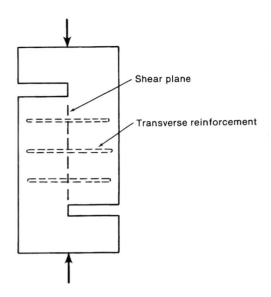

Fig. 17–1
Shear transfer specimen.

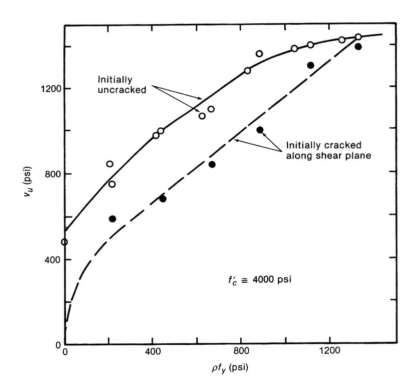

Fig. 17–2
Variation of shear strength
with reinforcement ratio, ρf_y.
(From Ref. 17–1).

In the uncracked specimens, a series of diagonal tension cracks occurred across the shear plane, as shown in Fig. 17–3. With further shear deformation, the compression struts between these cracks rotated at their ends (points A and B) such that point B moved downward relative to A. At the same time, the distance AC measured across the crack increased, stretching the transverse reinforcement. The tension in the reinforcement was equilibrated by an increase in the compression in the struts. Failure occurred when the bars yielded or the struts crushed.

Shear Friction, Horizontal Shear Transfer, and Composite Concrete Beams

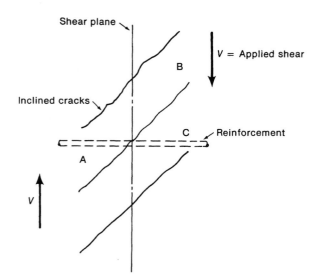

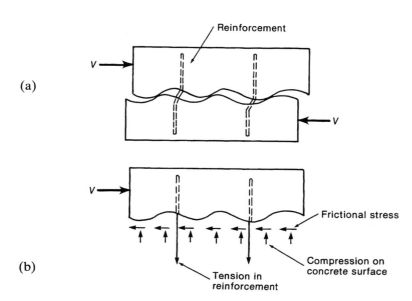

Fig. 17–3
Diagonal tension cracking along previously uncracked shear plane. (From Ref. 17–4.)

(a)

(b)

Fig. 17–4
Shear friction analogy. (From Ref. 17–4.)

When a shear is applied to an initially cracked surface, or a surface formed by placing one layer of concrete on top of an existing layer of hardened concrete, relative slip of the layers causes a separation of the surfaces as shown in Fig. 17–4a. If there is reinforcement across the crack, it is elongated by the separation of the surfaces and hence is stressed in tension. For equilibrium a compressive stress is needed as shown in Fig. 17–4b. Shear is transmitted across the crack by (1) friction resulting from the compressive stresses [17-5], and (2) by interlock of aggregate protrusions on the cracked surfaces combined with dowel action of the reinforcement crossing the surface. If the transverse reinforcement is perpendicular to the shear plane, Mattock [17-2] has suggested that the shear strength of a precracked surface is

$$V_n = 0.8A_{vf}f_y + A_cK_1 \qquad (17\text{--}1)$$

where A_{vf} is the area of reinforcement crossing the surface, A_c is the area of the concrete surface resisting shear friction, and $K_1 = 400$ psi for normal-weight concrete, 200 psi for "all-lightweight" concrete, and 250 psi for "sand-light weight" concrete.

The first term in Eq. 17–1 represents the friction, the coefficient of friction taken as 0.8 for concrete sliding on concrete. The second term represents the shear transferred by shearing off surface protrusions and by dowel action. Equation 17–1 is plotted in Fig. 17–5 for comparison with the measured strengths. Since this equation applies only for $\rho f_y = A_{vf} f_y / A_c$ greater than 200 psi, the curve is not plotted below this value. For Grade 60 reinforcement, this limit requires a minimum reinforcement ratio, $\rho = 0.0033$.

Design Rules in the ACI Code

ACI Sec. 11.7 presents design rules for cases "where it is appropriate to consider shear transfer across a given plane, such as an existing or a potential crack, an interface between dissimilar materials, or an interface between two concretes cast at different times." Typical examples are shown in Fig. 17–6.

In design, a crack is assumed along the shear plane and reinforcement is provided across that crack. The amount of reinforcement is computed using (ACI Sec. 11.7.4)

$$\phi V_n \geq V_u \qquad\qquad (6\text{–}14)$$
$$\text{(ACI Eq. 11–1)}$$

$$V_n = A_{vf} f_y \mu \qquad\qquad (17\text{–}2)$$
$$\text{(ACI Eq. 11–25)}$$

where $\phi = 0.85$ for shear and μ is the coefficient of friction taken equal to:

1. Concrete placed monolithically: 1.4λ

2. Concrete placed against hardened concrete with the surface intentionally roughened to a full amplitude of approximately $\frac{1}{4}$ in.: 1.0λ

3. Concrete placed against hardened concrete not intentionally roughened: 0.6λ

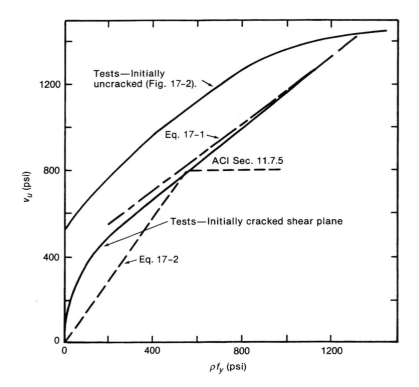

Fig. 17–5
Comparison of test results and design equations.

Shear Friction, Horizontal Shear Transfer, and Composite Concrete Beams

4. Concrete anchored to as-rolled structural steel by headed studs or reinforcing bars: 0.7λ

where $\lambda = 1.0$ for normal-weight concrete, 0.85 for "sand-lightweight" concrete, and 0.75 for "all-lightweight" concrete. In cases 2 and 3 the surface must be clean and free of laitance (a weak layer on the top surface of a pour due to bleed water collecting at the surface). ACI Sec. 11.7.10 requires that in case 4, the steel must be clean and free of paint. Case 2 specifies a "full amplitude" (wave height) of approximately $\frac{1}{4}$ in. but does not specify a "wave length" for the roughened surface. This was done to allow some freedom in satisfying this requirement. It was intended, however, that the wave length be on the same order of magnitude as the full amplitude, say $\frac{1}{4}$ to $\frac{3}{4}$ in.

The values of $\mu = 1.4$ and 1.0 are larger than the true coefficient of friction, which is given by Eq. 17–1 as about 0.8. The reason for this, as shown in Fig. 17–5, is the use of a friction model for a more complex response.

ACI Sec. 11.7.5 sets the upper limit on V_n from Eq. 17–2 as $0.2f_c'A_c$ or $800A_c$ pounds, whichever is smaller. Equation 17–2 and its upper limits are plotted in Fig. 17–5 for comparison with the test data. For low amounts of ρf_y, Eq. 17–2 is conservative.

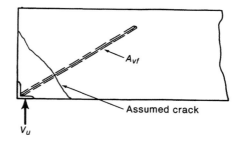

(a) Precast beam bearing.

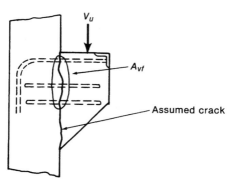

(b) Corbel

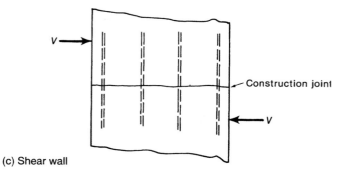

(c) Shear wall

Fig. 17–6
Examples of shear friction.
(From Ref. 17–6.)

17–2 Shear Friction

Because the reinforcement is assumed to yield in order to develop the necessary forces, the yield strength of the steel is limited to 60,000 psi. Each bar must be anchored on both sides of the crack to develop the bar. The steel must be placed approximately uniformly across the shear plane so that all parts of the crack are clamped together.

If tensile forces, N_u, act across the shear plane, they must be equilibrated by reinforcement, A_n, in addition to the shear friction reinforcement:

$$A_n = \frac{N_u}{\phi f_y} \tag{17-3}$$

Permanent net compressive forces across the crack, C_u, can be considered as directly additive to the force $A_{vf} f_y$ in Eq. 17-2:

$$V_n = (A_{vf} f_y + C_u)\mu \tag{17-4}$$

When shear friction reinforcement is inclined to the shear plane, the shear strength becomes

$$V_n = A_{vf} f_y (\mu \sin \alpha_f + \cos \alpha_f) \tag{17-5}$$
$$\text{(ACI Eq. 11-26)}$$

where α_f is the angle between the assumed crack plane and the shear friction reinforcement, as illustrated in Fig. 17-7. The origin of the two terms in parentheses in Eq. 17-5 is also illustrated. Only the normal component is multiplied by μ since it causes friction. Only those bars that are stressed in tension by the sliding motion (as is the case in Fig. 17-7) can be included in A_{vf}. Bars that are inclined so they are stressed in compression tend to kink and force the crack surfaces apart, reducing the shear friction.

Finally, it should be noted that in unusual cases, ACI Sec. 11.7.3 allows the use of Eq. 17-1 to compute V_n. This can be used if ρf_y exceeds 200 psi.

EXAMPLE 17-1   Design of the Reinforcement in the Bearing Region of a Precast Beam

Figure 17-8 shows the support region of a precast concrete beam. The factored beam reactions are 70 kips vertical force and a horizontal tension force of 15 kips. The horizontal force arises from restraint of shrinkage of the precast beam. Use $f_y = 60,000$ psi and normal-weight concrete.

1.   **Assume the cracked plane.** One possible assumed crack is shown in Fig. 17-1. It is necessary to provide enough reinforcement across this plane to prevent possible sliding.

2.   **Compute the area of steel required.** From Eq. 17-2,

$$V_n = A_{vf} f_y \mu$$

where $v_n = \phi V_n$. Therefore,

$$A_{vf} = \frac{V_u}{\phi f_y \mu}$$

where $V_u = 70$ kips and $\mu = 1.4\lambda$ since the crack plane is in monolithically placed concrete, and $\lambda = 1.0$ since it is normal-weight concrete.

$$A_{vf} = \frac{70}{0.85 \times 60 \times 1.4} = 0.98 \text{ in.}^2$$

The tensile force must also be transferred across the crack by reinforcement (Eq. 17-3):

$$A_n = \frac{15}{0.85 \times 60} = 0.29 \text{ in.}^2$$

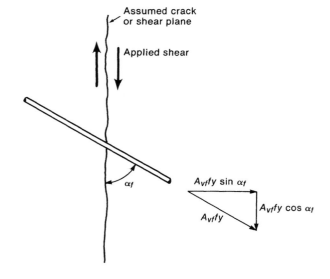

Fig. 17–7
Force components in a bar inclined to the shear plane.
(From Ref. 17–6.)

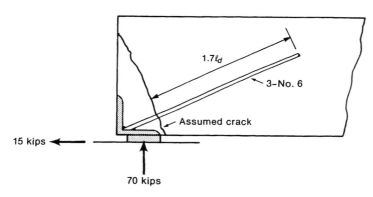

Fig. 17–8
Example 17–1.

Therefore, the total steel across the crack must be $0.98 + 0.29$ in.^{2} $= 1.27$ in.^{2}

Provide 3 No. 6 bars across the assumed crack. These must be anchored on both sides of the crack. This is done by welding them to the bearing angle and by extending them $1.7\ell_d$, as recommended in Ref. 17–7. ∎

17–3   COMPOSITE CONCRETE BEAMS

Frequently, precast beams or steel beams have a slab cast on top of them and are designed assuming that the slab and beam act as a monolithic unit to support loads. Such a beam and slab combination is referred to as a *composite beam*. This discussion will deal only with composite beams where the beam is precast concrete or other concrete cast at an earlier time.

Shored or Unshored Construction

When the slab concrete is placed, the precast beam can either be shored or unshored. If shored with the shores supporting the dead load of the precast beam, the dead load of the beam and slab is initially supported by the shores. When the strength of the concrete is high enough to resist the forces induced, the shores are removed and the dead load is resisted by the composite beam and slab. If the precast beam is not shored when the slab is placed, it supports its own weight plus the weight of the slab and slab forms. ACI Sec. 17.2 allows either construction process and requires that each element be strong enough to support all loads which it supports by itself. If the beam is shored, ACI Sec. 17.3 requires that the shores be left in place until the composite section has adequate strength to support all loads and to limit deflections and cracking.

Tests have shown that the ultimate strength of a composite beam is the same whether the member was shored or unshored during construction. For this reason, ACI Sec. 17.2.4 allows strength computations to be made considering only the final member.

Horizontal Shear

In the beam shown in Fig. 17–9a, there are no horizontal shear stresses transferred from the slab to the beam. It acts as two independent members. In Fig. 17–9b, horizontal shear stresses act on the interface and the slab and beam act in a composite manner. The ACI Code provisions for horizontal shear are given in ACI Sec. 17–5. Although the mechanism of horizontal shear transfer and that of shear friction are similar, if not identical, there is a considerable difference between the two sets of provisions, as shown in Fig. 17–10. The difference results from the fact that Eq. 17–1, ACI Sec. 17.5.2.3, and ACI Sec. 11.7 are all empirical attempts to fit test data.[17–1 to 17–3, 17–8 to 17–10] Equation 17–1 is valid for relatively short shear planes with lengths up to several feet, but is believed to give too high shear strengths for long shear transfer regions where the maximum stress is localized. It is also unconservative for low values of $\rho_v f_y$.

The tests reported in Ref. 17–3 included members with and without shear keys along the interface. The presence of shear keys stiffened the connection at low slips but had no significant effect on its strength.

From strength of materials, the horizontal shear stresses, v_h, on the contact surface between an uncracked elastic precast beam and a slab can be computed from

$$v_h = \frac{VQ}{I_c b_v} \tag{17–6}$$

where

$V =$ shear force acting on the section in question

$Q =$ first moment of the area of the slab or flange about the neutral axis of the composite section

$I_c =$ moment of inertia of the composite section

$b_v =$ width of the interface between the precast beam and the cast-in-place slab

Equation 17–6 applies to uncracked elastic beams and is only an approximation for cracked concrete beams. The ACI Code gives two ways of calculating the horizontal shear stress.

ACI Sec. 17.5.2 defines the horizontal shear force, V_{nh}, to be transferred as

$$\phi V_{nh} \geq V_u \tag{17–7}$$
$$\text{(ACI Eq. 17–1)}$$

This gives

$$v_{nh} = \frac{V_u / \phi}{b_v d} \tag{17–8}$$

This is based on the observation that in an element directly over the beam web, $v_{nh} = v_n$ and $v_n = V_n / b_v d$.

Alternatively, ACI Sec. 17.5.3 allows horizontal shear to be computed from the change in compressive or tensile force in the slab in any segment of its length.

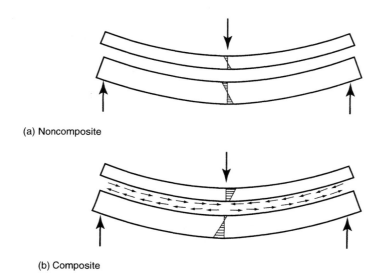

Fig. 17–9
Horizontal shear transfer in a
composite beam.

(a) Noncomposite

(b) Composite

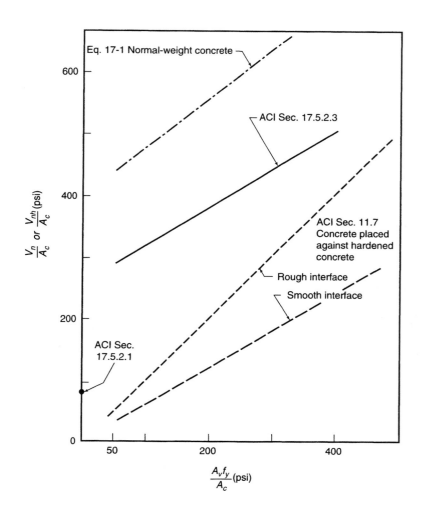

Fig. 17–10
Comparison of ACI Sec.
11.7, shear friction; ACI Sec.
17.5, horizontal shear
strength; and Eq. 17–1.

Figure 17–11 illustrates this clause. At midspan, the force in the compression zone is C as shown in Fig. 17–11a. All of this force acts above the interface. At the end of the beam, the force in the flange is zero. Thus the horizontal shear force to be transferred across the interface between midspan and the support is

$$V_{nh} = C \qquad (17\text{–}9)$$

A similar derivation could be made if the flange were in tension. ACI Sec. 17.5.3.1 says that when ties are provided to resist the horizontal shear calculated using Eq. 17–9, their distribution should approximately reflect the distribution of shear forces in the member. This implies that the horizontal shear stresses should be calculated from

$$v_{nh} = \frac{K V_{nh}}{A_c} \qquad (17\text{–}10)$$

where A_c is the contact area, and K is a factor to account for the distribution of the shear forces along the member. K is equal to the shear at a point divided by the average shear. For the beam in Fig. 17–11, K would vary from 2 at the end of the beam to zero at midspan. Thus the distribution of the horizontal shear stresses from Eqs. 17–9 and 17–10 would be as shown in Fig. 17–11b. For a constant shear force, K would be 1.

The two procedures give similar results, as will be seen in Example 17–2. The limits on V_{nh} from ACI Secs. 17.5.2.1 to 17.5.2.3 are given in Table 17–1.

In all cases the contact surfaces must be clean and free from laitance. The words "intentionally roughened" imply that the surface has been roughened with a "full amplitude" of $\frac{1}{4}$ in., where "full amplitude" refers to the total height (twice the amplitude) of the roughness. The "wave length" of the roughness is intended to be of the same magnitude as the

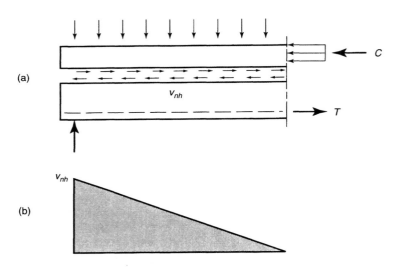

Fig. 17–11
Horizontal shear stresses in a composite beam.

TABLE 17–1 CALCULATION OF V_{nh}

ACI Section	Contact Surfaces	Ties	V_{nh}
17.5.2.1	Intentionally roughened	None	$80 b_v d$
17.5.2.2	Not roughened	Minimum from 17.6	$80 b_v d$
17.5.2.3	Intentionally roughened	$A_v f_y$	$\left(260 + \dfrac{0.6 A_v f_y}{b_v s}\right) \lambda b_v d$

Shear Friction, Horizontal Shear Transfer, and Composite Concrete Beams

height, say $\frac{1}{4}$ to $\frac{3}{4}$ in. When the factored shear force, $V_u = \phi V_{nh}$, at the section exceeds $\phi(500b_v d)$, where 500 is in psi, ACI Sec. 17.5.2.4 requires design using shear friction in accordance with ACI Sec. 11.7.4. This limit reflects the range of test data used to derive ACI Sec. 17.5.2.3.

ACI Sec. 17.6 requires that the ties provided for horizontal shear be not less than the minimum stirrups required for shear given by

$$A_v = \frac{50b_w s}{f_y}$$

(17–11)
(ACI Eq. 11–13)

The tie spacing shall not exceed 4 times the least dimension of the supported element, which is usually the thickness of the slab, but not more than 24 in. The ties must be fully anchored in both the beam stem and the slab.

Deflections

The beam cross section considered when calculating deflections depends on whether the beam was shored or unshored when the composite slab is placed. If it is shored so that the full dead load of both the precast beam and the slab is carried by the composite section, ACI Sec. 9.5.5.1 allows the designer to consider the loads to be carried by the full composite section when computing deflections. The modulus of elasticity should be based on the strength of the concrete in the compression zone, while the modulus of rupture should be based on the strength of the concrete in the tension zone. For nonprestressed beams constructed using shores, it is not necessary to check deflections if the overall height of the composite section satisfies ACI Table 9.5(a).

ACI Sec. 9.5.5.2 covers the calculation of deflections for unshored construction of nonprestressed beams. If the thickness of the precast member satisfies ACI Table 9.5(a), it is not necessary to consider deflections. If the thickness of the composite section satisfies the table but the thickness of the precast member does not, it is not necessary to compute deflections occurring after the section becomes composite, but it is necessary to compute the instantaneous deflections and that part of the sustained load deflections occurring prior to the beginning of effective composite action. This can be assumed to occur when the modulus of elasticity of the slab reaches 70 to 80% of its 28-day value, which occurs about 4 to 7 days after the slab is placed.

ACI Sec. 9.5.5.1 states that if deflections are computed, they should account for the curvatures induced by the differential shrinkage between the slab and the precast beam. Shrinkage of the slab relative to the beam, causes the slab to shorten relative to the beam. Because the slab and beam are joined together, this relative shortening causes the beam to deflect downward, adding to the deflections due to loads. Some of the shrinkage of the concrete in the beam will have occurred before the beam is erected in the structure. All of the slab shrinkage occurs after the slab is cast. As the slab shrinks relative to the beam, tensile stresses are induced in the slab and compressive stresses in the beam. These are redistributed to some degree by creep of the concrete in the slab and beam. This effect can be modeled using an *age-adjusted effective modulus, E_{caa}*, and an *age-adjusted transformed section* in the calculations as discussed in Sec. 3–4 and in Refs. 17–11 and 17–12.

EXAMPLE 17–2   Design of a Composite Beam

Precast simply supported beams that span 24 ft and are spaced 10 ft on centers are composite with a slab that supports an unfactored live load of 100 psf, a partition load of 20 psf, and a superimposed dead load of 10 psf. Design the beams and the composite beam and slab. Use $f_c' = 3000$ psi for the slab and 5000 psi for the precast beams and $f_y = 60,000$ psi.

1. **Select the trial dimensions.** For the end span of the slab, ACI Table 9.5(a) gives the minimum thickness of a one-way slab as $h = \ell/24 = 120$ in./24 = 5 in. For a simply supported beam, the table gives $h = \ell/16 = (24 \times 12)/16 = 18$ in. Deflections of the composite beam may be a problem if the overall depth is less than 18 in. However, for unshored construction ACI Sec. 9.5.5.2 requires that deflections of the precast member be considered if its overall depth is less than given by Table 9.5(a). To avoid this we shall try a 18-in.-deep precast beam with width 12 in. to allow the steel to be in one layer plus a 5-in. slab. For the precast beam, $d = 18 - 2.5 = 15.5$ in.

2. **Compute the factored loads on the precast beam.** Since the floor will be constructed in an unshored fashion, the precast beam must support its own dead load, the dead load of the slab, the weight of the forms for the slab, assumed to be 10 psf, and some construction live load, assumed to be 50 psf.

Dead loads:

Beam stem $w = \dfrac{12 \times 18}{144} \times 0.150 \text{ kcf} = 0.225 \text{ kip/ft}$

Slab $w = 5 \text{ in.}/12 \times 10 \text{ ft} \times 0.150 \text{ kcf} = 0.625 \text{ kip/ft}$

Forms $w = 10 \text{ ft} \times 0.010 \text{ ksf} = 0.100 \text{ kip/ft}$

Total $= 0.950 \text{ kip/ft}$

Live load: $w = 10 \text{ ft} \times 0.050 \text{ ksf} = 0.500 \text{ kip/ft}$

The ACI Code does not specifically address the load factors for this construction load case. We shall take $U = 1.4 D + 1.7L$, giving

$$w_u = 1.4 \times 0.950 + 1.7 \times 0.500 = 2.18 \text{ kips/ft}$$

3. **Compute the size of precast member required for flexure.**

$$M_u = \frac{2.18 \times 24^2}{8} = 157.0 \text{ ft-kips}$$

We shall select a steel percentage close to but less than the tension-controlled limit in the precast beam so that $\phi = 0.9$, and later check if that provides enough steel for flexure in the composite section. From Table A–3 for 5 ksi, the value of ρ corresponding to the tension-controlled limit is 0.021. We shall try $\rho = 0.02$. For this ρ, ρ, $\phi k_n = 927$ and $j = 0.858$.

$$\frac{bd^2}{12,000} = \frac{M_u}{\phi k_n} = \frac{150.5}{927}$$
$$= 0.162$$

From Table A–9, for $b = 12$ in. we require that $d = 13$ in. and $h = 13 + 2.5 = 15.5$ in. Therefore, the size chosen for deflection control is adequate.

For the 12 by 18 in. beam we require that

$$A_s = \frac{M_u}{\phi f_y jd} = \frac{157.0 \times 12,000}{0.9 \times 60,000 \times 0.858 \times 15.5}$$
$$= 2.62 \text{ in.}^2$$

Try a 12 by 18 in. precast beam with 2 No. 8 bars pus 2 No. 7 bars, $A_s = 2.78 \text{ in.}^2$.

$$a = \frac{2.78 \times 60,000}{0.85 \times 5000 \times 12} = 3.27 \text{ in.}$$

$$\frac{a}{d_t} = \frac{3.27}{15.5} = 0.211$$

This is less than a/d_t for the tension-controlled limit of 0.300; therefore, $\phi = 0.9$.

$$\phi M_n = \frac{0.9 \times 2.78 \times 60,000(13 - 3.27/2)}{12,000} = 173.4 \text{ ft-kips}$$

Therefore, OK.

4. Check the capacity of the composite member in flexure. Factored loads on composite member:

Dead load

	Precast stem	$w = 0.225$ kip/ft
	Slab	$w = 0.625$ kip/ft

Superimposed dead load

$w = 0.100$ kip/ft

Total $w = 0.950$ kip/ft   Factored $= 1.4 \times 0.950 = 1.33$ kips/ft

Live load

	Floor load	$w = 10$ ft $\times$ 0.100 ksf $= 1.0$ kip/ft
	Partitions	$w = 10$ ft $\times$ 0.020 ksf $= 0.2$ kip/ft

Total $w = 1.20$ kips/ft   Factored $= 1.7 \times 1.20 = 2.04$ kips/ft

Total factored load $= 3.37$ kips/ft

$$M_u = \frac{3.37 \times 24^2}{8} = 243 \text{ ft-kips}$$

Compute ϕM_n. From ACI Sec. 8.10.2, the effective flange width is 72 in. Overall height $= 18 + 5 = 23$ in.; $d = 23 - 2.5 = 20.5$ in. Assuming rectangular beam action where f_c' in the slab is 3000 psi,

$$a = \frac{2.78 \times 60,000}{0.85 \times 3000 \times 72} = 0.909 \text{ in.}$$

Since a is less than the flange thickness, rectangular beam action exists and $a = 0.909$ in. Also, a/d_t is much less than a/d_t for the tension-controlled limit and thus $\phi = 0.9$ for flexure.

$$\phi M_n = \frac{0.9 \times 2.78 \times 60,000(20.5 - 0.909/2)}{12,000}$$
$$= 251 \text{ ft-kips}$$

Therefore, the steel chosen is adequate to resist the moments acting on the composite section. **Use a 12 by 18 in. precast section with 2 No. 8 plus 2 No. 7 longitudinal bars and a 5-in. cast-in-place slab.**

5. Check vertical shear. V_u at d from the support $= 3.37$ kips/ft $\times (12 - 20.5/12)$ ft $= 34.7$ kips.

$$V_c = 2\sqrt{f_c'}b_w d \quad \text{where we shall use the smaller of the two concrete strengths}$$
$$= 2\sqrt{3000} \times 12 \times 20.5 = 26.9 \text{ kips} \quad \phi V_c = 22.9 \text{ kips}$$

Since $V_u > \phi V_c/2$, we need stirrups.

$$V_s = \frac{V_u}{\phi} - V_c = \frac{34.7}{0.85} - 26.9$$
$$= 13.9 \text{ kips at } d \text{ from the support}$$

Try No. 3 U stirrups, $A_v = 0.22$ in.^{2}.

$$s = \frac{A_v f_y d}{V_s} = \frac{0.22 \times 60 \times 20.5}{13.9}$$
$$= 19.5 \text{ in. on-centers}$$

Maximum spacing $= d/2 = 10.25$ in., say 10 in.

Minimum stirrups: $s = \dfrac{A_v f_y}{50 b_w} = \dfrac{0.22 \times 60,000}{50 \times 12}$

$$= 22.0 \text{ in. on-centers.}$$

We will select the stirrups after considering horizontal shear.

6. Compute the horizontal shear. Horizontal shear may be computed according to ACI Sec. 17.5.2 or 17.5.3. We shall do the calculations both ways and compare the results. We shall assume that the interface is intentionally roughened.

ACI Sec. 17.5.2: From Eq. 17–7 (ACI Eq. 17–1), $\phi V_{nh} \leq V_u = 34.7$ kips at d from the support. From ACI Sec. 17.5.2.2, an intentionally roughened surface without ties is adequate for

$$\phi V_{nh} = \phi 80 b_v d = 0.85 \times 80 \times 12 \times 20.5 = 16.7 \text{ kips}$$

Therefore, ties are required.

From ACI Sec. 17.5.2.3, if minimum ties are provided according to ACI Sec. 17.6 and the interface is intentionally roughened,

$$\phi V_{nh} = \phi(260 + 0.6\rho_v f_y)\lambda b_v d$$

The maximum tie spacing allowed by ACI Sec. 17.6.1 is 4×5 in. $= 20$ in. but not more than 24 in. The maximum spacing for vertical shear governs. Assume that the ties are No. 3 two leg stirrups at the maximum spacing allowed for shear $= 10$ in.

$$\rho_v = \frac{A_v}{b_v s} = \frac{0.22}{12 \times 10}$$
$$= 0.00183$$

and

$$\phi V_{nh} = 0.85(260 + 0.6 \times 0.00183 \times 60{,}000) \times 1.0 \times 12 \times 20.5$$
$$= 68.1 \text{ kips}$$

Thus minimum stirrups provide more than enough ties for horizontal shear. We shall use closed stirrups to better anchor them into the top slab. **Use No. 3 closed stirrups at 10 in. on-centers throughout the length of the beam.**

ACI Sec. 17.5.3: Alternatively, horizontal shear could be considered using ACI Sec. 17.5.3. The compression zone is entirely in the slab. Therefore, the force transferred across the interface is

$$V_{nh} = C = T = A_s f_y = 2.78 \text{ in.}^2 \times 60 \text{ ksi} = 166.8 \text{ kips}$$

From Eq. 17–11,

$$v_{nh} = \frac{KV_{nh}}{A_c}$$

where K is 2 at the support since the shear force diagram is a triangle, and $A_c = 12$ *in.* $\times 144$ in. $= 1728$ in.^{2} . Thus

$$v_{nh} = \frac{2 \times 166.8 \times 1000 \text{ lb}}{1728 \text{ in.}^2}$$
$$= 193 \text{ psi}$$

Since this exceeds 80 psi, ACI Sec. 17.5.2.1 requires ties. Minimum ties according to ACI Sec. 17–6 are equivalent to No. 3 two-legged stirrups at 10 in. on-centers with $\rho_v = 0.00183$. For minimum ties, ACI Sec. 17.5.2.3 gives

$$v_{nh} = (260 + 0.6\rho_v f_y) = (260 + 0.6 \times 0.00183 \times 60{,}000)$$
$$= 326 \text{ psi}$$

Thus minimum ties are satisfactory. We shall use closed stirrups to better anchor them into the top slab. **Use No. 3 closed stirrups at 10 in. on centers throughout the length of the beam.** ∎

Discontinuity Regions, Corbels, Deep Beams, and Joints

18–1 INTRODUCTION

Definition of Discontinuity Regions

Structural members may be divided into portions called *B-regions*, in which beam theory applies, including linear strains and so on, and other portions called *discontinuity regions*, or *D-regions*, adjacent to discontinuities or disturbances, where beam theory does not apply. In D-regions, a significant portion of the load is carried by inplane forces. Examples of D-regions are regions near concentrated loads and reactions, corbels, deep beams, joints, dapped ends, abrupt changes in cross section, holes and other discontinuities. Up to this point, most of this book has dealt with B-regions. For many years, D-region design has been by "good practice," rule of thumb, or empirical. Three landmark papers by Professor Schlaich of the University of Stuttgart and his co-workers[18-1 to 18-3] have changed this. This chapter will present rules and guidance for the design of D-regions based largely on these and other recent papers.

St. Venant's principle suggests that the localized effect of a disturbance dies out in about one member depth from the point of the disturbance. On this basis, D-regions are assumed to extend one member depth each way from the discontinuity. Figure 18–1 shows B-regions and D-regions in several structural members.

Behavior of D-Regions

Prior to cracking, an elastic stress field exists which can be determined using an elastic analysis such as a finite element analysis. Cracking disrupts this stress field, causing a major reorientation of the internal forces. After cracking, the internal forces can be modeled using a *strut-and-tie model* consisting of concrete compression struts, steel tension ties, and joints referred to as nodal zones. If the compression struts are narrower at their ends than they are at midsection, the struts may, in turn, crack longitudinally. For unreinforced struts this may lead to failure. On the other hand, struts with transverse reinforcement to restrain the cracking can carry further load and will fail by crushing as shown in Fig. 6–19. Failure may also occur by yielding of the tension ties or failure of the nodal zones. As always, failure initiated by yield of the steel tension ties tends to be more ductile and is desirable.

Basic Method of Solution of D-Regions

We shall start with an overview and an example, followed in subsequent sections by a discussion of the assumptions made in using strut-and-tie models. Most discontinuity regions can be solved by the following procedure:

1.   Isolate the D-regions. The discontinuity can be assumed to extend approximately a distance d from the discontinuity. The beam in Fig. 18–1a has been divided into four different D-regions plus a B-region. The region around the hole in the beam in Fig. 18–1c is represented by six D-regions. The beam–column joint in Fig. 18–2a can either be thought of as extending d into the members or terminating at the faces of the column and beam as shown by the shaded area. The latter corresponds more closely with American design practice and will be followed here.

2.   Compute the internal stresses on the boundaries of the element. This can be done using reinforced concrete theory or elastically ($\sigma = My/I + P/A$), as shown in Fig. 18–2a and b. To include the strength reduction factor, the stresses should be computed for M_u/ϕ, V_u/ϕ, or P_u/ϕ, where ϕ is generally taken equal to the value for shear, 0.85.

3.   Subdivide the boundary and compute the force resultants on each sublength. See Fig. 18–2c. Several strategies can be followed in doing this. The simplest is to take all the sublengths the same length. Alternatively, in Fig. 18–2c, the lengths of the two right-hand portions have been chosen so that the tension force resultant in the farthest-right sublength is the equal in magnitude to the compression force resultant in the sublength second from the right. For vertical equilibrium, the compressive force resultants in the remaining two sublengths must equal the applied load in the column. This subdivision makes the strut-and-tie model easier to draw, since the two right-hand forces directly cancel each other out. The problem illustrated in Fig. 18–2c is solved in Example 18–2.

4.   Draw a truss to transmit the forces from boundary to boundary of the D-region. In Fig. 18–3c (discussed more fully in Example 18–1) the compression members in the truss are shown dashed and the tension ties by solid lines. Generally speaking, the truss should be chosen to minimize the steel volume.[18-2]

When two noncollinear forces meet at a point, a third force is necessary for equilibrium. This is a fundamental rule of concrete detailing. Thus when the two compression struts meet at an angle at point B in Fig. 18–3c, a tension tie (BC) is needed for equilibrium. Similarly, the strut DE is needed to maintain equilibrium at joints D and E. If a reinforcing bar stressed in tension is bent around a corner a compression force into the apex of the bend or a tension force out of the apex is required for equilibrium.

5.   Check the stresses in the individual members in the truss. The steel ties can be assumed to be stressed to f_y and the concrete struts to $f_{ce} = vf'_c$, where v is taken from Table 18–1, discussed later. The allowable stresses on the faces of the nodes will also be discussed later.

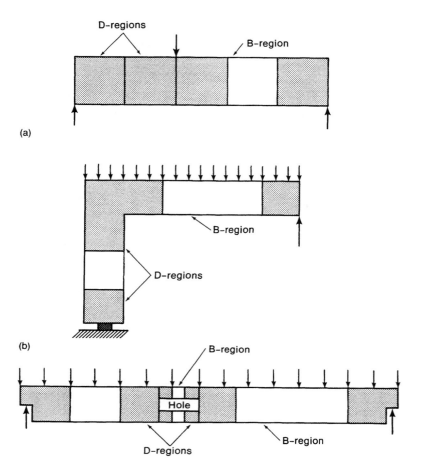

Fig. 18–1
B-regions and D-regions.

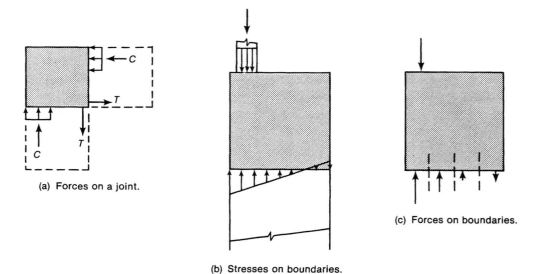

(a) Forces on a joint.

(c) Forces on boundaries.

(b) Stresses on boundaries.

Fig. 18–2
Forces on boundaries of
D-elements.

18-1   Introduction

755

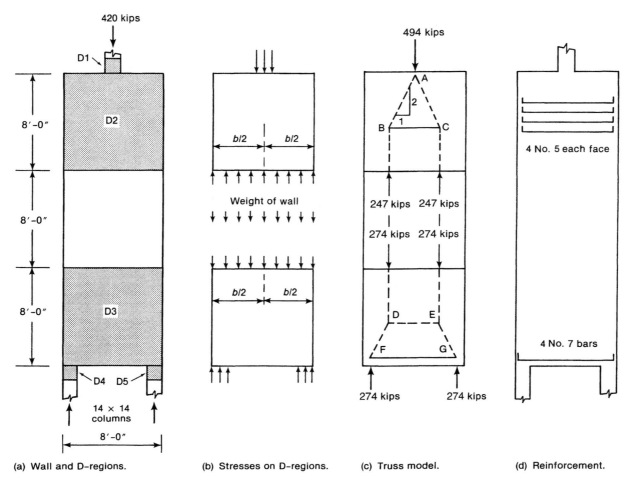

(a) Wall and D-regions. (b) Stresses on D-regions. (c) Truss model. (d) Reinforcement.

Fig. 18–3
D-regions in a wall—Example 18–1.

EXAMPLE 18–1 Design of D-Regions

The structure shown in Fig. 18–3a consists of a B-region and five D-regions. Design the reinforcement in regions $D2$ and $D3$. The wall is 14 in. thick and is prevented from buckling out of plane by floor slabs. Use $f'_c = 3000$ psi and $f_y = 60,000$ psi.

REGION D2

1. Isolate the D-regions. See Fig. 18–3a.

2. Compute the internal stresses on the boundaries of the region. We shall assume the stresses can be computed by $\sigma = P/A$ on each boundary. To include the strength reduction factor, we shall compute stresses for $P_n = P_u/\phi = 420/0.85 = 494$ kips. The weight of the wall shall be taken as $46/0.85 = 54$ kips and will be assumed to act at midheight of the wall.

3. Subdivide the boundary and compute the force resultants. We shall represent the top boundary by a single force at the middle of the column. The bottom boundary will be divided into two equal lengths, each with its resultant force of 247 kips.

4. Draw a truss. See Fig. 18–3c. Compression struts will always be shown by dashed lines and tension ties by solid lines. In drawing this truss it is necessary to have a value of the angle θ. This can either be obtained from stress trajectory plots such as Fig. 18–15a, which will be discussed later, or it can be assumed directly. In most cases a 2:1 slope can be assumed, provided this can be accommodated within the D-region (see Fig. 18–3c).

$$\theta = \arctan 2 = 63.4°$$

5. Check the stresses in the truss members.

Tension tie *BC*:

$$\text{force} = T_n = \frac{247}{\tan \theta}$$

$$= 123.7 \text{ kips}$$

$$A_s = \frac{T_n}{f_y} = \frac{123.7}{60}$$

$$= 2.06 \text{ in.}^2$$

Transverse steel having this area should be provided across the full width of the wall in a band extending about $0.3d$ above and below the position of tie *BC* (see Fig. 18–3d).
Use 8 No. 5 bars, 12 in. on centers, hooked both ends, half in each face.
A_s provided $= 2.48$ in.^{2} The top bar will be at 30 in. below the top of the wall.
Node *A*: Since the concrete struts fan out, the most critical concrete stresses are at node *A*. Since this node is compressed on all faces, we shall see later that $f_{ce} = 0.85 f_c' = 2.55$ ksi.

$$\text{Nominal force in column} = 494 \text{ kips}$$

$$\text{Maximum stress} = \frac{494}{14 \times 14} = 2.52 \text{ ksi—therefore OK}$$

REGION D3

This region is shown in Fig. 18–3a and c along with the resultant forces on the boundaries and a strut-and-tie model. The self-weight of the wall is idealized as a factored load of 46 kips applied at mid-height. This gives a nominal load of $46/0.85 = 54$ kips. Half of this acts in each vertical strut. Again $\tan \theta$ has been taken as 2.

Tension tie *FG*

$$\text{Force} = T_n = \frac{274}{\tan \theta} = 137 \text{ kips}$$

$$A_s = \frac{137}{60} = 2.28 \text{ in.}^2$$

Use 4 No. 7 bars in one layer with 90° hooks inside the column cages.

Nodes *F* and *G*: Since these each anchor a tension tie, we shall use a lower limiting stress than for node *A*. In Sec. 18–2 the limit for this case is given as $0.75 f_c' = 2.55$ ksi.

$$\text{Stress} = \frac{274}{14 \times 14} = 1.40 \text{ ksi} \quad \text{OK}$$

In addition to the steel shown in Fig. 18–3d, provide minimum wall steel according to ACI Sec. 14.3 and extend the column steel at least a compression development length into the wall. ■

18–2   COMPONENTS OF STRUT-AND-TIE MODELS

A strut-and-tie model for a deep beam is shown in Fig. 18–4. It consists of concrete compressive struts, reinforcing bars as tension ties, and joints or nodal zones. The strengths of these elements are discussed in this section and the layout of strut-and-tie models in the

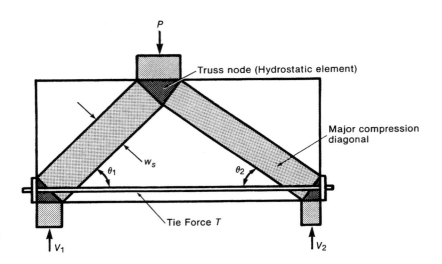

Fig. 18–4
Strut-and-tie model of a deep beam.

next. As discussed later, the strut-and-tie model will be analyzed for the internal forces due to a load of P_u/ϕ, where ϕ will normally be taken as 0.85 for shear. The forces in the two compression struts and the tie are C_{n1}, and C_{n2}, and T_n, respectively, where n stands for nominal.

A strut-and-tie model is a system of forces in equilibrium with a given set of loads. The lower bound theorem of plasticity states that the capacity of such a system of forces is a lower bound on the strength of the structure, provided that no element is loaded beyond its capacity. This assumes that the deformation capacity is not exceeded at any point before the assumed system of forces is reached. For this reason, the resultant forces in the members of the strut-and-tie model should be close to the final set of internal forces.

Compression Struts

In a strut-and-tie model the struts represent concrete compression stress fields with the prevailing compression in the direction of the strut. Struts are frequently idealized as prismatic or uniformly tapering members, but often vary in cross section along their length, as shown in Fig. 18–5b, because the concrete is wider at midlength of the strut than at the ends. Struts that vary in width are sometimes idealized as *bottle-shaped* as shown in Fig. 18–5b,[18–1 to 18–3] or are idealized using local truss models as shown in Fig. 18–5c. The spreading of the compression forces gives rise to transverse tensions, which may cause the strut to crack longitudinally. If the strut has no transverse reinforcement, it may fail after this cracking occurs. If adequate transverse reinforcement is provided, the strut will fail by crushing. In strut-and-tie models, the compression struts are shown by dashed lines along the axes of the struts.

Strut Failure by Longitudinal Cracking

Figure 18–6a shows one end of a bottle-shaped strut. The width of the bearing area is a, and the thickness of the strut is t. At midlength the strut has an effective width b_{ef}. Reference 18–1 assumes that a bottle shaped region at one end of a strut extends approximately $1.5b_{ef}$ from the end of the strut and in examples used $b_{ef} = \ell/3$ but not less than a, where ℓ is the length of the strut from face to face of the nodes. For short struts, the limit that b_{ef} not be less than a often governs. We shall assume that in a strut with bottle-shaped regions at each end

$$b_{ef} = a + \ell/6 \quad \text{but not more than the available width} \tag{18–1}$$

Discontinuity Regions, Corbels, Deep Beams, and Joints

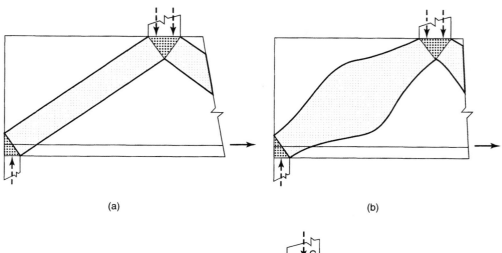

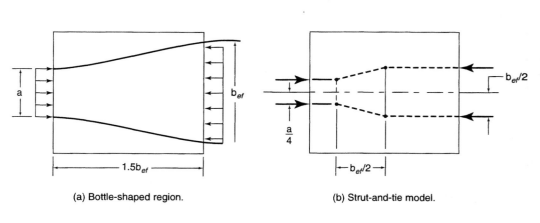

Fig. 18-5
Compression struts.

(a) Bottle-shaped region.

(b) Strut-and-tie model.

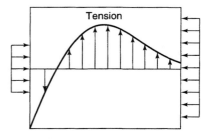

Fig. 18-6
Cracking of compression
struts.

(c) Transverse tensions and compressions.

18–2   Components of Strut-and-Tie Models   **759**

Figure 18–6b shows a strut-and-tie model for the bottle-shaped region. It is based on the assumption made in Ref. 18–4, that the longitudinal projection of the inclined struts is equal to $b_{ef}/2$. The transverse tension force T at one end of the strut is

$$T = \frac{C}{2}\left(\frac{b_{ef}/4 - a/4}{b_{ef}/2}\right)$$

or

$$T = \frac{C}{4}\left(1 - \frac{a}{b_{ef}}\right) \tag{18–2}$$

The force T causes transverse stresses in the concrete which may cause cracking. These are distributed as shown by the curved line in Fig. 18–6c. Analyses by Adebar and Zhou[18–5] suggest that the tensile stress distributions at the two ends of a strut are completely separate when ℓ/a exceeds about 3.5 and overlap completely when ℓ/a is 1.5 to 2. Assuming a parabolic distribution of transverse tensile stresses spread over a length of $1.6b_{ef}$ in a strut of length $2b_{ef}$, and equilibrating a tensile force of $2T$, indicates that the minimum load, C, at cracking is $0.57atf_c'$ for a strut with $a/b_{ef} = 2$. This corresponds to the value given in Refs. 18–2 and 18–3. From this analysis or from Refs. 18–2 and 18–3, we may conclude that the longitudinal cracking in the strut may be a problem if the bearing pressures on the ends of a strut exceed $0.55f_c'$. This is given in Table 18–1 as $f_{ce} = 0.65\nu_2 f_c'$ where ν_2 varies from 0.85 for 2500-psi concrete to 0.70 for 10,000-psi concrete. In tests of cylindrical specimens loaded axially through circular bearing plates with diameters less than that of the cylinders, failure occurred at 1.2 to 2 times the cracking loads.[18–5]

The maximum load on an unreinforced strut in a wall-like member such as the deep beam in Fig. 18–4, as governed by cracking of the concrete in the strut, is given by Eq. 18–2. This assumes that the compression force spreads in only one direction. If the bearing area does not extend over the full width of the member, there will also be transverse tensile stresses through the width of the strut which will require reinforcement through the thickness as shown in Fig. 18–7.

Compression Failure of Strut

The crushing strength of the concrete in a strut is referred to as the *effective strength*,

$$f_{ce} = \nu_1\nu_2 f_c' \tag{18–3}$$

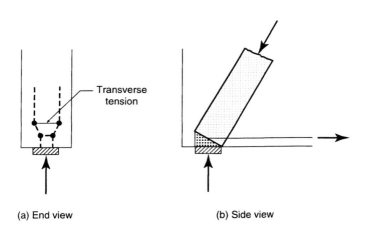

(a) End view   Transverse tension   (b) Side view

where the product $\nu_1\nu_2$ is an *efficiency factor* between 0 and 1.0. Various sources give differing values of the efficiency factor.[18-2, 18-6 to 18-11] The ACI Code does not address this problem. The major factors affecting the effective compressive strength are:

1. The concrete strength. Concrete becomes more brittle as the strength increases. This will be accounted for by the factor ν_2 from Ref. 18–11:

$$\nu_2 = 0.55 + \frac{15}{\sqrt{f_c'}} \qquad (18\text{--}4)$$

ν_2 is roughly equivalent to α_1 in the rectangular stress block.

2. The direction of cracking, whether parallel to the strut or at an angle to it.

3. Tension strains in the concrete transverse to the strut resulting from forces in reinforcement crossing the cracks. Collins and Mitchell[18-10] have shown that such strains reduce the compressive strength of uniformly strained concrete panels. See Sec. 3–2.

Recommended values of f_{ce} are given in Table 18–1. These are compatible with the values given in Sec. 6–4 and the load and resistance factors in the ACI Code.

Longitudinal cracking of the strut may prevent it from reaching its full compression capacity. To prevent longitudinal splitting failures of struts, horizontal and vertical reinforcement should be provided to resist the entire tensile forces, ΣT, in the transverse ties at each end of the strut-and-tie model in Fig. 18–6b when the compressive force in the strut,

TABLE 18–1   Recommended Values of Effective Compression Strength, f_{ce}

$$f_{ce} = \nu_1\nu_2 f_c' \text{ where } \nu_2 = \left(0.55 + \frac{15}{\sqrt{f_c'}}\right)^a$$

Structural Member	ν_1
Truss nodes[b]	
Joints bounded by compression struts and bearing plates	1.0
Joints anchoring one tension tie	0.85
Joints anchoring more than one tension tie	0.75
Struts[c]	
Uncracked uniaxially stressed struts or fields	1.0
Struts cracked longitudinally due to bottle shaped stress fields, containing transverse reinforcement based on Eq. 18–5	0.80
Struts cracked longitudinally due to bottle-shaped stress fields without transverse reinforcement	0.65
Struts in cracked zone with transverse tensions from transverse reinforcement	0.60
Severely cracked webs of slender beams[d]	
$\theta = 30°$	0.30
$\theta = 45°$	0.55

[a]From Ref. 18–11.

[b]From Ref. 18–9, modified to include ν_2.

[c]From Ref. 18–7, modified to fit ACI load and ϕ factors.

[d]From Ref. 18–10, modified to include ν_2.

C, is at its maximum value, where T is given by Eq. 18–2. The strut should therefore be crossed by approximately perpendicular reinforcement with a yield strength

$$A_s f_y \geq \Sigma \left[\frac{C}{4}\left(1 - \frac{a}{b_{ef}}\right)\right] \qquad (18\text{--}5)$$

where Σ implies the sum of the values at the two ends of the strut. If the reinforcement is at an angle θ to the strut, $A_s f_y$ should be divided by $\sin \theta$. This reinforcement will be referred to as *confining reinforcement*.

Design of Compression Struts

Once the strut-and-tie model has been laid out, the strength of a compression strut is computed as follows:

    1. If there is no transverse reinforcement in the strut, the strength should be taken as the compression causing cracking, which will equal or exceed $0.55 a t f_c'$, where a is the width of the node and t is the thickness of the element.

    2. If the strut is crossed by reinforcement satisfying Eq. 18–5, the strength of the strut should be calculated based on the smallest cross-sectional area of the strut and the effective concrete strengths given in Table 18–1

Tension Ties

The second major component of a strut-and-tie model is the tension tie. This represents one or several layers of steel in the same direction designed with $A_s f_y \geq T_n$, where $T_n = T_u / \phi$ is the force to be resisted by the tie.

    Tension ties may fail due to lack of end anchorage. The anchorage of the ties in the nodal zones is a major part of the design of a D-region using a strut-and-tie model. Tension ties are shown as solid lines in strut-and-tie models.

Nodal Zones

The joints in the strut-and-tie model are known as *nodal zones*. Three or more forces meet at a node. The forces meeting at a node must be in equilibrium. That is, $\Sigma F_x = 0$, $\Sigma F_y = 0$, and $\Sigma M = 0$ about the nodal point. The third condition implies that the lines of action of the forces must pass through a common point, or must be able to be resolved into forces that act through a common point. The two compressive forces shown in Fig. 18–8a meet at an angle and are not in equilibrium unless a third force is added, as shown in Fig. 18–8b or c. In the nodal zones at middepth of Fig. 18–3c, the three forces

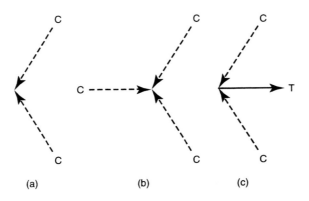

Fig. 18–8
Forces at nodal zones.

        (a)   (b)   (c)

Discontinuity Regions, Corbels, Deep Beams, and Joints

are compressive. At the reactions in this figure, the inclined compressive force and the compressive reaction need a tensile force for equilibrium. Nodal zones are classified as CCC if three compressive forces meet, as in Fig. 18–8b and as CCT if one of the forces is tensile as shown in Fig. 18–8c. CTT joints and TTT joints may also occur.

Two common ways of laying out nodal zones are illustrated in Figs.18–9 and 18–10. The prismatic compression struts in Fig. 18–4 are assumed to be stressed in uniaxial compression. A section perpendicular to the axis of a strut is acted on only by compression stresses, while sections at any other angle have combined compression and shear stresses. One way of laying out nodal zones is to place the sides of the nodes at right angles to the axes of the struts or ties meeting at that node, as shown in Fig. 18–9, and to have the same bearing pressure on each side of the node. When this is done for a CCC node, the ratio of the lengths of the sides of the node, $a_1 : a_2 : a_3$, is the same as the ratio of the forces in the three members meeting at the node, $C_1 : C_2 : C_3$, as shown in Fig. 18–9a. If one of the forces is tensile, the width of that side of the node is calculated from a hypothetical bearing plate on the end of the tie, which is assumed to exert a bearing pressure on the node equal to the compressive stress in the struts at that node, as shown in Fig. 18–9b. Nodes laid out in this fashion are sometimes referred to as *hydrostatic elements* since the in-plane stresses in the node are the same in all directions. In such a case, the Mohr's circle for the in-plane stresses reduces to a point.

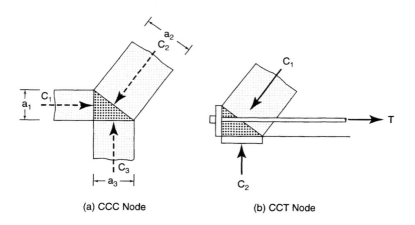

Fig. 18–9
Hydrostatic nodal zones.

(a) CCC Node

(b) CCT Node

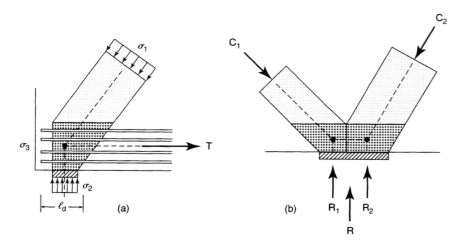

Fig. 18–10
Nodal zones within the inter-sections of members.

(a)

(b)

The use of hydrostatic elements can be tedious in design, except possibly for CCC nodes. More recently, the design of nodal zones has been simplified by considering the nodal zone to comprise that concrete lying within extensions of the members meeting at the joint as shown in Fig. 18–10.[18-3, 18-7] This allows different stresses to be assumed in the struts and over bearing plates, for example. Two examples are given in Fig. 18–10. Figure 18–10a shows a CTT node. The bars must be anchored within or to the left of the node. The length, ℓ_d, in which the bottom layer of bars must be developed is shown. The vertical face of the node is acted on by a stress σ_3 equal to the tie force T divided by the area of the vertical face. The stresses σ_1, σ_2, and σ_3 can all be different, provided that:

1. The resultants of the three forces coincide.
2. The stresses are within the limits given in Table 18–1.
3. The stress is constant on any one face.

Another example is shown in Fig. 18–10b. This may be subdivided into two subnodes. It is necessary to ensure that the stresses in the members entering the node, the stress over the bearing plate, and the stress on the vertical line dividing the two subnodes are within the limits in Table 18–1.

Nodal zones are assumed to fail by crushing. Anchorage of the tension ties is also a design consideration. If a tension tie is anchored in a nodal zone there is an incompatibility between the tensile strains in the bars and the compressive strain in the concrete of the node. This tends to weaken the node. The Canadian Code[18-9] states that unless special confinement is provided, the calculated stresses in the node regions shall not exceed the following:

1. $085f_c'$ in node regions bounded by compressive struts and bearing areas (CCC nodes)
2. $0.75f_c'$ in node regions anchoring a tension tie in only one direction (CCT nodes)
3. $0.65f_c'$ in node regions anchoring tension ties in more than one direction (CTT or TTT nodes)

The values given in Table 18–1 have been modified to include the ν_2 term reflecting the concrete strength (Eq. 18–4). Tests of CCT and CTT nodes reported in Ref. 18–12 indicate that $0.80f_c'$ could be developed in such nodes if properly detailed.

The Canadian Code gives two other requirements for the design of nodal zones. The bearing stresses on the sides of the nodes cannot exceed the values given above. Second, the tension tie reinforcement must be uniformly distributed over an effective area of concrete at least equal to the tie force divided by the concrete stress limits for the node.

Compression Fans

A compression fan is a series of compression struts that radiate out from a concentrated applied force to distribute that force to a series of localized tension ties such as the stirrups in a beam. An example is given in Fig. 18–11. Fans are shown over the reaction and under the load in Fig. 6–19b. The failure of a compression fan is shown in Fig. 6–21.

Compression Fields

A compression field is a series of parallel compression struts combined with appropriate tension ties and compression chords as shown in Fig. 18–11. A compression field is shown between the compression fans in Fig. 6–19b.

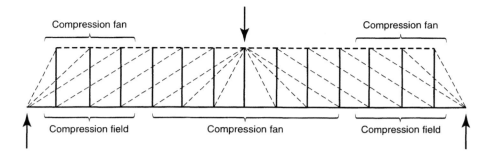

Fig. 18–11
Compression fans and compression fields.

Force Whirls, U Turns

From time to time situations arise where two adjacent forces are equal in magnitude but opposite in direction, as shown by the two forces on the right-hand side of the lower boundary of the D-region in Fig. 18–2c. In the strut-and-tie model for this case, these forces form a *force whirl* or a *U turn* consisting of tension ties and compression struts. Such a case is illustrated in Example 18–2.

Validity of Strut-and-Tie Models

The validity of strut-and-tie models was discussed in Sec. 6–4, and the measured and computed forces in the longitudinal reinforcement of a continuous beam are compared in Fig. 6–26. The comparison is seen to be very close, suggesting that the strut-and-tie model is a very good model in this case.

The validity of a strut-and-tie model for a given problem depends on whether the model represents the true situation. Concrete beams can undergo a limited amount of redistribution of internal forces as these change from the elastic uncracked state, through the elastic cracked state, to the plastic cracked state. If the truss that is chosen requires excessive deformation to reach the fully plastic state, it may fail prematurely. This is discussed more fully in Sec. 18–3.

EXAMPLE 18–2   Strut-and-Tie Model Containing a Force Whirl

Figure 18–12 shows a 12 in. square column supported on one side of a 12 in. by 96 in. wall. The total load in the column is 204 kips. Draw a strut-and-tie model for a nominal load of $204/\phi = 204/0.85 = 240$ kips. The concrete strength is 3000 psi and the yield strength of the steel is 60,000 psi.

   1. Isolate the D-regions. The D-region is assumed to extend vertically for one times the width of the wall.

   2. Compute the internal stresses on the boundaries of the element. The stresses on the lower edge of the D-region (Section A–I) have been computed using $\sigma = P/A + M_y/I$ and range from 677 psi at A to 261 psi at I.

   3. Subdivide the boundary and compute the force resultants on each sublength. The force in the column will be divided into two equal forces of 120 kips at the quarter points of the column width as shown in Fig. 18–12. The tensile force resultant in part G–I on the bottom of the D-region is

$$\frac{261 \text{ psi} \times 12 \text{ in.} \times 26.7 \text{ in.}}{2} = 41.8 \text{ kips}$$

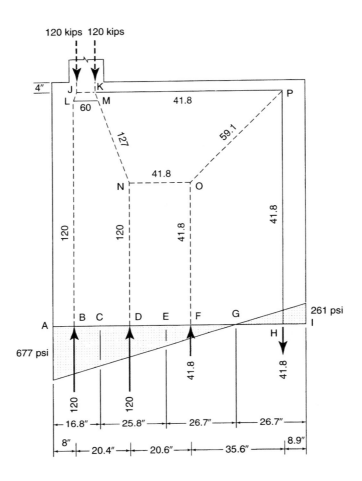

Fig. 18–12
Column supported on a
wall—Example 18–2.

We shall subdivide the length A–G to give a compressive force of 41.8 kips to counteract the tensile force, plus two 120-kip forces to equilibrate the two 120-kip forces in the column. The forces and their locations are given in Fig. 18–12.

4. **Draw a strut-and-tie model to transmit the forces from boundary to boundary of the D-region.** Compression struts will be shown by dashed lines and tension ties by solid lines. The forces at *F* and *H* offset each other. This is accomplished by the members *F–O*, *O–P*, and *P–H*. This is referred to as a force whirl or a U turn. Member *O–P* has been taken at 45°. At joint *P*, a horizontal tension force of 41.8 kips exists in member *K–P*. Similarly, a horizontal compression force of 41.8 kips is needed at *O* for equilibrium.

The column force at *J* is 9 in. from the edge of the wall. The compression force at *B* is 8 in. from the edge. The transition will be accomplished by member *J–L*, which will be assumed to have a slope of 2:1. This gives rise to a compressive force of 60 kips in member *J–K* and a tensile force of 60 kips in member *L–M*. Consideration of each of the other joints in turn leads to the member forces shown in Fig. 18–12.

5. **Check the stresses in the individual members of the truss.**

Compression struts: Figure 18–13 is an enlarged view of joints *J*, *K*, *L*, and *M*. This is the most congested area in the strut-and-tie model and if the stresses are acceptable in this region of the model they will be in all regions. The struts are shown in the light shading and the nodes in darker shading. The widths of the struts have been drawn using a stress of $0.8 \left(0.55 + 15/\sqrt{f_c'}\right) f_c' = 0.66 f_c'$, given in Table 18–1 for bottle shaped struts. It can be seen that none of the struts or nodal zones overlap. This indicates that the concrete stresses are satisfactory.

Minimum wall steel by ACI Sec. 14.3 is No. 4 at 18 in. o.c. vertical and No. 4 at 16 in. o.c. horizontal.

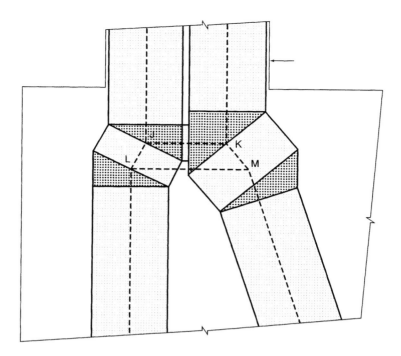

Fig. 18–13
Struts and nodal zones under column.

Member P–H:

$$A_s(\text{reqd}) = \frac{41.8}{60} = 0.70 \text{ in.}^2$$

Use 2 No. 4 vertical bars in each face of the wall at 4 in. o.c., first bar at 2 in. from edge.
Member K–P:

$$A_s(\text{reqd}) = 0.70 \text{ in.}^2$$

Provide the same steel as in *P–H*. Extend the bars past *J* and hook to anchor them. Bend the other end and lap splice with the steel in *P–H*.
Member L–M:

$$A_s(\text{reqd}) = \frac{60}{60} = 1.00 \text{ in.}^2$$

Use 3 No. 4 horizontal bars in each face, hooked at node *L* and extended ℓ_d past node *M*. The final reinforcement layout is shown in Fig. 18–14. ■

18–3 LAYOUT OF STRUT-AND-TIE MODELS

Strut-and-Tie Models

Strut-and-tie models can be drawn at two levels. In many cases it is sufficiently accurate to represent the struts and ties by lines along their centerlines, as shown in Fig. 18–3. In congested areas, however, the direction of the struts and their slopes and intersections are strongly affected by their widths as can be seen from Fig. 18–4 or 18–13. In such cases, it is necessary to consider the widths of the struts and draw all, or some part of, the strut-and-tie model to scale.

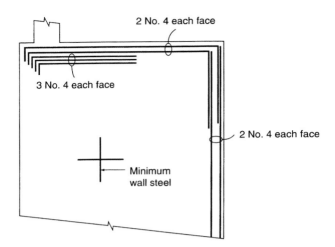

Fig. 18–14
Reinforcement in wall.

A strut-and-tie model should be in equilibrium with the applied forces on the boundaries of the D-region. If several load cases must be considered, it is necessary to develop a separate strut-and-tie model for each load case. The strut-and-tie model requiring the most reinforcement in any given location will govern the selection of steel at that point in the structure.

Several empirical rules that aid in laying out strut-and-tie models are given in the following sections.

Elastic Stress Trajectories

From an elastic analysis, such as a finite element analysis, it is possible to derive the stress trajectories in an uncracked D-region, shown in Fig. 18–15a for a deep beam. Principal compression stresses act parallel to the dashed lines, which are known as *compressive stress trajectories*. Principal tensile stresses act parallel to the solid lines, which are called *tensile stress trajectories*. Such a diagram shows the flow of internal forces and is a useful, but by no means essential, step in laying out a strut-and-tie model. The compressive struts should roughly follow the direction of the compressive stress trajectories, as shown by the refined and simple strut-and-tie models in Fig. 18–15c and e. Generally, the strut direction should be within $\pm 15°$ of the direction of the compressive stress trajectories.

Because a tie consists of a finite arrangement of reinforcing bars which are usually placed orthogonally in the member, there is less restriction on the conformance of ties with the tensile stress trajectories. However, they should be in the general direction of the tension stress trajectories.

Minimum Steel Content

The loads will try to follow the path involving the least forces and deformations. Since the tensile ties are more deformable than the compression struts, the model with the least and shortest ties is the best. Thus the strut-and-tie model in Fig. 18–16a is a better model than the one in Fig. 18–16b because Fig. 18–16a more closely approaches the elastic stress trajectories. Schlaich et al.[18-2, 18-3] propose the following criterion for guidance in selecting a good model:

$$\Sigma F_i \ell_i \epsilon_{mi} = \text{minimum}$$

Discontinuity Regions, Corbels, Deep Beams, and Joints

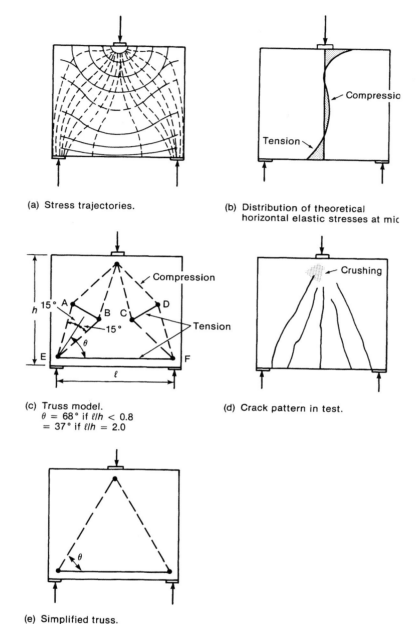

(a) Stress trajectories.

(b) Distribution of theoretical horizontal elastic stresses at mic

(c) Truss model.
$\theta = 68°$ if $\ell/h < 0.8$
$= 37°$ if $\ell/h = 2.0$

(d) Crack pattern in test.

(e) Simplified truss.

Fig. 18–15
Single-span deep beam.
(Adapted from Ref 18—1.)

where F_i, ℓ_i, and ϵ_{mi} are the force, length, and mean strain in strut or tie i, respectively. Because the strains in the concrete are small, the struts can be ignored in the summation.

An example of an unsuitable truss is given in Fig. 18–17. This shows one half of a simply supported beam with the flexural steel and one layer of "horizontal web reinforcement" at middepth. A possible plastic truss model for this beam consists of two trusses, one utilizing the lower steel as its tension tie, the other using the upper steel. For an ideally plastic material, the capacity would be the sum of the shears transmitted by the two trusses, $V_1 + V_2$. Tests[18-6] show, however, that the upper layer of steel has little, if any, effect on the strength. When this beam is loaded, the bottom tie yields first. Large deformations are

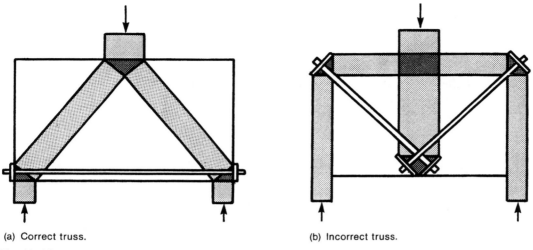

(a) Correct truss. (b) Incorrect truss.

Fig. 18–16
Correct and incorrect trusses.

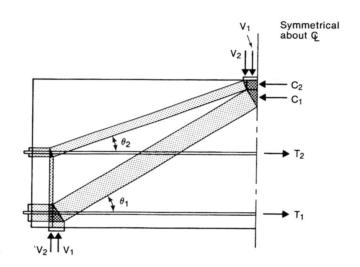

Fig. 18–17
Plastic truss-model for beam
with horizontal web rein-
forcement. (From Ref. 18–6.)

required before the upper tie can yield. Before these can fully develop, the lower truss will normally fail.

Agreement with Crack Pattern

If photographs of test specimens are available, the crack pattern may assist one in selecting the best strut-and-tie model. Figure 18–24a (which will be discussed later) shows the crack pattern in a dapped end at the support of a precast beam. Figure 18–24b, c, and d show possible models for this region. Compression strut *B–D* in Fig. 18–24d crosses a zone of cracking in the test specimen, which suggests that this is not a correct location for a compression strut.

Strength Reduction Factors

Because the ACI Code has strength reduction factors ϕ, of 0.9 for flexure, 0.85 for shear, and 0.7 for tied columns, it is sometimes difficult to select which value to use in the

Discontinuity Regions, Corbels, Deep Beams, and Joints

proportioning of a D-region representing a beam–column joint, for example. Since the most common D-region problems and corbels and deep beams, which are generally considered to be shear problems, we will normally use $\phi = 0.85$ in solving D-regions. The strut-and-tie models are drawn for loads of P_u/ϕ that give nominal compression forces, C_n, in the struts and nominal tie forces, T_n, in the ties. The resistances of the struts and ties are calculated from $A_c(v_1 v_2 f_c')$ for compression struts and $A_s f_y$ for tension ties.

18–4 BRACKETS AND CORBELS

A *bracket* or *corbel* is a short member that cantilevers out from a column or wall to support a load. The corbel is generally built monolithically with the column or wall, as shown in Fig. 18–18. The term "corbel" is generally restricted to cantilevers having shear span-to-depth ratios, a/d, less than or equal to 1.

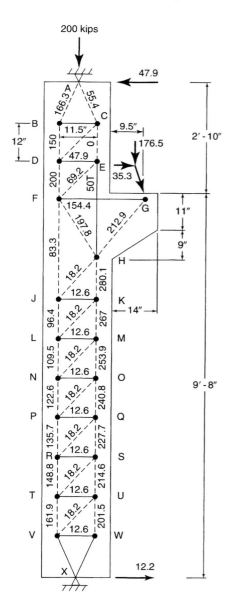

Fig. 18–18
Strut-and-tie model of a column and a corbel

Structural Action

A strut-and-tie model for a corbel and column is shown in Fig. 18–18. Within the corbel itself the structural action consists of an inclined compression strut, *G–H*, and a tension tie, *F–G*. Shears induced in the columns above and below the corbel are resisted by tension in the column ties and compression forces in struts between the ties.

In tests, corbels display several typical modes of failure, the most common of which are: yielding of the tension tie; failure of the end anchorages of the tension tie, either under the load point or in the column; failure of the compression strut by crushing or shearing; and local failures under the bearing plate. If the tie reinforcement is hooked downward, as shown in Fig. 18–19a, the concrete outside the hook may split off, causing failure. The tie should be anchored by welding it to a cross bar or plate. Bending the tie bars in a horizontal loop at the outer face of the corbel is also possible, but may be difficult to do and may require extra cover. If the corbel is too shallow at the outside end, there is a danger that cracking may extend through the corbel as shown in Fig. 18–19b. For this reason ACI Sec. 11.9.2 requires the depth of the corbel to be $0.5d$ at the outside edge of the bearing plate.

Design of Corbels

Two closely related design procedures for corbels will be presented, design using strut-and-tie models, and design according to ACI Sec. 11.9. The strut-and-tie method is a little more versatile than the ACI method, but both give essentially the same results within the range of application of the ACI Code.

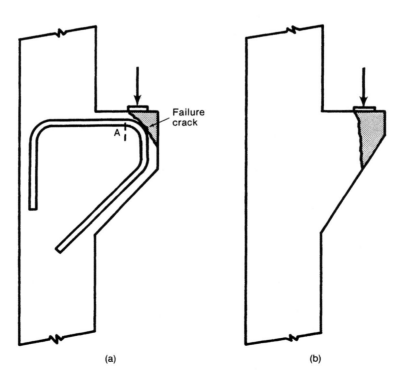

Fig. 18–19
Failures of corbels due to
poor detailing.

(a)

(b)

Discontinuity Regions, Corbels, Deep Beams, and Joints

EXAMPLE 18–3   Design of a Corbel Using a Strut-and-Tie Model

Design a corbel to transfer a precast beam reaction to a supporting column. The end of the beam is 14 in. wide. The factored shear, V_u, to be transferred is 150 kips. The column is 16 in. square and is supported as shown in Fig. 18–18. The beam is restrained against longitudinal shrinkage. Use $f'_c = 5000$ psi and $f_y = 60,000$ psi.

1.   Compute the distance, a, from the column to V_u. Assume a 12-in.-wide bearing plate. From ACI Sec. 10.17.1, the allowable bearing stress is

$$\phi 0.85 f'_c = 0.70 \times 0.85 \times 5 \text{ ksi} = 2.98 \text{ ksi}$$

The required width of the bearing plate is

$$\frac{150 \text{ kips}}{2.98 \text{ ksi} \times 12 \text{ in.}} = 4.2 \text{ in.}$$

Use a 12 in. $\times$ 5 in. bearing plate. Assuming that the beam overhangs the bearing plate by 6 in. and there is a 1 in. space between the face of the column and the end of the beam, the distance a is 9.5 in., as shown in Fig. 18–18.

2.   Compute the minimum depth. We shall base this on ACI Sec. 11.9.3.2.1: max. $V_n = 0.2 f'_c b_w d$ but not more than $V_n = 800 b_w d$ lb. $\phi V_n \geq V_u$. For 5000-psi concrete, $0.2 f'_c$ exceeds 800 psi; therefore, the second equation governs. Thus

$$\text{minimum } d = \frac{V_u}{\phi \times 800 b_w}$$

$$= \frac{150,000 \text{ lb}}{0.85 \times 800 \times 16 \text{ in.}} = 13.8 \text{ in.}$$

Thus the smallest corbel we could use is a corbel with $b = 16$ in., $h = 16$ in., and $d = 14$ in. For conservatism we shall use $h = 20$ in. and $d = 20$ in. $- (1\frac{1}{2}$ in. cover $+ \frac{1}{2}$ bar diameter), which equals 18 in. The corbel will be the same width as the column ($b = 16$ in.) to simplify the forming.

3.   Compute the forces on the corbel. The factored shear is 150 kips. Since the beam is restrained against shrinkage, we shall assume the normal force to be (ACI Sec. 11.9.3.4)

$$N_{uc} = 0.2 V_u = 30 \text{ kips}$$

We shall analyze the strut-and-tie model for the nominal forces, V_n and N_n. Since this is primarily a shear problem, we shall use $\phi = 0.85$

$$V_n = \frac{V_u}{\phi} = \frac{150}{0.85}$$

$$= 176.5 \text{ kips}$$

$$N_n = 35.3 \text{ kips}$$

4.   Lay out the strut-and-tie model. In laying out the strut-and-tie model, three assumptions were made. The portions of the column above and below the corbel were divided into 12-in.-high segments as shown in Fig. 18–18. Nodes B through F and H through W were assumed to be at the center of the longitudinal steel layers, taken as 2.25 in. from the surfaces of the column. The load acting on the corbel was represented by an inclined force acting through a point on the top of the corbel at 9.5 in. from the face of the column. Node G is at the intersection of this inclined force and the top steel in the corbel (2 in. below the top) and hence is at (9.5 in. $+$ 0.2 $\times$ 2 in.) $=$ 9.9 in. from the face of the column.

5.   Solve for the reactions. Summing moments about the hinge at node X gives the horizontal reaction at node A as

$$R_A = \frac{176.5 \text{ kips} \times (9.9 \text{ in.} + 16/2 \text{ in.}) + 35.3 \text{ kips} \times (9.5 \text{ ft} \times 12 \text{ in./ft})}{12.5 \text{ ft} \times 12 \text{ in./ft}}$$
$$= 47.9 \text{ kips to left in Fig. 18–18}$$

$$R_X = 12.6 \text{ kips to right in Fig. 18–18}$$

18–4   Brackets and Corbels

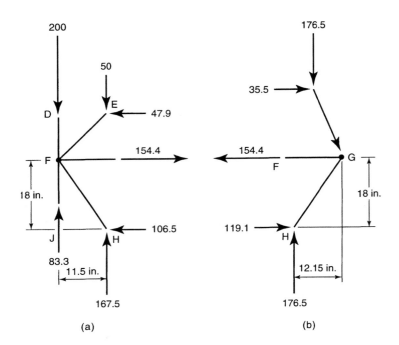

Fig. 18–20
Forces in members
meeting at nodes F and G—
Example 18–3.

(a)

(b)

6. Solve for the strut and tie forces. Starting at node *A*, solve the forces in the struts and ties using the method of joints. Figure 18–20 illustrates the solution of two joints in the corbel region.

Joint *G*: The applied loads, V_n and N_n, are 176.5 and 35.3 kips, as shown in Fig. 18–20b. The vertical force is equilibrated by the vertical component in strut *G–H*, which is thus 176.5 kips. Strut *G–H* has vertical and horizontal projections of 18 in. and 12.15 in., respectively, and a length of 21.7 in. From similar triangles, the horizontal component of the strut force is 119.1 kips and the axial force is 212.9 kips. Summing horizontal forces gives the tension in tie *F–G* as 154.4 kips.

Joint *F*: The forces in struts *D–F* and *E–F* have been solved by the time this joint is reached and are as shown in Fig. 18–20a. Summing horizontal forces gives the horizontal component of the force in strut *F–H* as 106.5 kips. Strut *F–H* has vertical and horizontal projections of 18 in. and 11.5 in., respectively, and a length of 21.4 in. From similar triangles, the vertical and axial components in strut *F–H* are 166.7 kips and 197.8 kips, respectively. Summing vertical forces gives the force in strut *F–J* as 83.3 kips.

The strut and tie forces are computed progressively joint by joint. When joint *X* is reached, the bar forces should add up to the computed reactions.

7. Compute the widths of the struts and see if they will fit in the space available. The struts are bottle-shaped stress fields that probably will crack longitudinally. They do not have tension forces transferred across them by reinforcement, except stresses induced by the opening of the cracks. From Eq. 18–3 and Table 18–1, the effective concrete strength of the struts is

$$f_{ce} = \nu_1 \nu_2 f_c' \quad \text{where } \nu_1 = 0.80 \text{ and } \nu_2 = 0.55 + \frac{15}{\sqrt{f_c'}} = 0.762$$

$$= 0.61 f_c' = 3049 \text{ psi}$$

The largest strut force is in *H–K*. We shall check the width of this strut first:

$$\text{width} = \frac{280.1 \text{ kips}}{3.05 \text{ ksi} \times 16 \text{ in.}}$$
$$= 5.74 \text{ in.}$$

Discontinuity Regions, Corbels, Deep Beams, and Joints

The axis of strut H–K is at 2.25 in. from the face of the column, which allows this strut to have a maximum width of 4.5 in. The node is too close to the edge of the column to allow the full width of the strut within the concrete.

Check if there is enough width for node H. The bottom face of node H anchors the 280.1-kip force in strut H–K. In addition, the tension in tie E–H must be developed by bond below node H, and hence this tie force applies a compressive force on the bottom of node H. Node H anchors one tie. From Eq. 18–3 and Table 18–1, the effective concrete strength in this node is

$$f_{ce} = \nu_1 \nu_2 f_c' \qquad \text{where } \nu_1 = 0.85 \text{ and } \nu_2 = 0.762$$

$$= 3240 \text{ psi}$$

$$\text{width of nodal zone required} = \frac{280.1 \text{ kips} + 50.0 \text{ kips}}{3.24 \text{ ksi} \times 16 \text{ in.}}$$

$$= 6.37 \text{ in.}$$

Only 4.5 in. is available. Node H must be moved farther from the column face. We shall move node H and the nodes below it so that they are located 4 in. from the face of the column. Because tie E–H is formed by the vertical steel in the column, we shall not move node E. There will be a slight inconsistency at node H since E–H does not act through H. This will be ignored.

8. Reanalyze the strut and tie forces. The strut and tie forces for the new geometry are shown in Fig.18–21. The forces in the critical struts have all gone up as has the force in tie F–G. The computed widths of the struts are also shown, based on $f_{ce} = 3049$ psi. The width of the bottom of the nodal zone at H needed to anchor H–K and E–H is

$$\text{width} = \frac{339.4 + 50}{3.24 \times 18} = 7.98 \text{ in.}$$

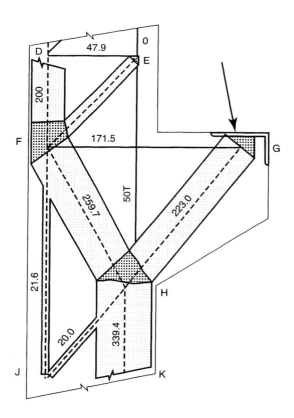

Fig. 18–21
Forces in members—revised geometry—Example 18–3

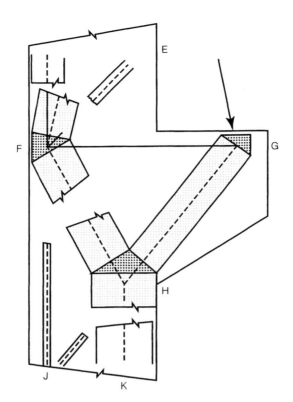

Fig. 18–22
Nodes resulting from re-
solved member forces—
Example 18–3.

There is just enough width for this node. The strut widths all fit into the outline of the corbel region. This solution will be accepted.

The nodal zones shown in Figs. 18–9 and 18–10 were all triangular, because three forces came together at each node. The nodes at F and H are not triangular and the intersection of H–J and H–K is awkward. In Fig. 18–22 the forces in struts D–F and E–F are resolved into one strut as are the forces in F–H and F–J. When this is done, the nodal zone becomes triangular, as does the one at H.

9. Select the reinforcement. The strut and tie forces in Fig. 18–21 are an acceptable solution for the corbel region. We shall now compute the amount of reinforcement required. Note that ϕ has already been included in the analysis of the bar forces.

Tie F–G:

$$A_s = \frac{171.6 \text{ kips}}{60 \text{ ksi}} = 2.86 \text{ in.}^2$$

Could use 5 No. 7 bars, $A_s = 3.00$ in.^{2} or 4 No. 8 bars, $A_s = 3.16$ in.^{2}.

Steel in the column: At the bottom of the corbel the column resists an axial force of $P_u/\phi = 200$ kips $+ 176.5$ kips $= 376.5$ kips, where ϕ was taken as 0.85. The moment about the centroid of the column at the bottom of the corbel is

$$M_u/\phi = 176.5 \times 17.9 + 35.3 \times 18 - 47.9 \times 54$$

$$= 1208 \text{ in.-kips}$$

Entering interaction diagrams with $P_u = 376.5 \times \phi = 376.5 \times 0.85$ kip and $M_u = 1208 \times 0.85$, we find that $\rho_t = 0.01$ is adequate. A similar check above the corbel also indicates that $\rho_t = 0.01$ is adequate. Thus $A_{st} = 2.56$ in.^{2} is required. **Use 4 No. 8 bars as the column reinforcement and 4 No. 8 bars with 90° hooks in the column as the corbel tie reinforcement.**

Confining reinforcement: The effective concrete strengths of the struts were taken from Table 18–1 for the case of bottle-shaped struts. These must have transverse reinforcement given by Eq. 18–5 to restrain the longitudinal cracking that may occur in such struts. Within the depth of the corbel this is governed by strut F–H, which has an axial force of 259.7 kips. Strut F–H has a length,

Discontinuity Regions, Corbels, Deep Beams, and Joints

ℓ, of 16.25 in. The widths, a, of the nodes at F and H are 259.7 kips/(3.24 ksi × 16 in.) = 5.01 in. The effective width of strut F–H, $b_{ef} = a + \ell/6 = 5.01 + 16.25/6 = 7.72$ in. From Eq. 18–5

$$A_s f_y \geq \Sigma\left[\frac{C}{4}\left(1 - \frac{a}{b_{ef}}\right)\right]$$

$$\geq 2\left[\frac{259.7}{4}\left(1 - \frac{5.01}{7.72}\right)\right]$$

$$= 45.6 \text{ kips}$$

$$A_s = 0.76 \text{ in.}^2$$

Use 2 horizontal closed No. 4 stirrups in the depth of the corbel. Below the corbel, strut H–K can be considered to act as the compression zone in the column. This is a prismatic stress field and does not need confinement to resist cracking due to bottle-shaped stress fields. We shall, however, provide two closed column ties immediately below the corbel.

10. Establish the anchorage of tie F–G at node G. The discontinuous end of tie F–G must be anchored to develop $A_s f_y$ at node G. This can be done by welding the bars to a transverse plate, angle, or bar. A welded cross-angle will be provided as shown in Fig. 18–23. If the horizontal force, N_{uc}, is required for the equilibrium of the structure, some direct connection would be required from the baseplate to the tension tie. This is not the case here.

11. Consider all other details. To prevent cracks similar to those shown in Fig. 18–19b, ACI Sec. 11.9.2 requires that the depth at the outside edge of the bearing area be at least 0.5d. ACI Sec. 11.9.4 requires that the stirrups be placed within $\frac{2}{3}d$ of the tension tie. Strut-and-tie theory suggests that they should be in the bottle-shaped regions at each end of the struts. We shall satisfy the ACI rule as much as is possible. ACI Sec. 11.9.7 requires that the anchorage of the tension tie be outside the bearing area. Finally, 2 No. 4 bars are provided to anchor the front ends of the horizontal stirrups. All of these aspects are satisfied in the final corbel layout similar to that given in Fig. 18–23. (Figure 18–23 shows the corbel designed by the ACI Code procedure, which required less steel in tie F–G, and more horizontal stirrups.)

The column should also be checked for shear. ∎

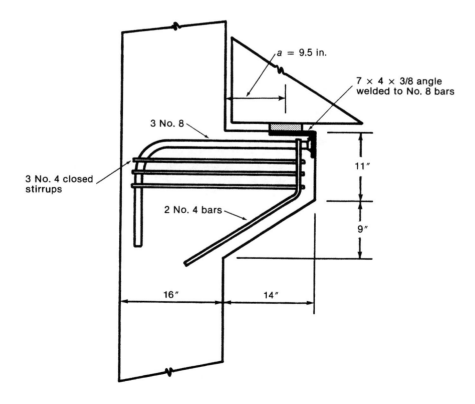

Fig. 18–23
Corbel design—
Example 18–4.

Design of Corbels by ACI Code Method

ACI Sec. 11.9 presents a design procedure for brackets and corbels. It is based in part on the strut-and-tie truss model, and in part on shear friction. The design procedure is limited to a/d ratios of 1.0 or less, because the strut action is assumed to be less efficient for longer corbels and because there is little test data for longer brackets.

The section at the face of the support is designed to resist the shear V_u, the horizontal tensile force N_{uc}, and a moment of $[V_u a + N_{uc}(h - d)]$, where the moment has been calculated relative to the tension steel at point G in Fig. 18–22. The maximum shear strength, V_n, shall not be taken greater than:

Normal-weight concrete: $0.2f'_c b_w d$, but not more than $800b_w d$ lb

All-lightweight or sand-lightweight concrete: $(0.2-0.07a/d)f'_c b_w d$, but not more than $(800-280a/d)b_w d$ lb

In design, the size of the corbel is selected so that $V_u \le \phi V_n$, based on the maximum shear strength. If a high value of V_n is used, cracking at service loads may lead to serviceability problems. The designer then calculates:

1. The area, A_{vf}, of shear friction steel required using Eq. 18–6:

$$V_n = A_{vf} f_y \mu \tag{18–6}$$

$$\text{(ACI Eq. 11–25)}$$

2. The area, A_f, of flexural reinforcement required to support a moment of $[V_u a + N_{uc}(h - d)]$ based on ACI Chap. 10 (Chap 4 of this book), and

3. The area, A_n, of direct tension reinforcement required by Eq. 18–7 to resist the tension force N_{uc}, where

$$A_n = \frac{N_{uc}}{\phi f_y} \tag{18–7}$$

In all these calculations, ϕ is taken equal to 0.85 since the behavior is dominated by shear.

The resulting area of tensile steel, A_s, and the placement of the reinforcement within the corbel is specified in ACI Secs. 11.9.3.5 and 11.9.4. In the corbel tests reported in Refs. 18–13 and 18–14, the best behavior was obtained in corbels that had some horizontal stirrups in addition to the tension tie shown in Fig. 18–18. Accordingly, ACI Sec. 11.9.3.5 requires that two reinforcement patterns be considered and the one giving the greatest area, A_s, be used:

1. A tension tie with an area $A_s = A_f + A_n$, plus horizontal stirrups having an area $A_f/2$

2. A tension tie with an area $A_s = (2A_{vf}/3) + A_n$, plus horizontal stirrups having an area $A_{vf}/3$

The horizontal stirrups are to be placed within $\frac{2}{3}d$ below the tension tie.

Because the tension tie in the truss in Fig. 18–18 is assumed to be stressed to f_y in tension between the loading plate and the column, it must be anchored outside the loading plate for that tension. This is done in one of several ways. It can be anchored by welding to an angle or bar at right angles to the tie (Fig. 18–23), or by welding to a transverse reinforcing bar of the same diameter as the tie, or by bending the bar in a horizontal loop. Although the tension tie could also be anchored by bending the bar in a vertical hook, as shown in Fig. 18–19a, this is discouraged since failures have occurred, as shown in that figure. ACI Sec. 11.9.7 requires that the outer edge of the bearing plate be inside the cross-bar or, in the case of the detail shown in Fig. 18–19a, inside the start of the bend (inside point A). The use of a welded transverse reinforcing bar requires special welding techniques (see ACI Sec. 3.5.2).

For corbels supporting precast beams, shrinkage and creep of the beams will generally cause a horizontal force N_{uc} (Fig. 18–18). Unless provisions are made to avoid such tensions, ACI Sec. 11.9.3.4 requires that corbels be designed for N_{uc} not less than $0.2V_u$. Ways of calculating N_{uc} are given in Ref. 18–15.

EXAMPLE 18–4   Design of a Corbel-ACI Code Method

Design of a corbel to transfer a precast beam reaction to a supporting column. The factored shear to be transferred is 150 kips. The column is 16 in. square. The beam being supported is restrained against longitudinal shrinkage. Use f'_c = 5000 psi and f_y = 60,000 psi.

1.   Compute the distance, a, from the column to V_u. Assume a 12-in.-wide bearing plate. From ACI Sec. 10.15.1, the allowable bearing stress is

$$\phi 0.85 f'_c = 0.70 \times 0.85 \times 5 \text{ ksi} = 2.98 \text{ ksi}$$

The required width of the bearing plate is

$$\frac{150 \text{ kips}}{2.98 \text{ ksi} \times 12 \text{ in.}} = 4.2 \text{ in.}$$

Use a 12 in. × 5 in. bearing plate. Assuming that the beam overhangs the bearing plate by 6 in. and a 1-in. gap is left between the end of the beam and the face of the column, the distance a is 9.5 in., as shown in Fig. 18–23.

2.   Compute the minimum depth, d. Base this calculation on ACI Sec. 11.9.3.2.1:

$$\phi V_n \geq V_u$$

$$\text{max. } V_n = 0.2 f'_c b_w d \text{ but not more than } V_n = 800 b_w d \text{ lb}$$

For 5000-psi concrete, $0.2f'_c$ exceeds 800 psi; therefore, the second equation governs. Thus

$$\text{minimum } d = \frac{V_u}{\phi \times 800 b_w}$$

$$= \frac{150,000 \text{ lb}}{0.85 \times 800 \times 16} = 13.8 \text{ in.}$$

Thus the smallest corbel we could use is a corbel with b = 16 in., h = 16 in., and d = 14 in. For conservatism we shall use h = 20 in. and d = 20 in. − ($1\frac{1}{2}$ in. cover + $\frac{1}{2}$ bar diameter), which equals 18 in. The corbel will be the same width as the column (16 in. wide) to simplify forming.

3.   Compute the forces on the corbel. The factored shear is 150 kips. Since the beam is restrained against shrinkage we shall assume the normal force to be (ACI Sec. 11.9.3.4)

$$N_{uc} = 0.2V_u = 30 \text{ kips}$$

The factored moment is

$$M_u = V_u a + N_{uc}(h - d)$$

$$= 150 \text{ kips} \times 9.5 \text{ in. } + 30 \text{ kips}(20 \text{ in.} - 18 \text{ in.})$$

$$= 1485 \text{ in.-kips}$$

4.   Compute the shear friction steel, A_{vf}. From Eqs. 6–13 and 18–6,

$$\phi V_n \geq V_u$$

$$A_{vf} = \frac{V_n}{\mu f_y} = \frac{V_u}{\phi \mu f_y}$$

where μ = 1.4λ for a shear plane through monolithic concrete and λ = 1.0 for normal-weight concrete. Therefore,

$$A_{vf} = \frac{150,000 \text{ lb}}{0.85(1.4 \times 1.0)60,000 \text{ psi}} = 2.10 \text{ in.}^2$$

5. **Compute the flexural reinforcement, A_f.** A_f is computed using Eq. 4–12b (with A_s replaced by A_f):

$$M_u = \phi A_f f_y \left(d - \frac{a}{2} \right)$$

where $\phi = 0.85$ (ACI Sec. 11.9.3.1) and

$$a = \frac{A_f f_y}{0.85 f_c' b}$$

As a first trial we shall assume that $(d - a/2) = 0.9d$. Thus

$$A_f = \frac{M_u}{\phi f_y (0.9d)}$$

$$= \frac{1485 \text{ in.-kips}}{0.85 \times 60 \times 0.9 \times 18} = 1.80 \text{ in.}^2$$

Since this is based on a guess for $(d - a/2)$, we shall compute a and recompute A_f:

$$a = \frac{1.80 \times 60}{0.85 \times 5 \times 16} = 1.59 \text{ in.}$$

$$A_f = \frac{1485 \text{ in.-kips}}{0.85 \times 60 \text{ ksi} (18-1.59/2) \text{ in.}}$$
$$= 1.69 \text{ in.}^2$$

Check if $\rho \leq 0.75\rho_b : a/d = 1.59/18 = 0.09$. From Table A–4, this is less than the value of a/d corresponding to $\rho = 0.75\rho_b$. Therefore, use $A_f = 1.69 \text{ in.}^2$.

6. **Compute the reinforcement, A_n, for direct tension.** From Eq. 18–7 and ACI Sec. 11.9.3.4,

$$A_n = \frac{N_{uc}}{\phi f_y} = \frac{30 \text{ kips}}{0.85 \times 60 \text{ ksi}}$$

$$= 0.59 \text{ in.}^2$$

7. **Compute the area of the tension tie reinforcement, A_s.** From ACI Sec. 11.9.3.5, A_s shall be the larger of

$$(A_f + A_n) = 1.69 + 0.59 = 2.28 \text{ in.}^2, \text{ or}$$

$$\left(\frac{2A_{vf}}{3} + A_n \right) = 1.40 + 0.59 = 1.99 \text{ in.}^2$$

Minimum A_s (ACI Sec. 11.9.5):

$$A_{s(\min)} = \frac{0.04 f_c'}{f_y} b_w d = 0.96 \text{ in.}^2$$

Therefore, $A_s = 2.28 \text{ in.}^2$. Try 3 No. 8 bars, giving $A_s = 2.37 \text{ in.}^2$.

8. **Compute the area of horizontal stirrups.**

$$0.5(A_s - A_n) = 2.28 - 0.59 = 0.85 \text{ in.}^2$$

Possible choices are 4 No. 3 double-leg stirrups, area = 0.88 in.², or 3 No. 4 double-leg stirrups, area = 1.20 in.². ACI Sec. 11.9.4 requires that these be placed within $(2/3)d$ measured from the tension tie.

9. **Establish the anchorage of the tension tie into the column.** The column is 16 in. square. Try a 90° standard hook. From ACI Sec. 12.5.1,

$$\ell_{dh} = \frac{1200 d_b}{\sqrt{f_c'}} \times 0.7 \times 0.8 = 9.50 \text{ in.}$$

measured from the face of the column. Therefore, **use 3 No. 8 bars hooked into the column.** The hooks are inside the column cage.

10. Establish the anchorage of the outer end of the bars. The outer end of the bars must be anchored to develop $A_s f_y$. This can be done by welding the bars to a transverse plate, angle, or bar. If the horizontal force, N_{uc}, is required for equilibrium of the structure, some direct connection would be required from the beam base plate to the tension tie. This is not the case here and a welded cross-angle will be provided, as shown in Fig. 18–23.

11. Consider all other details. To prevent cracks similar to those shown in Fig. 18–19b, ACI Sec. 11.9.2 requires that the depth at the outside edge of the bearing area be at least $0.5d$. ACI Sec. 11.9.4 requires that the stirrups be placed within $\frac{2}{3}d$ of A_s. ACI Sec. 11.9.7 requires that the anchorage of the tension tie be outside the bearing area. Finally, 2 No. 4 bars are provided to anchor the front ends of the stirrups. All of these aspects are satisfied in the final corbel layout given in Fig. 18–23. ∎

Comparison of the Strut-and-Tie Method and the ACI Method

The strut-and-tie method required more steel in the tension tie and less confining reinforcement than the ACI method. The strut-and-tie method explicitly considered the effect of the corbel on the forces in the column. The strut-and-tie method could also be used for corbels which have a/d greater than the limit of 1.0 given in ACI Sec. 11.9.1. For $a/d > 1$, the confining stirrups would be more efficient in restraining the splitting of the strut if they were vertical.

18–5 DAPPED ENDS

The ends of precast beams are sometimes supported on an end projection that is reduced in height as shown in Fig. 18–24. Such a detail is referred to as a *dapped-end*. Although several design procedures exist, the best method of design is by means of strut-and-tie models. Tests of such regions are reported in Refs. 18–12, 18–16, and 18–17.

Four common strut-and-tie models for dapped end regions are compared to the crack pattern observed in tests in Fig. 18–24. Cracking originates at the reentrant corner of the notch, point *A* in Fig. 18–24a. The strut-and-tie models in Fig. 18–24b to d all involve a vertical tie *B–C* at the end of the notched portion of the beam and an inclined strut *A–B* over the reaction. In tests, specimens with tie *B–C* composed of closed vertical stirrups with 135° bends around longitudinal bars in the top of the beam performed better than specimens with open-topped stirrups.[18-17] The horizontal component of the compressive force in A–B is equilibrated by the tension tie A–D. The three strut-and-tie models differ in the manner in which the horizontal tie is anchored at *D*. The model in Fig. 18–24c has the advantage that the force in tie *C–E* is lower, and hence easier to anchor, than the corresponding force in tie *C–F* in Fig. 18–24b. In Fig. 18–24d, tie *A–D* is anchored by strut *B–D*, which will be crossed by cracks, as shown in Fig. 18–24a. This suggests that Fig. 18–24d is not a feasible model.

The strut-and-tie model in Fig. 18–24e has an inclined hanger tie *B–C* and a vertical strut over the reaction. Care must be taken to anchor the tie *B–C* at its upper end. It is customary to provide a horizontal tie at *A* to resist any tensile forces due to restrained shrinkage of the precast beam. In tests,[18-17] dapped ends designed using the models in Fig. 18–24b or c performed equally well as ends designed using the model in Fig. 18–24e. A compound model, designed assuming that half the reaction was resisted by each of these two types of strut-and-tie models, also performed well in tests.

In laying out a dapped end support, it is good practice to have the depth of the extended part of the beam at least half of the overall height of the beam. The extended part of the beam should be deep enough that the inclined compression strut A–B at the support is

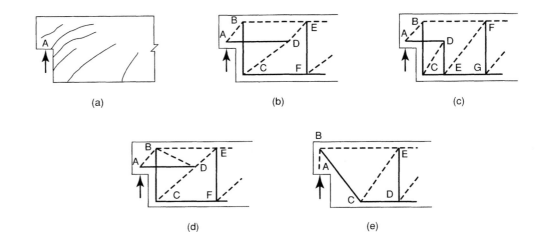

Fig. 18–24
Strut-and-tie models for a
dapped end.

(a) (b) (c)

(d) (e)

no flatter than 45°. Otherwise, the forces in this strut and the tie that meets it at the support become too large to deal with in a simple manner. Great care must be taken to anchor the bars in the vicinity of the dapp.

EXAMPLE 18–5   Design of a Dapped End Support

A precast beam supports a dead load of 2 kips/ft and a live load of 2.5 kips/ft on a 20-ft span. The beam is 30 in. deep by 15 in. wide and is made from 3000-psi concrete and Grade 60 reinforcement. The longitudinal steel is 4 No. 8 bars. Design the reinforcement in the support region.

1.   Isolate the D-region and compute the forces on the boundaries of the D-region. The D-region includes the extended part plus 30 in. of the beam length. The reaction is 70.5 kips vertically, and although no horizontal loads are present, we shall assume a horizontal reaction of 0.2 times the vertical reaction, equal to 14.1 kips, in keeping with ACI Sec. 11.9.3.4. Because this region can be considered primarily a shear dominated region, we shall take $\phi = 0.85$. Dividing the reactions by ϕ gives the reactions shown in Fig. 18–25. For simplicity we shall neglect the uniform dead and live loads acting on the D-region. Assuming that the reactions act 2 in. from the end, the moment at the right end is $34 \times 82.9 = 2819$ in.-kips. This moment can be represented as a couple with compressive and tensile forces of $2819/(0.9 \times 27) = 116.0$ kips. The axial tensile force is resolved into two forces of 8.3 kips. The forces on the edges of the D-region are shown in Fig. 18–25.

2.   Select a strut-and-tie model. The strut-and-tie model in Fig. 18–24c will be used.

3.   Estimate the locations of the nodes.
Node A: The size of the bearing at the support will be based on the effective concrete strengths, f_{ce}, given by Eq. 18–3 and Table 18–1. This node anchors one tension tie and from Table 18–1, $f_{ce} = 0.85 \nu_2 f_c'$, where

$$\nu_2 = 0.55 + \frac{15}{\sqrt{f_c'}} = 0.824$$

$$f_{ce} = 2101 \text{ psi}$$

The required bearing area is

$$\text{bearing area} = \frac{82,900}{2101} = 39.5 \text{ in.}^2$$

Assuming that the bearing will be an angle extending across the width of the end of the beam, we need an angle of leg width 2.63 in. We shall use a $4 \times 4 \times \frac{1}{2}$ in. angle, 15 in. long. We shall assume that the reaction acts 2 in. from the end of the member. Node A is located at the intersection of

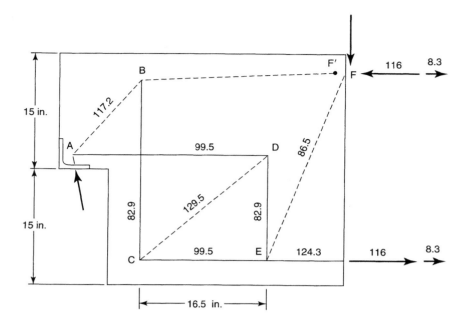

Fig. 18–25
Strut-and-tie model—
Example 18–5.

the inclined reaction and the bar A–D, which we shall assume is located 2 in. above the bottom of the extended part of the beam.

Node B: Node B is at the centerline of tie B–C. Cutting an inclined section inward from the reentrant corner, we find that the force in tie B–C is equal to the vertical reaction, 82.9 kips. The area of steel required in tie B–C is 82.9 kips/60 ksi = 1.38 in^2. We shall make tie B–C out of 4 No. 4 closed stirrups. Because the entire region depends on the anchorage of these stirrups, we shall hook them 135° around the 2 No. 4 bars provided in the upper corners of the stirrups. The stirrups will be placed with minimum $1\frac{1}{2}$ in. cover and a spacing of 1 in. on centers. The centerline of the group of stirrups is thus 4 in. from the end of the deeper portion of the beam.

We shall assume that the node is enclosed on the top by the top of the closed stirrups that make up tie B–C. This puts the top of the node at 2 in. below the top of the beam. Estimating that the node is 3 in. in overall depth puts the center of node B at 3.5 in. below the top of the beam.

Node C: Node C is at the intersection of the axis of tie B–C and the longitudinal steel. We will take it at 4 in. horizontally and 3 in. vertically from the lower corner.

Nodes D and E: Nodes D and E will be located during the solution of the strut and tie forces.

4. Compute the strut and tie forces. We shall ignore the slight slope in strut B–F. This will lead to a slightly higher force in tie B–C and a slight lack of closure in the force diagram. Use will be made of a scale drawing of the strut-and-tie model in Fig. 18–25 when computing the internal forces. The drawing should be large enough to scale lengths from.

Joint B: Strut A–B is at a 45° angle. The force in tie B–C is 82.9 kips tension. The vertical force in A–B is 82.9 kips, giving an axial compressive force of 117.2 kips. The force in B–F is 82.9 kips compression.

Joint A: The force in A–B has horizontal and vertical components of 82.9 kips. The reaction has vertical and horizontal components of 82.9 and 16.6 kips, respectively. Summing horizontal forces gives the force in A–D as 82.9 + 16.6 = 99.5 kips.

Joint D: The force in A–D is 99.5 kips and the tension in tie D–E is 82.9 kips. The force in C–D is $\sqrt{99.5^2 + 82.9^2} = 129.5$ kips. From similar triangles, D is located 16.5 in. horizontally from C.

Joint C: The horizontal component of the force in C–D is 99.5 kips. The horizontal force needed for equilibrium is 99.5 kips tension in C–E.

Joint E: Tie C–E has a tensile force of 99.5 kips and tie E–G a tensile force of 124.3 kips. Tie D–E has a tensile force of 82.9 kips. For equilibrium, strut E–F must have a vertical component of 82.9 kips and a horizontal component of 24.8 kips, giving an inclined compressive force of 86.5 kips. When the location of node F is worked out, it falls at F' rather than F (see Fig. 18–25). This lack of closure arises from ignoring the slope of B–F and measurement and round off errors.

Joint F: Summing horizontal and vertical forces show that joint F is in equilibrium.

5. Compute the strut widths and check whether they will fit. The first estimate of strut-and-tie forces is shown in Fig. 18–25. It is now necessary to compute the widths of the struts to see whether they will fit into the available space without overlapping. The effective concrete strength will be taken from Table 18–1 as $f_{ce} = 0.8v_2 f_c' = 1978$ psi. The struts will be assumed to have a thickness equal to the thickness of the beam, 15 in. except for A–B. In tests, the cover over the sides of the stirrups spalled off at node B. As a result, we shall assume that strut A–B has a width of 12 in.

$$\text{Strut } A\text{–}B \quad \text{width} = \frac{117.2}{1.98 \times 12} = 4.93 \text{ in.}$$

$$\text{Strut } B\text{–}F \quad \text{width} = \frac{82.9}{1.98 \times 15} = 2.79 \text{ in.}$$

$$\text{Strut } C\text{–}D \quad \text{width} = \frac{129.5}{1.98 \times 15} = 4.36 \text{ in.}$$

$$\text{Strut } E\text{–}F \quad \text{width} = \frac{86.5}{1.98 \times 15} = 2.91 \text{ in.}$$

Struts having these widths are shown in Fig. 18–26.

6. Compute the steel required in the ties.
Tie A–D:

$$A_s = \frac{\text{nominal tie force}}{f_y} = \frac{99.5 \text{ kips}}{60 \text{ ksi}}$$
$$= 1.66 \text{ in.}^2$$

Use 4 No. 6 bars welded to the angle. Development length $\ell_d = 32.9$ in. Theoretically, the bars should be anchored toward midspan from node D. Extend the bars 33 in. past D.

Tie B–C: In step 3 we selected 4 No. 4 closed stirrups for tie B–C. These will be closed stirrups with 2 No. 4 bars inside the top corners. These four closed stirrups will be anchored with 135° bends alternately around one or the other of these bars.

Tie C–E: At midspan, the bottom steel is 4 No. 8 bars for flexure. This is enough for the 99.5 kip force in C–E. However, it is necessary to develop this force within the node at C. The length available for development of the bars within the node is 7 in., as shown in Fig. 18–26. The development length of a No. 8 bar is 54.8 in. The force that can be developed in the No. 8 bars is

$$\frac{7 \text{ in.}}{54.8 \text{ in.}} \times 4 \times 0.79 \text{ in.}^2 \times 60 \text{ ksi} = 24.2 \text{ kips}$$

This is not enough. Either weld a plate to the ends of the bars or provide horizontal U bars to anchor the force. We shall provide horizontal U bars. The area required is

$$\frac{99.5}{60} = 1.66 \text{ in.}^2$$

Use 2 No. 6 U bars with bends adjacent to the end of the beam. Place these above the No. 8 bars with clear spaces of 1 in. between the bars. Lap splice these $1.3\ell_d = 42.8$ in., say 3 ft 8 in. into the beam.

Tie D–E: Similar to tie B–C, this tie can also be 4 No. 4 double leg stirrups. We shall arbitrarily spread them out over a longer length of beam than done for tie B–C. Since these are provided primarily to anchor tie A–D, their upper anchorage is less critical, and normal U stirrups with 135° hooks can be used.

Tie F–G: We shall assume that tie F–G is provided by stirrups in the B-region to the right of F–G.

7. Check the stresses on the sides of the nodes. Table 18–1 indicates that the effective concrete strengths in the nodes are larger than the strengths of the struts entering them. As a result, it is generally not necessary to check bearing stresses on nodes. We shall check two cases.

Height of node C: The steel should be in a height approximately equal to the height of node to anchor the tie force based on concrete stressed at $f_{ce} = 0.75v_2 f_c'$, $= 1854$ psi, the value from Table 18–1 for a node anchoring more than one tension tie.

$$\text{Nodal area required} = \frac{124.3 \text{ kips}}{1.85 \text{ ksi}} = 67.2 \text{ in.}^2$$

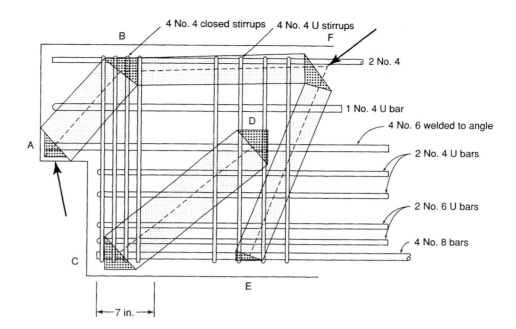

Fig. 18–26
Struts and reinforcement—
Example 18–5.

The width is 15 in. and a height of 4.48 in. is required. The No. 8 bars and No. 6 U bars take more than this, therefore OK.

Anchorage of strut A–B at node A: In tests,[18–17] the concrete outside the ties spalled off in the vicinity of node B. As a result we shall assume that strut A–B will have to be anchored on a thickness of 12 in. and a width equal to the width of the strut, 4.93 in. From Table 18–1 for a node anchoring one tie, $f_{ce} = 0.85v_2f'_c = 2100$ psi.

The force that can be anchored is $12 \times 4.93 \times 2.1$ ksi $= 124.3$ kips, which exceeds the force in A–B.

8. Provide the strut confining steel. In Sec. 18–2 the effective strength of the concrete in a strut was taken equal to the splitting strength of the strut unless reinforcement is provided transverse to the strut to restrain the splitting. The amount of this steel is given by Eq. 18–5:

$$A_s f_y \geq \Sigma\left[\frac{C}{4}\left(1 - \frac{a}{b_{ef}}\right)\right]$$

Strut A–B: Length of strut A–B = 13 in., the width, a, of the strut at A = 3.72 in. and at B = 4.93 in., the effective width is $b_{ef} = a + l/6 = 3.72 + 13/6 = 5.89$ in., the axial force in A–B = 117.2 kips:

$$A_s f_y \geq \frac{117.2}{4}\left(1 - \frac{3.72}{5.89}\right) + \frac{117.2}{4}\left(1 - \frac{4.93}{5.89}\right) = 10.8 + 4.78$$

$$\geq 15.6 \text{ kips}$$

$$A_s = \frac{15.6}{60 \text{ ksi}} = 0.26 \text{ in.}^2$$

Provide 1 No. 4 U bar 36 in. long.

Strut C–D: Length of strut C–D = 21.5 in., the width, a, of the strut at $C = 129.5/(1.85 \times 15) = 4.67$ in. and at D the same, $b_{ef} = a + l/6 = 4.67 + 21.5/6 = 8.25$ in., the axial force in A–B = 129.5 kips:

$$A_s f_y \geq 2\left[\frac{129.5}{4}\left(1 - \frac{4.67}{8.25}\right)\right] = 28.1 \text{ kips}$$

$$A_s = \frac{28.1}{60} = 0.47 \text{ in.}^2$$

18–5 Dapped Ends

785

This should be "approximately perpendicular" to the strut. We shall provide 2 No. 4 horizontal U bars with a total area of 0.80 in.[2].

Provide 2 No. 4 U bars 36 in. long. These are in addition to the two U bars provided to anchor tie C–E.

9. Design the stirrups for shear. Outside of the D-regions, shear is carried by concrete and stirrups.

The final reinforcement is shown in Fig. 18–26. ■

18–6 DEEP BEAMS

ACI Sec. 10.7.1 specifies that *deep beam action* must be considered when designing for *flexure* if ℓ_n/d is less than 5/2 for continuous spans or 5/4 for simple spans. Shorter members must be designed "taking into account the non-linear distribution of strain. . . ." ACI Sec. 11.8.1 specifies that deep beam action must be considered when designing for *shear* if ℓ_n/d is less than 5 and the beam is loaded at the top or compression face. These two definitions are somewhat arbitrary. A better definition is: A deep beam is a beam in which a significant amount of the load is carried to the supports by a compression thrust joining the load and the reaction. This occurs if a concentrated load acts closer than about $2d$ to the support, or for uniformly loaded beams with a span-to-depth ratio, ℓ_n/d, less than about 4 to 5.

Most typically, deep beams occur as *transfer girders*, which may be single span (Fig. 18–27) or continuous (Fig. 18–28). A transfer girder supports the load from one or more columns, transferring it laterally to other columns. Deep beam action also occurs in some walls and in pile caps. Although such members are not uncommon, no completely satisfactory method of design exists.

Fig. 18–27
Single-span deep beam.
(Photograph by J. G. MacGregor.)

Analyses and Behavior of Deep Beams

Elastic analyses of deep beams in the uncracked state are only meaningful prior to cracking. In a deep beam, cracking will occur at one-third to one-half of the ultimate load. After cracks develop, a major redistribution of stresses is necessary since there can be no tension across the cracks. The results of elastic analyses are of interest primarily because they show the distribution of stresses which cause cracking and hence give guidance as to the direction of cracking and the flow of forces after cracking. In Figs. 18–15a and 18–29a to 18–31a, the dashed lines are *compressive stress trajectories* parallel to the directions of the principal compressive stresses, and the solid lines are *tensile stress trajectories* parallel to the principal tensile stresses. Cracks would be expected to occur perpendicular to the solid lines (i.e., parallel to the dashed lines).

In the case of a single span beam supporting a concentrated load at midspan (Fig. 18–15), the principal compressive stresses act roughly parallel to the lines joining the load and the supports and the largest principal tensile stresses act parallel to the bottom of the beam. The horizontal tensile and compressive stresses on a vertical plane at midspan are shown in Fig. 18–15b. Although it cannot be seen from the figure, it is important to note that the flexural stress at the bottom is constant over much of the span. The stress trajectories in Fig. 18–15a can be simplified to the pattern given in Fig. 18–15c. Again, dashed lines represent compression struts and solid lines, tension ties. The angle θ varies approximately linearly from 68° (2.5:1 slope) for $\ell/d = 0.80$ or smaller, to 40° (0.85:1 slope) at $\ell/d = 1.8$. If such a beam were tested, the crack pattern would be as shown in Fig. 18–15d. Note that each of the three tension ties in Fig. 18–15c (*AB, CD,* and *EF*) has cracked. At

Fig. 18–28
Three-span deep beam, Brunswick building, Chicago. (Photograph by J. G. MacGregor.)

failure the shaded region in Fig. 18–15d would crush, or the anchorage zones at E and F would fail. The truss model in Fig. 18–15c could be simplified further to the model shown in Fig. 18–15e. This model does not explain the formation of the inclined cracks, however.

An uncracked, elastic, single-span beam supporting a uniform load has the stress trajectories shown in Fig. 18–29a. The distribution of horizontal stresses on vertical sections at midspan and the quarter point are plotted in Fig. 18–29b. The stress trajectories can be represented by the simple truss in Fig. 18–29c, or the slightly more complex truss in Fig. 18–29d. In the first case, the uniform load is divided into two parts, each represented by its resultant. In the second case, four parts were used. The angle θ varies from about $68°$ for $\ell/d = 1.0$ or smaller to about $55°$ for $\ell/d = 2.0$.[18-1] The crack pattern in such a beam is shown in Fig. 18–29e.

Figure 18–30a shows the stress trajectories for a deep beam supporting a uniform load acting on a ledge at the lower face of the beam. The compression trajectories form an arch with the loads hanging from it, as shown in Fig. 18–30b and c. The crack pattern in Fig. 18–30d clearly shows that the load is transferred upward by reinforcement until it acts on the compression arch, which then transfers the load down to the supports.

Similar diagrams are presented in Fig.18–31 for a multispan continuous deep beam. Here the angle θ is similar to that in a single-span deep beam. The crack pattern in a two-span continuous beam is shown in Fig. 18–31d.

The distribution of the flexural stresses in Fig. 18–29b and the truss models in Figs. 18–15 and 18–27 to 18–31 all suggest that the force in the longitudinal tension ties will be constant along the length of the deep beam. This implies that this force must be anchored at the joints over the reactions. Failure to do so is a major cause of distress in deep beams. ACI Sec. 12.10.6 alludes to this.

Strut-and-Tie Models for Deep Beams

Figure 18–4 shows a simple strut-and-tie model for a single-span deep beam. The loads, re-actions, struts, and ties in Fig. 18–4 are all laid out such that the centroids of each truss member and the lines of action of all externally applied loads coincide at each joint. This is necessary for joint equilibrium. In Fig. 18–4 the bars are shown with external end anchors. In a reinforced concrete beam, the anchorage would be accomplished with horizontal or vertical hooks, or in extreme cases, with an anchor plate as shown.

The truss shown in Fig. 18–4 can fail in one of three ways: (1) the tie could yield, (2) one of the struts could crush when the stress in the strut exceeded f_{ce}, or (3) a node could fail by being stressed greater than its effective compressive strength. Frequently, this involves a bearing failure at the loads or reactions. Since a tension failure of the steel will be more ductile than either a strut failure or a node failure, the beam should be pro-portioned so that the strength of the steel governs.

A second example consisting of a simple span beam with vertical stirrups subjected to a concentrated load at midspan is shown in Fig. 18–32. This is the sum of several trusses. One truss uses a direct compression strut running from the load to the support. This truss carries a shear V_c. The other truss uses the stirrups as vertical tension members and has compression fans under the load and over the reactions. The vertical force in each stirrup is computed assuming that the stirrup has yielded. The vertical force component in each of the small compression struts must be equal to the yield strength of its stirrup for the joint to be in equilibrium. The farthest left stirrup is not used, since one cannot draw a compression diagonal from the load point to the bottom of this stirrup without encroaching on the direct compression strut.

The compression diagonals radiating from the load point intersect the stirrups at the level of the centroid of the bottom steel, because a change in the force in the bottom steel is required to equilibrate the horizontal component of the force in the compression diago-

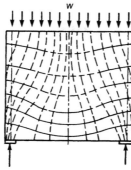

(a) Stress trajectories.

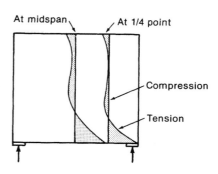

(b) Distribution of theoretical
horizontal elastic stresses.

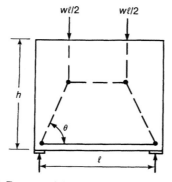

(c) Truss model.
$\theta = 68°$ if $\ell/h \le 1$
$= 54°$ if $\ell/h = 2$

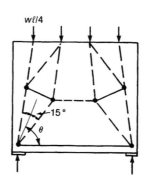

(d) Refined truss model.

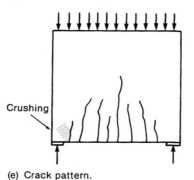

Fig. 18–29
Uniformly loaded deep beam.
(Adapted from Refs. 18–1
and 18–18.)

(e) Crack pattern.

nal. The force in the bottom steel is reduced at each stirrup by the horizontal component of the compression diagonal intersecting at that point. This is illustrated in Fig. 18–32b, where the stepped line shows the resulting tensile force in the bottom steel. The tensile force computed from beam theory, M/jd, is shown by a dashed line in the same figure. Note that the tensile force from the plastic truss analogy exceeds that from beam theory. This is in accordance with test results (see Fig. 6–26).[18–6]

Figure 18–33a shows a plastic truss model for a two-span continuous beam. Again the struts are shown in light shading and the hydrostatic elements in darker shading. At the interior support, two trusses carry the load. The upper truss shown in Fig. 18–33b utilizes the top reinforcement with a tie force, T_2, and the lower truss shown in Fig. 18–33c uses the

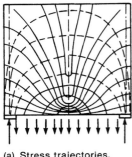

(a) Stress trajectories.

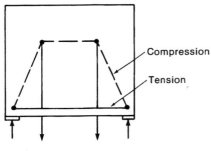

Compression

Tension

(b) Truss model.

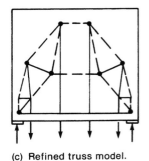

(c) Refined truss model.

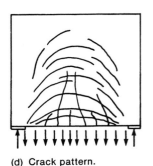

(d) Crack pattern.

Fig. 18–30
Deep beam loaded on the
bottom edge. (Adapted from
Refs. 18–1 and 18–18.)

bottom reinforcement which has a force T_1. The capacity of each truss can be computed from the geometry of the triangles and the $A_s f_y$ of the tension chord. The capacity of the beam is found by adding them together. The forces T_2 and T_1 are shown anchored at the load points and at the exterior support with bearing plates. Actually, the bars would be anchored by development or hooks beyond the locations of the bearing plates. Note that the tension forces, T_1 and T_2, are assumed to be constant between the bearing anchorage plates.

Design of Deep Beams

The ACI Code deals with deep beams in three sections. ACI Sec. 10.7.1 deals with flexural design and requires that the nonlinear distribution of stress (See Fig. 18–15b) and the possibility of lateral buckling be considered for deep beams. A continuous beam is considered to be deep for flexural design if the clear span-to-depth ratio, ℓ_n/d, is less than 2.5. The corresponding ratio is 1.25 for a simple span.

ACI Sec. 11.8 deals with the design of deep beams for shear and applies for members having ℓ_n/d less than 5.0 which are loaded on the top surface and supported on the bottom surface. Finally, ACI Sec. 12.10.6 requires that adequate end anchorage be provided for the tension steel in members where the reinforcement stress is not directly proportional to the moment, such as in deep beams.

At first glance, the ACI Code definitions of deep beams appear to conflict, but actually they do not. Special shear requirements are needed if ℓ_n/d is less than 5.0, but the beam can still be designed for flexure in the normal manner if ℓ_n/d exceeds 1.25 or 2.5 for simple or continuous spans, respectively. For ℓ_n/d less than these limits, the design should reflect the reduced lever arm indicated in elastic calculations (Fig. 18–15b).

A problem arises, however, in the case of a beam with a span-to-depth ratio, ℓ_n/d, of, for example 6, with a single concentrated load at d from one end. According to ACI Sec. 11.8.1,

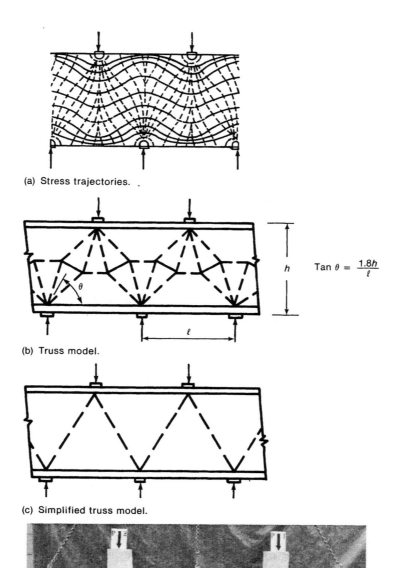

(a) Stress trajectories.

$$\text{Tan } \theta = \frac{1.8h}{\ell}$$

(b) Truss model.

(c) Simplified truss model.

Fig. 18–31
Multispan deep beam.
(Adapted from Ref. 18–1:
photograph courtesy of J.
G.MacGregor.)

this is not a deep beam. However, the short shear span acts as a deep beam. The other shear span is not "deep" and should be designed as a shallow beam.

The 1983 ACI Code provisions for deep beams for shear were amended in the 1986 Code Supplement. Two design procedures are presented in the supplement. For simple beams, design is based on ACI Eqs. 11–1, 11–2, 11–27, 11–29, and 11–30. The last three were based on a shear-friction analysis and are not borne out by more recent tests, although the design value of V_n is conservative for simple beams. For continuous beams the revised ACI Sec. 11.8 requires that design be based on slender beam design procedures or: ". . . on methods satisfying equilibrium and strength requirements. . . ." This procedure involves the use of strut-and-tie models such as those in Figs. 18–15e, 18–29, and so on. This method is recommended for design since a clear load path is provided. Regardless of which procedure is used, however, the minimum reinforcement requirements of ACI Secs. 11.8.9 and 11.8.10 must be satisfied.

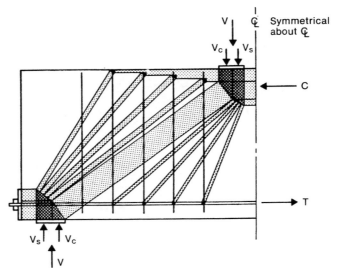

(a) Plastic truss model

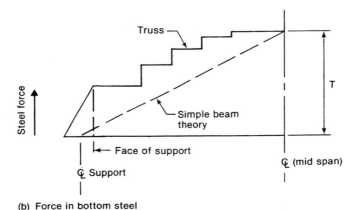

(b) Force in bottom steel

Fig. 18–32
Strut-and-tie model for a
deep beam with stirrups.

Design Using Strut-and-Tie Models

The design of a deep beam using a strut-and-tie model involves laying out a truss that will transmit the necessary loads. Once a satisfactory truss has been found, the joints and members of the truss are detailed to transmit the necessary forces. The overall dimensions of the beam must be such that the entire truss fits within the beam and has adequate cover.

Continuous deep beams are very stiff elements and, as such, are very sensitive to differential settlement of their supports due to foundation movements, or differential shortening of the columns supporting the beam. The first stage in the design of such a beam is to estimate the range of reactions and use this to compute shear and moment envelopes. Although some redistribution of moment and shear may occur, the amount will be limited.

ACI Sec. 11.8.4 limits ϕV_n in deep beams to $\phi(8 \text{ to } 10)\sqrt{f_c'}\, b_w d$, depending on the span-to-depth ratio of the beam. One can obtain an initial trial section on the basis of limiting $V_u = \phi V_n$ to $(6 \text{ to } 8)\sqrt{f_c'}\, b_w d$.

The crucial phase is the selection of the truss model to support the loads. The direction of the compression struts in each shear span must agree with the general direction of the principal compressive stresses in that shear span. The simplified stress trajectories in

Discontinuity Regions, Corbels, Deep Beams, and Joints

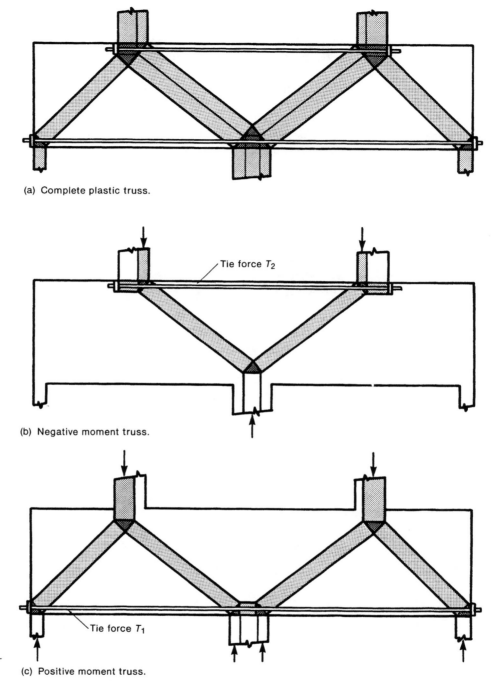

(a) Complete plastic truss.

Tie force T_2

(b) Negative moment truss.

Tie force T_1

(c) Positive moment truss.

Fig. 18–33
Strut-and-tie model for a two-span continuous beam.

Figs. 18–15 and 18–29 to 18–31 will be useful in establishing the truss model. Reference 18–1 suggests that the struts should be oriented within $\pm\ 15°$ of the angle θ shown in these figures. Values of θ were given in the description of these figures.

When several different trusses could be used, as shown in Fig. 18–16, the one that requires the least volume of steel (Fig. 18–16a) will most closely model the behavior of a concrete beam.[18-3] This truss will approach the elastic stress trajectories (Fig. 18–15a), due to the different stiffnesses of the concrete and the steel.

18–6   Deep Beams   793

If the beam is sufficiently slender that the compression fan regions at the load and the support do not overlap, no major compression struts will exist. Instead, there will be a compression field. In this case, the angle of the compression field, θ, should be determined using Eq. 6–13a, with the further limit that $25 \leq \theta \leq 65°$.

Once the geometry of the truss is established, a first estimate of the member forces can be computed. In many cases, the trusses will appear to be highly indeterminate. Such trusses can be solved easily, however, by assuming that the stirrups yield and the longitudinal steel yields at the points of maximum moments. (These assumptions are made in normal beam design in the ACI Code.) It is then possible to compute the force in each member and from this, the size required for each of the compression members to support the required compression forces. The compression struts will be stressed at f_{ce}, where f_{ce} is given in Table 18–1. Normally, $f_{ce} = 0.5f_c'$ in deep beams.

At this stage it is *essential* that the truss be *drawn to scale* to establish the size of the hydrostatic elements at the intersections of the truss members. When this is done the slopes of some of the struts will have changed and it is necessary to recompute the strut forces and sizes and the tie forces for the new geometry. The new strut sizes will require that the truss be *drawn to scale again*. This process is repeated until convergence is obtained, generally in one or two cycles. Once this has occurred, reinforcement is selected to provide $A_s f_y$ equal to, or greater than, the tensile forces in each of the stirrups and ties.

EXAMPLE 18–6   Design of a 36-ft-Span Transfer Girder

A transfer girder is supported at one end on a column and is fixed at the other end as shown in Fig. 18–34a. It supports the unfactored loads shown in the figure, including a column load of 1446 kips. The column above the beam is 28 in. square. Use the plastic truss analogy with θ between 25 and 65° and $f_{ce} = 0.5f_c'$. Use $f_c' = 5000$ psi and $f_y = 60,000$ psi.

1. Select the first trial size for the beam. Select the size of the beam on the basis of $V_u = \phi V_n$, where V_n is (6 to 8) $\sqrt{f_c'} b_w d$ and $\phi = 0.85$. This gives $b_w d$ between 3490 and 2610 in.². For $b_w = 28$ in. we require d between 125 and 93 in.

We shall try $b_w = 28$ in., h = 144 in., and d about 130 in.

2. Compute V_u/ϕ and so on. Since the design is governed by shear we shall use $\phi = 0.85$ throughout. Therefore, the loading diagram becomes that shown in Fig. 18-34b.

3. Draw the idealized truss. To simplify the drawing, we shall idealize the uniform load as a series of concentrated loads at a regular spacing. To reduce the number of such loads to be considered, we will assume they are at 2 ft on centers. Similarly, idealize the stirrups as being at 2 ft on centers and in this way compute the stirrup force per 2-ft interval. We will then select a stirrup size and spacing that will give a stirrup capacity at least equal to the required stirrup force.

A final truss is shown in Fig. 18–35. This truss was arrived at after several iterations based on the following calculations.

4. Draw a truss for the right end of the beam.

(a) Select the stirrups. Assume stirrups at 2 ft on centers to coincide with the loading points. Although stirrups will be provided at I–II, J–JJ, and T–TT, these will be disregarded in the calculations since they are loaded by compression struts which are steeper than 65°. This leaves 9 stirrups to transmit the shear.

Cut section 1–1 from I–J to SS–TT, as shown at the right side of Fig. 18–35.

$$V_u/\phi \text{ on this section} = 1471 - 11 \times 18.9 = 1263 \text{ kips}$$

Force per stirrup $= \frac{1263}{9} = 140.3$ kips/stirrup assuming that the stirrups are at 2 ft o.c., or 70.2 kips/ft

$$A_v/\text{ft} = \frac{70.2 \text{ kips/ft}}{60 \text{ ksi}} = 1.17 \text{ in.}^2/\text{ft}$$

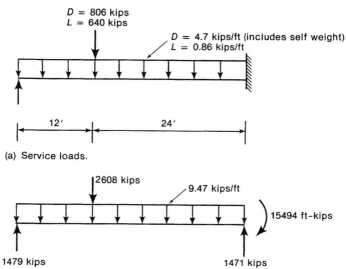

(a) Service loads.

$D = 806$ kips
$L = 640$ kips

$D = 4.7$ kips/ft (includes self weight)
$L = 0.86$ kips/ft

12' 24'

2608 kips

9.47 kips/ft

15494 ft-kips

1479 kips 1471 kips

(b) Ultimate loads and reactions, U/ϕ.

Fig. 18–34
Transfer girder loads. (From
Ref. 18–6.)

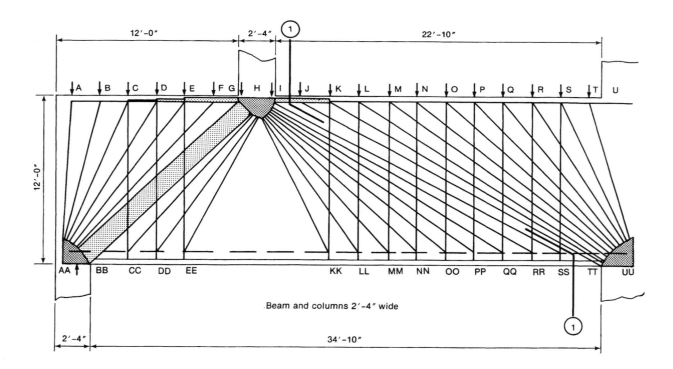

Beam and columns 2'-4" wide

Fig. 18–35
Transfer girder and strut-and-tie model—Example 18–6. (From Ref. 18–6.)

Use No. 7 double-leg stirrups at 12 in. o.c. This gives $A_v = 1.20$ in.^{2}/ft with a stirrup capacity of 144 kips every 2 ft. For simplicity, we shall continue to compute the forces in the truss, assuming that the stirrups carry 140.3 kips every 2 ft.

(b) **Check the diagonal struts.** The most highly stressed diagonal strut in the right end of the beam is the flattest strut engaging a stirrup. This is K–UU. Measuring from the drawing of the truss, it has a slope of 28°. The equilibrium of joint K is shown in Fig. 18–36a. The vertical component of the force in the strut is $140.3 + 18.9 = 159.2$ kips. The diagonal compression force D is:

$$D = \frac{140.3 + 18.9}{\sin 28°} = 339.1 \text{ kips}$$

From Eq. 18–3 and Table 18–1 the effective concrete strength, f_{ce}, is

$$f_{ce} = \nu_1 \nu_2 f_c'$$

where

$\nu_1 = 0.80$ because the deep beam will have horizontal and vertical steel in the web

$$\nu_2 = 0.55 + \frac{15}{\sqrt{f_c'}} = 0.762$$

$$f_{ce} = 0.80 \times 0.762 \times 5000 \text{ psi} = 3050 \text{ psi}$$

The area, $b_w w_s$, required is

$$\text{area} = \frac{339.1}{3.05} = 111.2 \text{ in.}^2$$

$$= 28 \text{ in. thick (through the beam web) by } 3.97 \text{ in. wide}$$

In a similar manner, the narrowest strut S–UU is found to be 2.06 in. wide. Thus the average width of the struts reaching UU is about 3 in.

At this point it is possible to refine the layout of the struts and fans in the right end. This gives the truss shown in Fig. 18–35. In drawing this it was assumed that the struts were all 4 in. wide where they meet at $U–U$ and H. The centroid of the lower tension tie was located at midheight of the truss node at UU. The support at UU is large enough to enclose the nodal element.

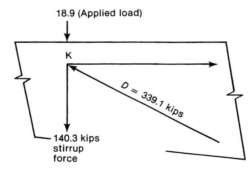

(a) Joint K.

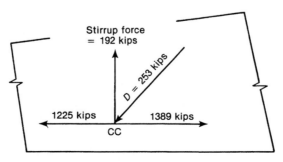

(b) Joint CC.

Fig. 18–36
Equilibrium of joints—
Example 18–6

Discontinuity Regions, Corbels, Deep Beams, and Joints

Before continuing on to compute the forces in the top and bottom chords, we shall lay out a truss for the left end of the beam.

5. Draw a truss for the left end of the beam.

(a) Select the stirrups. Several design philosophies are possible for the left end: (1) all the load could be carried by a major diagonal strut, (2) all the load could be carried by the stirrups, or (3) the load could be divided between the strut and the stirrups and their associated compression fans.

The ductility increases as the load transferred by stirrups and compression fans is increased. On the other hand, the shear capacity increases relatively little when stirrups are added, until all the shear is transferred by the stirrups. Based on test observations,[18-6] it appears that a minimum of 25 to 35% of the shear should be transmitted by the stirrups to ensure ductility.

As a first trial, use No. 7 double-leg stirrups at 12 in. o.c. in this span. This was arbitrarily chosen equal to the area selected in the right span. Referring to Fig. 18–35, we shall disregard the effectiveness of stirrup B–BB because it interferes with the main compression strut and F–FF since it is loaded by a strut steeper than 65°. This leaves three idealized stirrups C–CC, D–DD, and E–EE to transmit shear. For each 2-ft spacing $A_v f_y = 144$ kips.

V transmitted by the stirrups = $3 \times 144 = 432$ kips. Thus the stirrups transmit $432/1479 = 29\%$ of the end reaction. This is within the desired minimum range (25 to 35% minimum). If No. 7 double-leg stirrups were used at 9 in. on centers, $A_v f_y$ would increase to 192 kips/2 ft. Now roughly 40% of the shear is carried by stirrups. In view of the importance of this member, we shall **try No. 7 double-leg stirrups at 9 in. o.c. in the left shear span.**

$$V \text{ transmitted by the stirrups} = 3 \times 192 = 576 \text{ kips}$$

$$V \text{ transmitted by strut } H\text{–}AA = 1479 - 576 - 6 \times 18.9$$

$$= 790 \text{ kips}$$

$$\text{slope of strut } H\text{–}AA \text{ (measured from preliminary sketch)} = 43.5°$$

$$D = \frac{790}{\sin 43.5} = 1147 \text{ kips}$$

$$\text{width of strut } H\text{–}AA = \frac{1147}{3.05 \times 28}$$

$$= 13.4 \text{ in.}$$

Strut E–AA transmits a vertical force of $18.9 + 192 = 211$ kips. From the drawing, it has a slope of 53°.

$$D = \frac{211}{\sin 53} = 264 \text{ kips}$$

$$\text{width of strut } E\text{–}AA = \frac{264}{3.05 \times 28}$$

$$= 3.09 \text{ in.}$$

At this point it is possible to refine the layout of the struts and fans in the left shear span. This gives the truss layout drawing shown on the left side of Fig. 18–35. The joint at AA is shown in Fig. 18–37. The centroid of the tension tie should be located at midheight of the hydrostatic element at AA.

6. Compute the horizontal and vertical components of the forces in the struts. The horizontal and vertical forces in each of the diagonal struts are calculated in Table 18–2. In this calculation, the slope of the struts was measured from the final scale drawing of the beam (Fig. 18–35). The vertical force component in a strut is the sum of the stirrup force and the applied loads at the end of the strut. The horizontal component of the force in a strut is $V/\tan \theta$.

7. Compute the forces in the lower chord. The horizontal force in the lower chord is zero at the left of the truss node at AA. Struts A–AA, B–AA through H–AA all exert horizontal forces on this joint (Fig. 18–37). As a result, the force in the lower chord between AA and CC is (from Table 18–2)

$$T_{AA\text{–}CC} = (1.0 + 4.0 + 85.2 + 126.7 + 158.9 + 17.0) + 832.1$$

$$= 1225 \text{ kips}$$

The equilibrium of the joint at CC is shown in Fig. 18-36b. The diagonal compressive strut is at 49.6° and has a horizontal component of 164 kips.

$$T_{CC\text{-}DD} = 1225 + 164$$

$$= 1389 \text{ kips}$$

The forces in the lower chord are calculated in Table 18-3. The shaded portion of Fig. 18-38c represents the lower chord bar forces. The force in the lower chord increases from AA to EE. Because the struts change direction between EE and KK, the chord force decreases from KK to UU. Note that the lower chord is stressed in tension from end to end.

As a check on the computations, compute $T = M/jd$ based on the moment at the center of the column.

$$\frac{M_u}{\phi} = (1479 \text{ kips} \times 12.0 \text{ ft}) - \frac{9.45 \times 12.0^2}{2} \text{ ft-kips} - \left(\frac{2608}{2} \times \frac{7}{12}\right) \text{ ft-kips}$$

$$= 16{,}310 \text{ ft-kips}$$

From Fig. 18-35, $jd = 10.2$ ft. Therefore, the tension in the lower chord should be $T = 16{,}990/10.2 = 1599$ kips. This agrees well with the 1616 kips calculated in Table 18-3.

8. Compute the forces in the upper chord. The variation of forces along the top chord is calculated in Table 8-4 in the same way as the forces in the bottom chord were calculated. The top forces are plotted in Fig. 18-38a with the shaded diagram.

The force in the top steel at the right support can be computed as

$$T = \frac{M}{jd}$$

where jd is 10.08 ft from Fig. 18-35

$$T = \frac{15{,}494 \text{ ft-kips}}{10.08 \text{ft}}$$

$$= 1537 \text{ kips}$$

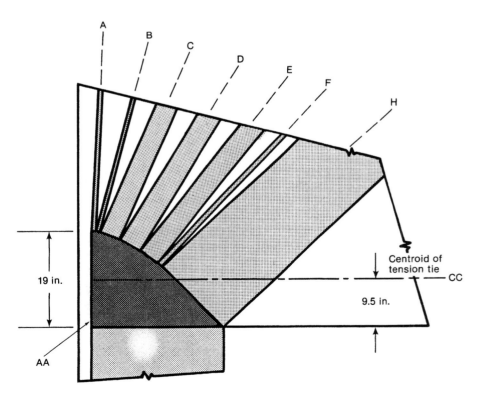

Fig. 18-37
Geometry of joint A-A—
Example 18-6.

Discontinuity Regions, Corbels, Deep Beams, and Joints

TABLE 18–2 Calculation of Strut Forces—Example 18–6

Strut	Slope (deg)	Force Components	
		Vertical (kips)	Horizontal (kips)
A–AA	87	18.9	1.0
B–AA	78	18.9	4.0
C–AA	68	192 + 18.9 = 210.9	85.2
D–AA	59	210.9	126.7
E–AA	53	210.9	158.9
F–AA	48	18.9	17.0
H–AA	43.5	789.6	832.1
H–CC	49.5	192	164.0
H–DD	56	192	129.5
H–EE	63	192	97.8
H–KK	64	140.3	68.4
H–LL	55	140.3	98.2
H–MM	48.5	140.3	124.1
H–NN	44	140.3	145.3
H–OO	39	140.3	173.3
H–PP	35	140.3	200.4
H–QQ	32.5	140.3	220.2
H–RR	30	140.3	243.0
H–SS	28	18.9	34.1
J–UU	29	18.9	34.1
K–UU	31	18.9 + 40.3 = 159.2	265.0
L–UU	33.5	159.2	240.5
M–UU	36	159.2	219.1
N–UU	38	159.2	203.8
O–UU	42	159.2	176.8
P–UU	46	159.2	153.7
Q–UU	51.5	159.2	126.6
R–UU	57	159.2	103.4
S–UU	65	159.2	74.2
T–UU	73	18.9	5.8

This compares with the 1524 kips computed in Table 18–4. The difference is less than 1% and is attributable to inaccuracies in drawing the truss and measuring the angles. If this difference is too large, a still more accurate drawing of the truss should be made and the calculations repeated.

9. **Select the longitudinal reinforcement.**

(a) **Top chord.** The maximum tension force is 1524 kips at the right support. Therefore,

TABLE 18–3   Calculation of Lower Chord Forces—Example 18–6

Panel Point	Load from Struts	Force Added (kips)	Force in Chord (kips)
End of beam			0
AA	A–AA through F–AA	392.8	393 tension (T)
AA	H–AA	832.1	1225 T
CC	H–CC	164.0	1384 T
DD	H–DD	129.5	1518 T
EE	H–EE	97.8	1616 T
KK	H–KK	−68.4	1548 T
LL	H–LL	−98.2	1449 T
MM	H–MM	−124.1	1325 T
NN	H–NN	−145.3	1180 T
OO	H–OO	−173.3	1007 T
PP	H–PP	−200.4	806 T
QQ	H–QQ	−220.2	586 T
RR	H–RR	−243.0	343 T
SS	H–SS	−263.9	79 T

$$A_s = \frac{1524}{60} = 25.40 \text{ in.}^2$$

Possible choices are:

17 No. 11, $A_s = 26.5$ in.^{2}

20 No. 10, $A_s = 25.4$ in.^{2}

26 No. 9, $A_s = 26.0$ in.^{2}

Try 20 No. 10 bars in five layers, 4 per layer, to allow lots of room for placing concrete through the bars.

The selection of the top bar cutoff points is illustrated in Fig. 18–38a and b. The required tensile force is plotted as the shaded diagram in Fig. 18–38a. The bar cutoff points are selected so that the diagram of the developed bar capacity always lies 12 bar diameters (15 in.) outside the diagram of the required tensile force in the chord. The developed capacity of a given layer of bars is assumed to vary linearly from zero at the cutoff point, to the full capacity $A_s f_y = 4 \times 1.27 \times 60 = 305$ kips, at ℓ_d from the cutoff point. The development length, ℓ_d, of a No. 10 top bar with case 1 confinement is 69 in. Arrange the bars with 2 in. clear vertical space between the layers. Because the truss calculations allowed for the effect of shear on the bar forces, it is not necessary to extend the bars further to satisfy ACI Sec. 12.12.3.

The cutoff points determined in Fig. 18–38a are shown in Fig. 18–38b. The top bars must be developed in the right-hand support.

(b) Bottom chord. The maximum tensile force is 1616 kips under the column. Therefore,

$$A_s = \frac{1616}{60} = 26.9 \text{ in.}^2$$

Possible choices are:

22 No. 10, $A_s = 27.9$ in.^{2}

18 No. 11, $A_s = 28.1$ in.^{2}

27 No. 9, $A_s = 27.0$ in.^{2}

TABLE 18–4 Calculation of Upper Chord Forces—Example 18–6

Panel Point	Load from Struts	Force Added (kips)	Force in Chord (kips)
End of beam			0
A	A–AA	1.0	
			1.0 compression (C)
B	B–AA	4.0	
			5.0 C
C	C–AA	85.2	
			90.2 C
D	D–AA	126.7	
			216.9 C
E	E–AA	158.9	
			367 C
F	F–AA	17.0	
			393 C
H	H–AA, H–CC H–DD, H–EE	1223	
			1616 C
H	H–KK, to H–SS	−1537	
			79 C
J	J–UU	−34.1	
			45 C
K	K–UU	−265.0	
			220 tension (T)
L	L–UU	−240.5	
			460 T
M	M–UU	−219.1	
			680 T
N	N–UU	−203.8	
			883 T
O	O–UU	−176.8	
			1060 T
P	P–UU	−153.7	
			1214 T
Q	Q–UU	−126.6	
			1340 T
R	R–UU	−103.4	
			1444 T
S	S–UU	−74.2	
			1518 T
T	T–UU	−5.8	
			1524 T

Two things must be considered choosing the bar size: (1) the bars must be anchored for almost their full yield strength within the column at the left end, and (2) the centroid of the tension chord at the left end must coincide with the midheight of the truss node element at this joint (see Fig. 18–37). The truss node element is 19 in. in height; therefore, the centroid of the tension chord should be 9.5 in. above the base of the beam.

Use 27 No. 9 bars in three layers of 7 bars and one layer of 6 bars. Due to the cover and the stirrup size, the first layer will be placed $3\frac{1}{2}$ in. above the bottom, the remaining layers will be 4 in. apart, center to center, placing the centroid of the group close to 9.5 in. above the soffit. Because this is not a standard bar spacing, a detail of the bar arrangement must be shown on the drawings. The anchorage at AA is shown in Fig. 18–39. To reduce the required length of the hooks, the hooked bars will be placed inside the column cage and column ties.

The selection of the bottom bar cutoff points is illustrated in Fig. 18–38b and c. Again, the diagram of the developed bar capacity extends $12d_b = 13.5$ in. outside the diagram of required tensile force. The basic development length, ℓ_d, of a No. 9 bottom bar with case 1 confinement is 47.7 in. The capacities of 7, 14, 21, and 27 No. 9 bars are 420, 840, 1260, and 1620 kips, respectively.

Because the truss model included the effects of shear on the bar forces it is not necessary to extend the bars further to satisfy ACI Sec. 12.10.3. Because the diagram of developed bar capacity is always outside the diagram of required bar force, ACI Sec. 12.11.3 is automatically satisfied.

10. Select the horizontal web reinforcement. Horizontal web reinforcement is not necessary as part of the truss, but does help to control crack widths. ACI Sec. 11.8.9 requires A_{vh} not less than $0.0025b_w s_2$, where s_2 is the spacing of the horizontal steel. If s_2 is taken as 12 in., we require 0.84 in.^{2}/12 in. **Provide No. 6 bars at 12 in. o.c. horizontal steel, each face.**

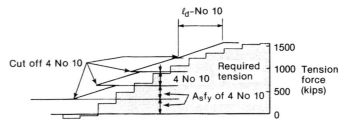

(a) Top chord bar force

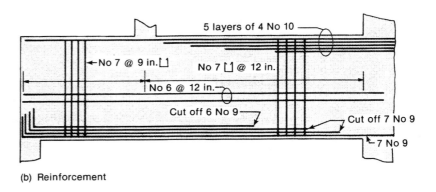

(b) Reinforcement

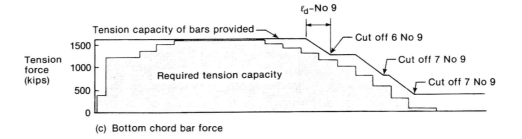

(c) Bottom chord bar force

Fig. 18–38
Forces in top and bottom
chords and reinforcement—
Example 18–6. (From Ref.
18–6.)

    11. Check the confinement of strut *AA–H*. The only major compression strut is strut *AA–H*, which has an axial force, *C*, of 1147 kips. From Eq. 18–5, this strut should be confined by reinforcement having a yield force of

$$A_s f_y \geq \Sigma\left[\frac{C}{4}\left(1 - \frac{a}{b_{ef}}\right)\right]$$

at each end. Because struts *AA–F* and *CC–H* limit the degree to which this strut can spread in a bottle-shaped fashion, we shall take *a* as 13.4 in., from step 5, and shall assume that b_{ef} is twice this, or 26.8 in. The required steel force is

$$A_s f_y \geq 2\left[\frac{1147 \text{ kips}}{4}\left(1 - \frac{13.4}{26.8}\right)\right] = 1147 \text{ kips}$$

$$A_s \geq \frac{1147 \text{ kips}}{60 \text{ ksi}} = 19.12 \text{ in.}^2$$

The stirrups and horizontal steel are more than adequate.   ■

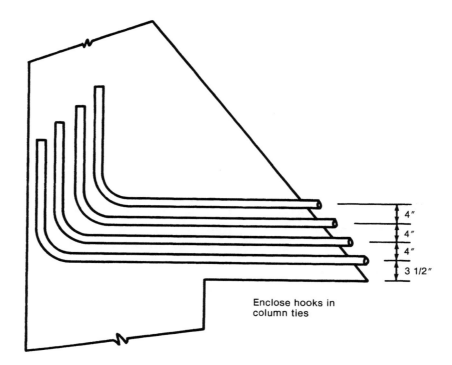

Fig. 18–39
Anchorage of tension tie at
joint A–A—Example 18–6.

4″
4″
4″
3 1/2″

Enclose hooks in
column ties

Deep Beam Design for Shear by the ACI Code

The ACI Code requirements for shear in deep beams distinguish between simply supported beams and continuous deep beams. The code requires that continuous deep beams be designed using a method such as the strut-and-tie model illustrated in the previous sections.

On the other hand, the code presents the traditional semiempirical design method for simple (single span) deep beams. The basic design equation for simple span deep beams is

$$V_u \le \phi(V_c + V_s) \tag{18–8}$$

where V_c is the shear carried by the concrete and V_s is the shear carried by the vertical and horizontal web reinforcement. The maximum shear that can be supported varies between $8\sqrt{f_c'}b_w d$ and $10\sqrt{f_c'}b_w d$ as a function ℓ_n/d. Paradoxically, the shorter the shear span, the smaller the maximum shear is. It appears that this requirement grew out of bearing failures in deep beam tests rather than shear failures.

Design for shear is carried out at a critical section located at $0.15\ell_n$ from the face of the support in uniformly loaded beams, and at the middle of the shear span for beams with concentrated loads. For both cases, however, the critical section shall not be farther than d from the face of the support (ACI Sec. 11.8.5). The shear reinforcement required at this section is used throughout the span. If a beam has one deep shear span and one ordinary shear span, a better interpretation would be that the shear reinforcement selected in the deep shear span would be used throughout that shear span.

The term V_c in Eq. 18–8 is given by

$$V_c = 2\sqrt{f_c'}b_w d \tag{18–9}$$

or by

$$V_c = \left(3.5 - 2.5\frac{M_u}{V_u d}\right)\left(1.9\sqrt{f_c'} + 2500\rho_w\frac{V_u d}{M_u}\right)b_w d \tag{18–10}$$

$$\text{(ACI Eq. 11–29)}$$

18–6   Deep Beams

The term $(3.5 - 2.5 M_u / V_u d)$ in Eq. 18–10 shall not be taken greater than 2.5 and V_c shall not be taken greater than $6\sqrt{f_c'}\, b_w d$. The terms M_u and V_u in Eq. 18–10 are the factored moment and shear at the critical section defined in the preceding paragraph.

Equation 18–10 breaks down in the case of a continuous deep beam supporting concentrated loads because M_u approaches zero at the critical section at $0.5a$ from the support. Partly for this reason, the ACI Code does not allow the use of Eq. 18–10 for continuous deep beams.

The term V_s in Eq. 18–8 is defined in ACI Sec. 11.8.8 as

$$V_s = \left[\frac{A_v}{s}\left(\frac{1 + \ell_n/d}{12}\right) + \frac{A_{vh}}{s_2}\left(\frac{11 - \ell_n/d}{12}\right) \right] f_y d \qquad \begin{array}{l} (18\text{–}11) \\ (\text{ACI Eq. } 11\text{-}30) \end{array}$$

where A_v and s are the area and spacing of the vertical shear reinforcement, and A_{vh} and s_2 refer to the horizontal shear reinforcement. This equation was derived on the basis of shear friction theory. It has two very serious shortcomings. First, there is a major discontinuity in strength as ℓ_n/d goes from below 5 to above 5. Second, and more critical, tests[18-6] show that horizontal web reinforcement is not effective for a/d ratios greater than about 0.75. Equation 18–11 suggests that horizontal web reinforcement is always more effective than vertical web reinforcement. Although a plastic truss model is a much better design basis than Eqs. 18–10 and 18–11 this method is not mentioned in ACI Sec. 11.8.2 for simple beams.

ACI Secs. 11.8.9 and 11.8.10 require minimum reinforcement in both the horizontal and vertical directions. This should be provided, regardless of the method of design followed, and should be placed in layers close to each face of the beam. This reinforcement will serve as skin reinforcement to limit crack widths.

18–7   BEAM–COLUMN JOINTS

In Chaps. 4, 5, 10, and 11, beams and columns were discussed as isolated members on the assumption that they can somehow be joined together to develop continuity. The design of the joints requires a knowledge of the forces to be transferred through the joint and the likely ways in which this transfer can occur. The ACI Code touches on joint design in several places:

1. ACI Sec. 7.9 requires enclosure of splices of continuing bars and of the end anchorages of bars terminating in connections of primary framing members, such as beams and columns.

2. ACI Sec. 11.11.2 requires a minimum amount of lateral reinforcement (ties or stirrups) in beam–column joints if the joints are not restrained on all four sides by beams or slabs of approximately equal depth. The amount required is the same as the minimum stirrup requirement for beams (ACI Eq. 11–14).

3. ACI Sec. 12.12.1 requires negative moment reinforcement in beams to be anchored in, or through, the supporting member by embedment length, hooks or mechanical anchorage.

4. ACI Sec. 12.11.2 requires that in frames forming the primary lateral load resisting system, a portion of the positive moment steel should be anchored in the joint to develop the yield strength, f_y, in tension at the face of the support.

None of these sections give specific guidelines for design. Design guidance can be obtained from Refs. 18–1 and 18–19. Extensive tests of beam column joints have been reported in Refs. 18–20 and 18–21.

Corner Joints: Opening

When considering joints at the intersection of a beam and column at a corner of a rigid frame, it is necessary to distinguish between joints that tend to be *opened* by the applied moments (Fig. 18–40) and those that tend to be *closed* by the applied moments (Fig. 18–42). Opening joints occur at the corners of frames and in L-shaped retaining walls. In bridge abutments, the joint between the wing-walls and the abutment is normally an opening joint.

The elastic distribution of stresses before cracking is illustrated in Fig. 18–40b. Large tensile stresses occur at the reentrant corner and in the middle of the joint. As a result, cracking develops as shown in Fig. 18–40c. A free-body diagram of the portion outside the diagonal crack is shown in Fig. 18–40d. The force T is necessary for equilibrium. If the reinforcement is not provided to develop this force, the joint will fail almost immediately after the development of the diagonal crack. A truss model of the joint is shown in Fig. 18–40e.

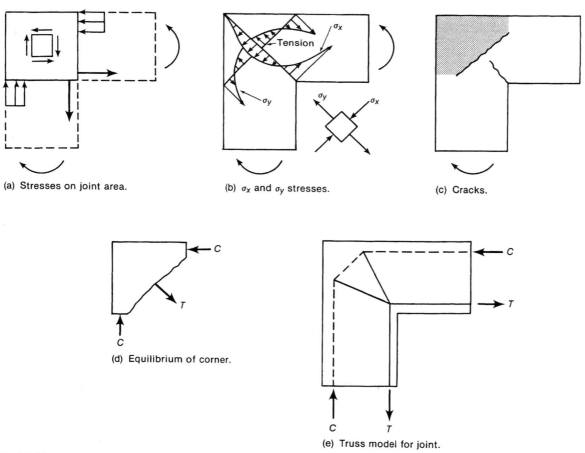

(a) Stresses on joint area.

(b) σ_x and σ_y stresses.

(c) Cracks.

(d) Equilibrium of corner.

(e) Truss model for joint.

Fig. 18–40
Stresses in an opening joint.

Figure 18–41a compares the measured efficiency of a series of corner joints reported in Refs. 18–20 and 18–21. The *efficiency* is defined as the ratio of the failure moment of the joint to the moment capacity of the members entering the joint. The reinforcement was detailed as shown in Fig. 18–41b to e. The solid curved line corresponds to the computed moment at which diagonal cracking is expected to occur in such a joint. Typical beams have reinforcement ratios of about 1%. At this reinforcement ratio, the very common joint details shown in Fig. 18–41d and e can transmit only 25 to 35% of the moment capacity of the beams.

Nilsson and Losberg[18–20] have shown experimentally that a joint reinforced as shown in Fig. 18–41b will develop the needed moment capacity without excessive deformations. The joint consists of two hooked bars enclosing the corner and diagonal bars having a total cross-sectional area of half that of the beam reinforcement. The tension in the hooked bars has a component across the diagonal crack, helping to provide the T force in Fig. 18–40d. The inclined bar limits the growth of the crack at the reentrant corner, slowing the propagation of cracking in the joint. The open symbols in Fig. 18–41 show the efficiency of the joints shown in Fig. 18–41a and b, both with and without the diagonal corner bar. It can be seen that the corner bar is needed to develop the full efficiency in the joint.

Corner Joints: Closing

The elastic stresses in a closing corner joint are exactly opposite to those in an opening corner joint. The forces at the ends of the beams load the joint in shear as shown in Fig. 18–42a. As a result, cracking of such a joint occurs as shown in Fig. 18–42b, with a major crack on the diagonal. Such joints generally have efficiencies between 80 and 100%. Problems arise due to bearing inside the bent bars in the corner, since these bars must transmit a force of $\sqrt{2}\,A_s f_y$ to the concrete on the diagonal of the joint. For this reason it may be desirable to increase the radius of this bend.

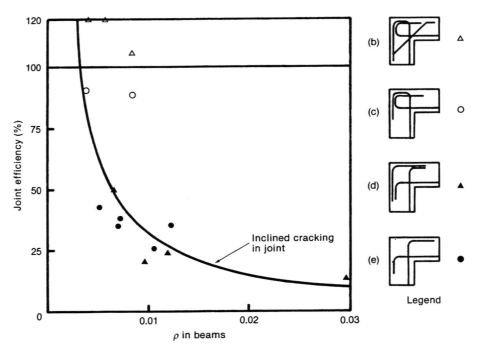

Fig. 18–41
Measured efficiency of opening joints.

(a) Test data.

Discontinuity Regions, Corbels, Deep Beams, and Joints

Frequently, the depth of the beam will be greater than that of the column, as shown in Fig. 18–43a. In such a case, the internal lever arm in the beam is larger than that in the column and, as a result, the tension force in the column steel will be larger than that in the beam. In the case shown, T_2 is three times T_1. For simplicity the effect of the shears in the beam and column have been omitted in drawing Fig. 18–43. Although the strut-and-tie model in Fig. 18–43a is in overall equilibrium, the bar force suddenly jumps by a factor of 3 at A. A strut-and-tie model that accounts for the change in bar force in such a region is shown in Fig. 18–43b. Stirrups are required in the joint to achieve the increase in tension force in the column. It can be seen from this strut-and-tie model that the stirrups in the joint region must provide a tie force

$$\Sigma T_3 = T_2 - T_1$$

The reinforcement is detailed as shown in Fig. 18–43c.

T Joints

T joints occur at exterior column–beam connections, at the base of retaining walls, or where roof beams are continuous over columns. The forces acting on such a joint can be idealized

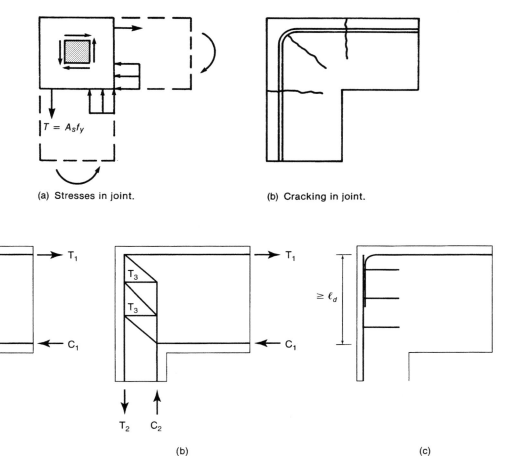

Fig. 18–42
Closing joints.

(a) Stresses in joint.

(b) Cracking in joint.

(a)

(b)

(c)

Fig. 18–43
Closing joint, beam deeper than column.

as shown in Fig. 18–44a. In this figure, compression struts are shown by dashed lines and tension ties by solid lines. Two different reinforcement patterns for column-to-roof beam joints are shown in Fig. 18–44b and c, and their measured efficiencies are shown in Fig. 18–45. The most common detail is that shown in Fig. 18–44b. This detail produces unacceptably low joint efficiencies. Joints reinforced as shown in Fig. 18–44c and d had much better performance in tests.[18-20] The hooks in these two patterns act to restrain the opening of the inclined crack and to anchor the diagonal compressive strut in the joint (see Fig. 18–44a).

In the case of a retaining wall, the detail shown in Fig. 18–44d is satisfactory to develop the strength of the wall, provided that the toe is long enough to develop bar A–B. The diagonal bar, shown in dashed lines, can be added if desired, to control cracking at the base of the wall at C.

Beam–Column Joints in Frames

The function of a beam–column joint in a frame is to transfer the loads and moments at the ends of the beams into the columns. Again, force diagrams can be drawn for such joints. The exterior joint in Fig. 18–46 has the same flow of forces as the T joint in Fig. 18–44a and cracks in the same way. An interior joint under gravity loads transmits the tensions and compressions at the ends of the beams and columns directly through the joint, as shown in Fig. 18–47a. An interior joint in a laterally loaded frame requires diagonal tensile and compressive forces within the joint as shown in Fig. 18–47b. Cracks develop perpendicular to the tension diagonal, A–B, in the joint and at the faces of the joint where the beams frame into the joint.

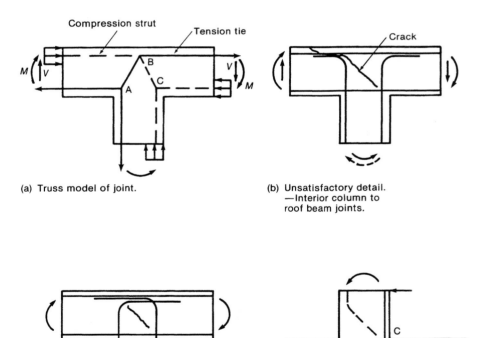

(a) Truss model of joint.

(b) Unsatisfactory detail.
—Interior column to roof beam joints.

(c) Satisfactory detail.
—Interior column to roof beam joint.

(d) Base of retaining wall.

Fig. 18–44
T joints.

Discontinuity Regions, Corbels, Deep Beams, and Joints

Design of Joints According to ACI 352

The ACI Committee 352[18-19] report on joint design divides joints into (1) *Type 1* joints for structures in nonseismic areas, and (2) *Type 2* joints where large inelastic deformations must be tolerated. The report further subdivides joints into *interior, exterior,* and *corner* joints, as illustrated in Fig. 18–48.

  The forces acting on the joints are computed from free-body diagrams of the joints, as shown in Figs. 18–49a and b. Figure 18–49b is an enlargement of the shaded part of Fig. 18–49a. The force T_n is the tensile force in the beam at its *nominal moment capacity.* Thus T_n is equal to $A_s \alpha f_y$ for the flexural steel in the beam, where $\alpha = 1.0$ for Type 1 joints and 1.25 for Type 2 joints. The column shears, V_{col}, are obtained from a frame analysis, or

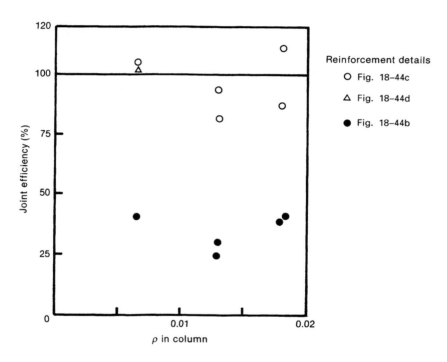

Fig. 18–45
Measured efficiency of T joints.

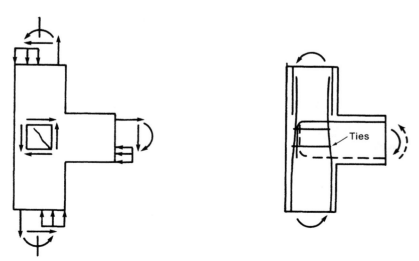

Fig. 18–46
Exterior beam column joint.

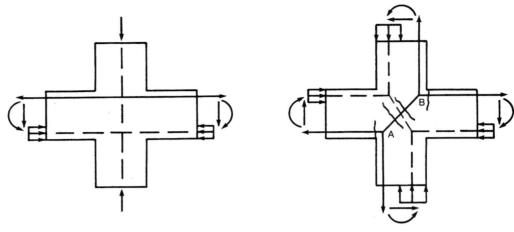

Fig. 18–47
Interior beam column joint.

(a) Forces due to gravity loads.

(b) Forces due to lateral loads.

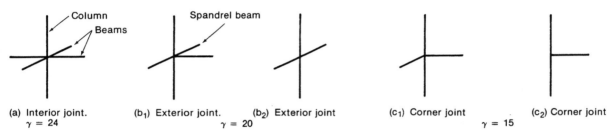

(a) Interior joint.
$\gamma = 24$

(b₁) Exterior joint.
$\gamma = 20$

(b₂) Exterior joint

(c₁) Corner joint
$\gamma = 15$

(c₂) Corner joint

Fig. 18–48
Classification of joints—ACI 352.
(γ Values are for type 1 joints.)

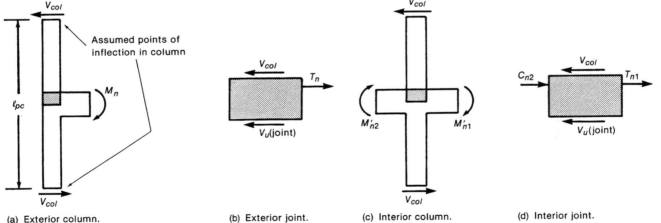

(a) Exterior column.

(b) Exterior joint.

(c) Interior column.

(d) Interior joint.

Fig. 18–49
Calculation of shear in joints.

for most practical cases can be estimated from the free bodies shown in Fig. 18–49a and c, where points of contraflexure are assumed at the midheight of each story.

The ACI Committee 352 design procedure for Type 1 (nonseismic) joints, consists of three main stages:

1. Providing confinement to the joint region by means of beams framing into the sides of the joint, or a combination of confinement from the column bars and ties in

Discontinuity Regions, Corbels, Deep Beams, and Joints

the joint region. The confinement allows the compression diagonal to form within the joint and intercepts the inclined cracks. For the joint to be properly confined, the beam steel must be inside the column steel.

2. Limiting the shear in the joint.

3. Limiting the bar size in the beams to a size that can be developed in the joint.

For best joint behavior, the longitudinal column reinforcement should be uniformly distributed around the perimeter of the column core. For Type 1 joints, ACI Committee 352 recommends that at least two layers of transverse reinforcement (ties) be provided between the top and the bottom levels of the longitudinal reinforcement in the deepest beam framing into the joint. The vertical center-to-center spacing of the transverse reinforcement should not exceed 12 in. in frames resisting gravity loads and shall not exceed 6 in. in frames resisting nonseismic lateral loads. In nonseismic regions the transverse reinforcement can be closed ties or closed ties formed by U-shaped ties and cap ties, or by U-shaped ties lap spliced within the joint.

The hoop reinforcement can be omitted within the depth of the shallowest beam entering an interior joint provided that at least three-fourths of the column width is masked by the beams on each side of the column and provided that the projection of the column outside the beam does not exceed 4 in. at any corner.

Transverse reinforcement is required in all seismic (Type 2) joints. ACI Secs. 21.6.2.2 and 21.6.2.3 specify the amount required.

The shear on a horizontal plane at midheight of the joint is limited to

$$V_n = \gamma \sqrt{f_c'}\, b_j h \tag{18–12}$$

where b_j and h are the width and thickness of the joint area, respectively, as defined in Fig. 18–50.

For joints in nonseismic areas (Type 1 joints), ACI Committee 352 recommends that γ be 24 for interior joints, 20 for exterior joints, and 15 for corner joints as illustrated in Fig. 18–48. ACI 352 and ACI Sec. 21.6.3.1 give $\gamma = 20$, 15, and 12, respectively, for interior, exterior, and corner Type 2 joints. For a joint to quality as an interior joint, the beams must cover at least three-fourths of the width and depth of each joint face, where the depth of the joint is taken as the depth of the deepest beam framing into the joint. To qualify as an exterior joint, the width of each beam must be at least three-fourths the width of that face of the column and no beam shall have an overall depth of less than three-fourths of the depth of the deepest beam at the joint. Joints that do not meet the requirements of an interior joint must be classified as exterior or corner joints.

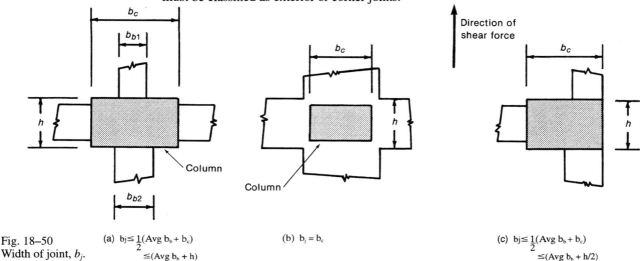

Fig. 18–50
Width of joint, b_j.

(a) $b_j \leq \frac{1}{2}(\text{Avg } b_b + b_c)$
$\leq (\text{Avg } b_b + h)$

(b) $b_j = b_c$

(c) $b_j \leq \frac{1}{2}(\text{Avg } b_b + b_c)$
$\leq (\text{Avg } b_b + h/2)$

Beam reinforcement terminating in a Type 1 joint should have 90° standard hooks with a development length, ℓ_{dh}, calculated in accordance with ACI Sec. 12.5, where ℓ_{dh} is measured from the beam–joint interface. If ℓ_{dh} is too large to fit into the joint, it may be necessary to reduce the diameter of the bars.

Rules for developing bars in Type 2 (seismic) joints are given in ACI Secs. 21.6.1.3 and 21.6.4. If hooks are used, ACI Sec. 21.6.4.1 requires that they may be embedded in confined concrete. ACI Eq. 21–5 for the ℓ_{dh} of hooks in joints is based on ACI Sec. 12.5.2 (Eq. 8–10 of this book) modified to include the effects of cover and ties (ACI Secs. 12.5.3.2 and 12.5.3.3). The resulting hook length was then increased by about 35% to allow for the effect of load reversals. The anchorage of straight bars in Type 2 joints is covered in ACI Secs. 21.6.4.2 and 21.6.4.3.

EXAMPLE 18–7   Design of Joint Reinforcement

An exterior joint in a braced frame is shown diagrammatically in Fig. 18–51a. The concrete and steel strengths are 3000 psi and 60,000 psi, respectively. The story-to-story height is 12 ft 6 in.

1.   Check the distribution of the column bars and lay out the joint ties. For Type 1 joints, no specific column bar spacing limits are given by ACI Committee 352. The column bars should be well distributed around the perimeter of the joint. Figure 18–51b shows an acceptable arrangement of column bars and ties.

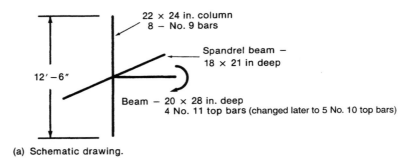

(a) Schematic drawing.

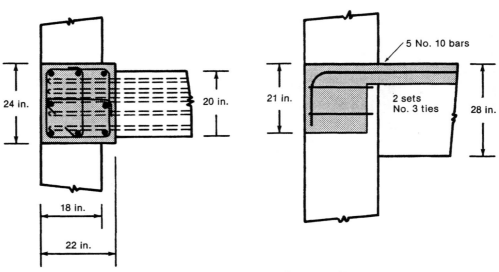

(b) Plan of joint.

(c) Elevation of joint.

Fig. 18–51
Joint design—Example 18–7.

Since the frame is braced, the frame is not the primary lateral load resisting mechanism. Hence the spacing of the joint ties can be $s \leq 12$ in., with at least two sets of ties between the top and bottom steel in the deepest beam. The required area of these ties will be computed in step 3.

2. Calculate the shear force on the joint.

We will check this in the direction perendicular to the edge. Since the frame does not resist lateral loads, there is no possiblity of a sway mechanism due to lateral loads parallel to the edge. A free body cut through the joint is similar to Fig. 18-49b. The column loads have been omitted to simplify Fig. 18-49. The shear in the joint is

$$V_{u(\text{joint})} = T_n - V_{col} \tag{18-13}$$

where

$$T_n = A_s \alpha f_y \text{ and } \alpha = 1.0 \text{ for a Type 1 joint}$$

$$= 4 \times 1.0 \times 1.56 \text{ in.}^2 \times 60 \text{ ksi} = 374 \text{ kips}$$

To compute V_{col} we must consider the free-body diagram in Fig. 18-49a where $\ell_{pc} = 12.5$ ft. The nominal moment capacity of the beam is

$$M_n = A_s f_y \left(d - \frac{a}{2} \right) = 4 \times 1.56 \times 60 \left(25 - \frac{7.34}{2} \right)$$

$$= 7986 \text{ in.-kips} = 665 \text{ ft-kips}$$

$$V_{col} = \frac{M_n}{12.5} = 53.2 \text{ kips}$$

Therefore,

$$V_{u(\text{joint})} = 374 - 53.2 = 321 \text{ kips}$$

3. Check the shear strength of the joint. From Fig. 18-50 the width of the joint

$$b_j \leq \tfrac{1}{2}(20 + 24) = 22 \text{ in.}$$

$$\leq 20 + 24 \text{ in.}$$

Use $b_j = 22$ in. The thickness of the joint, h, is equal to the column thickness. Use $h = 22$ in.

$$V_n = \gamma \sqrt{f_c'} \, b_j h$$

Check the joint classification: This will be an exterior joint, provided that all of the beams are at least three-fourths as wide as the corresponding column face and the shallowest beam is at least three-fourths of the depth of the deepest beam. Therefore, this is an exterior joint and $\gamma = 20$.

$$V_n = \frac{20\sqrt{3000} \times 22 \times 22}{1000} = 530 \text{ kips}$$

$$\phi V_n = 451 \text{ kips}$$

Since this exceeds $V_{u(\text{joint})} = 321$ kips the joint is acceptable in shear.

Using No. 3 ties, the required spacing of the joint ties to satisfy ACI Sec. 11.11.2 is

$$\text{area per set} = 3 \times 0.11 \text{ in.}^2 = 0.33 \text{ in.}^2$$

From ACI Eq. 11-13 (Eq. 6-22 of this book),

$$s = \frac{0.33 \times 60,000}{50 \times 20} = 19.8 \text{ in.}$$

but not more than 12 in. (to satisfy ACI 352). Provide two sets of ties in the joint.

4. Check the bar anchorages. From ACI Sec. 12.5, the basic development length of a hook is

$$\ell_{hb} = \frac{1200 d_b}{\sqrt{f_c'}} = \frac{1200 \times 1.41 \text{ in.}}{\sqrt{3000}}$$

$$= 30.9 \text{ in.}$$

18-7 Beam–Column Joints

813

If the beam bars are inside the column bars and have 2 in. of tail cover, ACI Sec. 12.5.3.2 allows

$$\ell_{dh} = 0.7 \times 30.9 = 21.6 \text{ in.}$$

The development length available $= 22 \text{ in.} - 1\frac{1}{2} \text{ in. (cover)} - \frac{3}{8} \text{ in. (tie)} = 20.125 \text{ in.}$ but not more than $22 \text{ in.} - 2 \text{ in. (cover on tail)} = 20 \text{ in.}$

Since ℓ_{dh} exceeds the space available, we must either increase the column size to 24 in., or reduce the bar size in the beam to No. 10 or smaller. Try 5 No. 10 bars.

$$\ell_{dh} = 0.7\left(\frac{1200 \times 1.27}{\sqrt{3000}}\right) = 19.5 \text{ in.}$$

therefore OK. Note that changing the bar area will change T_n, and hence change $V_{u(\text{joint})}$ to 326 kips, which is still less than V_n. **Use the joint as detailed in Fig. 18–51b and c with two sets of ties in the joint.** ∎

Joints between Wide Beams and Narrow Columns

Occasionally, a beam will be considerably wider than the column supporting it. The detailing of such a joint must provide a clear force path.

At an exterior beam column joint the tensile force T_n (see Figs. 18–46a and 18–44a) must be transferred to the joint core. This can be done by grouping the top bars inside the column bars, or by providing closely placed stirrups around the entire joint to create a horizontal "truss" action to transmit the tension into the joint.

If a wide deep beam is supported on a narrow column, the tensile force in the longitudinal bars outside the column will not be equilibrated by a compression strut since this strut will only exist over the column. Hence these bars will tend to shear the overhanging portions off the beam. Again, the region should be confined with stirrups to provide a horizontal truss to anchor these bars.

Finally, the truss analogy for shear strength suggests that only the portion of the web over the column is effective in transmitting diagonal compression stresses to the support. In a deep beam, this may overstress the region right over the column.

Use of Strut-and-Tie Models in Joint Design

Although strut-and-tie models can be used to design joint regions, the models sometimes become complex when the effects of shears in the beams and columns are included. For some purposes it is adequate to consider only the flexural forces in the steel and concrete when drawing the strut-and-tie models as done in Figs. 18–40e, 18–43, and 18–47b. If it is necessary to include the effects of beam and column shears, it is generally necessary to consider a strut-and-tie model that includes portions of the beams between the zero shear locations and portions of the columns, much as done in Fig. 18–18.

18–8   BEARING STRENGTH

Frequently, the load from a column or beam reaction acts on a small area on the top of a wall or a pedestal. As the load spreads out in the wall or pedestal, transverse tensile stresses develop, which may cause splitting and possibly failure of the wall or pedestal.

This section starts with a review of the internal stresses and forces developed in such a region, followed by a review of the ACI design requirements for bearing strength. The discussion of internal forces is based in large part on the excellent design manual by Schlaich.[18–1]

Internal Forces near Bearing Areas

Figure 18–52a shows the stress trajectories due to a concentrated load, F, acting at the center of the top of an isolated wall of length ℓ, height d, and thickness t, where $\ell = d$ and t is small compared to ℓ and d. As before, compressive stress trajectories are shown by dashed lines and tensile by solid lines. The horizontal stresses across a vertical line under the load are shown in Fig. 18–52b. These tend to cause a splitting crack directly under the load. For the case shown where the width of the load is 0.1ℓ, the maximum tensile stress is about $0.4p$, where $p = F/\ell t$.

For calculation purposes, this state of stress can be idealized as shown in Fig. 18–52c. Here the uniform compressive stress on the bottom surface is replaced by two concentrated forces, $F/2$ at the quarter points of the base. The inclined struts act at an angle θ, which for a load size of $w/d = 0.1$ is $65°$ for $\ell/d = 1.0$, or smaller. Flatter angles will occur if ℓ/d increases, reaching about $55°$ for a load with $w/d = 0.10$ on a wall having $\ell/d = 2.0$. For design, a slope of 2 vertical to 1 horizontal may be assumed, as done in Example 18–1 (Fig. 18–3c).

For equilibrium, the truss needs a horizontal tension tie, as shown by the solid line B–C in Fig. 18–52c. This is made up of well-anchored horizontal steel placed over the zone of horizontal tensile stress (Fig. 18–52b) and having $A_s f_y$ sufficient to serve as the horizontal tie in the truss.

A similar situation occurs when a series of concentrated loads act on a continuous wall, as shown in Fig. 18–52d. The horizontal force in the diagonal struts must be equilibrated by the tension tie B–C and a compression strut, C–B'. For force equilibrium on a vertical plane between C and B', the force in A–A' must be equal and opposite to C–B'. Elastic

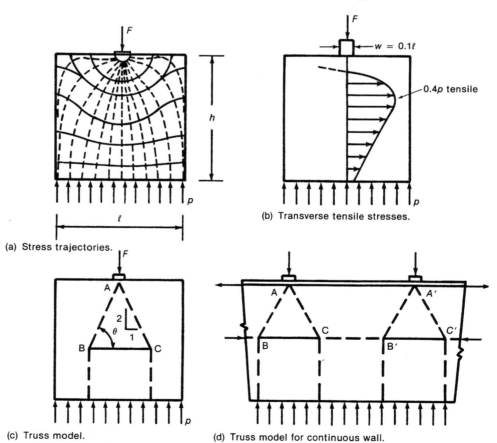

Fig. 18–52
Internal forces near bearing areas. (Adapted from Ref. 18–1.)

(a) Stress trajectories.

(b) Transverse tensile stresses.

(c) Truss model.

(d) Truss model for continuous wall.

analyses suggest that the tension in *B–C* is roughly twice the compression in *C–B′*. Reinforcement should be placed in the direction of the tension ties.

The two cases shown in Fig. 18–52 involved concentrated loads acting on thin walls. If the concentrated loads were to act on a square pedestal, similar stresses would develop in a three-dimensional fashion. Although the stresses would not be as large, it may be necessary to reinforce for them.

ACI Code Requirements for Bearing Areas

The ACI Code treats bearing on concrete in ACI Sec. 10.15 when dealing with normal situations, and ACI Sec. 18.13.1 when dealing with prestress anchorage zones. Section 18.13.1 requires consideration of the spread of forces in the anchorage zone and requires reinforcement where this leads to large internal stresses.

ACI Sec. 10.15 is based on tests by Hawkins[18-22] on unreinforced concrete blocks, supported on a stiff support, and loaded through a stiff plate. A section through such a test is shown in Fig. 18–53a. As load is applied, the crack labeled 1 occurs in the center of the block at a point under the load. This then progresses to the surface, as cracks 2. The resulting conical wedge is forced into the body, causing circumferential tension in the surrounding concrete. When this occurs, the radial cracks 3 form (Fig. 18–53b), the block breaks, and a "bearing failure" occurs. Two solutions are available: (1) provide reinforcement to replace the tension lost when crack 1 formed as done in Example 18–1, or (2) limit the bearing stresses so that internal cracking does not occur. ACI Sec. 10.15 follows the latter course.

The permissible bearing stress is set at $0.85f_c'$ if the bearing area is equal to the area of the supporting member and can be increased to

$$f_b = 0.85f_c'\sqrt{\frac{A_2}{A_1}} \text{ but not more than } 1.7\sqrt{f_c'} \qquad (18\text{--}14)$$

if the support has a larger area than the actual bearing area. In Eq. 18–14, A_1 refers to the actual bearing area and A_2 is the area of the base of a frustrum of a pyramid or cone with its upper area equal to the actual bearing area and having sides extending at 2 horizontal to 1 vertical until they first reach the edge of the block, as illustrated in Fig. 16–10.

The maximum bearing load, B_{max}, is computed from

$$B_{max} = \phi f_b A_1 \qquad (18\text{--}15)$$

where ϕ is 0.70, f_b is from Eq. 18–14, and A_1 is the bearing area.

The 2:1 rule used to define A_2 does not imply that the load spreads at this rate; it is merely an empirical relationship derived by Hawkins.[18-22] It should not be confused with the 2 vertical to 1 horizontal slope assumed in the truss model in Fig. 18–52c.

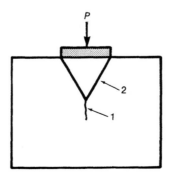

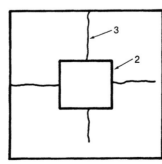

Fig. 18–53
Failure in a bearing test.

(a) Elevation.

(b) Plan.

In Fig. 5–5 a rudimentary strut-and-tie model was used to illustrate the spread of forces in the compression flange of a T beam. The forces in T-beam flanges will be examined more closely using strut-and-tie models of the beam web and flange. Figure 18–54a shows a strut-and-tie model of half of a simply supported beam loaded with a concentrated load at midspan (*J*). The horizontal components of the compression forces in the diagonal struts in the web apply loads to the top flange at *B*, *D*, and so on, as shown in Fig. 18–54b. Compression struts and transverse tension ties in the flange act to spread this compression across the width of the flange. Assuming that the web struts are inclined at 45° except at the reaction and the concentrated load, the horizontal components of the forces acting on the flange at *D*, *F*, and *H* are all equal to the shear V_n. Assuming that the flange struts are at a 2:1 slope, the transverse tensions arising from the forces acting on the flange at *D*, *F*, and *H* are $T = V_n/4$ and require transverse reinforcement with a capacity of

$$A_s f_y = V_n/4$$

For the force at *F*, for example, this steel would be distributed over the length of flange extending about *jd*/2 each way from the locations of the transverse tie resulting from the force acting on the flange at *F*. If transverse steel was needed for cross bending of the flanges due to loads on the overhanging flanges, it would be added to the steel needed to spread the compressive forces.

The strut-and-tie model in Fig. 18–54b indicates that there will not be any compression in the flange to the left of *B* and there will be a concentration of horizontal compres-

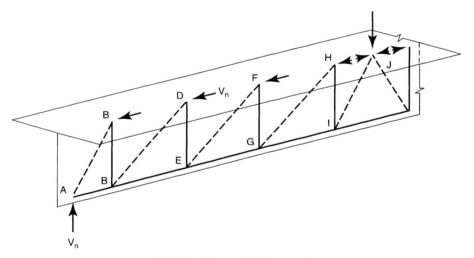

(a) Strut-and-tie model of beam web.

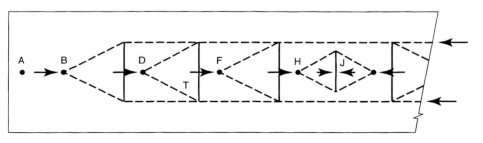

Fig. 18–54
Strut-and-tie models of a
T beam with the flange in
compression.

(b) Strut-and-tie model of compression flange.

sive force in the vicinity of the load at J because the horizontal component in strut I–J cannot spread across the flange.

The situation in a tension flange is more extreme as shown in Fig. 18–55. Here a cantilever T beam is loaded with a single concentrated load equal to V_n. Again assuming that the web struts are at 45° except at the concentrated load and the support, the horizontal components of the strut forces acting on the flange at C, E, G, and I are equal to V_n. Figure 18–55b is drawn assuming that the flexural tensile reinforcement in the flange is spread evenly over the width of the flange and hence can be represented by two bars at the quarter points of the width of the flange. The force acting on the flange at G is spread by compression struts to engage longitudinal steel at G' and G''. If the struts are at 2:1, a transverse tension tie G'–G'' is required to resist a transverse tension of $V_n/4$. The longitudinal forces acting on the flange at A and C are too close to the free end to be spread by 2:1 struts. As a result, the transverse tie forces, and hence the amounts of transverse reinforcement, are larger in this region than in the rest of the beam.

Figure 18–55b was drawn assuming that the longitudinal tension steel in the flange was evenly spread over the width of the flange. By using steel close to the web to resist the longitudinal forces introduced into the flange at A and C, the amounts of transverse steel needed at the end of the beam may be reduced.

Figures 6–22a and 8–17c illustrated the effects of shear in displacing the longitudinal steel force diagram away from the point of maximum moment in a rectangular beam. This effect is even more pronounced in the tension flange of a T beam. The forces in the flange steel at G' and G'' in Fig. 18–55b are controlled by the horizontal force applied to the flange at G. The horizontal in the flange at G would be computed by cutting a section through the web and summing moments about H.

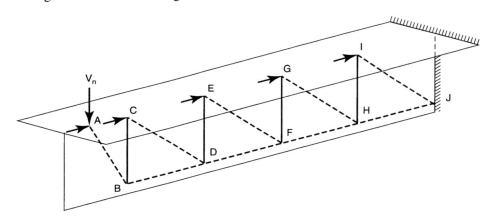

(a) Strut-and-tie model of beam web.

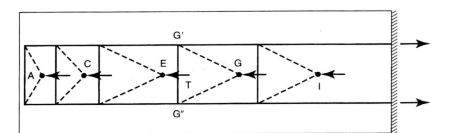

Fig. 18–55
Strut-and-tie models of a
T beam with the flange in
tension.

(a) Strut-and-tie model of tension flange.

Discontinuity Regions, Corbels, Deep Beams, and Joints

PROBLEMS

18–1 The deep beam shown in Fig. P18–1 supports a factored load of 1450 kips. The beam and columns are 24 in. wide. Draw a truss model neglecting the effects of stirrups and the dead load of the wall. Check the strength of the nodes and struts and design the tension tie. Use $f_c' = 4000$ psi and $f_y = 60,000$ psi.

18–2 Repeat Problem 18–1, including the dead load of the wall. Assume that stirrups crossing the lines AB and CD have a capacity $\phi \Sigma A_v f_y$ equal to one-third or more of the shear due to the column load.

18–3 Design a corbel to support a factored vertical load of 120 kips acting at 5 in. from the face of a column. The column and corbel are 14 in. wide. The

concrete in the column and corbel was cast monolithically. Use 5000-psi normal-weight concrete and $f_y = 60,000$ psi.

18–4 Repeat Problem 18–3 with a factored vertical load of 100 kips and a factored horizontal load of 40 kips.

18–5 Figure P18–5 shows the dapped support region of a simple beam. The factored reaction is 100 kips, $f_c' = 5000$ psi, and $f_y = 60,000$ psi (weldable). The beam is 16 in. wide.

(a) Isolate the D-region.

(b) Draw a truss to support the reaction.

(c) Detail the reinforcement.

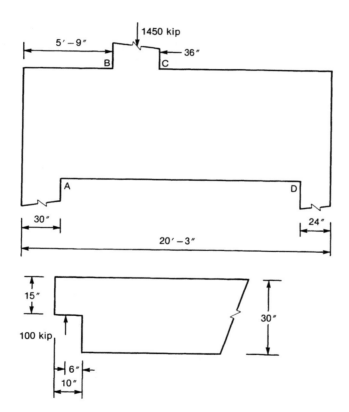

Fig. P18–1

Fig. P18–5

19

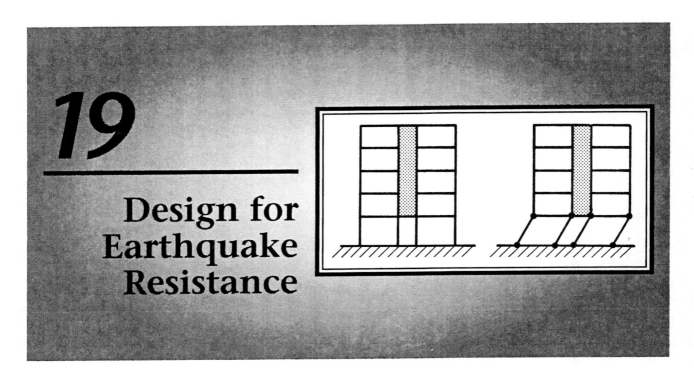

Design for
Earthquake
Resistance

19–1   EARTHQUAKES AND SEISMIC RESPONSE SPECTRA

An earthquake causes the ground under a structure to move rapidly back and forth imparting accelerations, a, to the base of the structure. If the structure were completely rigid, forces of magnitude $f = ma$ would be generated in it, where m is the mass of the structure. Since real structures are not rigid, the actual forces generated will differ from this value depending on the matching of the period of the building and the dominant periods of the earthquake. The determination of the force, f, is made more complicated because any given earthquake contains a wide and unpredictable range of frequencies and intensities of base acceleration.

The effect of a given earthquake on a wide range of structures is represented by a *response spectrum*. If a series of structures of varying periods of vibration, represented by the damped inverted pendulums in Fig. 19–1a, are subjected to a given earthquake, the maximum acceleration of each structure during the earthquake can be computed and plotted against the period of the structure as shown in Fig. 19–1b. Such a graph is known as an *acceleration response spectrum*. The vertical axis is called the *spectral acceleration*. In plotting these curves, $S_a = 1$ represents the maximum ground acceleration during the particular earthquake being studied. If $S_a = 2$ for a structure of a given period, the structure experiences an acceleration twice that of the ground. Each of the curves refers to a particular degree of damping. Damping is a measure of the dissipation of energy in the structure. Typically, a reinforced concrete building will have 1 to 2% damping prior to an earthquake. As cracking and structural and nonstructural damage develop during the earthquake, this increases to about 5%. As the damping increases, the spectral acceleration decreases as shown in Fig. 19–1b.

Figure 19–1b indicates that for the buildings having short fundamental periods, the maximum acceleration may be several times the ground acceleration. Since severe earthquakes may have maximum ground accelerations on the order of $0.2g$ to $0.4g$, this implies that the horizontal earthquake forces could be as large or larger than the weight of the

820

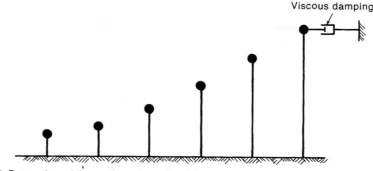

Viscous damping

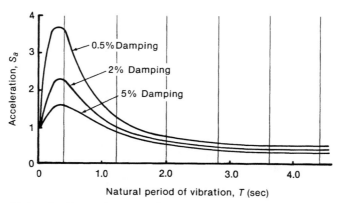

(a) Damped pendulums of varying natural frequencies.

(b) Acceleration response spectrum.

Fig. 19–1
Earthquake response spectrum (From Ref. 19–1.)

building if the building remains elastic. The forces actually used in the design of buildings for earthquakes are smaller than indicated in this figure because inelastic action in the structure tends to dissipate the earthquake forces.

As an undampened elastic pendulum is deflected to the right, energy is stored in it in the form of strain energy. The stored energy is equal to the shaded area under the load deflection diagram shown in Fig. 19–2a. When the pendulum is suddenly allowed to move back to its original position, this energy reenters the system as velocity energy and helps drive the pendulum to the left. This pendulum oscillates back and forth along the load deflection diagram shown.

If the pendulum were to develop a plastic hinge at its base, the load–deflection diagram for the same lateral deflection would be as shown in Fig. 19–2b. When this pendulum is suddenly allowed to move back to its original position, only the energy indicated by the triangle a–b–c reenters the system as velocity energy, the rest being dissipated by friction, heat, crack development, and so on.

Studies of hypothetical elastic and elastic–plastic buildings subjected to a number of different earthquake records suggest that the maximum lateral deflections of the elastic and elastic–plastic structures are roughly the same. Figure 19–3 compares the load deflection diagrams for an elastic structure and an elastic–plastic structure subjected to the same lateral deflection, Δ_u. The ratio of the maximum deflection, Δ_u, to the deflection at yielding Δ_y, is called the *displacement ductility ratio* μ:

$$\mu = \frac{\Delta_u}{\Delta_y}$$

(19–1)

19–1 Earthquakes and Seismic Response Spectra

821

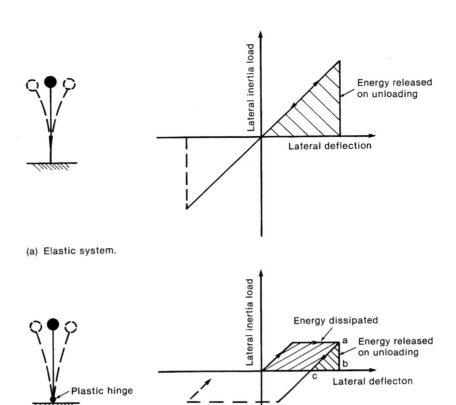

(a) Elastic system.

Fig. 19–2
Energy in vibrating pendulums. (From Ref. 19–2 copyright John Wiley & Sons, Inc.)

(b) Elastic–plastic system.

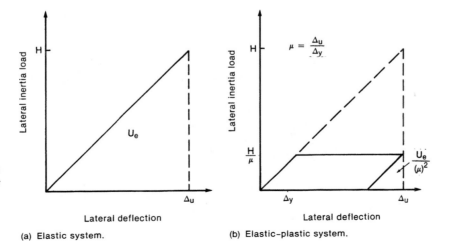

Fig. 19–3
Effect of ductility ratio, μ, on lateral force and strain energy in structures deflected to the same Δ_μ. (From Ref. 19–2; copyright John Wiley & Sons, Inc.)

(a) Elastic system.

(b) Elastic–plastic system.

Design for Earthquake Resistance

From Fig. 19–3 it can be seen that for a ductility ratio of 4, the lateral load acting on the elastic–plastic structure would be $\frac{1}{\mu} = \frac{1}{4}$ of that on the elastic structure, and the energy recovered in each cycle would be $\frac{1}{16}$ as great. Thus if a structure is ductile, it can be designed for lower seismic forces.

19–2   SEISMIC DESIGN PHILOSOPHY

The design philosophy embodied in American design codes is that buildings should remain in use and be essentially undamaged under earthquakes with a return period of 5 years or so, and buildings should not collapse under the strongest earthquake anticipated at the building location. This earthquake is taken as one with a 10% chance of excedance in 50 years. In a high seismic region such an earthquake will cause significant inelastic action in the structure. This will often be accompanied by extensive damage to nonstructural items such as cladding, partitions and ceilings.

Effect of Building Configuration

Perhaps most important in the design of a building for seismic loads is the choice of the building configuration, that is, the distribution of masses and stiffnesses in the building and the choice of load paths by which lateral loads will eventually reach the ground. The Uniform Building Code (UBC)[19–3] defines buildings as *regular* or *irregular*. Irregular buildings require special analysis or design features. Examples of irregularities are:

1. Large concentrated masses or abrupt changes in mass, stiffness, or strength from story to story. These should be avoided because they attract large forces and localized damage. The *soft story* created by terminating or greatly reducing the stiffness of the shear walls in the ground story (Fig. 19–4) concentrates forces and deformations at this level.

2. Discontinuous shear walls. The structural system should provide a number of continuous load paths to transmit horizontal loads to the foundations. The building shown in Fig. 19–5a has discontinuous shear walls at each end which rely on the floor members *AB* and *CD* in the second floor to transmit the shears to the walls in the ground floor. The rocking of the upper part of the building puts excessive loads on these floor members. Shear walls which have in-plane discontinuities (Fig. 19–5c) should be avoided also.

3. Large eccentricities. The building should be symmetrical or, if not, the distance between the center of mass, the point through which the seismic forces act on a given floor, and the center of resistance should be minimized. If there is an eccentricity, as shown in Fig. 19–6a, the building will undergo torsional deflections as shown. Due

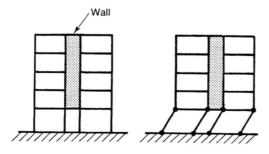

Fig. 19–4
Soft story due to discontinuous shear walls.

to this, the columns at *A* in Fig. 19–6a will experience larger shears than those at *B*. The location of the center of resistance is affected by the presence of both structural and "nonstructural" stiffening elements.

4. Low torsional stiffness. The building should have a significant torsional resistance. Because the individual walls in Fig. 19–6b are farther from the center of rotation than those in Fig. 19–6c, they provide more torsional resistance. The plan in Fig. 19–6d is particularly unsuitable. It has a large eccentricity and very little torsional resistance.

5. Vertical or horizontal geometric irregularities. The two parts of each of the buildings shown in Fig. 19–7 have different periods of vibration and severe damage can occur where they are joined together. For the building on the left, one solution is to separate the two wings with a joint that is wide enough so that the two wings can vibrate separately without banging together. If this is not practical, the corner region must be strengthened to resist the tendency to pull apart.

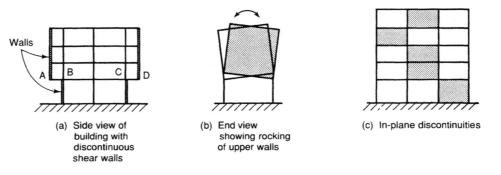

(a) Side view of building with discontinuous shear walls

(b) End view showing rocking of upper walls

(c) In-plane discontinuities

Fig. 19–5
Discontinous shear walls.

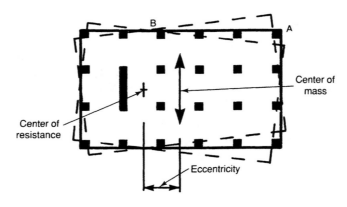

Center of resistance

Center of mass

Eccentricity

(a) Eccentricity of earthquake forces

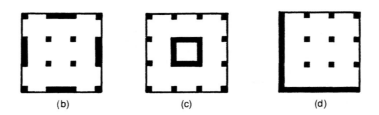

(b)   (c)   (d)

Fig. 19–6
Eccentricities and torsional deformations.

Design for Earthquake Resistance

6. Variations in column stiffnesses attract forces to the stiffer columns. Thus column *D* in Fig. 19–8 would be four times as stiff as column *A* for the same cross section and would initially be called on to resist four times the shear. Frequently, such a column will fail in shear above the wall. Sometimes the change in column stiffness is caused by nonstructural elements such as the masonry walls shown shaded in the figure.

7. Diaphragm discontinuities. Figure 19–9 shows a plan view of a floor diaphragm transmitting seismic forces to shear walls at the ends. The diaphragm acts as a wide flat beam that develops tension and compression on its two sides. Abrupt discontinuities or changes of stiffness of floor diaphragms such as the notch in the diaphragm in Fig. 19–9 lead to a concentration of damage in the weak area.

8. Nonparallel lateral force resisting systems. If the frames or walls that resist the lateral loads are not parallel to, or symmetric about, the major orthogonal axes of the lateral force resisting system, major torsional effects may be introduced.

19–3   CALCULATION OF SEISMIC FORCES ON STRUCTURES

The ACI Code does not give any guidance as to how structures should be analyzed for seismic actions. This is done in the general building code adopted in the area. Two different but

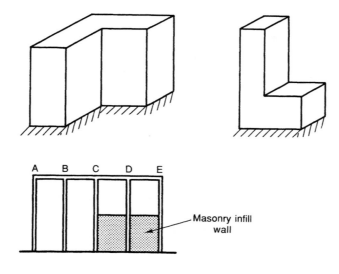

Fig. 19–7
Geometric irregularities.

Fig. 19–8
Differences in column stiffnesses.

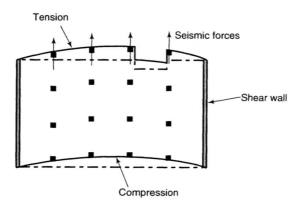

Fig. 19–9
Diaphragm discontinuities.

closely related sets of provisions exist, the *Uniform Building Code*[19–3] and ASCE 7–95, *Minimum Design Loadings for Buildings and Other Structures*[19–4]. The latter has just been updated based on a major reevaluation of seismic design rules, and will be used here.

The analysis of structures for earthquake motions according to ASCE 7–95 consists of three main steps:

(a) Determining the level of ground shaking at the site. This is a function of the effective peak acceleration, A_a, the effective peak velocity related acceleration, A_v, the soil type, and the corresponding acceleration seismic coefficient, C_a, or velocity seismic coefficient, C_v.

(b) Selecting the seismic performance category for the structure, and

(c) Carrying out the appropriate level of structural analysis to determine member forces and deformations.

Level of Ground Shaking at the Site

ASCE 7–95 presents maps of the earthquake ground motions across the United States, based on a 90% chance of not being exceeded in a 50-year period, which is equivalent to earthquakes having about a 1 in 475 chance of being exceeded in any given year. Two maps are given, one for the *effective peak acceleration, A_a*, defined as a coefficient representing ground motion at a period of 0.1 to 0.5 sec, the other for the *effective peak velocity-related acceleration, A_v*, defined as a coefficient representing ground motion at a period of about 1.0 sec. Figure 19–10 is a reproduction of the map of A_v.

The values of A_a and A_v in the maps are given for sites on rock. Different types of soils amplify the ground motions to varying degrees, and A_a and A_v are multiplied by site coefficients ranging from 0.8 to 3.5, depending on the type of subsoil, to get the *seismic coefficients, C_a and C_v*, which represent the ground motion acting on the structure at a given site.

Seismic Performance Category

ASCE 7–95 classifies buildings based on the nature of occupancy for the purposes of selecting wind, snow, and earthquake loads. The *use categories* range from I to IV, where

Use category I includes buildings with a low hazard to human life in the event of failure.

Use category II includes all buildings not included in use categories I, III, and IV.

Use category III includes buildings that represent a substantial hazard to human life in the event of collapse, including assembly buildings, schools, and so on.

Use category IV includes essential facilities intended to remain functional during and after a major earthquake, snow, or wind event.

Based on the use category and the value of the peak effective velocity-related acceleration, A_v, a *seismic performance category* is specified, ranging from A for low seismic hazard, corresponding to low values of A_v, to E for use category IV buildings in highly seismic areas as specified in Table 19–1. The type of structural analysis required and special provisions for irregular buildings get progressively more complex as the seismic performance category increases from A to E. The types of reinforced concrete structural systems and the allowable building heights for various systems are given in Table 19–2.

In Table 19–2, a *bearing wall system* is a system of floors supported by walls that resist both the gravity and lateral loads. A *building frame system* has a space frame (column and beam frame) which supports gravity loads, with lateral loads resisted by shear walls. A

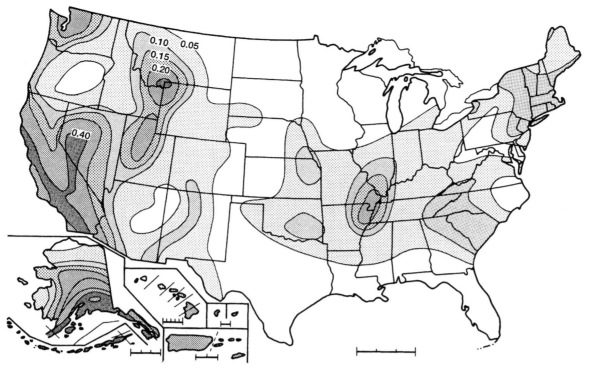

Fig. 19–10
Seismic zoning map. (From "Minimum Design Loads for Buildings and Other Structures" ASCE Standard ASCE 7-95, reproduced with the permission of the publisher, the American Society of Civil Engineers.)

TABLE 19–1 Seismic Performance Categories

Value of A_v	Use Category		
	I or II	III	IV
$A_v < 0.05$	A	A	A
$0.05 \leq A_v < 0.10$	B	B	C
$0.10 \leq A_v < 0.15$	C	C	D
$0.15 \leq A_v < 0.20$	C	D	D
$0.20 \leq A_v$	D	D	E

moment resisting frame resists both gravity and lateral loads. A *dual system* is a combination of a space frame resisting both gravity and lateral loads and a shear wall resisting lateral loads. An *inverted pendulum structure* has a large portion of its mass concentrated at the top of the structure.

Structural Analysis

ASCE 7–95 specifies two levels of structural analysis as a function of the seismic performance category and the type of structural irregularities, if any. For seismic performance

TABLE 19–2 Reinforced Concrete Structural Systems for Seismic Resistance

Basic Structural System and Seismic Force Resisting System	Response Modification Coefficient, R	Deflection Amplification Factor, C_d	Structural System Limitations and Building Height Limitations (ft) Seismic Performance Category: A&B		C	D	E
Bearing wall system							
Reinforced concrete shear walls	4.5	4	NL[a]		NL	160	100
Building frame system							
Reinforced concrete shear walls	5.5	5	NL		NL	160	100
Moment-resisting frame system							
Special moment frames (SMF)	8	5.5	NL		NL	NL	NL
Intermediate moment frames (IMF)	5	4.5	NL		NL	NL	NP[a]
Ordinary moment frames (OMF)	3	2.5	NL		NP	NP	NP
Dual system with a SMF capable of resisting at least 25% of prescribed seismic forces							
Reinforced concrete shear walls	8	6.5	NL		NL	NL	NL
Dual system with an IMF capable of resisting at least 25% of prescribed seismic forces							
Reinforced concrete shear walls	6	5	NL		NL	160	100
Inverted pendulum structures							
Special moment frames	2.5	2.5	NL		NL	NL	NL

Source: Abridged from Ref. 19–4.
[a]NL, not limited; NP, not permitted.

categories A, B, and C the *equivalent lateral force procedure* is permitted. It is also permitted for regular buildings up to about 20 stories and irregular buildings up to 5 stories in seismic performance categories D and E. For other buildings, the *modal analysis procedure* is required, sometimes with a site-specific response spectra. The modal analysis procedure considers the superimposed lateral forces resulting from several modes of vibration. A discussion of the modal analysis procedure is beyond the scope of this book. See Reference 19–6.

The equivalent lateral force procedure approximates the effects of the earthquake with a set of lateral forces distributed much as they would be in the first mode of vibration. A few aspects of the equivalent lateral force procedure given in ASCE 7–95 will be reviewed.

The seismic base shear in a given direction is

$$V = C_s W \qquad (19\text{-}2)$$

where

V = total lateral shear at the base of the building

C_s = seismic response coefficient, defined by Eq. 19–3 or 19–4

W = 1.0 dead load,

plus 0.25 times storage loads,

plus, in areas where a partition load is used in design, the larger of the actual partition loads or 10 psf,

plus the operating weight of permanent equipment and the contents of vessels, plus 20% or more of the snow load

$$C_s = \frac{1.2C_v}{RT^{2/3}} \qquad (19\text{--}3)$$

but need not be more than

$$C_s = \frac{2.5C_a}{R} \qquad (19\text{--}4)$$

where

C_v = seismic coefficient based on the soil profile and the value of A_v

C_a = seismic coefficient based on the soil profile and the value of A_a

R = response modification factor given in Table 19–2 (R is a function of the ductility and energy-dissipating characteristics of the structure; it is roughly equivalent to μ in Fig. 19–3b or Eq. 19–1)

T = fundamental period of the structure given by

$$T = C_T h_n^{3/4}$$

where

C_T = 0.030 for moment-resisting frames of reinforced concrete, or 0.020 for all other concrete buildings

h_n = height above the base to the top level in the building in feet

The variation of C_s with changing periods is shown in Fig. 19–11 for a reinforced concrete intermediate moment frame with a shear wall ($R = 5$), founded on stiff soil in a region where A_a and A_v are both 0.2g. The curve is similar in shape to the acceleration response spectrum for 5% damping in Fig. 19–1. The ordinates are different because Fig. 19–11 was drawn for a particular building, soil type, and ground movement, while Fig. 19–1 has been normalized to the ground acceleration. If the building had a period of 1 sec, the base shear would be about 0.1W.

For design purposes, the lateral load is converted to lateral forces at each story level as shown in Fig. 19–12. These are selected to simulate the distribution from a modal analysis. For building periods of 0.5 sec or less, buildings of up to 6 or 7 storys, first-mode vibration dominates and the lateral force distribution approaches an inverted triangle as shown in Fig. 19–12a. For very tall buildings higher vibration modes are significant and the

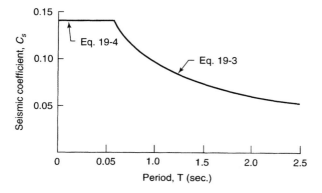

Fig. 19–11
Variation of seismic response coefficient, C_s, with period, T.

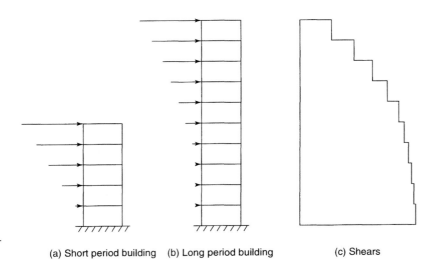

Fig. 19–12
Distribution of equivalent lateral forces and shears.

(a) Short period building   (b) Long period building   (c) Shears

equivalent lateral force distribution approaches a parabola with apex at the base of the building as shown in Fig. 19–12b. The shear, V_x, in any story x is the sum of the lateral forces acting above that story:

$$V_x = \sum_{i=x}^{n} F_i \qquad (19\text{–}5)$$

where n refers to the top level. This gives a shear distribution as shown in Fig. 19–12c.

ASCE 7–95 requires that the equivalent lateral forces be applied at some distance from the center of mass to increase the eccentricity (see Fig. 19–6a), thereby allowing for an accidental increase in the torsional effects. Such an increase may occur, for example, if a corner column such as A in Fig. 19–6a cracked and lost some of its stiffness before the other columns cracked. When this occurs, the center of resistance moves to the left, increasing the eccentricity and the torsional effects.

For buildings classified as seismic performance category B, the earthquake loads are assumed to act independently along the two orthogonal axes of the buildings. For seismic performance category C buildings having nonparallel lateral load resisting systems, and for all seismic performance category D and E buildings, 100% of the forces for one direction are added to 30% of the forces in the perpendicular direction, the directions chosen to give the worst effect for the member being designed.

Specific details on any of these calculation procedures are available in the applicable building codes and in books on earthquake engineering.[19–5,19–6] Two concepts are important here, however. First, the force developed in the structure does not have a fixed value, but instead results from the stiffness of the structure and its response to a ground vibration. Second, if a structure is detailed so that it can respond in a ductile fashion to the ground motion, the earthquake forces are reduced from the elastic values.

ACI Sec. 21.2.2 requires that all structural and nonstructural elements which materially affect the response of the structure be considered in the analysis.

19–4 DUCTILITY OF REINFORCED CONCRETE

Factors affecting the ductility of reinforced concrete beams under monotonically applied loadings have been discussed in Secs. 4–2, 4–3, 5–3, and 11–2 (see Figs. 4–16, 5–15, 5–16, and 11–5). The ductility of a beam increases as the ratio ρ/ρ_b goes down and as ρ'/ρ goes up, where ρ_b is the reinforcement ratio for balanced failure and ρ' is the ratio of compression reinforcement.

When a reinforced concrete member is subjected to load, flexural and shear cracks develop as shown in Fig. 19–13a. When the load is reversed, these cracks close and new cracks form. Following several cycles of loading the member will resemble Fig. 19–13b. The left end of the beam is divided into a series of blocks of concrete held together by the reinforcing cage. If the beam cracks through, as shown in Fig. 19–13b, shear is transferred across the crack by dowel action of the longitudinal reinforcement and grinding friction along the crack. After the concrete outside the reinforcement crushes, the longitudinal bars will buckle unless restrained by closely spaced stirrups or hoops. The hoops also provide confinement of the core concrete, increasing its ductility.

It should be noted that the ductility ratio μ is defined in Eq. 19–1 in terms of the deflection Δ at the end of the beam shown in Fig. 19–13 or at the top of the building. Since most of the deformation is concentrated in the cracked and hinging regions, the curvature ductility, ϕ_u/ϕ_y, must be several times the required deflection ductility, μ.

In Sec. 3–3 it was shown that concrete subjected to triaxial compressive stresses increases in both strength and ductility (Fig. 3–15). In a spiral column the lateral expansion of the concrete inside the spiral stresses the spiral in tension and this, in turn, causes a confining pressure on the core concrete leading to an increase in the strength and ductility of the core (Fig. 11–5). ACI Chap. 21 requires that beams, columns, and the ends of shear walls have *hoops* in regions where the reinforcement is expected to yield. Hoops are closely spaced closed ties or continuously wound ties or spirals, the ends of which have 135° hooks with 6 bar diameter (but not less than 3 in.) extensions. The hoops must enclose the longitudinal reinforcement and give lateral support to those bars in the manner required for column ties in ACI Sec. 7.10.5.3. Although hoops can be circular, they most often are rectangular as shown in Fig. 19–14 since most beams and columns have rectangular cross sections. In addition to confining the core concrete, the hoops restrain the buckling of the longitudinal bars and act as shear reinforcement.

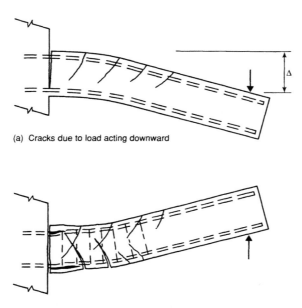

(a) Cracks due to load acting downward

(b) Cracks due to load acting upwards

Fig. 19–13
Beam subjected to cyclic loads.

19–4   Ductility of Reinforced Concrete

831

Frames with the ductile detailing required by ACI Chap. 21 can achieve deflection ductilities in excess of 5 and flexural walls about 4, compared to 1 to 2 for conventional concrete frames.

The response modification coefficients, R, given in Table 19–2 are a measure of the deflection ductilities various types of structures can attain.

19–5 GENERAL ACI CODE PROVISIONS FOR SEISMIC RESISTANCE

Seismic design provisions are presented in ACI Chap. 21, which covers only cast-in-place reinforced concrete structures. ACI Sec. 21.2.1.5 allows the use of precast or prestressed concrete structures provided that it can be demonstrated that the proposed system has strength and toughness equal or exceeding those for an equivalent reinforced concrete structure designed according to the applicable sections of ACI Chap. 21.

ACI Secs. 21.2.1.3 and 21.2.1.4 give the design requirements for regions of moderate seismic risk and high seismic risk, respectively, without defining "moderate" or "high." ACI Commentary Sec. R21.2.1 states that moderate seismic risk corresponds to Zone 2 and high seismic risk to Zones 3 and 4 in the 1988 edition of ASCE 7 and the 1991 edition of the *Uniform Building Code*. ASCE 7–95 does not use these terms. Instead, it defines the class of construction in terms of the seismic performance categories A to E, which, as seen earlier, are selected as a function of the ground acceleration, A_v, and the use category of the building. Table 19–2, abridged from ASCE 7–95, sets limits on the structural systems that can be used for each seismic performance category and ASCE 7–95 Secs. A.9.6.4 to A.9.6.7 set out those ACI Code sections that must be satisfied for each seismic performance category as summarized in Table 19–3.

ASCE 7–95 refers to a moment-resisting frame designed using ACI Chaps. 1 to 17 as an *ordinary moment frame* (OMF), and a moment-resisting frame designed using ACI Chaps. 1 to 17 plus ACI Sec. 21.8, which requires localized special detailing, as an *intermediate moment frame* (IMF). A moment-resisting frame designed using ACI Chaps. 1 to 17 plus ACI Secs. 21.2 to 21.7 is called a *special moment frame* (SMF).

Materials

The compressive strength of the concrete shall not be less than 3000 psi (ACI Sec. 21.2.4.1). Because some high-strength lightweight concretes display brittle crushing failures (see Fig. 3–25) the strength of lightweight concrete shall not exceed 4000 psi unless good behavior is documented.

Reinforcement resisting earthquake-induced stresses in frame members and the boundary elements of walls shall comply with ASTM A706. *Specification for Low-Alloy Steel Deformed Bars for Concrete Reinforcement*. Specially graded A615 steel may also be used.

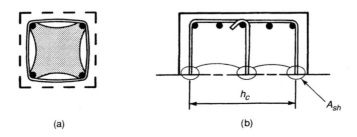

Fig. 19–14
Confinement by hoops.

(a)   (b)

TABLE 19-3 Sections of ACI Chapter 21 to Be Satisfied

	Seismic Performance Category			
	A	B	C	D and E
Frame members resisting earthquake effects in moment-resisting frames	None (OMF)	None (OMF) ASCE 7–95 requires some continuous beam bars Moment frames on soil other than stiff soil: 21–8	21.8 (IMF) Columns supporting discontinuous walls must have closely spaced ties	21.2 to 21.7 (SMF)
Walls and diaphragms resisting earthquake effects	None	None	None	21.2, 21.6
Frame members not resisting earthquake effects	None	None	None	21.2, 21.7, 21.8.2

Load Factors and Strength Reduction Factors

ACI Sec. 9.2.3 defines the load combinations to be used as

$$U = 0.75(1.4D + 1.7L + 1.7 \times 1.1E) \tag{19–6}$$
$$= 1.05D + 1.28L + 1.40E$$

and

$$U = 0.9D + 1.3 \times 1.1E \tag{19–7}$$
$$= 0.9D + 1.43E$$

ASCE 7–95 Sec. A9.6.1.1 states that if seismic design is carried out using the seismic loads from ASCE 7–95, the load factors for earthquake loads should be

$$U = 1.1(1.2D + 1.0E + \alpha_L L + 0.2S) \tag{19–8a}$$

and

$$U = 1.1(0.9D + 1.0E) \tag{19–9}$$
$$= 0.99D + 1.1E$$

where S = snow and $\alpha_L = 0.5$ except that it shall be equal 1.0 for garages, areas occupied as places of public assembly, and all areas where the live load is greater than 100 psf. For buildings with $\alpha_L = 0.5$, Eq. 19–8a becomes

$$U = 1.32D + 1.1E + 0.55L + 0.22S \tag{19–8b}$$

All of these load combinations are to be used with the strength reduction factors, ϕ, from ACI Sec. 9.3 (Sec. 2–6 of this book). Thus, for design, we shall use the ASCE 7–95 earthquake loads factored using Eqs. 19–8b and 19–9 in combination with the strength reduction factors from ACI Sec. 9.3.

ACI Sec. 9.3.4.1 presents a special strength reduction factor, ϕ, for a brittle type of member sometimes encountered in seismic designs. For any structural member with a nominal shear strength, V_n, less than the shear corresponding to the development of the nominal flexural strength ACI Sec. 9.3.4.1 requires ϕ for shear to be reduced from 0.85 to 0.60. The shear corresponding to the development of the flexural strength of the members is computed in accordance with Sec. 19–6 (see Fig. 19–18) for beams and Eq. 19–14 for columns except that in both cases M_n is substituted for M_{pr}. This requirement applies in the case of low-rise walls or portions at walls between openings. It is not applied to calculations of the shear strength of joints.

Strong Column–Weak Beam Design

If plastic hinges form in columns, the axial force causes a rapid degradation of the ability of the hinge to absorb energy while undergoing cyclic motions. As a result, the design of ductile moment-resisting frames attempts to force the structure to respond in what is referred to as strong column–weak beam action in which the plastic hinges induced by the seismic forces form at the ends of the beams, as shown in Fig. 19–15. The hinging regions are detailed to allow the plastic hinges to undergo yielding in both positive and negative moment.

19–6 FLEXURAL MEMBERS IN FRAMES DESIGNED FOR SEISMIC PERFORMANCE CATEGORIES D AND E

ACI Sec. 21.3 defines a flexural member as one proportioned to resist primarily flexure with a factored axial compressive force less than $(A_g f_c' / 10)$. Geometric limitations are

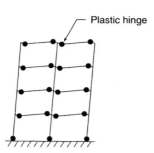

Plastic hinge

Fig. 19–15
Strong-column weak-beam
behavior.

placed on the span-to-depth ratio $(\ell_n \geq 4d)$ to avoid deep beam action. This should be borne in mind in designing coupling beams in shear walls (see Sec. 19–10). In addition, the width-to-depth ratio is limited $(b/d \geq 0.3)$. Finally, the width of the member should not be either (a) less than 10 in., or (b) more than the width of the supporting member plus $3/4d$ on either side. All of these limitations are intended to provide members that perform adequately in flexure.

Longitudinal Reinforcement

Seismic loads cause the moment diagram shown by the solid line in Fig. 19–16b when the frame is swaying to the right and an opposite diagram when the frame is swaying to the left. To this must be added the dead and live load moments shown in Fig. 19–16c giving the moment envelope in Fig. 19–16d. The maximum moments in the span normally occur at the face of a column. In addition to providing adequate moment resistance, flexural reinforcement must satisfy detailing requirements in ACI Sec. 21.3.2 to provide adequate ductility.

1. At least two bars must be provided continuously top and bottom.

2. The areas of each of the top and bottom reinforcement at every section shall not be less than given by Eq. 4–31 (ACI Eq. 10–3), nor less than $200b_w d/f_y$. The reinforcement ratio, $\rho = A_s/bd$, shall not exceed 0.025 for either the top or bottom reinforcement. Normally, ρ would not exceed about 0.015.

3. The positive moment strength at the face of the beam-column joint shall not be less than half the negative moment strength (see Fig. 19–16e). This provides $\rho' \simeq 0.5\rho$ and greatly improves the ductility of the ends of the beams.

4. At every section the positive and negative moment capacity shall not be less than one-fourth the maximum moment capacity provided at the face of either joint. This is also plotted in Fig. 19–16e.

The upper limit on ρ of 0.025 in item 2 is greater than $0.75\rho_b$ for Grade 60 steel and most concrete strengths. It is set this high taking into account the fact that there will always be confinement and compression steel with ρ' equal to at least 0.25ρ.

Development and Splicing of Flexural Reinforcement

The development lengths and splice lengths specified in ACI Secs. 12.2 and 12.15 apply in frames resisting seismic forces except as altered in ACI Sec. 21.5.4, which deals with development of bars in beam–column joints. ACI Sec. 21.3.2.3 *prohibits* lap splices

1. Within joints

2. Within $2d$ from the faces of joints

3. In locations where flexural yielding can occur due to lateral deformations of the frame

19–6   Flexural Members in Seismic Performance Categories D and E   **835**

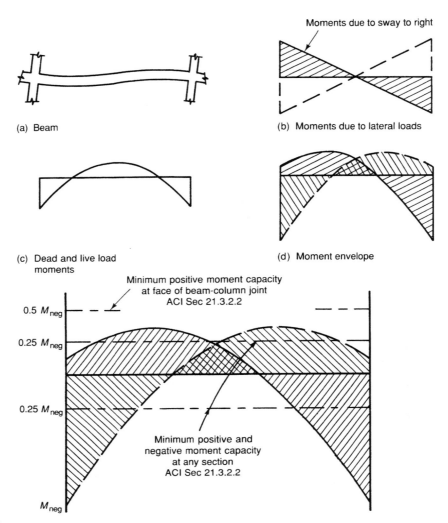

(a) Beam

Moments due to sway to right

(b) Moments due to lateral loads

(c) Dead and live load moments

(d) Moment envelope

Minimum positive moment capacity at face of beam-column joint ACI Sec 21.3.2.2

$0.5\ M_{neg}$

$0.25\ M_{neg}$

$0.25\ M_{neg}$

Minimum positive and negative moment capacity at any section ACI Sec 21.3.2.2

M_{neg}

(e) ACI Sec 21.3.2 detailing requirements

Fig. 19–16
Moment diagram due to gravity loads and seismic loads.

Lap splices must be enclosed by hoops or spirals at the smaller of 4 in. or $d/4$. Welded splices can be used as limited by ACI Sec. 21.3.2.4. Tack welding of bars for assembly purposes embrittles the bars locally and is not permitted.

Transverse Reinforcement

Transverse reinforcement is required to confine the concrete and prevent buckling of the compression bars in the hinging areas (ACI Sec. 21.3.3), to provide adequate shear strength (ACI Sec. 21.3.4), and as mentioned in the preceding section, to confine lap splices.

Confinement Reinforcement

Hoops for confinement and to control buckling of the longitudinal reinforcement are required

1. Over a length equal to $2d$ from the face of supports
2. Within $2d$ on each side of other locations where hinging can result due to lateral deformations of the frame
3. At lap splices

Design for Earthquake Resistance

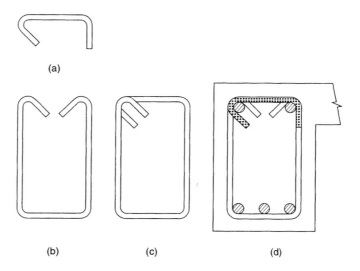

Fig. 19–17
Hoops and crossties.

(a)

(b)   (c)   (d)

The spacing of the hoops is specified in ACI Sec. 21.3.3.2. In the rest of the beam either stirrups or hoops are required at a maximum spacing of $d/2$.

ACI Sec. 21.1 defines a *seismic hook* as a hook on a stirrup, hoop, or cross-tie having a bend not less than 135° with a six-diameter (but not less than 3 in.) extension that engages the longitudinal reinforcement and projects into the concrete in the interior of the stirrup or hoop.

A *cross-tie* is defined as a continuous reinforcing bar having a seismic hook at one end and a hook not less than 90° with at least a six-diameter extension at the other end as shown in Fig. 19–17a. Both hooks shall engage peripheral longitudinal bars. The 90° hooks of two successive crossties engaging the same longitudinal bars are alternated end for end except as allowed in ACI Sec. 21.3.3.6.

A *hoop* is a closed tie as shown in Fig. 19–17c, or a continuously wound tie. A closed tie can be made up of several reinforcing elements each having seismic hooks at each end. This allows the use of a number of interlocking sheets of welded-wire fabric to be used to make up a cage of hoops and longitudinal bars for a beam or column. In flexural members ACI Sec. 21.3.3.6 allows hoops to be made up of a cross-tie as shown in Fig. 19–17a plus a stirrup with seismic hooks at each end as shown in Fig. 19–17b. If the longitudinal bars secured by the cross-ties are confined by a slab on only one side of the beam as shown in Fig. 19–17d, the 90° hooks on the cross-ties are placed on that side.

Shear Reinforcement

When the frame is displaced laterally through the inelastic deformations required to develop the ductility of the structure, the reinforcement at the ends of the beam will yield unless the moment capacity is several times the moment due to seismic loads. The yielding of the reinforcement sets an upper limit on the moments that can be developed at the ends of the beam. The design shear forces, V_e, are based on the shears due to factored dead and live loads (Fig. 19–18c) plus the shears due to hinging at the two ends of the beam for the frame swaying to the right or to the left as shown in Fig. 19–18b, where M_{pr} is the probable moment capacity of the members based on the dimensions and reinforcement at the joint assuming a tensile strength of $1.25f_y$ and $\phi = 1.0$. For a rectangular beam without axial loads,

$$M_{pr} = 1.25f_y A_s \left(d - \frac{1.25 A_s f_y}{2 \times 0.85 f_c' b} \right) \qquad (19\text{–}10)$$

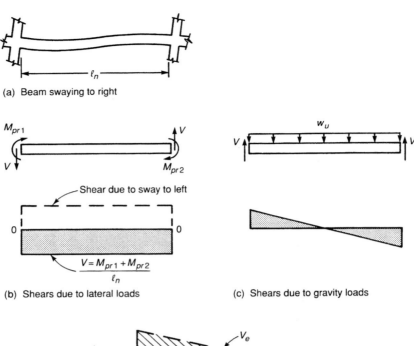

(a) Beam swaying to right

(b) Shears due to lateral loads

(c) Shears due to gravity loads

$$V = \frac{M_{pr1} + M_{pr2}}{\ell_n}$$

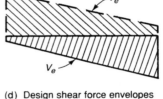

(d) Design shear force envelopes

Fig. 19–18
Shear force diagrams due to
gravity loads and seismic
loads.

The steel stress at the hinge is taken as $1.25f_y$ because the actual yield strength will normally exceed the specified value and furthermore, there may be limited strain hardening of the bars at the hinges.

The beam is then designed for the resulting shear force envelope with $V_u = V_e$ in the normal way except that if

(a) The shear due to the moments M_{pr1} and M_{pr2} is half or more of the total shear, V_e, *and*

(b) The factored axial compressive force (if any) including earthquake effects is less than $(A_g f'_c / 20)$,

then V_c is taken equal to zero. The damage to the hinging area due to repeated load reversals greatly reduces the ability of the cross section to resist shear, requiring more transverse reinforcement.[19-7] Hoops and stirrups provided to satisfy ACI Sec. 21.3.3 can also serve as shear reinforcement.

EXAMPLE 19–1 Design of a Flexural Member

The beam shown in Fig. 19–19a and b is in a moment resisting frame of an office building in a region where $A_v = 0.30g$. It supports a uniform unfactored dead load of 1.8 kips/ft and a uniform unfactored live load of 0.6 kip/ft. Earthquake loads cause unfactored end moments of $\pm$ 250 ft-kips at the exterior end and $\pm$ 240 ft-kips at the interior end. The concrete and steel strengths are 4000 psi and 60,000 psi. The layout of the building satisfies the definitions of a regular frame. Design the reinforcement.

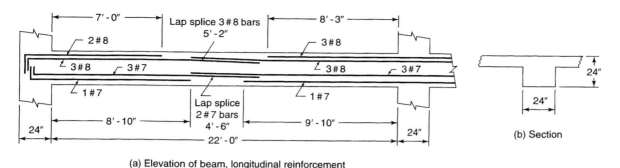

(b) Section

(a) Elevation of beam, longitudinal reinforcement

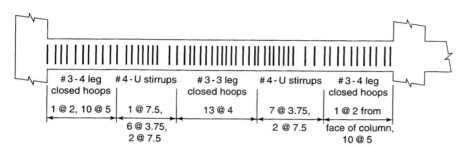

#3-4 leg closed hoops	#4-U stirrups	#3-3 leg closed hoops	#4-U stirrups	#3-4 leg closed hoops
1 @ 2, 10 @ 5	1 @ 7.5, 6 @ 3.75, 2 @ 7.5	13 @ 4	7 @ 3.75, 2 @ 7.5	1 @ 2 from face of column, 10 @ 5

(c) Hoops and stirrups

Fig. 19–19
Beam—Example 19–1.

1. **Select the seismic performance category and the level of seismic design.** An office building is use category II. From Table 19–1, for use category II and $A_v = 0.30g$, the structure must be designed for seismic performance category D. From Table 19–3, design must satisfy ACI Secs. 21.2 to 21.7.

2. **Compute the factored moments.**

Load combination 1—ACI Eq. 9–1: $U = 1.4D + 1.7L$.

Factored uniform load: $w_u = 1.4 \times 1.8 + 1.7 \times 0.6 = 3.54$ kips/ft

$$\text{Exterior negative gravity load moment} = \frac{w_u \ell_n^2}{16} = \frac{3.54 \times 22^2}{16}$$
$$= 107.1 \text{ ft-kips}$$

Midspan positive gravity load moment = 122.4 ft-kips

Interior negative gravity load moment = 171.3 ft-kips

Load combination 2—Eq. 19–8b: $U = 1.32D + 1.1E + 0.55L + 0.22S$

Factored uniform load: $w_u = 1.32 \times 1.8 + 0.55 \times 0.6 + 0.22 \times 0$
$$= 2.71 \text{ kips/ft}$$

Exterior negative gravity load moment = 81.9 ft-kips

Midspan positive gravity load moment = 93.6 ft-kips

Interior negative gravity load moment = 130.9 ft-kips

Load combination 3—Eq. 19–9: $U = 0.99D + 1.1E$

Factored uniform load: $w_u = 0.99 \times 1.8 = 1.78$ kips/ft

Exterior negative gravity load moment = 53.9 ft-kips

Midspan positive gravity load moment = 61.6 ft-kips

Interior negative gravity load moment = 86.2 ft-kips

Moments including earthquake effects from an elastic frame analysis are (positive moments cause compression in top fibers)

	Exterior Moment (ft-kips)	Midspan Moment (ft-kips)	Interior Moment (ft-kips)
Load combination 1	−107.1	**+122.4**	−171.3
Load combination 2			
Sway to right	+193.1	+93.6	**−395**
Sway to left	**−357**	+93.6	+ 133.1
Load combination 3			
Sway to right	**+221**	+61.6	−350
Sway to left	−329	+61.6	**+177.8**

3. Does the beam satisfy the definition of a flexural member? ACI Sec. 21.3.1 requires flexural members to have:

Factored compression force less than $A_g f_c'$. There is no axial load—OK

Clear span not less than 4 times the effective depth. $\ell_n/d = 22 \times 12/21.5 = 12.3$—OK

Width not: (a) less than 10 in.—OK

(b) more than width of column plus $\frac{3}{4}h$ on each side of column—OK

Thus the beam satisfies the requirements of a beam. If it did not, it would be necessary to change the dimensions of the beam.

4. Calculate the steel required for flexure.

Interior support, negative moment: Maximum negative moment from step 2 = −395 ft-kips. Assume one layer of steel, $d = 24 − 2.5 = 21.5$ in.

$$A_s = \frac{M_u}{\phi f_y jd}$$

Assume that $j = 0.90$.

$$A_s = \frac{395 \times 12,000}{0.9 \times 60,000 \times 0.9 \times 21.5}$$

$$= 4.54 \text{ in.}^2$$

Try 6 No. 8 bars, $A_s = 4.74$ in.² These will fit in one layer. ACI Sec. 21.5.1.4 requires the column dimension parallel to the bars to be at least $20d_b$. The columns are 24 in. square. This sets the maximum bar size as $24/20 = 1.2$ in. Thus a No. 8 bar is OK.

$$a = \frac{A_s f_y}{0.85 f_c' b} = \frac{4.74 \times 60,000}{0.85 \times 4000 \times 24}$$

$$= 3.49 \text{ in.}$$

$$\phi M_n = \phi A_s f_y (d − a/2)$$

$$= \frac{0.9 \times 4.74 \times 60,000(21.5 − 3.49/2)}{12,000}$$

$$= 421 \text{ ft-kips}$$

ACI Sec. 21.3.2.1: Check if $A_s \geq A_{s,min} = \dfrac{3\sqrt{f_c'}}{f_y} b_w d \geq \dfrac{200 b_w d}{f_y}$

$$= 1.63 \geq 1.72 \text{ in.}^2\text{—OK}$$

Check if $\rho = \dfrac{4.74}{24 \times 21.5} = 0.0092 \leq 0.025$—OK

ACI Sec. B.9.3: Check if section is tension controlled.

$$\frac{a}{d_t} = \frac{3.49}{21.5} = 0.162 \qquad \frac{a_{tc\ell}}{d_t} = 0.375\beta_1 = 0.391 \quad \text{(Eq. 4–23)}$$

Since $a/d_t = 0.162$ is less than $a_{tc\ell}/d_t = 0.319$, the section is tension controlled and $\phi = 0.9$. **Use 6 No. 8 top bars at the interior support.**

Interior support, positive moment: Maximum positive moment = 177.8 ft-kips, but ACI Sec. 21.3.2.2 requires that the positive moment capacity at the face of the joint not be less than 0.5 times the negative moment capacity = $0.5 \times 421 = 210.5$ ft-kips. Therefore, design $\phi M_n = 211$ ft-kips.
 Use 4 No. 7 bars, $A_s = 2.40$ in.^{2}, $\phi M_n = 223$ ft-kips. A_s satisfies the minimums.
Exterior support, negative moment: Maximum negative moment = -357 ft-kips.
 Use 5 No. 8 bars, $A_s = 3.95$ in.^{2}, $\phi M_n = 356$ ft-kips. A_s satisfies the minimums.
Exterior support, positive moment: Maximum positive moment = $+221$ ft-kips but not less than 0.5 times ϕM_n for the negative moment steel chosen = $0.5 \times 356 = 178$ ft-kips.
 Use 4 No. 7 Bars, $A_s = 2.40$ in.^{2}, $\phi M_n = 223$ ft-kips.
Midspan, positive moment: Maximum positive moment at midspan = 122.4 ft-kips.
 Use 3 No. 7 bars, $A_s = 1.80$ in.^{2}, $\phi M_n = 168.8$ ft-kips.
Minimum positive and negative moment capacity: ACI Sec. 21.3.2.2 requires that the minimum positive and negative moment capacities at any section along the beam not be less than 0.25 times the maximum negative moment capacity provided at either joint = 0.25×421 ft-kips = 105 ft-kips. 2 No. 7 bars are adequate as minimum steel.
 The steel chosen for flexure and to suit the detailing requirements is shown in Fig. 19–19a.

5. Compute the probable moment capacities, M_{pr}. The shears in the beam are computed assuming that plastic hinges form at each end of the beam with the reinforcement stressed to $1.25 f_y$ and $\phi = 1.0$ (see Eq. 19–10).

Exterior negative moment:

$$M_{pr} = \frac{1.25 \times 60,000 \times 3.95\left(21.5 - \dfrac{1.25 \times 60,000 \times 3.95}{1.7 \times 4000 \times 24}\right)}{12,000}$$

$$= 486 \text{ ft-kips}$$

Exterior positive moment: $M_{pr} = 306$ ft-kips
Interior negative moment: $M_{pr} = 572$ ft-kips
Interior positive moment: $M_{pr} = 306$ ft-kips

6. Compute the shear force envelope and design the stirrups. Figure 19–20a shows the moments and uniform load for load combination 2 acting on the beam with the frame swaying to the right. The reactions consist of two parts:

Reactions due to gravity loads = $w_u \ell_n / 2 = 29.8$ kips upward at each end, plus

$$\text{Reactions due to } M_{pr} \text{ at each end} = \frac{306 \text{ ft-kips} + 572 \text{ ft-kips}}{22 \text{ ft}}$$

$$= 39.9 \text{ kips downward at } A \text{ and upward at } B$$

R_A = 29.8 kips upward, due to gravity load + 39.9 kips downward, due to earthquake load
 = 10.1 kips downward

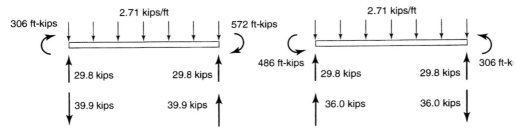

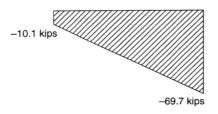 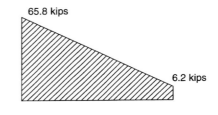

(a) Loads and end moments, frame swaying to right   (c) Loads and end moments, frame swaying to left

(b) Shear force from (a)   (d) Shear force from (c)

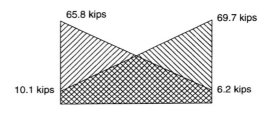

Fig. 19–20
Computation of shear forces—Example 19–1.

R_B = 29.8 kips upward, due to gravity load + 39.9 kips upward, due to earthquake load
= 69.7 kips upward

As a check: $\Sigma V = -10.1 - 2 \times 29.8 + 69.7 = 0$ —OK. The resulting shear force diagram is shown in Fig. 19–20b.

Similarly for the frame swaying to the left, the moments and loads are shown in Fig. 19–20c. The reactions are:

R_A = 29.8 kips upward, due to gravity load + 36.0 kips upward, due to earthquake load

= 65.8 kips upward

R_B = 29.8 kips upward, due to gravity load + 36.0 kips downward, due to earthquake load

= 6.2 kips downward

The shear force diagram is plotted in Fig. 19–20d. In Fig. 19–20e the absolute values of the two shear force diagrams are superimposed. The absolute values have been plotted to show directly the largest shear at any point in the span.

Stirrups for shear: ACI Sec. 21.3.4.2 states that V_c shall be taken equal to zero if

(a) Earthquake shear represents more than half the shear: it does at both ends, and

(b) Factored axial load including earthquake effect is less than $(A_g f_c' / 20)$: it is.

Therefore, $V_c = 0$.
End A: Maximum shear = 65.8 kips.

$$V_s = \frac{V_u}{\phi} - V_c = \frac{65.8}{0.85} - 0 = 77.4 \text{ kips}$$

ACI Sec. 11.5.6.8 sets the maximum $V_s = 8\sqrt{f_c'}b_w d = 261$ kips—OK

$$\frac{A_v}{s} = \frac{V_s}{f_y d} = \frac{77.4 \text{ kips}}{60 \text{ ksi} \times 21.5 \text{ in.}}$$

$$= 0.0600 \text{ in.}^2/\text{in.}$$

Try No. 3 four-leg stirrups, $A_v = 0.44$ in.^{2}, $s = \dfrac{0.44}{0.0600} = 7.33$ in. The final spacing will be chosen to satisfy both shear and confinement.

End *B*: Maximum shear $= 69.7$ kips.

$$V_s = \frac{V_u}{\phi} - V_c = \frac{69.7}{0.85} - 0 = 82.0 \text{ kips}$$

$$\frac{A_v}{s} = \frac{V_s}{f_y d} = \frac{82.0 \text{ kips}}{60 \text{ ksi} \times 21.5 \text{ in.}}$$

$$= 0.0636 \text{ in.}^2/\text{in.}$$

Try No. 3 four-leg stirrups, $A_v = 0.44$ in.^{2}, $s = \dfrac{0.44}{0.0636} = 6.92$ in.

Hoops for confinement: ACI Sec. 21.3.3.1 requires hoops over a distance of $2h = 2 \times 24$ in. $= 48$ in. from face of columns. Every corner and alternate longitudinal bars must be at the corner of a stirrup in accordance with ACI Sec. 7.10.5.3. ACI Sec. 21.3.3.2 requires the first hoop at 2 in. from the face of the column and a maximum spacing of hoops of:

    (a) $d/4 = 21.5$ in.$/4 = 5.38$ in.

    (b) 8 times the maximum longitudinal bar diameter $= 8 \times 0.875 = 7$ in.

    (c) $24 \times$ diameter of hoop bars $= 24 \times 0.375$ in. $= 9$ in. for No. 3 hoops, or

    (d) 12 in.

Therefore, **place first No. 3 four-leg hoop at 2 in. from the face of the columns at each end, plus 10 at 5 in.,** total 52 in. The hoops are shown in Figs. 19–19c and 19–21a. They consist of two U stirrups and a cross-tie.

Shear at 52 in. from the face of the column at end *B*

$$= 69.7 - \frac{69.7 \text{ kips}-10.1 \text{ kips}}{22 \text{ ft}} \times \frac{52 \text{ in.}}{12} = 58.0 \text{ kips}$$

$$\frac{A_v}{s} = \frac{58.0/0.85}{60 \times 21.5} = 0.0529 \text{ in.}^2/\text{in.}$$

Try No. 3 two-leg stirrups, $A_v = 0.22$ in.^{2}, $s = 4.16$ in. Try No. 4 two-leg stirrups, $A_v = 0.40$ in.^{2}, $s = 7.57$ in. Maximum spacing by ACI Sec. 21.3.3.4 is $d/2 = 21.5$ in.$/2 = 10.75$ in.

Use No. 4 double-leg stirrups with seismic hooks at both ends at 7.5 in. on centers in the rest of the beam. $V_s = 68.8$ kips. The stirrups are shown in Figs. 19–19c and 19–21b.

    7. Compute the cutoff points for the flexural reinforcement. The bar cutoffs are calculated assuming the ends of the beam are hinging at $\pm M_{pr}$ and the moment at the cutoff point is ϕM_n for the bars remaining after the cutoff point. For a given set of end moments, the bending

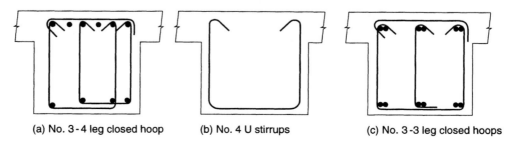

Fig. 19–21
Hoops and stirrups—
Example 19–1.

(a) No. 3 - 4 leg closed hoop   (b) No. 4 U stirrups   (c) No. 3 - 3 leg closed hoops

moment diagram will be a straight line if the factored gravity load on the beam is zero, as shown by the solid line in Fig. 19–22 for the frame swaying to the right. The moment diagrams for $w_u = 0.99D = 1.78$ kips/ft and $w_u = 1.32D + 0.55L = 2.71$ kips/ft are shown by dashed lines in Fig. 19–22. It can be seen that when cutting off the negative moment bars at the interior support, the distance B–C to the cutoff will be greater if $w_u = 1.78$ kips/ft. On the other hand, when cutting off the positive moment bars, the distance A–D to the cutoff will be longer if $w_u = 2.71$ kips/ft.

Negative moment steel, interior support. There are 6 No. 8 bars for negative moment at the interior support. We could cut off two at a time, leaving two at midspan, but the need to develop the continuing bars a distance ℓ_d past the cutoff points of the terminated bars would give problems. We shall cut off 3 No. 8 bars. For the remaining 3 No. 8 bars, $\phi M_n = 220$ ft-kips. A free-body diagram of the beam in a frame swaying to the right is shown in Fig. 19–23a. The bars will be cut off when the moment is -220 ft-kips. A free-body diagram of the right end of the beam is shown in Fig. 19–23b. Summing moments about the centroid at the cut section gives

$$\Sigma M_0 = 0 = -220 + 1.78x^2/2 - 59.5x + 572$$

or

$$0.89x^2 - 59.5x + 352 = 0$$

where

$$x = \frac{-b \pm \sqrt{b^2 - 4ac}}{2a}$$

$$= \frac{59.5 \pm \sqrt{(-59.5)^2 - 4 \times 0.89 \times 352}}{2 \times 0.89} = 6.56 \text{ ft}$$

ACI Sec. 12.10.3 requires that the bars extend the larger of $d = 21.5$ in. $= 1.79$ ft or $12d_b = 12$ in. past this point. Therefore, we could cut off 3 No. 8 bars at $6.56 + 1.79$ ft $= 8.35$ ft, say 8 ft 3 in. from the face of the interior column.

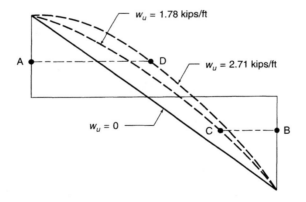

Fig. 19–22
Effect of uniform load on cutoff points.

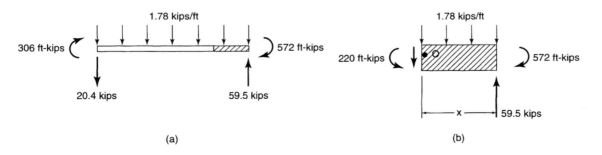

(a)   (b)

Fig. 19–23
Calculation of cutoff points—Example 19–1.

Design for Earthquake Resistance

The cutoff bars must extend ℓ_d from the face of the column, where, for No. 8 top bars,

$$\frac{\ell_d}{d_b} = \frac{f_y \alpha \beta \lambda}{20\sqrt{f'_c}} = \frac{60{,}000 \times 1.3 \times 1.0 \times 1.0}{20\sqrt{4000}}$$

$$= 61.7$$

and

$$\ell_d = 61.7 \times 1.0 \text{ in.} = 61.7 \text{ in., say 5 ft 2 in.}$$

Since 8 ft 3 in. exceeds 5 ft 2 in., OK. **Cut off 3 No. 8 top bars at 8 ft 3 in. from the face of the interior column.**

ACI Sec. 12.10.5 requires extra stirrups at the cutoff point unless the shear at the cutoff point is less than or equal to two-thirds of the shear capacity, which we shall take equal to the shear capacity of the stirrups, $V_s = 68.8$ kips, because over half of the shear at this location is due to earthquake loads, and the cutoff point will be a region of inclined cracking.

$$\tfrac{2}{3}\phi V_s = \tfrac{2}{3} \times 0.85 \times 68.8 \text{ kips} = 39.0 \text{ kips}$$

The shear at the cutoff point is 47.6 kips, which exceeds two-thirds of the shear capacity. Therefore, extra stirrups are required in the amount (ACI Sec. 12.10.5.2) of

$$\frac{A_v}{s} = \frac{60 b_w}{f_y} = 0.024 \text{ in.}^2/\text{in.}$$

at the maximum spacing of

$$s = \frac{d}{8\beta_b} = \frac{21.5}{8 \times 0.5} = 5.38 \text{ in.}$$

for a distance of $0.75d = 16.12$ in. along the bar being cut off, measured from the end of the bar. There are already No. 4 double-leg stirrups at 7.5 in. o.c. in this region. We shall provide extra stirrups at 7.5 in. o.c., halfway between the existing stirrups. These must each have $A_v = 0.024 \times 7.5 = 0.18$ in.^{2}. **Provide 3 extra No. 4 double-leg stirrups at 4.5 in. on centers adjacent to the cutoff point.**

Negative moment steel, exterior support. There are 5 No. 8 top bars at the exterior support. We shall cut off 2 of these. ϕM_n of the remaining bars is 220 ft-kips. In the same way, the negative moment is found to drop to 220 ft-kips at $x = 5.22$ ft from the left end. Extend the bar 1.79 ft and cut it off at $5.22 + 1.79 = 7.01$ ft, say 7 ft 0 in. This exceeds ℓ_d from the face of the support. **Cut off 2 No. 8 top bars at 7 ft 0 in. from the face of the exterior column. Provide 3 extra No. 4 double-leg stirrups at 4.5 in. on centers adjacent to the cutoff point.**

Positive moment, exterior support. There are 4 No. 7 bars at the exterior and interior ends. Three No. 7 bars are needed at midspan, with $\phi M_n = 168.8$ ft-kips. Figure 19–24a is a free-body diagram of the span with the frame swaying to the right. The uniform load is taken equal to 2.71 kips/ft because we are considering the positive moment steel. Figure 19–24b is a free-body diagram of the left end of the span. Solving gives $x = 7.00$ ft plus $d = 1.79$ ft $= 8.79$ ft, say 8 ft 10 in. as the actual cutoff point. This exceeds $\ell_d = 41.5$ in. for a No. 7 bottom bar. V_u at the cutoff point is -34.0 kips for this loading. $\tfrac{2}{3}\phi V_s = 45.9$ kips. Therefore, no additional stirrups are required at the cutoff point. **Cut off 1 No. 7 bar at 8 ft 10 in. from the face of the exterior column.**

Positive moment, interior support. In the same way, the actual cutoff point is found to be at 9.82 ft, say 9 ft 10 in. from the face of the interior column. **Cut off 1 No. 7 bar at 9 ft 10 in. from the face of the interior column.**

8. Design the lap splices for the continuous top and bottom steel. ACI Sec. 21.3.2.3 describes the lap splices required. These will be at midspan for both the top and bottom steel.

Top steel. Three No. 8 bars are lap spliced. The moment is low here and hence the steel stress is low. ACI Sec. 12.15.2 allows a Class A lap splice length $= 1.0\ell_d$, where $\ell_d = 6.17$ in. $= 5$ ft 2 in. Provide a 5 ft 2 in. lap splice enclosed by hoops at the smaller of

(a) $d/4 = 21.5$ in./4 $= 5.38$ in., or
(b) 4 in.

Lap splice 3 No. 8 top bars 5 ft 2 in. at midspan. Provide No. 3 three-leg closed hoops at 4 in. on centers along the length of the lap splice. The hoops consist of one U stirrup plus two cross-ties, as shown in Fig. 19–21c.

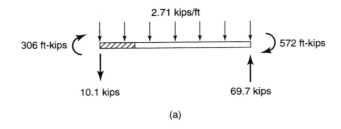

(a)

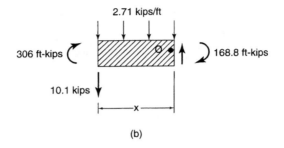

Fig. 19–24
Calculation of cutoff
points—Example 19–1.

(b)

Bottom steel. Three No. 7 bars are lap spliced at a point of maximum positive moment under gravity loads only. A Class B lap splice of length = $1.3\ell_d$ is required, where $1.3\ell_d = 1.3 \times 41.5 = 54$ in. **Lap splice 3 No. 7 bottom bars 4 ft 6 in. at midspan. Enclose the splices in the hoops provided for the top bar splices.**

The final design is shown in Figs. 19–19 and 19–21. ■

19–7 COLUMNS IN FRAMES DESIGNED FOR SEISMIC PERFORMANCE CATEGORIES D AND E

ACI Sec. 21.4 applies to columns in frames resisting earthquake forces and supporting a factored axial force exceeding $(A_g f_c'/10)$. Columns in frames in regions of high seismic risk must satisfy two geometric requirements. The smallest dimension through the centroid of the column must be at least 12 in. and the ratio of the shortest to longest cross-sectional dimensions shall not be less than 0.4. These limits ensure a minimum robustness and prevent lateral-torsional buckling which might occur with highly rectangular columns.

Required Capacity and Longitudinal Reinforcement

It is highly desirable that plastic hinges form in the beams rather than in the columns. Because the dead load must always be transferred down through the columns, the damage to the columns should be minimized. ACI Sec. 21.4.2.1 strongly encourages the use of a strong column–weak beam design. In the event that this is not possible, the columns in question shall be disregarded in the structural analysis (i.e., be assumed to have failed) if they add to the stiffness and strength of the building. If inclusion of such columns in the analysis has a negative effect on the stiffness or strength, they should be included.

Strong column–weak beam behavior is assured by requiring (Fig. 19–25) that

$$\Sigma M_e \geq 1.2 \Sigma M_g \qquad (19\text{–}11)$$
$$(\text{ACI Eq. 21–1})$$

where M_e is the *design* flexural capacity of the columns corresponding to the factored seismic load combination leading to the lowest axial load and hence lowest flexural strength;

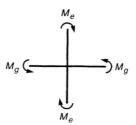

Fig. 19–25
Moments at a beam-column joint.

and M_g is the design flexural capacity of the girders at that joint. The words *design strength* are defined in ACI Sec. 9.3.1 as ϕM_n, ϕP_n, and so on.

Columns that do not satisfy Eq. 19–11 must have transverse reinforcement satisfying ACI Sec. 21.4.4 over their entire length.

Longitudinal reinforcement is designed for the axial loads and moments in the same way as in a nonseismic column. It may range from $\rho = 0.01$ to 0.06. Generally, it is difficult to place and splice much more than 2 to 3% reinforcement in a column.

Since there is a possibility of the cover concrete spalling in the regions near the ends of the column, lap splices are permitted only in the center half of the column height. Such splices must be designed as tension splices because the alternating moments due to the cyclic loads alternately stress the bars on each side of the column in tension and compression. Furthermore, there is frequently a possibility of uplift forces. The considerable length required for tension lap splices may require small diameter bars or welded or mechanical splices.

Transverse Reinforcement

Confinement Reinforcement

Transverse reinforcement in the form of spirals or hoops must be provided over a height of ℓ_o from each end of the column to confine the concrete and restrain the longitudinal bars from buckling. The height ℓ_o is the greater of (ACI Sec. 21.4.4.4)

   (a) The depth of the column d at the face of the joint,

   (b) One-sixth of the height of the column, or

   (c) 18 in.

Within the length ℓ_o ACI Sec. 21.4.4.2 requires that the spacing of the transverse reinforcement shall not exceed (a) one-fourth of the minimum thickness of the column, and (b) 4 in. Transverse reinforcement also serves as shear reinforcement.

If spirals are used, they are designed as outlined in Sec. 11–5 using ACI Eq. 10–6. An additional lower limit on the ratio of spiral reinforcement is given by ACI Eq. 21–2. This will govern if A_g/A_c is less than 1.27, which for $1\frac{1}{2}$ in. cover, will occur for columns larger than 24 in. in diameter.

Because the pressure on the sides of the hoops causes the sides to deflect outward, hoops are less efficient than spirals at confining the core concrete (Fig. 19–14). The equation for the required area of hoops, ACI Eq. 21–3, was based on the equation for spirals, ACI Eq. 10–6, but the constant was selected to give hoops with about one-third or more cross-sectional area than required for spirals:

$$A_{sh} = 0.3 \frac{s h_c f_c'}{f_{yh}} \left(\frac{A_g}{A_{ch}} - 1 \right)$$

(19–12)
(ACI Eq. 21–3)

but not less than

$$A_{sh} = 0.09 \frac{s h_c f'_c}{f_{yh}}$$

(19–13)
(ACI Eq. 21–4)

where

A_{ch} = cross-sectional area of the core of the column measured out-to-out of the hoops

A_g = gross area of the section

A_{sh} = total cross-sectional area of all the legs of the hoops and crossties within a spacing s and perpendicular to the dimension h_c (see Fig. 19–14b)

h_c = cross-sectional dimension of the column core measured center to center of outer legs of the hoops

s = spacing of the hoops measured parallel to the axis of the column

A_{sh} is calculated separately for each direction.

Figure 19–14 shows typical hoop arrangement for a column. The maximum distance between hoop or crosstie legs in the plane of the cross section is 14 in. The hoops must also satisfy ACI Sec. 7.10.5.3, which requires that every corner bar and alternate side bars be at the corner of a tie. Equation 19–13 gives a lower limit on the amount of confining reinforcement for columns larger than about 24 in. square.

Columns supporting discontinued shear walls are extremely susceptible to seismic damage. ACI Sec. 21.4.4.5 requires hoops or spirals over the full height of such members. These hoops must extend into the wall over the column and the footing or other member under the column.

Shear Reinforcement

The transverse reinforcement must also be designed for shear. The design shear force V_e is computed, assuming inelastic action in either the columns or the beams, and is given by

(a) the shear corresponding to plastic hinges at each end of the column given by

$$V_e = \frac{M_{prc\ top} + M_{prc\ btm}}{\ell_u}$$

(19–14)

where $M_{prc\ top}$ and $M_{prc\ btm}$ are the probable moment capacities at the top and bottom of the column and ℓ_u is the clear height of the column. These are obtained from an interaction diagram for the probable strength, $P_{pr} - M_{pr}$ of the column, for the range of factored loads on the member for the load combination under consideration.

(b) but need not be more than

$$V_e = \frac{\Sigma M_{prb\ top} DF_{top} + \Sigma M_{prb\ btm} DF_{btm}}{\ell_u}$$

(19–15)

where $\Sigma M_{prb\ top}$ and $M_{prb\ btm}$ are the sum of the probable moment capacities of the beams framing into the joints at the top and bottom of the column for the frame swaying to the left or right, and DF_{top} and DF_{btm} are the moment distribution factors at the top and bottom of the column being designed. This reflects the strong-column weak-beam philosophy and Eq. 19–11, which makes the beams weaker than the columns.

(c) but not less than the factored shear from a frame analysis.

Transverse reinforcement is designed for shear according to ACI Sec. 11.1.1 and V_c may be increased to allow for the effect of axial loads (see Sec. 6–8), except that within the length ℓ_0, defined in the discussion of confinement reinforcement, V_c shall be taken equal to zero when the earthquake induced shear force makes up half or more of the maximum shear force in the lengths ℓ_0 *if* the factored compression force is less than $A_g f'_c / 20$ (ACI Sec. 21.4.5.2). This clause needs revision since all of ACI Sec. 21.4 applies only when the factored axial force exceeds $A_g f'_c / 10$. As the code stands, it implies that V_c is never taken equal to zero for columns. We shall neglect the second part of this requirement and take V_c equal to zero when the earthquake-induced shear force makes up half or more of the maximum shear force in the lengths ℓ_0.

It should be noted that although the axial load increases V_c, it also increases the rate of shear degradation.[19-8] For this reason, V_c is ignored when a major portion of the shear results from earthquake loads.

EXAMPLE 19–2 Design of a Column

The column supporting the interior end of the beam designed in Example 19–1 is 24 in. square and is constructed of 4000-psi concrete and 60,000-psi steel. The floor-to-floor height is 12 ft with 24-in.-deep beams in each floor, giving a clear column height of 10 ft. The column size and the floor to floor heights are the same in the stories over and under the column being designed. The unfactored moments, shears, and axial loads from an elastic analysis for earthquake loads are given in Table 19–4. Design the reinforcement in the column.

1.   Select the seismic performance coefficient and the level of seismic design. From step 1 of Example 19–1, the structure must be designed for seismic performance coefficient D and must satisfy ACI Secs. 21.2 to 21.7.

2.   Compute the factored loads and moments. Again, we shall consider three load combinations. We shall assume that the effects of earthquakes in the two orthogonal directions are included in the moments given in Table 19–4.

Load combination 1—ACI Eq. 9–1: $U = 1.4D + 1.7L$
Load combination 2—Eq. 19–8b: $U = 1.32D + 1.1E + 0.55L + 0.22S$
Load combination 3—Eq. 19–9: $U = 0.99D + 1.1E$

The calculations are summarized in Table 19–5.

3.   Does the column satisfy the definition of a column? ACI Sec. 21.4.1 lists three requirements for a member to be designed as a column under ACI Sec. 21.4:

(a) Column resists earthquake-induced forces—OK
(b) Factored axial force exceeds $A_g f'_c / 10 = 24 \times 24 \times 4/10 = 230$ kips—OK
(c) Shortest cross-sectional dimension not less than 12 in.—OK

TABLE 19–4 Unfactored Axial Forces, Moments, and Shears in Column

	Dead Load	Live Load	Earthquake
Axial load, kips			
Column in story over	510	140	± 5
Column being designed	560	154	± 5
Column in story under	610	168	± 6
Moments, ft-kips[a]			
Top of column	−4	−1	± 195
Bottom of column	−4	−1	± 210
Shears, kips	0	0	40

[a]Counterclockwise moment on the end of a member is positive.

TABLE 19–5 Factored Forces and Moments on Columns

	Axial Load (kips)	Top Moment[a] (ft-kips)	Bottom Moment (ft-kips)	Shear (kips)
Column in story over				
Load combination 1	952			
Load combination 2	756			
Load combination 3	505			
Column being designed				
Load combination 1	1046	−7	−7	1.4
Load combination 2				
Sway to right	829	209	225	44
Sway to left	818	−220	−237	44
Load combination 3				
Sway to right	560	210	227	44
Sway to left	548	−219	−235	44
Column in story under				
Load combination 1	1140			
Load combination 2	904			
Load combination 3	610			

[a]Counterclockwise moment on the end of the column is positive.

(d) Ratio of cross-sectional dimensions not less than 0.4—OK

Therefore, design the column according to ACI Sec. 21.4. If these were not satisfied, it would be necessary to modify the column dimensions.

4. Initial selection of column steel. As a first trial we shall select a 24 × 24 in. column with 12 No. 8 bars, $A_{st} = 9.48$ in.^{2}:

$$\rho_g = \frac{9.48}{24 \times 24} = 0.0165$$

ACI Sec. 21.4.3.2 limits ρ_g to not less than 0.01 or more than 0.06—OK. No. 8 bars were chosen to avoid excessive splice lengths.

5. Check if the column strength satisfies $\Sigma M_e \geq \frac{6}{5} \Sigma M_g$. ACI Sec. 21.4.2.2 requires that the flexural strengths, ϕM_n, of the columns satisfy

$$\Sigma M_e \geq \frac{6}{5} \Sigma M_g$$

(19–11)
(ACI Eq. 21–1)

where ΣM_e is the sum of the ϕM_n strengths for the two columns meeting at a floor joint corresponding to the factored axial loads in the columns, and ΣM_g is the sum of the ϕM_n strengths of the beams meeting at the joint.

For the frame swaying to the right and left, respectively, the moments ϕM_n at the ends of the beams meeting at the top of the column, from step 4 of Example 19–1, are as shown in Fig. 19–26a and b. For the joint shown, the ΣM_g is the same for both cases.

$$\frac{6}{5} \Sigma M_g = 1.2 \times (421 \text{ ft-kips} + 223 \text{ ft-kips})$$

$$= 773 \text{ ft-kips}$$

Fig. 19–26
Moments at a beam column joint—Example 19–2.

(a) Frame swaying to right   (b) Frame swaying to left

For load combination 2, the axial load in the column in the story over the column being designed is 756 ft-kips. From the interaction diagram for $\phi P_n - \phi M_n$ in Fig. 19–27, the moment capacity corresponding to $\phi P_n = 756$ kips is $\phi M_n = 535$ ft-kips. The axial load in the column being designed is 829 kips, corresponding to a moment capacity, $\phi M_n = 515$ ft-kips. Thus at the top of the column,

$$\Sigma M_e = 535 \text{ ft-kips} + 515 \text{ ft-kips} = 1050 \text{ ft-kips}$$

This column cross section satisfies the requirement that $\Sigma M_e \geq \frac{6}{5} \Sigma M_g$ at the top of the column.

Assuming that the beams at the bottom of the column are the same as those at the top, $\frac{6}{5} \Sigma M_g = 773$ ft-kips. The moment capacity of the column being designed is 515 ft-kips. The axial load in the column in the story below the column being designed is 904 kips, corresponding to a moment capacity, $\phi M_n = 495$ ft-kips. Here, $\Sigma M_e = 515$ ft-kips $+ 495$ ft-kips $= 1010$ ft-kips, which is still greater than $\frac{6}{5} \Sigma M_g$—OK.

6. Design the confinement reinforcement. ACI Sec. 21.4.4.1(2) requires that the total cross-sectional area of hoop reinforcement not be less than the larger of

$$A_{sh} = 0.3 \left(\frac{s h_c f_c'}{f_{yh}} \right) \left(\frac{A_g}{A_{ch}} - 1 \right) \qquad \text{(19–12)} \\ \text{(ACI Eq. 21–3)}$$

or

$$A_{sh} = \frac{0.09 s h_c f_c'}{f_{yh}} \qquad \text{(19–13)} \\ \text{(ACI Eq. 21–4)}$$

where h_c is the cross-sectional dimension of the core, measured from center to center of hoops (see Fig. 19–14) $= 24$ in. $- 2 \times (1.5 + 0.5/2)$ in. $= 20.5$ in. and $A_{ch} =$ cross-sectional area of the core of the column, measured out-to-out of the transverse reinforcement $= (24 - 2 \times 1.5)^2$ in.$^2 = 441$ in.^{2}. Rearranging Eqs. 19–12 and 19–13 and solving gives

$$\frac{A_{sh}}{s} = 0.3 \left(\frac{20.5 \times 4000}{60,000} \right) \left(\frac{24 \times 24}{441} - 1 \right)$$

$$= 0.126 \text{ in.}^2/\text{in.}$$

and

$$\frac{A_{sh}}{s} = \frac{0.09 \times 20.5 \times 4000}{60,000}$$

$$= 0.123 \text{ in.}^2/\text{in.}$$

ACI Sec. 21.4.4.2 sets the maximum spacing as

    (a) 0.25 times the minimum cross-sectional dimension $= 0.25 \times 24$ in. $= 6$ in. or

    (b) 4 in.

For $s = 4$ in., the required $A_{sh} = 0.126 \times 4 = 0.504$ in.^{2}. Use No. 4 hoops with three legs in each direction as shown in Fig. 19–28, giving $A_{sh} = 0.60$ in.^{2} in each direction. This layout satisfies ACI Sec. 21.4.4.3.

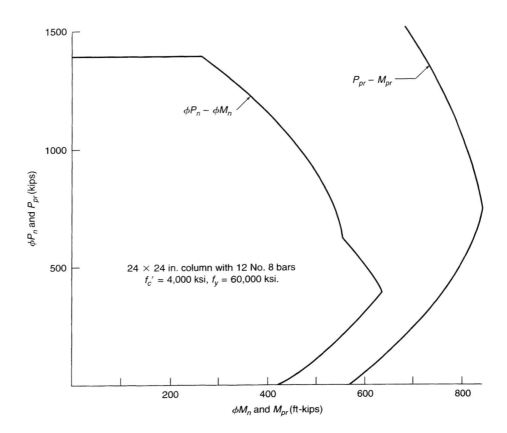

Fig. 19–27
Interaction diagrams—
Example 19–2.

ACI Sec. 21.4.4.4 requires hoop reinforcement over a length ℓ_0 adjacent to each end of the column, where ℓ_0 is the larger of

(a) The depth of the member at the joint face = 24 in.,
(b) One sixth of the clear height of the column = 120 in./6 = 20 in., or
(c) 18 in.

Thus ℓ_0 = 24 in. Throughout the rest of the height of the column, ACI Sec. 21.4.4.6 requires hoops at 6 in.

7. **Design the shear reinforcement.** The design shear force V_e shall be:

(a) The shear corresponding to plastic hinges at each end of the column given by

$$V_e = \frac{M_{\text{prc top}} + M_{\text{prc btm}}}{\ell_u} \qquad (19\text{–}14)$$

(b) but need not be more than

$$V_e = \frac{\Sigma M_{\text{prb top}} DF_{\text{top}} + \Sigma M_{\text{prb btm}} DF_{\text{btm}}}{\ell_u} \qquad (19\text{–}15)$$

(c) but not less than the factored shear from a frame analysis.

For load combination 2, the factored axial loads in the column being designed are 818 and 829 kips for sway to the left and right, respectively. For load combination 3, the factored axial loads in the column being designed are 548 and 560 kips for sway to the left and right, respectively. From the interaction diagram for P_{pr}–M_{pr} in Fig. 19–27, the maximum value of M_{pr} for the column is 835 ft-kips. Substituting into Eq. 19–14 gives

$$V_e = \frac{835 \text{ ft-kips} + 835 \text{ ft-kips}}{10 \text{ ft}} = 167 \text{ kips}$$

From step 5 of Example 19–1, the probable moment capacities of the beams framing into the joints at the top and bottom of the column are ± 572 ft-kips and ± 306 ft-kips. Since the columns in the stories over and under, and the column being designed all have the same stiffness, DF_{top} and DF_{btm} are both 0.5. Substituting into Eq. 19–15 gives

$$V_e = \frac{(572 + 306) \text{ ft-kips} \times 0.5 + (572 + 306) \text{ ft-kips} \times 0.5}{10 \text{ ft}} = 87.8 \text{ kips}$$

The shear V_e shall not be less than the factored shear from the analysis = 44 kips. Therefore, the design shear, $V_e = 87.8$ kips.

As stated earlier, we shall assume that ACI Sec. 21.4.5.2 states that V_c shall be taken equal to zero if the earthquake-induced shear represents half or more of the total design shear without the limit on axial loads. For the column being designed this is true; hence, V_c is zero.

$$V_s = \frac{V_u}{\phi} - V_c$$

$$V_u = V_e = 87.8 \text{ kips} \qquad V_c = 0$$

Therefore,

$$V_s = \frac{87.8}{0.85} = 103.3 \text{ kips}$$

$$\frac{A_v}{s} = \frac{V_s}{f_y d} = \frac{103.3}{60 \times 21.5}$$

$$= 0.0800 \text{ in.}^2/\text{in.}$$

For $s = 4$ in., $A_v = 0.36$ in.^{2}. The hoops for confinement have $A_v = 0.60$ in.^{2}—OK.

Outside of the lengths ℓ_0, V_c is given by Eq. 11–4:

$$V_c = 2 \left(1 + \frac{N_u}{2000 A_g} \right) \sqrt{f_c'} b_w d \qquad\qquad (6\text{--}17a)$$
$$\text{(ACI Eq. 11--4)}$$

$$= 2 \left(1 + \frac{548 \text{ kips} \times 1000}{2000 \times 576 \text{ in.}^2} \right) \sqrt{4000} \times 24 \text{ in.} \times 21.5 \text{ in.}$$

$$= 96.3 \text{ kips}$$

Since V_c exceeds V_u outside the length ℓ_0, stirrups are not needed for shear and instead will be provided for confinement.

Provide No. 4 hoops as shown in Fig. 19–28 at 2 in. from end of column, and 5 at 4 in. on centers at each end; provide similar No. 4 hoops at 6 in. on centers over the rest of the height.

8. Design lap splices for the column bars. ACI Sec. 21.4.3.2 requires that splices be in the middle of the column and should be designed as tension splices. At midheight of the column, the most tensile bar stress will be close to zero and hence will be less than $0.5f_y$ in tension. ACI Sec. 12.17.2.2 requires a Class B tension lap splice if all the bars are spliced at the same location. For a vertical No. 8 bar

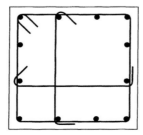

Fig. 19–28
Hoops—Example 19–2.

$$\ell_d = d_b \left(\frac{f_y \alpha \beta \lambda}{20\sqrt{f_c'}} \right) = 1.0 \left(\frac{60,000 \times 1.0 \times 1.0 \times 1.0}{20\sqrt{4000}} \right)$$ (8–10)

$$= 47.4 \text{ in.}$$

A Class B splice has a length of $1.3\ell_d = 1.3 \times 47.4$ in. $= 61.7$ in. ACI Sec. 12.17.2.4 allows this length to be multiplied by 0.83 if the ties throughout the splice length have an effective area of not less than $0.0015hs$, which for the $s = 6$ in. of the hoops away from the ends of the columns is 0.22 in.^{2}. The hoops have an area of 0.60 in.^{2}; therefore, they are adequate to allow this reduction. The lap length becomes 0.83×61.7 in. $= 51.2$ in., say 4 ft 4 in.

Lap splice all vertical bars with a 4 ft 4 in. lap splice at midheight of the column.

Use a 24-in.-square column with 12 No. 8 bars, $f_c' = 4000$ psi and $f_y = 60,000$ psi. Provide hoops as in step 7 and lap splice as in step 8. The reinforcement is shown in Fig. 19–28. ∎

19–8 JOINTS OF FRAMES

The flow of forces within beam–column joints and the design of such joints has been discussed in Section 18–7 and an example of the design of an exterior nonseismic joint is given in Example 18–7. Code provisions for joints in ductile moment-resisting frames (SMFs) are given in ACI Sec. 21.5. These differ from the design recommendations for nonseismic joints in a number of areas.

ACI Sec. 21.5.1.1 requires that joint forces be calculated taking the stress in the flexural reinforcement in the beams as $1.25f_y$. This is analogous to using the probable strength in the calculations of shear in columns and beams in ductile frames.

ACI Sec. 21.5.1.4 limits the diameter of the longitudinal beam reinforcement that passes through a joint to $\frac{1}{20}$ of the width of the joint parallel to the beam bars. When hinges form in the beams, the beam reinforcement is stressed to the actual yield strength of the bar on one side of the joint and is stressed in compression on the other side. This results in very large bond stresses in the joint, possibly leading to slipping of the bar in the joint. The minimum bonded length of such a bar in a joint is thus $20d_b$, which is considerably less than required by the development-length equations. The minimum bonded length was selected from test results of joints tested under cyclic loads.

ACI Sec. 21.5.2.1 requires hoop reinforcement around the column reinforcement in all joints in ductile moment-resisting frames. In joints confined on all four sides by beams satisfying ACI Sec. 21.5.2.2, the amount of hoop reinforcement is reduced and its spacing is liberalized within the depth of the shallowest beam entering the joint.

ACI Sec. 21.5.3.1 gives upper limits on the shear strength of joints. As indicated in Sec. 18–7, these are lower than the joint shear strengths recommended in nonseismic joints. This reflects the damage to joints resulting from cyclic loads.

ACI Sec. 21.5.4 gives special development lengths for hooks and straight bars in joints. These are shorter than the development lengths given in ACI Chap. 12 because the effects of the joint confinement by hoops have already been included.

EXAMPLE 19–3 Design an Interior Beam–Column Joint

Design the interior beam–column joint connecting the beams and columns from Examples 19–1 and 19–2. Beams, which are 24 in. by 24 in. in section frame into the 24 in. by 24 in. column on all four sides.

1. **Define the size of the joint.** The joint has width, depth, and vertical height of 24 in. The area of a horizontal section through the joint, A_j (see definition in ACI Sec. 21.0), is $A_j = 24 \times 24 = 576$ in.^{2}.

ACI Sec. 21.5.1.4 requires the length of the joint measured parallel to the flexural steel causing the joint shear to be at least 20 times the diameter of those bars $= 20 \times 1$ in.—OK.

Design for Earthquake Resistance

2. **Determine the transverse reinforcement for confinement.** ACI Sec. 21.5.2.1 requires confinement steel within the joint. Because the joint has beams on all four sides ACI Sec. 21.5.2.2 sets the amount of confinement steel as half of the confinement steel required in the ends of the columns, given by Eqs. 19–12 and 19–13 (ACI Eqs. 21–3 and 21–4). In the column, Eq. 19–12 (ACI Eq. 21–3) governed (see step 6 of Example 19–2) and required that $A_{sh}/s = 0.126$ in.²/in. Within the height of the joint, we require that

$$\frac{A_{sh}}{s} = 0.5 \times 0.126 \text{ in.}^2/\text{in.} = 0.063 \text{ in.}^2/\text{in.}$$

The vertical spacing of the hoops from ACI Secs. 21.4.4.2 and 21.5.2.2 is the smaller of

(a) $0.25 \times$ the least dimension of $A_j = 0.25 \times 24$ in., or
(b) 6 in.

The clear distance between the top and bottom beam steel is 18 in. Provide three sets of hoops, the first at 3 in. below the top steel. The required $A_{sh} = 6 \times 0.063 = 0.378$ in.². Use No. 3 three-legged hoops with the arrangement shown in Fig. 19–28.

3. **Compute the shear on the joint and check the shear strength.** Figure 19–29 is a free-body diagram of the joint for the frame swaying to the right. The beams entering the joint have probable moment capacities of –572 ft-kips and +306 ft-kips. At the joint the stiffnesses of the columns above and below the joint are the same, giving distribution factors of $DF = 0.5$ for each column. Thus the moment in the column over is

$$M_e = 0.5(572 + 306) = 439 \text{ ft-kips}$$

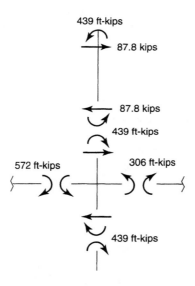

(a) Joint and columns

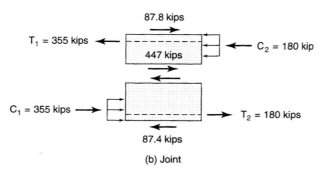

(b) Joint

Fig. 19–29
Freebody diagrams of joint—
Example 19–3

The shear in the column over is

$$V_e = \frac{439 + 439}{10 \text{ ft}} = 87.8 \text{ kips} \qquad \text{(see step 7 of Example 19–2.)}$$

The force in the steel in the beam on the left of the joint is

$$T_1 = 1.25 A_s f_y = 1.25 \times 6 \times 0.79 \text{ in.}^2 \times 60 \text{ ksi}$$
$$= 355 \text{ kips}$$

The compression force in the beam to the left is $C_1 = T_1 = 355$ kips.
Similarly, T_2 and C_2 in the beam to the right of the joint are $= 1.25 \times 4 \times 0.60 \times 60 = 180$ kips.
Summing horizontal forces gives the shear in the joint as

$$V_j = V_e - T_1 - C_2 \qquad\qquad\qquad (19\text{–}16)$$

$$= 87.8 \text{ kips to the right} - 355 \text{ kips to the left} - 180 \text{ kips to the left}$$

$$= 447.2 \text{ kips to the left}$$

From ACI Sec. 21.5.3.1 the nominal shear strength of a joint confined on all four sides is

$$V_n = 20\sqrt{f_c'}\, A_j = 20 \times \sqrt{4000} \text{ psi} \times 576 \text{ in.}^2$$

$$= 729 \text{ kips}$$

$$\phi V_n = 0.85 \times 729 = 619 \text{ kips}$$

Therefore, the joint has adequate shear strength.
Provide 3 No. 3 three-legged hoops at 6 in. on centers in the joint. ∎

19–9   DIAPHRAGMS

Floors and roofs serve as diaphragms to distribute the horizontal seismic or wind forces to vertical elements such as walls or frames which resist lateral loads as shown in Fig. 19–9. In effect, they act as deep flexural members lying in horizontal planes. ACI Sec. 21.6.4 gives minimum thicknesses and requirements for cast-in-place and composite diaphragms.

Flexural Strength

ACI Sec. 21.6.7 points out the need to provide tension and compression chords in diaphragms. The chord forces are equal to the factored axial force in the diaphragm, if any, plus the force obtained by dividing the factored moment at the section by the distance between the boundary elements (chords) of the diaphragm. If the edge of a diaphragm is notched as shown in Fig. 19–9, special details are required to transmit the tension chord force around the notch.

Shear Strength

ACI Secs. 21.6.5.2 and 3 give expressions for the nominal shear resistance of walls and diaphragms. These are expressed as a function of h_w/ℓ_w, where for a diaphragm, h_w is the length of the diaphragm or segment of diaphragm measured perpendicular to the direction of the seismic loads and ℓ_w is the dimension of the diaphragm segment under consideration, measured parallel to the earthquake forces. For small h_w/ℓ_w, the expressions allow higher shear strengths, recognizing D-region behavior.

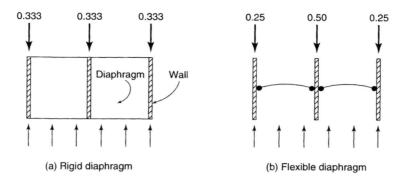

Fig. 19–30
Effect of diaphragm stiffness
on distribution of lateral
loads to walls in a building.

(a) Rigid diaphragm   (b) Flexible diaphragm

Effect of Diaphragm Stiffness on Lateral Load Distribution

Figure 19–30 illustrates the effect of diaphragm stiffness on the distribution of lateral loads to the lateral load resisting elements. The building shown in the figure has three walls of equal lateral stiffness. If the diaphragm is essentially rigid in plane and there is no torsion the three walls will displace the same amount and each wall will resist one-third of the total lateral load, as shown in Fig. 19–30a. On the other hand, if the diaphragm is flexible relative to the walls, the two end walls will each resist a quarter of the lateral shear and the center wall will resist half of it, as shown in Fig. 19–30b.

The following derivations are presented to give an idea of the factors affecting the relative stiffnesses of the walls and diaphragms. The lateral stiffness, K_ℓ, of a cantilever of height ℓ which is fixed at the base is

$$K_\ell = \frac{V}{\Delta} \tag{19–17}$$

where Δ is the lateral deflection on the top of the cantilever due to the load V at the top, equal to

$$\Delta = \frac{V\ell^3}{3EI} + \frac{1.2V\ell}{AG} \tag{19–18}$$

The first term is flexural deflections, the second is shear deflections. Substituting Eq. 19–18 into Eq. 19–17 and taking $G = E/2$ gives

$$K\ell = \frac{3EI}{\ell^3} + \frac{AE}{2.4\ell} \tag{19–19}$$

A similar expression can be derived for the lateral stiffness of a piece of diaphragm between two walls.

Benjamin[19–9] has shown that if the stiffness, $K_{\ell d}$, of the diaphragm exceeds about two times that of the walls, $K_{\ell w}$, the diaphragm will act as a rigid diaphragm in transmitting loads to the walls. If $K_{\ell d}/K_{\ell w} = 0$, the diaphragm is fully flexible.

19–10   STRUCTURAL WALLS

Structural walls or *shearwalls* are frequently used to resist a major fraction of the design shears. These are designed according to ACI Sec. 21.6. The factored shears, moments, and axial forces to be considered in the design of the wall are obtained from the frame analysis.

The first step in designing a wall is to determine whether boundary elements are needed. Using the design forces, the wall is analyzed using

$$\sigma = \frac{P}{A} \pm \frac{My}{I} \qquad (19\text{--}20)$$

where A and I are based on the gross cross section. If the maximum compressive fiber stress exceeds $0.2f_c'$ at any point, boundary elements are required over that part of the height of the wall where the maximum fiber stress exceeds $0.15f_c'$. This analysis is simply a convenient way to establish a limit value and is not intended to be representative of the true wall response.

Boundary elements (Fig. 19–31) are regions at the ends of the cross section of the wall which are reinforced as columns with the vertical reinforcement enclosed in hoop reinforcement. Boundary elements armor the edge of the walls for stress reversals and prevent buckling of the edge of the wall. Ideally, the wall should be designed using a strain compatibility solution such as given in Sec. 5–5 or 11–4. Because such a solution is tedious, a simpler method of design is presented in ACI Sec. 21.6.6.3. It is assumed that when the wall is displaced laterally, only the boundary element transmits compression to the next lower level as shown in Fig. 19–32. The boundary element at A is designed for a compressive force of $(W_u + M_u/z)$ as shown in Fig. 19–32 and a tensile force of (M_u/z) corresponding to sway in the opposite direction. It is assumed to act as an axially loaded short column or tension member and design is based on ACI Eq. 10–2 with $\phi = 0.70$ if the boundary element has hoop reinforcement.

The design of structural walls for shear is given in ACI Sec. 21.6.5. The basic design equation is essentially the same as the $\phi(V_c + V_s)$ procedure used in beam design. The basic shear v_c, stress carried by the concrete, is $2\sqrt{f_c'}$ psi except for short stubby walls, for which a higher stress is allowed. The horizontal reinforcement in the wall must be anchored in the boundary elements as specified in ACI Sec. 21.6.6.4.

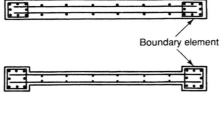

Fig. 19–31
Plan views of structural walls with boundary elements.

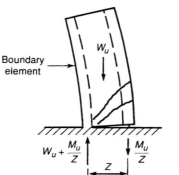

Fig. 19–32
Forces in boundary elements.

The ϕ factor is 0.85 for shear unless the nominal shear strength is less than the shear corresponding to development of the nominal flexural strength of the wall. In such a case ϕ is taken as 0.6 (ACI Sec. 9.3.4.1).

Frequently, two shear walls are *coupled* by beams or slabs spanning across a doorway or similar opening. Depending on the stiffness of the coupling beams, the walls act as two independent cantilevers as the coupling beam stiffness approaches zero, or one solid cantilever if the coupling beam stiffness is high. The coupling beams transmit shear from one cantilever to the other and undergo large shearing deformations as shown in Fig. 19–33a. As a result, these beams degrade rapidly in shear in an earthquake. Paulay[19-2] has shown experimentally that the reinforcement pattern shown in Fig. 19–33c transmits cyclic shear load much better than conventional top and bottom steel and stirrups. This acts as a truss with forces T_u and C_u, which transmit a moment and shear.

$$T_u = C_u = \phi A_s f_y \tag{19–21}$$

$$V_u = 2T_u \sin \alpha = 2\phi A_s f_y \sin \alpha \tag{19–22}$$

$$M_u = (\phi A_s f_y \cos \alpha)(h - 2d') \tag{19–23}$$

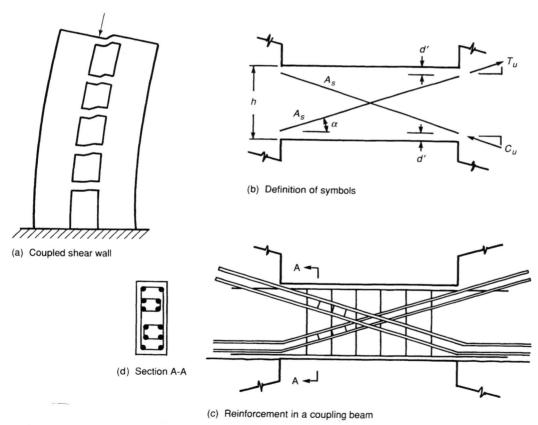

(a) Coupled shear wall

(b) Definition of symbols

(d) Section A-A

(c) Reinforcement in a coupling beam

Fig. 19–33
Coupled shear walls and coupling beams.

The diagonal bars should be tied to form intersecting column cages. Nominal top and bottom steel and stirrups are provided to prevent large pieces of concrete from dropping out during cyclic deformations.

Coupled walls may need boundary elements at the outside ends only or at both ends of both walls, depending on the stiffness of the coupling beams. This would be determined by an analysis according to ACI Sec. 21.6.6.1 and Eq. 19–20 using the moments and axial forces assigned to each part of the coupled wall.

19–11 FRAME MEMBERS NOT PROPORTIONED TO RESIST FORCES INDUCED BY EARTHQUAKE MOTIONS IN FRAMES SUBJECTED TO HIGH SEISMIC RISK

ACI Sec. 21.7 provides less stringent design requirements for members that are not part of the designated lateral load resisting system in structures subjected to severe earthquakes. Such members must be able to resist the factored axial forces due to gravity loads, and the moments and shears induced in them when the frame is deflected laterally through twice the elastically calculated lateral deflections under factored lateral loads. Traditionally, the so-called *building frame system* defined in Table 19–2 was designed assuming that the frame supported only gravity loads while shear walls resisted all of the lateral loads. In the 1994 Northridge earthquake, columns in a number of this type of building failed when forced through the lateral displacements imposed by that earthquake. In the 1995 code ACI Sec. 21.7 was made considerably more stringent.

19–12 FRAMES IN REGIONS OF MODERATE SEISMIC RISK

Portions of the United States have been designated as regions of moderate seismic risk. In these regions the relaxed seismic design provisions given in ACI Sec. 21.8 are applied along with ACI Chaps. 1 to 18. See Table 19–3.

In these regions the shear V_u in beams, columns, and two-way slabs is calculated as the larger of

(a) The shear resulting from reaching the nominal moment capacity $(\phi = 1.0, f_s = 1.0 f_y)$ at each end of the member in question, as shown in Fig. 19–18b, except M_n is substituted for M_{pr}, plus in the case of beams and slabs, the shear resulting from a gravity load of $0.75(1.4D + 1.7L)$. The factor 0.75 comes from ACI Eq. 9–2.

(b) The shear resulting from the load combination in ACI Sec. 9.2.3 with E replaced with $2E$. This gives

$$U = 0.75(1.4D + 1.7L + 1.1 \times 1.7 \times 2E) \qquad (19–24)$$

Special details are given in ACI Sec. 21.8.6 for the design of two-way slabs without beams. Such structures are not permitted as part of the lateral load resisting frame in regions of high seismic risk but can be used in regions of moderate seismic risk. It should be noted, however, that the lateral stiffness of unbraced two-way slab–column frames is very low and the lateral deflections under even moderate earthquakes may be very large. It should also be noted that under some circumstances there may be a moment reversal at the ends of slab spans which will require special detailing.

APPENDIX A

Design Aids

Fig. A–11   Nondimensional interaction diagram for tied columns with bars in four faces, $\gamma = 0.90$.

Fig. A–12   Nondimensional interaction diagram for circular *spiral* columns, $\gamma = 0.60$.

Fig. A–13   Nondimensional interaction diagram for circular *spiral* columns, $\gamma = 0.75$.

Fig. A–14   Nondimensional interaction diagram for circular *spiral* columns, $\gamma = 0.90$.

TABLE A–1   Areas, Weights, and Dimensions of Reinforcing Bars

Bar Size Designation No.[b]	Grades	Weight (lb/ft)	Nominal Dimensions[a]	
			Diameter (in.)	Cross-Sectional Area (in.²)
3	40, 60	0.376	0.375	0.11
4	40, 60	0.668	0.500	0.20
5	40, 60	1.043	0.625	0.31
6	40, 60, 75	1.502	0.750	0.44
7	60, 75	2.044	0.875	0.60
8	60, 75	2.67	1.000	0.79
9	60, 75	3.40	1.128	1.00
10	60, 75	4.30	1.270	1.27
11	60, 75	5.31	1.410	1.56
14	60, 75	7.65	1.693	2.25
18	60, 75	13.60	2.257	4.00

[a]The nominal dimensions of a deformed bar are equivalent to those of a plain round bar having the same weight per foot as the deformed bar.

[b]Bar numbers are based on the number of eighths of an inch included in the nominal diameter.

TABLE A–1M   Areas, Weights, and Dimensions of Reinforcing Bars—SI Units

Bar Size Designation No.[b]	Grades	Nominal Mass (kg/m)	Nominal Dimensions[a]	
			Diameter (mm)	Cross-Sectional Area (mm²)
10	300, 400	0.785	11.3	100
15	300, 400	1.570	16.0	200
20	300, 400, 500	2.355	19.5	300
25	400, 500	3.925	25.2	500
30	400, 500	5.495	29.9	700
35	400, 500	7.850	35.7	1000
45	400, 500	11.775	43.7	1500
55	400, 500	19.625	56.4	2500

[a]The nominal dimensions of a deformed bar are equivalent to those of a plain round bar having the same mass per metre as the deformed bar.

[b]Bar designation numbers are the nominal diameter rounded to the nearest 5 or 10 mm.

TABLE A–2   Welded-Wire Fabric

(a) Wires

Wire Size Number[a]		Nominal Diameter (in.)	Area (in.² per ft of width for center-to-center spacing, in.)			
Smooth	Deformed		4	6	10	12
W31	D31	0.628	0.93	0.62	0.372	0.31
W11	D11	0.374	0.33	0.22	0.132	0.11
W10	D10	0.356	0.30	0.20	0.12	0.10
W9	D9	0.338	0.27	0.18	0.108	0.09
W8	D8	0.319	0.24	0.16	0.096	0.08
W7	D7	0.298	0.21	0.14	0.084	0.07
W6	D6	0.276	0.18	0.12	0.072	0.06
W5.5		0.264	0.165	0.11	0.0066	0.055
W5	D5	0.252	0.15	0.10	0.06	0.05
W4	D4	0.225	0.12	0.08	0.048	0.04
W3.5		0.211	0.105	0.07	0.042	0.035
W2.9		0.192	0.087	0.058	0.035	0.029
W2.5		0.178	0.075	0.05	0.03	0.025
W2.1		0.162	0.063	0.042	0.025	0.021
W1.4		0.135	0.042	0.028	0.017	0.014

[a]Wire size number is 100 times the wire area in in.².

(b) Common Stock Welded-Wire Fabric

Style Designation[a]	Steel Area (in.²/ft)		Approximate Weight (lb/100 ft²)
	Longitudinal	Transverse	
6 × 6—W2.9 × W2.9	0.058	0.058	42
4 × 4—W2.1 × W2.1	0.062	0.062	44
6 × 6—W4 × W4	0.080	0.080	58
4 × 4—W2.9 × W2.9	0.087	0.087	62
6 × 6—W5.5 × W5.5	0.110	0.110	80
4 × 4—W4 × W4	0.120	0.120	85
4 × 4—W5.5 × W5.5	0.165	0.165	119

[a]The numbers in the style designation refer to: longitudinal wire spacing × transverse wire spacing − longitudinal wire size × transverse wire size.

TABLE A-3 Values of ϕk_n and j [a]

$$\omega = \frac{\rho f_y}{f'_c} \qquad \phi k_n = \phi[f'_c \omega(1 - 0.59\omega)] \qquad \frac{M_n}{\phi k_n} = \frac{bd^2}{12,000} \qquad j = 1 - 0.59\omega \qquad A_s = \frac{M_u}{\phi f_y jd}$$

| | $f_y = 40,000$ psi | | | | $f_y = 60,000$ psi | | | | | | | | | |
| | $f'_c = 3000$ psi | | $f'_c = 3750$ psi | | $f'_c = 3000$ psi | | $f'_c = 3750$ psi | | $f'_c = 4000$ psi | | $f'_c = 5000$ psi | | $f'_c = 6000$ psi | |
ρ	ϕk_n	j	ϕk_n	j	ϕk_n	j	ϕk_n	j	ϕk_n	j	ϕk_n	j	ϕk_n	j
0.0033					171	0.961	173	0.969	173	0.971	174	0.977	175	0.981
0.004					206	0.953	208	0.962	208	0.965	210	0.972	211	0.976
0.005	173	0.961	174	0.969	254	0.941	257	0.953	258	0.956	260	0.965	262	0.971
0.006	206	0.953	208	0.962	301	0.929	306	0.943	307	0.947	310	0.958	313	0.965
0.007	238	0.945	241	0.956	347	0.917	353	0.934	355	0.938	359	0.950	362	0.959
0.008	270	0.937	274	0.950	391	0.906	399	0.924	401	0.929	408	0.943	412	0.953
0.009	301	0.929	306	0.943	434	0.894	445	0.915	447	0.920	455	0.936	460	0.947
0.010	332	0.921	337	0.937	476	0.882	489	0.906	492	0.912	502	0.929	508	0.941
0.011	362	0.913	369	0.931	517	0.870	532	0.896	536	0.903	548	0.922	555	0.935
0.012	391	0.906	399	0.924	556	0.858	575	0.887	579	0.894	593	0.915	602	0.929
0.013	420	0.898	430	0.918	594	0.847	616	0.877	621	0.885	637	0.908	648	0.923
0.014	448	0.890	460	0.912			656	0.868	662	0.876	681	0.901	694	0.917
0.015	476	0.882	489	0.906			695	0.858	702	0.867	724	0.894	738	0.912
0.016	504	0.874	518	0.899			734	0.849	742	0.858	766	0.887	782	0.906
0.017	530	0.866	547	0.893					780	0.850	808	0.880	826	0.900
0.018	556	0.858	575	0.887					817	0.841	848	0.873	869	0.894
0.019	582	0.851	602	0.880							888	0.865	911	0.888
0.020	607	0.843	629	0.874							927	0.858	953	0.882
0.021			656	0.868							965	0.851	993	0.876
0.022			682	0.862									1030	0.870
0.023			708	0.855									1070	0.864
0.024			734	0.849										
0.025			758	0.843										

[a]Upper line in each column is below the entry for $\rho = 0.35\rho_b$; lower line is below the entry for $\rho = 0.5\rho_b$; bottom entry is the tension-controlled limit. Larger values of ρ can be used but require ϕ to be evaluated.

TABLE A–3M Values of ϕk_n and j—SI Units[a]

$$\omega = \frac{\rho f_y}{f'_c} \qquad \phi k = \phi[f'_c\omega(1-0.59\omega)] \qquad \frac{M_u}{\phi k_n} = \frac{bd^2}{10^6} \qquad A_s = \frac{M_u}{\phi f_y j d}$$

$$j = 1 - 0.59\omega$$

	$f_y = 300$ MPa				$f_y = 400$ MPa									
	$f'_c = 20$ MPa		$f'_c = 25$ MPa		$f'_c = 20$ MPa		$f'_c = 25$ MPa		$f'_c = 30$ MPa		$f'_c = 35$ MPa		$f'_c = 40$ MPa	
ρ	ϕk_n	j	ϕk_n	j	ϕk_n	j	ϕk_n	j	ϕk_n	j	ϕk_n	j	ϕk_n	j
0.0033					1.14	0.961	1.15	0.969	1.16	0.974	1.16	0.978	1.16	0.981
0.004	1.29	0.956			1.37	0.953	1.39	0.962	1.39	0.969	1.40	0.973	1.41	0.976
0.005	1.53	0.947	1.30	0.965	1.69	0.941	1.72	0.953	1.73	0.961	1.74	0.966	1.75	0.970
0.006	1.77	0.938	1.55	0.958	2.01	0.929	2.04	0.943	2.06	0.953	2.07	0.960	2.08	0.965
0.007	2.01	0.929	1.80	0.950	2.31	0.917	2.35	0.934	2.38	0.945	2.40	0.953	2.42	0.959
0.008	2.24	0.920	2.04	0.943	2.61	0.906	2.66	0.924	2.70	0.937	2.72	0.946	2.74	0.953
0.009	2.46	0.911	2.28	0.936	2.90	0.894	2.96	0.915	3.01	0.929	3.04	0.939	3.07	0.947
0.010	2.68	0.903	2.51	0.929	3.18	0.882	3.26	0.906	3.32	0.921	3.36	0.933	3.39	0.941
0.011	2.90	0.894	2.74	0.922	3.45	0.870	3.55	0.896	3.62	0.913	3.67	0.926	3.70	0.935
0.012	3.11	0.885	2.96	0.915	3.71	0.858	3.83	0.887	3.91	0.906	3.97	0.919	4.01	0.929
0.013	3.31	0.876	3.19	0.908	3.96	0.847	4.11	0.877	4.20	0.898	4.27	0.912	4.32	0.923
0.014	3.51	0.867	3.41	0.901			4.37	0.868	4.48	0.890	4.56	0.906	4.62	0.917
0.015	3.71	0.858	3.62	0.894			4.64	0.858	4.76	0.882	4.85	0.899	4.92	0.911
0.016	3.90	0.850	3.83	0.887			4.89	0.849	5.04	0.874	5.14	0.892	5.22	0.906
0.017	4.09	0.841	4.04	0.880					5.30	0.866	5.42	0.885	5.51	0.900
0.018			4.24	0.873					5.56	0.858	5.69	0.879	5.79	0.894
0.019			4.44	0.865					5.82	0.851	5.96	0.872	6.07	0.888
0.020			4.64	0.858					6.07	0.843	6.23	0.865	6.35	0.882
0.021			4.83	0.851							6.49	0.858	6.62	0.876
0.022			5.01	0.844							6.75	0.852	6.89	0.870
0.023													7.16	0.864
0.024													7.42	0.858

[a]Upper line in each column is below the entry for $\rho = 0.35\rho_b$; lower line is below the entry for $\rho = 0.5\rho_b$; bottom entry is the tension-controlled limit. Larger values of ρ can be used but require ϕ to be evaluated.

TABLE A–4 Ratio of Depth of Rectangular Stress Block for Balanced Failure (a_b), Compression-Controlled Limit ($a_{cc\ell}$), and Tension-Controlled Limit ($a_{tc\ell}$) to Effective Depth (d) or Depth to Extreme Tension Steel Layer (d_t)[a]

f_y (psi)		f_c' (psi)			
		Less than or equal to 4000	5000	6000	8000
40,000	a_b/d, $a_{cc\ell}/d_t$	0.582	0.548	0.514	0.445
	$0.75a_b/d$	0.437	0.411	0.385	0.334
	$a_{tc\ell}/d_t$	0.319	0.300	0.281	0.244
	$0.50a_b/d$	0.291	0.274	0.257	0.223
	$0.35a_b/d$	0.204	0.192	0.180	0.156
60,000	a_b/d, $a_{cc\ell}/d_t$	0.503	0.474	0.444	0.385
	$0.75a_b/d$	0.377	0.355	0.333	0.288
	$a_{tc\ell}/d_t$	0.319	0.300	0.281	0.244
	$0.50a_b/d$	0.252	0.237	0.222	0.192
	$0.35a_b/d$	0.176	0.166	0.155	0.135
	β_1	0.85	0.80	0.75	0.65

[a] a_b/d from Eq. 4–21; desirable range for beams a/d from 0.35 to 0.50 a_b/d; $a_{cc\ell}/d_t$ and $a_{tc\ell}/d_t$ from Eqs. 4–22 and 4–23. ACI Appendix B requires $\phi < 0.9$ if $a/d_t > a_{tc\ell}/d_t$.

TABLE A–4M Ratio of Depth of Rectangular Stress Block for Balanced Failure (a_b), Compression-Controlled Limit ($a_{cc\ell}$), and Tension Controlled Limit ($a_{tc\ell}$) to Effective Depth (d) or Depth to Extreme Tension Steel Layer (d_t)—SI Units[a]

f_y (MPa)		f_c' (MPa)			
		Less than or equal to 30	35	40	50
300	a_b/d, $a_{cc\ell}/d_t$	0.567	0.540	0.513	0.460
	$0.75a_b/d$	0.425	0.405	0.385	0.345
	$a_{tc\ell}/d_t$	0.319	0.304	0.289	0.259
	$0.50a_b/d$	0.283	0.270	0.256	0.230
	$0.35a_b/d$	0.198	0.189	0.180	0.161
400	a_b/d, $a_{cc\ell}/d_t$	0.510	0.486	0.462	0.414
	$0.75a_b/d$	0.382	0.364	0.346	0.310
	$a_{tc\ell}/d_t$	0.319	0.304	0.289	0.259
	$0.50a_b/d$	0.255	0.243	0.231	0.207
	$0.35a_b/d$	0.178	0.170	0.162	0.145
	β_1	0.85	0.80	0.77	0.69

[a] a_b/d from Eq. 4–21M; desirable range of a/d for beams from 0.35 to 0.50 a_b/d; $a_{cc\ell}/d_t$ and a_{tcl}/d_t from Eqs. 4–22M and 4–23. ACI Appendix B requires $\phi < 0.9$ if $a/d_t > a_{tc\ell}/d_t$.

TABLE A-5 Steel Ratios at Balanced Condition (ρ_b), Compression-Controlled Limit (ρ_{ccl}) and Tension-Controlled Limit (ρ_{tcl}) for Rectangular Beams with Tension Reinforcement Only[a]

f_y (psi)		f'_c (psi)						
		3000	3750	4000	5000	6000	8000
40,000	ρ_b, ρ_{ccl}	0.0371	0.0464	0.0495	0.0582	0.0655	0.0703
	$0.75\,\rho_b$	0.0278	0.0348	0.0371	0.0437	0.0491	0.0527
	ρ_{tcl}	0.0203	0.0254	0.0271	0.0319	0.0359	0.0414
	$0.50\rho_b$	0.0186	0.0232	0.0247	0.0291	0.0328	0.0352
	$0.35\rho_b$	0.0130	0.0162	0.0173	0.0204	0.0229	0.0246
60,000	ρ_b, ρ_{ccl}	0.0214	0.0267	0.0285	0.0335	0.0377	0.0405
	$0.75\rho_b$	0.0161	0.0200	0.0214	0.0251	0.0283	0.0307
	ρ_{tcl}	0.0135	0.0169	0.0181	0.0213	0.0239	0.0276
	$0.50\rho_b$	0.0107	0.0134	0.0143	0.0168	0.0189	0.0202
	$0.35\rho_b$	0.0075	0.0094	0.0100	0.0117	0.0132	0.0142
	β_1	0.85	0.85	0.85	0.80	0.75	0.65

[a] $\rho = A_s/bd$, ρ_b from Eq. 4-25; desirable steel ratio for beams, $\rho = 0.35$ to $0.50\rho_b$.

TABLE A-5M Steel Ratios at Balanced Condition (ρ_b), Compression-Controlled Limit ($\rho_{cc\ell}$), and Tension-Controlled Limit ($\rho_{tc\ell}$) for Rectangular Beams with Tension Reinforcement Only—SI Units

f_y (MPa)		f_c' (MPa)				
		20	25	30	35	40
300	$\rho_b, \rho_{cc\ell}$	0.0321	0.0401	0.0482	0.0535	0.0582
	$0.75\rho_b$	0.241	0.0301	0.0361	0.0401	0.0436
	$\rho_{tc\ell}$	0.0181	0.0226	0.0271	0.0301	0.0327
	$0.5\rho_b$	0.0160	0.0200	0.0241	0.0267	0.0291
	$0.35\rho_b$	0.0112	0.0140	0.0169	0.0187	0.0204
400	$\rho_b, \rho_{cc\ell}$	0.0217	0.0271	0.0325	0.0361	0.0393
	$0.75\rho_b$	0.0163	0.0203	0.0244	0.0271	0.0295
	$\rho_{tc\ell}$	0.0136	0.0169	0.0203	0.0226	0.0245
	$0.5\rho_b$	0.0109	0.0135	0.0162	0.0180	0.0196
	$0.35\rho_b$	0.0076	0.0095	0.0114	0.0126	0.0138
	β_1	0.85	0.85	0.85	0.81	0.77

[a] $\rho = A_s/bd$, ρ_b from Eq. 4-25M; desirable ρ for beams, $\rho = 0.35$ to $0.50\rho_b$; ACI Appendix B requires $\phi < 0.9$ for $\rho > \rho_{tc\ell}$.

TABLE A–6 Minimum Beam Web Widths, b_w for Various Bar Combinations, Interior Exposure, Minimum Bar Spacing (in.)[a,b]

No. of Bars	A Bar No.	A 0	A 5	B Bar No.	B 1	B 2	B 3	B 4	B 5	C Bar No.	C 1	C 2	C 3	C 4	C 5
1		5.5	13.0		7.0	8.5	9.5	11.0	12.5						
2		7.0	14.5		8.5	9.5	11.0	12.5	14.0						
3	4	8.5	16.0	3	10.0	11.0	12.5	14.0	15.5						
4		10.0	17.5		11.5	12.5	14.0	15.5	17.0						
5		11.5	19.0		13.0	14.0	15.5	17.0	18.5						
1		5.5	13.5		7.0	8.5	10.0	11.5	13.0		7.0	8.5	9.5	11.0	12.5
2		7.0	15.0		8.5	10.0	11.5	13.0	14.5		8.5	10.0	11.0	12.5	14.0
3	5	8.5	17.0	4	10.0	11.5	13.0	14.5	16.0	3	10.0	11.5	13.0	14.0	15.5
4		10.5	18.5		12.0	13.5	15.0	16.5	18.0		11.5	13.0	14.5	16.0	17.0
5		12.0	20.0		13.5	15.0	16.5	18.0	19.5		13.5	14.5	16.5	17.5	19.0
1		5.5	14.0		7.0	9.0	10.5	12.0	13.5		7.0	8.5	10.0	11.5	13.0
2		7.0	16.0		9.0	10.5	12.0	13.5	15.5		8.5	10.0	11.5	13.0	14.5
3	6	9.0	17.5	5	10.5	12.0	14.0	15.5	17.0	4	10.5	12.0	13.5	15.0	16.5
4		10.5	19.5		12.5	14.0	15.5	17.0	19.0		12.0	13.5	15.0	16.5	18.0
5		12.5	21.0		14.0	15.5	17.5	19.0	20.5		14.0	15.5	17.0	18.5	20.0
1		5.5	15.0		7.5	9.0	11.0	12.5	14.5		7.0	9.0	10.5	12.0	13.5
2		7.5	16.5		9.0	11.0	12.5	14.5	16.0		9.0	10.5	12.0	14.0	15.5
3	7	9.0	18.5	6	11.0	12.5	14.5	16.0	18.0	5	11.0	12.5	14.0	15.5	17.5
4		11.0	20.5		13.0	14.5	16.5	18.0	20.0		12.5	14.5	16.0	17.5	19.0
5		13.0	22.5		14.5	16.5	18.0	20.0	21.5		14.5	16.0	18.0	19.5	21.0
1		5.5	15.5		7.5	9.5	11.0	13.0	15.0		7.5	9.0	11.0	12.5	14.5
2		7.5	17.5		9.5	11.0	13.0	15.0	17.0		9.0	11.0	12.5	14.5	16.0
3	8	9.5	19.5	7	11.5	13.0	15.0	17.0	19.0	6	11.0	13.0	14.5	16.5	18.0
4		11.5	21.5		13.5	15.0	17.0	19.0	21.0		13.0	15.0	16.5	18.5	20.0
5		13.5	23.5		15.5	17.0	19.0	21.0	23.0		15.0	17.0	18.5	20.5	22.0
1		5.5	17.0		7.5	9.5	11.5	13.5	15.5		7.5	9.5	11.5	13.0	15.0
2		8.0	19.0		10.0	12.0	14.0	16.0	18.0		9.5	11.5	13.5	15.5	17.0
3	9	10.0	21.5	8	12.0	14.0	16.0	18.0	20.0	7	12.0	14.0	15.5	17.5	19.5
4		12.5	23.5		14.5	16.5	18.5	20.5	22.5		14.0	16.0	18.0	20.0	21.5
5		14.5	26.0		16.5	18.5	20.5	22.5	24.5		16.5	18.5	20.0	22.0	24.0
1		5.5	18.0		8.0	10.0	12.5	14.5	17.0		8.0	10.0	12.0	14.0	16.0
2		8.0	20.5		10.5	12.5	15.0	17.0	19.5		10.0	12.0	14.0	16.0	18.0
3	10	10.5	23.5	9	13.0	15.0	17.5	19.5	22.0	8	12.5	14.5	16.5	18.5	20.5
4		13.0	26.0		15.5	17.5	20.0	22.0	24.5		15.0	17.0	19.0	21.0	23.0
5		15.5	28.5		18.0	20.0	22.5	24.5	27.0		17.5	19.5	21.5	23.5	25.5
1		5.5	19.5		8.0	10.5	13.0	15.5	18.0		8.0	10.5	12.5	15.0	17.0
2		8.5	22.5		11.0	13.5	16.0	18.5	21.0		10.5	13.0	15.0	17.5	19.5
3	11	11.0	25.0	10	13.5	16.0	19.0	21.5	24.0	9	13.5	15.5	18.0	20.0	22.5
4		14.0	28.0		16.5	19.0	21.5	24.0	26.5		16.0	18.5	20.5	23.0	25.0
5		17.0	31.0		19.5	22.0	24.5	27.0	29.5		19.0	21.5	23.5	26.0	28.0

[a]Clear cover, $1\frac{1}{2}$ in.; No. 3 double-leg stirrup; $\frac{3}{4}$ in.-maximum-size aggregate.

[b]This table consists of three basic parts: Part A lists the web widths required for 1 to 5 bars of the sizes given in the left margin of part A. *Example:* 3 No. 5, min b_w = 8.5 in. Part A also lists the minimum b_w for 1 to 5 bars of the size given in the left margin plus 5 bars of the same size. *Example:* 3 No. 5 plus 5 No. 5, min b_w = 17.0 in. Part B lists the minimum b_w of 1 to 5 bars of the size given in the left margin of part A plus 1 to 5 bars of the size listed in the left margin of part B. *Example:* 3 No. 5 plus 2 No. 4, min b_w = 11.5 in. Part C is similar to part B.

Source: Based on a table from Ref. 4–11, used with the permission of the American Concrete Institute.

TABLE A–6M Minimum Beam Web Widths, b_w, for Various Bar Combinations, Interior Exposure, Minimum Bar Spacing—SI Units (mm)[a,b]

No. of Bars	A — Bar No.	A: 0	A: 5	B — Bar No.	B: 1	B: 2	B: 3	B: 4	B: 5	C — Bar No.	C: 1	C: 2	C: 3	C: 4	C: 5
1	15	140	340	10	180	210	250	280	340						
2		180	380		220	250	290	320	380						
3		220	420		260	290	330	360	420						
4		260	460		300	330	370	400	460						
5		300	500		340	370	410	440	500						
1	20	140	365	15	180	220	260	300	340	10	175	210	245	280	315
2		185	410		225	265	305	345	385		220	255	290	325	360
3		230	455		270	310	350	390	430		265	300	335	370	405
4		275	500		310	355	395	435	475		310	345	380	415	450
5		320	545		360	400	440	480	520		355	390	425	460	495
1	25	140	390	20	185	230	275	320	365	15	180	220	260	300	340
2		190	440		235	280	325	370	415		230	270	310	350	390
3		240	490		285	330	375	420	465		280	320	360	400	440
4		290	540		335	380	425	470	515		330	370	410	450	490
5		340	590		385	430	475	520	565		380	420	460	500	540
1	30	140	440	25	195	250	305	360	415	20	190	240	290	340	390
2		200	500		255	310	365	420	475		250	300	350	400	450
3		260	560		315	370	425	480	535		310	360	410	460	510
4		320	620		375	430	485	540	595		370	420	470	520	570
5		380	680		435	490	545	600	655		430	480	530	580	630
1	35	140	490	30	205	270	335	400	465	25	200	260	320	380	440
2		210	560		275	340	405	470	535		270	330	390	450	510
3		280	630		345	410	475	540	605		340	400	460	520	580
4		350	700		415	480	545	610	675		410	470	530	590	650
5		420	770		485	550	615	680	745		480	540	600	660	720

[a]Clear cover, 40 mm; No. 10 double-leg stirrup; 19 mm maximum-size aggregate; spacing based on the diameter of the largest bars.

[b]For directions on how to use this table, see Table A-6. footnote b.

TABLE A–7 Values of $bd^2/12{,}000$ for Use in Choosing Beam Sizes

$$\frac{\phi M_n}{\phi k_n} = \frac{bd^2}{12{,}000} \quad \text{or} \quad \frac{M_u}{\phi k_n} = \frac{bd^2}{12{,}000} \quad \text{where } \phi k_n \text{ is from Table A–3}$$

d	6	7	8	9	10	11	12	14	16	18	20	22	24	26	28	30	36	48
															b (in.)			
5	0.013	0.015	0.017	0.019	0.021	0.023	0.025	0.027	0.033	0.037	0.042	0.046	0.050	0.054	0.058	0.062	0.075	0.100
6	0.018	0.021	0.024	0.027	0.030	0.033	0.036	0.042	0.048	0.054	0.060	0.066	0.072	0.078	0.084	0.090	0.108	0.144
7	0.025	0.029	0.033	0.037	0.041	0.045	0.049	0.057	0.065	0.073	0.082	0.090	0.098	0.106	0.114	0.123	0.147	0.196
8	0.032	0.037	0.043	0.048	0.053	0.059	0.064	0.075	0.085	0.096	0.107	0.117	0.128	0.139	0.149	0.160	0.192	0.256
9	0.041	0.047	0.054	0.061	0.068	0.074	0.081	0.095	0.108	0.122	0.135	0.149	0.162	0.176	0.189	0.203	0.243	0.324
10	0.050	0.058	0.067	0.075	0.083	0.092	0.100	0.117	0.133	0.150	0.167	0.183	0.200	0.217	0.233	0.250	0.300	0.400
11	0.061	0.071	0.081	0.091	0.101	0.111	0.121	0.141	0.161	0.181	0.202	0.222	0.242	0.262	0.282	0.303	0.363	0.484
12	0.072	0.084	0.096	0.108	0.120	0.132	0.144	0.168	0.192	0.216	0.240	0.264	0.288	0.312	0.336	0.360	0.432	0.576
13	0.085	0.099	0.113	0.127	0.141	0.155	0.169	0.197	0.225	0.253	0.282	0.310	0.338	0.366	0.394	0.423	0.507	0.676
14	0.098	0.114	0.131	0.147	0.163	0.180	0.196	0.229	0.261	0.294	0.327	0.359	0.392	0.425	0.457	0.490	0.588	0.784
15	0.113	0.131	0.150	0.169	0.188	0.206	0.225	0.263	0.300	0.338	0.375	0.413	0.450	0.488	0.525	0.563	0.675	0.900
16	0.128	0.149	0.171	0.192	0.213	0.235	0.256	0.299	0.341	0.384	0.427	0.469	0.512	0.555	0.597	0.640	0.768	1.02
18	0.162	0.189	0.216	0.243	0.270	0.297	0.324	0.378	0.432	0.486	0.540	0.594	0.648	0.702	0.756	0.810	0.972	1.30
20	0.200	0.233	0.267	0.300	0.333	0.367	0.400	0.467	0.533	0.600	0.667	0.733	0.800	0.867	0.933	1.00	1.20	1.60
22		0.282	0.323	0.363	0.403	0.444	0.484	0.565	0.645	0.726	0.807	0.887	0.968	1.05	1.13	1.21	1.45	1.94
24		0.336	0.384	0.432	0.480	0.528	0.576	0.672	0.768	0.864	0.960	1.06	1.15	1.25	1.34	1.44	1.73	2.30
26			0.451	0.507	0.563	0.620	0.676	0.789	0.901	1.01	1.13	1.24	1.35	1.46	1.58	1.69	2.03	2.70
28			0.523	0.588	0.653	0.719	0.784	0.915	1.04	1.18	1.31	1.44	1.57	1.70	1.83	1.96	2.35	3.14
30				0.675	0.750	0.825	0.900	1.05	1.20	1.35	1.50	1.65	1.80	1.95	2.10	2.25	2.70	3.60
32				0.768	0.853	0.939	1.02	1.19	1.37	1.54	1.71	1.88	2.05	2.22	2.39	2.56	3.07	4.10
34					0.963	1.06	1.16	1.35	1.54	1.73	1.93	2.12	2.31	2.50	2.70	2.89	3.47	4.62
36					1.08	1.19	1.30	1.51	1.73	1.94	2.16	2.38	2.59	2.81	3.02	3.24	3.89	5.18
38						1.32	1.44	1.68	1.93	2.17	2.41	2.65	2.89	3.13	3.37	3.61	4.33	5.78
40						1.47	1.60	1.87	2.13	2.40	2.67	2.93	3.20	3.47	3.73	4.00	4.80	6.40
45							2.03	2.36	2.70	3.04	3.38	3.71	4.05	4.39	4.73	5.06	6.08	8.10
50							2.50	2.92	3.33	3.75	4.17	4.58	5.00	5.42	5.83	6.25	7.50	10.0
55							3.03	3.53	4.03	4.54	5.04	5.55	6.05	6.55	7.06	7.56	9.07	12.1
60							3.30	4.20	4.80	5.40	6.00	6.60	7.20	7.80	8.40	9.00	10.8	14.4

Source: Based on a Table in Ref. 4–12, used with the permission of the American Concrete Institute.

TABLE A–7M Values of $bd^2/10^6$ for Use in Choosing Beam Sizes—SI Units

$$\frac{\phi M_n}{\phi k_n} = \frac{bd^2}{10^6} \quad \text{or} \quad \frac{M_u}{\phi k_n} = \frac{bd^2}{10^6} \quad \text{where } \phi k_n \text{ is from Table A–3M}$$

d (mm)	b (mm) 150	175	200	225	250	275	300	350	400	450	500	600	700	800	900	1000	1500
125	2.34	2.73	3.13	3.52	3.91	4.30	4.69	5.47	6.25	7.03	7.81	9.38	10.94	12.50	14.06	15.63	23.44
150	3.38	3.94	4.50	5.06	5.63	6.19	6.75	7.88	9.00	10.13	11.25	13.50	15.75	18.00	20.25	22.50	33.75
175	4.59	5.36	6.13	6.89	7.66	8.42	9.19	10.72	12.25	13.78	15.31	18.38	21.44	24.50	27.56	30.63	45.94
200	6.00	7.00	8.00	9.00	10.00	11.00	12.00	14.00	16.00	18.000	20.00	24.00	28.00	32.00	36.00	40.00	60.00
225	7.59	8.86	10.13	11.39	12.66	13.92	15.19	17.72	20.25	22.78	25.31	30.38	35.44	40.50	45.56	50.63	75.94
250	9.38	10.94	12.50	14.06	15.63	17.19	18.75	21.88	25.00	28.13	31.25	37.50	43.75	50.00	56.25	62.50	93.75
275	11.34	13.23	15.13	17.02	18.91	20.80	22.69	26.47	30.25	34.03	37.81	45.38	52.94	60.50	68.06	75.63	113
300	13.50	15.75	18.00	20.25	22.50	24.75	27.00	31.50	36.00	40.50	45.00	54.00	63.00	72.00	81.00	90.00	135
325	15.84	18.48	21.13	23.77	26.41	29.05	31.69	36.97	42.25	47.53	52.81	63.38	73.94	84.50	95.06	106	158
350	18.38	21.44	24.50	27.56	30.63	33.69	36.75	42.88	49.00	55.13	61.25	73.50	85.75	98.00	110	123	184
375	21.09	24.61	28.13	31.64	35.16	38.67	42.19	49.22	56.25	63.28	70.31	84.38	98.44	113	127	141	211
400	24.00	28.00	32.00	36.00	40.00	44.00	48.00	56.00	64.00	72.00	80.00	96.00	112	128	144	160	240
450	30.38	35.44	40.50	45.56	50.63	55.69	60.75	70.88	81.00	91.13	101	122	142	162	182	203	304
500	37.50	43.75	50.00	56.25	62.50	68.75	75.00	87.50	100	113	125	150	175	200	225	250	375
550	45.38	52.94	60.50	68.06	75.63	83.19	90.75	106	121	136	151	182	212	242	272	303	454
600	54.00	63.00	72.00	81.00	90.00	99.00	108	126	144	162	180	216	252	288	324	360	540
650	63.38	73.94	84.50	95.06	106	116	127	148	169	190	211	254	296	338	380	423	634
700	73.50	85.75	98.00	110	123	135	147	172	196	221	245	294	343	392	441	490	735
800	96.00	112	128	144	160	176	192	224	256	288	320	384	448	512	576	640	960
900	122	142	162	182	203	223	243	284	324	365	405	486	567	648	729	810	1215
1000			200	225	250	275	300	350	400	450	500	600	700	800	900	1000	1500
1200					360	396	432	504	576	648	720	864	1008	1152	1296	1440	2160
1400					490	539	588	686	784	882	980	1176	1372	1568	1764	1960	2940
1500					563	619	675	788	900	1013	1125	1350	1575	1800	2025	2250	3375
2000						1100	1200	1400	1600	1800	2000	2400	2800	3200	3600	4000	6000

TABLE A–8 Cross-Sectional Areas, A_s, for Various Combinations of Bars (in.^{2})[a]

No. of Bars	A Bar No.	A 0	A 5	B Bar No.	B 1	B 2	B 3	B 4	B 5	C Bar No.	C 1	C 2	C 3	C 4	C 5
1		0.20	1.20		0.31	0.42	0.53	0.64	0.75						
2		0.40	1.40		0.51	0.62	0.73	0.84	0.95						
3	4	0.60	1.60	3	0.71	0.82	0.93	1.04	1.15						
4		0.80	1.80		0.91	1.02	1.13	1.24	1.35						
5		1.00	2.00		1.11	1.22	1.33	1.44	1.55						
1		0.31	1.86		0.51	0.71	0.91	1.11	1.31		0.42	0.53	0.64	0.75	0.86
2		0.62	2.17		0.82	1.02	1.22	1.42	1.62		0.73	0.84	0.95	1.06	1.17
3	5	0.93	2.48	4	1.13	1.33	1.53	1.73	1.93	3	1.04	1.15	1.26	1.37	1.48
4		1.24	2.79		1.44	1.64	1.84	2.04	2.24		1.35	1.46	1.57	1.68	1.79
5		1.55	3.10		1.75	1.95	2.15	2.35	2.55		1.66	1.77	1.88	1.99	2.10
1		0.44	2.64		0.75	1.06	1.37	1.68	1.99		0.64	0.84	1.04	1.24	1.44
2		0.88	3.08		1.19	1.50	1.81	2.12	2.43		1.08	1.28	1.48	1.68	1.88
3	6	1.32	3.52	5	1.63	1.94	2.25	2.56	2.87	4	1.52	1.72	1.92	2.12	2.32
4		1.76	3.96		2.07	2.38	2.69	3.00	3.31		1.96	2.16	2.36	2.56	2.76
5		2.20	4.40		2.51	2.82	3.13	3.44	3.75		2.40	2.60	2.80	3.00	3.20
1		0.60	3.60		1.04	1.48	1.92	2.36	2.80		0.91	1.22	1.53	1.84	2.15
2		1.20	4.20		1.64	2.08	2.52	2.96	3.40		1.51	1.82	2.13	2.44	2.75
3	7	1.80	4.80	6	2.24	2.68	3.12	3.56	4.00	5	2.11	2.42	2.73	3.04	3.35
4		2.40	5.40		2.84	3.28	3.72	4.16	4.60		2.71	3.02	3.33	3.64	3.95
5		3.00	6.00		3.44	3.88	4.32	4.76	5.20		3.31	3.62	3.93	4.24	4.55
1		0.79	4.74		1.39	1.99	2.59	3.19	3.79		1.23	1.67	2.11	2.55	2.99
2		1.58	5.53		2.18	2.78	3.38	3.98	4.58		2.02	2.46	2.90	3.34	3.78
3	8	2.37	6.32	7	2.97	3.57	4.17	4.77	5.37	6	2.81	3.25	3.69	4.13	4.57
4		3.16	7.11		3.76	4.36	4.96	5.56	6.16		3.60	4.04	4.48	4.92	5.36
5		3.95	7.90		4.55	5.15	5.75	6.35	6.95		4.39	4.83	5.27	5.71	6.15
1		1.00	6.00		1.79	2.58	3.37	4.16	4.95		1.60	2.20	2.80	3.40	4.00
2		2.00	7.00		2.79	3.58	4.37	5.16	5.95		2.60	3.20	3.80	4.40	5.00
3	9	3.00	8.00	8	3.79	4.58	5.37	6.16	6.95	7	3.60	4.20	4.80	5.40	6.00
4		4.00	9.00		4.79	5.58	6.37	7.16	7.95		4.60	5.20	5.80	6.40	7.00
5		5.00	10.00		5.79	6.58	7.37	8.16	8.95		5.60	6.20	6.80	7.40	8.00
1		1.27	7.62		2.27	3.27	4.27	5.27	6.27		2.06	2.85	3.64	4.43	5.22
2		2.54	8.89		3.54	4.54	5.54	6.54	7.54		3.33	4.12	4.91	5.70	6.49
3	10	3.81	10.16	9	4.81	5.81	6.81	7.81	8.81	8	4.60	5.39	6.18	6.97	7.76
4		5.08	11.43		6.08	7.08	8.08	9.08	10.08		5.87	6.66	7.45	8.24	9.03
5		6.35	12.70		7.35	8.35	9.35	10.35	11.35		7.14	7.93	8.72	9.51	10.30
1		1.56	9.36		2.83	4.10	5.37	6.64	7.91		2.56	3.56	4.56	5.56	6.56
2		3.12	10.92		4.39	5.66	6.93	8.20	9.47		4.12	5.12	6.12	7.12	8.12
3	11	4.68	12.48	10	5.95	7.22	8.49	9.76	11.03	9	5.68	6.68	7.68	8.68	9.68
4		6.24	14.04		7.51	8.78	10.05	11.32	12.59		7.24	8.24	9.24	10.24	11.24
5		7.80	15.60		9.07	10.34	11.61	12.88	14.15		8.80	9.80	10.80	11.80	12.80

[a]For directions on how to use this table, see Table A–6, footnote b.

Source: Based on a table from Ref. 4–12; used with the permission of the American Concrete Institute.

TABLE A–8M Cross-Sectional Areas, A_s, for Various Combinations of Bars—SI Units (mm²)[a]

A

No. of Bars	Bar No.	0	5
1	15	200	1200
2		400	1400
3		600	1600
4		800	1800
5		1000	2000
1	20	300	1800
2		600	2100
3		900	2400
4		1200	2700
5		1500	3000
1	25	500	3000
2		1000	3500
3		1500	4000
4		2000	4500
5		2500	5000
1	30	700	4200
2		1400	4900
3		2100	5600
4		2800	6300
5		3500	7000
1	35	1000	6000
2		2000	7000
3		3000	8000
4		4000	9000
5		5000	10000

B

No. of Bars	Bar No.	1	2	3	4	5
1	10	300	400	500	600	700
2		500	600	700	800	900
3		700	800	900	1000	1100
4		900	1000	1100	1200	1300
5		1100	1200	1300	1400	1500
1	15	500	700	900	1100	1300
2		800	1000	1200	1400	1600
3		1100	1300	1500	1700	1900
4		1400	1600	1800	2000	2200
5		1700	1900	2100	2300	2500
1	20	800	1100	1400	1700	2000
2		1300	1600	1900	2200	2500
3		1800	2100	2400	2700	3000
4		2300	2600	2900	3200	3500
5		2800	3100	3400	3700	4000
1	25	1200	1700	2200	2700	3200
2		1900	2400	2900	3400	3900
3		2600	3100	3600	4100	4600
4		3300	3800	4300	4800	5300
5		4000	4500	5000	5500	6000
1	30	1700	2400	3100	3800	4500
2		2700	3400	4100	4800	5500
3		3700	4400	5100	5800	6500
4		4700	5400	6100	6800	7500
5		5700	6400	7100	7800	8500

C

No. of Bars	Bar No.	1	2	3	4	5
1	10	400	500	600	700	800
2		700	800	900	1000	1100
3		1000	1100	1200	1300	1400
4		1300	1400	1500	1600	1700
5		1600	1700	1800	1900	2000
1	15	700	900	1100	1300	1500
2		1200	1400	1600	1800	2000
3		1700	1900	2100	2300	2500
4		2200	2400	2600	2800	3000
5		2700	2900	3100	3300	3500
1	20	1000	1300	1600	1900	2200
2		1700	2000	2300	2600	2900
3		2400	2700	3000	3300	3600
4		3100	3400	3700	4000	4300
5		3800	4100	4400	4700	5000
1	25	1500	2000	2500	3000	3500
2		2500	3000	3500	4000	4500
3		3500	4000	4500	5000	5500
4		4500	5000	5500	6000	6500
5		5500	6000	6500	7000	7500

[a]For directions on how to use this table, see Table A–6, footnote b.

TABLE A–9   Areas of Bars in a Section 1 ft Wide (in.^{2}/ft)

Bar Spacing (in.)	Bar No.				
	3	4	5	6	7
4	0.33	0.60	0.93	1.32	1.80
$4\frac{1}{2}$	0.29	0.53	0.83	1.17	1.60
5	0.26	0.48	0.74	1.06	1.44
$5\frac{1}{2}$	0.24	0.44	0.68	0.96	1.31
6	0.22	0.40	0.62	0.88	1.20
$6\frac{1}{2}$	0.20	0.37	0.57	0.81	1.11
7	0.19	0.34	0.53	0.75	1.03
$7\frac{1}{2}$	0.18	0.32	0.50	0.70	0.96
8	0.17	0.30	0.47	0.66	0.90
$8\frac{1}{2}$	0.16	0.28	0.44	0.62	0.85
9	0.15	0.27	0.41	0.59	0.80
$9\frac{1}{2}$	0.14	0.25	0.39	0.56	0.76
10	0.13	0.24	0.37	0.53	0.72
$10\frac{1}{2}$	0.13	0.23	0.35	0.50	0.69
11	0.12	0.22	0.34	0.48	0.65
$11\frac{1}{2}$	0.11	0.21	0.32	0.46	0.63
12	0.11	0.20	0.31	0.44	0.60
13	0.10	0.18	0.29	0.41	0.55
14	0.09	0.17	0.27	0.38	0.51
15	0.09	0.16	0.25	0.35	0.48
16	0.08	0.15	0.23	0.33	0.45
17	0.08	0.14	0.22	0.31	0.42
18	0.07	0.13	0.21	0.29	0.40

TABLE A–9M   Areas of Bars in a Section 1 m Wide—
                SI Units (mm²/m)

Bar Spacing (mm)	Bar No.			
	10	15	20	25
100	1000	2000	3000	5000
110	909	1818	2727	4545
120	833	1667	2500	4167
130	769	1538	2308	3846
140	714	1429	2143	3571
150	667	1333	2000	3333
160	625	1250	1875	3125
180	556	1111	1667	2778
200	500	1000	1500	2500
220	455	909	1364	2273
240	417	833	1250	2083
250	400	800	1200	2000
260	385	769	1154	1923
280	357	714	1071	1786
300	333	667	1000	1667
350	286	571	857	1429
400	250	500	750	1250
450	222	444	667	1111
500	200	400	600	1000

TABLE A–10 Limiting Values of d'/a for Checking if Compression Steel Yields[a]

f_y (psi	f_c' (psi)			
	≤ 4000	5000	6000	8000
40,000	0.636	0.675	0.720	0.831
50,000	0.500	0.532	0.567	0.654
60,000	0.365	0.388	0.414	0.477

[a]If d'/a exceeds the values given in this table, the compression steel will not yield before failure; calculated using Eq. 5–9.

TABLE A–10M Limiting Values of d'/a for Checking if Compression Steel Yields—SI Units[a]

f_y (MPa)	f_c' (MPa)			
	≤ 30	35	40	50
300	0.588	0.625	0.649	0.725
400	0.392	0.417	0.433	0.483

[a]If d'/a exceeds the values given in this table, the compression steel will not yield before failure; calculated using Eq. 5–9M.

TABLE A–11 Basic Tension Development Length Ratio, ℓ_{db}/d_b (in./in.)

$$\ell_d = \frac{\ell_{db}}{d_b} \times \beta\lambda \times d_b \text{, but not less than 12 in.}^{a}$$

Case 1: Clear spacing of bars being developed or spliced not less than d_b, clear cover not less than d_b, and stirrups or ties throughout ℓ_d not less than the Code minimum,

or

Case 2: Clear spacing of bars being developed or spliced not less than $2d_b$, and clear cover not less than d_b.

Bar No.	$f_c' = 3000$ psi Bottom Bar	Top Bar	$f_c' = 3750$ psi Bottom Bar	Top Bar	$f_c' = 4000$ psi Bottom Bar	Top Bar	$f_c' = 5000$ psi Bottom Bar	Top Bar	$f_c' = 6000$ psi Bottom Bar	Top Bar
$f_y = 60,000$ psi, uncoated bars, normal-weight concrete										
3 to 6	43.8	57.0	39.2	50.9	37.9	49.3	33.9	44.1	31.0	40.3
7 to 18	54.8	71.2	49.0	63.7	47.4	61.7	42.4	55.2	38.7	50.3
$f_y = 40,000$ psi, uncoated bars, normal-weight concrete										
3 to 6	29.2	38.0	26.1	34.0	25.3	32.9	22.6	29.4	20.7	26.9
Other Cases										
$f_y = 60,000$ psi, uncoated bars, normal-weight concrete										
3 to 6	65.7	85.5	58.8	76.4	56.9	74.0	50.9	66.2	46.5	60.5
7 to 18	82.2	106.8	73.5	95.6	71.1	92.6	63.6	82.8	58.1	75.5
$f_y = 40,000$ psi, uncoated bars, normal-weight concrete										
3 to 6	43.8	57.0	39.2	51.0	38.0	49.4	33.9	44.1	31.1	40.4

$^a\beta$, coating factor; λ, lightweight concrete factor.

TABLE A–11M Basic Tension Development Length Ratio, ℓ_{db}/d_b SI Units (mm/mm)

$$\ell_d = \frac{\ell_{db}}{d_b} \times \beta\lambda \times \ell_d \text{ , but not less than 300 mm}^{[a]}$$

Bar No.	$f_c' = 20$ MPa		$f_c' = 25$ MPa		$f_c' = 30$ MPa		$f_c' = 35$ MPa		$f_c' = 40$ MPa	
	Bottom Bar	Top Bar	Bottom Bar	Top Bar	Bottom Bar	Top Bar	Bottom Bar	Top Bar	Bottom Bar	Top Bar
Case 1: Clear spacing of bars being developed or spliced not less than d_b, clear cover not less than d_b, and stirrups or ties throughout ℓ_d not less than the Code minimum,										
or										
Case 2: Clear spacing of bars being developed or spliced not less than $2d_b$ and clear cover not less than d_b.										
$f_y = 400$ MPa, uncoated bars, normal-weight concrete										
10 to 20	42.9	55.8	38.4	49.9	35.1	45.6	32.5	42.2	30.4	39.5
25 to 45	53.7	69.8	48.0	62.4	43.8	57.0	40.6	52.7	37.9	49.3
$f_y = 300$ MPa, uncoated bars, normal-weight concrete										
10 to 20	32.2	41.9	28.8	37.4	26.3	34.2	24.3	31.6	22.8	29.6
Other Cases										
$f_y = 400$ MPa, uncoated bars, normal-weight concrete										
10 to 20	64.4	83.7	57.6	74.9	52.6	68.4	48.7	63.3	45.5	59.2
25 to 45	80.5	104.6	72.0	93.6	65.7	85.4	60.9	79.1	56.9	74.0
$f_y = 300$ MPa, uncoated bars, normal-weight concrete										
10 to 20	48.3	62.8	43.2	56.2	39.4	51.3	36.5	47.5	34.2	44.4

[a]β, coating factor; λ, lightweight concrete factor.

TABLE A–12 Basic Compression Development Length, ℓ_{dbc} (in.)[a]

	$\ell_d = \ell_{dbc} \times$ (Factors in ACI Sec. 12.3.3) f_c' (psi)		
Bar No.	3000	4000	5000 psi and up
$f_y = 60,000$ psi			
3	8	8	8
4	11	9	9
5	14	12	11
6	16	14	14
7	19	17	16
8	22	19	18
9	25	21	20
10	28	24	23
11	34	27	25
14	37	32	30
18	49	43	41
$f_y = 40,000$ psi			
3	8	8	8
4	8	8	8
5	9	8	8
6	11	9	9

[a]Lengths may be reduced if excess reinforcement is anchored or if the splice is enclosed in a spiral. See ACI Sec. 12.3.3. Reduced length shall not be less than 8 in.

TABLE A–12M Basic Compression Development Length, ℓ_{dbc}—SI Units (mm)[a]

	$\ell_d = \ell_{dbc} \times$ (Factors in ACI Sec. 12.3.3)				
	f_c' (MPa)				
Bar No.	20	25	30	35	40
$f_y = 400$ (MPa)					
10	224	200	200	200	200
15	335	300	274	254	240
20	447	400	365	338	320
25	559	500	456	423	400
30	671	600	548	507	480
35	783	700	639	592	560
45	1006	900	822	761	720
55	1261	1128	1030	953	892
$f_y = 300$ (MPa)					
10	200	200	200	200	200
15	252	225	205	200	200
20	335	300	274	254	240

[a]Lengths may be reduced if excess reinforcement is anchored or if the splice is enclosed in a spiral. See ACI Sec. 12.3.3. Reduced length shall not be less than 200 mm.

TABLE A–13 Basic Development Lengths for Hooked Bars, ℓ_{hb} (in.)

$\ell_{dh} = \ell_{hb} \times$ (Factors in 12.5.3)[a]
Normal-weight concrete, $f_y = 60,000$ psi
Standard 90° or 180° Hooks

Bar No.	f_c' (psi)			
	3000	4000	5000	6000
3	8.2	7.1	6.4	5.8
4	11	9.5	8.5	7.8
5	13.7	11.9	10.6	9.7
6	16.4	14.2	12.7	11.6
7	19.2	16.6	14.9	13.6
8	22	19	17	15.5
9	25	21	19	17.5
10	28	24	22	20
11	31	27	24	22
14	37	32	29	26
18	49	43	38	35

[a]ℓ_{dh} is defined in Fig. 8–12a. The development length of a hook, ℓ_{dh}, is the product of ℓ_{hb} from this table and factors relating to bar yield strength, cover, presence of stirrups, and type of concrete given in ACI Sec. 12.5.3. The resulting length ℓ_{dh} shall not be less than the larger of 8 bar diameters or 6 in.

TABLE A–13M Basic Development Lengths for Hooked Bars, ℓ_{hb}—SI Units (mm)

$\ell_{dh} = \ell_{hb} \times$ (Factors in 12.5.3)[a]
Normal-weight concrete, $f_y = 400$ MPa
Standard 90° or 180° hooks

Bar No.	f_c' (MPa)				
	20	25	30	35	40
10	253	226	206	191	179
15	358	320	292	270	253
20	436	390	356	330	308
25	563	504	460	426	398
30	669	598	546	505	473
35	798	714	652	603	564
45	977	874	798	739	691
55	1261	1128	1030	953	892

[a]ℓ_{dh} is defined in Fig. 8–12a. The development length of a hook, ℓ_{dh}, is the product of ℓ_{hb} from this table and factors relating to bar yield strength, cover, presence of stirrups, and type of concrete given in ACI Sec. 12.5.3. The resulting length ℓ_{dh} shall not be less than the larger of 8 bar diameters or 150 mm.

TABLE A–14 Minimum Thicknesses of Non-Prestressed Beams or One-Way Slabs Unless Deflections Are Computed

Exposure	Member	Minimum Thickness, h				Source
		Simply Supported	One End Continuous	Both Ends Continuous	Cantilever	
Not supporting or attached to partitions or other construction likely to be damaged by large deflections	Solid one-way slabs	$\ell/20$	$\ell/24$	$\ell/28$	$\ell/10$	ACI Table 9.5(a)
	Beams or ribbed one-way slabs	$\ell/16$	$\ell/18.5$	$\ell/21$	$\ell/8$	
Supporting or attached to partitions or other construction likely to be damaged by large deflections	All members: $\omega \leq 0.12$[a] and $\dfrac{\text{sustained load}}{\text{total load}} < 0.5$	$\ell/10$	$\ell/13$	$\ell/16$	$\ell/4$	Ref. 9–20
	All members: $\dfrac{\text{sustained load}}{\text{total load}} > 0.5$	$\ell/6$	$\ell/8$	$\ell/10$	$\ell/3$	

[a] $\omega = \rho f_y / f_c'$

TABLE A–15 Maximum Allowable Spiral Pitch, s (in.), for Circular Spiral Columns, Grade 60 Spirals[a]

Column Diameter (in.)	Core Diameter (in.)	f_c' (psi)						
		4000 Spiral Size		5000 Spiral Size		6000 Spiral Size		
		No. 3	No. 4	No. 3	No. 4	No. 3	No. 4	No. 5
12	9	2	$3\frac{1}{2}$	$1\frac{1}{2}$*	$2\frac{3}{4}$	—	$2\frac{1}{4}$	$3\frac{1}{2}$
14	11	2	$3\frac{1}{2}$	$1\frac{1}{2}$*	3	—	$2\frac{1}{4}$	$3\frac{1}{2}$
16	13	2	$3\frac{1}{2}$	$1\frac{1}{2}$*	3	—	$2\frac{1}{2}$	$3\frac{1}{2}$
18	15	2	$3\frac{1}{2}$	$1\frac{1}{2}$*	3	—	$2\frac{1}{2}$	$3\frac{1}{2}$
20	17	2	$3\frac{1}{2}$	$1\frac{3}{4}$	3	—	$2\frac{1}{2}$	$3\frac{1}{2}$
22	19	2	$3\frac{1}{2}$	$1\frac{3}{4}$	3	—	$2\frac{1}{2}$	$3\frac{1}{2}$
24	21	2	$3\frac{1}{2}$	$1\frac{3}{4}$	3	—	$2\frac{1}{2}$	$3\frac{1}{2}$
26	23	$2\frac{1}{4}$	$3\frac{1}{2}$	$1\frac{3}{4}$	$3\frac{1}{4}$	$1\frac{1}{2}$*	$2\frac{3}{4}$	$3\frac{1}{2}$
28	25	$2\frac{1}{4}$	$3\frac{1}{2}$	$1\frac{3}{4}$	$3\frac{1}{4}$	$1\frac{1}{2}$*	$2\frac{3}{4}$	$3\frac{1}{2}$
30	27	$2\frac{1}{4}$	$3\frac{1}{2}$	$1\frac{3}{4}$	$3\frac{1}{4}$	$1\frac{1}{2}$*	$2\frac{3}{4}$	$3\frac{1}{2}$
32	29	$2\frac{1}{4}$	$3\frac{1}{2}$	$1\frac{3}{4}$	$3\frac{1}{4}$	$1\frac{1}{2}$*	$2\frac{3}{4}$	$3\frac{1}{2}$
34	31	$2\frac{1}{4}$	$3\frac{1}{2}$	$1\frac{3}{4}$	$3\frac{1}{4}$	$1\frac{1}{2}$*	$2\frac{3}{4}$	$3\frac{1}{2}$
36	33	$2\frac{1}{4}$	$3\frac{1}{2}$	$1\frac{3}{4}$	$3\frac{1}{4}$	$1\frac{1}{2}$*	$2\frac{3}{4}$	$3\frac{1}{2}$
38	35	$2\frac{1}{4}$	$3\frac{1}{2}$	$1\frac{3}{4}$	$3\frac{1}{4}$	$1\frac{1}{2}$*	$2\frac{3}{4}$	$3\frac{1}{2}$
40	37	$2\frac{1}{4}$	$3\frac{1}{2}$	$1\frac{3}{4}$	$3\frac{1}{4}$	$1\frac{1}{2}$*	$2\frac{3}{4}$	$3\frac{1}{2}$

[a] The pitch is measured center to center of consecutive turns. Cover $1\frac{1}{2}$ in. to spiral. The tabulated values can be used with 1-in.-maximum-size aggregate, except that values marked with an asterisk require $\frac{3}{4}$-in.-maximum aggregate.

Source: From Ref. 11–4. Ref. 11–4 is the Design Handbook, reprinted with permission of the American Concrete Institute.

Table A-16 Maximum Number of Bars That Can Be Placed in Square
Columns with the Same Number of Bars in Each Face, Based on
Normal (Radial) Lap Splices, Minimum Bar Spacing[a]

b (in.)	A_g (in.²)		Bar No.						
			5	6	7	8	9	10	11
10	100	n_{max}	8	4	4	4	4		
		A_{st}	2.48	1.76	2.40	3.16	4.00		
		ρ_t	0.025	0.018	0.024	0.032	0.040		
12	144	n_{max}	12	8	8	8	4	4	4
		A_{st}	3.72	3.52	4.80	6.32	4.00	5.08	6.24
		ρ_t	0.026	0.024	0.033	0.044	0.028	0.035	0.043
14	196	n_{max}	16	12	12	12	8	8	4
		A_{st}	4.96	5.28	7.20	9.48	8.00	10.16	6.24
		ρ_t	0.025	0.027	0.037	0.048	0.041	0.052	0.032
16	256	n_{max}	—	16	16	12	12	8	8
		A_{st}		7.04	9.60	9.48	12.00	10.16	12.48
		ρ_t		0.028	0.038	0.037	0.047	0.040	0.049
18	324	n_{max}	—	20	20	16	16	12	12
		A_{st}		8.80	12.00	12.64	16.00	15.24	18.72
		ρ_t		0.027	0.037	0.039	0.049	0.047	0.058
20	400	n_{max}	—	—	20	20	16	16	12
		A_{st}			12.0	15.80	16.00	20.32	18.72
		ρ_t			0.030	0.039	0.040	0.051	0.047
22	484	n_{max}	—	—	24	24	20	16	16
		A_{st}			14.40	18.96	20.00	20.32	24.96
		ρ_t			0.030	0.039	0.041	0.042	0.052
24	576	n_{max}	—	—	28	28	24	20	16
		A_{st}			16.80	22.12	24.00	25.40	24.96
		ρ_t			0.029	0.038	0.042	0.044	0.043
26	676	n_{max}	—	—	32	28	24	20	20
		A_{st}			19.20	22.12	24.00	25.40	31.20
		ρ_t			0.028	0.033	0.036	0.038	0.046
28	784	n_{max}	—	—	36	32	28	24	20
		A_{st}			21.60	25.28	28.00	30.48	31.20
		ρ_t			0.028	0.032	0.036	0.039	0.040
30	900	n_{max}	—	—	—	36	32	28	24
		A_{st}				28.44	32.00	35.56	37.44
		ρ_t				0.032	0.036	0.039	0.042
32	1024	n_{max}	—	—	—	40	32	28	28
		A_{st}				31.60	32.00	35.56	43.68
		ρ_t				0.031	0.031	0.035	0.043

[a]Based on 1-in.-maximum-size aggregate.

Source: From Ref. 11–4. Ref. 11–4 is the Design Handbook, reprinted with permission of
the American Concrete Institute.

TABLE A–17 Maximum Number of Bars That Can Be Placed in Circular Columns, Based on Normal (Radial) Lap Splices, Minimum Bar Spacing[a]

Diameter (in.)	A_g (in.²)		Bar Size						
			5	6	7	8	9	10	11
12	113	n_{max}	8	7	6	6	—	—	—
		A_{st}	2.48	3.08	3.60	4.74			
		ρ_t	0.022	0.027	0.032	0.042			
14	154	n_{max}	11	10	9	8	7		
		A_{st}	3.41	4.40	5.40	6.32	7.00		
		ρ_t	0.022	0.029	0.035	0.041	0.046		
16	201	n_{max}	14	13	12	11	9	7	6
		A_{st}	4.34	5.72	7.20	8.69	9.00	8.89	9.36
		ρ_t	0.022	0.029	0.036	0.043	0.045	0.044	0.047
18	254	n_{max}	—	16	14	13	11	9	8
		A_{st}		7.04	8.40	10.27	11.00	11.43	12.48
		ρ_t		0.028	0.033	0.040	0.043	0.045	0.049
20	314	n_{max}	—	—	17	16	13	11	10
		A_{st}			10.20	12.64	13.00	13.97	15.60
		ρ_t			0.033	0.040	0.041	0.044	0.050
22	380	n_{max}	—	—	20	18	16	13	12
		A_{st}			12.00	14.22	16.00	16.51	18.72
		ρ_t			0.032	0.037	0.042	0.043	0.049
24	452	n_{max}	—	—	22	21	18	15	13
		A_{st}			13.20	16.59	18.00	19.05	20.28
		ρ_t			0.029	0.037	0.040	0.042	0.045
26	531	n_{max}	—	—	25	23	20	17	15
		A_{st}			15.00	18.17	20.00	21.59	23.40
		ρ_t			0.028	0.034	0.038	0.041	0.044
28	616	n_{max}	—	—	28	26	22	19	17
		A_{st}			16.80	20.54	22.00	24.13	26.52
		ρ_t			0.027	0.033	0.036	0.039	0.043
30	707	n_{max}	—	—	—	28	25	21	19
		A_{st}				22.12	25.00	26.67	29.64
		ρ_t				0.031	0.035	0.038	0.042
32	804	n_{max}	—	—	—	31	27	23	21
		A_{st}				24.29	27.00	29.21	32.76
		ρ_t				0.031	0.034	0.036	0.041

[a]Based on No. 4 spirals or ties, 1-in.-maximum-size aggregate, and $1\frac{1}{2}$-in. clear cover to spirals.

Source: This table is an abridged version of a table in Ref. 11–4 and is printed with the permission of the American Concrete Institute.

TABLE A–18 Number of Bars Required to Provide a Given Area of Steel[a]

Area (in.²)	Bar No. 5 (0.31)	Bar No. 6 (0.44)	Bar No. 7 (0.60)	Bar No. 8 (0.79)	Bar No. 9 (1.00)
1.24	**4**				
1.76		**4**			
1.86	6				
2.17	7				
2.40			**4**		
2.48	**8**				
2.64		6			
3.08		7			
3.16				**4**	
3.41	11				
3.52		**8**			
3.60			6		
3.72	**12**				
3.96		9			
4.00	13				**4**
4.20			7		
4.40		10			
4.65	15				
4.74				6	
4.80			**8**		
5.28		**12**			
5.40			9		
5.53				7	
5.72		13			
6.00			10		6
6.32				**8**	
7.00					7
7.04		**16**			
7.20			**12**		
7.48		17			
7.80			13		
7.90				10	
8.00					**8**
8.40			14		
8.69				11	
8.80		**20**			
9.00			15		9
9.48				**12**	
9.60			**16**		
10.00					10
10.80			18		

Area (in.²)	Bar No. 8 (0.79)	Bar No. 9 (1.00)	Bar No. 10 (1.27)	Bar No. 11 (1.56)	Bar No. 14 (2.25)
10.16			**8**		
10.27	13				
11.00		11			
11.06	14				
11.43			9		
11.85	15				
12.00		**12**			
12.48				**8**	
12.64	**16**				
12.70			10		
13.00		13			
13.43	17				
13.50					6
13.97			11		
14.00		14		9	
14.22	18				
15.00	19	15			
15.24			**12**		
15.60				10	
15.75					7
15.80	**20**				
16.00		**16**			
16.51			13		
16.59	21				
17.00		17			
17.16				11	
17.78			14		
18.00		18			**8**
18.72				12	
19.00		19	15		
20.00		**20**			
20.25				13	9
20.32			**16**		
21.00		21			
21.59			17		
21.84				14	
22.00		22			
22.50					10
22.86			18		
24.00		**24**			
24.96				**16**	

[a]Bold figures denote combinations that will give an equal number of bars in each side of a square column.

TABLE A–19 Lap Splice Lengths for Grade 60 Bars in Columns (in.)

f_c' (psi)	Bar No.						
	5	6	7	8	9	10	11

Compression lap splices

Lap splice length = (length from table) × (factors in note a)

	5	6	7	8	9	10	11
< 3000	26	31	35	40	46	51	56
≥ 3000	19	23	26	30	34	38	42

Tension lap splices

Lap splice length = (length from table) × $\beta\lambda^b$

Class A tension lap splice: half or fewer of the bars spliced at any location *and* $0 \le f_s \le 0.5f_y$ in tension, (ACI Sec. 12.17.2.2)

	5	6	7	8	9	10	11
3000	27.4	32.9	48.0	54.8	61.8	69.6	77.3
4000	23.7	28.4	41.5	47.4	53.5	60.2	66.8
5000	21.2	25.4	37.1	42.4	47.8	53.8	59.8
6000	19.4	23.3	33.9	38.7	43.7	49.1	54.6

Class B tension lap splices: more than half of the bars spliced at any section and/or f_s greater than $0.5f_y$ in tension, (ACI Sec. 12.17.2.2)

	5	6	7	8	9	10	11
3000	35.6	42.7	62.3	71.2	80.4	90.5	100.4
4000	30.8	37.0	53.9	61.6	69.5	78.3	86.9
5000	27.5	33.0	48.2	55.1	61.2	70.0	77.7
6000	25.2	30.2	44.0	50.3	56.7	63.9	70.9

ªCompression lap splices may be multiplied by 0.83 or 0.75 if enclosed by ties or spirals satisfying ACI Secs. 12.17.2.4 or 12.17.2.5.

ᵇβ = coating factor, λ = lightweight concrete factor.

TABLE A–20   Moment Distribution Factors for Slabs
without Drop Panels[a]

FEM (uniform load w) $= Mw\ell_2\ell_1^2$ $\qquad$ K (stiffness) $= kE\ell_2 t^3/12\ell_1$
$\qquad\qquad$ Carryover factor $=$ COF

c_1/ℓ_1		c_2/ℓ_2					
		0.00	0.05	0.10	0.15	0.20	0.25
0.00	M	0.083	0.083	0.083	0.083	0.083	0.083
	k	4.000	4.000	4.000	4.000	4.000	4.000
	COF	0.500	0.500	0.500	0.500	0.500	0.500
0.05	M	0.083	0.084	0.084	0.084	0.085	0.085
	k	4.000	4.047	4.093	4.138	4.181	4.222
	COF	0.500	0.503	0.507	0.510	0.513	0.516
0.10	M	0.083	0.084	0.085	0.085	0.086	0.087
	k	4.000	4.091	4.182	4.272	4.362	4.449
	COF	0.500	0.506	0.513	0.519	0.524	0.530
0.15	M	0.083	0.084	0.085	0.086	0.087	0.088
	k	4.000	4.132	4.267	4.403	4.541	4.680
	COF	0.500	0.509	0.517	0.526	0.534	0.543
0.20	M	0.083	0.085	0.086	0.087	0.088	0.089
	k	4.000	4.170	4.346	4.529	4.717	4.910
	COF	0.500	0.511	0.522	0.532	0.543	0.554
0.25	M	0.083	0.085	0.086	0.087	0.089	0.090
	k	4.000	4.204	4.420	4.648	4.887	5.138
	COF	0.500	0.512	0.525	0.538	0.550	0.563
$x = (1 - c_2/\ell_2^3)$		1.000	0.856	0.729	0.613	0.512	0.421

[a]c_1 and c_2 are the widths of the column measured parallel to ℓ_1 and ℓ_2.
Source: Ref. 14–8.

TABLE A-21   Moment Distribution Factors for Slabs with Drop
Panels, $h_1 = 1.25h$[a]

FEM (uniform load w) $= Mw\ell_2\ell_1^2$ K (stiffness) $= kE\ell_2t^3/12\ell_1$
Carryover factor $= COF$

c_1/ℓ_1		c_2/ℓ_2						
		0.00	0.05	0.10	0.15	0.20	0.25	0.30
0.00	M	0.088	0.088	0.088	0.088	0.088	0.088	0.088
	k	4.795	4.795	4.795	4.795	4.795	4.795	4.795
	COF	0.542	0.542	0.542	0.542	0.542	0.542	0.542
0.05	M	0.088	0.088	0.089	0.089	0.089	0.089	0.090
	k	4.795	4.846	4.896	4.944	4.990	5.035	5.077
	COF	0.542	0.545	0.548	0.551	0.553	0.556	0.558
0.10	M	0.088	0.088	0.089	0.090	0.090	0.091	0.091
	k	4.795	4.894	4.992	5.039	5.184	5.278	5.368
	COF	0.542	0.548	0.553	0.559	0.564	0.569	0.573
0.15	M	0.088	0.089	0.090	0.090	0.091	0.092	0.092
	k	4.795	4.938	5.082	5.228	5.374	5.520	5.665
	COF	0.542	0.550	0.558	0.565	0.573	0.580	0.587
0.20	M	0.088	0.089	0.090	0.091	0.092	0.093	0.094
	k	4.795	4.978	5.167	5.361	5.558	5.760	5.962
	COF	0.542	0.552	0.562	0.571	0.581	0.590	0.600
0.25	M	0.088	0.089	0.090	0.091	0.092	0.094	0.095
	k	4.795	5.015	5.245	5.485	5.735	5.994	6.261
	COF	0.542	0.553	0.565	0.576	0.587	0.598	0.609
0.30	M	0.088	0.089	0.090	0.092	0.093	0.094	0.095
	k	4.795	5.048	5.317	5.601	5.902	6.219	6.550
	COF	0.542	0.554	0.567	0.580	0.593	0.605	0.618

[a]h, Slab thickness; h_1, total thickness in drop panel.

Source: Ref. 14-8.

TABLE A–22 Moment Distribution Factors for Slabs with Drop Panels, $h_1 = 1.5h$[a]

FEM (uniform load w) = $M w \ell_2 \ell_1^2$ K (stiffness) = $kE\ell_2 t^3/12\ell_1$
Carryover factor = COF

c_1/ℓ_1		c_2/ℓ_2					
		0.00	0.05	0.10	0.15	0.20	0.25
0.00	M	0.093	0.093	0.093	0.093	0.093	0.093
	k	5.837	5.837	5.837	5.837	5.837	5.837
	COF	0.589	0.589	0.589	0.589	0.589	0.589
0.05	M	0.093	0.093	0.093	0.093	0.094	0.094
	k	5.837	5.890	5.942	5.993	6.041	6.087
	COF	0.589	0.591	0.594	0.596	0.598	0.600
0.10	M	0.093	0.093	0.094	0.094	0.094	0.095
	k	5.837	5.940	6.024	6.142	6.240	6.335
	COF	0.589	0.593	0.598	0.602	0.607	0.611
0.15	M	0.093	0.093	0.094	0.095	0.095	0.096
	k	5.837	5.986	6.135	6.284	6.432	6.579
	COF	0.589	0.595	0.602	0.608	0.614	0.620
0.20	M	0.093	0.093	0.094	0.095	0.096	0.096
	k	5.837	6.027	6.221	6.418	6.616	6.816
	COF	0.589	0.597	0.605	0.613	0.621	0.628
0.25	M	0.093	0.094	0.094	0.095	0.096	0.097
	k	5.837	6.065	6.300	6.543	6.790	7.043
	COF	0.589	0.598	0.608	0.617	0.626	0.635

[a]h, Slab thickness; h_1, total thickness in drop panel.

Source: Ref. 14–8.

TABLE A–23 Stiffness and Carryover Factors for Columns

$$K_c = k\,\frac{EL_c}{\ell_c}$$

		ℓ_c/ℓ_u								
t_a/t_b		1.05	1.10	1.15	1.20	1.25	1.30	1.35	1.40	1.45
0.00	k_{AB}	4.20	4.40	4.60	4.80	5.00	5.20	5.40	5.60	5.80
	C_{AB}	0.57	0.65	0.73	0.80	0.87	0.95	1.03	1.10	1.17
0.2	k_{AB}	4.31	4.62	4.95	5.30	5.65	6.02	6.40	6.79	7.20
	C_{AB}	0.56	0.62	0.68	0.74	0.80	0.85	0.91	0.96	1.01
0.4	k_{AB}	4.38	4.79	5.22	5.67	6.15	6.65	7.18	7.74	8.32
	C_{AB}	0.55	0.60	0.65	0.70	0.74	0.79	0.83	0.87	0.91
0.6	k_{AB}	4.44	4.91	5.42	5.96	6.54	7.15	7.81	8.50	9.23
	C_{AB}	0.55	0.59	0.63	0.67	0.70	0.74	0.77	0.80	0.83
0.8	k_{AB}	4.49	5.01	5.58	6.19	6.85	7.56	8.31	9.12	9.98
	C_{AB}	0.54	0.58	0.61	0.64	0.67	0.70	0.72	0.75	0.77
1.0	k_{AB}	4.52	5.09	5.71	6.38	7.11	7.89	8.73	9.63	10.60
	C_{AB}	0.54	0.57	0.60	0.62	0.65	0.67	0.69	0.71	0.73
1.2	k_{AB}	4.55	5.16	5.82	6.54	7.32	8.17	9.08	10.07	11.12
	C_{AB}	0.53	0.56	0.59	0.61	0.63	0.65	0.66	0.68	0.69
1.4	k_{AB}	4.58	5.21	5.91	6.68	7.51	8.41	9.38	10.43	11.57
	C_{AB}	0.53	0.55	0.58	0.60	0.61	0.63	0.64	0.65	0.66
1.6	k_{AB}	4.60	5.26	5.99	6.79	7.66	8.61	9.64	10.75	11.95
	C_{AB}	0.53	0.55	0.57	0.59	0.60	0.61	0.62	0.63	0.64
1.8	k_{AB}	4.62	5.30	6.06	6.89	7.80	8.79	9.87	11.03	12.29
	C_{AB}	0.52	0.55	0.56	0.58	0.59	0.60	0.61	0.61	0.62
2.0	k_{AB}	4.63	5.34	6.12	6.98	7.92	8.94	10.06	11.27	12.59
	C_{AB}	0.52	0.54	0.56	0.57	0.58	0.59	0.59	0.60	0.60
2.2	k_{AB}	4.65	5.37	6.17	7.05	8.02	9.08	10.24	11.49	12.85
	C_{AB}	0.52	0.54	0.55	0.56	0.57	0.58	0.58	0.59	0.59
2.4	k_{AB}	4.66	5.40	6.22	7.12	8.11	9.20	10.39	11.68	13.08
	C_{AB}	0.52	0.53	0.55	0.56	0.56	0.57	0.57	0.58	0.58
2.6	k_{AB}	4.67	5.42	6.26	7.18	8.20	9.31	10.53	11.86	13.29
	C_{AB}	0.52	0.53	0.54	0.55	0.56	0.56	0.56	0.57	0.57
2.8	k_{AB}	4.68	5.44	6.29	7.23	8.27	9.41	10.66	12.01	13.48
	C_{AB}	0.52	0.53	0.54	0.55	0.55	0.55	0.56	0.56	0.56
3.0	k_{AB}	4.69	5.46	6.33	7.28	8.34	9.50	10.77	12.15	13.65
	C_{AB}	0.52	0.53	0.54	0.54	0.55	0.55	0.55	0.55	0.55
3.5	k_{AB}	4.71	5.50	6.40	7.39	8.48	9.69	11.01	12.46	14.02
	C_{AB}	0.51	0.52	0.53	0.53	0.54	0.54	0.54	0.53	0.53
4.0	k_{AB}	4.72	5.54	6.45	7.47	8.60	9.84	11.21	12.70	14.32
	C_{AB}	0.51	0.52	0.52	0.53	0.53	0.52	0.52	0.52	0.52
4.5	k_{AB}	4.73	5.56	6.50	7.54	8.69	9.97	11.37	12.89	14.57
	C_{AB}	0.51	0.52	0.52	0.52	0.52	0.52	0.51	0.51	0.51
5.0	k_{AB}	4.75	5.59	6.54	7.60	8.78	10.07	11.50	13.07	14.77
	C_{AB}	0.51	0.51	0.52	0.52	0.51	0.51	0.51	0.50	0.49
6.0	k_{AB}	4.76	5.63	6.60	7.69	8.90	10.24	11.72	13.33	15.10
	C_{AB}	0.51	0.51	0.51	0.51	0.50	0.50	0.49	0.49	0.48
7.0	k_{AB}	4.78	5.66	6.65	7.76	9.00	10.37	11.88	13.54	15.34
	C_{AB}	0.51	0.51	0.51	0.50	0.50	0.49	0.48	0.48	0.47
8.0	k_{AB}	4.78	5.68	6.69	7.82	9.07	10.47	12.01	13.70	15.54
	C_{AB}	0.51	0.51	0.50	0.50	0.49	0.49	0.48	0.47	0.46
9.0	k_{AB}	4.80	5.71	6.74	7.89	9.18	10.61	12.19	13.93	15.83
	C_{AB}	0.50	0.50	0.50	0.49	0.48	0.48	0.47	0.46	0.45

Source: Ref. 14–7. Courtesy of the Portland Cement Association.

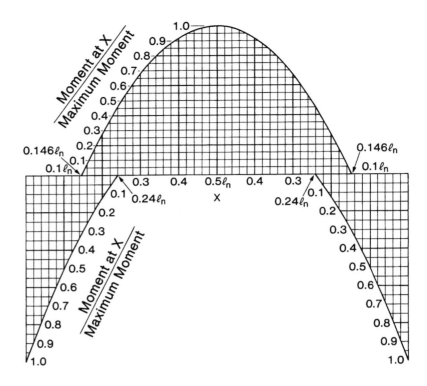

Fig. A–1
Bending moment envelope for typical interior span (moment coefficients: $-1/11, +1/16, -1/11$).

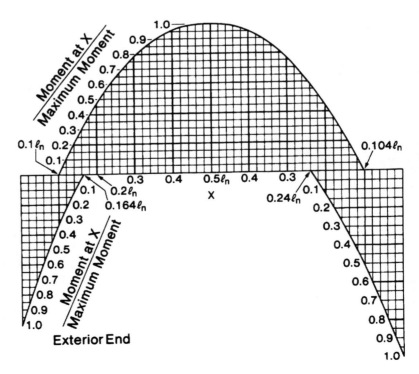

Fig. A–2
Bending moment envelope for exterior span with exterior support built integrally with a column (moment coefficients: $-1/16, +1/14, -1/10$).

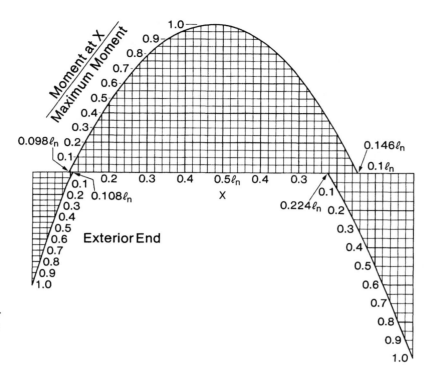

Fig. A–3
Bending moment envelope
for exterior span with exterior
support built integrally with a
spandrel beam or girder (moment coefficients:
$-1/24, +1/14, -1/10$).

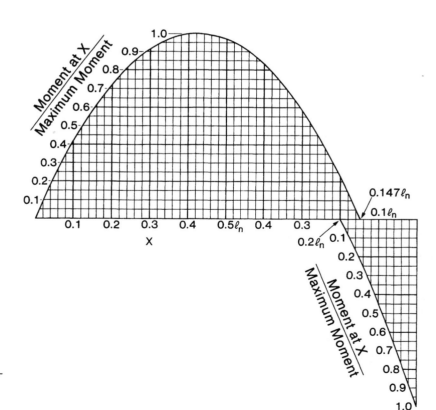

Fig. A–4
Bending moment envelope
for exterior span with discontinuous end unrestrained
(moment coefficients:
$0, +1/11, -1/10$).

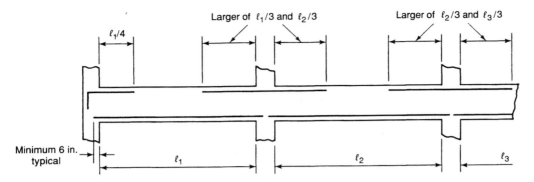

(a) Beam with closed stirrups.
If closed stirrups are not provided, see ACI Sec 7.13

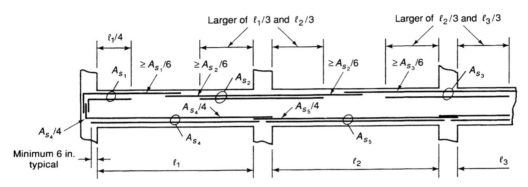

(b) Perimeter beam

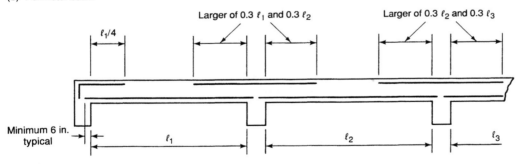

(c) One-way slab

Fig. A–5
Standard bar details.

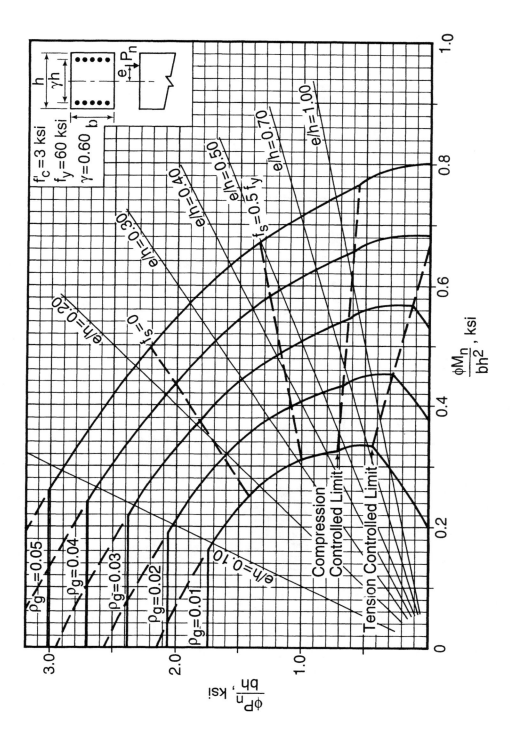

Fig. A–6
Nondimensional interaction
diagram for tied columns
with bars in two faces,
$\gamma = 0.60$.

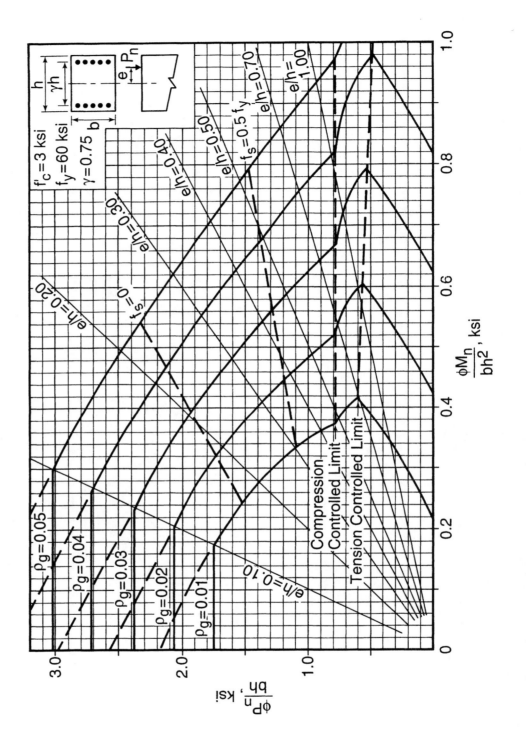

Fig. A–7
Nondimensional interaction
diagram for tied columns
with bars in two faces,
γ = 0.75.

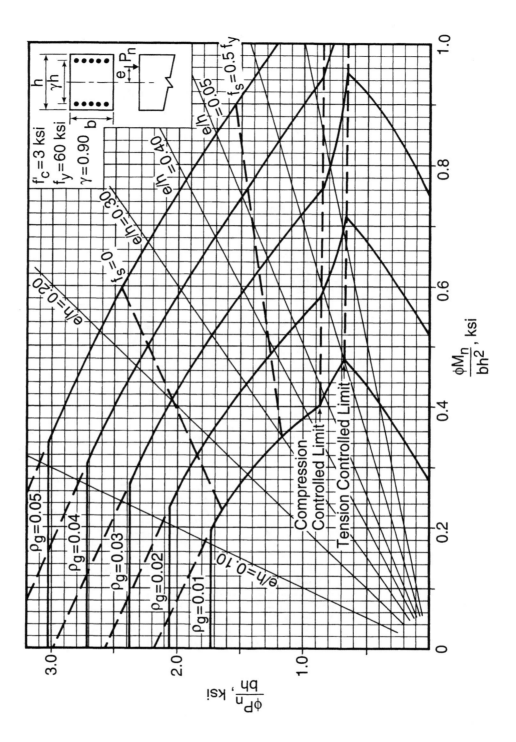

Fig. A–8
Interaction diagram for tied columns with bars in two faces, $\gamma = 0.90$.

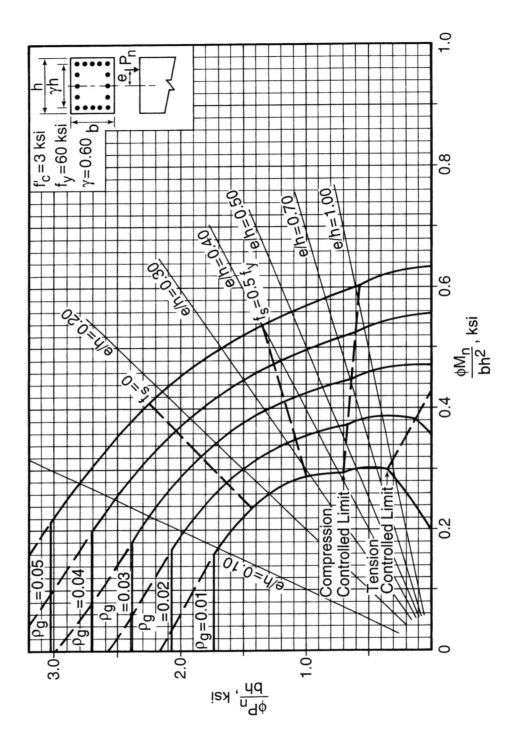

Fig. A–9
Interaction diagram for tied columns with bars in four faces, $\gamma = 0.60$.

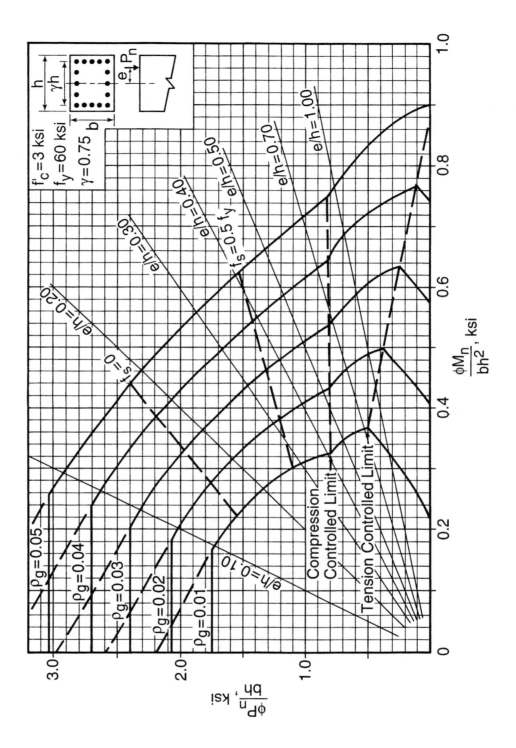

Fig. A–10
Interaction diagram for tied
columns with bars in four
faces, $\gamma = 0.75$.

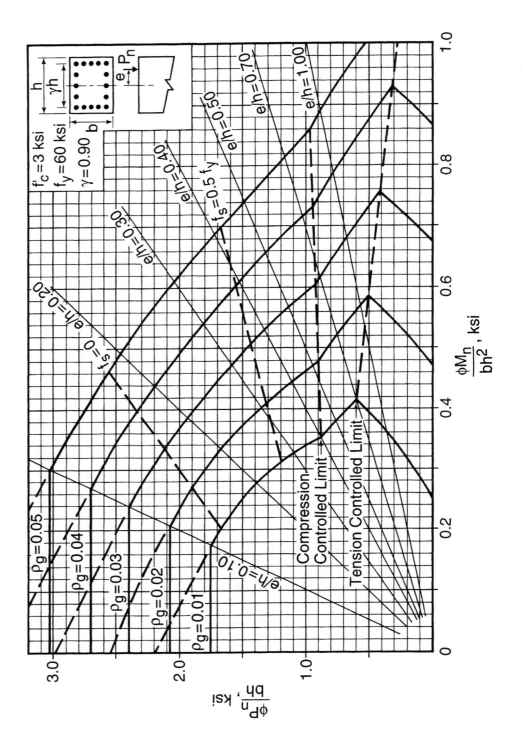

Fig. A–11
Interaction diagram for tied
columns with bars in four
faces, $\gamma = 0.90$.

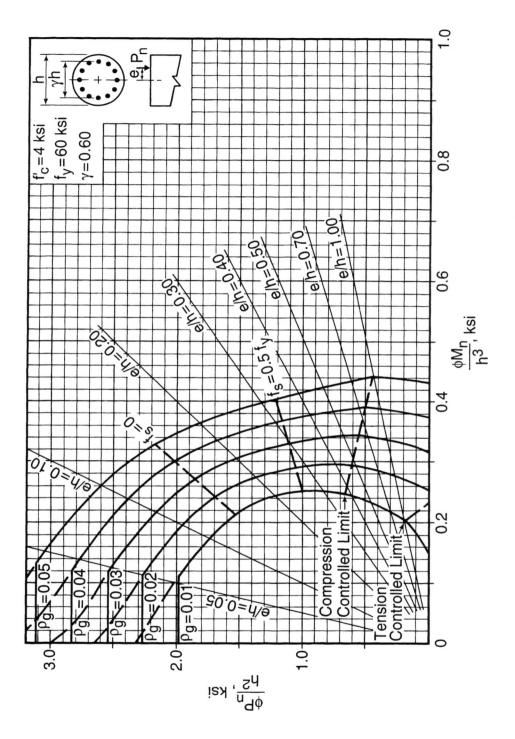

Fig. A–12
Interaction diagrams for *spiral* columns, $\gamma = 0.60$.

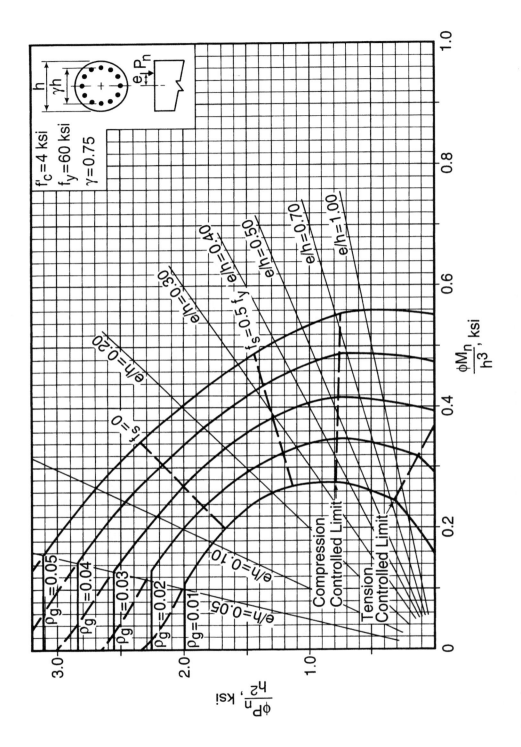

Fig. A–13
Interaction diagrams for *spiral* columns, $\gamma = 0.75$.

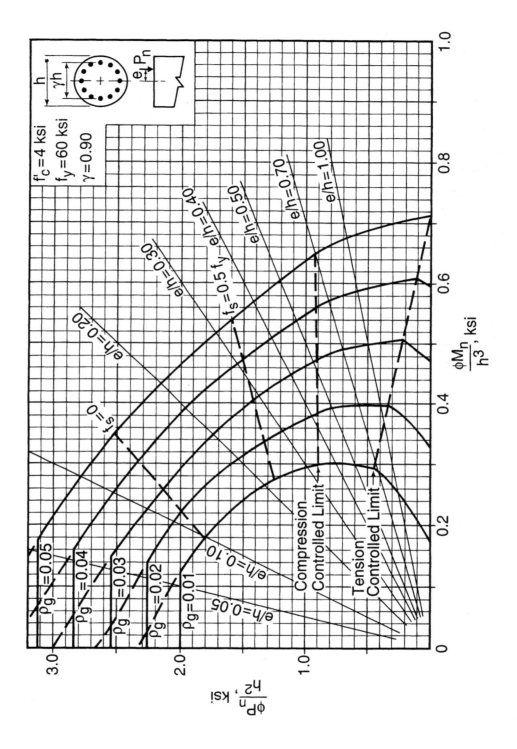

Fig. A–14
Interaction diagrams for *spiral* columns, $\gamma = 0.90$.

APPENDIX *B*

Notation

a = depth of equivalent rectangular stress block.

a = shear span, distance between concentrated load and face of support.

a_b = depth of rectangular stress block corresponding to balanced strain conditions.

$a_{cc\ell}$ = depth of rectangular stress block at the compression-controlled limit, in.

$a_{tc\ell}$ = depth of rectangular stress block at the tension-controlled limit, in.

A = effective tension area of concrete surrounding the flexural tension reinforcement and having the same centroid as that reinforcement, divided by the number of bars, in.^{2} (see Sec. 9–3).

A_b = area of an individual bar, in.^{2}.

A_c = area of core of spirally reinforced compression member measured to outside diameter of spiral, in.^{2}.

A_{cp} = area enclosed by outside perimeter of concrete cross section, in.^{2}

A_g = gross area of section, in.^{2}.

A_I = influence area (see Sec. 2–7).

A_ℓ = total area of longitudinal reinforcement to resist torsion, in.^{2}.

A_o = gross area enclosed by shear flow path, in.^{2}

A_{oh} = area enclosed by centerline of the outermost closed transverse torsional reinforcement, in.^{2}

A_s = area of nonprestressed tension reinforcement, in.^{2}.

A_s' = area of compression reinforcement, in.^{2}.

A_{sf} = area of tension reinforcement balancing the compression force in the overhanging flanges of a T beam.

A_{sk} = area of skin reinforcement per unit height in one side face, in.^{2}/ft.

A_{sw} = area of tension reinforcement balancing the compression force in the web of a T beam.

A_{s1} = area of tension reinforcement balancing the compression force in the compression reinforcement.

A_{s2} = area of tension reinforcement balancing the compression force in the concrete in a beam with compression reinforcement.

A_t = area of one leg of a closed stirrup resisting torsion, in.^{2}.

A_{tr} = total cross-sectional area of all transverse reinforcement which is within the spacing s and which crosses the potential plane of splitting through the reinforcement being developed, in.^{2}

A_v = area of shear reinforcement within a distance s, in.^{2}.

A_{vf} = area of shear-friction reinforcement, in.^{2}.

A_1 = loaded area in bearing.

A_2 = area of the lower base of the largest frustrum of a pyramid, cone, or tapered wedge contained wholly within the support and having for its upper base the loaded area, and having side slopes of 1 vertical to 2 horizontal (see Fig. 16–10).

b = width of compression face of member, effective compressive flange width of a T beam, in.

b_o = perimeter of critical section for two-way shear in slabs and footings, in.

b_w = web width, or diameter of circular section, in.

b_1 = length of critical shear perimeter for two-way shear, measured parallel to the span ℓ_1, in.

b_2 = length of critical shear perimeter for two-way shear perpendicular to b_1, in.

c = spacing or cover dimension, in.

c = distance from extreme compression fiber to neutral axis, in.

c_b = distance from extreme compression fiber to neutral axis corresponding to balanced strain conditions, in.

$c_{cc\ell}$ = distance from extreme compression fiber to the neutral axis when the strain in the extreme tension steel is ϵ_y in tension, in.

$c_{tc\ell}$ = distance from extreme compression fiber to the neutral axis when the strain in the extreme tension steel is 0.005 in tension, in.

c_1 = size of rectangular or equivalent rectangular column, capital, or bracket measured parallel to the span ℓ_1, in.

c_2 = size of rectangular or equivalent rectangular column, capital, or bracket measured parallel to ℓ_2, in.

C = compressive force in cross section, subscripts c = concrete, s = steel.

C = cross-sectional constant to define torsional properties.

C_m = factor relating the actual moment diagram of a slender column to an equivalent uniform moment diagram (Chap. 12).

C_m = moment coefficient (Chap. 10).

C_v = shear coefficient.

d = effective depth = distance from extreme compression fiber to centroid of tension reinforcement, in.

d' = distance from extreme compression fiber to centroid of compression reinforcement, in.

d_b = nominal diameter of bar, wire, or prestressing strand, in.

d_c = thickness of concrete cover measured from extreme tension fiber to center of the closest bar, in.

d_t = distance from extreme compression fiber to extreme tension steel, in.

D = dead loads, or related internal moments and forces.

D = diagonal compression force in the web of a beam or in a compression strut in a D-region, lb.

e = eccentricity of axial load on a column = M/P.

E_c = modulus of elasticity of concrete, psi.

EI = flexural stiffness of compression member.

E_s = modulus of elasticity of reinforcement, psi.

f'_c = specified compressive strength of concrete, psi.

$\sqrt{f'_c}$ = square root of specified compressive strength of concrete, psi.

f_{cd} = diagonal compressive stress in the web of a beam, psi.

f_{ce} = effective compressive strength of concrete in the web of a beam,
= vf'_c, psi.

f_{cr} = required average compressive strength for f'_c to meet statistical acceptance criteria.

f_{ct} = splitting tensile strength of concrete, psi.

f_r = modulus of rupture of concrete, psi.

f_s = calculated stress in reinforcement at service loads, ksi.

f_y = specified yield strength of nonprestressed reinforcement, psi.

h = overall thickness of member, in.

I = moment of inertia of section. Subscripts: b = beam, c = column, s = slab.

I_{cr} = moment of inertia of cracked section transformed to concrete.

I_e = effective moment of inertia for computation of deflection.

I_g = moment of inertia of gross concrete section, neglecting reinforcement.

I_{gt} = moment of inertia of gross transformed uncracked section.

I_{se} = moment of inertia of the reinforcement in a column about centroidal axis of member cross section.

jd = distance between the resultants of the internal compressive and tensile forces on a cross section.

J_c = property of assumed critical section for two-way shear, analogous to polar moment of inertia.

k = effective length factor for compression members.

k_n = factor used in calculating flexural capacity of a section.

K = flexural stiffness; moment per unit rotation. Subscripts: b = beam, c = column, ec = equivalent column, s = slab.

K_t = torsional stiffness of torsional member; moment per unit rotation.

K_{tr} = transverse reinforcement index, see Chap. 8.

ℓ = span length of beam or one-way slab, generally center to center of supports; clear projection of cantilever, in.

ℓ_a = additional embedment length at support or at point of inflection, in.

ℓ_c = length of compression member in a frame, measured from center to center of the joints in the frame.

ℓ_d = development length, in.

ℓ_{db} = basic development length in tension before modification.

ℓ_{dbc} = basic development length in compression.

ℓ_{dh} = development length of standard hook in tension, measured from critical section to outside end of hook, in.

ℓ_{hb} = basic development length of standard hook in tension, in.

ℓ_n = clear span measured face to face of supports:

    = clear span for positive moment, negative moment at exterior support, or shear;

    = average of adjacent clear spans for negative moment.

ℓ_u = unsupported length of compression member.

ℓ_1 = length of span of two-way slab in direction that moments are being determined, measured center-to-center of supports.

ℓ_2 = length of span of two-way slab transverse to ℓ_1, measured center-to-center of supports.

L = live loads, or related internal moments and forces.

m = moment per unit width, subscript xy = twisting moment, ft-kips/ft.

m_r = resisting moment per unit width, ft-kips/ft.

M_a = maximum moment in member at the stage for which deflections are being computed, see Sec. 9–4.

M_c = factored moment to be used for design of a slender compression member.

M_{cr} = cracking moment.

M_n = nominal moment strength, in.-lb.

M_o = total factored static moment.

M_{pr} = probable flexural moment strength, see Eq. 19–10.

M_s = moment in column due to loads causing appreciable sidesway.

M_u = moment due to factored loads.

M_1 = smaller factored end moment on a compression member, positive if the member is bent in single curvature, negative if bent in double curvature.

M_{1ns} = factored end moment on a compression member at the end at which M_1 acts, due to loads that cause no appreciable sidesway.

M_{1s} = factored end moment on a compression member at the end at which M_1 acts, due to loads that cause appreciable sidesway, calculated using a first-order elastic frame analysis.

M_2 = larger factored end moment on a compression member, always positive.

M_{2ns} = factored end moment on a compression member at the end at which M_2 acts, due to loads that cause no appreciable sidesway.

M_{2s} = factored end moment on a compression member at the end at which M_2 acts, due to loads that cause appreciable sidesway, calculated using a first-order elastic frame analysis.

n = modular ratio = E_s / E_c.

n = number of bars in a layer being spliced or developed at a critical section.

N_u = factored axial load normal to cross section occurring simultaneously with V_u; positive for compression, negative for tension, and to include effects of tension due to creep and shrinkage.

N_v = axial tension force due to shear, lb.

p_{cp} = outside perimeter of the concrete cross section, in.

p_h = perimeter of the centerline of the outermost closed transverse torsional reinforcement, in.

P_b = nominal axial load strength at balanced strain conditions.

P_c = critical load.

P_E = buckling load of an elastic, hinged-end column.

P_n = nominal axial load strength at given eccentricity.

P_o = nominal axial load strength at zero eccentricity.

P_u = axial force due to factored loads.

Q = stability index for a story; see Chap. 12.

r = radius of gyration of cross section of a compression member.

s = standard deviation.

s = spacing of shear or torsion reinforcement measured along the longitudinal axis of the structural member, in.

T_n = nominal torsional moment strength.

T_s = nominal torsional moment strength provided by torsion reinforcement.

T_u = factored torsional moment at section.

U = required strength to resist factored loads or related internal moments and forces.

v = shear stress.

v_c = nominal shear stress carried by concrete, psi.

v_n = nominal shear stress, psi.

V_c = nominal shear force carried by concrete.

V_e = design shear force in beams or columns due to dead, live, and seismic loads.

V_n = nominal shear strength.

V_{nh} = nominal horizontal shear strength.

V_s = nominal shear strength provided by shear reinforcement.

V_u = factored shear force at section.

w_c = weight of concrete, lb/ft^3.

w_D = factored dead load per unit area of slab or per unit length of beam.

w_L = factored live load per unit area of slab or per unit of length of beam.

w_u = total factored load per unit length of beam or per unit area of slab.

x = shorter overall dimension of rectangular part of cross section.

y = longer overall dimension of rectangular part of cross section.

y_t = distance from centroidal axis of cross section, neglecting reinforcement, to extreme fiber in tension.

z = quantity limiting distribution of flexural reinforcement.

α = angle between inclined stirrups and longitudinal axis of member.

(alpha)

α = ratio of flexural stiffness of beam section to flexural stiffness of a width of slab bounded laterally by centerlines of adjacent panels (if any) on each side of the beam = $E_{cb}I_b/(E_{cs}I_s)$.

α = reinforcement location factor for development length calculation.

α_m = average value of α for all beams on edges of a panel.

α_s = coefficient used to compute V_c in slabs.

α_1 = α in direction of ℓ_1.

α_2 = α in direction of ℓ_2.

β = coating factor for development length calculation.

(beta)

β = ratio of clear spans in long to short direction of two-way slabs.

β = ratio of long side to short side of footing.

β_a = ratio of dead load per unit area to live load per unit area (in each case without load factors).

β_b = ratio of area of reinforcement cut off to total area of tension reinforcement at section.

β_c = ratio of long side to short side of column or concentrated load.

β_d = (a) for nonsway frames, β_d is the ratio of the maximum factored axial dead load to the total factored axial load,

(b) for sway frames, except as required in (c), β_d is the ratio of the maximum factored sustained shear in a story to the total factored shear in that story,

(c) for stability checks of sway frames carried out in accordance with ACI Sec. 10.13.6, β_d is the ratio of the maximum factored axial dead load to the total factored axial load.

β_t = ratio of torsional stiffness of edge beam section to flexural stiffness of a width of slab equal to span length of beam, center-to-center of supports
$= E_{cb}C/(2E_{cs}I_s)$.

β_1 = ratio of depth of rectangular stress block, a, to depth to neutral axis, c.

γ = ratio of the distance between the outer layers of reinforcement in a column to the overall depth of the column (see Fig. 11–30).

(gamma)

γ = reinforcement size factor for development length calculations.

γ_f = fraction of unbalanced moment transferred by flexure at slab-column connections.

γ_v = fraction of unbalanced moment transferred by eccentricity of shear at slab–column connections $= 1 - \gamma_f$.

δ_{ns} = moment magnification factor for frames braced against sidesway, to reflect effects of member curvature between ends of compression member.

(delta)

δ_s = moment magnification factor for frames not braced against sidesway, to reflect lateral drift resulting from lateral and gravity loads.

ϵ = strain.

(epsilon)

ϵ_c = strain in concrete.

ϵ_{cu} = compressive strain at crushing of concrete.

ϵ_s = strain in steel.

ϵ_t = net tensile strain in extreme tension steel at nominal strength.

λ = multiplier for additional long-term deflection.

(lambda)

λ = correction factor related to unit weight of concrete.

μ = coefficient of friction.

ν = ratio of effective concrete strength in web of beam or compression strut to f'_c.

ξ = time-dependent factor for sustained load deflections.

ρ = ratio of nonprestressed tension reinforcement $= A_s/bd$.

ρ' = ratio of nonprestressed compression reinforcement A'_s/bd.

ρ_b = reinforcement ratio corresponding to balanced strain conditions.

ρ_g = ratio of total reinforcement area to cross-sectional area of column.

ρ_s = ratio of volume of spiral reinforcement to total volume of core (out-to-out spirals) of a spirally reinforced compression member.

$\rho_w = A_s/b_w d$.

ϕ = strength reduction factor.

θ = angle of compression struts in web of beam.

ω = mechanical reinforcement ratio $= \rho f_y/f'_c$.

References

Chapter 1

1-1 Hedley E. H. Roy, "Toronto City Hall and Civic Square," *ACI Journal, Proceedings,* Vol. 62, No. 12, December 1965, pp. 1481–1502.

1-2 CRSI Subcommittee on Placing Reinforcing Bars, *Reinforcing Bar Detailing,* Concrete Reinforcing Steel Institute, Chicago, 1971, 280 pp.

1-3 Robert Mark, *Light, Wind, and Structure: The Mystery of the Master Builders,* MIT Press, Boston, 1990, pp. 52–67.

1-4 Committee on Concrete and Reinforced Concrete, "Standard Building Regulations for the Use of Reinforced Concrete," *Proceedings, National Association of Cement Users,* Vol. 6, 1910, pp. 349–361.

1-5 Special Committee on Concrete and Reinforced Concrete, "Progress Report of Special Committee on Concrete and Reinforced Concrete," *Proceedings of the American Society of Civil Engineers,* 1913, pp. 117- 135.

1-6 Special Committee on Concrete and Reinforced Concrete, "Final Report of Special Committee on Concrete and Reinforced Concrete," *Proceedings of the American Society of Civil Engineers,* 1916, pp. 1657 – 1708.

1-7 Frank Kerekes and Harold B. Reid, Jr., "Fifty Years of Development in Building Code Requirements for Reinforced Concrete," *ACI Journal, Proceedings,* Vol. 50, No. 8, February 1954, pp. 441–472.

1-8 *Uniform Building Code,* International Conference of Building Officials, Whittier, Calif., various editions.

1-9 *Standard Building Code,* Southern Building Code Congress, Birmingham, Ala., various editions.

1-10 *Basic Building Code,* Building Officials and Code Administrators International, Chicago, various editions.

1-11 ACI Committee 318, *Building Code Requirements for Structural Concrete* (ACI 318-95) and *Commentary* (ACI 318R-95), American Concrete Institute, Detroit, 1995, 369 pp.

1-12 *Standard Specifications for Highway Bridges,* American Association of State Highway and Transportation Officials, Washington, D.C., various editions.

1-13 *CEB-FIP Model Code 1990,* Thomas Telford Services Ltd., London, for Comité Euro-International du Béton, Laussane, 1993, 437 pp.

Chapter 2

2–1 Donald Taylor, "Progressive Collapse," *Canadian Journal of Civil Engineering,* Vol. 2, No. 4, December 1975, pp. 517–529.

2–2 *Minimum Design Loads for Buildings and Other Structures,* ASCE 7-95, American Society of Civil Engineers, New York, 1995, 214 pp.

2–3 C. Allan Cornell, "A Probability Based Structural Code," *ACI Journal, Proceedings,* Vol. 66, No. 12, December 1969, pp. 974–985.

2–4 Alan H. Mattock, Ladislav B. Kriz, and Eivind Hognestad, "Rectangular Concrete Stress Distribution in Ultimate Strength Design," *ACI Journal, Proceedings,* Vol. 57, No. 8, February 1961, pp. 875–928.

2–5 Jong-Cherng Pier and C. Allan Cornell, "Spatial and Temporal Variability of Live Loads," *Proceedings ASCE, Journal of the Structural Division,* Vol. 99, No. ST5, May 1973, pp. 903–922.

2–6 James G. MacGregor, "Safety and Limit States Design for Reinforced Concrete," *Canadian Journal of Civil Engineering,* Vol. 3, No. 4, December 1976, pp. 484–513.

2–7 Bruce Ellingwood, Theodore V. Galambos, James G. MacGregor, and C. Allan Cornell, *Development of a Probability Based Load Criterion for American National Standard A58,* NBS Special Publication 577, National Bureau of Standards, U.S. Department of Commerce, Washington, D.C., June 1980, 222 pp.

2–8 James G. MacGregor, "Load and Resistance Factors for Concrete Design," *ACI Journal, Proceedings,* Vol. 80, No. 4, July–August 1983, pp. 279–287.

2–9 *Uniform Building Code,* International Conference of Building Officials, Whittier, Calif., various editions.

2–10 *Standard Building Code,* Southern Building Code Congress, Birmingham, Ala., various editions.

2–11 *Basic Building Code,* Building Officials and Code Administrators International, Chicago, various editions.

2–12 Donald A. Sawyer, "Ponding of Rainwater on Flexible Roof Systems," *Proceedings ACSE, Journal of the Structural Division,* Vol. 93, No. ST1, February 1967, pp. 127–147.

2–13 Allan G. Davenport, "Gust Loading Factors," *Proceedings ASCE, Journal of the Structural Division,* Vol. 93, No. ST6, June 1967, pp. 12–34

2–14 *Material and Cost Estimating Guide for Concrete Floor and Roof Systems,* Publication PA 136.01B, Portland Cement Association, Skokie, Ill., 1974, 14 pp.

2–15 Percival E. Pereiru, *McGraw-Hill's Dodge Construction Systems Costs 1980,* McGraw-Hill Information Systems Company, New York, 1980, 281 pp.

2–16 ACI Committee 340, *Design Handbook in Accordance with the Strength Design Method of ACI 318-89,* Vol. 1, *Beams, One-Way Slabs, Brackets, Footings, and Pile Caps,* ACI Publication SP-17(91), American Concrete Institute, Detroit, 1991.

2–17 ACI Committee 340, *Design Handbook in Accordance with the Strength Design Method of ACI 318-89,* Vol. 2, *Columns,* ACI Publication SP-17A(90), American Concrete Institute, Detroit, 1990.

2–18 ACI Committee 340, *Design Handbook in Accordance with the Strength Design Method of ACI 318-89,* Vol. 3, *Two-Way Slabs,* ACI Publication SP-17(91)(S), American Concrete Institute, Detroit, 1991.

2–19 *CRSI Handbook, 1992,* Concrete Steel Reinforcing Steel Institute, Schaumberg, Ill., 1992, 710 pp.

2–20 ACI Committee 315, *ACI Detailing Manual—1994,* ACI Publication SP-66 (94), American Concrete Institute, Detroit, 1994, 244 pp.

2–21 ACI Committee 301, *Specifications for Structural Concrete for Buildings,* ACI 301-95, American Concrete Institute, Detroit, 1995, 36 pp.

2–22 Mary K. Hurd and ACI Committee 347, *Formwork for Concrete,* 6th ed., ACI Publication SP-4, American Concrete Institute, Detroit, 1995, 500 pp.

2–23 *Manual of Concrete Practice,* Vol. 1 to 5, American Concrete Institute, Detroit, published annually.

2–24 ACI Committee 117, "Standard Specifications for Tolerances for Concrete Construction and Materials (ACI 117-90) and Commentary (ACI 117R-90)," *ACI Manual of Concrete Practice,* American Concrete Institute, Detroit, 1991 and later editions.

Chapter 3

3–1 Thomas T. C. Hsu, F. O. Slate, G. M. Sturman, and George Winter, "Micro-cracking of Plain Concrete and the Shape of the Stress-Strain Curve," *ACI Journal, Proceedings,* Vol. 60, No. 2, February 1963, pp. 209–224.

3–2 K. Newman and J. B. Newman, "Failure Theories and Design Criteria for Plain Concrete," Part 2 in M. Te 'eni (ed.), *Solid Mechanics and Engineering Design,* Wiley-Interscience, New York, 1972, pp. 83/1–83/33.

3–3 F. E. Richart, A. Brandtzaeg, and R. L. Brown, *A Study of the Failure of Concrete under Combined Compressive Stresses,* Bulletin 185, University of Illinois Engineering Experiment Station, Urbana, Ill., November 1928, 104 pp.

3–4   Hubert Rüsch, "Research toward a General Flexural Theory for Structural Concrete," *ACI Journal, Proceedings,* Vol. 57, No. 1, July 1960, pp. 1–28.

3–5   Llewellyn E. Clark, Kurt H. Gerstle, and Leonard G. Tulin, "Effect of Strain Gradient on Stress–Strain Curve of Mortar and Concrete," *ACI Journal, Proceedings,* Vol. 64, No. 9, September 1967, pp. 580–586.

3–6   Comité Euro-International du Béton, *CEB-FIP Model Code 1990,* Thomas Telford Services Ltd. London, 1993, 437 pp.

3–7   ACI Committee 214, *Recommended Practice for Evaluation of Strength Test Results of Concrete,* ACI 214–77, American Concrete Institute, Detroit, 1977, 14 pp.

3–8   Sher Ali Mirza, Michael Hatzinikolas, and James G. MacGregor, "Statistical Descriptions of the Strength of Concrete," *Proceedings ASCE, Journal of the Structural Division,* Vol. 105, No. ST6, June 1979, pp. 1021–1037.

3–9   H. F. Gonnerman and W. Lerch, *Changes in Characteristics of Portland Cement as Exhibited by Laboratory Tests over the Period 1904 to 1950,* ASTM Special Publication 127, American Society for Testing and Materials, Philadelphia, 1951.

3–10   ACI Committee 211, "Standard Practice for Selecting Proportions for Normal, Heavyweight, and Mass Concrete" (ACI 211.1-91), *ACI Manual of Concrete Practice,* American Concrete Institute, Detroit, 1992 and later editions, pp. 211.1-1 to 211.1-38.

3–11   ACI Committee 226, "Use of Fly Ash in Concrete" (ACI 226.3R-87), *ACI Manual of Concrete Practice,* American Concrete Institute, Detroit, 1988 and later editions, pp. 226.3R-1 to 226.3R-38.

3–12   ACI Committee 226, "Ground Granulated Blast Furnace Slag as a Cementitious Constituent in Concrete" (ACI 226.1R-87), ACI *Manual of Concrete Practice,* American Concrete Institute, Detroit, 1988 and later editions, pp. 226.1R-1 to 226.1R-16.

3–13   Walter H. Price, "Factors Influencing Concrete Strength," *ACI Journal, Proceedings,* Vol. 47, No. 6, December 1951, pp. 417–432

3–14   Paul Klieger, "Effect of Mixing and Curing Temperature on Concrete Strength," *ACI Journal, Proceedings,* Vol. 54, No. 12, June 1958, pp. 1063–1081.

3–15   ACI Committee 209, "Prediction of Creep, Shrinkage and Temperature Effects in Concrete Structures," *Designing for Creep and Shrinkage in Concrete Structures,* ACI Publication SP-76, American Concrete Institute, Detroit, 1982, pp. 193–300.

3–16   V. M. Malhotra, "Maturity Concept and the Estimation of Concrete Strength: A Review," *Indian Concrete Journal,* Vol. 48, No. 4, April 1974, pp. 122–126 and 138; No. 5, May 1974, pp. 155–159 and 170.

3–17   H. S. Lew and T. W. Reichard, "Prediction of Strength of Concrete from Maturity," *Accelerated Strength Testing,* ACI Publication SP-56, American Concrete Institute, Detroit, 1978, pp. 229–248.

3–18   F. Michael Bartlett and James G. MacGregor, "Equivalent Specified Concrete Strength from Core Test Data," *Concrete International,* Vol. 17, No. 3, March 1995, pp. 52–58.

3–19   F. Michael Bartlett and James G. MacGregor, "Statistical Analysis of the Compressive Strength of Concrete in Structures," *ACI Materials Journal,* Vol. 93, in press.

3–20   Jerome M. Raphael, "Tensile Strength of Concrete," *ACI Journal, Proceedings,* Vol. 81, No. 2, March–April 1984, pp. 158–165.

3–21   *Proposed Complements to the CEB-FIP International Recommendations—1970,* Bulletin d'Information 74, Comité Européen du Béton, Paris, March 1972 revision, 77 pp.

3–22   H. S. Lew and T. W. Reichard, "Mechanical Properties of Concrete at Early Ages," *ACI Journal, Proceedings,* Vol. 75, No. 10, October 1978, pp. 533–542.

3–23   H. Kupfer, Hubert K. Hilsdorf, and Hubert Rüsch, "Behavior of Concrete under Biaxial Stress," *ACI Journal, Proceedings,* Vol. 66, No. 8, August 1969, pp. 656–666.

3–24   Frank J. Vecchio and Michael P. Collins, *The Response of Reinforced Concrete to In-Plane Shear and Normal Stresses,* Publication 82-03, Department of Civil Engineering, University of Toronto, Toronto, March 1982, 332 pp.

3–25   Frank J. Vecchio and Michael P. Collins, "The Modified Compression Field Theory for Reinforced Concrete Elements Subjected To Shear," *ACI Journal, Proceedings,* Vol. 83, No. 2, March–April 1986, pp. 219–231.

3–26   J. A. Hansen, "Strength of Structural Lightweight Concrete under Combined Stress," *Journal of the Research and Development Laboratories, Portland Cement Association,* Vol. 5, No. 1, January 1963, pp. 39–46.

3–27   ACI Committee 363, "State-of-the-Art Report on High-Strength Concrete," *ACI Manual of Concrete Practice,* Vol. 1, American Concrete Institute, Detroit, 1993 and later editions, pp. 363R-1 to 363R-55.

3–28   Eivind Hognestad, Norman W. Hanson, and Douglas McHenry, "Concrete Stress Distribution in Ultimate Strength Design," *ACI Journal, Proceedings,* Vol. 52, No. 4, December 1955, pp. 475–479.

3–29   Paul H. Kaar, Norman W. Hanson, and H. T. Capell, "Stress-Strain Characteristics of High-Strength Concrete," *Douglas McHenry International Symposium on Concrete and Concrete Structures,* ACI Publication SP-55, American Concrete Institute, Detroit, 1978, pp. 161–186.

3–30 Adrian Pauw, "Static Modulus of Elasticity as Affected by Density," *ACI Journal, Proceedings,* Vol. 57, No. 6, December 1960, pp. 679–683.

3–31 Eivind Hognestad, *A Study of Combined Bending and Axial Load in Reinforced Concrete Members,* Bulletin 399, University of Illinois Engineering Experiment Station, Urbana, Ill., November 1951, 128 pp.

3–32 Claudio E. Todeschini, Albert C. Bianchini, and Clyde E. Kesler, "Behavior of Concrete Columns Reinforced with High Strength Steels," *ACI Journal, Proceedings,* Vol. 61, No. 6, June 1964, pp. 701–716.

3–33 S. H. Ahmad and Surendra P. Shah, "Stress–Strain Curves of Concrete Confined by Spiral Reinforcement," *ACI Journal, Proceedings,* Vol. 79, No.6. November–December 1982, pp. 484–490.

3–34 B. P. Sinha, Kurt H. Gerstle, and Leonard G. Tulin, "Stress–Strain Relations for Concrete under Cyclic Loading," *ACI Journal, Proceedings,* Vol. 61, No. 2, February 1964, pp. 195–212.

3–35 Surendra P. Shah and V. S. Gopalaratnam, "Softening Responses of Plain Concrete in Direct Tension," *ACI Journal, Proceedings,* Vol. 82, No. 3, May–June 1985, pp. 310–323.

3–36 Zedenek P. Bazant, "Prediction of Concrete Creep Effects Using Age-Adjusted Effective Modulus Method," *ACI Journal, Proceedings,* Vol. 69, No. 4, April 1972, pp. 212–217.

3–37 Walter H. Dilger, "Creep Analysis of Prestressed Concrete Structures Using Creep-Transformed Section Properties," *PCI Journal,* Vol. 27, No. 1, January–February 1982, pp. 99–118.

3–38 Amin Ghali and Rene Favre, *Concrete Structures: Stresses and Deformations,* Chapman & Hall, New York, 1986, 348 pp.

3–39 *Structural Effects of Time-Dependent Behaviour of Concrete,* Bulletin d'Information, 215, Comité Euro-International du Béton, Laussane, March 1993, pp. 265–291.

3–40 ACI Committee 216, "Guide for Determining the Fire Endurance of Concrete Elements," *Concrete International: Design and Construction,* Vol. 3, No. 2, February 1981, pp. 13–47.

3–41 Boris Bresler, "Lightweight Aggregate Reinforced Concrete Columns," *Lightweight Concrete,* ACI Publication SP-29, American Concrete Institute, Detroit, 1971, pp. 81–130.

3–42 ACI Committee 201, "Guide to Durable Concrete," (ACI 201.2R-92), *ACI Manual of Concrete Practice,* American Concrete Institute, Detroit, 1993 and later editions, pp. 201R-1 to 201.2R-41.

3–43 ACI Committee 222, "Corrosion of Metals in Concrete" (ACI 222R–85), *ACI Journal, Proceedings,* Vol. 82, No. 1, January–February 1985, pp. 3–32.

3–44 PCI Committee on Durability, "Alkali–Aggregate Reactivity—A Summary," *PCI Journal,* Vol. 39, No. 6, November–December, 1994, pp. 26–35.

3–45 ACI Committee 515, "A Guide to the Use of Waterproofing, Dampproofing, Protective, and Decorative Barrier Systems for Concrete" (ACI 515.R-79), *ACI Manual of Concrete Practice,* American Concrete Institute, Detroit, 1993 and later editions, pp. 515.1R-1 to 515.1R-44.

3–46 Sher Al Mirza and James G. MacGregor, "Variability of Mechanical Properties of Reinforcing Bars," *Proceedings ASCE, Journal of the Structural Division,* Vol. 105, No. ST5, May 1979, pp. 921–937.

3–47 T. Helgason and John M. Hanson, "Investigation of Design Factors Affecting Fatigue Strength of Reinforcing Bars—Statistical Analysis," *Abeles Symposium on Fatigue of Concrete,* ACI Publication SP-41, American Concrete Institute, Detroit, 1974, pp. 107–137.

3–48 ACI Committee 215, "Considerations for Design of Concrete Structures Subjected to Fatigue Loading," (ACI 215R-74, revised 1992), *ACI Manual of Concrete Practice,* American Concrete Institute, Detroit, 1993 and later editions, Detroit, pp. 215R-1 to 215R-24.

Chapter 4

4–1 Eivind Hognestad, *A Study of Combined Bending and Axial Load in Reinforced Concrete Members,* Bulletin 399, University of Illinois Engineering Experiment Station, Urbana, Ill., November 1951, 128 pp.

4–2 Hubert Rüsch, "Research toward a General Flexural Theory for Structural Concrete," *ACI Journal, Proceedings,* Vol. 57, No. 1, July 1960, pp. 1–28; Discussion, Vol. 57, No. 9, March 1961, pp. 1147–1164.

4–3 Alan H. Mattock, Ladislav B. Kriz, and Eivind Hognestad, "Rectangular Concrete Stress Distribution in Ultimate Strength Design," *ACI Journal, Proceedings,* Vol. 57, No. 8, February 1961, pp. 875–926.

4–4 *CEB-FIP Model Code 1990,* Thomas Telford Services Ltd., London, for Comité Euro-International du Béton, Laussane, 1993, 437 pp.

4–5 Dudley Charles Kent and Robert Park, "Flexural Members with Confined Concrete," *Proceedings ASCE, Journal of the Structural Division,* Vol. 97, No. ST7, July 1971, pp. 1969–1990; Closure to Discussion, Vol. 98, No. ST12, December 1972, pp. 2805–2810.

4–6 J. Stephen Ford, D. C. Chang, and John E. Breen, "Design Indications from Tests of Unbraced Multipanel Concrete Frames," *Concrete International; Design and Construction,* Vol. 3, No. 3, March 1981, pp. 37–47.

4–7 Paul H. Kaar, Norman W. Hanson, and H. T. Capell, "Stress–Strain Characteristics of High Strength Concrete," *Douglas McHenry International Symposium on Concrete Structures,* ACI Publication SP-55, American Concrete Institute, Detroit, 1978, pp. 161–185.

4–8 Robert F. Mast, "Unified Design Provisions for Reinforced and Prestressed Concrete Flexural and Compression Members," *ACI Structural Journal, Proceedings,* Vol. 89, No. 2, March–April 1992, pp. 185–199.

4–9 Charles S. Whitney, "Plastic Theory of Reinforced Concrete Design," *Proceedings ASCE,* December 1940; *Transactions ASCE,* Vol. 107, 1942, pp. 251–326.

4–10 P. W. Birkeland and L. J. Westhoff, "Dimensional Tolerance—Concrete," State-of-Art Report 5, Technical Committee 9, *Proceedings of International Conference on Planning and Design of Tall Buildings,* Vol. Ib, American Society of Civil Engineers, New York, 1972, pp. 845–849.

4–11 Sher-Ali Mirza and James G. MacGregor, "Variations in Dimensions of Reinforced Concrete Members," *Proceedings ACSE, Journal of the Structural Division,* Vol. 105, No. ST4, April 1979, pp. 751–766.

4–12 ACI Committee 340, *Design Handbook,* Vol. 1, *Beams, One-Way Slabs, Brackets, Footings, and Pile Caps, in Accordance with the Strength Design Method of ACI 318–91,* 4th ed., ACI Publication SP-17(91), American Concrete Institute, Detroit, 1991, 374 pp.

4–13 *CRSI Handbook 1992,* Concrete Reinforcing Steel Institute, Schaumberg, Ill., 1992, 710 pp.

Chapter 5

5–1 CRSI Subcommittee on Placing Reinforcing Bars, *Placing Reinforcing Bars,* Concrete Reinforcing Steel Institute, Chicago, 1971, 201 pp.

5–2 Stephen Timoshenko and J. N. Goodier, *Theory of Elasticity,* 2nd ed., McGraw-Hill, New York, 1951, pp. 171–177.

5–3 Gottfried Brendel, "Strength of the Compression Slab of T-Beams Subject to Simple Bending," *ACI Journal, Proceedings,* Vol. 61, No. 1, January 1964, pp. 57–76.

5–4 G. W. Washa and P. G. Fluck, "Effect of Compressive Reinforcement on the Plastic Flow of Reinforced Concrete Beams," *ACI Journal, Proceedings,* Vol. 49, No. 4, October 1952, pp. 89–108.

5–5 Mircea Z. Cohn and S. K. Ghosh, "Flexural Ductility of Reinforced Concrete Sections," *Publications,* International Association of Bridge and Structural Engineers, Zurich, Vol. 32-II, 1972, pp. 53–83.

Chapter 6

6–1 Boyd G. Anderson, "Rigid Frame Failures," *ACI Journal, Proceedings,* Vol. 53, No. 7, January 1957, pp. 625–636.

6–2 Howard P. J. Taylor, *Investigation of Forces Carried across Cracks in Reinforced Concrete Beams in Shear by Interlock of Aggregate,* TRA 42.447, Cement and Concrete Association, London, 1970, 22 pp.

6–3 Robert Park and Thomas Paulay, *Reinforced Concrete Structures,* A Wiley-Interscience Publication, Wiley, New York, 1975, 769 pp.

6–4 ACI-ASCE Committee 426, "The Shear Strength of Reinforced Concrete Members—Chapters 1 to 4," *Proceedings ASCE, Journal of the Structural Division,* Vol. 99, No. ST6, June 1973, pp. 1091–1187.

6–5 Jörg Schlaich, Kurt Schaefer, and Mattias Jennewein, "Towards a Consistent Design of Reinforced Concrete Structures," *Journal of the Prestressed Concrete Institute,* Vol. 32, No. 3, May–June 1987.

6–6 ACI-ASCE Committee 426, *Suggested Revisions to Shear Provisions for Building Codes,* American Concrete Institute, Detroit, 1978, 88 pp; abstract published in *ACI Journal, Proceedings,* Vol. 75, No. 9, September 1977, pp. 458–469; Discussion, Vol. 75, No. 10, October 1978, pp. 563–569.

6–7 Michael P. Collins and Denis Mitchell, *Prestressed Concrete Structures,* Prentice Hall, Englewood Cliffs, N. J., 1991, p. 366.

6–8 Peter Marti, "Basic Tools of Beam Design," *ACI Journal, Proceedings,* Vol. 82, No. 1, January–February 1985, pp. 46–56.

6–9 Peter Marti, "Truss Models in Detailing," *Concrete International Design and Construction,* Vol. 7, No. 12, December 1985, pp. 66–73.

6–10 David M. Rogowsky and James G. MacGregror, "Design of Reinforced Concrete Deep Beams," *Concrete International: Design and Construction,* Vol. 8. No. 8, August 1986, pp. 49–58.

6–11 *Bruchwiderstand und Bemessung von Stahlbeton-und Spannbetontragwerken (Ultimate Limit States and Design of Reinforced Concrete and Prestressed Concrete Structures),* Schweizerischer Ingenieur-und Architekten-Verein (SIA), Zurich, 1976.

6–12 *CEB-FIP Model Code 1990,* First Draft, Bulletin d'Information 195 and 196, Comité Euro-International du Béton, Lausanne, March 1990, 348 pp.

6–13 Michael P. Collins and Denis Mitchell, "Design Proposals for Shear and Torsion," *Journal of the Prestressed Concrete Institute,* Vol. 25, No. 5, September–October 1980, 70 pp.

6–14 Frank J. Vecchio and Michael P. Collins, "The Modified Compression Field Theory for Reinforced Concrete Elements Subjected to Shear," *ACI Journal, Proceedings,* Vol. 83, No. 2, March–April 1986, pp. 219–231.

6–15 ACI-ASCE Committee 326, Shear and Diagonal Tension," *ACI Journal, Proceedings,* Vol. 59, Nos. 1–3, January–March 1962, pp. 1–30, 277–344, and 352–396.

6–16 Theodore C. Zsutty, "Shear Strength Prediction for Separate Categories of Simple Beam Tests," *ACI Journal, Proceedings,* Vol. 68, No. 2 February 1971, pp. 138–143.

6–17 Fritz Leonhardt and Rene Walther, *The Stuttgart Shear Tests, 1961,* Translation 111, Cement and Concrete Association, London, 1964, 110 pp.

6–18 A. G. Mphonde and Gregory C. Frantz, "Shear Tests for High- and Low-Strength Concrete Beams without Stirrups," *ACI Journal, Proceedings,* Vol. 81, No. 4, July–August 1984, pp. 350–357.

6–19 A. H. Elzanaty, Arthur H. Nilson, and Floyd O. Slate, "Shear Capacity of Reinforced Concrete Beams Using High Strength Concrete," *ACI Journal, Proceedings,* Vol. 83, No. 2, March–April 1986, pp. 290–296.

6–20 CSA Technical Committee on Reinforced Concrete Design, *Design of Concrete Structures for Buildings,* CAN3-A23.3-M84, Canadian Standards Association, Rexdale, Ontario, December 1984, 281 pp.

6–21 Alan H. Mattock and J. F. Shen, "Joints between Reinforced Concrete Members of Similar Depth," *ACI Structural Journal, Proceedings,* Vol. 89, No. 3, May–June 1992, pp. 290–295.

Chapter 7

7–1 Stephen P. Timoshenko and J. N. Goodier, *Theory of Elasticity,* McGraw-Hill, New York, 1951, pp. 275–288.

7–2 E. P. Popov, *Mechanics of Materials, SI Version,* 2nd ed., Prentice Hall, Englewood Cliffs, N. J., 1978, 590 pp.

7–3 Thomas T. C. Hsu, "Torsion of Structural Concrete—Behavior of Reinforced Concrete Rectangular Members," *Torsion of Structural Concrete,* ACI Publication SP-18, American Concrete Institute, Detroit, 1968, pp. 261–306.

7–4 Ugor Ersoy and Phil M. Ferguson, "Concrete Beams Subjected to Combined Torsion and Shear—Experimental Trends," *Torsion of Structural Concrete,* ACI Publication SP-18, American Concrete Institute, Detroit, 1968, pp. 441–460.

7–5 N. N. Lessig, *Determination of the Load Carrying Capacity of Reinforced Concrete Elements with Rectangular Cross-Section Subjected to Flexure with Torsion,* Work 5, Institute Betona i Zhelezobetona, Moscow, 1959, pp. 4–28; also available as Foreign Literature Study 371, PCA Research and Development Labs, Skokie, Ill.

7–6 Paul Lampert, *Torsion und Biegung von Stahlbetonbalken (Torsion and Bending of Reinforced Concrete Beams),* Bericht 27, Institute für Baustatik, Zurich, January 1970.

7–7 Paul Lampert and Bruno Thürlimann, "Ultimate Strength and Design of Reinforced Concrete Beams in Torsion and Bending," *Publications,* International Association for Bridge and Structural Engineering, Zurich, Vol. 31-I, 1971, pp. 107–131.

7–8 Paul Lampert and Michael P Collins, "Torsion, Bending, and Confusion—An Attempt to Establish the Facts," *ACI Journal, Proceedings* , Vol. 69, No. 8 August 1972, pp. 500–504.

7–9 *CEB-FIP Model Code 1990,* First Draft, Bulletin d'Information 195 and 196, Comité Euro-International du Béton, Lausanne, March 1990, 348 pp.

7–10 CAN3-A23.3-M84, *Design of Concrete Structures for Buildings,* Canadian Standards Association, Rexdale, Ontario, 1984, 280 pp.

7–11 Thomas T. C. Hsu, "Torsion of Structural Concrete—Plain Concrete Rectangular Sections," *Torsion of Structural Concrete,* ACI Publication SP-18, American Concrete Institute, Detroit, 1968, pp. 203–238.

7–12 James G. MacGregor and Mashour G. Ghoneim, "Design for Torsion," *ACI Structural Journal,* Vol. 92, No. 2, March–April 1995, pp. 211–218.

7–13 Thomas T. C. Hsu, "Shear Flow Zone in Torsion of Reinforced Concrete," *Journal of Structural Engineering, ASCE,* Vol. 116, No. 11, November 1990, pp. 3206–3226.

7–14 Michael P. Collins and Denis Mitchell, "Design Proposals for Shear and Torsion," *PCI Journal,* Vol. 25, No. 5, September–October 1980, 70 pp.

7–15 Michael P. Collins and Paul Lampert, "Redistribution of Moments at Cracking—The Key to Simpler Torsion Design?" *Analysis of Structural Systems for Torsion,* ACI Publication SP-35, American Concrete Institute, Detroit, 1973, pp. 343–383.

7–16 Denis Mitchell and Michael P. Collins, "Detailing for Torsion," *ACI Journal, Proceedings,* Vol. 73, No. 9, September 1976, pp. 506–511.

7–17 ACI Committee 318, *Building Code Requirements for Structural Concrete* (ACI 318–95) and *Commentary* (ACI 318R-95), American Concrete Institute, Detroit, 1995, 369 pp.

7–18 ACI Committee 315, *ACI Detailing Manual,* ACI Publication SP-66(88), American Concrete Institute, Detroit, 1988, 218 pp.

Chapter 8

8–1 ACI Committee 408, "Bond Stress—The State of the Art," *ACI Journal, Proceedings,* Vol. 63, No. 11, November 1966, pp. 1161–1190; Discussion, pp. 1569–1570.

8–2 C. O. Orangun, J. O. Jirsa, and J. E. Breen, "A Reevaluation of Test Data on Development Length and Splices," *ACI Journal, Proceedings,* Vol. 74, No. 3, March 1977, pp. 114–122; Discussion, pp. 470–475.

8–3 J. O. Jirsa, L. A. Lutz, and P. Gergely, "Rationale for Suggested Development, Splice and Standard Hook Provisions for Deformed Bars in Tension," *Concrete International: Design and Construction,* Vol. 1, No. 7, July 1979, pp. 47–61.

8–4 *Bond Action and Bond Behaviour of Reinforcement, State-of-the-Art Report,* Bulletin d'Information 151, Comité Euro-International du Béton, Paris, April 1982, 153 pp.

8–5 Paul R. Jeanty, Denis Mitchell, and Saeed M. Mirza, "Investigation of Top Bar Effects in Beams," *ACI Structural Journal,* Vol. 85, No. 3, May–June 1988, pp. 251–257.

8–6 Robert A. Treece and James O. Jirsa, "Bond Strength of Epoxy-Coated Reinforcing Bars," *ACI Materials Journal,* Vol. 86, No. 2, March–April 1989, pp. 167–174.

8–7 J. L. G. Marques and J. O. Jirsa, "A Study of Hooked Bar Anchorages in Beam-Column Joints," *ACI Journal, Proceedings,* Vol. 72, No. 5, May 1975, pp. 198–209.

8–8 G. Rehm, "Kriterien zur Beurteilung von Bewehrungsstäben mit hochwertigem Verbund (Criteria for the Evaluation of High Bond Reinforcing Bars)" *Stahlbetonbau-Berichte aus Forschung und Praxis-Hubert Rüsch gewidmet,* Berlin, 1969, pp. 79–85

8–9 Joint Committee on Standard Specifications for Concrete and Reinforced Concrete, *Recommended Practice and Standard Specifications for Concrete and Reinforced Concrete,* American Concrete Institute, Detroit, June 1940, 140 pp.

8–10 M. Baron, "Shear Strength of Reinforced Concrete Beams at Point of Bar Cutoff," *ACI Journal, Proceedings,* Vol. 63, No. 1, January 1966, pp. 127–134.

8–11 Y. Goto and K. Otsuka, "Experimental Studies on Cracks Formed in Concrete around Deformed Tension Bars," *Technological Reports of Tohoku University,* Vol. 44, June 1979, pp. 49–83.

8–12 J. F. Pfister and A. H. Mattock, "High Strength Bars as Concrete Reinforcement," Part 5, "Lapped Splices in Concentrically Loaded Columns," *Journal of the Research and Development Laboratories,* Portland Cement Association, Vol. 5, No. 2, May 1963, pp. 27–40.

8–13 F. Leonhardt and K. Teichen, "Druck-Stoesse von Bewehrungsstaeben (Compression Splices of Reinforcing Bars)," *Deutscher Ausschuss fuer Stahlbeton,* Bulletin 222, Wilhelm Ernst, Berlin, 1972, pp. 1–53.

8–14 ACI Committee 439, "Mechanical Connections of Reinforcing Bars," *Concrete International: Design and Construction,* Vol. 5, No. 1, January 1983, pp. 24–35.

Chapter 9

9–1 Zedenek P. Bazant, "Prediction of Concrete Creep Effects Using Age-Adjusted Effective Modulus Method," *ACI Journal, Proceedings,* Vol. 69, No. 4, April 1972, pp. 212–217.

9–2 Walter H. Dilger, "Creep Analysis of Prestressed Concrete Structures Using Creep-Transformed Section Properties," *PCI Journal,* Vol. 27, No. 1, January–February 1982, pp. 99–118.

9–3 Amin Ghali and Rene Favre, *Concrete Structures: Stresses and Deformations,* Chapman & Hall, New York, 1986, 348 pp.

9–4 Fritz Leonhardt, "Crack Control in Concrete Structures," *IABSE Surveys,* IABSE Periodica, 3/1977, International Association for Bridge and Structural Engineering, Zurich, 1977, 26 pp.

9–5 ACI Committee 224, "Control of Cracking in Concrete Structures" (ACI 224R-90), *ACI Manual of Concrete Practice,* American Concrete Institute, Detroit, 1991 and later editions, pp. 224R-1 to 224R-43.

9–6 ACI Committee 224, "Causes, Evaluation, and Repair of Cracks in Concrete Structures" (ACI 224.1R-90), *ACI Manual of Concrete Practice,* American Concrete Institute, Detroit, 1991 and later editions, pp. 224.1R-1 to 224.1R-20.

9–7 Comité Euro-International du Béton, *CEB Manual—Cracking and Deformations,* Ecole Polytechnique Fédérale de Lausanne, 1985, 232 pp.

9–8 Peter Gergely and Leroy A. Lutz, "Maximum Crack Width in Reinforced Concrete Flexural Members," *Causes, Mechanism and Control of Cracking in Concrete,* ACI Publication SP-20, American Concrete Institute, Detroit, 1973, pp. 87–117.

9–9 Andrew W. Beeby, "Cracking, Cover, and Corrosion of Reinforcement," *Concrete International: Design and Construction,* Vol. 5, No. 2, February 1983, pp. 35–41.

9–10 ACI Committee 222, "Corrosion of Metals in Concrete" (ACI 222R-89), *ACI Manual of Concrete Practice,* American Concrete Institute, Detroit, 1990 and later editions, pp. 222R-1 to 222R-30.

9–11 *CEB-FIP Model Code 1990,* Thomas Telford Service Ltd., London, for Comité Euro-International du Béton, Lausanne, 1993, 437 pp.

9–12 Gregory C. Frantz and John E. Breen, "Design Proposal for Side Face Crack Control Reinforcement for Large Reinforced Concrete Beams," *Concrete International: Design and Construction,* Vol. 2, No. 10, October 1980, pp. 29–34.

9–13 ACI Committee 435, Deflections of Reinforced Concrete Flexural Members," *ACI Journal, Proceedings,* Vol. 63 No. 6 June 1966, pp. 637–674.

9–14 Dan E. Branson, "Compression Steel Effect on Long-Time Deflections," *ACI Journal, Proceedings,* Vol. 68, No. 8, August 1971, pp. 555–559.

References 917

9–15 Amin Ghali, "Deflection of Reinforced Concrete Members: A Critical Review," *ACI Structural Journal,* Vol. 90, No. 4, July–August 1993, pp. 364–373.

9–16 H. Mayer and Hubert Rüsch, "Damage to Buildings Resulting from Deflection of Reinforced Concrete Members," *Deutscher Ausschuss für Stahlbeton,* Bulletin 193, Wilhelm Ernst, Berlin, 1967, 90 pp.; English translation: Technical Translation 1412, National Research Council of Canada, Ottawa, 1970, 115 pp.

9–17 D. A. Sawyer, "Ponding of Rainwater on Flexible Roof Systems," *Proceedings ASCE, Journal of the Structural Division,* Vol. 93, No. ST1, February 1967, pp. 127–147.

9–18 ACI Committee 435, "Allowable Deflections," *ACI Journal, Proceedings,* Vol. 65, No. 6, June 1968, pp. 433–444.

9–19 Robert Ramsey, Sher-Ali Mirza, and James G. MacGregor, "Variability of Deflections of Reinforced Concrete Beams," *ACI Journal, Proceedings,* Vol. 76, No. 8, August 1979, pp. 897–918.

9–20 Jacob S. Grossman, "Simplified Computations for Effective Moment of Inertia I_e and Minimum Thickness to Avoid Deflection Computations," *ACI Journal, Proceedings,* Vol. 78, No. 6, November–December 1981, pp. 423–440.

9–21 "Motion Perception and Tolerance," Chapter PC-13, *Planning and Environmental Criteria for Tall Buildings,* Monograph on Planning and Design of Tall Buildings, Vol. PC, American Society of Civil Engineers, New York, 1981, pp. 805–862.

9–22 Fazlur R. Khan and Mark Fintel, "Effects of Column Exposure in Tall Structures," *ACI Journal, Proceedings,* Vol. 62, No. 12, December 1965, pp. 1533–1556; Vol. 63, No. 8, August 1966, pp. 843–864; Vol. 65, No. 2, February 1968, pp. 99–110.

9–23 Mark Fintel and S. K. Ghosh, *Column Shortening in Tall Structures—Prediction and Compensation,* EB108.01D, Portland Cement Association, Skokie, Ill., 1986, 35 pp.

9–24 David E. Allen, J. Hans Rainer, and G. Pernica, "Vibration Criteria for Long-Span Concrete Floors," *Vibrations of Concrete Structures,* ACI Publication SP-60, American Concrete Institute, Detroit, 1979, pp. 67–78.

9–25 "Commentary A, Serviceability Criteria for Deflections and Vibrations," *Supplement to the National Building Code of Canada 1990,* NRCC 30629, National Research Council of Canada, Ottawa, 1990, pp. 134–140.

9–26 ACI Committee 215, "Considerations for the Design of Concrete Structures Subjected to Fatigue Loading" (ACI 215R-74, revised 1992), *ACI Manual of Concrete Practice,* American Concrete Institute, Detroit, 1993 and later editions, pp. 215R-1 to 215R-24.

9–27 ACI Committee 343, "Analysis and Design of Reinforced Concrete Bridge Structures" (ACI 343R-88), *ACI Manual of Concrete Practice,* American Concrete Institute, Detroit, 1989 and later editions, pp. 343R-1 to 343R-162.

9–28 *Bond Action and Bond Behavior of Reinforcement,* Bulletin d'Information 151, Comité Euro-International du Béton, Paris, April 1982, 153 pp.

Chapter 10

10–1 ACI Committee 216, "Guide for Determining the Fire Endurance of Concrete Elements," *Concrete International: Design and Construction,* Vol. 3, No. 2, February 1981, pp. 13–47.

10–2 R. Ian Gilbert, "Shrinkage Cracking in Fully-Restrained Concrete Members," *ACI Structural Journal,* Vol. 89, No. 2, March–April 1992, pp. 141–150.

10–3 *Building Movements and Joints,* Engineering Bulletin EB 086.10B, Portland Cement Association, Skokie, Ill., 1982, 64 pp.

10–4 *Minimum Design Loads for Buildings and Other Structures,* ASCE 7-95, American Society of Civil Engineers, New York, 1995, 214 pp.

10–5 ACI-ASCE Committee 428, "Progress Report on Code Clauses for Limit Design," *ACI Journal, Proceedings,* Vol. 65, No. 9, September 1968, pp. 713–720.

10–6 Richard W. Furlong, "Design of Concrete Frames by Assigned Limit Moments," *ACI Journal, Proceedings,* Vol. 67, No. 4, April 1970, pp. 341–353.

10–7 Alan H. Mattock, "Redistribution of Design Bending Moments in Reinforced Concrete Continuous Beams," *Proceedings, Institution of Civil Engineers, London,* Vol. 13, 1959, pp. 35–46.

Chapter 11

11–1 Eivind Hognestad, *A Study of Combined Bending and Axial Load in Reinforced Concrete Members,* Bulletin 399, University of Illinois Engineering Experiment Station, Urbana, Ill., June 1951, 128 pp.

11–2 ACI Committee 105, "Reinforced Concrete Column Investigation," *ACI Journal, Proceedings,* Vol. 26, April 1930, pp. 601–612; Vol. 27, February 1931, pp. 675–676; Vol. 28, November 1931, pp. 157–578; Vol. 29, September 1932, pp. 53–56; Vol. 30, September–October 1933, pp. 78–90; November–December 1933, pp. 153–156.

11–3 ACI-ASCE Committee 327, "Report on Ultimate Strength Design," *Proceedings ASCE,* Vol. 81, October 1955, Paper 809. See also *ACI Journal, Proceedings,* Vol. 2, No. 7, January 1956, pp. 505–524.

11–4 ACI Committee 340, *Design Handbook Volume 2—Columns* (ACI 340.2R-91), ACI Publication SP-17A(90), American Concrete Institute, Detroit, 250 pp.

11–5 *CRSI Handbook 1992,* Concrete Reinforcing Steel Institute, Schaumberg, Ill., 1992, 710 pp.

11–6 Albert C. Bianchini, R. E. Woods, and Clyde E. Kesler, "Effect of Floor Concrete Strength on Column Strength," *ACI Journal, Proceedings,* Vol. 56, No. 11, May 1960, pp. 1149–1169.

11–7 ACI Committee 363, "State-of-the-Art Report on High-Strength Concrete," *ACI Manual of Concrete Practice,* American Concrete Institute, Detroit, 1993 and later editions, pp. 363R-1 to 363R-55.

11–8 Charles S. Whitney, "Plastic Theory of Reinforced Concrete Design," *Transactions ASCE,* Vol. 107, 1942, pp. 251–326.

11–9 Troels Brondiem-Nielsen, "Ultimate Limit States of Cracked Arbitrary Concrete Sections under Axial Load and Biaxial Bending," *Concrete International: Design and Construction,* Vol. 4, No. 11, November 1982, pp. 51–55.

11–10 James G. MacGregor, "Simple Design Procedures for Concrete Columns," Introductory Report, Symposium on Design and Safety of Reinforced Concrete Compression Members, *Reports of the Working Commissions,* Vol. 15, International Association of Bridge and Structural Engineering, Zurich, April 1973, pp. 23–49.

11–11 Albert J. Gouwens, "Biaxial Bending Simplified," *Reinforced Concrete Columns,* ACI Publication SP-50, American Concrete Institute, Detroit, 1975, pp. 233–261.

11–12 Boris Bresler, "Design Criteria for Reinforced Concrete Columns under Axial Load and Biaxial Bending," *ACI Journal, Proceedings,* Vol. 57, No. 5, November 1960, pp. 481–490; Discussion, pp. 1621-1638.

11–13 ACI Committee 318, *Building Code Requirements for Structural Concrete* (ACI 318-95) and *Commentary* (ACI 318R-95), American Concrete Institute, Detroit, 1995, 369 pp.

Chapter 12

12–1 James G. MacGregor, John E. Breen, and Edward O. Pfrang, "Design of Slender Columns," *ACI Journal, Proceedings,* Vol. 67, No. 1, January 1970, pp. 6–28.

12–2 Bengt Broms and Ivan M. Viest, "Long Reinforced Concrete Columns—A Symposium," *Transactions ASCE,* Vol. 126, Part 2, 1961, pp. 308–400.

12–3 *Commentary on Specification for the Design, Fabrication and Erection of Structural Steel for Buildings,* American Institute of Steel Construction, New York, 1970.

12–4 Walter J. Austin, "Strength and Design of Metal Beam-Columns," *Proceedings ASCE, Journal of the Structural Division,* Vol. 87, No. ST4, April 1961, pp. 1–34.

12–5 James G. MacGregor, Urs H. Oelhafen, and Sven E. Hage, "A Reexamination of the *El* Value for Slender Columns," *Reinforced Concrete Columns,* ACI Publication SP-50, American Concrete Institute, Detroit, 1975, pp. 1–40.

12–6 A. A. Gvozdev and E. A. Chistiakov, "Effect of Creep on Load Capacity of Slender Compressed Elements," State-of-Art Report 2, Technical Committee 23 ASCE-IABSE International Conference on Tall Buildings, *Proceedings,* Vol. III-23, August 1972, pp. 537–554.

12–7 Roger Green and John E. Breen, "Eccentrically Loaded Columns under Sustained Load," *ACI Journal, Proceedings,* Vol. 66, No. 11, November 1969, pp. 866–874.

12–8 Adam M. Neville, Walter H. Dilger, and J. J. Brooks, *Creep of Plain and Structural Concrete,* Construction Press, New York, 1983, pp. 347–349.

12–9 Robert F. Warner, "Physical-Mathematical Models and Theoretical Considerations," *Introductory Report, Symposium on Design and Safety of Reinforced Concrete Compression Members,* International Association on Bridge and Structural Engineering, Zurich, 1974, pp. 1–21.

12–10 Sher Ali Mirza, P. M. Lee, and D. L. Morgan, "ACI Stability Resistance Factor for RC Columns," *ASCE Structural Engineering,* Vol. 113, No. 9, September 1987, pp. 1963–1976.

12–11 Richard W. Furlong and Phil M. Ferguson, "Tests of Frames with Columns in Single Curvature," *Symposium on Reinforced Concrete Columns,* ACI Publication SP-13, American Concrete Institute, Detroit, 1966, pp. 55–73.

12–12 William B. Cranston, *Analysis and Design of Concrete Columns,* Research Report 20, Paper 41.020, Cement and Concrete Association, London, 1972, 54 pp.

12–13 Robert F. Manuel and James G. MacGregor, "Analysis of Restrained Reinforced Concrete Columns under Sustained Load, *ACI Journal, Proceedings,* Vol. 64, No. 1, January 1967, pp. 12–23.

12–14 Shu-Ming Albert Lai, James G. MacGregor, and Jostein Hellesland, "Geometric Non-linearities in Non-sway Frames, *Proceedings ASCE, Journal of the Structural Division,* Vol. 109, No. ST12, December 1983, pp. 2770–2785.

12–15 Thomas C. Kavenagh, "Effective Length of Framed Columns," *Transactions ASCE,* Vol. 127, Part 2, 1962, pp. 81–101.

12–16 William McGuire, *Steel Structures,* Prentice Hall, Englewood Cliffs, N.J., 1968, 1110 pp.

12–17 *Code for the Structural Use of Concrete,* CP 110, Part 1, British Standards Institution, London, 1972, 154 pp.

12–18 ACI Committee 340, *Design Handbook in Accordance with the Strength Design Method of ACI 318-77,* Vol. 2, *Columns,* ACI Publication SP-17A(78), American Concrete Institute, Detroit, 1978, 214 pp.

12–19 *PCI Design Handbook,* 3rd ed., Prestressed Concrete Institute, Chicago, 1985, 521 pp.

12–20 James G. MacGregor and Sven E. Hage, "Stability Analysis and Design of Concrete Frames," *Proceedings ASCE, Journal of the Structural Division,* Vol. 103, No. ST10, October 1977, pp. 1953–1970.

12–21 Noel D. Nathan, "Rational Analysis and Design of Prestressed Concrete Beam Columns and Wall Panels," *Journal of the Prestressed Concrete Institute,* Vol. 30, No. 3, May–June 1985, pp. 82–133.

12–22 J. Steven Ford, D. C. Chang, and John E. Breen, "Design Implications from Tests of Unbraced Multipanel Concrete Frames," *Concrete International: Design and Construction,* Vol. 3, No. 3, March 1981, pp. 37–47.

12–23 Shu-Ming Albert Lai and James G. MacGregor, "Geometric Non-Linearities in Multi-story Frames," *Proceedings ASCE, Journal of the Structural Division,* Vol. 109, No. ST11, November 1983, pp. 2528–2545.

12–24 James G. MacGregor, "Design of Slender Columns—Revisited," *ACI Structural Journal,* Vol. 90, No. 3, May–June 1993, pp. 302–309.

12–25 *Minimum Design Loads for Buildings and Other Structures,* ASCE 7-95, American Society of Civil Engineers, New York, 1995, 214 pp.

12–26 W. G. Godden, *Numerical Analysis of Beam and Column Structures,* Prentice Hall, Englewood Cliffs, N. J., 1965, 320 pp.

12–27 Randal W. Poston, John E. Breen, and Jose M. Roesset, "Analysis of Nonprismatic or Hollow Slender Concrete Bridge Piers," *Journal of the American Concrete Institute,* Vol. 82, No. 5, September–October 1985, pp. 731–739.

Chapter 13

13–1 Mete A. Sozen and Chester P. Siess, "Investigation of Multiple Panel Reinforced Concrete Floor Slabs: Design Methods—Their Evolution and Comparison," *ACI Journal, Proceedings,* Vol. 60, No. 8, August 1963, pp. 999–1027.

13–2 John R. Nichols, "Statical Limitations upon the Steel Requirements in Reinforced Concrete Flat Slab Floors," *Transactions ASCE,* Vol. 77, 1914, pp. 1670–1681.

13–3 Discussions of Ref. 13–2, *Transactions ASCE,* Vol. 77, 1914, pp. 1682–1736.

13–4 H. M. Westergaard and N. A. Slater, "Moments and Stresses in Slabs," *ACI Proceedings,* Vol. 17, 1921, pp. 415–538.

13–5 ACI Committee 318, *Building Code Requirements for Structural Concrete* (ACI 318-95) and *Commentary* (ACI 318R-95), American Concrete Institute, Detroit, 1995, 369 pp.

13–6 *Notes on ACI 318–95, Building Code Requirements for Reinforced Concrete, with Design Applications,* Portland Cement Association, Skokie, Ill. 1996, 860 pp.

13–7 David P. Thompson and Andrew Scanlon, "Minimum Thickness Requirements for Control of Two-Way Slab Deflections," *ACI Structural Journal,* Vol. 85, No. 1, January–February 1988, pp. 12–22.

13–8 James O. Jirsa, Mete A. Sozen, and Chester P. Siess, "Pattern Loadings on Reinforced Concrete Floor Slabs," *Proceedings ASCE, Journal of the Structural Division,* Vol. 95, No. ST6, June 1969, pp. 1117–1137.

13–9 R. E. Woodring and Chester P. Siess, *An Analytical Study of the Moments in Continuous Slabs Subjected to Concentrated Loads,* Structural Research Series 264, Department of Civil Engineering, University of Illinois, Urbana, May 1963, 151 pp.

13–10 Scott D. B. Alexander and Sidney H. Simmonds, "Ultimate Strength of Column–Slab Connections," *ACI Structural Journal, Proceedings,* Vol. 84, No. 3, May–June 1987, pp. 255–261.

13–11 ACI-ASCE Committee 426, "The Shear Strength of Reinforced Concrete Members," Chapter 5, "Shear Strength of Slabs," *Proceedings ASCE, Journal of the Structural Division,* Vol. 100, No. ST8, August 1974, pp. 1543–1591.

13–12 Paul E. Regan and Michael W. Braestrup, *Punching Shear in Reinforced Concrete,* Bulletin d'Information 168, Comité Euro-International du Béton, Lausanne, January 1985, 232 pp.

13–13 Johannes Moe, *Shearing Strength of Reinforced Concrete Slabs and Footings under Concentrated Loads,* Development Department Bulletin D47, Portland Cement Association, Skokie, Ill., April 1961.

13–14 ACI-ASCE Committee 326, "Shear and Diagonal Tension, Slabs," *ACI Journal, Proceedings,* Vol. 59, No. 3, March 1962, pp. 353–396.

13–15 ACI-ASCE Committee 426, "Suggested Revisions to Shear Provisions for Building Codes," *ACI Journal, Proceedings,* Vol. 75, No. 9, September 1978, pp. 458–469.

13–16 Neil M. Hawkins, H. B. Fallsen, and R. C. Hinojosa, "Influence of Column Rectangularity on the Behavior of Flat Plate Structures," *SP-30 Cracking, Deflection and Ultimate Load of Concrete Slab Systems,* American Concrete Institute, Detroit, 1971, pp. 127–146.

13–17 M. Daniel Vanderbilt, "Shear Strength of Continuous Plates, *Proceedings ASCE, Journal of the Structural Division,* Vol. 98, No. ST5, May 1972, pp. 961–973.

13–18 W. Gene Corley and Neil M. Hawkins, "Shearhead Reinforcement for Slabs," *ACI Journal, Proceedings,* Vol. 65, No. 10, October 1968, pp. 811–824.

13–19 Walter H. Dilger and Amin Ghali, "Shear Reinforcement for Concrete Slabs," *Proceedings ASCE, Journal of the Structural Division,* Vol. 107, No. ST12, December 1981, pp. 2403–2420.

13–20 Adel E. Elgabry and Amin Ghali, "Design of Stud-Shear Reinforcement for Slabs," *ACI Structural Journal,* Vol. 87, No. 3, May–June 1990, pp. 350–361.

13–21 ACI Committee 421, "Shear Reinforcement for Slabs" (ACI 421.1R), *ACI Structural Journal,* Vol. 89, No. 5, September–October 1992, pp. 587–589.

13–22 Neil M. Hawkins, "Shear Strength of Slabs with Moments Transferred to Columns," *Shear in Reinforced Concrete,* Vol. 2, ACI Publication SP-42, American Concrete Institute, Detroit, 1974, pp. 817–846.

13–23 Sidney H. Simmonds and Scott D. B. Alexander, "Truss Model for Edge Column–Slab Connections," *ACI Structural Journal,* Vol. 84, No. 4, July–August 1987, pp. 296–303.

13–24 Walter H. Dilger and Alaa Sherif, Discussion of "Proposed Revisions to Building Code Requirements for Reinforced Concrete," *Concrete International,* Vol. 17, No. 7, July 1995, pp. 70–73.

13–25 Jack P. Moehle, "Strength of Slab–Column Edge Connections," *ACI Structural Journal,* Vol. 85, No. 1, January–February 1988, pp. 89–98; Discussions and Closure, Vol. 85, No. 6, November–December 1988, pp. 703–709.

13–26 Joseph DiStasio and M. P. Van Buren, "Transfer of Bending Moment between Flat Plate Floor and Column," *ACI Journal, Proceedings,* Vol. 57, No. 3, September 1960, pp. 299–314.

13–27 Norman W. Hanson and John M. Hanson, "Shear and Moment Transfer between Concrete Slabs and Columns," *Journal of the Research and Development Laboratories, Portland Cement Association,* Vol. 10, No. 1, January 1968, pp. 2–16.

13–28 Amin Ghali and Adel Elgabry, Discussion of "Proposed Revisions to Building Code Requirements for Structural Concrete," *Concrete International,* Vol. 17, No. 7, July 1995, pp. 75–82.

13–29 ACI Committee 352, "Recommendations for Design of Slab-Column Connections in Monolithic Reinforced Concrete Structures," *ACI Structural Journal,* Vol. 85, No. 6, November–December 1988, pp. 675–696.

13–30 Denis Mitchell and William D. Cook, "Preventing Progressive Collapse of Slab Structures," *Proceedings ASCE, Journal of the Structural Division,* Vol. 110, No. ST7, July 1984, pp. 1513–1532.

13–31 R. K. Agerwal and Noel J. Gardner, "Form and Shore Requirements for Multistory Flat Plate Type Buildings, *ACI Journal, Proceedings,* Vol. 71, No. 11, November 1974, pp. 559–569.

13–32 John A. Sbarounis, "Multistory Flat Plate Buildings," *Concrete International: Design and Construction,* Vol. 81, No. 2, February 1984, pp. 70–77; No. 4, April 1984, pp. 62–70; No. 8, August 1984, pp. 31–35.

Chapter 14

14–1 Dean Peabody, Jr., "Continuous Frame Analysis of Flat Slabs," *Journal, Boston Society of Civil Engineers,* January 1948.

14–2 Chester P. Siess and Nathan M. Newmark, "Rational Analysis and Design of Two-Way Concrete Slabs," *ACI Journal, Proceedings,* Vol. 45, 1949, pp. 273–315.

14–3 Chester P. Siess and Nathan M. Newmark, *Moments in Two-Way Concrete Floor Slabs,* Bulletin 385, University of Illinois Engineering Experiment Station, Urbana, Ill., 1950, 124 pp.

14–4 W. Gene Corley and James O. Jirsa, "Equivalent Frame Analysis for Slab Design," *ACI Journal, Proceedings,* Vol. 67, No. 11, November 1970, pp. 875–884.

14–5 M. Daniel Vanderbilt, "Equivalent Frame Analysis of Unbraced Concrete Frames," *Significant Developments in Engineering Practice and Research—A Tribute to Chester P. Siess,* ACI Publication SP-72, American Concrete Institute, Detroit, 1981, pp. 219–246.

14–6 C. K. Wang, *Statically Indeterminate Structures,* McGraw-Hill, New York, 1953.

14–7 *Notes on ACI-318-95, Building Code Requirements for Reinforced Concrete with Design Applications,* Portland Cement Association, Skokie, Ill., 1996, 860 pp.

14–8 Sidney H. Simmonds and Janko Misic, "Design Factors for the Equivalent Frame Method," *ACI Journal, Proceedings,* Vol. 68, No. 11, November 1971, pp. 825–831.

14–9 Sidney H. Simmonds, *Effects of Supports on Slab Behavior,* meeting preprint 1697, presented at ASCE National Structural Engineering Meeting, Cleveland, Ohio, April 24–28, 1972.

14–10 ACI Committee 318, *Building Code Requirements for Reinforced Concrete,* (ACI 318–89), and *Commentary* (ACI 318R–89), American Concrete Institute, Detroit,1989, 353 pp.

Chapter 15

15–1 Arne Hillerborg, *Strip Method of Design,* E. & F., N. Spon, London, 1975, 258 pp.

15–2 Randal H. Wood and G. S. T. Armer, "The Theory of the Strip Method for Design of Slabs," *Proceedings, Institution of Civil Engineers, London,* Vol. 41, October 1968, pp. 285–313.

15–3 Kurt W. Johansen, *Yield-Line Theory,* Cement and Concrete Association, London, 1962, 181 pp.

15-4 Leonard L. Jones and Randal H. Wood, *Yield Line Analysis of Slabs,* Elsevier, New York, 1967.

15-5 Roger P. Johnson, *Structural Concrete,* McGraw-Hill, London, 1967, 271 pp.

15-6 Robert Park and William L. Gamble, *Reinforced Concrete Slabs,* Wiley-Interscience, New York, 1980, 618 pp.

15-7 Arne Hillerborg, "Equilibrium Theory for Concrete Slabs" (in Swedish), *Betong,* Vol. 41, No. 4, 1956, pp. 171–182.

15-8 Arne Hillerborg, "The Advanced Strip Method—A Simple Design Tool," *Magazine of Concrete Research,* Vol. 34, No. 121, December 1982, pp. 175–181.

Chapter 16

16-1 Joseph E. Bowles, *Foundation Analysis and Design,* 3rd ed., McGraw-Hill, New York, 1982, 816 pp.

16-2 Arthur N. Talbot, *Reinforced Concrete Wall Footings and Column Footings,* Bulletin 67, University of Illinois Engineering Experiment Station, Urbana, Ill., 1913, 96 pp.

16-3 Frank E. Richart, "Reinforced Concrete Wall and Column Footings," *ACI Journal, Proceedings,* Vol. 45, October 1948, pp. 97–128; November 1948, pp. 327–360.

16-4 ACI-ASCE Committee 326, "Shear and Diagonal Tension," Part 3, "Slabs and Footings," *ACI Journal, Proceedings,* Vol. 59, No. 3 March 1962, pp. 353–396.

16-5 ACI Committee 436, "Suggested Design Procedures for Combined Footings and Mats," *ACI Journal, Proceedings,* Vol. 63, No. 10, October 1966, pp. 1041–1058.

16-6 F. Kramrisch and P. Rogers, "Simplified Design of Combined Footings," *Proceedings ASCE, Journal of the Soil Mechanics Division,* Vol. 87, No. SM5, October 1961, pp. 19–44.

16-7 Perry Adebar, Daniel Kuchma, and Michael P. Collins, "Strut-and-Tie Models for the Design of Pile Caps: An Experimental Study" *ACI Structural Journal,* Vol. 87, No. 1, January-February 1990, pp. 81–92.

16-8 CRSI Handbook, 1992, Concrete Steel Reinforcing Steel Institute, Schaumberg, Ill., 1992, 710 pp.

Chapter 17

17-1 J. A. Hofbeck, I. A. Ibrahim, and Alan H. Mattock, "Shear Transfer in Reinforced Concrete," *ACI Journal, Proceedings,* Vol. 66, No. 2, February 1969, pp. 119–128.

17-2 Alan H. Mattock and Neil M. Hawkins, "Shear Transfer in Reinforced Concrete–Recent Research," *Journal of the Prestressed Concrete Institute,* Vol. 17, No. 2, March-April 1972, pp. 55–75.

17-3 Norman W. Hanson, "Precast-Prestressed Concrete Bridges," Part 2, "Horizontal Shear Connections," *Journal of the Research and Development Laboratories, Portland Cement Association,* Vol. 2, No. 2, May 1960, pp. 38–58.

17-4 ACI-ASCE Committee 426, "The Shear Strength of Reinforced Concrete Members, *Proceedings ASCE, Journal of the Structural Division,* Vol. 99, No. ST6, June 1973, pp. 1148–1157.

17-5 P. W. Birkeland and H. W. Birkeland, "Connections in Precast Concrete Construction, *ACI Journal, Proceedings,* Vol. 63, No. 3, March 1966, pp. 345–368.

17-6 ACI Committee 318, *Building Code Requirements for Structural Concrete* (ACI 318-95 and Commentary (ACI 318R–95), American Concrete Institute, Detroit, 1995, 369 pp.

17-7 *PCI Design Handbook–Precast and Prestressed Concrete,* 3rd ed., Prestressed Concrete Institute, Chicago, 1985, 521 pp.

17-8 Paul H. Kaar, L. B. Kriz, and Eivind Hognestad, "Precast-Prestressed Bridges: (1) Pilot Tests of Continuous Girders," *Journal of the Research and Development Laboratories, Portland Cement Association,* Vol. 2, No. 2, May 1960, pp. 21–37.

17-9 J. C. Saemann and George W. Washa, "Horizontal Shear Connections Between Precast Beams and Cast-in-Place Slabs," *ACI Journal, Proceedings,* Vol. 61, No. 11, November 1964, pp. 1383–1409.

17-10 Robert E. Loov and A. K. Patnaik, "Horizontal Shear Strength of Composite Concrete Beams with a Rough Interface," *PCI Journal,* Vol. 39, No. 1, January–February 1994, pp. 48–69.

17-11 Amin Ghali and Rene Favre, *Concrete Structures: Stresses and Deformations,* Chapman & Hall, New York, 1986, 348 pp.

17-12 Walter H. Dilger, "Creep Analysis of Prestressed Concrete Structures Using Creep-Transformed Section Properties," *PCI Journal,* Vol. 27, No. 1, January–February 1982, pp. 99–118.

Chapter 18

18-1 Jörg Schlaich and Dieter Weischede, *Detailing of Concrete Structures* (in German), Bulletin d'Information 150, Comité Euro-International du Béton, Paris, March 1982, 163 pp.

18-2 Jörg Schlaich, Kurt Schäfer, and Mattias Jennewein, "Toward a Consistent Design of Structural Concrete," *Journal of the Prestressed Concrete Institute,* Vol. 32, No. 3, May–June 1987, pp. 74–150.

18–3 Jörg Schlaich and Kurt Schäfer, "Design and Detailing of Structural Concrete Using Strut-and-Tie Models," *The Structural Engineer,* Vol. 69, No. 6, March 1991, 13 pp.

18–4 David M. Rogowsky and Peter Marti, "Detailing for Post-Tensioning," *VSL Report Series,* No. 3, VSL International Ltd., Bern, 1991, 49 pp.

18–5 Perry Adebar and Zongyu Zhou, "Bearing Strength of Compressive Struts Confined by Plain Concrete," *ACI Structural Journal,* Vol. 90, No. 5, September–October 1993, pp. 534–541.

18–6 David M. Rogowsky and James G. MacGregor, "Design of Deep Reinforced Concrete Continuous Beams," *Concrete International: Design and Construction,* Vol. 8, No. 8, August 1986, pp. 49–58.

18–7 *CEB-FIP Model Code 1990,* Thomas Telford Services, Ltd., London, for Comité Euro-International du Béton, Lausanne, 1993, 437 pp.

18–8 M. P. Nielsen, M. N. Braestrup, B. C. Jensen, and F. Bach, *Concrete Plasticity, Beam Shear—Shear in Joints—Punching Shear,* Special Publication of the Danish Society for Structural Science and Engineering, Technical University of Denmark, Lyngby/Copenhagen, 1978, 129 pp.

18–9 CSA Technical Committee on Reinforced Concrete Design, *A23.3-94 Design of Concrete Structures,* Canadian Standards Association, Rexdale, Ontario, December 1994, 199 pp.

18–10 Michael P. Collins and Denis Mitchell, "Design Proposals for Shear and Torsion," *Journal of the Prestressed Concrete Institute,* Vol. 25, No. 5, September-October 1980, 70 pp.

18–11 Konrad Bergmeister, John E. Breen, and James O. Jirsa, "Dimensioning of the Nodes and Development of Reinforcement," *Structural Concrete, IABSE Colloquium, Stuttgart 1991, Report,* International Association for Bridge and Structural Engineering, Zurich, 1991, pp. 551–556.

18–12 James O. Jirsa, Konrad Bergmeister, Robert Anderson, John E. Breen, David Barton, and Hakim Bouadi, "Experimental Studies of Nodes in Strut-and-Tie Models," *Structural Concrete IABSE Colloquium Stuttgart, 1991, Report,* International Association for Bridge and Structural Engineering, Zurich, 1991, pp. 525–532.

18–13 Ladislav Kriz and Charles H. Raths, "Connections in Precast Concrete Structures—Strength of Corbels," *Journal of the Prestressed Concrete Institute,* Vol. 10, No. 1, February 1965, pp. 16–47.

18–14 Alan H. Mattock, K. C. Chen, and K. Soongswany, "The Behavior of Reinforced Concrete Corbels," *Journal of the Prestressed Concrete Institute,* Vol. 21, No. 2, March-April 1976, pp. 52–77.

18–15 *PCI Design Handbook—Precast and Prestressed Concrete,* 3rd ed., Prestressed Concrete Institute, Chicago, 1985, 521 pp.

18–16 Alan H. Mattock and T. Theryo, *Strength of Members with Dapped Ends,* Research Project 6, Prestressed Concrete Institute, 1980, 25 pp.

18–17 William D. Cook and Denis Mitchell, "Studies of Disturbed Regions near Discontinuities in Reinforced Concrete Members," *ACI Structural Journal,* Vol. 85, No. 2, March–April 1988, pp. 206–216.

18–18 Fritz Leonhardt and Rene Walther, "Wandartige Trager (Wall-like Beams)," *Deutscher Ausschuss für Stahlbeton,* Vol. 178, Wilhelm Ernst, Berlin, 1966, 159 pp.

18–19 ACI-ASCE Committee 352, "Recommendations for Design of Beam-Column Joints in Monolithic Reinforced Concrete Structures," *ACI Journal, Proceedings,* Vol. 82, No. 3, May–June 1985, pp. 266–283.

18–20 Ingvar H. E. Nilsson and Anders Losberg, "Reinforced Concrete Corners and Joints Subjected to Bending Moment," *Proceedings ASCE, Journal of the Structural Division,* Vol. 102, No. ST6, June 1976, pp. 1229–1254.

18–21 P. S. Balint and Harold P. J. Taylor, *Reinforcement Detailing of Frame Corner Joints with Particular Reference to Opening Corners,* Technical Report 42.462, Cement and Concrete Association, London, February, 1972, 16 pp.

18–22 Neil M. Hawkins, "The Bearing Strength of Concrete Loaded through Rigid Plates," *Magazine of Concrete Research,* Vol. 20, No. 62, March 1968, pp. 31–40.

Chapter 19

19–1 Henry J. Degenkolb, *Earthquake Forces on Tall Structures,* Booklet 2028, Bethlehem Steel, Bethlehem, Pa., 1964, 25 pp.

19–2 Robert Park and Thomas Paulay, *Reinforced Concrete Structures,* Wiley-Interscience, New York, 1975, 768 pp.

19–3 *Uniform Building Code,* International Conference of Building Officials, Whittier, Calif., 1988, 926 pp.

19–4 *Minimum Design Loads for Buildings and Other Structures* (ASCE 7-95), American Society of Civil Engineers, New York, 1995, 214 pp.

19–5 Nathan M. Newmark and Emilio Rosenblueth, *Fundamentals of Earthquake Engineering,* Prentice Hall, Englewood Cliffs, N. J., 1971, 639 pp.

19–6 Anil K. Chopra, *Dynamics of Structures–A Primer,* Earthquake Engineering Research Institute, Berkeley, Calif., 1981, 126 pp.

19–7 Egor P. Popov, Vitelmo V. Bertero and H. Krawinkler, *Cyclic Behavior of Three R/C Flexural Members with High Shear,* EERC Report 72–5, Earthquake Engineering Research Center, University of California, Berkeley, October 1972.

19–8 James K. Wight and Mete A. Sozen, "Shear Strength Decay of RC Columns under Shear Reversals," *Proceedings ASCE, Journal of the Structural Division,* Vol. 101, No. ST5, May 1975, pp. 1053–1065.

19–9 Jack R. Benjamin, *Statically Indeterminate Structures,* McGraw-Hill, New York, 1959, 350 pp.

19–10 ACI Committee 318, *Building Code Requirements for Structural Concrete* (ACI 318–95) and *Commentary* (ACI 318R–95), American Concrete Institute, Detroit, 1995, 369 pp.

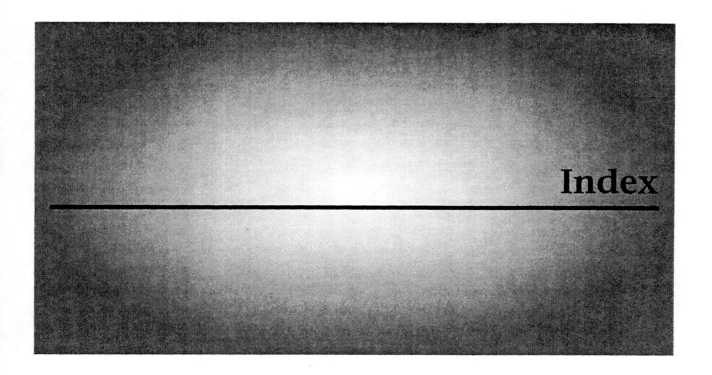

Index

S